S0-DUV-675

The
Alkaline Rocks

Intrusive contact between granitic basement overlain by sandstone with sheets of dolerite (left) and agpaitic nepheline syenites (right). The latter consist of poikilitic sodalite syenites (naujaites) at the bottom and of foyaitic rocks at the top. Head of the Kangerdluarssuk fjord, the Ilímaussaq intrusion, South Greenland.

The Alkaline Rocks

Edited by

H. Sørensen
Institut for Petrologi
Universitetets Mineralogisk-Geologiske Instituter, Copenhagen

A Wiley–Interscience Publication

JOHN WILEY & SONS
LONDON · NEW YORK · SYDNEY · TORONTO

Library of Congress catalog card number 72-5725

ISBN 0 471 81383 4

Printed in Great Britain by
The Garden City Press Limited
Letchworth, Hertfordshire SG6 1JS

List of Contributors

D. K. Bailey, *Department of Geology, Reading University, Whiteknights, Reading RG6 2AB, England*

D. S. Barker, *Department of Geological Sciences, The University of Texas, Austin 12, Texas 78712, U.S.A.*

T. Yu. Bazarova, *Institute of Geology and Geophysics, Siberian Branch Ac., Sci., U.S.S.R., Novosibirsk 90, U.S.S.R.*

K. Bell, *Department of Geology, Carleton University, Ottawa 1, Ontario, Canada*

G. D. Borley, *Department of Geology, Imperial College of Science and Technology, Prince Consort Road, London, S.W.7, England*

L. S. Borodin, *Institute of Mineralogy, Geochemistry and Crystal Chemistry of Rare Elements, The Academy of Sciences of the U.S.S.R., Sadovnicheskaja 71, Moskva 113127, U.S.S.R.*

P. Bowden, *University of St. Andrews, Department of Geology, St. Andrews, Fife, Scotland*

E. L. Butakova, *All-Union Geological Research Institute, Vasilievsky Ostrov, Sredny Prospekt 72-b, Leningrad, U.S.S.R.*

A. D. Edgar, *Department of Geology, The University of Western Ontario, London 72, Ontario, Canada*

P. Floor, *Geologisch- en Mineralogisch Instituut der Rijksuniversiteit — Leiden, Garenmarkt, Leiden, Holland*

V. I. Gerasimovsky, *Vernadsky Institute of Geochemistry and Analytical Chemistry of the Academy of Sciences of the U.S.S.R., Vorobievskoye shosse 47a, Moskva V-334, U.S.S.R.*

P. G. Harris, *Department of Earth Sciences, The University of Leeds, Leeds LS2 9JT, England*

L. N. Kogarko, *Vernadsky Institute of Geochemistry and Analytical Chemistry of the Academy of Sciences of the U.S.S.R., Vorobievskoye shosse 47a, Moskva V-334, U.S.S.R.*

V. P. Kostyuk, *Institute of Geology and Geophysics, Siberian Branch Ac., Sci., U.S.S.R., Novosibirsk 90, U.S.S.R.*

W. C. Luth, *Department of Mineral Engineering, Stanford University, Stanford, California 94305, U.S.A.*

R. MacDonald, *Department of Environmental Sciences, University of Lancaster, Bailrigg, Lancaster, Lancashire, England*

M. Mathias, *9 Third Street, Marandellas, Rhodesia*

A. S. PAVLENKO, *Vernadsky Institute of Geochemistry and Analytical Chemistry of the Academy of Sciences of the U.S.S.R., Vorobievskoye shosse 47a, Moskva V-334, U.S.S.R.*

A. R. PHILPOTTS, *The University of Connecticut, Department of Geology and Geography, Storrs, Connecticut 06268, U.S.A.*

A. I. POLYAKOV, *Vernadsky Institute of Geochemistry and Analytical Chemistry of the Academy of Sciences of the U.S.S.R., Vorobievskoye shosse 47a, Moskva V-334, U.S.S.R.*

J. L. POWELL, *Department of Geology, Oberlin College, Oberlin, Ohio 44074, U.S.A.*

I. D. RYABCHIKOV, *Institute of Geology of Ore Deposits, Petrography, Mineralogy and Geochemistry of the Academy of Sciences of the U.S.S.R., Staromonetni per. 35, Moskva V-17, U.S.S.R.*

TH. G. SAHAMA, *The University of Helsinki, Institute of Geology and Mineralogy, Snellmaninkatu 5, Helsinki 17, Finland*

E. I. SEMENOV, *Institute of Mineralogy, Geochemistry and Crystal Chemistry of Rare Elements, The Academy of Sciences of the U.S.S.R., Sadovnicheskaja 71, Moskva 113127, U.S.S.R.*

V. S. SOBOLEV, *Institute of Geology and Geophysics, Siberian Branch Ac., Sci., U.S.S.R., Novosibirsk 90, U.S.S.R.*

H. SØRENSEN, *Institut for Petrologi, Universitetets Mineralogisk-Geologiske Instituter, Øster Voldgade 5, Dk-1350 Copenhagen K, Denmark*

D. C. TURNER, *Department of Geology, Ahmadu Bello University, Zaria, Nigeria*

B. G. J. UPTON, *Grant Institute of Geology, King's Buildings, West Mains Road, Edinburgh EH9 3JW, Scotland*

V. P. VOLKOV, *Vernadsky Institute of Geochemistry and Analytical Chemistry of the Academy of Sciences of the U.S.S.R., Vorobievskoye shosse 47a, Moskva V-334, U.S.S.R.*

J. F. G. WILKINSON, *Department of Geology, University of New England, Armidale, New South Wales 2351, Australia*

W. WIMMENAUER, *Mineralogisches Institut der Universität Freiburg i. Br., 78 Freiburg i. Br., Hebelstrasse 40, Germany*

P. J. WYLLIE, *Department of the Geophysics Sciences, The University of Chicago, 1101 East 58th Street, Chicago, Illinois 60637, U.S.A.*

Preface

The 'alkaline rocks' constitute with regard to areal distribution an insignificant part of the igneous rocks. Daly (1933) estimated that in North America less than 0·05% of the area occupied by igneous rocks is made up of quartz-free or quartz-poor alkaline rocks. Barker (1969) has pointed out that this may be underestimated by several orders, but even then these rocks contribute very little to the volume of igneous rocks available for study. Alkaline rocks have, nevertheless, more than any other group of igneous rocks burdened petrographic nomenclature with rock names. This is due to the 'exotic' mineral composition of many alkaline rocks which invites the invention of new names.

Apart from these purely petrographical and mineralogical aspects, the petrology of the alkaline rocks is most complex and elucidates, furthermore, a number of petrological principles. These rocks have therefore attracted vivid attention throughout the history of modern petrology. Last, but not least, alkaline rocks are gaining an increasing economic importance, which is another cause of an intensified study. For these reasons it is felt justified to compile a collective volume comprising reviews of various aspects of the petrology of the alkaline rocks.

During the last few years a number of monographs devoted to major rock groups has appeared: E. Raguin (1965) on granites, O. F. Tuttle and J. Gittins (1966) and E. Wm. Heinrich (1966) on carbonatites, P. J. Wyllie (1967) on ultramafic and related rocks, H. H. Hess and A. Poldervaart (1967, 1969) on basalts and K. R. Mehnert (1969) on migmatites. A number of corresponding books and collective works have been published in the USSR on various rock groups, including the alkaline rocks. When planning this book on the alkaline rocks care has been taken to avoid unnecessary overlap with these earlier monographs. For this reason, carbonatites, melilite-bearing rocks and alkaline ultramafic rocks are not described as such but are mentioned and discussed in their proper associations with the types of alkaline rocks covered by the present book. Spilites and lamprophyres are not treated.

The book is divided into seven sections: I. Introduction; II. Petrography and Petrology; III. Regional Distribution and Tectonic Relations; IV. Alkaline Provinces; V. Conditions of Formation; VI Petrogenesis; and VII. Economic Mineralogy.

During the work on the book the editor realized that due to the limitations on space it would not be possible to cover all aspects of the petrology of alkaline rocks. Thus it was decided not to include a special section on the mineralogy of alkaline rocks and also not to compile a list of all alkaline complexes of the world corresponding to the valuable catalogues of carbonatites compiled by Tuttle and Gittins (1966) and Heinrich (1966). Instead it has been endeavoured to select examples which illustrate the major rock types, their occurrence and association. For the same reason the references do not present a complete bibliography, but may serve as an entrance into the literature for those interested in further information.

I wish to extend my thanks and appreciation to all the contributors to the book for their friendly and understanding collaboration. I also wish to acknowledge the valuable advice received when planning the book from many colleagues all over the world. The names are too many to be listed here, only one will be mentioned, that of the late professor T. F. W. Barth who took a vivid interest also in this project.

Gitte Sjørring has offered invaluable assistance in all stages of the work on the book from the first correspondence to the preparation of the indexes. Vibeke Knudsen has prepared the figures of Chapters II.1 and II.2.

I am extremely grateful to Agnete Steenfelt, Lotte Melchior Larsen, my wife Helle and my daughter Lise for their assistance in the compilation of the indexes.

Figures 9 to 13 in Chapter II.2 have been reproduced with the permission of the Geological Survey of Greenland.

Professor Albert Streckeisen, Berne, chairman of the Commission on Systematics in Petrology, the International Union of the Geological Sciences, has critically read the part of the glossary dealing with plutonic alkaline rocks and has offered much invaluable advice. Dr. V. P. Volkov, Moscow, who in all stages of preparation of the book, has assisted with information about Sovietic contributions to the study of alkaline rocks, has also helped me with data on new rock names. I am most grateful to these two colleagues for their friendly collaboration but wish to emphasize that I alone remain responsible for the presentation of the data in this book. Dr C. K. Brooks, Copenhagen, has kindly given advice on the introduction to the glossary.

H. SØRENSEN

Contents

'The alkaline rocks constitute a group that is difficult to mark off sharply from their more abundant sub-alkaline relatives.'

(*N. L. Bowen, 1928, p. 234*)

I. Introduction

H. Sørensen

(with contributions by V. P. Volkov)

This chapter reviews the various definitions proposed for 'alkaline rocks' and the history of the study of these rocks.

I.1. Definition of the Term Alkaline Rock

In 'Dictionary of Geological Terms', prepared under the direction of the American Geological Institute, it is recommended that 'alkalic' be used instead of 'alkaline' in definitions of rocks containing alkalis in excess of the amount needed to form feldspar with the available silica (*Dolphin Reference Book*, 1962, pp. 11, 12). This is because geologic usage gives 'alkaline' so many different meanings, e.g. also as a synonym for basic, that it is ambiguous without further qualification. However, a survey of the literature leaves the impression that alkaline is preferred by most authors, now and in the earlier literature. Alkaline is therefore used in the title of the book. Alkali is used as prefix in names of rocks, e.g. alkali syenite.

The term alkaline rock has been used in at least the following meanings:

1. Igneous rocks of Atlantic or alkaline series (branch, group, facies).
2. Igneous rocks with alkali feldspar as predominant feldspar, that is with more alkalis than average for their clans.
3. Igneous rocks with feldspathoids.
4. Igneous rocks with alkali-lime index less than 51.
5. Igneous rocks with feldspathoids and/or soda-pyroxenes and/or -amphiboles.

I.1.1.

Iddings (1892) introduced the terms alkali group and subalkali group for the two widely developed rock series: basalt–trachyte–phonolite and basalt–andesite–rhyolite. Harker in 1896 pointed out that this 'two-fold division of Iddings is of very wide application and that there is a very general correspondence of the areas of the alkali and subalkali group respectively with the areas of the Atlantic and Pacific types of coast line as defined by Suess. The one type is found around the Atlantic and part of the Indian Ocean and in the Polar basins, the other, generally speaking, around the Pacific' (Harker, 1909, p. 93). Harker therefore suggested that one may distinguish an Atlantic and a Pacific facies of eruptive rocks corresponding with distinct phases in crust-movements of a large order. The Atlantic rock series is step-by-step richer in alkalis and poorer in Ca and Mg than the rocks of the Pacific branch.

Becke (1903) on the basis of a comparative study of the eruptive rocks of the Bohemian Mittelgebirge (České středohoří) and the Andes mountains independently arrived at this two-fold division of the igneous rocks and distinguished an Atlantic 'Sippe' (tephritic) and a Pacific 'Sippe' (andesitic). He warned against using Atlantic and alkaline as synonyms since other rocks also occur in Atlantic provinces.

The Atlantic branch is characterized by: (1) alkali feldspar abundant in acid and intermediate rocks and in many rocks of low acidity; (2) micro- and cryptoperthites common, zoning rare in feldspars; (3) feldspathoids common; (4) quartz only in acid and not in intermediate rocks; (5) pyroxene and amphibole often sodic, orthorhombic pyroxene not present; (6) mica and garnet common; (7) plagioclase only in mafic rocks (Harker, 1909, p. 91).

Niggli (1920, p. 162) introduced the term Mediterranean and considered the following provinces:

I. Pazifischer Provinzialtypus = Kalk-Alkalireihe = gabbrodioritische Reihe.

II. Atlantischer Provinzialtypus = Natronreihe = foyaitisch-theralitische Reihe.

III. Mediterraner Provinzialtypus = Kalireihe = syenitisch-(monzonitisch-)-shonkinitische Reihe.

As emphasized by Zavaritsky (1950) and Barth (1962) the geographical names for the main petrographical provinces have gradually become obsolete. Pacific is replaced by terms as subalkalic, calc-alkalic or calcic, Atlantic by alkalic (Barth, 1962, p. 170). Zavaritsky (1941, 1950) distinguished the alkaline and calc-alkaline rock associations on the basis of their petrochemical affinities as indicated in variation diagrams. He especially emphasized the 'pantelleritic trend' in alkaline rock associations.

The concept of Atlantic and Pacific provinces has in the recent literature been partly superseded by that of the major associations of (basaltic) rocks (see Chapter II.4):

1. alkali olivine basalt series: alkali olivine basalt–hawaiite–mugearite–trachyte $<$ phonolite / pantellerite
2. tholeiitic series: tholeiitic basalt–tholeiitic andesite–dacite–rhyolite
3. calc-alkali series: tholeiite–andesite–dacite–rhyolite.

I.1.2.

Rosenbusch (1910, p. 227) established two series of igneous rocks based on mineralogical and chemical criteria:

I. 'Alkali-Reihe': alkali granite, alkali syenite and nepheline syenite forming a foyaitic series; and essexite, theralite, shonkinite, missourite, ijolite forming a theralitic series including also alkali pyroxenites.

II. 'Alkalikalk-Reihe': granite, diorite, gabbro, peridotite.

The alkaline series was characterized as follows: 'Der molekulare Alkaligehalt übersteigt häufig den der Tonerde, so dass letztere nicht zur Bindung der Alkalien in Alkalitonerdesilikate ausreicht, ein Teil von ihnen ist mit Fe_2O_3 im Aegirinmolekul verbunden' (Rosenbusch, 1923, p. 40). The alkaline rocks are characterized by: alkalifeldspar (only feldspar in acid rocks, often together with plagioclase in silica-poor rocks); feldspathoids common; pyroxenes and amphiboles often sodic, no rhombic pyroxene; accessories such as eudialyte, melanite, låvenite, astrophyllite, etc. (Rosenbusch, 1923, p. 51).

The two main series of alkaline rocks were, however, distinguished by the presence or absence of plagioclase (albite excluded as it was considered an alkali feldspar). Alkali granite (rhyolite) and -syenite (-trachyte) were thus characterized by the presence of alkali feldspar and the absence of plagioclase.

This distinction has also been used in many subsequent petrographical systems based mainly on qualitative or quantitative modal compositions (Daly, 1914; Tröger, 1935; Niggli, 1931, 1935; Johannsen, 1939; Rittmann, 1952; Jung and Brousse, 1959; Ronner, 1963; Streckeisen, 1967, etc.).

Daly (1914) grouped five of the ten families of eruptive rocks recognized by Rosenbusch as the alkaline clans: nepheline and leucite syenites; essexites; shonkinites and theralites; missourites and fergusites; ijolites and bekinkinites. The term alkaline was not used as a chemical description. The 'designation "alkaline clans" will do no harm if it is remembered that it is used merely as the most available, brief name for a great syngenetic group of rock families' (Daly, 1914, p. 410). Daly emphasized that alkaline and subalkaline species are often associated and are not independent, neither with regard to space nor time.

Niggli (1931, 1935) followed Rosenbusch's main division of rocks, but alkali granites containing soda pyroxene or -amphibole were termed arfvedsonite granite, etc.; syenites containing riebeckite, etc. natron syenite. Niggli (1931, p. 323), however, when speaking of syenodiorites, stated that alkali diorite and alkali syenodiorite contain soda pyroxene and -amphibole. He thus also used 'alkali' to indicate a content of sodic pyriboles.

Lacroix (1933) subdivided granites and syenites into alkaline and calc-alkaline families. Rocks containing soda pyroxene and -amphibole were termed hyperalkaline (hyperalkaline alkali granites, alkali syenites and feldspathoidal syenites). Alkaline rocks were defined as follows: 'Les roches alcalines sont, par principe, celles riches en alkalis, c'est-à-dire . . . syénites et granites alcalins et *à fortiori* hyperalcalins. Il faut y ajouter toutes les roches feldspathoidiques' (Lacroix, 1933, p. 186).

Alkali granites and -syenites (in the sense of Rosenbusch) containing soda pyroxene and -amphibole are termed peralkaline by Hatch, Wells and Wells (1961) and by Nockolds (1954); natron alkali granite, etc. by Ronner (1963) and soda granite, etc. by Rittmann (1952) and Streckeisen (1967).

In the above-mentioned petrographical systems alkaline igneous rocks are first of all defined by their predominance of alkali feldspar (in most systems comprising albite) and lack of or poverty in plagioclase. Feldspathoidal rocks also fall in the alkaline magma series.

I.1.3

The 'alkaline clans' considered by Daly (1914) only comprise feldspathoidal rocks. This is an example of the use of the term alkaline to describe igneous rocks containing feldspathoids.

Zavaritsky (1955) emphasized not only mineralogical and chemical data, but also the type of natural rock association when characterizing alkaline rocks. He distinguished seven groups of igneous rocks, of which only two were considered to be 'typically alkaline'. These are the group of feldspathoidal syenites and phonolites including the urtite-monmouthite subgroup, and the group of alkali gabbroid and -basaltic rocks including the subgroup of feldspar-free, feldspathoidal rocks. The members of the latter group belong to a single rock association. Calc-alkaline and alkaline types were distinguished in the granite-rhyolite group and the syenite-trachyte group. Alkali pyroxenites, including jacupirangites, were considered as members of the calc-alkali group of ultrabasic rocks.

Eliseev (1957) classified nepheline-bearing rocks by means of the ternary diagram—nepheline–total of mafic minerals–alkali feldspar. Four groups of nepheline-bearing rocks were distinguished: (1). urtite–ijolite–melteigite; (2) juvite–malignite; (3). nepheline syenites; (4). alkali gabbroids. This classification only covers a restricted part of the alkaline rocks.

I.1.4.

Alkaline rocks are in many petrographical systems defined on the basis of chemical criteria (see for instance Ronner, 1963, p. 307, etc.).

Lang (1891) based his petrographical system on the contents (weight per cent) of alkalis and lime, $Na_2O + K_2O \gtrless CaO$ ('Alkalien-Vormacht' or 'Kalk-Vormacht').

Loewinson-Lessing (1899) subdivided his four main groups of igneous rocks (acid, neutral, basic, ultrabasic) by means of the relation between $Na_2O + K_2O$, $CaO + MgO + FeO$ and $Al_2O_3 + Fe_2O_3$. His alkalic subgroups are: urtite (ultrabasic), nepheline syenite, phonolite and tinguaite (basic), tephrite, orthophyre and trachyte (neutral), nordmarkite, pantellerite, granite and liparite (acid).

Michel-Lévy (1897) considered two primary magmas, an alkaline and a ferro-magnesian. Subdivision was undertaken according to the contents of lime, potash and soda in feldspars and feldspathoids

Cross, Iddings, Pirsson and Washington (1902) also considered the relation between alkalis and lime. Peralkalic rocks have the total of salic alkalis distinctly higher than salic lime. Femic soda is not used to classify alkaline rocks. Domalkalic rocks have salic alkalis dominant over salic lime. The ratio

$$\frac{Na_2O + K_2O}{CaO} \text{ (sal)}$$

is used to define 'rangs'.

Niggli (1923) defined three series of 'magma types': Natron-Reihe: alk $>$ al or al — alk small; k and mg low. Kali-Reihe: al — alk small; high k, also in si-poor members. Kalkalkali-Reihe: al $>$ alk.

Rittmann (1960) proposed a similar classification on the basis of a 'serial index'

$$\frac{(Na_2O + K_2O)^2}{SiO_2 - 43} \text{ (weight per cent),}$$

and the proportion of alkalis ($Na_2O \sim K_2O$; $Na_2O > K_2O$, $Na_2O < K_2O$).

Peacock (1931) suggested a four-fold division of igneous rock series according to their alkali-lime indices. The alkali-lime index is the percentage of SiO_2 at which the curves for CaO and $Na_2O + K_2O$ intersect in an ordinary variation diagram. The alkalic series have the index lower than 51, the alkali-calcic series between 51 and 56, the calc-alkalic series between 56 and 61 and the calcic series higher than 61.

Wright (1969) has proposed an 'alkalinity' ratio

$$\left(\frac{Al_2O_3+CaO + \text{total alkalis}}{Al_2O_3+CaO - \text{total alkalis}}\right) \text{(weight per cent)},$$

in order to distinguish alkaline and calc-alkaline rock series over the full range of silica contents.

Most of these classifications are based on the relation between alkalis and lime and thus mainly reflect the nature of the feldspar present.

I.1.5.

Shand (1922, 1933) pointed out that in the more common igneous rocks alkalis are associated with alumina and silica to form feldspar and mica. In alkali feldspar the 'molecular' ratio $Na_2O + K_2O : Al_2O_3 : SiO_2 = 1 : 1 : 6$. In muscovite this ratio is 1 : 3 : 6.

'An alkaline rock, then, if names are to mean anything, should be one in which the alkalis are in the excess of 1 : 1 : 6, either alumina or silica or both being deficient' (Shand, 1922, p. xix). This means that contents of alkali feldspar and/or mica do not entitle a rock to be termed alkaline.

The thus-defined alkaline rocks may then be divided into three subgroups:

I. Silica adequate or excessive, alumina deficient: the rocks are composed of alkali feldspar, sodic pyroxene and/or amphibole, quartz may be present (alkali granite, pantellerite, nordmarkite).

II. Alumina adequate or excessive, silica deficient: the rocks are composed of feldspar, feldspathoids, mica, hornblende, augite, corundum, etc. (mica foyaite, etc.).

III. Both silica and alumina deficient: the minerals are: feldspathoids, sodic pyroxene and/or amphibole, eudialyte, alkali feldspar, etc. (agpaitic nepheline syenites).

Shand's definition of alkaline rocks is advocated by Polánski (1949) using terms introduced by Ussing (1912), Fersman (1929) and Goldschmidt (1930). He distinguishes:

(1. plumasitic rocks, $K + Na < Al$; $K + Na < 1/6\ Si$; subalkaline)

2. miaskitic rocks, $K + Na < Al$; $K + Na > 1/6\ Si$;
3. agpaitic rocks, $K + Na > Al$, $K + Na \gtrless 1/6\ Si$.
} alkaline

Barth (1962) also advocates Shand's definition of alkaline rocks and distinguishes ekeritic, miaskitic and agpaitic types (see Chapter II.2).

Harker (1954, pp. 35 and 41) and Moorhouse (1959, p. 262) emphasize that alkali pyroxene and amphibole are the distinguishing marks of alkali granites and syenites.

Turner and Verhoogen (1960, p. 194) state that 'nowadays rocks are usually classed as alkaline when the content of $(K_2O + Na_2O)$ is sufficiently high, as compared with SiO_2 or Al_2O_3, for specially alkaline (usually sodic) minerals such as feldspathoids and aegirine to appear'.

In an official report of the Petrographic Committee of the Government of the USSR distributed by Professor A. Streckeisen in connection with a re-evaluation of his report of 1967 (Streckeisen, 1967) it is pointed out that alkaline and calc-alkaline rocks are not to be mixed in diagrams of classification. According to Soviet use, only rocks containing feldspathoids, or when these are absent, alkali amphiboles and/or alkali pyroxenes are considered as alkaline rocks. Granites and syenites enriched in alkali feldspar are not classified as alkali granite and syenite. They are termed syenite, alaskite, monzosyenite and granite. The discussion of the detailed classification of alkaline rocks is, however, still vivid in the USSR.

I.1.6.

This survey of the literature on the alkaline rocks leaves the impression that the term 'alkaline rock' has been used differently by different petrographers and sometimes in so vague a way that it is hard to know what is covered by the term. The editor of this volume has found that the definition given by Shand is the one most feasible when planning a book of limited size and entitled 'the alkaline rocks'. The definition used in the preparation of the book may be expressed most briefly by stating that alkaline igneous rocks are characterized by the presence of feldspathoids and/or alkali pyroxenes and -amphiboles.

In the succeeding chapters it will be shown that alkaline rocks also are characterized by special geological environments and that they form members of a number of characteristic rock associations.

1.2. History of Study of the Alkaline Rocks

It will not be possible to review the history of study of the alkaline rocks without repetition and without overlap with succeeding chapters. The present chapter will therefore only list some of the more important 'milestones' in the study of alkaline rocks. Other accounts can be found in Daly (1914 and 1933) Shand (1922, 1949), Smyth (1927), Fischer (1961) and in the standard petrological textbooks. The studies of carbonatites will not be mentioned as they are so well covered by the carbonatite memoirs by Tuttle and Gittins (1966) and by Heinrich (1966).

The 'pre-microscopic' ages of the history of petrology will not be treated here. It should only be mentioned that some of the more common alkaline rocks were given names in these periods. Examples are: rhomb porphyry (v. Buch, 1810), syenite (Plinius; Werner, 1788), trachyte (Haüy; Roth, 1861), foyaite (Blum, 1861), ditroite (Zirkel, 1866), phonolite (Haüy), leucitophyre (von Humboldt, 1837), miaskite (Rose, 1839), basanite (Plinius), teschenite (Hohenegger, 1861), nephelinite (Naumann, 1849), and tephrite (Plinius; Brongniart, 1813).

During the last two decades of the nineteenth century and the first three decades of the present century most of the alkaline rocks considered in this book were described and named, first of all by Rosenbusch, Brøgger, Lacroix and Washington. After 1930 very few names have been added to the list of rock names; on the contrary, many names have been found to be completely unnecessary.

During the period from *c.* 1880 to *c.* 1930 a number of the now classic occurrences of alkaline rocks were described in great detail and the discussions of the origin and geological relations of the alkaline rocks presented in these classic papers have contributed much to the development of the then young science of petrology.

Examples are: the Oslo Province (Brøgger, 1890, and a number of subsequent memoirs published during the years 1894 to 1933); Arkansas, also Magnet Cove (Williams, 1891); Khibina (Umptek) and Lovozero (Lujaur-Urt) (Ramsay, 1890, etc.); Serra de Monchique (Kraatz-Kochlau and Hackman, 1897); Leucite Hills (Cross, 1897); the Highwood Mountains (Weed and Pirsson, 1895); the alkaline province of northern Bohemia (Hibsch, 1926); Puy-de-Dôme and Mont Dore (Michel-Lévy, 1890). A. Lacroix should be especially mentioned for a series of memoirs on alkaline provinces: Pouzac (1890), Montreal (1890), Northern Madagascar (1903, 1922, 1923), Puy-de-Dôme (1908) and Archipel de Los (1911). Further examples are: the Monteregian Hills (Adams, 1903 etc.); Haliburton–Bancroft (Adams and Barlow, 1910); Laacher See (Brauns, 1922); Transvaal (Brouwer, 1910–17); Ilímaussaq and Igaliko (Ussing, 1898, 1912); Mount Ascutney, Ascension Island and Saint Helena (Daly, 1903, 1925, and 1927); Khibina and Lovozero (Fersman, Kupletsky and many others from *c.* 1922); Magnet Cove and the Roman Province (Washington, 1900 and 1896–1906); the differentiated sills of Scotland (Tyrrell, 1917 and 1948); Mariupol (Morozewicz, 1930); the Ilmen Mountains (Belijankin, 1909–10, 1926, etc.); Dunedin (Marshall, 1906); the nepheline gneisses of Portugal (Osann, 1907); Shands long row of papers on Loch Borolan (1906, 1939), Pilanesberg (1928), Spitzkop or Sekukuniland (1922), the Fransport Line (1923), Leeuwfontein (1922); the western Rift Valley of East Africa (Holmes and Harwood, 1932), etc.

During this period Harker (1896) and Becke (1903) demonstrated that alkaline rocks are associated with foundering of fault blocks in non-orogenic areas (Atlantic coast lines). Daly (1910) presented his limestone-syntexis hypothesis and Smyth (1913 and 1927) and Gilson (1928) emphasized the role of volatiles (mineralizers) in the formation of alkaline rocks. Bowen (1928) discussed the origin of alkaline rocks in the light of experimental evidence.

The early examiners of alkaline rocks explored a practically virgin field and had not only to name minerals and rocks, but also to describe and explain the geological relations and the genesis of the rocks. As little or nothing was given the authors could apply all their imaginative power with the result that many of these old papers are still illuminating reading.

In the years after 1930 many of the classic localities were re-examined in the light of new field and laboratory experience, Examples are: the Oslo Province (the new series on The Igneous Rock Complex of the Oslo Region, 1943, etc.); Khibina and Lovozero (see IV.2)*; Leucite Hills, the Roman Province and the Western Rift Valley (see II.5 and III.2); the Highwood Mountains (Larsen *et al.*, 1940); Puy-de-Dôme, Pouzac, Mont Dore and Northern Bohemia (IV.4); Serra de Monchique (Czygan, 1969); the Monteregian Hills (IV.6); oceanic islands (IV.7); Haliburton–Bancroft (Tilley, 1957, II.7); South Africa (III.5); Ditró (II.2); Ilímaussaq (IV.3, Ferguson, 1964 and 1970, the series: 'Contributions to the Mineralogy of Ilímaussaq', 1965, etc.).

A number of provinces of alkaline rocks were discovered and studied in the USSR; Meimecha-Kotuj (III.4); Aldan (Bilibin, 1941, 1947, Kravchenko and Vlasova, 1962); Tuva (IV.5); Baikal (Andreev, *et al.*, 1969); Kuznetsk Alatau (Andreeva, 1968); Turkestano-Alai (Perchuk, 1964, Schinkarev, 1966, Molchanova, 1966), etc.

During this period the quantitative petrographical aspects have gained an increasing interest and the mineralogical–petrographical data have been supplemented to a very large extent with information on trace elements (see V.3) and strontium isotopes (see V.4). The petrological discussions are at least partly based on experimental data (see V.1. and many other chapters of this book). Fluid inclusions in the minerals of alkaline rocks have yielded information about the physico-chemical conditions of formation of these minerals (see V.2). Isotopic age determinations are of ever increasing importance for the interpretation of alkaline provinces and intrusions (see III.3, III.4, IV.2, IV.3, IV.4, etc.). They show that alkaline igneous activity in a given region may persist during longer periods than other types of igneous activity (cf. IV.2, IV.3, and II.3). Also tectonic and geophysical methods are used in studying location, internal structures and mode of emplacement of alkaline massifs (see for instance III.2, III.3, III.4, and IV.2).

Historical aspects are stressed in a number of the chapters of the book. The changing ideas of derivation of alkaline magmas are reviewed in Chapter VI.8.

* Readers are referred to the relevant Chapters of this book where similarly cited.

I. REFERENCES

Adams, F. D., 1903. The Monteregian Hills: A Canadian petrographical province. *J. Geol.*, **11**, 239–82.

Adams, F. D., and Barlow, A. E., 1910. Geology of the Haliburton and Bancroft areas, Province of Ontario. *Mem. Geol. Surv. Can.*, **6**.

Andreev, G. V., Scharakshinov, A. O., and Litvinovsky, B. A., 1969. *Intrusions of Nepheline Syenite of the Western Zabaikal'ye* (in Russian). Izd 'Nauka'.

Andreeva, E. D., 1968. *Alkaline Magmatism of the Kuznetsk Alatau* (in Russian). Izd 'Nauka'.

Barth, T. F. W., 1962. *Theoretical Petrology*. John Wiley and Sons, 2nd ed., 1–416.

Becke, F., 1903. Die Eruptivgebiete des böhmischen Mittelgebirges und der amerikanischen Andes. *Tschermaks miner. petrogr. Mitt. Neue Folge*, **22**.

Belijankin, D. S., 1909, 1910. Outlines of the petrography of the Ilmen Mountains (in Russian). *Izv. S-Peterb. Politechn. Inst.*, **12** and **13**.

Belijankin, D. S., 1926. On the interpretation of the Ilmen complex (in Russian). *Geol. Vestnik*, **5**.
Bilibin, Yu. A., 1941. 'Post-Jurassic intrusions of the Aldan district' (in Russian). In *Petrografiya SSSR*, **I**, 10.
Bilibin, Yu. A., 1947. Petrography of the Yllymakh Intrusion (in Russian). *Gosgeoltekhizdat*.
Bowen, N. L., 1928. *The Evolution of the Igneous Rocks*. Princeton Univ. Press, N.J., 332 pp.
Brauns, R., 1922. Die phonolitischen Gesteine des Laacher-See-Gebietes und ihre Beziehungen zu anderen Gesteinen dieses Gebietes. *Neues Jb. Miner.*, **46**, 1–116.
Brøgger, W. C., 1890. Die Mineralien der Syenitpegmatitgänge der südnorwegischen Augit- und Nephelinsyenite. *Z. Kristallogr. Miner.*, **16**, 1–663.
Brouwer, H. A., 1910. *Oorsprong en samenstellung der Transvaalsche nepheliensyeniten*. Boek- en Kuntstdrukkerij v/h Mouton & Co.'s Gravenhage, 1–180.
Cross, W., 1897. Igneous rocks of the Leucite Hills and Pilot Butte, Wyoming. *Am. J. Sci.*, **4**.
Cross, W., Iddings, J. P., Pirsson, L. V., and Washington, H. S., 1902. A quantitative chemico-mineralogical classification and nomenclature of igneous rocks. *J. Geol.*, **10**, 2, 555–690.
Czygan, W., 1969. Petrographie und Alkali-Verteilung im Foyait der Serra de Monchique, Süd-Portugal. *Neues Jb. Miner. Abh.*, **111**, 32–73.
Daly, R. A., 1903. The geology of Ascutney Mountain, Vermont. *Bull. U.S. Geol. Surv.*, **209**.
Daly, R. A., 1910. Origin of alkaline rocks. *Bull. geol. Soc. Am.*, **21**, 87–118.
Daly, R. A., 1914. *Igneous Rocks and Their Origin*. McGraw-Hill, 1–563.
Daly, R. A., 1925. The geology of Ascension Island. *Proc. Am. Acad. Arts Sci.*, **60**, 3–80.
Daly, R. A., 1927. The geology of Saint Helena Island. *Proc. Am. Acad. Arts Sci.*, **62**, 2, 31–92.
Daly, R. A., 1933. *Igneous Rocks and the Depths of the Earth*. McGraw-Hill, New York.
Dolphin Reference Book, 1962. Dolphin Books, New York.
Eliseev, N. A., 1957. On the problem of nepheline-bearing rock classification (in Russian). *Zapiski Vses. min. ob-va*, **86**, 629–31.
Ferguson, J., 1964. Geology of the Ilímaussaq alkaline intrusion, South Greenland. *Meddr. Grønland*, **172**, 4, 1–82.
Ferguson, J., 1970. The significance of the kakortokite in the evolution of the Ilímaussaq intrusion, South Greenland. *Meddr. Grønland*, **190**, 1, 1–193.
Fersman, A., 1929. Geochemische Migration der Elemente und deren wissenschaftliche und wirtschaftliche Bedeutung, erläutert an vier Mineralvorkommen: Chibina–Tundren, Smaragdgruben, Uran–grube Tuja–Mujun, Wüste Karakumy. *Abh. prakt. Geol. BergwLehre*, **18**, 1–116.
Fischer, W., 1961. *Gesteins- und Lagerstättenbildung im Wandel der wissenschaftlichen Anschauung*. E. Schweizerbart'sche Verlagsbuchhandlung, Stuttgart, 1–592.
Gilson, I. C., 1928. On the origin of the alkaline rocks. *J. Geol.*, **26**, 471–4.
Goldschmidt, V. M., 1930. Elemente und Minerale pegmatitischer Gesteine. *Nachr. Gesellsch. Wiss. Göttingen. Math. Phys. kl.*, 370–8.
Harker, A., 1896. The natural history of igneous rocks. I. Their geographical and chronological distribution. *Sci. Progress*, **6**.
Harker, A., 1909. *The Natural History of the Igneous Rocks*. Macmillan, London.
Harker, A., 1954. *Petrology for Students*. Cambridge University Press, 1–283.
Hatch, F. H., Wells, A. K., and Wells, M. K., 1961. *Petrology of the Igneous Rocks*. Thomas Murby and Co., London, 1–515.
Heinrich, E. Wm., 1966. *The Geology of Carbonatites*. Rand McNally and Company, Chicago, 1–555.
Hibsch, I.E., 1926. Erläuterungen zur geologischen Übersichtskarte des Böhmischen Mittelgebirges. *Heimatkunde des Elbegaues Tetschen*, **3**. Lieferung, 1–139.
Holmes, A., and Harwood, H. F., 1932. Petrology of the volcanic fields east and southeast of Ruwenzori, Uganda. *Q. J. geol. Soc. Lond.*, **88**, 370–442.
Iddings, J. P., 1892. The origin of igneous rocks. *Bull. phil. Soc. Washington*, **12**.
Johannsen, A., 1939. *A Descriptive Petrography of the Igneous Rocks. Vol 1: Introduction, Textures, Classifications and Glossary*. Chicago University Press, 1–318.
Jung, J., and Brousse, R., 1959. *Classification modale des roches éruptives*. Masson and Cie, Paris, 1–122.
Kraatz-Kochlau, K. von, and Hackman, V., 1897. Der Elaeolithsyenit der Serra de Monchique, seine Gang- und Contactgesteine. *Tschermaks miner. petrogr. Mitt., Neue Folge*, **16**, 197–307.
Kravchenko, S. M., and Vlasova, E. V., 1962. Alkaline rocks of the Central Aldan (in Russian). *Trudy IMGRE*, **14**.
Lacroix, A., 1890. Description des syénites néphéliniques de Pouzac (Hautes Pyrénées). *C. R. Acad. Sci. France*, **106**, 1011–13.
Lacroix, A., 1908. Le mode de formation du Puy-de-Dome et les roches qui les constituent. *C.R. Acad. Sci. France*, **147**, 826–31.
Lacroix, A., 1911. Les syénites néphéliniques de l'Archipel de Los et leurs minéraux. *Nouv. Archs. Mus. Paris*, **5**, 3, 1–132.
Lacroix, A., 1922. *Minéralogie de Madagascar. II Minéralogie Appliquée, Lithologie*. Paris, 1–694.
Lacroix, A., 1933. Contribution à la connaissance de la composition chimique et minéralogique des

roches éruptives de l'Indochine. *Bull. Serv. géol. Indochine*, **20**, 3, 1–208.

Lang, H. O., 1891. Das Mengenverhältnis von Calcium, Natrium und Kalium als Vergleichspunkt und Ordnungsmittel der Eruptivgesteine. *Bull. Soc. géol. Belge*, **5**.

Larsen, E. S., Hurlbut, C. S., jr., Griggs, D., Buie, B. F., and Burgess, C. H., 1939–41. Igneous rocks of the Highwood Mountains, Montana. *Bull. geol. Soc. Am.*, **50**, 1043–112; **52**, 1733–868,

Loewinson-Lessing, F., 1899, 1900. Kritische Beiträge zur Systematik der Eruptivgesteine. I, II, III. *Tschermaks miner. petrogr. Mitt. Neue Folge*, **18, 19**.

Michel-Lévy, A., 1890. Situation stratigraphique des régions volcaniques de l'Auvergne, la Chaine des Puys. Le Mont Dore es ses alentours. *Bull. Soc. géol. France*, 3.Série, **18**, 688–814.

Michel-Lévy, A., 1897. Note sur la classification des roches éruptives. *Bull. Soc. géol. France*, **25**, 326–77.

Molchanova, T. V., 1966. Structural position, petrology and origin of the potassium basic-alkaline rocks (in Russian). *Trudy IGN*, **159**.

Moorhouse, W. W., 1959. *The Study of Rocks in Thin Section*. Harper and Brothers, New York, 1–514.

Morozewicz, J., 1930. Der Mariupolit und seine Blutsverwandten. *Mineralog. petrogr. Mitt.*, **40**, 335–436.

Niggli, P., 1920. Systematik der Eruptivgesteine. *Zentbl. Miner. Geol. Paläont.*, 161–74.

Niggli, P., 1923. 'Gesteins- und Mineralprovinsen' I. Niggli, P. and Beger, P. J., *Einführung*. Berlin, Verlag von Gebrüder Borntraeger, 1–602.

Niggli, P., 1931. Die quantitative mineralogische Klassifikation der Eruptivgesteine. *Schweiz. miner. petrogr. Mitt.*, **11**, 296–364.

Niggli, P., 1935. Zur mineralogischen Klassifikation der Eruptivgesteine. *Schweiz. miner. petrogr. Mitt.*, **15**, 295–318.

Nockolds, S. R., 1954. Average chemical composition of some igneous rocks. *Bull. geol. Soc. Am.*, **65**, 1007–32.

Peacock, M. A., 1931. Classification of igneous rock series. *J. Geol.*, **39**, 54–67.

Perchuk, L. L., 1964. *Physico-Chemical Petrology of Granitoid and Alkaline Intrusions of the Central Turkestan-Alay* (in Russian). Izd. 'Nauka'.

Polánski, A., 1949. The alkaline rocks of the East-European Plateau. *Bull. Soc. Amis Sci. Lett. Poznán*, Serie B, **10**, 119–84.

Ramsay, W., 1890. Geologische Beobachtungen auf der Halbinsel Kola. *Fennia*, **3**, 7, 22–50.

Rittmann, A., 1952. Nomenclature of volcanic rocks proposed for the use in the catalogue of volcanoes, and keytables for the determination of volcanic rocks. *Bull. Volcan.*, **12**, 75–102.

Rittmann, A., 1960. *Vulkane und ihre Tätigkeit*. Ferdinand Enke Verlag, Stuttgart, 1–336.

Ronner, F., 1963. *Systematische Klassifikation der Massengesteine*. Springer-Verlag, Wien, 1–380.

Rosenbusch, H., 1910. *Elemente der Gesteinslehre*. 3rd. ed. Stuttgart.

Rosenbusch, H. 1923. *Elemente der Gesteinslehre*. (ed.: by A. Osann). Schweizerbart'sche Verlagsbuchhandlung, Stuttgart, 1–779.

Schinkarev, N. F., 1966. *Upper-Palaeozoic Magmatism of the Turkestan-Alay* (in Russian). Izd. Leningrad Gos. Univers.

Shand, S. J., 1906. Ueber Borolanite und die Gesteine des Croc-na-Sroine-Massivs in Nord-Schottland. *Neues Jb. Beil.Bd.* **22,** 413–52.

Shand, S. J., 1921. The igneous rocks of Sekuniland. *Trans. geol. Soc. S. Afr.*, **24,** 111–49.

Shand, S. J., 1921. The igneous complex of Leeuwfontein, Pretoria District. *Trans. geol. Soc. S.Afr.*, **24,** 232–49.

Shand, S. J., 1922. The problem of the alkaline rocks. *Proc. geol. Soc. S.Afr.*, **25,** xix-xxxiii.

Shand, S. J., 1923. The alkaline rocks of the Franspoort line, Pretoria district. *Trans. geol. Soc. S.Afr.*, **24,** 81–100.

Shand, S. J., 1928. The geology of Pilandsberg (Pilaan's Berg) in the Western Transvaal. *Trans. geol. Soc. S.Afr.*, **31,** 97–156.

Shand, S. J., 1933. Zusammensetzung und Genesis der Alkaligesteine Sudafrikas. *Mineralog. petrogr. Mitt.*, **44,** 211–16.

Shand, S. J., 1939. Loch Borolan laccolith, Northwest Scotland. *J. Geol.*, **47,** 408–20.

Shand, S. J., 1949. *Eruptive Rocks*. Thomas Murby and Co., 1–488.

Smyth, C. H., 1913. Composition of the alkaline rocks and its significance as to their origin. *Am. J. Sci.*, **36,** 1–36.

Smyth, C. H., 1927. The genesis of alkaline rocks. *Proc. Am. phil. Soc.*, **66,** 535–80.

Streckeisen, A., 1967. Classification and nomenclature of igneous rocks. *Neues Jb. Minerl. Abh.*, **107,** 144–240.

Tilley, C. E., 1957. Problems of alkali rock genesis. *Q. J. geol. Soc. Lond.*, **113,** 323–60.

Tröger, W. E., 1935. *Spezielle Petrographie der Eruptivgesteine. Ein Nomenklatur-Kompendium*. Schweizerbart'sche Verlagsbuchhandlung, Stuttgart, 1–360.

Turner, F. J., and Verhoogen, J., 1960. *Igneous and Metamorphic Petrology*. McGraw-Hill, 2nd edn. 1–694.

Tuttle, O. F. and Gittins, J., eds., 1966. *Carbonatites*. Interscience, New York—London, 591 pp.

Tyrrell, G. W., 1917. The picrite-teschenite sill of Lugar (Ayrshire). *Q. J. geol. Soc. Lond.*, **72,** 84–131.

Tyrrell, G. W., 1948. A boring through the Lugar sill. *Trans. geol. Soc. Glasgow*, **21,** 157–202.

Ussing, N.V., 1898. Mineralogisk-petrografisk Undersøgelse af grønlandske Nefelinsyeniter og beslægtede Bjergarter. *Meddr. Grønland,* **14,** 1–220.

Ussing, N. V., 1912. Geology of the country around Julianehaab, Greenland. *Meddr. Grønland,* **38,** 1–376.

Washington, H. S., 1906. *The Roman Comagmatic Region.* Carnegie Institution of Washington, Publication 57, 1–199.

Weed, W. H., and Pirsson, L. V., 1895. Highwood Mountains of Montana. *Bull. geol. Soc. Am.,* **6,** 389–422.

Williams, J. F., 1891. The igneous rocks of Arkansas. *Rep. Ark. geol. Surv.,* 1890, 1–457.

Wright, J. B., 1969. A simple alkalinity ratio and its application to questions of non-orogenic granite genesis. *Geol. Mag.,* **106,** 4, 370–84.

Zavaritsky, A. N., 1941. *Recalculation of Chemical Analyses of Igneous Rocks* (in Russian). Izd. Akad. Nauk SSSR.

Zavaritsky, A. N., 1950. *Introduction to the Petrochemistry of the Igneous Rocks* (in Russian). Izd. Akad. Nauk SSSR.

Zavaritsky, A. N., 1955. *Igneous Rocks* (in Russian). Izd. Akad. Nauk SSSR.

II. Petrography and Petrology

II.1. INTRODUCTION

H. Sørensen

As reviewed in Chapter I the term 'alkaline' has been used differently by the authorities of petrography. This is reflected in the petrographical systems which classify rocks of alkaline affinity in different, partly contradictory, ways. Thus 'theralite' and 'essexite' change position in the recent systems of Ronner (1963) and Streckeisen (1967). This confusion can in most cases be traced back to insufficient or ambiguous primary descriptions of the rocks.

Surveys of the petrographical systems have been published by Johannsen (1939), Ronner (1963) and Streckeisen (1967, pp. 207–11) and reference has been made to a number of these systems in Chapter I. A comparative and critical discussion of the systems is therefore considered passé and unnecessary. Only the petrographical skeleton of the book will be discussed here.

Some contributors to the book refer to the recent system of Streckeisen (1967) which subdivides igneous rocks with M (colour index) less than 90 according to the modal contents of their light-coloured minerals (see Fig. 1). This traditional way of classification is based on feldspar ratios and contents of quartz or foids. It does not reflect the importance of the mafic minerals.

The alkaline rocks considered in this book are first of all rocks characterized by the presence of feldspathoids and/or alkali pyroxenes or -amphiboles (see I.1.6).* Contents of alkali feldspar or mica do not according to this definition, proposed by Shand (1922), entitle rocks to be called alkaline. For this reason the modal contents and the types of mafic minerals are of the greatest systematic importance, as is emphasized by the Soviet petrographic committee (Fig. 2) and by most petrographic systems (i.e. Ronner, 1963, and Jung and Brousse, 1959). Unfortunately there is no agreement on the division of rocks according to colour index (cf. Streckeisen, 1967, p. 193).

Polanski (1949) following a proposal by Smulikowski (1934) has to the $QAPF$-diagram added an orthogonal diagram in which the An-content of the plagioclase and the total percentage of dark minerals are plotted along the ordinate axis and the ratio of $P + Q$ to $A + F$ on the abscisse axis. This method was made for the plotting of rock series but is not convenient for classificatory purposes.

* Readers are referred to the relevant Chapters of this book where similarly cited.

II.1.1. Classification

A schematic classification of the main types of alkaline rocks is presented in Table 1. The divisions used are those suggested by Streckeisen (1967) on the basis of a thorough survey of the literature. The table is not to be considered as a contribution to the standing discussion of a universal petrographic system but is only designed as a practical tool to be used in connection with this book.

Some calc-alkalic rocks commonly associated with alkaline rocks are also listed in the table as they are mentioned in the book and as they may contain subordinate amounts of feldspathoids and/or alkali pyriboles.

The following few remarks may explain the main divisions of the table:

1. The table is based on modal compositions indicated as volume percentages (cf. Streckeisen, 1967).

2. The contents of quartz and feldspathoids indicated are volume percentages of total amount of light-coloured minerals.

3. In some cases rocks containing nepheline

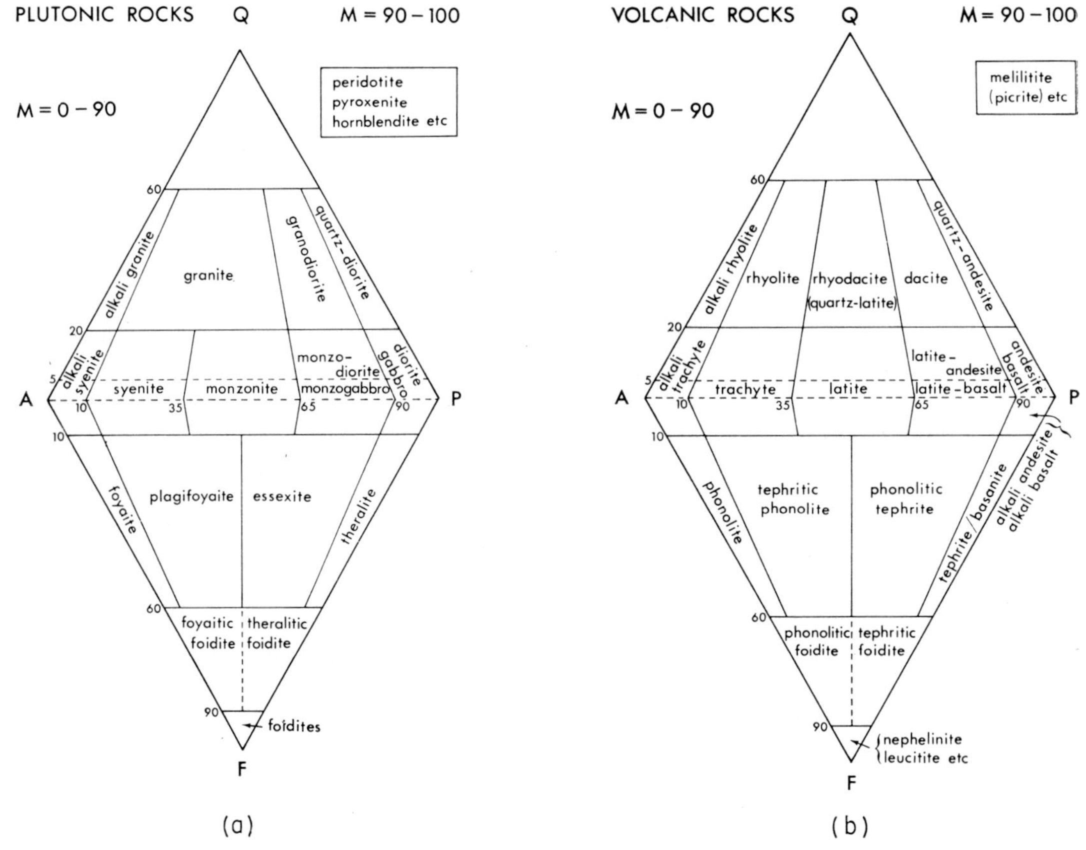

Fig. 1 The *Q–A–P–F* double-triangles of Streckeisen (1967) for plutonic and volcanic rocks. The diagrams are based on modal compositions.

Q = Quartz;
A = Alkali feldspars (including albite An $_{00-05}$);
P = Plagioglase An $_{05-100}$;
F = Feldspathoids;
M = Mafic minerals

and other Na-feldspathoids are separated from those containing K-feldspathoids including pseudoleucite.

4. In the case that several feldspathoids occur in a specific petrographic type, only one name is indicated in most cases. Thus phonolites containing nepheline, analcime, hauyne, etc. are not distinguished.

5. The feldspar ratios indicated are the fractions that plagioclase makes up of the total feldspar of the rocks (volume per cent).

6. Albite $(An)_{0-5}$ is counted as alkali feldspar; plagioclase then covers the range An_5–An_{100}.

7. The systematic importance of the anorthite contents of the plagioclase is not considered in the table.

8. The type of mafic mineral is considered, transitions between groups of rocks containing Na-poor mafic minerals and Na-rich mafic minerals are indicated.

9. Many rock names are given to a certain range of modal compositions. This is indicated in the more prominent cases (foyaite, syenite, etc.).

10. A number of the names listed in the table are obsolete. They are mentioned in order to show that rocks have been described which fall

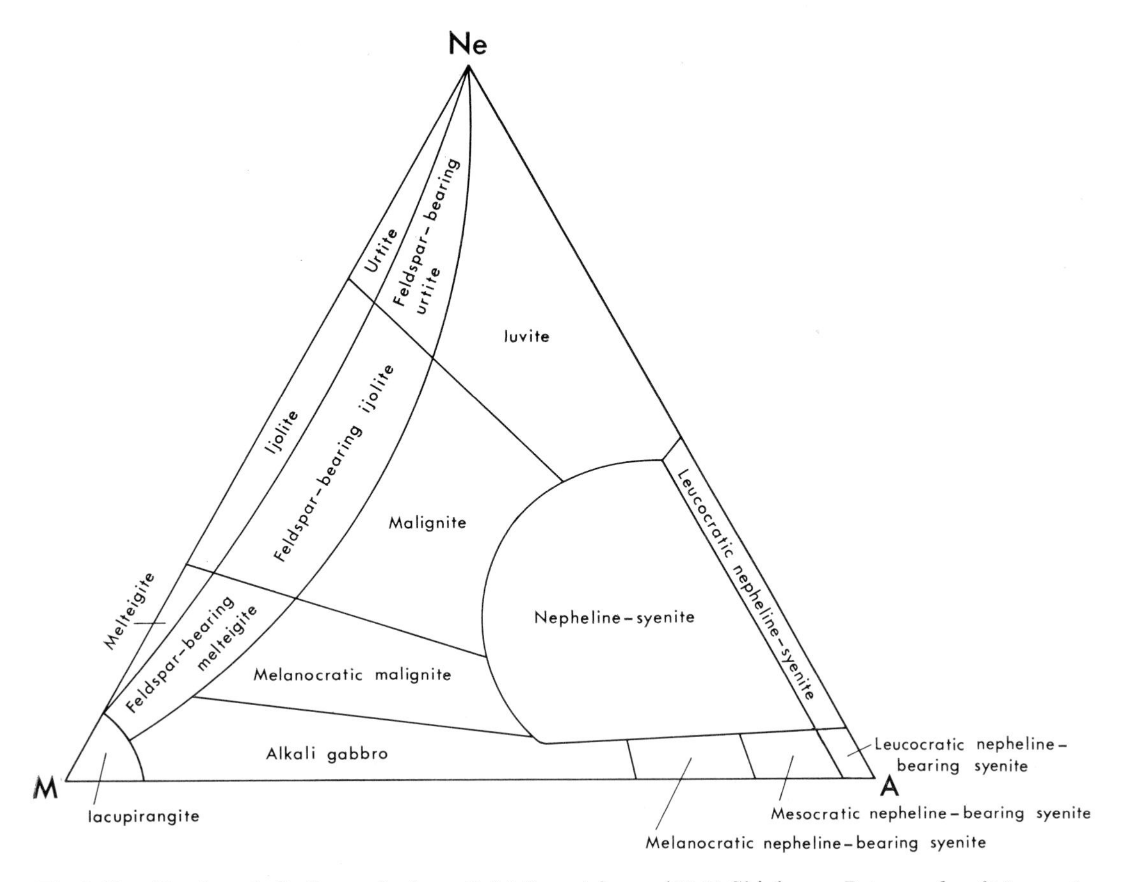

Fig. 2 Classification of alkaline rocks from G. M. Sarantsina and N. F. Shinkarev. *Petrography of Magmatic and Metamorphic Rocks* (in Russian). 'Nedra', Leningrad, 1967 (Made available by Professor A. Streckeisen, Bern)

in practically all 'pigeon holes' of the table. Most of these names should be replaced by general names like those proposed by Streckeisen (1967) or Ronner (1963). A simplified classification is presented in Table 2 (see also II.1.2).

II.1.2. Nomenclature

A restricted number of the many names given to alkaline rocks are listed in Table 1. The table shows that a wide range of alkaline rocks have been described but also that many of the fields of the table are occupied by unfamiliar names.

The reader is referred to the works of Johannsen (1939), Tröger (1935 and 1938), Jung and Brousse (1959) and Ronner (1963) for definitions of the many names attached to alkaline rocks.

Many of the rocks in question are very rare, indeed, as pointed out by Streckeisen (1967, Figs. 63 and 64). The most common alkaline rocks are alkali syenites (trachytes) and feldspathoidal syenites (phonolites), cf. II.2 and II.5; foidites (ijolites, nephelinites, etc.), cf. II.3 and II.5; alkali basaltic rocks (tephrites, essexites, etc.), cf. II.4; and alkali granites and rhyolites, cf. II.6. Gneissic varieties of alkali granites, syenites and nepheline syenites are discussed in II.7.

TABLE 1 Modal Classification

see text for explanation	>20% quartz	5–20% quartz	0–5% quartz, 0–10% foids						10–60% foids Na-foids	K-foids
feldspar ratio	<10% plagioclase				10–35% plagioclase	35–65% plagioclase	65–90% plagioclase	90–100% plagioclase		
mafics / colour index (vol. %)	Na-pyriboles								olivine augite hornblende biotite	
0–35	alkali granite ekerite *paisanite* *alkali rhyolite* *pantellerite* *comendite* grorudite	alkali quartz syenite *alkali quartz trachyte*	alkali syenite umptekite pulaskite *alkali trachyte*	------syenite------------ albitite nordmarkite bostonite ------larvikite----------- ------*trachyte*----------- ------*vulsinite*----------- ------*rhombporphyry*------	plauenite akerite *domite*	monzonite ---syenodiorite--- ---*trachyandesite*-- *latite* *doreite* *shoshonite*	monzodiorite *latite-andesite* kjelsåsite plumasite	anorthosite diorite *andesite*	nepheline diorite craigmontite raglanite *nepheline andesite*	
35–65	rockallite lindinosite		lusitanite ordosite	covite	durbachite vogesite	kentallenite *absarokite* ---syenogabbro---- ---*trachybasalt* ----	monzogabbro kauaiite *latite basalt* sörkedalite	gabbro *basalt* *hawaiite* camptonite monchiquite	theralite teschenite *tephrite* *basanite*	*leucite tephrite* *leucite basanite* *ottajanite* *vesuvite* *kivite*
65–90			shonkinite tveitåsite		shonkinite			montrealite *madeirite* ouachitite *ankaramite* *oceanite*	*atlantite*	
>90			Na-pyroxenite							

of Alkaline rocks*

10–60% foids		10–60% K-foids	10–60% Na-foids	60–90% Na-foids		60–90% K-foids		90–100% K-foids		90–100% Na-foids
50–90% plagioclase	10–50% plagioclase		0–10% plagioclase	10–40% feldspar				0–10% feldspar (absolute amount of feldspar)		
				plagioclase 0–50%	plagioclase 50–100%	plagioclase 50–100%	plagioclase 0–50%			
				Na-pyriboles				augite hornblende biotite olivine Na-pyriboles		
nepheline syenodiorite *vicoite (leucite)* *orvietite (leucite)* *ordanchite* *nepheline trachyandesite*	miaskite nepheline monzonite *nepheline latite* *latitic phonolite* *tephritic phonolite* *tristanite* *viterbite (leucite)* *tautirite*	borolanite *orendite* *leucite phonolite* *leucitophyre*	----- nepheline syenite– ----- foyaite-------- ----- juvite--------- ----- mariupolite---- ----- *phonolite*------- litchfieldite khibinite naujaite *kenyite* tinguaite lardalite ditroite canadite lujavrite	feldspar urtite *phonolitic nephelinite*		*leucitophyre*	*phonolitic leucitite*	italite *leucitophyre*	arkite *leucitophyre*	urtite congressite monmouthite
glenmuirite essexite *phonolitic tephrite* *phonolitic basanite*	nosykombite *sommaite (leucite)*	*jumillite* *wyomingite* *lamproite*	ledmorite malignite covite *murite* lujavrite		theralitic foidite *tephritic nephelinite*	*tephritic leucitite*		fergusite *leucitite* *olivine leucitite* *madupite*	*leucite-nephelinite* *niligongite*	ijolite *nephelinite* *olivine nephelinite* *ngurumanite*
	bekinkinite		----- shonkinite------		feldspar melteigite			missourite shonkinite	*ugandite*	melteigite fasinite *ankaratrite*
			Na-pyroxenite							jacupirangite

* Volcanic rocks indicated by italics.

TABLE 2 Simplified Modal Classification

	>20% quartz	5–20% quartz	0–5% quartz 0–10% foids		10–60% foids	
				Not alkaline		
Feldspar ratio		0–10% plagioclase		10–35% plagioclase	90–100% plagioclase	65–90% plagioclase
mafics / colour index		Na-pyroxene Na-amphibole			olivine augite	hornblende
0–35	alkali granite (alkf) sodic granite (ab)	alkali quartz syenite (alkf)	alkali syenite (alkf) —— pulaskite (ne) —— nordmarkite (q) —— albite syenite (ab) albitite (ab)	syenite (alkf, kf, plag) larvikite (alkf + ternf)	feldspathoidal diorite (<An_{50}) feldspathoidal leucogabbro (>An_{50})	feldspathoidal leucomonzogabbro (<An_{50}, alkf, ne) leucoessexite
	alkali rhyolite	*alkali quartz trachyte*	*alkali trachyte*	*trachyte*	*feldspathoidal andesite*	*leucotephrite*
35–65			lusitanite (alkali melasyenite)	melasyenite	feldspathoidal gabbro (>An_{50}) theralite	feldspathoidal monzogabbro essexite (<An_{50}, alkf, ne) essexite gabbro (>An_{50})
				melatrachyte	*alkali basalt (tephrite, basanite)*	*phonolitic tephrite and basanite*
65–90			shonkinite (alkali melasyenite)			
90–100			Na-pyroxenite			

of Alkaline Igneous Rocks

10–60% foids					60–90% foids		90–100% foids
					10–40% feldspar		
35–65% plagioclase	10–35% plagioclase	0–10% plagioclase			0–50% plagioclase	50–100% plagioclase	
biotite			Na-pyroxenes Na-amphiboles		augite hornblence	and/or	Na-pyroxene Na-amphibole
	FELDSPATHOIDAL SYENITES						
	Miaskitic nepheline syenites		**Intermediate ('khibinitic') nepheline syenites**	**Agpaitic nepheline syenites**			
feldspathoidal monzonite (plagimiaskite)	miaskite (alkf, plag) lardalite (alkf, ternf)	miaskite (alkf,ab) litchfieldite (alkf, ab) sodalite syenite (sod, alkf, (ne))	—foyaite (alkf)— juvite (kf) mariupolite (ab) tinguaite (alkf/ab, kf) ———	foyaite (alkf) sodalite-nepheline-syenite (alkf, sod, ne) sodalite syenite (alkf, sod)	syenitic foidite	dioritic and gabbroic leucofoidite	urtite (ne) italite (lc)
feldspathoidal latite *tephritic phonolite*	*miaskitic phonolite*		*phonolite*	*agpaitic phonolite*	*phonolitic nephelinite* (*leucitophyre*)	*tephritic leuconephelinite*	*leuconephelinite* *leucoleucitite*
	mesomiaskite	———	canadite (ab,alkf) malignite (alkf) ——— tinguaite (alkf/ab, kf) ———	lujavrite (alkf/ab, kf)	syenitic melafoidite (feldspar ijolite)	gabbroic foidite	ijolite (ne) fergusite (lc) tavite (sod)
					phonolitic melanephelinite	*tephritic nephelinite* *tephritic leucitite*	*nephelinite* *leucitite*
		———	shonkinite (alkf) ———				melteigite (ne) missourite (lc)
			Na-pyroxenite				jacupirangite (ne)

Abbreviations: ne = nepheline; lc = leucite or pseudoleucite; sod = sodalite; alkf = alkali feldspar, often microperthite; ab = albite (An_{0-5}); kf= potash feldspar; alkf/kf,ab = either alkf or kf + ab (subsolvus syenites); ternf = ternary feldspar; plag = plagioclase.

II.1. REFERENCES

Johannsen, A., 1939. *A Descriptive Petrography of the Igneous Rocks. Vol. 1: Introduction, textures, classifications and glossary*. Univ. Chicago Press, 1–318.

Jung, J., and Brousse, R., 1959. *Classification modale des roches éruptives*. Masson & Cie, Paris, 1–122.

Polanski, A., 1949. The alkaline rocks of the East-European Plateau. *Bull. Soc. Amis Sci. Lett. Poznán*. Serie B, **10,** 119–84.

Ronner, F., 1963. *Systematische Klassifikation der Massengesteine*. Springer-Verlag, Wien, 1–380.

Shand, S. J., 1922. The problem of the alkaline rocks. *Proc. geol. Soc. S.Afr.*, **XXV,** xix–xxxiii.

Smulikowski, K., 1934. Les roches éruptives des Andes de Bolivia. *Arch. Min. Soc. Warsaw*, **10,** 162–234.

Streckeisen, A., 1967. Classification and nomenclature of igneous rocks. *Neues Jb. Miner. Abh.*, **107,** 144–240.

Tröger, W. E., 1935. *Spezielle Petrographie der Eruptivgesteine. Ein Nomenklatur-Kompendium.* Schweizerbart'sche Verlagsbuchhandlung, Stuttgart, 1–360.

Tröger, W. E., 1938. Spezielle Petrographie der Eruptivgesteine. Eruptivgesteinsnamen. *Fortsch. Miner., Kristall., und Petrogr.*, **23,** 1, 41–90.

II.2. ALKALI SYENITES, FELDSPATHOIDAL SYENITES AND RELATED LAVAS

H. Sørensen

This chapter deals with the groups of alkaline rocks in which alkali feldspar is the predominant or only feldspar and which contain feldspathoids and/or soda pyriboles (cf. I* and II.1).

II.2.1. Classification

A classification of alkali syenites and feldspathoidal syenites (and their volcanic equivalents) based on modal contents of quartz, feldspathoids and mafic minerals, on feldspar ratio and on types of mafic minerals is presented in Tables 1 and 2, Chapter II.1. These rock groups have been further subdivided in a number of ways:

(*a*) According to type of mafic minerals or important accessories: augite syenite, eudialyte-nepheline syenite, etc.

(*b*) According to type of alkali feldspar: Nepheline syenites having microperthitic alkali feldspar are often termed foyaite, those having orthoclase—juvite, albite—mariupolite and albite-potash feldspar—litchfieldite.

A petrologically more satisfactory way of classifying syenitic rocks is to group them into hyper- and subsolvus types which are characterized by respectively high- and low-temperature mineral assemblages (cf. V.1.2).

(*c*) Nepheline syenites are now generally divided into agpaitic and miaskitic types. The term agpaitic was introduced by Ussing (1912) in order to characterize the peralkaline nepheline syenites of the Ilímaussaq intrusion: 'if *na*, *k* and *al* are the relative amounts of Na-, K-, and Al-atoms in the rock, the agpaites may be characterized by the equation

$$\frac{na + k}{al} \geqslant 1{\cdot}2,$$

whereas in ordinary nepheline syenites this ratio does not exceed 1·1' (Ussing, 1912, p. 341). Fersman (1929) introduced the division of nepheline syenites into the agpaitic and miaskitic groups, which are characterized by having the ratio, $(Na_2O + K_2O)/Al_2O_3$ (molecular proportions) higher or lower than one, respectively. This ratio is termed the agpaitic coefficient or index.

* Readers are referred to the relevant Chapters of this book where similarly cited.

TABLE 1 Distinction of Agpaitic and Miaskitic Nepheline Syenites

	Agpaitic Nepheline Syenites	Intermediate Types	Miaskitic Nepheline Syenites
The agpaitic coefficient:			
1. $\frac{Na_2O + K_2O}{Al_2O_3}$ (molecular proportions):			
a. Ussing (1912):	$\geqslant$ 1·2	<1·1 in other nepheline syenites	
b. Fersman (1929):	>1		<1
c. Goldschmidt (1930):	>1 (includes also silica-oversaturated rocks)		<1 (= plumasitic rocks including over-saturated types)
2. $\frac{(Na,K)_2O+(Ca,Mg,Fe)O}{(Al,Fe)_2O_3 + SiO_2}$ (Gerasimovsky, 1941, 1956)	> ca. 0·22		< ca. 0·22
3. $\frac{Na}{Al–K}$ (Zlobin, 1959) (rel. no. of atoms)	>1·1	0·85–1·1	<0·85
4. $\frac{K + Na + Ca + Sr}{Al + Fe^{3+} + Ti}$ (Ginzburg and Portnov, 1966)	No boundaries indicated, the ratio changes with crystallization of magma		
Polanski, 1949 (based on Shand, 1922)	$K_2O + Na_2O > Al_2O_3$ $K_2O + Na_2O \gtrless 1/6\ SiO_2$ (molecular proportions) also oversaturated rocks		$K_2O + Na_2O < Al_2O_3$ a. $K_2O + Na_2O > 1/6SiO_2$ (miaskitic) b. $K_2O + Na_2O < 1/6SiO_2$ (plumasitic)
Semenov, 1967 Division based on type of amphibole (and on F/Ca ratio in amphibole formula)	1. Agpaitic rocks s.str. = F-arfvedsonite or Ilímaussaq type (F/Ca∼10)	2. Mg-arfvedsonite or Lovozero type (F/Ca∼5) 3. Al-arfvedsonite or Khibina type (F/Ca ∼2) 4. Catophorite or Langesundfjord type (F/Ca∼1)	5. miaskitic rocks s.str. ∼ hastingsite or Ilmen type (F/Ca∼0·1)
Division favoured by the author			
a. chemical features: (cf. Chapter V.3)	Na + K > Al Na > K Ca, Mg low or subordinate Fe high (often $Fe_2O_3 >$ FeO) High Zr, Nb, Be, Li, Zn, RE, Th, U, Cl, F Low CO_2	Na + K $\geqslant$ Al Mixed chemistry	Na + K < Al often K > Na Ca, Mg often high lower Fe Generally low contents of rare elements and volatiles, except CO_2
b. mineralogical features: (see further table 2)	Eudialyte, F-arfvedsonite, villiaumite, chkalovite typomorphic	Mixed mineral association	Zircon, titanite, apatite, hornblende, biotite typomorphic

The agpaitic coefficient has been discussed by a number of authors, see Table 1. The characteristic features of the agpaitic rocks are, however, caused not only by the peralkalinity of these rocks, but also by other chemical parameters (cf. Tables 1 and 2) and by the conditions of crystallization (Sørensen, 1960). This is evident from the existence of nepheline syenites, for instance most tinguaites, which are peralkaline without containing characteristic agpaitic minerals. There are also *leucocratic* rocks which according to the modal compositions are agpaitic but which have agpaitic coefficients lower than one (Azambre and Girod, 1966). This is because the feldspar and nepheline have non-stoichiometric compositions resulting in lower (Na + K)/Al-ratios than indicated by the formulae, and because Na and K may be substituted by Ca, Ba, Sr, Rb and Cs (Zlobin, 1959; Ginzburg and Portnov, 1966).

Many nepheline syenites have modal compositions which are intermediate between those

TABLE 2 Mineralogical Characteristics of Nepheline Syenites (partly based on Gerasimovsky, 1956, Ginzburg and Portnov, 1966, and Semenov, 1967)

	Amphibole (Semenov, 1967)	Pyroxene Mica	Characteristic Felsic Minerals		Characteristic Rare-Metal Minerals and Accessories	
Agpaitic Nepheline Syenites	F-arfvedsonite	Aegirine/acmite Polylithionite	Nepheline Microcline Albite Sodalite (hackmanite) Analcime Natrolite Ussingite	Eudialyte Lovozerite Epistolite Steenstrupine Chkalovite Rinkite Villiaumite Sphalerite Aenigmatite		
Intermediate Types of Nepheline Syenites	Mg-arfvedsonite	Aegirine/acmite Polylithionite	Nepheline Alkali feldspar Sodalite Analcime	Eudialyte Rinkolite Catapleiite Loparite	Murmanite Epididymite Sphalerite	
	Al-arfvedsonite	Aegirine/acmite Aegirine-augite Diopside Biotite			Lamprophyllite Leucophane	
	Catophorite Barkevikite	Aegirine-augite Biotite		Eudialyte Låvenite Britholite Leucophane Chevkinite	Zircon	Apatite Titanite Ilmenite Fluorite
Miaskitic Nepheline Syenites	Hastingsite	Diopside-hedenbergite Augite Biotite	Nepheline Alkali feldspar Albite Cancrinite Plagioclase Sodalite Noseane	Magnetite Allanite Calcite Corundum Melanite Aeschynite	Pyrochlore	

characteristic for agpaitic and miaskitic rocks (cf. Gerasimovsky, 1956; Ginzburg and Portnov, 1966). Semenov (1967), emphasizing the importance of the chemical composition of the amphiboles, especially the contents of Na, Ca and F, distinguishes accordingly five types of nepheline syenites of which three are intermediate between the agpaitic and miaskitic types (cf. Table 1). The agpaitic rocks are dominated by sodium-bearing minerals: arfvedsonite, aegirine, sodalite, eudialyte, villiaumite, etc. The minerals of the miaskitic rocks have low contents of sodium and fluorine: hastingsite, aegirine-augite, biotite, zircon, titanomagnetite, titanite, etc. The intermediate types have combinations of sodium-rich and sodium-free minerals, for instance eudialyte, titanite, apatite and ilmenite in type 3 (cf. Tables 1 and 2).

II.2.2. Petrography

II.2.2.1. Introduction

Most essential and accessory minerals found in igneous rocks are present in syenites and trachytes. Only minerals requiring special conditions of formation, e.g. strong silica-undersaturation, are absent. The feldspathoidal rocks show some restrictions with regard to mineralogical composition. Orthorhombic pyroxene is absent (cf. V.1.2). Plagioclase and forsteritic olivine are rare or lacking and do not occur together with sodic pyroxenes and amphiboles. Quartz is of course lacking.

II.2.2.2. Hypersolvus Syenites and Feldspathoidal Syenites

These alkali feldspar-rich rocks are closely related. The widespread silica-saturated augite syenites (larvikites) pass into lardalitic nepheline syenites; the slightly undersaturated pulaskites into foyaitic nepheline syenites. Augite syenitic and pulaskitic rocks (alkali syenites) may both pass into quartz-bearing syenites: nordmarkites.

Augite syenites are only slightly, if at all, alkaline and pass into 'normal' calc-alkaline syenites (orthosyenites), which have sodic plagioclase, biotite, hornblende and diopsidic augite as characteristic minerals.

II.2.2.2.1. Larvikites and lardalites are hypidiomorphic or xenomorphic granular and generally of massive appearance. The larvikites of the Oslo region have two discrete feldspars, namely rhomb-shaped grains of antiperthitic oligoclase and rectangular grains of perthitic potash feldspar (Barth, 1944). Larvikite may have interstitial quartz or feldspathoids. Lardalite contains large subhedral grains of nepheline and may also contain rhomb-shaped grains of unmixed ternary feldspar.

These rocks are more basic than the average alkali syenite and nepheline syenite and are chemically first of all characterized by fairly high contents of Ca (reflected in their contents of An-bearing feldspars, augitic pyroxenes and the frequent accessories apatite and titanite) and Mg.

II.2.2.2.2. Pulaskites and foyaites are characterized by rectangular grains of alkali feldspar. In the former rock, which contains less than 5% nepheline and less than 10% mafic minerals, the feldspar tablets are densely packed. In the latter rock, which contains a higher proportion of nepheline and mafic minerals, respectively 25–50% and more than 10%, the feldspar tablets form a framework, the interstices being occupied by nepheline, mafic minerals and accessories (Fig. 1). This intergranular texture is often termed 'foyaitic'; feldspathoidal syenites exhibiting this texture are, vice

Fig.1 Foyaite. Southern part of the Ilímaussaq intrusion, South Greenland

versa, termed foyaites. The writer finds it therefore unfortunate to use 'foyaite' as the general term for feldspathoidal syenites, as for instance proposed in Streckeisen's petrographical system (cf. Streckeisen, 1964, p. 218; 1967, p. 172).

The nepheline is generally anhedral in pulaskites, while euhedral grains or clusters of grains are more common in foyaites. Sodalite is an important constituent of some foyaites and forms interstitial euhedral to anhedral grains. Noseane and hauyne are more rare in foyaitic rocks (cf. IV.4).

The khibinites of the Khibina intrusion are foyaite-type rocks of massive or laminated appearance (IV.2).

II.2.2.2.3. Quartz-bearing syenites, nordmarkites, generally have less than 5% mafic minerals. They are hypidiomorphic granular and are mainly made up of tabular alkali feldspar. The quartz is interstitial or forms crystals in miarolitic cavities.

II.2.2.2.4. Mineralogy of hypersolvus syenites. The Ca-poor alkali feldspars of hypersolvus syenitic rocks are tabular parallel to {010}. The range in bulk composition is represented by the formula $Or_{25-45}Ab_{50-70}An_{0-5}$ (wt. %). The tablets may have homogeneous cores but are generally developed as crypto- and microperthites or antiperthites, sometimes with rims of albite. Anorthoclase may be present, but orthoclase and microcline are more common. Microcline grid-twinning is, however, rarely seen and may as in Serra de Monchique (Czygan, 1969) be present only in spots, patches of varying triclinicity being found within one grain.

Hypersolvus syenites and nepheline syenites often have interstitial albite formed by unmixing of the primary high-temperature feldspar or by late crystallization, perhaps as a result of late-magmatic albitization.

The rhomb-shaped feldspars of larvikites (and rhomb porphyries) are zoned, having cores of micro- to crypto-antiperthitic oligoclase and perthitic An-poorer rims. These feldspars are interpreted as stages in the inversion and unmixing of primary high-temperature ternary feldspars (Muir and Smith, 1956; Upton, 1964).

Galakhov (1969) distinguishes two groups of rock-forming nepheline in the Khibina intrusion by means of the triangular diagram Si–Al + Fe^{3+}–Na + K + (Ca,Mg). The nephelines of the foyaites and khibinites plot close together while those of the rischorrite, urtite-ijolite, and apatite-nepheline ore form another group (cf. IV.2). Apart from this there is little information in the literature about variation in chemistry of nepheline within intrusions (see further II.2.7).

The nephelines (and also the feldspar and primary sodalite) of agpaitic rocks are often crowded with microlites of aegirine (or arfvedsonite). These microlites are interpreted as either products of unmixing of originally iron-rich high-temperature minerals, or as crystal embryos caught by crystals floating in a consolidating magma. Galakhov (1969) has pointed out that the nepheline of the leucocratic trachytoid khibinite of Khibina crystallized earlier than the mafic minerals, but contains numerous aegirine microlites. In the melanocratic khibinite the nepheline is later than the mafic minerals and free of aegirine microlites.

Nepheline, primary sodalite and feldspar are in agpaitic rocks often substituted by late sodalite, analcime, natrolite or ussingite; in miaskitic rocks by cancrinite, white mica, analcime and natrolite. These minerals also fill the interstices between the early minerals.

Augite and ferroaugite (sometimes titanaugite) are the pyroxenes of larvikitic syenites, while salite-hedenbergite, aegirine-augite and aegirine, often in zoned grains, are predominant in alkali syenites. Aegirine-augite and aegirine (acmite when brown in thin section) are the most common pyroxenes of feldspathoidal syenites (especially agpaitic types), but augite, ferroaugite and salite-hedenbergite are found in miaskitic types. Aegirine is in agpaitic rocks often formed during deuteric or post-magmatic processes and may then form radiating groups of long thin needles or felty crusts and interstitial masses.

Augite syenites and lardalites have kaersutitic or barkevikitic (ferrohastingsitic) amphiboles which are respectively rich and poor in Ti. Pulaskites, foyaites and nordmarkites have catophorite and barkevikite in the least alkaline types, Mg-Al-arfvedsonitic amphibole in the more alkaline types, and F-arfvedsonite in agpaitic types (Semenov, 1967). Amphiboles are also deuteric products in syenitic rocks and form mantles around early olivine, pyroxene and iron ore.

Olivine and biotite in syenites and miaskitic and intermediate nepheline syenites are practically always rich in iron being represented by almost pure fayalite and lepidomelane-annite.

The typical accessories are zircon, apatite, titanite and titanomagnetite. Fluorite is commonly present and may in cases be a primary mineral. More sodic types have aenigmatite, astrophyllite, låvenite, pyrochlore and melanite; agpaitic types eudialyte-eucolite, rinkite-mosandrite, villiaumite, etc. (cf. Table 2).

II.2.2.3. Shonkinites and Malignites

The shonkinites of the Highwood Mountains, Montana (Hurlbut and Griggs, 1939; Nash and Wilkinson, 1970) consist of euhedral Ca-rich augite (salite), olivine (Fa_{22-40}), biotite, apatite and titanomagnetite with interstitial sanidine, aegirine-augite, and zeolites. The zeolites are probably secondary after nepheline. The colour index is 40–65. The type shonkinite is thus a melanocratic syenite with alkaline affinities. The term now covers silica-saturated and undersaturated melanocratic syenites of calc-alkalic as well as of alkalic type.

Malignite is the term for nepheline syenites containing more than about 50% mafic minerals.

II.2.2.4. Lujavrites and Tinguaites

Lujavrites and tinguaites have 20–50% mafic minerals. Lujavrites are fine- to course-grained rocks which are characterized by a pronounced feldspar lamination. They contain laths of perthite or in the case of the lujavrites from Ilímaussaq, South Greenland (Ussing, 1912) separate laths of microcline and albite. The needles or prisms of aegirine and arfvedsonite, stout crystals of nepheline and platy crystals of eudialyte are oriented with their longest directions in the plane of lamination in such a way that the nepheline crystals, and sometimes crystals of sodalite, are wrapped by the other minerals. Arfvedsonite and aegirine also form felty interstitial aggregates or large poikilitic anhedra. Analcime and natrolite are widespread matrix minerals.

The microclines of lujavrites and other agpaitic nepheline syenites do not show cross-hatching, but an intricate penetration twinning according to the albite law (Fig. 2). They are practically maximum microcline and non-perthitic varieties are chemically almost pure microcline (cf. Ussing, 1898; Sørensen, 1962, p. 224, Semenov, 1969, p. 126, Vlasov *et al.*, 1959, p. 248). Lujavrites are rich in rare minerals, first of all eudialyte, which is an essential mineral.

Tinguaites are fine-grained or dense, sometimes porphyritic, rocks characterized by a matrix rich in aegirine, either as parallel needles or as felty interstitial aggregates. Alkali feldspar, nepheline, leucite, pseudoleucite, sodalite, cancrinite, analcime, aegirine or aegirine-augite, and biotite form phenocrysts. The groundmass is, in addition to aegirine and the phenocrystal minerals, composed of soda amphibole, natrolite, apatite, zircon, titanite, melanite, aenigmatite, låvenite, mosandrite and pectolite. The alkali feldspar varies from sanidine to microcline + low albite ($Or_{88}Ab_{12}$ and $Ab_{95}An_5$ in natrolite tinguaite from Toror Hills, Hytönen, 1959).

Tinguaites may be considered as dyke (or low pressure) equivalents of lujavrites but are of more widespread occurrence. They differ from lujavrites in only rarely containing complex minerals such as eudialyte (Azambre and Girod, 1966).

II.2.2.5. Sodalite Syenites and Sodalite-nepheline Syenites

Primary sodalite is described from a variety of syenitic rocks. Examples are the sodalite syenite of the upper part of the Square Butte laccolith, Montana (Hurlbut and Griggs, 1939) and the zone of sodalite foyaite of the upper part of the Ilímaussaq intrusion (Sørensen, 1969), which both have sodalite interstitial to the framework of alkali feldspar. The sodalite syenite of Square Butte contains barkevikitic hornblende and grades downwards into shonkinite containing augite, olivine and biotite. The Ilímaussaq sodalite foyaite grades upwards into sodalite-free foyaite. In both examples the sodalite-bearing syenites appear to be products of more or less *in situ* crystallization of magma progressively enriched in Cl, SO_3 and Na (cf. VI.4).

The sodalite foyaite of Ilímaussaq is underlain by poikilitic sodalite-nepheline syenite (naujaite, Ussing, 1912) which contains up to 70% sodalite (hackmanite). This rock, in which small crystals of sodalite are enclosed in coarse interlocking grains of alkali feldspar, aegirine, arfvedsonite, eudialyte and aenigmatite (Fig. 3), is interpreted by all investigators as a flotation cumulate (cf. IV.3; Sørensen, 1969), crystals of sodalite having floated in a volatile-rich magma. Similar rocks occur in the Lovozero intrusion (cf. IV.2; VI.5).

The sodalite syenites of Bohemia (Hibsch, 1902, 1926) occur in stocks, laccoliths and dykes and are composed of alkali feldspar, diopside, aegirine-augite, barkevikite, magnetite, titanite and apatite. Sodalite, and in cases also hauyne, occur as large crystals and also form small crystals enclosed in the alkali feldspar. These rocks correspond chemically to the widespread essexites of this province apart from their higher contents of volatiles. The early crystallization of sodalite in the volatile-rich magma appears thus to have had a pronounced influence on the crystallization.

Fig. 2 Microcline from agpaitic vein, Ilímaussaq. × 30, + nic. In the lower half of the photo the characteristic albite twinning is distinct, in the upper part the twinning is 'gritty'

Sodalite syenites are often termed ditroites, but this is unfortunate, because the sodalite of the Ditró rocks occupies fractures and zones of crushing.

The primary, grey, green or black sodalite of sodalite syenites is often distinguished by its content of microlites of aegirine, arfvedsonite or ore dust, while late, often blue, sodalite is clear and free from such inclusions.

II.2.2.6. Albite-Rich Nepheline Syenites and Nepheline Gneisses

Albite-rich nepheline syenites are described under the names: litchfieldite (potash feldspar-albite) canadite (albite-potash feldspar, normative anorthite) and mariupolite (albite). To these may be added miaskites s.str., which are leucocratic biotite-nepheline syenites containing albite or oligoclase. These rocks are often of gneissic appearance (cf. II.7).

Fig. 3 Poikilitic sodalite syenite (naujaite), the Ilímaussaq intrusion. The match is 4·5 cm long. Small crystals of sodalite (grey) are enclosed in anhedra of microcline (white), arfvedsonite or aegirine (black) and eudialyte (grey)

Some of these rocks have laths of perthitic alkali feldspar set in a granular matrix of albite and cross-hatched microcline (cf. Barker, 1965; Streckeisen, 1954; Appleyard, 1967, 1969) and there may also be grains of oligoclase. Other rocks have only albite and microcline. The nepheline forms phenocrysts (porphyroblasts?) or is part of the matrix. It often contains inclusions of feldspar, pyroxene, etc. indicating a late origin. The dominant mafic minerals are biotite, hastingsitic or barkevikitic hornblende, diopside-hedenbergite or aegirine-augite and magnetite or titanomagnetite. The most common accessories are titanite, zircon, calcite, cancrinite, apatite, fluorite, corundum, scapolite, vesuvianite, muscovite, zeolites and pyrochlore. Blue sodalite is widespread on fractures.

Nepheline gneisses are interpreted as protoclastically foliated igneous rocks (Ditró, Roumania, Streckeisen, 1960); as forcefully injected nearly crystalline feldspar mush lubricated by thinly fluid nepheline syenite melt (Litchfield, Maine, Barker, 1965); as products of nephelinization (Haliburton–Bancroft, Ontario, Tilley, 1957; Appleyard, 1967, 1969; Stjernöy, Northern Norway, Sturt and Ramsay, 1965); as metasomatic or migmatitic covers of anatectic diapirs (Ditró, Codarcea *et al.*, 1958, and Ianovichi *et al.*, 1968); or as deformed and recrystallized igneous rocks (Malawi, Bloomfield, 1970; Vishnevogorsk, V. P. Volkov, personal information), see also II.7.

In addition to the above-mentioned nepheline gneisses, the origin of which is under discussion, there are gneissic syenites and nepheline syenites located within undeformed igneous rocks. Examples are the 'ditroites' of the Oslo region (Brøgger, 1890) and the gneissoid rischorrites of Khibina (Galakhov, 1959). In the former example the 'ditroites' may represent metasomatically altered zones of deformation in larvikite (Oftedahl, 1960, p. 331). The gneissoid rischorrite may have been protoclastically deformed during emplacement of the rischorrite ring (cf. IV.2).

II.2.2.7. Nepheline Monzonites and -Diorites

Nepheline monzonites and -diorites (in part plagifoyaite and essexite) are described from a number of provinces, but play a very subordinate rôle.

Parts of the nepheline gneisses of the Haliburton–Bancroft belt have plagioclase as the only feldspar. These dioritic rocks are associated in the field and are described under the names dungannonite, raglanite and craigmontite (Adams and Barlow, 1910).

Nepheline monzonite and diorite are, for instance, described from the Monteregian province (IV.6), from northern Madagascar (Lacroix, 1922) and Pouzac in the Pyrenees

(Azambre, 1967). In these localities the rocks are made up of oligoclase-andesine, nepheline, potash feldspar, titanaugite, barkevikitic or kaersutitic hornblende and titanomagnetite. Biotite, sodalite, titanite, olivine (rich in forsterite) and apatite also occur.

II.2.2.8. Trachytes and Phonolites

The terms trachyte and phonolite are used here in the sense of Rosenbusch (cf. Streckeisen, 1967) to distinguish alkali feldspathic lavas containing less than, respectively more than 10 vol. % feldspathoids. The term phonolite designates rocks having nepheline as the only feldspathoid. Sodalite phonolite contains one, sodalite-nepheline phonolite two feldspathoids. Trachytes and phonolites have in most cases less than 10–15% mafic minerals.

Most trachytes and phonolites are dense (aphanitic) holocrystalline rocks, often porphyritic. Glass and vesicles are generally of subordinate importance, but characterize some occurrences, e.g. the froth flows of Kenya (see II.2.4). The groundmass is dominated by alkali feldspar which may form slender laths or microlites in parallel orientation, giving a trachytic texture, or form more stout laths giving an orthophyric texture.

Trachytes vary from quartz- to feldspathoid-bearing types and from 'normal' or calc-alkalic trachyte to alkaline types. Phonolites vary from miaskitic to agpaitic types.

'Normal' trachytes and miaskitic phonolites show similarities on a number of points. They have phenocrysts of alkali feldspar (and sometimes oligoclase-andesine), clinopyroxenes (augite, in part titaniferous, diopside-hedenbergite, sometimes with rims of sodic pyroxene), hornblende (kaersutite or oxykaersutite, barkevikite or oxyhornblende), iron-rich biotite and iron-rich olivine. The biotite and hornblende phenocrysts are commonly partly or completely resorbed and have mantles of clinopyroxene and iron ore ('opacite'). The groundmass is made up of the same minerals, but biotite and hornblende are rare.

In normal trachytes the clinopyroxene phenocrysts may have cores of orthopyroxene and there may be interstitial quartz or feldspathoids in the matrix.

Miaskitic phonolites contain, in addition to the above-mentioned minerals, nepheline, sodalite, noseane, hauyne, analcime, leucite and pseudoleucite.

Alkali trachytes and agpaitic phonolites have phenocrysts of alkali feldspar, aerigine-augite to aegirine and soda amphibole (arfvedsonite, riebeckite, catophorite, barkevikite). Fayalite and iron-rich biotite may also occur in alkali trachytes, while agpaitic phonolites have nepheline, sodalite and analcime.

The groundmasses of alkali trachytes and agpaitic phonolites are made up of the above-mentioned minerals. The soda pyroxenes and -amphiboles occur as interstitial needles, as felty aggregates and as pseudopoikilitic masses enclosing feldspar, etc. Amphiboles are more widespread than in normal trachytes and miaskitic phonolites.

The alkali feldspars of trachytes and phonolites are homogeneous or cryptoperthitic sanidine, soda sanidine and anorthoclase, generally in the range $Or_{20-40}Ab_{55-80}An_{0-5}$ (wt. %). The sanidine phenocrysts may be enriched in barium as is the case in the melaphonolites of the Highwood Mountains (Larsen, 1941).

The miaskitic phonolites of Cantal, Auvergne have phenocrysts of noseane and sodalite (colourless in thin section), while the agpaitic ones have black sodalite and analcime, but no noseane (Varet, 1969a). The content of iron in the sodalite-group minerals decreases from phenocrysts to the groundmass minerals in the miaskitic types, while the opposite relation is seen in agpaitic phonolites. This may be caused by the different order of crystallization in these two groups, iron minerals being early in miaskites, but late in agpaites (Brousse, *et al.*, 1969; Varet, 1969a).

Yagi (1966) has demonstrated that the clinopyroxenes of trachytes and phonolites, which have crystallized at low partial oxygen pressures, are low in the acmite component. The agpaitic phonolites of Cantal have, however, acmite-rich pyroxenes (Varet, 1969b). The groundmass pyroxenes of trachytes and phonolites are richer in acmite than the phenocrysts (Hytönen, 1959; Nash *et al.*, 1969; Tyler and King, 1967).

The most common accessories of trachytes and miaskitic phonolites are: titanite, titano-magnetite, zircon and apatite. In miaskitic phonolites melanite and aenigmatite are quite widespread, cancrinite is described from few occurrences. Pectolite and wollastonite have been found in some regions, the latter may be due to assimilation of calcareous rocks. One or more minerals such as eudialyte, mosandrite, låvenite and perhaps rosenbuschite have been found in agpaitic types of phonolites (Azambre and Girod, 1966; Varet, 1969a).

II.2.2.9. Rhomb Porphyries and Domites

Rhomb porphyries and domites are examples of oligoclase-bearing trachytes which should most properly be named latite or trachyandesite.

The rhomb porphyries from the type region around Oslo (Oftedahl, 1960) are regarded as the effusive equivalents of larvikite. The rhomb-shaped phenocrysts have cores of micro- to crypto-antiperthite ($Or_{10-15}Ab_{45-75}An_{10-45}$) mantled by An-poor alkali feldspar. The cores of the phenocrysts are structurally disordered or intermediate and partly unmixed, the matrix feldspar is Ca-poor, strongly unmixed and structurally ordered or intermediate (Harnik, 1969). Some trachytes have two coexisting ternary feldspars—soda-sanidine and potash-oligoclase (Carmichael, 1965).

Domite is an oligoclase trachyte rich in biotite and with minor tridymite.

II.2.3. Petrochemistry

Syenites and trachytes do not deviate much from the average igneous rock (Clarke and Washington, 1924) with regard to SiO_2, TiO_2, Fe_2O_3, FeO and MnO, but have generally higher contents of Al_2O_3, Na_2O and K_2O and lower contents of MgO and CaO (Table 3).

Nepheline syenites and phonolites are, when compared with the average igneous rock, characterized by lower amounts of SiO_2, MgO and CaO; and higher amounts of Al_2O_3, Na_2O and K_2O. FeO, Fe_2O_3 and TiO_2 vary within wide limits, partly depending on the type of nepheline syenite (cf. V.3).

The trace elements of syenitic and felds-pathoidal syenitic rocks are treated in Chapter V.3. Here it should only be pointed out that some varieties of miaskitic rocks are extremely poor in rare elements (Heier, 1964, 1965; Semenov, 1967). These rocks may have high contents of Ba and Sr, and low contents of Rb and Cs in spite of high contents of Na and K (Heier and Taylor, 1964). This depletion of Rb and Cs may be a result of escape of the gas phase in equilibrium with the crystallizing magma (Heier, 1964, p. 213). In this connection it is worth emphasizing that rare elements are especially concentrated in nepheline syenites rich in Na, K, and volatiles, especially F.

Agpaitic rocks are characterized by higher than average contents of Cl, F and S, while CO_2 appears to be a minor component (see VI.4). The gas phase of high-temperature gas–liquid inclusions in nepheline, also of agpaitic rocks, are however, dominated by CO_2 (see V.2). Gas inclusions along healed fractures in nepheline and sodalite from a number of agpaitic, intermediate and miaskitic occurrences are dominated by hydrocarbons (Petersilie, 1963; Petersilie *et al.*, 1965; Ikorskii, 1967). The carbon of these hydrocarbon gases is enriched in C^{13} and supposed to be of inorganic origin (Petersilie and Sørensen, 1970).

II.2.4. Mode of Occurrence of Trachytes and Phonolites

Magmas of trachytic and phonolitic composition are characterized by rather high to moderate contents of silica and alumina, but are, on the other hand, often rich in alkalis and volatiles. This is clearly displayed in the contrasted mode of occurrence of rocks formed from such magmas.

Trachytes and phonolites as those of Auvergne (IV.4) and oceanic islands (IV.7) form short thick flows or domes (IV.4, Figs. 3, 4). The domes have 'endogeneous' structures: they have grown by consolidation of very viscous material in and immediately around the orifice.

TABLE 3 Chemical Composition of Selected Syenites, Nepheline Syenites, Trachytes and Phonolites

	Average Igneous Rock (Clarke and Washington, 1924)	Larvikite, Oslo (Oftedahl, 1960)	Average Syenite, Gardar Province (Watt, 1966)	Average (Alkali) Syenite (Nockolds, 1954)	Average Nepheline Syenite (Gerasimovsky, 1963)	Average Foyaite, Khibina (Galakhov, 1967)	Mafic Phonolite (Lower Shonkinite Chill) Shonkin Sag, Montana Nash and Wilkinson, 1970)	Average, Aegirine Lujavrite, Ilímaussaq (Gerasimovsky, 1969)
SiO_2	59·12	57·80	58·26	61·86	53·34	55·07	47·00	53·15
TiO_2	1·05	1·15	1·31	0·58	0·83	0·94	0·80	0·28
ZrO_2					0·30	0·04		1·12
Al_2O_3	15·34	18·82	16·45	16·91	20·10	21·66	12·91	16·03
Fe_2O_3	3·08	1·60	1·52	2·32	3·58	2·53	1·30	9·39
FeO	3·80	3·50	5·70	2·63	2·34	1·36	7·20	1·18
MnO	0·12	0·14	0·15	0·11	0·23	0·15	0·15	0·23
MgO	3·49	1·48	0·87	0·96	0·78	0·64	7·75	0·41
CaO	5·08	3·72	3·48	2·54	2·40	1·35	9·70	0·72
Na_2O	3·84	6·48	5·52	5·46	8·46	9·13	1·85	11·13
K_2O	3·13	3·97	5·01	5·91	5·77	5·50	6·45	3·45
H_2O+	1·15	0·64	0·49	0·53	} 1·22	0·55	1·74	} 3·15
H_2O-		0·02				0·21	0·40	
P_2O_5	0·30	0·55	0·37	0·19	0·25	0·27	0·91	0·07
CO_2	0·10	0·10	0·87			0·11	1·01	
Cl		0·05			0·13	0·07		0·07
F		0·04			0·12	0·07		0·05
S		0·03			0·22		0·06	
SO_3						0·15		0·21
No. of Analyses	5159	1	13	25	285	16	1	6

C.I.P.W. Weight Norms

q	10·18	0·00	0·00	1·85	0·00	0·00	0·00	0·00
c	0·00	0·00	0·00	0·00	0·00	0·00	0·00	0·00
z	0·00	0·00	0·00	0·00	0·45	0·06	0·00	1·67
or	18·51	23·48	29·63	34·96	34·13	32·53	29·03	20·41
ab	32·45	49·23	46·65	46·15	25·42	31·95	0·00	22·12
an	15·33	10·70	5·27	4·13	0·30	2·63	7·82	0·00
lc	0·00	0·00	0·00	0·00	0·00	0·00	7·15	0·00
ne	0·00	2·80	0·00	0·00	24·45	23·69	8·47	22·22
hl	0·00	0·08	0·00	0·00	0·22	0·12	0·00	0·12
th	0·00	0·00	0·00	0·00	0·00	0·27	0·00	0·37
ac	0·00	0·00	0·00	0·00	0·00	0·00	0·00	25·61
ns	0·00	0·00	0·00	0·00	0·00	0·00	0·00	0·00
di	5·83	2·92	3·52	5·94	4·24	0·99	22·72	2·33
wo	0·00	0·00	0·00	0·00	1·60	0·00	0·00	0·00
hy	8·63	0·00	4·79	1·43	0·00	0·00	0·00	0·00
ol	0·00	3·96	1·98	0·00	0·00	0·80	13·81	0·84
mt	4·47	2·32	2·20	3·36	4·34	1·66	1·88	0·75
il	1·99	2·18	2·49	1·10	1·58	1·79	1·52	0·53
hm	0·00	0·00	0·00	0·00	0·59	1·39	0·00	0·00
ap	0·71	2·01	0·88	0·45	0·59	0·64	2·15	0·17
fr	0·00	0·00	0·00	0·00	0·20	0·09	0·00	0·09
pr	0·00	0·06	0·00	0·00	0·41	0·00	0·11	0·00
cc	0·24	0·24	2·10	0·00	0·00	0·26	2·43	0·00

	Average Naujaite, Ilímaussaq (Gerasimovsky, 1969)	Miaskite, Vishnevogorsk (Ronenson, 1966)	Nepheline Syenite, Stjernöy (Heier, 1961)	Average (Alkali) Trachyte (Nockolds, 1954)	Average Phonolite (Nockolds, 1954)	Average Miaskitic Phonolite, Cantal (Varet, 1969 a)	Average Agpaitic Phonolite; Cantal (Varet, 1969 a)	Sodalite Phonolite, Mt. Suswa (Nash *et al.*,) 1969)	Sodalite Phonolite, Mt. Suswa, (Residual Glass) Nash *et al.*, 1969)
SiO_2	46·82	56·96	52·73	61·95	56·90	57·76	58·53	57·58	57·1
TiO_2	0·30	0·42	0·51	0·73	0·59	0·16	0·11	1·02	1·1
ZrO_2	0·41							0·09	
Al_2O_3	22·42	22·29	23·71	18·03	20·17	18·15	17·34	16·83	17·7
Fe_2O_3	3·00	1·27	1·89	2·33	2·26	3·93	3·35	3·25	3·2
FeO	2·10	1·77	1·04	1·51	1·85	0·28	0·12	4·39	5·2
MnO	0·13	0·08	0·06	0·13	0·19	0·09	0·11	0·33	0·4
MgO	0·16	0·51	0·24	0·63	0·58	0·12	0·11	0·74	0·6
CaO	1·24	1·41	2·54	1·89	1·88	3·67	2·94	2·19	0·9
Na_2O	15·93	6·52	7·78	6·55	8·72	7·22	8·53	7·86	8·6
K_2O	3·61	7·07	8·08	5·53	5·42	5·36	5·63	4·95	5·5
H_2O +	} 1·52	} 0·66	0·26	0·54	0·96	1·35	1·51	0·21	
H_2O −			0·05			0·95	0·39	0·13	
P_2O_5	0·03	0·09	0·05	0·18	0·17	0·27	0·06	0·25	
CO_2		0·66	0·77					0·13	
Cl	2·90				0·23	0·28[x)]	0·33[xx)]	0·16	
F	0·29								
S									
SO_3	0·15				0·13	0·99[x)]	1·22[xx)]		
No. of Analyses	8	9	1	15	47	7([x)]3)	4([xx)]2)	1	1

	Average Naujaite	Miaskite	Nepheline Syenite	Average (Alkali) Trachyte	Average Phonolite	Average Miaskitic Phonolite	Average Agpaitic Phonolite	Sodalite Phonolite	Sodalite Phonolite (Residual Glass)
q	0·00	0·00	0·00	0·00	0·00	0·00	0·00	0·00	0·00
c	0·00	3·07	0·00	0·00	0·00	0·00	0·00	0·00	0·00
z	0·61	0·00	0·00	0·00	0·00	0·00	0·00	0·13	0·00
or	21·35	41·82	47·79	32·71	32·06	31·70	33·30	29·28	32·53
ab	7·83	34·00	7·68	53·80	36·45	44·19	40·88	38·64	29·61
an	0·00	2·24	5·84	3·42	1·21	5·79	0·00	0·00	0·00
lc	0·00	0,00	0·00	0·00	0·00	0·00	0·00	0·00	0·00
ne	47·27	11·44	31·48	0·85	18·79	4·47	9·12	10·96	16·62
hl	4·81	0·00	0·00	0·00	0·38	0·46	0·55	0·27	0·00
th	0·27	0·00	0·00	0·00	0·23	1·76	2·17	0·00	0·00
ac	8·66	0·00	0·00	0·00	0·00	0·00	3·47	5·61	9·24
ns	1·64	0·00	0·00	0·00	0·00	0·00	0·00	0·00	0·45
di	3·38	0·00	1·23	3·40	4·19	0·66	0·61	7·20	3·88
wo	0·00	0·00	0·00	0·17	0·76	4·10	5·61	0·00	0·00
hy	0·00	0·00	0·00	0·00	0·00	0·00	0·00	0·00	0·00
ol	1·52	2·05	0·02	0·00	0·00	0·00	0·00	2·60	5·48
mt	0·00	1·84	1·87	2·75	3·28	0·44	0·07	1·89	0·00
il	0·57	0·80	0·97	1·39	1·12	0·30	0·21	1·94	2·09
hm	0·00	0·00	0·60	0·43	0·00	3·63	2·10	0·00	0·00
ap	0·07	0·21	0·12	0·43	0·40	0·64	0·14	0·59	0·00
fr	0·59	0·00	0·00	0·00	0·00	0·00	0·00	0·00	0·00
pr	0·00	0·00	0·00	0·00	0·00	0·00	0·00	0·00	0·00
cc	0·00	1·59	1·86	0·00	0·00	0·00	0·00	0·31	0·00

Bubble vesicles are of subordinate importance, flow structures may be present, as for instance at Drachenfels (H. and E. Cloos, 1927). Many effusions of trachyte and phonolite are accompanied by pyroclastic material and tuff pipes.

In Cantal, France, flows of agpaitic phonolites are thinner than flows of miaskitic composition, indicating a less viscous agpaitic magma (Varet, 1969a).

Flood lavas of trachyte and phonolite in Kenya, which reach lengths of more than a hundred kilometres, may be erupted from fissures related to the early Rift faulting (Wright, 1965).

Mount Suswa, Kenya is an example of a shield volcano composed of sodalite phonolite (Johnson, 1969). The area covered is 270 km^2, the slope of the volcano forms an angle of 3°.

MacCall (1964) and Locardi (1965) have described extensive fragmental effusions of respectively sodic trachytes and phonolites from Kenya and leucite-bearing trachytes from the Roman volcanic province. Welding and vitroclastic structures are absent, or less conspicuous than in rhyolitic ignimbrites. Both authors consider surface vesiculation of magma erupted with irregularly distributed dissolved gases as the cause of the 'frothing' and this type of flows is termed 'froth flows' by MacCall, or foam lavas.

The froth flows of Kenya often have vesicular crystalline sodalite-containing patches enclosed in glassy, less vesicular material without sodalite. This shows that crystals and gas bubbles grew most easily in the least viscous, volatile-rich parts of the flows. In the Highwood Mountains, Montana, thin flows of phonolite are similarly vesicular, while thick flows are dense (Larsen, 1941).

Ignimbritic trachytes and phonolites of Gran Canaria (Schmincke, 1970) have hydroxyl-bearing phenocrysts (amphiboles and biotite) while the 'normal' lavas have hydroxyl-free phenocrysts.

This discussion shows that nearly identical alkaline magmas in some cases loose their gas phase explosively, while in other cases they retain the gases during eruption so that volatile-bearing minerals may crystallize, high mobility be obtained and agpaitic phonolites be formed. The causes of this retention of volatiles may be sought in different temperatures of the erupting magmas, in slight chemical differences, in differences in contents and composition of volatiles, in volume of magma, in the proportion of gases, liquid and crystals, and in swiftness of eruption (cf. Tazieff, 1969, and VI.4).

II.2.5. Types of Intrusions

The size of syenitic bodies varies from less than 1 km^2 to more than 1500 km^2, the majority of the masses being small in size. The bodies are generally discordant; concordant relations are shown by gneissic syenites and nepheline syenites—and by sills and laccoliths.

Evidence of forceful intrusion is rarely noted around syenitic intrusions; the litchfieldite at Litchfield, Maine may be an example (Barker, 1965). Most discordant bodies of syenitic rocks may consequently be classified as permissive intrusions. They have generally rounded or elliptical outlines and vertical or steep contacts.

1. Sills and laccoliths: Syenitic and feldspathoidal syenitic lenses are common in differentiated alkali basalt sills (II.4). Laccoliths of trachyte and phonolite occur for instance in northern Bohemia; they consolidated either near the surface or pierced to the surface (Kopecký, 1966, p. 575). The Shonkin Sag laccolith, the Highwood Mountains, Montana (Hurlbut and Griggs, 1939), provides an excellent example of *in situ* differentiation being composed of shonkinite and residual alkali syenite (Fig. 4).

2. Dykes of microsyenite, tinguaite, etc. are abundant in and around syenite and feldspathoidal syenite intrusions. The dykes are generally thin and of limited extent. They are nevertheless of great petrological interest as they in many cases are products of *in situ* crystallization and thus provide direct information about the primary magmas (cf. VI.2).

Giant dykes having marginal zones of alkali gabbro and central zones of over- or undersaturated syenites occur in the Gardar Province (IV.3, Bridgwater and Harry, 1968). They appear

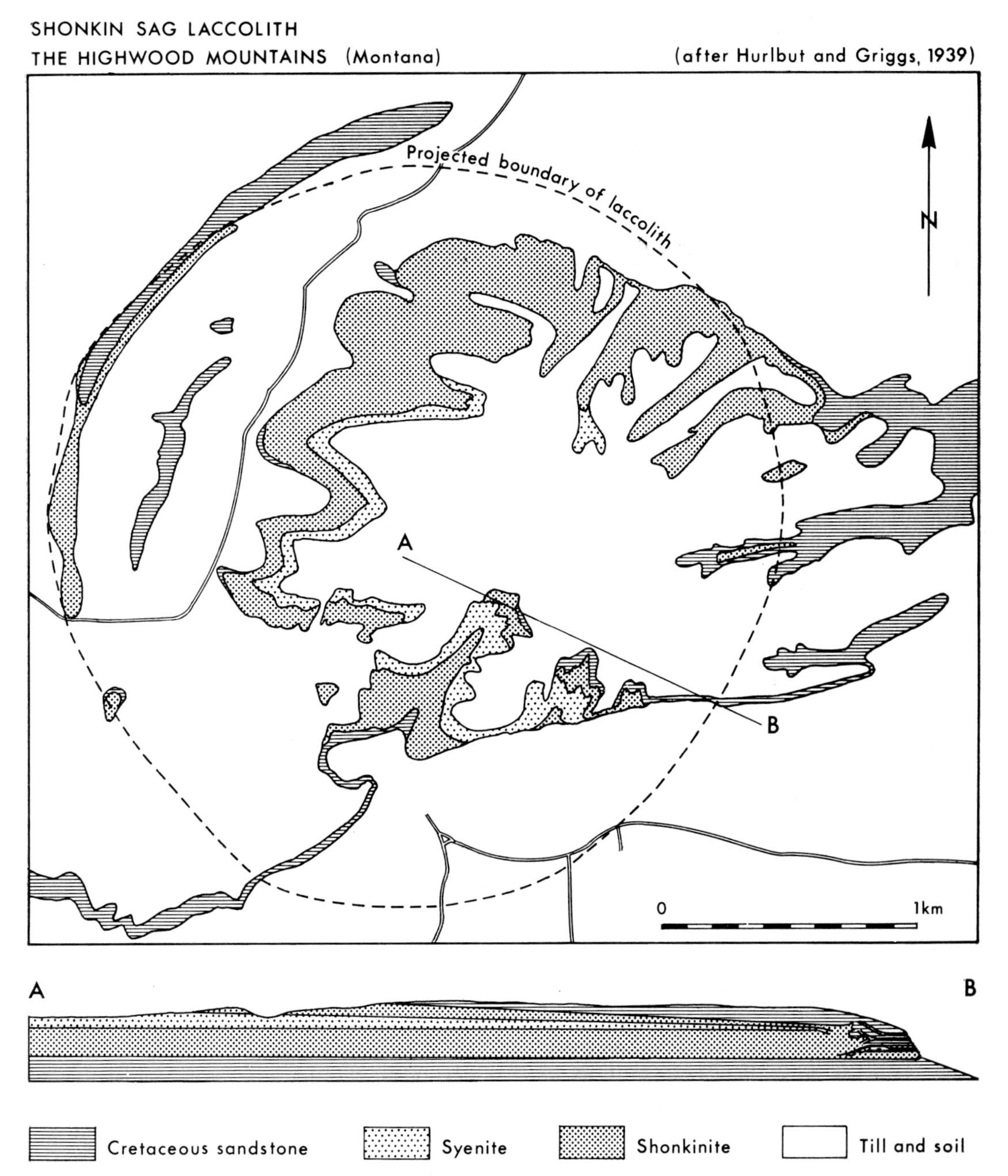

Fig. 4 The Shonkin Sag laccolith. Cross section exaggerated

to have been emplaced by stoping and to have been formed by successive injection of a differentiated alkali basaltic magma, syenite being later than gabbro (Bridgwater and Coe, 1969, p. 74).

The 'headed' dyke of the Highwood Mountains (Buie, 1941) is composed of an upper bulbous body and a lower thin feeder dyke. The arched upper zone consists of fergusite on top of shonkinite, and is a product of gravitative differentiation after intrusion of the shonkinite magma.

3. Central instrusions: Syenites and feldspathoidal syenites are members of several types of central intrusions.

In necks, plugs and stocks having diameters from a few tenths of metres up to several kilometres the rocks are often more or less concentrically arranged, or small plugs of the late members intrude the main body of the stock. The Monteregian intrusions (IV.6) and the alkaline ultramafic intrusions of the Kola Peninsula (IV.2) illustrate this type of occurrence (IV.2, Fig. 3; IV.6, Figs. 1,2). In the Mount Johnson intrusion of the Monteregian province (see IV.6.6.5) a marginal zone of pulaskite grades into a core of essexite, (IV.6, Fig. 1), otherwise intermediate rocks are rare in this province and in similar intrusions elsewhere. This speaks against *in situ* differentiation of one magma injection and suggests a repeated injection of magma differentiated at depth.

The plug- and stock-like intrusions of syenitic rocks are generally believed to be emplaced by stoping (cf. IV.6 and Philpotts, 1970). However, the annular arrangement of the rocks of many stock-like intrusions may also indicate emplacement of magma along ring fractures.

Some bodies of syenitic rocks show internal gneissic structures. At Almunge, Central Sweden, canaditic nepheline syenite forms an incomplete ring of mostly schistose en-echelon dykes surrounding a central mass of umptekite which in places intersects the canadite. The umptekite is of migmatitic appearance and grades into fenitic migmatites (Gorbatchev, 1960).

The Octjabr' (Mariupol) massif, Ukraine (1740–1950 m.y.) is composed of annular zones which grade into each other (Fig. 5). A marginal zone of coarse-grained quartz syenite and granosyenite surrounds trachytoid ferrohastingsite syenites and gneissic hastingsite-biotite nepheline syenites. Albite-nepheline syenite, in part gneissic (mariupolite), containing aegirine and Na-amphibole forms bodies of different shape which are confined to tectonic zones (V. P. Volkov, personal information; Eliseev *et al.*, 1965).

Norra Kärr, Southern Sweden, is made up mainly of foliated eudialyte-catapleiite-nepheline syenites, the adjoining gneisses are fenitized (Adamson, 1944; von Eckermann, 1968). The foliated rocks present an igneous mineral assemblage of agpaitic type, but the textures are of crystalloblastic type.

The geological setting of Almunge, Norra Kärr and Mariupol indicate that they have been formed deep in the crust. All three occurrences are post-orogenic, but Norra Kärr may have been subjected to late tectonic processes (Koark, 1960). The foliated rocks of Norra Kärr may, however, also be compared with zones of foliated amphibolite occurring in undeformed doleritic dykes in West Greenland. These amphibolites may have formed from a hydrous magma at high water pressures (Ramberg, 1948; Windley, 1969). As discussed by Rast (1969, p. 357) schistosity in igneous rocks can originate by crystal growth in a medium transmitting stress. Also injection of magma under initial pressure and subsequent relaxation of the pressure will allow the walls of the magma to compress the crystal mush. Mechanisms of this type may explain the contradictory mineralogical and textural peculiarities of the rocks of Norra Kärr.

4. Ring intrusions: The White Mountain magma series intrusions, New Hampshire, are notable examples (Billings, 1945, Chapman, 1968). The ring dykes are commonly made up of syenites, while granites form the central stocks of these complexes.

Foyaitic rocks occupy the cores of many ring intrusions and tinguaites commonly occur as late dykes (III.5, Figs. 3, 4). Some syenitic ring

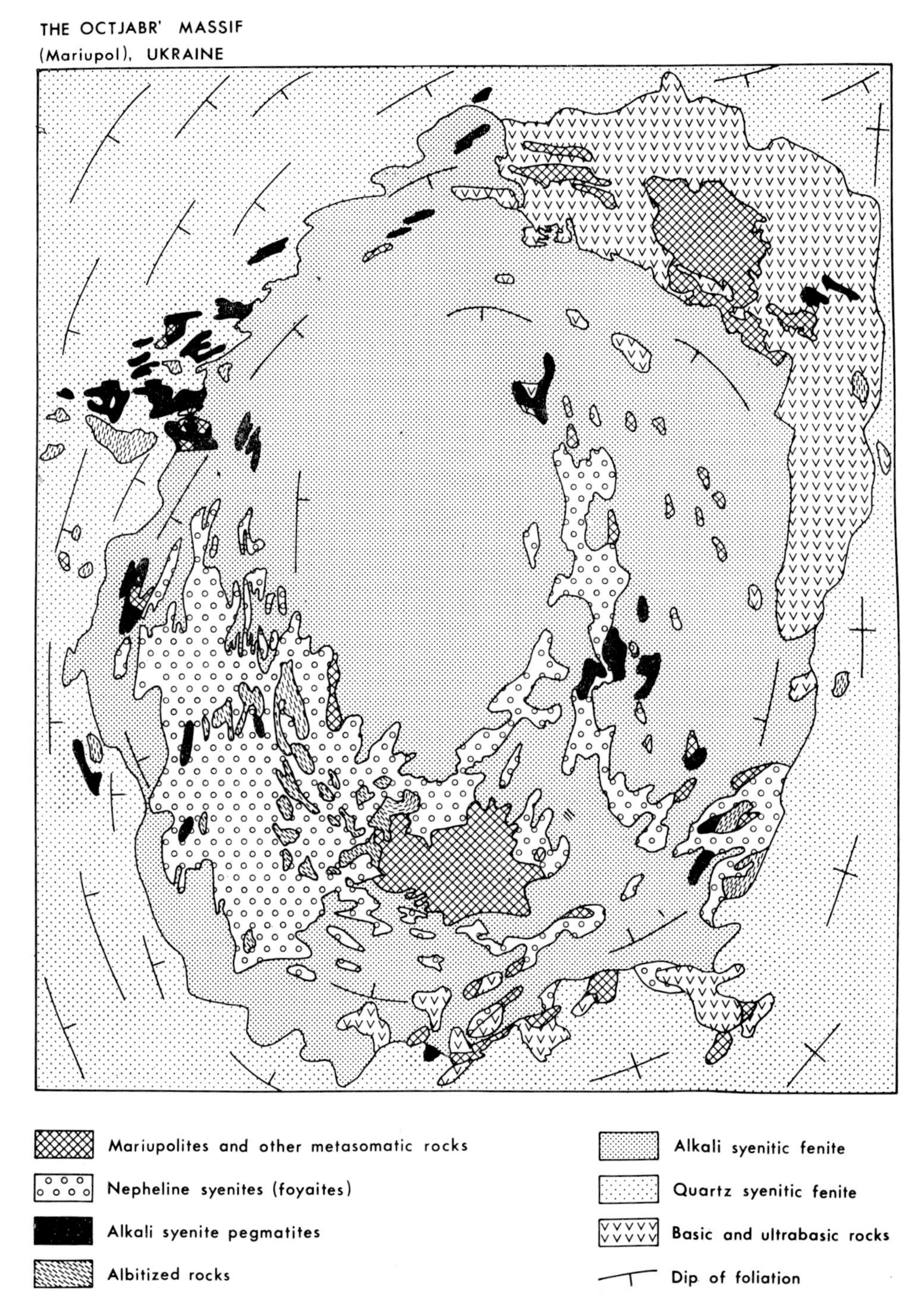

Fig. 5 The Octjabr' massif (Mariupol), Ukraine (after Eliseev *et al.*, 1965; V. P. Volkov, personal information)

intrusions, as for instance Kalkfeld (III.5), have cores of carbonatite.

The largest of all nepheline syenite intrusions, Khibina, the Kola Peninsula (IV.2), is composed of alternating horseshoe-shaped cone and ring intrusions which are steep or dip inwards. (IV.2, Fig. 4).

Ring intrusions consist mainly of granitic and syenitic rocks, while gabbroic rocks are rare. This indicates that passive dilational emplacement connected with subsidence of country rocks into less dense magmas prevails in the formation of ring intrusions (cf. Chapman, 1968).

5. Layered intrusions: Marginal uptilting of layered rocks in the downfaulted central block in ring intrusions indicates that the ring fractures are funnel-shaped, the funnel opening upwards. Chapman (1968) suggests that saucer-shaped structures are formed by crystallization of melts resting on the top of subsided blocks within ring faults, blocks which again rest on the floor existing in the magma chamber before subsidence took place. This observation may connect ring intrusions with layered alkaline complexes such as Lovozero, the Kola Peninsula, Ilímaussaq, South Greenland and Kangerdlugssuaq, East Greenland.

Geophysical data indicate that the visible layered part of Lovozero is underlain by a concentrically zoned stem (IV.2). The magmas forming Lovozero may therefore, like those forming the adjacent Khibina intrusion, have ascended along ring fractures, perhaps accompanied by (underground?) cauldron subsidence or roof spalling.

Stoping is clearly at least partly responsible for the emplacement of the visible saucer-shaped rock units of the Ilímaussaq intrusion (IV.3; Sørensen, 1970).

The Kangerdlugssuaq intrusion, East Greenland (Wager, 1965; Kempe, Deer and Wager, 1970) is constructed as a pile of saucers which show descreasing contents of SiO_2 inwards and upwards and vary from quartz nordmarkite over nordmarkite and pulaskite to foyaite (Fig. 6). There are gradational contacts and sporadic lamination and rhythmic layering which dip at angles of 30–60° inwards. This indicates a formation by *in situ* differentiation.

6. Conformable bodies: The two large Hercynian nepheline syenite massifs, Ilmen and Vishnevogorsk (Fig. 7) are together with a number of small sheet-like bodies and dykes located in the core of an anticlinorium of the Ural mountains. Their emplacement (268 m.y.) was preceded by granitoid magmatism (286 m.y.). The massifs are conformably enclosed by crystalline schists and amphibolites and wedge out, partly as dykes, parallel to the schistosity of the country rocks. The massifs were formed during two phases: (1) miaskitic nepheline syenites; (2) vein miaskites, lamprophyres and miaskitic pegmatites. The endocontact zones are made up of miaskitic nepheline syenites containing plagioclase, amphibole, corundum or scapolite, syenites and nearly monomineralic nepheline rocks. The miaskitic magma is believed to have been generated anatectically by metasomatic infiltration of alkaline abyssal solutions into aluminosilicate sediments within a narrow zone in the axial part of the geosyncline at depths of 8–12 km (Ronenson, 1966; V. P. Volkov, personal information). Metasomatic recrystallization of the adjacent crystalline schists resulted in the formation of alkali migmatites and fenites.

7. Simple homogeneous intrusions: In addition to the necks of syenite, sodalite syenite, etc. found in volcanic districts and probably representing vent-fillings, there are larger bodies, sometimes subvolcanic, sometimes without visible connection to volcanic rocks, which consist of one rock type only. The Serra de Monchique foyaite, Portugal is an example (Fig. 8). This elongated body is located in folded schists and sandstone with its longest axis parallel to the structures of the country rocks. The body is formed from one injection of magma which did not differentiate after emplacement (Cyzgan, 1969). Pulaskitic rocks near the contacts may be due to assimilation of country rocks.

Also the bodies of nordmarkite and larvikite of the Oslo Province appear to be made up of mainly one rock type (Oftedahl, 1960). Inwardly dipping lamination together with sporadic steep mineral layering indicate that the large

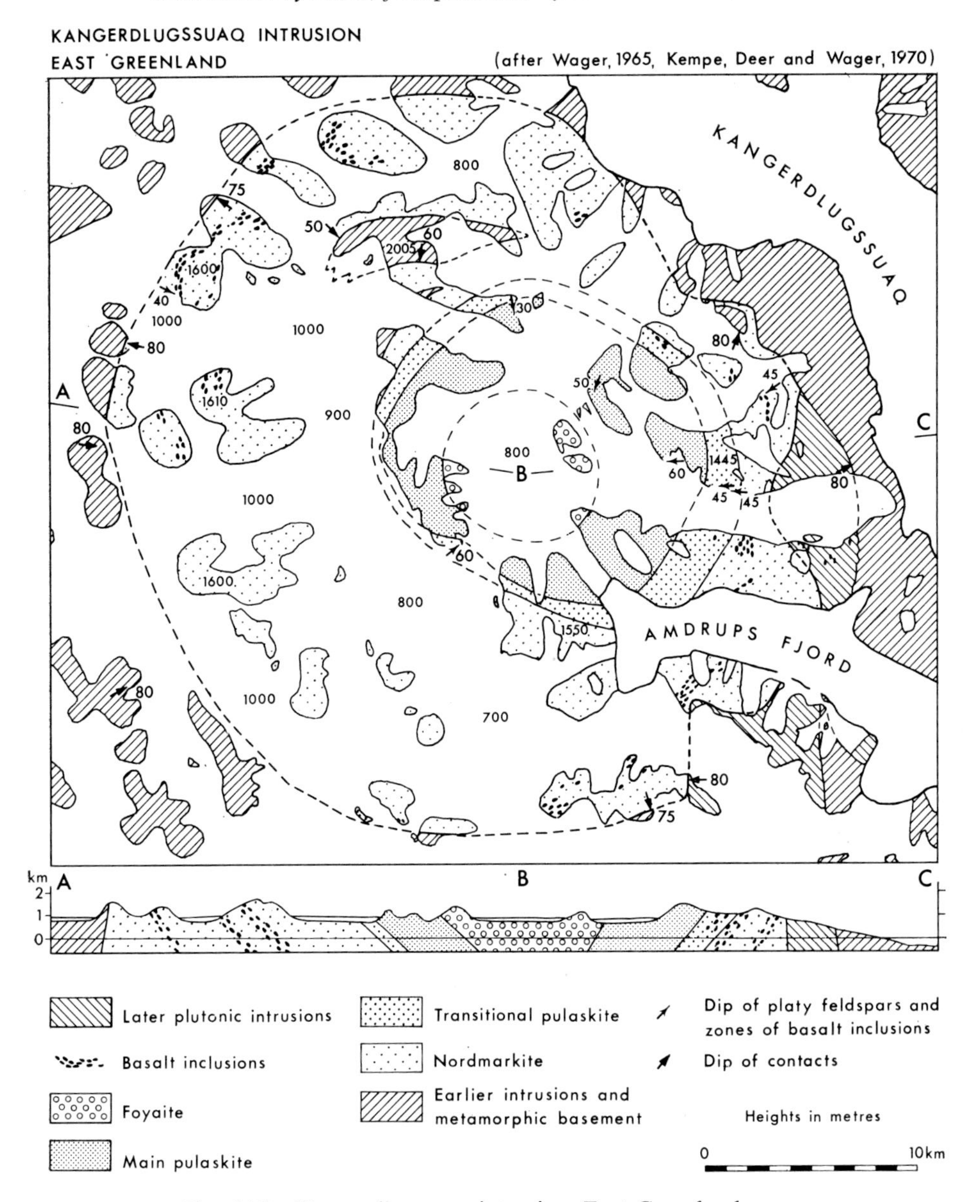

Fig. 6 The Kangerdlugssuaq intrusion, East Greenland

southern body of larvikite constitutes one large intrusion (Meighan, 1968).

The Nunarssuit complex, South Greenland (Harry and Pulvertaft, 1963) is made up of separate bodies of syenite and granite which show opposite or indefinite contact relations at nearby localities. No chilling is seen. This indicates a rapid succession of pulses of magma and variations in speed of consolidation at different places (Harry and Richey, 1963).

8. Marginal syenite in granite massifs: The Beverly alkali syenite is an example of alkali syenite forming a marginal facies of alkali granite (Cape Ann, Massachusetts, Toulmin, 1964). The contact between syenite and granite is sharp and marked by an abrupt increase of

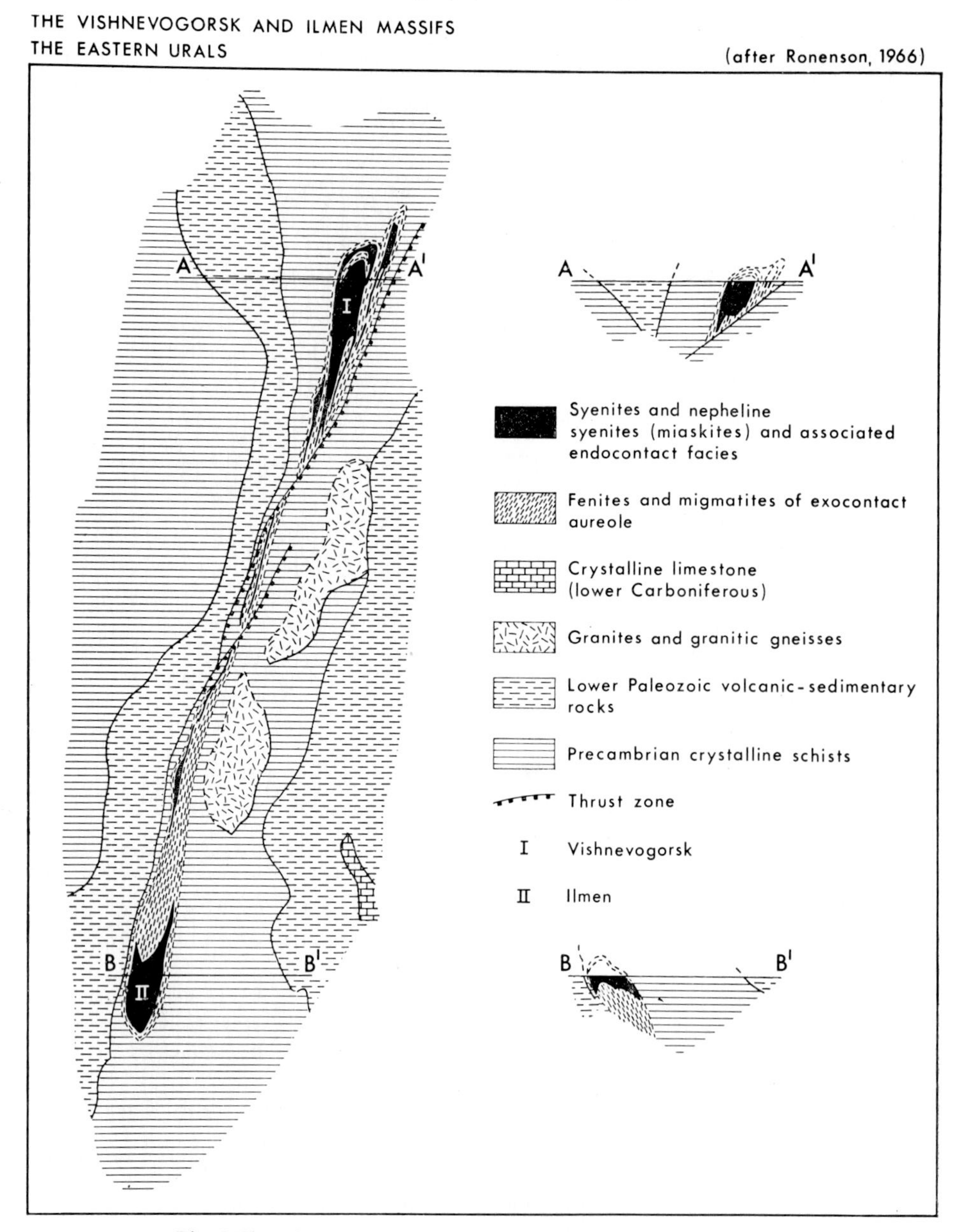

Fig. 7 The Vishnevogorsk and Ilmen massifs, the Eastern Urals

quartz in an otherwise continuous framework of feldspar crystals. This syenite appears to be formed by accumulation of feldspar crystals in a part of the granite magma having a shallow floor during a period of release of volatiles.

9. Emplacement: The emplacement of magmas of alkali syenitic and nepheline syenitic composition is facilitated by their high initial temperatures and in some cases also by long intervals of crystallization (cf. V.1). The magmas are generally less dense than the crustal rocks and in the case of volatile-rich magmas highly mobile, and thus show a tendency to ascend to higher levels than the basic magmas from which

they may have been derived. This may explain the preponderance of syenitic rocks in some provinces, as for instance the Kola and Gardar alkaline provinces (IV.3; VI.2; Sørensen, 1970, p. 331).

10. Contact metamorphism: The country rocks adjacent to syenitic intrusions are modified to varying extent in zones which may be several hundred metres wide.

Fenitization of the country rocks associated with carbonatites and alkaline rocks has been treated recently by McKie (1966) and Verwoerd (1966), see also VI.7. Information about fenitic zones around agpaitic nepheline syenites is collected by Vlasov, ed. (1968). This subject will therefore not be treated further here, it should only be emphasized that fenitization is evidence of the high gas contents of some syenitic magmas.

Hornfels zones containing minerals such as cordierite, andalusite and hypersthene are described from a number of alkaline intrusions, such as Khibina (cf. Sørensen, 1970), the Monteregian Hills (IV.6) and Ditró (Streckeisen, 1960 and 1968). Rheomorphism of the country rocks is also seen (IV.6). These indications of high temperatures of intrusion are consistent with experimental data (see V.1) and information obtained from fluid inclusions (see V.2).

II.2.6. Textures and Structures of Syenitic Rocks

Ussing (1912) pointed out that the agpaitic nepheline syenites of Ilímaussaq do not display the order of separation of minerals emphasized by Rosenbusch for igneous rocks, the felsic minerals being earlier than the mafic ones. Fersman (1929) distinguished subsequently agpaitic and miaskitic orders of separation, the miaskitic nepheline syenites generally obeying the Rosenbusch rule.

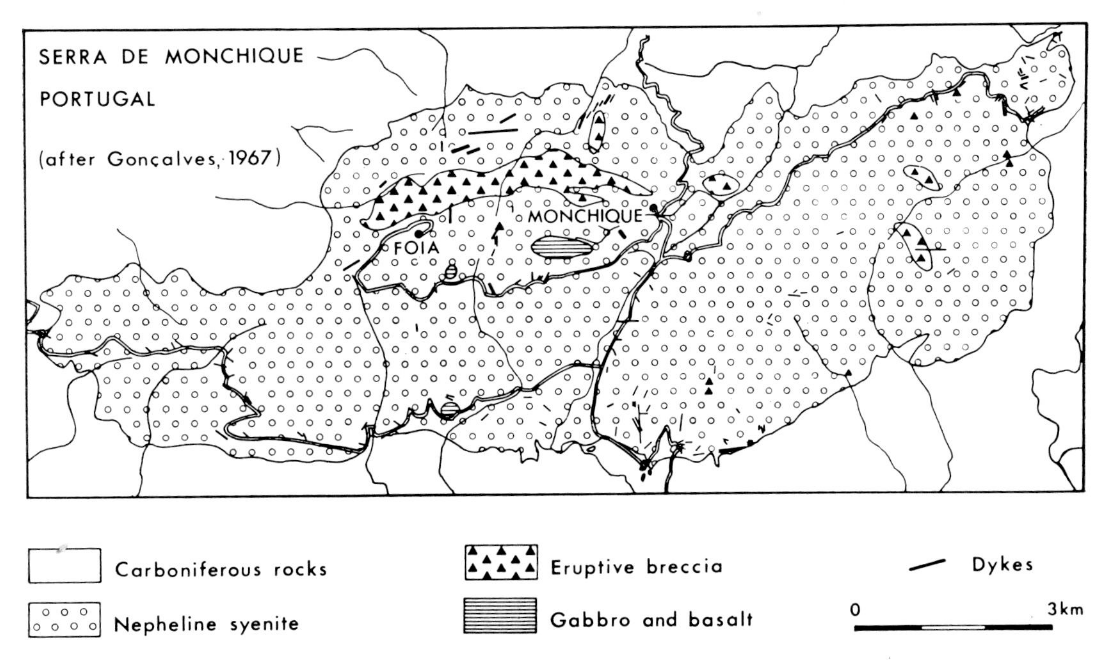

Fig. 8 The Serra de Monchique massif, Portugal (after Gonçalves, 1967)

The order of separation estimated on the basis of degree of idiomorphism should, however, not be overemphasized in the discussion of alkaline rocks for the following reasons: (1) in some alkaline rocks the essential minerals all begin to separate within a very narrow temperature interval; (2) in other rocks, especially the peralkaline ones, the interval of crystallization is very wide (cf. V.1), which favours late-magmatic reactions; (3) many alkaline rocks are cumulates and composed of minerals which before accumulation crystallized in a great volume of magma.

Many augite syenites, pulaskites and foyaites may be regarded as products of *in situ* crystallization formed by plastering of crystals on the walls of the magma chamber or crystallizing as a suspension of feldspar plates when the feldspar growth of successive layers of crystals on the walls of stagnant magmas in which concentration gradients in the front of crystallization give rise to rhythmic layering (Hess, 1960).

Syenites and nepheline syenites formed by accumulation of crystals are characterized by igneous lamination and sometimes also by rhythmic layering. The lamination is generally supposed to be a result of current activity in the magma (cf. Upton, 1960 and 1961). Steep lamination and layering of the pulaskites and essexites of Mount Johnson in the Monteregian

Fig. 9 Augite syenite, Ilímaussaq intrusion. Left: Synneusis texture caused by lumping together of augite, fayalite and titanomagnetite. Right: Steeply inclined inch-scale layering caused by alternation of layers enriched in felsic and mafic minerals

and the magma have nearly identical densities or the magma is very viscous and consolidation rapid. These rocks often display a characteristic synneusis texture having the mafic minerals, fayalite, augite and iron ore, clustered together (Fig. 9). This spotted texture is enhanced by late-magmatic reactions, rims of biotite or amphibole surrounding primary grains of especially fayalite and iron ore.

Steeply inclined 'inch-scale' layering in augite syenites (Fig. 9) may be taken as evidence of province provide strong evidence of vertical magma currents (IV.6, Figs. 4, 7).

Fine examples of rhythmic layering ascribed to convection current activity are found in the augite syenites of the Nunarssuit and Kûngnât intrusions, South Greenland (IV.3, Harry and Pulvertaft, 1963; Upton, 1960) in which gravity stratification, cross-bedding, wash-out channels (Fig. 10) and slumping phenomena are well-developed in certain zones. In the western centre of the Kûngnât intrusion the dip of the layering is up to 70° inwards in the lower fayalite-ferroaugite syenites, but only 10–25° in the overlying more alkalic syenites. This indicates that the volatile-rich magma, from which the latter rocks formed, was less viscous than the magma in which the lower augite syenites formed by bottom accumulation of crystals. The layered and laminated alkali syenites are overlain by laminated alkali syenites and these again by unlaminated and unlayered alkali syenites. This succession is interpreted as a result of the establishment of stagnant conditions in the magma due to accumulation of volatiles in higher levels, so that the upper volatile-rich parts of the magma are less dense and crystallize at lower temperatures than the lower volatile-poor parts (Upton, 1960, p. 116). In volatile-poor magmas crystallization may take place from the top which promotes convective currents; in volatile-rich magmas crystallization takes place at lower levels under nearly stagnant conditions (cf. Sørensen, 1969).

Fig. 10 Erosion channel (trough banding) in gravity layered syenite, Nunarssuit, the Gardar Province, South Greenland. The hammer handle is 40 cm long. Vertical dark bands caused by water seeping down the face. (photo T. C. R. Pulvertaft)

Rhythmic layering in nepheline syenites corresponds closely to that seen in syenites. The layering in miaskitic foyaites dips at steep angles, for instance 30–70° in the Grønnedal-Íka foyaites of South Greenland (Emeleus, 1964). The rocks of this intrusion have also developed pronounced feldspar lamination parallel to the layering. The steep dips are explained by high viscosity, by slight density contrast between crystals and magma and by rapid consolidation of the intercumulus melt (Emeleus, 1964, p. 26, see also Wager and Brown, 1968, p. 500). The rocks in question are poor in volatile-bearing minerals.

Parts of the agpaitic nepheline syenites of Lovozero and Ilímaussaq display almost horizontal, although wavy, layering (Fig. 11) and a pronounced igneous lamination conformable with the layering. The layers show mineral-grading; cross-bedding and trough banding are absent or rare. This suggests a formation of the layering by sedimentation of crystals in a volatile-rich mainly stagnant magma, possibly as a result of intermittent crystallization (Sørensen, 1969). In the upper part of the Ilímaussaq intrusion poikilitic sodalite syenites (naujaites) were formed by flotation of sodalite crystals (IV.3; II.2.2.5).

Fig. 11 Rhythmic igneous layering in eudialyte-arfvedsonite-nepheline syenite (kakortokite). The Ilímaussaq intrusion

It should finally be mentioned that planar structures are formed in some rocks as a result of emplacement of the magma during periods of deformation. The lujavrites of Ilímaussaq (II. 2.2.4) may serve as an example (Ussing, 1912, p. 330; Sørensen, 1969, 1970).

II.2.7. Crystallization of Nepheline Syenites

The agpaitic and miaskitic nepheline syenites form two major paragenetical groups characterized by different orders of crystallization of the minerals. In the agpaitic rocks the felsic minerals have crystallized earlier than the mafic ones; crystallization appears to take place over a considerable range of temperatures; at 1000 bars P_{H_2O} the liquidus and solidus of agpaitic melts are separated by 400–500 °C (cf. V.1). In the miaskitic rocks the mafic minerals are first to form, but the pyroxene, alkali feldspar and nepheline appear within an interval of 30–40 °C below the liquidus and the range of crystallization is 100–200 °C (cf. V.1).

The nepheline coexisting with high-temperature feldspar varies widely in chemical composition, while that of 'plutonic' assemblages exhibits a narrow range of compositions (cf. Tilley, 1957; Barth, 1963). The ratio K/(Na + K) (atomic proportions) varies thus from 7 to 37 in the nephelines of volcanic rocks, and from 14–23 in plutonic rocks (Tröger, 1969, p. 201).

A theoretical analysis of the equilibrium relations in nepheline-feldspar parageneses has been carried out by Perchuk and Ryabchikov (1968) and the data compared with nepheline-feldspar assemblages from igneous intrusions. The whole compositional range of coexisting nepheline and alkali feldspars is represented in their material, most assemblages correspond to equilibrium temperatures of 400–600 °C. Assemblages formed below 500 °C have a narrow compositional range, the nepheline approaches the Morozevicz composition—$Ne_{75\cdot0}Ks_{20\cdot5}Qz_{4\cdot5}$—closely. The results indicate that adjustment of the compositions of the minerals takes place during cooling.

In the subsolvus nepheline syenites, which are formed by slow cooling of hypersolvus assemblages, by low temperature consolidation, or by low temperature recrystallization, the nepheline is close in composition to the Morozevicz–Buerger convergence field (Tilley, 1957) and coexists with discrete grains of potash feldspar, often microcline, and low-albite. The lujavrites of Ilímaussaq have discrete laths of pure low-albite and pure maximum-triclinic microcline associated with nepheline of the ideal Morozevicz composition (Sørensen, 1962; Piotrowski and Edgar, 1970). From textural relations the albite and microcline are estimated to have been formed in mutual equilibrium and later than the nepheline.

Nepheline gneisses are also examples of subsolvus assemblages formed during metamorphism and recrystallization (II.2.6; II.7; Tilley, 1957; Heier, 1965; Piotrowski and Edgar, 1970).

Hytönen (1959), Sørensen (1962) and Wilkinson (1965) have pointed out that analcime and natrolite may accompany nepheline and alkali feldspars in rocks consolidated at high water-vapour pressures and at temperatures in the range of 500–400 °C. The analcime and natrolite of these rocks, for instance some lujavrites and tinguaites, are then to be regarded as primary igneous minerals.

A series of nepheline syenite parageneses may be established on the basis of the available data (cf. Tilley, 1957; Hytönen, 1959; Wilkinson, 1965; Welman, 1970):

A. Hypersolvus assemblages:
  1. Phonolitic: nepheline–(sodalite)–sanidine or anorthoclase;
  2. lardalitic: nepheline–(sodalite)–cryptoperthitic alkali feldspar;
  3. foyaitic: nepheline–(sodalite)–microperthitic alkali feldspar;

B. Subsolvus assemblages:
  1. Primary igneous assemblages:
     (i) nepheline syenite pegmatites;
     (ii) litchfielditic: nepheline–microperthite–albite;
     (iii) lujavritic and tinguaitic: nepheline–sodalite–microcline–albite;
     (iv) analcime tinguaitic: nepheline–alkali feldspar–analcime–sodalite;
     (v) analcime-natrolite tinguaitic: microcline–albite–analcime–natrolite;
     (vi) natrolite tinguaitic: microcline–natrolite;
  2. Recrystallized assemblages:
     nepheline gneisses: nepheline–microcline or microperthite–albite
     mariupolitic: nepheline–albite.

The mafic minerals also offer valuable information about the conditions of crystallization. The importance of the chemical composition of pyroxenes in determining the trend of evolution of syenitic rocks has been emphasized by Edgar and Nolan (1966). The substitution of $(OH)^-$ by F^- elevates the high-temperature stability limit of amphiboles and accounts for the presence of primary alkali amphiboles in igneous rocks. Riebeckite-arfvedsonite solid solutions are further stable at magmatic temperatures at relatively low oxidation states, even in the absence of fluorine (Ernst, 1968, p. 100).

In the case of sodic rocks the primary arfvedsonitic amphibole may be substituted by late acmite ($\pm$ biotite) which indicates varying vapour and oxygen pressures, or changes in temperature of crystallization.

II.2.8. Pegmatites

The nepheline syenite pegmatites of Langesundfjord, Ilímaussaq and Lovozero are well-known for the wealth of minerals they contain; namely respectively about 80, 120 and 150 different species.

Nepheline syenite pegmatites may be divided according to their chemical and mineralogical characteristics into agpaitic and miaskitic types and types intermediate between these 'end members' (Semenov, 1967).

Agpaitic pegmatites occur not only in and around agpaitic nepheline syenites but also in and around miaskitic (or intermediate) types. Some of the Langesundfjord nepheline syenite pegmatites occurring in larvikite, gneissic nepheline syenite and in the exocontact zone have, for instance, agpaitic mineral assemblages (eucolite, mosandrite, catapleiite, etc.). Also the pegmatites of the layered Kangerdluagssuaq intrusion (Kempe *et al.*, 1970), which is composed of intermediate (catophorite-arfvedsonite) nepheline syenites, have agpaitic mineral assemblages (eudialyte, astrophyllite).

Agpaitic pegmatites are often strikingly zoned (Vlasov *et al.*, 1959; Sørensen, 1962). Replacement phenomena are widespread and result in the formation of bodies or veins of albite, late potash feldspar, fine-grained natrolite, ussingite, analcime and a number of beryllium, rare earth, niobium, thorium, and lithium minerals (Figs. 12, 13).

Fig. 12 Phase pegmatite forming horizon in poikilitic sodalite syenite, Ilímaussaq. The white crystals bridging the pegmatite consist of microcline and sodalite

The pegmatites of miaskitic nepheline syenites are generally simple and undifferentiated, but zoned pegmatites and replacement bodies and veins occur locally, for instance in the Vishnevogorsk massif, the Urals (cf. Vlasov, ed., 1968, p. 280). The cores and replacement bodies of these pegmatites contain albite, zeolites, hydrargillite, cancrinite and calcite. Miaskitic pegmatites are generally poor in minerals containing Nb, RE, Be, Th, etc.

II.2.9. Pneumatolytic-Hydrothermal Veins

The replacement bodies of the agpaitic pegmatites of Ilímaussaq are of the same mineralogical composition as the numerous pneumatolytic-hydrothermal veins of this intrusion (Sørensen, 1962). These late veins are made up of one or more of the following minerals: albite, analcime, natrolite, ussingite and aegirine and/or arfvedsonite-riebeckite and besides of numerous rare minerals containing beryllium, rare earths, etc. Carbonates are practically lacking, but fluid inclusions in the minerals contain CO_2 and hydrocarbons.

The pneumatolytic-hydrothermal veins associated with miaskitic rocks are dominated by carbonates, cancrinite, albite, sericite, and further contain zircon, pyrochlore, ilmenite, aegirine-augite, hedenbergite, hastingsite, diaspor, fluorite, allanite and britholite (cf. Semenov, 1967).

Hydrothermal veins containing calcite, dolomite, baryte, hematite, bästnäsite, sulphides, thorite, etc. are associated with syenitic rocks, alkalic as well as calc-alkalic, in a number of regions, as for instance Iron Hill, Colorado (Hedlund and Olsen, 1961); Gallinas Mountains, New Mexico (Perhac and Heinrich, 1964), and Mountain Pass, California (Olson *et al.*, 1954). Vlasov, ed. (1968, p. 343) suggests that this type of deposit is associated mainly with biotite and barkevikite granosyenites (cf. Section VII).

Ca-poor syenites and nepheline syenites have crusts of felt-like aegirine on fractures, the occurrence of which recalls the crusts of epidote found on fractures in calc-alkalic rocks.

II.2.10. Association and Petrogenesis

Alkali syenites and feldspathoidal syenites and their volcanic equivalents are members of a number of petrological associations (cf. IV.1) and appear to be products of a variety of processes. These processes are treated exhaustively in Chapters VI.1 to VI.7, and will be reviewed only briefly here.

1. The trachytes and phonolites of oceanic islands appear from geological data to be products of crystallization of residual melts derived from larger volumes of alkaline olivine basalts (cf. IV.7), but this origin may be contradicted by isotopic data (cf. Gast, 1967 and V.4). An origin by magmatic differentiation is also proposed for some continental regions (cf. IV.2; IV.3; IV.4).

Shand has repeatedly pointed out that there are two main stages in the development of feldspathoidal alkaline rocks from common magmas: (a) desilication, (b) introduction of alkalis. Fersman (1939) added oxidation as another important process.

2. The fact that alkali basaltic and trachytic-phonolitic magmas alternate in some volcanic provinces, as for instance the Kaiserstuhl volcano (IV.4), has led to the conclusion that the basaltic and trachytic-phonolitic magmas of such provinces may be intermittent eruptions from different levels of layered magma chambers (IV.3) or that they are of independent origin formed by upper mantle or deep crustal anatexis (VI.1). The huge effusions of trachytic and phonolitic rocks in some provinces may also be products of anatexis (VI.1).

Anatectic processes in orogenic zones have been considered to be responsible for the formation of syenitic rocks, as for instance the miaskitic nepheline syenites of the Urals (Ronenson, 1966; cf. II.2.5). Vorobieva (1963) has pointed out that these nepheline syenites have nepheline: feldspar ratios of 35:65 and that feldspar and nepheline may be in micrographic intergrowth which suggest a consolidation from anchieutectic melts (see also Payne, 1968). Anatexis is believed to have been preceded by metasomatic alteration (fenitization) of the original gneisses and metasediments. Thus Kotina and Yaroshevsky (1970) suggest that incongruent melting of albite (or alkali feldspar) during the interaction of primary granitoid rocks with silica-undersaturated alkaline solutions resulted in nephelinization (cf. also Saha, 1961; Currie, 1968, 1970;

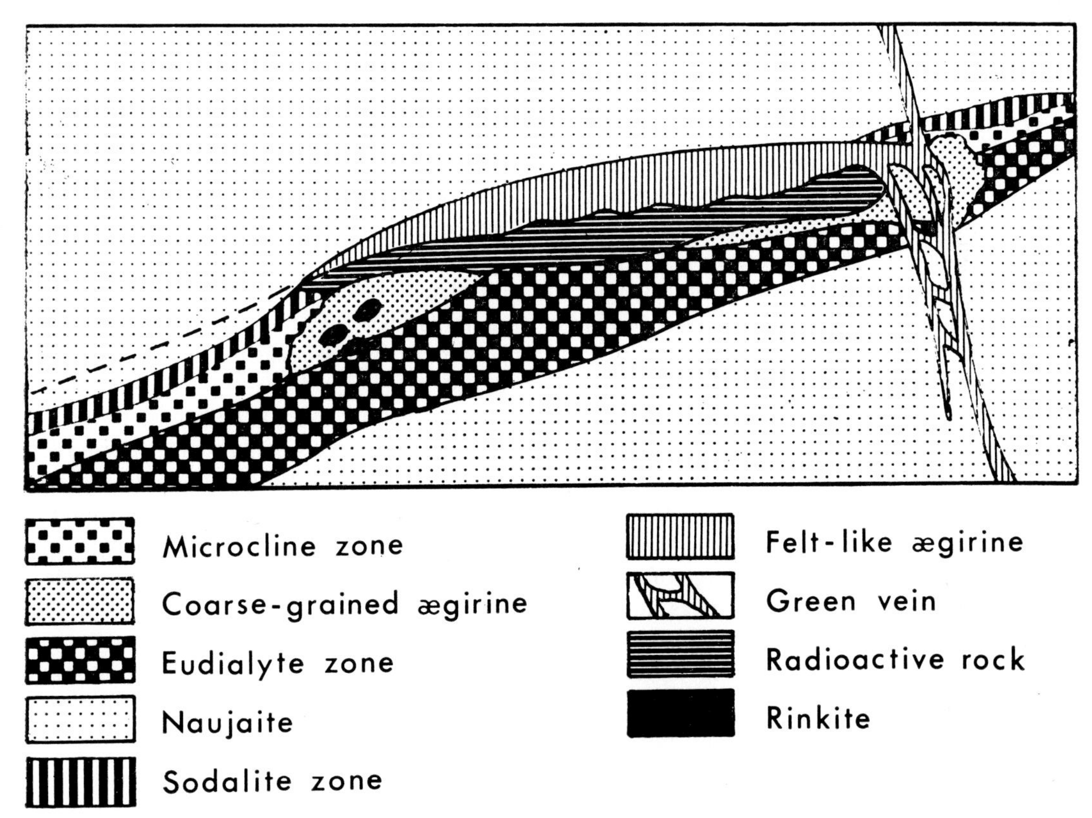

Fig. 13 Sketch of facies pegmatite in poikilitic sodalite syenite, Ilímaussaq. The pegmatite is approximately 50 cm thick. It is asymmetrically zoned. The radioactive rock is a replacement body of natrolite, analcime and albite

V.1). This process may precede the generation of palingenic nepheline-syenite melts. Kopecký *et al.* (1970) discuss the importance of fenitization along deep fault zones prior to generation of alkaline melts (cf. IV.4.4.2).

A formation by partial melting is in accordance with the fact that most granites and nepheline syenites plot in the low-temperature regions of the 'residua system' (V.1; VI.2), but as pointed out by Hamilton and MacKenzie (1965) the lack of sediments of suitable composition restricts the application of this mechanism with regard to nepheline syenites and phonolites (cf. V.1).

3. The limestone assimilation hypothesis of Daly–Shand was long the favoured explanation of the origin of feldspathoidal rocks, but now gains little support (VI.3). Currie (1970) has pointed out, however, that desilication of magmas and solid crustal rocks may be brought about by a flux of an aqueous phase through them.

4. Resorption of silicate minerals such as biotite and hornblende may be responsible for the formation of some types of syenitic rocks (see VI.6).

5. Syenites forming the marginal parts of granite (and monzonite) massifs are variously explained as products of assimilation of basic country rocks in granitic magmas or as derivatives of the granitic magmas. Both explanations have, for instance, been advanced in the interpretation of syenitic bodies associated with the Keivy alkali granites of the Kola Peninsula (IV.2), the former by Ginzburg (1958), the latter by Chumakov (1958).

6. The (calc-alkalic) syenites associated with anorthosites are variously considered as formed from liquid left after removal of plagioclase crystals from a gabbroic magma (Yoder, 1969); as results of regional anatexis around intrusive anorthosite (cf. Isachsen, 1969; Hodge *et al.*, 1970), or to be of independent origin (cf. Buddington, 1969). In the Gardar alkaline province there is a clear genetic relation between anorthosite and alkali syenitic rocks (IV.3).

7. The role of volatiles in the genesis of feldspathoidal and alkaline rocks has been emphasized repeatedly since Smyth (1913), cf. VI.4; VI.5; VI.8. Volatiles are undoubtedly of prime importance in the genesis of many types of feldspathoidal rocks.

8. The association of feldspathoidal and granitic rocks within one intrusion presents a problem of its own which is not yet solved (cf. VI.2.5).

9. Finally it should be pointed out, that metasomatic processes may have been involved in the formation of the gneissic and some massive types of alkali syenites and nepheline syenites, cf. VI.7; VI.8.6. Thus Eliseev *et al.* (1965) consider the mariupolites of the Oktjabr'sky (Mariupol) massif, Ukraine, as products of transformation of primary magmatic calc-alkaline syenites or nepheline syenites due to interaction with alkaline metasomatic solutions.

II.2. REFERENCES

Adams, F. D., and Barlow, A. E., 1910. Geology of the Haliburton and Bancroft areas, Province of Ontario. *Mem. Geol. Surv. Can.*, **6.**

Adamson, O. J., 1944. The petrology of the Norra Kärr district. An occurrence of alkaline rocks in southern Sweden. *Geol. För. Stockh. Förh.*, **66,** 113–255.

Appleyard, E. C., 1967. Nepheline gneisses of the Wolfe belt, Lyndoch Township, Ontario. I. Structure, stratigraphy, and petrography. *Can. J. Earth Sci.*, **4,** 371–95.

Appleyard, E. C., 1969. Nepheline gneisses of the Wolfe belt, Lyndoch Township, Ontario. II. Textures and mineral paragenesis. *Can. J. Earth Sci.*, **6,** 689–717.

Azambre, B., 1967. *Sur les roches intrusives soussaturées du Crétacé des Pyrénées.* Thèse Faculté des Sciences de Paris, 1–147.

Azambre, B., and Girod, M., 1966. Phonolites agpaïtiques. *Bull. Soc. fr. Minér. Christallogr.*, **89,** 514–20.

Barker, D. S., 1965. Alkalic rocks at Litchfield, Maine. *J. Petrology*, **6,** 1–27.

Barth, T. F. W., 1944. Studies on the igneous rock

complex of the Oslo region. II. Systematic petrography of the plutonic rocks. *Skr. norske Vidensk-Akad. Oslo l. Mat.-naturv. Kl.*, **9,** 1–104.

Barth, T. F. W., 1963. The composition of nepheline. *Schweiz. miner. petrogr. Mitt.*, **43,** 153–64.

Billings, M. P., 1945. Mechanics of igneous intrusion in New Hampshire. *Am. J. Sci.*, **243 A,** Daly volume, 40–68.

Bloomfield, K., 1970. 'Orogenic and post-orogenic plutonism in Malawi', in T. N. Clifford and I. G. Gass, eds. *African Magmatism and Tectonics.* Oliver & Boyd, 119–56.

Bridgwater, D., and Coe, K., 1969. 'The rôle of stoping in the emplacement of the giant dykes of Isortoq, South Greenland'. In G. Newall and N. Rast, eds.: Mechanism of Igneous Intrusion. *Geol. J. Spec. Iss.*, **2,** 67–78.

Bridgwater, D., and Harry, W. T., 1968. Anorthosite xenoliths and plagioclase megacrysts in Precambrian intrusions of South Greenland. *Meddr Grønland*, **185,** 2, 1–243.

Brøgger, W. C., 1890. Die Mineralien der Syenitpegmatitgänge der südnorwegischen Augit- und Nephelinsyenite. *Z. Kristallogr. Miner.*, **16,** 1–663.

Brousse, R., Varet, J., and Bizouard, H., 1969. Iron in the minerals of the sodalite group. *Contr. Miner. Petrogr.*, **22,** 169–84.

Bryan, W. B., 1970. 'Alkaline and peralkaline rocks of Socorro Island, Mexico', in P. H. Abelson, ed. *Annual Rep. of the Director Geophys. Lab. Yb. Carnegie Insts.*, **68,** 194–200.

Buddington, A. F., 1969. 'Adirondack anorthositic series', in Y. W. Isachsen, ed., Origin of Anorthosite and Related Rocks. *Mem. N.Y. St. Mus. Sci. Serv.*, **18,** 215–31.

Buie, B. F., 1941. Igneous rocks of the Highwood Mountains, Montana. Part III. Dikes and related intrusives. *Bull. Geol. Soc. Am.*, **52,** 1753–808.

Carmichael, I. S. E., 1965. Trachytes and their feldspar phenocrysts. *Mineralog. Mag.*, **34,** 107–25.

Chapman, C. A., 1968. 'A comparison of the Maine coastal plutons and the magmatic central complexes of New Hampshire', in E–An Zen *et al.*, eds. *Studies of Appalachian Geology: Northern and Maritime.* Interscience, 385–98.

Chumakov, A. A., 1958. 'The origin of the alkali granites of Keivy' (in Russian). In *Alkali Granites of the Kola Peninsula*, 308–68. Izd. Akad. Nauk. SSSR.

Clarke, F. W., and Washington, H. S., 1924. The composition of the Earth's crust. *Prof. Pap. U.S. geol. Surv.*, **127,** 1–117.

Cloos, H., and Cloos, E., 1927. Die Quellkuppe des Drachenfels am Rhein, ihre Tektonik und Bildungsweise. *Z. Vulkanol.*, 11, 33–40.

Codarcea, Al., Codarcea-Dessila, M., and Ianovici, V., 1958. Structure géologique du massif des roches alcalines de Ditrau. *Rev. Géol.-Geogr. Acad. R.P.R. II*, **1,** Bucarest.

Currie, K. L., 1968. On the solubility of albite in supercritical water in the range 400 to 600 °C and 750 to 3500 bars. *Am. J. Sci.*, **266,** 321–41.

Currie, K. L., 1970. An hypothesis on the origin of alkaline rocks suggested by the tectonic setting of the Monteregian Hills. *Can. Mineral.*, **10,** 411–20.

Czygan, W., 1969. Petrographie und Alkali-Verteilung im Foyait der Serra de Monchique, Süd-Portugal. *Neues Jb. Miner. Abh.*, **111,** 32–73.

Eckermann, H. von, 1968. New contributions to the interpretation of the genesis of the Norra Kärr alkaline body in southern Sweden. *Lithos*, **1,** 76–88.

Edgar, A. D., and Nolan, J., 1966. Phase relations in the system $NaAlSi_3O_8$ (albite) –$NaAlSiO_4$ (nepheline) –$NaFeSi_2O_6$ (acmite) –$CaMgSi_2O_6$ (diopside) –H_2O and its importance in the genesis of alkaline undersaturated rocks. *Miner. Soc. India, IMA volume, Fourth General Meeting*, 176–81.

Eliseev, N. A., Kushev, V. G., and Vinogradov, D. P., 1965. *The Proterozoic Intrusive Complex of the Eastern Priazovye* (in Russian). Izd. 'Nauka'.

Emeleus, C. H., 1964. The Grønnedal-Íka alkaline complex, South Greenland. *Meddr. Grønland*, **172,** 3, 1–75.

Ernst, W. G., 1968. *Amphiboles.* Springer-Verlag Berlin, Heidelberg, New York, 1–125.

Fersman, A., 1929. Geochemische Migration der Elemente und deren wissenschaftliche und wirtschaftliche Bedeutung, erläutert an vier Mineralvorkommen: Chibina-Tundren—Smaragdgruben—Uran-Grube Tuja-Mujun—Wüste Karakumy. *Abh. prakt. Geol. BergwLehre*, **18,** 1–116.

Fersman, A., 1939. *Geochemistry III* (in Russian). Leningrad.

Galakhov, A. V., 1959. *The Rischorrites of the Khibina Alkaline Massif* (in Russian). Izd. Akad. Nauk., 1–169.

Galakhov, A. V., 1967. Chemical composition of rocks in the Khibiny alkalic massif. *Dokl. Acad. Sci.* (English translation), **171,** 225–8.

Galakhov, A. V., 1969. 'Rock-forming nepheline as indicator of magmatic differentiation (based on the Khibina alkaline massif)' (in Russian), in I. V. Bel'kov, ed. *Material on the Mineralogy of the Kola Peninsula*, **7,** Izd. 'Nauka', 120–5.

Gast, P. W., 1967. 'Isotope geochemistry of volcanic rocks'. In Hess, H. H. and Poldervaart, A., eds.: *Basalts. I.* Interscience, 325–58.

Gerasimovsky, V. I., 1941. On the role of zirconium in minerals of nepheline-syenite massifs. *Comptes Rendus (Dokl. Acad. Sci.) USSR*, **30,** 820–1.

Gerasimovsky, V. I., 1956. Geochemistry and mineralogy of nepheline syenite intrusions. *Geochemistry*, 494–510.

Gerasimovsky, V. I., 1963. 'Geochemical features of agpaitic nepheline syenites', in A. P. Vinogradov, ed. *Chemistry of the Earth's Crust, I*, 104–18. Israel Translation Program. Jerusalem.

Gerasimovsky, V. I., 1969. *Geochemistry of the Ilímaussaq Alkaline massif* (in Russian). Izd. 'Nauka', Moskva, 1–174.

Ginzburg, I. V., 1958. 'Geological position and internal tectonics of the alkali granites on the Kola Peninsula' (in Russian), in *Alkali Granites of the Kola Peninsula*, 213–24. Izd. Akad. Nauk SSSR.

Ginzburg, A. I., and Portnov, A. M., 1966. Mineral associations in the alkalic rocks (in Russian). *Geokhimiya*, 28–36.

Goldschmidt, V. M., 1930. Elemente und Minerale pegmatitischer Gesteine. *Nachr. Gesellsch. Wiss. Göttingen. Math. Phys. Kl.*, 370–8.

Gonçalves, F., 1967. Subsídios para o conchecimento geológico do maçio eruptivo de Monchique. *Com. Serv. Geol. Portugal*, **52**, 169–84.

Gorbatschev, R., 1960. On the alkali rocks of Alumunge. A preliminary report on a new survey. *Bull. geol. Inst. Univ. Uppsala*, **39**, 1–69.

Hamilton, D. L., and MacKenzie, W. S., 1965. Phase-equilibrium studies in the system $NaAlSiO_4$ (nepheline)–$KAlSiO_4$ (kalsilite)–SiO_2–H_2O. *Mineralog. Mag.*, **34** (Tilley volume), 214–31.

Harnik, A. B., 1969. Strukturelle Zustände in den Anorthoklasen der Rhombenporphyre des Oslogebietes. *Schweiz. miner. petrol. Mitt.*, **49**, 509–67.

Harry, W. T., and Pulvertaft, T. C. R., 1963. The Nunarssuit intrusive complex South Greenland. Part I. General Description. *Meddr Grønland*, **169**, 1, 1–136.

Harry, W. T., and Richey, J. E., 1963. Magmatic pulses in the emplacement of plutons. *Liverpool Manchester geol. J.*, **3**, 254–68.

Hedlund, D. C., and Olson, J. C., 1961. Four environments of thorium-, niobium-, and rare-earth-bearing minerals in the Powderhorn district of southwestern Colorado. *Prof. Pap. U.S. geol. Surv.*, **424–B**, 283–6.

Heier, K. S., 1961. Layered gabbro, hornblendite, carbonatite and nepheline syenite on Stjernøy, North Norway. *Norsk Geol. Tidsskr.*, **41**, 109–56.

Heier, K. S., 1964. Geochemistry of the nepheline syenite on Stjernøy, North Norway. *Norsk Geol. Tidsskr.*, **44**, 205–15.

Heier, K. S., 1965. A geochemical comparison of the Blue Mountain (Ontario, Canada) and Stjernoy (Finmark, North Norway) nepheline syenites. *Norsk Geol. Tidsskr.*, **45**, 41–52.

Heier, K. S., and Taylor, S. R., 1964. A note on the geochemistry of alkaline rocks. *Norsk Geol. Tidsskr.*, **44**, 197–204.

Hess, H. H., 1960. Stillwater igneous complex, Montana. *Mem. Geol. Soc. Am.*, **80**, 1–230.

Hibsch, I. E., 1902. Ueber Sodalithaugitsyenit im böhmischen Mittelgebirge und über die Beziehungen zwischen diesem Gestein und dem Essexit. *Tschermak's miner. petrogr. Mitt.*, **21**, 157–70.

Hibsch, I. E., 1926. Erläuterungen zur geologischen Übersichtskarte des Böhmischen Mittelgebirges. *Heimatkunde des Elbegaues Tetschen*, **3.** Lieferung, 1–139.

Hodge, D. S., Smith, B. D., and Smithson, S. B., 1970. Quantitative geophysical study of petrogenesis of syenites related to Laramie anorthosite, Wyoming, U.S.A. *Lithos*, **3**, 237–50.

Hurlbut, C. S., jr., and Griggs, D., 1939. Igneous rocks of the Highwood Mountains Montana. Part I. The laccoliths. *Bull. geol. Soc. Am.*, **50**, 1043–112.

Hytönen, K., 1959. On the petrology and mineralogy of some alkaline volcanic rocks of Toror Hills, Mt. Moroto, and Morulinga in Karamoja, Northeastern Uganda. *Bull. Comm. géol. Finl.*, **184**, 75–137.

Ianovici, V., Rădulescu, D., Rădulescu, I., and Săndulescu, M., 1968. Crystalline, Mesozoic complexes and volcanism in the East Carpathians (central sector). *23rd Intern geol. Cong. Prague, Guide to Excursion*, **47 AC**, 1–25.

Ikorskii, S. V., 1967. *The Organic Substances in the Minerals of Igneous Rocks* (in Russian), 1–120. Izd. 'Nauka', Leningrad.

Isachsen, Y. M., 1969. 'Origin of anorthosite and related rocks—a summarization', in Y. W. Isachsen, ed., Origin of Anorthosite and Related Rocks. *Mem. NY St. Mus. Sci. Serv.*, **18**, 435–45.

Johnson, R. W., 1969. Volcanic geology of Mount Suswa, Kenya. *Phil. Trans. R. Soc. ser. A*, **265**, 383–412.

Kempe, D. R. C., Deer, W. A., and Wager, L. R., 1970. Geological investigations in East Greenland. VIII. The petrology of the Kangerdlugssuaq alkaline intrusion, East Greenland. *Meddr Grønland*, **190**, 2, 1–49.

Koark, H. J., 1960. Zum Gefügeverhalten des Nephelins in zwei Vorkommen alkaliner kristalliner Schiefer, 'Grennait von Norra Kärr und canaditischer Gneis von Almunge'. *Bull. geol. Instn Univ. Uppsala*, **39**, 1–31.

Kopecký, L., 1966. 'Tertiary volcanics', in J. Svoboda, ed. *Regional Geology of Czechoslovakia. I. Bohemian Massif*, 554–81. Publ. House Czech. Acad. Sciences, Prague.

Kopecký, L., Doběs, M., Fiala, J., and Šťovičková, N., 1970. Fenites of the Bohemian massif and the relations between fenitization, alkaline volcanism and deep fault tectonics. *Sb. geol. ved. Praha*, **G. 16**, 51–107.

Kotina, R. P. and Yaroshevsky, A. A., 1970. On the possibility of desilication in the palingenic-metasomatic process of alkaline complexes settling within the granitoid formation (in Russian). *Geokhimiya*, **2**, 199–209.

Lacroix, A., 1922. *Minéralogie de Madagascar. II. Minéralogie Appliquée, Lithologie*. Paris, 1–694.

Larsen, E. S., 1941. Igneous rocks of the Highwood Mountains, Montana. Part II. The extrusive rocks. *Bull. geol. Soc. Am.*, **52,** 1733–52.

Locardi, E., 1965. Tipi di ignimbrite di magmi mediterranei. Le ignimbriti del vulcano di Vico. *Atti Soc. tosc. Sci. nat. Ser. A*, **72,** 55–173.

McCall, G. J. H., 1964. Froth flows in Kenya. *Geol. Rundschau*, **54,** 1148–95.

McKie, D., 1966. 'Fenitization'. in O. F. Tuttle and J. Gittins, eds. *Carbonatites*. 261–94. Interscience.

Meighan, I. G., 1968. Some aspects of the alkaline rocks of the Oslo Province. Unpublished Abstr. of Symposium: *Petrology of peralkaline rocks*, Reading.

Muir, I. D. and Smith, J. V., 1956. Crystallization of feldspars in larvikites. *Z. Kristallogr.*, **107,** 182–95.

Nash, W. P., Carmichael, I. S. E., and Johnson, R. W., 1969. The mineralogy and petrology of Mount Suswa, Kenya. *J. Petrology*, **10,** 409–39.

Nash, W. P., and Wilkinson, J. F. G., 1970. Shonkin Sag laccolith, Montana. I. Mafic minerals and estimates of temperature, pressure, oxygen fugacity and silica activity. *Contr. miner. petrol.*, **25,** 241–69.

Nockolds, S. R., 1954. Average chemical compositions of some igneous rocks. *Bull. geol. Soc. Am.*, **65,** 1007–32.

Oftedahl, C., 1960. Permian rocks and structures of the Oslo region. *Norg. geol. Unders.*, **208,** 298–343.

Olson, J. C., Shawe, D. R., Pray, L. C., and Sharp, W. N., 1954. Rare-earth mineral deposits of the Mountain Pass district San Bernardino County California. *Prof. Pap. U.S. geol. Surv.*, **261,** 1–75.

Payne, J. G., 1968. Geology and geochemistry of the Blue Mountain nepheline syenite. *Can. J. Earth Sci.*, **5,** 259–73.

Perhac, R. M. and Heinrich, E. Wm., 1964. Fluorite–bastnaesite deposits of the Gallinas Mountains, New Mexico and bastnaesite paragenesis. *Econ. Geol.*, **59,** 226–39.

Perchuk, L. L., and Ryabchikov, I. D., 1968. Mineral equilibria in the system nepheline-alkali feldspar-plagioclase and their petrological significance. *J. Petrology*, **9,** 123–67.

Petersilie, I. A., 1963. 'Organic matter in igneous and metamorphic rocks of the Kola Peninsula' (in Russian), in A. P. Vinogradov, ed. *Chemistry of the Earth's Crust* **1**, 48–62. Izd. Ak. Nauk., Moskva (English translation: Israel Program for Scientific Translations. Jerusalem, 1966).

Petersilie, I. A., Andreeva, E. D., and Sveshnikova, E. V., 1965. Organic matter in the rocks of some alkaline massifs of Siberia (in Russian). *Izv. Akad. Nauk. SSSR Geol. Ser.*, **6,** 26–38.

Petersilie, I. A., and Sørensen, H., 1970. Hydrocarbon gases and bituminous substances in rocks from the Ilímaussaq alkaline intrusion, South Greenland. *Lithos*, **3,** 59–76.

Philpotts, A. R., 1970. Mechanism of emplacement of the Monteregian intrusions. *Can. Mineral.*, **10,** 395–410.

Piotrowski, J. M., and Edgar, A. D., 1970. Melting relations of undersaturated alkaline rocks from South Greenland compared to those of Africa and Canada. *Meddr Grønland*, **181,** 9, 1–62.

Polanski, A., 1949. The alkaline rocks of the East-European Plateau. *Bull. Soc. Amis Sci. Lett. Poznán. Serie* **10**, 119–84.

Ramberg, H., 1948. On sapphirine-bearing rocks in the vicinity of Sukkertoppen (West Greenland). *Meddr Grønland*, **142,** 5, 1–32.

Rast, N., 1969. 'The initiation, ascent and emplacement of magmas', in G. Newall and N. Rast, eds., Mechanism of Igneous Intrusion. *Geol. J. Spec. Iss.*, **2,** 339–62.

Ronenson, B. M., 1966. The origin of miaskites and the associated rare-metal mineralizations (in Russian). *Geologiya mestorozhdeniy redkikh elementov*, **28,** 1–174. Izd. 'Nedra', Moskva.

Saha, P., 1961. System nepheline–albite. *Am. Miner.*, **46,** 859–85.

Schmincke, H. U., 1970. Ignimbrite sequence on Gran Canaria. *Bull. volcan.*, **33,** 1199–218.

Semenov, E. I., 1967. 'The mineralogical-geochemical types of derivatives of nepheline syenites. The activity of alkalis and volatiles' (in Russian), in R. P. Tichonenkova and E. I. Semenov, eds. *Mineralogy of Pegmatites and Hydrothermalites from Alkaline Massifs*, 52–71.

Semenov, E. I., 1969. *Mineralogy of the Ilímaussaq Alkaline Massif* (in Russian). Izd. 'Nauka', Moskva, 1–165.

Sheynmann, Yu. M., Apel'tsin, F. R., and Nechayeva, Ye. A., 1961. *Alkalic Intrusions, Their Mode of Occurrence and Associated Mineralization* (in Russian). Geologiya Mestorozhdenii Redkikh Elementov, 12–13, Moskva (Gosgeoltekhizdat), 177 pp. (Review in *Int. Geol. Rev.*, **5,** 451–8).

Smyth, C. H., 1913. Composition of the alkaline rocks and its significance as to their origin. *Am. J. Sci.*, **36,** 1–36.

Sørensen, H., 1960. On the agpaitic rocks. *Rep. 21st Intern Geol. Congr. Sess. Norden*, **13,** 319–27.

Sørensen, H., 1962. On the occurrence of steenstrupine in the Ilímaussaq massif, Southwest Greenland. *Meddr Grønland*, **167,** 1, 1–251.

Sørensen, H., 1969. Rhythmic igneous layering in peralkaline intrusions. *Lithos*, **2,** 261–83.

Sørensen, H., 1970. Internal structures and geological setting of the three agpaitic intrusions—Khibina, and Lovozero of the Kola Peninsula and Ilímaussaq, South Greenland. *Can. Mineral.*, **10,** 299–334.

Streckeisen, A., 1952 and 1954. Das Nephelinsyenit-Massiv von Ditro (Siebenbürgen). *Schweiz. miner. petrogr. Mitt.*, **32** and **34,** 251–308, 335–409.

Streckeisen, A., 1960. On the structure and origin of the nepheline-syenite complex of Ditro (Transylvania, Roumania). *Rep. 21st Intern Geol. Congr. Sess. Norden*, **13,** 228–38.

Streckeisen, A., 1964. Zur Klassifikation der Eruptivgesteine. *Neues Jb. Miner. Mh.*, **7,** 195–222.

Streckeisen, A., 1967. Classification and nomenclature of igneous rocks. *Neues Jb. Miner. Abh.*, **107,** 144–240.

Streickeisen, A., 1968. Stilpnomelan im Kristallin der Ostkarpathen. *Schweiz. miner. petrogr. Mitt.*, **48,** 751–80.

Sturt, B. A., and Ramsay, D. M., 1965. The alkaline complex of the Breivikbotn area, Sørøy, Northern Norway. *Norg. geol. Unders.*, **231,** 1–164.

Tazieff, H., 1969. 'Mechanism of ignimbrite eruption', in G. Newall and N. Rast, eds., Mechanism of Igneous Intrusion. *Geol. J. Spec. Iss.*, **2,** 157–64.

Tilley, C. E., 1957. Problems of alkali rock genesis. *Q. Jl. geol. Soc. Lond.*, **113,** 323–60.

Toulmin, P. 3rd., 1964. Bedrock geology of the Salem quadrangle and vicinity, Massachusetts. *Bull. U.S. geol. Surv.*, **1163–A,** 1–79.

Tröger, W. E., 1969. *Optische Bestimmung der gesteinsbildenden Minerale. Teil 2. Auflage 2.* Publ. by: H. U. Bambauer, F. Taborszky and H. D. Trochim. E. Schweizerbart'sche Verlagsbuchhandlung Stuttgart, 1–822.

Tyler, R. C. and King, B. C., 1967. The pyroxenes of the alkaline igneous complexes of eastern Uganda. *Mineralog. Mag.*, **36,** 5–21.

Upton, B. G. J., 1960. The alkaline igneous complex of Kûngnât Fjeld, South Greenland. *Meddr Grønland*, **123,** 4, 1–145.

Upton, B. G. J., 1961. Textural features of some contrasted igneous cumulates from South Greenland. *Meddr Grønland*, **123,** 6, 1–31.

Upton, B. G. J., 1964. The geology of Tugtutôq and neighbouring Islands, South Greenland. Part II. Nordmarkitic syenites and related alkaline rocks. *Meddr Grønland*, **169,** 2, 1–62.

Ussing, N. V., 1898. Mineralogisk-petrografisk Undersøgelse af grønlandske Nefelinsyeniter og beslægtede Bjergarter. *Meddr Grønland*, **14,** 1–220.

Ussing, N. V., 1912. Geology of the country around Julianehaab Greenland. *Meddr Grønland*, **38,** 1–376.

Varet, J., 1969 a. Les phonolites agpaïtiques et miaskitiques du Cantal septentrional (Auvergne, France). *Bull. volcan.*, **33,** 621–56.

Varet, J., 1969 b. Les pyroxènes des phonolites du Cantal (Auvergne, France). *Neues Jb. Miner. Mn.*, **4,** 174–84.

Verwoerd, W. J., 1966. 'Fenitization of basic igneous rocks', in O. F. Tuttle and J. Gittins, eds. *Carbonatites*. Interscience, 295–310.

Vlasov, K. A., Kuz'menko, M. V., and Es'kova, E. M., 1959. *The Lovozero Alkaline Massif* (in Russian). Izd. Acad. Sci., Moskva, 1–624 (English translation: Oliver & Boyd, Edinburgh, 1966).

Vlasov, K. A. (ed), 1968. *Geochemistry and Mineralogy of Rare Elements and Genetic Types of their Deposits. III. Genetic Types of Rare-Element Deposits.* Israel Program for Scientific Translations. Jerusalem, 1–916.

Vorobieva, O. A., 1963. 'Problems of alkaline magmatism' (in Russian), in G. D. Afanasiev, ed. *Problems of Magmas and Genesis of Igneous Rocks*. Izd. Akad. Nauk, Moskva, 76–83.

Wager, L. R., 1965. The form and internal structure of the alkaline Kangerdlugssuaq intrusion, East Greenland. *Mineralog. Mag.*, **34,** 487–97.

Wager, L. R., and Brown, G. M., 1968. *Layered Igneous Rocks*. Oliver & Boyd, Edinburgh and London, 1–588.

Watt, W. S., 1966. Chemical analyses from the Gardar igneous province, South Greenland. *Rap. Grønlands geol. Unders.*, **6,** 1–92.

Welman, T. R., 1970. The stability of sodalite in a synthetic syenite plus aqueous chloride fluid system. *J. Petrology*, **11,** 49–71.

Wilkinson, J. F. G., 1965. Some feldspars, nephelines and analcimes from the Square Top intrusion, Nundle, N.S.W. *J. Petrology*, **6,** 420–44.

Windley, B. F., 1969. 'Primary quartz ferro-dolerite/garnet amphibolite dykes in the Sukkertoppen region of West Greenland', in G. Newall and N. Rast, eds., Mechanism of Igneous Intrusion. *Geol. J. Spec. Iss.*, **2,** 79–92.

Wright, J. B., 1965. Petrographic sub-provinces in the Tertiary to recent volcanics of Kenya. *Geol. Mag.*, **102,** 541–57.

Yagi, K., 1966. The system acmite–diopside and its bearing on the stability relations of natural pyroxenes of the acmite–hedenbergite–diopside series. *Am. Miner.*, **51,** 976–1000.

Yoder, H. S., jr., 1969. 'Experimental studies bearing on the origin of anorthosite.' in Y. W. Isachsen, ed., Origin of Anorthosite and Related Rocks, 8. *Mem. NY St. Mus. Sci. Serv.*, **18,** 13–55.

Zlobin, B. I., 1959. Parageneses of dark minerals in alkalic rocks and a new expression for the agpaitic coefficient. *Geochemistry*, 507–18.

II.3. NEPHELINITES AND IJOLITES

D. K. Bailey

Probably the most perplexing groups of alkaline rocks are those composed predominantly of feldspathoid and clinopyroxene, the melteigite-ijolite-urtite intrusives and the nephelinite lavas. Not only do these rocks pose severe problems for any process of differentiation from a basaltic parent, but they show abundant evidence of abnormal amounts of associated volatiles which, in the intrusions, superimpose an additional problem of complex metasomatism. Consequently, extreme caution is needed in interpreting the field, mineralogical and chemical relations.

II.3.1. Definitions

Ijolite Series. In the most recent description of the type area for ijolite, Lehijärvi (1960) follows the nomenclature of Johannsen (1938), namely: *urtite* when nepheline exceeds 70%; *ijolite* when nepheline is 70–50%; and *melteigite* when nepheline is less than 50%. Other authors, e.g. Pulfrey (1950), King and Sutherland (1960), have preferred to extend the ijolite range as far as rocks with colour indices of 70 at the expense of melteigite (cf. II.1). In view of the nomenclatural jungle that has grown up around the alkaline rocks it would seem preferable to retain Johannsen's division, which is close to that of Brøgger's definition of melteigite from the type area (1921).

Lava Series. Nomenclature ambiguities are troublesome in the lavas, too. *Nephelinite* seems to be generally accepted as an essentially feldspar-free nepheline-pyroxene lava, but when any olivine is present Johannsen calls the rock nephelite-basalt. The term nephelinite is not particularly apt, but nephelite-basalt is wholly illogical. Quite apart from debasing the term 'basalt' it requires the use of yet another name when olivine becomes an essential constituent, e.g. tannbuschite (Johannsen, 1938) or ankaratrite (Lacroix, 1916: quoted by Johannsen, 1938). Fortunately many authors now tend to use *olivine nephelinite* for those lavas in which olivine is an essential constituent. This is the terminology preferred in this account.

II.3.2. Mineralogy

Ijolite Series. The principal constituents are nepheline and pyroxene with various amounts of andradite-melanite garnet (vars. schorlomite and iivaarite) and wollastonite, and typically accessory amounts of perovskite, titanite, apatite, biotite and iron-titanium oxides. Late-stage or secondary minerals frequently include cancrinite, sodalite, analcime, pectolite and various zeolites. Calcite is variously described as primary or secondary and the rocks may grade towards carbonatite, e.g. pyroxene-sövite (von Eckermann, 1948).

Most analysed nephelines from these rocks have high potash values, those from the ijolite type area showing almost constant alkali values, $100K/(K + Na) \sim 19$. They may also contain gas and liquid inclusions and oriented pyroxene needles, which from the analyses may be presumed to be aegirine. Oriented aegirine needles appear to be a fairly common feature of plutonic nephelines, cf. II.2*; Vlasov *et al.* (1966). These needles may be an exsolution feature, because nepheline shows extensive solution in the system nepheline-acmite (Bailey and Schairer, 1966).

Clinopyroxenes in the ijolite series have a wide range of composition from titanaugite in the cafemic members, through aegirine-diopside and aegirine-hedenbergite to aegirine. Zoning is common, with a pronounced tendency for enrichment in aegirine in the margins, corresponding with a late crystallization of aegirine needles in the groundmass. Its appearance in pegmatites

* Readers are referred to the relevant Chapters of this book where similarly cited.

and metasomatized country rocks (fenites) confirms its residual character. In a very general sense the increase of aegirine in the pyroxene correlates with increasing nepheline content in the rocks, but some pyroxene-rich rocks have high aegirine contents (see Lehijärvi, 1960, Table 1). This anomalous composition is probably attributable to metasomatism (the aegirine-rich rocks typically have acicular, radiating green pyroxenes) and contributes to the confusion in this rock series. Melanite is more common, and may be an essential mineral, in the pyroxene-rich rocks; it may grade towards andradite in the lighter coloured rocks. It is commonly strongly zoned and may be euhedral or irregular and poikilitic (poikiloblastic?). Wollastonite is more common in the nepheline-rich members of the series, and presents pitfalls to a field classification based on colour index. Melilite is not recorded from ijolites proper, but melilite-rich rocks such as uncompahgrite and turjaite are part of the association and normally carry some pyroxene and nepheline.

Lava Series. There are no lava equivalents of urtites and ijolites. Nephelinites correspond most closely with the cafemic melteigites, with pyroxenes ranging from titanaugite phenocrysts (in olivine-nephelinites) to acicular groundmass aegirine. Like pyroxene, nepheline seems to form over the whole range of crystallization, though when nepheline phenocrysts are abundant, sanidine usually appears in the groundmass, and the lava becomes a feldspathic or phonolitic nephelinite. Accessory and secondary minerals are similar to those of the plutonic rocks, including melanite but with the addition of melilite and the notable exception of wollastonite. Glass may be present in small amounts.

The appearance of olivine in the lavas is the most remarkable difference from the plutonic rocks. King (1965) infers that this difference is due to crystallization of the plutonic rocks in the presence of volatiles. When olivine occurs as occasional phenocrysts in nephelinites it is commonly corroded; in the olivine-nephelinites, however, it forms euhedral phenocrysts and may appear as a groundmass mineral. In the latter rocks nepheline phenocrysts are absent. In composition the olivines are normally forsterite-rich, but from some lavas they are exceptionally rich in Ca_2SiO_4 (Sahama and Hytönen, 1958).

II.3.3. Textures

Rocks of the ijolite series are characterized by extreme variability of grain size and texture, from their appearance in outcrop down to the microscopic scale. Mineral segregation (in clots and bands) pegmatoid-breccia and vein-structures are common, making these rocks particularly intractible to mapping or to generalizations about their composition. Published descriptions suggest a tendency for the more leucocratic members to vein and intrude the pyroxene-rich members.

Nepheline is commonly euhedral in the leucocratic rocks and less frequently in the ijolites and melteigites. Pyroxene tends to be euhedral in the melteigites, but acicular aegirine is common throughout and underlines the fruitless nature of any textural generalizations about a series of rocks in which metasomatism has frequently determined the final crystallization.

The sequence of crystallization in the lavas is more clearly related to their composition. Olivine is characteristically early among the phenocrysts, followed by pyroxene, closely succeeded by nepheline; the two latter minerals continue to crystallize throughout. Porphyritic melanite may be present in olivine-free nephelinite. Melilite occurs as a phenocryst only in melilite-rich lavas that are properly termed melilitites. In the more melanocratic lavas nepheline may be entirely interstitial and difficult to determine optically. The final residue in the average nephelinite may be glassy, or may take the form of analcime and zeolites filling interstices and vesicles.

II.3.4. Chemistry

Ijolite Series. With most rock series it is difficult to make useful generalizations about the bulk chemistry without regard to the mode and the compositions of the individual minerals, but in the case of the ijolite series any generalizations can be positively misleading. This springs partly from the variability of grain size, texture and fabric, but is largely due to a classification based on nepheline content, regardless of the proportions of other minerals and, most particularly the compositions of the pyroxenes. A rock composed largely of aegirine is quite different from one composed largely of titanaugite, but both are called melteigite. A glance at the published analyses of melteigites alone shows

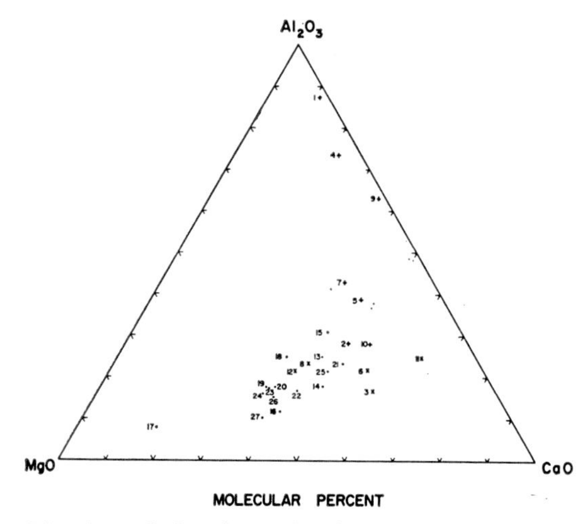

Fig. 1 Variation in molecular CaO–MgO–Al_2O_3 in the analyses listed in Tables 1 and 2 (numbers 1–24). Numbers 25, 26 and 27 are 'average melilite nephelinite', 'average olivine melilite nephelinite' and 'average olivine melilitite' of Nockolds (1954). Lavas are shown as points; urtites and ijolites as plus signs; melteigites as crosses

for example: SiO_2 values ranging from 38–52%; Al_2O_3 from 3–19%; CaO from 8–21%; and Na_2O from 3–11%. Some of the variation is due to minerals such as biotite and melanite assuming importance, but considerable variation is due simply to the composition range of the pyroxene. Some of the confusion about ijolitic rocks could be cleared if the proportions of the minerals other than nepheline, and the composition of the pyroxene, were plainly stated. A new classification is needed but is probably best deferred, pending a sounder knowledge of the rocks. For the present, perhaps, a growing awareness of the problem among petrologists may be the best remedy. In Table 1 a selection of analyses are presented, chiefly to illustrate the scope of the ijolite series. Fig. 1 represents an attempt to show ijolitic and nephelinitic rocks in terms of three variables that are more informative than mere nepheline content. Inspection of nephelinite and ijolite analyses indicates that Al_2O_3, MgO and CaO tend to vary most significantly. The variation between the volcanic and plutonic suites is distinct. The ankaratrite (17) would seem to be olivine cumulitic.

A statistical test of the chemical variation among nephelinites and associated lavas of the West Eifel and South-west Uganda (F. E. Lloyd, pers. comm. 1970) using principal component analysis, has revealed that the first principal component, which accounts for 67% of the total chemical variation involves CaO + Fe_2O_3 versus MgO + FeO. In Fig. 2 these two

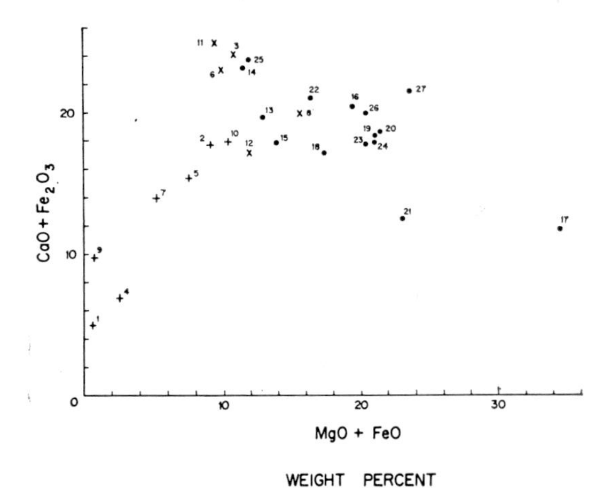

Fig. 2 Diagram showing CaO + Fe_2O_3 versus MgO + FeO in the analyses of Tables 1 and 2 (numbers 1–24). Numbers 25, 26 and 27 are 'average melilite nephelinite', 'average olivine melilite nephelinite', and 'average olivine melilitite' (Nockolds, 1954). Symbols as in Fig. 1

factors have been plotted for the analyses given in Tables 1 and 2, resulting in a remarkable discrimination between the lavas and the intrusives. The distribution in Fig. 2 could be partly ascribed to variations in olivine and nepheline contents but this is not a wholly adequate explanation. Research now in progress by F. E. Lloyd should shed more light on this variation pattern.

TABLE 1 Analyses of Rocks of the Ijolite Series

	1	2	3	4	5	6	7	8	9	10	11	12
SiO_2	45·43	46·15	40·64	42·59	42·58	41·90	40·85	40·90	41·73	37·29	42·05	51·12
Al_2O_3	28·77	15·70	10·58	27·42	18·46	12·20	20·46	13·65	25·13	14·41	12·05	10·41
Fe_2O_3	3·10	3·55	4·18	2·49	4·01	6·41	3·86	6·67	1·92	4·23	7·93	4·72
FeO	0·40	3·40	4·18	1·89	4·19	4·32	5·08	7·14	0·58	6·10	5·06	4·47
MnO		0·18	0·28	0·09	0·20	0·22	0·09		0·13	0·32	0·96	0·20
MgO	0·22	5·52	6·47	0·69	3·22	5·45	3·56	8·35	0·18	4·15	2·18	7·38
CaO	1·86	14·16	19·91	4·38	11·38	16·60	10·06	13·28	7·84	13·69	17·01	10·43
BaO			0·11						0·05	0·14		
Na_2O	16·16	7·24	4·75	14·12	9·55	5·10	8·71	3·06	11·74	5·61	4·95	6·77
K_2O	3·38	2·61	1·86	3·82	2·55	2·66	2·98	1·65	5·72	4·22	3·15	2·31
H_2O			0·14						0·40	0·04		0·11
H_2O+		0·93	0·27	0·42	0·55	0·87	0·98	0·50	1·86	0·69	0·67	1·05
TiO_2		0·38	2·24	0·35	1·41	2·21	2·44	3·99	0·08	3·46	2·36	0·66
CO_2			2·08	1·30	0·38	0·82	0·40		2·06	3·29		0·25
P_2O_5		0·77	1·91	0·44	1·52	1·24	0·53	0·81	0·28	0·77	1·66	0·14
Cl			0·03						tr.	0·33		
F			0·12						tr.	0·32		
S			0·05						0·03	2·43	0·54	
ZrO_2			0·10									
Total	99·32	100·59	99·90						99·73	101·68	100·54	100·02
CIPW Norm												
or		5·54		8·3	10·0	6·7		9·4		3·89	18·40	13·64
ab								1·6			0·52	1·49
an		2·67		0·6		2·5	8·1	18·6		3·89		
lc		7·75		10·9	3·9	7·4	13·9		26·60	16·57	1·39	
ne		33·15		64·5	43·7	23·3	39·8	13·1	52·54	23·28	22·55	21·24
kp												
hl										0·59		
th										0·43		
ac									2·77			13·67
ns												0·24
di									1·83	22·46	15·13	
wo		26·12		4·3	18·5	27·9	11·2	17·6	0·12	3·13	22·06	20·56
en		13·74		1·7	8·0	13·6	8·1	14·7				18·37
fs		3·02		0·9	2·4		2·1	0·7				7·48
ol							0·8	4·5				
cs							2·7		7·05			
mt		5·15		3·7	5·8	8·1	5·6	9·7	1·39	1·86	7·41	
hm						0·8				2·88	2·88	
il		0·71		0·6	2·7	4·3	4·6	7·6	0·15	6·54	4·51	1·26
pf												
ap		1·81		1·0	3·6	2·9	1·3	1·9	0·62	2·02	3·90	0·34
pr										4·48	1·12	
cc				3·0	0·9	1·8	0·9		4·70	7·50		0·57

Legend:

1. Type urtite. Ramsay (1896) from Johannsen (1938).
2. Ijolite from the type area. Lehijärvi, (1960, p. 33).
3. Type melteigite. Brögger (1921) from Johannsen (1938).
4. Average urtite (6 analyses). Nockolds (1954).
5. Average ijolite (11 analyses). Nockolds (1954).
6. Average melteigite (9 analyses). Nockolds (1954).
7. Ijolite with titanaugite (3 analyses). Nockolds (1954).
8. Melteigite with titanaugite (3 analyses). Nockolds (1954).
9. Urtite from Usaki, Kenya. Pulfrey, (1950, p. 428).
10. Fine-grained ijolite, Magnet Cove, Arkansas, Erikson and Blade, (1963, p. 33).
11. Melanite-melteigite-porphyrite, Alnö, Sweden, von Eckermann, (1948, p. 92).
12. Melteigite, Iivaara. Lehijärvi, (1960, p. 39).

Lava Series. Chemically, nephelinites present a more coherent group, though the range of composition is still daunting to any attempt at generalization. Variation in the cafemic mineralogy, and especially in the pyroxene composition, is not such an acute problem as in the plutonic rocks; the chemical variations in nephelinites are due largely to the proportions of olivine and nepheline, and gradations in the rocks towards melilitites on the one hand, and phonolitic nephelinites on the other. Consequently, if nephelinite and olivine-nephelinite were defined as rocks composed essentially of nepheline, pyroxene, and olivine, generalizations about their chemical relations could be valid, and possibly even meaningful! It seems likely that the petrology of nepheline-pyroxene rocks will be elucidated only by using the undoubted magmatic rocks as a starting point. Judging from the mineralogy and chemistry of the lavas, many of the rocks of the ijolite series appear to be differentiates in which volatile activity and metasomatism have played significant, if not essential, rôles.

Table 2 gives some illustrative analyses of nephelinite lavas and dykes, and two sets of averages for nephelinite and olivine-nephelinite, which show some interesting contradictions. The two 'average' olivine-nephelinites are basically similar in analysis and norm, but the two 'average' nephelinites show differences that, if they are not significant, must render all four averages invalid. Analysis 20, a melilite olivine nephelinite from Honolulu, has been included as a representative of the melilite-bearing lavas commonly associated with nephelinites, and because, together with the nephelinite of Analysis 19, it is currently being used in experimental studies by Yoder.

II.3.5. Associations

Some of the rocks associated with ijolitic and nephelinitic suites have been mentioned in the previous discussion, but will be summarized here.

With the incoming of plagioclase, melteigite grades towards theralite or nepheline gabbro and nephelinite towards basanite and tephrite, but this is not a common association. Ijolite and urtite sometimes carry small amounts of alkali feldspar, and as this increases in quantity they pass into malignite and then nepheline syenites. An analogous gradation is found in nephelinite, phonolitic nephelinite, and phonolite. Nephelinite also shows a gradation towards melilitite with increasing melilite content, but although melilite rich rocks are found in many plutonic complexes, there is little evidence of a gradational series from ijolite. This is perhaps another facet of the oddly divergent trends of the ijolite and nephelinite series noted by King and Sutherland (1960).

Dunite, peridotite and pyroxenite are less common associates of the plutonic series (cf. IV.2; VI.7). The lava series, at the olivine melilitite end, shows affinities with kimberlite composition, which is represented by late dykes of alnöite-kimberlite in the plutonic complexes.

The rocks of the plutonic suite commonly contain carbonate, and CO_2 appears in small amounts in many nephelinite analyses. In many ijolite complexes carbonatite occurs as a separate rock, and this obvious enrichment in volatiles is manifested at the surface in the predominantly fragmental nature of nephelinite volcanic piles. The massive pile of Mt. Elgon, probably the largest nephelinite volcano, is estimated to contain 30–40% of zeolites and calcite (Davies, 1952). This richness in volatiles and carbonates must not be overlooked in any consideration of nephelinite genesis.

II.3.6. Distribution

All the clearly defined nephelinite volcanoes are in the anorogenic sectors of the crust—either in the ocean basins or the stable continents—and past nephelinite volcanicity appears to conform to this pattern.

Most ijolitic complexes were emplaced in tectonically 'quiet' or epeirogenic regions, and are typically unconnected with any orogenic movement. Exceptions to this may occur, notably in the eastern Rocky Mountains, Ice River in British Columbia (Rapson, 1966) and

TABLE 2 Analyses of Rocks of the Nephelinite Series

	13	14	15	16	17	18	19	20	21	22	23	24
SiO_2	41·51	44·08	42·0	43·6	41·44	38·60	39·14	35·83	39·07	41·85	39·40	40·29
Al_2O_3	12·66	10·07	14·1	8·3	7·01	15·09	12·23	10·86	12·82	10·35	11·50	11·32
Fe_2O_3	7·72	8·85	5·5	4·9	3·16	5·39	5·28	6·43	8·75	6·89	5·14	4·87
FeO	6·32	3·37	8·7	6·1	7·89	7·83	8·02	10·17	6·39	6·23	8·05	7·69
MnO	0·18	0·36	0·30	0·28	0·18	0·20	0·20	0·23	0·26	0·16	0·20	0·22
MgO	6·49	7·96	5·1	13·2	26·48	9·41	12·86	11·09	6·14	10·13	12·20	13·28
CaO	11·91	14·31	10·3	15·6	8·23	11·72	13·11	12·21	14·20	14·03	12·61	12·99
Na_2O	4·63	3·17	7·4	3·1	1·83	4·52	3·95	5·45	4·09	3·55	3·43	3·14
K_2O	1·65	2·10	2·1	0·8	0·94	1·56	1·23	1·82	2·07	1·44	1·20	1·44
H_2O	0·41	0·99	2·6	1·5	0·35	1·00	0·16	0·27		2·48	2·55	
H_2O+	3·51	2·80			1·57	1·20	0·48	0·72	1·59			1·08
TiO_2	2·25	1·49	2·1	2·3	0·72	2·50	2·54	2·64	3·86	2·36	3·04	2·90
CO_2	0·10	0·03					0·13					
P_2O_5	1·05	0·96	0·16	0·22	0·26	0·93	0·58	1·38	0·76	0·71	0·88	0·78
Total	100·39	100·54	100·4	99·9	100·06	99·95	99·91	99·78		100·18	100·20	
CIPW Norm												
or	9·4	12·2					4·11	6·00	1·7	5·56		
ab	5·2	5·8										
an	9·2			6·1	8·1	16·0	11·95		10·3	8·06	12·51	12·8
lc		7·2	10·0	3·9	4·4	2·8			8·3	2·18	5·67	6·5
ne	18·5	11·4	32·7	14·5	8·5	24·6	18·18	24·99	18·7	16·47	15·62	14·2
kp						3·6						
ac			1·8									
ns												
di	34·1	43·0	39·6	52·0	22·8	29·2	30·60	27·17		44·54	33·01	
wo		0·9							23·0			17·2
en									15·3			13·1
fs												2·2
ol			1·6	8·9	46·5	12·8	17·16	16·90		4·61	14·83	16·8
cs			0·5		1·2						0·86	1·6
mt	11·1	7·7	6·7	7·2	4·4	5·7	7·66	9·28	10·4	9·98	7·42	7·2
hm		3·5							1·4			
il	4·3	2·9	4·0	4·3	1·4	3·4	4·86	5·01	7·3	4·56	5·78	5·5
ap	2·4	2·4	0·3	1·5	0·7	1·9	1·34	3·23	1·8	1·68	2·02	1·8
cc	0·2											
hy	2·2							5·11				

Legend:

13. Nephelinite ('melanephelinite'), Napak, Uganda. King, p. 77. Table 2, Analysis 8, 1965.
14. Nephelinite ('melanephelinite'), Mt. Elgon, Uganda. King, p. 76, Table 1, Analysis 9, 1965.
15. Olivine melanephelinite, Moroto Mt., Uganda, Varne, p. 175, Table 2, Analysis 4, 1968.
16. Olivine melanephelinite, Moroto Mt., Uganda. Varne, p. 175, Table 2, Analysis 1, 1968.
17. Ankaratrite, Mt. Elgon, Uganda. King, p. 76, Table 1, Analysis 8, 1965.
18. Tannbuschite, Trinidade, W. Atlantic. de Almeida, p. 167, Table XVIII, Analysis 2, 1961.
19. Olivine nephelinite, Oahu, Hawaiian Islands, Yoder and Tilley, p. 362, Table 2, Analysis 24, 1962.
20 Melilite olivine nephelinite, Honolulu. Tilley, Yoder and Schairer, p. 71, Table 2, Analysis 0, 1965.
21. Average nephelinite (8 analyses). Nockolds (1954).
22. Average nephelinite (15 analyses). Wood, p. 165, Table 8, Analysis 3, 1969.
23. Average olivine nephelinite (39 analyses). Wood, p. 165, Table 8, Analysis 1, 1969.
24. Average olivine nephelinite (21 analyses). Nockolds (1954).

the various complexes of Colorado (Tuttle and Gittins, 1966), and in the High Atlas of Morocco (Agard, 1960), but although these are fold mountain belts the tectonic condition at the time of emplacement of the alkaline complexes is not yet clear (cf. III.3).

In the ocean basins the distribution of nephelinite volcanicity is varied (cf. IV.7). In the Atlantic, nephelinites are most strongly in evidence on those islands furthest from the median rise: Bermuda, Fernando de Noronha, Martim Vas and Trinidade in the western Atlantic and the Canary and Cape Verde Islands in the east. Some of these volcanoes rise from the abyssal plains and consist predominantly, if not wholly (Trinidade), of nephelinite and associated rocks. In the Pacific the distribution shows no such simple relationship. Nephelinites and their associates are reported from several islands in the Micronesia–Polynesia archipelagos (Truk, Ponape, Samoa, Cook, and Tahiti) forming a south-east trending arc in the south-western portion of the Pacific basin. In the mid-Pacific, the waning stages of activity in the Hawaiian chain have provided the well-known nephelinitic association of the Honolulu Series (Winchell, 1947).

As might be expected these volcanics may carry xenoliths of their coarse-grained equivalents, such as melteigite; and deeper erosion on Tahiti (Lacroix, 1929), Cape Verde (Part, 1950) and the Canaries has revealed the intrusive members of the association.

The nephelinite association and the intrusive ijolite suite are represented in all the major continental areas, where they appear to be much more common than in the oceans.

Present records appear to show only sparse representation of nephelinites and ijolites in India, S. America (Tuttle and Gittins, 1966) and Australasia (Joplin, 1964). N. American localities are listed by Barker (cf. III.3).

The rocks are well represented in Europe, in the Carboniferous of the British Isles; in the Tertiary–Recent volcanism of France and Germany (IV.4; Jung and Brousse, 1961). The distribution in the USSR has been summarized by Vorobieva (1960), see also III.4 and IV.2.

Undoubtedly the outstanding developments of nephelinites and the plutonic suite are in the African continent, particularly along the rift zones. A comprehensive survey of the distribution of alkaline rocks in eastern and southern Africa was presented by King and Sutherland (1960), and this has been elaborated, both in terms of distribution and time sequence, in various contributions to the UNESCO Seminar on the East African Rift System (1965). A large number of new analyses of African nephelinites has been made by Wood (1968) who draws particular attention to the fact that olivine-poor nephelinites are especially abundant in Africa and other continental provinces, in contrast with the preponderance of olivine nephelinites in the oceanic provinces.

II.3.7. Form

Ijolitic rocks are present in varying amounts in the unusually large alkaline plutons, such as Khibina and Lovozero (IV.2), and Pilanesberg (cf.III.5) (Pilaansberg) (pers. comm. E. A. Retief, 1968) but the most common type of ijolitic complex is subvolcanic. This is typically cylindrical in form, a few kilometres in diameter, and made up of several intrusions, giving rise to concentric arcuate outcrops when seen in plan. Carbonatite is frequently the final intrusive phase (cf. IV.2).

There is little doubt that in many instances there was an overlying nephelinitic volcano. The classic and oft-figured example is Napak, in E. Uganda (King, 1949), where a plug of ijolite-carbonatite forms the eroded core of a large nephelinite volcano, the remnants of which consist of 97% pyroclastics. The central plug has a diameter of 2 km, and the original volcano had a basal diameter of over 30 km and a possible height greater than 4 km. A recent detailed account of the ijolites at Napak and other complexes in E. Uganda is provided by King and Sutherland (in Tuttle and Gittins, 1966), and a petrological survey by King (1965).

In most ijolite complexes metasomatism is much in evidence (cf. VI.7), even where there is good evidence of igneous intrusion. The extreme case is reached in some complexes where the investigator has concluded that the ijolitic rocks are entirely metasomatic, either formed *in situ* or subsequently emplaced as a breccia (Parsons, 1961). The two origins, magmatic and metasomatic, are not mutually exclusive: they may, in some cases, be stages in the same process. If, however, some ijolites are rheomorphic fenites (von Eckermann, 1948) and others are true igneous differentiates from a nephelinite magma, it should be possible eventually to discern the difference in their mineralogy and petrochemistry.

II.3.8. Nephelinite Synthesis

In recent years considerable attention has been given to synthetic studies at atmospheric pressure, on joins in the system Na_2O–CaO–MgO–Al_2O_3–SiO_2 (Schairer and Yoder, 1964; Schairer, Tilley and Brown, 1968), the join diopside–akermanite–nepheline (Onuma and Yagi, 1967) and various other joins involving melilite compositions. Most of this work has followed from earlier studies on basalts (Yoder and Tilley, 1962) defined in terms of a composition tetrahedron, forsterite–diopside–nepheline–silica (the 'basalt tetrahedron'). This model was found inadequate to describe melilite-bearing lavas, which must be referred to an enlarged system forsterite–larnite–nepheline–silica, which takes account of the calcium-rich composition of most melilites (cf. V.1). This is the 'expanded basalt tetrahedron'. A crystallization 'flow-diagram' was deduced from the first studies but this has been modified by later experiments. Modifications were found when anorthite was added to the system (Schairer, Tilley and Brown, 1968) where it seemed that one important crystallization pattern led to a single eutectic (L) at which nepheline–plagioclase–diopside–melilite crystallize simultaneously. The crystallization sequences are such that diopside and nepheline are never seen to crystallize together without either melilite or plagioclase. The same result was found by Onuma and Yagi (1967) on *di-aker-ne*, and by Schairer and Yoder (1960a) on *fo-ne-di* (V.1, Fig. 3) and was implicit in the earlier work of Bowen (1922) on *ne-di*. In fact, none of the new joins has intersected the assemblage forsterite + diopside + nepheline + liquid which would approximate to olivine nephelinite. So far as I can ascertain this assemblage has been found only in the join diopside–nepheline–SiO_2 (Schairer and Yoder, 1960b; V.1, Fig. 2). Forsterite composition does not lie in this join and olivine must react out under equilibrium crystallization. Strangely enough, the assemblage forsterite + diopside + nepheline + liquid does not appear in the composition-plane forsterite–diopside–nepheline (Schairer and Yoder, 1960a; V.1, Fig. 3). Thus, the 'olivine–nephelinite' which figures in the recent flow diagrams can be regarded only as a device to fit the proposed model of crystallization. Olivine-free nephelinites (without melilite) cannot be fitted to the experimental model at all, and are consequently even more enigmatic! Most of the apparent conflicts between the experimental products and the rocks have arisen from the use of a quaternary model (flow sheet) to describe what is essentially a quinary system. Distinct quinary behaviour in the expanded basalt tetrahedron has since been recorded by Schairer and Yoder (1969 and 1970) and forced them to resort to experiments in the system CaO–MgO–Al_2O_3–SiO_2 as a truly quaternary model for the melilitite–nephelinite association. O'Hara and Biggar (1969) also base their hypotheses about these rocks on the same quaternary system. Relationships in the system CaO–MgO–Al_2O_3–SiO_2 at one atmosphere pressure are complex, but still far removed from a natural system in which the additional components Na_2O, K_2O, Fe, O, CO_2 and H_2O are *evidently* important. The analogies drawn from CaO–MgO–Al_2O_3–SiO_2 lack conviction and do not seem justified. Nephelinites obviously belong in a much more complex chemical system, and it is probable also that the interrelations between the magmas of the nephelinite association require explanation in terms of pressure variation. This is seen when the melting relations of nephelinitic lavas are examined.

Compared with basalts the nephelinitic lavas tend to have high liquidus temperatures (Tilley,

Yoder and Schairer, 1965) but one of the most notable features in their crystallization, compared with basalts, is the large range of temperature between the crystallization of the major phases. An olivine-nephelinite from Hawaii has olivine on the liquidus at 1305 °C but nepheline does not start to crystallize until 1085 °C. If the narrow crystallization temperature span shown by the basalts is taken to indicate their proximity to a univariant condition (Yoder and Tilley, 1962) then one must infer that nephelinites are considerably removed from this condition at atmospheric pressure. Such a wide temperature span is also inconsistent with the 1 atmosphere synthetic results (Schairer, Tilley and Brown, 1968) which seem to show the major 'nephelinite' invariant points close together in composition and temperature.

One-atmosphere experiments in the system Na_2O–Al_2O_3–Fe_2O_3–SiO_2 (Bailey and Schairer, 1966) have shed some light on the lower-temperature, end-stages of nephelinite crystallization. A quaternary reaction point was located, acmite + hematite + nepheline + albite + liquid, which was close in composition to the join nepheline–acmite, i.e. the liquid at the invariant point was relatively rich in the acmite and nepheline components. This liquid is similar to the composition of the type-malignite (Johannsen, 1938). It is interesting because the final stages of nephelinite crystallization commonly involve aegirine, and nepheline, and, in many instances, alkali feldspar (or aegirine + analcime). The synthetic invariant point is thus the analogue of the nephelinite residium and, of course, malignite. Furthermore, it is linked by a univariant crystallization path, acmite + nepheline + albite + liquid, to a quaternary eutectic which is the equivalent of a peralkaline nepheline syenite or phonolite. It illustrates a possible lineage from nephelinite to phonolite, and malignite to nepheline syenite. A curious consequence of the proximity of the quaternary reaction point to the join nepheline–acmite is that small fluctuations in conditions could alter the subsequent path of crystallization, so that it follows a trend of increasing acmite content in the liquid. In the natural environment, fluctuations in the volatile phase might well initiate such a trend and give rise to the aegirine-rich 'melteigites' described from many localities. It may be reasonable to extrapolate these findings to consideration of the subvolcanic conditions, because Nolan (1966) located a similar point in the system acmite–nepheline–albite–H_2O at 1 kilobar pressure.

II.3.9. Origin

A general review of the origin of nephelinites and ijolites, if it is to be kept within reasonable bounds, must be selective, and even then has to be restricted to outlining the main hypotheses. Given these terms of reference, such a review must be biased, no matter how well-intentioned. I have therefore chosen to focus attention on those hypotheses that appear most amenable to test. It must be stated at the outset that a general *primary* origin by crustal metasomatism, possibly followed by rheomorphism, cannot be considered even though some ijolites may have formed by this process (cf. VI.7). The eruption of nephelinite magmas, particularly in the ocean basins, must eliminate this as a serious general explanation. For the same reason, a general origin involving assimilation of sialic crust must be ruled out, and also because the available trace element and isotopic data indicate a mantle origin for the association (cf. V.4). Hypotheses involving reaction of normal basic magmas with sedimentary limestone (Wyllie, cf. VI.3) or of carbonatite with sialic material (Holmes, 1950; von Eckermann, 1948; Dawson, 1966) will not be considered, even though they might warrant consideration for specific complexes.

Suggested primary magmas, from which the nephelinite and ijolite associations are derived, cover what appears to be a wide range; alkali pyroxenite (Davies, 1952); alkali peridotite (carbonated) (King and Sutherland, 1960); peridotite (Strauss and Truter, 1951); and kimberlite (Saether, 1957; von Eckermann, 1961). The common feature of these primary magmas is their ultramafic character and, in most cases, involvement of uncommon amounts of alkalis and volatiles. This point is made clear when King (1965) suggests that the 'immediate parental magma' of the E. Uganda province was of 'nephelinite/melteigite composition'. The question that remains is, how is this magma formed, or what special conditions are required to produce nephelinite magma rather than the more common basaltic magmas? No feasible differentiation scheme has yet been proposed that could derive nephelinites from basaltic magma at low pressures. Part of the answer to the question must lie in magma generation at higher pressures, but before delving into this complex subject the rôle of two heteromorphs of nephelinite must be examined.

Following the discovery of pargasitic hornblende xenoliths, of nephelinitic composition, in the nephelinites of the Moroto volcano, Varne (1968) has proposed that the nephelinite lava series derives from the incongruent melting of pargasite. He envisages that the two series of lavas represented at Moroto, nephelinite and alkali basalt, are both derived by partial melting of hydrated peridotite: the nephelinites from the hornblendic portion and the olivine basalts from the volatile-poor fraction. This idea has been questioned by Wood (1968) because of the incompatibility between the trace elements in the hornblende and those in the nephelinites.

Tilley and Yoder (1968) in a series of experiments on natural nephelinites have shown that under pressure these are converted to an essentially pyroxenite composition, with accessory biotite, hornblende, olivine and magnetite. In the presence of excess water, *especially at lower pressures*, there is a prominent development of hornblende along with clinopyroxene. These results, and the natural pargasite from Moroto, show the possibility of *nephelinite heteromorphs* at depth, but they do not bring us much nearer to the problem of how the *nephelinite composition* originates as a discrete phase (or assemblage) in the Earth. Although the authors themselves do not comment on it, their experimental results have possibly a direct bearing on the occurrence of melteigite as the plutonic equivalent of nephelinite. Experimentally, it seems that pressure, with or without added volatiles, is sufficient to suppress the stability of olivine in nephelinite compositions. Experiments carried out at different times, on different nephelinitic starting materials and in different laboratories, all show pressure ranges over which no olivine appears below the solidus (Yoder and Tilley, 1962; Tilley and Yoder, 1968; Bultitude and Green, 1970). This must be a significant factor in nephelinite petrogenesis.

The most recent attempts to fit the development of nephelinite magmas into a general framework of magma generation in the mantle are as follows:

1. O'Hara (1968) suggests that nephelinitic magma is a product of special crystal fractionation from an initial picritic liquid.
2. Kushiro (1968), and Kushiro and Kuno (1963), suggest that nephelinite represents a special condition of partial melting of mantle peridotite.
3. Wood (1968) favours this idea of special partial melting, but finds it necessary to add zone refining and crystal fractionation to produce the final nephelinite liquid.
4. Bultitude and Green (1968) suggest that nephelinitic magmas are the products of crystal fractionation or partial melting in the presence of H_2O.

II.3.9.1. Anhydrous High-Pressure Crystal Fractionation

O'Hara's mechanism involves the separation, from a picritic liquid, of a mixture of garnet and clinopyroxene ('eclogite fractionation') at pressures around 30 kb. Protracted eclogite fractionation is claimed to produce highly undersaturated liquids resembling kimberlite. The nepheline normative composition of the residual liquid may be enhanced (to give nephelinites) by the separation of a spinel-lherzolite assemblage (spinel + olivine + clinopyroxene + orthopyroxene) at intermediate pressures. According to this latter mechanism the spinel-lherzolite nodules in nepheline-bearing lavas are cognate xenoliths (accumulates). O'Hara suggests that eclogite fractionation causes such a large reduction in the volume of liquid that other processes, such as zone-refining, are unnecessary to explain the high trace and minor element contents of nepheline-bearing lavas.

II.3.9.2. Anhydrous High-Pressure Melting

The basis of Kushiro's suggestion is that, given a relatively homogeneous peridotite mantle, partial melting at successively higher pressures yields increasingly undersaturated liquids, which become basanitic or nephelinitic if the degree of melting is limited. Kushiro and Kuno (1963) derive a figure of only 2% melting of peridotite to yield a liquid of basanitic composition, so presumably even less melting is involved in the production of nephelinitic liquids.

II.3.9.3. Subsequent Zone-Refining and Fractionation

Wood (1968) adopts the hypothesis of Kushiro and Kuno, but is not satisfied about its ability to cater for his finding that nephelinites have a generally higher tenor of trace elements than alkali basalts and tholeiites. He appeals to the mechanism of zone-refining (Harris, 1957; cf. VI.1a) during ascent of the magma, involving continuous solution of the roof and wall-rocks and precipitation in the magma, to provide the necessary trace element pattern. Unfortunately this process is inadequate because the magma at higher levels would have to correspond to the appropriate melt-product for the lower-pressure environment, i.e. it would not be nephelinitic according to Kushiro and Kuno's scheme. Indeed, if Kushiro and Kuno's scheme is applicable, zone-refining should not be necessary because the minute amounts of liquid involved would incorporate high concentrations of trace and minor elements, derived from the early melting of accessory and minor phases in the mantle peridotite (cf. O'Hara, 1968, Fig. 8).

Wood is the only author who examines the relationship between olivine-nephelinite and nephelinite. On the basis of his average analyses (quoted in Table 2) he proposes that nephelinite may be derived from a parental olivine-nephelinite by separation of olivine and spinel, but he recognizes the difficulty that spinel is not recorded as a phenocryst in nephelinites. Yoder and Tilley (1962, p. 363) have in fact recorded spinel inclusions in the olivine phenocrysts of the olivine nephelinite of the Pali dike, Oahu. (Analysis 19, Table 2), but any obvious spinel in the lavas is probably confined to lherzolite nodules, which Wood considers, from other evidence, to be accidental, *not* cognate, xenoliths. If Wood's mechanism has operated it requires that the spinel has been cleanly separated from the liquid, *without* separation of the co-precipitated olivine.

An alternative to the mechanism proposed by Wood would be that the two magmas, olivine nephelinite and nephelinite, are different partial melts, the former involving more melting of the olivine and spinel constituents in the source region. Kushiro and Kuno (1963) suggest, in passing, that melting of spinel might be involved in the production of deep-seated magma, but they do not substantiate this view and the later experiments of Kushiro (1968) show decreasing spinel-contents in liquids at higher pressures.

II.3.9.4. High Pressure Crystallization or Melting in the Presence of H_2O

The experiments of Bultitude and Green (1968) are focused on the relationship between olivine nephelinite and olivine melilite nephelinite and, like all the experimental work to date, do not shed any light on the olivine-poor nephelinites. They claim that in the presence of H_2O, at high pressure, orthopyroxene is the dominant liquidus phase in an olivine nephelinite composition. The result of orthopyroxene fractionation would be to drive the liquid composition towards olivine melilite nephelinite. O'Hara (1968) deems their conclusions 'unacceptable at present', and questions whether their final run products represent the original starting material. Furthermore Kushiro's results (1969, 1970) using sealed capsules, are in direct conflict with those of Bultitude and Green, and Kushiro concludes that their unsealed experiments have lost water and alkalis during the runs. In view of the typically explosive character of nephelinite volcanism, some nephelinites at least would be expected to contain orthopyroxene phenocrysts if nephelinite genesis is controlled by orthopyroxene separation, but this does not seem to be the case.

II.3.9.5. Melting in an Open System

One of the basic assumptions in all the above hypotheses is that the melting and crystallization takes place in a closed system, or one of constant bulk composition. But when the volatile-rich nature of the nephelinite association is considered, this assumption becomes highly questionable, if not wholly implausible. Examination of the analyses of rocks of the nephelinite association shows variations in most of the important elements that are inexplicable in terms of a straightforward system of crystal $\rightleftharpoons$ liquid equilibria. On these bases, *any* hypothesis of nephelinite development must give adequate consideration to the rôle of volatile and mobile elements. Admittedly, this invokes an agency of unexplored potential, but high concentrations of volatiles are an integral part of this association and it would be illogical to ignore their rôle in petrogenesis.

Two assumptions about the volatiles can be made. Firstly, melting must result from a disturbance in the energy distribution in the region of magma generation, and there is no *a priori* reason why this region should be a closed system with respect to volatiles if it is open to energy: in fact, the *a priori* assumption should be the opposite. Secondly, in the presence of volatiles, melting should start at considerably lower temperatures than under anhydrous conditions (see Kushiro, 1969) and result in a slight reduction of the total volume of the system. The first result will be the generation of a small volume of volatile-saturated melt at temperatures much lower than those required for large-scale partial melting: this first melt will contain higher than normal concentrations of trace and minor elements, and consequently will appear to be strongly 'differentiated'. The formation and migration (or eruption) of this initial melt could produce a further influx of volatiles from the region surrounding the melt-zone, and indeed, the whole process may be initiated, and propagated by a localization of mantle de-gassing (cf. III.2; VI.1a, b).

It is also possible to make reasonable deductions about the nature of the gaseous phase of nephelinite volcanism. Carbonatites are a characteristic part of the nephelinite–ijolite association; explosive volcanism is typical and the pyroclastic deposits are rich in carbonates. The only active volcanoes in East Africa are nephelinitic, Oldonyo Lengai in the Eastern Rift and Nyiragongo and Nyamalgira in the Western Rift. The latest phases of activity on Lengai have been alkali carbonate ash showers and lavas. Gas is continuously issuing from the lava lake at Nyiragongo and the predominant species is CO_2 (Chaigneau *et al.*, 1960). Combusted fume from fissures around the Nyiragongo lava lake is also rich in CO_2 (A. T. Huntingdon, personal communication, 1972.) Tazieff points out the ratio of CO_2 to solid products from this volcano is difficult to estimate but must be extraordinarily high (pers. comm., 1969). The conclusion is inescapable that we are not dealing with a simple silicate system, but one in which CO_2 and carbonates have an important (if not paramount) rôle.

In all these phenomena water appears to be playing a subsidiary rôle to carbon dioxide. Experimental studies also suggest that this must be so. Kushiro (1969 and 1970) shows that the addition of H_2O to mantle compositions at high pressures enlarges the liquidus stability field of olivine, whereas several high-pressure studies of nephelinitic compositions (Yoder and Tilley, 1962; Tilley and Yoder, 1968; Bultitude and Green, 1970) show that olivine stability is suppressed at high pressures. Olivine is also absent in melteigite, the plutonic equivalent of nephelinite, suggesting that H_2O, which might be expected to enhance the stability of olivine, plays a subordinate rôle. The experiments of Yoder (1970) on phlogopite–H_2O–CO_2 indicate that the stability of this mineral, which is important in the higher pressure equivalents of the nephelinite–melilitite association, is enhanced by high partial pressures of carbon dioxide.

In looking for a petrogenic control of the three magma types, nephelinite, olivine nephelinite and olivine-melilite nephelinite, perhaps the most promising candidate is carbon dioxide.

Examination of the analyses, and the variation diagrams of Figs. 1 and 2, suggests that the olivine nephelinites could be represented as nephelinite + olivine ± spinel. Some of these with abundant olivine phenocrysts are presumably low-pressure accumulates, and this is supported by the high temperature of the olivine liquidus in such rocks. The others, typified by oceanic olivine nephelinites, may represent mantle melting in the presence of limited amounts of CO_2 and H_2O (possibly with H_2O dominant) such that the stabilities of olivine and spinel are not too seriously curtailed. The olivine-poor nephelinites, which are so characteristic of the continental volcanoes, and their deep-seated equivalents, the melteigites, represent melting in a more volatile-rich environment where CO_2 is dominant, and the olivine stability is strongly suppressed. This condition is to be expected below a stable continental plate where there is impeded degassing of the mantle, and this in itself leads to extensive metasomatism and the development of alkaline ultramafic mineralogies poor in olivine (see III.2).

As the melilite content of nephelinite increases it grades towards melilitite, with a decrease in silica and a relative increase in metal oxides, especially CaO. The changes involved are so subtle that more analyses of fresh, and totally unaltered lavas are probably needed to obtain a clear picture. With so much volatile activity in these melts it is probably unnecessary to invoke any special crystal fractionation or partial melting to relate nephelinite to melilitite. Sahama (1962) has pointed out that the melilite-rich magmas of Nyiragongo may represent a highly carbonated silicate melt, produced essentially by volatile-streaming in the upper parts of a nephelinite magma chamber. Perhaps the relationship could be summarized most simply by saying that the association of nephelinite + carbonatite is equivalent to melilitite + CO_2. Carbonatite thus represents the condition whereby a discrete carbonate fluid has separated from the silicate melt.

II.3. REFERENCES

Agard, J., 1960. Les carbonatites associeés du massif de roches alcalines du Tamazert. *XXI*[st] *Int. Geol. Cong. Part XIII*, 293–303.

Almeida, F. F. M. de, 1961. Geologia e petrologia da Ilha da Trinidade. *Brazil Div. Geol. Min. Monographia*, **18.**

Bailey, D. K., and Schairer, J. F., 1966. The system $Na_2O–Al_2O_3–Fe_2O_3–SiO_2$ at 1 atms., and the petrogenesis of alkaline rocks. *Jour. Petrol.*, **7,** 114–70.

Bowen, N. L., 1922. Genetic features of the alnöitic rocks of Isle Cadieux, Quebec. *Am. Jour. Sci.*, **3,** 1–34.

Brøgger, W. C., 1921. Das Fengebiet in Telemark, Norway. *Vidensk. selsk. Skrifter, I. Mat.–Naturv. Klasse*, **9.**

Bultitude, R. J., and Green, D. H., 1968. Experimental study at high pressures on the origin of olivine nephelinite. *Earth Planetary Sci. Letters*, **3,** 325–37.

Bultitude, R. J., and Green, D. H., 1970. Highly undersaturated rocks in upper mantle conditions. *Nature*, **226,** 748–9.

Chaigneau, M., Tazieff, H., and Fabre, R., 1960. Composition des gaz volcaniques du lac de lave permanant du Nyiragongo. *Acad. Sci. (Paris). C.R.*, **250,** 2482–5.

Davies, K. A., 1952. The building of Mt. Elgon, E. Africa. *Geol. Surv. Uganda, Mem. 7.*

Dawson, J. B., 1966. Oldoinyo Lengai. In *The Carbonatites*, Eds. Tuttle and Gittins. John Wiley and Sons, New York, 155–68.

von Eckermann, H., 1948. The alkaline district of Alnö Island. *Sver. Geol. Undersök. Ser. Ca. No. 36.*

von Eckermann, H., 1961. The petrogenesis of the Alnö alkaline rocks. *Bull. Geol. Inst. Univ. Upsala*, **40,** 25–36.

Erikson, R. L., and Blade, L. V., 1963. Geochemistry and petrology of the alkalic igneous complex at Magnet Cove, Arkansas. *U.S. Geol. Surv. Prof. Paper 425.*

Harris, P. G., 1957. Zone refining and the origin of potassic basalts. *Geochim. Cosmochim. Acta*, **12,** 195–208.

Holmes, A., 1950. Petrogenesis of katungite and its associates, *Am. Mineral.*, **35,** 772–92.

Johannsen, A., 1938. *A Descriptive Petrography of the Igneous Rocks.* U. of Chicago Press, Chicago.

Joplin, G. A., 1964. *A Petrography of Australian Igneous Rocks.* Angus & Robertson, Sydney.

Jung, J., and Brousse, R., 1961. *Bull. Serv. Carte Geol. France*, **58,** 569–629.

King, B. C., 1949. The Napak area of Karamoja, Uganda. *Geol. Surv. Uganda, Mem. 5.*

King, B. C., 1965. Petrogenesis of the alkaline igneous rock suites of the volcanic and intrusive centres of E. Uganda. *Jour. Petrol.*, **6,** 67–100.

King, B. C., and Sutherland, D. S., 1960. Alkaline rocks of Eastern and Southern Africa. *Science Progress*, **48,** 298–321; 504–24; 709–20.

Kushiro, I., 1968. Compositions of magmas formed by partial zone melting of the Earth's upper mantle. *Jour. Geophys. Res.*, **73,** 619–34.

Kushiro, I., 1969. Discussion of the paper 'The origin of basaltic and nephelinitic magmas in the Earth's mantle' by D. H. Green. *Tectonophys.*, **7,** 427–36.

Kushiro, I., 1970. Systems bearing on melting of the upper mantle under hydrous conditions. *Carn. Inst. Wash. Yr. Bk. 68*, 240–45.

Kushiro, I., and Kuno, H., 1963. Origin of primary basalt magmas and classification of basaltic rocks. *Jour. Petrol.*, **4,** 75–89.

Lacroix, A., 1929. La constitution lithologique des îles de la Polynésie australe. *Mem. Acad. Sci. France*, **59,** 1–80.

Lehijärvi, M., 1960. The alkaline district of Iivaava, Kuusamo, Finland. *Bull. Comm. Geol. Fin.*, **185.**

Nockolds, S. R., 1954. Average chemical compositions of some igneous rocks. *Bull. Geol. Soc. Amer.* **65,** 1007–32.

Nolan, J., 1966. Melting-relations in the system $NaAlSi_3O_8 - NaAlSiO_4 - NaFeSi_2O_6 - CaMgSi_2O_6 - H_2O$. *Quart. Jour. Geol. Soc. Lond.*, **122,** 119–58.

O'Hara, M. J., 1968. The bearing of phase equilibria studies on the origin and evolution of basic and ultrabasic rocks. *Earth Sciences. Revs.*, **4,** 69–133.

O'Hara, M. J., and Biggar, G. M., 1969. Diopside + spinel equilibria, anorthite and forsterite reaction relations in silica-poor liquids in the system $CaO–MgO–Al_2O_3–SiO_2$ at atmospheric pressure. *Am. Jour. Sci.*, **267–A,** 364–90.

Onuma, K., and Yagi, K., 1967. The system diopside–akermanite–nepheline. *Am. Min.*, **52,** 227–43.

Parsons, G. E., 1961. Niobium-bearing complexes east of Lake Superior. *Ontario Dept. Mines, Geol. Rep.*, **3.**

Part, G. M., 1950. Volcanic rocks from the Cape Verde Islands. *Bull. Brit. Mus.* **Vol. 1,** *No. 2*, 27–72.

Pulfrey, W., 1950. Ijolitic rocks near Homa Bay, W. Kenya. *Quart. Jour. Geol. Soc. Lond.*, **105,** 425–59.

Rapson, J. E., 1966. Carbonatite in the alkaline complex of the Ice River area. *Min. Soc. India, I.M.A. Volume*, 9–22.

Saether, E., 1957. The alkaline rock province of the Fen area in southern Norway. *Norske Vidensk. selsk. skrifter. No. 1.*

Sahama, Th. G., 1962. Petrology of Mt. Nyiragongo. *Trans. Edin. geol. Soc.*, **19,** 1–28.

Sahama, Th. G., and Hytönen, K., 1958. Calcium-bearing magnesium-iron olivines. *Am. Mineral.*, **43,** 862–71.

Schairer, J. F., Tilley, C. E., and Brown, M. A., 1968. The join nepheline–diopside–anorthite and its relation to alkali basalt fractionation. *Carn. Inst. Wash. Yr. Bk. 66*, 467–71.

Schairer, J. F., and Yoder, H. S. Jn., 1960a. The system forsterite–nepheline–diopside. *Carn. Inst. Wash. Yr. Bk. 59*, 70–1.

Schairer, J. F., and Yoder, H. S. Jn., 1960b. The nature of residual liquids from crystallization, with data on the system nepheline–diopside–silica. *Am. Jour. Sci.*, **258–A,** 273–83.

Schairer, J. F., and Yoder, H. S. Jn., 1964. Crystal and liquid trends in simplified alkali basalts. *Carn. Inst. Wash. Yr. Bk. 63*, 65–74.

Schairer, J. F., and Yoder, H. S. Jn., 1969. The join albite–anorthite–akermanite. *Carn. Inst. Wash. Yr. Bk. 67*, 104–5.

Schairer, J. F., and Yoder, H. S. Jn., 1970. Pyroxenes and related systems. *Carn. Inst. Wash. Yr. Bk. 68*, 202–14.

Strauss, C. A., and Truter, F. C., 1951. The alkaline complex at Spitskop. *Trans. Geol. Soc. S. Africa*, **53,** 81–125.

Tilley, C. E., and Yoder, H. S. Jn., 1968. The pyroxenite facies conversion of melilite-bearing assemblages. *Carn. Inst. Wash. Yr. Bk. 66*, 457–60.

Tilley, C. E., Yoder, H. S. Jn., and Schairer, J. F., 1965. Melting relations of volcanic tholeiite and alkali-rock series. *Carn. Inst. Wash. Yr. Bk. 64*, 69–82.

Tuttle, O. F., and Gittins, J., 1966. *The Carbonatites.* John Wiley and Sons, New York.

UNESCO *Seminar on the East Africa Rift System.* 1965. University College, Nairobi.

Varne, R., 1968. The petrology of Moroto Mountain, Eastern Uganda, and the origin of nephelinites. *Jour. Petrol.*, **9,** 169–90.

Vlasov, K. A., Kuzmenko, M. V., and Eskova, E. M., 1966. *The Lovozero alkali massif.* Oliver and Boyd, Edinburgh.

Vorobieva, O. A., 1960. Alkali rocks of the U.S.S.R. *XXI*[st] *Int. Geol. Cong. Part XIII*, 7–17.

Winchell, H., 1947. Honolulu Series, Oahu, Hawaii. *Bull. Geol. Soc. Amer.*, **58,** 1–48.

Wood, C. P., 1968. *A geochemical study of E. African alkaline lavas and its relevance to the petrogenesis of nephelinites.* Unpublished Ph.D. thesis, U. of Leeds.

Yoder, H. S. Jn., 1970. Phlogopite–H_2O–CO_2: an example of the multicomponent gas problem. *Carn. Inst. Wash. Yr. Bk. 68*, 236–40.

Yoder, H. S. Jn., and Tilley, C. E., 1962. Origin of basalt magmas. *Jour. Petrol.*, **3,** 342–532.

II.4. THE MINERALOGY AND PETROGRAPHY OF ALKALI BASALTIC ROCKS

J. F. G. Wilkinson

II.4.1. Introduction

Basaltic rocks, essentially Ca-rich clinopyroxene-calcic plagioclase assemblages, can be broadly divided into alkaline and subalkaline types (Chayes, 1966; Wilkinson, 1967; Le Maitre, 1968). The subalkaline basalts include the mafic members of the tholeiitic and calc-alkali series i.e. the P and H series as defined by Kuno (1959). Alkali basalts and basanites and their intrusive equivalents (collectively designated in this account as alkali basaltic rocks) are the most common members of the alkali olivine basalt magma type (Tilley, 1950; *see* Chayes, 1964, p. 1584), originally defined in a restricted rôle in the Mull volcanic succession (Bailey and others, 1924), and later elevated to world status as the olivine basalt magma type (Kennedy, 1933).

The iron- and potash-free system diopside–forsterite–nepheline–quartz—the so-called 'basalt tetrahedron'—conveniently portrays the fundamental mineralogical relationships between the principal basaltic types, five groups of basalts being recognized according to their normative components (Fig. 1; Yoder and Tilley, 1962, pp. 349–53).

1. Tholeiite (oversaturated): normative quartz and hypersthene.
2. Tholeiite (saturated; hypersthene basalt): normative hypersthene.
3. Olivine tholeiite (undersaturated): normative hypersthene and olivine.
4. Olivine basalt: normative olivine.
5. Alkali basalt: normative olivine and nepheline.

Many petrologists would classify rocks belonging to Groups 1–3 as subalkaline. Alkali basalts, projecting on the Ne side of the critical plane of undersaturation and adjacent to it, are composed of olivine, Ca-rich clinopyroxene, plagioclase (more calcic than An_{50}), and opaque oxides; if present, nepheline and/or analcime have only accessory status. With increasing undersaturation, expressed by an increase in modal nepheline or analcime, alkali basalts merge into basanites, whose content of normative nepheline (*ne*) exceeds 5% (Macdonald and Katsura, 1964; Green and Ringwood, 1967; Coombs and Wilkinson,

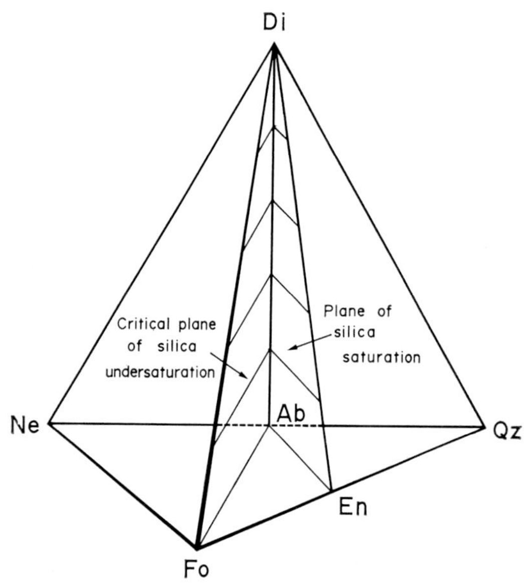

Fig. 1 Diagrammatic representation of the system Di–Fo–Ne–Qz showing the plane of silica saturation Di–En–Ab, and the critical plane of silica undersaturation Di–Fo–Ab. (After Yoder and Tilley, 1962, Fig. 1)

1969). High-alumina variants of alkaline and subalkaline basalts, either aphyric or porphyritic in plagioclase, have also been described; these rocks contain more than 17% Al_2O_3. Additional chemical variation among individual lineages may stem from differences in ratios such as $(FeO + Fe_2O_3)/MgO$ and K_2O/Na_2O; nevertheless a spectrum of basaltic compositions between

TABLE 1 Mineralogical Differences Between Alkaline and Subalkaline Mafic Rocks

Alkaline	Subalkaline
Pyroxene	
Titaniferous salite or Ca-rich augite; pigeonite and/or hypersthene typically absent. Exsolution lamellae absent from Ca-rich pyroxenes, even in slowly cooled rocks.	Augite (generally TiO_2-poor), which may be zoned to subcalcic augite, with pigeonite or hypersthene or both. Hypersthene is the common Ca-poor pyroxene in calc-alkali mafic rocks. With appropriate cooling exsolution lamellae are often present in Ca-rich and Ca-poor pyroxenes.
Olivine	
Olivine is generally a phenocryst *and* groundmass phase; normal zoning common. Ca-poor pyroxene coronas absent but in some rocks the olivines may be mantled by diopsidic clinopyroxene.	When present, olivine is typically only a phenocryst phase and is usually unzoned. Olivine often partially resorbed or mantled by Ca-poor pyroxenes.
Amphibole	
TiO_2-rich kaersutitic amphibole may be present.	Except in dolerite pegmatites, amphibole is typically absent from tholeiites Calc-alkali gabbros often carry a green or green-brown calciferous amphibole, frequently mantling augite.
Felsic minerals	
Plagioclase is labradorite-sodic bytownite, which may be zoned to anorthoclase. Interstitial alkali feldspar may be present. Variable modal nepheline and analcime, or other 'undersaturated' zeolites.	Plagioclase is labradorite-sodic bytownite. A pigmented vitreous acid residuum is common in tholeiites; in intrusive rocks this is represented by a quartzo-feldspathic intergrowth; modal quartz in the more oversaturated variants.

The general applicability of certain criteria in Table 1 requires critical reappraisal. Thus some tholeiitic lavas contain a violet-tinted (presumably titaniferous) clinopyroxene, and groundmass olivine may also be present, without evidence of a reaction relation (Carmichael, 1964; Chayes, 1966, pp. 129–30; Wilkinson, 1968b). Alkali feldspar is a potential groundmass phase in relatively potassic tholeiitic rocks, and in olivine-normative types, this will not be accompanied by modal quartz.

the basanites and quartz tholeiites has been demonstrated and this is fundamental to basalt classification and to many petrogenetic concepts.

II.4.2. Differences between Alkaline and Subalkaline Basalts

II.4.2.1. Mineralogical and Petrographic Criteria

Mineralogical differences between alkaline and subalkaline basalts have been discussed and evaluated by Kennedy (1933), Tilley (1950), Wilkinson (1956a; 1967; 1968b), Kuno and others (1957), Kuno (1959, 1960) and Macdonald and Katsura (1964). Some of the more widely applied criteria are listed in Table 1. Alkaline and subalkaline rocks can generally be discriminated petrographically if they are medium- or coarse-grained, or sufficiently removed from the critical plane of undersaturation to permit the appearance of mineral phases which reflect their degree of saturation, e.g. nepheline or Ca-poor pyroxenes. A petrographic distinction is more difficult or impossible, for fine grained rocks whose compositions fall in the transition zone between the alkaline and subalkaline types, represented, for example, by the assemblage olivine–clinopyroxene–plagioclase–opaque oxide–(glass). In many instances a petrographic assessment of the alkaline or subalkaline affinities of this assemblage may not be possible, since there are limitations in determining optically the nature of the groundmass clinopyroxenes, especially if these are Al- or Ti-rich.

II.4.2.2. Chemical Criteria

Chemical analyses, combined with normative compositions, provide strong evidence for the alkaline or subalkaline character of basaltic rocks. However, quite apart from the uncertainties introduced by the quality of the analyses themselves (cf. Yoder and Tilley, 1962, p. 352), chemical studies require the selection of fresh material. In practise it is often difficult to satisfy this condition, since basaltic rocks, particularly the alkaline types, are very susceptible to low temperature alteration, often associated with selective leaching (Wilshire, 1959). As a result of oxidation and hydration of olivine and oxidation of titanomagnetite, the norms of alkali basaltic rocks acquire a more subalkaline character. Thus an analcime-bearing olivine basalt containing a nepheline-normative clinopyroxene may display substantial *hy* due to alteration of the olivine (*see* Tilley and Muir, 1962, Table 2).

The Harker-type variation diagram (Na_2O + K_2O *vs.* SiO_2 wt. %), incorporating the boundary line between the Hawaiian alkaline and tholeiitic suites (Macdonald and Katsura, 1964, Fig. 1), has been widely applied to differentiate members of the two volcanic series. With the exception of melanocratic picritic or ankaramitic variants, the alkaline mafic lavas contain more (Na_2O + K_2O than tholeiitic rocks with similar silica contents. This diagram provides a useful preliminary guide to magmatic affinities but it possesses distinct limitations for rocks close to the Hawaiian datum line (Wilkinson, 1968b), since this should not be expected to be valid for all basaltic provinces (cf. Aoki, 1967, Fig. 4).

An unambiguous two-fold division of basaltic rocks into alkaline or subalkaline can be based on normative compositions alone, depending on the presence of *ne* or *qz* respectively. Chayes (1966) has estimated that rocks containing neither *ne* nor *qz* include probably less than 25% of analyses of Cainozoic mafic volcanics. Many *hy*-bearing basalts which project close to the transitional zone in Figure 1 include types which cannot be accommodated in the basalt classification of Yoder and Tilley (1962), despite their intrinsic alkaline characters. Not infrequently these basalts are associated in the field with oversaturated alkaline salic lavas such as pantellerites, comendites, etc. These 'mildly alkaline' *hy*-bearing basalts have been distinguished from basalts transitional to typical tholeiites by their indicator ratio (I.R.), expressed by Hy/(Hy + 2 Di) in molecular proportions of C.I.P.W. normative minerals (Coombs, 1963). The mildly alkaline basalts of Réunion, Easter Island, Galapagos, Tutuila and Ascension (I.R. = 0·38–0·00) possess a lower degree of metasilicate saturation than tholeiites (I.R. = 0·65 ± 0·20). Poldervaart (1964) has separated *hy*-bearing 'alkali' basalts from olivine tholeiites according to the positive or negative value of Ab − 1·95886 En_{hy} − 1·49064 Of_{hy}, where Ab, En_{hy}, Of_{hy} represent the normative albite, enstatite in hypersthene, and ferrosilite in hypersthene, respectively. Chayes (1966) has classified basalts as alkaline or subalkaline depending on the negative (alkaline) or positive (subalkaline) value of $D^* - 32{\cdot}26$, where $D^* = (x_i + 1{\cdot}119\, x_j) - (0{\cdot}006\, x^2_i + 0{\cdot}014\, x_i x_j + 0{\cdot}011\, x^2_j)$, $x_i = hy'$, $x_j = ol'$, and $hy' + ol' + di' = 100$. This discriminant correctly classifies more than 96% of analyses to which it is applicable (*see* Chayes, 1966, Fig. 3). Le Maitre (1968) has also applied discriminant functions to delineate alkaline and subalkaline series.

II.4.3. Classification of Alkali Basaltic Rocks

Alkali basaltic rocks are composed essentially of Ca-rich clinopyroxene and calcic plagioclase and they may be specifically classified by the relative proportions of plagioclase and alkali feldspar, and feldspathoid contents. The modal plagioclase is more calcic than An_{50}; normative plagioclase compositions (An × 100)/(Ab + An) although commonly in the labradorite or sodic bytownite range, are partly conditioned by the

TABLE 2 Classification of Alkali Basaltic Rocks

	Intrusive	Extrusive	Remarks
Alkali feldspar < 10% of total feldspar content; Feldspathoid < 10% of the rock	Alkali gabbro; Alkali dolerite	Alkali basalt	Analcime-bearing alkali dolerite ≡ crinanite whose usage is not recommended (Wilkinson, 1955)
Alkali feldspar 10% to 40% of total feldspar content; Feldspathoid < 10% of the rock	Monzogabbro; Trachydolerite	Trachybasalt	Alkali feldspar an essential constituent; monzogabbro (Streckeisen, 1967) ≡ alkali mangerite of Hurum type (Nockolds, 1954) ≡ syenogabbro of other classifications (cf. Johannsen, 1937)
Alkali feldspar < 10% of total feldspar content; Feldspathoid > 10% of rock	Theralite	Nepheline basanite (nepheline-olivine tephrite)	Essential nepheline; type Duppau theralite relatively rich in alkali feldspar and close to essexite
	Teschenite	Analcime basanite (analcime-olivine tephrite)	Essential analcime; analcimization of feldspars common
Alkali feldspar 10% to 40% of total feldspar content; Feldspathoid > 10% of rock	Essexite	Nepheline trachybasalt	Essential nepheline and alkali feldspar; ideally the modal plagioclase should be more calcic than An_{50}, in which case rock may be designated gabbro-essexite; the plagioclase in many essexites may be less calcic than An_{50}. The Oslo and Mount Royal 'essexites' misnamed (Streckeisen, 1967)
	Glenmuirite	Analcime trachybasalt	Essential analcime and alkali feldspar; many 'teschenites' are probably closer to glenmuirite in modal composition

(i) An essential feature of all these assemblages is the pair Ca-rich clinopyroxene + calcic plagioclase (>An_{50}); olivine is generally present, often as an essential constituent (> 10% of the rock)

(ii) Gabbroic rocks are coarse-grained (average grainsize > 5 mm), doleritic rocks medium-grained (average grainsize 1–5 mm), basaltic rocks fine-grained (average grainsize < 1 mm)

alumina content of the clinopyroxene, by Ab in alkali feldspar, and by modal analcime.

This account includes only two classifications of alkali basaltic rocks. One of these (Table 2), based on the classification of Nockolds (1954), has previously been adopted in a generalized petrography of basaltic rocks (Wilkinson, 1967). The other follows the recommendations of a committee on igneous rock nomenclature (Streckeisen, 1967); see II.1*. Differences between the two systems depend mainly on the amount of feldspathoid considered 'essential', on the slightly differing feldspar ratios of some rocks, e.g. essexite, and on the provision in the Streckeisen classification of a group of foidites at $F = 60$ which link moderately undersaturated alkali basaltic types and ultra-alkaline rocks, ideally devoid of essential feldspar, and which include the nephelinites and analcimites.

The practical difficulties in the microscopic assessment of modal alkali feldspar and feldspathoid obviously impose several limitations on 'pigeon-hole' type classifications; furthermore ternary feldspars tend to obscure the simplified

* Readers are referred to the relevant Chapters of this book where similarly cited.

two-fold feldspar grouping (plagioclase *vs.* alkali feldspar) that such classifications demand. That igneous rock nomenclature should be based primarily on modal composition is widely accepted, but few classifications co-ordinate the mode with rock chemistry. A degree of internal consistency between mode and chemical composition when applied to nomenclature is highly desirable but rarely achieved. This problem is illustrated by the modal and chemical implications of the term 'alkali basalt'.

Although widely applied to designate mildly undersaturated fine-grained basalts, the term alkali basalt has seldom been defined in petrographic accounts of alkali volcanic successions. Many generalized modal classifications of mafic rocks place the upper f.r. of alkali basalt close to 90 and similar rocks whose modal and normative plagioclase is more calcic than An_{50} and f.r. between 90 and 65 have been termed trachybasalt or latite-basalt (Streckeisen). Although criticized because of its derivation, the term trachybasalt has been more widely employed than latite-basalt for the mildly potassic basalts which may be prominent in some alkali basaltic provinces (Aoki, 1959; Le Maitre, 1962; Uchimuzu, 1966; Streckeisen, 1967, Fig. 64).

Since the andesitic member of sodic alkaline lineages, namely hawaiite, has been defined partly by a K_2O/Na_2O ratio of less than 1:2 (Macdonald and Katsura, 1964), it is appropriate to examine this alkali ratio as a possible parameter in the nomenclature of its mafic associates. Uchimuzu (1966) and Kurasawa (1967), following Kuno (1954) in part, have defined trachybasalts as mafic volcanic rocks whose normative (Or $\times$ 100)/(Or + Ab + An) exceeds 15. Assuming (An $\times$ 100)/(Ab + An) = 50, this value implies a K_2O/Na_2O ratio close to 1:2. As a means of distinguishing alkali basalts from trachybasalts, the limiting 1:2 alkali ratio probably results in excessive modal anorthoclase or soda sanidine in alkali basalts if the f.r. required by the modal classifications is to be retained. If $A + P = 50$, the upper limit of modal alkali feldspar should then approximate to only 5% of the rock.

In view of very limited data on the chemical *and* modal compositions of fine-grained alkali basaltic rocks and lack of detailed information on the nature and chemistry of strongly zoned plagioclase (where ternary solid solution is likely to be extensive), a correlation of modal alkali feldspar and alkali contents of the host rocks is not possible at present. The normative feldspar compositions of Hebridean and Hawaiian alkali basalts project in the one feldspar field in the Ab–An–Or system and much of the K_2O revealed by the rock analyses (Table 4, Nos. 3 and 4) is presumably present in potassic plagioclase. More potassic normative feldspar compositions, e.g. $Ab_{42}An_{43}Or_{15}$, plot in the two feldspar field, and soda sanidine then becomes a potential phase. Mildly undersaturated alkaline rocks (either *hy*- or *ne*-bearing) with K_2O/Na_2O ratios less than 1:3 probably conform in feldspar ratios to the modal definitions of alkali basalt presented herein. Lavas with alkali ratios between 1:3 to 1:2 may be transitional between sodic alkali basalts and trachybasalts.

Macdonald and Katsura (1964), Green and Ringwood (1967) and Coombs and Wilkinson (1969) have suggested that 5% *ne* delineate mildly undersaturated sodic alkali basalts and trachybasalts from the more undersaturated basanites (basanitoids if modal feldspathoid is not apparent) and feldspathoidal trachybasalts; 10% modal nepheline (vol. %) reduces to less *ne* in the norm, much of the Ks appearing as *or*. However some *ne* derives from the contained clinopyroxene, a lesser amount is consequent on Fe_2TiO_4 in titanomagnetite, and hence the problem of internal consistency again arises. Coombs and Wilkinson (1969) have pointed out that the limiting value of 5% *ne* handles more rationally the rôle of essential analcime (appearing in the norm as *ab* + *ne*); at basaltic compositions the parameter ($\gtrless$ 5% *ne*) distinguishes rocks which lie either on an alkali trachytic or phonolitic trend. In the Streckeisen classification the division between alkali basalts and nepheline basanites at $F = 10$, equivalent to about 5% modal feldspathoid when $A + P + F = 50$, may approximate more closely to $ne = 5$.

In Table 2 no provision has been made for

melanocratic alkali basalts and related rocks, in which olivine and clinopyroxene usually exceed 60%. Such rocks include alkali picrite basalt (= oceanite) and ankaramite, in which the principal ferromagnesian mineral is olivine and clinopyroxene respectively. Limburgite (typically feldspar-free) contains phenocrysts of olivine and clinopyroxene in a vitric groundmass studded with microlites of olivine, clinopyroxene and opaque oxide; the *ne* contents of limburgites will indicate whether they are mildly or moderately undersaturated.

II.4.4. The Mineralogy of Alkali Basaltic Rocks

In recent years a considerable amount of data on the mineralogy of alkali basaltic rocks has become available. The following account is of necessity limited in its scope. For more detailed accounts of the mineral groups the reader is referred particularly to the excellent compilations of Deer, Howie and Zussman (1962a, 1962b, 1963a, 1963b) and Brown (1967). With the exception of analcime, which has close affinities with the feldspathoids, the zeolite mineral group has not been discussed (*see* Deer, Howie and Zussman, 1963b).

II.4.4.1. Olivines

Magnesium-rich members of the forsterite (Mg_2SiO_4)—fayalite (Fe_2SiO_4) solid solution series (Table 3, Nos. 1 and 2) are important phases in alkali basaltic and picritic rocks, the more fayalitic variants being confined, in rocks of basaltic composition, to the evolved facies of differentiated mafic intrusions. Microprobe analyses of olivines have been reported by Smith and Strenstrom (1965) and Smith (1966a). Fa-rich olivines are enriched in Ti, Mn and Zn; Cr and Ni are higher in Fo-rich samples. The Ca content increases with increasing temperature and higher Fa content. Roedder (1965) has described liquid CO_2 inclusions in olivines from basalts; these inclusions are not as common in this paragenesis as in olivines from peridotite nodules.

The compositions of olivine phenocrysts in alkali basaltic rocks generally range from Fa_{20} to Fa_{35}. Although olivines from picritic rocks may be more Mg-rich than Fa_{20}, there are usually measurable compositional differences between cognate olivines and xenocrystal olivines from ultramafic inclusions, whose compositions are usually close to Fa_{10}. Groundmass olivines tend to be slightly more Fa-rich than coexisting phenocrysts, and compared with the latter, are depleted in Ni (Forbes and Banno, 1966). Alkali basaltic olivines frequently display normal zoning (Tomkeieff, 1939), and in individual crystals compositional differences of 15–20% Fa between core and margin are not uncommon. Examples of extreme zoning are furnished by some olivines from the Garbh Eilean sill, which are zoned from Fa_{33} to Fa_{87} (Johnston, 1953).

The crystallization trend of olivines controlled by low-pressure fractionation of alkali basaltic magmas, traced by the compositions of successive crops of crystals and by zoning in individual crystals, is analogous to the trend in the Fo (M.Pt. 1890 °C)—Fa (M.Pt. 1205 °C) system (Bowen and Schairer, 1935), namely enrichment in Fa with decreasing temperature. This may be inhibited if fractionation takes place under oxidizing conditions, as in the Square Top intrusion in which olivines from rocks of wide compositional range retain compositions close to Fa_{30} (Wilkinson, 1966b). The continuous olivine crystallization sequence in the Garbh Eilean teschenite-picrite association (approximately Fa_{16} to Fa_{97}; Johnston, 1953) and the Black Jack teschenite sill (Fa_{21} to Fa_{60}; Wilkinson, 1956b) may be contrasted with the early disappearance of olivine (and subsequent reappearance upon strong fractionation) in most differentiated tholeiitic intrusions, consequent on the well-known reaction between olivine and quartz-normative liquid to yield Ca-poor pyroxene. This reaction does not take place during the low pressure crystallization of alkali basaltic rocks but the reaction, olivine + liquid → diopside, has been demonstrated in the system $NaAlSiO_4$–$CaMgSi_2O_6$–SiO_2 (Schairer and Yoder, 1960), and observed in certain New Zealand alkali basalts and nepheline basanites by Searle (1961, Figs. 15–17) and Coombs and Wilkinson (1969), and in alkali dolerites by Wilshire (1967).

In strong contrast to the coexisting clinopyroxenes, which are invariably fresh, olivines in alkali basaltic rocks are very susceptible to alteration, as a result of either deuteric or weathering processes. The green or greenish brown or red or brown olivine alteration products are usually designated 'bowlingite' or 'iddingsite' respectively (Wilshire, 1958; Smith, 1961; Haggerty and Baker, 1967; Baker and Haggerty, 1967). High temperature alteration of olivine, sometimes accompanied by reheating, yields either symplectic or microsymplectic magnetite, which may be subsequently oxidized to lamellar hematite, or exsolution of hematite associated with a more forsteritic olivine (Searle, 1961; Le Maitre, 1962; Haggerty and Baker, 1967). 'Bowlingite' (a complex phyllosilicate with smectite a major constituent) and 'iddingsite' (a mixture of goethite and interstratified phyllosilicates also containing a smectite) are the characteristic olivine alteration products at low to intermediate temperatures.

II.4.4.2. Pyroxenes

Alkali basaltic rocks contain only one principal pyroxene phase, namely a member of the diopside-hedenbergite series (Wilkinson, 1956a), more specifically a salite or Ca-rich augite, as defined by Poldervaart and Hess (1951, Fig. 1B). Although some basanitic rocks contain minor acmitic pyroxene, either in the groundmass or rimming more Ca-rich phenocrysts, alkali basaltic pyroxene assemblages characteristically are simpler than their subalkaline counterparts which may show several crystallization trends, and which may, under appropriate conditions, undergo subsolidus exsolution and inversion (*see* Brown, 1967, pp. 121–3). The Ca-rich pyroxenes from subalkaline rocks tend to contain less Al, Ti ($TiO_2 < 1\%$), Na ($Na_2O < 0{\cdot}4\%$), and to some extent, less Fe^{3+} than the salites or Ca-rich augites from alkaline rocks.

The general pyroxene formula is

$X_{1-p}Y_{1+p}Z_2O_6$ ($0 \leqslant p \leqslant 1$) where
X = Ca, Na; Y = Mg, Fe^{2+}, Mn, Ni, Al, Fe^{3+}, Cr, Ti;
Z = Si, Al, Fe^{3+}

Table 3 lists analyses of clinopyroxenes from a variety of alkali basaltic host rocks, and the numbers of ions per six oxygens (Nos. 3 to 6). When Al is insufficient to complete the ideal two tetrahedral ions per formula unit, Fe^{3+} is allocated to make Z = 2. It has been more or less conventional to allocate Ti to make up any deficiency in the Z-group, but more recent data do not favour the Ti → Si diadochy (Tiba, 1966; Yagi and Onuma, 1967; Hartman, 1969). Because most alkali basaltic clinopyroxenes are comparatively Ti-rich (Wilkinson, 1956a, Table 2; Le Bas, 1962, Table 1), and are characterized in thin section by purplish brown or lilac tints, they are frequently designated as titansalite or titanaugite. Chemical definition of titanaugite or titansalite is purely arbitrary. Huckenholz (1965a; *see* Deer, Howie and Zussman, 1963a, p. 113) chooses a lower limit of 3% TiO_2 for titanaugites or titansalites, and calls those with 2 to 3% TiO_2 titaniferous augites or titaniferous salites. Yagi and Onuma (1967) have placed the lower limit of titania in titanaugites at 2% TiO_2. It should be noted that synthetic titaniferous clinopyroxenes (Fe^{2+}- or Fe^{3+}-free) are colourless (Segnit, 1953; Yagi and Onuma, 1967), in contrast to natural titanaugites, whose violet tints have been ascribed to Ti^{3+} instead of Ti^{4+} (Goldschmidt, 1954, p. 417). Synthetic Fe^{3+}-rich diopsides display strong orange-brown colours (Segnit, 1953).

The norms of alkali basaltic clinopyroxenes reflect the degree of undersaturation of their host rocks. Clinopyroxenes from mildly alkaline olivine basalts contain minor *hy*, whereas the pyroxenes in nepheline-normative rocks contain minor *ne*, accompanied, in clinopyroxenes from moderately undersaturated rocks, by *lc* or *cs* (Tilley and Muir, 1962; Yoder and Tilley, 1962; Coombs, 1963; Muir and Tilley, 1964). The variable degree of saturation indicated by the normative compositions of Ca-rich clinopyroxenes from alkaline and subalkaline rocks (generally strongly *hy*-normative, sometimes with *qz*) is reflected in differing Si and Al contents. Kushiro (1960) and Le Bas (1962) have demonstrated that the amount of Al, most of which is tetrahedral Al_Z, is primarily a function

TABLE 3 Analyses and Optical Properties of Minerals from Alkali Basaltic Rocks

	1	2	3	4	5	6	7
SiO_2	38·25	34·08	51·05	48·61	46·51	45·52	49·52
TiO_2	0·09	0·04	1·30	1·91	2·72	1·90	0·77
Al_2O_3	0·99	0·00	4·46	4·80	6·61	8·50	9·10
Fe_2O_3	0·82	0·27	1·28	2·75	2·28	3·11	1·61
FeO	22·42	47·30	4·87	7·14	8·61	2·76	5·18
MnO	0·28	0·65	n.d.	0·20	0·17	0·10	0·15
MgO	36·66	17·83	15·89	13·42	11·15	14·32	16·06
CaO	0·09	0·00	21·04	20·38	21·40	21·80	16·28
Na_2O	0·04	—	0·44	0·63	0·55	0·74	1·35
K_2O	0·06	—	0·09	0·11	0·06	0·07	0·13
H_2O+	0·39	0·05	—	n.d.	n.d.	0·50	0·00
H_2O-	0·04	0·00	—	0·03	nil	—	0·00
Cr_2O_3	—	—	—	0·09	—	0·22	0·08
Total	100·13	100·22	100·42	100·07	100·06	99·54	100·31
α	—	1·747–1·754	—	—	1·697–1·702	—	1·689
β	—	1·775–1·783	1·698	—	—	1·698	1·694
γ	—	1·790–1·798	—	—	1·725–1·731	—	1·714
2V	—	71(−)	54 (+)	—	55 (+)	56 (+)	50 (+)

Numbers of ions

	4(O)		6(O)				
Si	0·999	1·008	1·866	1·816	1·758	1·701	1·796
Al	0·030	—	0·134	0·184	0·242	0·299	0·204
Al	—	—	0·058	0·027	0·053	0·076	0·185
Ti	0·002	0·001	0·036	0·054	0·077	0·053	0·021
Fe^{3+}	0·016	0·006	0·035	0·079	0·063	0·088	0·044
Cr_{2+}	—	—	—	0·004	—	0·006	0·002
Fe	0·490 } 1·98	1·170 } 1·98	0·149 } 2·00	0·223 } 2·01	0·270 } 2·01	0·086 } 2·04	0·157 } 2·01
Mn	0·006	0·016	—	0·007	0·005	0·003	0·005
Mg	1·428	0·786	0·865	0·751	0·633	0·798	0·868
Ca	0·003	—	0·824	0·816	0·866	0·873	0·632
Na	0·002	—	0·031	0·045	0·041	0·053	0·095
K	0·002	—	0·004	0·004	0·002	0·003	0·006
OH	—	—	—	—	—	—	—
Mg	74·4	40·2	Ca 44	44	47	47	37
Fe^{2+}	25·6	59·8	Mg 46	40	34	43	51
			Σ Fe 10	16	19	10	12

Legend:

1. Olivine from 'kylite', Benbeoch, Ayrshire (Drever and MacDonald, 1967, Table 2). Analyst J. G. MacDonald. (Rock analysis Table 5, No. 9).
2. Olivine from analcime-bearing trachydiabase, Mt. Moroto, Karamoja, Uganda (Hytönen, 1959, Table 15). Analyst A. Juurinen.
3. Ca-rich augite from picrite basalt (G 121), Gough Island, South Atlantic (Le Maitre, 1962, Table 4). Analyst R. W. Le Maitre. (Rock analysis Table 4, No. 15).
4. Ca-rich augite from alkali olivine basalt (65992), Hualalai, Hawaii (Yoder and Tilley, 1962, Table 3). Analyst J. H. Scoon. (Rock analysis Table 4, No. 4).
5. Heavy clinopyroxene fraction (salite) from basalt (4519–34; 67545), Mid-Atlantic Ridge (45° 44′ N, 27° 43′ W) (Muir and Tilley, 1964, Table 3). Analyst J. H. Scoon. (Rock analysis Table 4, No. 2).
6. Salite from basanitoid (HF 109), Hocheifel area, Western Germany (Huckenholz, 1965b, Table 2). Analyst H. G. Huckenholz.
7. Clinopyroxene megacryst in analcime basanite, Armidale, New South Wales (Binns, 1969, Table 2). Analyst G. I. Z. Kalocsai. (Total includes P_2O_5 0·08).
8. Kaersutite from kaersutite trachybasalt (60525), (Dōgo, Oki Islands (Uchimuzu, 1966, Table 11).

8	9	10	11	12	13	14	15
39·43	39·68	40·10	0·20	51·60	44·35	47·23	53·98
5·94	4·97	4·44	25·65	0·15	tr.	—	—
13·60	14·48	14·53	2·39	29·93	32·22	26·72	23·49
4·76	10·72	3·24	18·28	0·94	0·47	1·29	0·47
8·18	1·30	10·05	50·13	0·08	—	—	—
0·14	0·16	0·14	0·68	tr.	—	—	—
11·58	12·95	11·00	2·04	0·00	0·23	n.d.	—
10·94	11·92	11·06	0·16	12·46	0·54	1·40	0·77
2·32	2·11	2·99	—	4·19	19·69	14·70	13·33
1·65	1·29	1·62	—	0·54	1·93	0·73	0·32
1·18	0·33	0·73	—	0·10	0·53	8·23	8·11
0·20	0·00	0·10	—	0·01	nil	0·17	0·00
—	—	—	—	—	—	—	—
99·92	99·91	100·00	99·53	100·00	99·96	100·47	100·47
1·689	1·679–1·681	1·680	a 8·493Å	—	ε 1·527	n 1·493	n 1·480–1·484
1·704	1·721–1·723	1·700		—	ω 1·531	a 13·744Å	
1·719	1·750–1·752	1·715		—			
80–77(−)	76(−)	80(−)		—			

Number of Ions

24(O, OH)			32(O)	32(O)	32(O)	96(O)	
5·900	5·886	6·039	0·058	9·422	8·495	28·447	31·604
2·100	2·114	1·961	0·821	6·443	7·273	18·972	16·212
0·299	0·417	0·618	—	— } 16·02	— } 15·84	— } 48·01	— } 48·02
0·668	0·554	0·503	5·634	0·021	—	—	—
0·536	1·196	0·367	4·019	0·129	0·067	0·590	0·204
— } 5·13	— } 5·21	— } 5·24	— } 23·89	—	—	—	—
1·023	0·161	1·266	12·247	0·012	—	—	—
0·018	0·020	0·018	0·168	—	—	—	—
2·582	2·863	2·469	0·888	—	—	—	—
1·754	1·895	1·785	0·051	2·438 } 4·06	0·110	0·904	0·482
0·673 } 2·74	0·606 } 2·75	0·872 } 2·97		1·484	7·308	17·170	15·135
0·315	0·244	0·311		0·125	0·472	0·557	0·239
1·178	0·326	0·733				(H_2O)16·533	15·839
				Ab 35·3	Ne 90·9	79·6	69·4
				An 61·5	Ks 6·7	3·0	1·0
				Or 3·2	Qz 2·4	17·4	29·6

Analyst H. Haramura.

9. Oxykaersutite from oxykaersutite trachyandesite, Iki Islands, Japan (Aoki, 1964, Table 2). Analyst K. Aoki.
10. Kaersutite megacryst from basalt, San Carlos, Arizona (Mason, 1968, Table 1). Analyst J. Nelen. (Analysis recalculated to 100% after deducting 1% ilmenite.)
11. Titanomagnetite from analcime-olivine theralite (ST. 28), Square Top intrusion, Nundle, New South Wales (Wilkinson, 1965b, Table 1). Analyst J. H. Pyle. (Normative composition mt 30·4, usp 66·0, il 3·5; rock analysis Table 5, No. 6).
12. Plagioclase from tephritic trachybasalt, Mt. Etna (Tanguy, 1966, Table 2). Analyst A. Nétillard. (Analysis recalculated to 100% after deducting 1·5% apatite.)
13. Nepheline from analcime-olivine theralite (ST. 28), Square Top intrusion, Nundle New South Wales (Wilkinson, 1965a, Table 3). Analyst J. H. Pyle.
14. Analcime from analcime-olivine theralite (ST. 28), Square Top intrusion, Nundle, New South Wales (Wilkinson, 1963, Table 1). Analyst M. Chiba.
15. Analcime from analcime alkali dolerite (72103), Okabe, Takakusayama district, Japan (Tiba, 1966, Table 9). Analyst T. Tiba.

of the Si content of the parent melt, so that clinopyroxenes from the more undersaturated alkaline rocks contain a smaller proportion of Si and larger proportion of Al than clinopyroxenes from subalkaline rocks. The Al_z and Ti contents of titanaugites tend to vary sympathetically and Yagi and Onuma (1967, Fig. 6) have suggested that these ions are mainly present as the hypothetical titanpyroxene molecule $CaTiAl_2O_6$. The maximum solubility of $CaTiAl_2O_6$ in diopside at atmospheric pressure is about 11% (Yagi and Onuma, 1967), corresponding to 4% TiO_2, and therefore less than the reported titania contents of some natural titanaugites. At high pressures (10–25 kb.), the solubility of $CaTiAl_2O_6$ in diopside decreases markedly, suggesting that, despite high Al contents, most titanaugites are comparatively low pressure phases, a conclusion consistent with their textural relations in the host rocks.

Any Al in octahedral co-ordination is probably present mainly as Ca-Tschermak's molecule whose entry into diopside, according to the coupled substitution AlAl for MgSi, is favoured by high pressures (Clark, Schairer and de Neufville, 1962). It should be noted that $CaMgSi_2O_6$–$CaAl_2SiO_6$ solid solutions synthesized at atmospheric pressure may contain up to 11·7% Al_2O_3 (Hytönen and Schairer, 1961). Analyses of groundmass clinopyroxenes from alkali basaltic rocks recalculated according to the method of Kushiro (1962) frequently contain only relatively small amounts of $CaAl_2SiO_6$ (less than 2 mol %), in contrast to clinopyroxenes from ultramafic inclusions (up to 18 mol % $CaAl_2SiO_6$), which usually possess a small jadeite component as well (Aoki and Kushiro, 1968). Exceptions are provided by some clinopyroxenes (coexisting with kaersutite and Al-rich spinel) from the Iki Islands, Japan, alkaline rocks (Aoki, 1964), and by megacrysts of aluminian Ca-poor augite (coexisting with aluminian orthopyroxene megacrysts) in certain hawaiites and basanites (Table 3, No. 7; Kuno, 1964; Binns, 1969; Binns, Duggan and Wilkinson, 1970). These megacrysts have been interpreted as high-pressure cognate phases.

Some facies of the Atumi dolerite, Japan, are exceptional in that Ca-poor pyroxenes have been reported in nepheline-normative rocks (Kushiro, 1964). Yoder and Tilley (1962, Table 20) have demonstrated that the 'pigeonitic pyroxene' (CaO = 13·5%) in an alkali basalt from Hiva Oa, Marquesas Islands (Barth, 1931) is in fact a *ne*-normative high lime clinopyroxene, characteristic of critically undersaturated rocks.

Little is achieved by referring the optical properties of titaniferous clinopyroxenes (Wilkinson, 1956a, Table 2) to the well-known optical properties/composition diagrams (*see* Deer, Howie and Zussman, 1963a, pp. 131–2) in order to derive their compositions. The Al, Ti, and Na contents of these clinopyroxenes often exceed the minimum values assigned to the pyroxene standards on which such diagrams are based, and there is the additional problem of differing effects on optical properties, depending on the position of a particular ion in the pyroxene structure (Hori, 1954; Henriques, 1958; Winchell, 1961). Some titaniferous salites may have low 2V's (27°–40°), quite inconsistent with their Ca-rich nature (Huckenholz, 1965a, 1965b). Since clinopyroxenes in alkali basaltic rocks show a relatively limited range in composition (i.e. in terms of $Ca + Mg + \Sigma Fe = 100$), analytical data provide the only reliable means of determining compositional variation. Microprobe analyses of members of the Di-Hd series have been reported by Smith (1966b).

Clinopyroxene crystallization trends in differentiated alkaline mafic intrusives and alkali volcanic lineages are now comparatively well defined, although more detailed assessments of the controls of some elements, e.g. Al and Ti, are required. In differentiated intrusions, such as the Garbh Eilean sill, Shiant Isles (Murray, 1954), the Black Jack teschenite sill, Gunnedah, New South Wales (Wilkinson, 1957a), and the Square Top theralite-tinguaite intrusion, Nundle, New South Wales (Wilkinson, 1966b), and alkaline sequences such as those at Morotu, Sakhalin (Yagi, 1953b), Okonjeje, South Africa (Simpson, 1954) and Mount Dromedary, New South Wales (Boesen, 1964) and Shonkin Sag, Montana (Nash and Wilkinson, 1970), the clinopyroxenes belonging to the Di–Hd series trend parallel to the Di–Hd join in the pyroxene quadrilateral and experience only a limited degree of iron enrichment, compositions more iron-rich than Fe_{25} being exceptional. However, it should be noted that microprobe data on individual grains of Ca-rich pyroxenes in the more evolved Shonkin Sag soda syenites indicate substantial substitution of Mg by Fe^{2+} (Nash and Wilkinson, 1970, Fig. 10). With more advanced differentiation the clinopyroxenes trend away from the Ca–Mg–Fe plane towards aegirine $NaFe^{3+}Si_2O_6$, well illustrated by pyroxenes from the alkali dolerite–monzonite–syenite differentiation sequence at Morotu (Yagi, 1966) and the shonkinite–syenite series at Shonkin Sag (Nash and Wilkinson, 1970).

Various chemical definitions of the sodic pyroxenes have been proposed (cf. Deer, Howie and Zussman, 1963a, p. 80; Yagi, 1966). The initial stages of Na–Fe^{3+} enrichment are represented by violet tinted or pale green soda augites ($Na_2O > 1{\cdot}5\%$; Yagi, 1953b; Huckenholz, 1965a; Wilkinson, 1966b), either rimming more diopsidic cores, or forming discrete crystals which may be rimmed by aegirine augite or aegirine. Experimental studies in the system diopside–hedenbergite–acmite indicate a complete series of solid solutions in the subsolidus region (Yagi, 1966; Nolan and Edgar, 1963; Nolan, 1969). These data are contrary to Aoki's (1964) proposal that Ca-rich and Na-rich clinopyroxenes are separated by a wide immiscibility gap, and that aegirine forms by reaction between hematite (magnetite in nature) and liquid. This reaction is rare in natural assemblages (cf. Nolan, 1966) but has been recorded in certain soda syenites from Shonkin Sag.

The more evolved members of some alkaline volcanic series carry a member of the Di–Hd series in which substitution of Mg by Fe^{2+} is extreme, so that these clinopyroxenes, often coexisting with Fa-rich olivine or aenigmatite, trend through ferrosalite or Ca-rich ferroaugite to hedenbergite or soda hedenbergite (Muir and Tilley, 1961; Carmichael, 1962; Aoki, 1964; Uchimuzu, 1966). Only very rarely, as for example in a hortonolite trachyte from Kakarashima Island, Japan, is there the suggestion of an additional trend from Ca-rich augite to subcalcic ferroaugite (Aoki, 1964). Olivine is typically absent from natural assemblages of acmite-rich pyroxenes and the development of the latter as a significant phase is not conditioned by an increase in oxygen fugacity but by oxygen fugacities diminishing less rapidly with temperature than if olivine (and magnetite) were present (Nicholls and Carmichael, 1969; Nash and Wilkinson, 1970).

Olivines coexisting with Ca-rich clinopyroxenes have the higher Fe/Mg ratios, and compositional differences between the two minerals become more extreme in later differentiates (Murray, 1954, Fig. 2; Aoki, 1964, Fig. 1). Zoning in individual grains, or compositional trends between phenocryst and groundmass pyroxenes, may be consistent with an increase in Fe/Mg ratios (Wilkinson, 1957a, 1966b; Huckenholz, 1965b). An initial increase in Ti in the early stages of differentiation may later be followed by a decrease in this constituent (Yagi and Onuma, 1967). In some intrusions, such as Black Jack and Square Top, clinopyroxene Al_z decreases with differentiation, suggesting that, in addition to the composition of the melt, temperature also controls the Al $\rightarrow$ Si substitution, increased temperatures favouring the entry of Al in fourfold co-ordination (Thompson, 1947; Buerger, 1948).

II.4.4.3. Kaersutite

Kaersutite is a pargasitic amphibole in which Ti ranges from about 0·5 to 1 atom per formula unit; some Mg is replaced by Fe^{2+} and the ratio Fe^{3+}/Fe^{2+} may vary considerably (Table 3, Nos. 8 and 10). An approximate formula of many kaersutites can be written: $(Na,K)Ca_2(Mg,Fe^{2+})_{3-4}(Ti,Al,Fe^{3+})_{2-1}Al_2Si_6O_{22}(O,OH)_2$. In relatively unoxidized kaersutites $Mg > Fe^{2+}$. Brown calciferous amphiboles with $Fe^{2+} > Mg$, less than 0·5 Ti, and more akin to ferropargasite, may be termed barkevikite (cf. Wilkinson, 1961, Table 4). Intratelluric kaersutites, initially displaying low Fe^{3+}/Fe^{2+} ratios (Howie, 1963; Tiba, 1966), may be converted to oxykaersutite (analogous to the common hornblende $\rightarrow$ oxyhornblende transformation; Barnes, 1930), characterized by high Fe^{3+}/Fe^{2+} ratios, low OH, increased refractive indices and birefringence, and more intense absorption and pleochroism (Table 3, No. 9). The transformation to oxykaersutite, as indicated by a significant change in optical properties, has been assumed to have taken place when $Fe^{3+}/Fe^{2+} \simeq 2$ (Aoki, 1963). Chemical and optical data on kaersutites and oxykaersutites from a variety of parageneses have been listed by Wilkinson (1961), Deer, Howie and Zussman (1963a), Aoki (1963), and Mason (1968).

In extrusive rocks kaersutite frequently displays evidence of resorption and attendant breakdown to a variety of mineral phases, including Ca-rich clinopyroxene, olivine, opaque oxides, feldspar and nepheline (Benson, 1939; Aoki, 1959; Uchimuzu, 1966; Coombs and Wilkinson, 1969). The stability of kaersutite has not been investigated experimentally, but pargasite $NaCa_2Mg_4Al^{VI}\ Al_2^{IV}Si_6O_{22}(OH)_2$ dehydrates at 840 °C to 1025 °C for water pressures from 250 to 800 bars to aluminous diopside + forsterite + nepheline + anorthite + spinel + vapour (Boyd, 1956). The norms of many kaersutites are similar to those of olivine nephelinites.

Kaersutite has been recorded from a variety of alkaline mafic and intermediate rocks, both sodic (Tiba, 1966) and potassic (Aoki, 1959; Coombs and Wilkinson, 1969). Mason (1968) has directed attention to garnet-pyroxene xenoliths containing kaersutite in alkaline volcanic rocks and tuffs, and also to megacrysts, apparently cognate, coexisting with tschermakitic clinopyroxene, pyrope-rich garnet, anorthoclase, titanbiotite and spinel (Mason, 1966, 1968; Dickey, 1968; Binns, 1969). He has suggested that such kaersutites (even though some are comparatively Fe-rich) crystallized in the upper mantle (*see* Ernst, 1968, pp. 74–6; Le Maitre, 1969; Aoki, 1970; Best, 1970). There are apparently no definitive chemical criteria whereby kaersutites of high-pressure origin may be distinguished from those which clearly crystallized at low pressures, as indicated by the occurrence of kaersutite as a groundmass constituent of certain alkaline mafic rocks, as parallel intergrowths with titaniferous clinopyroxene, or in relatively leucocratic cross-cutting veins (cf. Yagi, 1953b). As yet crystallization trends of kaersutites have not been evaluated in detail but the principal chemical change is presumably an increase in Fe^{2+} at the expense of Mg, in accordance with optical measurements on zoned kaersutites in which the optic axial angle decreases towards the margins, consistent with a trend towards ferropargasite.

II.4.4.4. Iron–titanium Oxides

The most important series in the FeO–Fe_2O_3–TiO_2 system are: (i) the magnetite (Fe_3O_4)–ulvöspinel (Fe_2TiO_4) series, termed the β-series by Verhoogen (1962), with complete solid solution above approximately 600 °C (Vincent and others, 1957; Kawai, 1959); (ii) the trigonal hematite (α-Fe_2O_3)–ilmenite ($FeTiO_3$) series or α-series, with complete miscibility between the two-end members above 1000 °C (Carmichael, 1961). Members of the orthorhombic solid solution series pseudobrookite (Fe_2TiO_5)–'ferropseudobrookite' ($FeTi_2O_5$), much more restricted in paragenesis than the preceding series, may be minor constituents of some alkali gabbros, mugearites, etc. (Carmichael and Nicholls, 1967).

Titanomagnetites in extrusive rocks typically are homogeneous, in contrast to magnetites from more slowly cooled rocks which may contain lamellae of ulvöspinel, generally in the (100) plane of the host, and/or ilmenite parallel to (111).

Although iron–titanium oxides are ubiquitous accessories in alkali basaltic rocks, few detailed studies of these minerals have been carried out. The available data suggests the members of the β-series (or their oxidized equivalents) are more common than those belonging to the trigonal series; when it is present ilmenite may form phenocrysts but more often it tends to be confined to the groundmass as minute flakes and needles (Le Maitre, 1962; Uchimuzu, 1966; White, 1966; Wright, 1967a). The coexisting iron–titanium oxides can yield useful information on geothermometry and oxygen geobarometry (Buddington and Lindsley, 1964). The oxygen geobarometer may have to be slightly modified to take account of ferrous–ferric iron equilibria as a function of the alkali contents of the host rocks (Carmichael and Nicholls, 1967). There are presently no analytical data on the coexisting titanomagnetites and ilmenites from alkali basaltic rocks. Crystallization trends of titanomagnetites involve a decrease in the Fe_2TiO_4 component (Wilkinson, 1965b, Fig. 1; Huckenholz, 1965a, Fig. 19; Uchimuzu, 1966), usually accompanied by decreases in Al, Mg, Cr, V, Ni, and Co. Carmichael and Nicholls (1967) have shown that, compared with the coexisting groundmass β-phase, the titanomagnetite phenocrysts in a kenyte are enriched in Fe_2TiO_4. Minor amounts of chromiferous spinel may occur as inclusions to silicate phenocrysts or as cores to titanomagnetites (Wright, 1967a).

II.4.4.5. Feldspars

There are few detailed chemical and other data on the feldspar assemblages of alkali basaltic lineages, particularly the fine-grained variants, partly because of the difficulties in separating pure feldspar fractions for chemical and X-ray analysis, partly because plagioclase compositions are generally determined from optical properties only. Most feldspar analyses record only K_2O, Na_2O and CaO, and hence the possibility of systematic departures from stoichiometric compositions, conceivably controlled by host rock compositions, cannot at present be assessed.

The compositions of plagioclases in alkali basaltic rocks are usually in the labradorite range, less commonly sodic bytownite. Sodic labradorites containing up to 0·9% K_2O (Wilkinson, 1965a; Huckenholz, 1965a) represent an initial trend towards the potash andesines and potash oligoclases (Muir, 1962, Fig. 1) of more evolved alkaline rocks, and hence are more potassic than their subalkaline counterparts.

Alkali feldspars in extrusive alkali basaltic rocks and from shallow intrusives belong to the low sanidine–high albite series, and may be sanidine-cryptoperthites with bulk compositions ranging from approximately Or_{45} to Or_{65} (Coombs, in Brown, 1955; Hamilton and Neuerberg, 1956; Wilkinson, 1965a; Tanguy, 1966), or anorthoclase or anorthoclase cryptoperthites (e.g. Huckenholz, 1965a). Plagioclase crystallization trends involve enrichment in Ab, and, to a lesser extent Or, extending through potash andesine and potash oligoclase to lime anorthoclase (Muir and Tilley, 1961; Huckenholz, 1965a; Coombs and Wilkinson, 1969). It is inferred that most plagioclases in alkali basaltic rocks are structurally high- or intermediate-temperature types; rarely, as in an analcime-bearing trachydolerite from Mount Moroto, Uganda, low-temperature plagioclases have been recorded (Hytönen, 1959). Alkali feldspars in the theralites at Square Top, Nundle, New South Wales, display an initial decrease in their Or contents with fractionation, but in the tinguaitic differentiates, containing only one feldspar in contrast to the two-feldspar assemblages in the more mafic rocks, the alkali feldspars become enriched in Or with fractionation (Wilkinson, 1965a, Fig. 2). Some alkali basaltic lavas contain prominent megacrysts of anorthoclase, sharply delineated from the groundmass feldspar assemblage (Edwards, 1938; Vlodavetz and Shavrova, 1953; Binns, 1969); detailed appraisals of this paragenesis remain to be carried out.

II.4.4.6. Nepheline

The limited compositional data on nephelines from alkali basaltic rocks indicate highly sodic compositions, close to Ne_{90}; their crystallization trends mainly involve a slight enrichment in Ks (Wilkinson, 1965a; Wilkinson, 1966a; Tanguy, 1966; Coombs and Wilkinson, 1969).

II.4.4.7. Analcime

Depending on the substitution NaAl $\rightarrow$ Si, the compositions of analcimes in alkali basaltic rocks may depart significantly from the 'ideal' formula $NaAlSi_2O_6 \cdot H_2O$ ($a \sim 13{\cdot}72$ Å). Experimental data (Saha, 1959, 1961; Peters, Luth and Tuttle, 1966) indicate a wide range of analcime solid solutions, from $Na_2Al_2Si_3O_{10}$ ($a \sim 13{\cdot}75$ Å) to $NaAlSi_3O_8$ ($a \sim 13{\cdot}67$ Å), i.e. corresponding, apart from water content, to natrolite (cf. Table 3, No. 14) and albite; igneous analcimes containing more than 34 Si per formula unit (96 O) are relatively uncommon. K and Ca may replace

Na to a limited extent, approximately 2% K_2O defining the upper potash content of natural analcimes (Wilkinson, 1968a). A measure of the NaAl $\rightarrow$ Si substitution is obtained from unit cell dimensions or variation in the analcime (639) reflection, measured against the (331) peak of silicon as internal standard (Peters, Luth and Tuttle, 1966, Fig. 1; Coombs and Whetten, 1967, Fig. 1). Chemical, X-ray and optical data on analcimes from a variety of igneous parageneses have been listed by Deer, Howie and Zussman (1963b, Table 44), Coombs and Whetten (1967) and Wilkinson (1968a).

Analcime is a common accessory mineral in alkali basalts, and is an essential primary constituent of analcime basanites, teschenites and related rocks. Deuteric or secondary analcimes in vughs and gas cavities are often slightly birefringent (Coombs, 1955). Well-defined euhedral analcime phenocrysts in some analcimites provide the most convincing textural evidence for a primary origin (Larsen, 1941; Wilkinson, 1968a, Figs. 1 and 2; *see* Peters, Luth and Tuttle, 1966). There are comparatively few data defining analcime crystallization trends. Analcimes from the more differentiated facies of the Black Jack teschenite sill contain less K_2O than earlier analcimes (Wilkinson, 1958), and in the theralite–tinguaite differentiation sequence at Square Top, the substitution NaAl $\rightarrow$ Si decreases with decreasing temperature (Wilkinson, 1963).

II.4.5. Generalized Petrography of Alkali Basaltic Rocks

Alkali basaltic rocks and their cogeners dominate the volcanic islands of the oceanic basins. They are also prominent in a continental environment, as, for example, in north-eastern New South Wales (McDougall and Wilkinson, 1967). Alkali basalts tend to have higher contents of Fe_2O_3, TiO_2, Na_2O, K_2O and P_2O_5, and tend to be poorer in SiO_2 and Al_2O_3 than subalkaline basalts. Chayes (1964; 1965) has discussed chemical differences between 'circum-oceanic' and 'oceanic' basalts. With the exception of higher TiO_2 and 'alkalinity' relative to SiO_2 (reflected in alkali basalts by *ne* or minor *hy*) Chayes (1964, p. 1584) concluded that 'as far as content of what are ordinarily considered essential constituents is concerned, most basalts could belong as well to one group as to the other' (i.e. to either the oceanic or circumoceanic group). Most oceanic basalts contain considerably more than 1·75% TiO_2 (Chayes, 1965); high alumina alkaline variants (cf. Table 4, No. 6; Yoder and Tilley, 1962, Tables 23 and 24) appear to be less common than their subalkaline analogues. Chayes' generalization is applicable only to oceanic islands or seamounts since it does not take account of the low potash tholeiites of the deeper parts of the oceans (Engel and Engel, 1964; Engel, Engel and Havens, 1965).

The analyses listed in Table 4 illustrate the variation in the major oxides of some alkali basaltic lavas. The K_2O/Na_2O ratios of the mildly undersaturated types (Nos. 1 to 9) range from 0·16 (No. 5) to 0·78 (No. 9), and a more or less comparable range of alkali ratios is provided by the moderately undersaturated extrusives (Nos. 10 to 14). Sodic alkali basalts tend to have relatively high $(FeO + Fe_2O_3)/MgO$ ratios but low-iron sodic lineages are known. Although not so apparent in potassic mafic lavas, low Fe/Mg ratios tend to be characteristic of the intermediate members of potassic lineages (Table 6, Nos. 6 to 10). Moderately undersaturated variants (basanites and nepheline trachybasalts) are enriched in alkalis relative to SiO_2. The melanocratic picrite basalts (olivine-rich) and ankaramites (clinopyroxene-rich) contain high MgO and CaO contents respectively; enrichment in mafic minerals takes place mainly at the expense of plagioclase so that the alumina contents of these rocks are correspondingly low (Table 4, nos. 14 and 15).

Kuno (1959, Table 8), Chayes (1964, Table 1), Wilkinson (1967, Table 7) and Le Maitre (1968, Appendix B) have listed references to many alkali volcanic associations, particularly those of the oceanic islands. Green and Poldervaart (1955) and Coombs (1963) have presented averaged chemical compositions of some alkali basaltic suites. Sodic volcanic lineages include the Hebridean (Bailey and others, 1924; Tilley and Muir, 1962; Anderson and Dunham, 1966), the mildly undersaturated Hawaiian series (Macdonald and Katsura, 1964; White, 1966), the Hocheifel, West Germany, series (Huckenholz, 1965a), and the alkaline rocks of the Takakusayama district, Japan (Tiba, 1966). Relatively potassic suites, in which trachybasalts and trachyandesites figure prominently, have been described by Aoki (1959—Iki Islands, Japan), Le Maitre (1962—Gough Island, South Atlantic), Baker and others (1964—Tristan da Cunha), Uchimuzu (1966—Dōgo, Oki Islands, Japan), and Kurasawa (1967—north-west Kyushu). The volcanic rocks of Ascension Island (Daly, 1925), Pantelleria (Washington, 1914; Zies, 1962), Galapagos Islands (Richardson, 1933), Easter Island (Bandy, 1937), the Older Series of Mauritius (Walker and Nicolaysen, 1954), and Réunion (Upton and Wadsworth, 1966) include mildly alkaline *hy*-bearing basalts (*see* Coombs, 1963; Chayes, 1966). Moderately undersaturated volcanic associations include the Tahiti basanitoids (Williams, 1933), the East African Rungwe volcanics (Harkin, 1960), the basanites of the Honolulu series (Winchell, 1947), the Auckland basanites (Searle, 1960), the Hocheifel basanitoids (Huckenholz, 1965b), the recent lavas of Etna (Washington and others, 1926; Tanguy, 1966), and the moderately undersaturated members of the East Otago Volcanic Province (Coombs and Wilkinson, 1969).

The intrusive alkali basaltic rocks usually form comparatively small intrusions, discordant or plug-like, or concordant sheets and sills in which picritic facies may be developed; the felsic differentiates are represented by veins and schlieren of alkali trachytic or phonolitic composition, depending on the degree of undersaturation of the mafic host rocks. Alkali gabbros and alkali dolerites (Table 5, Nos. 1 to 3) occur in the Okenjeje Complex, South-West Africa (Simpson, 1954), in the Takakusayama series, Japan (Tiba, 1966), and the Prospect intrusion near Sydney, New South Wales (Wilshire, 1967). The following examples of teschenites (cf. Table 5, Nos. 4 and 5) may be noted: the Garbh Eilean sill, Shiant Isles (Walker, 1930; Drever and Johnston, 1959), the teschenitic and associated intrusions in the Midland Valley of Scotland (MacGregor and MacGregor, 1948, pp. 63–71), the Landywood sill, Staffordshire (Barton, 1963), the alkaline intrusives at Morotu, Sakhalin (Yagi, 1953b), the Circular Head laccolith, Tasmania (Edwards, 1941), the Mount Nebo sills, New South Wales (Edwards, 1953), and the Black Jack sill, near Gunnedah, New South Wales (Wilkinson, 1958). The mafic facies of the Lugar sill, Ayrshire (which also includes teschenitic types; Tyrrell, 1917, 1948, 1952; Phillips, 1968), the Waihola sill, East Otago (Benson, 1942; Coombs and Wilkinson, 1969) and the Square Top intrusion, Nundle, New South Wales (Wilkinson, 1965a) illustrate the more characteristic features of theralites (Table 5, No. 6).

Despite their comparatively simple gross mineralogy, alkali basaltic rocks vary considerably in grain size and textures. The textures of the intrusives range through basaltic, doleritic, and gabbroic, termed by Walker (1923) porphyritic, ophitic, and non-ophitic respectively. Porphyritic (phyric) and glomeroporphyritic textures are probably the most common textures of the extrusives but aphyric and microphyric textures are also important; variolitic and subvariolitic lavas are comparatively rare (*but* see Muir and Tilley, 1964, p. 412). The groundmass textures of lavas are ophitic, subophitic, intergranular or intersertal.

Olivine is a common phenocryst and groundmass phase and when euhedral typically displays well developed prism and dome faces. Elongated and skeletal olivines with deep embayments, or parallel growths of olivine have been attributed to rapid skeletal crystallization by Drever and Johnston (1957). Fayalitic olivines tend to be markedly subophitic with plagioclase. Inclusions of picotite or iron–titanium oxides are common in olivines from the more mafic rocks. Normal zoning in olivines is characteristically indicated by their interference colours which increase outwards in sections $\perp$ β and γ, and decrease in sections $\perp$ α. Alteration to 'bowlingite' and 'iddingsite' typically proceeds around crystal margins or along the imperfect cleavages; less commonly olivine alters to carbonates. Olivine cores may be surrounded by an iddingsitized

TABLE 4 Analyses and Norms of some Alkali Basaltic Lavas

	1	2	3	4	5	6
SiO_2	47·28	48·65	47·90	46·53	46·56	47·42
TiO_2	2·75	1·44	1·57	2·28	2·02	2·04
Al_2O_3	14·55	15·99	15·28	14·31	15·93	17·13
Fe_2O_3	2·45	2·18	1·70	3·16	1·61	2·06
FeO	9·70	6·19	9·10	9·81	10·32	7·26
MnO	0·18	0·15	0·17	0·18	0·20	0·16
MgO	8·74	9·66	7·30	9·54	9·00	8·34
CaO	10·03	11·52	12·07	10·32	10·51	9·71
Na_2O	2·78	2·71	2·81	2·85	3·21	3·66
K_2O	0·84	0·57	0·53	0·84	0·51	1·45
H_2O^+	0·23	0·75	1·27	0·08	0·02	0·48
H_2O^-	0·08	0·30	0·46	nil	nil	0·33
P_2O_5	0·44	0·21	0·16	0·28	0·26	0·48
Rest	—	—	—	0·06	0·06	—
Total	100·05	100·32	100·32	100·24	100·21	100·52
			Norms			
Or	4·96	3·37	3·13	4·96	3·01	8·57
Ab	23·52	22·93	21·78	20·49	19·68	20·09
An	24·75	29·79	27·52	23·78	27·56	26·04
Lc	—	—	—	—	—	—
Ne	—	—	1·08	1·96	4·05	5·89
Di	17·95	20·79	25·63	20·72	18·67	15·30
Hy	3·21	1·83	—	—	—	—
Ol	15·55	14·17	13·63	18·62	20·38	15·85
Mt	3·55	3·16	2·46	4·58	2·33	2·99
Il	5·22	2·73	2·98	4·33	3·84	3·87
Ap	1·03	0·49	0·37	0·65	0·61	1·12
Rest	0·31	1·05	1·73	0·14	0·08	0·81
Total	100·05	100·31	100·31	100·23	100·21	100·53
Differentiation index	28·5	26·3	26·0	27·4	26·7	34·6
Normative $\frac{An \times 100}{Ab + An}$	51·3	56·5	55·8	53·7	58·3	56·4
$\frac{(FeO + Fe_2O_3) \times 100}{MgO + FeO + Fe_2O_3}$	58·2	46·4	59·7	57·6	57·0	52·8

1. Basalt (Re 168), Piton de la Fournaise, Réunion (Upton and Wadsworth, 1966, Table 1). Analyst W. J. Wadsworth.
2. Basalt (4519–34; 67545), Mid-Atlantic Ridge (45° 44′ N, 27° 43′ W) (Muir and Tilley, 1964, Table 1). Analyst J. H. Scoon (Heavy clinopyroxene fraction $Ca_{47}Mg_{34}Fe_{19}$; olivine Fa_{16-30}; plagioclase An_{75-55}).
3. Alkali olivine basalt (63366), Hebridean alkali series, Fingal's Cave, Staffa (Tilley and Muir, 1962, Table 3). Analyst J. H. Scoon.
4. Alkali olivine basalt (65992), Hualalai, Hawaii (Yoder and Tilley, 1962, Table 2). Analyst J. H. Scoon. (Clinopyroxene $Ca_{44}Mg_{40}Fe_{16}$).
5. Alkali olivine basalt (S), first lava, April 1964, Surtsey, Iceland (Tilley, Yoder and Schairer, 1967, Table 4). Analyst J. H. Scoon.
6. Olivine basalt (basanitic) (GA 1253), Rodriguez Island, Indian Ocean, (Upton, Wadsworth and Newman, 1967, Table 1). Analysts Avery and Anderson.
7. Olivine trachybasalt, Wei-tschou Island, South China Sea (Yagi, 1953a; Table 2). Analyst K. Yagi. (Olivine Fa_{33-42}; plagioclase An_{54-50}).
8. Trachybasalt ('olivine basalt', C–20–57), Clarion Island, Mexico (Bryan, 1967, Table 3). Analyst H. B. Wiik.
9. Porphyritic trachybasalt (G 97), Gough Island, South Atlantic (Le Maitre, 1962, Table 10). (Olivine Fa_{23}; plagioclase An_{54-36}).
10. Nepheline basanite, 7 miles west of Inverell, New South Wales (Wilkinson, 1966a, Table 1). Analyst M. Chiba. (Olivine Fa_{32-33}; plagioclase An_{63}; nepheline $100K/(K + Na) = 9·2$. Mode: olivine 24,

7	8	9	10	11	12	13	14	15	16	17	18
48·98	46·89	48·89	42·63	41·65	46·36	43·03	43·61	46·57	44·44	47·06	47·65
2·88	3·03	3·07	2·11	1·96	3·54	2·66	2·43	1·85	3·00	3·44	1·13
14·12	15·78	16·15	13·07	15·17	16·19	13·82	10·28	8·20	13·24	17·14	18·13
2·25	2·53	1·53	2·02	3·84	3·66	3·70	4·45	1·20	5·44	3·29	2·63
7·40	8·68	8·15	10·78	9·59	6·94	8·95	8·95	9·75	6·01	6·65	6·48
0·15	0·14	0·13	0·20	0·21	0·18	0·21	0·17	0·14	0·16	0·18	n.d.
7·98	7·86	7·25	10·19	7·64	4·57	8·59	13·97	19·65	8·54	4·35	4·19
10·30	8·72	7·58	10·97	10·72	9·45	9·23	12·59	9·43	10·99	9·00	9·01
2·61	2·97	3·35	3·35	4·10	3·97	3·91	2·20	1·56	2·15	4·08	2·78
1·61	1·78	2·61	0·93	1·20	3·15	2·14	0·62	1·18	2·14	3·40	7·47
1·03	0·85	0·42	1·77	2·09	0·29	2·37	0·51	0·11	1·89	0·37	0·13
0·32	0·00	0·44	1·22	1·07	0·19	0·78	0·12	0·12	1·46	0·27	0·11
0·30	0·68	0·15	1·00	0·67	1·42	0·73	0·30	0·26	1·04	0·75	0·50
—	—	—	—	—	—	0·23	0·15	0·04	0·05	—	0·26
99·93	99·91	99·72	100·24	99·91	99·91	100·35	100·35	100·06	100·55	99·98	100·47
						Norms					
9·51	10·52	15·42	5·50	7·09	18·62	12·65	3·66	6·97	12·65	20·10	14·46
22·09	24·38	23·59	8·98	6·25	17·92	8·37	6·84	11·52	16·69	15·45	—
22·06	24·47	21·33	17·89	19·45	17·06	13·84	16·34	11·89	20·16	18·42	14·73
—	—	—	—	—	—	—	—	—	—	—	23·54
—	0·41	2·58	10·49	15·41	8·49	13·39	6·38	0·91	0·82	10·33	12·78
21·74	11·59	12·38	24·41	23·91	16·51	22·19	35·20	26·52	21·65	17·15	21·80
6·19	—	—	—	—	—	—	—	—	—	—	—
7·56	16·70	15·17	20·73	13·81	5·51	14·50	19·38	36·12	9·19	4·85	5·59
3·26	3·67	2·22	2·93	5·57	5·31	5·36	6·45	1·74	7·89	4·77	3·94
5·47	5·75	5·83	4·01	3·72	6·72	5·05	4·62	3·51	5·70	6·53	2·13
0·70	1·59	0·35	2·33	1·56	3·35	1·69	0·70	0·61	2·42	1·77	1·30
1·35	0·85	0·86	2.99	3·16	0·38	3·15	0·78	0·27	3·40	0·64	0·24
99·93	99·93	99·73	100·26	99·93	99·87	100·19	100·35	100·06	100·57	100·01	100·51
31·6	35·3	41·6	25·0	28·8	45·0	34·1	16·9	19·4	30·2	45·9	50·8
50·0	50·1	47·5	66·6	75·7	48·8	62·3	70·5	50·8	54·7	54·4	—
54·7	58·8	57·2	55·7	63·7	69·9	59·6	49·0	35·8	57·3	69·6	68·5

clinopyroxene 29, plagioclase 15, nepheline 14, glass 13, opaques 5).

11. Nepheline basanite (OU 20670), Omimi, East Otago (Coombs and Wilkinson, 1969, Table 2). Analyst G. I. Z. Kalocsai. (Olivine Fa_{21}; nepheline Ne_{88}. Mode: olivine 12, clinopyroxene 40, feldspar (dominantly plagioclase) 11, nepheline 22, zeolites + apatite 7, opaques 8).
12. Trachybasalt (347), Tristan da Cunha (Baker and others, 1964, Table 6). Analyst J. R. Baldwin. (Plagioclase phenocrysts An_{85-80}, groundmass An_{54}; minor leucite in groundmass).
13. Sanidine basanite (OU 5766) Siberia Hill, north-east Otago (Benson, 1942, Table 5; Brown, 1955, Table 1). Analyst F. T. Seelye.
14. Ankaramitic picrite basalt, Haleakala, Maui, Hawaiian Islands (Tilley, Yoder and Schairer, 1964, Table 1). Analyst J. H. Scoon.
15. Picrite basalt (G 121), Gough Island, South Atlantic (Le Maitre, 1962, Table 10). Analyst R. W. Le Maitre. (Olivine Fa_{15}; clinopyroxene $Ca_{44}Mg_{46}Fe_{10}$, plagioclase An_{67}).
16. Limburgite, parish of Lancefield, Victoria (Edwards, 1938, Table 9). Analyst A. B. Edwards.
17. Leucite-bearing trachybasalt (351), Tristan da Cunha (Baker *et al.*, 1964, Table 6). Analyst J. R. Baldwin.
18. Leucite tephrite, lava of 1872, below Observatory, Mount Vesuvius (Washington, 1906). Analyst H. S. Washington. (Calculated mode: labradorite 17·4, leucite 34·9, nepheline 9·0, augite 30·3, olivine 5·6, opaques 1·5, apatite 1·3).

TABLE 5 Analyses, Norms, Modes and Mineralogy of some Alkali Basaltic Intrusive Rocks

	1	2	3	4	5	6	7	8	9
SiO_2	46·01	43·94	48·21	44·78	46·41	44·90	46·88	43·85	43·15
TiO_2	2·46	2·45	1·45	2·49	2·43	2·13	2·81	1·12	0·92
Al_2O_3	13·13	14·03	13·29	14·03	15·66	14·97	17·07	11·12	8·67
Fe_2O_3	2·84	1·95	1·87	4·15	4·28	2·62	3·62	2·22	2·86
FeO	10·09	11·65	7·46	9·15	6·31	8·26	5·94	9·79	9·83
MnO	0·18	0·32	0·17	0·14	0·22	0·15	0·16	0·14	0·33
MgO	8·90	10·46	8·51	9·57	5·40	8·45	4·85	17·50	20·70
CaO	9·08	8·99	13·96	8·12	8·13	9·08	9·49	7·73	7·16
Na_2O	3·03	2·68	2·55	3·30	4·44	5·02	5·09	2·12	1·66
K_2O	0·96	0·33	1·34	1·77	2·47	1·72	2·64	0·98	0·73
H_2O^+	2·58	2·31	0·53	2·05	3·59	1·21	0·97	2·36	3·86
H_2O^-	0·30	0·85	0·24	0·14		0·38		0·18	0·26
P_2O_5	0·49	0·20	0·48	0·62	0·66	0·97	0·48	0·26	0·29
Rest	0·06	0·26	—	—	—	—	—	0·26	0·20
Total	100·11	100·42	100·06	100·31	100·00	99·86	100·00	99·63	100·62
Norms									
Or	5·67	1·95	7·92	10·46	14·60	10·16	15·60	5·79	4·31
Ab	24·74	20·09	12·71	18·76	21·74	13·29	14·23	12·70	12·95
An	19·40	25·29	20·87	18·25	15·52	13·24	15·94	17·94	14·05
Ne	0·48	1·40	4·80	4·96	8·57	15·81	15·63	2·84	0·59
Di	18·24	14·68	36·59	14·47	16·43	20·58	22·38	15·00	15·63
Ol	18·71	25·65	9·82	19·03	7·20	15·09	3·56	36·62	42·19
Mt	4·12	2·83	2·71	6·02	6·20	3·80	5·25	3·22	4·15
Il	4·67	4·65	2·75	4·73	4·62	4·05	5·34	2·13	1·75
Ap	1·14	0·47	1·12	1·45	1·54	2·26	1·12	0·61	0·68
Rest	2·94	3·42	0·77	2·19	3·59	1·59	0·97	2·80	4·32
Total	100·11	100·43	100·06	100·32	100·01	99·87	100·02	99·65	100·62
Differentiation index	30·9	23·4	25·4	34·2	44·9	39·3	45·5	21·3	17·9
Normative $\frac{An \times 100}{Ab + An}$	44·0	55·7	62·2	49·3	41·7	49·9	52·8	58·5	52·0
$\frac{(FeO + Fe_2O_3) \times 100}{MgO + FeO + Fe_2O_3}$	59·2	56·5	52·3	58·2	66·2	56·3	66·3	40·7	35·8
Modes									
Olivine (plus alteration products)	14·9	22	7·0	23·0	—	13·1	—	34·7	49·5
Clinopyroxene	33·2	19	41·0	19·7	—	23·0	—	24·5	20·2
Opaque minerals	7·7	5	3·2	10·2	—	8·7	—	1·6	0·8
Plagioclase	36·7	52	37·8	38·0	—	24·0	—	28·5	18·2
Alkali feldspar	—	—	2·4		—	2·0	—		
Nepheline	—	—	3·1	—	—	12·1	—	'others'	'others'
Zeolite (mainly analcime)	—	2	—	9·1	—	17·1	—	7·3	9·1
Mesostasis	1·9	—	—	—	—	—	—		
Biotite	4·7	—	4·9	—	—	—	—	3·2	2·2
Amphibole		—	tr.	—	—	—	—	—	—
Apatite	0·9	—	0·6	—	—	—	—	0·1	n.f.

TABLE 5 (*Continued*)

	Mineralogy								
Olivine (Fa)	(25)	—	27–46	25	—	30*	—	26*	—
Clinopyroxene									
Ca	46*	—	47*	48*	—	47*	—	48*	—
Mg	37	—	38	37	—	40	—	41	—
ΣFe	17	—	15	15	—	13	—	11	—
Plagioclase (An)	(54)	'labra-dorite'	80–70	69	—	56*	—	—	—

* Composition from analysis.

Legend:

1. Alkali dolerite, composite sample, upper chilled margin, Prospect intrusion, 18 miles west of Sydney, New South Wales (Wilshire, 1967, Table 5). Analyst J. H. Pyle. Generalised olivine and plagioclase compositions taken from Wilshire (1967, Fig. 12).
2. Crinanite (analcime-bearing alkali dolerite), 1 mile north of Inver Cottage, Jura, Argyllshire (Flett, 1911). Analyst E. G. Radley. Mode determined on original slice by W. Q. Kennedy (Johannsen, 1938, p. 72).
3. Alkali gabbro (158), Adams Shoulder, Okonjeje, South-West Africa (Simpson, 1954, Table 2). Analyst E. S. W. Simpson.
4. Teschenite Bl (approaching glenmuirite), 20 feet above lower contact, Black Jack sill, near Gunnedah, New South Wales (Wilkinson, 1958, Table 6). Analyst J. F. G. Wilkinson.
5. Average glenmuirite (13 analyses) (Nockolds, 1954, Table 11).
6. Analcime-olivine theralite (ST.28), close to lower contact, Square Top intrusion, near Nundle, New South Wales (Wilkinson, 1965a, Table 1). Analyst M. Chiba. (Alkali feldspar $Or_{55}Ab_{40}An_5$; nepheline $Ne_{91}Ks_7Qz_2$).
7. Average essexite (15 analyses) (Nockolds, 1954, Table 11).
8. Picroteschenite ('kylitic'), 76 feet above base of southern cliff, Benbeoch, Ayrshire (Drever and Macdonald, 1967, Table 3). Analyst J. G. Macdonald.
9. Alkaline picrite, Craigdonkey, Benquhat Hill, Ayrshire (Drever and Macdonald, 1967, Table 3).

zone, followed by a rind of more fayalitic olivine (Upton and others, 1967; Baker and Haggerty, 1967). In addition to granular coronas of clinopyroxene, olivines are sometimes rimmed by kaersutite (Aoki, 1959), and biotite may form discontinuous reaction rims to both phenocryst and groundmass olivine. Cognate olivines can often be distinguished from xenocrystal olivines from peridotite xenoliths (frequently coexisting with xenocrystal orthopyroxene and spinel) by their more euhedral habit, more fayalitic compositions and by the absence of undulose extinction and banded structures (cf. Hamilton, 1957; Muir and Tilley, 1964).

The Ca-rich clinopyroxene may be neutral coloured, grey green (in picrite basalts), or brownish or lilac tinted. The lilac tints are more pronounced in the clinopyroxenes from moderately undersaturated host rocks, which show strong inclined dispersion of the optic axes. Hourglass and sector zoning and colour zoning are common (Strong, 1969); the various zones often tend to be outlined by opaque oxide inclusions. Abundant inclusions of olivine, opaque oxide and groundmass material often impart a sieve-like appearance to the clinopyroxene. In more undersaturated rocks the Ca-rich pyroxene is rimmed by green acmitic pyroxene, especially where the crystals abut areas of nepheline, analcime or alkaline mesostasis. Clinopyroxene and opaque oxides may be intergrown in a graphic or subgraphic manner and similar structures involving clinopyroxene and nepheline or plagioclase have been described (Benson, 1942; (Coombs and Wilkinson, 1969).

Titanomagnetite of early crystallization forms euhedral cubes or octahedra, often arranged in the synneusis structure; markedly skeletal habits are displayed by titanomagnetite phenocrysts in the more evolved intrusive facies (cf. Wilkinson, 1957b, Fig. 2). Ilmenite may be common as rods and flakes in the groundmass; in slowly cooled rocks it forms lamellae parallel to (111) of the host spinel.

Kaersutite can be either a phenocryst or groundmass mineral; the phenocrysts in extrusive rocks usually display evidence of breakdown (cf. II.4.4.3). Biotite is typically confined to the groundmass or forms reaction rims to opaque oxides or other ferromagnesian minerals.

In extrusive and doleritic alkali basaltic rocks, the plagioclase most commonly has a lath-like habit but in the gabbroic types it tends to be tabular or short prismatic. The plagioclase is often mantled by anorthoclase, but in the more potassic intrusives, anorthoclase or sanidine form discrete crystals. The alkali feldspars of trachybasalts are often interstitial to plagioclase and show undulose extinction. In intrusives the plagioclase is commonly altered to clay minerals, analcime, natrolite or thomsonite, or rarely to prehnite, as in the Prospect intrusion (Wilshire, 1967). Nepheline forms rectangular or hexagonal crystals but in the more undersaturated basanites it tends to be coarsely granular and poikilitic. Primary analcime is generally interstitial but in rocks such as teschenites, it is also a deuteric alteration product of plagioclase. A variety of zeolites have been recorded from alkali basaltic rocks, occurring as vesicle fillings, or as alteration products of nepheline or feldspars; on occasion the zeolites appear to be primary (deuteric) phases. The zeolite species typically are those favoured by a SiO_2-poor environment (Coombs and others, 1959, Fig. 5), namely natrolite, chabazite, thomsonite, phillipsite and mesolite. Interstitial leucite occurs in some of the Tristan da Cunha trachybasalts (Baker and others, 1964). Strongly prismatic apatite, often referred to fluor-apatite, is an ubiquitous accessory mineral.

Alkali basaltic rocks from flows and shallow intrusives often display petrographic evidence of the production of low-melting residua, composed of alkali feldspars, sodic clinopyroxenes, opaque oxide, and variable amounts of nepheline, analcime or other zeolites. These assemblages are represented in volcanic rocks by 'trachytic' segregations (Le Maitre, 1962; Bryan, 1967; Wright, 1967b) and in teschenites, theralites and essexites by alkaline mesostasis (Wilkinson, 1958). Schlieren or pegmatoid segregations, analogous to tholeiitic dolerite pegmatites (Walker, 1953), are distinguished from the host rocks by their coarser grain size, more extensive deuteric alteration of olivine, plagioclase or nepheline, and increased zeolite or feldspathoid contents. The habits of the olivines, clinopyroxenes and opaque oxides are often indicative of free- and skeletal growth (Wilkinson, 1958; Wilshire, 1967; Coombs and Wilkinson, 1969). Rapidly cooled lavas may contain variable amounts of intersertal pale brown glass, trachytic or phonolitic in composition (Wilkinson, 1966a).

II.4.6. Alkaline Picrites

Alkaline picrites, the more mafic associates of alkali dolerites, teschenites etc., contain more than 40% modal olivine; when modal olivine ranges from 20–40%, the appropriate mafic rock term may be prefixed by picro- (Drever and Johnston, 1965). The Garbh Eilean sill, Shiant Isles (Walker, 1930; Drever and Johnston, 1965), the intrusions in the Nebo district, New South Wales (Edwards, 1953), the Midland Valley of Scotland (*see* Drever and Johnston, 1967a, p. 52) and north Skye (Simkin, 1967) are well-known examples of mafic–picritic associations. Alkaline picrites are composed of olivine (Fa_{15-23}), clinopyroxene (ca. $Ca_{46}Mg_{43}Fe_{11}$) and calcic plagioclase (An_{74-72}), with variable amounts of opaque oxide, analcime, kaersutitic amphibole and phlogopitic biotite. Poikilitic textures are distinctive, with olivines (typically unzoned) enclosed by clinopyroxene and bytownite (*see* Drever and Johnston, 1967a, Fig. 3.3).

The picritic minor intrusions which form relatively small dykes, sills and sheets unassociated with mafic rocks, and which are well developed in the Cuillins of Skye and on Ubekendt Ejland, West Greenland (Drever, 1956; Drever and Johnston, 1958; Wyllie and Drever, 1963; Drever and Johnston, 1967b) should also be mentioned. As yet the magmatic affinities of these rocks have not been clearly defined but inspection of their normative composition indicates mainly a subalkaline character, less commonly mildly alkaline, with minor *hy* (Drever and Johnston, 1958, Table 2).

II.4.7. Leucite-Bearing Basaltic Rocks

The appearance of leucite in mafic alkaline volcanics is conditioned partly by relatively high K_2O/Na_2O ratios, partly by an appropriate level of undersaturation. Leucite tephrite (Table 4, No. 18) and leucite basanite (= leucite-olivine tephrite), the leucite-bearing analogues of nepheline tephrite and nepheline basanite respectively (cf. Table 2), by definition should contain only accessory alkali feldspar. However modal data on lavas so designated would probably indicate the not infrequent role of alkali feldspar as an essential constituent, and hence these volcanics which are comparable with leucite-bearing absarokite (Iddings, 1895; Holmes and Harwood, 1937; Joplin, 1968), may be more appropriately termed leucite trachybasalts (cf. Baker *et al.*, 1964); more leucite-rich variants correspond with the vicoites of the Roman province (Washington, 1906). With decreasing modal feldspar and increasing undersaturation, leucite tephrites merge into the leucitites.

Leucite-bearing basaltic rocks mainly occur as variants within petrographic provinces displaying a wide range of leucite-bearing eruptives, e.g. the Roman province of Western Italy (Washington, 1906) and the Bufumbira area in Uganda (Holmes and Harwood, 1937) (see also Iddings, 1913, pp. 268–9). The leucite-bearing trachybasalts of Tristan da Cunha (Baker *et al.*, 1964) exemplify variants from a moderately potassic lineage characterized by only a sporadic development of leucite. The lava represented by Analysis 17 in Table 4 (see also Analysis 12) contains phenocrysts of clinopyroxene, kaersutite, and plagioclase (An_{70}–An_{80}) in an intergranular groundmass of plagioclase (An_{68}), pyroxene, opaque oxide, alkali feldspar and leucite.

The more leucite-rich tephrites often carry conspicuous phenocrysts of leucite, frequently displaying multiple twinning, and accompanied by a second generation of this feldspathoid in the groundmass. The pinkish or pale grey-green clinopyroxene is a diopside-salite (cf. Carmichael, 1967; Tournon, 1969). Alkali feldspar sometimes margins plagioclase and frequently is an interstitial groundmass constituent. Phlogopitic biotite is a potential phenocryst or groundmass phase and nepheline may also be a minor component of the groundmass.

II.4.8. Inclusions in Alkali Basaltic Rocks

In basaltic rocks mafic and ultramafic inclusions are almost exclusively confined to the alkali basaltic types (generally nepheline-normative) and are exceedingly rare in tholeiitic lavas (*see* White, 1966, p. 270; Babkine and others, 1966). The ultramafic inclusions have been most intensively studied (Ross and others, 1954; Wilshire and Binns, 1961; Talbot and others, 1963; Yamaguchi, 1964; White, 1966; Forbes and Kuno, 1967) but according to Forbes and Kuno (1965) the most ubiquitous inclusion type is gabbroic in composition. The most common ultramafic variant is lherzolite, composed of olivine (Fa_8–Fa_{15}), aluminian enstatite, aluminian chrome diopside and minor $MgAl_2O_4$-rich spinel. Other ultramafic inclusions include dunites, wehrlites, and various pyroxenites. In Hawaii, lherzolites are more common in moderately to strongly undersaturated hosts, and their component minerals possess relatively constant compositions. In contrast the dunites, wehrlites and pyroxenites are more iron-rich, compositional variation of olivine and pyroxene is more extensive, and these inclusions, with gabbroic types, preferentially occur in mildly undersaturated hosts (White, 1966). Lherzolite inclusions display evidence of deformation and cataclasis, indicated by undulose extinction, kink bands and lamellae parallel to (100) in olivine, undulose extinction in diopside, and kink banding in enstatite. Preferred orientation of olivine and pyroxenes tends to be characteristic (cf. Brothers, 1960; Collée, 1963). Evidence of disequilibria between ultramafic inclusions and the enclosing basaltic rock is widespread (Wilshire and Binns, 1961; White, 1966). Orthopyroxene may be mantled by fine granular olivine in a cryptocrystalline matrix, diopside grains have a porous outer zone, depleted in Na and Al relative to the core, and containing very minor feldspar and (?) nepheline, whilst the olivines at the boundaries of inclusions are enriched in Fe, Ca and Mn and depleted in Mg and Ni. An opaque margin, enriched in Fe and depleted in Al, is developed around spinel.

The gabbroic inclusions may be orthopyroxene-bearing (Le Maitre, 1965; White, 1966) or devoid of orthopyroxene and composed of olivine, clinopyroxene, plagioclase and kaersutitic amphibole in varying proportions (Baker and others, 1964; Upton and others, 1967; Kuno, 1967). All compositional gradations between anorthosite and pyroxenite exist, and individual suites of inclusions may show wide variation in grain size; layers with contrasting mineralogy are present in some specimens.

II.4.9. The Nomenclature of Some Evolved Alkaline Volcanic Rocks

Investigations of alkaline petrographic provinces such as the Hebridean and Hawaiian and the volcanic rocks of Gough Island and Tristan da Cunha have demonstrated a general coherence in mineralogy and chemistry in the members of specific lineages. These lineages may be broadly defined by parameters such as the ratios K_2O/Na_2O and $(FeO + Fe_2O_3)/MgO$, and by the degree of undersaturation, expressed as *ne*. The mildly undersaturated Hebridean and Hawaiian series alkali basalt → hawaiite → mugearite → benmoreite → alkali trachyte, and the trachybasalt → trachyandesite → tristanite → trachyte series of Gough Island provide examples of sodic and moderately potassic lineages. More undersaturated equivalents of these lineages have also been described e.g. a sodic series nepheline basanite → nepheline hawaiite → nepheline mugearite → nepheline benmoreite → phonolite in the rather heterogeneous East Otago Volcanic Province has been proposed by Coombs and Wilkinson (1969).

Hawaiite contains a modal and normative plagioclase whose composition is in the andesine range, the modal plagioclase being zoned from calcic andesine to lime anorthoclase (Muir and Tilley, 1961). As defined by Macdonald (1960) and Macdonald and Katsura (1964), hawaiite has a K_2O/Na_2O ratio less than 1:2 (Table 6, Nos. 1 and 2). Hawaiites may contain approximately equal proportions of plagioclase and lime anorthoclase, and hence the distinction between hawaiite and trachyandesite is based primarily on the sodic or potassic nature of the rock, and not on differing feldspar ratios. Trachyandesite ($K_2O/Na_2O > 1:2$), equivalent to doreite as defined by Nockolds (1954), contains a plagioclase (modal and normative) in the andesine–oligoclase range (Table 6, Nos. 6 and 7), and alkali feldspars (anorthoclase and sanidine) comprise approximately 10–40% of the total feldspar (cf. Streckeisen, 1967, Fig. 6). The differentiation index of trachyandesite ($\Sigma\, qz + ab + or + ne + lc$; Thornton and Tuttle, 1960) is less than 65. The differentiation indices of the more evolved potassic lavas, namely tristanites (D.I. = 65–75), span the so-called 'Daly gap' (Tilley and Muir, 1964; *see* Le Maitre, 1962; Baker and others, 1964). The K_2O/Na_2O ratios of mugearites are less than 1:2 (a corollary of the definition of hawaiite according to Macdonald and Katsura, 1964). The normative plagioclase compositions of mugearites are in the oligoclase range (Table 6, No. 4); lime anorthoclase exceeds potassic plagioclase and interstitial soda sanidine may also be present (Muir and Tilley, 1961). Benmoreite, the sodic member of mildly undersaturated lineages with a composition appropriate to the Daly gap, has a D.I. = 65–75, and (typically) $53 < SiO_2 < 57$; $3 < CaO < 6$; $5\frac{1}{4} < (FeO + Fe_2O_3) < 8\frac{1}{2}$ (Table 6, No. 5).

Typical hawaiites and mugearites display comparatively high $(FeO + Fe_2O_3)/MgO$ ratios, reflected in the Fe-rich compositions of the contained olivines and pyroxenes; lavas with similar alkali ratios but possessing distinctly lower $(FeO + Fe_2O_3)/MgO$ ratios are probably best termed low-iron hawaiites and low-iron mugearites. Trachyandesites often display lower Fe/Mg ratios than hawaiites and mugearites, a criterion employed by Uchimuzu (1966, Fig. 8) to distinguish the rocks by a 'mugearite–trachyandesite' boundary on an AMF diagram. This

TABLE 6 Analyses and Norms of some Evolved Alkaline Volcanics

	1	2	3	4	5	6	7	8	9	10
SiO_2	45·73	48·60	47·56	49·68	55·76	51·46	52·26	49·04	55·37	52·95
TiO_2	3·36	3·16	2·14	2·13	1·78	2·69	2·62	1·12	1·71	1·43
Al_2O_3	16·30	16·49	17·93	16·99	16·55	17·12	16·90	18·23	17·56	19·14
Fe_2O_3	3·87	4·19	4·64	3·45	3·10	2·96	3·84	2·36	2·83	3·25
FeO	11·45	7·40	6·20	8·99	6·02	6·05	4·69	6·01	4·59	2·86
MnO	0·20	0·18	0·21	0·27	0·22	0·15	0·14	0·15	0·11	0·20
MgO	5·25	4·70	3·13	2·79	1·08	4·03	3·15	4·54	2·22	2·02
CaO	7·42	7·79	7·13	5·46	3·23	5·94	6·25	6·52	4·64	5·33
Na_2O	4·33	4·43	6·26	5·78	6·28	4·06	4·87	5·35	4·65	6·55
K_2O	0·59	1·60	2·27	1·90	3·87	3·69	3·46	3·25	4·52	4·37
H_2O^+	0·97	—	0·66	1·77	0·95	1·09	0·66	1·67 }	1·40	1·12
H_2O^-	0·48	—	1·15	0·34	0·80	0·26	0·54	1·24 }		
P_2O_5	0·33	0·69	0·74	0·48	0·40	0·26	0·61	0·29	0·34	0·37
Cl	—	—	—	—	—	—	—	0·15	—	—
CO_2	—	—	0·24	—	0·03	—	—	—	—	—
Rest	—	—	—	0·37	0·07	—	—	—	0·10	0·43
Total	100·28	99·23	100·26	100·40	100·14	100·03	99·99	99·92	100·04	100·02

Norms

	1	2	3	4	5	6	7	8	9	10
Or	3·49	9·45	13·42	11·23	22·87	23·40	20·45	19·21	26·71	25·82
Ab	31·72	34·76	26·87	37·83	47·40	29·66	36·31	21·21	39·35	28·43
An	23·31	20·38	14·12	14·81	5·55	16·80	14·05	16·72	13·70	9·93
Ne	2·66	1·48	14·14	6·00	3·11	2·54	2·65	12·43	—	14·62
Di	9·38	11·10	12·50	7·67	6·52	8·85	10·27	11·21	5·78	11·10
Hy	—	—	—	—	—	—	—	—	1·32	—
Ol	15·52	8·38	4·45	10·22	4·00	7·42	3·10	9·73	3·54	0·28
Mt	5·61	6·08	6·73	5·00	4·49	4·29	5·57	3·42	4·10	4·71
Il	6·38	6·00	4·06	4·05	3·38	5·11	4·98	2·13	3·25	2·72
Ap	0·77	1·60	1·62	1·12	0·93	0·61	1·42	0·68	0·79	0·86
Cc	—	—	0·55	—	0·07	—	—	—	—	—
Rest	1·45	—	1·81	2·48	1·82	1·35	1·20	2·91	1·50	1·55
Total	100·29	99·23	100·27	100·41	100·14	100·03	100·00	99·90	100·04	100·02
Differentiation index	37·9	45·7	54·4	55·1	73·4	55·6	59·4	52·9	66·1	68·9
Normative $\frac{An \times 100}{Ab + An}$	42·3	37·0	34·4	28·1	10·5	36·2	27·9	44·1	25·8	25·9
$\frac{(FeO + Fe_2O_3) \times 100}{MgO + FeO + Fe_2O_3}$	74·5	71·1	77·6	81·7	89·4	69·1	73·0	64·8	77·0	75·2

Legend:

1. Hawaiite, Hebridean alkali series, Dun Hill, Skye (Tilley, Yoder and Schairer, 1967, Table 4). Analyst J. H. Socoon.
2. Average Hawaiian hawaiite (33 analyses) (Macdonald and Katsura, 1964, Table 10).
3. Nepheline hawaiite approaching nepheline mugearite (OU 22487), North Head section, near entrance to Otago Harbour, New Zealand (Coombs and Wilkinson, 1969, Table 7). Analyst M. Chiba.
4. Mugearite, Hebridean alkali series, Druim na Criche, near Mugeary, Skye (Muir and Tilley, 1961, Table 4). Analyst J. H. Scoon.
5. Benmoreite, Hebridean alkali series, so-called 'mugearite', E. of Kinloch Hotel, Mull (Bailey and others, 1924, p. 26; Tilley and Muir, 1964, Table 1).
6. Trachyandesite ('trachybasalt', G 164), Gough Island, South Atlantic (Le Maitre, 1962, Table 10). Analyst R. W. Le Maitre.
7. Trachyandesite, Gonoura-machi, Iki Islands, Japan (Aoki, 1959, Table 15). Analyst K. Aoki.
8. Porphyritic nepheline trachyandesite (OU 20658), Brinns Point, East Otago (Coombs and Wilkinson, 1969, Table 1). Analyst G. I. Z. Kalocsai. (Norm includes hl 0·25).
9. Tristanite (average of 3 analyses), Gough Island (Tilley and Muir, 1964, Table 1).
10. Feldspathoidal tristanite (= average nepheline latite, according to Nockolds, 1954, Table 10) (Tilley and Muir, 1964, Table 2).

distinction neglects the differing alkali ratios of lavas belonging to either the sodic or potassic series.

Nepheline hawaiite, nepheline mugearite and nepheline benmoreite are the feldspathoidal equivalents of hawaiite, etc. and contain essential feldspathoid (Coombs and Wilkinson, 1969). The corresponding potassic lavas are nepheline trachyandesite (= ordanchite; Nockolds, 1954) and feldspathoidal tristanite, equated by Tilley and Muir (1964) with the average nepheline latite of Nockolds (1954). The 1:2 potash/soda ratio is useful in distinguishing fine-grained sodic or potassic lavas of intermediate D.I. since it primarily reflects the nature of the respective feldspar assemblages. However the alkali ratios of moderately undersaturated lavas are partly controlled by modal feldspathoid, and consequently inspection of the alkali contents of these rocks does not necessarily indicate the dominantly sodic or potassic nature of the contained alkali feldspars.

The more evolved mildly alkaline lavas may be *ne*- or *hy*-bearing (Table 6, No. 9) or oversaturated, as illustrated by a Hebridean benmoreite with $qz = 6{\cdot}9\%$ (Tilley, Yoder and Schairer, 1967, Table 3). Olivine and Ca-rich clinopyroxenes are the principal ferromagnesian minerals in hawaiites, trachyandesites, etc. Although probably more common in potassic lineages, kaersutite also occurs as phenocrysts in low-iron sodic rocks (Tiba, 1966). High-iron hawaiites and mugearites typically are devoid of phenocrystal calciferous amphibole or biotite.

II.4. REFERENCES

Anderson, F. W., and Dunham, K. C., 1966. *The geology of Northern Skye*. Mem. geol. Surv. U.K., 216 pp.

Aoki, K., 1959. Petrology of alkali rocks of the Iki Islands and Higashi-Matsuura district, Japan. *Tôhoku Univ. Sci. Rept.*, Ser. III, **6**, 261–310.

Aoki, K., 1963. The kaersutites and oxykaersutites from alkali rocks of Japan and surrounding areas. *J. Petrology*, **4**, 198–210.

Aoki, K., 1964. Clinopyroxenes from alkaline rocks of Japan. *Am. Miner.* **49**, 1199–223.

Aoki, K., 1967. Petrography and petrochemistry of latest Pliocene olivine tholeiites of Taos area, northern New Mexico, U.S.A. *Contr. Mineral. Petrology*, **14**, 190–203.

Aoki, K., 1970. Petrology of kaersutite-bearing ultramafic and mafic inclusions in Iki Islands, Japan. *Contr. Mineral. Petrology*, **25**, 270–83.

Aoki, K., and Kushiro, I., 1968. Some clinopyroxenes from ultramafic inclusions in Dreiser Weiher, Eifel. *Contr. Mineral. Petrology*, **18**, 326–7.

Babkine, J., Conquéré, F., and Vilminot, J. C., 1966. Nodules de péridotite et cumulats d'olivine. *Bull. Soc. fr. Miner. Christallogr.*, **89**, 262–8.

Bailey, E. B., and others, 1924. *Tertiary and post-Tertiary geology of Mull, Loch Aline, and Oban.* Mem. geol. Surv. Scotland, 445 pp.

Baker, P. E., Gass, I. G., Harris, P. G., and Le Maitre, R. W., 1964. The volcanological report of the Royal Society expedition to Tristan da Cunha, 1962. *Trans. R. Soc. Philos.*, **256**, 439–575.

Baker, I., and Haggerty, S. E., 1967. The alteration of olivine in basaltic and associated lavas. Part II: Intermediate and low temperature alteration. *Contr. Mineral. Petrology*, **16**, 258–73.

Bandy, M. C., 1937. Geology and petrology of Easter Island. *Bull. geol. Soc. Am.*, **48**, 1589–610.

Barnes, V. E., 1930. Changes in hornblende at about 800 °C. *Am. Miner.* **15**, 393–417.

Barth, T. F. W., 1931. Pyroxen von Hiva Oa, Marquesas Inseln und die formel titanhaltiger augite. *Neues Jb. Miner. Abh*, **64**, 217–24.

Barton, M. E., 1963. The petrology of a teschenite sill at Landywood, Staffordshire. *Geol. Mag.*, **100**, 533–50.

Benson, W. N., 1939. Kaersutite and other brown amphiboles in the Cainozoic igneous rocks of the Dunedin district. *Trans R. Soc. New Zealand*, **69**, 283–308.

Benson, W. N., 1942. The basic igneous rocks of Eastern Otago and their tectonic environment. Part II. *Trans. R. Soc. New Zealand*, **72**, 85–110.

Best, M. G., 1970. Kaersutite-peridotite inclusions and kindred megacrysts in basanitic lavas, Grand Canyon, Arizona. *Contr. Mineral. Petrology*, **27**, 25–44.

Binns, R. A., 1969. High pressure megacrysts in basanitic lavas near Armidale, New South Wales. *Am. J. Sci.* (Schairer vol.), **267A**, 33–49.

Binns, R. A., Duggan, M. B., and Wilkinson,

J. F. G., 1970. High pressure megacrysts in alkaline lavas from northeastern New South Wales. *Am. J. Sci.*, **269,** 132–68.

Boesen, R. S., 1964. The clinopyroxenes of a monzonitic complex at Mount Dromedary, New South Wales. *Am. Miner.* **49,** 1435–57.

Bowen, N. L., and Schairer, J. F., 1935. The system MgO.FeO.SiO_2. *Am. J. Sci.*, 5th ser., **29,** 151–217.

Boyd, F. R., 1956. 'Hydrothermal investigations of amphiboles', in Abelson, P. H., Ed. *Researches in Geochemistry*, John Wiley & Sons, New York, 377–96.

Brothers, R. N., 1960. Olivine nodules from New Zealand. *Rep. 21st Intern. geol. Congr.*, Copenhagen, pt. 13, 68–81.

Brown, D. A., 1955. The geology of Siberia Hill and Mount Dasher, North Otago. *Trans. R. Soc. New Zealand*, **83,** 347–72.

Brown, G. M., 1967. 'Mineralogy of basaltic rocks', in Hess, H. H., and Poldervaart, A., Eds., *Basalts, vol. 1*, Interscience, New York, 103–62.

Bryan, W. B., 1967. Geology and petrology of Clarion Island, Mexico. *Bull geol. Soc. Am.*, **78,** 1461–76.

Buddington, A. F., and Lindsley, D. H., 1964. Iron–titanium oxide minerals and synthetic equivalents. *J. Petrology*, **5,** 310–57.

Buerger, M. J., 1948. The role of temperature in mineralogy. *Am. Miner.*, **33,** 101–21.

Carmichael, C. M., 1961. The magnetic properties of ilmenite-hematite crystals. *Proc. R. Soc. Lond.*, Ser. A, **263,** 508–30.

Carmichael, I. S. E., 1962. Pantelleritic liquids and their phenocrysts. *Mineralog. Mag.*, **33,** 86–113.

Carmichael, I. S. E., 1964. The petrology of Thingmuli, a Tertiary volcano in eastern Iceland. *J. Petrology*, **5,** 435–60.

Carmichael, I. S. E., 1967. The mineralogy and petrology of the volcanic rocks from the Leucite Hills, Wyoming. *Contr. Mineral. Petrology*, **15,** 24–66.

Carmichael, I. S. E., and Nicholls, J., 1967. Iron–titanium oxides and oxygen fugacities in volcanic rocks. *J. geophys. Res.*, **72,** 4665–87.

Chayes, F., 1964. A petrographic distinction between Cenozoic volcanics in and around the open oceans. *J. geophys. Res.*, **69,** 1573–88.

Chayes, F., 1965. Titania and alumina content of oceanic and circumoceanic basalt. *Mineralog. Mag.* (Tilley vol.), **34,** 126–31.

Chayes, F., 1966. Alkaline and subalkaline basalts. *Am. J. Sci.*, **264,** 128–45.

Clark, S. P., Schairer, J. F., and de Neufville, J., 1962. Phase relations in the system $CaMgSi_2O_6$–$CaAl_2SiO_6$ at low and high pressure. *Yb. Carnegie Inst. Wash.*, **61,** 59–68.

Collée, A. L. G., 1963. A fabric study of lherzolites with special reference to ultrabasic nodular inclusions in the lavas of Auvergne (France). *Leid. geol. Meded.*, **28,** 3–102.

Coombs, D. S., 1955. X-ray observations on wairakite and non-cubic analcime. *Mineralog Mag.*, **30,** 698–708.

Coombs, D. S., 1963. Trends and affinities of basaltic magmas and pyroxenes as illustrated by the diopside–olivine–silica diagram. *Miner. Soc. Am. Special Paper 1*, 227–50.

Coombs, D. S., Ellis, A. J., Fyfe, W. S., and Taylor, A. M., 1959. The zeolite facies, with comments on the interpretation of hydrothermal syntheses. *Geochim. Cosmochim. Acta*, **17,** 53–107.

Coombs, D. S., and Whetten, J. T., 1967. Composition of analcimes from sedimentary and burial metamorphic rocks. *Bull. Geol. Soc. Am.* **78,** 269–82.

Coombs, D. S., and Wilkinson, J. F. G., 1969. Volcanic lineages and fractionation trends in some undersaturated volcanic rocks and shallow intrusives from the East Otago Volcanic Province (New Zealand) and related rocks. *J. Petrology*, **10,** 440–501.

Daly, R. A., 1925. The geology of Ascension Island. *Proc. Am. Acad. Arts Sci.*, **60,** 3–80.

Deer, W. A., Howie, R. A., and Zussman, J., 1962a. *Rock-Forming Minerals, Vol. 1. Ortho- and Ring Silicates.* 333 pp., Longmans, London.

Deer, W. A., Howie, R. A., and Zussman, J., 1962b. *Rock-Forming Minerals, Vol. 5. Non-silicates.* 371 pp., Longmans, London.

Deer, W. A., Howie, R. A., and Zussman, J., 1963a. *Rock-Forming Minerals, Vol. 2. Chain Silicates.* 379 pp., Longmans, London.

Deer, W. A., Howie, R. A., and Zussman, J., 1963b. *Rock-Forming Minerals, Vol. 4. Framework Silicates.* 435 pp., Longmans, London.

Dickey, J. S., 1968. Eclogitic and other inclusions in the Mineral Breccia Member of the Deborah Volcanic Formation at Kakanui, New Zealand. *Am. Miner.*, **53,** 1304–19.

Drever, H. I., 1956. The geology of Ubekendt Ejland, West Greenland: Part II. The picritic sheets and dykes of the east coast. *Meddr Grønland*, **137,** 1–41.

Drever, H. I., and Johnston, R., 1957. Crystal growth of forsteritic olivine in magmas and melts. *Trans. Royal Soc. Edinburgh*, **63,** 289–315.

Drever, H. I., and Johnston, R., 1958. The petrology of picritic rocks in minor intrusions—a Hebridean group. *Trans. Royal Soc. Edinburgh*, **63,** 459–99.

Drever, H. I., and Johnston, R., 1959. The lower margin of the Shiant Isles sill: *Q. J. geol. Soc. Lond.*, **114,** 343–65.

Drever, H. I., and Johnston, R., 1965. New petrographical data on the Shiant Isles picrite. *Mineralog. Mag.* (Tilley vol.), **34,** 194–203.

Drever, H. I., and Johnston, R., 1967a. 'The ultrabasic facies in some sills and sheets', in Wyllie, P. J., Ed. *Ultramafic and Related Rocks*. John Wiley and Sons, New York, 51–63.

Drever, H. I., and Johnston, R., 1967b. 'Picritic minor intrusions', in Wyllie, P. J., Ed. *Ultramafic and Related Rocks*. John Wiley and Sons, New York, 71–82.

Drever, H. I., and MacDonald, J. G., 1967. Some new data on 'kylitic' sills and associated picrites in Ayrshire, Scotland. *Proc. R. Soc. Edinburgh*, sect. B, **70**, 31–48.

Edwards, A. B., 1938. The Tertiary volcanic rocks of central Victoria. *Q. J. geol. Soc. Lond.*, **94**, 243–320.

Edwards, A. B., 1941. The crinanite laccolith of Circular Head, Tasmania. *Proc. R. Soc. Vict.*, **53**, 403–15.

Edwards, A. B., 1953. Crinanite-picrite intrusions in the Nebo district of New South Wales. *Proc. R. Soc. Vict.*, **65**, 9–29.

Engel, A. E. J., and Engel, C., 1964. Composition of basalts from the Mid-Atlantic Ridge. *Science*, **144**, 1330–3.

Engel, A. E. J., Engel, C., and Havens, R. G., 1965. Chemical characteristics of oceanic basalts and the upper mantle. *Bull. geol. Soc. Am.*, **76**, 719–34.

Ernst, W. G., 1968. *Amphiboles: Crystal Chemistry, Phase Relations and Occurrence*, 125 pp., Springer-Verlag, New York Inc.

Flett, J. S., 1911 in Peach, B. N., and others, *The geology of Knapdale, Jura and North Kintyre*. Mem. Geol. Surv. Scotland, **28**, 149 pp.

Forbes, R. B., and Banno, S., 1966. Nickel–iron content of peridotite inclusion and cognate olivine from an alkali olivine basalt. *Am. Miner.*, **51**, 130–40.

Forbes, R. B., and Kuno, H., 1965. The regional petrology of peridotite inclusions and basaltic host rocks. *Upper Mantle Symposium* (*New Delhi, 1964*), 161–79.

Forbes, R. B., and Kuno, H., 1967. 'Peridotite inclusions and basaltic host rocks', in Wyllie, P. J., Ed. *Ultramafic and Related Rocks*. John Wiley and Sons, New York, 328–37.

Goldschmidt, V. M., 1954, *Geochemistry*. 730 pp., Oxford, University Press.

Green, D. H., and Ringwood, A. E., 1967. The genesis of basaltic magmas. *Contr. Mineral. Petrology*, **15**, 103–90.

Green, J., and Poldervaart, A., 1955. Some basaltic provinces. *Geochim. Cosmochim. Acta*, **7**, 177–88.

Haggerty, S. E., and Baker, I., 1967. The alteration of olivine in basaltic and associated lavas. Part I. High temperature alteration. *Contr. Mineral. Petrology*, **16**, 233–57.

Hamilton, J., 1957. Banded olivines in some Scottish Carboniferous olivine basalts. *Geol. Mag.*, **94**, 135–9.

Hamilton, W. B., and Neuerberg, G. J., 1956. Olivine-sanidine trachybasalt from the Sierra Nevada, California. *Am. Miner.* **41**, 851–73.

Harkin, D. A., 1960. *The Rungwe volcanics at the northern end of Lake Nyasa*. Mem. geol. Surv. Tanganyika, **2**, 172 pp.

Hartman, P., 1969. Can Ti^{4+} replace Si^{4+} in silicates? *Mineralog. Mag.*, **37**, 366–9.

Henriques, Å., 1958. The influence of cations on the optical properties of clinopyroxenes. Part I and II. *Ark. Miner. Geol.*, **2**, 341–8, 381–4.

Holmes, A., and Harwood, H. F., 1937. The petrology of the volcanic field of Bufumbira, south-west Uganda. Mem. geol. Surv. Uganda, **3**, 291 pp.

Hori, F., 1954. Effects of constituent cations on the optical properties of clinopyroxenes. *Univ. Tokyo Coll. General Educ., Sci. Papers*, **4**, 71–83.

Howie, R. A., 1963. Kaersutite from the Lugar sill, Ayrshire, and from alnöite breccia, Alnö Island, Sweden. *Mineralog. Mag.*, **33**, 718–20.

Huckenholz, H. G., 1965a. Der petrogenetische Werdegang der Klinopyroxene in den tertiären Vulkaniten der Hocheifel. I. Die Klinopyroxene der Alkalinolivinbasalt-Trachyt-Assoziation. *Beitr. Miner. Petrogr.*, **11**, 138–95.

Huckenholz, H. G., 1965b. Der petrogenetische Werdegang der Klinopyroxene in den tertiären Vulkaniten der Hocheifel. II. Die Klinopyroxene der Basanitoide. *Beitr. Miner. Petrogr.*, **11**, 415–48.

Hytönen, K., 1959. On the petrology and mineralogy of some alkaline volcanic rocks of Toror Hills, Mt. Morotu and Morolinga in Karamoja, north-eastern Uganda. *Bull. Comm. géol. Finl.*, **184**, 75–132.

Hytönen, K., and Schairer, J. F., 1961. The plane enstatite-anorthite-diopside and its relation to basalts. *Yb. Carnegie Inst. Wash.* **60**, 125–39.

Iddings, J. P., 1895. Absarokite–shoshonite–banakite series. *J. Geol.* **3**, 935–59.

Iddings, J. P., 1913, *Igneous Rocks, Vol II*, 685 pp., John Wiley & Sons, New York.

Johannsen, A., 1937. *A Descriptive Petrography of the Igneous Rocks. Vol. III*, 360 pp., University of Chicago Press.

Johnston, R. 1953. The olivines of the Garbh Eilean sill, Shiant Isles. *Geol. Mag.*, **90**, 161–71.

Joplin, G. A., 1968. The shoshonite association: a review. *J. Geol. Soc. Australia*, **15**, 275–94.

Kawai, N., 1959. Subsolidus phase relation in titanomagnetite and its significance in rock magnetism. *Rep. 20th Intern. Geol. Congr., Mexico*, pt. II–A, 103–20.

Kennedy, W. Q., 1933. Trends of differentiation in basaltic magmas. *Am. J. Sci.*, **25**, 239–56.

Kuno, H., 1954. *Volcanoes and Volcanic Rocks* (in Japanese). Iwanami and Co., Tokyo.

Kuno, H., 1959. Origin of Cenozoic petrographic provinces of Japan and surrounding areas. *Bull. volcan.*, **20,** 37–76.

Kuno, H., 1960. High-alumina basalt. *J. Petrology*, **1,** 121–45.

Kuno, H., 1964. 'Aluminian augite and bronzite in alkali olivine basalt from Taka-Sima, North Kyushu, Japan', in *Advancing Frontiers in Geology and Geophysics*, Osmania Univ. Press, Hyderabad, 205–20.

Kuno, H., 1967. 'Mafic and ultramafic nodules from Itinome-Gata, Japan', in Wyllie, P. J., Ed. *Ultramafic and Related Rocks.* John Wiley and Sons, New York, 337–42.

Kuno, H., and others, 1957. Differentiation of Hawaiian magmas. *Jap. J. Geol. Geogr.*, **28,** 179–218.

Kurasawa, H., 1967. *Petrology of the Kita-matsuura basalts in the northwest Kyushu, southwest Japan.* Geol. Surv. Japan, Rept. 217, 108 pp.

Kushiro, I., 1960. Si–Al relations in clinopyroxenes from igneous rocks. *Am. J. Sci.*, **258,** 548–54.

Kushiro, I., 1962. Clinopyroxene solid solutions. Part I. The $CaAl_2SiO_6$ component. *Jap. J. Geol. Geog.*, **33,** 213–20.

Kushiro, I., 1964. Petrology of the Atumi dolerite, Japan. *J. Fac. Sci. Univ. Tokyo*, **15,** 135–202.

Larsen, E. S., 1941. Igneous rocks of the Highwood Mountains, Montana. Part II. The extrusive rocks. *Bull. geol. Soc. Am.*, **52,** 1733–52.

Le Bas, M. J., 1962. The role of aluminium in igneous pyroxenes in relation to their parentage. *Am. J. Sci.*, **260,** 267–88.

Le Maitre, R. W., 1962. Petrology of volcanic rocks, Gough Island, South Atlantic. *Bull. geol. Soc. Am.*, **73,** 1309–40.

Le Maitre, R. W., 1965. The significance of gabbroic xenoliths from Gough Island, South Atlantic. *Mineralog. Mag.* (Tilley vol.), **34,** 303–17.

Le Maitre, R. W., 1968. Chemical variation within and between volcanic rock series—a statistical approach. *J. Petrology*, **9,** 220–52.

Le Maitre, R. W., 1969. Kaersutite-bearing plutonic xenoliths from Tristan da Cunha, South Atlantic. *Mineralog. Mag.*, **37,** 185–97.

Macdonald, G. A., 1960. Dissimilarity of continental and oceanic rock types: *J. Petrology*, **1,** 172–7.

Macdonald, G. A., and Katsura, T., 1964. Chemical composition of Hawaiian lavas. *J. Petrology*, **5,** 82–133.

MacGregor, M., and MacGregor, A. G., 1948. *The Midland Valley of Scotland.* Br. reg. Geol. 95 pp.

Mason, B., 1966. Pyrope, augite and hornblende from Kakanui. *New Zealand J. Geol. Geophys.*, **9,** 474–80.

Mason, B., 1968. Kaersutite from San Carlos, Arizona, with comments on the paragenesis of this mineral. *Mineralog. Mag.*, **36,** 997–1002.

McDougall, I., and Wilkinson, J. F. G., 1967. Potassium–argon dates on some Cainozoic volcanic rocks from northeastern New South Wales. *J. Geol. Soc. Australia*, **14,** 225–34.

Muir, I. D., 1962. The paragenesis and optical properties of some ternary feldspars. *Norsk. Geol. Tidsskr.*, **42,** 477–92.

Muir, I. D., and Tilley, C. E., 1961. Mugearites and their place in alkali igneous rock series. *J. Geol.*, **69,** 186–203.

Muir, I. D., and Tilley, C. E., 1964. Basalts from the northern part of the rift zone of the Mid-Atlantic Ridge. *J. Petrology*, **5,** 409–34.

Murray, R. J., 1954. The clinopyroxenes of the Garbh Eilean sill, Shiant Isles. *Geol. Mag.*, **91,** 17–31.

Nash, W. P., and Wilkinson, J. F. G., 1970. Shonkin Sag laccolith, Montana. I Mafic minerals and estimates of temperature, pressure, oxygen fugacity and silica activity. *Contr. Mineral. Petrol.*, **25,** 241–69.

Nicholls, J., and Carmichael, I. S. E., 1969. Peralkaline acid liquids: A petrological study. *Contr. Mineral. Petrol.*, **20,** 268–94.

Nockolds, S. R., 1954. Average chemical compositions of some igneous rocks. *Bull. geol. Soc. Am.*, **65,** 1007–32.

Nolan, J., 1966. Melting relations in the system $NaAlSi_3O_8$–$NaAlSiO_4$–$NaFeSi_2O_6$–$CaMgSi_2O_6$–H_2O and their bearing on the genesis of alkaline undersaturated rocks. *Q. J. geol. Soc. Lond.*, **122,** 119–57.

Nolan, J., 1969. Physical properties of synthetic and natural pyroxenes in the system diopside–hedenbergite–acmite. *Mineralog. Mag.*, **37,** 216–29.

Nolan, J., and Edgar, A. D., 1963. An X-ray investigation of synthetic pyroxenes in the system acmite–diopside–water at 1,000 kg/cm^2 water vapour pressure. *Mineralog. Mag.*, **33,** 625–34.

Peters, Tj., Luth, W. C., and Tuttle, O. F., 1966. The melting of analcite solid solutions in the system $NaAlSiO_4$–$NaAlSi_3O_8$–H_2O. *Am. Miner.*, **51,** 736–53.

Phillips, W. J., 1968. The crystallization of the teschenite from the Lugar sill, Ayrshire. *Geol. Mag.*, **105,** 23–34.

Poldervaart, A., 1964. Chemical definition of alkali basalts and tholeiites, *Bull. geol. Soc. Am.*, **75,** 229–32.

Poldervaart, A., and Hess, H. H., 1951. Pyroxenes in the crystallization of basaltic magma. *J. Geol.*, **59,** 472–89.

Richardson, C., 1933. 'Petrology of the Galapagos Islands', in Chubb, L. J., Geology of the Galapagos, Cocos, and Eastern Islands. *Bull. Bernice P. Bishop Mus.*, **110,** 45–67.

Roedder, E., 1965. Liquid CO_2 inclusions in olivine-bearing nodules and phenocrysts from basalts. *Am. Miner.*, **50,** 1746–82.

Ross, C. S., Foster, M. D., and Meyers, A. T., 1954. Origin of dunites and of olivine-rich inclusions in basaltic rocks. *Am. Miner.*, **39,** 693–737.

Saha, P., 1959. Geochemical and X-ray investigations of natural and synthetic analcites. *Am. Miner.*, **44,** 300–13.

Saha, P., 1961. The system $NaAlSiO_4$ (nepheline)–$NaAlSi_3O_8$ (albite)–H_2O. *Am. Miner.*, **46,** 859–84.

Schairer, J. F., and Yoder, H. S., 1960. The nature of residual liquids from crystallization, with data on the system nepheline–diopside–silica. *Am. J. Sci.*, **258A** (Bradley vol.), 273–83.

Searle, E. J., 1960. Petrochemistry of the Auckland basalts. *New Zealand J. Geol. Geophys.*, **3,** 23–40.

Searle, E. J., 1961. The petrology of the Auckland basalts. *New Zealand J. Geol. Geophys.*, **4,** 165–204.

Segnit, E. R., 1953. Some data on synthetic aluminous and other pyroxenes. *Mineralog. Mag.*, **30,** 218–26.

Simkin, T., 1967. 'Flow differentiation in the picritic sills of north Skye' in Wyllie, P. J., Ed., *Ultramafic and Related Rocks*, John Wiley and Sons, New York, 64–9.

Simpson, E. S. W., 1954. The Okonjeje igneous complex, South-West Africa. *Trans. geol. Soc. S. Afr.*, **57,** 124–72.

Smith, J. V., 1966a. X-ray-emission microanalysis of rock-forming minerals. II. Olivines. *J. Geol.*, **74,** 1–16.

Smith, J. V., 1966b. X-ray-emission microanalysis of rock-forming minerals. VI. Clinopyroxenes near the diopside-hedenbergite join. *J. Geol.*, **74,** 463–77.

Smith, J. V., and Stenstrom, R. C., 1965. Chemical analyses of olivines by the electron microprobe. *Mineralog. Mag.* (Tilley vol.), **34,** 436–59.

Smith, W. W., 1961. Structural relationships within pseudomorphs of olivine. *Mineralog. Mag.*, **32,** 823–25.

Streckeisen, A., 1967. Classification and nomenclature of igneous rocks. *Neues Jb. Miner. Abh.*, **107,** 144–240.

Strong, D. F., 1969. Formation of the hour-glass structure in augite. *Mineralog. Mag.*, **37,** 472–9.

Talbot, J. L., and others, 1963. Xenoliths and xenocrysts from lavas of the Kerguelen Island Archipelago. *Am. Miner.*, **48,** 159–79.

Tanguy, J. C., 1966. Les laves récentes de l'Etna. *Bull. Soc. géol. France*, **8,** 201–17.

Thompson, J. B., 1947. Role of aluminium in the rock-forming silicates. *Bull. geol. Soc. Am.*, **58,** 1232.

Thornton, C. P., and Tuttle, O. F., 1960. Chemistry of igneous rocks. I. Differentiation index. *Am. J. Sci.*, **258,** 664–84.

Tiba, T., 1966. Petrology of the alkaline rocks of the Takakusayama district, Japan. *Tohoku Univ. Sci. Rept.*, Ser. III, **9,** 541–610.

Tilley, C. E., 1950. Some aspects of magmatic evolution. *Q. J. geol. Soc. Lond.*, **106,** 37–61.

Tilley, C. E., and Muir, I. D., 1962. The Hebridean Plateau magma type. *Trans. geol. Soc. Edinburgh*, **19,** 208–15.

Tilley, C. E., and Muir, I. D., 1964. Intermediate members of the oceanic basalt–trachyte association. *Geol. För. Stockh. Förh.*, **85,** 434–43.

Tilley, C. E., Yoder, H. S., and Schairer, J. F., 1964. New relations on melting of basalts. *Yb. Carnegie Inst. Wash.*, **63,** 92–7.

Tilley, C. E., Yoder, H. S., and Schairer, J. F., 1967. Melting relations of volcanic rock series. *Yb. Carnegie Inst. Wash.*, **65,** 260–9.

Tomkeieff, S. I., 1939. Zoned olivines and their petrogenetic significance. *Mineralog. Mag.*, **25,** 229–51.

Tournon, J., 1969. Les roches basaltiques de la province de Gerone (Espagne): basanites à leucite et basanites à analcime. *Bull. Soc. franç. Miner. Cristallogr.*, **92,** 376–82.

Tyrrell, G. W., 1917. The picrite-teschenite sill of Lugar (Ayrshire). *Q. J. geol. Soc. Lond.*, **72,** 84–131.

Tyrrell, G. W., 1948. A boring through the Lugar sill. *Trans. geol. Soc. Glasg.*, **21,** 157–202.

Tyrrell, G. W., 1952. A second boring through the Lugar sill. *Trans. geol. Soc. Edinburgh*, **15,** 374–92.

Uchimuzu, M., 1966. Geology and petrology of alkali rocks from Dogo, Oki Islands. *J. Fac. Sci. Univ. Tokyo*, Sec. II, **16,** 85–159.

Upton, B. G. J., and Wadsworth, W. J., 1966. The basalts of Réunion Island, Indian Ocean. *Bull. volcan.*, **29,** 7–24.

Upton, B. G. J., Wadsworth, W. J., and Newman, T. C., 1967. The petrology of Rodriguez Island, Indian Ocean. *Bull. geol. Soc. Am.*, **78,** 1495–506.

Verhoogen, J., 1962. Oxidation of iron–titanium oxides in igneous rocks. *J. Geol.*, **70,** 168–81.

Vincent, E. A., and others, 1957. Heating experiments on some natural titaniferous magnetites. *Mineralog. Mag.*, **31,** 624–55.

Vlodavetz, V. I., and Shavrova, N. N., 1953. On anorthoclase from a lava in the Darigan volcanic region. *Acad. Sci. USSR*, **2,** 71–6.

Walker, F., 1923. Notes on the classification of Scottish and Moravian teschenites. *Geol. Mag.*, **60**, 242–9.

Walker, F., 1930. The geology of the Shiant Isles (Hebrides). *Q. J. Geol. Soc. Lond.*, **86,** 355–98.

Walker, F., 1953. The pegmatitic differentiates of basic sheets. *Am. J. Sci.*, **251,** 41–60.

Walker, F., and Nicolaysen, L. O., 1954. The petrology of Mauritius. *Colon. Geol. Miner. Resour.*, **4,** 3–43.

Washington, H. S., 1906. *The Roman comagmatic region*. Publs. Carnegie Inst. Wash., 57, 199 pp.

Washington, H. S., 1914. The volcanoes and rocks of Pantelleria. Part III. Petrology. *J. Geol.*, **22,** 16–27.

Washington, H. S., Aurousseau, M., and Keyes, M. G., 1926. The lavas of Etna. *Am. J. Sci.*, **12,** 371–408.

White, R. W., 1966. Ultramafic inclusions in basaltic rocks from Hawaii. *Contr. Miner. Petrology*, **12,** 245–314.

Wilkinson, J. F. G., 1955. The terms teschenite and crinanite. *Geol. Mag.*, **92,** 282–90.

Wilkinson, J. F. G., 1956a. Clinopyroxenes of alkali olivine-basalt magma. *Am. Miner.*, **41,** 724–43.

Wilkinson, J. F. G., 1956b. The olivines of a differentiated teschenite sill near Gunnedah, New South Wales. *Geol. Mag.*, **93,** 441–55.

Wilkinson, J. F. G., 1957a. The clinopyroxenes of a differentiated teschenite sill near Gunnedah, New South Wales. *Geol. Mag.*, **94,** 123–34.

Wilkinson, J. F. G., 1957b. Titanomagnetites from a differentiated teschenite sill. *Mineralog. Mag.*, **31,** 443–54.

Wilkinson, J. F. G., 1958. The petrology of a differentiated teschenite sill near Gunnedah, New South Wales. *Am. J. Sci.*, **256,** 1–39.

Wilkinson, J. F. G., 1961. Some aspects of the calciferous amphiboles, oxyhornblende, kaersutite and barkevikite. *Am. Miner.*, **46,** 340–54.

Wilkinson, J. F. G., 1963. Some natural analcime solid solutions. *Mineralog. Mag.*, **33,** 498–505.

Wilkinson, J. F. G., 1965a. Some feldspars, nephelines and analcimes from the Square Top intrusion, Nundle, N.S.W. *J. Petrology*, **6,** 420–44.

Wilkinson, J. F. G., 1965b. Titanomagnetites from a differentiation sequence, analcime-olivine theralite to analcime tinguaite. *Mineralog. Mag.*, **34,** (Tilley vol.), 528–41.

Wilkinson, J. F. G., 1966a. Residual glasses from some alkali basaltic lavas from New South Wales. *Mineralog. Mag.*, **35,** 847–60.

Wilkinson, J. F. G., 1966b. Clinopyroxenes from the Square Top intrusion, Nundle, New South Wales. *Mineralog. Mag.*, **35,** 1061–70.

Wilkinson, J. F. G., 1967. 'The petrography of basaltic rocks', in Hess, H. H., and Poldervaart, A., Eds. *Basalts, Vol. 1.* Interscience, New York, 163–214.

Wilkinson, J. F. G., 1968a. Analcimes from some potassic igneous rocks and aspects of analcime-rich igneous assemblages. *Contr. Mineral. Petrology*, **18,** 252–69.

Wilkinson, J. F. G., 1968b. The magmatic affinities of some volcanic rocks from the Tweed Shield Volcano, S.E. Queensland—N.E. New South Wales. *Geol. Mag.*, **105,** 275–89.

Williams, H., 1933. Geology of Tahiti, Moorea, and Maio. *Bull. Bernice P. Bishop Mus.*, **105.**

Wilshire, H. G., 1958. Alteration of olivine and orthopyroxene in basic lavas and shallow intrusions. *Am. Miner.*, **43,** 120–47.

Wilshire, H. G., 1959. Deuteric alteration of some volcanic rocks. *Proc. R. Soc. New South Wales*, **93,** 105–20.

Wilshire, H. G., 1967. The Prospect alkaline diabase-picrite intrusion, New South Wales, Australia. *J. Petrology*, **8,** 97–163.

Wilshire, H. G., and Binns, R. A., 1961. Basic and ultrabasic xenoliths from volcanic rocks of New South Wales. *J. Petrology*, **2,** 185–208.

Winchell, H., 1947. Honolulu series, Oahu, Hawaii. *Bull. geol. Soc. Am.*, **58,** 1–48.

Winchell, H., 1961. Regressions of physical properties on the compositions of clinopyroxenes. II. Optical properties and specific gravity. *Am. J. Sci.*, **259,** 295–319.

Wright, J. B., 1967a. The iron–titanium oxides of some Dunedin (New Zealand) lavas, in relation to their palaeomagnetic and thermomagnetic character. *Mineralog. Mag.*, **36,** 425–35.

Wright, J. B., 1967b. Contributions to the volcanic succession and petrology of the Auckland Islands. II. Upper parts of the Ross Volcano. *Trans. R. Soc. New Zealand*, **5,** 71–87.

Wyllie, P. J., and Drever, H. I., 1963. The petrology of picritic rocks in minor intrusions—a picritic sill on the island of Soay (Hebrides). *Trans. R. Soc. Edinburgh*, **65,** 155–77.

Yagi, K., 1953a. Olivine trachybasalt from Weitschou Island, South China Sea. *Tôhoku Univ. Sci. Rept.*, Ser. III, **4,** 201–8.

Yagi, K., 1953b. Petrochemical studies on the alkalic rocks of the Morotu district, Sakhalin. *Bull. geol. Soc. Am.*, **64,** 769–810.

Yagi, K., 1966. The system acmite–diopside and its bearing on the stability relations of natural pyroxenes of the acmite–hedenbergite–diopside series. *Am. Miner.*, **51,** 976–1000.

Yagi, K., and Onuma, K., 1967. The join $CaMgSi_2O_6$ –$CaTiAl_2O_6$ and its bearing on the titanaugites. *J. Fac. Sci. Hokkaido Univ.*, Ser. IV, **13,** 463–83.

Yamaguchi, M., 1964. Petrogenetic significance of ultrabasic inclusions in basaltic rocks from southwest Japan. *Mem. Fac. Sci. Kyushu Univ.*, **15,** 163–219.

Yoder, H. S., and Tilley, C. E., 1962. Origin of basalt magmas: an experimental study of natural and synthetic rock systems. *J. Petrology*, **3,** 342–532.

Zies, E. G., 1962. A titaniferous basalt from the island of Pantelleria. *J. Petrology*, **3,** 177–80.

II.5. POTASSIUM-RICH ALKALINE ROCKS

Th. G. Sahama

II.5.1. Introduction

Alkali rocks of the potassic (mediterranean) suite show a wt. % of K_2O in a marked excess over that of Na_2O. The excess of potassium is mainly contained in leucite (rarely kalsilite), in potassium feldspar, in the glass phase or, in some rocks, in phlogopite (biotite). Because leucite is not stable in deep-seated conditions, the K-rich feldspathoidal rocks are restricted to volcanic or subvolcanic environments. Also saturated potassic to perpotassic rocks of clear alkaline affinity are not found in a plutonic environment.

A plot of the alkali ratio against the degree of undersaturation (Fig. 1) characterizes the K-rich alkaline rocks. The perpotassic rocks can be conveniently subdivided into two distinct series: rocks of orenditic affinity (original orendite from Leucite Hills, Wyoming; Cross, 1897) and those of kamafugitic affinity (KAtungite–MAFurite–UGandite from Toro–Ankole, Uganda; Holmes, 1950).

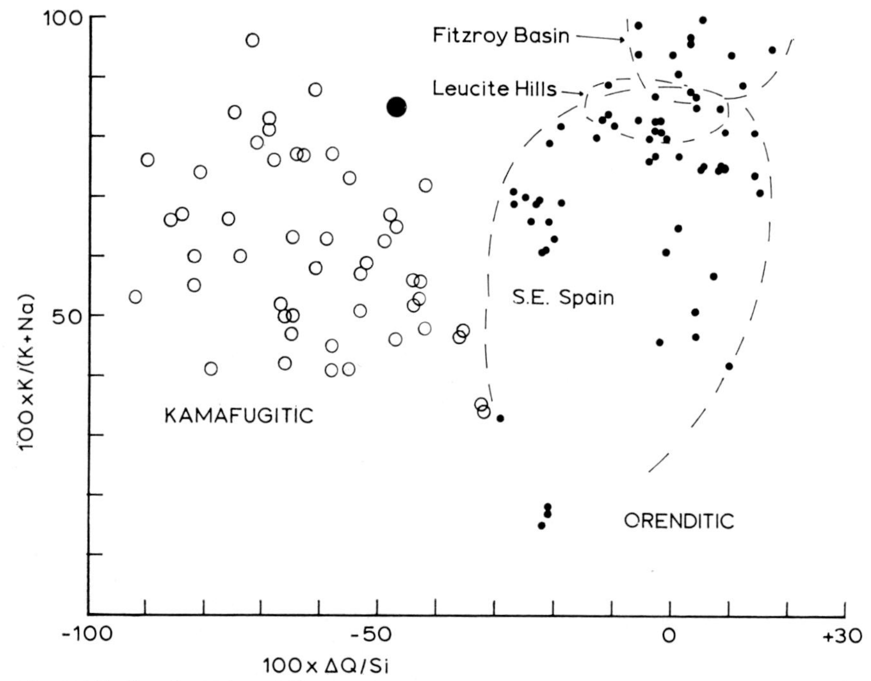

Fig. 1 Alkali ratio 100 × K/(K + Na) plotted against the degree of undersaturation 100 × △Q/Si. K, Na, Si: atomic numbers of these cations. △Q: excess or deficiency in silica of the saturated norm. Large black circle indicates madupite

The orenditic rocks cluster around the point of silica saturation and represent lamproites (Mg- and K-rich lamprophyres) of varying mineralogical composition. The kamafugitic rocks are strongly undersaturated and cover both chemically and mineralogically a considerable diversity of rock types.

The nomenclature of the K-rich alkaline

rocks is complex. The rock names used are derived from locality names and mostly apply to the rocks of that area only. This circumstance is due to the fact that the petrographic character of these rocks is variable to such an extent that many areas display their own distinct rock species. Being based solely on the modal light-coloured minerals, the Streckeisen (1967) scheme is not sufficiently specific to characterize these rocks. In that scheme the orenditic rocks would fall into the fields of alkali trachyte, phonolite or even of feldspathoidite (cf.II.1)*. The kamafugitic rocks would all be included in the feldspathoidites. Therefore, the original nomenclature used in literature will be retained in this chapter.

It is not possible to review here all the potassic rocks known. The discussion will be restricted to a few selected rock suites of perpotassic character.

II.5.2. Rocks of Orenditic Affinity

II.5.2.1. Field Occurrence and Petrography

The following areas exhibiting lamproites of orenditic affinity were selected for discussion:

a. Fitzroy River basin in the West Kimberley area of Western Australia (Wade and Prider, 1940; Prider, 1960). Surface flows, intrusive plugs and sills or flat dipping sheets scattered over an area of some 150 km in diameter. The gas-drilled lamproite pipes are of Jurassic age cutting through the Precambrian granitic basement and a sequence of various sedimentary formations including limestones, sandstones and shales.

The lamproites are characterized by the following mineral assemblages:

Fitzroydite: leucite + phlogopite.
Cedricite: leucite + diopside.
Mamilite: leucite + magnophorite.
Wolgidite: leucite + magnophorite + diopside + phlogopite.

The rocks are strongly autometasomatically altered. Altered olivine is constantly present. The abundant leucite is the only felsic constituent and no primary K-feldspar occurs. Leucite is mostly altered to turbid material which contains secondary orthoclase (Prider and Cole, 1942) indicating that the crystallization of leucite from a silica-rich residuum took place at temperatures close to that of the leucite–orthoclase reaction.

The lamproite types encountered are subdivided into: (i) the undersaturated wolgidite group with SiO_2 less than 46% (coarse-grained wolgidites) and (ii) the oversaturated orendite group with SiO_2 over 51% (the rest of the lamproite types including the fine-grained wolgidites).

Field evidence indicates that the orendite group rocks crystallized at or near the surface whereas the crystallization of the wolgidite group rocks took place at considerable depth. The lamproitic magma was subjected to chemical differentiation during ascent: a silicic and more alkaline orenditic fraction on the top, a more subsilicic and less alkaline wolgiditic one lower down in the reservoir. The differentiation was effected by enrichment of the alkalis on top through gaseous transfer and by sinking of olivine phenocrysts. Assimilation of the rocks perforated did not notably contribute to the development of the orenditic and wolgiditic magma fractions.

The various orendite group lamproites display a virtually constant chemical composition and differ from each other in the mineralogical composition only, just reflecting successive stages of crystallization with the general sequence: fitzroydite → cedricite → mamilite → wolgidite.

b. Leucite Hills, Wyoming (Cross, 1897; Kemp, 1897; Kemp and Knight, 1903; Schultz and Cross, 1912; Carmichael, 1967). Miocene to Pliocene volcanic cones with surface flows, isolated necks and dykes, intruded sheets. The lamproites occur in an area of some 40 × 50 km cutting through early Tertiary sediments which rest on a large dome of underlying Cretaceous strata.

The two main rock types, wyomingite and orendite, show phenocrysts of phlogopite and olivine (mostly altered). Magnophorite and

* Readers are referred to the relevant Chapters of this book where similarly cited.

diopside are constant constituents of the groundmass which contains abundant leucite in wyomingite and leucite + sanidine in orendite. Some orendites are interbanded with layers of wyomingite. The third rock type, madupite from Pilot Butte, occurs as a surface flow showing phenocrysts of diopside and poikilitic phlogopite in a largely glassy groundmass.

The field data available do not disclose the genetic relationships between the various Leucite Hills rock types.

Besides various sedimentary rocks, granite and gabbro are found as xenoliths.

c. South-eastern Spain, extending from the province of Albacete through Murcia to Almeria (Fúster *et al.*, 1967, reviewing previous literature; Borley, 1967). Mostly breccia-mantled pipes or dykes and only rarely subaerial volcanics scattered over an area of 100 × 120 km. The Miocene lamproites cut through the sedimentary bedrock at the oriental edge of the Betical and Subbetical orogenic belts of the Alpine folding.

The lamproites contain olivine and phlogopite (both partly altered), diopside, leucite (analcime), magnophorite and sanidine. Leucite and phlogopite appear stable at the beginning of crystallization, sanidine and magnophorite at the later stages of consolidation.

Three crystalline rock types are distinguished: jumillite (with potential olivine and leucite), cancalite (with potential olivine) and fortunite (saturated). The distinction cannot be made solely on the basis of a thin section study but requires chemical analysis. The chemical composition of the vitrophyric varieties, verites, is variable and is partly affected by hydrothermal alteration.

The rocks are mainly potassic to perpotassic, but sodopotassic to sodic varieties are known among the jumillites and fortunites.

Field data relating the various lamproite types to each other are not available.

II.5.2.2. Chemistry

Table 1 summarizes the average chemical composition of the various lamproites of orenditic affinity (neglecting water and most trace elements).

Chemical characteristics are: high K and Mg combined with moderate Si, relatively low Al and very low Na. Marked enrichment of Ti (most pronounced in Fitzroy basin) and P.

An important feature is the alkali excess over Al in most orenditic lamproites. In the Fitzroy basin rocks and in the wyomingites and orendites of the Leucite Hills this excess is shown by K alone.

Trace element data available (Wade and Prider, 1940; Carmichael, 1967) indicate remarkable enrichment of Ba with high ratio Ba/K, Sr (Ba predominating over Sr), Zr, F, S, Cr, Ni.

II.5.2.3. Mineral Composition

The order of crystallization of the main constituents is: olivine, phlogopite, diopside, leucite (analcime), sanidine, magnophorite, quartz (if present). Some Spanish lamproites (core of the Cerro Negro plug, Fortuna) contain orthopyroxene (bronzite to enstatite) instead of olivine. In most cases olivine and phlogopite are partially resorbed, olivine being altered to montmorillonite, nontronite, saponite, etc., or mantled by a reaction rim of phlogopite or sanidine. The relative order of crystallization of leucite and sanidine is variable (Leucite Hills).

When fresh, olivine shows a composition of 10± mol % Fa (Carmichael, 1967; Fúster *et al.*, 1967) with a marked Ni content (Carmichael).

The pale-coloured phlogopite, most characteristic of the orenditic rocks, has been analysed from all lamproite areas concerned (Prider, 1939; Wade and Prider, 1940; Velde, 1967; Cross, 1897; Carmichael, 1967; Fúster *et al.*, 1967; Borley, 1967). It is highly titanian (least pronounced in Leucite Hills). Compared with the bulk rock, the mineral is highly enriched in Cr, less so in Ba and shows a marked content of Ni. Al is mostly deficient to fill the four-fold sites and K is in excess over Al (especially in Fitzroy basin).

The composition of the clinopyroxene (Carmichael, 1967; Fermoso, 1967a; Borley, 1967) is not too far from the Mg end of the diopside–hedenbergite series.

TABLE 1 Average Chemical Composition of Some K-Rich Alkaline Rocks of Orenditic Affinity

	Fitzroy River basin[1]		Corsica[2]	Leucite Hills[3]			S.E. Spain[4]		
	Wolgidite group	Orendite group	Lamproite	Madupite	Wyomingite	Orendite	Jumillite, Jumilla	Cancalite, Cancarix	Fortunite, Fortuna
Number of analyses	2	8	1	2	4	4	13	13	6
SiO_2	44·92	52·57	56·23	43·11	52·40	54·12	47·72	55·39	56·68
A_2lO_3	6·58	9·81	12·06	8·50	10·57	9·74	7·72	9·30	10·81
Fe_2O_3	6·03	5·47	1·91	5·35	3·05	3·42	3·01	2·33	1·85
FeO	2·00	1·52	2·91	0·96	1·44	0·80	3·40	2·88	4·09
MnO	0·08	0·06	0·07	0·14	0·06	0·06	0·10	0·08	0·04
MgO	11·44	6·31	6·90	10·96	6·78	7·74	16·27	12·01	10·17
CaO	4·66	2·84	3·48	12·13	4·57	3·69	7·11	3·71	2·72
BaO	1·91	0·88		0·78	0·78	0·53			
Na_2O	0·56	0·37	1·03	0·82	1·32	1·24	1·71	1·49	2·15
K_2O	7·71	10·80	10·00	7·59	11·03	11·54	4·99	8·77	6·90
TiO_2	6·96	4·90	1·14	1·98	2·28	2·44	1·43	1·74	1·41
P_2O_5	1·69	0·96	0·79	1·51	1·74	1·40	1·68	1·00	0·73

Source of data:
1. Calculated from the analyses by Wade and Prider (1940) and Prider (1960).
2. Velde (1967).
3. Calculated from the analyses summarized by Carmichael (1967).
4. Fúster *et al.* (1967).

Leucite is twinned except in the Leucite Hills where it is strictly isotropic. The difference in twinning between Leucite Hills and the other areas concerned may be due to differences in thermal history or in composition. The leucite from Leucite Hills displays a non-stoichiometric composition with a marked excess of Si and K over Al and with a fairly high content of the iron leucite component (Carmichael, 1967). The only analysis of leucite from Fitzroy basin (Carmichael) reveals a strictly stoichiometric composition with a notable Fe content.

Sanidine is relatively rich in the iron component (Velde, 1967; Carmichael, 1967; Fermoso, 1967b), in Leucite Hills even more so than the classical Madagascar iron-rich K-feldspar. The mineral is expectedly poor in the sodium component averaging 6 mol % Ab in the Spanish lamproites and less than 2 mol % Ab in Leucite Hills rocks.

Magnophorite (potassian analogue to richterite) is restricted to the orenditic lamproite suite. This amphibole with its distinct pleochroism from colourless or pale yellow to reddish was detected by Cross (1897) in Leucite Hills and studied by Prider (1939) from Fitzroy basin. Later it has been described from south-eastern Spain by Hernández-Pacheco (1965) and from Corsica by Velde (1967). The mineral shows high wt. % of K_2O equalling or even exceeding that of Na_2O and is highly magnesian (ferroan in Corsica). It is relatively poor in Al and enriched in Sr.

Priderite (Norrish, 1951) and wadeite (Prider, 1939) are accessory constituents specific of the orenditic lamproites. Both minerals were found in the Fitzroy basin rocks and were later shown to be common in the Leucite Hills (Carmichael, 1967). The occurrence of priderite in the Corsica lamproite was discovered by Velde (1968). Priderite represents a tetragonal KBaTi-oxide of a non-stoichiometric composition (Dryden and Wadsley, 1958) and was originally mistaken as rutile. Fúster *et al.* (1967) report rutile in the

Spanish lamproites; the presence of priderite has not been tested.

Apatite is a common accessory. Spinel is reported from Leucite Hills (chromian) and S.E. Spain, baryte from Leucite Hills (Sr-bearing and Fitzroy basin. Perovskite has been found in some wolgidites (Fitzroy basin) and in madupite (Leucite Hills). Madupite is the only rock of the orenditic lamproite areas concerned which contains magnetite (titanian). Ilmenite is absent.

II.5.3. Rocks of Kamafugitic Affinity

II.5.3.1. Field Occurrence and Petrography

The following two areas exhibiting perpotassic rocks of kamafugitic affinity will be considered:

a. Provinces of Toro and Ankole in the southwestern Uganda, east and south-east of the Ruwenzori horst. A suite of middle-late Pleistocene volcanics penetrate the Precambrian bedrock and the overlying Kaiso series of Pleistocene lacustrine-fluviatile deposits. The isolated volcanic fields are: Rusekere, Fort Portal, Ndale, Katwe-Kikorongo, Bunyaruguru and Katunga (Holmes and Harwood, 1932; Holmes, 1936, 1937, 1942a, 1945, 1950, 1952, 1956; Combe, 1937; Combe and Holmes, 1945; Neuvonen, 1956; Sahama and Wiik, 1952; Sahama, 1954).

The rocks cover an extensive range of petrographic variation. The critical constituents of the most conspicuous rock types are:

Ugandite: Olivine, augite, leucite.
Mafurite: Olivine, augite, kalsilite.
Katungite: Olivine, melilite, glass phase.

The rare extreme variety of katungite, called proto-katungite, is free from olivine. The potash ankaratrite–melaleucitite series represents a sodopotassic analogue to the kamafugitic rocks proper and is characterized by the predominance of augite instead of olivine. The felsic constituents are: nepheline in potash ankaratrite proper, nepheline + leucite in leucite ankaratrite and leucite in melaleucitite.

Rocks intermediate between all these rock types are found and the entire suite is summarized in Fig. 2. Mica does not belong to the critical minerals of the suite though biotite (phlogopite)-bearing varieties are known.

The eruptions of the small volcanoes encountered were largely explosive. Pyroclastics are abundant and true lavas occur to a minor extent. Although good representatives of the rock types mentioned have been found in lava flows, the entire range of variation is essentially based on ejected blocks. The scarcity of exposures *in situ* makes it impossible to establish the genetic relationships between the various rock types in the way presented by Prider (1960) for the Fitzroy basin rocks.

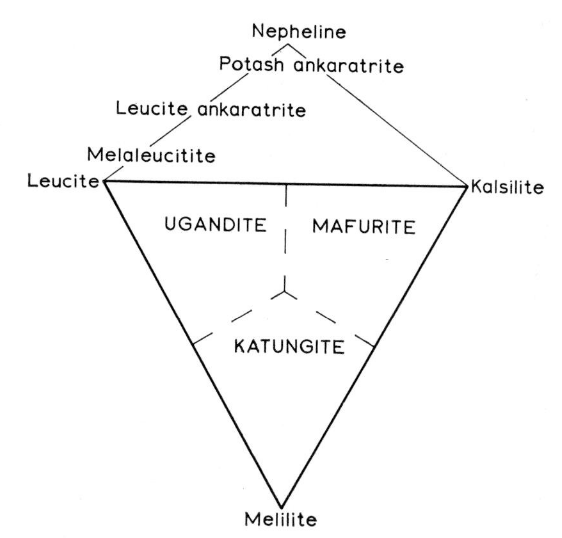

Fig. 2 Petrographic classification of the Toro–Ankole volcanics. Modified from Holmes

More or less carbonated volcanics are common and true carbonatites as ejected blocks or even as surface outflows have been found (v. Knorring and Du Bois, 1961).

Ejected blocks of kamafugitic rocks and xenoliths of cognate subvolcanic rocks occur in abundance in the pyroclastics and in the lavas. After their dominant minerals, olivine–biotite–pyroxene, the subvolcanic rocks were called the O.B.P. series by Holmes (1950, p. 776). This rock series contains pyroxenite, biotite-pyroxenite, glimmerite and peridotite. Minerals of the O.B.P. series occur in the tuffs as xenocrysts.

Accidental xenoliths derived from the sialic basement (Karagwe–Ankolean metasedimentary rocks, granites) are common. Transfused xenoliths of quartz (Holmes, 1936) and partially leucitized granite (Holmes, 1945; Combe and Holmes, 1945) have been recorded.

b. Umbria, Italy. Two small close-lying volcanoes, San Venanzo (essentially a pyroclastic cone) and Pian di Celle (a stratovolcano, explosive in the begin and effusive at the end) and a third similar one of Cupaello (effusive) (Rodolico, 1937; Mittempergher, 1965a,b).

These Pleistocene volcanoes represent peripheral manifestations of the Vulcinian volcanism in the Roman volcanic province and cut through late Tertiary clayey sediments with underlying Mesozoic limestones.

The Pian di Celle lava, called venanzite, is essentially an olivine-melilitite with some leucite and kalsilite. Diopside and, as a last crystallization, phlogopite are found (Holmes, 1942b; Mittempergher, 1965a). Mineralogically, venanzite corresponds to the katungite with augite of Toro–Ankole. Pegmatoid varieties with abundant phlogopite occur. The tuffs of the San Venanzo volcano, erupted prior to the lavas and representing the top of the magma pile, are considerably more rich in silica, containing even some sanidine and plagioclase, but no kalsilite and only rare leucite and melilite (Sartori, 1966). The main constituent is clinopyroxene. Remnants of carbonatic bombs are abundant.

Also the Cupaello lava represents melilitite (Rodolico, 1937), called coppaelite, in which the presence of kalsilite has been verified (Mittempergher, 1965 a, p. 468).

TABLE 2 Chemical Composition of Some Rocks Representative of the Kamafugitic Affinity

	Toro–Ankole, Uganda				Umbria, Italy	
	Katungite. Katunga[1]	Mafurite. Mafuru, Bunyaruguru[2]	Ugandite. Katwe[3]	Potash ankaratrite. Nabugando, Katwe–Kikorongo[4]	Venanzite. Pian di Celle[5]	Coppaelite. Cupaello[6]
Si_2O	35·37	39·06	43·85	38·58	40·52	41·45
Al_2O_3	6·50	8·18	7·32	9·27	10·43	7·56
Fe_2O_3	7·23	4·61	3·63	5·56	4·66	4·41
FeO	5·00	4·98	6·84	7·02	2·92	2·96
MnO	0·24	0·26	0·20	0·30	0·11	0·14
MgO	14·08	17·66	15·37	10·78	12·65	11·20
CaO	16·79	10·40	11·13	15·72	16·23	15·99
BaO	0·25	0·32	0·17			
Na_2O	1·32	0·18	2·50	2·34	1·11	0·55
K_2O	4·09	6·98	3·28	2·50	7·41	5·33
TiO_2	3·87	4·36	3·12	5·08	0·74	1·20
P_2O_5	0·74	0·61	0·52	0·67	0·32	1·21

Source of data:
1. Holmes, (1937, p. 205).
2. Holmes (1942a, p. 212).
3. Holmes (1956, p. 15).
4. Holmes, (1952, p. 196).
5. Mittempergher (1965 a, p. 460).
6. Mittempergher (personal communication).

II.5.3.2. Chemistry

Table 2 summarizes the chemical composition of some typical examples of the kamafugitic rocks (in a way analogous to Table 1). Because these rocks display virtually a continuous series between the type rocks (Fig. 2), no average compositions have been calculated. Reviews of the existing chemical analyses of the Toro–Ankole volcanics have been compiled by

Denaeyer *et al.*, (1965) and El-Hinnawi (1965). The trace elements have been studied by Higazy (1954).

Some of the chemical characteristics of the orenditic lamproites apply also to the kamafugitic rocks. The most conspicuous differences between the two rock groups will be discussed here.

The kamafugitic rocks represent alkaline ultramafics, strongly to even extremely undersaturated with respect to silica (Fig. 1). Although they are generally less alkalic than the orenditic rock groups the ratio Fe/Mg being higher in the kamafugitic rocks.

In contrast to the orenditic lamproites in which the alkali excess over Al is a rule rather than an exception, the kamafugitic rocks mostly show no normative Ac.

The general trend in the trace element contents of the kamafugitic rocks follows that of the orenditic lamproites. Probably the most significant difference between the two rock groups is found in the ratio Ba/Sr. In the orenditic lamproites this ratio is decidedly greater than unity.

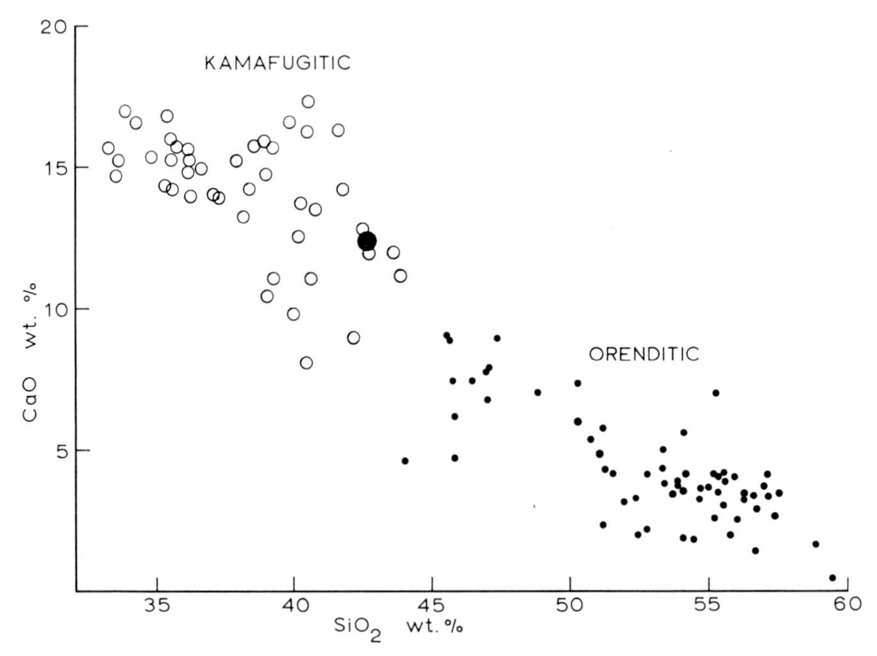

Fig. 3 CaO plotted against SiO_2. Large black circle indicates madupite

lamproites, the alkali ratio varies within the same range. Another feature of importance is the systematically higher Ca content of the kamafugitic rocks (Fig. 3). In the orenditic lamproites Mg displays a clear negative correlation with Si, numerically specific to each of the orenditic suites concerned. No correlation can be traced between Mg and Si in the kamafugitic rocks (Fig. 4). Total iron (as FeO in Fig. 5) is fairly independent of the Mg content in both The data given by Higazy (1954) indicate this ratio to be close to unity in mafurite and ugandite, but considerably less in katungite and in the potash ankaratrite–melaleucitite series.

II.5.3.3. Mineral Composition

Chemical data for the constituent minerals of the kamafugitic rocks are not ample.

Olivine (monticellite in the coppaelite of Table 2) is fresh and occurs preferably as

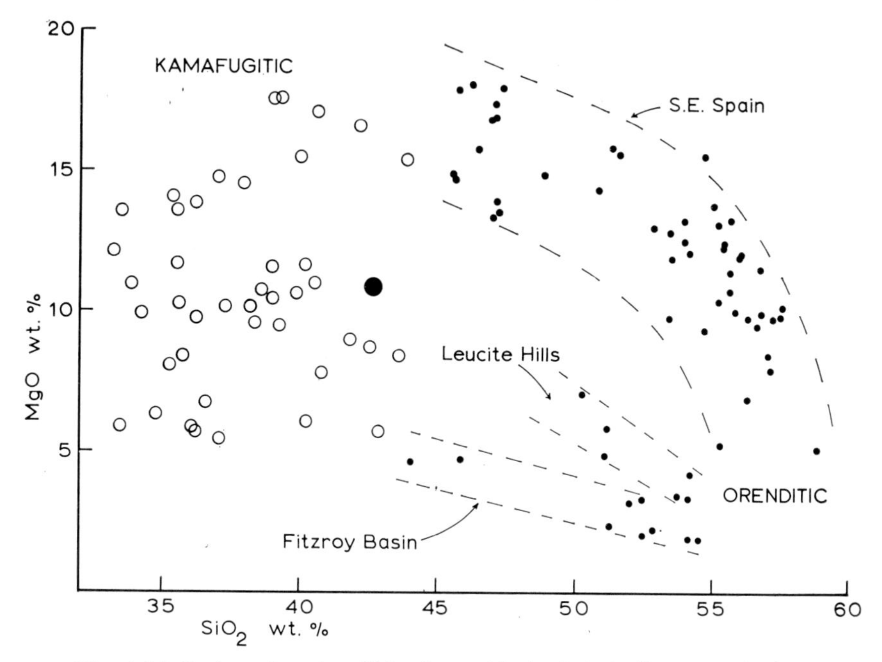

Fig. 4 MgO plotted against SiO_2. Large black circle indicates madupite

phenocrysts. The two existing chemical analyses (Neuvonen, 1956; Sahama, 1954) indicate a composition 10± mol % Fa. The clinopyroxene is rarely colourless (except in melilite-rich rocks) but shows a slightly greenish (acmitic) or purple brownish (titanian) tint. The only analysis (from potash ankaratrite: Sahama and Wiik, 1952) indicates a titanian salite.

Melilite (Neuvonen, 1956; Mittempergher, 1965a; Sahama, 1967) is essentially a solid solution of åkermanite (iron åkermanite) and alkali melilite. The gehlenite component is negligible.

Leucite is unaltered and twinned. The two analyses (Sahama and Wiik, 1952; Bannister, Sahama and Wiik, 1952) indicate no remarkably high Fe content nor any non-stoichiometric composition.

Nepheline (potassian: Sahama and Wiik, 1952) is reported from the rocks of the potash ankaratrite–melaleucitite series.

Because nepheline and kalsilite as separate grains of the same rock cannot be distinguished from each other in thin section, their joint occurrence in the kamafugitic rocks has not been tested. Only in San Venanzo small amounts of nepheline have been shown to occur in the kalsilite-bearing rock (Smith in: Sahama, Neuvonen and Hytönen, 1956).

Kalsilite was originally detected in the mafurite of the Bunyaruguru field (Bannister, 1942) and has been shown to be a common

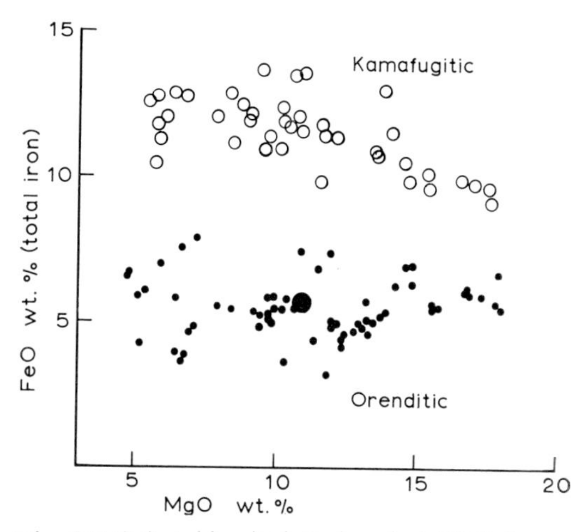

Fig. 5 FeO (total iron) plotted against MgO. Large black circle indicates madupite

constituent also in the melilite-bearing rocks of both Toro–Ankole and Umbria. Additional kalsilite analyses have been published from Toro–Ankole by Neuvonen (1956), by Sahama (1954) and from San Venanzo by Bannister, Sahama and Wiik (1952).

The only analysis of mica (called biotite: Combe and Holmes, 1945, p. 377) indicates the composition of a relatively iron-rich Ba- and Ti-bearing phlogopite. Apart from xenocrysts, feldspars and amphiboles are absent in the lavas.

Apatite is constantly present. Perovskite (analysis: Neuvonen, 1956) is common especially in the melilite-bearing rocks. Opaque iron oxide minerals occur in all types of the kamafugitic rocks. The presence of any alkali-TiZr-minerals, corresponding to priderite etc., of the orenditic rocks, has not been tested.

II.5.4. Petrogenesis

II.5.4.1. Retrospect

The petrogenesis of the K-rich rocks considered offers two separate problems: (*a*) genetic interrelationship between the various rock types of a specific area, and (*b*) derivation of the K-rich magma from a parent source.

a. The chemical and mineralogical characteristics displayed by the rocks of a particular area make it evident that the individual flows, plugs, sheets, etc., represent manifestations of the same co-magmatic province and must be petrogenetically derivable from each other or from a common K-rich magma. Any theory dealing with the interrelationships between the various rock types of a given province should be based on field observations. Unfortunately, indicative field data are available only from the Fitzroy basin (Prider, 1960) illustrating the chemical differentiation of the magma into an orenditic top part and a wolgiditic fraction deeper down.

A chemical differentiation caused by a gaseous upward transfer of more volatile elements, rather than fractional crystallization, represents an instrumental agent in the chemical differentiation of the lamproite magma in the Fitzroy basin (separation into wolgiditic-orenditic: Prider, 1960) and in S.E. Spain (jumillite-cancalite-fortunite: Fúster *et al.*, 1967). This view is in accordance with the common occurrence of volcanic breccias mantling and capping the lamproite pipes. The orendite-wyomingite liquid can hardly be derived from a madupitic parent solely by crystal fractionation (Carmichael, 1967). Madupite is intermediate between the rocks of the orenditic and kamafugitic affinities and is chemically not too different from coppaelite. Could it be that madupite just represents a local manifestation of a kamafugitic genesis (interaction of a restricted carbonatite pocket in the magma mass) in an otherwise orenditic magma development?

b. In discussing the origin of highly K-rich igneous rocks, many authors extend their arguments to cover both the orenditic and the kamafugitic rock groups. The rocks of these two groups certainly exhibit fundamental parallelism in chemistry applying to the general trace element pattern and to several features in the contents of the main elements. As has been illustrated above (Tables 1 and 2; Figs. 1, 3–5), the contents of some other elements or their ratios display, however, marked differences between the two rock groups. These differences should not be overlooked.

The authors who have studied the orenditic rocks do not consider a crustal contamination to be significant in modifying the chemistry of the lamproite magma. This statement holds true despite the fact that a large part of the lamproite masses have ascended through various sedimentary beds including limestones.

On the other hand, according to Mittempergher (1965a), an assimilation of limestone stoped by the K-rich magma was instrumental in producing undersaturation of the Umbria kamafugitic rocks. A limestone syntexis is rejected by Holmes (1950) in the evolution of the Toro–Ankole rocks. According to his latest hypothesis, this rock series results from an interaction between a carbonatite magma of deep-seated origin and the granitic basement illustrated by the schematical reaction equation:

$$\text{granite} + \text{carbonatite} = \text{kamafugite} + \text{O.B.P. series.}$$

The composition postulated for the carbonatite is somewhat uncommon for such rocks and the rocks of the O.B.P. series may possibly be interpreted as subvolcanic crystallizations from the kamafugitic magma itself. Based on *n*-dimensional statistics of a number of analyses of the leucite-bearing Birunga rocks, Cundari and Le Maitre (1970) consider these magmas as having been generated by an incomplete fractionation of a parent with biotite-pyroxenite affinity.

Chemically speaking, it makes no difference whether the carbonate material is derived from limestone assimilation or is of a deep-seated magmatic origin. The main point is that strongly undersaturated K-rich alkaline rocks, like those of the kamafugitic series, can form only if a K-rich magma becomes desilicated through interaction of carbonatic material (Marinelli and Mittempergher, 1966). Thus, it could be inferred that the difference in the degree of undersaturation between the orenditic and the kamafugitic rocks (Fig. 1) is due to the extent in which carbonate material interacts with the K-rich magma.

The desilication of an alkaline magma is beautifully demonstrated by the permanently molten lava lake of the Nyiragongo volcano (North Kivu). The solidified parts of the lava lake and the ejected blocks lying around, often sprinkled with kalsilite phenocrysts, contain abundant vesicles filled with quartz (Sahama, 1962). The Si evidently escaped through gaseous transfer from the deeper parts of the magma basin where possibly a contemporary carbonatite is in action (Meyer, 1958).

The orenditic lamproite magma is considered as of mantle origin by Carmichael (1967). Prider (1960) derives the Fitzroy basin lamproites from a mica peridotite magma generated in small cupolas in the roof of an extended peridotite magma mass. His interpretation implies that the chemical characteristics of the lamproite material were developed essentially prior to the ascent of the magma. The small Umbria volcanoes concerned are too isolated to be connected with any definite stage of magmatic evolution in the Roman province (Mittempergher, 1965a,b). The potassic volcanics of northern Latium in general representing differentiates from anatectic magmas (Marinelli and Mittempergher, 1966), the possibility of a crustal origin, or at least of a strong crustal syntexis of the magma, must be taken into account. Holmes' hypothesis dealing with the Toro–Ankole volcanics implies a mixed crustal–mantle origin.

II.5.4.2. Chemical Differentiation

Irrespectively of the origin attributed to the K-rich magmas treated in this section, the following statements are valid: K-rich rocks of pronounced orenditic or kamafugitic affinity are extremely rare and occur in a few restricted areas only. Such rocks are confined to volcanic or subvolcanic environments and lack plutonic counterparts. The chemical composition including the trace element pattern shows features common to these rocks despite the fact that the areas concerned are widely separated from each other. Quite a variety of potassic to sodopotassic alkaline rocks display a varying tendency to approach the chemical composition of the orenditic rocks (example: Highwood Mountains, Montana; Larsen *et al.*, 1939 and 1941) or that of the kamafugitic rocks proper (example: Nyiragongo volcano, North Kivu; Sahama, 1962). These facts indicate that the processes capable of producing such magmas are of a universal application, but will only exceptionally find conditions sufficiently favourable to result in typically orenditic or kamafugitic rocks.

The main point in the petrology of the orenditic-kamafugitic rocks is the nature of the chemical differentiation which results in the composition found. The hypothesis of zone refining put forward by Harris (1957) is capable of explaining the peculiar chemistry of the rocks (except the marked enrichment of Ti: Prider, 1960), but implies a rather complicated mechanism of melting and precipitation during ascent. Such a mechanism is not required in the differentiation by filtration (Marinelli and Mittempergher, 1966) which could be illustrated as vaguely analogous to a chromatographic process. In a melt moving in direction of the pressure gradient, the more mobile elements are selectively enriched in the top portion of the ascending magma. The filtration process may be strengthened by a gaseous transfer of the alkalis etc., by assimilation of the rocks perforated and, during periods of rest in the magma ascent, by fractional crystallization.

The differentiation by filtration as outlined by Marinelli and Mittempergher (1966) is a major mechanism in the evolution of the potassic igneous suite of northern Latium. The chemistry of this suite is still far from an orenditic-kamafugitic composition. It seems, however, not impossible that, under exceptionally favourable conditions (narrow subvertical magma column with steep pressure gradient; volcanic to subvolcanic) a chemical differentiation could advance far enough to generate even an orenditic-kamafugitic magma. Such a magma would represent an extreme end product in the evolution and, therefore, would be so rarely acquired.

This hypothesis of differentiation by filtration is essentially a parallel to the view of the upward diffusion of more volatile components by Saether (1950) and Prider (1960), as well as to the principle of a selective enrichment of the incompatible elements as developed by Green and Ringwood (1967) (see also VI.5). Carbonatic material may or may not become enriched to an extent sufficient to modify the process.

II.5.4.3. Selective Alkali Enrichment

A selective enrichment of K with respect to Na is known locally, or even regionally, from many alkaline rocks of volcanic to subvolcanic environment. In its extreme development, such as is displayed in the orenditic-kamafugitic rocks proper, the enrichment cannot reasonably be attributed to fractional crystallization.

The only mineral which is common in the volcanic–subvolcanic alkaline magmas and which would be capable of extracting K in a selective manner is leucite. Being a mineral of a very low density, leucite phenocrysts, under favourable conditions, can become accumulated locally in the top part of the magma column. Convincing examples of such leucite accumulations have been given by Buie (in Larsen *et al.*, 1939 and 1941) and by Sahama and Meyer (1958). Further examples are offered by the classical leucite-rich tuffs of Villa Senni and Rochamonfina in Central Italy. Such extreme accumulations do not reach extensive dimensions. Wherever sizable leucite phenocrysts occur, the possibility of crystal accumulation should, however, be taken into consideration. The effect of the accumulation process upon the bulk composition of the magma will mostly not be significant.

On the other hand, numerous examples are known in which a selective enrichment either of K or of Na by processes other than fractional crystallization (differentiation by filtration, gaseous transfer, upward migration etc.) is evident. A few of such examples will be listed below.

Rittmann (1933) has shown on Vesuvius that a selective migration of Na has caused a passive enrichment of K in the series of undersaturated leucitic rocks. The excess of Na illustrated by the normative Ns is characteristic of the soda rhyolites (hyalopantellerites) of the Island of Pantelleria. A series of chemical analyses of the crystalline core and of the largely glassy crust on the same spot of the flow were carried out by Romano (1968) and have shown that the soda excess is confined to the crust. After eruption and during solidification of the lava flow, Na migrated through gaseous transfer from the core to the overlying glassy part of the lava mass. A most spectacular example of a large-scale enrichment of Na by gaseous transfer is manifested by the Oldoinyo Lengai volcano in Tanzania (Dawson, 1962, 1966; Dawson *et al.*, 1968).

The selective enrichment of K with respect to Na in the top part of the magma column of the Katunga volcano was illustrated by Holmes (1950, p. 778). A similar phenomenon in the deeply dissected ancient lava lake of the Radicofani volcano in south-eastern Tuscany was demonstrated by Innocenti (1967). Here the enrichment of K through gaseous transfer in the top part of the column is evident.

Such marked shifts in the ratio K/Na being manifested by subaerial volcanics, even in post-eruptional environment, more radical shifts of the ratio can well be expected to occur during the long continued ascent of the magma column. Which one of the alkalis is apt to become enriched, depends on a bewildering number of factors and is hardly predictable. In the scale of a hand specimen, the crystallization of some zeolites in the vesicles of a solidifying lava through 'sweating' of the magma may be taken as an analogue to the selective gaseous transfer of the alkalis in a magma column.

II.5.4.4. Summary

In conclusion, the generation of the potassic to perpotassic magmas is most probably complex and can hardly be attributed to the operation of just one single process of a more universal application. The following points are to be considered:

Ascent of mantle material already somewhat enriched in the more volatile components. This enrichment may have been affected by upward diffusion in the magma mass prior to the ascent and/or by partial melting of the roof of the extended magma basin.

Chemical differentiation within the magma column caused by selective diffusion, gaseous transfer and crystal fractionation. The effect increases with increasing pressure gradient.

Crustal development of anatectic magma with subsequent chemical differentiation.

Stoping and assimilation of the bedrock perforated.

Interaction of carbonatitic differentiates (syntectic or deep-seated) of the magma. A carbonatic magma portion may not have played any significant role (rocks of orenditic affinity) or its influence may have been strong (rocks of kamafugitic affinity). In Toro–Ankole, mafurite and ugandite may have resulted from a less extended participation of the carbonatite magma whereas katungite and potash ankaratrite may have originated from more strongly carbonated K-rich magma portions.

Evidently, the extent to which these factors contribute to the evolution of the K-rich magmas is variable and must be tested separately for each province or occurrence. Isotopic and trace element studies may offer important data. The ratio Sr^{87}/Sr^{86} in the Toro–Ankole volcanics seems to favour a mixed mantle–crustal origin (Bell and Powell, 1969). This interpretation of the data presupposes that the ascending mantle material was originally depleted in potassium and in radiogenic strontium. Because such an assumption may possibly not hold true, the use of the strontium isotopic data as genetic indicators must be excercised with care (Powell and Bell, 1970).

A comparison of the orenditic-kamafugitic rocks with kimberlites certainly is tempting, but is as debatable as the kimberlite problem itself.

II.5. REFERENCES

Bannister, F. A., 1942. Kalsilite, a polymorph of $KAlSiO_4$, from Uganda. *Mineralog. Mag.*, **26,** 218–24.

Bannister, F. A., Sahama, Th. G., and Wiik, H. B., 1952. Kalsilite in venanzite from San Venanzo, Umbria, Italy. *Mineralog. Mag.*, **30,** 46–8.

Bell, K., and Powell, J. L., 1969. Strontium isotopic studies of alkalic rocks: The potassium-rich lavas of the Birunga and Toro–Ankole regions, East and Central Equatorial Africa. *J. Petrology*, **10,** 536.

Borley, G. D., 1967. Potash-rich volcanic rocks from southern Spain. *Mineralog. Mag.*, **36,** 364–79.

Carmichael, I. S. E., 1967. The mineralogy and petrology of the volcanic rocks from the Leucite Hills, Wyoming. *Contr. Miner. Petrology*, **15,** 24–66.

Combe, A. D., 1937. The Katunga volcano, South-West Uganda. *Geol. Mag.*, **74,** 195–200.

Combe, A. D., and Holmes, A., 1945. The kalsilite-bearing lavas of Kabirenge and Lyakauli, South-West Uganda. *Trans. R. Soc. Edinburgh*, **61,** 359–79.

Cross, W., 1897. Igneous rocks of the Leucite Hills and Pilot Butte, Wyoming. *Am. J. Sci.*, **4,** 115–41.

Cundari, A., and Le Maitre, R. W., 1970. On the petrogeny of the leucite-bearing rocks of the Roman and Birunga volcanic regions. *J. Petrology*, **11,** 33.

Dawson, J. B., 1962. The geology of Oldoinyo Lengai. *Bull. volcan.*, **24,** 349–87.

Dawson, J. B., 1966. 'Oldoinyo Lengai—an active volcano with sodium carbonate flows', in O. F.

Tuttle and J. Gittins (Eds.), *Carbonatites*, John Wiley and Sons, New York, 155–68.

Dawson, J. B., Bowden, P., and Clarke, G. C., 1968. Activity of the carbonatite volcano Oldoinyo Lengai, 1966. *Geol. Rundschau*, **57**, 865–79.

Denaeyer, M.-E., Schellinck, F., and Coppez. A., 1965. Recueil d'analyses des laves du fossé tectonique de l'Afrique Centrale (Kivu, Rwanda, Toro–Ankole). *Ann. Musée Roy. de l'Afrique Centrale Ser. IN 8º, Sci. Geol.*, **49**, 1–234.

Dryden, J. S., and Wadsley, A. D., 1958. The structure and dielectric properties of compounds with the formula Ba_x $(Ti_{8-x}Mg_x)O_{16}$. *Trans. Faraday Soc.*, **54**, 1574–80.

El-Hinnawi, E. E., 1965. Petrochemical characters of African volcanic rocks. Part III: Central Africa. *Neues Jahrb. Mineral. Abh.*, **103**, 126–46.

Fermoso, M. L., 1967a. El diópsido de las rocas volcánicas de Jumilla (S.E. de Espana). *Estud. Geol.*, **23**, 31–3.

Fermoso, M. L., 1967b. Composición química de las sanidinas de las rocas lamproíticas espanolas. *Estud. Geol.*, **23**, 29–30.

Fúster, J. M., Gastesi, P., Sagredo, J., and Fermoso, M. L., 1967. Las rocas lamproiticas del S.E. de Espana. *Estud. Geol.*, **23**, 35–69.

Green, D. H., and Ringwood, A. E., 1967. The genesis of basaltic magmas. *Contr. Mineral. Petrology*, **15**, 103.

Harris, P. G., 1957. Zone refining and the origin of potassic basalts. *Geochim. Cosmochim. Acta*, **12**, 195–208.

Henshaw, D. E., 1955. The structure of wadeite. *Mineral. Mag.*, **30**, 585–95.

Hernández-Pacheco, A., 1965. Una richterita potásica de rocas volcánicas alcalinas, Sierra de las Cabras (Albacete). *Estud. Geol.*, **20**, 265–70.

Higazy, R. A., 1954. Trace elements of volcanic ultrabasic potassic rocks of southwestern Uganda and adjoining part of the Belgian Congo. *Bull. Geol. Soc. Am.*, **65**, 39–70.

Holmes, A., 1936. Transfusion of quartz xenoliths in alkali basic and ultrabasic lavas, South-West Uganda. *Mineral. Mag.*, **24**, 408–21.

Holmes, A., 1937. The petrology of katungite. *Geol. Mag.*, **74**, 200–19.

Holmes, A., 1942a. A suite of volcanic rocks from south-west Uganda containing kalsilite (a polymorph of $KAlSiO_4$). *Mineral. Mag.*, **26**, 197–216.

Holmes, A., 1942b. A heteromorph of venanzite. *Geol. Mag.*, **79**, 225–32.

Holmes, A., 1945. Leucitized granite xenoliths from the potash-rich lavas of Bunyaruguru, South-West Uganda. *Am. J. Sci.*, **243–A**, Daly Volume, 313–32.

Holmes, A., 1950. Petrogenesis of katungite and its associates. *Am. Miner.*, **35**, 772–92.

Holmes, A., 1952. The potash ankaratrite-melaleucitite lavas of Nabugando and Mbuga craters, South-West Uganda. *Trans. Edinburgh Geol. Soc.*, **15**, 187–213.

Holmes, A., 1956. The ejectamenta of Katwe crater, South-West Uganda. *Verhandel. Koninkl. Ned. Geol. Mijnbouwk. Genoot.*, **16**, 1–28.

Holmes, A., and Harwood, H. F., 1932. Petrology of the volcanic fields east and south-east of Ruwenzori, Uganda. *Q. J. Geol. Soc. Lond.*, **88**, 370–442.

Innocenti, F., 1967. Studio chimico-petrografico delle vulcaniti di Radicofani. *Rendiconti della Societa Mineralogica Italiana*. **23**, 99–128.

Kemp, J. F., 1897. The Leucite Hills of Wyoming. *Bull. Geol. Soc. Am.*, **8**, 169–82.

Kemp, J. F., and Knight, W. C., 1903. Leucite Hills of Wyoming. *Bull. Geol. Soc. Am.*, **14**, 305–36.

v. Knorring, O., and Du Bois, C. G. B., 1961. Carbonatitic lava from Fort Portal area in Western Uganda. *Nature*, **192**, 1064–5.

Larsen, E. S., Hurlbut Jr., C. S., Griggs, D., Buie, B. F., and Burgess, C. H., 1939 and 1941. Igneous rocks of the Highwood Mountains, Montana. *Bull. Geol. Soc. Am.*, **50**, 1043–112, and **52**, 1733–868.

Marinelli, G., and Mittempergher, M., 1966. On the genesis of some magmas of typical Mediterranean (potassic) suite. *Bull. volcan.*, **29**, 113–40.

Meyer, A., 1958. Carbonatites—quelques grands traits. *Commission de Cooperation Technique en Afrique au Sud du Sahara, Réunion Conjointe, Léopoldville*, 295–301.

Mittempergher, M., 1965a. Vulcanismo e petrogenesi nella zona di San Venanzo (Umbria). *Atti Soc. Toscana Sci. Nat.*, Serie A, **72**, 437–79.

Mittempergher, M., 1965b. Volcanism and petrogenesis in the S. Venanzo area (Italy). *Bull. volcan.*, **28**, 1–12.

Neuvonen, K. J., 1956. Minerals of the katungite flow. *Compt. Rend. Soc. Géol. Finlande.*, **29**, 1–7.

Norrish, K., 1951. Priderite, a new mineral from the leucite-lamproites of the West Kimberley area, Western Australia. *Mineral. Mag.*, **24**, 496–501.

Powell, J. L., and Bell, K., 1970. Strontium isotopic studies of alkalic rocks: Localities from Australia, Spain and The Western United States. *Contr. Miner. Petrology*, **27**, 1.

Prider, R. T., 1939. Some minerals from the leucite-rich rocks of the West Kimberley area, Western Australia. *Mineral. Mag.*, **25**, 373–87.

Prider, R. T., 1960. The leucite lamproites of the Fitzroy Basin, Western Australia. *J. Geol. Soc. Australia*, **6**, 71–118.

Prider, R. T., 1965. Noonkambahite, a potassic batisite from the lamproites of Western Australia. *Mineral. Mag.*, **34**, 403–5.

Prider, R. T., and Cole, W. F., 1942. The alteration

products of olivine and leucite in the leucite-lamproites from the West Kimberley area, Western Australia. *Am. Miner.*, **27**, 373–84.

Rittmann, A., 1933. Die geologisch bedingte Evolution und Differentiation des Somma-Vesuv Magmas. *Z. Vulkanol.*, **15**, 8–94.

Roedder, E. W., 1951. The system K_2O–MgO–SiO_2. *Am. J. Sci.*, **249**, 81–130 and 224–48.

Rodolico, F., 1937. Le zone vulcaniche di San Venanzo a di Cupaello. *Boll. Soc. Geol. Ital.*, **56**, 33–66.

Romano, R., 1968. Sur l'origine de l'excés de soude (Ns) dans certaines laves de l'ile de Pantelleria. *Symposium International de Volcanologie, Canary Islands*. To be published in *Bull. Volcanol.*

Saether, E., 1950. On the genesis of peralkaline rock provinces. *International Geological Congress* **18**, *London, Part II*, 123–30.

Sahama, Th. G., 1954. Mineralogy of mafurite. *Compt. Rend. Soc. Géol. Finlande*, **27**, 21–8.

Sahama, Th. G., 1962. Petrology of Mt. Nyiragongo. *Trans. Edinburgh Geol. Soc.*, **19**, 1–28.

Sahama, Th. G., 1967. Iron content of melilite. *Compt. Rend. Soc. Géol. Finlande*, **39**, 17–28.

Sahama, Th. G., and Meyer, A., 1958. Study of the volcano Nyiragongo, a progress report. *Institut des Parcs Nationaux du Congo Belge, Exploration du Parc National Albert, Missions d'etudes vulcanologiques*, Fasc. 2, 1–85.

Sahama, Th. G., Neuvonen, K. J., and Hytönen, K., 1956. Determination of the composition of kalsilites by an X-ray method. *Mineral. Mag.*, **31**, 200–8.

Sahama, Th. G., and Wiik, H. B., 1952. Leucite, potash nepheline, and clinopyroxene from volcanic lavas from southwestern Uganda and adjoining Belgian Congo. *Am. J. Sci.*, Bowen Volume, 457–70.

Sartori, F., 1966. Su di una tufite della zone di San Venanzo (Umbria). *Atti Soc. Toscana Sci. Nat.*, Serie A, **73**, 25–48.

Schultz, A. R. and Cross, W., 1912. Potash-bearing rocks of the Leucite Hills, Sweetwater County, Wyoming. *U.S. Geol. Surv.*, Bulletin 512, 1–39.

Streckeisen, A., 1967. Die Klassifikation der Eruptivgesteine. *Geol. Rundschau*, **55**, 478–91.

Velde, D., 1967. Sur un lamprophyre hyperalcalin potassique: la minette de Sisco (ile de Corse). *Bull. Soc. Franc. Mineral. Crist.*, **90**, 214–23.

Velde, D., 1968. A new occurrence of priderite. *Mineral. Mag.*, **36**, 867–70.

Wade, A., and Prider, R. T., 1940. The leucite-bearing rocks of the West Kimberley area, Western Australia. *Q. J. Geol. Soc. Lond.*, **96**, 39–98.

II.6. OVERSATURATED ALKALINE ROCKS: GRANITES, PANTELLERITES AND COMENDITES

P. Bowden

II.6.1. Introduction

This chapter is principally concerned with a discussion on silicic rocks which, according to Shand's definition, have a molecular excess of alkalis over alumina, and are termed *peralkaline*. The numerical value of this molecular ratio is called *peralkalinity*, *agpaitic index* or *agpaitic coefficient* (cf.II.2.1).*

Murthy and Venkataraman (1964) have discussed at length the world-wide distribution of peralkaline granites. Consequently greater emphasis has been placed in this chapter on silicic volcanics belonging to the *comendite* and *pantellerite* clans. Comendites and pantellerites are the peralkaline equivalents of rhyolites with lower Al and more Fe, Ti and Na than their calc-alkali or alkali equivalents. In this chapter the term *pantellerite* refers to the more peralkaline and iron-rich volcanic group and is consistent with the description of specimens from the type locality on the Isle of Pantelleria, between Tunisia and Sicily. *Comendite*, which is less peralkaline and iron-rich, was first described from Commende, San Pietro Island, S.W. Sardinia. Although the distinction between comendite and pantellerite is often treated arbitrarily by many workers, Lacroix (1927) recognized an apparent continuity of the comendite–pantellerite series and proposed that silicic volcanic rocks in this series containing more than 12·5%

* Readers are referred to the relevant Chapters of this book where similarly cited.

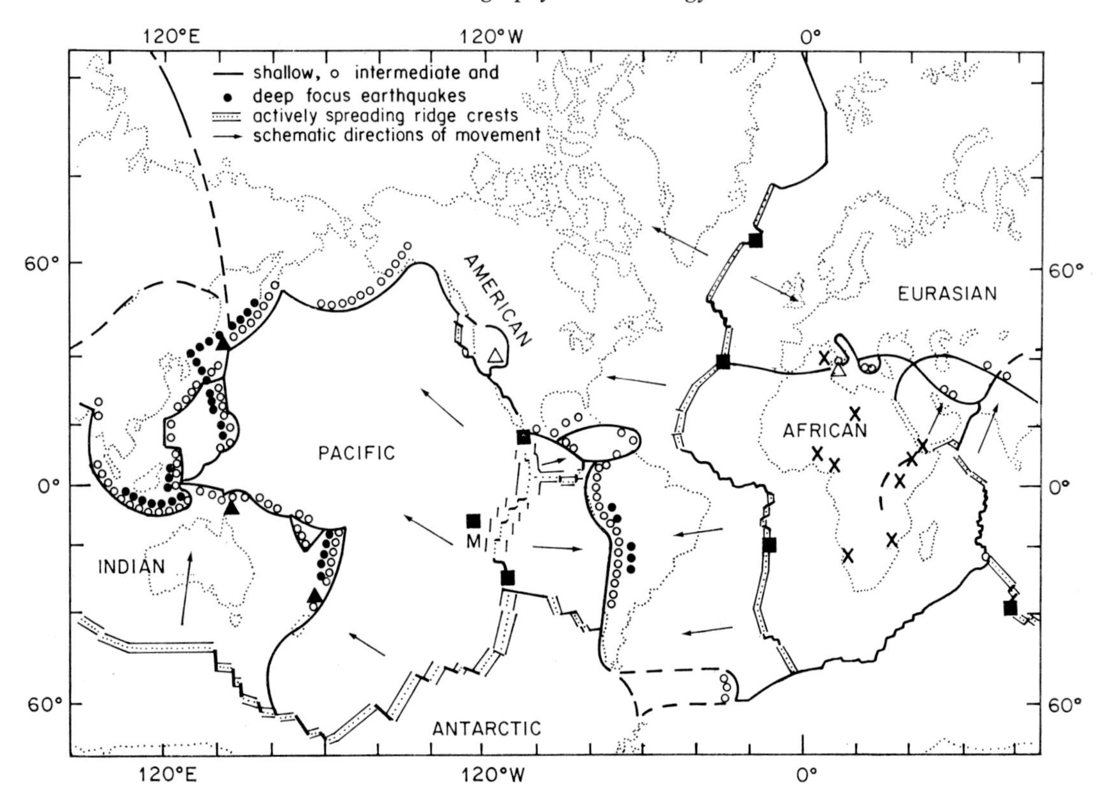

Fig. 1 The relationship between plate tectonics and sites of peralkaline magmatism

Selected examples collated from various literature sources directly and indirectly referenced in the text. The basic diagram containing a summary of the seismicity, and the extent of lithospheric plates is taken from Vine (1970).

Key: Crosses refer to some continental occurrences in Africa related to epeirogenic doming and rifting.
Closed squares are sites of peralkaline silicic vulcanism above actively spreading ridge crests.
Open triangles represent peralkaline silicic activity at continental plate margins. Pantelleria has been tentatively assigned to this group (McKenzie, 1970).
Closed triangles are sites near oceanic trenches referred to as 'behind island arcs' in the text (cf. Vine, 1970).

total normative femic constituents should be termed pantellerites, whilst those with correspondingly less femics should be called comendites. With more data available this criterion is not entirely satisfactory (Ewart *et al.*, 1968; Nicholls and Carmichael, 1969). To overcome this problem, the writer, in co-operation with Dr. W. E. Stephens, has utilized principal component analysis and cluster analysis (Parks, 1966). This approach has shown that SiO_2, Al_2O_3, FeO, Na_2O and MnO values can be used to discriminate between comendites and pantellerites. Where anomalies of classification have arisen in the literature, this statistical treatment places the rocks in question positively into the comendite group.

II.6.2. Modes of Occurrence

Peralkaline silicic rocks are located in areas of alkaline and calc-alkaline magmatism in both continental and oceanic environments. An attempt has been made in Fig. 1 to plot representative occurrences of some peralkaline silicic rocks on a world map in relation to the seismicity of the earth and the position of lithospheric plates (Vine, 1970). There appears to be several tectonic settings for peralkaline magmatism.

II.6.2.1. Peralkaline Silicic Magmatism Related to Continental Doming and Rifting, e.g. Africa

Peralkaline silicic rocks are commonly found in non-orogenic continental regions which have been subjected to crustal swelling and rifting (Le Bas, 1971). Peralkaline magmatism can occur during *pre-rifting* (epeirogenic doming), *initial rifting* (development of linear fractures and beginning of crustal attenuation) and *continued rifting* (extensive crustal attenuation leaving little or no sialic crust on the rift floor). Each event appears to be characterized by particular suites and varying proportions of peralkaline silicic and associated rocks.

Comendites frequently occur in subvolcanic ring complexes associated with non-peralkaline volcanics. Comenditic associations are typical products of epeirogenic doming exemplified by the elevated domes of vitreous to granophyric peralkaline and alkaline lavas in Tibesti (Vincent, 1970). In rift valleys comendites are more closely linked with peralkaline trachytes and pantellerites. Examples of these associations are found in Kenya (III.2.3). However, the proportion of comendite becomes diminished as the rift evolves and in well-developed rift systems (Ethiopia) where crustal attenuation is more extensive comendites are virtually absent (cf. Mohr, 1970).

Pantellerites are related to the development of rifting. For example, there is a Tertiary to Recent site of crustal swelling in the Cameroun which possesses distinct protofractures, and significantly pantellerites have been recorded at Mount Mba Nsché (Koch, 1955). The Ethiopian Rift System consisting of the Main Ethiopian Rift and the Danakil Depression, a N.N.W.–S.S.E. trending graben structure en echelon to the Red Sea rift (Tazieff *et al.*, 1969; Barberi *et al.*, 1970) provides good examples of pantelleritic volcanism and its relationship with the extent of rifting. Within the Main Ethiopian Rift alkali trachyte-alkali rhyolite-pantellerite volcanic complexes (e.g. Fantale, Gibson, 1967) are located on the rift floor overlying older alkali basalts. On the margins of the Danakil Depression near the western scarp of the Ethiopian plateau pantellerites associated with non-peralkaline silicic rocks are found in volcanic massifs, e.g. P. Pruvost, Boina. Further to the north in Afar rhyolites of pantelleritic affinity occur at several volcanic centres, but in the region centred along the Erta'Ale range where crustal attenuation is extensive the principal rock types are basalts intermediate between alkali and tholeiitic and trachytes with only small quantities (~0·5 % by volume) of slightly peralkaline glassy rhyolites.

Peralkaline granites are common on continental regions of epeirogenic doming frequently displayed as subvolcanic ring structures. During the period of gradual upwarping of the country rocks prior to rifting, the composition of the associated plutonic rocks may vary. For instance, peralkaline and associated non-peralkaline granitic ring complexes in the Nigeria and South Niger regions (IV.8.2.4) could represent early doming features. In contrast, but probably indicating a later stage of pre-rift development, are associations of peralkaline granites, syenites, syenogabbros, gabbros and anorthosites found in Aïr (IV.8.2.3). Similar rock types in the Gardar province (IV.3.5.1; VI.2.4) may also be connected with protograben formation. It is clear that peralkaline plutonism is an essential part of *pre-rift* tectonics with peralkaline granites usually restricted to this environment. For example, peralkaline granites are located on the Western plateau of the Danakil Depression and as a remnant horst in the centre of the rift at Affara Dara (Barberi *et al.*, 1970). According to J. Varet (personal communication, 1971) the granites have identical Tertiary isotopic ages (25 m.y.) and therefore just predate the beginning of rifting. However, xenoliths of peralkaline granite and quartz syenite have also been recorded in peralkaline lavas erupted on oceanic islands (IV.7) above actively spreading ridge crests, and hence peralkaline plutonism may be also a minor but important early stage in most areas of tensional tectonics.

II.6.2.2. Peralkaline Silicic Volcanism on Oceanic Islands

Such examples occur on Soccorro and Easter Island (East Pacific Rise), Iceland, Azores and Ascension Island on the Mid-Atlantic Ridge, and St. Paul situated above the Mid-Indian Rise. All these occurrences are located on actively spreading ridge crests (see Fig. 1). It is also tempting to suggest that the Pacific islands of Marquesas (M) have been carried westwards by the Pacific plate from the Mid-Pacific Rise, and were therefore generated in a similar environment to Easter Island and Soccorro. According to Bryan (1969), Soccorro exhibits virtually identical rock types to those described from Pantelleria. However, the occurrence of pantellerite on oceanic islands is very rare. Comendites are the dominant silicic rock type but they are of minor proportions compared with the volumes of alkali basalt, trachybasalt, trachyandesite, trachytes and phonolites at each centre (cf. IV.7.4.1). In the Atlantic there are also islands on the basin and continental slope regions where phonolites are more abundant than trachytes but whose end products may still be comenditic (IV.7.4).

II.6.2.3. Peralkaline Silicic Volcanism on Island Arcs

Examples such as Mayor Island (New Zealand), New Guinea and Hokkaido N. E. Japan, are situated in regions of seismicity associated with shallow and intermediate focus earthquakes, and close to active plate subduction zones. The association of compression tectonics with peralkaline silicic volcanism seems at first to be anomalous, for oceanic peralkaline magmatism is normally associated with tensional tectonics and spreading at ridge crests. However a process equivalent to crustal extension may be occurring on a limited scale 'behind island arcs' (Vine, 1970). Such regions are the sites of comendite volcanism.

II.6.2.4. Peralkaline Silicic Volcanism on Continental Plate Margins

The only good example of this type is found in the Great Basin, Nevada. Here peralkaline silicic rocks, both ash-flow tuffs and lavas of principally comenditic affinity, with minor pantellerites, appear to be restricted to the marginal parts of the Great Basin (Noble *et al.*, 1968). In contrast, quartz latites rich in strontium and potassium outcrop in the centre of the Basin. Peralkaline silicic volcanism in this region has been attributed to the overriding of the northern part of the East Pacific Rise by the north American plate during the Tertiary resulting in pervasive crustal extension and rifting. An alternative hypothesis has been expressed by Christiansen and Lipman (1972) who attribute the development of Cenozoic peralkaline volcanism to changes in plate behaviour after the collision of the East Pacific Rise with a mid-Tertiary continental-margin trench causing direct contact of the American and western Pacific plates along a right-lateral transform fault system.

II.6.3. Petrography

II.6.3.1. Granites

Most of the peralkaline granites are leucocratic, holocrystalline, hypidiomorphic and medium-grained, with potash-feldspar, albite, quartz, aegirine and arfvedsonite-riebeckite as the principal minerals. Aenigmatite, astrophyllite and biotite are minor but important accessory minerals; fluorite, pyrochlore, cryolite and elpidite are rare but occasionally rise to major proportions. There are also assemblages of hastingsite, catophorite, magnesioriebeckite, sodic hedenbergite, and fayalite pseudomorphs in the less peralkaline types. Typical modal data for a selection of peralkaline granites from various regions are given in Table 1.

Two types of perthites are usually present: a low albite-orthoclase perthite which may represent earlier formed feldspar, and a low albite-microcline perthite partly or completely replacing the orthoclase perthite. In general, peralkaline granites contain abundant patchy replacement microcline perthite with microcline obliquity values in the range 0·75 to 1·00. There is a variable amount of antiperthite with rims of

TABLE 1 Modal Analyses of some Peralkaline Granites

	1	2	3	4	5	6
Quartz	32·9	93·6 (Quartz + Feldspar)	31·8	40	25	39·4
Feldspar	57·0		62·5	50	37	54·0
Alkali amphibole	8·70 (Alkali amphibole + Pyroxene)	3·9	3·0	9	9	5·5
Pyroxene		—	0·5	1	25	—
Zircon		0·1	tr	—		—
Astrophyllite	2·41 (Zircon + Astrophyllite + Ore, etc.)	2·4	2·0	—	4 (Zircon + Astrophyllite + Ore, etc.)	1·1
Ore, etc.		—	—	—		—

dashed line—mineral absent. All values in vol. %.

Legend:
1. Mungeria, W. Rajasthan India, Murthy and Venkataraman (1964. Table 3, column 2).
2. Kungnat, S. Greenland, Upton (1960, p. 79).
3. Ekerite, Oslo, Norway. Dietrich *et al.*, (1965, Table 2, analysis E2).
4. Corsica, Murthy and Venkataraman (1964. Table 3, column 12, original analysis by Quin, 1962).
5. Rockall, North Atlantic, Sabine (1960, Table 3, specimen 2).
6. North Conway, New Hampshire, U.S.A., Chapman and Williams (1935, Table 1).

coarsely twinned albite (An~5) between adjacent crystals. Albite twinning may also be developed in the sodic phase forming the host.

Late stage albitization may be locally severe with a large proportion of deuteric albite (An ~ 4) replacing earlier formed feldspar and amphibole. Working with the system albite–water at pressures of 1 to 10 Kb and temperatures of 200–700 °C, Martin (1969) found that in the presence of sodium silicate low albite might be expected to form most readily in the temperature interval 300–400 °C. Thus the occurrence of low albites in peralkaline granites may be used to suggest that hydrothermal modification, by solutions of albitic-acmitic composition, has been an important process.

The coarse grained peralkaline granites contain up to about 40% of modal quartz as clear anhedral and sometimes strained crystals. In most granites of this type quartz is obviously of late formation and frequently appears to have replaced microcline perthite. However, in the associated less peralkaline types the proportion of free quartz is lower but a high temperature-bipyramidal quartz is often present.

II.6.3.2. Volcanics

In hand specimen comendites vary from coarse grained, pale blue to greenish-grey banded lavas with individual bands up to 3 cm across, to fine grained dark greenish-grey glassy rocks with alternating yellowish-green and dark grey layers between 1 mm and 10 mm thick. Eutaxitic texture is very common. Comenditic obsidian is also locally present occurring as frozen skins covering comendite flows, interbedded with lavas, and as glassy lenses in comenditic ash-flow tuffs, and welded vitroclastic tuffs. Typical modal analyses for three comenditic porphyritic obsidians are presented in Table 2.

In thin section the textural features of comendites range from hyaline and hypohyaline to microgranular, and may be spherulitic, porphyritic or aphyric. Sodic-sanidine and quartz are major phenocrysts, but they may exhibit partial corrosion and resorption. Fayalite (Fa 98–100), hedenbergite-aegirine, arfvedsonite-riebeckite, and ferrohastingsite are only rarely recorded as microphenocrysts in comendites. The groundmass frequently contains a second generation of the phenocryst assemblage as microlites associated with one or more of the following: aenigmatite, cristobalite, tridymite, brown hornblende (? catophorite, barkevikite), magnetite or hematite, often set in a vitreous to microfelsitic or granophyric matrix. Comenditic obsidians which have not devitrified are extremely rare (e.g. modal analyses 2 and 3), for instance almost all the comendites from the type area in Sardinia

TABLE 2 Modal Analyses of some Peralkaline Porphyritic Obsidians

	1	2	3	4	5	6
Phenocrysts						
Quartz	6·0	—	—	0·8	—	0·7
Feldspar	9·3	0·6	<1	10·7	3·3	5·5
Pyroxene	—	tr.	tr.	0·2	tr.	0·3
Aenigmatite	—	tr.	—	0·8	—	0·3
Amphibole	—	—	—	—	—	1·8
Opaques	—	—	tr.	—	tr.	—
Tridymite	0·3	—	—	—	—	—
Vitreous* groundmass	84·3	99·4	>99	87·5	96·7	91·7

* Refers to major part of groundmass.
tr. trace, dashed line—mineral absent. All values in vol. %

Legend:
1. Comendite, Le Fontane quarry, San Pietro Island, Sardinia, Chayes and Zies (1962, Table 6 sample No. 40B5).
2. Comendite, Mayor Island, New Zealand, Ewart *et al.*, (1968, Table 4 sample No. P29532). (Some confusion exists over nomenclature of Mayor Island obsidians—see text).
3. Comendite, Midhyma, Iceland, Bailey and Macdonald (1970, Table 1 sample No. 295).
4. Pantellerite, Gelkhamar. Isle of Pantelleria, Chayes and Zies (1962, Table 6 sample No. PRC 2000).
5. Pantellerite, Fantale, Ethiopia, Dickinson and Gibson (1972, Table 4, sample No. Y 486).
6. Pantellerite, Lake Naivasha, Kenya, Nicholls and Carmichael (1969, Table 1, sample No. 121R).

(Garbarino and Maccioni, 1968) have a partially or completely microcrystalline groundmass. Occasionally anorthoclase is recorded instead of sodic sanidine, with aenigmatite as major phenocrysts in some comendites. Such an assemblage appears to be more characteristic of pantellerites than of comendites.

The pantellerite spectrum also embraces lavas, tuffs and obsidians. In hand specimen pantellerites may therefore exhibit similar features to the comendites, except that pantellerites are often darker in colour, ranging from green to black, and trachytic flow textures are more common. The type specimens from Pantelleria (e.g. Carmichael, 1962) and similar volcanics from other regions are porphyritic obsidians with phenocrysts of lime-poor anorthoclase. Frequently corroded quartz, sodic hedenbergite and aenigmatite are associated with anorthoclase in the phenocrystic assemblage. These minerals may also be accompanied by fayalite (Fa 98–100) and arfvedsonite-riebeckite. Ilmenite is a rare microphenocryst. The groundmass may show a variable proportion of brown glass ($N \sim 1{\cdot}510$) with a second generation of the phenocryst assemblage as microlites, and associated with brown hornblende (? catophorite, barkevikite) and rare magnetite, biotite and apatite. Ferrorichterite has also been reported as a phenocryst and groundmass mineral in pantellerite from Lake Naivasha, Kenya (Nicholls and Carmichael, 1969).

The frequency of occurrence of pantelleritic obsidians is far greater than for the comenditic obsidians. This observation may be related to the higher proportion of normative sodium metasilicate in the pantellerites which inhibits devitrification. Typical modal analyses of these volcanics are presented in Table 2.

It is obvious that feldspar is the principal precipitating early mineral phase in peralkaline silicic volcanic rocks. On the whole the feldspar phenocrysts are not zoned and have a restricted range in composition. For the comendites, values range from $Or_{60}Ab_{40}$ to $Or_{35}Ab_{65}$ (Garbarino and Maccioni, 1968) but the range narrows to $Or_{39}Ab_{61}$–$Or_{33}Ab_{67}$ for some pantellerites (Nicholls and Carmichael, 1969), with the feldspar microlites slightly more enriched in potassium than the associated phenocrysts in the quartz-bearing assemblage. It should be noted that quartz may be absent as a phenocrystic mineral in pantellerites, but it is ubiquitous in the comendites with cristobalite and more frequently tridymite in the devitrified groundmass. Chalcedony has also been found in a comendite which contained zeolite-filled cavities.

II.6.4. Geochemistry

II.6.4.1. Major Elements

The interpretation of chemical data from peralkaline silicic rocks is subject to many problems. For instance if we accept that peralkaline silicic glasses (obsidians) could represent samples of magmatic liquids, then we must remember that such glasses can readily hydrate, losing alkalis and other petrogenetically important constituents (VI.2.4; Noble, 1968; Macdonald, 1969). According to Taylor (1968) hydrated obsidians and perlites from western U.S.A. have D/H and O^{18}/O^{16} isotopic compositions that indicate that the glasses have not simply absorbed water but have exchanged with large quantities of it. A similar problem must exist with epizonal granites (Taylor and Forester, 1971), where circulating groundwater may enrich, deplete and amend their chemical and isotopic constitution. Thus only by stringent scrutiny of analyses, and careful examination of thin sections under the microscope, for evidence of water interaction, hydration, devitrification, etc., can data on peralkaline silicic rocks give meaningful geochemical trends.

All available evidence suggests that non-hydrated glassy rocks best represent the composition of peralkaline silicic liquids immediately prior to solidification. Accordingly six selected analyses are presented in Table 3, two of which are of separated non-hydrated glasses. It should however be emphasized that analyses of glass separates cannot be truly equated with the original magmatic liquid, for the phenocryst composition is ignored. Nevertheless, provided chemical comparisons are made between the glasses themselves, meaningful trends can be established which can then be correlated with the porphyritic obsidians. Significant features of peralkaline silicic glassy rocks are extreme depletion in MgO and a somewhat erratic range of low CaO values (Noble *et al.*, 1969). Examination of Table 3 reveals that pantellerites are more enriched in Fe_2O_3, FeO, TiO_2 and MnO than comendites, and that they have a markedly higher agpaitic index due to an absolute increase in Na_2O and decrease in Al_2O_3, for K_2O stays reasonably constant. Silica averages 69·8% in the pantellerites whilst comendites yield a mean of 73·5%. The analysis of comendite from Sardinia (Table 3, column 1) suggests that the sample may not be completely fresh, and may have interacted with ground water.

For completeness, CIPW norms calculated for the six analyses in Table 3 are tabulated in Table 4. But the conventional way of plotting normative salic compositions in the Q–Or–Ab projection of the experimental system (Tuttle and Bowen, 1958) is of limited value for peralkaline compositions for it cannot show the degree of peralkalinity. Since SiO_2, Al_2O_3, $Na_2O + K_2O$ constitute more than 90% by weight of all peralkaline silicic rocks, Bailey and Macdonald (1969) have placed special emphasis on a triangular diagram using these components (Fig. 2) which may reveal trends consistent with feldspar fractionation (cf. VI.2.4).

When analyses of comendite obsidians (Bailey and Macdonald, 1970) and selected pantelleritic obsidian analyses from Macdonald and Bailey (1972) are plotted in Fig. 2, they separate into two distinct groups shown by the dotted regions. Further the comendite obsidians of undisputed continental origin and some pantellerite obsidians fall within a quartz-feldspar cotectic zone defined by the experimental boundaries Or–Silica and Ab–silica at 1 atmosphere (Schairer and Bowen, 1955 and 1956). The points group around quartz-feldspar minima, A, B and C determined in the $NaAlSi_3O_8$–$KAlSi_3O_8$–SiO_2–H_2O experimental system with added amounts of acmite and sodium metasilicate (Carmichael and Mackenzie, 1963). It is significant that comendites *sensu stricto* trend towards minimum A (least peralkaline) whilst type pantellerites group around minimum C (most peralkaline with

TABLE 3 Analyses of some Peralkaline Porphyritic Obsidians and Non-Hydrated Glass Separates

	1	2	3	4	5	6
SiO_2	75·31	75·8	75·25	69·25	69·81	70·13
ZrO_2	0·18	n.d.	0·09	0·78	0·25	0·22
TiO_2	0·21	0·09	0·13	0·25	0·45	0·30
Al_2O_3	10·43	12·14	11·99	9·10	8·59	7·97
Fe_2O_3	3·22	0·80	0·90	3·70	2·28	2·77
FeO	0·80	0·94	1·25	2·61	5·76	5·27
MgO	0·10	0·01	0·02	0·01	0·10	0·07
CaO	0·13	0·30	0·27	0·06	0·42	0·55
Na_2O	3·99	5·22	4·79	7·00	6·46	7·46
K_2O	4·65	4·48	4·67	4·29	4·49	4·24
H_2O+	0·51	0·12	0·15	0·14	0·14	0·54
H_2O-	0·38	0·14	0·05	0·01	0·05	0·06
P_2O_5	0·03	—	—	0·01	0·13	0·04
MnO	0·09	0·06	0·08	0·19	0·28	0·26
SO_3	tr.	n.d.	n.d.	0·08	0·06	n.d.
Cl	0·05	0·25	0·18	0·78	0·76	0·37
F	0·18	0·29	0·25	1·30	0·30	n.d.
	100·26	100·64	100·07	99·56	100·33	100·25
less O for F, Cl	0·09	0·18	0·14	0·72	0·30	0·08
Total	100·17	100·46	99·93	98·84	100·03	100·17
Agpaitic index	1·11	1·10	1·08	1·79	1·81	2·12
SiO_2	85·30	82·42	83·67	82·31	83·11	82·74
Al_2O_3	6·96	7·87	7·85	6·37	6·02	5·54
$Na_2O + K_2O$	7·74	8·71	8·47	11·31	10·86	11·72

tr. trace. n.d. not determined.

Legend:

1. Comendite, Sardinia, for details see Table 2, analysis 1, and Chayes and Zies (1962, Table 7).
2. Comendite, Iceland, for details see Table 2, analysis 3.
3. Glass separate, from porphyritic comendite obsidian, Nevada, Spearhead member of Thirsty Canyon tuff, Noble (1968, Table 1, sample No. D100737).
4. Glass separate from porphyritic pantellerite obsidian, Nevada, Gold Flat member of Thirsty Canyon tuff, Noble 1965 (Table 1, sample No. D100126).
5. Pantellerite, Isle of Pantelleria, for details see Table 2, analysis 4. Fluorine data taken from Noble and Haffty (1969, Table 2, column B), remainder of data from Chayes and Zies, (1962, Table 7).
6. Pantellerite, Kenya, for details see Table 2, analysis 6.

8·3% *ac* and 8·3% *ns*). Certain samples such as the Mayor Island obsidians, fall near minimum B (4·5% *ac* and 4·5% *ns*). Since statistical treatment has indicated that these obsidians are more closely linked to comendites it is suggested than comendites *sensu lato* may divide within the cotectic zone into a subgroup which became more peralkaline, denoted by those points near minimum B, and into a less peralkaline group near minimum A. Such associations are found in the type locality on San Pietro Island Sardinia (Garbarino and Maccioni, 1968).

Continental comendites fall within the cotectic zone indicating that they are quartz-feldspar minima compositions of increasing peralkalinity. In contrast oceanic comendites from spreading ridge crests, and from New Guinea, an island arc situation, lie between the quartz-feldspar cotectic zone and trachytic compositions. No such division can be made for the pantellerites since, although some pantellerites plot in Fig. 2 outside the cotectic zone, the majority of the pantelleritic obsidians used are from evolving continental rift systems and from Pantelleria.

TABLE 4 CIPW Norms for some Peralkaline Porphyritic Obsidians and Non-Hydrated Glass Separates

		1	2	3	4	5	6
Q		35·80	30·77	30·77	27·52	29·08	29·08
or		27·47	26·47	27·59	25·35	26·53	25·05
ab		27·76	37·50	35·67	22·92	19·19	17·39
ac		4·63	2·31	1·93	10·70	6·60	8·01
ns		—	0·08	—	2·93	3·90	7·25
di	wo	—	0·18	0·18	—	0·06	1·03
	en	—	—	—	—	—	0·02
	fs	—	0·20	0·19	—	0·06	1·14
hy	en	0·25	0·02	0·05	0·02	0·25	0·15
	fs	—	1·49	1·84	4·73	10·29	8·52
mt		2·26	—	0·34	—	—	—
hm		0·05	—	—	—	—	—
il		0·40	0·17	0·25	0·47	0·85	0·57
ap		—	—	—	—	0·31	0·09
Z		0·27	—	0·13	1·16	0·37	0·33
fl		0·18	0·60	0·51	0·08	0·62	—
hl		0·08	0·41	0·30	1·29	1·25	0·61

For key to samples 1–6 see Table 3.

Such samples analysed by Zies (1960) and Carmichael (1962) may lie within the cotectic zone surrounding minimum C, or trend away outside the cotectic zone towards trachytic compositions. Apart from the obvious differences in the relative amounts of SiO_2 and Al_2O_3, the most significant feature of these analyses is that the Pantellerian samples within the cotectic zone have higher contents of normative halite (*hl*), normative fluorite (*fl*) and normative sodium metasilicate (*ns*). Analyses of pantellerite obsidians and glass separates from other regions exhibit a similar relationship for *ns*, but the differences in *hl* and *fl* are not always diagnostic. Obviously the pantellerite trend in Fig. 2 requires a different interpretation.

II.6.4.2. Minor and Trace Elements

Fluorine and chlorine are important minor elements in peralkaline silicic rocks, but there are distinctive anomalies in their relative abundances. A study of the available analytical data reveals that fluorine is more common and widely distributed than chlorine. The chlorine content is higher in the type specimen of pantellerite (Lovering, 1966) than fluorine, but this observation is repeated in only a few other similar specimens. Romano (1969) has noted from his analyses of one pantellerite flow on Pantelleria that there was a concomitant increase in both sodium and chlorine from the nearly holocrystalline inner part to the outer vitreous crust. The effect of sodium chloride in the experimental system albite–water has been studied by Koster van Groos and Wyllie (1968) at approximately 900 °C and 4 kb pressure. It was found that although sodium chloride has little effect on the albite liquidus, the solubility of water is greatly increased. Further sodium chloride fractionates preferentially into the aqueous fluid instead of the silicate phase. Thus the high chlorine values recorded by Lovering may be due to an immiscible aqueous saline phase (Roedder and Coombs, 1967) distributed submicroscopically in pantellerite glass (cf. VI.5). However, fluorine is the more important constituent in peralkaline granites and in the majority of peralkaline obsidians. Significantly, sodium fluoride fractionates into the silicate phase in contrast to the trend shown by sodium chloride (VI.4; Kogarko, 1961). Such partitioning of fluorine and immiscibility of chlorine may

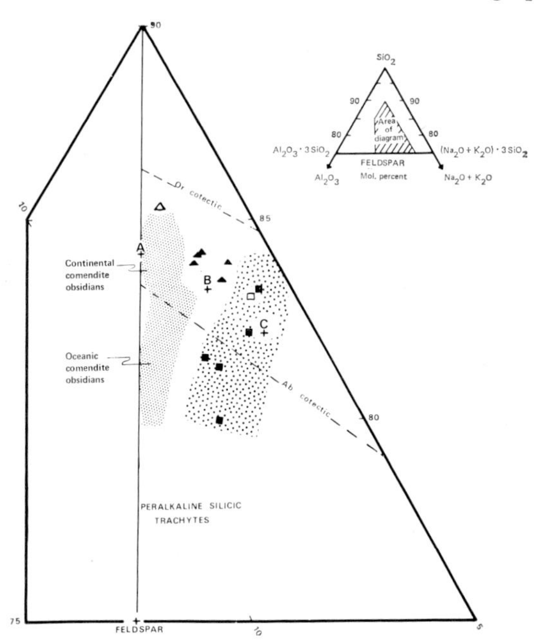

Fig. 2 Molecular SiO_2, Al_2O_3 and $Na_2O + K_2O$ for suites of comenditic and pantelleritic porphyritic obsidians and separated glasses

Area of enlarged portion of the triangular diagram is shown in the upper right corner.

Zone of widely spaced dots represents the region of pantellerites, selected from Macdonald and Bailey (1972). Comendite analyses published by Bailey and Macdonald (1970) plot within the zone of closely spaced dots. Mayor Island obsidians (Ewart *et al.*, 1968) are shown as closed triangles; pantellerite PRC 2000 from Pantelleria (Chayes and Zies, 1962) is represented as an open square, and comendite from Sardinia (Chayes and Zies, 1962) by an open triangle. Other analyses from Pantelleria (Carmichael, 1962; Zies, 1960; Noble and Haffty, 1969) are indicated by closed squares.

The quartz–feldspar cotectic zone is defined by the broken lines. The quartz–feldspar minima compositions A, B, and C are taken from Carmichael and Mackenzie (1963). Basic diagram is modified from Bailey and Macdonald (1970, Fig. 1, p. 345).

have an important bearing on the distribution of trace elements, for the most abundant trace elements in peralkaline silicic rocks are those which readily form fluoride complexes (e.g. Nb, Zr, Sn). These elements are also important economically, concentrating as pyrochlore and elpidite, and as columbite and cassiterite in associated non-peralkaline silicic rocks.

A selection of trace element data for some porphyritic obsidians and glass separates are presented in Table 5. Most elements, such as B, Ga, La, Nb, Sn, Y, Yb, and Zr are believed by many workers to represent rest concentrations developed through precipitation and separation of crystallizing phases. The depletion in Sr, Ba, V, Sc, Co and Ni testifies to the efficient removal of these elements by mafic and salic phenocrysts. For instance the wide variation in Ba values is related to the efficiency of feldspar removal. The contrast between the Ba values in the porphyritic obsidians (anal. 5 and 6) and an equivalent glass separate (anal. 4) strengthens this observation. A similar mechanism can be proposed for the Sr depletion.

Rare earth distributions have not been extensively studied in peralkaline silicic rocks but Ewart *et al.* (1968) have shown that the rare earth fractionation patterns in Mayor Island obsidians are dominated by a relative depletion in Eu, whilst in the trachybasalt inclusions there is a correspondingly relative enrichment in Eu (Fig. 3). Since Haskin *et al.* (1966) have shown that feldspars exhibit a preferential concentration of Eu (2+) compared with coexisting mafic

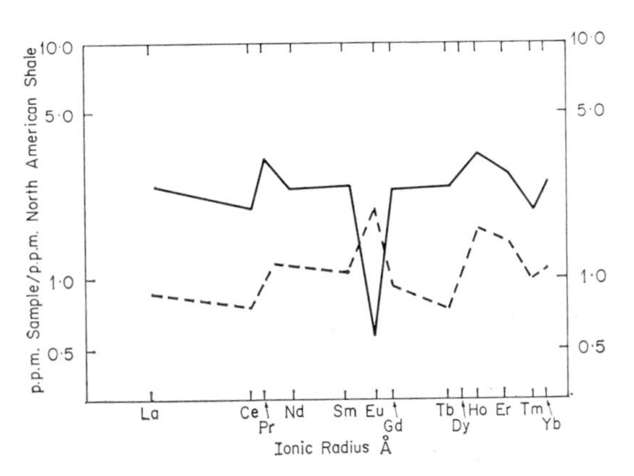

Fig. 3 Rare earth patterns relative to North American composite shale (Haskin *et al.*, 1966) for a Mayor Island obsidian P.29532, and a trachybasalt xenolith. Diagram adapted from Ewart *et al.* (1968, Fig. 9, p. 134)

Continuous line with europium depletion—peralkaline obsidian.

Dotted line with europium enhancement—trachybasalt.

TABLE 5 Trace Element Data for some Peralkaline Porphyritic Obsidians and Non-Hydrated Glass Separates

	1	2	3	4	5	6
B	40	n.d.	30	100	30	n.d.
Ba	10	8·6	<5	5	50	780
Be	10	8·2	6	56	11	11
Co	<5	n.d.	<2	<5	<5	<10
Cr	<1	n.d.	<1	3	<1	<10
Cs	n.d.	5·2	<5	10	n.d.	n.d.
Cu	8	5·6	<1	3	5	2·4
Ga	28	32	30	44	45	30
La	170	95	110	580	360	100
Li	n.d.	59	78	260	n.d.	30
Mo	2	12·6	5	9	22	3
Nb	150	69	70	670	450	145
Ni	2	n.d.	<2	5	3	<10
Pb	70	21	40	210	30	23
Rb	n.d.	134	390	950	n.d.	105
Sc	2	0·7	<2	7	4	<10
Sn	16	5·7	6	52	24	<10
Sr	6	0·3	2	1·5	7	19
Th	n.d.	17·4	$23{\cdot}5 \pm 0{\cdot}5$	n.d.	n.d.	n.d.
U	n.d.	5·1	$5{\cdot}31 \pm 0{\cdot}9$	n.d.	n.d.	n.d.
V	<5	n.d.	<5	<5	<5	<3
Y	140	145	73	490	220	110
Yb	14	9·2	6	55	20	10
δO^{18}	n.d.	n.d.	n.d.	7·0	n.d.	n.d.

All values in parts per million except for δO^{18} which is recorded in parts per mil. n.d. not determined.

Legend:
1. Comendite, Sardinia, see Tables 2, 3 and 4. Data from Noble and Haffty (1969, Table 1).
2. Comendite, New Zealand, see Table 2 analysis 2 for details.
3. Comendite glass separate. Spearhead member of Thirsty Canyon tuff, trace data determined on sample No. D100437, *ref.* Noble (1965, Table 2): uranium and thorium data from Rosholt and Noble (1969, sample No. Ttsu-035).
4. Glass separate from pantellerite, Nevada, see Table 3 and 4. Data from Noble and Haffty (1969, Table 1): δO^{18} value reported by Taylor (1968).
5. Pantellerite, Isle of Pantelleria, see Tables 2, 3 and 4. Data from Noble and Haffty (1969, Table 1).
6. Pantellerite, Fantale, see Table 2, analysis 5 for details.

phenocrysts, Ewart *et al.* (1968) consider that feldspar removed from the obsidian parent magma has been concentrated in the associated trachybasalt. Trace element data therefore appear to suggest that feldspar fractionation has been an important process in the generation of peralkaline silicic liquids.

II.6.5. Petrogenesis

When discussing the origin of peralkaline silicic liquids it first must be established whether peralkaline granites are the plutonic equivalents of peralkaline silicic lavas. Certainly the phenocryst and groundmass assemblages of peralkaline silicic volcanics can be equated mineralogically and chemically with the quartz syenites and associated granites of variable peralkalinity. The close association of some plutonic activity with explosive ignimbritic and ash-flow tuff eruptions is sufficient to suggest that similar mechanisms and similar sources are involved (Harris *et al.*, 1970). There is also a reported transition from peralkaline granite to comendite at Jebel Khariz, a dissected volcano in south Arabia (Gass and Mallick, 1968). Peralkaline granites may therefore be the plutonic equivalents of continental comendites.

Looking at the schematic spectrum of peralkaline magmatism (Fig. 4) there appears to be a continuous evolutionary process from early epeirogenic doming through to crustal separation, rifting of the continents, spreading at active ridge crests, and crustal extension near plate subduction zones. Any consistent petrogenetic model must explain the various peralkaline silicic products and their associated rock types. The models must also account for the petrochemical variations.

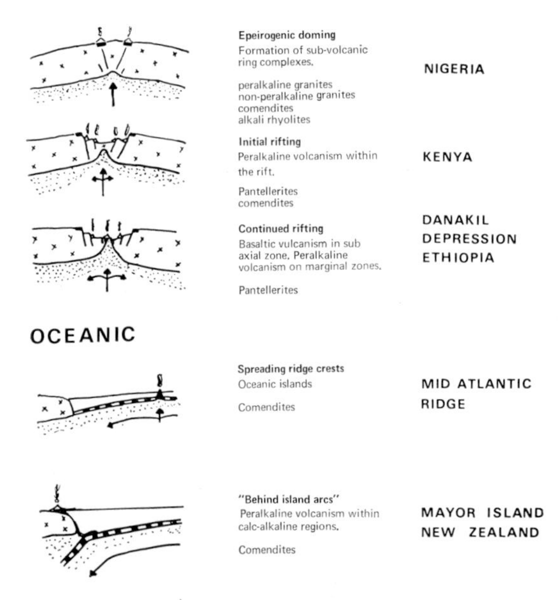

Fig. 4 Schematic evolution of peralkaline silicic magmatism

The sequential diagrams are modified from Tarling and Tarling (1971, pp. 102–3). Crosses represent continental crust, dotted zone the upper mantle; oceanic crust is shown as black and white stripes, and arrows indicate direction and intensity of doming and spreading.

Examples of various stages in the schematic evolution are given in the right-hand column.

If current models are accepted then the generation of peralkaline silicic liquids, in areas of crustal attenuation such as in the floors of continental rift systems, at the edges of oceanic rifts, and at spreading ridge crests, must be closely linked with events in the upper mantle. In contrast, regions of thicker continental crust may provide a ready source for peralkaline silicic rocks intruded into elevated plateaux and domes. The end products are similar, the sources are different. So too might be the mechanism of generation, for in areas of crustal attenuation fractionation from a mildly alkaline basalt magma is probably more likely, whereas progressive melting within a thick crust could be the dominant control for producing peralkaline silicic liquids in a continental environment (cf. VI.1b). Both processes are not entirely exclusive and may operate in regions, with a crust of intermediate thickness, and type.

Some indication of this division has been obtained by Bailey and Macdonald (1970) who studied a suite of comenditic obsidians collected both from oceanic and continental environments (Fig. 2). They concluded that the oceanic peralkaline specimens had a compositional spread consistent with a derivation via a trachytic stem from a basaltic parent, but that continental comendites were generated by partial melting within the continental crust (cf. VI.2). Peralkaline granites are believed to have been derived in a similar way to the continental comendites.

From both field and laboratory studies it is assumed by many workers that feldspar fractionation is a critical process in the origin of pantelleritic liquids. However Macdonald *et al.* (1970) have found that obsidians ranging from soda trachyte to pantellerite, erupted in the central part of the East African rift, cannot be explained in terms of simple fractional crystallization nor in terms of crystal $\rightleftharpoons$ liquid equilibria. But if there was participation of an alkali-bearing vapour phase this might allow a cooling path from trachyte to rhyolite to be obtained.

There are also a number of anomalies in both the field observations and the geochemistry which cannot be entirely reconciled with the concept of feldspar fractionation as a differentiation process for the genesis of pantellerites. At Pantelleria, Villari (1970) has clearly stated that two eruptive silicic cycles began with pantellerite and ended with soda trachytes. Peralkaline ash-flow tuff units examined by Noble (1965) are strongly peralkaline at their base and become more soda trachytic at the top of the unit. The ash-flow tuff at Fantale (Gibson, 1970) is another similar example, all of which strongly suggest that pantelleritic liquids may have arisen simply by the concentration of volatiles, causing enrichment in *ns*, near the top of the magma chamber. There is no need to envoke fractional crystallization or any of the current models available in the literature. Pantellerites may simply be sodium silicate volatile-concentrates of an original soda trachytic magma. (Upton, 1960; cf. IV.3). If eruption is quiet and slow then pantelleritic lavas could be continually reproduced by volatile rise. However, should the magma unit empty quickly then the transitions recorded by Villari, Noble and Gibson of pantellerite towards sôda trachyte will be produced. Such sequential events will be shown diagrammatically in Fig. 2 as a trend outside the cotectic zone towards trachytic compositions, depending on the degree of volatile concentration. An elucidation of this concept might be achieved by a deeper understanding of the behaviour of various isotopes.

For instance, there are isotopically and chemically zoned ash-flow tuff units (Noble and Hedge, 1969; Gibson, 1970) which in certain regions appear to be an important rapidly erupted phase associated with the development of large calderas. Also the relative enrichment of anorthoclase phenocrysts in Sr^{87}/Sr^{86} compared with the glass matrix at Fantale (Dickinson and Gibson, 1972) is further possible proof of isotope zonation in an evolving peralkaline magma chamber. It should however be emphasised that isotopes of the same element do not fractionate. This has been proved by a study of boron isotopes B^{11}/B^{10} which have a large mass difference by which any fractionation would be detected. Exhaustive measurements on samples from Fantale yielded a constant isotopic abundance ratio of 4·06 (Bowden, unpublished data). Nevertheless peralkaline lavas which become increasingly pantelleritic with time do show progressive enrichment in Rb/Sr and Sr^{87}/Sr^{86}. A similar relationship appears to hold for obsidians from Pantelleria (Barberi *et al.*, 1969) and Ethiopia (Barberi, *et al.*, 1970) presented in Table 6. Although the variations are attributed to inhomogenieties in the mantle source rocks, to

TABLE 6 Rb–Sr Isotope Data for some Recent Pantellerites and Comendites

Sample and Locality	Rb(ppm)	Sr(ppm)	Rb/Sr	Rb^{87}/Sr^{86}	Sr^{87}/Sr^{86}	Reference
Pantellerite (P8) Porto di Nika, Pantelleria	173	13	13·3	38	0·7027	Barberi *et al.* (1969).
Pantellerite (P9) Cala Tramontana, Pantelleria	198	14	14·1	41	0·7067	
Pantellerite (CH15) P. Pruvost, Ethiopia	139	10·5	13·2	38·2	0·7110	Barberi *et al.* (1970).
Pantellerite (G15) P. Pruvost, Ethiopia	130	6·4	20·3	58·5	0·7158	
Pantellerite (Y408) Fantale, Ethiopia	119	13·9	8·6	24·8	0·7048	Dickinson and Gibson (1972).
Pantellerite (Y360) Fantale, Ethiopia	151	11·0	13·7	39·6	0·7060	
Pantellerite (Y335) Fantale, Ethiopia	167	6·2	26·9	78·2	0·7074	
Comendite (K87) Jebel Khariz, S. Arabia	85·0	82·5	1·0	1·36	0·7045	Dickinson *et al.* (1969).
Comendite (JB331) Jebel umm Birka, S. Arabia	100	129	0·8	2·23	0·7045	

crustal contamination or to progressive melting in the crust, the variations may be related to the degree of volatile concentration prior to eruption.

To summarize, it is obvious that there is an urgent need for a consistent petrogenetic model to explain all the observed facts concerning peralkaline silicic rocks. However, the concept of volatile transfer and concentration is not new and may be the mechanism responsible for the variations seen over the whole peralkaline spectrum.

II.6. REFERENCES

Bailey, D. K., and Macdonald, R., 1969. Alkali-feldspar fractionation trends and the derivation of peralkaline liquids. *Am. J. Sci.*, **267,** 242–8.

Bailey, D. K., and Macdonald, R., 1970. Petrochemical variations among mildly peralkaline (comendite) obsidians from the oceans and continents. *Contr. Mineral. Petrology*, **28,** 340–51.

Barberi, F., Borsi, S., Ferrara, G., and Innocenti, F., 1969. Strontium isotopic composition of some recent basic volcanites of the southern Tyrrhenian sea and Sicily channel. *Contr. Mineral. Petrology*, **23,** 157–72.

Barberi, F., Borsi, S., Ferrara, G., Marinelli, G., and Varet, J., 1970. Relations between tectonics and magmatology in the northern Danakil Depression (Ethiopia). *Phil. Trans. R. Soc. Lond.*, **A 267,** 293–311.

Bryan, W. B., 1969. Alkaline and peralkaline rocks of Soccorro Island, Mexico. *Ann. Rept. Director Geophys. Lab. Carnegie Inst. Washington*, Yearbook, **68,** 194–200.

Carmichael, I. S. E., 1962. Pantelleritic liquids and their phenocrysts. *Mineral. Mag.*, **33,** 86–113.

Carmichael, I. S. E., and Mackenzie, W. S., 1963. Feldspar–liquid equilibria in pantellerites: an experimental study. *Am. J. Sci.*, **261,** 382–96.

Chapman, R. W., and Williams, C. R., 1935. The evolution of the White Mountain magma series. *Am. Miner.*, **20,** 502–30.

Chayes, F., and Zies, E. G., 1962. Sanidine phenocrysts in some peralkaline volcanic rocks. *Ann. Rept. Director Geophys. Lab. Carnegie Inst. Washington*, Yearbook **61,** 112–18.

Christiansen, R. L., and Lipman, P. W., 1972. Cenozoic volcanism and plate-tectonic evolution of the Western United States. II. Late Cenozoic. *Phil. Trans. R. Soc. Lond.*, **A 271,** 249–84.

Dickinson, D. R., Dodson, M. H., Gass, I. G., and Rex, D. C., 1969. Correlation of initial $^{87}Sr/^{86}Sr$ with Rb/Sr in some late tertiary volcanic rocks of South Arabia. *Earth Planet Sci. Lett.*, **6,** 84–90.

Dickinson, D. R., and Gibson, I. L. (1972). Feldspar fractionation and anomalous $^{87}Sr/^{86}Sr$ ratios in a suite of peralkaline silicic rocks. *Bull. Geol. Soc. Am.*, **83,** 231–40.

Dietrich, R. V., Heier, K. S., and Taylor, S. R., 1965. Petrology and geochemistry of ekerite: *Studies on the Igneous Rock Complex of the Oslo region*, **20,** 33 pp.

Ewart, A., Taylor, S. R., and Capp, A. C., 1968. Geochemistry of the pantellerites of Mayor Island, New Zealand. *Contr. Mineral. Petrology*, **17,** 116–40.

Garbarino, C., and Maccioni, L., 1968. Volcanic rocks of San Pietro Island (South west Sardinia) I Comendites. *Period. Mineral. Roma*, **37,** 895–983.

Gass, I. G., and Mallick, D. I. J., 1968. Jebel Khariz: an Upper Miocene strato-volcano of comenditic affinity on the South Arabian coast. *Bull. volcanol.*, **32,** 33–88.

Gibson, I. L., 1967. Preliminary account of the volcanic geology of Fantale, Shoa, Ethiopia. *Bull. Geophys. Obs. Addis Ababa*, **10,** 59–67.

Gibson, I. L., 1970. A pantelleritic welded ash-flow tuff from the Ethiopian rift valley. *Contr. Mineral. Petrology*, **28,** 89–111.

Harris, P. G., Kennedy, W. Q., and Scarfe, C. M., 1970. Volcanism versus plutonism—the effect of chemical composition. *Geol. J.* (Special Issue), **2,** 'Mechanism of Igneous Intrusion', edited by G. Newall and N. Rast, The Seal House Press, Liverpool, 187–200.

Haskin, L. A., Frey, F. A., Schmitt, R. A., and Smith, R. H., 1966. Meteoritic, solar and terrestrial rare-earth distributions. *Phys. Chem. Earth*, **7,** 167–321.

Koch, P., 1955. Pantellerites of Mount Mba Nsché (Camerouns). *Compt. Rend. Acad. Sci. Paris*, **241,** 893–5.

Kogarko, L. N., 1961. Conditions of formation of villiaumite in nepheline syenites (Lovozero massif). *Geochemistry*, **1961,** 84–7.

Koster Groos, A. F. K., van., and Wyllie, P. J., 1968. Melting relationships in the system, $NaAlSi_3O_8$–NaF–H_2O to 4kb pressure, *J. Geol.*, **76,** 50–70.

Lacroix, A., 1927. Les rhyolites et les trachytes hyperalcalins quartzifères a propos de ceux de la Corée. *Compt. Rend. Acad. Sci. Paris*, **185,** 1410–15.

Le Bas, M. J., 1971. Peralkaline volcanism, crustal swelling and rifting. *Nature*, **230,** 85–6.

Lovering, J. F., 1966. Electron microprobe analysis of chlorine in two pantellerites. *J. Petrology*, **7,** 65–7.

Macdonald, R., 1969. The petrology of alkaline dykes from the Tugtutoq area, South Greenland. *Bull. Geol. Soc. Denmark*, **19,** 257–82.

Macdonald, R., Bailey, D. K., and Sutherland, D. S., 1970. Oversaturated peralkaline glassy trachytes from Kenya. *J. Petrology*, **11,** 507–17.

Macdonald, R., and Bailey, D. K., 1972. The chemistry of peralkaline oversaturated obsidians. *U.S. Geol. Surv. Prof. paper*, **440–N,** part 1.

Martin, R. F., 1969. The hydrothermal synthesis of low albite. *Contr. Mineral. Petrology*, **23,** 323–39.

McKenzie, D. P., 1970. Plate tectonics of the Mediterranean Region. *Nature*, **226,** 239–43.

Mohr, P. A., 1970. Volcanic composition in relation to tectonics in the Ethiopian rift system: a preliminary investigation. *Bull. volcan.* **34,** 141–57.

Murthy, M. V. N., Venkataraman, P. K., 1964. Petrogenetic significance of certain platform peralkaline granites of the world. *The Upper Mantle Symposium, New Delhi*, 127–49.

Nicholls, J., and Carmichael, I. S. E., 1969. Peralkaline acid liquids: a petrological study. *Contr. Mineral. Petrology*, **20,** 268–94.

Noble, D. C., 1965. Gold Flat member of the Thirsty Canyon tuff—a pantellerite ash-flow sheet in Southern Nevada. *U.S. Geol. Surv. Prof. paper* **525–B,** B85–B90.

Noble, D. C., 1968. Systematic variation of major elements in comendite and pantellerite glasses. *Earth Planet Sci. Lett.*, **4,** 167–72.

Noble, D. C., Chipman, D. W., and Giles, D. L., 1968. Peralkaline silicic volcanic rocks in Northwestern Nevada. *Science*, **160,** 1337–8.

Noble, D. C., and Haffty, J., 1969. Minor-element and revised major-element contents of some Mediterranean pantellerites and comendites. *J. Petrology*, **10,** 502–9.

Noble, D. C., Haffty, J., and Hedge, C. E., 1969. Strontium and magnesium contents of some natural peralkaline silicic glasses and their petrogenetic significance. *Am. J. Sci.*, **267,** 598–608.

Noble, D. C., and Hedge, C. E., 1969. Sr^{87}/Sr^{86} variations within individual ash-flow sheets. *U.S. Geol. Surv. Prof. paper* **650–C,** C133–C139.

Parks, J. M., 1966. Cluster analysis applied to multivariate geologic problems. *J. Geol.*, **74,** 703–15.

Quin, J. P., 1962. La lindinosite (granite mésocrate à riebeckite) du massif d'Evisa (Corse). *Bull. Soc. Géol. France*, 7th ser, **4,** 380–3.

Roedder, E., and Coombs, D. S., 1967. Immiscibility in granitic melts, indicated by fluid inclusions in ejected granitic blocks from Ascension Island. *J. Petrology*, **8,** 417–51.

Romano, R., 1969. Sur l'origine de l'excès de sodium (*ns*) dans certaines laves de l'Île de Pantelleria. *Bull. volcan.*, **33,** 694–700.

Rosholt, J. N., and Noble, D. C., 1969. Loss of uranium from crystallized silicic volcanic rocks. *Earth Planet. Sci. Lett.*, **6,** 268–70.

Sabine, P. A., 1960. The geology of Rockall, North Atlantic. *Bull. Geol. Surv. Gt. Br.*, **16,** 156–78.

Schairer, J. F., and Bowen, N. L., 1955. The system $K_2O–Al_2O_3–SiO_2$. *Am. J. Sci.*, **253,** 681–746.

Schairer, J. F., and Bowen, N. L., 1956. The system $Na_2O–Al_2O_3–SiO_2$. *Am. J. Sci.*, **254,** 129–95.

Sundius, N., 1946. The classification of the hornblendes and the solid solution relations in the amphibole group. *Arsbok. Sveriges Geol. Undersok.*, **40,** no. 4, 36 pp.

Tarling, D. H., and Tarling, M. P., 1971. *Continental Drift*, G. Bell and Sons Ltd., London, 112 pp.

Taylor, H. P. Jr., 1968. The oxygen isotope geochemistry of Igneous rocks. *Contr. Mineral. Petrology*, **19,** 1–71.

Taylor, H. P. Jr., and Forester, R. W., 1971. Low–O^{18} igneous rocks from the intrusive complexes of Skye, Mull and Ardnamurchan, Western Scotland. *J. Petrology*, **12,** 465–97.

Tazieff, H., Marinelli, G., Barberi, F., and Varet, J., 1969. Géologie de l'Afar Septentrional. Première expedition du CNRS-France et du CNR-Italie (Décembre 67–Février 68). *Bull. volcan.*, **33,** 1039–72.

Tuttle, O. F., and Bowen, N. L., 1958. Origin of granite in the light of experimental studies in the system $NaAlSi_3O_8–KAlSi_3O_8–SiO_2–H_2O$. *Geol. Soc. Am. Mem.*, **74,** 1–153.

Upton, B. G. J., 1960. The alkaline igneous complex of the Kungnat Fjeld, South Greenland. *Medd. Grønland*, **123,** No. 4. 145 pp.

Villari, L., 1970. The caldera of Pantelleria. *Bull. volcan.*, **34,** 758–66.

Vincent, P. M., 1970. 'The evolution of the Tibesti volcanic province, eastern Sahara', in *African Magmatism and Tectonics*, T. N. Clifford and I. G. Gass, eds., Oliver and Boyd, Edinburgh, 301–19.

Vine, F. J., 1970. The Geophysical Year. *Nature*, **227,** 1013–17.

Zies, E. G., 1960. Chemical analyses of two pantellerites. *J. Petrology*, **1,** 304–8.

II.7. ALKALINE GNEISSES

P. Floor

II.7.1. Introduction

A discussion of alkaline gneisses in this book is justified since they reflect the behaviour of alkaline rocks under metamorphic and/or metasomatic conditions quite different in certain respects from those of magmatic crystallization.

II.7.1.1

In order to cover all alkaline rocks called *gneiss* in the literature, the following definition had to be adopted: gneiss in this chapter is a leucocratic metamorphic, metasomatic or igneous rock with feldspar as an essential component and characterized by one or more of the following structural features:

1. a banded appearance, caused by alternations of rock bands with different mineralogical compositions;
2. a foliated appearance, caused by alternations of thin laminae of dark and light minerals within one rock type;
3. (sub)parallel orientation of one or more mineral species.

II.7.1.2

Depending on their alumina content *alkaline* gneisses may be alumina-undersaturated (peralkaline, with Na-amphibole, Na-pyroxene) or alumina-oversaturated (with biotite, Ca-amphibole, Ca-pyroxene).

Metasediments may have peralkaline characteristics, containing, e.g., aegirine and a sodic amphibole. Generally, however, these rocks are feldspar-poor and therefore not covered by the definition of gneiss given above. Such *peralkaline schists* will not be dealt with in this chapter (see Fischer and Nothaft, 1954, and Miyashiro, 1967, for this subject).

II.7.1.3

The *first record* of alkaline gneiss known to the author is that of Macpherson (1881), describing 'glaucophane-bearing gneissic syenite' in the Vigo area, Spain. It would appear later that the glaucophane is actually osannite, the optical variety of riebeckite, first described by Hlawatsch (1906) from Cevadais, Portugal. The latter occurrence of alkaline gneiss has been called 'the first known representative of the alkaline plutonic rocks in the facies of the crystalline schists' by Rosenbusch (1898; p. 484), though he mentions the gneisses near Vigo as well.

II.7.1.4

The following explanations of the *mode of origin* of alkaline gneisses were found in the literature:

1. Penetrative deformation and metamorphic recrystallization of alkaline igneous rocks. Most rocks were gneissified after their consolidation (e.g. Floor, 1966); in a few instances synmetamorphic intrusion has been envisaged (e.g. Richardson, 1968).
2. Metasomatic transformation of pre-existing (meta)sedimentary and (meta)igneous rocks (Sturt and Ramsay, 1965).
3. Flow-orientation in igneous rocks (e.g. Payne, 1968).
4. Syn- to late-kinematic crystallization of igneous rocks (e.g. Marmo *et al.*, 1966). The available data are insufficient to decide whether the gneissic appearance was caused by deformation of already crystallized minerals before complete consolidation of an intrusive melt (protoclasis) or by oriented crystallization of the whole melt under stress conditions.
5. Isochemical metamorphic recrystallization of appropriately composed sediments (e.g. Bloomfield, 1968).

In most alkaline gneisses recrystallization has largely or completely obliterated deformation textures and formed granoblastic textures instead. The recrystallized minerals often have preferred optical or shape orientations with respect to the deformation pattern of the rocks, thus demonstrating that the recrystallization was actually metamorphic. Fabric analyses of metamorphic nepheline, albite, aegirine and biotite were made by Koark (1961), of nepheline and scapolite by Sturt and Appleyard (in Appleyard, 1969), of nepheline by Sturt (1961) and Gellatly (1964) and of quartz by Phadke (1967). Gellatly discusses the differences between the orientations of magmatic albite and nepheline and of metamorphic nepheline.

Subsequent deformations, with or without metamorphic recrystallization, occasionally destroyed prior metamorphic recrystallization structures indicative of the origin of the gneisses to such an extent that they have lost much of their diagnostic value.

II.7.1.5

The *nomenclature* adopted in this contribution depends on the origin of the rock. The name of the original igneous rock with the suffix gneiss is given to rocks of proven igneous origin. If a gneiss does not derive from an igneous rock or is of uncertain origin, it is called two-feldspar, perthite or plagioclase gneiss depending on the kind of feldspar assemblage, with the addition of 'quartz' or 'nepheline' when present.

II.7.1.6

The *age* of alkaline gneisses has not been especially considered. Since their formation, as a rule, did not take place at very shallow depths, it is to be expected that they are more frequently found in relatively old, more deeply eroded, environments. The occurrences mentioned in this chapter are of Paleozoic or older age. Peralkaline schists, frequently formed under glaucophanitic facies conditions, may also be younger.

II.7.1.7

Most alkaline gneisses discussed have no *economic value*; only the Blue Mountain complex, Eastern Ontario, Canada, is of considerable economic interest (Allen and Charsley, 1968).

II.7.1.8

The occurrences known to the author are listed in Table 1. Unfortunately, it was impossible to acquire a sufficiently detailed knowledge of gneisses in the Soviet Union and China within the time available for the preparation of this contribution. According to Dr. A. S. Pavlenko (written communication, 1968), the three principal areas in the U.S.S.R. are: the Kola peninsula, the Ukrainian shield and the Aldan shield.

Peralkaline gneisses are distinguished by locality names in italics in Table 1.

In order to facilitate selection of literature by the reader according to his special interests, a classification is given after each reference. Not all the literature on each occurrence is given in Table 1. In case of differing opinions concerning the modes of origin of alkaline gneisses those given in the corresponding column are based on the most recent references, unless stated otherwise.

II.7.1.9

The *chemical compositions* of alkaline gneisses strongly depend on their history. Those of metamorphosed igneous alkaline rocks are identical to their unmetamorphic analogues (compare Tables 2, 3 with those elsewhere in this book). The compositions of alkaline gneisses derived from (meta)sediments with or without metasomatism or from (meta)igneous rocks with the participation of metasomatic processes vary widely since they are determined by the original rock compositions and the amount and characteristics of metasomatic transformation. A few examples are given in Table 4.

II.7.1.10

In the subsequent sections only orthogneisses and metasomatic gneisses will be separately dealt with, as they alone provide data of special interest within the scope of this book.

TABLE 1. Occurrences of Alkaline Gneisses
(for explanation see sections II.7.1.5, II.7.1.8 and II.7.4)

	Locality	Rock type
Europe		
Austria	*Gloggnitz*	aeg–rieb rhyolite-gneiss (perthite relics present)
Finland	*Kiihtelysvaara*	?catapleite?–astrophyllite-aeg-alk.amph-neph syenite-gneiss
	Otanmäki	aeg–rieb granite-gneiss
Norway	Sørøy	(aeg.aug-hast-) bi-neph two-feldspar gneiss, bi two-feldspar gneiss; occasionally bi-neph syenite-gneiss (perthite relics present)
Portugal	*Cevadais*, *Arronches*, *Vaiamonte–Monforte* (Prov. Alto Alentejo)	rieb–aeg (quartz-) syenite- and granite-gneisses; aeg-neph syenite-gneiss (a few outcrops at Cevadais only; perthite relics present)
	Alter Pedroso, *various occurrences W and S of Elvas* (Prov. Alto Alentejo)	gneissic parts in predominantly massive (aeg-) rieb (-quartz) syenite to granite (perthite relics present in gneissic parts)
Scotland	*Carn Chuinneag* (Ross and Cromarty)	aeg–rieb granite-gneiss
	Glen Dessarry (Invernesshire)	zoned body of bi syenite-gneiss with aeg.aug (core), aeg.aug and hbl (intermediate zone) and hbl (marginal zone) (perthite relics present)
	Glen Lui (Aberdeenshire)	quartz-free aeg two-feldspar gneiss
Spain	*Sierra de Galiñeiro* (Vigo)	(astrophyllite-) aeg–rieb granite-gneisses and bi-ferrohast granite-gneiss; bi-ferrohast two-feldspar gneiss

Enclosing Rocks (Grade of Regional Metamorphism)	Published Mode of Origin	Authors	
Metasediments (greenschist facies)	Metamorphic peralkaline rhyolite	Phadke (1967) Zemann (1951) Cornelius (1951) Keyserling (1903)	12457 12457 7 12347
Boulder	Movement during crystallization (protoclasis)	Eskola and Sahama (1930)	23457
Granite-gneisses	Late-kinematic peralkaline granite	Marmo *et al.* (1966)	123456
Metagabbro and metasediments (mainly in almandine amphibolite facies)	Mainly metasomatic transformations caused by highly fluid neph syenite pegmatite magma; a few deformed and recrystallized pegmatite bodies. Transformation and intrusion took place during Caledonian orogeny	Sturt and Ramsay (1965) Sturt (1961)	123456 28
Metasediments (greenschist and amphibolite facies)	Lower Paleozoic intrusives, deformed and recrystallized by Hercynian orogeny. Locally postcrystalline deformation	Teixeira and Torre de Assunção (1958) Aires Barros (1958) Teixeira and Torre de Assunção (1956) Osann (1907) Hlawatsch (1906) Gonçalves (1971) Torre de Assunção and Gonçalves (1970)	1248 189 189 123458 35 127 12347
Metagabbro and metasediments (greenschist and amphibolite facies)	Lower Paleozoic intrusives only locally deformed with or without recrystallization by Hercynian orogeny	Neves Correia (1959) Serralheiro (1957) Teixeira and Torre de Assunção (1956) Burri (1928)	245 189 189 123456
Biotite granite-gneiss (almandine amphibolite facies)	Pre- or syn-metamorphic Caledonian intrusion	Long (1964) Harker (1962)	6 123458
Metasediments (almandine amphibolite facies)	Pre- or syn-metamorphic intrusion; metamorphism of pre-Caledonian (Moine) age	Richardson (1968)	123456
Metasediments	Moine schists transformed by Na-metasomatism from Cairngorm granite pegmatites	McLachlan (1951)	123456
Metasediments (cordierite amphibolite facies)	Ordovician peralkaline and alkaline granites deformed and metamorphosed by Hercynian orogeny; contact-metasomatic paragneisses	Floor (1966)	123456

TABLE 1. Occurrences of Alkaline Gneisses (*continued*)

	Locality	Rock type
	Sierra de Corzón (Prov. La Coruña)	(astrophyllite-) aeg–rieb granite-gneisses
	Several occurrences between Malpica and Noya (Prov. La Coruña)	bi-ferrohast granite-gneisses
Sweden	Almunge alkaline complex (Uppsala area)	schistose 'canadite': (hast-) bi (-cancr-) neph syenite
	Mölndal (Gothenburg area)	quartz-rich alk.amph. two-feldspar gneiss
	Norra Kärr district (Gränna, Småland)	'grennaite': (eudialyte-) catapleiite-aeg-neph syenite-gneiss; 'kaxtorpite': eckermannite-aeg syenite-gneiss ± pectolite and neph
Africa		
Algeria	*In Hihaou* (Ahaggar, W of Tamanrasset)	aeg–rieb granite-gneiss
Angola–Zair	*Noqui–Matadi*	aeg–rieb granite-gneiss (perthite relics present)
Cameroons	*Mount Bollo* (S of Garoua)	(aeg-) rieb granite-gneiss
Ghana	Kpong–Somanya (NNE of Accra) and Dufo–Jirawde (NE of Accra)	banded rieb-bi-aeg.aug-ab-neph gneisses
Malagasy Republic	Ianakafy (Betroka)	(di-) hbl perthite gneiss
	Makaraingobe massif (Morafenobe)	aeg.aug-hast-neph two-feldspar gneiss
Malawi	*Mchinji–Lilongwe road–Ngara Hill*	discontinuous belt of outcrops of banded quartz-rich aeg two-feldspar gneisses
	Ncheu	bi-neph-oligoclase[1] gneisses ± hast

Enclosing Rocks (Grade of Regional Metamorphism)	Published Mode of Origin	Authors	
Metasediments (amphibolite facies)	Prehercynian peralkaline granites deformed and metamorphosed by Hercynian orogeny; postcrystalline second deformation	Parga–Pondal (1967) Parga–Pondal (1956)	49 89
Metasediments and granite-gneisses (amphibolite facies)	Prehercynian alkaline granites deformed and metamorphosed by Hercynian orogeny; postcrystalline second deformation	den Tex and Floor (1967) Parga–Pondal (1967)	89 49
Metasediments, migmatites and granites, partly fenitized and mobilized into umptekite	Metamorphic deformation of nepheline syenite (ring) dykes	Gorbatschev (1961) Koark (1961)	12346 29
Metasedimentary series between granites and migmatites	Meta-arkose, metasomatically enriched in sodium	Lundegårdh (1953)	179
Granite to granite-gneiss, schistose gneiss and quartzdioritic hornblende gneiss	Metamorphic deformation of (nepheline) syenites	Koark (1969) Koark (1961) Adamson (1944)	1246 26 123456
Metasediments (amphibolite facies)	Metamorphic deformation of peralkaline granite	Le Fur (1966)	89
Metasediments	Recrystallized sheared peralkaline granite (Holmes); the gneissic texture (of uncertain origin) has been observed only very locally in the Noqui peralkaline granite (Korpershoek)	Korpershoek (1964) Mortelmans (1948) Holmes (1915)	12348 1247 12348
Biotite-amphibole gneisses (amphibolite facies)	Orthogneiss	Koch (1959) Koch (1955)	89 89
Acid gneisses	Not stated	Allen and Charsley (1968)	489
Together with oriented syenite and migmatic syenite in gneiss and migmatite	Metapelites enriched in K, Na and P during sedimentation	Megerlin (1968)	128
Migmatites	Metasomatic transformation of older migmatic rocks along dislocation zones	Welter (1964)	12347
Metasediments (almandine amphibolite facies)	Metamorphic glauconite-rich feldspathic sandstones	Bloomfield (1968) Bloomfield (1967)	1247 1247
Metasediments	Metamorphic analcime-bearing sediments	Bloomfield (1968) Bloomfield (1965)	1247 489

TABLE 1. Occurrences of Alkaline Gneisses (*continued*)

	Locality	Rock type
	Nsanje	bi-hast-neph syenite gneisses with igneous texture and perthite bi-neph syenite gneisses
	South Vipya plateau (NE of Mzimba)	quartz-rich rieb–aeg two-feldspar gneiss
	Tambani (NW of Chikwawa)	bi- (ms-) neph syenite gneiss
Nigeria	Shaki	zoned two-feldspar gneiss-body with di + bi (core), di + amph + bi (intermediate zones) and amph (marginal zones)
Somali Republic	Darkainle complex (NW of Hargeisa)	A. (alk.amph-aeg.aug-) green bi (-neph) syenite gneiss
		B. Banded brown bi-neph-albite gneisses + bi-rich (cancr-) plag. gneisses; homogeneous bi-rich cancr/neph plag. or two-feldspar gneisses
Tanzania	*Kilonwa* (Zoissa area, NE of Dodoma)	(Mg.arfv–aeg-) bi syenite-gneiss (perthite relics present)
	Lungolo (Handeni area)	'mariupolite gneiss': bi-calcite-neph-albite gneiss
	Mbozi (NE of Tunduma)	aeg.aug-amph two-feldspar gneiss (± quartz or neph)
Upper Volta	Several occurrences near *Yabo* (N of Ouagadougou)	aeg granite- and syenite-gneiss
Asia		
India	Kishangarh (Rajasthan)	banded amph-neph two-feldspar gneisses
North Korea	*Pockchinsan district* (S of Wonsan)	rieb, aeg and bi-bearing perthite + two-feldspar gneisses ± quartz or nepheline
North Vietnam	*Pia-Ma massif* (NW of Bac Kan)	(aeg-) hast-neph syenite-gneiss (perthite relics present)
South Yemen	Socotra island	rieb[2] granite-gneiss (perthite relics present)

Enclosing Rocks (Grade of Regional Metamorphism)	Published Mode of Origin	Authors	
Metasediments (amphibolite facies)	Presumedly igneous origin	Bloomfield (1968) Bloomfield (1956)	1247 127
Metasediments against nepheline syenite	Contact Na-metasomatic transformation of psammite	Bloomfield (1967) Bloomfield (1965)	89 9
Metasediments (amphibolite facies?)	Regionally metamorphosed nepheline syenite body, possibly synkinematic in age	Bloomfield (1968) Cooper and Bloomfield (1961)	1247 12457
Biotite–muscovite-bearing basement gneisses	Transformation of basement gneiss by potash feldspathization and Ca, Fe, Mg metasomatism	Oyawoye (1967)	123456
Predominantly metarhyolites and (semi) pelitic biotite schists	A. Intrusive mass with tectonic foliation; undeformed primary foliation only locally preserved. B. Contact-metasomatic transformation of metasediments; solutions mainly formed through hydrothermal alteration of nepheline in magmatic syenite, possibly partly as residuum of nepheline syenite magma.	Gellatly and Hornung (1968)	1246
Sheared synorogenic granite and migmatites	Recrystallized sheared peralkaline syenite	Kempe (1968)	12457
Granulites	Presumedly igneous origin	Kempe (1968)	89
Associated with Mbozi syenite-gabbro complex in metasediments	Metasomatic transformation of quartzo-feldspathic metasediments, probably by volatile-rich solutions from deeper-seated neph syenite	Brock (1968)	12346
Metasediments	Oriented border facies of confined post-tectonic intrusives; in one case syn-tectonic granite	Ducellier (1963)	89
Metasediments and diorite-gneiss (greenschist facies and almandine amphibolite facies)	Syntectonic metasomatic transformation of gabbro; alkaline fluids probably derived from granitizing source	Niyogi (1966)	123456
Metasediments (greenschist facies and almandine amphibolite facies)	Not settled; arguments in favour of both igneous and metasomatic origin	Miyashiro and Miyashiro (1956)	123456
Sediments, metasediments, garnet-bearing amphibolites	Recrystallized deformed nepheline syenites	Lacroix (1928) Bourret (1922)	248 89
Metasediments	Not stated Granite-gneiss	Bichan (1968) Bichan (written comm; 1968)	9

TABLE 1. Occurrences of Alkaline Gneisses (*continued*)

	Locality	Rock type
North America		
Canada	Haliburton–Bancroft, Peterborough, etc. areas (eastern Ontario)	A. banded neph plag gneisses and some two-feldspar gneisses with amph, bi, clino-pyroxene and garnet as predominant dark minerals
		B. Bi–ms (-neph) syenite-, bi-neph syenite- and (aeg.aug-) hbl-neph syenite-gneisses (perthite present)

[1] oligoclase partly antiperthitic, considered by Bloomfield (1968; p. 133) as a hypersolvus assemblage.
[2] according to observations by the present author on a specimen kindly put at his disposal by Dr. R. Bichan and the Dept. of Earth Sciences, the University of Leeds, probably a Mg-riebeckite with γ = b: violet, β : yellow, α : greenish blue, and a small $\alpha \Lambda$ c.

Key to literature classification:
1. Field data
2. Petrographical data ⎰ other than mere mention
3. Mineralogical data ⎱ of constituent minerals
4. Chemical analyses of rocks
5. Chemical analyses of minerals
6. Well-documented analysis of rock-history
7. History discussed but not documented with many data
8. History only slightly dealt with or not mentioned
9. Only short mention of alkaline gneiss occurrence

II.7.2. Orthogneisses

II.7.2.1 Peralkaline Orthogneisses

The major province of peralkaline orthogneisses, taken into consideration in this chapter, is situated in the western part of the Iberian peninsula. The related age, composition and situation within the Hercynian orogen of the Portuguese and Spanish occurrences of this province, listed separately in Table 1, render a similar or even common parentage highly probable. Chemical analyses of peralkaline orthogneisses from the Iberian and some other occurrences are given in Table 2.

Most *Portuguese* occurrences are located in the province of Alto Alentejo (Fig. 1). One has recently been discovered in Trás-os-Montes (A. Ribeiro, personal communication). Their position within the Hercynian orogen is certainly a more superficial one than that of the N.W. Spanish gneisses, for several of them are less deformed, and the deformed ones less recrystallized than those in Spain. The grade of regional metamorphism, not precisely known, probably ranges into the amphibolite facies, but is generally weaker (Teixeira, 1956; pp. 12–14). Fig. 1 shows that the less deformed peralkaline rocks (Alter Pedroso and occurrences S and W of Elvas) intruded into gabbro complexes, which are themselves intrusive into Cambrian limestones. It is tentatively suggested here that the gabbros protected the peralkaline rocks against penetrative deformation.

Syenites predominate, but quartz syenites and

Enclosing Rocks (Grade of Regional Metamorphism)	Published Mode of Origin	Authors	
High grade Grenville metasediments	A. Most authors: metasomatic recrystallization of Grenville metasediments caused by highly fluid nepheline syenite magma; often posterior metamorphism. According to Appleyard, the transformations are synchronous with the orogenic period that caused deformation and metamorphic recrystallization of the metasomatic rocks.	Appleyard (1969)	1236
		Appleyard (1967)	126
		Gittins (1961)	123456
		Tilley (1958)	123456
		Moyd (1949)	126
		Gummer and Burr (1946)	1246
		Adams and Barlow (1910)	123458
	B. Intrusive masses with flow foliation and cataclastic deformation; metamorphic recrystallization in cases only limited	Payne (1968)	1246
		Fairbairn (1941)	28

Legend:

ab	albite	ferrohast	ferrohastingsite
aeg	aegirine	hast	hastingsite
aeg. aug	aegirine-augite	hbl	hornblende
alk. amph	alkaline amphibole	mg. arfv	magnesio-arfvedsonite
bi	biotite	ms	muscovite
cancr	cancrinite	neph	nepheline
di	diopside	rieb	riebeckite

granites are not uncommon at Vaiamonte–Monforte, near Elvas and at Cevadais. Nepheline gneisses are known from Cevadais only. They carry albite, nepheline, microcline, aegirine, ore, cancrinite, biotite and (locally) sodalite. Alkali amphibole, present in all other rocks, is notably lacking in this nepheline gneiss. Most occurrences show a very wide variation in colour index, grain size, structure and—in the ocurrences W and S of Elvas—dark mineral content. Peralkaline pegmatites are not rare. Special mention should be made of the veins up to 30 cm wide, consisting almost entirely of riebeckite, in the Alter Pedroso complex (Burri, 1928).

The syenites from Alter Pedroso offer good examples of the gradual transition of igneous rocks via cataclastic ones (Fig. 2a) into metamorphic gneisses. Mesoperthites of the syenites are locally bent, granulated along their margins, and broken. In some rocks the granulated mass has recrystallized into fine-grained mosaics of albite and microcline, in others recrystallization has been only weak. The gneisses at Cevadais, representing more advanced stages of deformation and recrystallization, are granoblastic two-feldspar gneisses. Their foliation is seen to bend around the scarce fragments of perthite that escaped granulation.

The peralkaline granite-gneisses in the *Vigo area* are much more homogeneous. Their bulk composition, colour index, grain size and structure vary only little. They are well-foliated and fine- to medium-grained with a granoblastic texture throughout the complex.

In the same area, biotite and biotite-ferrohastingsite gneisses with similar textures are intimately associated with a hybrid series of melanocratic rock fragments enclosed in a great variety of meso- and leucocratic plagioclase-rich, generally quartz-poor rocks. This series has been

TABLE 2 Chemical Analyses of Peralkaline Orthogneisses

	A	B	C	D	E	F	G	H	I	K
SiO_2	76·03	74·66	73·80	75·52	71·03	65·88	63·12	59·52	57·46	64·28
TiO_2	0·10	0·32	0·23	0·28	0·37	tr	0·10	tr	0·10	0·38
Al_2O_3	11·74	8·85	11·90	10·24	11·38	16·03	16·40	21·24	22·06	17·33
Fe_2O_3	2·44	3·26	1·90	1·44	3·56	2·56	3·33	2·71	2·15	2·63
FeO	0·65	3·54	1·91	3·20	1·49	1·84	2·70	0·48	1·24	1·16
MnO	0·04	tr	0·12	0·07	0·05		0·08	tr	0·03	0·03
MgO	0·04	0·09	0·33	0·13	1·19	0·29		0·12	0·12	0·22
CaO	0·11	0·53	0·30	0·45	0·83	0·25	1·14	0·48	1·68	0·03
Na_2O	4·74	3·68	5·05	3·95	4·55	7·44	7·11	10·72	9·75	7·36
K_2O	4·07	4·46	4·93	4·30	4·40	4·66	5·68	3·92	4·80	5·47
H_2O^+	0·28	0·67	0·13	0·38	0·69	0·34	0·57	0·50	0·45	0·73
H_2O^-	0·04	0·08	0·04							0·11
P_2O_5				0·01	0·05	0·02	0·17		0·30	0·18
ZrO_2		0·51	0·04			0·45		0·16		
CO_2								0·21		
Li_2O				0·24	0·26					
	100·28	100·65	100·68	100·21	100·30	99·76	100·40	100·06	100·14	99·91

Legend:

A. Strongly mylonitized aegirine-riebeckite gneiss, Gloggnitz, Austria (Zemann, 1951).
B. Gneissic riebeckite-aegirine granite, Noqui, Angola (Holmes, 1915).
C. Gneissic aegirine-riebeckite granite, Carn Chuinneag, Scotland (in Harker, 1962).
D. Astrophyllite- and lepidomelane-bearing riebeckite gneiss, Vigo, Spain (Floor, 1966).
E. Astrophyllite-bearing aegirine-riebeckite gneiss, Vigo, Spain (Floor,1966).
F. Quartz-bearing Na-amphibole gneiss, Cevadais, Portugal (Osann, 1907).
G. Riebeckite-aegirine orthogneiss, Cevadais, Portugal (Texeira and Torre de Assunção, 1958).
H. Aegirine-bearing nepheline gneiss, Cevadais, Portugal (Osann, 1907).
I. Gneissic nepheline syenite, Cevadais, Portugal (Teixeira and Torre de Assunção, 1958).
K. Sheared aegirine-arfvedsonite-biotite syenite, Kilonwa, Tanzania (Kempe, 1968).

interpreted by Floor (1966; pp. 181–3) as gabbro, intruded and fragmented by peralkaline granite and subsequently metamorphosed.

Metasediments in the area generally have a greywacke composition and a low-pressure amphibolite facies mineral assemblage with andalusite or cordierite. Metasediments overlying the peralkaline, the biotite and the biotite-ferrohastingsite gneisses, however, contain ferrohastingsite, biotite and some microcline. Floor (1966; pp. 177–9) supposes them to be normal metasediments, transformed by Na (and K) metasomatic emanations from the peralkaline rocks at the time of their Hercynian recrystallization (and possibly also of their igneous crystallization).

Calc-alkaline granite gneisses abound in a narrow zone from the Vigo area northwards. They are roughly coeval with the alkaline gneisses, which reappear in this zone in the province of La Coruña (Fig. 1). Here, however, a late Hercynian deformation acted strongly, destroying all prior metamorphic recrystallization textures (den Tex and Floor, 1967).

II.7.2.2. The Behaviour of Peralkaline Rocks During Metamorphism

Data on the behaviour of peralkaline rocks during metamorphism are scanty and not very precise. The chemical compositions of igneous and metamorphic amphiboles and pyroxenes, including trace elements, have not yet been studied in a sufficiently critical way to provide reliable conclusions. Redistribution of certain elements during metamorphic recrystallization may have taken place, but has only rarely been

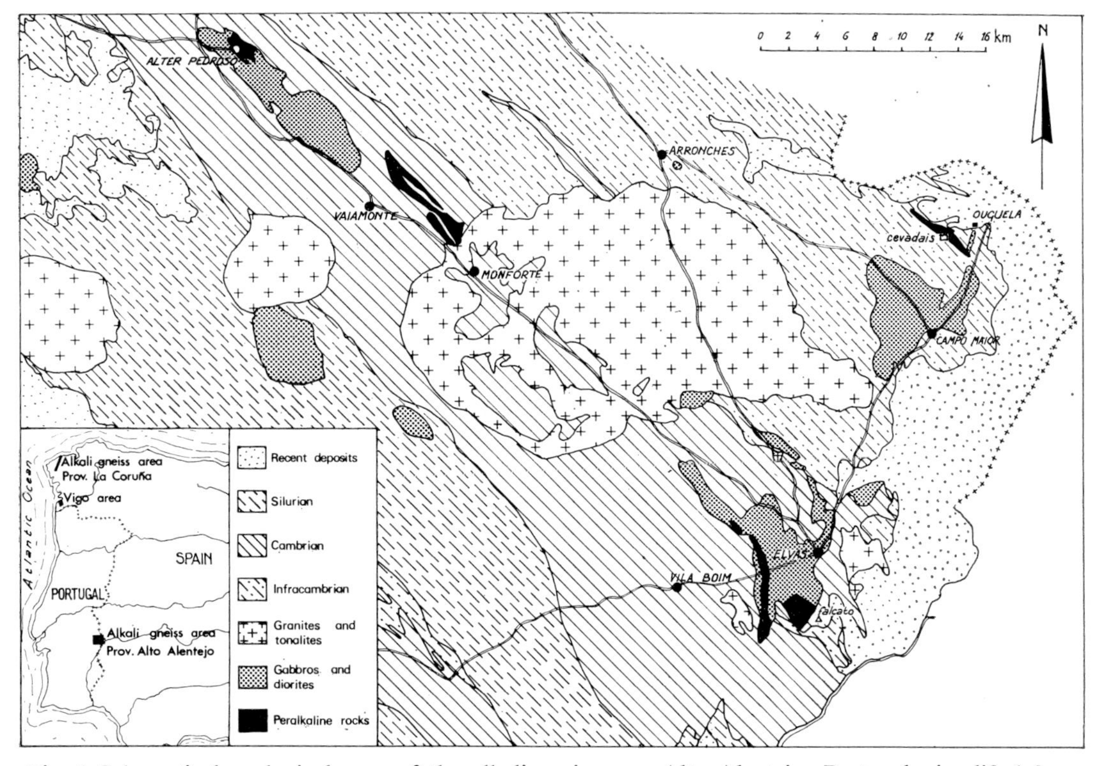

Fig. 1 Schematical geological map of the alkali gneiss area Alto Alentejo, Portugal, simplified from Teixeira and Torre de Assunçao (1956). Location of the occurrences of the Iberian alkaline gneiss province in inset

demonstrated beyond doubt. The following data are available:

1. *Feldspar*. As exemplified by the Portuguese syenites and orthogneisses, the perthites of peralkaline igneous rocks are bent, broken and marginally or completely granulated during their gneissification. When metamorphism accompanies the cataclastic destruction, the granulated masses recrystallize into separate phases of albite and little or non-perthitic potash feldspar, whereas perthite relics (present where specifically mentioned in Table 1) remain perthitic (Figs. 2 a and b; see also Section II.7.4). The microcline from the Vigo area, Spain, has been shown to be a nearly pure, maximum triclinic microcline, so in that area the separation into the albite and microcline phases has been almost complete (Floor, 1966; pp. 41–2). The Vigo albite has a notable porphyroblastic to poikiloblastic habit. Local occurrence of albite of similar habit has been described by Zemann (1951; p. 12) in more massive varieties of the gneiss near Gloggnitz, Austria. Floor (1966; pp. 154–5) suggests that fluorine (abundantly present as fluorite in the Spanish gneisses) has influenced both size and crystallization temperature of albite.

2. *Dark constituents*. Structural and textural relations demonstrate that riebeckite, aegirine, biotite and astrophyllite may (re)crystallize under regional metamorphic conditions. Undeformed astrophyllite in gneiss near Vigo even occurs at right angles to the foliation (Floor, 1961). Ernst's study of the stability of artificial riebeckite and riebeckite-arfvedsonite solid solutions suggests that metamorphic riebeckitic amphiboles may turn out to be purer end-member riebeckites than igneous ones (Ernst, 1962). Before using analyses given in the

Fig. 2(a) Incipient granulation of perthitic alkali feldspar with only limited recrystallization. Cataclastic riebeckite syenite, N. of Alter Pedroso, Portugal. Nearly crossed polarizers, 30×

Fig. 2(b) Granular albite and microcline surrounding perthite relics. Grains to the right of the upper perthite clearly have an elongated shape, demonstrating the cataclastic origin of the present recrystallized texture. Kilonwa syenite-gneiss, Tanzania. Crossed polarizers, 35×

literature to test this suggestion, or before analyzing new material, it is essential to verify, among other things, whether the amphibole is metamorphic, an igneous relic, or whether both generations occur together. Amphibole in joints cutting the gneissic foliation is certainly metamorphic but may have crystallized well after the culmination of regional metamorphism. The first 'osannite' described (Hlawatsch, 1906) occurs in that way. Riebeckite in coarse quartz-riebeckite veins at Carn Chuinneag has about the same trace element composition as the country-rock riebeckite. According to Harker (1962; p. 228) this evidences 'that the coarser material segregated from the bulk of the rock during total recrystallization in the period of regional metamorphism'.

3. *The mobility of elements during regional metamorphism.* In the Vigo area, the presence of albite porphyroblasts (and some microcline) in paragneisses adjacent to peralkaline gneiss demonstrates the migration of sodium (and some potassium) away from the peralkaline rocks, probably after the complete separation of the albite and microcline phases (Floor, 1966; p. 176). The peralkaline gneiss near the contacts became impoverished in Na and contains magnetite instead of riebeckite or aegirine (Floor, 1966; p. 160). Magnetite-bearing gneisses were also noted near Gloggnitz

(Zemann, 1951; p. 11) and Alter Pedroso (Burri, 1928; p. 398), but sodium enrichment of adjacent rocks has not been recorded from these localities.

Again in the Vigo area, some indications of the mobility of zirconium have been found. Partial analyses of aegirine and riebeckite suggest that their ZrO_2 contents are lower than in igneous aegirine and riebeckite. It is possible that metamorphic alkali amphiboles and pyroxenes cannot accommodate as much Zr (and probably also other elements) as can igneous ones. Newly formed zircons often have a bipyramidal habit, just as in igneous peralkaline rocks. Apparently the alkaline environment strongly influences both habit and crystallization temperature of zircon (Floor, 1966; p. 156).

II.7.2.3. Alumina-Oversaturated Orthogneisses

The alumina-oversaturated syenite complex at Glen Dessarry, Scotland, consists of two syenite types, the older of which is richer in dark constituents. Both types originally contained one alkali feldspar, aegirine-augite, biotite, magnetite, titanite, apatite, (calcite, allanite and zircon). Richardson (1965; pp. 159–62) argues that the syenites intruded more or less synchronously with, rather than before, Moine folding and metamorphism. The petrology of the original rocks and the modification of the Glen Dessarry complex under the influence of deformation and regional metamorphism after its intrusion has been thoroughly discussed and well illustrated by Richardson (1968).

The igneous textures of the syenites were destroyed by granulation of the original mesoperthites and crystallization of separate orthoclase and plagioclase; linear gneisses resulted. The orthoclase phases in the perthites and in the unmixed granular aggregates both have a monoclinic structure and the same composition, though in the surrounding staurolite and kyanite or sillimanite-bearing schists muscovite is stable. Pyroxene has been altered into hornblende under the influence of a fluid phase entering the complex. This fluid phase, possibly containing hydrogen in excess, also effected an overall reduction of the original syenites. There is a clear relation between the extent to which the modifications took place and the contours of the complex, the least affected rocks occurring in the centre. Some chemical analyses of Glen Dessarry syenites and syenite gneisses are given in Table 3.

TABLE 3 Chemical Analyses of Alumina-Oversaturated Orthogneisses

	A	B	C	D
SiO_2	53·63	54·97	61·30	60·43
TiO_2	1·33	1·22	0·71	0·69
Al_2O_3	13·75	13·96	17·68	17·45
Fe_2O_3	7·15	2·45	1·97	1·85
FeO	4·18	5·92	2·21	2·60
MnO	0·22	0·19	0·08	0·10
MgO	2·74	4·92	1·51	1·27
CaO	7·72	6·97	2·11	3·98
Na_2O	4·25	3·37	5·70	4·31
K_2O	4·15	3·87	6·13	6·47
H_2O^+	0·35	0·85	0·63	0·35
H_2O^-	0·05	0·01	0·04	0·02
P_2O_5	0·88	0·81	0·34	0·26
CO_2	0·24	0·11	0·16	0·04
	100·64	99·62	100·57	99·82

Legend:

A. Mafic biotite-bearing pyroxene syenite with mainly igneous texture and perthite $>$ granular K. feldspar and plagioclase, Glen Dessarry, Scotland (Richardson, 1968).

B. Mafic biotite-hornblende syenite, mainly granulated with perthite $\ll$ granular K. feldspar and plagioclase, Glen Dessarry, Scotland (Richardson, 1968).

C. Leucocratic pyroxene-bearing hornblende-biotite syenite with mainly igneous texture and perthite $>$ granular K. feldspar and plagioclase, Glen Dessarry, Scotland (Richardson, 1968).

D. Leucocratic biotite-hornblende syenite, mainly granulated with K.feldspar and plagioclase, Glen Dessarry, Scotland (Richardson, 1968).

The Glen Dessarry complex is very similar to the Shaki complex, Nigeria, although a completely different origin has been proposed for the latter (see Table 1). Unlike Glen Dessarry, however, clinopyroxene at Shaki overgrew amphibole.

TABLE 4 Chemical Analyses of Metasomatic Nepheline Gneisses

	A	B	C	D	E	F	G
SiO_2	58·97	57·53	49·59	48·52	44·19	44·17	39·45
TiO_2	0·11	0·36	0·47	0·15	0·35	0·58	0·70
Al_2O_3	23·58	21·74	23·32	28·62	27·09	15·26	19·48
Fe_2O_3	0·60	1·33	1·44	0·51	1·74	2·73	2·37
FeO	1·58	2·12	4·32	1·30	5·37	11·77	6·13
MnO	0·12	0·11	0·11	0·02	0·19	0·32	0·13
MgO	0·14	0·46	1·18	0·35	0·32	1·20	2·95
CaO	0·51	1·88	5·68	8·93	3·90	12·75	13·64
Na_2O	10·76	8·91	9·92	7·59	11·99	7·62	9·00
K_2O	3·00	3·73	2·35	2·92	2·73	2·19	1·96
H_2O^+	0·50	0·75	0·72	0·67	0·97	0·55	0·93
H_2O^-	0·07	0·13	0·03	0·11	nil	0·08	0·10
P_2O_5	0·00	0·12	0·38	0·04	0·13	0·21	0·81
CO_2	0·06	0·79	0·59	0·06	1·05	1·08	2·20
rest						0·18	
	100·00	99·96	100·10	99·79	100·02	100·69	99·85

Legend:

A. Nepheline gneiss in metasediment, Sørøy, Norway (Sturt and Ramsay, 1965).

B. Nepheline gneiss in metagabbro, Sørøy, Norway (Sturt and Ramsay, 1965).

C. Hornblende-nepheline-albite gneiss, York River, Ontario, Canada (Tilley, 1958).

D. Nepheline - microcline - andesine - scapolite - biotite - garnet gneiss, Monmouth township, Ontario, Canada (Gittins, 1961).

E. Hornblende-(biotite)-plagioclase-nepheline-(calcite) gneiss, Monmouth township, Ontario, Canada (Gittins, 1961).

F. Hedenbergite-nepheline gneiss, York River, Ontario, Canada (Tilley, 1958).

G. Hornblende - nepheline - (plagioclase) - (calcite) gneiss, Monmouth township, Ontario, Canada (Gittins, 1961).

C, D, E, F and G are considered to represent nephelinized limestones.

II.7.3. Metasomatic Gneisses

Some alkaline gneisses are merely the product of metasomatism acting on already existing gneissic or banded rocks. Their gneissic character is not the result of the processes that made them into alkaline rocks. Well-known examples are the nepheline gneisses of the Haliburton–Bancroft area, Canada, though even some of these also underwent modification through subsequent regional metamorphic recrystallization (Gittins, 1961). Appleyard (1967a,b; 1969), however, considers some Canadian occurrences, like the Norwegian one (see below), as syn-orogenic.

Other alkaline gneisses, e.g., at Sørøy, Norway, and Kishangarh, India, were formed as a result of metasomatic transformations accompanying deformation and regional metamorphism, but in some cases their gneissic structure may be inherited as well. At Sørøy, a great variety of rocks has been formed due to the fact that the transformations took place prior to, synchronous with or later than the second phase of Caledonian deformation.

The metasomatic processes are often seen to take place along zones of weakness (contacts of contrasting rock types, shear zones, faults, etc.). Emanations from nepheline syenite magma, very rich in volatiles, are thought to have caused metasomatism in Norway, Canada, Somali Republic and Tanzania. In these cases intrusive nepheline syenite is found in the same area, but a direct relation could not always be demonstrated. At Sørøy and Darkainle, carbonatites are present as well. Sturt and Ramsay (1965; p. 103) and Gellatly and Hornung (1968; pp. 688–9) have clearly demonstrated that the metasomatic processes outlined above are not related with the carbonatite. Some chemical analyses of metasomatic alkaline gneisses are listed in Table 4.

II.7.4. General Characteristics of Alkaline Gneisses

An alkaline rock recrystallizing into a gneiss with or without the participation of metasomatic processes, forms minerals of which species and composition are determined by the bulk composition of the rock at the moment of recrystallization and the conditions of metamorphism (e.g., T, P_{total}, P_{H_2O}, P_{O_2}, etc.). This causes a certain convergence in the production of alkaline gneisses as far as their mineralogical composition is concerned. The following features have been noted in most alkaline gneisses:

1. They usually carry a subsolvus feldspar assemblage: separate plagioclase and potash feldspar. The latter is only slightly or non-perthitic.
2. Hydrated dark minerals (amphiboles, biotite) are more common than anhydrous ones (pyroxenes).

Enclosing rocks show that many alkaline gneisses underwent amphibolite facies metamorphism, the remainder greenschist facies metamorphism. Their unfavourable chemical composition renders determination of metamorphic grade in the rocks themselves difficult. Most peralkaline gneisses are too low in calcium to permit crystallization of plagioclase more calcic than albite. Alumina-saturated or over-saturated alkaline rocks, on the other hand, generally contain more Ca; the anorthite content of plagioclase may then be used as an indicator of metamorphic grade. The Kishangarh alkaline complex, India, is an interesting example in having a small part, containing albite, in a greenschist-facies environment, whereas the remainder carries more basic plagioclase (An_{18-55}) and is surrounded by almandine and staurolite-bearing schists (Niyogi, 1966; pp. 74–5).

The subsolvus feldspar assemblage indicates that the metamorphic recrystallization took place under rather high water pressures; not rarely does this assemblage contrast with an original hypersolvus alkaline feldspar, formed under relatively dry conditions and conserved as a relic. Textural evidence clearly shows that the complete separation can only be effected by a combination of cataclastic deformation and subsequent granoblastic recrystallization.

Growth of perthitic feldspar under wet amphibolite-facies conditions was advocated by Megerlin (1967); according to Bloomfield (1968; pp. 89–92) stress is the critical factor in the crystallization of perthite under amphibolite facies conditions in Malawi. The experimental data available do not seem to support these hypotheses very strongly.

The common occurrence of biotite and Ca-amphibole in the gneisses is in accordance with rather high water pressures during metamorphic recrystallization. The preponderance of Na-amphibole over Na-pyroxene cannot be ascribed to the same cause, since crystallization of these iron-rich minerals also depends on the partial pressure of oxygen. Ernst (1962) pointed out that riebeckitic amphiboles are stable at relatively low oxygen pressures only, while acmite and hematite or magnetite crystallize at higher P_{O_2}. Floor (1966; pp. 157–9) tried to deduce the metamorphic facies using Ernst's stability data and those of annite + quartz (Eugster and Wones, 1962), but this procedure badly needs confirmation and calibration on other natural occurrences before it can be advocated as a useful tool.

It is hoped that the present contribution somewhat weakens Miyashiro's (1967; p. 152) statement that 'aegirine and riebeckite are known rarely to occur in rocks that suffered non-glaucophanitic metamorphism'.

II.7.5. Distinction Between the Various Modes of Origin of Alkaline Gneisses

Criteria to distinguish between meta-igneous and metasomatic gneisses were reviewed by Gittins (1961; pp. 304–6) and applied to some Canadian examples. In most cases it will not be possible to prove the origin of a single specimen. A careful study of field relations and microscopical features, combined with a great number of chemical analyses may provide sufficient data for an unambiguous choice. In the reviewed literature the following criteria proved especially useful:

1. Field relations: (in)homogeneity, nature of the contacts (sharp, diffuse, fragmented, etc.), comparison of number and intensity of deformative phases within and around the alkaline gneisses.
2. Microscopial features: habits of minerals, presence of minerals with igneous properties, characteristics and frequency of replacement textures, crystallization order, orientation of minerals.
3. Chemical composition: trace element contents of rocks and minerals, zoning of minerals, (non)linearity of plots of chemical analyses in variation diagrams, comparison with adjacent non-alkaline rocks.

II.7. REFERENCES

Adams, F. D., and Barlow, A. E., 1910. Geology of the Haliburton and Bancroft areas, province of Ontario. *Mem. geol. Surv. Branch*, **6,** 1–419.

Adamson, O. J., 1944. The petrology of the Norra Kärr district. *Geol. För. Stockholm Förh.*, **66,** 113–255.

Aires-Barros, L., 1958. Contribução para o cohecimento da petrografia da região de Vaiamonte-Monte da Torre das Figueiras (Alto-Alentejo): *Bol. Mus. Lab. Miner. geol. Univ. Lisb.*, 7 ser, **26,** 255–67.

Allen, J. B., and Charsley, T. J., 1968. *Nepheline-Syenite and Phonolite*, 169 pp., H.M.S.O., London.

Appleyard, E. C., 1967a. Syn-orogenic nepheline rocks in eastern Ontario and northern Norway. *Can. Mineralogist*, **9,** 238.

Appleyard, E. C., 1967b, Nepheline gneisses of the Wolfe belt, Lyndoch township, Ontario. I. Structure, stratigraphy, and petrography. *Can. J. Earth Sci.*, **4,** 371–95.

Appleyard, E. C., 1969. Nepheline gneisses of the Wolfe Belt, Lyndoch township, Ontario. II. Textures and mineral paragenesis. *Can. J. Earth Sci.*, **6,** 689–717.

Bichan, R., 1968. The igneous and metamorphic rocks of Socotra. *12th ann. rept. scientific results 1966–7, Res. Inst. Afr. Geol. Univ. Leeds*, 33–5.

Bloomfield, K., 1956. Nepheline gneisses of southern Nyasaland. *20th Int. Geol. Congress, Mexico City, Ass. Serv. Geol. Africanos*, 291–301.

Bloomfield, K., 1965. A reconnaissance survey of alkaline rocks in the northern and central regions. *Rec. Geol. Surv. Malawi*, **5,** 17–64.

Bloomfield, K., 1967. Aegirine-gneisses in central Malawi. *Q.J. Geol. Soc. London*, **123,** 93–8.

Bloomfield, K., 1968. The pre-Karroo geology of Malawi. *Mem. Geol. Surv. Malawi*, **5,** 166 pp.

Bourret, R., 1922. Etudes géologiques sur le nord-est du Tonkin. *Bull. Serv. Geol. Indochine*, **11,** Fasc. 1, 1–326.

Brock, P. W. G., 1968. Metasomatic and intrusive nepheline-bearing rocks from the Mbozi syenite-gabbro complex, southwestern Tanzania. *Can. J. Earth Sci.*, **5,** 387–419.

Burri, C., 1928. Zur Petrographie der Natronsyenite von Alter Pedroso, und ihrer basischen Differentiate. *Schw. Min. Petr. Mitt.*, **8,** 374–436.

Cooper, W. G. G., and Bloomfield, K., 1961. The geology of the Tambani-Salambidwe area. *Bull. Geol. Surv. Nyasaland*, **13,** 1–63.

Cornelius, H. P., 1951. Bemerkungen zur Geologie der Riebeckitgneise in der Grauwackenzone des Semmeringgebietes. *Tsch. Min. Petr. Mitt.*, III, **2,** 24–6.

Ducellier, J., 1963. Contribution à l'étude des formations cristallines et métamorphiques du centre et du Nord de la Haute-Volta. *Mém. B.R.G.M.*, **10,** 1–320.

Ernst, W. G., 1962. Synthesis, stability relations, and occurrence of riebeckite and riebeckite-arfvedsonite solid solutions. *J. Geol.*, **70,** 689–736.

Eskola, P., and Sahama, Th. G., 1930. On astrophyllite-bearing nephelite syenite gneis, found as a boulder in Kiihtelysvaara, E. Finland. *Bull. Com. Geol. Finlande*, **92,** 77–88.

Eugster, H. P., and Wones, D. R., 1962. Stability relations of the ferruginous biotite, annite. *J. Petrol.*, **3,** 82–125.

Fairbairn, H. W., 1941. Petrofabric relations of nepheline and albite in litchfieldite from Blue Mountain, Ontario. *Am. Miner.*, **26,** 316–20.

Fischer, G., and Nothaft, J., 1954. Natronamphibol-(Osannit)-Aegirinschiefer in den Tarntaler Bergen. *Tsch. Min. Petr. Mitt.*, **4,** (Festband Bruno Sander), 396–419.

Floor, P., 1961. Astrofilita, un mineral nuevo en España. *Notas Comuns. Inst. Geol. Miner. España*, **62,** 59–72. Errata: **65,** 155.

Floor, P., 1966. Petrology of an aegirine-riebeckite gneiss-bearing part of the Hesperian Massif: The Galiñeiro and surrounding areas, Vigo, Spain. *Leidse Geol. Med.*, **36,** 1–203.

le Fur, Y., 1966. *Nouvelles observations sur la structure de l'Antécambrien du Hoggar nord occidental* (*Région d'In Hihaou*), Thèse dipl. dr. 3me cycle, Nancy, 116 pp.

Gellatly, D. C., 1964. Nepheline and feldspar orientations in nepheline syenites from Darkainle, Somali Republic. *Am. J. Sci.*, **262**, 635–42.

Gellatly, D. C., and Hornung, G., 1968. Metasomatic nepheline-bearing gneisses from Darkainle, Somali Republic. *J. Geol.*, **76**, 678–91.

Gittins, J., 1961. Nephelinization in the Haliburton–Bancroft district, Ontario, Canada. *J. Geol.*, **69**, 291–308.

Gonçalves, F. A., 1971. Subsidios para o conhecimento geológico do Nordeste Alentejano. *Mem. Serv. geol. Portugal, nov. ser.*, **18**, 1–62.

Gorbatschev, R., 1961. On the alkali rocks of Almunge. A preliminary report on a new survey. *Bull. Geol. Inst. Uppsala*, **39**, nr. 5, 1–69.

Gummer, W. K., and Burr, S. V., 1946. Nephelinized paragneisses in the Bancroft area, Ontario. *J. Geol.*, **54**, 137–68.

Harker, R. I., 1962. The older ortho-gneisses of Carn Chuinneag and Inchbae. *J. Petrol.*, **3**, 215–37.

Hlawatsch, C., 1906. Über den Amphibol von Cevadaes (Portugal). *Rosenbusch-Festschrift, Stuttgart*, 68–76. (Abstr. in *N. Jb. Min.*, 1908, **1**, 24–5.)

Holmes, A., 1915. A contribution to the petrology of north-western Angola. (4. Aegirine-riebeckite granite from the lower Congo). *Geol. Mag.*, **52**, 267–72.

Kempe, D. R. C., 1968. The Kilonwa syenite, Tanzania. *Q. J. Geol. Soc. Lond.*, **124**, 91–100.

Keyserling, H. Graf, 1903. Der Gloggnitzer Forellenstein, ein feinkörniger ortho-riebeckitgneis. *Tsch. Min. Petr. Mitt.*, neue Folge, **22**, 109–58.

Koark, H. J., 1961. Zum Gefügeverhalten des Nephelins in zwei Vorkommen alkaliner kristalliner Schiefer, 'Grennait' von Norra Kärr und canaditischer Gneiss von Almunge (Schweden). *Bull Geol. Inst. Uppsala*, **39**, nr. 4, 1–31.

Koark, H. J., 1969. Zu Hülle, Inhalt, Gefüge und Alter des Alkaligesteinsvorkommen von Norra Kärr in südlichen Mittelschweden. *Geol. För. Stockholm Förh.*, **91**, 159–84.

Koch, P., 1955. Sur un gneiss à riebeckite du Nord-Cameroun. *C.R. Somm. S. Geol. France*, 274–5.

Koch, P., 1959. Le précambrien de la frontière occidentale du Cameroun central. *Bull. Dir. Mines Géol., Territoire du Cameroun*, **3**, 1–300.

Korpershoek, H. R., 1964. The geology of degree sheet Sul B–33/H–N (Noqui-Tomboco). *Bol. Serv. Geol. Min. Angola*, **10**, 5–105.

Lacroix, A., 1916. Les syénites à riebeckite d'Alter Pedroso (Portugal), leurs formes mésocrates (lusitanites) et leur transformation en leptynites et en gneiss. *C. R. Ac. Sci. Paris*, **163**, 279–83.

Lacroix, A., 1928. La syénite néphélinifère de Haut-Tonkin et le gneiss qui en dérive. *Fennia*, **50**, nr. 37, 1–9.

Long, L. E., 1964. Rb–Sr chronology of the Carn Chuinneag intrusion, Ross-shire, Scotland. *J. Geoph. Res.*, **69**, 1589–97.

Lundegårdh, P. H., 1953. Petrology of the Mölndal-Styrsö-Vallda region in the vicinity of Gothenburg. *Sver. geol. unders.*, C, **531**, 1–58.

Macpherson, J., 1881. Apuntes petrográficos de Galicia. *An. Soc. Esp. Hist. Nat.*, **10**, 49–87.

Marmo, V., Hoffrén, V., Hytönen, K., Kallio, P., Lindholm, O., and Siivola, J., 1966. On the granites of Honkamäki and Otanmäki, Finland, (with special reference to the mineralogy of accessories). *Bull. Com. Geol. Finlande*, **221**, 1–34.

McLachlan, G. R., 1951. The aegirine granulites of Glen Lui, Braemar, Aberdeenshire. *Miner. Mag.*, **29**, 476–95.

Megerlin, N., 1967. Sur les syénites de la région de Ianakafy (Sud de Madagascar). *C. R. Sem. Geol. Madagascar*, 39–41.

Miyashiro, A., and Miyashiro, T., 1956. Nepheline syenites and associated alkalic rocks of the Fukushin-zan district, Korea. *J. Fac. Sci., Tokyo*, II, **10**, 1–64.

Miyashiro, A., 1967. Chemical composition of rocks in relation to metamorphic facies. *Jap. J. Geol. Geogr.*, **38**, 149–57.

Mortelmans, G., 1948. Le granite de Noqui et ses phénomènes de contact. *Bull. Soc. Belge Géol. Pal. Hydrol.*, **57**, 519–40.

Moyd, L., 1949. Petrology of the nepheline and corundum rocks of southeastern Ontario. *Am. Miner.*, **34**, 736–51.

Neves Correia, J. M., 1959, Roches sodiques des régions d'Alter-Pedroso et d'Elvas (Alentejo, Portugal). *20th Int. Geol. Congress, Mexico City, 1956, seción XI–A: Pétrologia y Mineralogia*, 239–52.

Niyogi, D., 1966. Petrology of the alkalic rocks of Kishangarh, Rajasthan, India. *Geol. Soc. Amer Bull.*, **77**, 65–82.

Osann, A., 1907. Ueber einen Nephelinreichen Gneis von Cevadaes, Portugal. *N. Jb. Min.*, **II**, Abh., 109–28.

Oyawoye, M. O., 1967. The petrology of a potassic syenite and its associated biotite pyroxenite at Shaki, western Nigeria. *Contr. Min. Petrology*, **16**, 115–38.

Parga-Pondal, I., 1956. Nota explicativa del mapa geológico de la parte N.O. de la provincia de la Coruña. *Leidse Geol. Med.*, **21**, 468–84.

Parga-Pondal, I., 1967. 'Datos geológico-petrográficos de la provincia de La Coruña', in *Estudio agrobiólógico de la Provincia de La Coruña*, Vigo, Artes gráficas de Faro de Vigo, 46 pp.

Payne, J. G., 1968. Geology and geochemistry of the Blue Mountain nepheline syenite. *Can. J. Earth Sci.*, **5**, 259–73.

Phadke, A. V., 1967. Petrology and structure of the riebeckite gneiss from the area near Gloggnitz in the graywacke zone of Austria. *Jb. Geol. Bundesanstalt*, **110,** 199–216.

Richardson, S. W., 1965. *The petrology of the Glen Dessarry complex, Inverness-shire*, Ph.D. thesis, University of Oxford, 183 pp.

Richardson, S. W., 1968. The petrology of the metamorphosed syenite in Glen Dessarry, Inverness-shire. *Q. J. Geol. Soc. Lond.*, **124,** 9–51.

Rosenbusch, H., 1898. *Elemente der Gesteinslehre*, E. Schweizerbart, Stuttgart, 546 pp.

Serralheiro, A., 1957. Esboço geológico da região de Alter Pedroso. *Bol. Soc. Geol. Portugal*, **12,** fasc III, 3–12.

Sturt, B. A., 1961. Preferred orientation of nepheline in deformed nepheline syenite gneisses from Sørøy, northern Norway. *Geol. Mag.*, **98,** 464–6.

Sturt, B. A., and Ramsay, D. M., 1965. The alkaline complex of the Breivikbotn area, Sørøy, northern Norway. *Norges Geol. Unders.*, **231,** 1–142.

Teixeira, C., 1956, *Notas sobre geologia de Portugal; O complexo cristalofilino antigo*, Empr. Lit. Flumineuse, Lisboa, 20 pp.

Teixeira, C., and Torre de Assunção, C. F., 1956. Novos elementos para o conhecimento das rochas hiperalcalinas sódicas do Alto Alentejo. *Rev. Fac. Ci., Univ. Lisboa*, 2 ser., C, **5,** 173–208.

Teixeira, C., and Torre de Assunçao, C. F., 1958. Sur la géologie et la pétrographie des gneiss à riebeckite et aegyrine et des syénites à néphéline et sodalite de Cevadais, près d'Ouguela (Campo Maior), Portugal. *Com. Serv. Geol. Portugal*, **48,** 31–56.

den Tex, E., and Floor, P., 1967. 'A blastomylonitic and polymetamorphic "graben" in western Galicia (N.W. Spain)', in *Etages tectoniques*, La Baconnière, Neuchâtel, 169–78.

Tilley, C. E., 1957. Problems of alkali rock genesis. *Q. J. Geol. Soc. Lond.*, **113,** 323–60.

Torre de Assunção, C. F., and Gonçalves, F. A., 1970. Contribuiçao para o conhecimento das rochas hiperalcalinas e alcalinas (gnaisses hastingsíticos) do Alto Alentejo (Portugal). *Bol. Soc. Geol. Portugal*, **17,** 187–228.

Welter, C., 1964. Contribution à la pétrographie et à la genèse du complexe des gneiss a néphéline du Makaraingobe (Ouest de Madagascar). *C.R. Sem. Geol. Madagascar*, 57–62.

Zemann, J., 1951. Zur Kenntnis der Riebeckitgneise des Ostendes der nordalpinen Grauwackenzone. *Tsch. Min. Petr. Mitt.*, III, **2,** 1–23.

III. Regional Distribution and Tectonic Relations

III.1. INTRODUCTION

H. Sørensen

Since Harker (1896) it has been generally accepted that alkaline rocks are formed under relative tectonic quiescence and that the *mise-en-place* of the magmas was facilitated by block faulting and foundering. The tectonic quiescence allowed that magmas could differentiate without escape of volatiles and also prevented mixing of magma fractions. In accordance with this most continental provinces of alkaline rocks are located in tectonically stable regions of the crust, including regions of completed folding. Alkaline rocks are less conspicuous in the active orogenic belts. The oceanic occurrences of alkaline rocks are mainly found in oceanic islands situated away from the midocean ridges.

Kuznetsov (1958, 1964, V. P. Volkov, personal communication) emphasizing the correlation of magmatism and tectonism introduced the term 'magmatic formation'. This term covers rock associations which are mutually related with respect to assemblage of rocks, petrochemical and geochemical affinities, and location in similar geological structures. He considered the following magmatic formations: (1) geosynclinal, (2) orogenic, and (3) in tectonically stable regions. Two of the orogenic formations, the trachyandesitic volcanic formation and its intrusive equivalent, the gabbro-monzonite–syenite formation, are confined to the rigid blocks of the mobile belts, as for instance median masses, and to major fault zones cutting these blocks. The magmatic formations of the stable regions, platforms and shields, are generally made up of alkaline types, when the tholeiitic–basaltic (trapp) formation is excluded. Kuznetsov emphasized that abyssal differentiation of the material of the interior zones of the Earth resulted in regeneration of tectonic activity of the platforms after long periods of quiescence. During these periods of tectonic activity the trapp and alkali basaltic associations precede alkaline intrusions of central type. The diversity of magmatic formations in the tectonically stable regions is explained by vertical migration of magma chambers generated at different depths within the Earth's mantle.

There is generally a close connection between alkaline igneous activity and major tectonic structures, first of all fault zones. The most prominent examples are the rift zones (see III.2),* such as the East African ones, the Rhine–Oslo graben (IV.4), the Gardar rift, South Greenland (IV.3), the Monteregian province, Quebec (IV.6).

The occurrences of alkaline rocks often are located at the intersections of fault zones, examples are found in the Kola and Gardar Provinces (IV.2, IV.3, Sørensen, 1970), and in Northern Bohemia (Kopecký *et al.*, 1970). The lines of occurrences of alkaline rocks and the elongation of individual intrusions are commonly transverse to the predominant regional structures of the country rocks, this is for instance the case in the Gardar province (IV.3) and in the Monteregian province (IV.6; Kumarapeli, 1970).

The reticulate arrangement of individual intrusions on nodular points in a regional lattice has been especially emphasized by Chapman (1968) for the White Mountain Magma Series Intrusions, New Hampshire.

Occurrences of alkaline rocks are also situated at bends of the strike of monoclines or flexures, examples are the Kangerdlugssuaq intrusion, East Greenland (Kempe *et al.*, 1970) and the Nuanetsi Province, South East Africa (Cox, 1970).

In some provinces there is no evident relation

* Readers are referred to the relevant Chapters of this book where similarly cited.

between tectonic lines and occurrences of alkaline rocks (cf. III.3; III.4). The alkaline rocks may here be associated with concealed fault zones or zones of crustal weakness.

Alkaline rocks are frequently found in updomed or arched structures as is the case in the rift valley occurrences (III.2). The relation between arching and location of alkaline igneous occurrences is for instance also seen in the Siberian alkaline provinces (III.4), in the Nigeria–Niger province, West Africa (Black and Girod, 1970) and in the alkaline provinces along the eastern edge of the North American Rocky Mountain system (e.g. Cripple Creek, Lovering and Goddard, 1950 and Bear Paw Mountains, Montana, Bryant *et al.*, 1960).

The provinces of alkaline rocks often show a distinct zonal or linear variation with regard to mineralogy, chemistry and mode of occurrence. The prominent examples are the potassic western rift and the sodic eastern rift of East Africa, the western United States showing an eastward increasing alkalinity, and some Siberian occurrences (III.4). The Monteregian line of intrusions is a small scale example having carbonatites, ijolites and melilite-bearing rocks in the west, quartz- and plagioclase-bearing rocks in the east (IV.6; Philpotts, 1970; Kumarapeli, 1970). Also in the northern Bohemia province there is a regional zonation, leucite being the predominant feldspathoid in the west in Dupovsky hory, sodalite in the east in České středohoří (Kopecký, 1966). Sheynmann *et al.* (1961) point out that the content of SiO_2 in occurrences of the gabbroid alkaline association (see IV.1) increases from areas of slight mobility towards such mobile zones as geanticlines of fold belts.

In some provinces, such as the Gardar province, parts of the East African rift and the Tertiary alkaline province of Central Europe, there is no pronounced regularity in distribution of the alkaline rocks and often two parent magmas appear to have been constantly available, for instance basaltic and nephelinitic magmas in the eastern rift, which also contains huge amounts of trachyte and phonolite in regions of upwarping and structural culmination, perhaps formed by partial crustal melting (III.2; King, 1970).

Oceanic and continental alkaline rocks differ in a number of respects. Thus leucite is most widespread in the continental provinces, which also show the highest proportion of the more evolved derivatives of basaltic magmas such as rhyolite and phonolite. Gorshkov (1970) has pointed out that continental and oceanic alkaline igneous rocks in spite of similar mineralogy and geochemistry show different trends of evolution, the oceanic rocks forming the series tholeiitic basalt–olivine basalt–alkaline derivatives, the continental phonolites being derived from primary alkaline olivine basalts. These two series of rocks show different rates of increase in alkalinity during differentiation.

The provinces of alkaline rocks often show a long duration of the alkaline igneous activity, for instance about 200 m.y. in the Gardar province (IV.3), 600 m.y. in the St. Lawrence Valley system (Currie, 1970), and about 1400 m.y. in the Kola–Karelia–Scandinavia zone of alkaline rocks (Kukharenko, 1967).

Backlund (1932) emphasized that the emplacement of alkaline magmas in stable regions (platforms or forelands) was simultaneous with orogenic processes elsewhere. The emplacement by perforation of stable continental areas was termed *epeirodiastresis*. This synchronism of spatially separated orogenesis and alkaline igneous activity is especially emphasized for the Siberian provinces of alkaline rocks (III.4) and for the Caledonian and Hercynian alkaline intrusions of the Kola Peninsula (IV.2).

Even if alkaline rocks are found mainly in stable regions, quite a few occurrences are described from mobile belts. Thus in the East Otago Pliocene Province of New Zealand the alkaline rocks erupted during a period of rather mild folding and compressional faulting (cf. review in Turner and Verhoogen, 1960). Barker (1969, III.2) has pointed out that 21% of the feldspathoidal rocks of North America occur in orogenic belts. The miaskites of the Urals are also believed to be synorogenic (II.2.5.6), and the nepheline syenites of Söröy, Northern Norway and of Blue Mountain, Ontario were emplaced before the latest phases of folding (Sturt and Ramsay, 1965 and Payne, 1968). Even if it is not always clear if the nepheline syenites in question are syntectonic or emplaced during intervening periods of relative tectonic quiescence, nepheline syenites are beyond doubt members of the igneous suites of some orogenic belts and may have been present in others, where, however, the chemical environments have been unfavourable for survival of undersaturated rocks during the late- and post-orogenic processes.

The volcanic activity of the oceanic ridges and the circum-Pacific belt is now generally considered in the light of the new 'global plate tectonics'. The continental rift zones may be results of the breaking up of warped rigid continental plates (cf. III.2.6), of displacement along irregular strike-slip faults (Knopoff, 1970), or of tensional stretching of the upper part of the crust at places where mantle-derived material is moving upwards to form wedges, dyke swarms or subcrustal domes (cf. III. 2.6; Harris, 1970; Baker and Wohlenberg, 1971 and Illies, 1970). In the oceans the shallow basaltic activity results in the formation of subordinate quantities of alkaline melts, while there is no alkaline activity in the deep-seated andesitic igneous activity of the circum-Pacific belt. The igneous activity originating at intermediate depths under continental regions results in magmas, which are predominantly alkaline in regions of crustal arching.

The distribution and tectonic setting of alkaline rocks are in this part of the book illustrated by chapters describing the alkaline rocks of rift zones (III.2), and the alkaline rocks of selected regions. Barker (III.3) has compiled a catalogue of all known occurrences in North America; Butakova (III.4) a thorough analysis of the geological relations of the alkaline rocks of Siberia with special emphasis on their relation to contemporaneous mobile belts; and Mathias (III.5) the geological setting of the classical occurrences of alkaline rocks in Southern Africa. The geological relations of alkaline rocks are also discussed in chapters of part IV of the book, e.g. the suite of alkaline rocks of the oceanic islands (IV.7).

III.1. REFERENCES

Backlund, H. G., 1932. On the mode of intrusion of deep-seated alkaline bodies. *Bull. geol. Instn. Univ. Uppsala*, **24,** 1–24.

Baker, B. H., and Wohlenberg, J., 1971. Structure and evolution of the Kenya rift valley. *Nature*, **229,** 538–42.

Barker, D. S., 1969. North American feldspathoidal rocks in space and time. *Bull. Geol. Soc. Amer.*, **80,** 2369–72.

Black, R., and Girod, M., 1970. Late Palaeozoic to Recent igneous activity in West Africa and its relationship to basement structure, in *African Magmatism and Tectonics*, eds. T. N. Clifford and I. G. Gass. Oliver and Boyd, Edinburgh, 185–210.

Bryant, B., Schmidt, R. G., and Pecora, W. T., 1960. Geology of the Maddux quadrangle, Bearpaw Mountains, Blaine County, Montana. *Bull. U.S. geol. Surv.*, **1081–C,** 91–116.

Chapman, C. A., 1968. 'A comparison of the Maine coastal plutons and the magmatic central complexes of New Hampshire', in E–An Zen *et al.*, eds., *Studies of Appalachian Geology: Northern and Maritime*. Interscience, 385–98.

Cox, K. G., 1970. 'Tectonics and vulcanism of the Karroo period and their bearing on the postulated fragmentation of Gondwanaland', in T. N. Clifford and I. G. Gass, eds., *African Magmatism and Tectonics*. Oliver and Boyd, Edinburgh, 211–36.

Currie, K. L., 1970. An hypothesis on the origin of alkaline rocks suggested by the tectonic setting of the Monteregian Hills. *Can. Miner.*, **10,** 411–20.

Gorshkov, G. S., 1970. Two types of alkaline rocks—two types of upper mantle. *Bull. volcan.*, **33,** 1186–97.

Harker, A., 1896. The natural history of igneous rocks. I. Their geographical and chronological distribution. *Sci. Progress*, **6.**

Harris, P. G., 1970. 'Convection and magmatism with reference to the African Continent', in T. N. Clifford and I. G. Gass, eds., *African Magmatism and Tectonics*. Oliver and Boyd, Edinburgh, 419–38.

Illies, J. H., 1970. 'Graben tectonics as related to crust-mantle interaction', in J. H. Illies and St. Mueller, eds., *Graben Problems*. Int. Upper Mantle Project, Sci. Rep. 27. E. Schweizerbart'sche Verlagsbuchhandlung, 4–27.

Kempe, D. R. C., Deer, W. A., and Wager, L. R., 1970. Geological investigations in East Greenland. VIII. The petrology of the Kangerdlugssuaq alkaline intrusion East Greenland. *Meddr Grønland*, **190**, 2, 1–49.

King, B. C., 1970. 'Vulcanicity and rift tectonics in East Africa'. In T. N. Clifford and I. G. Gass, eds., *African Magmatism and Tectonics*. Oliver and Boyd, Edinburgh, 263–84.

Knopoff, L., 1970. 'The Rhinegraben: a segment of the world rift system. Problems of continental rift structures', in H. J. Illies and St. Mueller, eds., *Graben Problems*. Int. Upper Mantle Project, Sci. Rep. 27. E. Schweizerbart'sche Verlagsbuchhandlung, 1–4.

Kopecký, L., 1966. 'Tertiary volcanics', in J. Svoboda, ed., *Regional Geology of Czechoslovakia. I. The Bohemian Massif.* Publ. House Czech. Acad. Sciences, Prague, 554–81.

Kopecký, L., Dobes, M., Fiala, J., and Stovicková, N., 1970. Fenites of the Bohemian massif and the relations between fenitization, alkaline volcanism and deep fault tectonics. *Sb. geol. Věd. Praha*, **G. 16**, 51–107.

Kukharenko, A. A., 1967. The alkaline magmatism of the Eastern part of the Baltic Shield (in Russian) *Zap. Vses. Miner. Obsch.*, **96**, 547–66.

Kumarapeli, P. S., 1970. Monteregian alkalic magmatism and the St. Lawrence rift system in space and time. *Can. Miner.*, **10**, 421–31.

Kuznetsov, Yu. A., 1958. 'Magmatic formations', in *Zakonomernosti rasmeshcheniya poleznykh iskopaemykh* (in Russian), vol. I, Izd. Akad. Nauk SSSR.

Kuznetsov, Yu. A., 1964. *Chief Types of Magmatic Formations* (in Russian). Izd. 'Nedra'.

Lovering, T. S., and Goddard, E. N., 1950. Geology and ore deposits of the Front Range, Colorado. *Prof. Pap. U.S. geol. Surv.*, **223**, 1–139.

Mohr, P. A., 1970. Plate tectonics of the Red Sea and East Africa. *Nature*, **228**, 547–8.

Payne, J. G., 1968. Geology and geochemistry of the Blue Mountain nepheline syenite. *Can. J. Earth Sci.*, **5**, 259–73.

Philpotts, A. R., 1970. Mechanism of emplacement of the Monteregian intrusions. *Can. Miner.*, **10**, 395–410.

Sheynmann, Yu. M., Apel'tsin, F. R., and Nechayeva Ye. A., 1961. 'Alkalic Intrusions, Their Mode of Occurrence and Associated Mineralization' (in Russian). *Geologiya Mestorozhdenii Redkikh Elementov*, 12–13, Moscow (Gosgeoltekhizdat) 177 pp. (Review in *Int. Geol. Rev.*, **5**, 451–8).

Sørensen, H., 1970. Internal structures and geological setting of the three agpaitic intrusions—Khibina and Lovozero of the Kola Peninsula and Ilímaussaq, South Greenland. *Can. Miner.*, **10**, 299–334.

Sturt, B. A., and Ramsay, D. M., 1965. The alkaline complex of the Breivikbotn area, Sørøy, Northern Norway. *Norg. geol. Unders.*, **231**, 1–164.

Turner, F. J., and Verhoogen, J., 1960. *Igneous and Metamorphic Petrology*. McGraw-Hill, 2nd edition 1–694.

III.2. CONTINENTAL RIFTING AND ALKALINE MAGMATISM

D. K. Bailey

'Storied of old in high immortal verse
Of dire chimeras and enchanted isles
And rifted rocks whose entrance leads to Hell—
For such there be, but unbelief is blind.'
John Milton. *Comus.*

III.2.1. Introduction

The terms 'rift' and 'rift zone', have become common parlance in recent discussions of mid-ocean ridges and continental drift, and with this usage their earlier meanings have been extended. The old concept of a graben, or zone of trough faulting, is too restricted to cover all those features which are now popularly described as rifts, and before discussing the associated magmatism it is essential first to compare oceanic and continental 'rifts'.

On the continents, rift zones are essentially cratogenic structures: they vary in width, depth, length and complexity, being expressed as fault troughs, faulted monoclines, belts of block faulting, and transverse faults, in various combinations. All these elements, for instance, appear in the East African Rift zone. The northern limb of this great continental rift is linked by the Gulf of Aden to the Carlsberg Ridge of the Indian Ocean, but its other limbs are entirely continental. Other continental rifts, with the possible exception of the Basin and Range area of the western U.S.A, appear to be unconnected with the mid-ocean ridge system (see Fig. 2).

Largely on the basis of physiographic analogy the oceanic 'rifts' are correlated with the rifts on the continents, but the available evidence and argument suggest that they are fundamentally different structures! The mid-ocean ridges are generally regarded as the sites of crustal *separation*, and the formation of new crust at rates of 1 to 10 cm/year, whereas the continental rifts are manifestly floored and walled by ancient crust. The differences are illustrated in Fig. 1, which shows the oceanic rift floor being continually renewed, with the pre-existing floor being displaced laterally, and *upwards*, to form the new rift margins. Estimates of crustal distension or attenuation across the Rhine rift (4·8 km since middle Eocene: McConnell, 1969) and the Kenya rift (Baker and Wohlenburg, 1971) give separation rates of 0·01 cm/year and 0·05 cm/year. To obtain an objective view, therefore, the continental and oceanic features should not be prematurely lumped together. It is the current vogue, for instance, to depict the East African Rifts as an off-shoot of the oceanic ridge-system, and Wilson (1967, p. 32) asserts 'they appear to be incipient ocean that has failed to develop'. This kind of assumption can only serve to obscure the real nature of both kinds of structure. Having made this assertion about the East African Rift, Wilson is then forced to conclude (p. 33) that the other continental rifts 'form a separate class' from the oceanic rifts.

There is another vital feature which distinguishes oceanic and continental 'rifts', the characteristic magmatism is different in each case. The magmatism of the mid-ocean ridges is dominantly basaltic, with tholeiites and oversaturated basalts in the crestal, or 'rift' zones

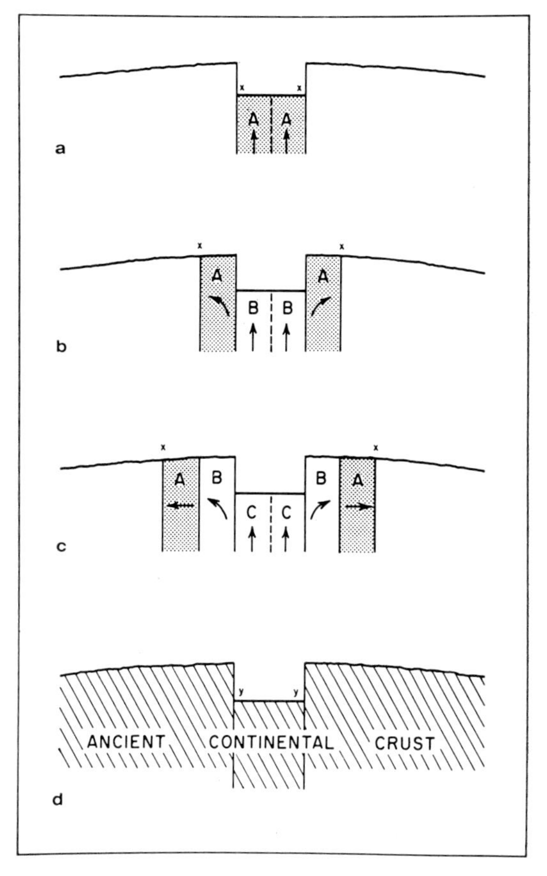

Fig. 1 Model diagrams to illustrate the fundamentally different mechanics involved in mid-ocean and continental 'rift' formation. The oceanic 'rift' is the site of new crust formation and crustal separation: (a), (b) and (c) show three stages, the arrows indicate the overall sense of movement in each stage. Material in the floor of the rift is continually transferred, sideways and upwards, to be replaced by new material from below. Note especially the movements of points X which start as the bottom of the rift wall, and contrast these with the points Y in Fig. 1(d). In the continental rift the floors and margins are in pre-existing crust and are not being continually replaced

(McBirney and Gass, 1967). Magmatism in the continental rift zones is almost the converse, alkali enrichment and silica undersaturation are two of its hall marks.

III.2.2. Continental Rifts and Up-Warps

Inspection of the distribution of alkaline

igneous activity is sufficient to show that this is not exclusively restricted to rift zones, but where continental rifting is known, alkaline magmatism is strongly in evidence. It is, therefore, important to realize that commonly the rift zones are merely the fractured crests of crustal arches, or upwarps, which are the major tectonic features (Bailey, 1964). When the distribution of alkaline activity is compared with regions of cratogenic uplift the correlation is stronger still, and it is hard to avoid the conclusion that the two phenomena, uplift and alkaline magmatism, are related. Many examples of this may be seen in the Western U.S.A. (Eardley, 1961, pp. 563–604), and the same relationship has been demonstrated for the major West African provinces by Black and Girod (1968 and 1970). Therefore, although there is a strong correlation between rifting and alkaline magmatism, it should be recognized that both are generally related to uplift.

Examples of the association of alkaline magmatism and recognized rifts are numerous. In Europe, the Carboniferous activity of the Midland Valley of Scotland, the Permian of the Oslo graben, and the Cainozoic of the Rhine graben are self-evident, and the faulted uplift of the French Massif Central should also be noted (cf. IV.4)*. With increasing age, and depth of erosion, it becomes increasingly difficult to relate the magmatism to the earth movements. This problem arises with the E–W line of older alkaline complexes of the Transvaal, typified by Pilanesberg (cf. III.5). But Innes (1960) and Innes *et al.*, (1967) have indicated that in Canada a line of alkaline complexes in Ontario coincides with the Kapuskasing gravity high, which they interpret as an ancient rift zone. Similarly, the Gardar alkaline activity of Greenland was related to rifting (cf. IV.3; Sørensen, 1970). Other continental rift zones are shown in Fig. 2. One of these, the Basin and

* Readers are referred to the relevant Chapters of this book where similarly cited.

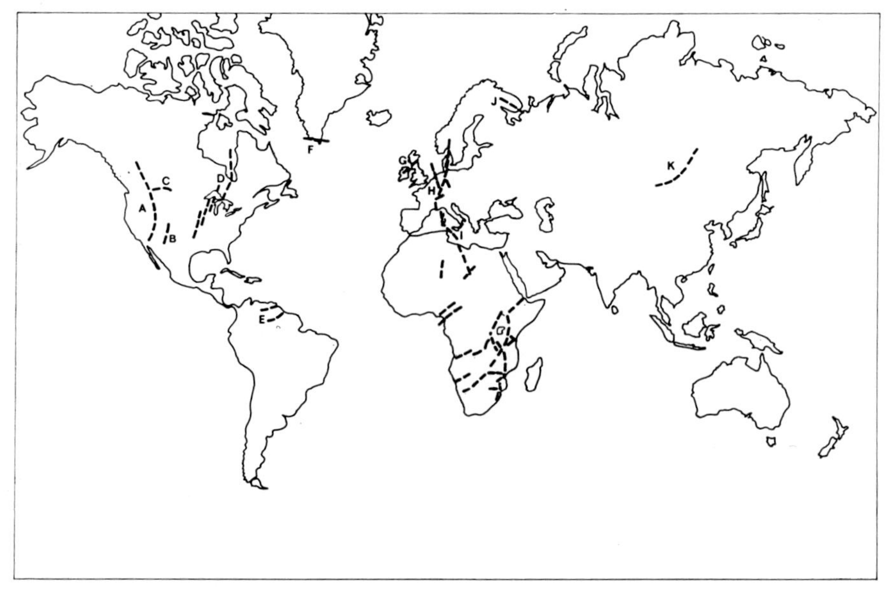

Fig. 2 Mercator's projection, showing the world distribution of known and postulated continental rifts (shown by heavy broken lines). A—Basin and Range, and B—Rio Grande (Cook, 1969); C—Buried rift in W. Canada (Kanesewich *et al.*, 1969); D—Kapuskasing and Mid-continental (Innes, 1960); E—Guiana and Venezuela (McConnell, 1969); F—Gardar (Sørensen, 1970); G—Midland Valley; H—Rhine–Oslo system, and I—Sardinia–Hon graben (Illies, 1969); J—Kola uplift and fracture zones (1 : 5,000,000 Tectonic map of Eurasia. Moscow, 1966); K—Baikal (Florensov, 1969). For details of African rift pattern see Fig. 3

Range and the 'System of Great Trenches' (Eardley, 1961), calls for comment. It has many of the attributes, such as block faulting and uplift, of a typical continental rift zone (although it is exceptionally wide) and much of the associated magmatism is also characteristic. It has been described as an active rift zone by Cook (1969) but he argues further that it is the landward extension of the East Pacific Rise. It is difficult to see the physical reality embodied in this concept, and such an extension is not shown in the plate tectonic maps of Oliver *et al.* (1969). In terms of the new tectonic model proposed by Morgan (1971) the whole concept of an active mid-ocean rise overridden by a continent becomes meaningless.

In this chapter attention will be focussed on the African craton in an attempt to develop a coherent synthesis of the relationships between arching, rifting and alkaline magmatism. This is not merely because the greatest continental rifts are found in Africa, but also because they display the greatest range of magma types, in composition and in time. Erosion levels range from very deep Precambrian complexes to active volcanoes. Above all, the African shield has escaped orogenic deformation since the late Precambrian and has not suffered the ravages of Quaternary continental glaciation, so that ancient structural features, and high-level erosion surfaces dating as far back as the Mesozoic, have been preserved.

III.2.3. Magmatism and Tectonics in the East African Rift Zone

Figure 3 shows those regions of the African continent of uplift and rifting where there is a known association of alkaline magmatism. By far the most spectacular is the East and Central African rift zone, extending from the Red Sea to Mozambique and Zambia, possibly through Botswana into S. W. Africa (Bailey, 1961). This zone may be conveniently divided into two parts; a northern sector, extending as far south as south-west Tanzania, in which Tertiary to Recent volcanism is strongly in evidence; and a southern sector of older rifting (Jurassic–Cretaceous) with associated plutonic and subvolcanic complexes. The alkaline activity along the rift can be most conveniently reviewed by starting in the north and working southwards.

In the most northerly region, Ethiopia, the extensive lava fields extend far beyond the boundaries of the rift. According to Mohr (1963; 1965) the earliest activity was the eruption of fissure basalts within and along the margins of the proto-rift during Eocene times. The subsequent Magdala lavas are thickly developed over the plateaux away from the rifts, having been erupted over the gently sloping surface of the Arabo-Ethiopian swell. This activity was in two major episodes: Oligocene–Miocene fissure basalts with abundant silicic and pyroclastic flows towards the top; followed by the main phase of uplift and rifting and the building of great Miocene–Pliocene basalt shields, with associated rhyolites and phonolites. During the Pleistocene violent silicic eruptions occurred within the rift, whilst basalts were erupted in the

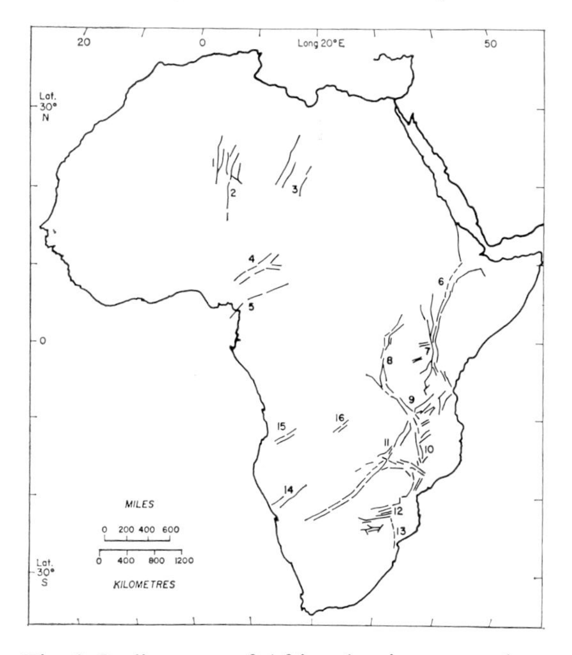

Fig. 3 Outline map of Africa showing areas where uplift and rifting are associated with alkaline magmatism. (1) Atakor; (2) Aïr; (3) Tibesti; (4) Benue; (5) Cameroun; (6) Ethiopia; (7) Eastern Rift; (8) Western Rift; (9) Rungwe; (10) Malawi; (11) Luangwa-mid-Zambezi; (12) Limpopo; (13) Lebombo monocline; (14) Damaraland; (15) Angola; (16) Alto Zambezi

neighbouring plateaux. The latest eruptions within the rift are scoriaceous olivine basalts, or peralkaline rhyolites.

The Ethiopian sector contrasts with the East Africa rift further south in its paucity of nephelinite–carbonatite activity, and regional trachyte/phonolite magmatism. Alkali basalts are developed on a far more extensive scale than elsewhere, but it is noteworthy that the proportion of rhyolites (as flows and tuff deposits) is high, about 25% of the volcanics (Mohr, pers. comm. 1967).

The lava fields of Ethiopia extend across the border into Kenya as the extensive, but not well-known, area of basalt and basanite east of Lake Rudolf. In Kenya, Uganda and Northern Tanzania generally, however, basalts are much less abundant. The sequence, distribution, and relation to tectonics have been summarized recently by Wright (1963; 1965), Baker (1965), Williams (1965), King (1965; 1970) and Baker and Wohlenberg (1971). Table 1 has been adapted from that of Baker (1965) to provide a convenient general summary, and to emphasize what seem to be the important magmatic episodes. It is consistent with the more detailed successions recorded by the other authors, and Baker himself, but I have taken the liberty of omitting the local terminology.

Table 1 refers specifically to Kenya and the earliest nephelinite–carbonatite activity of the Kavirondo region is only the southern end of a chain of nephelinite volcanoes extending northwards through Elgon and Napak in E. Uganda. These represent the earliest volcanism on the crustal swell which was subsequently broken by rifting further to the east. Nephelinite activity on a smaller scale occurred along the line of the

TABLE 1 Generalized Volcanic Sequence in the Kenya Rift Valley

		Volcanic Events	*Tectonic Events*
	Recent and Pleistocene	Minor basaltic cones	Minor renewals of movement in rift floor
		Comendite and Rhyolite plugs	Grid faulting—collapse and breaking of rift floor
		Phonolitic-trachytic calderas in rift	Graben faulting—early grid faults in rift floor
		Mt. Kenya and Kilimanjaro volcanoes	
3.	Pliocene	TRACHYTIC IGNIMBRITES FLOOD TRACHYTES IN RIFT FLOOR	Down flexing of rift floor, especially in northern sector
		Basalts and Basanites of rift floor	
		Nephelinite-trachyte-pyroclastic volcanoes of rift floor	Uplift of rift shoulders
2.	Upper and Middle Miocene	EXTENSIVE FISSURE ERUPTIONS OF PHONOLITE OVER A WIDE AREA	Major faulting on west side of rift. Local faulting on E. side. Monoclinal down flexuring of rift floor
		Eruption of alkali basalts and tuffs from many small centres.	Gentle down-warping of rift zone
1.		ERUPTION OF NEPHELINITES, MELILITITES, EMPLACEMENT OF ALKALINE IGNEOUS ROCKS AND CARBONATITES IN KAVIRONDO GULF AREA.	Up-doming of the West-central part of Kenya
			Deposition of Miocene sediments. Stability—formation of sub-Miocene erosion surface.

proto-rift but there was no corresponding activity east of the rift (Wright, 1963), so that the nephelinite magmatism is assymmetrically distributed along the western side of the rift axis. A feature that requires more investigation in this context arises from the common observation that the western margin of the rift is a zone of strong faulting, while the eastern margin is predominantly a monoclinal flexure. Clearly the assymmetry of the magmatism finds a counterpart in the tectonic expression (see Fig. 5).

Uplift of the axial region continued, and culminated with the outpouring of flood-phonolites. Major rifting along the axis of the upwarp coincided with, or immediately followed, the phonolite eruptions. Central volcanoes then developed within the collapsed area, but the next regional magmatic event was the eruption of floods of trachyte lavas and thick ash deposits. Like the earlier phonolites, these trachytes spread beyond the rift zone, and again the outpouring was followed by further collapse, forming the central graben of the present day. In the main rift, Pleistocene–Recent activity has been confined to trachytic calderas, and central volcanoes emitting basalt or peralkaline rhyolite. But further south, in northern Tanzania, where the graben begins to lose its identity the Recent volcanoes (e.g. Lengai and Meru) are of phonolite–nephelinite–carbonatite character. Some of the volcanoes in northern Tanzania are deeply eroded, and although they appear to be post-rifting it is clear that nephelinite activity has persisted in this sector of the rift for a considerable period. Subvolcanic carbonatite–ijolite complexes, such as Oldonyo Dili, may be as old as Miocene or even Cretaceous (McKie, 1966).

The Western Rift is well-known for its Pleistocene–Recent explosive potash-rich volcanicity around the eastern and southern margins of the Ruwenzori horst (cf. II.5). Further south, the Birunga field, north of Lake Kivu, displays a range of Pliocene-active volcanoes, of various types, examplified by the great active volcanoes of Nyamalgira (leucite-basanite) and Nyiragongo (nephelinite) only 15 kilometres apart.

Even more anomalous is the volcanic field south of Lake Kivu, where, as pointed out by Holmes (1940), tholeiites are mixed with alkali basalts and trachytes, and some unusual potash-rich rhyolites. Of the thirty-eight basalt analyses listed by Denaeyer and Schellinck (1965) and plotted here in Fig. 4, ten fall in the tholeiitic field defined by Macdonald and Katsura (1964). More information is needed on the sequence and age of these volcanics because they are a rare example of tholeiites in the continental part of the East African Tertiary rifts.

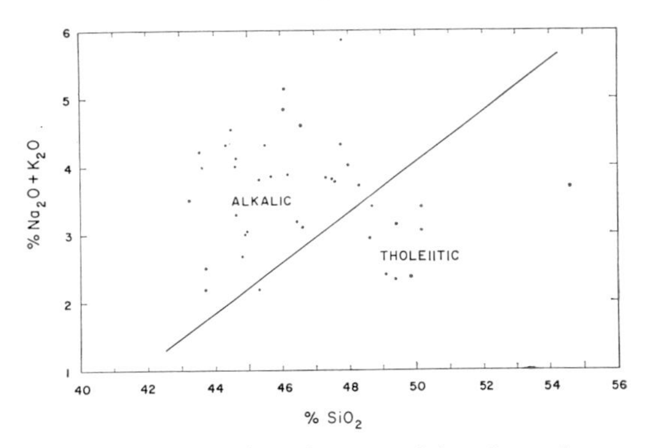

Fig. 4 Alkali–Silica diagram of basalt analyses from S. Kivu province (Denaeyer and Schellinck, 1965). The line separating 'alkaline' from 'tholeiitic' is taken from Macdonald and Katsura (1964)

Five carbonatite complexes, Bingu, Lueshe, Kawezi (Zaire), Karonge (Urundi) and Sangu (Tanzania) are listed by Tuttle and Gittins (1966) from the Western Rift. From the available descriptions they appear to be deep-seated deposits, pre-dating the Tertiary rifting.

Further south, the Western Rift rejoins the Eastern Rift at Mbeya (Tanzania). In the region of intersection there is a notable concentration of carbonatites and alkaline intrusions ranging in age from Precambrian to Cretaceous (Bailey, 1961; Tuttle and Gittins 1966). In Pliocene–Recent times this rift intersection has also been the locus of the most southerly volcanism in the Tertiary rifts—the Rungwe volcanics (Harkin, 1960). The association is alkali basalt–trachyte–phonolite, with some nephelinite. Because of the abnormally large amounts of felsic lavas Harkin is forced to conclude that fractional crystallization of a basic parent magma is an inadequate petrogenetic mechanism for this province. He invokes contamination of an alkali basalt parent magma by syenitic and carbonatitic material at depth.

South of the Rungwe intersection, the line of the Eastern Rift continues as the Luangwa–mid-Zambezi rift in Zambia, while the line of the Western Rift is continued by the Lake Malawi (Nyasa) rift in Malawi. These southern rifts were formed almost entirely by post-Karroo earth-movements (U. Jurassic–L. Cretaceous) with only minor rejuvenation in Tertiary–Recent times. The magmatism that accompanied the Mesozoic rift formation is now revealed in plutonic and subvolcanic complexes. These are magnificently displayed in the type area of the Chilwa Alkaline Province, in southern Malawi (Dixey *et al.*, 1955; Garson, 1965; Bloomfield, 1965). In addition to the classic ijolite-carbonatite vents, there are slightly older syenite, syeno-granite, and nepheline syenite plutons, which are the deep-seated equivalents of the great trachytic volcanoes of the Tertiary rifts to the north. Whatever the relationship between rifting and alkaline magmatism, the Chilwa Province shows beyond doubt that it is *not accidental* but has been *reproduced* at different periods and in different places. Indeed, the pre-Tertiary alkaline complexes of the rifts to the north indicate that the process has also been repeated at different periods in the same region. Repeated activity of this type emphasizes the need for a petrogenetic hypothesis of general application, which must eliminate such processes as assimilation and magma mixing. Singular, or unique, thermal events must also be ruled out as inadequate.

This argument is further strengthened by consideration of other rifts. In the Rhine–Oslo zone, for instance, the Tertiary–Recent alkaline volcanism of the Rhine rift has its image in the Permian alkaline complexes of the Oslo graben, and presumably even in the older L. Palaeozoic complexes such as Fen.

III.2.4. The Relationship Between Rifting and Alkaline Magmatism

In recent years the strong correlation of continental rifting and alkaline magmatism seems to have become widely accepted, and it may be assumed that the two phenomena are expressions of the same fundamental process. Before considering possible processes, however, the critical features must be set down (see Bailey, 1964 for the evidence and the argument).

1. In general rifts are merely the broken crests of crustal arches, which are the major tectonic features.
2. These arches have been positively uplifted with respect to the surrounding broad basinal areas: they are not lag-areas that have failed to keep pace with subsidence.
3. Uplift took place in distinct stages, indicated by flights of erosion surfaces which show increasing vertical separation as they rise towards the rift shoulders.
4. Preservation of a series of uplifted erosion surfaces requires a structurally competent crust.
5. Magmatism frequently seems to be most intense where uplift is greatest, especially at culminations such as rift intersections.

In an earlier paper (Bailey, 1964) the point was made that the Tertiary rift zones have been uplifted, and then peneplained, several times since the Jurassic. This requires that approximate isostatic equilibrium had been achieved at the end of each erosion cycle. To achieve this, less dense material must be added at depth, to compensate and provide for each succeeding uplift. A recent gravity profile across East Africa by Sowerbutts (1969) has shown the need for a model with a mass deficiency in the upper mantle under the rifts, and the adjacent plateaux. Continual uplift of this region requires that lighter material has been successively added to the upper mantle layers. Mere thermal expansion of the upper mantle (Harris, 1969; Gass, 1970) is not enough, because there would then have to be collapse of the uplift as the heat was dissipated.

As far as the magmatism is concerned the following points need emphasis.

1. Although alkali basalt is a typical member of the association, there are vast amounts of felsic magmas, even in dominantly basaltic

provinces such as Ethiopia. In some provinces, such as Chilwa, basaltic rocks are a very minor part of the association.

2. At certain periods trachytic and phonolitic magmas have been developed on a regional scale. King (1970) has estimated the volume of trachytic volcanics in the central portion of the Kenya rift as exceeding 300,000 cubic kilometres.

3. Peralkalinity is a strongly characteristic property of the felsic magmas. Alkali metasomatism is typical around the intrusions.

4. Rocks of the nephelinite-carbonatite associations are present on a scale unmatched by any other region of the crust. This could be regarded as the critical association of the rifts.

5. Perhaps the most important feature of the volcanism is its highly explosive, fragmental nature. *Abnormal enrichment in volatiles* and alkalis is the keynote of rift magmatism. The volatiles cannot be re-cycled oceanic or crustal material, they must be largely mantle-derived.

Consider the magmatism first. It is characterized by large quantities of what are normally considered 'low' temperature, or 'residual' fluids—felsic magma and volatiles. Indeed, the lavas of the East African Rift were the prime example used by Bowen (1937) to demonstrate the applications of Petrogeny's Residua System (nepheline–kalsilite–silica). Bowen clearly considered the rhyolites, trachytes and phonolites of the rift to be residua from the fractional crystallization of basalt magma (see, however, Macdonald, *et al.*, 1970). This was challenged (Bailey, 1964; Bailey and Schairer, 1966) on the grounds that the volume of felsic magma would require vast quantities of basaltic magma at depth, and the alternative proposed was that of partial melting of material of basic composition. Basaltic magma, after all, is generally considered to be the product of partial melting of material of peridotite composition, at even higher temperatures. It is hardly logical to postulate the generation of vast quantities of unseen basalt magma, only to have it crystallize in order to provide the observed lower temperature residua. It is proposed to examine the partial melting hypothesis in the light of the observed magmatism, and then to consider how this might be related to possible mechanisms of uplift and rifting.

III.2.5. Partial Melting

The general requirements for melting within the Earth will be found in the part of this book dealing with anatexis (VI.1a, b) and the detailed arguments will be found there on the possible controls and their influence on possible source materials. From the actual magmatism in the rifts it will be assumed only that we are witnessing the effects of a melting regime operating on the upper mantle and deep crust.

Clearly, if we are looking for a regional or general control of magmatism, two conditions should be present. Firstly, certain types of magma must be shown to be highly characteristic of the rift setting (e.g. carbonatites). Secondly, particular magmas must be generated in large volumes on a regional scale—either in concentrations of central volcanoes (e.g. the chain of large nephelinite volcanoes of E. Uganda) or as flood eruptions. The last category, flood eruption, is especially important in the hypothesis of partial melting, because, while the products of a central volcano can possibly be ascribed to differentiation in a local, subjacent magma chamber, such a process *cannot account for floods of magma of constant composition over a large region*. It is asserted, therefore, that regional floods of alkali basalt (Ethiopia), and trachyte and phonolite (Kenya) are the products of pervasive melting in the mantle and deep crust. Regional development of the unusual magma, nephelinite (E. Uganda and N. Tanzania), can also be described only in terms of a general, pervasive process. Within a cycle of regional melting, individual central volcanoes, emitting their own specialized products, can develop and persist: they represent local hotspots superimposed on a regional thermal pattern.

Given that regional eruptions are the result of pervasive melting, three episodes in the magmatic history of the central part of the East African rift are of paramount importance. These are emphasized in Table 1, and correlate with crucial tectonic phases; they are:

3. Flood trachytes (Kenya: main rifting);
2. Flood phonolites (Kenya: main doming);
1. Regional nephelinites (E. Uganda: early uplift).

These episodes can be linked to progressive heating of the mantle and deep crust, or gradual uprise of geo-isotherms below the rift zones. The early nephelinites may be correlated with incipient melting in the upper mantle (see II.3), and represent the response to a newly initiated heating cycle below the uplifted rift zone (Fig. 5). They were followed by more localized basalt volcanicity—representing local hot-zones along which more extensive melting took place in the mantle. With the passage of time the heating cycle affected the uppermost mantle and deep crust, where conditions were such that the early melt products were phonolitic; the major uplift of the Kenya dome culminated in a regional outpouring of phonolite lavas.

Development of the rift along the axis of the uplifted region, meant that a segment of the crust was lowered into the heated region. Re-establishment of the isotherms at higher levels in this segment resulted in pervasive partial melting at higher levels in the crust and outpouring of regional trachytes. Subsequent local development of calderas and central volcanoes in the rift floor represent slowing down of the heating cycle, and persistence of heat channels at local points in the crust. Along the eastern flanks of the dome dominantly phonolitic activity persisted, at centres such as Mt. Kenya and Kilimanjaro.

The scheme outlined above depicts the operation of a cycle of pervasive melting passing from the mantle into the crust. It raises the question of why no regional floods of basalt were developed during the cycle. But if it is accepted that nephelinite is the incipient melt-product of the deep upper mantle (see II.3 for discussion) there is no problem. It means, simply, that the heat-flux through the mantle generally had not been sufficient to cause the more advanced degree of melting involved in regional basalt generation. A local basaltic centre thus represents a local high in the thermal regime of the mantle below this sector of the rift. As the rift is followed northwards into Ethiopia flood basanites and basalts come in, and we are witnessing the overwhelming effects of a greater heat flux through the mantle. In this situation, the lower melting products are bound to appear diminished in relation to the basalts. But in addition, the great heat loss involved in large-scale basalt effusion may so disturb the heating cycle that sufficient heat is not available to effect pervasive melting in the deep crust.

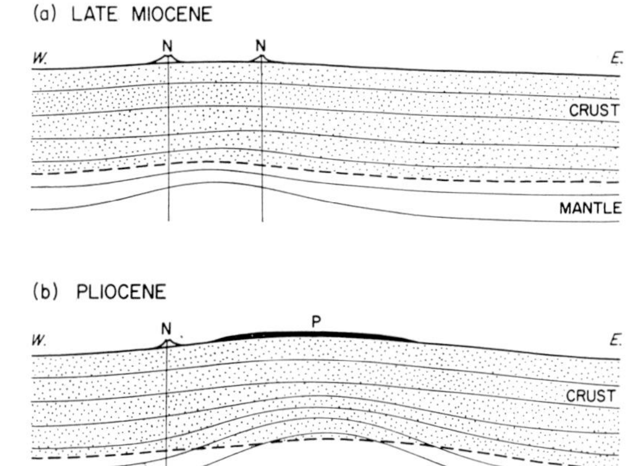

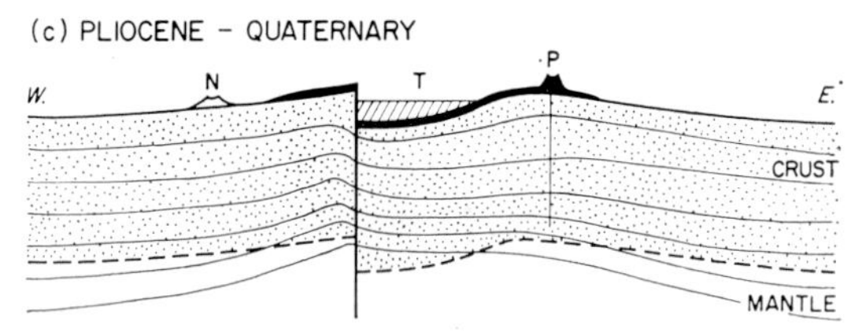

Fig. 5 Schematic sections across the Uganda–Kenya section of the East African Rift (adapted from Baker, 1965). Showing three phases of regional magmatism associated with the development of the rift arch, and the progressive uprise of the geo-isotherms (fine lines)

(a) Early nephelinite–carbonatite phase (N). Rising isotherms in the upper mantle causing incipient melting.

(b) Major uplift of the Kenya dome. Rising isotherms in the uppermost mantle and deep crust, resulting in pervasive partial melting and phonolite floods (P).

(c) Rift zone lowered, causing further uprise of isotherms in this crustal segment. Pervasive partial melting and regional trachyte magmatism (T) along rift. Central phonolitic volcanoes (P) (e.g. Mt. Kenya) on the eastern rift shoulder–residual arch structure.

Finally it should be pointed out that nephelinites and carbonatites may also appear last in the sequence, during the waning stages of the heating cycle as the isotherms collapse. The present day activity in N. Tanzania, and the late stages on Kilimanjaro, would fit this part of the pattern. Nephelinites and/or carbonatites may also occur alone, as in the Rufunsa province of Zambia (Bailey, 1966), where the heating cycle fails after incipient melting and eruption from the mantle. By the same token, the kimberlite–carbonatite association of the non-rifted areas would represent an aborted cycle involving only the eruption of gas-transported solid material from the mantle.

III.2.6. Mechanisms of Crustal Arching and Heating

Heating of the zone below the continental arches is amply testified in the magmatism—the outstanding question is how are the two processes, uplift and heating, related?

In a recent discussion of the basaltic magmatism of the East African Rift, Harris (1969) proposes that the volcanism and the uplift are both reflections of the thermal state of the underlying mantle. In other words heating of the mantle causes expansion, and thence uplift of the overlying crust, accompanied by volcanism. A similar mechanism is proposed by Gass (1970). It is difficult to fit the characteristic alkaline magmatism, and the rift tectonics into this simple picture. Moreover, non-volcanic sectors of the rift, such as Lake Tanzania, are not accounted for in this scheme. But perhaps the biggest discrepancies in the thermal expansion model are, (1) persistence of the rifts and alkaline magmatism from Cretaceous (and possibly Precambrian) times; (2) failure of uplifts to collapse as the heat is dissipated; and, (3) the source of the heat, and its repeated generation in long narrow belts. Harris appears to favour upward mantle convection as the ultimate driving force, but if this is the mechanism it has failed to produce any crustal separation over a long period of time, and it requires a complex pattern of small convection cells. On the basis of the gravity data Sowerbutts (1969) has concluded that the mantle convection alone is inadequate as the heating agent, and appeals to heat focussing in the rift zones by volatile transport as proposed by Bailey (1970).

Some years ago (Bailey, 1964), I took the then unfashionable view that the basin and swell pattern of the African continent might be the response of an anistropic rigid plate to radial compressive forces, originating in the girdle of mid-ocean ridges around Africa. Now, all the major features of the Earth are being described in terms of rigid plate tectonics (Morgan, 1968; Le Pichon, 1968; Isacks, Oliver and Sykes, 1968). So far little attention has been given to internal deformation in the large rigid plates, but already the suggestion has been made (Morgan, 1968, p. 1960) that the Macquarie Ridge in the Pacific is a compression feature, complementary to a colinear tensional ridge to the north. The East African Rift might be an analogous structure, continuing as it does into the tensional zone of the Gulf of Aden. In any case, if it is accepted that arching may be essentially a mechanical response to forces transmitted through a rigid plate, then convective wedging below the rift zones is no longer necessary to their formation. If an arch forms in a competent rigid plate, partial melting by decompression may take place in the underlying zone of reduced pressure. Furthermore, volatiles from the surrounding mantle will flow into the zone of pressure relief, bringing additional heat and reducing the melting range of the rocks therein (Bailey, 1970). In a subsequent extension of these ideas (Bailey, 1972), it is pointed out that initial bending and fracturing of the lithosphere is all that is needed to set a cycle of uplift, rifting, and alkaline magmatism in motion. Once warping and crustal fracturing are established the zone will to some extent be self-perpetuating—acting as a heat and volatile drain on the underlying mantle. In addition to carrying heat, the volatile flux will bring with it mobile elements, especially alkalis; and, under favourable conditions, phlogopite, amphibole and carbonates will be formed in the mantle. These less dense minerals will cause expansion and uplift, and provide parental materials for the later alkaline magmatism. This mechanism is in harmony with the abundance of volatile-rich, alkaline magmas so characteristic of the rifts. It explains the intermittent uplift of the rift zones as local responses to major tectonic events at the margins of the rigid plate.

This model has the further advantage that the type, and the amount of magmatism, depends on the interplay of several factors, chief of which are: the thermal condition of the mantle and crust prior to uplift; the nature of the volatiles and their level of derivation; and, the rate and method of volatile release. Those parts of the rift where magmatism is feeble, or absent, may have been regions of abnormally low heat-flow prior to uplift; starting from this condition melting by decompression may be impossible; influx of volatiles is not likely to induce melting either, but it would produce some heating and light-element metasomatism of the underlying mantle, causing expansion and satisfying the density requirements.

Variation in the nature of the volatiles, and their source-levels in the mantle might explain some of the variability seen in the alkaline magmatism of the rifts. For instance, the unusual potassic magmatism around the strongly uplifted Ruwenzori horst in S.W. Uganda, may be a direct consequence of the peculiar crust/mantle relations that must exist in this sector.

Rate and method of volatile release will influence the expression and possibly the character of the resulting magmatism. Release through a relatively small number of 'open' channels will favour central type volcanicity, possibly with higher temperature magmas generated on a local scale. Waves of volatiles diffusing through a broad region will be conducive to eventual flood eruptions.

Probably the most important feature of this hypothesis is that it provides its own specific answer to a problem that is fundamental to all magmatism—concentration of the heat necessary for melting. Warping and fracturing, with resultant pressure relief, localize the degassing of the underlying mantle, forming a zone into which volatiles migrate. Heat and alkalis carried by the volatiles, are thus focussed along these zones and their combined effects can set in motion a cycle of alkaline magmatism in the mantle and deep crust.

III.2. REFERENCES

Bailey, D. K., 1961. The mid-Zambezi–Luangwa rift and related carbonatite activity. *Geol. Mag.*, **98,** 277–84.

Bailey, D. K., 1964. Crustal warping—a possible tectonic control of alkaline magmatism. *Jour. Geophys. Res.*, **69,** 1103–11.

Bailey, D. K., 1966. 'Carbonatite volcanoes and shallow intrusions in Zambia', in *The Carbonatites*, eds. O. F. Tuttle and J. Gittins. John Wiley and Sons, New York.

Bailey, D. K., 1970. Volatile flux, heat focussing and the generation of magma. *Geol. J., Spec. Iss.*, **2,** 177–86.

Bailey, D. K., 1972. Uplift, rifting and magmatism in continental plates. *Leeds Univ. J. Earth Sci.*

Bailey, D. K., and Schairer, J. F., 1966. The system Na_2O–Al_2O_3–Fe_2O_3–SiO_2 at 1 atms., and the petrogenesis of alkaline rocks. *J. Petrology*, **7,** 114–70.

Baker, B. H., 1965. The rift system in Kenya (Pt. 1. pp. 82–4) and, An outline of the geology of the Kenya rift valley (Pt. II. pp. 1–19). *UNESCO Seminar*. East African Rift Systems, University College, *Nairobi*.

Baker, B., and Wohlenberg, J., 1971. Structure and evolution of the Kenya Rift Valley. *Nature*, **229,** 538–42.

Black, R., and Girod, M., 1968. Contrôle structural du volcanisme ancient et récent dans les régions du Hoggar, Aïr, Nigéria et Cameroun. *Proc. Geol. Soc. Lond.*, **1644,** 263–6.

Black, R., and Girod, M., 1970. 'Late Palaeozoic to Recent activity in West Africa and its relationship to Basement structure', in *African Magmatism and Tectonics*, eds. T. N. Clifford and I. G. Gass. Oliver and Boyd, Edinburgh, 185–210.

Bloomfield, K., 1965. The geology of the Zomba area. *Bull. geol. Surv. Malawi*, **16,** 193 pp.

Bowen, N. L., 1937. Recent high-temperature research on silicates and its significance in igneous geology. *Am. J. Sci.*, **33,** 1–21.

Cook, K. L., 1969. Active rift system in the Basin and Range Province. *Tectonophysics*, **8,** 469–511.

Denaeyer, M. E., and Schellinck, F., 1965. Recueil

d'analyses des laves du fossé tectonique de l' Afrique Centrale. *Annales Mus. Roy. l'Afrique Centrale*, **49.**

Dixey, F., Campbell Smith, W., and Bissett, C. B., 1955. The Chilwa Series of southern Nyasaland. *Bull. geol. Surv. Nyasaland*, **5** (2nd edn. revd.).

Eardley, A. J., 1961. *Structural geology of North America*. 2nd edn., Harper and Row, New York.

Florensov, N. A., 1969. Rifts of the Baikal mountain region. *Tectonophysics*, **8**, 443–56.

Garson, M. S., 1965. Carbonatites in southern Malawi. *Bull. geol. Surv. Malawi*, **15.**

Gass, I. G., 1970. 'Tectonic and magmatic evolution of the Afro-Arabian dome', in *African Magmatism and Tectonics*, eds. T. N. Clifford and I. G. Gass. Oliver and Boyd, Edinburgh, 285–300.

Harkin, D. A., 1960. The Rungwe volcanics at the northern end of Lake Nyasa. *Mem. geol. Surv. Tanganyika*, **2.**

Harris, P. G., 1969. Basalt type and rift valley tectonism. *Tectonophysics*, **8**, 427–36.

Holmes, A., 1940. Basaltic lavas of S. Kivu, Belgian Congo. *Geol. Mag.*, **77**, 89–101.

Illies, J. H., 1969. An intercontinental belt of the world rift system. *Tectonophysics*, **8**, 5–29.

Innes, M. J. S., 1960. Gravity and isostasy in N. Ontario and Manitoba. *Publ. Dom. Obs. Ottawa*, **21**, 263–338.

Innes, M. J. S., Goodacre, A. K., Weber, J. R., and McConnell, R. K., 1967. Structural implications of the gravity field in Hudson Bay and vicinity. *Can. J. Earth Sci.*, **4**, 1–17.

Isacks, B., Oliver, J., and Sykes, L. R., 1968. Seismology and the new global tectonics. *J. Geophys. Res.*, **73**, 5855–900.

Kanesewich, E. R., Clowes, R. M., and McCloughlan, C. H., 1969. A buried Precambrian rift in Western Canada. *Tectonophysics*, **8**, 513–28.

King, B. C., 1965. Volcanism in eastern Africa and its structural setting. *Proc. geol. Soc. Lond.*, **1629**, 16–19.

King, B. C., 1970. 'Vulcanicity and rift-tectonics in East Africa', in *African Magmatism and Tectonics*, eds. T. N. Clifford and I. G. Gass. Oliver and Boyd, Edinburgh, 263–83.

Le Pichon, X., 1968. Sea-floor spreading and continental drift. *J. Geophys. Res.*, **73**, 3661–97.

Macdonald, G. A., and Katsura, T., 1964. Chemical composition of Hawaiian lavas. *J. Petrology*, **5**, 82–133.

Macdonald, R., Bailey, D. K., and Sutherland, D. S., 1970. Oversaturated peralkaline glassy trachytes from Kenya. *J. Petrology*, **11**, 507–17.

McBirney, R., and Gass, I. G., 1967. Relations of oceanic volcanic rocks to mid-oceanic rises and heat flow. *Earth Planet. Sci. Lett.*, **2**, 265–76.

McConnell, R. B., 1969. Fundamental fault zones in the Guiana and West African Shields in relation to presumed axes of Atlantic spreading. *Bull. geol. Soc. Am.*, **80**, 1775–82.

McKie, D., 1966. 'Fenitization', in *The Carbonatites*, eds. O. F. Tuttle and J. Gittins. John Wiley and Sons, New York, 261–94.

Mohr, P. A., 1963. The Ethiopian Cainozoic lavas. *Bull. geophys. Obs. Addis Ababa*, **3**, 103–44.

Mohr, P. A., 1965. Re-classification of the Ethiopian Cainozoic volcanic succession. *Nature*, **208**, 177–8.

Morgan, W. J., 1968. Rises, trenches, great faults and crustal blocks. *J. Geophys. Res.*, **73**, 1959–82.

Morgan, W. J., 1971. Convection plumes in the lower mantle. *Nature*, **230**, 42–3.

Oliver, J., Sykes, L., and Isacks, B., 1969. Seismology and the new global tectonics. *Tectonophysics*, **7**, 527–41.

Sørensen, H., 1970. Internal structures and geological setting of the three agpaitic intrusions—Khibina and Lovozero of the Kola peninsula and Ilímaussaq, South Greenland. *Can. Miner.*, **10**, 299–334.

Sowerbutts, W. T. C., 1969. Crustal structure of the East Africa plateau and rift valleys from gravity measurements. *Nature*, **223**, 143–6.

Tuttle, O. F., and Gittins, J., 1966. *The Carbonatites*. John Wiley and Sons, New York.

Williams, L. A. J., 1965. Petrology of the volcanic rocks associated with the rift system in Kenya. *UNESCO Seminar*. East African Rift System. University College, *Nairobi*. Pt. II, 33–9.

Wilson, J. Tuzo., 1967. Rift valleys and continental drift. *Trans. Leicester Lit. Phil. Soc.*, **61**, 22–35.

Wright, J. B., 1963. A note on possible differentiation trends in Tertiary to Recent lavas of Kenya. *Geol. Mag.*, **100**, 164–80.

Wright, J. B., 1965. Petrographic sub-provinces in the Tertiary to Recent volcanics of Kenya. *Geol. Mag.*, **102**, 541–77.

III.3. ALKALINE ROCKS OF NORTH AMERICA

D. S. Barker

III.3.1. Introduction

Table 1 lists North American occurrences of feldspathoidal rocks and silica-saturated or oversaturated rocks containing sodic pyroxenes or amphiboles. The widespread lamprophyres, where not known to be associated with feldspathoid-feldspar rocks, are arbitrarily excluded from this compilation. For seven occurrences in Table 1, the interpretation as carbonatite-alkaline rock complexes has been based on intense circular or elliptical aeromagnetic anomalies (Satterly, 1968). Calcalkaline plutons can produce the same aeromagnetic patterns (Card, 1965); these localities are therefore listed with reservation and preceded by asterisks.

TABLE 1 North American Occurrences of Alkaline Rocks.

For each occurrence, this format is followed: name, longitude W, latitude N; form; estimated area (km^2); rock types; age in millions of years (geochronologic method); references. For abbreviations, see footnote at end of table.

Cape Richards, 79° 20′, 83°; R; 40; sy, qz monz, a–r gr; 390 ± 18 (2), 347 ± 15 (1); Frisch (in press).
Selawik Hills, 160°, 66° 15′; I; 200; ne sy, mal, sy, shonk, um, nep; 107 ± 2.8 (1); Patton and Miller (1968).
Spotted Fawn Creek, 138° 32′, 64° 17′; ?; ?; pslc phon; Paleozoic or younger; K. L. Currie (personal communication, 1968).
Chichagof Island, 135°, 57° 45′; I; 125; ne sy, sod sy, di, monz, gd, gr, gb; > 406 ± 16 (2); Lanphere *et al.* (1965).
Bokan Mountain, 132° 10′, 54° 55′; I; 7; sy, a–r gr; 431 ± 21 (2), 446 ± 22 (2); Lanphere *et al.* (1964).
Lonnie, 124° 30′, 56°; C; ?; carb, sy; ?; Rowe (1958).
Verity, 119°, 52° 30′; ?; ?; carb, sod sy; ?; Rowe (1958).
Rogers Pass-Big Bend, 118°, 52°; D, C; ?, ne sy, sy; Paleozoic?; Wheeler (1962, 1964).
Ice River, 116°, 51°; I; 125; carb, ne sy, sod sy, nep, um; 340 (1); Rapson (1966).
Crowsnest Pass, 114° 30′, 49° 45′; P, F, S, D; 1500; anl phon; Cretaceous; Mackenzie (1956).
Coryell Intrusions, 118°, 49° 30′; I, Pl; ?; sy, qz sy, gr, monz; Eocene or later; Little (1960).
Shasket Creek, 118° 33′, 48° 58′; Pl, D; 2; ne sy, shonk, sy, monz; Cretaceous?; Parker and Calkins (1964).
Mount Kruger, 119° 37′, 49°; I, D; 25; ne sy, mal, carb; Oligocene; Campbell (1939); Rinehart and Fox (1968).
Rock Creek, 119° 05′, 49°; I, F; 12; anl phon; post Early Oligocene; Daly (1912).
Big Spruce Lake, 115° 57′, 63° 33′; R; 12; ne sy, sy; 1785 (1); Leech *et al.* (1963).
Kaminak Lake, 94° 50′, 62° 15′; R; 10; ne sy, sy, nep; ?; Davidson (1968).
Carb Lake, 92°, 54° 45′; R; ?; carb; ?; Satterly (1968).
*Conifer-Sumach, 94° 10′, 50° 30′; R; ?; ?; ?; Satterly (1968).
*Falcon Island, 94° 45′, 49° 23′; R; ?; ?; ?; Satterly (1968).
Poohbah Lake, 91° 40′, 48° 20′; I; 35; mal; ?; Lawson (1896).
Wausau, 89° 45′, 44° 50′; D, I; 7; ne sy, sy, qz sy; ?; Emmons (1953).
Big Beaverhouse, 89° 55′, 52° 55′; R; 2; carb, um; 1005 (1); Duffell *et al.* (1963).
Schryburt Lake, 89° 35′, 52° 35′; R; 2; carb; ?; Duffell *et al.* (1963).
Prairie Lake, 86° 44′, 49° 02′; R; ?; carb, ne sy, nep; 1112 (1); Gittins (1966).
Coldwell, 86° 30′, 48° 45′; R; 300; ne sy, sy; 1065 (1), 1070 (5); Fairbairn *et al.* (1959).
Killala Lake, 86° 30′, 49° 10′; R; 5; ne sy, sy, gb; ?; Coates (1968).
Chipman Lake, 86° 15′, 49° 55′; R; ?, carb; ?; Gittins (1966).
*Nagagami River, 84° 15′, 50° 10′; R; ?; ?; ?; Satterly (1968).
*Squirrel River, 84° 15′, 50° 20′; R; ?; ?; ?; Satterly (1968).
*Mammamattawa, 84° 20′, 50° 25′; R; ?; ?; ?; Satterly (1968).
*Kingfisher River, 84° 30′, 50° 35′; R; ?; ?; ?; Satterly (1968).
Martison Lake, 83° 25′, 50° 20′; R; ?; carb; ?; Satterly (1968).

Firesand River, 84° 40′, 48°; R; 2; carb, um, qz sy; 1048 (1); Gittins *et al.* (1967).
French River, 80° 40′, 46° 05′; C; 2; gn ne sy, sy; ?; J. G. Payne, pers. comm. (1968).
Township 107, 81° 45′, 46° 40′; R; ?; carb, ne sy; 1560 (1); Gittins (1966).
Manitou Islands, 79° 35′, 46° 15′; R; 2; carb, ne sy, qz sy; 560 (1); Lowdon *et al.* (1963).
Callander Bay, 79° 25′, 46° 15′; R; ?; carb, ne sy, shonk; ?; Gittins (1966).
Seabrook Lake, 83° 19′, 47°; R; 1; carb, nep; 1103 (1); Parsons (1961).
Lackner and Portage, 83° 05′, 47° 50′; R, I; 20; carb, ne sy, um; 1090 (1), 1069 (1); Rowe (1958).
Nemegosenda Lake, 83° 05′, 48°; R; 20; carb, ne sy, um; > 1010 (3); Gittins *et al.* (1967).
*Shenango, 82° 49′, 48° 25′; R; ?; ?; ?; Satterly (1968).
Cargill, 82° 50′, 49° 20′; R; ?; carb, um; 1740 (1); Gittins (1966).
Clay-Howells, 82° 05′, 49° 50′; R; ?; carb; 1010 (1); Gittins (1966).
Goldray, 81°, 50° 30′; R; ?; carb; 1695 (1); Satterly (1968).
Argor, 80° 40′, 50° 50′; R; ?; carb; 1655 (1); Satterly (1968).
Blue Mountain, 77° 58′, 44° 40′; C; 7; ne sy; 1285 ± 41 (6), 900 (1); Krogh and Hurley (1968); Payne (1968).
Haliburton–Bancroft, 78°, 45°; C; 50; ne sy, gn ne sy, sy, nep, ne gb; 800–1000 (1, 3); Gittins (1967); Appleyard (1967).
Kipawa, 79°, 46° 45′; C; 2, gn ne sy, gn nep; 1290 (3), 904 (1); Macintyre *et al.* (1969).
Booth, 78° 30′, 46° 50′; C; ?; ne sy, gn ne sy; ?; A. F. Laurin (pers, comm., 1969).
Cabonga Reservoir, 76° 29′, 47° 12′; C; ?; gn ne sy; 894 (1); Macintyre *et al.* (1969).
Meach Lake, 75° 30′, 45° 30′; R; ?; carb, qz sy; 920 (1); Hogarth (1966).
Mt. Laurier, 75° 30′, 46° 30′; ?; ?; ne sy, sy, qz sy; 802–908 (1); Doig and Barton (1968).
St. Veronique, 75° 01′, 46° 31′; R; ?; ne sy, sy, um; 859 (1), 967 ± 30 (1); Doig and Barton (1968); Wanless *et al.* (1967).
Obedjiwan (Gouin Reservoir), 74° 55′, 48° 40′; R; 50; ne sy, nep; 988 (3), 931 (1); Macintyre *et al.* (1969).
Albanel, 73°, 50° 50′; C; ?; gn ne sy, gr; ?; Neilson (1953).
Chicoutimi, 71°, 48° 20′; D, R?; ?; carb, sy; 560, 563 (1), 568 (4); Doig and Barton (1968).
Lac de la Brèche, 78° 45′, 56° 30′; Pl, D, P; ?; carb, kimb; ?; Dimroth (1967).
Letitia Lake, 62° 25′, 54° 15′; C; 2; gn ne sy, sy; ?; Brummer and Mann (1961).
Mutton Bay, 59°, 50° 50′; R; 25; sy, um, carb; 566–577 (1), 630 ± 35 (5); D. P. Gold (pers. comm., 1968); Doig and Barton (1968).
Aillik, 59° 15′, 55° 15′; D; ?; carb, lamp; 570 (1); Leech *et al.* (1963).
Oka, 74° 30′, 45° 30′; R; 7; carb, nep, um, lamp; 114 ± 7 (5); Gold *et al.* (1967); Philpotts, cf. IV.6.
Monteregian Hills West (Mt. Royal, St. Bruno, St. Hilaire, Mt. Johnson, Rougemont, and satellitic bodies), 73–74°, 45° 30′; R, D, S, Pl; 30; ne sy, sod sy, ne gb; 113–124 (6); Gold (1967); Philpotts, cf. IV.6.
Monteregian Hills East (Mt. Yamaska, Shefford, Brome, Megantic), 71° 30′–73°, 45° 30′; R, D; 70; ne sy, qz sy, gb; 84–122 (1, 5, 6); Gold (1967); Philpotts, cf. IV. 6.
White Mountains, 72°, 43° 45′; D, I, R; 1250; sy, qz sy, a–r gr, gb; 110–185 (1); Foland *et al.* (1970).
Red Hill, 71° 30′, 43° 45′; R; 15; ne sy, sod sy, sy, qz sy, gr; ?; Quinn (1937).
Cuttingsville, 72° 50′, 43° 30′; I; 4; ne sy, sod sy, sy, ne gb; ?; Eggleston (1918).
Pleasant Mountain, 70° 50′, 44°; R; 7; anl sy, sy, qz sy, trach; ?; Jenks (1934).
Litchfield, 69° 50′, 44° 15′; R; 6; ne sy, sy; 242 ± 5 (3), 234 ± 5 (1); Barker (1965); Burke *et al.*, (1968).
Agamenticus, 70° 40′, 43° 12′; R; 12; sy, a–r gr; 227 ± 3 (6), 216 (1); Hoefs (1967); Foland *et al.* (1970).
Essex County, 70° 50′, 42° 33′; I, D; 300; ne sy, sy, a–r gr; 440 (6, 8); Toulmin (1964); R. E. Zartman (pers. comm., 1968).
Cashes Ledge, 68° 56′, 42° 54′; ?; ?; a–r gr; ?; Toulmin (1957).
Southern Massachusetts, Rhode Island, eastern Connecticut, 71–72°, 41° 30′–42°; I, C; ?; a–r gr, gn a–r gr; ?; Chute (1950); Quinn *et al.* (1949); Goldsmith (1961); Lundgren (1963).
Beemerville, 74° 42′, 41° 14′; S, D; 2; ne sy, phon; 437 ± 22 (1), 436 ± 41 (5); Zartman *et al.* (1967).
Brookville, 74° 57′, 40° 24′; S; 0·001; ne sy, anl sy; Late Triassic; Barker and Long (1969).
Augusta County, 79°, 38° 15′; D; 2; ne sy, sy, gb; 149 (1, 2), 114 ± 12 (5); Zartman *et al.* (1967).
Mount Rogers, 81° 30′, 36° 30′; I, F; ?; a–r gr, rhy; Late Precambrian; Rankin (1967).
Hicks Dome, 88° 30′, 37° 30′; D; ?; phon, lamp, kimb; 260 (1, 5); Grogan and Bradbury (1968).
Bateman No. 1 Well, 90°, 35° 15′; ?; ?; ne sy; Late Cretaceous; ?; Kidwell (1951).
Robroy-MacGregor No. 1 Well, 89° 45′, 35° 30′; ?; ?; ne sy; Late Cretaceous: Kidwell (1951).
Monroe Uplift, 91°, 33°; F, P; ?; ne sy?, phon; Late Cretaceous; Moody (1949); Kidwell (1951).
Jackson Dome, 90°, 32° 15′; F, P; ?; ne sy?, phon; Late Cretaceous; Moody (1949); Kidwell (1951).

Perkins and White, Lee No. 1 Well, 91°, 35°; F, P; ?; phon; Kidwell (1951).

Little Rock, 92° 05′, 34° 35′; I; 600; ne sy, sy, qz sy; 88 (1); Zartman *et al.* (1967).

Brazil Branch, 92° 45′, 35°; Pl; < 0·001, ne sy fragments in intrusive breccia; ?; Croneis and Billings (1929).

Magnet Cove, 93°, 34° 30′; R; 12; carb, ne sy, pslc sy, um, nep, phon, trach; 97 (1, 5); Erickson and Blade (1963); Zartman *et al.* (1967).

Potash Sulfur Springs, 93° 05′, 34° 30′; R; 2; carb, ne sy, sy, um, phon, trach; ?; Hollingsworth (1967).

Uvalde County–Balcones fault zone, 97°–100° 30′, 28° 30′–31° 15′; S, Pl, Lac, F, P, D; 25, nep, um, anl phon; 70 ± 5 (2, 3, 4); Spencer (1969); Burke *et al.* (1968).

Headquarters Mountains and Wichita Mountains, 99°, 34° 45′; I; ?; a–r gr; 525 ± 25 (1); Ham *et al.* (1964).

Rainy Creek, 115° 30′, 48° 30′; R, D, I; 15; ne sy, sy, um; 94 (5); Boettcher (1967).

Margin of Boulder Batholith, 112° 10′, 46° 35′; D; < 0·001; shonk; 68–78 (1); Knopf (1957).

Sweet Grass Hills, 111° 30′, 49°; I, S, Lac, D; 250; sy, phon, trach, di; ?; Kemp and Billingsley (1921)

Bearpaw Mountains, 109° 30′, 48° 10′; I, D, S, Lac, F; 750; carb, phon, ne sy, pslc sy, sy, shonk, um, qz monz; 52 ± 2 (1); Hearn *et al.* (1964).

Little Rocky Mountains, 108° 30′, 48°; I, S, Lac, D; ?; phon, sy, qz sy; ?; Weed and Pirsson (1896b).

Highwood Mountains, 110° 30′, 47° 30′; I, S, Lac, D, F; 400; ne sy, phon, anl phon, lc phon, sy, shonk, monz; ?; Larsen *et al.* (1941).

Big Belt Mountains, 112°, 47° 10′; I, S, Lac, D, F; ?; anl trachybasalt, qz latite; ?; Lyons (1944).

Little Belt Mountains, 110° 45′, 47°; I, S, Lac, D; ?; anl–ne sy, shonk, qz sy; 41 ± 1 (1), 45 ± 2 (2); Weed and Pirsson (1899); McDowell (1966).

Castle Mountains, 110° 45′, 46° 30′; I, S, Lac, D; ?; a–r trach; ?; Weed and Pirsson (1896a).

Moccasin Mountains, 109° 40′, 47° 10′; I, S, Lac, D; ?; a–r qz sy; ?; Blixit (1933).

Judith Mountains, 109° 10′, 47° 10′; I, S, Lac, D; 25; anl sy, phon, sy, gr; 49 ± 2 (2); Weed and Pirsson (1898); McDowell (1966).

Crazy Mountains, 110° 15′, 46°; I, S, Lac, D; 25; ne sy, phon, sy, gb, a–r trach, qz di; ?; Wolff (1938); Simms (1965).

Missouri Buttes–Devil's Tower–Black Hills, 103° 45′–104°, 44° 20′–44° 35′; Pl, Lac, S; 25; ne sy, pslc sy, phon, nep, sy, qz sy; 40·5 (3), 50 ± 1·5 (2), 49 ± 1·5 (2), 54 ± 2 (2), 57 ± 1·5 (1); Darton and Paige (1925); Basset (1961); McDowell (1966).

Leucite Hills, 109°, 42°; F, P, Pl, D; 25; lc–phlogopite–iron sanidine–diopside–glass rocks, 1·1 ± 0·4 (1); Carmichael (1967); McDowell (1966).

Mount Rosa, 104° 45′, 38° 45′; S; 10; a–r gr, a–r sy, lamp; 1040 (2); Gross and Heinrich (1965).

Iron Hill, 107° 10′, 38° 30′; funnel; 30; carb, ne sy, um, nep; 1481 ± 59 (6); Temple and Grogan (1965); Fenton and Faure (1969).

McClure Mountain, 105° 30′, 38° 20′; R, D; 50; carb, ne sy, sy, um; 704 ± 70 (6); Shawe and Parker (1967); Fenton and Faure (1969).

Cripple Creek, 105° 08′, 38° 44′; Pl, F, P; 25; anl sy, sod sy, phon; 34 ± 1 (2); Koschmann (1949); McDowell (1966).

South Park, 105° 45′, 39°; S, D; ?; anl sy, anl diabase; post-Cretaceous, pre-Oligocene; Jahns (1938).

San Rafael Swell, 111°, 39°; D, S, Pl; ?; anl sy, anl diabase; ?; Gilluly (1927).

Moon Canyon, 111° 10′, 40° 45′; Pl, F; ?; phlogopite–diopside–anl–sanidine rocks; 38·5 ± 2 (4); Best *et al.* (1968).

La Plata, 108° 06′, 37° 25′; D, S; ?; sy; 120 ± 4 (2); Eckel (1949); McDowell (1966).

Hopi Buttes and Navajo Country, 110°, 35° 20′; Pl, S, D, F; ?; um, anl basalt, ne trachybasalt, olivine leucitite; Pliocene; Williams (1936).

La Sal Mountains, 109° 10′, 38° 30′; Lac, D; ?; sod sy, sy, gr, di, monz; 23, 26 ± 3 (2), 32 ± 2 (9); Hunt (1958); Stern *et al.* (1965).

Pleasant Valley, 104°, 36° 30′; F, S, D; 60; phon, alkali-olivine basalt; Holocene; Baldwin and Muehlberger (1959).

Pajarito Mountain, 105° 26′, 33° 14′; ?; 2; sod? sy, qz sy, sy; 1170 ± 24 (2), 1190 ± 25 (2), 1155 ± 25 (5); Kelley (1968).

Cornudas Mountains and Diablo Plateau, 105° 30′, 32°; Lac, D, S; 20; anl–ne sy, phon, trach, qz sy; 43–28 (1,5,6,); Gross (1965); Barker and Long (in preparation).

Davis Mountains–Barrilla Mountains, 104° 15′, 31°; F, D, S, Pl, Lac; 25; ne sy, sy, phon, a–r rhy, a–r gr; post-Oligocene; McAnulty (1955); Eifler (1951).

Christmas Mountains, 103° 27′, 29° 25′; I, D, S; 10; anl sy, um, a–r rhy, alkali–olivine basalt; Tertiary; Swadley (1958); Jenkins (1959).

Terlingua–Big Bend, 103° 40′, 29° 20′; Lac, D, S, Pl, F; 500; anl sy, a–r rhy, a–r gr, trach, nep, mal, alkali–olivine basalt; post-Oligocene; Yates and Thompson (1959); Maxwell *et al.* (1967); Gunn *et al.* (1968).

Solitario, 102° 45′, 29° 30′; Pl, S, D; 0·2; anl sy, rhy, trach; ?; Lonsdale (1940).

Northern Coahuila, 101° 30′–102° 35′, 29° 30′; Pl, S, D, Lac; 7; ne–anl sy, anl gb, nep, um, a–r rhy, qz monz; 35–43 (1); Daugherty (1963, 1965); Sewell (1968).

Sierra de Picachos, 100°, 26° 15′; Lac, S, D; 35; anl–ne sy, anl gb, um; ?; McKnight (1963).

San Carlos Mountains, 98° 30′, 24° 40′; Lac, I, Pl, S, D; 125; ne sy, ne monz, phon, qz sy, nep; ?; Watson (1937).
Mountain Pass, 115° 30′, 35° 30′; I, D; 2; carb, pslc? shonk, qz sy, gr; 1400 ± 50 (1, 5); Olson *et al.* (1954); Lanphere (1964).
Tin Mountain, 117° 30′, 37°; I; ?; ne sy, qz sy; ?; McAllister (1952).
Deep Spring Valley, 118°, 37° 15′; Pl; 0·1; olivine melaleucitite and olivine–lc trachybasalt; ?; Nash and Nelson (1964).
New Idria, 120° 34′, 36° 18′; I; 0·1, anl sy, sy; ?; Eckel and Myers (1946).
Lincoln County, 123° 50′, 44° 30′; S, I; 250; ne–anl sy; late Eocene; Snavely and Wagner (1961).

Abbreviations for form of occurrence:

R = ring dike or zoned stock
Lac = laccolith
D = dykes
S = sills
F = flows
P = pyroclastic rocks
C = concordant foliated body
I = irregular discordant pluton
Pl = plugs, breccia pipes, diatremes

Abbreviations for minerals:

ne = nepheline
lc = leucite
pslc = pseudoleucite
anl = analcime
sod = sodalite
a–r = aegirine and/or riebeckite
qz = quartz

Abbreviations for rock names:

gn = gneissic
sy = syenite
mal = malignite
shonk = shonkinite
nep = nephelinite, urtite, ijolite, melteigite
um = pyroxenite, jacupirangite, limburgite
lamp = lamprophyres, alnoite
phon = phonolite, tinguaite
carb = carbonatites
gb = gabbro, essexite, theralite, teschenite
monz = monzonite
gr = granite
gd = granodiorite
di = diorite
trach = trachyte
rhy = rhyolite
kimb = kimberlite

Notation for age determination methods:

1 = K–Ar, micas
2 = K–Ar, amphibole or pyroxene
3 = K–Ar, nepheline or alkali feldspar
4 = K–Ar, whole rock
5 = Rb–Sr, biotite or feldspar
6 = Rb–Sr, whole rock and isochron
7 = Lead–alpha, zircon
8 = Pb^{207}–Pb^{206}, zircon
9 = Pb^{206}–U^{238}, zircon

Poor exposures, combined with the wide use of drilling, geophysical and geochemical techniques, have resulted in a reversal of the normal situation; the isotopic ages of twelve alkaline complexes are known before their field relations and petrology.

The table is necessarily incomplete, yet it lists all occurrences known to the writer as of December 1969. The tabulation of silica-undersaturated rock occurrences is more complete than the listing of silica-saturated and oversaturated ones; many accounts of riebeckite-bearing syenites and granites have very probably escaped this writer's search.

The references cited are not necessarily the definitive works on given areas; citations are confined to the most recent papers that can serve as guides to the earlier literature.

Some conclusions specifically concerning the feldspathoidal rocks have already been published (Barker, 1969) without the documentation provided here.

III.3.2. Abundance and Distribution

By a conservative estimate, alkaline rocks crop out over at least 8,000 square kilometres of the North American continent. Alkaline rocks are more widespread, and more abundant, than formerly acknowledged. Daly (1933, p. 39) stated that 'the alkaline clans, though including more than half of the named igneous species, are only incidental products of a planet whose eruptions have been overwhelmingly of the subalkaline kind'. This is true, but Daly's often-quoted estimate (Daly, 1933, p. 38) that alkaline rocks make up less than a tenth of 1 % by volume of all igneous rocks needs revision upward, perhaps by a factor of ten.

Of the occurrences listed, one-third lie in the Precambrian shield and only four in the craton (Fig. 1). 'Craton' is here used for that portion of the shield veneered by sedimentary rocks that are undeformed or only gently folded. The

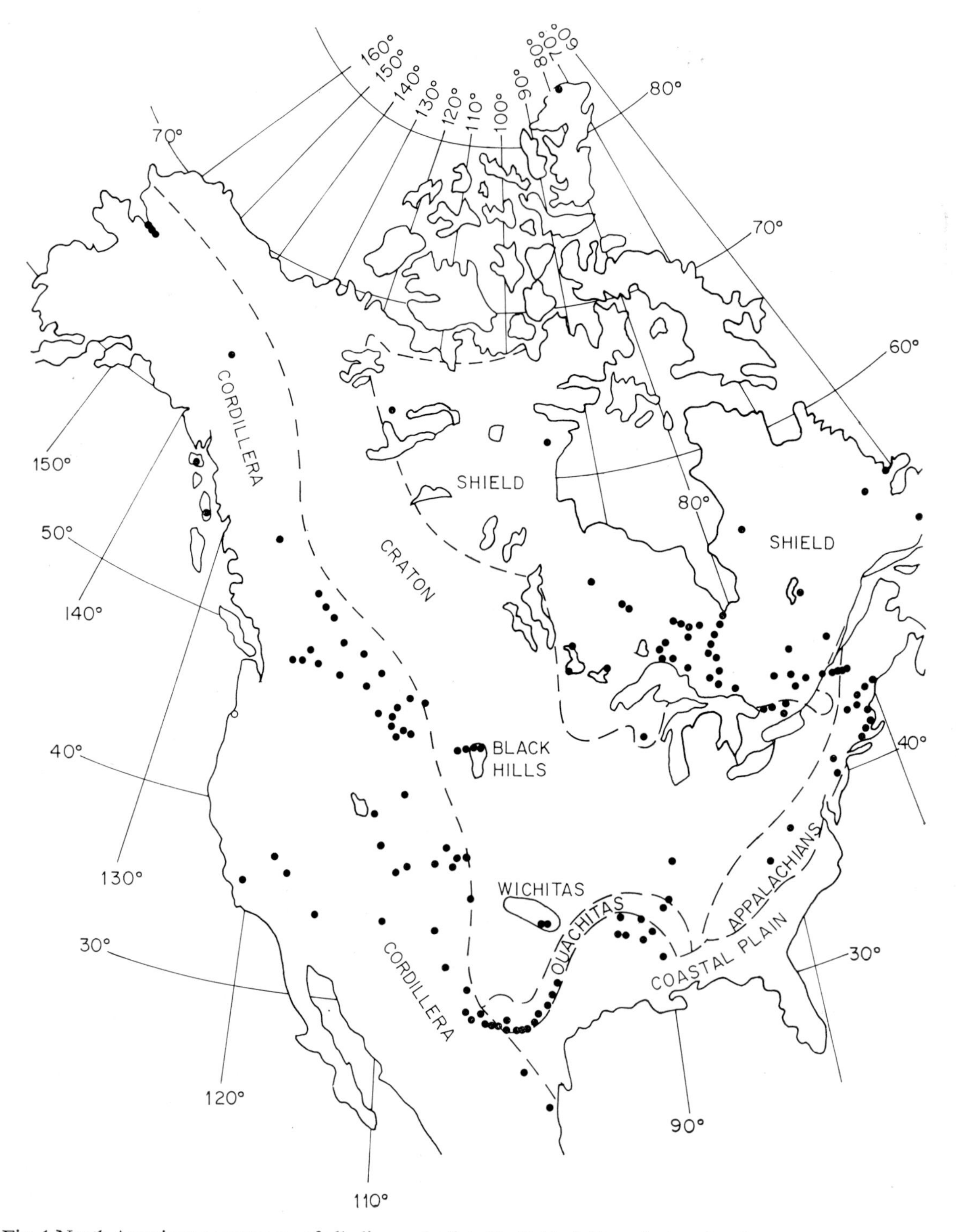

Fig. 1 North American occurrences of alkaline rocks listed in Table 1. Boundaries of major tectonic units of the continent are generalized from King (1959)

remainder are located in linear belts within, and generally parallel or perpendicular to the axes of, Phanerozoic orogens. The occurrences of 'alkaline' rocks in the midcontinent region discussed by Zartman *et al.*, (1967) have not been listed, with one exception, because they consist of kimberlites or lamprophyres without feldspathoid-feldspar rocks.

The exception is the Hicks Dome structure, Illinois, in which tinguaite dykes have recently been found. Some midcontinental 'cryptovolcanic' structures may be alkaline intrusions that have not been breached by erosion; furthermore, Currie and Shafiqullah (1968) reported alkaline ultramafic rocks, rich in K and Mg, forming dykes and the groundmass of intrusive breccias in five circular structures in Canada that have been considered meteorite impact craters. These five occurrences are not listed in Table 1.

III.3.3. Time of Emplacement

Alkaline rocks are far from being exclusively post-tectonic. At least 24 of the 127 occurrences listed were emplaced before or during major uplift, folding, faulting, or granite emplacement (see Bass, 1970; Barker, 1970). These include Selawik Hills, Chichagof Island, Bokan Mountain, Lonnie, Rogers Pass-Big Bend, Crowsnest Pass, Mount Kruger, French River, Blue Mountain, Haliburton–Bancroft, Kipawa, Booth, Cabonga Reservoir, Albanel, Letitia Lake, Essex County, Southern Massachusetts–Rhode Island–Connecticut, Beemerville, Headquarters and Wichita Mountains, shonkinite marginal to the Boulder Batholith, Big Belt and Crazy Mountains, northern Black Hills, and Lincoln County.

The Selawik Hills plutons, Alaska and the sills of Lincoln County, Oregon are particularly instructive as examples of alkaline magmatism contemporaneous with tectonism in eugeosynclines.

In many regions, alkaline magmas were active for much longer times than they could conceivably be stored or generated within the crust. Along the Balcones Fault Zone, Texas, the time of alkaline intrusive and pyroclastic activity was stratigraphically demonstrated by Spencer (1969) to range from Lower Cenomanian to Maestrichtian, a span of approximately 40 million years.

Toulmin (1961, p. 778) predicted that 'more detailed age studies may conceivably show a complete gradation in age among all the "alkalic" rocks of New England'. This has been amply confirmed, with ages ranging from 185 to 440 million years.

The great diversity of isotopic and stratigraphic ages for rocks of similar composition within restricted areas (for example, British Columbia–Washington; Colorado; New Mexico; eastern Ontario and Quebec; New England) indicates that time correlation based upon petrologic similarity and geographic proximity is as unjustified for alkaline igneous rocks as it is for sedimentary rocks. Strongly implied is the necessity for obtaining precise isotopic ages as frequently as one obtains chemical analyses.

Because of the spatial and temporal overlap of alkaline rock groups in North America, the writer is reluctant to define 'petrographic provinces'; the occurrences are arranged in Table 1 in an attempt to group presumably cogenetic bodies, but provinces cannot be circumscribed in Fig. 1.

Times of emplacement fall in intervals centred around 1700, 1100, 560, 440, 250, 190, 100, and 40 million years, the same times deduced from other data for widespread orogenic events in North America (Barker, 1969).

III.3.4. Form of Alkaline Rock Bodies

In the Canadian Shield, alkaline rocks occur most commonly in subcylindrical zoned stocks and ring complexes averaging 5km in diameter. Ring complexes are also characteristic of occurrences in the Monteregian Hills (cf. IV.6)*, northern New England, central Arkansas (Fig. 2), and northern Coahuila, but are by no means the exclusive mode of occurrence.

Particularly intriguing are extensive belts of concordant bodies of gneissic nepheline syenite

* Readers are referred to the relevant Chapters of this book where similarly cited.

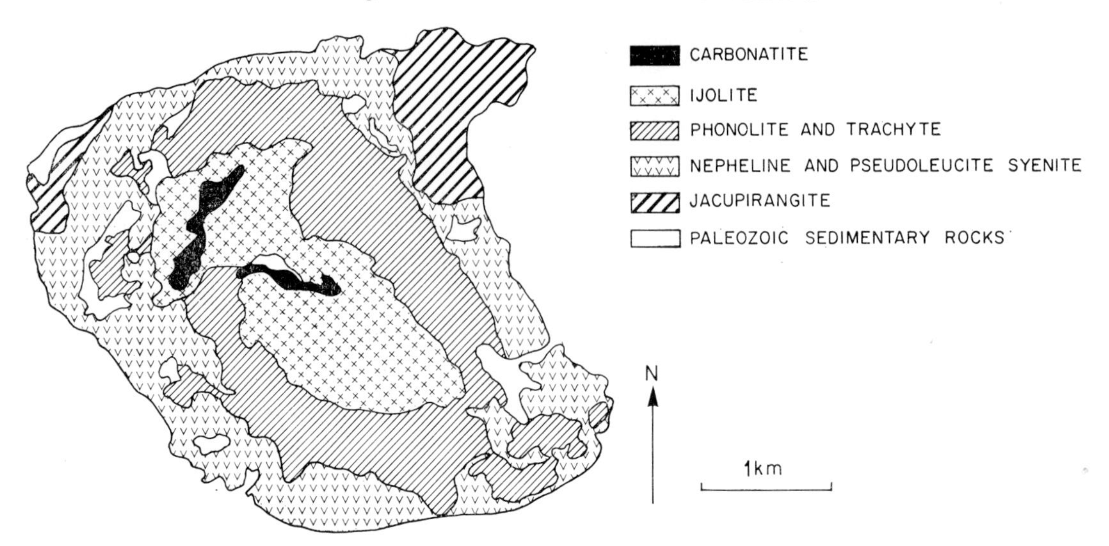

Fig. 2 Geologic map of Magnet Cove Complex, Arkansas, generalized from Erickson and Blade (1963)

(cf. II.2 and II.7). One belt, stretching from Ontario to Labrador, consists of French River (Fig. 3), Blue Mountain, Haliburton–Bancroft, Kipawa, Booth, Cabonga Reservoir, Albanel and Letitia Lake. Another, less well known, belt in the Canadian Rockies contains the Rogers Pass-Big Bend and Lonnie occurrences.

Laccoliths, small stocks, sills and dyke swarms are the dominant style of emplacement in undeformed or gently folded sedimentary rocks (Fig. 4). Mineralogically, these shallow intrusions are characterized by the presence of analcime in place of some or all nepheline, and by the apparent absence of carbonatites except in diatremes.

It should be possible to set up a depth classification of alkaline complexes, based upon comparison of varying levels of exposure, as Buddington (1959) did for granites.

III.3.5. Summary

The long time spans during which alkaline magmas are emplaced within small areas, the disposition of alkaline rocks in linear belts, the presence of essentially identical alkaline rocks in ocean basins (Borley, cf. IV.7) and on continents, the low initial Sr^{87}/Sr^{86} ratios (cf. V.4) and the

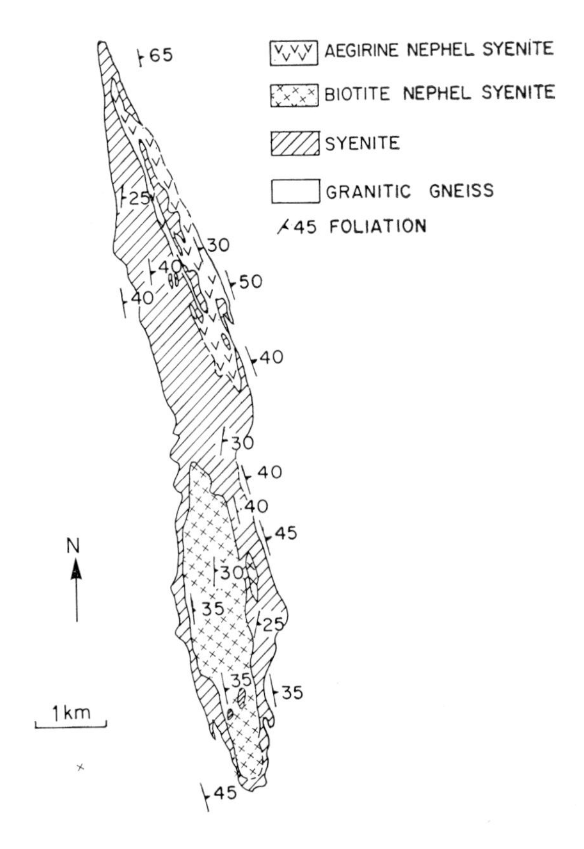

Fig. 3 Geologic map of French River Complex, Ontario, generalized from J. G. Payne (manuscript in preparation)

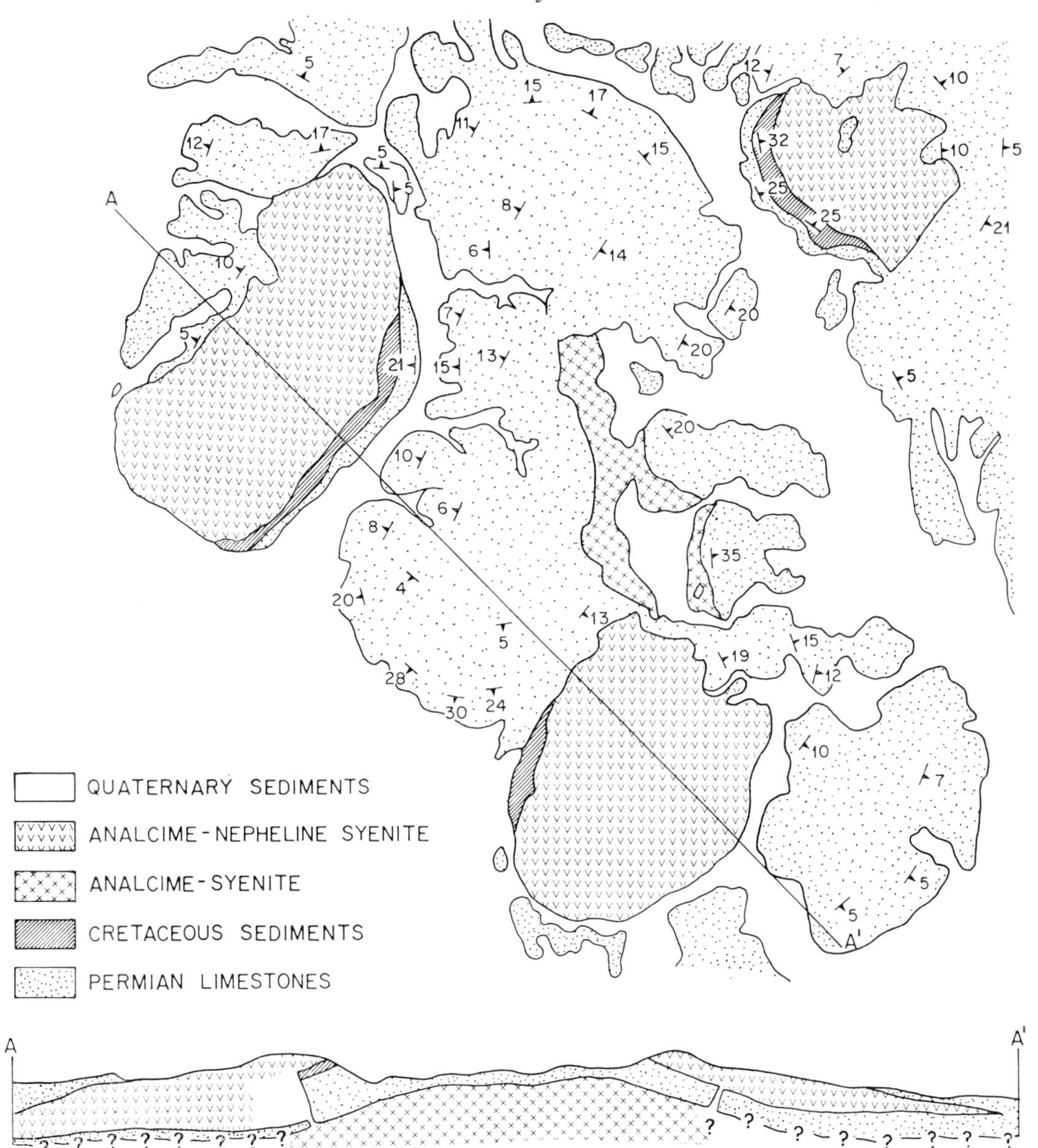

Fig. 4 Geologic map of Sierra Tinaja Pinta, Hudspeth County, Texas, generalized from Gross (1965)

common low-melting compositions (Bailey and Schairer, 1966 and V.1) all suggest that alkaline magma is generated within the mantle and moves up into the crust along recurrently opened fractures. In North America, most of these fractures are not obvious at the surface.

Geochronologic studies of the half of known occurrences that remain undated should receive highest priority.

ACKNOWLEDGMENTS

This research was supported by the U.S. National Science Foundation, Grants GA-652 and GA-11154. K. L. Currie, T. Frisch, J. Gittins, D. P. Gold, B. C. Hearn, Jr., A. F. Laurin, T. P. Miller, R. L. Parker and R. E. Zartman provided much new information but are not responsible for its interpretation here. Many others also contributed data and insight.

III.3. REFERENCES

Appleyard, E. C., 1967. Nepheline gneisses of the Wolfe Belt, Lyndoch Township, Ontario. I. Structure, stratigraphy, and petrography. *Can. J. Earth Sci.*, **4**, 371–95.

Bailey, D. K., and Schairer, J. F., 1966. The system Na_2O–Al_2O_3–Fe_2O_3–SiO_2 at 1 atmosphere, and the petrogenesis of alkaline rocks. *J. Petrology*, **7**, 114–70.

Baldwin, B., and Muehlberger, W. R., 1959. Geologic studies of Union County, New Mexico. *New Mexico Inst. Min. Tech., State Bur. Mines Min. Res., Bull.*, **63**.

Barker, D. S., 1965. Alkalic rocks at Litchfield, Maine. *J. Petrology*, **6**, 1–27.

Barker, D. S., 1969, North American feldspathoidal rocks in space and time. *Bull. geol. Soc. Am.*, **80**, 2369–72.

Barker, D. S., and Long, L. E., 1969. Feldspathoidal syenite in a quartz diabase sill, Brookville, New Jersey. *J. Petrology*, **10**, 202–21.

Barker, D. S., 1970. North American feldspathoidal rocks in space and time: Reply. *Bull. geol. Soc. Am.*, **81**, 3501–2.

Bass, M. N., 1970. North American feldspathoidal rocks in space and time: Discussion and Rebuttal. *Bull. geol. Soc. Am.*, **81**, 3493–500.

Basset, W. A., 1961. Potassium–argon age of Devil's Tower, Wyoming. *Science*, **134**, 1373.

Best, M. G., Henage, L. F., and Adams, J. A. S., 1968. Mica peridotite, wyomingite, and associated potassic igneous rocks in northeastern Utah. *Am. Miner.*, **53**, 1041–8.

Blixit, J. E., 1933. Geology and gold deposits of the North Moccasin Mountains, Fergus County, Montana. *Montana Bur. Mines Geol., Mem.* **8**, 25 pp.

Boettcher, A. L., 1967. The Rainy Creek alkaline-ultramafic igneous complex near Libby, Montana. I. Ultramafic rocks and fenite. *J. Geol.*, **75**, 526–53.

Brummer, J. J., and Mann, E. L., 1961. Geology of the Seal Lake area, Labrador. *Bull. geol. Soc. Am.*, **72**, 1361–82.

Buddington, A. F., 1959. Granite emplacement with special reference to North America. *Bull. geol. Soc. Am.*, **70**, 671–747.

Burke, W. H., Jr., Otto, J. B., and Denison, R. E., 1968. Potassium–argon dating of basaltic rocks. *J. Geophys. Res.*, **74**, 1082–6.

Campbell, C. D., 1939. The Kruger alkaline syenites of southern British Columbia. *Am. J. Sci.*, **237**, 527–49.

Card, K. D., 1965. The Croker Island complex, north channel of Lake Huron. *Ontario Dep. Mines Geol. Circ.*, **14**, 11 pp.

Carmichael, I. S. E., 1967. The mineralogy and petrology of the volcanic rocks from the Leucite Hills, Wyoming. *Contr. Miner. Petrogr.*, **15**, 24–66.

Chute, N. E., 1950. Bedrock geology of the Brockton quadrangle, Massachusetts. *U.S. Geol. Surv., Geol. Quad. Map* **GQ-5**.

Coates, M. E., 1968. Geology of Stevens-Kagiano Lake area, District of Thunder Bay. *Ontario Dept. Mines. Geol. Rep.*, **68**.

Croneis, C., and Billings, M. P., 1929. New areas of alkaline igneous rocks in central Arkansas. *J. Geol.*, **37**, 542–61.

Currie, K. L., and Shafiqullah, M., 1968. Geochemistry of some large Canadian craters. *Nature*, **218**, 457–9.

Daly, R. A., 1912. Geology of the North American Cordillera at the forty-ninth parallel. *Mem. geol. Surv. Can.* **38**.

Daly, R. A., 1933. *Igneous Rocks and the Depths of the Earth.* McGraw-Hill, New York.

Darton, N. H., and Paige, S., 1925. Central Black Hills folio, South Dakota. *U.S. Geol. Surv., Atlas, Folio* **219**.

Daugherty, F. W., 1963. La Cueva intrusive complex and dome, northern Coahuila, Mexico. *Bull. Geol. Soc. Am.*, **74**, 1429–38.

Daugherty, F. W., 1965. Aligned intrusive complexes in northern Coahuila, Mexico (abstract). *Geol. Soc. Am., Spec. Pap.* **87**, 41–2.

Davidson, A., 1968. An occurrence of alkali syenite, Kaminak Lake Map-area (55L), District of Keewatin. *Geol. Surv. Can., Pap.* **68-1A**, 127–9

Dimroth, E., 1967. Lac de la Brèche carbonatites. *Bull. Can. Geophys.*, **20**, 225–6.

Doig, R., and Barton, J. M., Jr., 1968. Ages of carbonatites and other alkaline rocks in Quebec. *Can. J. Earth Sci.*, **5**, 1401–7.

Duffell, S., MacLaren, A. S., and Holman, R. H. C., 1963. Red Lake, Lansdowne House Area, northwestern Ontario. *Geol. Surv. Can., Pap.* **63–65**.

Eckel, E. B., 1949. Geology and ore deposits of the La Plata District, Colorado. *Prof. Pap. U.S. geol. Surv.*, **219**, 179 pp.

Eckel, E. B., and Myers, W. B., 1946. Quicksilver deposits of the New Idria district, San Benito and Fresno Counties, California. *Calif. Div. Mines, Rep.*, **42**, 81–124.

Eggleston, J. W., 1918. Eruptive rocks at Cuttingsville, Vermont. *Am. J. Sci.*, 4th ser., **45**, 377–410.

Eifler, G. K., Jr., 1951. Geology of the Barrilla Mountains, Texas. *Bull. geol. Soc. Am.*, **62**, 339–54.

Emmons, R. C., 1953. 'Petrogeny of the syenites and nepheline syenites of central Wisconsin', in Emmons, R. C., Ed., Selected petrogenic relationships of plagioclase. *Geol. Soc. Am., Mem.* **52**, 71–87.

Erickson, R. L., and Blade, L. V., 1963. Geochemistry and petrology of the alkalic rock complex at Magnet Cove, Arkansas. *Prof. Pap., U.S. geol. Surv.* **425**, 95 pp.

Fairbairn, H. W., Bullwinkel, H. J., Pinson, W. H., and Hurley, P. M., 1959. Age investigation of syenites from Coldwell, Ontario. *Proc. geol. Assoc. Can.*, **11**, 141–4.

Fenton, M. D., and Faure, G., 1969. The isotopic evolution of terrestrial strontium (abstract). *Geol. Soc. Am., Abstracts with Programs for 1969*, part **7**, 64.

Foland, K. A., Quinn, A. W., and Giletti, B., 1970. Jurassic and Cretaceous isotopic ages of the White Mountain magma series (abstract). *Geol. Soc. Am., Abstracts with programs for 1970*, part **1**, 19–20.

Frisch, T., in press. Metamorphic and plutonic rocks of northernmost Ellesmere Island, Arctic Archipelago. *Bull. geol. Surv. Can.*

Gilluly, J., 1927. Analcite diabase and related alkaline syenite from Utah. *Am. J. Sci.*, 5th ser., **14**, 199–211.

Gittins, J., 1966. 'Summaries and bibliographies of carbonatite complexes', in Tuttle, O. F., and Gittins, J., Eds., *Carbonatites*, Wiley–Interscience, New York, 417–541.

Gittins, J., 1967. 'Nepheline rocks and petrological problems of the Haliburton–Bancroft area', in Jenness, S. E., Ed., Geology of parts of eastern Ontario and western Quebec, 1967, *Geol. Assoc. Can., Guidebook, Kingston, Ontario meeting*, 31–57.

Gittins, J., MacIntyre, R. M., and York, D., 1967. The ages of carbonatite complexes in eastern Canada. *Can. J. Earth Sci.*, **4**, 651–5.

Gold, D. P., 1967. 'Alkaline ultrabasic rocks in the Montreal area, Quebec', in Wyllie, P. J., Ed., *Ultramafic and Related Rocks*, John Wiley and Sons, New York, 288–302.

Gold, D. P., Vallee, M., and Perrault, G., 1967. 'Field guide to the mineralogy and petrology of the Oka area, Quebec', in Jenness, S. E., Ed., Geology of parts of eastern Ontario and western Quebec, *Geol. Assoc. Can., Guidebook, Kingston, Ontario meeting*, 147–65.

Goldsmith, R., 1961. Axial-plane folding in southeastern Connecticut. *Prof. Pap. U.S. geol. Surv.*, **424-C**.

Grogan, R. M., and Bradbury, J. C., 1968. 'Fluorite–zinc–lead deposits of the Illinois–Kentucky mining district', in Ridge, J. D., Ed., *Ore Deposits in the United States, Volume 1*, AIME, New York, 370–99.

Gross, E. B., and Heinrich, E. Wm., 1965. Petrology and mineralogy of the Mount Rosa area, El Paso and Teller Counties, Colorado: I. The granites. *Am. Miner.*, **50**, 1273–95.

Gross, R. O., 1965. *Geology of Sierra Tinaja Pinta and Cornudas Station areas, northern Hudspeth County, Texas.* Unpublished M.A. thesis, The University of Texas at Austin, 119 pp.

Gunn, B. M., Carman, M. F., and Cameron, K. L., 1968. Some characteristics of igneous rocks from the Big Bend region, Texas, and their relation to the Trans-Pecos igneous province (abstract). *Geol. Soc. Am., Program for Annual Meeting, Mexico City*, 121.

Ham., W. E., Denison, R. E., and Merritt, C. A., 1964. Basement rocks and structural evolution of Southern Oklahoma. *Oklahoma geol. Surv., Bull.* **95**, 302 pp.

Hearn, B. C., Jr., Pecora, W. T., and Swadley, W. C., 1964. Geology of the Rattlesnake quadrangle, Bearpaw Mountains, Blaine County, Montana. *Bull. U.S. geol. Surv.*, **1181-B**, 66 pp.

Hoefs, J., 1967. A Rb–Sr investigation in the southern York County area, Maine. M.I.T.-1381-15, *Fifteenth Annual Progress Report for 1967, U.S. Atomic Energy Commission Contract AT(30-1)-1381, Massachusetts Institute of Technology*, 127–9.

Hogarth, D. D., 1966. Intrusive carbonate rock near Ottawa, Canada. *Miner. Soc. India*, I.M.A. Volume, 45–53.

Hollingsworth, J. S., 1967. 'Geology of the Wilson Springs Vanadium deposits, Garland County, Arkansas', in *Guidebook, Geol. Soc. America Field Conference, Central Arkansas Economic Geology and Petrology*, Arkansas Geol. Comm., Little Rock, 22–8.

Hunt, C. B., 1958. Structural and igneous geology of the La Sal Mountains, Utah. *Prof. Pap. U.S. geol. Surv.*, **294-I**, 305–64.

Jahns, R. H., 1938. Analcite-bearing intrusives from South Park, Colorado. *Am. J. Sci.*, 5th ser., **36**, 8–26.

Jenkins, E. C., 1959. *Hypabyssal rocks associated with the Christmas Mountains gabbro, Brewster County, Texas.* Unpublished M.A. thesis, The University of Texas at Austin, 117 pp.

Jenks, W. F., 1934. Petrology of the alkaline stock at Pleasant Mountain, Maine. *Am. J. Sci.*, 5th ser., **28**, 321–40.

Kelley, V. C., 1968. Geology of the alkaline Precambrian rocks at Pajarito Mountain, Otero County, New Mexico. *Bull. geol. Soc. Am.*, **79**, 1565–72.

Kemp, J. F., and Billingsley, P., 1921. Sweet Grass Hills, Montana. *Bull. geol. Soc. Am.*, **32**, 437–78.

Kidwell, A. L., 1951. Mesozoic igneous activity in the northern Gulf Coastal Plain. *Trans. Gulf Coast Assoc. geol. Soc.*, **1**, 182–99.

King, P. B., 1959. *The Evolution of North America.* Princeton University Press, 190 pp.

Knopf, A., 1957. The Boulder bathylith of Montana. *Am. J. Sci.*, **255**, 81–103.

Koschmann, A. H., 1949. Structural control of the gold deposits of the Cripple Creek district, Teller County, Colorado. *Bull. U.S. geol. Surv.*, **955b**.

Krogh, T. E., and Hurley, P. M., 1968, Strontium isotopic variations and whole-rock isochron studies, Grenville Province of Ontario. *J. Geophys. Res.*, **73**, 7107–25.

Lanphere, M. A., 1964. Geochronologic studies in the eastern Mojave Desert, California. *J. Geology*, **72**, 381–99.

Lanphere, M. A., Loney, R. A., and Brew, D. A., 1965. Potassium–argon ages of some plutonic rocks, Tenakee area, Chichagof Island, Southeast Alaska. *Bull. U.S. geol. Surv.*, **525-B**, 108–11.

Lanphere, M. A., MacKevett, E. M., Jr., and Stern, T. W., 1964. Potassium–argon and lead-alpha ages of plutonic rocks, Bokan Mountain area, Alaska. *Science*, **145**, 705–7.

Larsen, E. S., Hurlbut, C. S., Jr., Burgess, C. H., and Buie, B. F., 1941. Igneous rocks of the Highwood Mountains: Part VII. Petrology. *Bull. geol. Soc. Am.*, **52**, 1857–68.

Lawson, A. C., 1896. On malignite, a family of basic plutonic orthoclase rocks rich in alkalies and lime intrusive in the Coutchiching schists of Poohbah Lake. *Univ. Calif., Dept. Geol., Bull.*, **1**, 337–62.

Leech, G. B., Lowdon, J. A., Stockwell, C. H., and Wanless, R. K., 1963. Age determinations and geological studies. *Geol. Surv. Can., Pap.* **63–17**.

Little, H. W., 1960. Nelson Map-area, west half, British Columbia. *Mem. geol. Surv. Can.*, **308**, 205 pp.

Lonsdale, J. T., 1940. Igneous rocks of the Terlingua–Solitario region, Texas. *Bull. geol. Soc. Am.*, **51**, 1539–626.

Lowdon, J. A., Stockwell, C. H., Tipper, H. W., and Wanless, R. K., 1963. Age determinations and geological studies. *Geol. Surv. Can., Pap.* **62-17**.

Lundgren, L., Jr., 1963. The bedrock geology of the Deep River quadrangle. *Conn. Geol. Nat. Hist. Surv., Quad. Rep.* **13**, 40 pp.

Lyons, J. B., 1944. Igneous rocks of the northern Big Belt range, Montana. *Bull. geol. Soc. Am.*, **55**, 445–72.

Macintyre, R. M., York, D., and Gittins, J., 1969. The K–Ar characteristics of nepheline. *Earth Planet Sci. Lett.*, **7**, 125–31.

MacKenzie, H. N. S., 1956. Crowsnest volcanics. *J. Alberta Soc. Petrol. Geol.*, **4**, 70–4.

Maxwell, R. A., Lonsdale, J. T., Hazzard, R. T., and Wilson, J. A., 1967. Geology of Big Bend National Park, Brewster County, Texas. *Univ. Texas Bur. econ. Geol. Pub.*, **6711**, 320 pp.

McAllister, J. F., 1952. Rocks and structure of the Quartz Spring area, northern Panamint Range, California. *Calif. Div. Mines, Spec. Rep.* **25**, 38 pp.

McAnulty, W. N., 1955. Geology of Cathedral Mountain quadrangle, Brewster County, Texas. *Bull. geol. Soc. Am.*, **66**, 531–78.

McDowell, F. W., 1966. *Potassium–argon dating of Cordilleran intrusives.* Unpublished Ph.D. dissertation, Columbia University, 242 pp.

McKnight, J. F., 1963. *Igneous rocks of Sombreretillo area, northern Sierra de Picachos, Nuevo Leon, Mexico.* Unpublished M. A. thesis, the University of Texas at Austin, 89 pp.

Moody, C. L., 1949. Mesozoic igneous rocks of the northern Gulf Coastal Plain. *Bull. Am. Assoc. Petrol. Geologists*, **33**, 1410–28.

Nash, D. B., and Nelson, C. A., 1964. Leucite-bearing volcanic plugs, Deep Spring Valley California (abstract). *Geol. Soc. Am., Spec. Pap.* **82**, 267.

Neilson, J. M., 1953. Albanel area, Mistassini Territory, Quebec. *Quebec Dept. Mines, Geol. Rep.* **53**.

Olson, J. C., Shawe, D. R., Pray, L. C., Sharp, W. N., and Hewett, D. F., 1954. Rare earth mineral deposits of the Mountain Pass district, San Bernadino County, California. *Prof. Pap. U.S. geol. surv.*, **261**, 75 pp.

Parker, R. L., and Calkins, J. A., 1964. Geology of the Curlew quadrangle, Ferry County, Washington. *Bull. U.S. Geol. Survey*, **1169**, 95 pp.

Parsons, G. E., 1961. Niobium-bearing complexes east of Lake Superior. *Ontario Dept. Mines, Geol. Rep.*, **3**, 73 pp.

Patton, W. W., Jr., and Miller, T. P., 1968. Regional geologic map of the Selawik and southeastern Baird Mountains quadrangles, Alaska. *U.S. geol. Surv., Misc. geol. Inv., Map* **I-530**.

Payne, J. G., 1968. Geology and geochemistry of the Blue Mountain nepheline syenite. *Can. J. Earth Sci.* **5**, 259–73.

Quinn, A., 1937. Petrology of the alkaline rocks at Red Hill, New Hampshire. *Bull. geol. Soc. Am.*, **48**, 373–402.

Quinn, A., Ray, R. G., and Seymour, W. L., 1949. Bedrock geology of the Pawtucket quadrangle, Rhode Island–Massachusetts. *U.S. geol. Surv., Geol. Quad. Map* **GQ-1**.

Rankin, D. W., 1967. Magmatic activity and orogeny in the Blue Ridge Province of the Southern Appalachian Mountain system in northwestern North Carolina and southwestern Virginia (abstract). *Geol. Soc. Am., Program for Annual Meeting, New Orleans*, 181.

Rapson, J. E., 1966. Carbonatite in the alkaline complex of the Ice River area, southern Canadian Rocky Mountains. *Miner. Soc. India*, I.M.A. Volume, 9–22.

Rinehart, C. D., and Fox, K. F., Jr., 1968. Alkalic plutons and diatremes of northern Okanagan Highlands, Washington. *Prof. Pap. U.S. Geol. Survey*, **600-A**, p. A–102.

Rowe, R. B., 1958. Niobium (columbium) deposits of Canada. *Geol. Surv. Can., Econ. Geol. Ser.*, **18**, 108 pp.

Satterly, J., 1968. Aeromagnetic maps of carbonatite-alkalic complexes in Ontario. *Ontario Dept. Mines, Prelim. Map* **P452**.

Sewell, C. R., 1968. The Candela and Monclova belts of igneous intrusions—a petrographic province in Nuevo Leon and Coahuila, Mexico (abstract). *Geol. Soc. Am., Program for Annual Meeting, Mexico City*, 273.

Shawe, D. R., and Parker, R. L., 1967. Mafic-ultramafic layered intrusion at Iron Mountain, Fremont County, Colorado. *Bull. U.S. geol. Surv.* **1251-A**.

Simms, F. E., 1965. Hypabyssal alkaline bodies and structure of part of the northwestern Crazy Mountains, Montana (abstract). *Am. Miner.*, **50**, 291–2.

Snavely, P. D., Jr., and Wagner, H. C., 1961. Differentiated gabbroic sills and associated alkalic rocks in the central part of the Oregon Coast Range. *Prof. Paper, U.S. geol. Surv.*, **424-D**, 156–61.

Spencer, A. B., 1969. Alkalic igneous rocks of the Balcones Province, Texas. *J. Petrology*, **10**, 272–306.

Stern, T. W., Newell, M. F., Kistler, R. W., and Shawe, D. R., 1965. Zircon uranium-lead and thorium-lead ages and mineral potassium–argon ages of La Sal Mountains rocks, Utah. *J. Geophys. Res.*, **70**, 1503–7.

Swadley, W. C., 1958. *Petrology of the Christmas Mountains gabbro, Brewster County, Texas*. Unpublished M.A. thesis, the University of Texas at Austin, 67 pp.

Temple, A. K., and Grogan, R. M., 1965. Carbonatite and related alkalic rocks at Powderhorn, Colorado. *Econ. Geol.*, **60**, 672–92.

Toulmin, P., 3rd, 1957. Notes on a peralkaline granite from Cashes Ledge, Gulf of Maine. *Am. Miner.*, **42**, 912–15.

Toulmin, P., 3rd, 1961. Geological significance of lead-alpha and isotopic age determinations of 'alkalic' rocks of New England. *Bull. geol. Soc. Am.*, **72**, 775–80.

Toulmin, P., 3rd, 1964. Bedrock geology of the Salem quadrangle and vicinity, Massachusetts. *Bull. U.S. geol. Surv.*, **1163-A**, 79 pp.

Wanless, R. K., Stevens, R. D., Lachance, G. R., and Edmonds, C. M., 1967. Age determinations and geological studies: K–Ar isotopic ages, Report 7. *Geol. Surv. Can., Pap.* **66-17**.

Watson, E. H., 1937. The geology and biology of the San Carlos Mountains, Tamaulipas, Mexico; Part 2, Igneous rocks of the San Carlos Mountains. *Michigan Univ. Studies Sci. series*, **12**, 99–156.

Weed, W. H., and Pirsson, L. V., 1896a. Geology of the Castle Mountains mining district, Montana. *Bull. U.S. geol Surv.*, **139**.

Weed, W. H., and Pirsson, L. V., 1896b. Geology of the Little Rocky Mountains, Montana. *J. Geol.*, **4**, 399–428.

Weed, W. H., and Pirsson, L. V., 1898. Geology and mineral resources of the Judith Mountains of Montana. *U.S. geol. Surv., Ann. Rep.* **18**, pt. 3, 437–616.

Weed, W. H., and Pirsson, L. V., 1899. Geology of the Little Belt Mountains, Montana. *U.S. geol. Surv., Ann. Rep.* **20**, pt. 3, 257–581.

Wheeler, J. O., 1962. Rogers Pass Map-area, British Columbia and Alberta. *Geol. Surv. Can., Pap.* **62-32**.

Wheeler, J. O., 1964. Big Bend Map-area, British Columbia. *Geol. Surv. Can., Pap.* **64-32**.

Williams, H., 1936. Pliocene volcanoes of the Navajo-Hopi country. *Bull. geol. Soc. Am.*, **47**, 111–72.

Wolff, J. E., 1938. Igneous rocks of the Crazy Mountains, Montana. *Bull. Geol. Soc. Am.*, **49**, 1569–626.

Yates, R. G., and Thompson, G. A., 1959. Geology and quicksilver deposits of the Terlingua district, Texas. *Prof. Pap. U.S. geol Surv.*, **312**.

Zartman, R. E., Brock, M. R., Heyl, A. V., and Thomas, H. H., 1967. K–Ar and Rb–Sr ages of some alkalic intrusive rocks from central and eastern United States. *Am. J. Sci.*, **265**, 848–70.

III.4. REGIONAL DISTRIBUTION AND TECTONIC RELATIONS OF THE ALKALINE ROCKS OF SIBERIA

E. L. Butakova

III.4.1. Introduction

Numerous major and minor provinces of alkaline rocks of various compositions are known from vast territories of Siberia (the Far East and the North-East included). They are mainly magmatic, but the rocks may be highly metasomatically altered. The Siberian associations of alkaline rocks may be subdivided into associations that have been formed: (1) within the Siberian platform that has a mainly Archean basement; (2) within the Proterozoic, Early Caledonian, Mesozoic and Cenozoic fold belts which border the platform.

III.4.2. Associations of Alkaline Rocks of the Siberian Platform

These occurrences are subdivided into two groups: (1) associations of ultramafic and alkaline rocks with carbonatites; (2) associations composed mainly of nepheline and alkali syenites.

In the first group the Permian-Triassic Maimecha–Kotuj petrographic province located at the northern margin of the platform is the largest, and most variable in petrographic composition and with regard to the facies of rocks. It is also the most interesting as to tectonic setting (Fig. 1). The carbonatite-bearing association of alkaline and ultramafic rocks located in the Aldan Shield is of similar type (Fig. 1), but is considered to be of Late Proterozoic age. Other petrographical provinces of alkaline-ultramafic rocks can only with considerable limitations be compared with these provinces. Examples are the Permian-Triassic Chadobets province represented by veins of picrite porphyrites, picrites and carbonatite-like rocks and located at the south-western margin of the platform; and the Mesozoic dykes of alkaline-ultramafic rocks (monticellite and nepheline-monticellite picrite porphyrites found at the north-eastern margin of the platform (Fig. 1).

The second group of alkaline provinces of the platform is presented by the Mesozoic intrusive-volcanic province of the Aldan Shield and by the early Paleozoic Udzhin province of nepheline and alkali syenites located in the north-east of the platform (Fig. 1).

III.4.2.1. The Maimecha–Kotuj Province

Described in many papers (Butakova and Egorov, 1962, etc.), this province is composed of a thick series of alkaline, subalkaline and ultrabasic extrusives (different alkaline basaltoids, trachybasalts, trachyandesites, alkali trachytes, and meimechites (picrite porphyrites); of later intrusions of ultramafic and alkaline rocks (dunite, peridotite, pyroxenite, ijolite, melteigite, jacupirangite, different rocks rich in melilite) and of carbonatite, as well as of dykes of various alkaline ultramafic and subalkaline rocks. All the individual massifs of the province, with the exception of the Gulin intrusion, are of insignificant dimensions, and are coarse-concentric, pipe-shaped, stock-shaped or cone-shaped intrusions of central type. The Gulin intrusion differs in many respects from the other intrusive bodies of the province. Its exposed parts dominated by dunites and peridotites, cover more than 500 km^2. On the basis of geophysical data some investigators hold that the area of the Gulin intrusion exceeds 2000 km^2 and that its major part is covered by thick Meso-Cenozoic deposits of the Lena–Enisei trough.

The Maimecha–Kotuj province is a fine example of the relation between alkaline magmatism and the large tectonic structures of

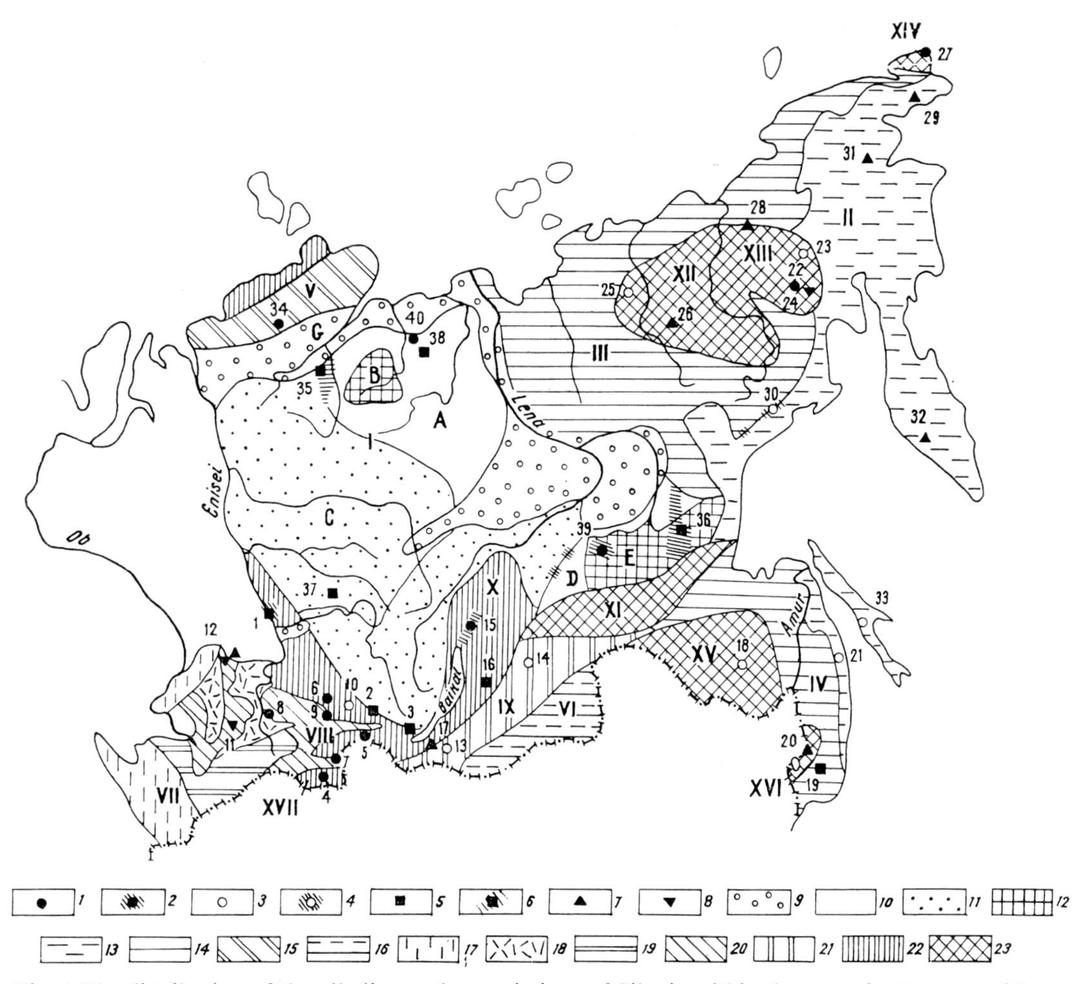

Fig. 1 The distribution of the alkaline rock associations of Siberia within the tectonic structures. (Compiled by E. L. Butakova.) Scale 1 : 42,000,000

Associations of alkaline rocks: (1) mainly nepheline and alkaline syenites; (2) the same showing a considerable territorial distribution; (3) alkali and subalkali granites, granosyenites and quartz syenites; (4) the same showing a considerable distribution; (5) alkaline and ultramafic rocks with carbonatites; (6) the same showing a considerable distribution; (7) alkaline basaltoids; (8) nepheline and alkali syenites and gabbroids.

The numeration of the alkaline rocks of the folded regions is that of the table (1–34). The alkaline rocks of the Siberian platform: (35) Maimecha–Kotuj; (36) Late Proterozoic province of the Aldan Shield; (37) Chadobets; (38) the north-east of the platform; (39) Aldan Mesozoic province; (40) Udzhin.

Tectonic structures of the Siberian platform: Structures of the platform cover: (9) Mesozoic troughs; (10) anteclises; (11) syneclises. Structures of the basement: (12) Archean folded structures of the shields. Tectonic structures of folded regions: (13) Cenozoic; (14) Mesozoic; (15) Early Mesozoic; (16) mainly Hercynian; (17) Hercynian; (18) Middle–Late Paleozoic troughs among the Caledonides; (19) Late Caledonian or Caledonian proper; (20) Early Caledonian; (21) Early Caledonian and Late Proterozoic (Baikal); (22) Baikalian; (23) Pre-Cambrian (Baikalian and more ancient) reworked during Paleozoic or Mesozoic movements (structures of median masses and the folded system of the Stanovoy ridge).

(I) Siberian platform: (A) Anabar (Anabara–Olenek) anteclise; (B) Anabar Shield; (C) Tunguska syneclise; (D) Aldan anteclise; (E) Aldan Shield; (G) Lena–Enisei trough. Folded regions and systems: (II) Koryak–Kamchatka region; (III) Verkhoyano–Chukotka region; (IV) Sikhote–Alin system; (V) Taimyr region; (VI) Mongolo–Okhotsk region; (VII) Ob–Zaisan region; (VIII) Altay–Sayan region; (IX) Transbaikal region; (X) Baikal region; (XI) Stanovoy system. Median masses: (XII) Kolyma; (XIII) Omolon; (XIV) Eastern Chukotka; (XV) Burein; (XVI) Khankay; (XVII) Sangilen.

the platform. The area of distribution of alkaline and ultramafic rocks is elongated submeridionally and it is thinning out towards the south (Fig. 2). The length of this wedge-shaped zone is about 300 km, and the maximum width (in the north—at the boundary of the platform) reaches 120 km. The alkaline and ultramafic rocks are situated in the zone of transition from one large tectonic platform structure, the Tunguska syneclise, to another structure of the first order, the Anabar anteclise. The wedging-out towards the south of the area occupied by alkaline and ultramafic rocks is in the direction away from the margin of the platform. The greatest masses of these rocks, as for instance the Gulin intrusion and all alkaline and ultrabasic extrusives, are furthermore concentrated near the margin of the platform. This emphasizes the primary significance of another large structural element, namely the boundary of the platform with the Lena–Enisei trough, for the spatial distribution of the alkaline and ultramafic rocks. Geophysical data indicate that this boundary in the greatest part of its extension is a fault zone of latitudinal to north-eastern strike. The submeridional and north-western faults traced in the basement of the trough conjugate with the faults bordering the platform. The area of distribution of the rocks of the Maimecha–Kotuj province is confined to the southern extension within the platform of these faults. The linear distribution of intrusions and dykes of alkaline and ultramafic rocks (Fig. 2), as well as trap dykes, indicates that the whole zone considered has been cut by numerous tectonic fissures which were channels for mainly alkaline and ultrabasic magmas of abyssal origin (Butakova 1961, Egorov *et al.*, 1968).

The main directions of the fractures that controlled the distribution of the bodies of alkaline and ultramafic rocks are the following: (1) submeridional, north-north-west or north-north-east, (2) latitudinal or almost latitudinal, (3) north-west. The large Gulin intrusion and also minor intrusive bodies are often found at the intersections of fractures of different trends (Fig. 2).

Parallel with the fact that the ascent of magma

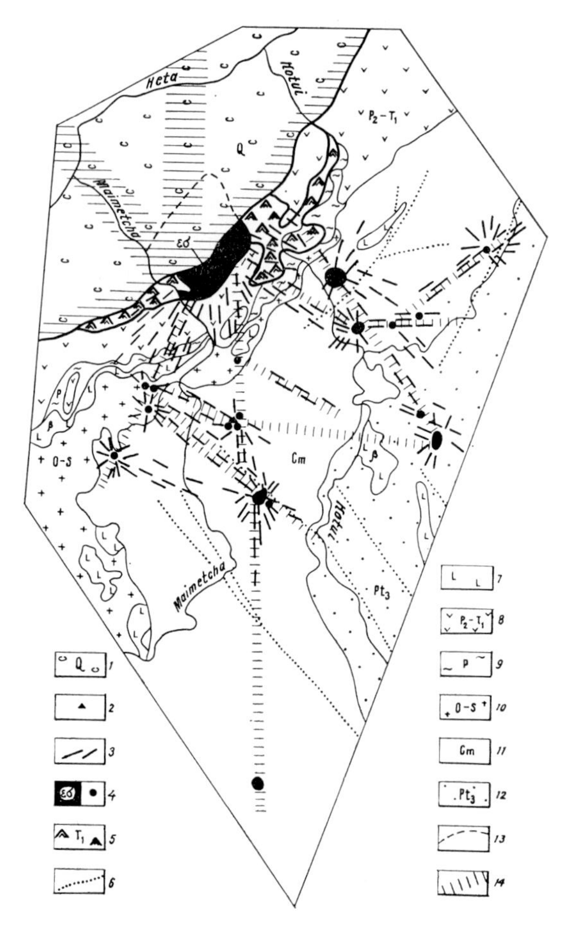

Fig. 2 The tectonic setting of intrusions and extrusives of the Maimecha–Kotuj province. (Compiled by E. L. Butakova according to the data by L. S. Egorov, N. P. Surina, E. L. Butakova and other investigators.) Scale 1 : 3,500,000

(1) Quaternary deposits (in the territory of the Lena–Enisei trough); (2) pipes of Mesozoic kimberlites. Rocks of the Maimecha-Kotuj complex: (3) dykes of alkaline rocks of Triassic age; (4) intrusions of alkaline and ultramafic rocks of Triassic age (small intrusive bodies shown without scale by black circles); (5) lavas, rarely tuffs of alkaline and subalkaline composition, mainly Early Triassic rocks of the trappean complex (Late Permian–Early Triassic); (6) basalt dykes; (7) dolerite sills; (8) basalts and their tuffs; (9) Permian terrigeneous deposits; (10) carbonate deposits of Ordovician and Silurian age, to a smaller degree Devonian; (11) Cambrian carbonate deposits; (12) Late Proterozoic carbonate deposits; (13) inferred northern boundary of the Gulin intrusion under the deposits of the trough; (14) abyssal fractures.

and the distribution of intrusive bodies and extrusives are clearly controlled by faults, a reverse relation is also observed on a smaller scale, namely the control of local tectonics on the location of the individual magmatic bodies of the Maimecha–Kotuj province. This influence is observed in the formation of dome-shaped structures associated with the intrusions, from several hundred of metres to 20–60 km in diameter, sharply protruding at the background of a very gentle monoclinal dip of the sedimentary and volcanic series. These domes are dissected by radial fissures (Fig. 2) occupied by dykes of alkaline and ultrabasic rocks (Surina, 1966).

On the basis of the present data it is possible to picture the structural environment of the origin of the alkaline and ultramafic rocks in the following way. The initial basaltic trap volcanism was especially intensive in Early Mesozoic times in a vast territory in the western part of the Siberian platform. The immense trap volcanism was undoubtedly associated with the highly significant tectonic movements that caused the downwarping of the large Tunguska syneclise and the breaking up of the Earth's crust in the parts of the platform which were tectonically most disturbed. The southern part of the Taimyr folded region was also involved in the trap volcanism. In the region considered two large fault zones have been formed: (1) a submeridional zone extending from the Taimyr region and dying out at a distance of about 300 km to the south of the present northern margin of the platform and (2) a latitudinal zone that later became the northern boundary of the platform as a result of the downwarping of the western part of the Lena–Enisei trough—the Khatanga basin. At the intersection of these zones alkaline basaltoid magma and ultrabasic and alkaline melts welled up from the upper mantle (Butakova and Egorov, 1962). The resulting alkaline, subalkaline and ultrabasic effusives of the Maimecha–Kotuj province are markedly different from the rather monotonous tholeiitic basalts of the remaining territory of the platform. The ultramafic lavas—meimechites—occurring at the intersection of these abyssal fault zones are unique with respect to their high contents of magnesia and their relatively large volumes. They were succeeded by the Gulin intrusion made up mainly of ultramafic rocks. This intrusion is exceptional with regard to size not only in this province but also when compared with the intrusions in other associations of alkaline ultramafic rocks throughout the world. At this focus, the site of intersection of two major abyssal fault zones, bodies of carbonatites, the largest of this province, are also found.

III.4.2.2. The Alkaline Ultramafic Rocks and Carbonatites of the Aldan Shield

Intrusions of central type are in this province made up of dunite, peridotite, pyroxenite, ijolite, melteigite, nepheline and alkali syenites, theralite, carbonatite and other rocks. The tectonic setting of this province has been insufficiently defined in the literature. Most of the intrusions are concentrated near the eastern and south-eastern margin of the Aldan anteclise. Some of them cluster towards the zone of abyssal faults, along which the platform conjugates with the Verkhoyansk folded system. Some intrusions are found in the uplifts which complicate the anteclise. A relation between location of intrusive bodies and the intersection of fractures is noted.

The information about the tectonic setting of the two other provinces of alkaline ultramafic rocks mentioned in II.4.2 are even poorer. The *Chadobets province* is a dome-shaped uplift located in the south-western margin of the Tunguska syneclise. The *Mesozoic dyke province* is found on the eastern slope of the Anabar anteclise and in its zone of transition to the Olenek uplift.

III.4.2.3. The Aldan Mesozoic Province of Alkaline Rocks

This province is first of all characterized by a considerable representation of rare types of pseudoleucitic and epileucitic rocks. The rocks of the province are rather widely distributed in the north-western and northern parts of the Aldan shield. Pseudoleucitic and epileucitic porphyries, pseudoleucitic basalts and tephrites, phonolitic porphyries, phonolites, alkali trachytes, tuffs and tuff breccias of these rocks are prominent among the extrusive alkaline rocks. Nepheline and alkali syenites associated with pseudoleucitic syenites and syenite-porphyries and alkaline porphyries dominate among the intrusive rocks. The rocks of the province are clustered in several districts which are all characterized by a shattered Earth's crust. The major regional trends of fractures are north-west and north-east. Most of the districts of alkaline rocks, the largest ones included, are located in the north where the Archean and Early Proterozoic series of the Aldan Shield dive under the platform cover. The Aldan province is suggested to have been formed during arched uplift and repeated revival of abyssal fractures (Bilibina *et al.*, 1967).

III.4.2.4. The Udzhin Province

This consists of two intrusions of central type made up mainly of nepheline and alkali syenite, with minor melteigite and urtite. These intrusions are found at the extreme north-eastern margin of the Anabar anteclise where Late Proterozoic and Cambrian formations are covered by the Mesozoic rocks of the Lena–Enisei trough. Thus, this province is located almost symmetrically with respect to the Maimecha–Kotuj province, which is found at the north-western margin of the Anabar anteclise. The outcrops of the alkaline rocks of the Udzhin are confined to a minor uplift complicating the north-eastern margin of the anteclise and to a zone of almost meridional fractures shattering this uplift (Erlikh, 1966, 1971).

III.4.2.5. Tectonic Relations

The present review has demonstrated some common features of the relation between the associations of alkaline rocks and the tectonic structures of the platform. Such common features of tectonic setting are for almost all provinces of alkaline rocks: (1) the location of most of the provinces, including the largest and most complicated ones, at the limbs of large arched uplifts, anteclises or shields (the Maimecha Kotuj province, the late Proterozoic and Mesozoic provinces of the Aldan Shield, the Udzhin province); (2) a close relation of the associations of alkaline rocks with deep (abyssal) faults and, particularly, with their intersections. Some provinces are located in relatively minor dome-shaped uplifts (Chadobets, Udzhin and a part of the masses of alkaline and ultramafic rocks of the Aldan Shield).

Most of the provinces of alkaline rocks of the Siberian platform were formed in early or late Mesozoic times. There are also provinces of late Proterozoic or early Paleozoic age. Both types of provinces were formed under platform conditions after a period of relative tectonic quietness. The disturbance of this quietness in the late Proterozoic, early Paleozoic, and in Mesozoic times resulted in arched uplifts, in formation or rejuvenation of deep fractures and in magmatism of different types, including alkaline activity.

III.4.3. Associations of Alkaline Rocks of Folded Regions

The associations of alkaline rocks of the folded regions may be petrographically subdivided into five groups mentioned in the order from those most widely distributed to those that occur more rarely:

1. Associations dominated by nepheline and alkali syenites, but also containing non-feldspathic feldspathoidal rocks (ijolites, melteigites, urtites) and alkali granites. The associations of this group are subdivided into sodic (predominant) and potassic series. The potassic series is characterized by a considerable development of pseudoleucitic or epileucitic rocks.
2. Associations dominated by alkaline (and calc-alkaline) granites, granosyenites and quartz syenites.
3. Associations of alkaline basaltoids (analcimite, nephelinite and leucitite, limburgite, teschenite, crinanite, ankaratrite and others).
4. Associations of non-feldspathic feldspathoidal rocks (ijolites and melteigites), nepheline and alkali syenites and ultramafic rocks (pyroxenites, jacupirangites) with carbonatites.

5. Associations of alkali and nepheline syenites with gabbroids, sometimes alkaline (essexites and theralites).

These groups of associations of alkaline rocks of the folded regions correspond in many respects to those distinguished by, for instance, Sheynmann *et al.* (1961).

I. The first group comprises most alkaline associations of the Altay–Sayan region, the Synnyr and Dezhnew associations, the alkali and nepheline syenites of the Abkit association and the rocks of Taimyr (see Table 1).

II. The second group comprises the Gutar and the Transbaikalian associations, and also the alkali granites and syenites of the Altakhin and Pribrezhny associations and all alkali granites of the Verkhoyano–Chukotka region.

III. The third group consists of the alkaline basaltoids of the Sikhote–Alin, Baikal, Verkhoyano–Chukotka and Koryak–Kamchatka regions.

IV. The fourth group comprises Kij–Tatarsky, Belozimin, Bolshezhidoj, Saizhin and Kokshary associations.

V. The rock associations of Kuznetsk Alatau and Omolon belong to the fifth group.

III.4.3.1. Age of the Country Rocks

Most of the associations of alkaline rocks (about 60%) distributed in folded regions of different ages are associated with Proterozoic (Baikal and more ancient) or sometimes Archean tectonic structures (see the table). In the Post-Proterozoic folded regions alkaline rocks are mainly confined to median masses of Proterozoic (and Archean) folded structures: Sangilen, Burein, Khankay, Kolyma, Omolon, Eastern-Chukotka. A number of provinces of alkaline rocks are located in folded structures of early Caledonian age. Only two or three alkaline associations are found in the territory occupied by Late Paleozoic–Early Mesozoic, Mesozoic and Cenozoic folded structures. This decrease in number of associations from the more ancient folded structures to younger ones is quite evident in spite of the debatable character of the knowledge of the age of some tectonic structures enclosing complexes of alkaline rocks and of the age of some complexes.

III.4.3.2. Relationship Between Age of Country Rocks and Type of Alkaline Rock

With regard to the distribution of alkaline rocks of different petrographical composition it is evident that in the Baikalides and other Pre-Cambrian structures the following types are concentrated: (1) most associations of nepheline and alkali syenites; (2) almost all associations of alkaline ultramafic rocks and carbonatites; (3) more than half of the associations of alkali granites and related rocks; and (4) more than half of the associations of alkaline basaltoids. Nearly all the remaining associations of nepheline and alkali syenites and alkaline ultramafic rocks and carbonatites, as well as the associations of leucocratic alkaline rocks and gabbroids (type V), are found in folded structures of Paleozoic age. It is highly characteristic that the alkaline rocks situated in folded structures of Mesozoic and Cenozoic age are made up almost exclusively of alkaline basaltoids and alkali granites.

Thus, it is seen that the Baikal and more ancient folded structures of Siberia are the hosts of the rocks richest in alkalis and that the younger folded structures accommodate associations of alkali rocks relatively poor in alkalis and also associations of rocks saturated with silica (alkali granites and granosyenites).

III.4.3.3. Age of the Alkaline Rocks

The alkaline rocks of the folded regions of Siberia are of presumably Paleozoic, Mesozoic, Tertiary and Quaternary ages (see Table 1). Alkaline rocks of Pre-Cambrian age are not known at all. About half of the associations of alkaline rocks are of Paleozoic—mainly, Middle and Late Paleozoic—age. Almost all associations of nepheline and alkali syenites, most associations of nepheline syenites, alkali syenites and gabbroids, and those of alkaline ultramafic rocks and carbonatites were formed in Middle and Late Paleozoic times. Most associations of

TABLE 1 Associations of Alkaline Rocks of the Folded Regions of Siberia

		Dominating Rocks	Age of Alkaline Rocks	Age of the Enclosing Tectonic Structures
		I. In the Altay–Sayan folded region (the Enisei ridge, Eastern Sayan, Eastern Tuva)		
1	Kij–Tatarsky	Nepheline syenites, ijolite–melteigite–jacupirangite, alkali syenites, carbonatites	Pz_2 or Pz_3	Archean or Late Proterozoic (Baikal)
2	Belozimin	Ijolites and melteigites, carbonatites, nepheline and alkali syenites, pyroxenites	P or T_1	Late Proterozoic
3	Bolshezhidoy	Pyroxenites, nepheline syenites	Pz_2 or Pz_3	Archean
4	Sangilen	Nepheline syenites, ijolites, melteigites and urtites, pyroxenites and carbonatites	D_{2-3} and $C_3 - T_1$	Late Proterozoic
5	Botogol	Nepheline and alkali syenites. Urtites are found	Pz_2 or Pz_1?	Late Proterozoic
6	Pezin	Nepheline and alkali syenites	D?	Late Proterozoic
7	Biykhem	Nepheline syenites, alkali syenites and quartz alkali syenites, alkali granites	P	Late Proterozoic
8	Saibar	Nepheline and alkali syenites, quartz alkali and subalkali syenites	Pz_2	Early Caledonian
9	Aksug	Alkali syenites, subalkali syenites, quartz syenites and granites. Nepheline syenites are found	Pz_2 or Pz_1	Early Caledonian
10	Gutar	Alkali and subalkali syenites, quartz syenites, granosyenites and alkali granites	Pz_2	Early Caledonian
11	Gabbro–syenite of Kuznetsk Alatau	Nepheline and alkali syenites and gabbro	O, S or D	Early Caledonian
12	Alkali gabbroids, nepheline and alkali syenites of Kuznetsk Alatau	Nepheline and alkali syenites and alkali gabbroids	D	Early Caledonian
		II. In the Transbaikal folded region		
13	Malokunaley (Kunaley)	Alkali and subalkali granites and granosyenites, alkali and nepheline syenites	Y_{1-2}	Late Proterozoic
14	Nerchugan	Alkali and subalkali granites, granosyenites, quartz syenites	Y_3?	Late Proterozoic
		III. In the Baikal folded region		
15	Synnyr	Nepheline, alkali and pseudoleucitic syenites, pseudoleucitites, quartz syenites and granosyenites	C	Late Proterozoic
16	Saizhin	Nepheline syenites, pyroxenites, gabbro, theralites, ijolites. Carbonatites are found	Mz?	Late Proterozoic
17	Alkaline basaltoids of the trachybasalt association	Monchiquites, crinanites, essexites, limburgites	Tr	Late Proterozoic

TABLE 1 (*continued*)

		Dominating Rocks	Age of Alkaline Rocks	Age of the Enclosing Tectonic Structures
		IV. In the Sikhote–Alin folded system (together with the Burein and Khankay median masses)		
18	Alkali syenites and granites of Altakhin	Alkali syenites and granites	Mz_1?	Proterozoic (of the Burein median masses)
19	Kokshary	Pyroxenites, jacupirangites, nepheline syenites, ijolites, melteigites. Carbonatites are found	J	Early Mesozoic
20	Alkaline basaltoids	Ankaratrites, ankaratrite-picrites, limburgites, picrites etc.	Pg	Archean (of the Khankay median masses)
21	Alkali granites of Pribrezny	Alkali granites	Pg	Meso-Cenozoic
		V. In the Verkhoyano–Chukotka folded region (together with the Omolon, Kolyma and Eastern–Chukotka median masses)		
22	Alkali and nepheline syenites of Abkitsk	Alkali and nepheline syenites	Late O or S	Pre-Cambrian (of the Omolon mass)
23	Alkali granites of granitoid associations	Alkali granites	Cr	Pre-Cambrian (of the Omolon mass)
24	Omolon	Crinanites, teschenites, essexites, alkali syenites, tinguaites, theralites	Cr–Pg	Pre-Cambrian (of the Omolon mass)
25	Alkali granites and related rocks of Tommot	Alkali granites, granosyenites and syenites	Pz_2 or Pz_3	Pre-Cambrian (of the Kolyma mass)
26	Alkaline basaltoids and gabbroids	Teschenite-picrites, essexites, crinanites, teschenites, analcime basalts	Cr_2–Pg	Pre-Cambrian (of the Kolyma mass)
27	Dezhnew	Nepheline and alkali syenites	Cr?	Proterozoic (of the Eastern-Chukotka mass)
28	Alkaline basaltoids of the Pezhen and other associations	Alkali basalts, leucite basalts, leucitites	Y_3	Mesozoic
29	Alkaline basaltoids	Nepheline, leucite and analcime basalts	Pliocene-Q	Mesozoic and Proterozoic
30	Alkali granites of the granitoid association	Alkali granites	Cr_2	Mesozoic
		VI. In the Koryak–Kamchatka folded region		
31	Alkaline basaltoids of Rarytkin	Teschenites, crinanites	Pg	Cenozoic
32	Alkaline basaltoids and gabbroids of Kamchatka	Teschenites, shonkinites, tephrites	Ng	Cenozoic
33	Syenite–monzonite complex of Sakhalin	Alkali syenites	Ng	Cenozoic
		VII. In the Taimyr folded region		
34	Alkali and nepheline syenites	Alkali and nepheline syenites	Mz_1?	Early Mesozoic

Pz_2 or Pz_3 = Middle or late Paleozoic; P = Permian; D_{2-3} = middle–late Devonian; C_3 = Late Carboniferous; T_1 = Early Triassic; O = Ordovician; S = Silurian; J_{1-2} = early–middle Jurassic; J_3 = late Jurassic; Mz = Mesozoic; Tr = Tertiary; Pg = Paleogene; Cr = Cretaceous; Q = Quaternary; Ng = Neogene.

alkali granites and related rocks are of Mesozoic age, and almost all associations of alkaline basaltoids are Cenozoic.

Most alkaline rocks are significantly separated in age from the time of formation of the enclosing folded structures. Thus, most intrusions located in Proterozoic (and Early Paleozoic) folded structures are of Middle and Late Paleozoic age (see Table 1). Most Mesozoic and a part of the Cenozoic associations are also found in the Pre-Cambrian folded structures.

III.4.3.4. Synchronism with Nearby Orogenic Processes

The development of associations of alkaline rocks coincides in time with the maximum activity of tectonic and magmatic processes (folding and calc-alkali magmatism) in neighbouring orogenic zones. The author concludes therefore, that the alkaline magmatism 'was excited' in the more ancient blocks of the Earth's crust under the influence of tectonic–magmatic processes taking place in neighbouring active mobile belts. The alkaline rocks of Middle and Late Paleozoic age may thus be considered to be results of 'Hercyninian activation' of Proterozoic and Early Paleozoic tectonic structures. With different degrees of certainty all or almost all associations of alkaline rocks of the Altay–Sayan region and the Synnyr intrusions of the Baikal region may be considered to be formed as results of 'Hercynian activation'. The Altay–Sayan region is fringed by Hercynian fold belts for long distances in the south-west, south and south-east (the Ob–Zaisan and Mongolo–Okhotsk folded regions). A Mesozoic activation of Proterozoic tectonic structures may be responsible for the formation of the alkaline rocks of Transbaikal, the Saizhin association of the Baikal region, the alkaline rocks of the Burein mass, and apparently, most of the alkaline rocks of the Kolyma and Omolon median masses and the Dezhnev intrusion of the Eastern Chukotka mass. The ancient folded structures enclosing these alkaline complexes are bordered by the Mesozoic Mongolo–Okhotsk and Verkhoyano–Chukotka folded regions.

III.4.3.5. Relation to Type of Geosyncline

The mode of distribution of alkaline rocks in zones of different types of development, i.e. the more mobile zones of eugeosynclinal type and the relatively stable zones of miogeosynclinal type has been considered by the author in the Altay–Sayan folded region. These two types of development were distinguished mainly by means of the presence of geosynclinal volcanism in the former. The Altay–Sayan folded region is comprised of Proterozoic (Baikal), and Caledonian folded structures. Alkaline rocks are known only in Proterozoic and Early Caledonian folded structures. In these structures the alkaline rocks are distributed in zones of different tectonic mobility (Butakova, 1969).

In contiguous parts of the Proterozoic and Early Caledonian folded structures of the eastern Altay–Sayan region there are seven complexes of alkaline rocks which are separated in space and which differ from one another with respect to mineralogical, petrographical and chemical composition (see Table 1 and Fig. 3). The enclosing folded rocks form a series of zones of increasing mobility. This is displayed in a gradual increase of the role of geosynclinal volcanism from zone to zone, in a progressively later folding, and in an increasing significance of granitoid magmatism. The Sangilen median mass was the most stable of these zones. A continuous Proterozoic formation of carbonatic and terrigeneous deposits indicates a stable regime of miogeosynclinal character. The zone of the Kolbin–Udinsk Cambrian trough located among the Baikalides of Eastern Sayan closes the considered series of zones. Being a zone of abyssal faulting it remained mobile for a long time.

The associations of alkaline rocks enclosed in the zones of this series display a gradual decrease of alkalinity and increasing contents of silica with increasing intensity of tectonic action seen in the enclosing rocks. This is manifested in a decreasing amount of nepheline syenites and non-feldspathic feldspathoidal rocks and in an accompanying increase in amount of at first alkali syenites, and then of alkaline and sub-alkaline granites and granosyenites. The alkaline

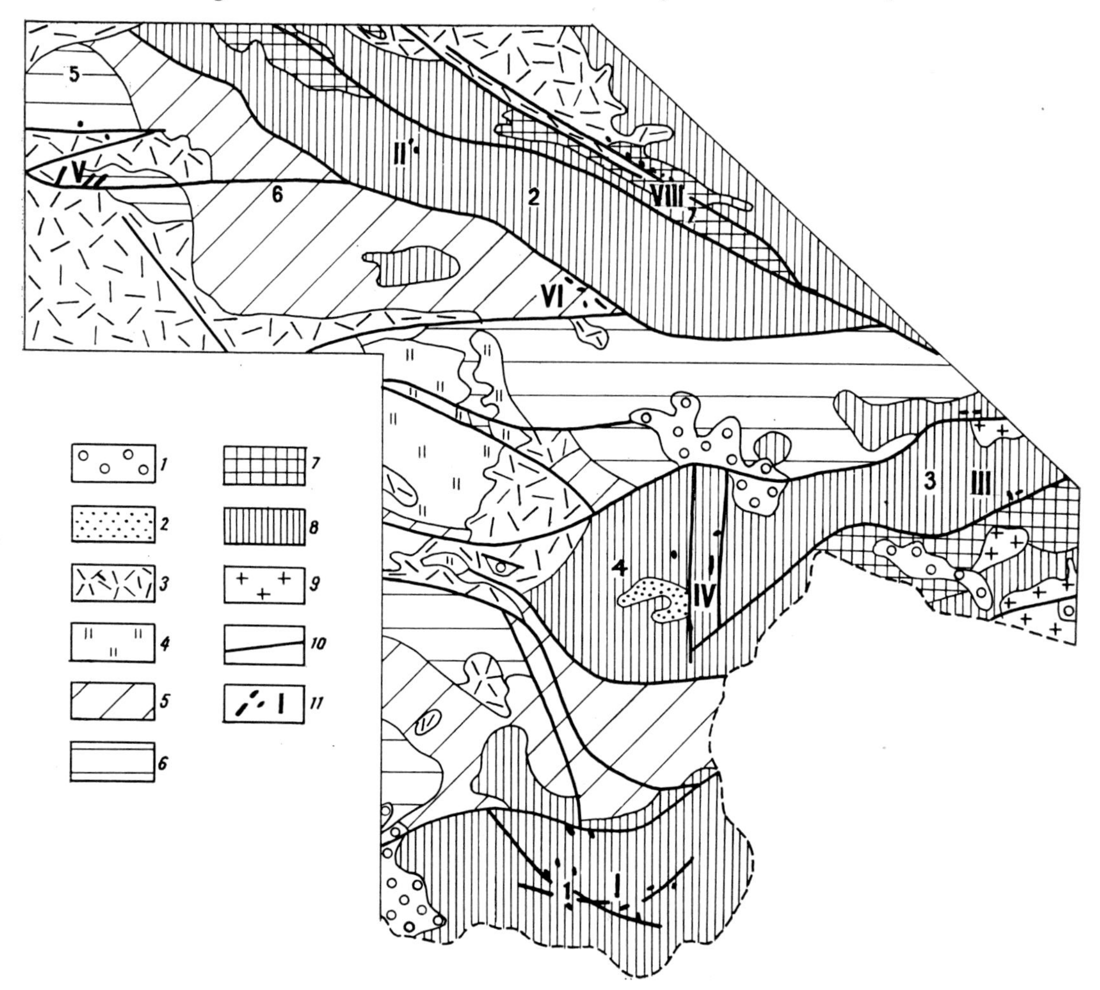

Fig. 3 The distribution of alkaline rock associations within the tectonic structures of the Altay–Sayan folded region. (Compiled by E. L. Butakova on the basis of the scheme of the tectonic zonation by E. N. Yanov *et al.*) (1) Cenozoic intermontane basins; (2) Jurassic–Cretaceous intermontane depressions; (3) Middle–Late Paleozoic intermontane depressions; (4) tectonic structures of Late Caledonian or of the Caledonian system proper. Tectonic structures of the Early Caledonian system: (5) synclinoria and (6) anticlinoria in the zones of eugeosynclinal type; (7) external geosynclinical troughs; (8) tectonic structures of the Baikal folded system in zones mainly of miogeosynclinal type; (9) tectonic structures of the basement of the folded region; (10) abyssal fractures; (11) intrusions of alkaline rocks. Scale 1:4,600,000.

magmatism of the most stable Sangilen zone has the most sharply defined alkaline character; nepheline syenites dominate and non-feldspathic feldspathoidal rocks are significantly distributed. There is very little alkali syenite and granite. The Gutar association, which is situated in the most mobile zone, the Kolbin–Udinsk zone, is characterized by a complete lack of nepheline syenite and non-feldspathic feldspathoidal rocks, and by a minor distribution even of alkali syenites. This province is made up of alkaline and subalkaline granites, granosyenites and quartz syenites. Associations situated in regions of intermediate tectonic setting between Sangilen and Gutar are made up mainly of nepheline and alkali syenites in alternating ratios. The amount of nepheline syenite gradually decreases and the amount of alkali syenite increases in the direction from Pezin and Botogol to Aksug. The alkali syenites of the latter complex often contain quartz.

This comparison of the main features of the tectonic development of the Baikal and Early Caledonian zones of the eastern part of the Altay–Sayan region with the main features of the associations of alkaline rocks found in these zones shows that the character of the alkaline magmatism depends on the tectonic development of the enclosing rocks. Thus the initial geosynclinal tectonic activity has an important influence on the character of the alkaline igneous rocks which are products of the final igneous activity of the region.

III.4.3.6. Relation to Faults

Butakova (1965, 1969) has discussed the relation of alkaline magmatism with abyssal and other fault zones in the eastern part of the Altay–Sayan folded region. The location of the alkaline provinces (Fig. 3) and also individual intrusions is indicative of the relation with fault zones, as are the form, structure and dimensions of the intrusions. The largest polyphase intrusions with the most complicated structures are associated with complicated zones of abyssal fractures. They are located either directly in the main components of abyssal fault zones (most magmatic bodies of Sangilen and the largest intrusions of other provinces) or are confined to their secondary components (intrusions of the Botogol, Gutar, Pezin provinces). The complicated polyphase intrusions originated at places of the most intensive disruptive tectonics characterized by intersections of fractures of different strike. Most intrusions of Eastern Tuva and Eastern Sayan are fissure intrusions closely related morphologically with abyssal faults or with conjugate fractures of higher orders.

The relation between alkaline magmatism and abyssal faults in other folded regions of Siberia has not been studied in detail but may be indicated by a close territorial relation of the alkaline rocks with faults and by the elongation of the intrusive bodies. Thus, the masses of the Kij–Tatarsky province are found in the fault zones of the western margin of the Baikalides of the Enisei ridge. The Belozimin and Bolshezhidoy provinces are confined to the zone of the Main Sayan fault bordering the Siberian platform to the south (Dodin, 1966). The chain of intrusions of the Synnyr province is elongated along the abyssal fault zone (Fig. 4) which breaks the arch of the Baikal–Vitim uplift (Salop, 1967).

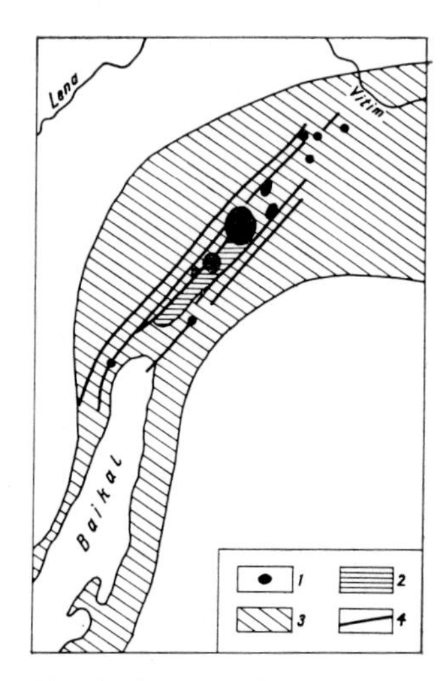

Fig. 4 The tectonic setting of the Synnyr province of alkaline rocks (according to L. I. Salop, 1967)

(1) Intrusions of alkaline rocks; (2) grabens filled with Cambrian deposits; (3) Baikal–Vitim arched uplifts of Middle Proterozoic age; (4) abyssal faults. Scale 1 : 9,500,000.

III.4.4. Conclusions

Harker (1909), Smyth (1927) and Sheynmann *et al.* (1961) have emphasized the importance of tectonic processes in the genesis of alkaline rocks This is confirmed and developed by the present analysis of the mode of occurrence of the alkaline rocks of Siberia.

The following main conclusions may be drawn:

1. The formation of thc alkaline rocks of the platforms and folded regions is preceded by long

periods of relative tectonic quietness. These periods are significantly longer than it was considered before. The formation of the alkaline rocks and especially those richest in alkalis is separated in time from the formation of the tectonic structures of the enclosing rocks. The dominant confinement of alkaline rocks to the parts of the folded regions, which were tectonically the most stable for a long time supports this statement.

2. The observed relation between the chemical, mineralogical and petrographical composition of alkaline rocks and the previous tectonic evolution of the enclosing rocks is an important evidence of the leading significance of the tectonic factor in the origin of alkaline rocks. This conclusion is supported by the dominant concentration of the most strongly alkaline rocks in the Proterozoic and Archean structures of the Siberian platform and of the folded regions of Siberia, which were tectonically stable for the longest periods (median masses and zones of miogeosynclinal type).

3. The origin of the alkaline rocks is related to the disturbance of the period of relative tectonic quietness succeeding folding and non-alkaline granitoid magmatism. This disturbance took place by the formation of large arched uplifts of some sections of the platforms and folded regions and in the downwarping of other sections, and in the formation and rejuvenation of abyssal and other faults.

4. The alkaline igneous activity in stabilized folded structures was in most cases triggered out by intensive tectonic and magmatic activity in neighbouring mobile belts.

5. The constant relation of the alkaline rocks of folded regions and platforms with abyssal faults indicates that the alkaline magmas (or their alkalis) are of abyssal origin.

III.4. REFERENCES (in Russian unless otherwise stated)

Bilibina, T. V., Dashkova, A. D., Donakov, V. I., Titov, V. K., and Shukin, S. I., 1967. *Petrology of the Alkaline Volcanic-Intrusive Complex of the Aldan Shield (Mesozoic)*. 'Nedra', Leningrad.

Butakova, E. L., 1961. Tectonic conditions of the origin of the complex of alkaline and ultrabasic rocks of the north of the Siberian platform. *Geol. and Geophys.*, **1**.

Butakova, E. L., 1965. Structural environment of formation of alkaline rocks of Eastern Tuva. *Geol. and Geophys.*, **5**.

Butakova, E. L., 1969. 'On the genesis of the alkaline formations of folded regions (the example of an alkaline formation of the Altay–Sayan region', in *Problems of Petrology and Genesis of Minerals*, Nauka, Moscow.

Butakova, E. L., and Egorov, L. S., 1962. 'The Maimecha–Kotuj complex of the formation of alkaline and ultrabasic rocks', in *Petrography of Eastern Siberia*. Izd. AN, Moscow.

Dodin, A. L., 1966. The Zima complex of ultrabasic-alkaline rocks,' in *Geology of the Siberian Platform*, Nedra, Moscow.

Egorov, L. S., Rudyachenok, V. M., and Surina, N. P., 1968. On the structural–geological setting of ultrabasic-alkaline rocks of the Maimecha–Kotuj province. *Dokl. AN*, **192**, 1.

Erlikh, E. N., 1966. 'Alkaline intrusions and basic extrusives of the Udzhin uplift,' in *Geology of the Siberian Platform*, Nauka, Moscow.

Erlikh, E. N., 1971. Magmatism of the early Paleozoic cycle, the Anabar anteclise, the Udzhin uplift. *Geology of the USSR*, **18**, 1, 2.

Harker, A., 1909. *The Natural History of the Igneous Rocks* (in English), Methuen, London.

Salop, L. I., 1967. *Geology of the Baikal Mountain Region*. Nedra, Moscow.

Sheynmann, Yu. M., Apeltsin, F. R., Nechaeva, E. A., 1961. Alkaline intrusions, their distribution and associated mineralizations. *Geol of Mineral Deposits of Rare Elements*, **12–13**, Moscow.

Smyth, C. H., 1927. The genesis of alkaline rocks (in English). *Proc. Am. phil. Soc.*, **66.**

Surina, N. P., 1966, Structural-tectonic setting of kimberlites and their relation with the rocks of the alkaline-ultrabasic formation in the Maimecha–Kotuj region (the north of the Siberian platform). *Sov. Geol.*, **3**.

BIBLIOGRAPHY ON THE ALKALINE ROCKS OF SIBERIA (in Russian unless otherwise stated)

I. SIBERIAN PLATFORM

1. The Maimecha–Kotuj province

Butakova, E. L., 1956. On the petrology of the Maimecha–Kotuj complex of ultramatic and alkaline rocks. *Trans. NIIGA*, **89,** 6.

Butakova, E. L., 1959. The role of metasomatism in the formation of alkaline rocks. *Intern. Geol. Rev.*, **3,** 3 (in English).

Butakova, E. L., 1961. Tectonic conditions of the origin of the complex of alkaline and ultramafic rocks of the north of Siberian platform. *Geol. and Geophys.*, **1.**

Butakova, E. L., 1962. The alkaline and ultramafic extrusives of the Maimecha–Kotuj magmatic complex of the north of the Siberian platform. *Trans. of I. All-Union Volcanolog. Conference*, Acad. Sci. USSR, Moscow.

Butakova, E. L., 1966. 'The Maimecha–Kotuj complex of ultramafic and alkaline rocks', in *The Geology of the Siberian Platform*, Nedra, Moscow.

Bukatova, E. L., and Egorov, L. S., 1962. 'The Maimecha–Kotuj complex of the formation of alkaline and ultrabasic rocks', in *The Petrography of East Siberia*, **1**. Acad. Sci. USSR, Moscow.

Egorov, L. S., 1964. On the problem of the origin of carbonatites. *Izv. Acad. Sci. USSR, ser. geol.*, **1.**

Egorov, L. S., 1969. The melilitic rocks of the Maimecha–Kotuj province. *Trans. NIIGA*, **159.**

Egorov, L. S., Goldburt, T. L., and Schichorina, K. M., 1961. Geology and petrography of the magmatic rocks of the Gulin intrusion. *Trans. NIIGA*, **122.**

Egorov, L. S., Surina, N. P., and Rudjachenok, V. M., 1968. On the structural–geological place of the ultramafic-alkaline rocks in the Maimecha–Kotuj province. *Dokl. Acad. Sci. USSR*, **182,** 1.

Epshtein, E. M., Anikeeva, L. I., and Michajlova, A. F., 1961. Metasomatic rocks and phlogopite content of the Gulin intrusion. *Trans. NIIGA*, **122.**

Evzikova, N. Z., and Ilcenko, L. N., 1965. New indications of a primary–sedimentary origin of the carbonatites of the Gulin intrusion. *Dokl. Acad. Sci. USSR*, **165.**

Gonshakova, V. I., and Egorov, L. S., 1968. *Petrochemical Features of the Ultramafic–Alkaline Rocks of the Maimecha–Kotuj Province*. Nauka, Moscow.

Ivanov, A. I., 1963. Stratigraphy of the volcanic formations of the east slope of the Tunguska syneclise. *Sci. Trans. NIIGA*, **1.**

Jabin, A. G., 1967. Geochemistry of alkalis in extrusive and dyke ultramafic alkaline magmatism. *Geochemistry*, **2.**

Kostyuk, V. P., and Panina, L. I., 1970. On the temperature conditions of crystallization of the Gulin intrusion's alkaline rocks. *Dokl. Acad. Sci. USSR*, **194,** 4.

Leontiev, L. N., Gladkich, V. S., Jabin, A. G., and Juk-Pochekutov, K. A., 1965. *Petrology and Geochemical Features of the Complex of Ultramafic and Alkaline Rocks and Carbonatites*. Nauka, Moscow.

Sheynmann, Iu. M., 1947. On the new petrographic province on the north of Siberian platform. *Izv. Acad. Sci. USSR, ser. geol.*, **1.**

Sheynmann, Iu. M., Apeltsin, F. R., and Nechaeva, E. A., 1961. Alkaline intrusions, their distribution and associated mineralizations. *Geol. of Mineral Deposits of Rare Elements*, **12–13,** Moscow.

Surina, N. P., 1968. Geology of the dykes of ultramafic–alkaline rocks of the Maimecha–Kotuj region (Siberian platform's north). *Geol. and Geophys*, **10.**

2. The ultramafic and alkaline rocks and carbonatites of the Aldan shield

Andreev, G. V., 1962. Petrography of the Konder intrusion. *Trans. Buryat Complex, Sci. Res. Inst.*, **9,** ser. geol.

Bogomolov, M. A., 1968. 'Some features of the petrology of central type intrusions with a dunite core on the Aldan shield', in *Metasomatism and Other Questions of Physico-Chemical Petrology*. Nauka, Moscow.

Elianov, A. A., and Andreev, G. V., 1960. New intrusion of central type on the Aldan shield. *Mineral Raw Material*, **1.**

Stoialov, S. P., 1961. The Arbarastah intrusion of ultramafic and alkaline rocks. *Trans. VAGT*, **7.**

Zlenko, N. L., Shpak, N. S., and Elianov, A. A., 1966. 'Ultramafic-alkaline intrusions of the Aldan anteclise', in *The Geology of the Siberian Platform*. Nedra, Moscow.

3. The Chadobets province

Bagdasarov, Iu. A., and Frolov, A. A., 1968. On rare-metal carbonatites of the Chadobets uplift. *Dokl. Acad. Sci. USSR*, **178,** 1.

Chubugina, V. L., 1964. 'On alkaline-ultramafic rocks of the Chadobets uplift and their connection', in *Materials on Geology and Mineral Products of Krasnojarsk Region*. Krasnojarsk book publ. office.

Polunina, L. A., 1966. 'The Chadobets complex of ultramafic–alkaline rocks', in *Geology of the Siberian Platform*. Nauka, Moscow.

Zaitsev, N. S., and Liachovich, V. V., 1955. Ultramafic dyke rocks of the Chadobets uplift. *Izv. Acad. Sci. USSR, ser. geol.*, **2.**

4. Dykes of alkaline–ultramafic rocks and carbonatites of north-eastern margin of the platform

Krutoiarski, M. A., 1966. 'Ultramafic–alkaline rocks of the Anabar–Olenek region', in *Geology of the Siberian Platform*. Nauka, Moscow.

5. The Aldan Mesozoic province of alkaline rocks

Biblibin, Iu., A., 1947. *Petrology of the Illimach Pluton*. State geological publ. office, Moscow.

Bilibina, T. V., Dashkova, A. D., Donakov, V. I., Titov, V. K., and Shukin, S. I., 1967. *Petrology of the Alkaline Volcanic–Intrusive Complex of the Aldan Shield* (*Mesozoic*). Nedra, Leningrad.

Kravchenko, S. M., and Vlasova, E. V., 1962. Alkaline rocks of the Central Aldan (on materials of study of accessory minerals). *Trans. IMGRE*, **14.**

6. The Udzhin province

Erlikh, E. N., 1966. 'Alkaline intrusions and basic extrusives of the Udzhin uplift', in *Geology of the Siberian Platform*. Nauka, Moscow.

II. FOLDED REGIONS (see Table 1)

1. The Altay–Sayan region (*and Enisei ridge*)

A. References for all or some associations of alkaline rocks of the region

Butakova, E. L., 1965. Structural environment of formation of alkaline rocks of the Eastern Tuva. *Geol. and Geophys.*, **5.**

Butakova, E. L., 1968. 'The Middle–Late Paleozoic alkaline formation of Eastern Tuva and Eastern Sayan', in *Geological Structure of USSR*. Nedra, Moscow.

Butakova, E. L., 1969. 'On the genesis of the alkaline formations of folded regions (the example of alkaline formation of the Altay–Sayan region)', in *Problems of Petrology and Genesis of Minerals*. Nauka, Moscow.

Dodin, A. L., 1966. 'The magmatic formations of Eastern Sayan and their metallogenic significance', in *The Theses of Papers on I Siberian Petrographical Conference*. Nedra, Moscow.

Ivanova, T. N., Polevaia, N. I., and others, 1961. Absolute ages of some magmatic and metamorphic rocks of the Altay–Sayan region's central part. *Trans. VSEGEI, new ser.*, **58.**

Klarovskii, V. M., and Kostyuk, V. P., 1965. On the age of the alkaline rocks of the Eastern Sayan's eastern part. *Dokl. Acad. Sci. USSR*, **162,** 2.

Pavlenko, A. S., 1963. 'Petrology and some geochemical features of the Eastern Tuva's middle Paleozoic complex of granitoides and alkaline rocks', in *The Problems of Magma and Genesis of Igneous Rocks*. Publ. Office Acad. Sci. USSR, Moscow.

Volbuev, M. I., Zikov, S. I., Stupnikova, N. I., Musatov, D. I., and Strizhov, V. P., 1966. Materials on absolute ages of magmatic complexes and polymetallic ore deposits of the Sayan–Altay folded region and the Enisei ridge. *Trans. XIII Session of Commission on Investigation of Absolute Ages of Geological Formations*, Nauka, Moscow.

Yashina, R. M., and Borisevich, I. V., 1966. Absolute age of the alkaline rocks of the Eastern Tuva. *Trans. of XIII Session of Commission on Investigation of Absolute Ages of Geological Formations*, Nauka, Moscow.

B. References for individual associations

Kij–Tatársky

Belov, V. P., 1968. Alkaline syenites of the Kij complex (Enisei ridge). *Bull. MOIP, part geol.*, **5.**

Belov, V. P., Volobuev, M. I., Zikov, S. I., and Stupnikova, N. I., 1969. Radiological ages of rocks of the Enisei ridges; the Kij alkaline intrusion. *Proc. MGU, geol. ser.*, **IV,** 6.

Samoilova, N. V., 1962. Petrochemical features of an association of ijolite-melteigite rocks and nepheline syenites (alkaline intrusion of Enisei ridge). *Trans. IGEM Acad. Sci. USSR*, **76.**

Sveshnikova, E. V., 1965. 'The nepheline–syenite complex of the Transangara region (Enisei ridge)', in *Alkaline Magmatism of the Folded Surroundings of the Siberian Platform's Southern Part*. Nauka, Moscow.

Belozimin

Dodin, A. L., 1966. 'The Zima complex of ultramafic–alkaline rocks', in *Geology of the Siberian Platform*. Nedra, Moscow.

Bolshezhidoy

Dodin, A. L., 1966. 'The Zima complex of ultrabasic–alkaline rocks', in *Geology of the Siberian Platform*. Nedra, Moscow.

Konev, A. A., 1960. Intrusion of ore perovskite pyroxenites in the Eastern Sayan. *Dokl. Acad. Sci. USSR*, 133, 4.

Sangilen

Kononova, V. A., 1961. Urtite–ijolite intrusions of the south-eastern Tuva and some questions of their genesis. *Trans. IGEM Acad. Sci. USSR.*, **60.**

Kononova, V. A., 1965. 'On the interaction of nepheline syenites and marbles after an example of the Tchahirtoi injection field (South-East of Tuva)', in *Alkaline Magmatism of the Folded Surroundings of the Siberian Platform.* Nauka, Moscow.

Yashina, R. M., 1957. Alkaline rocks of the south-eastern Tuva. *Izv. Acad. Sci. USSR, ser. geol.*, **5.**

Yashina, R. M., 1963. 'Magmatic metasomatism of dolomitic marbles and its role in alkaline petrogenesis of the south-eastern Tuva', in *Problems of Magma and Genesis of Igneous Rocks.* Publ. office Acad. Sci. USSR, Moscow.

Yashina, R. M., 1964. On the influence of the environment on the development of contact–reaction process at the magmatic stage of formation of nepheline–syenite intrusions (after an example of alkaline intrusions of the south-eastern Tuva). *Trans. of III All-Union Petrograph. Conference,* Nauka, Moscow.

Botogol

Kostyuk, V. P., and Bazarova, T. Yu., 1966. *Petrology of the Alkaline Rocks of the Eastern Part of Eastern Sayan.* Nauka, Moscow.

Sobolev, V. S., 1947. *Petrology of the Botogol Alkaline Intrusion.* Irkutsk region publ. office.

Solonenko, V. P., 1950. Genesis of alkaline rocks and graphite of the Botogol intrusion. *Izv. Acad. Sci. USSR, ser. geol.*, **6.**

Pezin

Krymskii, V. M., 1969. Petrographical features of the Pezin alkaline intrusion. *Trans. SNIIGGIMS*, **61,** Krasnojarsk book publ. office.

Biykhem

Butakova, E. L., 1963. The Dugdu alkaline intrusion in Eastern Tuva. *Trans. WSEGEI, new ser.*, **98,** 5.

Kovalenko, V. I., Okladnikova, L. V., Pavlenko, A. S., and Philippov, L. V., 1965. 'Petrology of the middle-Paleozoic complex of granitoids and alkaline rocks of the Eastern Tuva', in *Geochemistry and Petrology of Magmatic and Metasomatic Formations.* Nauka, Moscow.

Kudrin, V. S., 1962. Alkaline intrusions of the NE Tuva. *Sov. Geol.*, **4.**

Kudrin, V. S., and Kudrina, M. A., 1960. On alkali granitoids of the Eastern Tuva. *Mineral Raw Materials*, **1.**

Saibar

Edelstein, Ia. S., 1930. On new region of distribution of alkaline (nepheline–aegirine) rocks in South Siberia. *Geol. Proc.*, **7,** 1–3.

Fedorov, E. E., 1948. On the question of syenite intrusion of the Tuba-Sida region. *Materials of VSSGSY, gen. ser.*, **8.**

Kostyuk, V. P., and Guletskaia, E. S., 1967. On the mineralogy of the Saibar intrusion (Eastern Sayan). *Geol. and Geophys.*, **7.**

Luchitskii, I. V., 1959. Nepheline ores and alkaline nepheline-bearing rocks of the Krasnojarsk region's south. *Trans. Krasnojarsk Complex Expedition*, Publ. office of Acad. Sci. USSR, Moscow.

Saranchina, G. M., 1940. Alkaline rocks of the Saibar intrusion (West Siberia, Krasnojarsk region). *Sci. Trans. LGU*, **45,** 8.

Aksug

Dmitriev, L. V., Kotina, R. P., 1966. Form and structural position of the Katun intrusion of East Sayan. *Sov. Geol.*, **9.**

Gutar

Lin, N. G., and Morozov, L. N., 1963. Alkali granitoids of the Eastern Sayan's central part. *Theses of Papers on III All-Union Petrograph. Conference*, Publ. office of Siberian branch of Acad. Sci. USSR, Novosibirsk.

Rik, L. P., 1961. The Ognit intrusive complex. *Mater. on Geology and Mineral Product of Irkutsk Region*, **1,** 28.

Kuznetsk Alatau

Andreeva, E. D., 1968. *Alkaline Magmatism of the Kuznetsk Alatau.* Nauka, Moscow.

Bajenov, I. K., Scobelev, Iu. D., Kortusov, M. P., Vrublevskii, V. A., and others, 1963. Geological structure and petrography of nepheline rocks of the Kuznetsk Alatau. *Materials on Geology of Western Siberia*, **64.**

Dovgal, V. N., and Bognibov, V. I., 1965. 'Old gabbro–syenite complex of Kuznetsk Alatau', in *The Magmatic Formations of Altay–Sayan Folded Region.* Nauka, Moscow.

Kortusov, M. P., 1963. The Kij gabbro–syenite complex Mariin taiga (Kuznetsk Alatau). *Trans. Inst. Geol. Geophys. Siberian Branch Acad. Sci. USSR*, **33,** Novosibirsk.

Kortusov, M. P., 1964. 'Geological–petrographical features of nepheline-bearing rocks of the Kuznetsk Alatau's north-western part: Trans. III All-Union petrogr. conference', in *The Genesis of the Alkaline Rocks.* Nauka, Moscow.

Kortusov, M. P., and Makarenko, N. A., 1968. Nepheline-bearing rocks of the Mariin taiga and their genetical features. *Trans. Inter-High-School Sci. Conferention*, Publ. office of Tomsk University.

Luchitskii, I. V., 1961. *Volcanism and Tectonic Geology of the Devonian Depressions of the Minusinsk Intermountain Trough.* Publ. office of Acad. Sci. USSR, Moscow.

Michalev, V. G., 1962. Geological characteristic of the alkaline intrusions of the Goriachei mountain. *Geol. and Geophys.*, **5.**

Michalev, V. G., 1968. Some data on the geology and petrography of the nepheline rocks of the Kuznetsk Alatau. *Mater. on Geol. and Mineral Products of Krasnojarsk Region, 5* Krasnojarsk book publ. office.

Saranchina, G. M., 1936. Complex of nepheline–melilite–monticellite rocks of the Patin intrusion in district Mountain Shoria of Western Siberia. *Sci. Trans. LGU, 9, ser. geol., soil, geogr.*, **2.**

2. Transbaikal

Malokunaley (Kunaley)

Dvorkin-Samarskii, V. A., 1965. *Formations of Granitoids of the Sayan–Baikal Mountain Region.* Buryat book. publ. office, Ulan-Ude.

Jalbason, D. I., Skripkina, V. V., and Maksimova, E. A., 1967. The Mesozoic intrusive formations of the Buryatia's south (Western Transbaikal). *Mater on Geol. Conference*, Publ. office of Buryat ASSR, Ulan-Ude.

Kuznetsova, T. V., 1962. Nepheline syenites of the Borgoi alkaline intrusion. *Trans. East-Siberian Branch of All-Union Mineralog. Soc.*, **3.**

Ladaeva, V. M., 1960. Nepheline syenites of the Buryatia's southern district. *Materials on Geology and Mineral Deposits of Buryat ASSR*, **1 (IV),** Ulan-Ude.

Nechaeva, E. A., 1960. Alkali granitoids of the Transbaikal region. *Trans. II All-Union Petrogr. Conference*, State geological technical publ. office, Moscow.

Pleshanova, A. L., 1968. Accessory minerals of the Borgoi ridge's alkaline rocks (South-Western Transbaikal region). *Trans. of Conference of IMGRE*, Moscow.

Smirnov, G. V., 1963. 'Petrochemical Features of Nepheline Syenites of r. Jida's lower reaches', in Questions of Magmatism and Tectonic Geology of Buryatia. *Trans. of BKNII of Siberian Branch of Acad. Sci. USSR*, **12.**

Nerchugan

Alektrova, V. A., 1968. Features of the alkali and subalkali granitoids of the North-Eastern Transbaikal region. *Bull. MOIP, part geol.*, **5.**

3. Baikal

Synnyr

Andreev, G. V., 1965. *Petrology of the Synnyr Alkaline Pluton.* Buryat book publ. office, Ulan-Ude.

Andreev, G. V., and Dvorkin-Samarskii, A. V., 1966. The Akit alkaline intrusion. *Trans. BKNII, ser. geol.*, **22,** Ulan-Ude.

Archangelskaia, V. V., 1964. The Synnyr intrusion of alkaline rocks and its apatites. *Dokl. Acad. Sci. USSR*, **158,** 3.

Dvorkin-Samarskii, V. A., and Belov, I. V., 1962. 'Magmatic and metamorphic formations of north Baikal region', in *Petrography of Eastern Siberia*, **2.** Publ. office Acad. Sci. USSR, Moscow.

Jidkov, A. Ia, 1961. New North-Baikal alkaline province and some features of its rocks' nepheline content. *Dokl. Acad. Sci. USSR*, **140,** 1.

Jidkov, A. Ia., 1962. Complex Synnyr intrusion of syenites of the North-Baikal alkaline province. *Geol. and Geophys.*, **9.**

Jidkov, A. Ia., 1963. Unique mineral deposits of pseudoleucite ultrapotash syenites. *Dokl. Acad. Sci. USSR*, **152,** 2.

Kostyuk, V. P., Panina, L. I., and Guletskaia, E. S., 1970. 'Mineralogy of the high potash alkaline rocks of complex Synnyr pluton (North Baikal region)', in *Problems of Petrology and Genesis of Minerals*, **2.** Nauka, Moscow.

Pak, A. S., Zak, S. I., Gorstka, V. N., Filatov, V. G., and others, 1969. *Geological Structure and Apatite Content of the Synnyr Alkaline Intrusion, Collected Articles.* Nauka, Leningrad.

Panina, L. I., 1966. Some data on the temperature conditions of formation of the Synnyr alkaline intrusion. *Dokl. Acad. Sci. USSR*, **170,** 6.

Velikoslavinskii, D. A., Kazakov, A. N., Lobach-Iuchenko, L. B., Manuilova, M. M., Petrov, V. B., and Socolov, Iu. M., 1962. 'Magmatic and metamorphic formations of the North-Baikal upland', in *Petrology of Eastern Siberia*, **2.** Publ. office Acad. Sci. USSR, Moscow.

Saizhin

Andreev, G. V., Litvinovskii, B. A., and Sharakshinov, A. O., 1969. *Intrusions of Nepheline Syenites of the Western Transbaikal Region.* Nauka, Moscow.

Belov, I. V., 1963. *Trachybasalt Formation of the Baikal Region.* Publ. office. Acad. Sci. USSR, Moscow.

Konev, A. A., 1962. On new intrusions of nepheline rocks on the Vitim plateau. *Trans. East Siberian Branch All-Union Mineral. Soc.*, **3,** Irkutsk.

Konev, A. A., 1962. Petrography of alkaline, ultrabasic and basic rocks of the Saizhin and the Gulhen plutons (Vitim plateau). *Trans. East Siberian geol. Inst. Siberian Branch Acad. Sci. USSR*, **11.**

4. Sikhote–Alin (with Burein and Khankay median masses)

Putintsev, V. K., 1968. 'The Burein median mass', in *Geological Structure of the USSR, 3.* Nedra, Moscow.

Zalishak, B. L., 1969. *The Kokshary Intrusion of Ultramafic and Alkaline Rocks (South Sea Country).* Nauka, Moscow.

Gapeeva, G. M., 1961. Alkaline basaltoids of the Sichote–Alin ridge. *Trans. IGEM*, **45.**

Rub, M. G., and Zalishak, B. L., 1964. Alkaline intrusive rocks of the sea country. *Izv. Acad. Sci. USSR, ser. geol.*, **10.**

Russ, V. V., and Bikovskaia, E. V., 1968. 'The Sichote–Alin folded system', in *Geological Structure of the USSR, 3.* Nedra, Moscow.

5. Verkhoyano–Chukotka folded region (with the Omolon, Kolyma and Eastern-Chukotka median masses)

Alkali and nepheline syenites of the Abkitsk association

Spetnii, A. P., 1968. 'The Verkhoyano–Chukotka folded region. Kolyma, Omolon and Ohotsk median masses', in *Geological Structure of the USSR, 3.* Nedra, Moscow.

Alkali granites of the granitoid association (Omolon mass)

Spetnii, A. P., 1968. 'The Verkhoyano–Chukotka folded region. Kolyma, Omolon and Ohotsk median masses', in *Geological Structure of the USSR, 3.* Nedra, Moscow.

Omolon association

Bilibin, Iu. A., 1958. 'Essexite–teschenite complex of Omolon region', in *Selected Works of Iu. A. Bilibin, 1.* Publ. office Acad. Sci. USSR, Moscow.

Alkali granites and related rocks of the Tommot association

Nekrasov, I. Ia., 1962. *Magmatism and Ore Content in the North-Western Part of the Verkhoyano–Chukotka Folded Region.* Publ. office Acad. Sci. USSR, Moscow.

Alkaline basaltoids and gabbroids (Kolyma mass)

Nekrasov, I. Ia., 1962. *Magmatism and Ore Content in the North-Western Part of the Verkhoyano–Chukotka Folded Region.* Publ. office Acad. Sci. USSR, Moscow.

Dezhnev

Firsov, L. B., 1964. On the absolute age of the syenites of the Dezhnev cape. *Kolyma*, **11,** Magadan.

Perchuk, L. L., 1963. 'Magmatic replacement of carbonaceous rock series with the formation of nepheline syenites and other alkaline rocks after an example of the Dezhnev intrusion', in *Physico-Chemical Problems of Formation of Rocks and Ores, 2.* Publ. office Acad. Sci. USSR, Moscow.

Alkaline basaltoids of the Pezhen and other associations

Bilibin, Iu. A., 1958. 'Olivine and alkaline basalts of the Anui–Omolon region', in *Selected Works, 1.* Publ. office Acad. Sci. USSR, Moscow.

Alkaline basaltoids (Eastern-Chukotka median mass)

Rabkin, M. I., 1954. Alkaline, basic and ultrabasic extrusives of the Chukotka peninsula's south part. *Trans. NIIGA*, **43,** 3.

Alkali granites of the granitoid association

Gelman, M. L., 1964. Late-Mesozoic minor intrusions of Western Chukotka. *Mater. on Geology and Mineral Products of North-East of the USSR*, **17,** Magadan.

6. Koryak-Kamchatka folded region

Alkaline basaltoids of the Raritkin association

Mihailov, A. F., 1968. 'The Pekulney zone', in *Geological Structure of the USSR, 3.* Nedra, Moscow.

Alkaline basaltoids and gabbroids of Kamchatka

Mihailov, A. F., 1968. 'The West-Kamchatka–Koryak zone', in *Geological Structure of the USSR, 3.* Nedra, Moscow.

Alkaline syenites of the syenite-monzonite association of Sakhalin

Mihailov, A. F., 1968. 'Island Sakhalin', in *Geological Structure of the USSR, 3.* Nedra, Moscow.

7. Taimyr folded region

Alkaline and nepheline syenites

Ravich, M. G., 1968. 'The mountain's Taimyr', in *Geological Structure of the USSR, 3.* Nedra, Moscow.

Ravich, M. G., and Chaika, L. A., 1959. Minor intrusions of the Birranga range (Taimyr peninsula). *Trans. NIIGA*, **38.**

List of abbreviations of names of scientific-research institutes, scientific societies and other institutions.

BKNII—Buryat Complex Scientific-Research Institute
VAGT—All-Union Aero-Geological Trust
VIMS—All-Union Institute of Mineral Raw Material
VSEGEI—All-Union Scientific-Research Geological Institute
IMGRE—Institute of Mineralogy, Geochemistry and Crystallochemistry of Rare Elements
LGU—Leningrad State University
MGU—Moscow State University
MOIP—Moscow Society of Investigators of Nature
NIIGA—Scientific-Research Institute of Geology of the Arctic

III.5. ALKALINE ROCKS OF SOUTHERN AFRICA

M. Mathias

III.5.1. Introduction

The area described in this section comprises South-West Africa, Rhodesia, the Republic of South Africa and Botswana, thus having a northern boundary approximately aligned along the Kunene and Zambesi rivers but excluding Mozambique in the east. Countries lying farther north have not been included owing partly to limitation of space and partly to the excellent coverage of the alkaline rocks associated with carbonatites and the admirable summaries of B. C. King and D. S. Sutherland (1960).

Rock associations have been used as the basis for the major subdivision with a secondary subdivision into localities. Few classifications are wholly satisfactory. Some complexes have borderline characteristics and a subjective decision has to be made. Subsequent work may necessitate a regrouping.

III.5.2. Carbonatite Complexes

III.5.2.1. General

This field has been well covered by recent publications and the reader is referred particularly to the admirable general summaries by Tuttle and Gittins (1966), Heinrich (1966) and Verwoerd (1967). In view of this it will suffice to summarize the chief characteristics and comment on features of interest.

Throughout Southern Africa the carbonatite-bearing peralkaline complexes are typically subvolcanic. Goudini in the Transvaal and Osongombo in South-West Africa are exceptional in that they have been less eroded and represent respectively a volcanic complex and an intermediate type transitional between volcanic and subvolcanic (Verwoerd 1966, 1967).

With the exception of the Premier diamond mine kimberlite centre in the Transvaal, there is little correspondence between kimberlite and alkaline occurrences in Southern Africa. The main clusters of kimberlite and melilite basalt lie (*a*) near Gibeon in South-West Africa, 25° 10′ S, 17° 50′ E, (*b*) in Namaqualand, (*c*) in the northern Cape Province, (*d*) in the Orange Free State and (*e*) in Lesotho. These localities are all remote from alkaline complexes. Both groups, however, occur in stable areas or regions subject to vertical tectonics (Fig. 1).

Fenitization is characteristic of the carbonatite complexes and detailed general and comparative descriptions of the process have been given by Verwoerd (1966, 1967) and McKie (1966). Less commonly, fenitization has been described in connection with complexes which have no known carbonatite and into this category fall the Messum Complex of Damaraland, South-West Africa which shows fenitization of both acid and basic rocks (Mathias, 1956); the Granitberg foyaite, south of Lüderitz in South-West Africa (Kaiser, 1922; Kaiser and Beetz, 1926); the Semarule Hills syenite of Botswana

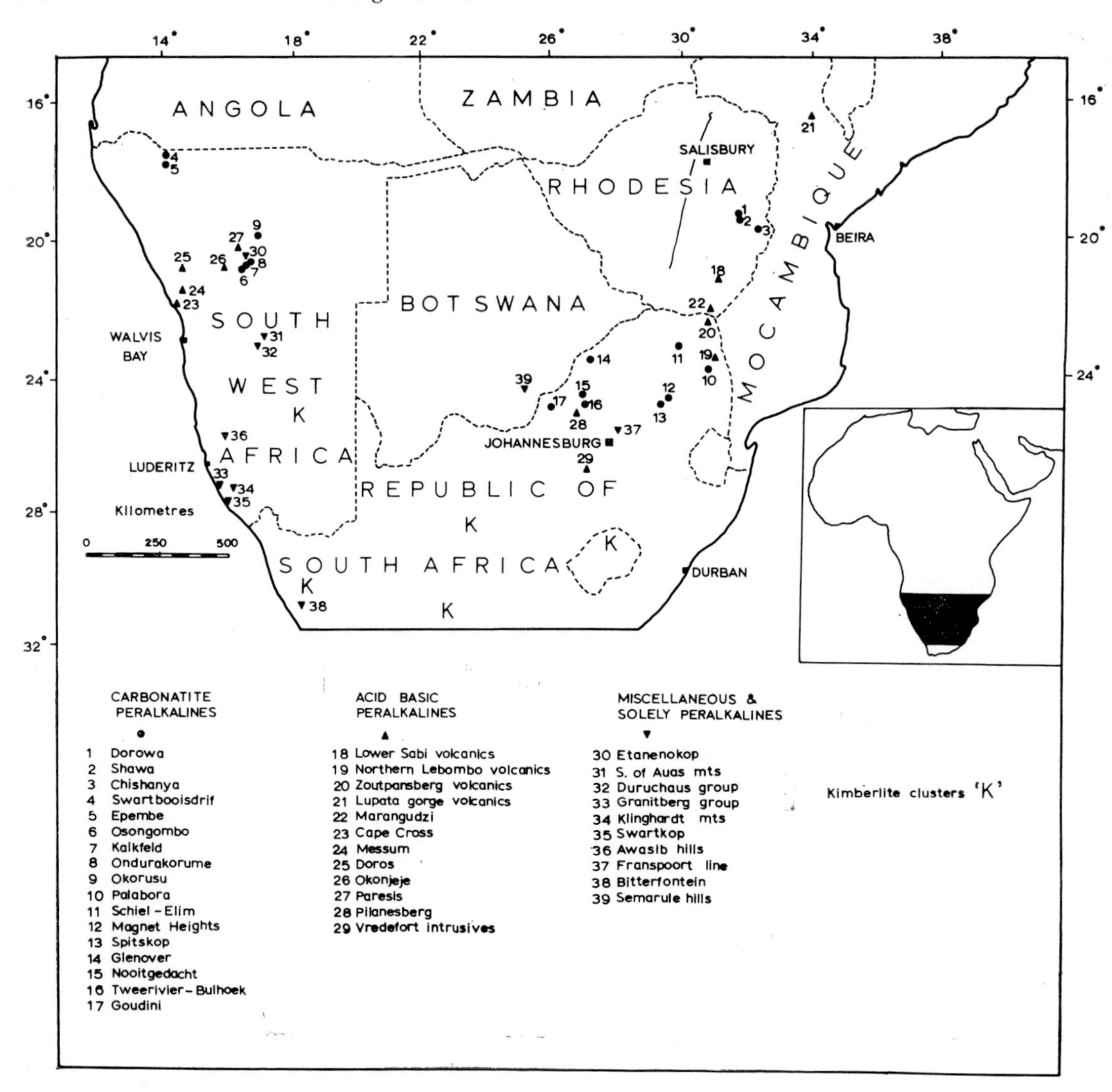

Fig. 1 Locality sketch map

(King, 1955, 1960) and an umptekite occurrence in Namaqualand (Jansen, 1960).

III.5.2.2. Rhodesia

The three complexes of Shawa, Dorowa and Chishanya are probably contemporary and belong to an early phase of Stormberg activity. (Mennell, 1946; Swift, 1952; Tyndale-Biscoe, 1956; Johnson, 1961, 1966; Gittins, 1966). The age of Shawa is given as 195 $\pm$15 m.y. by the Rb–Sr method using biotite from an ijolite (Nicolaysen *et al.*, 1962; Vail, 1968). The country rocks are Archaean granite-gneiss and intrusive dolerite. All have ijolitic rocks as the main alkalic intrusions and associated dyke swarms of nephelinite. At Shawa the ijolite is considered by Johnson (1961, 1966) as a likely parental magma. It is accompanied by serpentinized dunite. At Chishanya and Dorowa both ijolite and nepheline syenite may be mobilized fenites. All three complexes are surrounded by zones of fenitized Basement granite-gneiss and at Dorowa the earlier dolerite has been partially fenitized.

III.5.2.3. South West-Africa

The Damaraland complexes (Martin *et al.* 1960; Verwoerd, 1966, 1967) of Osongombo (Gittins, 1966), Kalkfeld (Stahl, 1930; Van Zijl, 1962), Ondurakorume (Gittins, 1966) and Okorusu (Stahl, 1930; Van Zijl, 1962; Gittins, 1966) lie along a N.E.-trending zone which includes also the acid–basic alkaline complexes of Cape Cross, Messum, Okonjeje, Paresis, the alkalic granite pluton of Brandberg and the agpaitic complex of Etanenokop (Fig. 2). They are characterized by the association of carbonatite and nepheline syenite as the major intrusives. All show fenitization of the country rocks which are Precambrian Damara System sediments and syntectonic Salem granite. Verwoerd (1966, 1967) believes that Osongombo, Kalkfeld and Ondurakorume correspond to a sequence of increasing erosion levels. The origin of the outer ring of Kalkfeld is in dispute. Van Zijl (1962) considers it to be an integral part of the complex whereas Verwoerd (1967) is of the opinion that it is a granitic fenite.

The Swartbooisdrif and Epembe (Gittins, 1966; Verwoerd, 1967; Toerien, private communication), occurrences lie farther north, near the border with Angola, and are situated respectively at the eastern extremity and south of the Kunene Basic Complex which is itself a southern continuation of the immense anorthosite mass of Southern Angola. Swartbooisdrif has unusual features which merit mention. A K/Ar age of 749 m.y. is reported by Verwoerd (1967) for biotite from a nepheline-syenite pegmatite at Swartbooisdrif. The complex has been studied by D. K. Toerien who has kindly permitted the following description

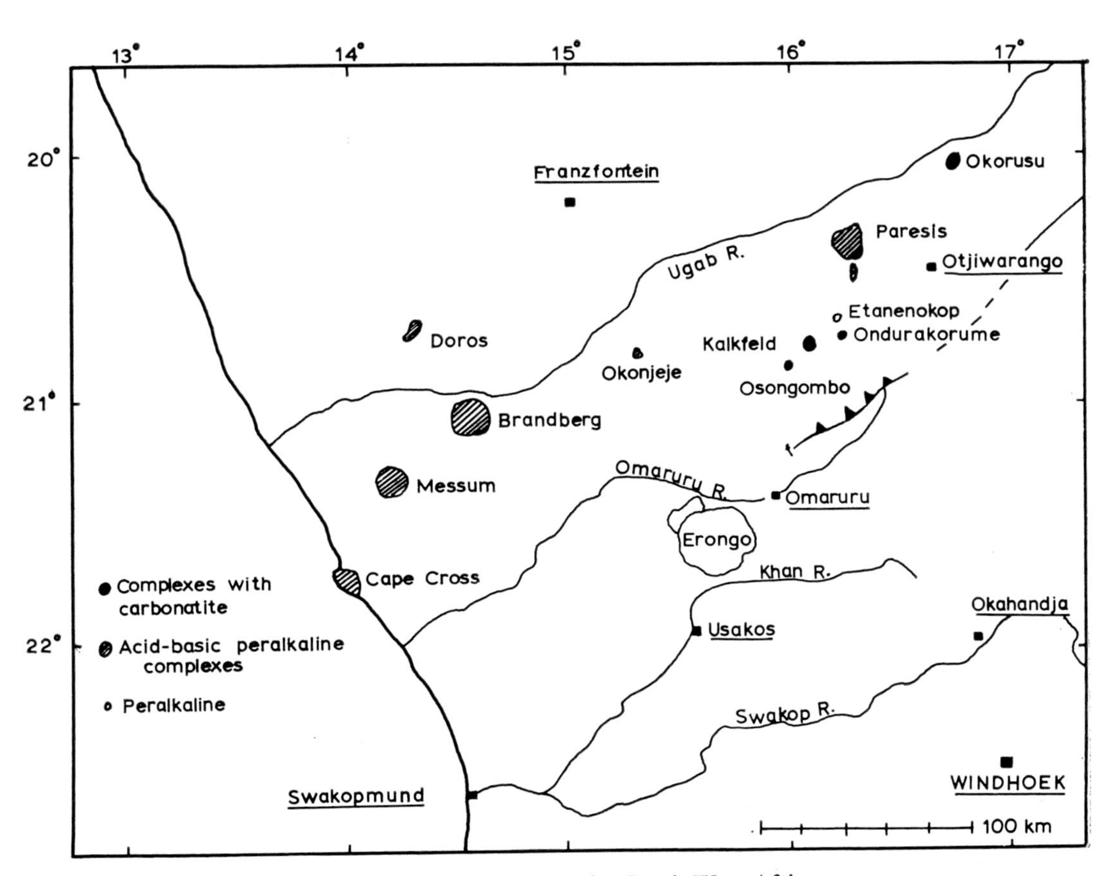

Fig. 2 Locality sketch map for South West Africa

which is published in terms of the Government Printers Copyright Concession No. 3947 of 5/1/68. The peralkaline rocks occur in the form of a fan-shaped anastomosing dyke-swarm some 30 kms long and located on a W.N.W.–E.S.E. trending fracture zone. A plug approximately 1·5 × 1 km occurs at the apex of the fan and consists of a carbonatite core surrounded by nepheline–cancrinite–sodalite syenite. The dykes consist of carbonatite, nepheline syenite and mica lamprophyre. Many of them are remarkable for their composite banded character with concentrations of alternating light and dark minerals parallel to the strike. Both alkaline rocks and carbonatites are believed to have resulted from metasomatism of the country rock anorthosite by CO_2-rich alkalic solutions penetrating along the fracture zone. Partial or complete segregation followed, giving bands of carbonatitic, nepheline-syenitic and mica lamprophyric compositions.

III.5.2.4. South Africa

These complexes all occur in the Transvaal, north of Pretoria (Gittins, 1966; Verwoerd, 1966, 1967). They comprise Palabora (Hall, 1912; Du Toit, 1931; Shand, 1931: Gevers, 1948; Russell *et al.*, 1954; Bouwer, 1957; Forster, 1958; Lombaard *et al.*, 1964; Hanekom *et al.*, 1965) and the related Schiel–Elim group (E. A. Viljoen, private communication); Glenover; Goudini; Spitskop (Shand, 1921a; Strauss and Truter, 1950a; Holmes, 1958; Verwoerd, 1964); Magnet Heights (Strauss and Truter, 1950b); Nooitgedacht and Tweerivier–Bulhoek. A lead isotope age determination on a U-bearing thorianite from the Palabora carbonatite yielded an age of 2060 m.y. (Holmes and Cahen, 1957). This makes it contemporary with the Bushveld Complex (Nicolaysen *et al.*, 1958). Pyroxenites and other basic rocks are associated with Palabora, Schiel–Elim, Glenover and Spitskop. Alkaline rocks are limited to small plugs, dykes and sills except at Spitskop where there are major intrusions of ijolite and foyaite. Minor intrusions of alkalic syenite associated with carbonatite injection-breccia are a feature of the Palabora–Schiel–Elim group and crop out over a wide area around the complexes and in two extensive east–west zones (Verwoerd, 1967).

III.5.3. Alkaline Acid-Basic Association

III.5.3.1. General

Besides the central type subvolcanic complexes which will be referred to under regional headings, alkaline rocks occur as lavas both early and late in the Karroo volcanic cycle. This concept of a volcanic cycle was proposed by Cox and co-authors (1965) and an excellent summary with the proposed genesis of the alkalic magma is given by Johnson (1966, pp. 219–22). Nephelinites and allied rocks are interbedded with olivine basalts and limburgites near the base of the lava succession in the lower Sabi region of Rhodesia (Swift *et al.*, 1953); the northern Lebombo (Du Toit, 1929; Lombaard, 1952) and north and east of the Zoutpansberg range in the Transvaal (Rogers, 1925; Du Toit, 1929). The late phase is represented by the association basalt→trachyte→phonolite and examples occur overlying the Karroo rhyolites in Mozambique (Young, 1920; Cox *et al.*, 1965) and in the Lupata gorge of the Zambesi river where analcime-bearing lavas dated at 115 $\pm$10 m.y. by K–Ar on feldspar are considered by Flores (1964) to belong to an upper division of the Lupata group which uncomformably overlies Karroo sandstones and basalts. Phonolites, kenyites and blairmorites have also been recognized in this area (Dixey *et al.*, 1955).

III.5.3.2. Rhodesia

The Marangudzi complex has been described in two unpublished Ph.D. theses of the University of London by A. C. Gifford (1961) and G. Rees (1960). It lies 80 kms east of Beit bridge and 32 kms north of the Limpopo river in the extreme south of Rhodesia. It covers 78 sq km and is intrusive into Basement rocks. It is an extremely fine example of an acid–basic alkaline complex with six intrusion centres which have migrated along a S.E.–N.W. line (Fig. 3). Successive emplacements are of olivine gabbro; a ring dyke of granite passing inwards into

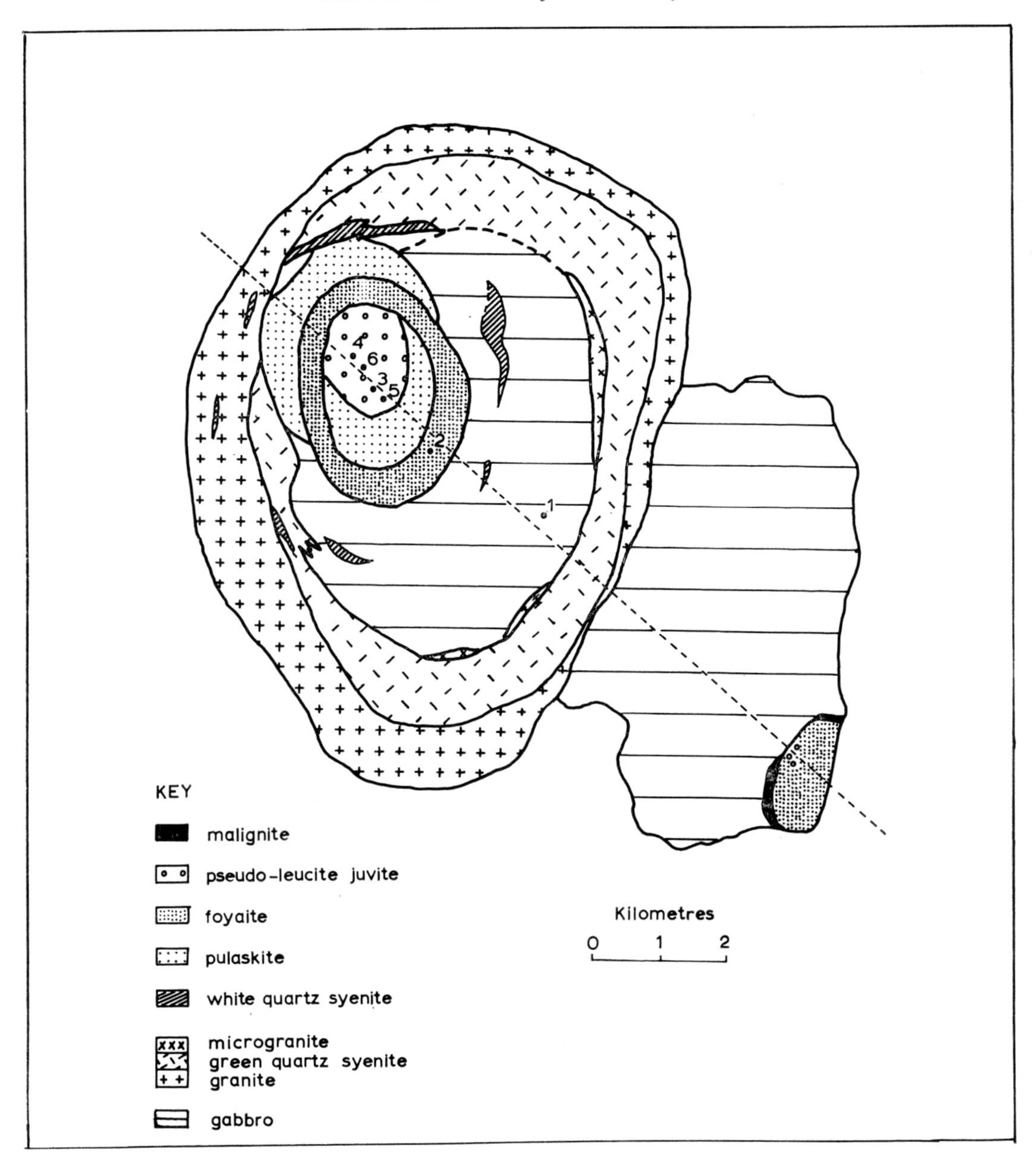

Fig. 3 Marangudzi complex, Rhodesia (after A. C. Gifford)

green quartz syenite and microgranite; a discontinuous ring of white quartz syenite; a pulaskite ring; a partial ring of foyaite and an innermost phase of pseudoleucite juvite. Alkaline rocks were also intruded at the Madawula centre in the south east. They consist mainly of nepheline syenite with marginal malignite in the north and west. The age of the complex is 190 $\pm$ 12 m.y. by K–Ar determinations (Gough *et al.*, 1964). Both radial and 'swarm' dykes are concentrated in a zone trending E.–W. and approximately eight kms wide.

III.5.3.3. South-West Africa

The ring complexes of Cape Cross (Reuning, 1929; Gevers, 1932; Linning, 1968), Messum (Korn and Martin, 1952, 1954; Mathias, 1956, 1957), Okonjeje (Korn and Martin, 1939; Simpson, 1954), and Paresis are aligned in a N.E. direction from Cape Cross on the Atlantic coast (Fig. 2). The Doros plug (Reuning, 1924, 1929; Reuning and Martin, 1957) is offset to the north. Since the review account by Martin and co-authors (1960) the Paresis volcano has been described by Siedner (1963, 1965a and b), Manton and Siedner (1967) and whole rock K–Ar age determinations have been carried out on regional basalts and dolerites and on basic rocks from four of the complexes with the following results: Doros 125 m.y., Messum 123 m.y., Okonjeje 164 m.y. and Paresis 136 m.y. (Siedner and Miller, 1968). Cape Cross, Messum and Paresis represent decreasing erosional levels.

Messum, Okonjeje and Marangudzi show striking similarities. All have the same general sequence basic → acidic → alkalic. All possess notable examples of contaminated rocks and highly calcic anorthosites. Okonjeje differs from the others in showing igneous lamination and cryptic layering with increase in iron and soda in a downward and outward direction. Messum alone shows fenitization effects.

The Brandberg granite stock has a fine-grained alkaline facies and thereby provides a link with the Damaraland granite plutons of Erongo and the Gross and Klein Spitzkoppe (Cloos, 1929; Chudoba, 1930; Cloos and Chudoba, 1931; Korn and Martin, 1952).

III.3.5.4. South Africa

Two intrusions belonging to this category occur in the Transvaal. They are quite different but both remarkable in their way. One is the Pilanesberg complex and the other the intrusive rocks associated with the Vredefort dome.

The Pilanesberg complex is one of the three largest alkaline intrusions known. It was des-described thus by Shand in 1928 and still holds this position today. This ring complex covers an area of 570 sq km in the western Transvaal and is intrusive into the Bushveld Complex at the junction of the norite and granite. In a classic paper Shand (1928) describes the Pilanesberg which he envisages as developing in the following sequence: residual magma from the Bushveld granite was entrapped at depth and there reacted with the Transvaal system dolomite, becoming desilicated and alkaline in composition. This alkaline magma, rich in volatiles, broke through forming a group of volcanoes at Pilanesberg which is located at the intersection of the E.–W. axis of the Bushveld complex with the N.N.W. zone of weakness delineated by the Franspoort line of alkaline intrusives. The alkaline lavas coalesced and formed a thick blanket of phonolitic tuffs. Additional magma consolidated as a laccolith of red foyaite between the overlying tuffs and underlying norite. Cauldron subsidence occurred and white foyaite welled up around the edge of the sinking lava 'cake'. This foyaite ring had an acidified marginal zone of red syenite and passed transitionally into the later green foyaites and lujavrites. Tinguaite was extruded as a sheet forming the Beacon Heights prior to the injection of the green foyaites and also later, both contemporaneously with the green foyaites and as a residual phase emplaced as a ring dyke.

Dolerites occurring in the north-west and north-central areas were interpreted as xenolithic in origin. The Pilanesberg intrusion was accompanied by a dyke swarm radiating from Pilanesberg mainly in south and south-easterly directions and comprising doleritic, monzonitic and syenitic rocks.

Subsequent dating gives an age range from 1290 $\pm$ 180 m.y. to 1330 $\pm$ 80 m.y. for the dyke swarm (Schreiner and Van Niekerk, 1958; Van Niekerk, 1962) and a K–Ar age of 1250 $\pm$ 50 m.y. for the Pilanesberg complex (Retief, 1963). As the Bushveld granite has been reliably dated at 1950 $\pm$ 150 m.y. (Nicolaysen *et al.*, 1958) Shand's theory of the generation of the alkaline magma from Bushveld granite is no longer acceptable.

More recent studies of Pilanesberg have been made by Cloete (1957) and Retief (1962, 1963).

Cloete concentrated on the dolerite occurrences and came to the conclusion that they are intrusive into, but coeval with, the alkaline rocks and do not occur as xenoliths of older dolerite as suggested by Shand (1928). Retief has remapped the complex and I am indebted to him for the accompanying sketch map (Fig. 4) and a summary of his detailed work. Apart from the major difference concerning the age and origin of the alkalic magma, additions and changes in interpretation include the following:

a. Two periods of vulcanicity separated by intrusion of early foyaite, syenite and quartz syenite are positively identified.

b. A plug of alkali granite was discovered by Cloete (1957) and a surrounding triangular area of alkali syenite by Retief (1963).

c. All tinguaites are thought to be post-white foyaite and pre-green foyaite.

d. The tinguaite which was considered by Shand (1928) to be a ring dyke is re-interpreted as a cone sheet.

e. No transition zone is recognized between the white and green foyaites, and the former southern transition zone is described as a separate intrusion to which the name 'Ledig-foyaite group' is given.

f. White foyaite is found to consist of three ring dykes and not a single intrusion separated by later green foyaite.

g. The relative age of the nepheline feldspar porphyry is fixed as later than the volcanics and white foyaite. This porphyry occurs as steeply inclined sheets.

The Vredefort dome is a remarkable tectonic feature, variously interpreted, which has a core of Archaean granite surrounded by steeply dipping uptilted and overfolded rocks belonging to the Dominion Reef, Witwatersrand, Ventersdorp and Transvaal Systems. Alkaline rocks occur in the north and north-west and have a similar age (2020 $\pm$ 70 m.y.) to that of the Bushveld Complex (L. O. Nicolaysen, private communication). Excellent petrographic descriptions are given by Hall and Molengraaff (1925); Nel (1927) and Tilley (1960). A recent study has been made by A. A. Bisschoff and the following is a brief summary extracted from information which he has kindly supplied.

The alkaline rocks include the following: (*a*) alkali granite of the Rietfontein complex intrusive into Transvaal dolomite and bordered to the south by intrusions of olivine gabbro, wehrlite and troctolite; (*b*) two plutons of alkali granite adjacent to the Vaal river and intrusive into the lower division of the Witwatersrand System; (*c*) dykes of mariupolite closely associated with (*b*); (*d*) dykes of alkali granite aplite in the north-eastern section of the dome. The plutons of alkali granite were forcefully injected along major strike faults which have a more or less concentric disposition to the core of the dome. They were emplaced after the main event which caused the formation of the dome. Dykes of mariupolite occur mainly in the sedimentary roof and hood zone of the northern pluton and, with two minor exceptions, are absent from the deeply eroded cores of the plutons. These dykes are considered to be a product of desilication, probably through the agency of escaping vapours. A later foyaitic dyke, which is a member of the Pilanesberg dyke swarm, cuts both mafic and alkalic intrusives.

III.5.4. Miscellaneous and Exclusively Alkaline Occurences

III.5.4.1. South-West Africa

Etanenokop lies within the line of Damaraland intrusives and is unusual in being entirely foyaitic in composition. It was mapped by W. J. Verwoerd in 1956 and has been studied by E. A. Retief (private communication) who describes it as consisting of three concentric rings and a slightly eccentric core.

Rocks of the foyaite family occur also in four further localities:

a. South of Windhoek, near the Auas mountains and between Aris and Krumhuk (Rimann, 1914; Gevers, 1933, 1934).

b. Near Duruchaus and approximately 60 km S.S.W. of Windhoek and 40 kms N.W. of Rehoboth (De Kock, 1934).

c. In the coastal region south of Lüderitz,

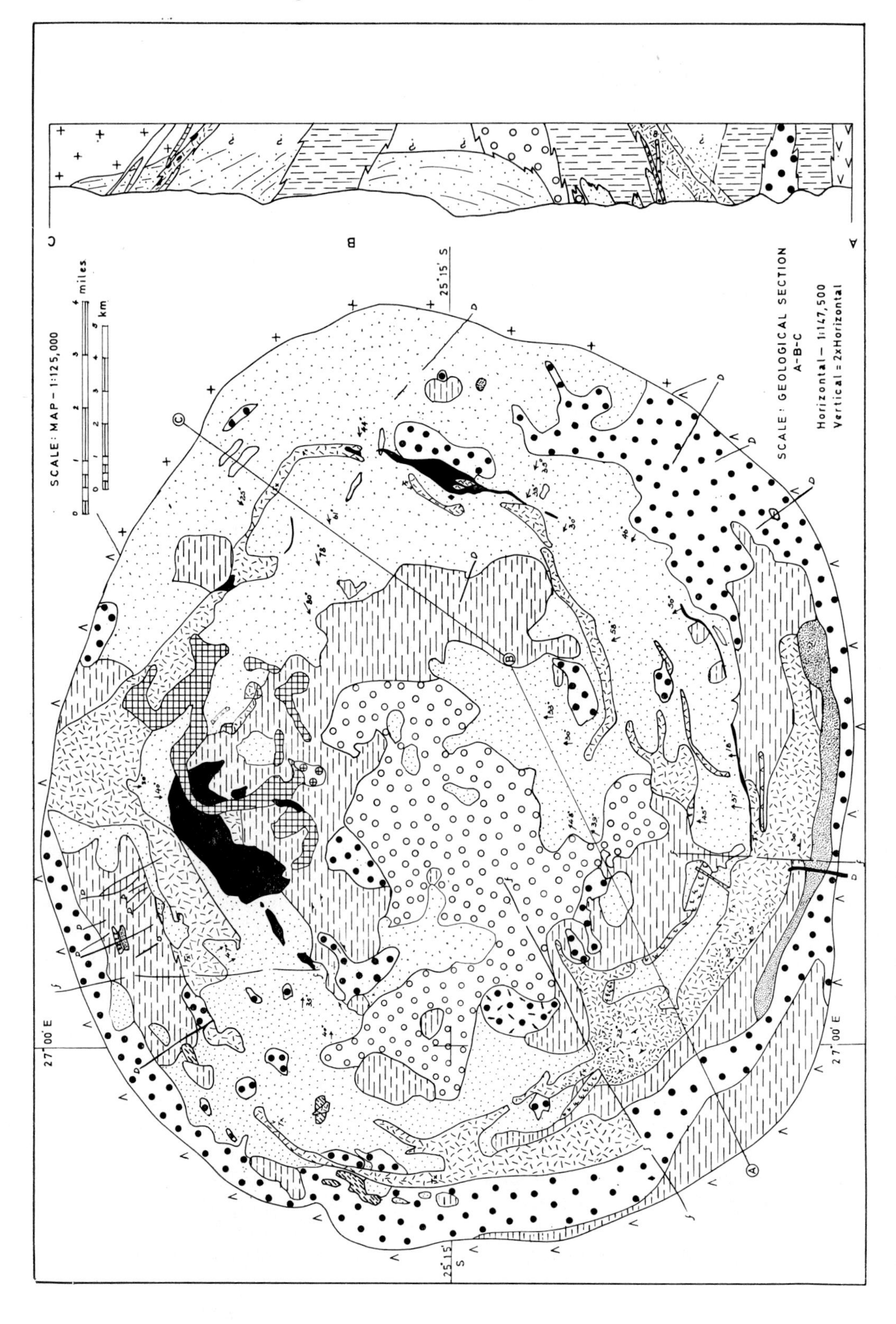
SCALE: MAP - 1:125,000
miles
km
SCALE: GEOLOGICAL SECTION
A-B-C
Horizontal - 1:147,500
Vertical = 2xHorizontal
25°15' S
27°00' E
A
B
C
D

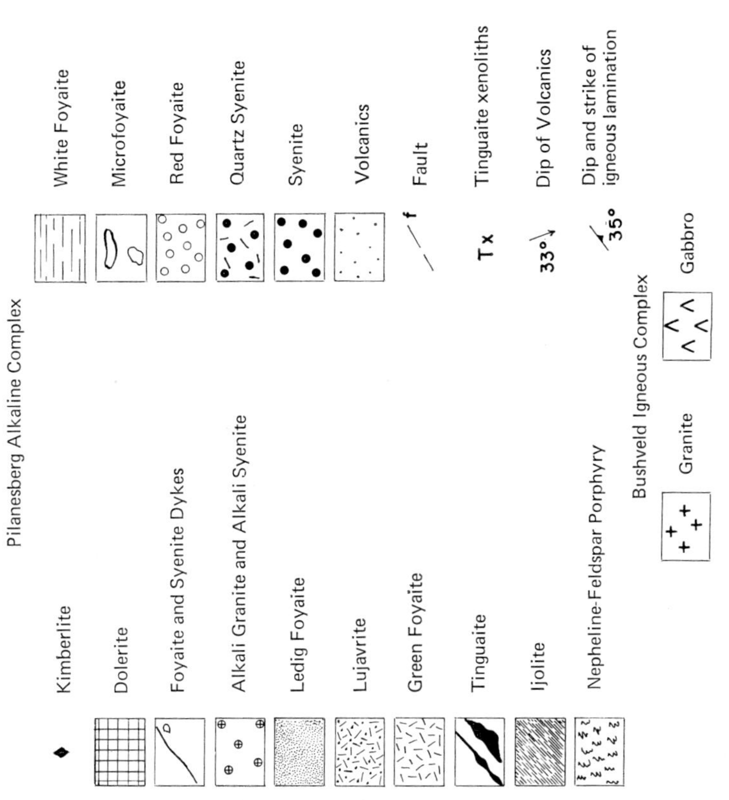

Fig. 4 Pilanesberg complex, western Transvaal (drawn by A. E. Retief)

where Granitberg is the largest and best known of a group of stocks and plugs (Wagner, 1910; Kaiser, 1922; Kaiser and Beetz, 1926).

d. In the area of the Klinghardt mountains some 40 km E.N.E. of Bogenfels (Kaiser and Beetz, 1926).

Phonolite plugs occur as far south as 28° 13′ (G. C. Stocken, private communication) and one plug, 'Swartkop' north of Chameis, has been dated by the K–Ar WR method at 37 m.y. (N. J. Snelling, private communication).

An occurrence, not hitherto recorded, of dykes of fresh phonolite is reported by H. Martin (private communication) on the western side of the Awasib hills approximately 130 km N.N.E. of Lüderitz.

III.5.4.2. South Africa

The Franspoort 'line' of alkaline intrusives as originally described by Shand (1921b, 1922) consisted of four complexes N.E. of Pretoria of which the Leeuwfontein complex is the largest. Verwoerd (1967) reports that, since 1950, a further five occurrences have been found within this zone.

Small outcrops of alkaline syenites and fenites have been described from Namaqualand, 350 km north of Cape Town, 31° S, 17° 55′ E (Jansen, 1960).

III.5.4.3. Botswana

Syenites which are partly alkaline occur in the Semarule Hills Complex 24° 28′ S, 25° 34′ E. They are considered by King (1955) to be fenitized granites showing local mobilization (cf. Poldervaart, 1952; King and Sutherland, 1960).

III.5.5. Distribution and Structural Control

No simple overall pattern of tectonic control is apparent for the alkaline rocks considered here. Rifting, so conspicuous to the north, dies out southward and although there undoubtedly is structural control in the localization of individual complexes, King and Sutherland (1960) and Verwoerd (1967) have rightly warned against over-simplification.

With regard to carbonatite emplacement Verwoerd (in press) states that if there is structural control then it is not directly related to major tectonic features in southern Africa.

A great deal more work on structural analysis needs to be done if a true pattern is to emerge. At present it is only possible to give the evidence of distribution and faulting where it is known and to draw attention to certain features of alignment which, as Garson and Campbell Smith (1958), James (1956), Verwoerd (1967) and others have suggested, may be expressions of abyssal fractures or zones of weakness.

Most conspicuous amongst these postulated fracture zones are those striking north-east from the western margin of the continent and the prominent parallel oceanic feature, the Walvis Ridge (Fig. 5). The northernmost of this thrice-recurrent continental lineament is the Angola line of fifteen alkaline and basic complexes, some with carbonatite. They occur in a zone of parallel faults 130 km long which may possibly extend to link with the Lucapa kimberlite-bearing graben. A Rb–Sr age determination on a mica from a nepheline syenite in the Chivara complex gives an age of 112 ± 8 m.y. (De Sousa Machado, 1958; Lapido Loureiro and de Vries, 1968). The next in the series is the

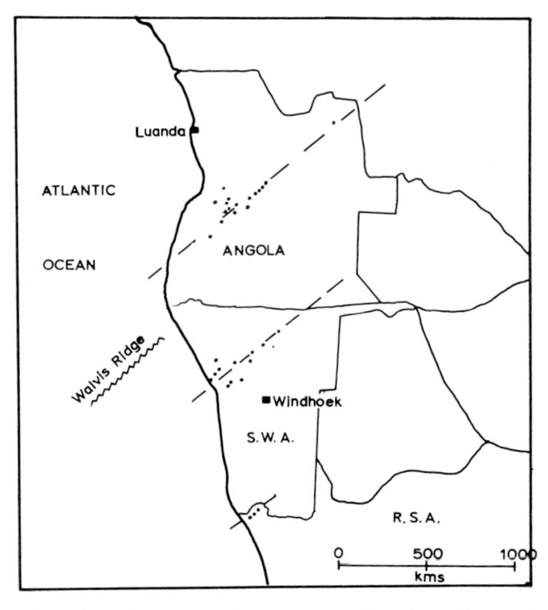

Fig. 5 Alignment of complexes in Angola and South West Africa

Damaraland line of intrusives stretching from Cape Cross to Okorusu which have K–Ar WR ages in the range 123–135 m.y. omitting the aberrantly high age of 164 m.y. for Okonjeje (Siedner and Miller, 1968). An approximately equal distance further south and with the same strike, but considerably older in age, is the Kuboos–Bremen line of acidic intrusions which has been dated on zircon at 550 m.y. (De Villiers and Burger, 1967). In between the Kuboos and Damaraland line are two structural features indicative of compression but these almost certainly predate the alkalic eruptives. The first is the thrust plane, with over-riding of the younger Karroo System sediments by those of the Damara System, which trends S.W. towards Omaruru. The second is the overfolding and thrusting from the north and development of S.E.-striking tear faults which Guj (1967) has reported. These latter, when rejuvenated as normal faults are thought to have caused the localization of the alkalic rocks on the southern side of the Auas mountains, south of Windhoek.

Of the remaining occurrences of alkaline rocks in South-West Africa, Swaartbooisdrif and Epembe are situated in E.S.E. and W.N.W. fractures, and the foyaites and phonolites of the littoral zone south of Lüderitz, lie parallel to the 'grain' of the country rocks.

Both King and Sutherland (1960) and Verwoerd (1967) have drawn attention to the fact that it is possible by extending the line originally proposed by Truter (1955), to form an arc extending from the Semarule Hills syenite complex in Botswana to Palabora in the east which will embrace nearly all the alkaline and carbonatitic centres. This, however, does not represent any known faulting. Another abyssal fracture postulated by Verwoerd (1967) extends south-west along the Great Dyke of Rhodesia through the group of complexes just north of Pretoria, to the Vredefort Dome.

In Rhodesia the Shawa and Dorowa complexes lie within a shield area and are not associated with any major fault although Dorowa lies on a shatter-belt which has a N.N.W. trend (Johnson, 1961, 1966). Marangudzi may be situated on an extension of the Messina line of copper mineralization north-eastward to the Mateke hills granophyre of the Nuanetsi area (Rees, 1960; Gifford, 1961).

Cox and co-authors (1965) advocate a tectonic control allied to high heat flow as the cause of magma generation in the Nuanetsi–Lebombo province. The tectonism is of the block-fault type which is classically associated with alkalic magmas and several primary magmas are produced under conditions of waxing and waning intensity. Alkalic magmas of nephelinite-type occur early in the cycle as products of partial fusion of the mantle, acidic magmas at the maximum due to melting of the sial and the basalt $\rightarrow$ trachyte $\rightarrow$ phonolite association at a late stage as products of crystallization differentiation. This is in contrast to the type Karroo area in South Africa where there was depression but no marked deformation and no alkalic or acidic magmas formed.

Recent high pressure experimental studies (Green and Ringwood, 1967; O'Hara and Yoder, 1967; Ito and Kennedy, 1967; Bultitude and Green, 1968) have shown that partial melting of the mantle and fractionation of the picritic product at depths of approximately 35–70 kms is governed by early precipitation of aluminous orthopyroxenes and clinopyroxenes. The resultant magmas are likely to be alkali olivine basalts or basanites with a few per cent of normative nepheline. Highly undersaturated alkaline magmas, it is suggested, require some process of enrichment in alkalis and fugitive constituents such as envisaged by Kennedy (1955).

III.5 REFERENCES

Bultitude, R. J., and Green, D. H., 1968. Experimental study at high pressures on the origin of olivine nephelinite and olivine melilite nephelinite magmas. *Earth Planet. Sci. Lett.*, **3,** 325–37.

Bouwer, R. F., 1957. *The carbonate member of the Palabora Igneous Complex*. Ph.D. thesis, Univ. of Cape Town, Rep, S. Africa.

Chudoba, K., 1930. *Brandbergit, ein neues aplitisches Gestein aus dem Brandberg* (*Südwestafrika*). Centralbl. f. Mineral. etc. A 389–95.

Cloete, D. R., 1957. *Die dolerietvoorkomste van die noordelike Pilanesberg*. M. Sc. thesis, Univ. of Pretoria, Rep. S. Africa.

Cloos, H., 1929. *Die jungen Plateaugranite in Südwestafrika*. Centralbl. f. Mineral. etc.

Cloos, H., and Chudoba, K., 1931. Der Brandberg Bau, Bildung und Gestalt der jungen Plutone in Südwestafrika. *Neues Jb. Min. Geol. Pal.*, Beit. Bd. **66,** Abt. B. 1–130.

Cox, K. G., Johnson, R. L., Monkman, L. J., Stillman, C. J., Vail, J. R., and Wood, D. N., 1965. The geology of the Nuanetsi igneous province. *Phil. Trans. Roy. Soc. Lond.*, ser. A., **257,** 71–218.

de Kock, W. P., 1934. The geology of the Western Rehoboth: An explanation of Sheet F. 33–W3. (Rehoboth). *Dept. Mines. South West Africa Mem.*, **1,** 148 pp.

de Sousa Machado, F. J., 1958. The volcanic belt of Angola and its carbonatites. *C.C.T.A. Pub.*, **44.** Leopoldville, 309–17.

de Villiers, J., and Burger, A. J. (1967). Note on the minimum age of certain granites from the Richtersveld area. *Geol. Surv. S. Africa Ann.*, **6,** 83–4.

Dixey, F., Campbell Smith, W., and Bissett, C. B., 1955, The Chilwa Series of Southern Nyasaland. *Bull. geol. Surv. Nyasaland*, **5,** (revised), 71 pp.

du Toit, A. L., 1929. The volcanic belt of the Lebombo—a region of tension. *Trans. Roy. Soc. S. Africa*, **18,** pt. 1, 189–217.

du Toit, A. L., 1931. The genesis of the pyroxenite-apatite rocks of Palabora, Eastern Transvaal. *Trans. geol. Soc. S. Africa.*, **34,** 107–27.

Flores, G., 1964. On the age of the Lupata rocks, lower Zambesi River, Mozambique. *Trans. geol. Soc. S. Africa*, **67,** 111–18.

Forster, I. F., 1958. Paragenetical ore mineralogy of the Loolekop–Phalaborwa carbonatite complex, Eastern Transvaal. *Trans. geol. Soc. S. Africa*, **61,** 359–63.

Garson, M. S., and Campbell Smith, W., 1958. Chilwa Island. *Geol. Surv. Nyasaland, Mem.*, **1,** 127 pp.

Gevers, T. W., 1932. Kaoko-eruptives and alkali-rocks at Cape Cross, S.W. Africa. *Trans. geol. Soc. S. Africa*, **35,** 85–96.

Gevers, T. W., 1933. Alkali-rocks in the Auas mountains, south of Windhoek, S. W. A. *Trans. geol. Soc. S. Africa*, **36,** 77–88.

Gevers, T. W., 1934. Jüngere Vulkanschlote in den Auasbergen Südlich von Windhuk in S.W.-Afrika. *Z. Vulkanol.*, **16,** 7–42.

Gevers, T. W., 1948. Vermiculite at Loolekop, Palabora, North East Transvaal. *Trans. geol. Soc. S. Africa*, **51,** 133–73.

Gifford, A. C., 1961. *The geology of Eastern Marangudzi, Southern Rhodesia*. Ph.D. thesis, Univ. of London, England.

Gittins, J., 1966. 'Summaries and bibliographies of carbonatite complexes', in Tuttle, O. F., and Gittins, J., Eds., *Carbonatites*, Interscience, New York and London, 417–541.

Gough, D. I., Brock, A., Jones, D. L., and Opdyke, N. D., 1964. The palaeomagnetism of the ring complexes at Marangudzi and the Mateke Hills. *J. Geophys. Res.*, **69,** 2499–507.

Green, D. H., and Ringwood, A. E., 1967. The genesis of basaltic magmas. *Contr. Miner. Petrology*, **15,** 103–90.

Guj, P., 1967. Structural geology of the Auas mountains, Windhoek district, S.W.A. *Geol. Surv. S. Africa Ann.*, **6,** 55–62.

Hall, A. L., 1912. The crystalline metamorphic limestone of Lulukop and its relationship to the Palabora plutonic complex. *Trans. geol. Soc. S. Africa*, **15,** 18–25.

Hall, A. L., and Molengraaff, G. A. F., 1925. The Vredefort Mountain Land in the Southern Transvaal and the Northern Orange Free State. *Verh. Koninkl. Ned. Akad. Wetensch.*, 2 sect., pt. 24, **3,** 183 pp.

Hanekom, H. J., van Staden, C. M. v. H., Smit, P. J., and Pike, D. R., 1965. The geology of the Palabora igneous complex. *Geol. Surv. S. Africa, Mem.*, **54,** 185 pp.

Heinrich, E. W., 1966. *The Geology of Carbonatites*. Rand McNally, Chicago, Illinois, 555 pp.

Holmes, A., 1958. Spitskop carbonatite, Eastern Transvaal. *Bull. geol. Soc. Am.*, **69,** 1515–26.

Holmes, A., and Cahen, L., 1957. Géochronologie Africaine 1956. *Mém. Acad. roy. Belg., Cl. Sci.*, **8, 5,** 1–169.

Humphrey, W. A., 1912. The geology of Pilandsberg. *Geol. Surv. S. Africa, Mines Dept., Ann. Rep. for 1911*, Pt. 3, 75 pp.

Ito, K., and Kennedy, G. C., 1967. Melting and phase relations in a natural peridotite to 40 kilobars *Am. J. Sci.*, **265,** 519–38.

James, T. C., 1956. Carbonatites and rift valleys in East Africa. *Abstract. Int. Geol. Congr. 20th sess. Mexico*, Assoc. de Serv. Geol. Africanos. 325 pp.

Jansen, H., 1960. The geology of the Bitterfontein area, Cape Province. *Expl. Sheet 253, (Bitterfontein), Geol. Surv. S. Africa*, 97 pp.

Johnson, R. L., 1961. The geology of the Dorowa and Shawa carbonatite complexes, Southern Rhodesia. *Trans. geol. Soc. S. Africa*, **64,** 101–45.

Johnson, R. L., 1966. 'The Shawa and Dorowa carbonatite complexes, Rhodesia', in Tuttle, O. F., and Gittins, J., Eds., *Carbonatites*, Interscience, New York and London, 205–24.

Kaiser, E., 1922. Über zwei verschiedenartige Injektionen syenitischer Magmen. *Sitsber. Bayer. Akad., Wiss. Math. Phys. Kl.*, 255–84.

Kaiser, E., and Beetz, W., 1926. *Die Diamantenwüste Südwest-Afrikas. Bd. 1*, Dietrich Reimer, Berlin, 321 pp.

Kennedy, G. C., 1955. Some aspects of the role of water in rock melts. *Geol. Soc. Am. Spec. Pap.* **62,** 489–504.

King, B. C., 1955. Syenitisation de granites à Semarule près de Molepolole, protectorat du Bechuanaland. *C.N.R.S. Coll.*, **68,** Paris, 1–16.

King, B. C., and Sutherland, D. S., 1960. Alkaline rocks of Eastern and Southern Africa. Pts. 1, 2 and 3. *Sci. Progr.*, **48,** 298–321, 504–24 and 709–20.

Korn, H., and Martin, H., 1939. Junge Vulkano-Plutone in Südwestafrika. *Geol. Rundschau*, **30,** 631–6.

Korn, H., and Martin, H., 1952. Der Intrusionsmechanismus der grossen Karroo-plutone in Südwestrafrika. *Geol. Rundschau*, **41,** 41–58.

Korn, H., and Martin, H., 1954. The Messum igneous complex in South-West Africa. *Geol. Soc. S. Africa*, **57,** 83–122.

Lapido-Loureiro, F. Eduado de Vries, 1968. Subvolcanic carbonatite structures of Angola. *Int. Geol. Congr. 23rd sess., Prague. Proc.*, Sec. 2, 147–61.

Linning, K., 1968. *Die stollingskompleks Kaap Kruis, Suidwes-Afrika.* M.Sc. thesis, University of Pretoria, Pretoria, Rep. of South Africa, 108 pp.

Lombaard, B. V., 1952. Karroo dolerites and lavas. *Trans. geol. Soc. S. Africa*, **55,** 175–98.

Lombaard, A. F., Ward-Able, N. M., and Bruce, R. W., 1964. 'The exploration and main geological features of the copper deposit in carbonatite at Loolekop, Palabora complex', in Haughton, S. H., Ed., *The Geology of some Ore Deposits in Southern Africa, v.2*, Geol. Soc. S. Africa, Johannesburg, 315–37.

McKie, D., 1966. 'Fenitization', in Tuttle, O. F., and Gittins, J., Eds., *Carbonatites*, Interscience, New York and London, 261–94.

Manton, W. I., and Siedner, G., 1967. Age of the Paresis complex, South-West Africa. *Nature*, **216,** 1197–8.

Martin, H., Mathias, M., and Simpson, E. S. W., 1960. The Damaraland sub-volcanic ring complexes in South West Africa. *Int. Geol. Congr., 21st sess.*, Pt. 13, 156–74.

Mathias, M., 1956. The petrology of the Messum igneous complex, South-West Africa. *Trans. geol. Soc. S. Africa*, **59,** 23–57.

Mathias, M., 1957. The geochemistry of the Messum igneous complex, South-West Africa. *Geochim. Cosmochim. Acta*, **12,** 29–46.

Mennell, F. P., 1946. Ring structures with carbonate cores in Southern Rhodesia. *Geol. Mag.*, **83,** 137–40.

Molengraaff, G. A. F., 1905. Preliminary note on the geology of the Pilandsberg and a portion of the Rustenburg district. *Trans. geol. Soc. S. Africa*, **8,** 108–9.

Nel, L. T., 1927. The geology of the country around Vredefort, an explanation of the geological map. *Geol. Surv. Union S. Africa*, 134 pp.

Nicolaysen, L. O., de Villiers, J. W. L., Burger, A. J., and Strelow, F. W. E., 1958. New measurements relating to the absolute age of the Transvaal System and of the Bushveld Igneous Complex. *Trans. geol. Soc. S. Africa*, **61,** 137–63.

Nicolaysen, L. O., Burger, A. J., and Johnson, R. L., 1962. The age of the Shawa carbonatite complex. *Trans. geol. Soc. S. Africa*, **65,** pt. 1, 293–4.

O'Hara, M. J., and Yoder, Jr., H. S., 1967. Formation and fractionation of basic magmas at high pressures. *Scottish J. Geol.*, **3,** 67–117.

Poldervaart, A., 1952. The Gaberones granite (South Africa). *Int. Geol. Congr. 19th sess. Algiers. Assoc. des Serv. Géol. Africains*, Pt. 20, 315–33.

Rees, G., 1960. *The geology of West Marangudzi.* Ph.D. thesis, Univ. of London, England.

Retief, E. A., 1962. Preliminary observations on the feldspars from the Pilanesberg alkaline complex, Transvaal, South Africa. *Norsk. Geol. Tidsskr., Feldspar*, **42,** 493–513.

Retief, E. A., 1963. *Petrological and mineralogical studies in the southern part of the Pilanesberg alkaline complex, Transvaal, South Africa.* D.Phil. thesis, Univ. of Oxford, England.

Reuning, E., 1924. Die Entwicklung der Karrooformation im südlichen Kaokofeld, Südwestafrika. *Neues. Jb. Min. Geol. Pal.*, Beil. Bd. **52,** Abt. B. 94–114.

Reuning, E., 1929. Differentiation der Karroo-Eruptiva im südlichen Kaokofeld Südwestafrika. *Int. Geol. Congr. 15th sess., S. Africa*, 28–36.

Reuning, E., and Martin, H., 1957, Die Prä-karroo-handschaft, die Karroo-Sedimente und Karroo-Eruptivgesteine des südlichen Kaokofeldes in Südwestafrika. *Neues. Jb. Miner. Abh.*, **91**, 193–212.

Rimann, E., 1914. Trachyt, Phonolith, Basalt in Deutsch–Südwestafrika. *Centralbl. Miner. etc.*, No. 2, 33–7.

Rogers, A. W., 1925. Notes on the North-Eastern part of the Zoutpansberg district. *Trans. geol. Soc. S. Africa*, **28**, 33–53.

Russell, H. D., Hiemstra, S. A., and Groeneveld, D., 1954. The mineralogy and petrology of the carbonatite at Loolekop, Eastern Transvaal. *Trans. geol. Soc. S. Africa*, **57**, 197–208.

Schreiner, G. D. L., and van Niekerk, C. B., 1958. The age of a Pilanesberg dyke from the Central Witwatersrand. *Trans. geol. Soc. S. Africa*, **61**, 197–203.

Shand, S. J., 1921a. The nepheline rocks of Sekukuniland. *Trans. geol. Soc. S. Africa*, **24**, 111–49.

Shand, S. J., 1921b. The igneous complex of Leeuwfontein, Pretoria district. *Trans. geol. Soc. S. Africa*, **24**, 232–49.

Shand, S. J., 1922. The alkaline rocks of the Franspoort line, Pretoria district. *Trans. Geol. Soc. S. Africa*, **25**, 81–100.

Shand, S. J., 1928. The geology of Pilandsberg (Pilaan's Berg) in the Western Transvaal: a study of alkaline rocks and ring-intrusions. *Trans. geol. Soc. S. Africa*, **31**, 97–156.

Shand, S. J., 1931. The granite–syenite–limestone complex of Palabora, Eastern Transvaal and the associated apatite deposits. *Trans. geol. Soc. S. Africa*, **34**, 81–105.

Siedner, G., 1963. *Geology and geochemistry of the Paresis igneous complex, South West Africa*. Ph.D. thesis, Univ. of Cape Town, Rep. of S. Africa.

Siedner, G., 1965a. Structure and evolution of the Paresis igneous complex, South West Africa. *Trans. geol. Soc. S. Africa*, **68**, 177–202.

Siedner, G., 1965b. Geochemical features of a strongly fractionated alkali igneous suite. *Geochim. Cosmochim. Acta*, **29**, 113–38.

Siedner, G., and Miller, J. A., 1968. K–Ar age determinations on basaltic rocks from South West Africa and their bearing on continental drift. *Earth Planet. Sci. Lett.*, **4**, 451–8.

Simpson, E. S. W., 1954. The Okonjeje igneous complex, South West Africa. *Trans. geol. Soc. S. Africa*, **57**, 125–72.

Stahl, A., 1930. Eisenerze im nördlichen Südwestafrika. *Neues. Jb. Min. geol. Pal.*, Beit. Bd. **64**, Abt. B, 165–200.

Strauss, C. A., and Truter, F. C., 1950a. The alkali complex at Spitskop, Sekukuniland, Eastern Transvaal. *Trans. geol. Soc. S. Africa*, **53**, 81–125.

Strauss, C. A., and Truter, F. C., 1950b. Post-Bushveld ultrabasic, alkali, and carbonatitic eruptives at Magnet Heights, Sekukuniland, Eastern Transvaal. *Trans. geol. Soc. S. Africa*, **53**, 169–90.

Swift, W. H., 1952. The geology of Chishanya, Buhera district, Southern Rhodesia. *Trans. Edinburgh geol. Soc.*, **15**, 346–59.

Swift, W. H., White, W. C., Wiles, J. W., and Worst, B. G., 1953. The geology of the lower Sabi coalfield. *Bull. geol. Surv. Southern Rhodesia*, **40**, 96 pp.

Tilley, C. E., 1960. Some new chemical data on the alkali rocks of the Vredefort Mountain Land, South Africa. *Trans. geol. Soc. S. Africa*, **63**, 65–70.

Truter, F. C., 1955. Modern concepts of the Bushveld igneous complex. *C.C.T.A., Salisbury*, 77–87.

Tuttle, O. F., and Gittins, J., Eds. 1966, *Carbonatites*. Interscience New York and London, 591 pp.

Tyndale-Biscoe, R., 1956. Alkali ring complexes in Southern Rhodesia. *Int. geol. Congr. 20th sess., Mexico, Assoc. de serv. Geol. Africanos*, 335–8.

Vail, J. R., 1968. The southern extension of the East African rift system and related igneous activity. *Geol. Rundschau*, **57**, 601–14.

van Niekerk, C. B., 1962. The age of the Gemspost dyke from the Venterspost gold mine. *Trans. geol. Soc. S. Africa*, **65**, Pt. 1, 105–11.

van Zijl, P. J., 1962. The geology, structure and petrology of the alkaline intrusions of Kalkfeld and Okorusu and the invaded Damara rocks. *Ann. Univ. Stellenbosch*, **37**, series A, 237–340.

Verwoerd, W. J., 1964. The significance of fenitised granite-pegmatites in the Spitskop complex. *Trans. geol. Soc. S. Africa*, **67**, 219–26.

Verwoerd, W. J., 1966. South African carbonatites and their probable mode of origin. *Ann. Univ. Stellenbosch*, **41**, Series A, 233 pp.

Verwoerd, W. J., 1967. The carbonatites of South Africa and South West Africa. *Geol. Surv. S. Africa, Handbook* **6**, 452 pp.

Wagner, P. A., 1910. About an occurrence of nepheline-syenite in Lüderitzland German South West Africa. *Centralbl. Mineral. etc.* 721–2.

Young, R. B., 1920. The rocks of a portion of Portuguese East Africa. *Trans. geol. Soc. S. Africa*, **23**, 98–113.

IV. Alkaline Provinces

IV.1. INTRODUCTION

H. Sørensen

It has commonly been pointed out, as for instance by Sheynmann *et al.*, (1961), that alkaline rocks have no single parental magma, but that they are products of differentiation of different magmas under conditions enabling enrichment in alkalis to take place. As, furthermore, nearly identical alkaline rocks intrude rocks of widely different structure and chemical composition, assimilation appears to be of subordinate importance in the formation of alkaline rocks. This indicates that alkaline rocks are formed from magmas generated in the deep crust or upper mantle or from derivatives of such magmas.

Sheynmann *et al.* (1961) distinguished three genetic formations or associations of alkaline rocks but admitted that a number of occurrences could not be classified under these headings. These three associations are:

1. *The ultrabasic alkaline association* dominated by ultramafic rocks rich in olivine and/or pyroxene. Later phases consist of ijolite-melteigite, syenite and nepheline syenite, melilite-bearing rocks and carbonatite. Fenitization of the exocontact zones is pronounced. In volcanic examples nephelinites and melilite-bearing rocks are associated with trachytes and phonolites. Sodic and potassic lineages are exemplified by respectively the Kola and Maimecha–Kotuj provinces (cf. IV.2 and III.4),* and by the western Rift valley of East Africa (cf. II.5). The Kander lineage is represented by a few complexes in the Aldan shield made up of ultramafic rocks, diorites, syenites and alkali granites.

2. *The gabbroid alkaline association* consists of alkali gabbro associated with syenites, nepheline syenites and occasionally alkali granites. Ijolites-urtites also occur. In some examples, as for instance many oceanic islands (IV.7), the felsic rocks appear to be derivatives of the alkali basaltic magmas; in other examples there is an intermittent eruption of alkali basaltic and trachytic-phonolitic magmas and in cases also nephelinite (cf. Williams, 1970; IV.4) or pantellerite (cf. Bryan, 1970).

3. *The granitoid alkaline association* is dominated by alkali granites and associated alkali syenites and nepheline syenites. It may be difficult to distinguish this association from the gabbroid association when syenites, nepheline syenites and granites are predominant and gabbroid rocks are of subordinate importance in the erosional section available for study (cf. IV.3).

These three associations appear to be products of differentiation of three main types of parental magmas, ultrabasic-nephelinitic, alkali basaltic and granitic. The two first-named groups of mother magmas appear to demand abyssal differentiation under relative tectonic quiescence in order to permit alkalis and volatiles to accumulate to such an extent that alkaline felsic rocks are formed. The third group may be a result of increasing alkalinity accompanying granitic activity in fold belts (Vorobieva, 1960).

King and Sutherland (1960) distinguished two main alkaline associations in East Africa, a nephelinite-carbonatite and an olivine basalt–trachyte–phonolite association, which correspond to the two first-named of the above-mentioned associations listed by Sheynmann *et al.* (1961). King (1970) distinguishes two genetic lava series in the eastern rift, a strongly alkaline, nepheline-bearing (ankaratrite, melanephelinite, nephelinite, phonolite) and a mildly alkaline,

* Readers are referred to the relevant Chapters of this book where similarly cited.

nepheline-free series (alkali basalt, trachybasalt, alkali-trachyte, soda rhyolite).

The three associations of Sheynmann *et al.* do not, as emphasized by the authors themselves, cover all occurrences of alkaline rocks. The main exceptions are the predominant trachytes and phonolites in some provinces as for instance Kenya (Nash *et al.*, 1969), independent masses of syenite or nepheline syenite as for instance Serra de Monchique and Norra Kärr, Sweden (II.2.9), and some miaskitic nepheline syenites as those of the Urals (cf. II. 2.5.6). In at least some of these examples an anatectic origin of the magmas is suspected (cf. VI.1).

When selecting examples of provinces of alkaline rocks, some classical regions like the Oslo province, the Midland Valley of Scotland, the East Otago province, and Hawaii, have been left aside, since these provinces are so well-covered by the petrological textbooks. The examples chosen are provinces studied in detail and in part not yet covered by review papers. The Kola province (IV.2) and the Monteregian Hills (IV.6) represent the ultrabasic alkaline association; the Gardar province (IV.3), Central Europe (IV.4), and the oceanic islands (IV.7) mainly the gabbroid alkaline association; the Mongol–Tuva (IV.5) and Nigeria–Niger (IV.8) provinces the granitoid alkaline association, but the latter province may also be considered a member of the gabbroid alkaline association (cf. Black and Girod, 1970). The chapters of parts II and III of the book present additional examples.

IV.1 REFERENCES

Black, R., and Girod, M., 1970. 'Late Palaeozoic to recent igneous activity in West Africa and its relationship to basement structure', in T. N. Clifford and I. G. Gass, eds., *African Magmatism and Tectonics.* Oliver and Boyd, 185–210.

Bryan, W. B., 1970. 'Alkaline and peralkaline rocks of Socorro Island, Mexico', in P. H. Abelson, ed,. *Annual rep. of the Director Geophys. Lab. Yb. Carnegie Instn. Wash.*, **68,** 194–200.

King, B. C., 1970. Structure of the East African rift system. *Proc. geol. Soc. Lond.*, **1663,** 150.

King, B. C., and Sutherland, D. S., 1960. Alkaline rocks of eastern and southern Africa. *Sci. Progr.*, **48,** I–III, 298–321, 504–24, 709–20.

Nash, W. P., Carmichael, I. S. E., and Johnson, R. W., 1969. The mineralogy and petrology of Mount Suswa, Kenya. *J. Petrology*, **10,** 409–39.

Sheynmann, Yu. M., Apel'tsin, F. R., and Nechayeva, Ye. A., 1961. Alkalic Intrusions, Their Mode of Occurrence and Associated Mineralization (in Russian). *Geologiya Mestorozhdenii Redkikh Elementov*, **12–13,** Moscow (Gosgeoltekhizdat) 177 pp. (review in *Int. Geol. Rev.*, **5,** 451–8).

Williams, L A. J., 1970. The volcanics of the Kenya rift Valley. *Proc. geol. Soc. Lond*, **1663,** 151.

Vorobieva, O. A., 1960. Alkali rocks of the USSR. *Rep. 21st Intern. geol. Congr. Norden*, **13,** 7–17.

IV.2. KOLA PENINSULA

V. I. Gerasimovsky, V. P. Volkov, L.N. Kogarko and A. I. Polyakov

Numerous occurrences of alkaline rocks of different age and composition belonging to various genetic types are observed within the Kola Peninsula and Northern Karelia.

The Kola Peninsula and Northern Karelia comprise a part of the Baltic Shield which consists of ancient intensively granitized metamorphic rocks. There are numerous bodies of granitoid composition and subordinate ultrabasic and basic intrusions. Acid and basic effusive rocks are characteristic for the igneous activity of the Lower Proterozoic geosynclinal stage. The beginning of consolidation of the folded rocks of the region is dated as the end of the

Middle Proterozoic. This period was characterized by calm tectonic activity with major block slips. By this time the formation of the main synclinoria and anticlinoria of north-western strike (Fig. 1) came to an end. At the same time the alkaline massifs began to form being confined to certain long-living mobile tectonic zones located between major consolidated crustal blocks (Tzirul'nikova *et al.*, 1968; Kukharenko, 1967).

The most ancient alkaline intrusions (Yelet'ozero and Gremjakha–Virmes) belonging to the alkaline–gabbroid association are of Middle Proterozoic age. The alkaline granitoid association of rocks and their accompanying granosyenites, alkali syenites and nepheline syenites (massifs of Western Keyvi and widely spread within the Kola Peninsula) were formed in the same period. The association of rocks comprising numerous Caledonian ultramafic alkaline massifs (Afrikanda, Khabozero, Kovdor, Turii mis, etc.), the basaltoid effusives of the Lovozero formation and the Hercynian nepheline syenites (Khibina and Lovozero) is of Paleozoic age.

IV.2.1. The Alkaline-Gabbroid Association

The Yelet'ozero massif (50 km^2) is located in Archean gneisses in Northern Karelia (Fig. 2). It is of elliptical shape and is concentrically zoned (Bogachev *et al.*, 1963).

The Yelet'ozero massif was formed during three intrusive phases. The rocks of the first intrusive phase have the widest distribution. They consist of gabbro and subordinate peridotite, pyroxenite and anorthosite and comprise a ring-like body within the marginal part of the massif. Rare dykes of porphyrite, diabase and spessartite cutting the gabbroids belong to the second intrusive phase. Alkali syenites and biotite nepheline syenites of miaskitic type comprise the third intrusive phase and form an elongated body in the central part of the massif.

The average chemical composition of the rocks of the first intrusive phase recalls that of olivine basalt (Table 1). During the evolution of the basic melt the contents of SiO_2, Al_2O_3 and alkalis increase while there is a simultaneous decrease in mafic components. The latest differentiates are represented by nepheline syenites.

The Gremjakha–Virmes massif (130 km^2) is located in the north-western part of the Kola Peninsula and is emplaced in lower Archean gneisses and gneiss-granites. The massif was formed during three intrusive phases (Polkanov *et al.*, 1967).

The most abundant rocks in the massif are those of the first intrusive phase mainly consisting of varieties of gabbro. The differentiation of the gabbroic melt resulted in two rock series: (*a*) peridotite–pyroxenite–gabbro–anorthosite and (*b*) peridotite–orthoclase gabbro–akerite–pulaskite.

The second intrusive phase forms a steeply dipping body in the central part of the massif and is dominated by biotite-aegirine nepheline syenite (foyaite). Ijolite-melteigite, urtite, juvite, malignite, and alkali syenite are found as subordinate, relatively rare varieties. The geological relationships of the alkaline rocks indicate their syngenetic character. Sections with primary pseudostratification are found within the alkaline rocks.

The third intrusive phase is composed of alkali aegirine-arfvedsonite granite, granosyenite (nordmarkite) and, rarely, alkali syenite. Granosyenite is found in the contact zones near the gabbroid rocks (Bergman, 1962).

The average chemical composition of the rocks of the first intrusive phase is, as seen in Table 1 (analysis no. 3), similar to that of alkaline olivine basalt (Kukharenko, 1967). Fractionation of olivine, pyroxene and plagioclase crystals during the formation of the rocks of the first intrusive phase resulted in an enrichment in silica and alkalis which gave rise to alkaline gabbroids and their syngenetic rocks and finally to ijolite-urtite and nepheline syenite. The syenite melt presumably underwent alternative evolutions to form either nepheline syenite or alkali granite as residual members of the trends of differentiation (Polkanov *et al.*, 1967).

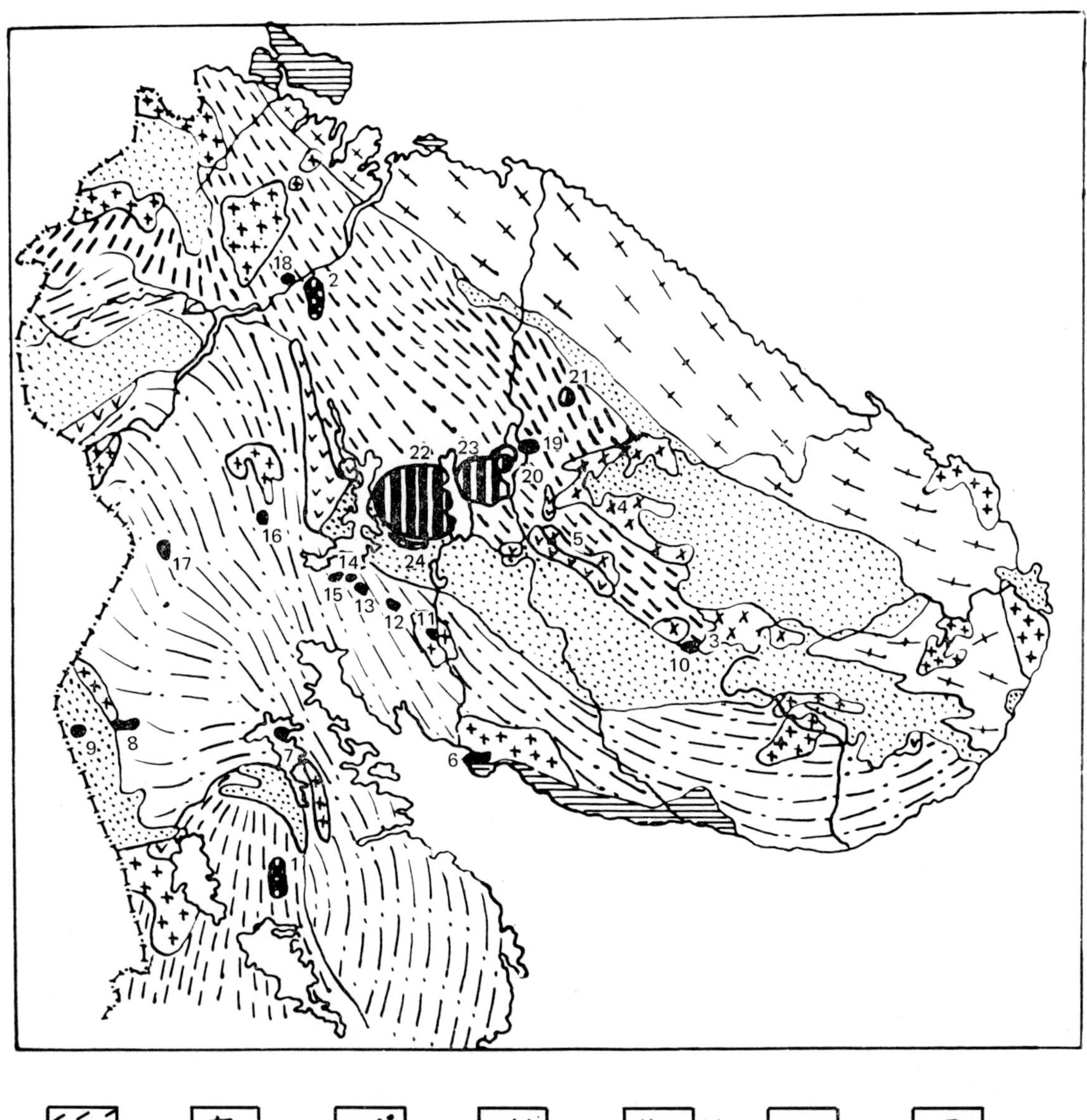

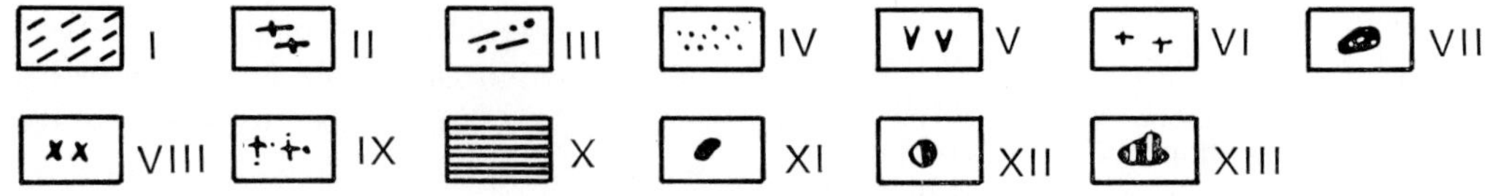

Fig. 1 Schematical map of the distribution of the alkaline massifs of the Kola Peninsula and Northern Karelia. Compiled from Kukharenko, 1967 and Shurkin, 1968

(I) Lower Archean complex of gneisses, granite-gneisses and migmatites; (II) undivided complex of Archean gneisses, granite-gneisses and granites; (III) Upper Archean Belomorsk gneisses, granite-gneisses, granulites and metabasites; (IV) Lower Proterozoic supracrustal and volcanogenic sequence; (V) gabbro and ultrabasites; (VI) granitoids; (VII) massifs of the alkaline–gabbroid association of rocks ((1) Yelet'ozero; (2) Gremjakha–Virmes); (VIII) massifs of the alkaline granitoid association ((3) River Ponoy massifs; (4) Western Keyvi; (5) Belye Tundri); (IX) porphyritic granites and rapakivi granites; (X) Upper Proterozoic sedimentary–metamorphic rocks; (XI) Caledonian ultramafic alkaline massifs ((6) Turii mis; (7) Kovdozero; (8) Vuoriyarvi; (9) Sallanlatvi; (10) Pesotchniy; (11) Ingozero; (12) Salmagorsky; (13) Lesnaya Varaka; (14) Ozernaya Varaka; (15) Afrikanda; (16) Mavrgubinsky; (17) Kovdor; (18) Sebl'yavr; (19) Kurga); (XII) Paleozoic sedimentary and volcanic rocks (alkaline basaltoids); (20) Lovozero Formation; (21) Kontozero; (XIII) Hercynian nepheline syenite massifs ((22) Khibina; (23) Lovozero; (24) Soustova massif).

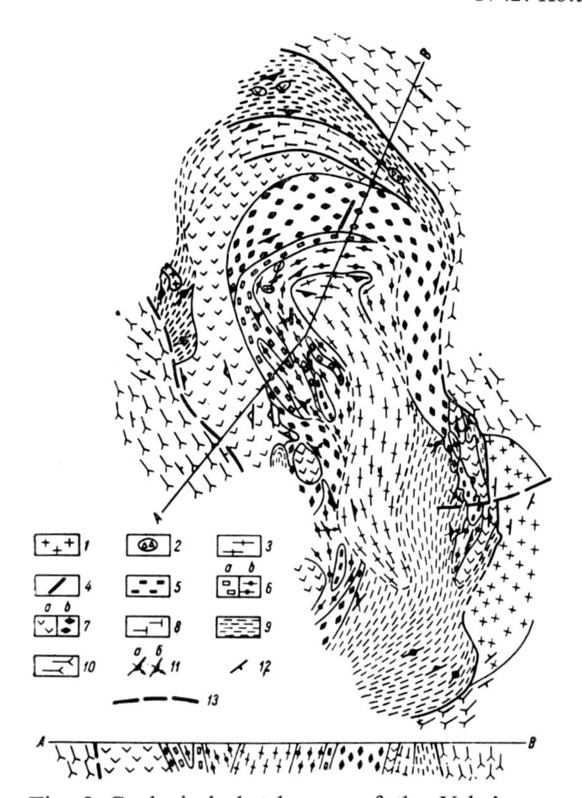

Fig. 2 Geological sketch map of the Yelet'ozero massif (after A. I. Bogachev, S. I. Zak *et al.*, 1963)
(1) plagioclase–microcline and microcline granites. Third intrusive phase: (2) explosive breccia; (3) alkali and nepheline syenites. Second intrusive phase: (4) porphyrite, diabase and spessartite dykes. First intrusive phase: (5) mica peridotite and mica orthoclase gabbro; (6) plagioclasite–peridotite ((a) plagioclasite, leucocratic gabbro; (b) banded olivine gabbro, melanocratic olivine gabbro, peridotite); (7) coarse-grained gabbro and medium-grained feldspathic gabbro ((a) coarse-grained gabbro, bytownitite; (b) medium-grained banded gabbro); (8) sideronitic pyroxenite; (9) fine-grained and medium-grained gabbro; (10) plagioclase and plagioclase–microcline gneiss-granites; (11) dip of layering ((a) higher than 45°; (b) less than 45°); (12) strike of foliation; (13) tectonic zones.

IV.2.2. The Alkaline Granitoid Association

Middle Proterozoic alkali granites are widespread within the Kola Peninsula (Batieva and Bel'kov, 1968; Tchumakov, 1958). These rocks form large sheet-like interformational bodies. The greatest area is covered by the Western Keyvi massif (2000 km²), minor ones are represented by the massifs of Belye Tundri, Rivers Ponoy and Strel'na, Kanozero, etc. The alkali granite massifs are confined to tectonic zones. Aenigmatite-aegirine-arfvedsonite granite is predominant. It has an approximate anchieutectic relationship between the quartz and feldspars (microcline and acid plagioclase). Biotite granite, alkali granosyenite, syenite and nepheline syenite are relatively rare members. The granosyenites appear as contact facies of the alkali granites adjacent to basic country rocks. Alkali syenite and miaskitic nepheline syenite form two massifs within the contact zone of the Western Keyvi alkali granite intrusion in the area of the Sakharyok river basin.

Most of the granites have an excess of alkalis over aluminium. I. D. Batieva and I. V. Bel'kov (1968) emphasize the decrease of silica and the simultaneous increase in alkalis during the process of formation of the rocks comprising the alkaline granitoid association from biotite granite to aegirine-arfvedsonite granite and finally to alkali syenitic and miaskitic rocks.

IV.2.3. The Association of Ultramafic Alkaline Rocks, Nepheline Syenites and Alkaline Basaltoids

These rocks are, because of similar geological position, similarity in geological structure of the massifs, and the genetical parentage, considered to form a Paleozoic association of ultramafic alkaline rocks, nepheline syenites and alkaline basaltoids (Gerasimovsky *et al.*, 1966; Kukharenko, 1967).

IV.2.3.1. The Caledonian Assemblage of Ultramafic Alkaline Rocks

The massifs comprising the ultramafic alkaline assemblage of rocks are confined to fractures of sublatitudinal strike which may be grouped into two belts (Fig. 1). The first belt includes the intrusions of Turii Peninsula, Kovdozero, Vuoriyarvi, etc., the second—Kovdor, the Khabozero group of intrusions, Salmagorsky, etc. About twenty massifs with areas from 0·8 to 35 km² are known. They form polyphase central type intrusions which have discordant steeply dipping contacts

TABLE 1 Average Chemical Composition of the Proterozoic Alkaline Rocks (Kola Peninsula)

Components	Alkaline–gabbroid association					Alkaline–granitoid association	
	Yelet'ozero		Gremjakha–Virmes				
	1	2	3	4	5	6	7
SiO_2	43·19	55·45	47·06	44·01	50·15	74·23	55·23
TiO_2	2·58	0·37	2·25	1·76	0·84	0·34	0·19
Al_2O_3	17·61	21·32	14·30	14·99	21·93	11·16	20·26
Fe_2O_3	4·57	1·54	4·48	6·15	2·20	2·22	4·08
FeO	7·10	2·68	9·78	7·80	5·32	1·84	1·32
MnO	0·17	0·06	0·14	0·22	0·10	0·07	0·16
MgO	6·78	0·18	6·96	4·21	1·23	0·08	0·30
CaO	12·98	0·95	9·32	9·15	2·72	0·46	1·48
Na_2O	1·63	6·99	3·54	6·26	9·76	4·18	8·56
K_2O	0·63	8·39	0·59	2·92	4·98	4·92	5·84
H_2O^+	0·20	0·15	0·70	1·07	0·98	0·22	0·50
H_2O^-			0·28	0·16	0·17	—	0·12
heat loss	0·77	0·06	—	—	—	0·41	0·22
P_2O_5	0·67	—	0·60	0·35	0·04	0·05	—
Number of analyses	11	1	10	4	6	16	Averaged sample
Data	Bogachev *et al.* (1963)		Polkanov *et al.* (1967) and authors			Batieva, Bel'kov (1968)	Authors

1. gabbro; 2. nepheline syenites; 3. gabbro; 4. ijolite–melteigites; 5. nepheline syenites; 6. aenigmatite–aegirine–arfvedsonite granites; 7. miaskites of Sakharyok massif, contain also ZrO_2, 0·51%; F, 0·57%.

against the Archean and Proterozoic country rocks (Fig. 3). The exocontact zones consist usually of fenite and tveitåsite; their thickness is frequently proportional to the areas of the intrusions. The massifs are concentrically zoned, the separate intrusive phases having the form of conical bodies (Kukharenko *et al.*, 1965).

According to geological and geochronological data the intrusions of the assemblage are of Caledonian age, their absolute ages being 300–450 m.y. (Polkanov and Gerling, 1961, 1964).

The geological sequence of the massifs is represented by three series of rocks: (1) ultramafic rocks—olivinite, pyroxenite, nepheline pyroxenite; (2) alkaline rocks—ijolite, melteigite, urtite and vein nepheline and cancrinite syenite; (3) postmagmatic melilite-, phlogopite-, apatite-, forsterite-, magnetite rocks and allied carbonatites.

The ultramafic rocks are the predominant rocks when seen on the presently exposed surface and make up about 60% of the area of the intrusions. Alkaline rocks play a subordinate rôle and the postmagmatic rocks with carbonatites are unessential (5%).

The central parts of the massifs consist generally of ultramafic rocks while the marginal zones are represented by alkaline rocks. From centre to margin the following succession is seen: olivinite, pyroxenite, ijolite-melteigite. All these types of rocks occur commonly as formations of different age and are characterized by their intrusive relationships and by primary magmatic textures.

The olivinite consists mainly of olivine (Fa_{10-15}) with essential quantities of titanomagnetite and the typomorphic accessory perovskite. Layering is not infrequent. The main rock-forming mineral of the pyroxenite is a

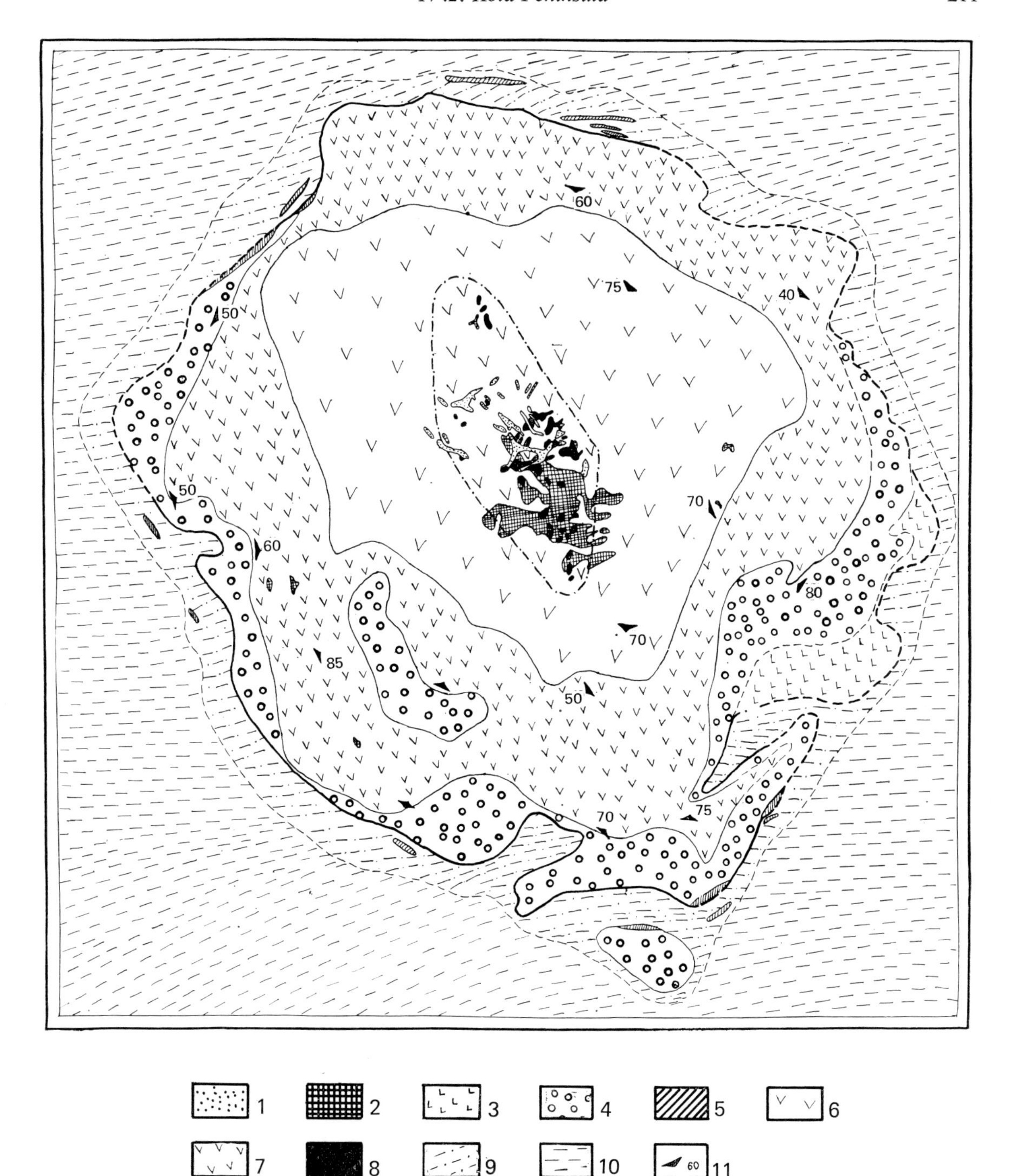

Fig. 3 Geological sketch map of the Afrikanda massif (after A. A. Kukharenko *et al.*, 1965)

(1) alkaline pegmatites; (2) calcite–amphibole–pyroxene rocks; (3) melteigites; (4) nepheline pyroxenites; (5) feldspathic pyroxenites; (6) coarse- and medium-grained pyroxenites; (7) fine-grained pyroxenites; (8) olivinites; (9) fenites; (10) gneisses; (11) strike and dip of trachytoid structures.

monoclinic pyroxene $Di_{79}Hd_{10}Aug_8Ac_3$ (Aug is the 'augitic' components such as $CaTi(Al_2O_6)$, $CaFe(AlSiO_6)$, etc remaining after subtraction of Hd + Di + Ac (Kukharenko *et al.*, 1965)). The local enrichment in titanomagnetite is peculiar. The marginal facies of the pyroxenite often contains some nepheline (nepheline pyroxenite).

The ijolite–melteigite–urtite series of rocks is represented by nepheline-pyroxene rocks varying from jacupirangite to urtite. The monoclinic pyroxene is more alkalic than that of the ultramafic rocks ($Di_{68}Hd_{15}Aug_7Ac_{10}$). Perovskite and melanite enriched in calcium are characteristic accessory minerals.

Nepheline and cancrinite syenites are usually represented by leucocratic vein rocks having miaskitic associations of mafic and accessory minerals (aegirine–diopside, biotite, zircon, pyrochlore, titanite).

According to Kukharenko *et al.* (1965) the postmagmatic rocks and the carbonatites are results of continuous metasomatic processes. During the first stage of alteration the melilite and phlogopite rocks were formed through contact interaction of magmatic ultramafic rocks and ijolite-melteigite having contrasting chemical compositions. The following stages of metasomatic alteration are displayed in the mobilization of calcium, magnesium and iron from the magmatic rocks by alkalic and carbonic acid solutions which resulted in the formation of apatite-forsterite-magnetite ores and of carbonatites.

The chemical compositions of the ultramafic-alkaline rocks are listed in Table 2. The main petrochemical peculiarity of the assemblage is the successive formation of undersaturated ultramafic rocks, the nepheline-pyroxene alkaline series of rocks and finally nepheline syenites. The petrochemical trends of the ultramafic as well as of the alkaline series of rocks have similar tendencies resulting in the enrichment in silica, alumina and alkalis and in the simultaneous decrease of calcium, magnesium and iron in the final differentiates.

All varieties of the Caledonian assemblage of rocks are characterized by high contents of CaO, TiO_2 and iron and low ratios of $MgO/(FeO + Fe_2O_3)$. The average composition of the assemblage is near to that of Daly's melilite basalt (Kukharenko *et al.*, 1965).

All these petrochemical peculiarities are inherent to the igneous rock series located within the consolidated parts of the Earth's crust (shields, platforms) and may serve as possible arguments in favour of the hypothesis of ultrabasic alkaline magma genesis due to partial melting of upper mantle material of eclogitic or peridotitic composition (Kukharenko *et al.*, 1965).

IV.2.3.2. The Alkaline Basaltoids

The Paleozoic volcanic-sedimentary sequence (the Lovozero Formation) formed the original roofs of the Khibina and Lovozero nepheline syenite massifs (Eliseev, 1946; Kirichenko, 1962). The effusives occur as xenoliths and are most abundant in the alkaline rocks of the Kontozero fault trough. In some places similar rocks occur in dykes and in volcanic necks. The volcanic rocks include picrite-porphyrite, augite porphyrite, olivine melteigite-porphyry, essexite-porphyrite, melilite basalt, trachyte, rhomb-porphyry, phonolite and their pyroclastic equivalents. The basic varieties are predominant. The formation of the effusives apparently went on for a long period of time. The ultrabasic and basic effusives are succeeded by the alkaline varieties at the final stages of volcanism.

As to chemical and mineralogical composition the series of effusives of the Kola Peninsula belongs to the undersaturated alkaline basaltoid group of rocks. The petrochemical evolution is similar to that of the ultramafic alkaline rocks. An increase in silica, alumina and alkalis is observed in the sequence: picrite–augite porphyrite – essexite porphyrite – rhombporphyry – phonolite.

IV.2.3.3. The Hercynian Nepheline Syenite Assemblage

The Khibina (1327 km^2) and Lovozero (650 km^2) massifs are the largest alkaline massifs in the world and are confined to a long-living tectonic zone of abyssal fractures cutting the folded structures of the Archean and Proterozoic

TABLE 2 The Average Chemical Composition of the Paleozoic Alkaline Rocks (Kola Peninsula)

Components	Caledonian ultramafic–alkaline association						Hercynian association of nepheline syenites: Khibina massif								Lovozero massif			
	1	2	3	4	5	6	7	8	9	10	11	12	13	14	15	16	17	18
SiO_2	33·82	36·52	40·76	41·30	51·91	36·38	53·01	52·90	54·08	51·59	49·92	55·03	55·07	53·22	51·39	53·94	52·89	53·62
TiO_2	1·71	5·30	2·54	2·84	0·58	3·22	0·87	0·94	0·91	1·01	2·31	1·01	0·94	1·05	1·13	1·05	1·37	1·12
ZrO_2	—	—	—	—	—	—	0·04	0·12	0·24	—	0·08	0·04	0·04	0·09	0·17	0·29	1·36	0·48
Al_2O_3	1·54	4·13	14·19	16·18	18·95	7·94	17·99	22·40	20·21	22·59	21·90	21·48	21·66	21·26	20·39	17·73	15·10	17·39
Fe_2O_3	11·34	10·92	5·66	4·59	3·33	8·37	5·67	2·77	3·59	2·51	4·00	2·84	2·53	2·59	3·45	4·61	6·69	4·93
FeO	10·80	7·83	4·74	4·25	1·80	6·88	2·14	1·29	1·58	1·76	3·04	1·70	1·36	1·58	1·49	0·94	1·18	1·01
MnO	0·20	0·09	0·13	0·14	0·11	0·12	0·29	0·25	0·23	0·18	0·19	0·13	0·15	0·18	0·22	0·30	0·53	0·34
MgO	36·90	12·41	7·01	6·89	0·85	14·95	0·59	0·60	0·51	0·75	1·64	0·44	0·64	0·65	1·39	0·92	1·13	0·98
CaO	1·33	18·97	15·99	13·25	4·90	14·46	1·95	1·38	1·55	1·55	4·88	1·35	1·35	1·80	2·00	1·07	1·66	1·22
Na_2O	0·07	0·89	5·13	6·74	9·63	3·17	7·48	10·00	9·57	7·22	10·42	9·12	9·13	9·81	10·61	11·16	10·28	10·97
K_2O	0·08	0·61	2·48	2·86	3·98	1·52	6·12	5·71	5·97	9·96	5·19	5·54	5·50	6·52	5·13	6·11	4·91	5·86
H_2O^-	1·80 (H_2O^- + H_2O^+)	1·94 (H_2O^- + H_2O^+)	0·86 (H_2O^- + H_2O^+)	0·68 (H_2O^- + H_2O^+)	1·83 (H_2O^- + H_2O^+)	0·66 (H_2O^- + H_2O^+)	0·20	0·30	0·22	0·29	0·10	0·26	0·21	0·15	1·18 (H_2O^- + H_2O^+)	0·15	0·06	0·14
H_2O^+							2·92	0·80	0·83	0·32	0·74	0·36	0·55	0·64		0·94	1·17	0·99
heat loss							—	—	—	—	—	—	—			0·20	0·64	0·27
P_2O_5	0·14	0·33	0·40	0·16	0·17	0·37	0·45	0·14	0·14	0·25	0·62	0·22	0·27	0·29	0·27	0·22	0·10	0·20
Number of analyses	45	21	14	23	9	130	3	8	9	8	30	4	16	averaged sample	17	12 averaged samples	5 averaged samples	Data columns 15,16,17
Data	Kukharenko *et al.* (1965)						Galakhov (1959, 1966)							Kukharenko *et al.* (1968)	Gerasimovsky *et al.* (1966)			

Legend:
1. Olivinites.
2. Pyroxenites.
3. Melteigites.
4. Ijolites.
5. Nepheline and cancrinite syenites.
6. Average composition of the ultramafic–alkaline complexes.
7. Nepheline and alkali syenites (intrusive phase I).
8. Massive khibinites (phase II).
9. Trachytoid khibinites (phase III).
10. Rischorrites (phase IV).
11. Ijolite-urtites (phase V).
12. Medium-grained aegirine-nepheline syenites (phase VI).
13. Foyaites (phase VII).
14. Average composition of the Khibina massif.
15. Nepheline and nosean syenites (phase I).
16. Average composition of the urtite–lujavrite–foyaite complex (phase II).
17. Eudialyte lujavrites (phase III).
18. Average composition of the Lovozero massif.

rocks. This tectonic zone has a north-eastern strike and continues within the territories of Sweden and Norway. Regeneration in Middle and Upper Paleozoic times resulted in block faulting within the Khibina–Lovozero–Kontozero graben (Volotovskaya, 1967). The presence of melilite-bearing ultrabasic alkaline rocks within the Khibina massif (Galakhov, 1966a) indicates the probability of a coincidence of Caledonian and Hercynian magmatism within the same fault system.

The Khibina and Lovozero intrusions were formed during the Hercynian tectonic–magmatic stage (290 $\pm$ 10 m.y.)

The Khibina massif is a polyphase intrusion having the central type of structure. It is of isometric shape; the steep contacts with the country rocks have been traced at depth for 7 km by geophysical methods (Shablinsky, 1963). The contact rim of the Khibina massif is composed of exocontact alkaline metasomatic rocks (fenite, tveitåsite, albitite, etc.) and of endocontact facies of alkali and nepheline syenite. The contact zone is several hundreds of metres thick.

The alkaline rocks of the Khibina intrusion break through Archean granite-gneisses and Proterozoic volcanic-sedimentary rocks. They were formed during seven intrusive phases. The most ancient rocks are substituted by younger ones in a centripetal direction (Eliseev *et al.*, 1939; Zak and Kamenev, 1964).

The first intrusive phase is represented by nepheline and alkali syenites and nepheline syenite-porphyries occurring in the form of small bodies and xenoliths within the peripheric zone of the massif (Volotovskaya, 1939). The massive khibinite (coarse-grained nepheline syenite, Table 2) was formed in phase II. Phase III is made up of trachytoid khibinites. In the very deepest part of the exposed vertical section these rocks are pseudostratified and occur as an alternating sequence of leucocratic nepheline syenite and melanocratic nepheline-pyroxene rocks of ijolitic type (Galakhov, 1966b). Phase IV is represented by the rischorrites which comprise a complex ring-like intrusive body. The rischorrites are characterized by poikilitic structures and by the occurrence of graphic, dactylotypic and micropegmatitic intergrowths of nepheline and alkali feldspar (Galakhov, 1959). According to Tikhonenkov (1963) autometasomatic and probably metasomatic processes played an important role in the formation of these rocks. Ijolite, melteigite and urtite are the main members of the fifth intrusive phase and comprise the well known stratified complex of rocks containing the world's largest apatite-nepheline ore deposit (Ivanova, 1963). The complex is of a complicated geological structure and was formed in several stages. It embraces the entire sequence of nepheline-pyroxene rocks from jacupirangite to urtite. Phase VI consists of medium-grained aegirine nepheline syenites and phase VII of foyaites.

The internal structure of the massif is characterized by alternating ring-like and conical intrusive bodies. The alkaline rocks having pseudostratified structure and trachytoid textures are confined to the conical fractures (phases III, V, VII) while the ring-like intrusive phases (I, II, IV, VI) are characterized by homogeneous structures and massive textures.

The final stages of igneous activity within the Khibina massif are represented by numerous alkaline pegmatites (Fersman *et al.*, 1937) and by dykes consisting of alkaline gabbroids and lamprophyres (theralite- and shonkinite-porphyry, nepheline syenite-porphyry, monchiquite, tinguaite, etc.). Several bodies of kimberlite-like picrite-porphyrite are located in the vicinity of the Khibina massif.

The satellite intrusion *Soustova massif* (38 km^2) is located to the south of the Khibina massif and is mainly composed of alkali syenite and subordinate nepheline and analcime syenites (Soustov, 1938).

The chemical composition of the nepheline syenites of the Khibina massif is very close to Daly's average nepheline syenite. There are no distinct fluctuations during the formation of the massif (Table 2) with the exception of the rischorrites with extremely high ratios of K_2O/Na_2O. The ijolite–melteigite–urtite series is highly undersaturated with respect to silica and is enriched in CaO, TiO_2, P_2O_5 and mafic

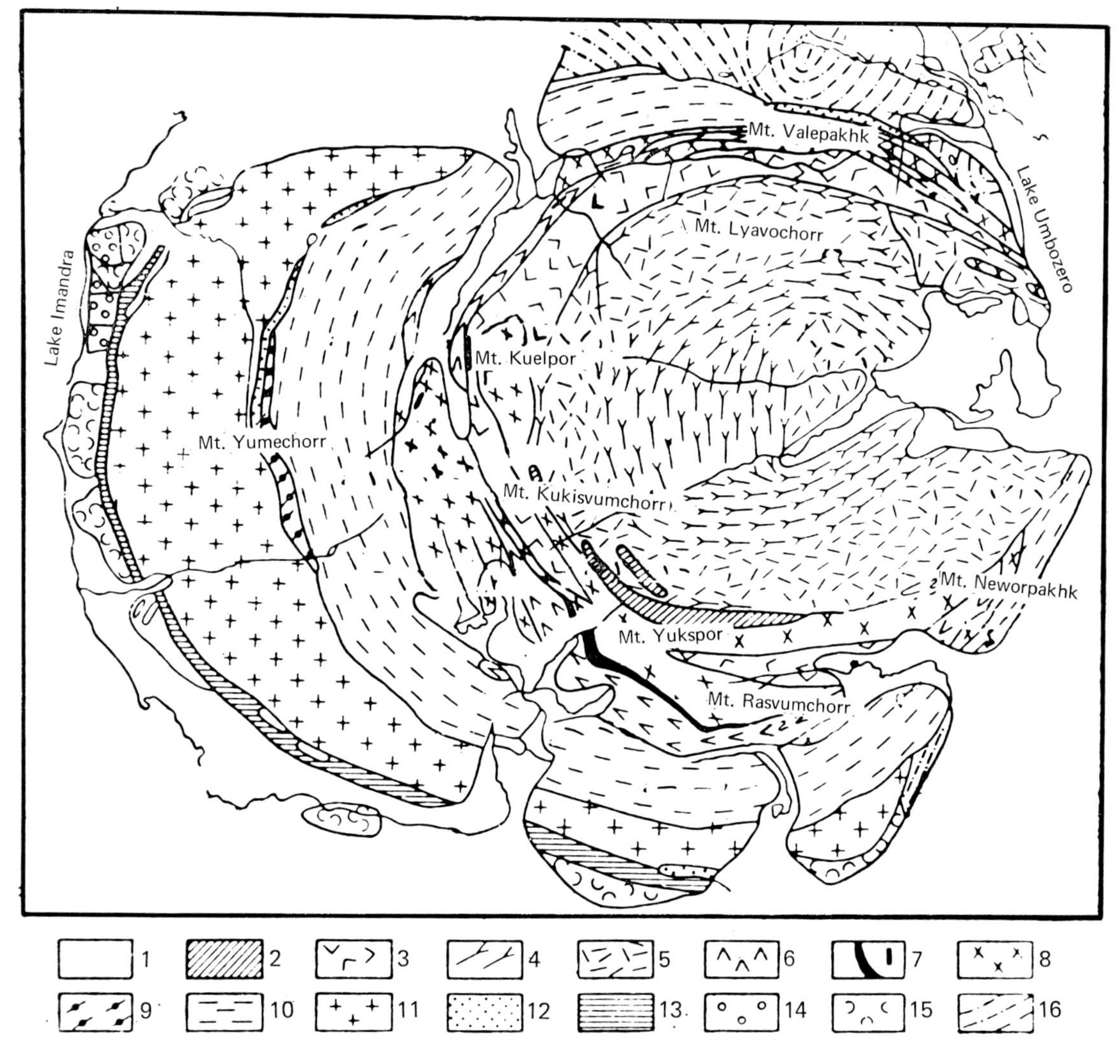

Fig. 4 Geological map of the Khibina massif (after E. N. Volodin and N. A. Eliseev, published in T. N. Ivanova, 1963)

(1) Quaternary rocks; (2) fine-grained mica–aegirine–amphibole nepheline syenites; (3) medium-grained aegirine–nepheline syenites; (4) trachytoid foyaites; (5) massive foyaites; (6) ijolites, urtites, malignites; (7) apatite–nepheline rocks; (8) rischorrites; (9) alkaline syenite–porphyries; (10) trachytoid khibinites; (11) massive khibinites; (12) alkali and nepheline syenites; (13) plagioclase–pyroxene hornfels; (14) quartz gabbro–diabases; (15) pillow lavas, meta-mandelstein lavas, greenschists, tuffitic–sedimentary rocks; (16) gneisses.

components. The absence of any regular geochemical trends in the evolution of the rock sequence of the Khibina massif leads to the hypothesis that there were several magma chambers at different depths in the Earth crust. The primary magma differentiated in these chambers which is the cause of the contrasting series of nepheline syenites and ijolite-melteigite (Galakhov, 1968).

The *Lovozero massif* is a complex body formed during four intrusive phases (Gerasimovsky *et al.*, 1966). The intrusion has steep contacts with the surrounding Archean granite-gneisses. The contact has been traced at depth

TABLE 3 Petrographic Composition of the Main Rock Types of the Khibina and Lovozero Alkaline Massifs

Massif	Rocks	Intrusive phase	Area of distribution %	Main rock-forming minerals	Characteristic accessory minerals
Khibin	Khibinites	II, III	43·9	Microcline, nepheline, aegirine, arfvedsonite	Eudialyte, lamprophyllite, titanite
	Rischorrites	IV	10·5	Microcline, orthoclase, nepheline, aegirine, aegirine–augite, arfvedsonite, biotite	Eudialyte, lamprophyllite, astrophyllite, titanite, apatite
	Ijolites, melteigites, urtites	V	5·6	Nepheline, aegirine–augite	Apatite, titanite, titanomagnetite
	Foyaites and medium-grained aegirine-nepheline syenites	VI, VII	38·9	Microcline, nepheline, arfvedsonite, aegirine–augite, biotite	Eucolite, astrophyllite, titanite
	Nepheline and nosean syenites	I	3·2	K–Na feldspar, nepheline, nosean, aegirine–diopside ($Ac_{35}Hd_{15}Di_{50}$) magnesio-riebeckite	Låvenite, ilmenite, titanite, apatite
Lovozero	Urtites		3·8	Nepheline, microcline, sodalite	Apatite, titanates
	Foyaites		32·3	Microcline, nepheline, aegirine ($Ac_{85}Hd_{6}Di_{9}$) arfvedsonite, sodalite, analcime	Eudialyte, lamprophyllite, murmanite, villiaumite
	Aegirine lujavrites	II		Microcline, aegirine, nepheline ($Ac_{74}Hd_{10}Di_{16}$)	Eudialyte, titanates, lamprophyllite, rinkolite
	Amphibole lujavrites		40·8	Microcline, arfvedsonite, nepheline	Titanite, apatite
	Eudialyte lujavrites	III	18·0	Microcline, aegirine ($Ac_{82}Hd_{7}Di_{11}$), eudialyte, arfvedsonite, nepheline	Lamprophyllite, murmanite, lovozerite
	Poikilitic sodalite syenites	II, III	1·3	Microcline, sodalite, nepheline, aegirine	Eudialyte, villiaumite, tchinglusuite

for 5 km by geophysical methods (Shablinsky, 1963). According to seismic data the lower part of the massif (12 × 16 km in cross-section) has a stock-like shape and a concentrically zoned structure.

The upper part of the massif (20 × 30 km in area) is represented by a laccolith-like differentiated body. It is mainly composed of nepheline syenites of the II and III intrusive phases and of xenoliths of the rocks from phase I. Detailed petrographical descriptions of the alkaline rocks (Table 3) are published in a

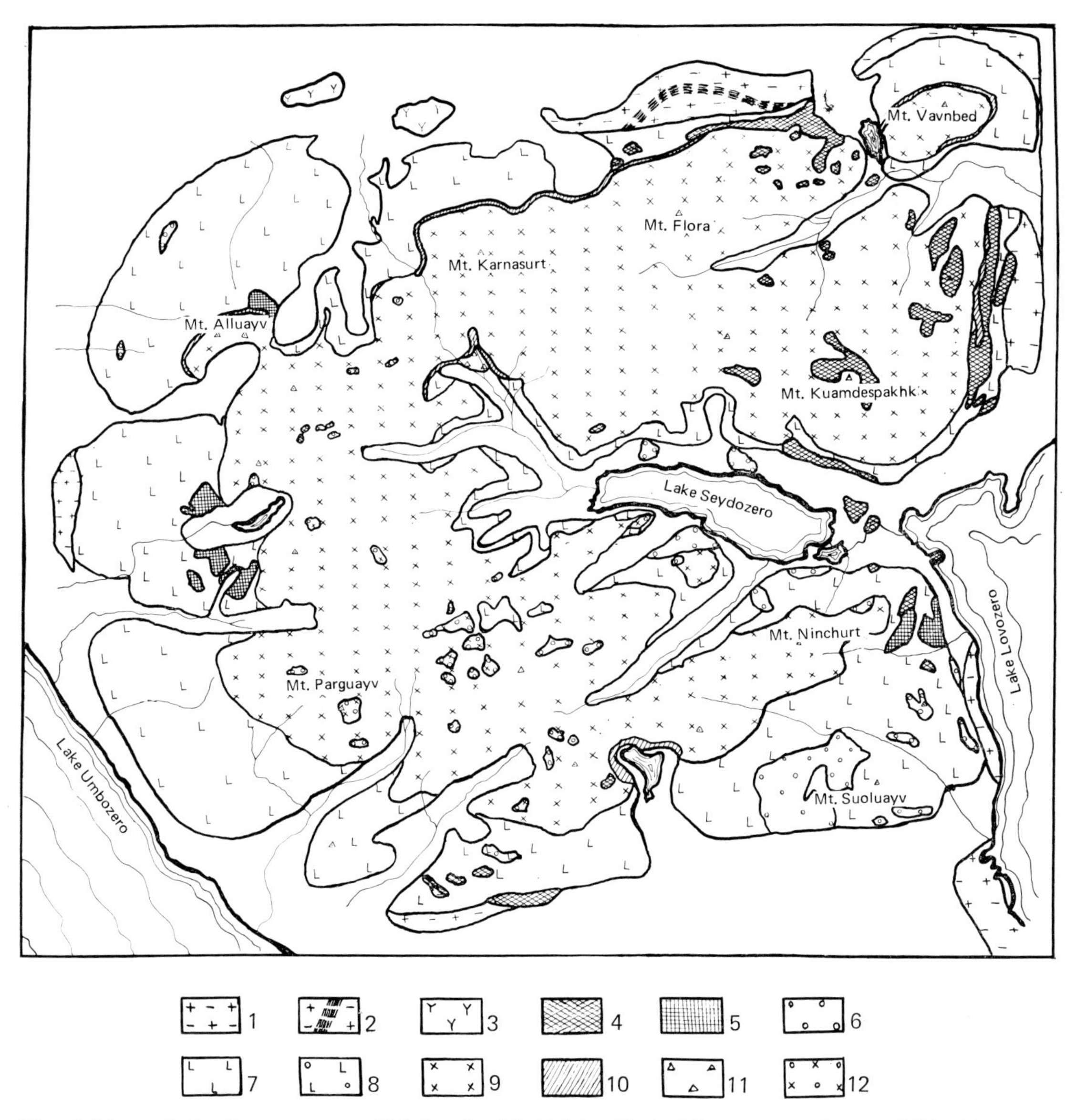

Fig. 5 Map of the Lovozero massif (after Ja. M. Feigin, N. A. Eliseev and others, published in V. I. Gerassimovsky *et al.*, 1966)

Lower Archean (A_1): (1) granite gneisses. Proterozoic (Prz): (2) sillimanite–andalusite schists; (3) ultrabasic rocks. Paleozoic magmatic complex (Pz): (4) effusive–sedimentary rocks of the Lovozero Formation. Intrusive rocks of the Lovozero massif: phase I: (5) metamorphosed nepheline syenites; (6) poikilitic nosean syenites, nepheline–nosean syenites and even-grained nepheline syenites; phase II: (7) urtites, foyaites, lujavrites; (8) poikilitic sodalite syenites; intrusive phase III: (9) eudialyte lujavrites; (10) porphyritic lujavrites; (11) porphyritic lovozerite–murmanite lujavrites; (12) poikilitic sodalite syenites and tawites.

number of publications (Vorob'eva, 1940; Eliseev and Fedorov, 1953; Gerasimovsky *et al.*, 1966).

Phase I consists of metamorphosed nepheline syenites, even-grained miaskitic nepheline syenites and poikilitic nepheline and nosean syenites, the latter probably being the products of autometasomatic replacement of the miaskitic nepheline syenites.

Phase II is represented by a rhythmically layered complex of alkaline rocks, the regular sequence of alternation of rocks in the vertical section is preserved throughout the area of the massif (from bottom to top: urtite → foyaite → lujavrite). There are generally gradual transitions between the rocks of phase II but the contacts between the urtite layers and the underlying lujavrite layers are commonly sharp.

The rocks of phase III comprise a coarsely layered complex of eudialyte lujavrite and associated porphyritic lujavrites and murmanite lujavrite. The rocks cut and rest on the differentiated complex (phase II). The eudialyte lujavrites are situated in the central part of the massif; Bussen and Sakharov (1967) have described the body as an ethmolith.

Poikilitic sodalite syenites occur as lensoid bodies within the rocks of the second and third intrusive phases and are presumably syngenetic with the surrounding lujavrites (Gerasimovsky *et al.*, 1966).

Highly differentiated alkaline pegmatites rich in rare minerals are found amongst the alkaline rocks of the Lovozero massif (Gerasimovsky, 1939; Vlasov *et al.*, 1959). The rocks comprising the dyke series (phase IV) are similar to the late dykes of the Khibina massif and are represented mainly by monchiquite, camptonite and tinguaite.

The rocks of the Lovozero intrusion are characterized by an extreme excess of alkalis in relation to alumina $(Na_2O + K_2O)/Al_2O_3 = 1{\cdot}40$ by the predominant rôle of sodium over potassium, by a high ratio of Fe_2O_3/FeO, and by the enrichment in rare elements and volatile components. The rocks of phase I are of miaskitic type ($(Na_2O + K_2O)/Al_2O_3 \leqslant 1$) whereas the rocks of phases II and III are of peralkaline agpaitic type. The petrochemical evolution of the agpaitic series of rocks resulted in a continuous increase in the rôle of the alkalis contained in the aluminosilicates, in an increasing amount of mafic components, as well as in a silica saturation in the final derivatives (Table 2). The compositional points of the agpaitic nepheline syenites probably fall in the low-temperature region of the hypothetical phase diagram Ne–Kp–Q–Ac which may indicate the residual character of the agpaitic alkaline magmas (Gerasimovsky *et al.*, 1966).

IV.2.4. On the Genesis of the Alkaline Rocks

The abyssal differentiation of the Proterozoic alkaline-gabbroid association and of the Paleozoic association of ultramafic alkaline rocks, basic effusives and nepheline syenites display many similar features. Presumably this was due to the platform (and subplatform) conditions of the alkaline magmatic activity within the Kola Peninsula. The compositions of the primary magmas of these alkaline assemblages are approximately analogous and near to the composition of alkaline olivine basalt or melilite basalt (the latter refers to the Caledonian assemblage, see Kukharenko *et al.*, 1965). The differences in chemical composition between the two types of basalt are not essential because of their probable comagmatic character (O'Hara, 1968). The pronounced alkalinity of the primary basaltic magma reflects according to experimental data an abyssal character (30–60 km) of its generation (Yoder and Tilley, 1962; Green and Ringwood, 1967).

It is convenient to consider the physicochemical evolution of an alkali basalt magma by means of the liquidus diagram of the system Fo–Di–Ne–Ab (Schairer and Yoder, 1960). The separation and partial fractionation of olivine and pyroxene at the first stages of crystallization differentiation resulted in the formation of peridotite and pyroxenite. These rocks occur as the first members of the geological sequence in all alkaline intrusive assemblages within the Kola Peninsula; for instance, the peridotites and pyroxenites of the Gremjakha–Virmes and

Yelet'ozero massifs, the olivinites and pyroxenites of the Caledonian alkaline assemblage, and the picrite-porphyrites among the effusives of the Lovozero Formation.

The alkaline nature of the original magma is the cause of its direct evolution into melts which plot near the Ne–Di cotectic line. The consumption of olivine takes place in accordance with the reaction Ol + L = Ne + Di. This stage of differentiation is represented by the ijolite–melteigite series of rocks of the alkaline complexes, for instance, in the Gremjakha–Virmes massif, the ultramafic alkaline intrusions, and the Khibina massif. If the primary magma was less alkaline and more saturated in silica the simultaneous crystallization of diopside, olivine and feldspars gives the gabbroid series of rocks (Gremjakha–Virmes, Yelet'ozero).

At the very last stages of differentiation of the alkaline basalt magma the liquids plot close to the anchieutectic composition (Aeg-Di + Ne + Fsp), the latter approximately corresponding to the nepheline syenites. Nepheline syenites are usually the final derivatives of the alkaline intrusions of the Kola Peninsula. The outlined differentiation sequence is due to a definite order of crystallization being observed in all the alkaline assemblages of rocks under consideration: Ol → Ol + Px → Px + Ne → Px + Ne + Fsp.

The rocks, which are products of the simultaneous crystallization of Ne and Px or of Ne + Px + Fsp, are characterized by a deficit in alkalis in relation to alumina (the agpaitic coefficient $\leqslant 1$). The agpaitic nepheline syenites may be the final result of a continuous and gradual crystallization differentiation of alkali basaltic magma due to fractionation of the alumina contained in the anorthite component of the plagioclases and in nepheline, as well as of the Tschermak molecule in the pyroxenes (Kogarko and Polyakov, 1967). The peculiarities of the differentiation of agpaitic melts are illustrated by the geological history of the Lovozero alkaline intrusion. The evolution of the alkaline magma may be considered using the simplified three-component diagram nepheline–potassium feldspar–acmite. Analogous trends of evolution are observed in the crystallization differentiation of the epigenetic series of rocks: II and III intrusive phases and in the syngenetic series of rocks of phase II (urtite, foyaite, lujavrite). This analogy is caused by the identical sequence of crystallization of the minerals in all the alkaline rocks of the massif: Ne → Ne + Fsp → Ne + Fsp + Aeg. The primary igneous layering is presumably a result of a rhythmic eutectic crystallization and fractionation under conditions of a very slow heat removal from the magma chamber. The chemical compositions of the lujavrites plot near the ternary eutectic in the system Ac–Di–Ne–Ab–H_2O with a low content of Di in the alkali pyroxene (Nolan, 1966). This mechanism of formation of magmatic layering is similar to that of rhythmic zoning in zone melting (Vinogradov, 1962).

Crystallization differentiation was obviously the main process responsible for the formation of the Proterozoic and Paleozoic alkaline assemblages (Polkanov *et al.*, 1967; Kukharenko *et al.*, 1965; Gerasimovsky *et al.*, 1966). At the same time Kukharenko *et al.* (1965) accept the essential rôle of the diffusion of alkalis towards the marginal zones of the Caledonian intrusions during the precrystallization stage and of gravitational differentiation of the ultrabasic alkaline melt during its ascent to the magma chamber. Some investigators propose a metasomatic formation of the ijolites and melteigites of the Caledonian assemblage of rocks (Borodin, 1958; VI.7).*

The alkaline granites of the Kola Peninsula are considered to be palingenetic rocks with highly abundant autometasomatic phenomena in contrast to the Proterozoic and Paleozoic ultramafic alkaline rocks, alkaline gabbroids and nepheline syenites.

* Readers are referred to the relevant Chapters of this book where similarly cited.

IV.2. REFERENCES

Batieva, I. D., and Bel'kov, I. V., 1968. 'The granitoid formations of the Kola Peninsula', in *Outlines of the petrology, mineralogy and metallogeny of the Kola Peninsula granites.* Izd-vo 'Nauka', Leningrad, 5–143 (in Russian).

Bergman, I. A., 1962. The accessory mineralization of the Gremjakha–Virmes pluton (in Russian). *Materialy po mineralogii Kol'skogo poluostrova*, 3, 20–36.

Bogachev, A. I., Zak, S. I., Safonova, G. P., and Inina, K. A., 1963. *The geology and petrology of Yelet'ozero gabbroid massif in Karelia.* (in Russian). 159 pp., Izd-vo Akademii Nauk SSSR, Moscow and Leningrad.

Borodin, L. S., 1958. On the nephelinization and aegirinization of pyroxenites concerning with the problem on the genesis of the alkaline rocks of the ijolite–melteigite type (in Russian). *Izv. Akad. Nauk SSSR, ser. geol.*, **6**, 48–57.

Bussen, I. V., and Sakharov, A. S., 1967. *The geology of the Lovozero tundras* (in Russian). 126 pp., Izd-vo 'Nauka', Moscow.

Eliseev, N. A., 1946. Devonian effusives of the Lovozero tundras (in Russian). *Zap. Vses Miner. Obsch.*, **75**, no. 2.

Eliseev, N. A., and Fedorov, E. E., 1953. The Lovozero pluton and its deposits (in Russian). *Mater. Lab. Geol. Dokembr.*, **1**, 1–306.

Eliseev, N. A., Ozhinsky, I. S., and Volodin, E. N., 1939. Geological Map of the Khibin tundras (in Russian). *Trudy Lengeolupr.*, **19**, 1–38.

Fersman, A. E., Smol'yaninov, N. A., and Bonshtedt, E. M., (Eds.) 1937. *Minerals of the Khibina and Lovozero tundras.* Izd-vo Akad. Nauk SSSR, 563 pp., Moscow and Leningrad (in Russian).

Galakhov, A. V., 1959. *The rischorrites of the Khibina alkaline massif* (in Russian). 169 pp., Izd-vo Akad. Nauk SSSR, Moscow and Leningrad.

Galakhov, A. V., 1966a. On the occurrence of the alkaline-ultrabasic magmatism in the Khibina tundras (Kola Peninsula) (in Russian). *Dokl. Akad. Nauk SSSR*, **170**, no. 3, 657–61.

Galakhov, A. V., 1966b. 'On the stratification of the trachytoid khibinite intrusion of the Khibina alkaline massif', in *The alkaline rocks of the Kola Peninsula.* Izd-vo 'Nauka', Moscow and Leningrad, 4–12 (in Russian).

Galakhov, A. V., 1966c. The chemical composition of the rocks of the Khibina alkaline massif (in Russian). *Dokl. Akad. Nauk. SSSR*, **171**, no. 5, 1179–82.

Galakhov, A. V., 1968. The petrochemistry of natural rock series and the alkaline magma evolution of the Khibina massif', in *Geological structure, evolution and ore contents of the Kola Peninsula.* Apatity, 97–114 (in Russian).

Gerasimovsky, V. I., 1939. Pegmatites of the Lovozero alkaline massif (in Russian). *Trudy Inst. Geol. Nauk*, **18**, ser. min.-geokhim., no. 5, 1–46.

Gerasimovsky, V. I., Volkov, V. P., Kogarko, L. N., Polyakov, A. I., Saprikina, T. V., and Balashov, Yu. A., 1966. *The geochemistry of the Lovozero alkaline massif* (in Russian). 395 pp., Izd-vo 'Nauka', Moscow. (English translation by D. A. Brown, Canberra, Australian National University Press, 1968.)

Green, D. H., and Ringwood, A. E., 1967. The genesis of basaltic magmas. *Contr. Miner. Petrology*, **15**, 103–90.

Ivanova, T. N., 1963. *Apatite deposits of the Khibina tundras* (in Russian). 287 pp. Gosgeoltekhizdat, Moscow.

Kirichenko, L. A., 1962. 'On Paleozoic sedimentary and effusive rocks of the Kola Peninsula', in *Materials of the geology and commercial minerals of the North-Western part of RSFSR* (in Russian). Gosgeoltekhizdat, 27–45.

Kogarko, L. N., and Polyakov, A. I., 1967. The problems on genesis of the agpaitic nepheline syenites (in Russian). *Geokhimiya*, **2**, 131–43.

Kukharenko, A. A., 1967. Alkaline magmatism of the eastern part of the Baltic shield (in Russian). *Zap. Vses. Miner. Obsch.*, **96**, no. 5, 547–66.

Kukharenko, A. A., *et al.*, 1965. *The Caledonian complex of ultrabasic-alkaline rocks and carbonatites of the Kola Peninsula and Northern Karelia* (in Russian). 772 pp., *Izd-vo 'Nedra', Moscow.*

Kukharenko, A. A., *et al.*, 1968. Average chemical composition of the Khibina alkaline massif (in Russian). *Zap. Vses. Miner. Obsch.*, **97**, no. 2, 133–50.

Nolan, J., 1966. Melting relations in the system $NaAlSi_3O_8$–$NaAlSiO_4$–$NaFeSi_2O_6$–$CaMgSi_2O_6$–H_2O and their bearing on the genesis of alkaline undersaturated rocks. *Q. J. Geol. Soc. Lond.*, **122**, Part 2, 119–57.

O'Hara, M. J., 1968. The bearing on phase equilibria studies in synthetic and natural systems on the origin and evolution of basic and ultrabasic rocks. *Earth Sci. Rev.*, **4**, no. 2, 69–135.

Polkanov, A. A., Eliseev, N. A., Eliseev, E. N., and Kavardin, G. I., 1967. *Gremjakha–Virmes massif at the Kola Peninsula* (in Russian). 235 pp., Izd-vo 'Nauka', Moscow and Leningrad.

Polkanov, A. A., and Gerling, E. K., 1961. Geochronology and the geological evolution of the Baltic shield and its folded marginal zones (in Russian). *Trudy Labor. Geol. Dokembr.*, **12**, 7–102.

Polkanov, A. A., and Gerling, E. K., 1964, A preliminary geological age scale of Pre-Cambrian—The Hercynides of the Baltic shield (in Russian). *Trudy Labor. Geol. Dokembr.*, **19**, 176–84.

Schairer, J. F., and Yoder, H. S., 1960. The nature of residual liquids from crystallization with data on the system nepheline–diopside–silica, *Am. J. Sci.*, **258A**, 273–83.

Shablinsky, G. N., 1963. The method of reflected waves in the investigations on the abyssal structures of the alkaline massifs on the Kola Peninsula (in Russian). *Zap. Leningr. Gorn. Inst.*, **XVI**, no.2.

Shurkin, K. A., 1968. 'The main features of geological structure and history of the eastern part of the Baltic shield', in *Geology and abyssal structure of the eastern part of the Baltic shield.* Izd-vo 'Nauka', Leningrad, 5–59 (in Russian).

Soustov, N. I., 1938. A new alkaline massif in the vicinity of the Khibina tundras on the Kola Peninsula (in Russian). *Trudy Petrogr. Inst.*, **12**, 89–106.

Tikhonenkov, I. P., 1963. *Nepheline syenites and pegmatites of the Northern part of the Khibina massif and the role of Postmagmatic phenomena in their formation* (in Russian). 244 pp. Izd-vo Akad. Nauk SSSR, Moscow.

Tchumakov, A. A., 1958. 'On the origin of the alkaline granites of Keyvi', in *The alkaline granites of the Kola Peninsula*, (in Russian). Izd-vo Akad. Nauk SSSR, Moscow and Leningrad, 308–68.

Tzirul'nikova, M. Ya., Chechel, E. K., Shustova, L. E., and Sokol, R. S., 1968. *The abyssal earth's crust structure of the eastern part of the Baltic shield* (in Russian). Izd-vo 'Nauka', Leningrad, 178–84.

Vinogradov, A. P., 1962. Zone melting as a method of study of some radial processes in the Earth (in Russian). *Geokhimiya*, **3**, 269–70.

Vlasov, K. A., Kuz'menko, M. V., and Es'kova, E. M., 1959. *The Lovozero alkaline massif* (in Russian). 623 pp. Izd-vo Akad. Nauk SSSR, Moscow.

Volotovskaya, N. A., 1939. The xenoliths of nepheline and alkaline syenites in massive khibinite of the northern part of the Khibina massif (in Russian). *Zap. Vses. Miner. Obsch.*, **68**, no. 1, 45–67.

Volotovskaya, N. A., 1967. New data on the regularities of the Hercynian alkaline rock location within the eastern part of the Baltic shield (in Russian). *Dokl. Akad. Nauk SSSR*, **173**, no. 3, 645–7.

Vorob'eva, O. A., 1940. 'On the origin of layering in the Lovozero alkaline massif', in *Productive forces of the Kola Peninsula, vol. 1*, (in Russian). Izd-vo Akad. Nauk SSSR, Moscow, 119–28.

Yoder, H. S., and Tilley, C. E., 1962. Origin of basalt magmas: an experimental study of natural and synthetic rock system. *J. Petrology.*, **3**, no. 3, 1–342.

Zak, S. I., and Kamenev, E. A., 1964. New data on the geology of the Khibina alkaline massif (in Russian). *Sov. Geol.*, **7**, 42–52.

IV.3. THE ALKALINE PROVINCE OF SOUTH-WEST GREENLAND

B. G. J. Upton

IV.3.1. Introduction

IV.3.1.1 General Statement

In south-west Greenland a suite of alkalic igneous complexes outcrops across a zone extending some 180 km from east to west and approximately 80 km from north to south. These are largely composed of foyaites, nordmarkites and granites and are, for the most part, deeply dissected and magnificently exposed. However, some are at least partially covered by glaciers and others by the waters of the Labrador Sea. The extraordinary variety of alkaline rocks displayed in these complexes has attracted the attentions of geologists since the early part of the nineteenth century and an astonishing number of new minerals, including such currently well-known species as arfvedsonite, sodalite, eudialyte, aenigmatite and cryolite, were first recorded from these rocks. Ivigtut, Kvanefjeld (Ilímaussaq intrusion) and Narssarssuk (N.W. margin of the Igdlerfigssalik intrusion) are among the more celebrated mineral localities (see Sørensen, 1967).

Most of the intrusive rocks of the region, post-dating the last regional metamorphic episode, together with the sediments and lavas comprising the Eriksfjord Formation (Poulsen, 1964), were formed during a cycle of cratogenic activity that Wegmann (1938) referred to as the Gardar cycle. Radiometric dating (Moorbath *et al.*, 1960, Bridgwater, 1965) has since shown that while the Gardar cycle may have extended over a total interval of some 400 m.y. prior to 1000 m.y., magmatism appears to have reached its peak between 1250 and 1150 m.y. ago (Bridgwater, 1967, van Breemen and Upton, 1972).

IV.3.1.2. The Nature of Gardar Activity

South-west Greenland consists very largely of granites, migmatites and metasedimentary rocks ranging from greenschist to high grade gneisses. This basement complex was subjected throughout the Gardar to severe faulting accompanied by the emplacement of regional dyke swarms and the central complexes alluded to in the introductory paragraph. The following generalizations are offered concerning the broad pattern of Gardar geology;

1. The regional fissuring and introduction of dyke swarms occurred intermittently during a long period of faulting and preceeded the majority of the central complexes.

2. The earlier dyke swarms are predominantly basic (olivine dolerites and gabbros for the most part) whereas the younger swarms are characteristically salic, with an abundance of trachytes.

3. The central complexes are composed, almost exclusively, of salic rocks (larvikitic and nordmarkitic syenites, granites and nepheline syenites).

4. Anorthositic inclusions are common within the basic intrusives (i.e. those with less than 53% silica).

5. Extensive surface volcanism is believed to have taken place. Although surface products of the inferred alkalic central volcanoes are virtually absent, early Gardar basaltic lavas are preserved in the eastern part of the province.

6. Strain release took place principally by fissuring and by movement along near vertical wrench-faults, some of which can be shown to have had large vertical displacements. Horizontal displacements terminated before the climax of alkaline magmatism although vertical adjustments persisted until much later.

IV.3.2. Regional Tectonics

IV.3.2.1. Faults and Dykes

Dykes to which a Gardar age has been assigned outcrop from Kap Farvel to Frederikshaab, and are also present in the basement of south-east Greenland (Bridgwater and Gormsen, 1969).

The Gardar faults affect the entire alkaline province. Steeply hading or vertical faults, with notable horizontal components of movement, fall into two main groups, namely (i) those trending between E.S.E. and E., and (ii) those trending between N.N.W. and N.N.E. Whereas the former show left-lateral displacements, most of the latter are right-lateral wrench faults. Consequently, there appears to have been a conjugate set of transcurrent faults, intermittently active throughout the Gardar, that may be explicable in terms of a tensional stress along a N.W.–S.E. axis. The left-lateral, easterly trending faults can be traced for long distances; some at least had long pre-Gardar histories of movement (Henriksen, 1960) and the total displacement associated with them is estimated at between 15–20 km. An important suite of dolerite dykes (the BDOs of Greenland Geological Survey nomenclature) was emplaced at an early stage in the Gardar along this same general direction. These dykes appear to have been intruded along shear-planes and there is frequently evidence for further shearing, after emplacement. The main dyke swarms trend between E.N.E. and N.E. and are seen in their greatest concentrations in the Nunarssuit–Isortoq and

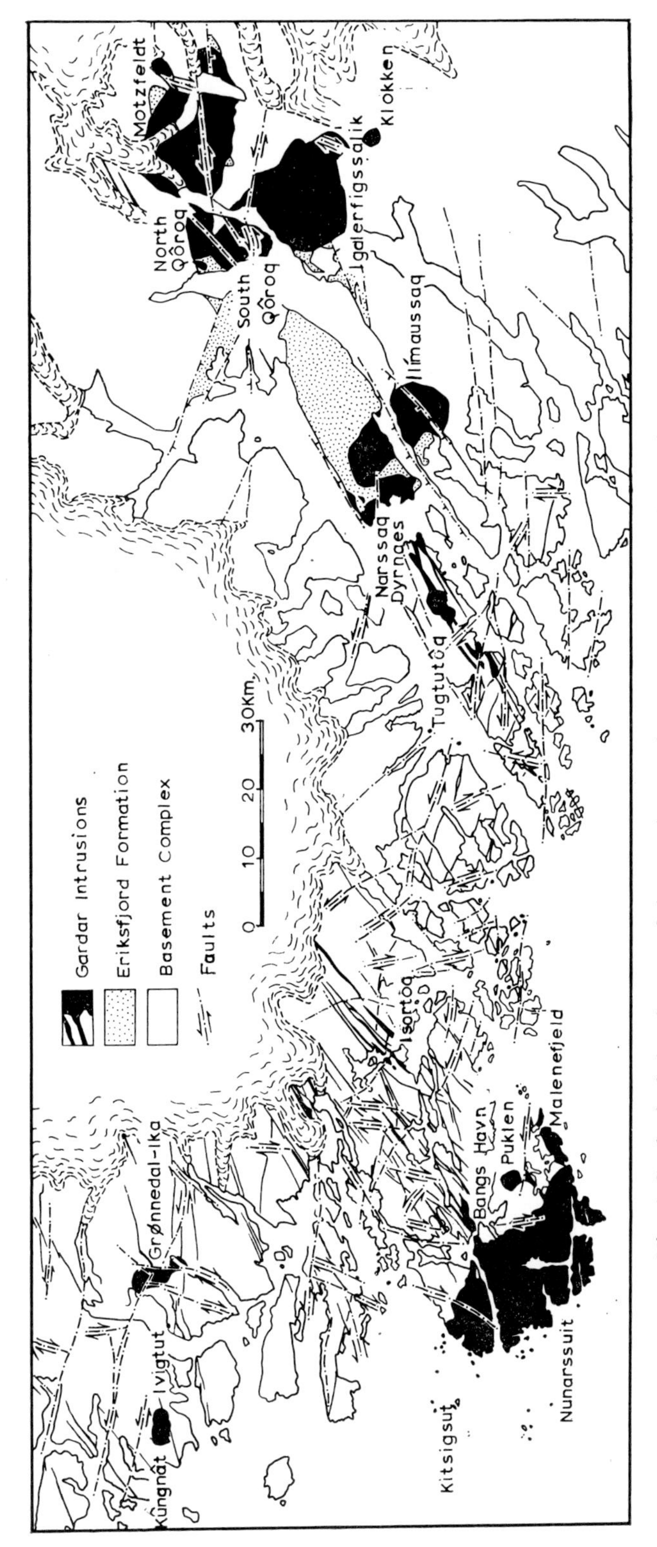

Fig. 1 Geological sketch map of the Gardar province showing principal intrusions and faults

Tugtutôq–Motzfeldt zones, where the alkali magmatism was also most intense. Many of these dykes too, at least in the Tugtutôq area, were intruded along shear-planes with small movements occurring subsequently within the consolidated dykes. In the two zones specified above, the main dyke swarms have been intruded parallel to the foliation in the gneisses and migmatites inherited from the earlier orogenic deformations. This behaviour is not seen however for the Gardar swarms of the Ivigtût region where the dykes are typically discordant to the basement structures.

IV.3.2.2. The Eriksfjord Formation

Most of the early E.N.E. and N.E. trending dykes are olivine dolerites and gabbros, and may reasonably be suspected of having supplied copious quantities of basaltic lava for fissure eruption across much of the province. The basaltic lava plateaux were probably developing while active faulting was in progress. Vertical displacements can be demonstrated for some of the easterly trending left-lateral faults and large vertical movements also apppear to have occurred along some of the E.N.E.–N.E. trending shears, e.g. in the vicinity of Ilímaussaq (Stewart, 1964). Stewart suggests that accumulation of the lavas was greatest on the subsiding fault-blocks. Such blocks provided troughs or grabens within which fluviatile, lacustrine and aeolian sediments were also being deposited (Wegmann, 1938 and Poulsen, 1964). According to Stewart (1964), of the 3 km sequence of basalts and sandstones preserved within the Ilímaussaq peninsula, approximately half is composed of lavas. The lavas and sediments comprise the Eriksfjord Formation, successions of which may have been notably thinner on neighbouring fault blocks. The evidence for the former presence of an extensive cover of lavas and sediments across much of the province comes (i) from relics preserved by local subsidence in the Gardar intrusions and diatremes, and (ii) from morainal material suggesting the possibility of further outcrops beneath the ice-cap.

IV.3.3. The Gardar Magmas

IV.3.3.1. Accumulitic Rocks

One of the most remarkable aspects of the province is the igneous layering that characterises most of the larger intrusions, irrespective of their composition. The principal features of the layered intrusions have been reviewed by Ferguson and Pulvertaft (1963) and also by Wager and Brown (1968).

Not only are such phenomena as rhythmic layering and igneous lamination displayed by the gabbroic bodies, but they are common in the larger larvikitic and nordmarkitic syenites, nepheline syenites and even in some of the granites (Harry and Emeleus, 1960). Cryptic layering and phase layering have been demonstrated in some of the intrusions and may be suspected in several others.

There is consensus of opinion among those who have worked on these intrusions that viscosities must have been sufficiently low in the salic magmas to permit highly effective gravitative separation of crystals from liquid. The presence of trough-banding and false-bedding in several of the intrusions, comparable to the structures described from Skaergaard, suggest also that convective overturn played an active rôle in the cooling histories and aided crystal fractionation. In brief it is suspected that the majority of the larger intrusives are composed of accumulitic rocks (cf. II.2.6).

IV.3.3.2. Magma Compositions

Since unmodified marginal chilled facies are not typically developed around the contacts of these intrusives and the lavas are invariably altered and oxidized, it is to the finer grained dyke-rocks that one must turn in search of more direct information concerning the composition of the magmas themselves. The main (E.N.E. and N.E.) dyke swarms involve feldspathic olivine basalts transitional to hawaiites, grading through mugearites to trachytes, and thence to both phonolites and comenditic rhyolites (Macdonald, 1969a and b). The phonolites and rhyolites are commonly peralkalic, containing

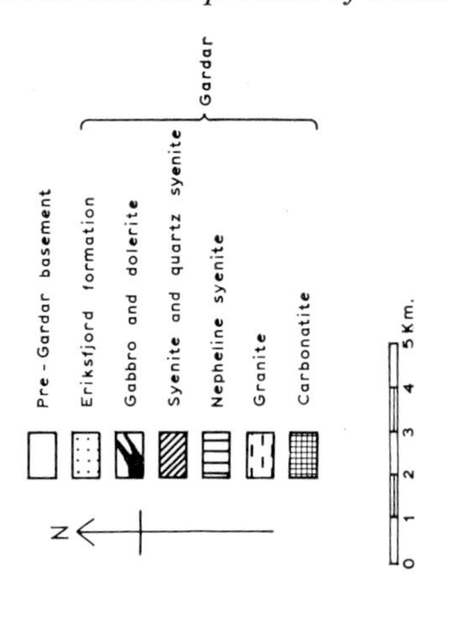

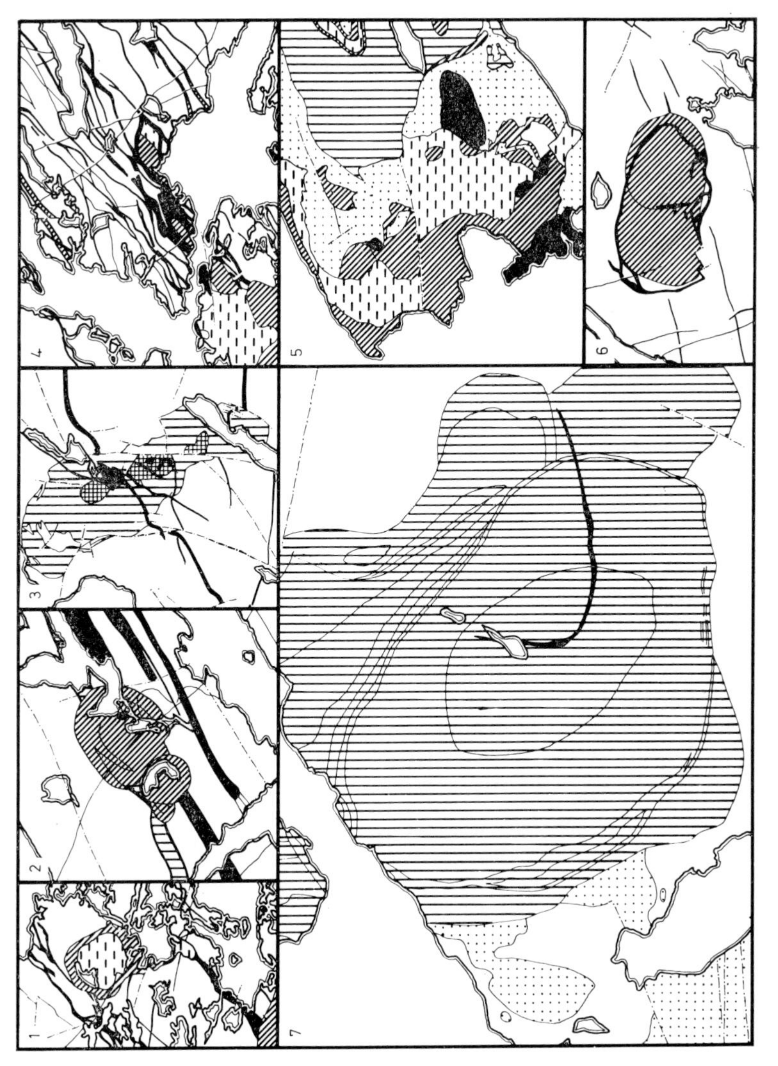

Fig. 2 Simplified geological maps of some of the Gardar intrusive centres
(1) Puklen; (2) Central Tugtutôq; (3) Grønnedal–Ika; (4) Bangs Havn; (5) Dyrnaes–Narssaq and part of Ilímaussaq; (6) Kûngnât Fjeld, (7) Igdlerfigssalik.

normative acmite ± sodium metasilicate. Since there is a fair degree of correlation between the composition and assemblages of the phenocrysts in the dyke rocks and those of the inferred cumulus minerals in the layered intrusions, it is assumed that there was a close identity between the magmas responsible in both situations (cf. Upton, Thomas and Macdonald, 1971).

IV.3.3.3. Alkali–Silica Diagram

The composition of some 300 analysed igneous rocks from the province are plotted on an alkali–silica diagram (Fig. 3) in which no attempt has been made to discriminate between accumulitic and non-accumulitic materials. The diagram has been contoured according to the amounts of normative nepheline or quartz (with a small field for hypersthene normative rocks with neither *ne* nor *qz*) and according to their differentiation-index (Thornton and Tuttle 1960). The data define an anvil-shaped field (outlined in Fig. 4), showing the low temperature divergence away from trachytic compositions towards the respective undersaturated and oversaturated cotectics of the 'residua system'. While magmatic evolution beyond the trachytic stage is believed to have been mainly dependent on fractionation of alkali feldspar in the range Or_{25-40}, the more extreme alkalic compositions reflect differing degrees of nepheline and/or sodalite accumulation.

IV.3.3.4. Basalt–Trachyte Succession

Some of the basic rocks are tholeiitic, including some at least of the BDO swarm and some N.E.-trending dykes occurring well away from the main alkaline centres. The basic rocks typically associated with the alkaline intrusions have relatively high alumina, titania, phosphorus, alkalis and Fe/Mg values, with MgO generally below 7%. The compositions plot close to the critical plane of undersaturation in the 'simple basalt system' *ne–ol–cpx–qz* (Yoder and Tilley, 1962) and near to the plagioclase–olivine join, i.e. the bulk of the basaltic magmas were of an aluminous mildly alkalic or 'transitional' (Coombs, 1963) variety. The sequence from basalt to trachyte is, however, typically alkalic in that there is evidence neither for an olivine/liquid reaction relationship nor (with rare exceptions) for any low calcium pyroxene. Olivines range from *c.* Fa_{30-100}, becoming unstable only in the more extreme differentiates. Since the assemblage fayalite + magnetite + alkali feldspar ± quartz is common in the syenitic rocks, the oxygen fugacity of the magmas was probably approximated by the fayalite–magnetite–quartz buffer curve (Morse, 1968).

Volatile concentration in the more extreme residual magmas (e.g. the Ilímaussaq agpaites and the peralkaline granites of Nunarssuit and Kûngnât Fjeld) allowed formation of complex pegmatite–aplite bodies and wall–rock metasomatism. With the exception of some of the volatile rich differentiates, however, the majority of the felsic magmas crystallized under hypersolvus conditions (with respect to the feldspars, cf. II.2).

IV.3.3.5. The 'Big Feldspar Dykes'

The so-called 'big feldspar dykes' constitute an important element in the E.N.E./N.E. dyke swarms. They post-date the main dolerite/gabbro intrusions while preceeding the late sinistral movements on the easterly transcurrent faults, the emplacement of the latest alkaline complexes and the main trachyte dyke swarms. The big feldspar dykes are of great interest in that their chilled marginal facies are typically trachytic or mugearitic, grading inwards towards more basic centres. Megacrysts of plagioclase and inclusions of anorthosite, absent from the alkalic marginal zones, occur in increasing concentration towards the most basic central zone, (Bridgwater, 1967, 1968, Bridgwater and Harry, 1968). Bridgwater concluded that the dykes were formed through the intrusion of an already differentiated magma column, tapped from the top downwards. He attributed the distribution of the feldspathic material through the different zones not to any form of flowage differentiation (Bridgwater 1968) but to gravitational equilibrium in a density stratified magma body, prior to the final intrusion.

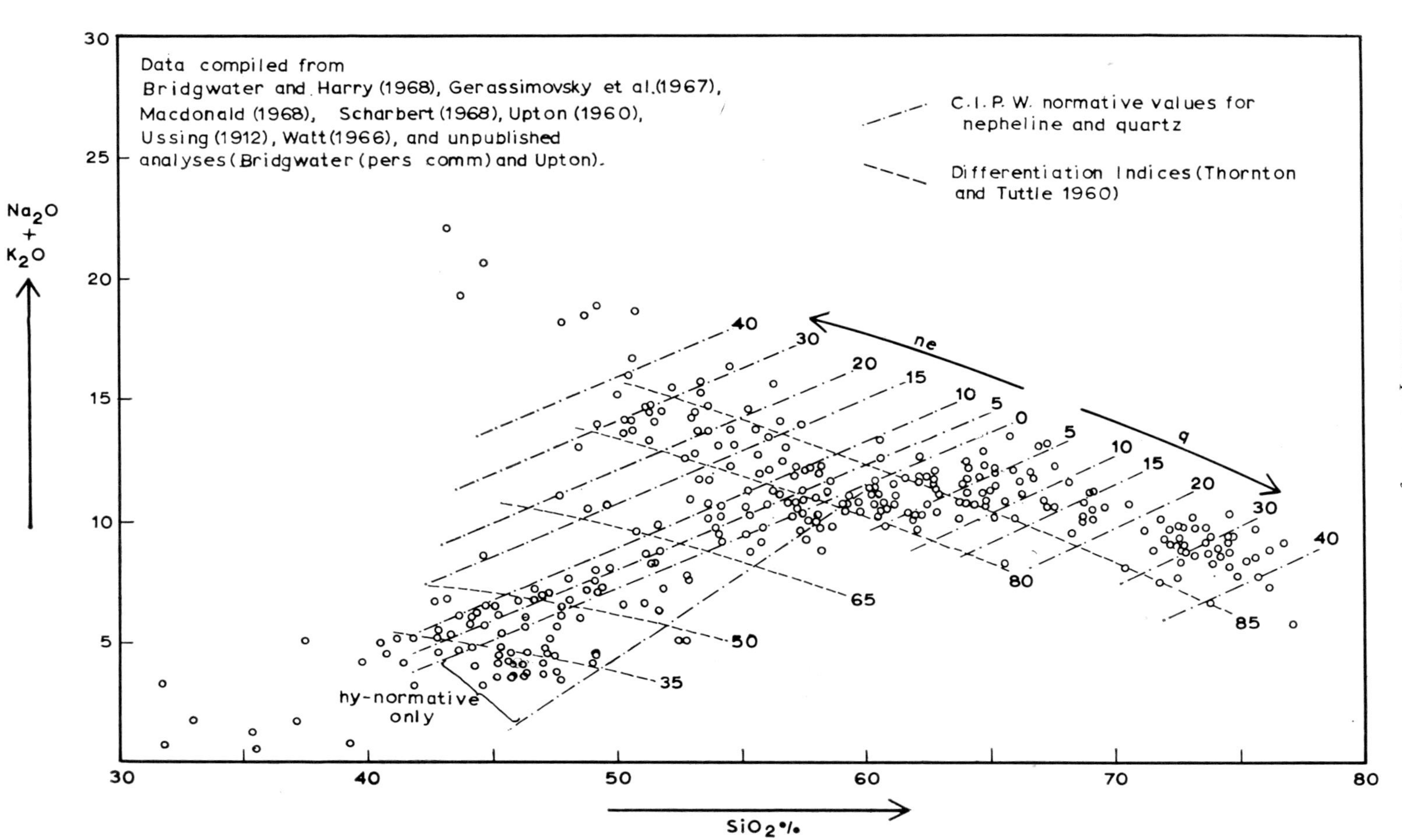

Fig. 3 Alkali–silica diagram for Gardar igneous rocks, contoured for differentiation indices and normative nepheline or quartz

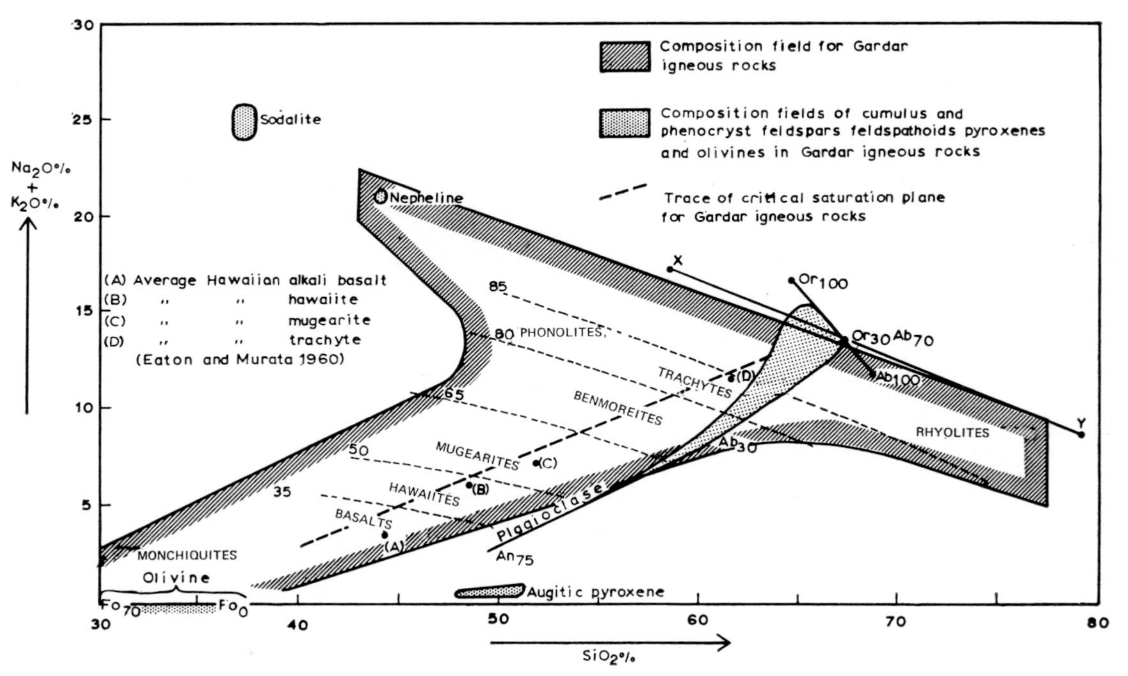

Fig. 4 Alkali–silica diagram indicating compositions of possible fractionating phases. Trace of critical undersaturation plane and differentiation index contours redrawn from Fig. 3. For comparative purposes points A–D, representing a basalt–trachyte extrusive sequence, are also shown. X and Y respectively represent the minima on the feldspar–nepheline and feldspar–quartz cotectics in the 'residua system' at 1000 kg/cm² (Hamilton and Mackenzie, 1965; Tuttle and Bowen, 1958)

IV.3.3.6. Alkaline Ultrabasic and Carbonatitic Rocks

Although quantitatively insignificant, volatile-rich ultrabasic magmas were also erupted during the Gardar magmatism. These gave rise to hypabyssal intrusions and lava flows and are associated with small gas-drilled diatremes in the eastern part of the province (Upton, 1962, Walton, 1965, Stewart, 1970). The hypabyssal intrusions include lamprophyric dykes (e.g. Ayrton, 1963, Ayrton and Burri, 1967) and miscellaneous small intrusions of mica peridotite, mica pyroxenite, monchiquite, alnoite and carbonatite (Emeleus, 1964, Walton, 1965, Upton, 1966, and Stewart, 1970). Extrusive flows of monchiquite and carbonatized melilite-rock have also been described by Stewart (1964 and 1970).

Relatively high K and Mg values in some of these Si and Al deficient rocks invite comparisons with the potassic ultramafic lavas of Uganda (cf. II.5).

IV.3.4. Emplacement of the Intrusions

IV.3.4.1. Giant Dykes

Widths of up to 800 m are attained by some of the N.E. and E.N.E. trending basic dykes of the Tugtutôq, Nunarssuit and Johann Dahl's Land swarms, (Upton 1964b, Pulvertaft 1965, and Walton 1965) whose consequent slow cooling produced coarse-grained troctolitic gabbro cumulates. Several of the dykes showing 'giant dimensions' (i.e. widths > 100 m) however also involve syenogabbros, augite syenites and foyaites or quartz syenites. These more alkalic associates commonly occur as central components within the dykes, enclosed by an envelope of slightly earlier gabbroic rocks. In some examples it appears that these alkaline rocks crystallized from relatively differentiated magmas that were injected along the median plane of a more basic precursor to produce composite dykes lacking internal chilled contacts. Nevertheless, in some of the giant-dykes, the possibility that the more differentiated rocks are the product of *in situ* fractionation cannot be excluded.

With transition from basic to less dense felsic magma, the intrusive mechanism was modified from a simple dilatational process to one in which piece-meal stoping became predominant. The importance of stoping in the emplacement of giant-dykes in the Isortoq area has been emphasized by Bridgwater and Coe (1970).

It is apparent at several localities (e.g. Isortoq, Puklen, Bangs Havn and Tugtutôq) that by continued stoping the more alkaline and felsic magmas were able to break through the sheath of gabbroic material within which they were initially confined.

IV.3.4.2. Permissive Emplacement of Larger Plutons

Further emplacement of the more differentiated magmas is believed to have taken place, at favourable localities, by persistent thermal shattering and stoping of the surrounding rocks, with the ultimate production of large ovoid chambers within which convection and crystal accumulation could occur. In short, it is proposed that a progression can be established from simple dilational basic dykes to the major, permissively emplaced, intrusions of syenite, foyaite and granite (Fig. 5). The conclusion is not to be drawn however, that all the major centres necessarily evolved in this manner, but rather that it is highly probable that some of them did and it is at least a plausible working hypothesis for the remainder.

A single large intrusion having been formed, a focus of structural weakness was automatically provided for the intrusion of further stocks or ring-dykes and, within limits, the larger the complex grew, the more likely was it that further magmatic accession would take place at that locality. Arrested stages in the progression from 'giant composite dyke' to central complex can be recognized in the Bangs Havn Complex (Harry and Pulvertaft, 1963, p. 17), the Puklen complex (Pulvertaft, 1961) and in the relationship of the Assorutit composite dyke on Tugtutôq (Upton, 1962) to the Dyrnaes–Narssaq complex. By far and away the most successful localities in this evolutionary process were those in the vicinity of Igaliko i.e., the Motzfeldt, Qôroq and Igdlerfigssalik centres (Emeleus and Harry, 1970) and

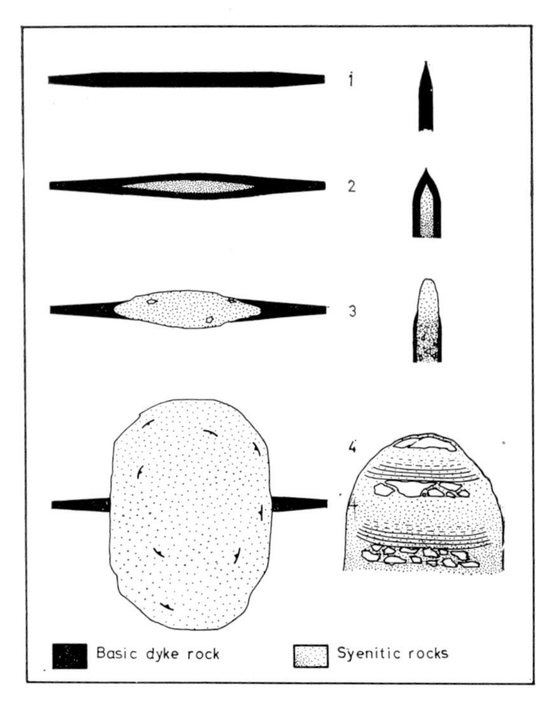

Fig. 5 Diagrammatic plans and sections indicating suggested progression from dilational dykes to major salic intrusions

Nunarssuit (Harry and Pulvertaft, 1963), where large and complex volcanoes persisted over long periods. The resulting intrusive complexes at these localities attained sizes between 30 and 45 km.

An unexplained structural peculiarity of many of the complexes is that the individual intrusions are ovoid in plan, with their long axes lying transverse to the dominant N.E./E.N.E. direction of dyke intrusion. This tendency to oriented elongation is demonstrated by the intrusions at Grønnedal–Ika, Nunarssuit, Puklen, Dyrnaes–Narssaq, Ilímaussaq and the Igdlerfigssalik and Qôroq centres, but not by those at Kûngnât, Klokken and central Tugtutôq. These last named are also anomalous in that their location does not seem to have been controlled by the intersection of the easterly left-lateral tear-faults with major dykes or dyke-swarms in the way that it does for the others.

As stated above, the general rule for the major alkaline complexes was permissive emplacement brought about by repeated cauldron-subsidence producing suites of ring-dykes (e.g. the early centres in central Tugtutôq, Upton, 1962) or by piece-meal stoping (e.g. the xenolithic foyaite of Grønnedal–Ika (Emeleus, 1963) and the main granite at Dyrnaes–Narssaq), or alternatively, by a process somewhat intermediate between these two concepts. This involved the spalling off of sub-horizontal sheets or rafts of roofing rock, with lateral dimensions several times their thickness. These sank into the underlying, lower-density, trachyte or phonolite magmas, either fragmenting as they did so (e.g. at Kûngnât Fjeld, central Tugtutôq, Nunarssuit and Motzfeld) or remaining essentially coherent (e.g. Grønnedal–Ika). Each roof-spalling event terminated one period of gravitative accumulation and convection and the sheets or rafts of inclusions provided a new base for a subsequent series of layered cumulates. In this way the larger alkaline magma bodies may have worked their way towards the surface, diminishing in volume and becoming increasingly fractionated as crystallization proceeded.

IV.3.4.3. Structural Level

The complexes occurring in the west of the province, namely Kûngnât Fjeld, Grønnedal–Ika and Nunarssuit, are seen at rather lower structural levels than those of the east. The main reason for this assertion is that the eastern complexes are either seen to be in contact with supracrustal rocks of the Eriksfjord Formation, or else contain relatively unmetamorphosed inclusions derived from this formation (as in the central Tugtutôq complex). Metavolcanic inclusions within the Kûngnât Fjeld and Nunarssuit syenites are also believed to be derived from Gardar lava sequences but the level at which these complexes are exposed must lie well below the unconformity that separated these sequences from the underlying migmatites and gneisses.

Perhaps the most informative, from a structural point of view, of the central complexes is Ilímaussaq. The salient features of this remarkable complex have been reviewed by Sørensen (1958), Wager and Brown (1968) and Ferguson (1970). The Ilímaussaq complex is crossed by the two north-easterly trending faults, both down-throwing towards the north-west so that erosion reveals it at three structural levels. The roof zone is seen to the north-west in the Ilímaussaq peninsula where part of the original lava cover is preserved (Sørensen *et al.*, 1969). The depth of the magma chamber is estimated by Sørensen (1969) to have lain between 2 and 5 km below the surface.

The dominant rock-type in the two north-western fault blocks is naujaite, (cf. II.2.2.5), an extraordinary cumulate whose formation has been attributed to uprise of sodalite (hackmanite) crystallizing from a peralkaline phonolitic magma, exceptionally enriched in chlorine. In the south-eastern block, where the erosion level lies below the basal Gardar unconformity, revealing the contacts against late-kinematic basement granites, over 400 m of an equally remarkable layered syenite (kakortokite) succession is exposed. The kakortolites are regarded as part of a cumulate succession that built up from the intrusion floor while the naujaites were forming beneath the roof. An intensely fractionated suite of lujavrites (again at least partly of accumulitic origin, pers. comm., J. Engell) represents the latest of the undersaturated alkaline rocks at this centre and is considered to have formed from residual magma following the separation of the naujaites and kakortokites.

Heat-flow data from Ilímaussaq (Sass *et al.*, 1972) indicate that the alkaline rocks (containing high concentrations of heat-producing isotopes) are superficial and are probably underlain at 1–2 km depth by more mafic rocks deficient in K, Th and U.

IV.3.4.4. Basic Ring-Dykes

The requisite conditions for cone-sheet formation appear to have been at no time satisfied in the emplacement of the complexes. Ring-dykes composed of basic rocks are rare but instances of thin gabbroic or syenogabbroic ring-dykes are to be found at Kûngnât Fjeld (Upton, 1960), Motzfeldt and Igdlerfigssalik (Emeleus and Harry, 1970). These late-stage ring-dykes are illuminating in that they provide evidence for the persistence of basic magma beneath these otherwise wholly syenitic complexes. Furthermore the Kûngnât Fjeld and Igdlerfigssalik cases indicate that this magma was still available at the very latest stages of the Gardar activity, when basaltic magma had long since ceased to ascend in the form of regional dykes.

IV.3.4.5. Forcible Intrusion

Clear-cut examples of forcibly intruded magmas are rare in the province. Possible examples include the carbonatite bodies, the largest of which is the søvite plug that cuts the Grønnedal–Ika foyaites (Emeleus, 1963), and the still smaller siderite–cryolite ore body at Ivigtût (Pauly, 1960). The latter is intrusive into a small peralkaline granite stock: joint patterns in the surrounding rocks suggest that room was made for the ore body by an expansion of explosive violence (Berthelsen, 1962).

IV.3.5. Magmatic Sequences of Abundances

IV.3.5.1. Order of Intrusion

With the important exception of the Ilímaussaq complex which includes both under- and over-saturated rocks, the magmas intruded at any one centre or group of centres were generally already fully committed to either the phonolitic or the rhyolitic side of the 'residua system' at the time of intrusion. Grønnedal–Ika and the coalescing complexes in the east of the province consist wholly of undersaturated rocks (mainly foyaites), while the Kûngnât Fjeld, Nunarssuit, Puklen, Central Tugtutôq, Klokken and Dyrnaes–Narssaq complexes are composed of saturated or over-saturated syenites and granites (although small quantities of nepheline may appear in the norms of some of the more basic syenites from these complexes).

With the principal exception of the 'big feldspar dykes', described in Section 3.5, and the late basic ring-dykes already cited, there is a persistent tendency throughout the province for more differentiated magmas to follow after the less differentiated. As already noted there is evidence from the dyke swarms that those of basaltic composition dominate the early swarms while trachytes, phonolites and rhyolites characterize the later ones. Superimposed on this 'long term drift', it is often possible to recognize shorter term cycles showing the same evolutionary pattern. It is only close to the top of the plateau lava series on the Ilímaussaq peninsula that trachytes appear (Stewart, 1964), and in the composite dykes (other than the 'big feldspar dykes') the younger components typically have higher alkalis and iron/magnesium ratios than the early components. The same behaviour is demonstrable in many of the intrusive complexes. Clear instances are provided at Dyrnaes–Narssaq and Ilímaussaq. At the former, oversaturated augite syenites and nordmarkites are succeeded by peralkaline granite. The syenites themselves are approximate age equivalents of the Assorutit syenite in the Tugtutôq 'giant-dyke' assemblage. This syenite post-dates a syenogabbro body which cuts the main gabbros (Upton, 1962, 1964b). At the adjacent and younger Ilímaussaq complex, an undersaturated intrusion that produced the early augite syenites and foyaites was followed by intrusion of magma that yielded the agpaite suite (i.e. naujaite, kakortokite and lujavrite). All the geochemical and mineralogical data indicate this agpaitic magma to have been richer in alkalis, volatiles and rare elements, i.e. to have been a more differentiated magma than its predecessor. According to Hamilton's interpretation (1964) crystallization of the agpaites was followed by successive pulses of magma which produced first the quartz syenites and then the Ilímaussaq arfvedsonite granite in what was essentially a repetition of the much earlier Dyrnaes–Narssaq sequence.

IV.3.5.2. Relative Abundances

Augite syenites of larvikitic affinities form important masses in almost all the complexes, other than Grønnedal–Ika and central Tugtutôq, and also figure prominantly in several of the composite giant dykes. With appearance of modal quartz or nepheline the augite syenites grade respectively into quartz syenites or foyaites. In spite of the comparative abundance of the augite syenites and more salic rocks, rocks compositionally intermediate between the augite syenites and the gabbros are rarely seen in the major complexes and, when they are, it is as relatively insignificant bodies. Trachybasaltic and syenogabbroic rocks are common among the dyke swarms but here too there is some suspicion that they are scarce in relation to the basaltic dykes on the one hand and the trachytes, phonolites and rhyolites on the other. The histograms in Fig. 6 relate frequency to (i) silica content and (ii) Thornton–Tuttle differentiation index, for analysed Gardar igneous rocks. The minima apparent in these at SiO_2 50–55% and T.T. Index 60–65 correspond to the trachybasaltic/syenogabbroic compositions and, following the arguments of Chayes (1963), the minima are believed to be of geological significance and not to merely reflect selective bias among the workers responsible for the data collection. Another minimum, in the histogram for silica, at SiO_2 65–70% is also apparent (cf. Watt, 1966). This relative shortage of compositions intermediate between quartz syenites and granites

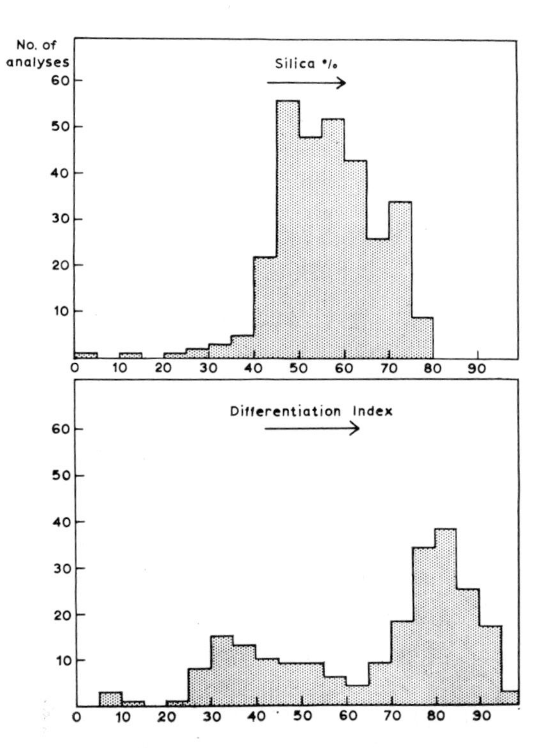

Fig. 6 Histograms relating number of analyses to (a) SiO_2% and (b) Thornton–Tuttle differentiation index. (Data from same sources as listed in Fig. 3)

appears to hold true for both the dyke swarms and the major complexes (Upton, Thomas and Macdonald, 1971). Other minima may be anticipated among the basic/ultrabasic compositions but existing data are too scanty for any definite conclusions to be drawn at present.

IV.3.6. Feldspathic Inclusions and Anorthosites

IV.3.6.1. Cognate Nature of Inclusions

Bridgwater, in Bridgwater and Harry (1968), noted that there is a correlation between the concentration of dykes carrying feldspar megacrysts and anorthositic inclusions and the areas showing strong alkalic activity. From a detailed study of the relationships between the megacrysts, xenoliths and the various Gardar intrusives, Bridgwater presented a case for an intimate genetic connection between the anorthosites and the Gardar magmas as a whole and he stressed the petrographic similarities linking the Gardar xenoliths with the anorthosite bodies outcropping in the Nain and Michikamau regions of Labrador.

IV.3.6.2. Suspected Anorthosite at Depth

The distribution of the xenoliths throughout the province led Bridgwater (Bridgwater and Harry, 1968) to postulate the existence at depth, of an elongate parental body, some 250–500 km long by 50–100 km broad. Furthermore, compositional studies of the xenolithic rocks indicated the principal component of this body to be granular anorthosite composed of labradorite (An_{60}), with subordinate olivine, Fe–Ti oxides, ortho- and clinopyroxenes. The presence of occasional well-layered xenoliths was taken as evidence that these granular anorthosites are locally associated with layered anorthositic gabbros which, by analogy with anorthosites in Labrador and Angola, may overlie the granular material. Bridgwater considered that the formation of the anorthosites may have been proceeding through the early and middle Gardar period and, in order to explain the virtual confinement of the granular anorthosite xenoliths to mid-Gardar dykes, he proposed that a change in the tectonic conditions at this time caused the brecciation of the parent body.

Bridgwater's principal deductions; (*a*) that a large anorthosite complex (or possibly complexes) underlies much of the province and (*b*) that any hypothesis purporting to explain the genesis of the Gardar magmas must be capable of embracing the formation of the anorthosite, appear irrefutable.

IV.3.7. Petrogenesis of Gardar Magmas

IV.3.7.1. Basalts and Anorthosites

During the period 1300–1000 m.y. B.P., a large-scale thermal event affected North America, Greenland and Northern Europe, resulting in widespread volcanism and crustal re-adjustments. The primary magma generated beneath the Gardar province underwent fractionation at depth to produce the mildly alkaline basalts that erupted as flows and high level intrusives and which were parental to most of the alkaline magmas. Variable degrees of clinopyroxene fractionation at depths between 60 and 15 km may have been the principal agency by which the high alumina, Fe/Mg and K + Na/Ca characteristics of the basalts were acquired (Upton, 1971). Periodic release of basaltic magma in response to crustal rupture, gave the main dyke swarms and the fissure-fed lava plateaux. Disruption and elutriation of the anorthositic cumulates attended the ascent of the magma. The granulation and cataclasis which distinguishes so much of the xenolithic Gardar anorthosite, may have resulted from tectonic disturbances operating within the uppermost mantle and lower crust while the anorthositic cumulates were forming.

IV.3.7.2. Alkaline Felsic Rocks

$Sr^{87/86}$ initial ratios of syenitic, granitic and foyaitic rocks from the Kûngnât Fjeld, central Tugtutôq and Hviddal intrusives fall within the range 0·684–0·709 (van Breemen and Upton, in press). Although rheomorphism or assimilation of crustal rocks may have played a significant rôle in the genesis of some of the Gardar felsic intrusives, the relatively low 87/86 values for the intrusives investigated so far suggest that these processes were relatively unimportant. On the basis of present data, generation of the Gardar trachytes, phonolites and comendites by crystal fractionation from parental basalt magmas remains the most plausible hypothesis.

It is suggested that the felsic magmas were the liquid residua complementary to the anorthositic cumulates referred to above. On this basis the peralkaline character of Ilímaussaq, for example, is the result of persistent operation of the 'plagioclase effect' to which Bowen (1945) drew attention. The theory that anorthosite formation may result in production of alkaline rocks has been discussed by Yoder (1970) and is strongly indicated by the close association of syenites and anorthosites in such complexes as Adirondacks (Silver, 1969) and Michikamau (Emslie, 1970). Although the supposition that the alkaline felsic magmas owe so much of their chemical constitution to plagioclase fractionation could be tested by investigation of the rare earth element distribution patterns, this work remains to be done.

Further modification at low pressures (< 10 kb) by fractionation along feldspar–olivine–liquid and feldspar–olivine–clinopyroxene–liquid (and lower temperature) equilibria may have produced the broad range of silica over- and under-saturated felsic magmas. The basaltic parental magmas tended to remain critically undersaturated throughout much of this low pressure evolution, becoming diverted towards phonolitic or comenditic termini only after the magmas had attained mugearitic or trachytic compositions.

Studies on the silica-oversaturated rocks of Kûngnât Fjeld, central Tugtutôq and the Tugtutôq dyke swarms (Macdonald and Edge, 1970, and Upton, Thomas and Macdonald, 1971) led to the conclusion that alkali loss (by metasomatism, volatile loss and devitrification of glasses) took place to variable extents. By implication, some of the highly fractionated Gardar granites and rhyolites containing riebeckite, aegirine, aenigmatite and astrophyllite, may have crystallized from thoroughly peralkaline magmas (comparable to modern glassy pantellerite), but with excess alkalis lost during, or subsequent to, crystallization.

As outlined in Section 3.5, the 'big feldspar dykes' provide evidence for the existence of differentiated magma bodies during the Gardar activity. These are construed as having been bodies within which alkalis and silica became concentrated upwards to give mugearitic or trachytic compositions in the apical parts. Whether this process operated through the agency of crystal fractionation in vertically extended magma bodies possessing strong temperature and pressure gradients, or whether a diffusive mechanism was involved, remains uncertain.

The relative scarcity of rocks with differentiation indices between 60 and 65 (Section 5.2 above) has a parallel in the 'Daly-gap' common in oceanic lava sequences (e.g. Chayes, 1963, Bryan, 1964, and Cann, 1968). Possibly the explanation lies in the magmas of this (mugearitic) composition range having had viscosities higher than (i) the hotter and less siliceous basalts on the one hand and (ii) the cooler trachytic magmas on the other, in which the effects of lower temperature and higher silica may have been effectively counteracted by the higher volatile contents. As noted in Section 3.1, the abundance of layered structures in the alkaline complexes indicates relatively low viscosities for the salic magmas.

IV.3.7.3. Alkaline Ultrabasic and Carbonatitic Magmas

Periodic small-scale eruption of highly undersaturated magmas produced the mica pyroxenites, alnöites, carbonatites etc. referred to in Section 3.6. These Si and Al deficient cafemic magmas appear to have been relatively enriched in K, P, Ti and other so-called incompatible elements, and to have had comparatively high K/Na ratios.

Such chemical characteristics could be explained on the supposition that the magmas represent initial partial melt liquids ($< 5\%$) in the mantle that were erupted with little or no further modification. Such an hypothesis would be in accord with the suggestions regarding genesis of olivine melilitite and kimberlite recently proposed by Green (1971) and Dawson (1972) respectively. Alternatively these characteristics might have been acquired through a process of wall-rock reaction such as envisaged by Green and Ringwood (1967). A third proposal, following O'Hara and Yoder (1967), is that such magmas were the produce of extended fractionation of basalt under eclogite facies conditions (>20 kb).

The 'initial melt' and 'wall-rock reaction' models would account for formation of such magmas at the beginning of the volcanism but would encounter problems in explaining their genesis after the main eruption of Gardar basalts and dolerites—a stage at which the upper mantle was presumably strongly depleted in incompatible elements. The eclogite-fractionation model avoids this objection and is attractive in that it offers a mechanism whereby magmas retained at depth (> 60 km) fractionate to variable extents so that incompatible-enriched residua were available for eruption at irregular intervals throughout the period of volcanism.

IV.3.8.1. Late Precambrian Continental Rifting

Similarities between the late Precambrian Gardar activity in Greenland, and the Tertiary events associated with the East African Rifts have been remarked upon by a number of authors. Stewart (1970) for example, has drawn attention to the similarity of volcanic activity around the Qagssiarssuk area with that around Ruwenzori in the African western rift. The narrow zone of intense olivine basalt–trachyte–comendite dyking in the Tugtutôq region may represent a deeply dissected rift-zone analogous to the Plio-Pleistocene rift-confined volcanics of similar composition seen in southern Kenya (Wright, 1965).

As outlined in Section 2.1, the region was subjected to tensional stresses involving (i) movements along conjugate faults and (ii) fissuring parallel or sub-parallel to the foliation of basement rocks. The fissuring was associated with magma ascent and a crustal dilation in excess of 10 km. It is not yet clear whether the Gardar rifting and volcanism was associated with the same type of up-arching as seen in younger continental rift environments.

Probably the alkaline province of south-west Greenland was but one sector of a greatly extended zone of penecontemporaneous rifting which included the Keweenawan basin of western Lake Superior (Smith *et al.*, 1966; Phinney, 1970). As suggested by Harris (1969) apropos of the East African Rift system, the degree of silica saturation of the basalts erupted may show a positive correlation with the amount of crustal extension involved. The tholeiitic mid-Keweenawan lavas may thus denote a zone of greater extension than that of southern Greenland. In both areas, crustal separation during the interlude 1300–1000 m.y. B.P. stopped short of creation of new oceanic crust.

IV.3.8.2. Geochemical Implications

The igneous rocks of the province tend to be not only relatively (and often absolutely) rich in iron, but also to be strongly titaniferous. (The average TiO_2 for 77 analyses with > 3% MgO is 3·25%.) While concentrations of Zr in both under- and oversaturated-felsic rocks are among the highest known for terrestrial rocks (cf. Macdonald and Parker, 1970), striking concentrations of Li, Be, F, Cl, Nb, Th and U have also been reported from various parts of the province.

Although crystal fractionation must have played a major rôle in producing these geochemical characteristics, the possibility that migration of volatiles and other 'mobile elements' occurred within the mantle to become concentrated within a zone of lower pressure or tension thereby facilitating magmatism (Bailey 1970), should not be overlooked. In other words the primary Gardar magmas themselves may already have been relatively enriched in these components. Whatever the mechanisms involved, it may be concluded that the upper mantle underwent a thorough and highly selective depletion of rare elements during the evolution of the south-west Greenland alkaline province.

ACKNOWLEDGMENTS

This chapter has been prepared with the permission of the Director of the Geological Survey of Greenland (Grønlands Geologiske Undersøgelse) which has been responsible for the mapping in south-west Greenland as well as much of the subsequent research. The author wishes to thank D. Bridgwater, K. G. Cox, C. H. Emeleus, M. J. O'Hara and T. C. R. Pulvertaft for their criticisms and cooperation in the preparation of the manuscript.

IV.3. REFERENCES

Ayrton, S. N., 1963. A contribution to the geological observations in the region of Ivigtût, S.W. Greenland. *Bull. Grønlands geol. Unders*, **37**, 139 pp. (also *Meddr. Grønland*, Bd. **167**, No. 3).

Ayrton, S. N., and Burri, M., 1967. L'évolution du socle Précambrien dans la région de Qagssimiut, Groëland Meridional. *Bull. Grønlands geol. Unders*, **66**, (also *Meddr. Grønland*, Bd. **115**, No. 2).

Bailey, D. K., 1970. 'Volatile flux, heat-focussing and the generation of magma', in *Mechanism of Igneous Intrusion* (ed. Newall and Rast), Liverpool, 177–86.

Berthelsen, A., 1962. On the geology of the country around Ivigtût, S.W. Greenland. *Geol. Rundschau*, **52**, 269–79.

Bowen, N. L., 1945. Phase equilibria bearing on the origins and differentiation of alkaline rocks. *Am. J. Sci.*, **243**, A., 75–89.

Bridgwater, D., 1965. Isotopic age determinations from South Greenland and their geological setting. *Bull. Grønlands geol. Unders*, **53**, 56 pp. (also *Meddr. Grønland*, Bd. **179**, No. 4).

Bridgwater, D., 1967. Feldspathic inclusions in the Gardar igneous rocks of south Greenland and their relevance to the formation of major anorthosites in the Canadian Shield. *Can. J. Earth Sci.*, **4**, 995–1014.

Bridgwater, D., 1968. Mechanics of flow differentiation in ultramafic and mafic sills: A discussion. *J. Geol.*, **76**, 569–99.

Bridgwater, D., and Coe, K., 1970. 'The role of stoping in the emplacement of the giant dykes of Isortoq, South Greenland', in *Mechanism of Igneous Intrusion* (ed. Newall and Rast), Liverpool, 67–78.

Bridgwater, D., and Gormsen, K., 1969. Geological reconnaissance of the Precambrian rocks of south-east Greenland. *Geol. Surv. Greenland, Rep.*, **19**, 43–50.

Bridgwater, D., and Harry, W. T., 1968. Anorthosite xenoliths and plagioclase megacrysts in Precambrian intrusions of south Greenland. *Bull. Grønlands geol. Unders*, **77**, 243 pp. (also *Meddr. Grønland*, Bd. **185**, No. 2).

Bryan, W. B., 1964. Relative abundance of intermediate members of the oceanic basalt–trachyte association; Evidence from Clarion and Socorro Islands, Revillagigedo Islands, Mexico. *J. Geol. Res.*, **69**, 14, 3047–9.

Cann, J. R., 1968. Bimodal distribution of rocks from volcanic islands. *Earth Planet. Sci. Lett.*, **4**, 479–80.

Chayes, F., 1963. Relative abundance of intermediate members of the oceanic basalt–trachyte association. *J. Geol. Res.*, **68**, No. 5, 1519–34.

Coombs, D. S., 1963. Trends and affinities of basaltic magmas and pyroxenes as illustrated on the diopside–olivine–silica diagram. *Min. Soc. Amer. Spec. Pap.*, **1**, 227–50.

Dawson, J. B., 1972. Kimberlites and their relation to the mantle. *Phil. Trans. Roy. Soc. Lond.*, A, **271**, 297–311.

Eaton, J. P., and Murata, K. J., 1960. How volcanoes grow. *Science*, **132**, No. 3432, 925–38.

Emeleus, C. H., 1964. The Grønnedal–Ika alkaline complex, south Greenland. *Bull. Grønlands geol. Unders*, **45**, 75 pp. (also *Meddr. Grønland*, Bd. **172**, No. 3).

Emeleus, C. H., and Harry, W. T., 1970. The Igaliko nepheline syenite complex, south Greenland. General description. *Bull. Grønlands geol. Unders*, **85**, 115 pp. (also *Meddr. Grønland*, Bd. **172**, No. 3).

Emslie, R. F., 1970. The geology of the Michikamau intrusion. *Geol. Surv. Can. Pub.*, paper 68–57.

Ferguson, J., 1964. Geology of the Ilímaussaq alkaline intrusion, south Greenland. *Bull. Grønlands geol. Unders*, **39**, 82 pp. (also *Meddr. Grønland*, Bd. **172**, No. 4).

Ferguson, J., 1970. The differentiation of agpaiitic magmas: The Ilímaussaq intrusion, south Greenland. *Can. Miner.*, **10**, 335–49.

Ferguson, J., and Pulvertaft, T. C. R., 1963. Contrasted styles of layering in the Gardar province of south Greenland. *Min. Soc. Amer. Spec. Pap.*, **I.** (I.M.A.) 10–21.

Gerassimovsky, V. I., and Kuznetsova, S. Ya., 1967. On the petrochemistry of the Ilímaussaq intrusion, south of Greenland. *Geochem. Int.*, **4**, No. 2, 236–46.

Gill, R. C. O., 1971. Chemistry of peralkaline phonolite dykes from the Grønnedal-Íka area, south Greenland. *Contr. Miner. Petrology*, **34**, 87–100.

Green, D. H., 1971. Composition of basaltic magmas as indicators of conditions of origin: application to oceanic volcanism. *Phil. Trans. Roy. Soc. Lond.*, A **268**, 707–25.

Green, D. H., and Ringwood, E. A., 1967. The genesis of basaltic magmas. *Contr. Miner. Petrology*, **15**, 103–90.

Hamilton, E. I., 1964. The geochemistry of the northern part of the Ilímaussaq intrusion, south Greenland. *Bull. Grønlands geol. Unders*, **42**, 104 pp. (also *Meddr. Grønland*, Bd. **162**, No. 10).

Hamilton, D. L., and MacKenzie, W. S., 1965. Phase-equilibrium studies in the system $NaAlSiO_4$ (nepheline)–$KAlSiO_4$ (kalsite)–SiO_2–H_2O. *Mineralog. Mag.*, **34**, (Tilley Vol.) 214–31.

Harris, P. G., 1969. Basalt type and African rift valley tectonism. *Tectonophysics*, **8**, 427–36.

Harry, W. T., and Emeleus, C. H., 1960. Mineral layering in some granite intrusions of S.W. Greenland. *Int. Geol. Congr. 21st. session. Norden*, 172–81.

Harry, W. T., and Pulvertaft, T. C. R., 1963. The Nunarssuit intrusive complex, south Greenland, Pt 1. General description. *Bull. Grønlands geol. Unders*, **36**, 136 pp. (also *Meddr. Grønland*, Bd. **169**, No. 1).

Henriksen, N., 1960, Structural analysis of a fault in south-west Greenland. *Bull. Grønlands geol. Unders*, **26**, 40 pp. (also *Meddr. Grønland*, Bd. **162**, No. 9).

Macdonald, R., 1969a. The petrology of alkaline dykes from the Tugtutôq area, south Greenland. *Bull. geol. Soc. Denmark*, **19**, 256–81.

Macdonald, R., 1969b. Mid-Gardar feldspathoidal dykes in the Tugtutôq region, south Greenland. *Bull. geol. Soc. Denmark*, **20**, 64–6.

Macdonald, R., and Edge, R. A., 1970. Trace element distribution in alkaline dykes from the Tugtutôq region, south Greenland. *Bull. geol. Soc. Denmark*, **20**, 38–58.

Macdonald, R. and Parker, A., 1970. Zirconium in alkaline dykes from the Tugtutôq region, south Greenland. *Bull. geol. Soc. Denmark*, **20**, 59–63.

Moorbath, S., Webster, R. K., and Morgan, J. W., 1960. Absolute age-determinations in south-west Greenland. *Bull. Grønlands geol. Unders*, **25**, 14 pp. (also *Meddr. Grønland*, Bd. **169**, No. 9).

Morse, S. A., 1968. Syenites. *Carnegie Inst. Wash. Yb.*, **67**, 112–20.

Phinney, W. C., 1970. Chemical relations between Keweenawan lavas and the Duluth Complex, Minnesota. *Bull. geol. Soc. Am.*, **81**, 2487–96.

Sass, J. H., Nielsen, B. L., Wollenberg, H. A., and Munroe, R. J. (1972). Heat flow and surface radioactivity at two sites in southern Greenland. *J. Geophy. Research*, **77**, 6435–44.

Silver, L. T., 1969. A geochronologic investigation of the anorthosite complex. Adirondack Mountains, New York. *N. Y. state Museum and Sci. Service Mem.*, **18**, 233–51.

Smith, T. J., Steinhart, J. S., and Aldrich, L. T., 1966. Crustal structure under Lake Superior, in *The Earth Beneath the Continents.* (ed. Steinhart and Smith). A. G. U. Geophys. Monograph **10**, 181–97.

Stewart, J. W., 1964. *The earlier Gardar igneous rocks of the Ilímaussaq area, south Greenland.* Unpub. Ph.D. thesis Durham.

Stewart, J. W., 1970. Precambrian alkaline–ultramafic/carbonatite volcanism at Qagssiarssuk, south Greenland. *Bull. Grønlands geol. Unders*, **84**, 70 pp. (also *Meddr. Grønland*, Bd. **186**, No. 4).

Sørensen, H., 1958. The Ilímaussaq batholith. A review and discussion. *Bull. Grønlands geol. Unders*, **19**, 48 pp. (also *Meddr. Grønland*, Bd. **162**, No. 3).

Sørensen, H., 1967. On the history of exploration of the Ilimaussaq alkaline intrusion, south Greenland. *Bull. Grønlands geol. Unders*, **68**, 32 pp. (also *Meddr. Grønland*, Bd. **181**, No. 3).

Sørensen, H., 1969. Rhythmic layering in peralkaline intrusions. An essay review on Ilímaussaq (Greenland) and Lovozero (USSR). *Lithos*, **2**, 261–83.

Thornton, C. P., and Tuttle. O. F., 1960. Chemistry of igneous rocks. I. Differentiation Index. *Am. J. Sci.*, **258**, 664–84.

Tuttle, O. F., and Bowen, N. L., 1958. Origin of granite in the light of experimental studies in the system $NaAlSiO_3O_8$–$KAlSiO_3O_8$–SiO_2–H_2O. *Geol. Soc. Am., Mem.*, **74**, 145 pp.

Upton, B. G. J., 1960. The alkaline igneous complex of Kûngnât Fjeld, south Greenland. *Bull. Grønlands geol. Unders*, **27**, 145 pp. (also *Meddr. Grønland*, Bd. **123**, No. 4).

Upton, B. G. J., 1962. Geology of Tugtutôq and neighbouring islands, south Greenland. Pt. I. *Bull. Grønlands geol. Unders*, **34**, 59 pp. (also *Meddr. Grønland*, Bd. **169**, No. 8).

Upton, B. G. J., 1964a. Geology of Tugtutôq and neighbouring islands, south Greenland, Pt. II. Nordmarkitic syenites and related igneous rocks. *Bull. Grønlands geol. Unders*, **44**, 62 pp. (also *Meddr. Grønland*, Bd. **169**, No. 2).

Upton, B. G. J., 1964b. The geology of Tugtutôq and neighbouring islands, south Greenland. Pt. III. Olivine gabbros, syeno-gabbros and anorthosites. Pt. IV. The nepheline syenites of the Hviddal composite dyke). *Bull. Grønlands geol. Unders*, **48**, 80 pp. (also *Meddr. Grønland*, Bd. **169**, No. 3).

Upton, B. G. J., 1966. Ultrabasic intrusives from Narssaq and Tugtutôq. *Geol. Surv. Greenland, Rep.*, **11**, 41–4.

Upton, B. G. J., 1971. Melting experiments on chilled gabbros and syenogabbros. *Carnegie Inst. Wash. Yb.*, **70**, 112–18.

Upton, B. G. J., Thomas, J. E., and Macdonald, R., 1971. Chemical variation within three alkaline complexes in south Greenland. *Lithos*, **4**, 163–84.

Ussing, N. V., 1912. Geology of the country around Julianehaab, Greenland. *Meddr. Grønland*, Bd. **138**, 426 pp.

Van Breemen, O., and Upton, B. G. J. (1972). The age of some Gardar Intrusive Complexes, south Greenland. *Bull. geol. Soc. Am.*, **83** 3381–90.

Wager, L. R., and Brown, G. M., 1968. *Layered Igneous Rocks*. 588 pp., Oliver and Boyd Ltd., Edinburgh.

Walton, B., 1965. Sanerutian appinitic rocks and Gardar dykes and diatremes, north of Narssarssaq, south Greenland. *Bull. Grønlands geol. Unders*, **57**, 66 pp. (also *Meddr. Grønland*, Bd. **179**, No. 9).

Watt, W. S., 1966. Chemical analyses from the Gardar igneous province, south Greenland. *Geol. Surv. Greenland, Rep.*, **6**, 92 pp.

Watt, W. S., 1968. Petrology and geology of the Precambrian Gardar dykes on Qaersuarssuk, south Greenland. *Geol. Surv. Greenland, Rep.*, **14**, 50 pp.

Wegmann, C. E., 1938. Geological investigations in southern Greenland. Pt. I. *Meddr. Grønland*, Bd. **113**, No. 2, 148 pp.

Wright, J. B., 1965. Petrographic sub-provinces in the Tertiary to Recent volcanics of Kenya. *Geol. Mag.*, **102**, 541–57.

Yoder, H. S., 1970. Experimental studies bearing on the origin of anorthosite. *N. Y. State Museum and Sci. Service, Mem.*, **18**, 13–22.

Yoder, H. S., and Tilley, C. E., 1962. Origin of basalt magmas: An experimental study of natural and synthetic rock systems. *J. Petrology*, **3**, Pt. 3, 342–532.

IV.4 THE ALKALINE PROVINCE OF CENTRAL EUROPE AND FRANCE

W. Wimmenauer

IV.4.1. Introduction

Central Europe and France have been the theatre of intense alkaline volcanism from the upper Cretaceous up to the Pleistocene. The main activity was in the Miocene; upper Cretaceous and Pleistocene volcanism is restricted to a few districts. The region considered here extends from the western Pyrenees over the French Central Massif, southern and western Germany and Bohemia to Silesia. The length of this belt is nearly 1700 km, the maximal width approximately 350 km.

Several isolated blocks of Hercynian and older formations and the platform sediments (Trias to Pleistocene) between them form the main elements of the geological structure. The principal centres of the Cenozoic volcanism are situated in those areas, where there are Tertiary grabens and graben-like depressions in or between the Hercynian blocks (Fig. 1): the grabens of the Limagne and the Forez in the French Central Massif; the Rhine graben between the Vosges and the Black Forest; in continuation of it the Hessian depression between the Rheinisches Schiefergebirge and the Thüringer Wald; the North Bohemian Basin within the Bohemian Massif. The bulk of the volcanic material issued at the surface lies mostly not in the grabens themselves, but on the adjoining horst blocks or near to the principal dislocations. The areas of the less disturbed platform sediments are usually poorer in volcanic phenomena; examples of major volcanic districts in this location are the Rhön (Germany) and the Urach volcanic field with its 300 diatremes. The zones of alpine folding are almost devoid of younger alkaline magmatism; the intrusives of the northern Pyrenees, the teschenite–picrite associations of the western Carpathians and the quantitatively very subordinate

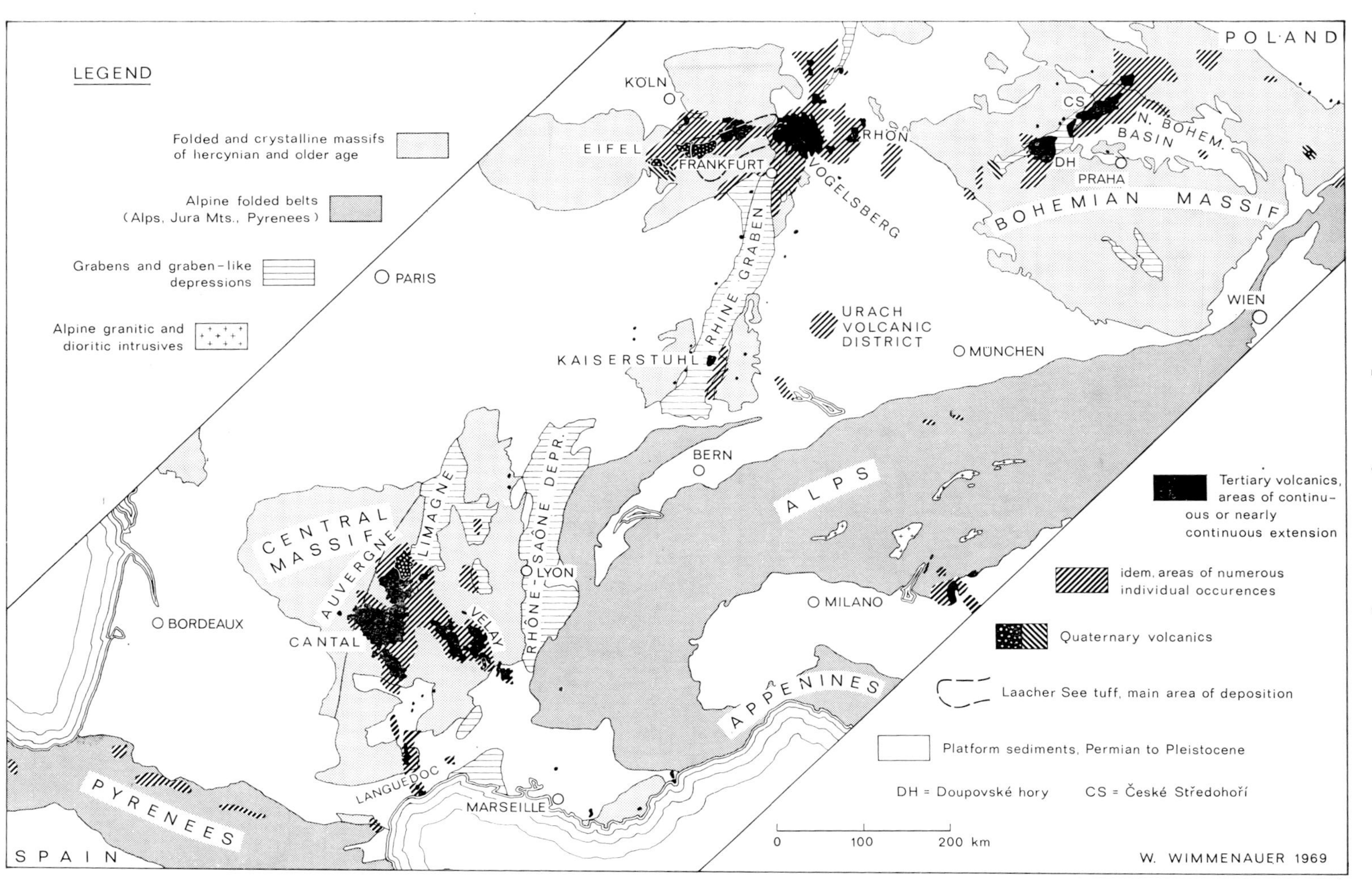

Fig. 1 Map of the Central European and French neovolcanic province

monchiquite dykes in the northern Calcareous Alps are the few exceptions to this rule.

The following descriptions of the three subprovinces will mainly concern the typical occurrences of the alkaline rocks and their interpretation. The more recent literature (from about 1950) has been consulted by preference. These publications, in turn, are based on the classic papers of the older authors, listed below:

Auvergne, Chaîne des Puys: A. Michel-Lévy (1890), A. Lacroix (1908), Ph. Glangeaud (1913), J. Jung (1946); Eifel: H. Laspeyres (1901), R. and A. Brauns (1911, 1925), G. Kalb (1934); Eifel and Westerwald: W. Ahrens (1931); Vogelsberg: W. Schottler (1937); Rhön: H. Bücking (1910); Urach volcanic field: W. Branco (1894), H. Cloos (1941); České Středohoří: F. Becke (1903), J. E. Hibsch (1926), K. H. Scheumann (1922); Younger volcanics of Central Europe: H. Jung (1928).

The last mentioned paper gives a large collection of chemical analyses and a first comprehensive view of the whole Central European volcanic province under the aspect of P. Niggli's system.

IV.4.2. The Alkaline Volcanics of France

IV.4.2.1. General Remarks

The Tertiary and Quaternary volcanic districts of France are concentrated on the Central Massif along three principal tectonic directions: N.W.–S.E., N.–S. and N.E.–S.W. (Jung and Brousse, 1962, p. 514). In the Auvergne and the Velay, volcanic rocks continuously cover several thousand square kilometres. The largest of the units, Cantal, occupies 2700 km^2 and consists of nearly 4000 km^3 of volcanic rocks.

Jung and Brousse (1962) distinguish several types of igneous rock associations, most of them of alkaline character. These 'Séries pétrographiques de réference' are:

I. calc-alkaline basalt–andesite–dacite–rhyolite;

II. olivine-poor basalt–trachyandesite–trachyte–rhyolite;

III. olivine-rich basalt–analcime tephrite or hauyne tephrite–phonolite;

IV. nepheline basanite–nepheline tephrite–phonolite;

V. nephelinite, melilitite.

Series I is developed only rarely and incompletely in France. In several districts, only one series or merely one of the basalt types are represented. In others, as Mont Dore and Cantal, two series and two basalt types occur together.

Jung and Brousse (1962) enumerate the following volcanic districts:

a. Velay (Mézenc, Mégal): Miocene to Villafranchian; alkaline olivine basalt type III–tephrite–phonolite; oceanite and ankaratrite as basic differentiates (Mergoil, 1968).

b. Basalt plateaus of Coirons, Aubrac, Devès, Margeride, Cézallier, W-Auvergne: Miocene to Villafranchian; predominant alkaline olivine basalts with subordinate tephritic differentiates (Colin, 1970).

c. Quaternary basalt flows of Cézallier and Vivarais; basalts type III.

d. Limagne: Oligocene to Villafranchian; the series alkaline olivine basalts (type III)–tephrite–phonolite is clearly developed. More undersaturated basanites and ankaramites also occur. The igneous constituents of the 'pépérites' are called trachyandesites by Michel (1953).

e. Forez and Bourgogne: Oligocene to Villafranchian; alkaline olivine basalts type III, basanites, ankaratrites and monchiquites.

f. Chaîne de la Sioule: Miocene to older Quaternary; alkaline olivine basalts type III and basanites.

g. Chaîne des Puys: Quaternary; olivine-poor basalt type II–labradorites (feldspars > mafites)–trachyandesite–trachyte (dômite).

h. Mont Dore: upper Miocene to Villafranchian; series II and III.

i. Cantal: Miocene to Villafranchian; series II and III.

k. Boutaresse: Villafranchian; basalt type II–trachyandesite–alkali trachyte.

l. Hérault (Causses, Escandorgue, Bas-Languedoc): Villafranchian to older Quaternary;

basalts of type II and III, leucite basanite (Capdevila, 1962), absarokite (Babkine, Conquéré and Vilminot, 1968), diatremes (Berger, Brousse and Causse, 1968).

The districts are of very different size. *a.* to *e.* belong to the N.W.–S.E. system, *f.* (Chaîne de la Sioule) to the N.E.–S.W. system and *g.* to *l.* to the N–S system.

m. Provence W: ankaratrites and trachybasalts (Coulon, 1967).

n. Provence E (district of Antibes): Calc-alkaline basalt–labradorite–dacite.

o. Pyrenees: upper Cretaceous; picrites, teschenites, monchiquites, nepheline syenites.

The districts *m.* to *o.* are situated outside of the Central Massif, *n.* and *o.* in close vicinity to the alpine orogenic belt.

The most prominent of the districts will be reviewed below.

For geochemistry of minor and trace elements in several rock associations of the French province see Coulomb and Goldstejn (1964), Dupuy (1965), Létolle and Kulbicki (1969, 1970) and Brousse (1967a).

IV.4.2.2. The Rock Associations of Mont Dore

The Mont Dore volcanic massif rests upon the granites and gneisses of Auvergne. It is elliptical and measures 35 km N.–S. and 16 km E.–W. The volcanics cover a surface of about 600 km^2, the volume is 200 to 250 km^3; the highest elevation attains 1886 m (Puy de Sancy). The rocks can be divided into two series (Brousse, 1961b, p. 245 f):

Saturated series (*A*): melabasalt (ankaramite)–olivine-poor basalt (type II)–mesocratic trachyandesite (doréite)–quartz-bearing latite (sancyite)–quartz-bearing trachyte–rhyolite.

Undersaturated series (*B*): olivine-rich basalt–basalt 'demi-deuil' (basalt with trachydoleritic structure)–leucocratic labradorite–leucocratic tephrite (ordanchite)–phonolite.

Chemical analyses and modes of the two basalt types, regarded as representing the parental magmas of the two series, and examples of the final differentiates, are given in Table 1. The degree of undersaturation, as also indicated by nepheline in the norm, is approximately equal in both types of basalt, but differentiation leads finally to rhyolites on the one hand, and phonolites on the other (Fig. 2). Ankaramites appear as melanocratic differentiates of series A.

The plagioclase-rich rocks of both series (basalts, 'basaltes demi-deuil', labradorites) form lava flows, whereas the phonolites and the more acidic rocks appear as intrusive plugs. Series B is comparatively poor in pyroclastics, on the other hand, ignimbritic sancyites, rhyolite tuffs and pumice flows are abundant in series A (Brousse and Lefèvre 1967). Among the trachytes, two divergent trends of evolution (pantelleritic and phonolitic) have been observed by Brousse and Varet (1967). According to Brousse (1961b, p. 199), the two series are completely independent in space and time. The main eruption centres are grouped around a central elliptic caldera of 8 by 12 km diameter.

Brousse (1961b, p. 200) has explained the origin of the two series in the following way: the general parent magma was an olivine-rich basalt, whose differentiation led to the undersaturated series B. The formation of the saturated series A is interpreted by assimilation of siliceous crustal matter and subsequent differentiation. The assimilation continued also during the prolonged differentiation process. Jung and Brousse (1962, p. 618 f) emphasize rather an early modification of the basalt magmas by a more or less effective fractionation of pyroxenes at depth. Early crystallization of pyroxene and its fractionation during a slow rise of the magma would form an undersaturated basalt melt, which differentiated to tephrites and phonolites. In the absence of fractionation of this kind, primitive, tholeiitic basalt magmas ascend quickly to the surface.

The rock associations of Mont Dore have been interpreted in a quite different way by Glangeaud and Létolle (1962, 1965). Considerable amounts of crustal sialic material have been melted by the introduction of heat connected with the intrusion of the basaltic magma. Hybrid intermediate magmas were formed by the mixture of the primary basaltic and the

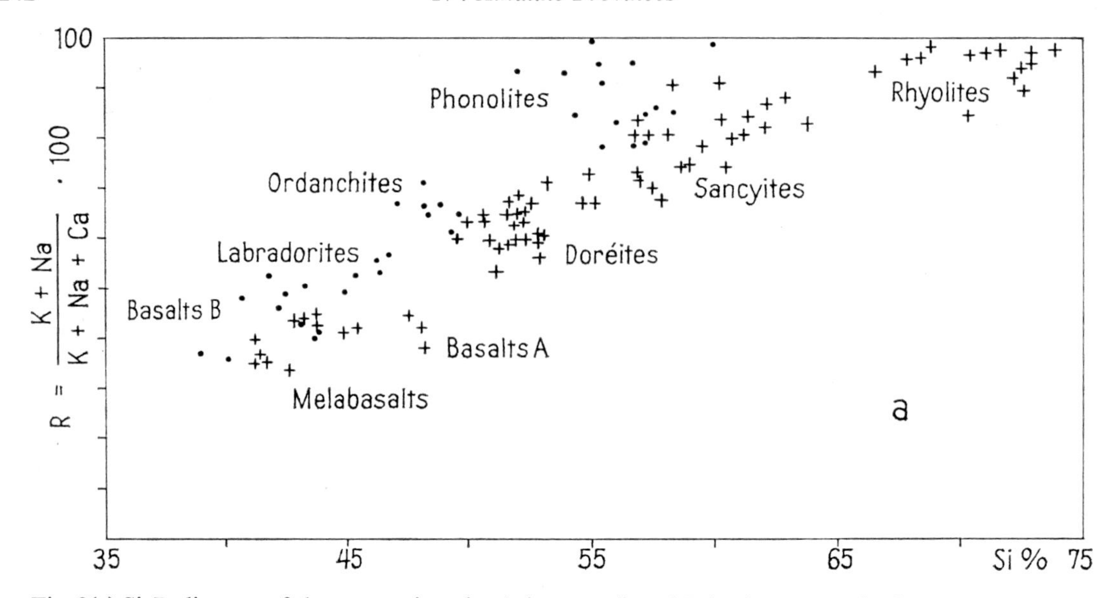

Fig. 2(a) Si–R-diagram of the magmatic series A (saturated) and B (undersaturated) of Mont Dore (after Jung and Brousse, 1962)

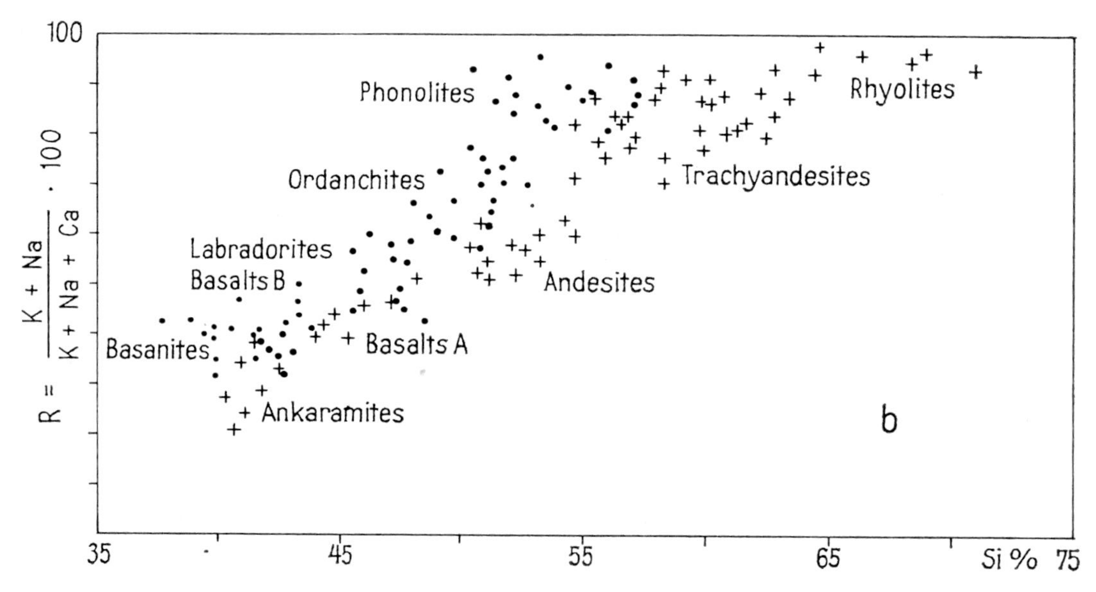

Fig. 2(b) Si–R-diagram of the magmatic series A (saturated) and B (undersaturated) of Cantal (after Vatin-Perignon, 1968)

palingenetic 'granitic' magma. They crystallized as 'sancyites' and 'doreites' at the surface. Chemical analyses of the rocks show a regular variation of the most important metal oxides against silica. Calculated mixtures of basalt and rhyolite in different proportions fit in very well with the average analyses of the sancyites and doréites (Table 2). Furthermore, the high participation of acidic lavas and ignimbrites speak against the classic concept of crystallization—differentiation in the case of Mont Dore. Not all of the rock types of Mont Dore can be

TABLE 1 Igneous Rocks of Mont Dore, Cantal and Chaîne des Puys

	Mont Dore					Cantal				Chaîne des Puys		
	1	2	3	4	5	6	7	8	9	10	11	12
SiO_2	43·60	42·20	71·95	59·60	59·76	47·76	45·31	54·30	56·34	47·20	47·05	65·20
TiO_2	2·70	2·85	—	0·20	0·80	2·71	2·79	0·28	0·06	3·06	3·62	0·41
Al_2O_3	13·90	11·60	14·20	20·53	19·08	15·75	14·44	21·91	17·76	16·43	15·40	18·35
Fe_2O_3	8·60	6·25	0·56	1·71	2·25	5·05	5·07	2·64	3·58	5·24	4·56	1·69
FeO	5·10	7·05	1·13	0·66	1·35	6·28	6·83	0·97	0·06	6·50	7·18	1·55
MnO	0·18	0·13	—	0·11	—	0·25	0·18	0·19	0·18	0·17	0·16	0·22
MgO	8·40	12·10	1·12	0·25	0·59	6·45	8·77	0·47	0·40	6·16	6·54	0·25
CaO	11·60	11·90	0·80	1·10	3·16	10·63	11·10	3·46	2·39	9·92	9·89	2·24
Na_2O	3·70	2·65	5·80	6·91	5·69	3·17	3·15	8·13	9·88	3·05	2·42	5·54
K_2O	1·80	1·80	4·42	6·00	5·01	1·35	1·61	5·65	4·97	1·71	2·12	3·86
P_2O_5	0·90	1·05	—	0·02	0·11	0·60	0·75	0·20	0·04	0·56	0·89	0·07
H_2O^+	0·25	0·70	0·83	1·63	1·63	—	—	0·09	3·19	—	0·13	0·37
H_2O^-	—	—	—	1·39	0·28	—	—	1·00	0·46	—	—	—
SO_3	—	—	—	—	—	—	—	0·69	0·99	—	—	—
Cl	—	—	—	—	—	—	—	0·39	0·24	—	—	—
	100·73	100·28	100·81	100·02	100·27	100·00	100·00	100·28	100·49	100·00	99·96	99·75

Legend:

1. Olivine-poor basalt, Cascade du Lac Guéry. Mode (vol. %): Augite 37·5; olivine 10·1; plagioclase (An 40–50) 44·0; titanomagnetite 6·6; biotite 1·0; apatite 0·8.
2. Olivine-rich basalt, lower lava flow of Charlannes. Mode (vol. %): Augite 26·7; olivine 14·4; plagioclase (An 55) 50·9; titanomagnetite 4·1; biotite 0·2; apatite 1·6; calcite 2·1.
3. Rhyolite, station moyenne du funiculaire de la Bourboule à Charlannes.
4. Phonolite, Le Piton.
5. Phonolite, La Roche Sanadoire.
6. Basalts of the saturated series, average.
7. Basalts of the undersaturated series, average.
8. Miaskitic phonolite, Vinsac.
9. Agpaitic phonolite, Repastils.
10. Average basalt of the Chaîne des Puys.
11. Essexibasalt, Puy de Dôme.
12. Dômite, summit of Puy de Dôme.

(1)–(5) and (12) from Brousse (1961a and 1961b); (6) and (7) from Vatin–Pérignon (1968); (8) and (9) from Varet (1969); (10) from Yamasaki and Brousse (1963); (11) from Bentor (1954).

TABLE 2 Average Analyses of Mont Dore Rocks and Calculated Mixtures of the Parent Magmas (after Glangeaud and Létolle, 1962)

	Average Basalt, Mt. Dore	Mixture Basalt–Rhyolite 2:1	Average Doréite Mt. Dore	Mixture Basalt–Rhyolite 1:1	Average Sancyite Mt. Dore	Mixture Basalt–Rhyolite 1:2	Average Rhyolite Mt. Dore
SiO_2	46·4	55·0	54	59	61	62	71
Al_2O_3	17	15·5	16	15·2	16	15	15
Fe_2O_3	5	4·2	4	3·5	3	2·9	1
FeO	7	5·2	4	1·6	2·5	1·5	1
MgO	7	5	4	3·6	2·5	2·7	0·5
CaO	10	7·3	7·5	6	4	4·6	2
Na_2O	3·5	3·5	4	3·5	4·5	3·5	4
K_2O	2	2·7	3·5	3	4	3·4	4·5

explained satisfactorily by this basalt–rhyolite hybridization hypothesis. The basic end members of the series, ankaramites and mareugites, are regarded as differentiates of the primary basalt magma. The tephrites and ordanchites, too, deviate from the main line of evolution by aberrant MgO and Na_2O contents. This is even clearer with the trachytes and phonolites. Their composition is characterized by strong enrichment in alkalis. Glangeaud and Létolle think of gas differentiation, which led to concentration of alkaline fluids in the upper part of the magma chamber. Crystallization differentiation with sinking of mafites and rising of feldspars may also have acted along with the hybridization processes. These additional assumptions explain the particular position of the alkaline rocks within the Mont Dore association.

IV.4.2.3. The Rock Associations of the Cantal

The large Cantal volcanic complex, about 50 km south of Mont Dore, has recently been described by Vatin-Pérignon (1968). It is a polygenic stratovolcano with multiple alternation of lava flows and tuff eruptions and a very variable sequence of rock types. The eruption centres are grouped around and above a large foundered area in the underground (17 × 20 km in diameter). The rocks are separated in a saturated series A and an undersaturated series B (see Fig. 2 and Table 1):

A. ankaramite–basalt–andesite–trachyandesite–hyalorhyolite;
B. basanite–basalt and labradorite–ordanchite–phonolite.

The two basalt types recall the corresponding basalts of Mont Dore, but are generally richer in SiO_2. Both basalts are clearly undersaturated; basalt A would not immediately yield the actual series A by simple crystallization–differentiation. Therefore Vatin-Pérignon (1968, p. 266) assumes —as Brousse does for Mont Dore—that the rocks of series A have been formed by assimilation of SiO_2-rich crustal material and subsequent differentiation. The ordanchites of series B have been studied and re-defined by Goer de Hervé and Vatin-Pérignon (1966).

In several publications, Varet (1967, 1969a and 1969b) discerns *miaskitic* and *agpaitic phonolites* in the northern parts of the Cantal (see II.2.1).* The miaskitic rocks have a composition close to the 'phonolitic minimum' in the system Ab–Or–Ne–Di, whereas the agpaitic rocks occupy 'apoeutectic' positions in the same system. Varet (1969b) explains the agpaitic differentiation as a continued evolution beyond the miaskitic stage, caused by gas differentiation and higher oxygen partial pressure. These conditions brought about further increase in alkalis and the late crystallization of aegirinic pyroxene.

IV.4.2.4. Rocks and Evolution of Puy de Dôme

Puy de Dôme is the most prominent among the nearly hundred Quaternary volcanoes of the Chaîne des Puys W of Clermont–Ferrand. This range is famous for the good preservation of the volcanic phenomena; it covers a N–S strip 3 to 5 km wide and 30 km long. Several lava flows extend 10 to 30 km into the foreland (Fig. 3).

* Readers are referred to the relevant Chapters of this book where similarly cited.

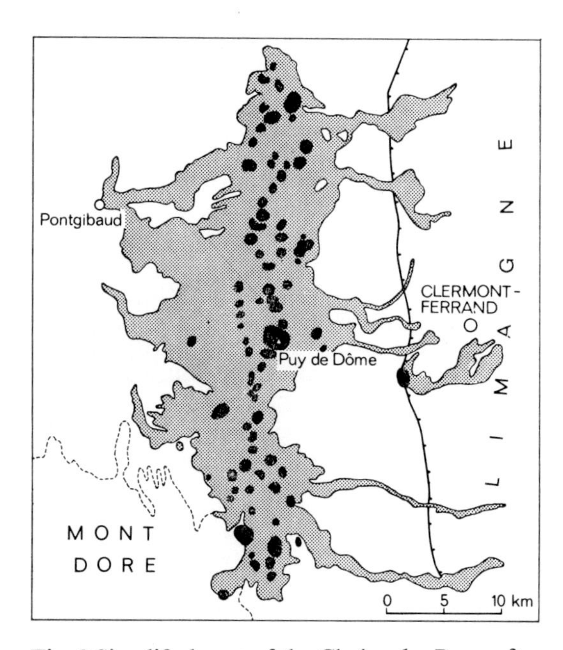

Fig. 3 Simplified map of the Chaîne des Puys, after Brousse (1963). Numerous volcanic cones, the 'puys' (black) and lava flows (shaded) of Quaternary age

Age determinations by the ^{14}C method applied to fragments of lignitic wood embedded in domitic scoriae of two localities gave values of respectively 8540 and 8730 ($\pm$ 300) years (Brousse, Delibrias, Labeyrie, and Rudel, 1966). More recent studies of Brousse *et al.* (1970) came to 12,800 and 5,700 years as oldest and youngest ages of the Chaîne des Puys.

Puy de Dôme (Fig. 4) rises to about 500 m above the crystalline basement; its absolute altitude is 1465 m. The rocks building Puy de Dôme are: lava flows of trachybasalt and essexibasalt; trachybasaltic, essexibasaltic, trachyandesitic, trachytic and rhyolitic pyroclastics, and a central intravolcanic intrusive plug of dômite, a cristobalite- and tridymite-bearing biotite trachyte (about 20 vol. % silica minerals). The composition of the parental basalt magma has been calculated by Yamasaki and Brousse (1963, p. 207) (see Table 1). Within the actual association of Puy de Dôme, the essexibasalt of Sigoine is closest to the calculated composition.

The evolution of Puy de Dôme and its magmas has been described by Bentor (1954) in the following way: The first phase of activity yielded trachybasaltic lava flows and a small cone of pyroclastics. During the following phase of rest, the magma differentiated in its chamber into an upper trachytic and a complementary lower 'ultrabasic' portion. Then, a phase of explosive activity built the Puy Lacroix, which today is partially covered under the younger products of Puy de Dôme proper (hornblende dômite, trachybasalt). In the third phase, the first pyroclastic cone of Puy de Dôme was formed by acidic biotite trachyte. This phase ended with rhyolitic tuffs, which overlie the trachytic tuffs. The viscosity of the trachytic magma was much augmented by the strong loss of gas during the continued explosive activity. The central plug of dômite within the tuff cone of the third phase rose almost to the surface. Then the trachytic magma was nearly used up. The subsequent so-called 'ultrabasic' magma reached the surface through a lateral additional crater, forming the youngest of the lava flows. To the last products belongs also the trachyandesite of Petit Puy de Dôme with labradorite and kaersutite; it is explained as a mixture of the essexibasaltic and a rest of the trachytic magma. Current investigations have lead to conclusions different from those of Bentor, and approaching, again, to the ideas of Lacroix (1908), (R. Brousse, personal communication).

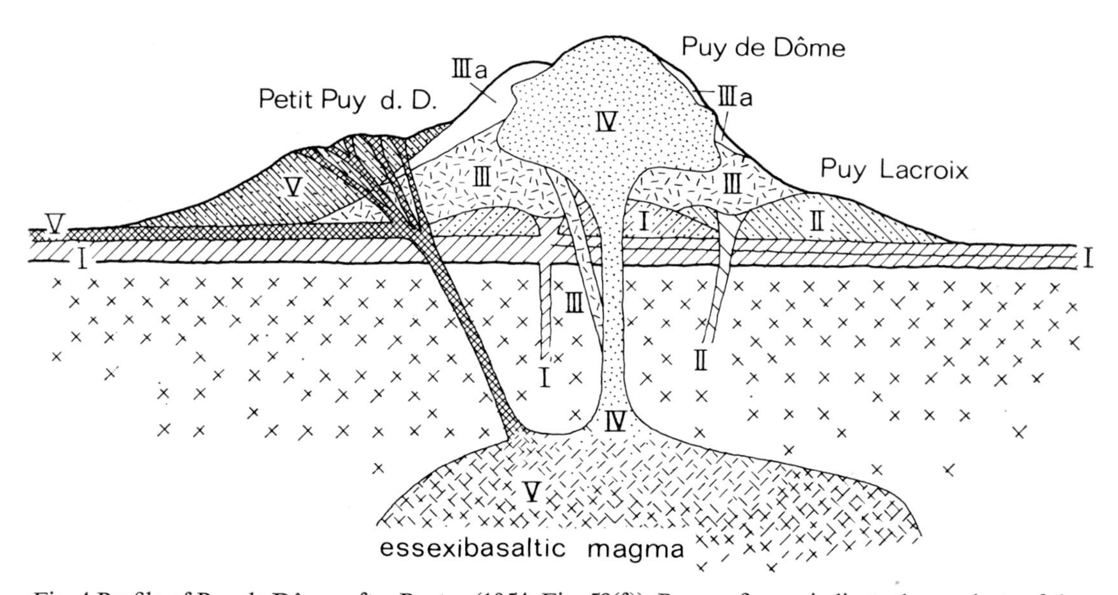

Fig. 4 Profile of Puy de Dôme, after Bentor (1954, Fig. 58(f)). Roman figures indicate the products of the phases of activity according to their temporal succession

IV.4.2.5. The Pépérites of the Limagne

Finally, the volcanic district of the Limagne shall be briefly dealt with because of its particular volcanic phenomena. Volcanic activity lasted from the Stampian (upper Oligocene) to the Pliocene; the variety of rocks goes from ultrabasic ankaratrites through basanites, basalts and trachyandesites to trachytes. (*NB.* This nomenclature used by Michel, 1953 is different from that chosen by Jung and Brousse, 1962.) In several individual volcanoes of the district, there occur typical 'pépérites', mixed volcanic-sedimentary rocks. The 'pépérites' consist of rounded (but not abraded), mostly vitreous lava particles in a marly or calcareous matrix. The size of the lava particles normally varies from 0·5 to 10 mm; exceptionally they reach 10 cm in diameter. The proportion of these lava fragments and the matrix is subject to great variations. The chemical and mineralogical investigation of the magmatic components led Michel (1953) to define them as trachyandesites. The accompanying lavas and dyke rocks of the same volcanoes are mostly basanites, basalts and mandchurites.

According to Michel (1953, pp. 70–6), the pépérites have been formed by violent intrusion of trachyandesite lava into unconsolidated marls and calcareous sediments of the Stampian. Rapid cooling and strong degassing caused the bursting into small round particles and the intensive interpenetration with the sediments. The mixed products so formed could spread out for more than one kilometre; they overlie or are interstratified with the unchanged sediments.

IV.4.2.6. The Intrusive Undersaturated Alkaline Rocks of the Pyrenees

These rocks are distributed over an area of 450 km E–W extension (Azambre, 1967). The occurrences are most frequent in three regions: south of San Sebastian (Spain), south of Pau and Tarbes and between Narbonne and Perpignan near the Mediterranean coast. The map of Azambre records 77 individual occurrences, which cut the Mesozoic rocks as dykes and small massifs and a few lava flows. The youngest strata traversed belong to the Senonian. The rocks vary from alkaline picrites through monchiquites and teschenites to nepheline syenites. The lavas are spilites and potassic spilites. The rocks of the two nepheline syenite massifs have been thoroughly described by Azambre (1967, pp. 16–51, and 1970).

a. Pouzac (about 100 m diameter): calc-alkalic nepheline syenite of miaskitic type ('monzonite néphélinique') (Table 3);

b. Fitou (1000 × 300 m) and Plâtrière de Fitou (300 × 200 m), miaskitic to agpaitic, with sodalite, cancrinite, eudialyte, mosandrite and catapleiite (Table 3).

TABLE 3 Nepheline Syenites of the Pyrenees (after Azambre, 1967).

	1	2	3
SiO_2	52·66	54·94	53·60
TiO_2	0·94	0·31	0·55
Al_2O_3	21·13	22·26	19·76
Fe_2O_3	2·11	0·90	1·68
FeO	1·60	0·72	1·20
MnO	0·15	0·15	0·19
MgO	1·22	0·40	0·80
CaO	4·03	1·77	3·36
Na_2O	7·10	8·11	8·10
K_2O	4·55	5·60	5·95
P_2O_5	0·32	0·39	0·24
H_2O^+	2·94	3·77	3·18
H_2O^-	0·53	0·61	0·64
	99·28	99·93	99·25

Legend:
1. Nepheline monzonite, Pouzac.
2. Nepheline syenite, Fitou.
3. Nepheline syenite, Plâtrière de Fitou.

The rocks are derived from an alkaline olivine basaltic magma which differentiated into a leucocratic (nepheline syenites) and a melanocratic branch (teschenites, monchiquites, picrites).

IV.4.3. The Alkaline Magmatites of Germany

IV.4.3.1. General Remarks

The neovolcanic subprovince of Germany is separated from the French subprovince by a gap of 250 km (see Fig. 1). The small volcanic districts of the Hassberge and the Oberpfalz (Bavaria) form the loose connection with the Bohemian subprovince.

The German subprovince can be subdivided into the following districts:

a. Vogelsberg, a complex shield volcano, about 75 km in diameter and 2500 km² in surface area; Miocene to Pliocene; alkaline olivine basalts, olivine tholeiites, rare phonolites (Schottler, 1937, Lehmann and Flörke, 1950, Ernst, 1961, 1968, Tilley, 1958, Schorer, 1970).

b. Westerwald; numerous smaller and larger occurrences; upper Oligocene to Pliocene; alkaline basalts, trachyandesites, trachytes, trachyte and basalt tuffs (Hentschel and Pfeffer, 1954, Ahrens, 1957, Ahrens and Villwock, 1964).

c. Hocheifel and northern Eifel, Siebengebirge; numerous small and medium sized occurrences; upper Miocene; alkaline basalts, hawaiites, latites, trachytes, trachyte tuffs (Frechen, 1962, Huckenholz, 1965a, b and c, 1966, Frechen and Vieten, 1970, Jasmund *et al.*, 1972).

d. Laacher See volcanic area; Quaternary; trachytic and other alkaline tuffs, selbergites, tephrites, ultrabasic alkaline basaltoids and their tuffs (Frechen, 1962).

e. Western Eifel (maar district); mostly Quaternary; numerous small and medium sized individual occurrences, lava flows, scoriaceous agglomerates and tuffs; nephelinites, leucitites and their varieties (Frechen 1962). The Dreiser Weiher volcano with its famous peridotite nodules belongs to this district (Frechen 1948, 1963). Jasmund and Schreiber (1965) emphasize the geochemical difference of the Tertiary olivine trachybasalts and the Quaternary tephritic nephelinites.

f. Rhine Graben; many, mostly small single occurrences in the adjoining horst blocks; uppermost Cretaceous to Miocene; basanites, olivine nephelinites, melilite ankaratrites, and pipe breccias of these rocks; shonkinite and sanidine nephelinite at the Katzenbuckel E of Heidelberg (Frenzel, 1967, Wimmenauer, 1967).

g. Kaiserstuhl; group of volcanoes (16 × 10 km), in the southern part of the Rhine Graben; subvolcanic centre exposed; Miocene; tephrites, essexites, limburgites, phonolites and carbonatites (Wimmenauer 1966 and 1967).

h. Hegau; several small single volcanoes; Miocene to lower Pliocene; melilite ankaratrites and their tuffs, phonolites (von Engelhardt and Weiskirchner, 1963, Weiskirchner, 1967a and 1967b).

i. Urach volcanic district; diatremes with rare magmatic dykes; melilite ankaratrites and olivine melilitites (Weiskirchner 1967b).

k. Rhön, probably upper Miocene; doleritic basalts, alkaline basalts, tephrites, basanites, olivine nephelinites, phonolitic tephrites, phonolites (Ficke, 1961, Eigenfeld, 1963, Martini *et al.*, 1970).

l. Hassberge; numerous dykes and a few vent fillings; probably Miocene; olivine nephelinites, alkaline basalts, phonolites (Schuster, 1927).

m. Volcanic district of northern Hessen and southern Niedersachsen; many small to medium sized occurrences, locally clustered (e.g. Habichtswald near Kassel); mostly Miocene; tholeiite basalts (rare), alkaline olivine basalts and their tuffs, basanites, limburgites, olivine nephelinites, melilite ankaratrites (Wedepohl, 1961 and 1968 a and b, Lohmann, 1964, Hentschel, 1968, Schneider, 1970).

IV.4.3.2. The Vogelsberg Basalt Volcano

The Vogelsberg is a complex of basaltic lava sheets and flows and subordinate basalt tuffs. In the centre of the hills, there are numerous younger vent filling and dykes exposed by erosion. Schottler (1937) subdivided the basaltic effusive rocks into 'trap' and basalt proper; the two rock types have been compared by Lehmann and Flörke (1950) with the tholeiitic basalts and the olivine basalts respectively, of the Island of Mull. The alkaline basalts are generally more variable than the tholeiitic ones; no true transition seems to exist between these two rock series (Schorer, 1970) (cf. Table 4). Phonolites appear at the beginning of the volcanic events; they play only a very subordinate quantitative rôle.

The two main basalt types, 'trap' and alkaline olivine basalt, alternate repeatedly in the sequence of effusions. This has recently been confirmed in the cores of the Rainrod I bore hole. Ernst and collaborators (1968) distinguish in this profile lavas poor in silica (38–44% SiO_2, average 42%) and lavas rich in silica (50–54% SiO_2, average 52%), which correspond to alkaline olivine basalts and tholeiitic basalts, respectively. Tuffs are associated with the alkaline basalt type. One of the basalt flows is remarkably rich in Al_2O_3, K_2O, Zr and other trace elements; a modification of the alkaline olivine basalt magma by assimilation of sialic material may be supposed. Ernst (1968, p. 37) assumes that the difference of the two main basalt types was effected by fractionation of one primitive parental magma at different depths. High-pressure fractionation is thought to be responsible for the variability of the alkali basalts, the tholeiitic ones may represent more primitive magma types (Schorer, 1970).

The observations in the bore hole Rainrod I allow the conclusion that the lava flows were erupted from at least two vents, one of which produced the tholeiitic, the other the alkaline basaltic magma.

The Miocene lava sheet of the Main 'trap' (after the river Main) adjoins the south-west of the Vogelsberg. It partly occupies the northernmost extremity of the Rhine graben. Today mostly eroded or covered, it is perhaps the largest individual lava body of Europe north of the Alps (30 × 35 km in surface area). The rock is clearly tholeiite basaltic; the analysis given in Table 4 is quartz normative (see IV.4.3.5.2).

IV.4.3.3. The Alkaline Olivine Basalt–Trachyte Association of the Hocheifel

A series of Miocene volcanic rocks of the Hocheifel has been studied in great detail by Huckenholz (1965–6). The rocks vary from ankaramites over basanites, alkaline olivine basalts s. str., hawaiites, mugearites and alkaline trachytes to leucotrachytes (Table 4). Huckenholz assumes an alkaline olivine basaltic parental magma, which yielded, by fractionated crystallization–differentiation, hawaiites, mugearites and trachytes. Basanites and ankaramites are explained by accumulation of clinopyroxene and olivine. The basaltoid rocks contain relics of high pressure–high temperature parageneses with Cr-bearing salite to augite, Ni-rich olivine, bronzite and chromian spinel. These minerals show distinct phenomena of corrosion and of reactions with the basaltic melt. The question remains open, if they be residuals ('restites') of the partial melting of mantle rock or relics of the primary crystallization of the basalt magma at great depth.

The main crystallization starts with a diopsidic augite, which gradually changes into hedenbergite- and acmite-bearing Ti-augite in the course of the differentiation of the magma. The acmite contents increase from about 5% in the salite phenocrysts up to 15% in the pyroxenes of the matrix. Aegirine augites proper occur subordinately. The olivines coexisting with the clinopyroxenes vary from chrysolite with 12 atomic % Fe to hyalosiderite with 40 atomic % Fe; the most magnesian olivines belong to the relictic high temperature–high pressure paragenesis, the most iron-rich to the groundmass of the mugearites (Table 5).

The anorthite contents of the plagioclases change from maximal 60% in the alkaline olivine basalts to 25% in the analcime trachyte. Alkali feldspars occur as the latest crystallizations in some basalts; they are more abundant in the hawaiites and mugearites and predominate in the trachytes (sodic sanidine $Ab_{54\cdot5}\,Or_{39\cdot2}An_{6\cdot3}$) The behaviour of the feldspars of the rock series studied speaks very much for the action of crystallization–differentiation (Huckenholz, 1965a, pp. 187–8 and Fig. 21).

The chemical evolution of the rocks of the Hocheifel is shown in Fig. 5 (from the data given by Huckenholz, 1965 and 1966). The distribution of niobium, a trace element characteristic for alkaline rocks, has been studied by Huckenholz (1965c).

In the Siebengebirge near Bonn, adjoining to the north of the Eifel, a subalkaline and a peralkaline rock series can be distinguished (Frechen and Vieten, 1970):

TABLE 4 Igneous Rock Types of the Vogelsberg, the Hocheifel, Northern Hessen and Niedersachsen

	Vogelsberg				Hocheifel							N.H., Ns.		
	1	2	3	4	5	6	7	8	9	10	11	12	13	14
SiO_2	51·84	48·6	42·6	41·2	42·8	42·6	45·6	46·6	50·0	55·3	61·6	47·8	42·4	38·6
TiO_2	2·55	2·1	2·3	2·6	2·4	2·8	2·6	2·8	1·9	0·85	0·8	2·53	2·32	2·78
Al_2O_3	13·04	14·6	13·1	14·0	12·6	13·8	13·7	15·3	17·8	19·2	17·9	13·1	11·56	11·49
Fe_2O_3	3·90	3·9	4·1	4·4	3·9	4·4	3·6	4·5	4·8	2·4	3·2	3·56	4·36	4·75
FeO	6·46	7·3	7·1	7·3	6·5	6·1	6·9	5·9	4·3	1·8	0·7	7·07	6·46	6·27
MnO	0·12	0·2	0·2	0·2	0·2	0·2	0·2	0·2	0·2	0·19	0·1	0·17	0·20	0·20
MgO	6·01	8·6	11·3	13·2	14·2	10·3	10·1	6·6	3·1	1·6	0·9	9·0	13·89	14·33
CaO	7·20	8·8	11·0	10·9	11·5	12·6	10·9	10·4	7·0	3·8	1·9	8·6	10·57	12·42
Na_2O	3·12	2·9	2·8	2·5	3·0	3·4	3·0	3·7	5·5	6·5	5·4	3·13	2·63	3·06
K_2O	0·74	1·1	1·5	1·6	0·8	1·0	1·1	1·3	2·4	4·5	4·3	1·97	1·64	1·85
P_2O_5	0·38	0·4	0·8	0·7	0·5	0·6	0·5	0·5	0·7	0·12	0·3	0·80	1·09	1·22
CO_2	tr.	0·1	0·8	0·5	0·5	0·3	0·2	0·2	1·3	0·2	0·4	—	—	—
H_2O^+	2·55	0·7	1·5	0·6	0·9	1·6	1·3	1·6	1·5	2·75	1·2	1·45	1·97	1·70
H_2O^-	2·05	0·7	0·9	0·3	0·2	0·3	0·3	0·4	0·5	0·75	1·3	0·58	0·40	0·82
	99·96	100·0	100·0	100·0	100·0	100·0	100·0	100·0	100·0	99·96	100·0	99·76	99·49	99·49

All analyses given are averages except nos. (1) and (10)

Legend:
1. Main 'trap', Dietesheim/Main.
2. 'Trap' basalts, 13.
3. Olivine basalts, 5.
4. Feldspar-free basaltoids, 3.
5. Ankaramites, 7.
6. Basanites, 7.
7. Alkaline olivine basalts, 5.
8. Hawaiites, 4.
9. Mugearites, 2.
10. Analcite alkali trachyte, 1.
11. Leukotrachytes, 5.
12. Alkaline olivine basalts, 25.
13. Nepheline basanites and limburgites, 9.
14. Olivine nephelinites, partly melilite-bearing, 18.

(1–4) calculated from the data of Schottler (1937); (5–10) from Huckenholz (1965–1966); (11) from Grünhagen (1964); (12–14) from Wedepohl (1968a).

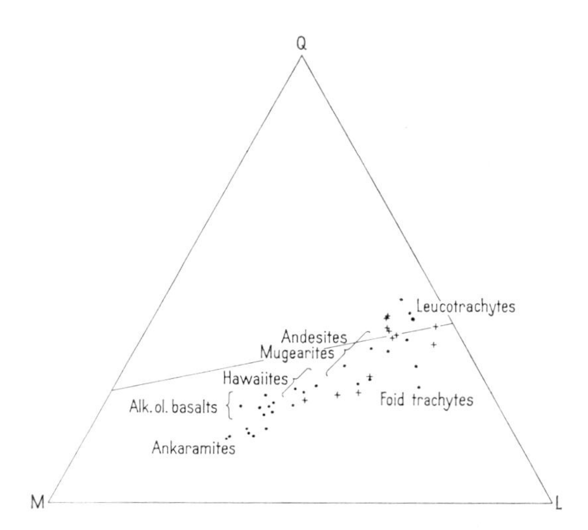

Fig. 5 QLM diagram (Niggli–Burri) of magmatic rocks of the Hocheifel (dots) and Siebengebirge (crosses). From the data of Huckenholz (1965, 1966) and Frechen (1962)

A. latite basalt–latite–quartz latite–trachytic latite–trachyte–quartz trachyte.

B. sanidine basanite–foidic latite–foidic trachyte–alkali trachyte.

Most of the rocks present belong to the less alkaline series A. The quartz trachyte plug of the Drachenfels is well known from the study of H. and E. Cloos (1927) of the sanidine orientation and their reconstruction of the intrusive dome on the basis of the feldspar fabric.

IV.4.3.4. The Quaternary Volcanism of the Laacher See Area

IV.4.3.4.1. General remarks. The most extreme alkaline rocks of the German subprovince occur in this area: selbergites with alkali oxides up to 18% (Table 6), schorenbergites, hauyne-nepheline leucitophyres. Various carbonatitic

TABLE 5 Clinopyroxenes of the Alkaline Olivine Basalt–Trachyte Association (after Huckenholz, 1965a and 1966)

	Relics of the HT–HP Phase	Phenocrysts	Groundmass
	Cr-salite	Ti-salite	Ti-salite
Ankaramites	$Ca_{47}Mg_{41}Fe_{12}$ ~0·45% Cr_2O_3 ~1·7% TiO_2	$Ca_{45}Mg_{42}Fe_{13}$ ~0·15% Cr_2O_3 ~2·5% TiO_2	$Ca_{45}Mg_{37}Fe_{18}$ ~0·15% Cr_2O_3 ~3–3·5% TiO_2
	Cr-salite	Ti-salite	Na–Ti-salite
Basanitoids	$Ca_{45}Mg_{43}Fe_{12}$ 0·2–0·5% Cr_2O_3 1·4–1·9% TiO_2	$Ca_{45}Mg_{41}Fe_{15}$ 0·17–0·28% Cr_2O_3 1·8–2·5% TiO_2	$Ca_{46}Mg_{37}Fe_{17}$ $>3\%$ TiO_2
	Cr-augite	Ti-augite	Na–Ti-augite
Alkaline Olivine Basalts	$Ca_{44}Mg_{44}Fe_{12}$ 0·3–0·6 Cr_2O_3 1·1–1·5 TiO_2	$Ca_{44}Mg_{42}Fe_{14}$ 0·12–0·30 Cr_2O_3 1·6–2·0 TiO_2	$Ca_{43}Mg_{40}Fe_{17}$ ~0·05% Cr_2O_3 2·2–2·9% TiO_2
	Cr-augite	Ti-augite	Na–Ti-augite
Hawaiites	$Ca_{44}Mg_{43}Fe_{13}$ 0·18–0·27 Cr_2O_3 1·15–2·20 TiO_2	$Ca_{44}Mg_{41}Fe_{15}$ 0·10–0·17 Cr_2O_3 1·8–2·3% TiO_2	$Ca_{44}Mg_{38}Fe_{18}$ 2·0–3·3% TiO_2
		Na–Ti-augite	Na–Ti-augite
Mugearites		$Ca_{44}Mg_{41}Fe_{15}$ ~0·1% Cr_2O_3 1·8–2·3% TiO_2	$Ca_{44}Mg_{38}Fe_{18}$ ~1·8% Ti
		Ti-augite	Na–Ti-salite
Trachytes		$Ca_{44}Mg_{40}Fe_{16}$ 1·3–2·6% TiO_2	$Ca_{45}Mg_{30}Fe_{15}$ ~3·3% TiO_2

rocks have been found among the ejectamenta.

Metasomatic transformation of the rocks surrounding the magma chamber (phyllites, mica schists) produced the famous *sanidinites*, which have been described preferably by Brauns (1911), Kalb (1934, 1936), Frechen (1947) and Schürmann (1960). The craters of the young Quaternary pumice tuffs are located in and near to the present Laacher See (Fig. 6). Originally, the tuffs covered an area approximately 120 km long and 40 km wide. Thin layers of fine pumice ash with the characteristic heavy minerals have been traced in young Quaternary peat moors in Germany and Switzerland at distances of up to 500 km.

The volcanic activity extended over a period of 420,000 years from the Waal stage (older Quaternary) up to the Alleröd stage (youngest Quaternary, about 11,000 years before the present day). One age determination (on the selbergite of the Schellkopf) is considerably higher (570,000 years). The dating of the different volcanic phases has been carried out by Frechen and co-workers by correlation with the loesses and fluviatile sediments of the Rhine (see Frechen, 1962, pp. 28–38). Quantitative age

TABLE 6 Igneous Rocks of the Laacher See District (after Frechen, 1962)

			Quaternary pumice tuffs						Trass
	1	2	3	4	5	6	7	8	9
SiO_2	50·41	47·90	55·56	55·14	55·49	57·40	56·93	52·18	51·21
TiO_2	0·40	1·70	0·15	0·23	0·25	0·41	1·15	1·32	0·23
Al_2O_3	22·15	16·75	21·68	22·52	21·63	23·09	20·04	17·59	20·24
Fe_2O_3	2·27	4·52	2·02	1·18	1·40	1·94	0·95	1·89	1·57
FeO	1·12	3·65	—	0·97	0·61	—	1·48	3·81	0·32
MnO	0·23	0·17	0·73	0·33	0·28	tr.	0·14	0·40	0·23
MgO	0·16	4·56	0·25	0·35	0·39	0·13	1·76	5·64	0·47
CaO	1·75	8·45	0·45	0·98	1·82	1·66	3·75	7·07	1·71
Na_2O	8·83	6·15	10·47	10·02	8·65	8·12	5·39	3·27	5·69
K_2O	9·10	4·21	4·73	5·69	5·37	5·70	6·67	4·98	7·02
CO_2	0·31	0·14	—	—	0·06	tr.	0·17	tr.	—
Cl	0·38	0·04	—	—	—	—	—	—	—
SO_3	0·61	0·58	0·08	—	0·11	0·57	0·36	0·26	—
P_2O_5	n.d.	0·43	0·03	—	tr.	—	—	0·16	—
H_2O^+	2·65	0·80	2·70	2·38	3·36	1·18	1·02	1·04	11·35
H_2O^-	n.d.	0·30	0·45	—	0·50	—	0·13	0·28	—
	100·37	100·55	99·56	99·79	99·92	100·20	99·94	99·93	100·04

Legend:
1. Selbergite, Schellkopf near Brenk. Mode (calculated vol %): Sanidine (ab 26) 27·9; leucite (Na–lc 5) 26·0; nepheline (ks 9) 21·6; noselite 9·4; sodalite 5·3; biotite tr; aegyrine augite 8·4; magnetite tr; titanite 0·7; calcite 0·7.
2. Tephrite lava of Niedermendig (with 0·20% BaO). Mode (calculated vol %): Plagioclase (an 22) 32·5; leucite (Na–lc 5) 21·7; nepheline (ks 9) 13·5; hauynite 4·8; augite 19·1; olivine (fa 15) 3·0; magnetite 4·2; apatite 0·9; calc te 0·3.
3. Meerboden pumice tuff (with 0·26% F).
4. Frauenkirch pumice tuff.
5. Laacher See pumice tuff II.
6. Laacher See pumice tuff III.
7. and 8. Laacher See pumice tuff V (8 with 0·04 % F).
9. Chabasitized trass, Kruft.

determinations by the K–Ar method have been published by Frechen and Lippolt (1965).

Petrographically, an older Quaternary foyaitic–foiditic series, a sequence of so-called alkaline 'basalts', and the series of younger Quaternary pumices and trasses can be discerned (Fig. 7).

IV.4.3.4.2. The older Quaternary foyaitic–foiditic series. The rocks vary from subvolcanic foyaites (ejected blocks only) through selbergites, leucitophyres and schorenbergites to nephelinites. Tuffs of selbergite are more abundant than the compact rocks. The foyaitic rocks (noseane foyaite, cancrinite foyaite, nepheline foyaite) are more feldspathic than the effusive rocks of the region. Among the less frequent subvolcanic ejectamenta, there are riedenite (noseane–biotite augitite), rodderite (noseane–sanidine–biotite augitite), boderite (sanidine–noseane–biotite–augitite), augitite and carbonatitic rocks (sövite with various silicate minerals such as sanidine, noseane, nepheline, cancrinite, scapolite, biotite, augite; cf. Brauns and Brauns, 1925). Isotopic studies of the carbon and oxygen of the subvolcanic rocks and the carbonatites (Taylor, Frechen and Degens, 1967) showed only little variation of the $\delta^{18}O$ and the $\delta^{13}C$ values. The results speak much in favour of a close petrogenetic relationship of all these rocks and particularly of a magmatic origin of the carbonatites. The isotopic composition of the carbon dioxide, issued today at Laacher See, fits very well to that of the carbonatites. The most frequent rocks of the series, the selbergites, are volcanic equivalents of khibinites; feldspathoids predominate over feldspars (Table 6). Schorenbergite is a leucocratic noseane leucitophyre

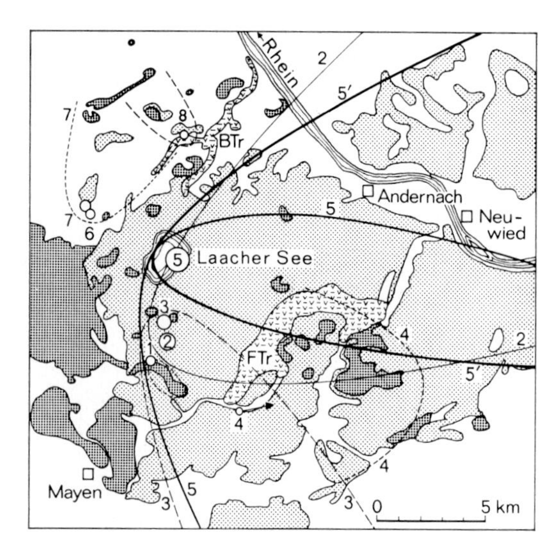

Fig. 6 Simplified map of the Laacher See area (from Frechen, 1962). Light shading: Pumice tuffs; heavy shading: Quaternary igneous rocks other than pumice tuffs; circles, hyperbolae and figures: situation of the eruption centres and areas of deposition of the different pumice tuffs (1 = Obermendig; 2 = Meerboden; 3 = Niedermendig; 4 = Frauenkirch; 5 and 5′ = Laacher See; 6 = Glees; 7 = Hüttenberg; 8 = Kahlenberg); BTr = Brohltal trass; FTr = Frauenkirch trass

with a very high alkali content (8·20% Na_2O, 8·50% K_2O, 47·12% SiO_2; for the complete analysis see Frechen, 1962, p. 49). The peralkaline character of these rocks, especially of the selbergite, cannot be satisfactorily explained, neither by normal crystallization–differentiation, nor by carbonate syntexis in the sense of Daly. Therefore, Frechen (1962, p. 26) assumes that the magma was additionally enriched in alkalis by gas differentiation. By a hypothetical reaction, schematically:

$$CaAl_2Si_2O_8 + 2Na \rightarrow 2NaAlSiO_4 + Ca,$$

potential plagioclase was suppressed; plagioclase does not occur, in fact, in these rocks. Considerable amounts of potassium must also have been introduced.

The assumption of the introduction of mobile alkalis is furthermore supported by the metasomatic transformations observed in the ejected rock fragments of the deeper substratum, the *sanidinites*. Frechen (1947; 1962, pp. 26–7) distinguishes an alkali-calcium metasomatism and an alkali metasomatism, which transformed, one after the other, the phyllites and mica schists of the basement. Quartz, biotite and muskovite are the main minerals of the original rocks. In the phase of alkali-calcium metasomatism, gneissose rocks with quartz, sanidine, plagioclase, actinolite and diopside were formed. The alkali metasomatism proper produced the sanidinites s. str. with sanidine, aegirine and alkali amphibole as main minerals. This sequence becomes evident by the observation, that the plagioclase of the metasomatic 'gneisses' and basites is ultimately transformed into sanidine.

IV.4.3.4.3. The 'alkaline basaltic' lavas and tuffs. The formation of 'alkaline basaltic' flows, cones of scoriae and complex volcanic buildings lasted during the entire period from the Waal stage (older Quaternary) up to the Wuerm II/III interstadial. The rocks are mostly tephrites, basanites, leucitites and nephelinites. The content of foids is so high in all the rocks quoted by Frechen (1962, pp. 61–75), that true basalts are not present. The best known rock is the tephritic lava of Niedermendig, which is exploited in large open-cut and underground quarries (analysis and mode see Table 6). Subvolcanic rocks of the series foyaite-foidite and alkaline syenite, monzonite, sanidine gabbro occur as xenoliths. The tephritic lava of the nearby Ettringer Bellerberg contains pyrometamorphic limestone xenoliths with wollastonite, garnet, diopside, gehlenite, larnite, portlandite, brownmillerite, hydrocalumite $Ca_2Al_2(OH)_7 . 3H_2O$, afwillite $Ca_2H_2[SiO_4]_2 . 2H_2O$, ettringite $Ca_6Al_2[(OH)_4|SiO_4)] . 26\,H_2O$, spinel and pyrrhotite (Hentschel, 1964; Jasmund and Hentschel 1964; see also Grünhagen and Mergoil, 1963).

IV.4.3.4.4. The younger Quaternary pumice tuffs and the trass. The pumice tuffs of the Laacher See area can be subdivided, corresponding to their eruption localities and sequence, into eleven individual layers, five of which belong to the Laacher See volcano proper (Frechen, 1962, pp. 83–112).

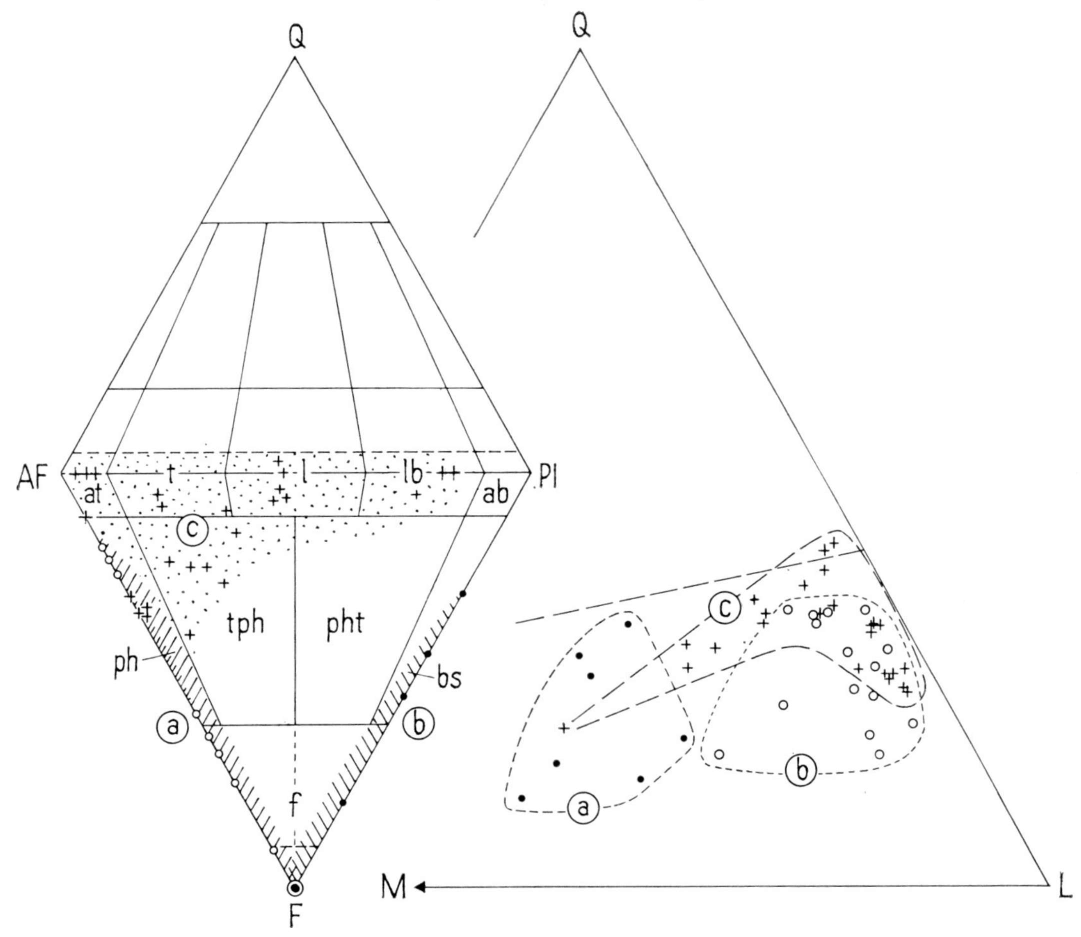

Fig. 7 Double triangle (Streckeisen) and QLM diagram (Niggli-Burri) of selected rocks of the Quaternary Laacher See volcanic district (from the data of Frechen, 1962)

a = alkali basaltic series, b = foyaitic–foiditic series, c = Laacher See pumice tuffs and comagmatic subvolcanic rocks. Abbreviations in the Streckeisen double triangle: Q = quartz, AF = alkali feldspars, Pl = plagioclase, F = feldspathoids, at = alkali trachyte, t = trachyte, l = latite, lb = latitic basalt, ab = alkali basalt, ph = phonolite, tph = tephritic phonolite, pht = phonolitic tephrite, bs = basanite, f = foidite.

The chemical compositions and the calculated mineral compositions of the pumice tuffs vary between those of phonolitic, foid-trachytic and foid-latitic rocks (Table 6). The analyses are arranged in the order of eruption of the rocks. The observations may be interpreted in such a way that, first, alkali- and gas-rich differentiates from the upper parts of the magma chamber were issued; later on followed the deeper parts of the magma with high contents of plagioclase and mafic minerals. The different tuffs are also distinguished by their heavy mineral content, which even allows their identification at distances of several hundred kilometres. The quantity of crystallized minerals is very low in all the pumice tuffs, even in the more basic members of the series. The predominant part of the tuffs is vitreous. Therefore the variation of the composition cannot simply be understood as the result of gravitational crystallization–differentiation. According to Frechen (1962, p. 92), the different fractions of the magma probably originated by gas differentiation in a molecularly dispersed state. The pumice tuffs contain frequent xenoliths of rocks of the underground, which have been partly transformed into sanidinites (see IV.4.3.4.2.). Ejectamenta of the older Quaternary igneous series occur, too. Several comagmatic subvolcanic rocks are described: syenite, alkali syenite, monzonite, mangerite, sanidine gabbro, noseane foyaite, hauyne foyaite, -khibinite, -syenite, -monzonite, and -mangerite, and sanidine gabbro. The hauyne rocks are also comprised as 'gleesites' after their main locality (Kalb, 1934–6; Schürmann, 1960). Jasmund and Seck (1964) regard the 'gleesites' as endogenous products of the trachyte magma rather than as metasomatic transformations of xenoliths.

At several places in the Laacher See area, 'trass' flows (pozzolanas) are interstratified with the pumice tuffs. They reach up to 60 m in thickness. The trasses follow topographical depressions in a way recalling lava flows. They seem to be deposits of very mobile suspensions of small rock and glass fragments in hot gases. According to the observation of carbonized wood inclusions, the temperature must have reached at least 350 to 400 °C. The unaltered magmatic particles of the trass consist of glass; chabazite and analcime are products of transformation under the subsurface water table. The unaltered rocks correspond largely to the pumice tuffs in chemical composition and heavy mineral content. Because of their hydraulic properties, the trasses have already been mined in underground galleries by the Romans.

Fig. 7 illustrates the difference of the three Quaternary magmatic series of the Laacher See area in the double triangle of Streckeisen. The marked divergence of the series is clearly visible. A normal alkaline olivine basalt with predominant plagioclase and little nepheline is not represented. At the present state of knowledge it is difficult to tell to which of the basalt types the parental magma belonged. In any case, the magmatic evolution led to extremely alkaline final differentiates. An important rôle may be ascribed to gas differentiation as is indicated by the violent explosivity of the Laacher See volcanism.

In a comparative petrochemical study of the volcanites of the Cape Verde Islands and the Rhenanian volcanic area (Eifel, Siebengebirge and Westerwald), Burri (1960) comes to the conclusion that the strongly undersaturated magmas were generated by syntexis of carbonate rocks. This should be particularly valid for the feldspar-free basaltoid rocks (leucitites, nephelinites) and the melilite-bearing rocks. Burri alleges, beyond the petrochemical criteria, the carbonatic-silicatic ejectamenta and the strong explosivity of the Quaternary volcanoes as arguments for his hypothesis.

IV.4.3.5. The Volcanic and Subvolcanic Rocks of the Rhine Graben and its Surroundings

IV.4.3.5.1. General remarks; distribution and age relationships. Tertiary igneous rocks occur in two separated areas (Fig. 8) of the Rhine graben region (cf. Wimmenauer, 1967). The Kaiserstuhl volcano lies in the heart of the southern magmatic province (Wimmenauer, 1966). Many small tuff pipes and narrow dykes of basic rocks are scattered over the neighbouring horst blocks (Black Forest and Vosges) and their border hills. The Hegau volcanic area (80–90 km S.E. of the Kaiserstuhl) consists of several prominent eroded necks, tuff pipes and tuff sheets ('Deckentuffe'). More than 300 tuff pipes are known in the Urach volcanic area, about 100 km E. of the graben.

Between the southern and northern volcanic region, there is a gap of nearly 130 km, which is practically devoid of Tertiary igneous rocks at the surface.

In the northern region many dykes, necks and tuffs pipes of predominantly basic rocks break through the lower Paleozoic, the Permian and the lower Triassic sandstone (Buntsandstein) of the Odenwald, Spessart and Taunus Mountains. A lava effusion of about 30 × 40 km surface area, the Main 'trap', occupies the northeastern extremity of the graben near Frankfurt/Main.

The igneous activity in the Rhine graben region is extended from the uppermost Cretaceous (Katzenbuckel E. of Heidelberg) into the Pliocene (Hegau, Main 'trap'?) The maximum activity was in the Miocene (Kaiserstuhl, Vogelsberg); the many small volcanic intrusions

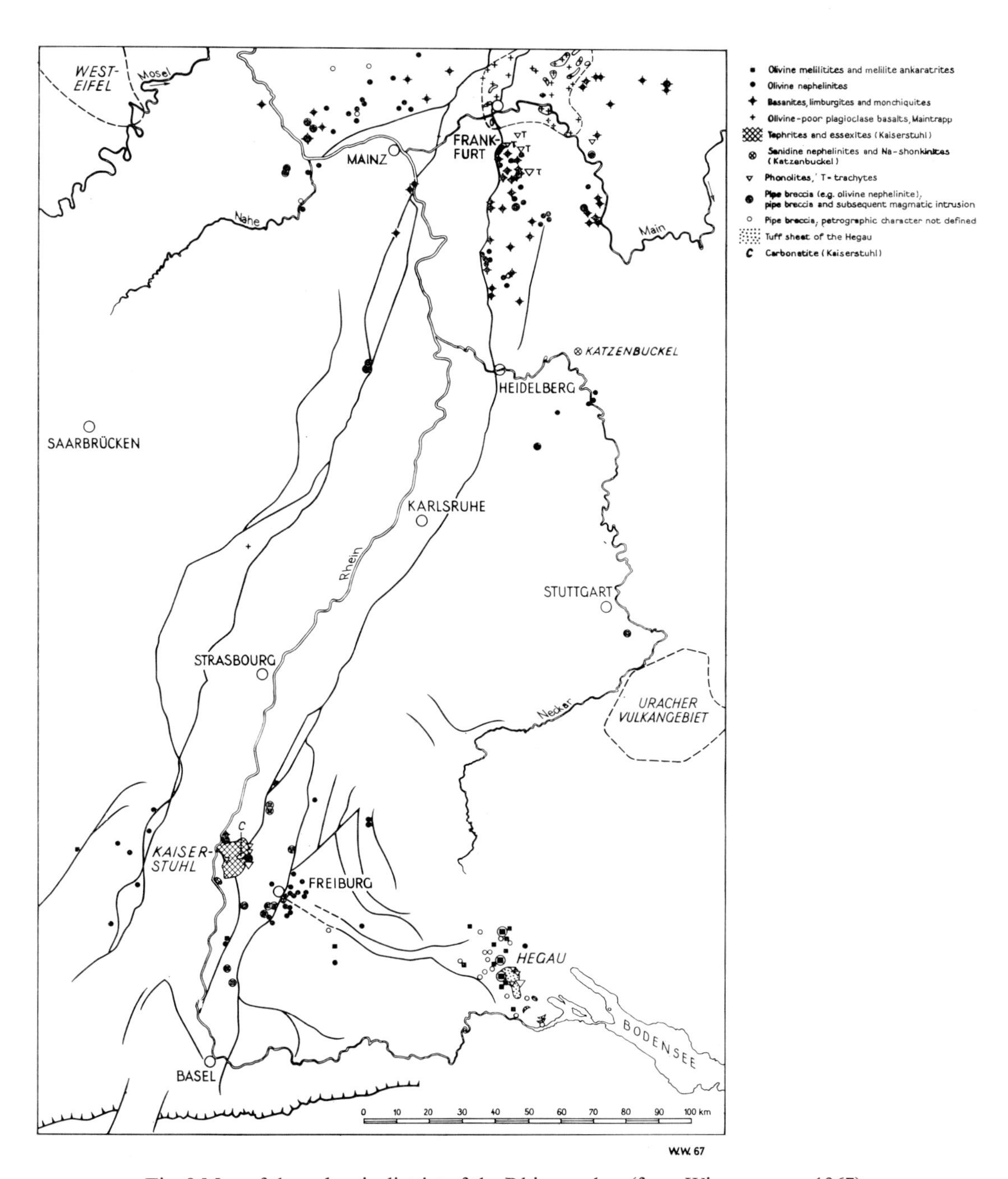

Fig. 8 Map of the volcanic district of the Rhine graben (from Wimmenauer, 1967).

in the horst blocks are not yet dated with sufficient certainty. There is still a controversy about the age of the Main 'trap' (middle Miocene or Pliocene). Absolute age determinations (K–Ar method) gave 66×10^6 Katzenbuckel $16–18 \times 10^6$ for the Kaiserstuhl, and (7?–)$12–14 \times 10^6$ y for the Hegau.

IV.4.3.5.2. Petrology. The igneous rocks of the Rhine graben region can be classified into eight groups:

a. Olivine-poor tholeiitic basalts (Main 'trap');

b. Basanites and limburgites (dykes and necks in the horsts; lava flows in the Kaiserstuhl);

c. Olivine nephelinites (dykes, necks and tuff pipes in the horst blocks; rare lava flows);

d. Melilite ankaratrites and olivine melilitites (several dykes and tuff pipes in the horsts and in the graben filling; necks and tuff pipes in the Hegau; tuff pipes and subsequent dykes in the Urach volcanic area);

e. Sanidine nephelinites and Na-shonkinites (Katzenbuckel);

f. Rocks of the essexite family in the Kaiserstuhl (tephrites, essexites, theralites, related dyke rocks and subordinate differentiates);

g. Phonolites (Hegau, Kaiserstuhl, Spessart), trachytes (northern Odenwald); differentiated dyke rocks of the phonolitic family (Kaiserstuhl);

h. Carbonatites (Kaiserstuhl).

The rocks of these groups will be briefly characterized below; their position in the *QLM*-triangle (Niggli–Burri) is shown in Fig. 9; for chemical analyses see Table 7.

a. The olivine-poor *tholeiitic basalts* of the Main 'trap' differ from the other basaltoids of the region by their comparatively high silica content ($>48\%$). Their field in the *QLM*-triangle is slightly below the saturation line. The structure is ophitic with a glassy mesostasis; there are two kinds of clinopyroxene: pigeonite and augite; some more silicic varieties also contain bronzite (Ernst, 1961). All these qualities place the Main 'trap' rocks near the tholeiite basalts.

b–d. Olivine nephelinites are widely distributed over the whole region adjoining the Rhine graben. The remarkable similarity of these rocks over large distances and their apparent independence from other basalts favour the assumption of a primary character of their magma. The frequent occurrence of peridotite inclusions suggests an ascent from great depth. In some places, nephelinitic-carbonatitic differentiates are produced.

There are continuous transitions into *basanites*, limburgites with glassy groundmasses, and melilite-bearing varieties. The latter grade into the strongly undersaturated melilite ankaratrites of the Hegau and the olivine *melilitites* of the Urach volcanic region and elsewhere. The preferred occurrence of these ultrabasic magmas in the more explosive type volcanoes should be mentioned.

e. The *sanidine nephelinites* and Na-shonkinites of the Katzenbuckel E. of Heidelberg differ greatly from the rocks of all the other groups. They are very rich in alkalis and phosphorus, fairly rich in iron, but poor in silica. This composition is most probably the result of a pronounced pneumatolytic differentiation, for which many indications are exhibited by the rocks and minerals themselves (Frenzel, 1967).

f. The main rocks of the *Kaiserstuhl* (*tephrites and essexites*) are distinguished from the basalts of similar SiO_2 content by lower fm and mg, but higher c, alk, k and al (Niggli numbers). This is mineralogically expressed by scarcity of olivine, predominance of Ca-rich augite, leucite in the lavas and potash feldspar in the subvolcanic rocks. Subordinate basanites and limburgites also occur as well as leucocratic differentiates. The origin of the typical Kaiserstuhl rocks is difficult to trace, as alkaline olivine basalts are not represented in the area. Wimmenauer (1966, pp. 201–2) assumed the evolution from an olivine nephelinitic parent magma by assimilation of sialic matter and differentiation.

g. Phonolites and rare trachytes are members of pronouncedly bimodal rock series in the Hegau, the Odenwald (and the Vogelsberg outside the Rhine graben region). Also at Kaiserstuhl no real convergence of the mineral compositions of the essexitic and phonolitic

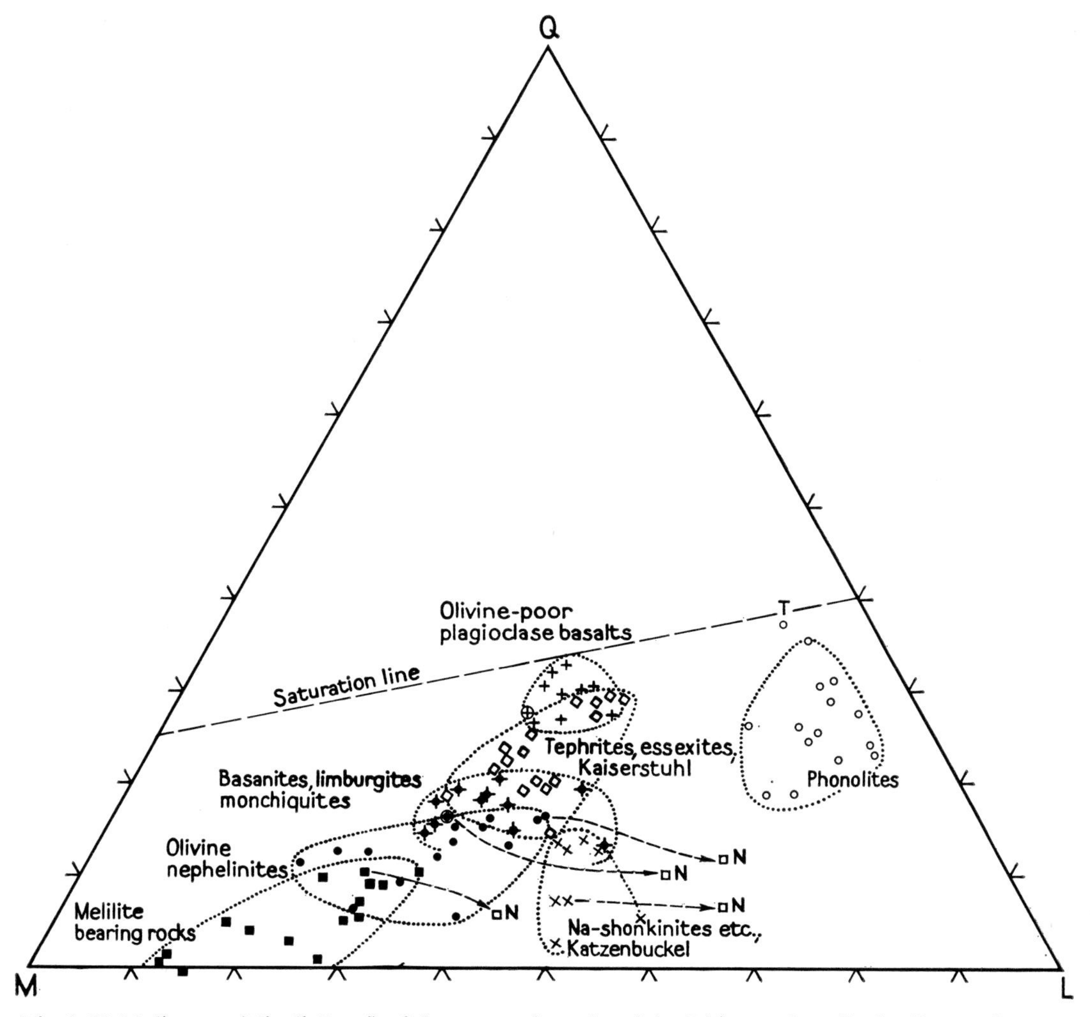

Fig. 9 QLM diagram (Niggli–Burri) of the magmatic rocks of the Rhine graben district (from Wimmenauer, 1967). Dashed lines connect nephelinitic differentiates (N) with their host rocks. Encircled cross and dot indicate the averages of the Vogelsberg 'trap' basalt and alkaline olivine basalt respectively. T= trachyte.

rocks is observed. It may be concluded, that the phonolites developed rather independently by intensified differentiation of the sial-contaminated magma. Many fenitized xenoliths of gneiss and granite support this hypothesis.

h. A *carbonatite* body of about 1 km^2 in surface area and many carbonatite dykes intrude into the subvolcanic rocks in the centre of the Kaiserstuhl. Most of the rocks are sövites, composed of calcite, apatite, magnetite, mica and forsterite. Pyrochlore and niobian perovskite are sparse accessories. The carbonatites are accompanied by subvolcanic breccias and in places by ultrapotassic feldspathic rocks, recently studied by Sutherland (1967). There are no intermediate rocks known between the carbonatites and the silicate rocks of the area; but it is noteworthy to state that the well known nephelinite–phonolite–carbonatite association (King and Sutherland, 1960) is also represented in the Kaiserstuhl, albeit obscured by the predominance of the essexitic-tephritic rocks.

TABLE 7 Main Rock Types of the Rhine Graben District and its Surroundings

	Kaiserstuhl						Katzenbuckel		Hegau			
	1	2	3	4	5	6	7	8	9	10	11	12
SiO_2	41·9	44·0	45·2	55·8	1·8	1·3	40·01	39·76	52·3	36·5	40·3	37·1
TiO_2	3·0	2·8	2·0	0·5	0·2	<0·1	3·46	3·51	0·5	2·9	2·4	2·4
Al_2O_3	13·8	14·1	15·3	20·2	1·1	0·4	13·58	13·96	19·6	9·1	12·8	7·9
Fe_2O_3	7·4	6·7	7·6	2·3	6·0	0·4	9·02	9·51	2·2	6·9	4·3	6·0
FeO	5·3	4·9	4·9	1·1	1·3	<0·1	6·68	5·73	0·7	6·5	7·3	4·3
MnO	—	—	0·2	0·1	0·7	0·2	0·15	0·16	0·2	0·2	0·3	0·2
MgO	7·6	5·7	5·2	1·2	1·6	0·2	5·70	5·05	1·2	14·7	10·2	17·7
CaO	13·0	13·0	11·3	4·2	48·7	54·8	8·49	8·53	3·2	14·7	13·6	17·3
Na_2O	2·9	2·6	3·1	5·9	0·2	⩽0·1	6·25	6·74	8·3	2·8	3·2	1·6
K_2O	1·9	1·8	2·6	4·9	0·1	0·1	3·06	2·83	4·9	1·5	1·0	0·5
P_2O_5	0·4	0·5	0·4	0·1	4·9	0·8	1·42	1·64	0·2	0·8	0·8	0·9
CO_2	0·3	0·2	0·3	0·7	32·8	41·2	—	—	1·0	0·4	0·4	<0·1
H_2O^+	1·6	2·6	1·4	2·9	0·3	0·6	2·18	2·58	4·8	3·2	3·2	3·3
H_2O^-	0·9	1·2	0·6	—	0·2	0·2			0·9			0·8
	100·0	100·0	100·0	100·0	99·9	100·4	100·00	100·00	100·0	100·0	100·0	100·0

Legend:

1. Limburgites, Kaiserstuhl, 5.
2. Tephrites, Kaiserstuhl, 7.
3. Essexites, Kaiserstuhl, 5.
4. Phonolite, Kirchberg, Kaiserstuhl, 2.
5. Carbonatite, Schelingen, Kaiserstuhl.
6. Carbonatite, Badloch, Kaiserstuhl.
7. Soda shonkinites, Katzenbuckel, 3 (after Frenzel, 1967).
8. Sanidine nephelinites, Katzenbuckel, 5 (after Frenzel, 1967).
9. Phonolites Hegau, 8.
10. Melilite ankaratrites, Hegau, 10.
11. Olivine nephelinites, whole Rhine graben region, 9.
12. Olivine melilitite, Hochbohl, Urach volcanic area.

Van Wambeke *et al.* (1964) have pointed out, that all Kaiserstuhl rocks are rich in Nb, the rare earths of the Ce group, Ba, Sr and Pb, as compared with the respective rock families in other alkaline provinces. Special concentrations of the elements named below are observed in the olivine nephelinites: Cu, V, Fe, Ni, Cr, Co; in the tephrites and essexites: Ti; in the phonolites: U, Th, Rb, Zr; in the carbonatites: the alkaline earths, Nb, RE, Mn, Zn and Pb.

Gehnes (unpublished communication) found remarkably higher contents of Li, Rb, Sr, Ba, Zr, V, Nb, Cr, Ni and Co in the olivine nephelinites of the Rhinegraben, than the averages of the alkaline olivine basalts (see also IV.4.5.).

IV.4.3.6. The Volcanites of Northern Hessen and Southern Niedersachsen

Many small single volcanoes and groups of volcanoes are distributed over an area about 100 km long and 50 km wide north of the Vogelsberg. The rocks have been studied in the last few years particularly by Wedepohl (1961, 1963a, 1963b, 1968a and b). Other petrographic and petrochemical papers have been published by Hentschel (1958), Schwarzmann (1957), Koritnig (1964) and Lohmann (1964). The most common basalt types are alkaline olivine basalts; in some subdistricts, basanites, limburgites and olivine nephelinites also occur. The most basic members of the basaltoid group contain melilite (36–9% SiO_2). More acid basalt varieties (tholeiite basalt, trachydolerite) are rare. The average analyses of the three most important types of basalts are shown in Table 4. The analyses of the alkaline olivine basalts are very similar to those of the average of the oceanic alkaline olivine basalts; but the potash content of this continental series is almost twice as high as that of the oceanic rocks.

The presence of melilite in the rocks and its origin have been discussed by Wedepohl (1961,

1963a) and Lohmann (1964). By reason of the particularly high Cr and Ni contents, Wedepohl arrives at the conclusion that the assimilation of limestone cannot be the cause of the undersaturation and the formation of melilite. Lohmann (1964) assumes that reactions of the alkali basaltic melt with early precipitates, such as olivine, augite and magnetite, led to the crystallization of melilite.

Geochemical studies of alkali basalts and their peridotite inclusions led Wedepohl (1963) and Herrmann and Wedepohl (1966) to the result that the peridotites were not accumulates of early minerals of the enclosing rocks. The origin of peridotite nodules, which are frequent in French and German alkaline basalts, will not be discussed here; attention may be drawn to the papers of Ernst, Frechen, de Roever, Den Tex and Wedepohl in *Neues Jahrbuch für Mineralogie*, Monatshefte 1963, and to the papers of Ernst (1936), Brousse (1967b) and Schütz (1967). $^{87}Sr/^{86}Sr$ ratios of the Tertiary basalts of all types range between 0·7031 and 0·7054; they are very similar to those obtained from Hawaii. The ratios found in phonolites and trachytes from the Westerwald and the Laacher See area are probably influenced by crustal material (Hoefs and Wedepohl, 1968).

IV.4.4. The Alkali Magmatites of the Bohemian Massif and its Surroundings

IV.4.4.1. General Remarks

The two main districts of this subprovince, the Doupovské horý and the České Středohoří, are situated within and on the margins of the North Bohemian basin, a graben-like depression in the Bohemian Massif, filled with Tertiary sediments. Many smaller groups of volcanoes and single volcanoes are distributed over a large area besides the two principal centres. According to the situation and the age of the volcanic formations, the subprovince may be subdivided into the following units:

a. Volcanic district of the Oberpfalz (Bavaria); Miocene–Pliocene; vent fills and several lava flows; alkaline olivine basalts and olivine nephelinites (Strunz, 1967).

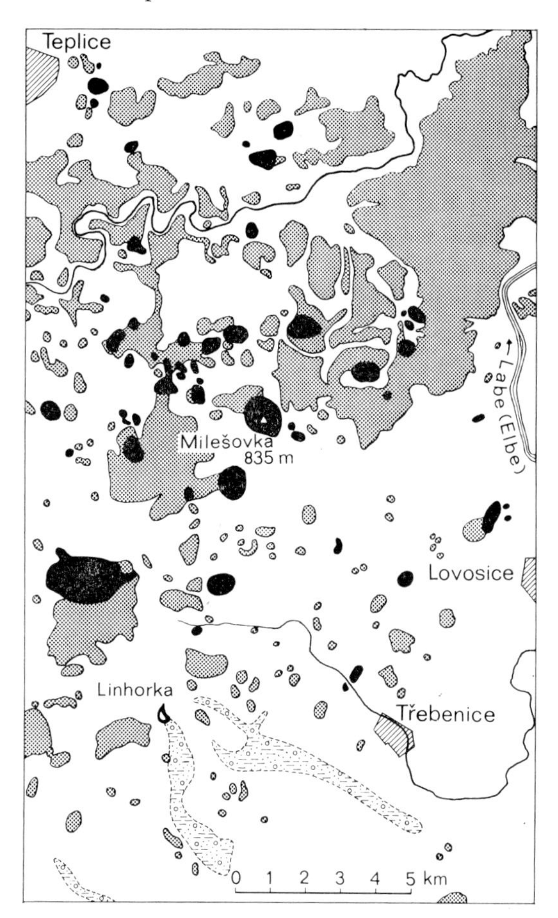

Fig. 10 Simplified map of the southwestern section of the České Středohoří, after Hibsch (1926)
Black: phonolite plugs; shaded: basaltic and tephritic lava sheets and eroded necks. Elongated areas in the southern part of the map are pyrope-bearing Quaternary sediments, originating from ultrabasic vent breccias, e.g. Linhorka

b. Doupovské horý, a stratovolcano approximately 30 km in diameter; Oligocene, Miocene (?); alkaline olivine basalts and their tuffs, tephrites, trachybasalts, trachytes, leucitites and olivine leucitites, olivine nephelinites, nephelinites, melilite-bearing rocks (Zartner, 1938, Kopecký, 1966a). Central essexite plug with dyke suite (Tröger, 1939).

c. České Středohoří (Böhmisches Mittelgebirge to the German writing authors); a very complex district with lava sheets, tuffs, plugs, small suvolcanic intrusions and many dykes

(Fig. 10); middle Oligocene to uppermost Miocene (eventually lowest Pliocene); great variation of the magmatites from ultrabasic melilite rocks and carbonatites to foidal trachytes (Kopecký 1966a). Many, mostly small single occurrences of Miocene basalts, olivine nephelinites and phonolites are strewn over two zones 40 km wide both north and south of the Doupovské horý and the České Středohoří. They stretch from the vicinity of Dresden in the N.E. down to Plzen in the S.W. (Pietzsch, 1963).

d. The volcanic districts of the Lausitz–Elbe (Labe) volcano-tectonic zone and the Bohemian Cretaceous, closely connected with *b*; upper Miocene; alkaline olivine basalt, olivine nephelinite, phonolite (Pacák, 1957, Pietzsch, 1963, Kopecký, 1966a).

e. Volcanic district of Silesia (Smulikovski, 1957, 1960; Kardymowicz, 1967, Birkenmajer and Nairn, 1969, Szpila, 1962). This belt runs from the northern end of the Lausitz district near Görlitz to the Annaberg (Swieta Anna Mt., Chodyniecka, 1967) near Oppeln east of the Oder river; mostly dispersed single occurrences, predominantly basalts, basanites and olivine nephelinites; Miocene.

f. Quaternary volcanoes near Cheb (Eger to the German writing authors); olivine nephelinite to olivine melilitite (Kettner, 1958).

g. Quaternary volcanoes near Bruntál (northern Moravia); alkaline olivine basalt to olivine nephelinite (Pacák, 1928).

h. Basic, partly alkaline magmatites in the Silesian nappe of the West Carpathians, beyond the eastern edge of the map, Fig. 1 (Roth and Hanzlíková, 1968). Submarine effusions, tuffs, tuffites, sills. Teschenites, picrites and diabases, often with strong deuteric alterations. Lower Cretaceous. The association is remarkably similar to the Cretaceous magmatites on the northern margin of the Pyrenees (see IV.4.2.6).

The volcanic activity in Northern Bohemia can be subdivided, after Kopecký (1966a), into three main phases. The first phase coincides approximately with the Savian movements of the Oligocene–Miocene limit. It was by far the most productive period; its volcanic buildings are now strongly eroded, so subvolcanic intrusions are frequently exposed at the surface. All kinds of alkaline rocks, the most basic types excepted, were formed during the first phase. Knorr (1932) made a more detailed distinction of an oldest basalt-phonolite stage, an essexite-tephrite stage, a younger trachyte-phonolite stage and an ultimate stage of basalts and nepheline-melilite rocks. The second phase, approximately contemporaneous with the Attic orogenic phase in the lowest Pliocene, yielded alkaline olivine basalts and more undersaturated basaltoids such as tephrites, olivine nephelinites and melilite-bearing rocks (polzenites, see IV.4.4.2). The third, Quaternary phase in western Bohemia and northern Moravia produced olivine basalts, olivine nephelinites and olivine melilitites. It may be parallelized with the Rhodanic or Wallachic orogenic phase.

IV.4.4.2. The České Středohoří

The alkaline rocks of this region vary widely from basalt through tephrite to phonolite and also include more uncommon types, such as leucitite, hauynitite, analcimite, picrite and olivine melilitite. Towards the more silicic side, foidal trachyte and trachyte are developed. Essexite (rongstockite) and sodalite syenite occur in small subvolcanic massifs (Table 8). Dyke rocks, *viz.*, monchiquite, gauteite, tinguaite and bostonite are very frequent around these massifs. In the Polzen area, at the border between the České Středohoří and the Lausitz–Elbe zone, alkali-ultrabasic dyke rocks occur. At the south-eastern margin of the České Středohoří there are several vent breccias with alkali-picrite igneous constituents (so-called 'kimberlitoids' (Strnad, 1962)). A speciality of the region is the abundance of minerals of the sodalite-hauyne group, which occur frequently along with or instead of nepheline. Chemical analyses of the rocks have been compiled by Jung (1928, pp. 160–99), Hibsch (1926, pp. 144–68), Kopecký (1966a, pp. 558–61) and Kavka (1968, pp. 106–48). Kavka's collection (220 analyses) comprises nearly all the rock types of the region except the basalts proper and the ultrabasic basaltoids. Many extreme alkaline rocks with

TABLE 8 Igneous Rocks of the České Středohoří

	1	2	3	4	5	6
SiO_2	42·1	44·43	54·44	55·94	50·27	30·05
TiO_2	1·5	4·09	0·30	0·14	1·54	1·62
Al_2O_3	15·5	12·50	22·68	21·97	17·94	9·64
Fe_2O_3	5·2	7·24	2·69	2·47	6·84	2·12
FeO	7·0	6·37	0·27	0·10	1·76	9·10
MnO	0·2	0·21	0·27	0·33	0·28	0·99
MgO	9·5	5·80	0·5	tr.	2·15	16·15
CaO	11·4	11·16	0·17	1·53	4·56	17·65
Na_2O	3·6	3·35	9·57	8·60	7·73	2·91
K_2O	1·7	2·50	5·99	5·54	4·09	1·94
P_2O_5	0·7	0·75	0·07	tr.	0·03	0·91
CO_2	0·3	—	1·26*	0·94*	—	3·58
H_2O^+	1·3	0·95			2·22	3·72
	100·0	99·35	98·80	99·40	99·41	100·94

Legend:
1. Average of alkaline olivine basalts ('Feldspatbasalte'), 8, calculated from the data of Hibsch, 1926.
2. Essexite, Leština E Ustí (Kavka, 1968, Anal. Holecková.)
3. Phonolite, used in the glass industry, Želenice.
4. Phonolite, used in the glass industry, Nestemice.
5. Sodalite syenite, Valtírov near Ustí.
6. Vesecite of Kleinhaida (from Scheumann, 1922), with 0·56% SO_3.

(3)–(5) from Kopecký (1966).

* = loss on ignition.

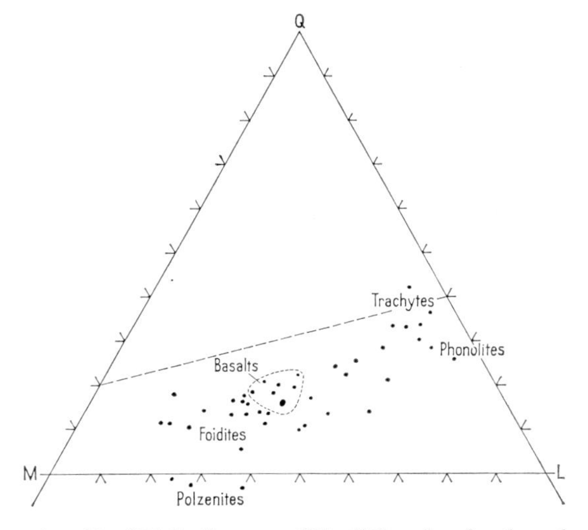

Fig. 11 QLM diagram (Niggli-Burri) of selected analyses from the Doupovske horý and České Středohoří (from the data of Kopecký, 1966). Large dot indicates the average of Hibsch's 'feldspar basalts', representing more than 60 vol-% of the volcanic rocks of the district

strong SiO_2-undersaturation and alkali contents almost reaching 17 wt.% are listed. Some of these rocks are used in the glass industry (Table 8). The *QLM*-values of the analyses listed by Kopecký (1966a, pp. 558–61) are shown in Fig. 11. Furthermore, the average of Hibsch's analyses 27–32, 34 and 36 is plotted. It represents the most frequent basalt type (62% of all exposed rocks in the České Středohoří).

Geochemical studies have been carried out by Macháček and Shrbený (1970a, b). There is a regular variation of the minor elements in the series olivine basalt–trachyte; the phonolites and melilite rocks appear to be products of special trends of evolution.

Kavka (1964–8) uses the term phonolite in a very wide sense and divides this rock family into 'foidal trachytes' and 'foidal trachyandesites'. The characteristic enrichment in alkalis and desilification go parallel with the decrease of the mafic minerals; a maximum is attained in a definite state of evolution (e.g. at colour indices of 12 to 14 in many phonolites of the Most, Bilin and Ustí regions). With a further decrease of the colour index, the foid content decreases, too. Kavka rejects simple differentiation of the parent magma by sinking of olivine and pyroxene as the probable cause of the alkali enrichment. He believes it to be the result of pneumatolytic processes, *viz.*, assimilation of alkaline fluids by a basaltic magma. These alkaline fluids are believed to originate from a 'crustal' source rather than from the magma itself.

Other indications of alkali migration between magma and crust have been described by Kopecký (1966b) and Kopecký, Fiala, Dobeš and Stovíčková (1970). Among the xenoliths of the pipe breccia near Kostal (on the southern margin of the České Středohoří), leucocratic and mafitic *fenites* are found. The gneisses, granulites, granodiorites, granites and gabbrodiorites of the basement have been affected by fenitization. The transformations correspond to those of the fenitization zones I to IV of von Eckermann (1948). Nepheline syenites (with zircon and

TABLE 9 Modal Composition of Polzenitic Dyke Rocks from the Polzen Area (Tröger, 1939) and of Fenites from Hůrky (Kopecký *et al.*, 1970), vol %.

	1	2	3	4
Melilite	33	34	22	—
Monticellite	10	—	—	—
Olivine	23	22	15	5
Augite	—	—	—	30
Ti-amphibole	—	—	—	20
Biotite	12	22	8	20
Ore minerals, perovskite, apatite	8	6	5	6
Nepheline	14	16	11	14
Hauynite, lasurite	—	—	5	5
Calcite	sec.	sec.	13	sec.

	5	6	7	8	
Plagioclase	27·1	—	—	—	relict minerals
K–feldspar	37·2	20	10	>2	
Biotite	7·5	—	—	—	
Muscovite	2·4	—	—	—	
Accessories	0·7	—	—	—	
Quartz	25·1	25	+	—	
K–feldspar	—	20	25	25	new-formed minerals
Albite	—	25	55	45	
Biotite	—	2	—	—	
Alkali pyroxene	—	—	7	5	
Alkali amphibole	—	7	—	—	
Nepheline	—	—	—	10	
Cancrinite	—	—	—	10	
Apatite, sphene	—	+	+	+	
Zircon, eudialyte	—	—	+	>1	

Legend:
1. Vesecite.
2. Modlibovite.
3. Luhite.
4. Wesselite.
5. Porphyritic biotite granite, Tis type.
6. Alkaline hornblende granitic fenite.
7. Alkali pyroxene syenitic fenite.
8. Cancrinite-nepheline syenite.

pyrochlore) and carbonate-bearing types are the most evolved rocks of this series. One single block of phlogopite sövite has been encountered. The mafitic fenites are partly alkali pyroxenites, some of them biotite-bearing, and partly alkali hornblendites. Calcite, apatite, titanite and pyrochlore occur as additional minerals. At Hurky, 70 km west of Prague, the Hercynian granite of Tis is fenitized in an area of 800 × 300 m (Klomínský, 1963, Kopecký *et al.*, 1970). Fenites of the stages I to IV are represented; they range from slightly transformed granites up to cancrinite-nepheline syenites (Table 9). The fenitization is connected with intense mylonitization of the granite. The geological situation indicates a pre-Tertiary age of the alkali metasomatism; it belongs, consequently, not to the Tertiary magmatism of the region. The same is held, too, for the xenoliths of the Kostal breccia.

The pyrope-bearing pipe breccias on the southern margin of the České Středohoří (Kopecký, Píšová and Pokorný, 1967) are more or less altered alkaline olivine basalts and picrites. In the breccias, there occur also fragments of garnet periodotite with pyrope, ortho- and clinopyroxene and chromian spinel, the source rocks of which have been detected in the crystalline basement in several bore holes at depths of only a few hundred metres. Therefore, these rocks are not fragments of the Earth's mantle, but constituents of the metamorphic assemblage of the Bohemian Massif. The pyropes released by weathering of these pipe breccias have been redeposited in Quaternary sediments (see Fig. 10) and have been exploited during several centuries for jewellery. In the same sediments, three diamonds were found in 1869, 1927 and 1958, but attempts at diamond extraction on an industrial scale gave no positive results. Among the numerous heavy minerals of the breccias, moissanite (SiC) should be mentioned (Bauer, Fiala and Hříchová, 1963).

At the eastern end of the České Středohoří, near to the Lausitz–Elbe (Labe) zone, there appear, in the Polzen area, numerous alkaline ultrabasic dykes up to 10 km in length. The rocks are partly melilite-ankaratrites and olivine nephelinites, partly the so-called *polzenites*, characterized by uncommon mineralogic compositions (Scheumann, 1922: vesecite, modlibovite, luhite and wesselite; see Table 9). As a consequence of extreme undersaturation in silica, some of these rocks lack pyroxene. Scheumann explains this by assimiliation of the cretaceous limestone, which is cut by the dykes. These dyke rocks are comagmatic members of a more comprehensive series, which contains trachybasalt (essexitic composition), phonolites hauynophyres, gauteites and camptonites. Scheumann (1922, p. 545) refrained from deriving the alkaline character of the series from limestone syntexis. After Kopecký, Fiala, Dobeš and Štovíčková (1970), the melilite rocks (and the phonolites) occur preferably in connection with deep-reaching fractures of the Krušné horý–Ohře tectonic–volcanic zone.

IV.4.4.3. The Doupovské horý

The rock assemblage of the Doupovské horý begins with a volcanoclastic series up to 100 m thick. The centre of the hills is a large stratovolcano, 80% of which consists of pyroclastics. The interstratified lavas are first leucitites, olivine leucitites and olivine nephelinites, later basalts, basanites and tephrites. The youngest lava effusions, again nephelinites, leucitites, tephrites, phonolites and trachytes, erupted particularly in the more peripheral parts of the stratovolcano. Essexite bodies with dykelets of nepheline syenite intruded comparatively late within the central vent and are accompanied by a suite of radially oriented dykes of monchiquite, camptonite and bostonite. As a whole, the rocks of the Doupovské horý are characterized by a more or less strong undersaturation with respect to silica and by the frequent presence of leucite. More details and references to the earlier literature have been given by Kopecký (1966a).

IV.4.5. Conclusions

Alkaline rocks, particularly olivine basalts, form the bulk of the rocks in the Tertiary and Quaternary magmatic province of France and Central Europe. The following rock associations are most frequent:

a. alkaline olivine basalt–tephrite–phonolite;
b. alkaline olivine basalt–trachyandesite–trachyte.

The nomenclature of the intermediate part of the series is applied very variably. Therefore a number of other names are often used, which designate at least partly rocks of similar composition to those chosen above (e.g. trachybasalt, trachydolerite, ordanchite, doréite, hawaiite, mugearite, labradorite, andesite and others).

Basaltoid rocks poorer in silica, such as basanite or limburgite occur also frequently in these associations. Ankaramites are mostly regarded as accumulative modifications of the alkaline olivine basalts. Feldspar-free basaltoids, especially olivine nephelinites and melilite ankaratrites, are not rarely members of the two main associations.

Complete examples of these associations are:

(*a*) Velay, Mont Dore series B, Cantal series B, Rhön, České Středohoří; (*b*) Chaîne des Puys, Hocheifel, Westerwald, České Středohoří.

Associations of a third type

c. alkaline olivine basalt–trachyandesite–sancyite–rhyolite

are very well represented in Mont Dore and the Cantal, whereas in Central Europe, the quartz-bearing end products of differentiation are very scarce.

Incomplete associations, the intermediate members of which are partly or entirely missing, are also fairly frequent. In such cases, basalts and phonolites or trachytes or even ultrabasic feldspar-free basaltoids and feldspar-rich phonolites are confronted with each other, for instance basalts and phonolites in Vogelsberg, melilite ankaratrites and phonolites in the Hegau, olivine nephelinites, basalts and phonolites in the Erzgebirge (Saxony) and in the Lausitz.

In rarer cases, e.g. in the Kaiserstuhl, basalts proper are almost nonexistent. The dominating rock association here is tephrite (and its subvolcanic equivalent essexite)-phonolite; olivine-bearing rocks occur only in the more peripheral parts of the volcano: limburgite, olivine nephelinite. In Kaiserstuhl, there also appear different varieties of carbonatite. Breccias with ultrabasic constituents and extremely potassic trachytic and phonolitic rocks are found in the immediate neighbourhood of the carbonatites (Wimmenauer, 1966, Sutherland, 1967).

Several districts of the French and Central European subprovinces consist almost exclusively of basalts, basanites, olivine nephelinites and other basaltoid rocks: Margeride, Devès, Chaîne de la Sioule, Auvergne occidentale, Forez, Bourgogne, Causses, Escandorgue, Bas-Languedoc; Vogelsberg, northern Hessen, Oberpfalz, Silesia and, to a certain degree, also the stratovolcano of Doupovské horý. Several of these districts have specific qualities, e.g. frequency of leucite-bearing rocks in Doupovské horý.

The olivine nephelinites and some melilite ankaratrites appear partly associated with other basalts and their differentiates, partly also independently in small, often isolated vents and dykes (Wimmenauer, 1963; Jung and Brousse, 1962, pp. 31 and 32).

This mode of occurrence proves the autonomous development of these ultrabasic basaltoids with regard to the more common basalt types. There are, of course, also transitional rocks (e.g. basanites).

Some districts show very uncommon rocks and rock associations, which indicate special ways of evolution:

The Quaternary volcanoes of the Laacher See district (ultrabasic-alkaline basaltoids, tephrites, peralkaline phonolites (= selbergites), foidal latites and foidal trachytes.

Katzenbuckel: Sanidine nephelinite and soda shonkinite.

Polzen district: ultrabasic melilite rocks ('polzenites'), melilite ankaratrites, trachybasalts and phonolites.

Several theories have been put forward during the last decades for the interpretation of the different alkaline associations of Central Europe and France. They are briefly summarized on the following pages.

a. 'Classic' fractional crystallization–differentiation, e.g.: alkaline olivine basalt–trachyte association of the Hocheifel (Huckenholz, 1965, 1966); alkaline olivine basalt–tephrite–phonolite associations of the Mont Dore and Cantal (Brousse, 1961b, Vatin–Pérignon, 1968); basalt–trachyte association of the Puy de Dôme (Bentor, 1954); association of the České Středohoří (Knorr, 1932); Eifel and Westerwald (Jasmund *et al.*, 1972).

b. Crystallization–differentiation with gas differentiation: Agpaitic phonolites of Cantal (Varet, 1969); Quaternary volcanics of the Laacher See district (Frechen, 1962); essexite → sodalite syenite in České Středohoří (Knorr, 1932); assimilation of alkaline fluids (phonolites of the same region: Kavka, 1965). Liquid immiscibility is envisaged by de Goër de Hervé (1968) for vesicular pegmatoids in dolerites from Cantal.

c. Crystallization–differentiation and assimilation of acidic, sialic material: Basalt–trachyandesite–sancyite–rhyolite associations of Mont Dore and Cantal (Brousse, 1961b, Vatin-Pérignon, 1968); phonolites of the Kaiserstuhl (Eigenfeld, 1954; Wimmenauer, 1966); some aberrant basalts of the Vogelsberg (Ernst, 1968).

d. Crystallization–differentiation and limestone syntexis: polzenites and other melilite rocks of the Polzen district (Scheumann, 1922); phonolites and other undersaturated alkaline rocks of the Kaiserstuhl and the Rhön (Eigenfeld, 1954, Ficke, 1961); undersaturated alkaline rocks of the Eifel (Burri, 1960); melilite ankaratrites of the Hegau (Leopold, 1940). For discussion of this theory see Wedepohl (1961, 1963).

e. Mixing of a sialic and a simic (basaltic) magma: Rock series of Mont Dore (Glangeaud and Létolle, 1962, 1965).

f. Metasomatic processes: Ultrapotassic trachytes in Kaiserstuhl (Sutherland, 1967); fenite ejectamenta in the Kaiserstuhl and in the České Středohoří (Wimmenauer, 1966; Kopecký, 1966b, Kopecký *et al.*, 1970).

In most of the quoted theories, primary *basaltic parent magmas* are assumed. The plotting of the average analyses of the basalts of the most important magmatic districts in an

Alk_2O/SiO_2 diagram shows clearly their close affiliation to the alkaline olivine basaltic type (Fig. 12). The calculation of the *CIPW*-norm of 251 basalts of France by Velde (1967) proves the strong predominance of nepheline-normative rocks (221). Already Jung (1928) calculated a theralite-gabbroic composition for the predominant rock type in Central Europe, which corresponds very well to the alkaline olivine basalt average. Only the 'trap' basalts of the Vogelsberg and the Main 'trap' have a more peripheral position. Ernst (1968) and Ernst *et al.* (1968) explain the difference between the alkaline olivine basalts and the olivine tholeiites of the Vogelsberg as the result of fractional crystallization or fractional fusion at different depths.

In addition to the alkaline olivine basalts with about 41–7% SiO_2, olivine nephelinites and melilite ankaratrites with 35–41% SiO_2 also exist at numerous places. Kushiro and Kuno (1963) and Bultitude and Green (1968) assume a particularly great depth of origin for these magmas. Recently, Varne (1968) envisaged the possible origin of olivine nephelinite magma by incongruent melting of pargasitic amphibole in the mantle (cf. II.3). Spencer (1969) regards olivine nephelinite as a parental magma type, capable of differentiating into either melilite-bearing or basanitic and tephritic daughter magmas. Kopecký *et al.* (1970) emphasize the close connection of the melilite rock–phonolite association with deep-reaching fractures in the basement of northern Bohemia. The mode of formation of the melilite rocks is now regarded as a purely magmatic process; for most of them, limestone syntexis as the cause of the undersaturation in silica can be excluded (Wedepohl 1963a). The same author (Wedepohl 1968b, p. 65) emphasizes the uniformity of the $^{87}Sr/^{86}Sr$- and Rb/Sr-ratios in all kinds of basalts of northern Hessen and Niedersachsen; these observations indicate a similar provenance of the three main basalt types of the region. Elements, which are specific for magmatic residuals (also volatiles) are increasingly concentrated in the rock sequence tholeiite basalt–alkaline olivine basalt–olivine nephelinite (see IV.4.3.5.2). A more detailed sequence of rock types was established

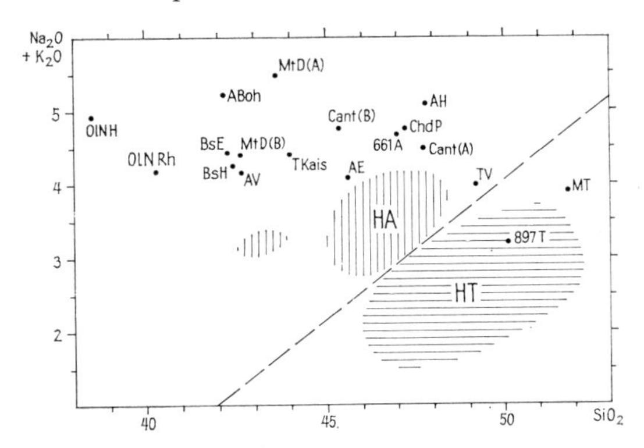

Fig. 12 $Alk_2O–SiO_2$ diagram of the main basalt types of the Central European and French province, compared with the Hawaiian basalts (Macdonald and Katsura, 1964). Abbreviations: ABoh = alkaline basalts, Bohemia; AE = alkaline basalts, Hocheifel; AH = alkaline basalts, northern Hessen and southern Niedersachsen; AV = alkaline basalts, Vogelsberg; BsE = basanites, Hocheifel; BsH = basanites northern Hessen and southern Niedersachsen; Cant(A) = basalt, saturated series, Cantal; Cant(B) = basalts, undersaturated series, Cantal; ChdP = basalts Chaîne des Puys; HA = Hawaiian alkaline olivine basalts (from Macdonald and Katsura, simplified); HT = Hawaiian tholeiitic basalts (from Macdonald and Katsura, simplified); MtD(A) = basalts, saturated series, Mont Dore; MtD(B) = basalts, undersaturated series, Mont Dore; MT = Main 'trap'; OlNH = olivine nephelinites, northern Hessen and southern Niedersachsen; OlNRh = olivine nephelinites, Rhine graben; TV = 'trap' basalts, Vogelsberg; 661 A = average of 661 alkaline basalts (Manson, 1967); 897 T = average of 897 tholeiitic basalts (Manson, 1967)

by Herrmann (1968) on the basis of the lanthanide distribution and the (La-Eu/Gd-Lu) ratio. Again, tholeiite basalt appears to be the primitive, olivine nephelinite the most developed basalt type.

Yamasaki and Brousse (1963) discuss the variability of the basalts with particular respect to their associations in France. They arrive at the conclusion, that the differences between the main types are the consequence of different depths of origin. The most undersaturated olivine nephelinites and melilite ankaratrites had the longest distance of ascent; they appear but in small quantities. With decreasing depth of formation, the composition of the magmas gradually passed over to that of the alkaline olivine basalt and finally to that of the tholeiitic basalt. Crystallization–differentiation and assimilation of sialic rocks contribute essentially to the further evolution of the primary magmas.

Ernst and Schwab (1971) drew the attention to the possible substitution of C^{4+} and $(H_4)^{4+}$ for Si^{4+} in silicates at very high pressures. Decomposition of such minerals would lead to undersaturated melts and release of important quantities of gases. The authors emphasize the bearing of this process on the well-known strong degassing of ultrabasic magmas, and on the origin of kimberlites and carbonatites.

Concluding, it may be stated that the common association of many types of alkaline rocks with abundant alkaline olivine basalt clearly evidences their close genetic relationship, namely by means of crystallization–differentiation. Olivine nephelinite may also locally assume the rôle of a parental magma. Several rock types, unusually rich in alkalis, require additional hypotheses, preferably gas differentiation or related processes.

ACKNOWLEDGEMENTS

The author is greatly indebted to Prof. Dr. R. Brousse, Prof. Dr. J. Frechen and Dr. L. Kopecký for critically reading parts of the manuscript, and to Mrs. Erica Jupe for editing the English of the text.

IV.4. REFERENCES

Ahrens, W., 1931. Altersfolge und Kennzeichnung der verschiedenen Trachyttuffe des Laacher-See-Gebietes. *Neues Jb. Miner., Beilage Bd.* **64 A**, 517–44.

Ahrens, W., 1957. Exkursion im südwestlichen Westerwald, I. *Fortschr. Miner.*, **35**, 109–16.

Ahrens, W., and Villwock, R., 1964. Exkursion in den Westerwald am 6. Sept. 1964. *Exkursionsführer der Deutschen Mineralogischen Gesellschaft*, **42**. Jahrestagung, 23 pp.

Azambre, B., 1967. *Sur les roches intrusives sous-saturés du Crétacé des Pyrénées (Picrites, teschénites, monchiquites, syénites néphéliniques).* Thesis, Paris, 147 pp.

Azambre, B., 1970. Les monchiquites et autres roches basiques intrusives accompagnant les syénites néphéliniques des Corbières. *C. R. Acad. Sci. Paris, Ser. D*, **271**, 641–3.

Babkine, J., Conquéré, F., and Vilminot, J. C., 1968. Les caractères particuliers du volcanisme au nord de Montpellier: l'absarokite du Pouget, la ferrisalite sodique de Grabels. *Bull. Soc. fr. Miner. Cristallogr.*, **91**, 141–50.

Bauer, J., Fiala, J., and Hřichová, R., 1963. Natural alpha silicon carbide. *Am. Miner.*, **48**, 620–34.

Becke, F., 1903. Die Eruptivgebiete des böhmischen Mittelgebirges und der amerikanischen Anden. *Tschermak's miner. petrogr. Mitteil.*, **22**, 209–65.

Bentor, Y. K., 1954. La chaîne des Puys (Massif Central Français). *Bull. Serv. Carte géol. France*, **52**, No. 242, 373–806.

Berger, E., Brousse, R. and Causse, Chr., 1968. Les pipes et les diatrèmes des Causses septentrionaux. *Bull. Soc. géol. France*, 7ème série, **10**, 588–600.

Birkenmajer, K., and Nairn, A. E. M., 1969. Paleomagnetic investigations of the Tertiary and Quaternary igneous rocks. V. The basic Tertiary basalts of Lower Silesia, Poland. *Geol. Rundschau*, **58**, 697–712.

Branco, W., 1894. *Schwabens 125 Vulkan-Embryonen und deren tufferfüllte Ausbruchsröhren.* Stuttgart, 816 pp.

Brauns, R., 1911. *Die kristallinen Schiefer des Laacher See-Gebietes und ihre Umbildung zu Sanidinit.* Schweizerbart, Stuttgart, 61 pp.

Brauns, R., 1921. Die phonolithischen Gesteine des Laacher-See-Gebietes und ihre Beziehungen zu anderen Gesteinen dieses Gebietes. *Neues Jb. Miner., Beilage-Band*, **46**, 1–116.

Brauns, R. and Brauns, A., 1925. Ein Carbonatit aus dem Laacher Seegebiet. *Cbl. Miner. etc.*, **A**, 97–101.

Brousse, R., 1961a. Analyses chimiques des roches volcaniques tertiaires et quaternaires de la France. *Bull. Serv. Carte géol. de la France*, **58**, No. 263, 1–140.

Brousse, R., 1961b. Minéralogie et pétrographie des roches volcaniques du Mont-Dore (Auvergne). *Bull. Soc. franç. Miner. Cristallogr.*, **84**, 131–86.

Brousse, R., 1963. La phonolite des Compains et les phonolites néogènes de France. *C. R. Congr. Soc. Savantes*, 88ème session, Clermont-Ferrand, 93–114.

Brousse, R., 1967a. Géochimie de l'uranium dans les roches volcaniques du massif du Mont Dore (Auvergne). *C. R. Congr. Soc. Savantes*, 91ème session, Rennes, 269–85.

Brousse, R., 1967b. La place des ultra-basites en France. *Geol. Rundschau*, **57**, 621–55.

Brousse, R., Delibrias, G., Labeyrie, J., and Rudel, A., 1966. Datation par la méthode du carbone 14, d'une éruption dômitique de la Chaîne des Puys. *C. R. Acad. Sci. Paris*, **263**, 1812–15.

Brousse, R., Delibrias, G., and Labeyrie, J., 1970. Utilisation des sols fossiles sous scories, pour la datation par le Carbone 14, du volcanisme. *Bull. Volcanol.*, **34**, 254–60.

Brousse, R., and Lefèvre, C., 1967. Nappes des ponces de Cantal et du Mont-Dore; leurs aspects volcanologiques, pétrographiques et minéralogiques. *Bull. Soc. géol. France*, 7ème sér., **8**, 223–45.

Brousse, R., and Varet, J., 1967. Les trachytes du Mont-Dore et du Cantal septentrional et leurs enclaves. *Bull. Soc. géol. France*, 7ème sér., **8**, 246–62.

Brousse, R., Varet, J., and Bizouard, H., 1969. Iron in the minerals of the sodalite group. *Contr. Miner. Petrology*, **22**, 169–84.

Bücking, H., 1910. Die Basalte und Phonolithe der Rhön, ihre Verbreitung und ihre chemische Zusammensetzung. *Sitzungsber. preuss. Akad. Wiss.*, **24**, 490–519.

Burri, C., 1960. Petrochemie der Capverden und Vergleich des Capverdischen Vulkanismus mit demjenigen des Rheinlandes. *Schweiz. mineral. petrogr. Mitt.*, **40**, 151–62.

Capdevila, R., 1962. Etude pétrographique et structurale du complexe volcanique de Saint-Thibéry (Herault). *Bull. Soc. géol. France*, 7ème sér., **4**, 13–17.

Chodyniecka, L., 1967. The basalt from Swieta Anna mountain. *Polska Akad. Nauk. Prace Miner.*, **8**, 56 pp.

Cloos, H., 1941. Bau und Tätigkeit von Tuffschloten. *Geol. Rundschau*, **32**, 6–108.

Cloos, H. and Cloos, E., 1927. Die Quellkuppe des Drachenfels am Rhein, ihre Tektonik und Bildungsweise. *Z. Vulkanol.*, **11**, 33–40.

Colin, F., 1970. Les basaltes alcalins sodiques de l'Aubrac (Massif Central Française). *Bull. Volcanol.*, **33**, (1969), 1237–45.

Coulomb, R., and Goldstejn, M., 1964. Comportement géochimique du Ta, Th, Co, Sc, U dans les laves de la Chaîne des Puys. *Bull. Soc. franç. Miner. Cristallogr.*, **87**, 163–5.

Coulon, C., 1967. Le volcanisme tertiaire de la région toulonnaise. (Var). *Bull. Soc. géol. France, 7ème sér.*, **9**, 691–700.

Dupuy, C., 1965. Variation des teneurs en Cu, Ni, Cr, Co et V dans quelques roches volcaniques des Causses et du Bas-Languedoc. *Bull. Soc. géol. France*, 7ème sér., **1**, 32–6.

v. Eckermann, H., 1948. The alkaline district of Alnö Island. *Sveriges geol. Unders.*, Ser. Ca., **36**, 176 pp.

Eigenfeld, R., 1954. Zur Genese von Alkaligesteinen. *Ber. phys. -med. Ges. Würzburg*, **66**, 95–114.

Eigenfeld, R., 1963. Über Feldspat- und Plagioklas–Augit–Knollen in Vulkaniten der Rhön und der Hassberge. *Fortschr. Miner.*, **40**, 61.

v. Engelhardt, W., and Weiskirchner, W., 1963. Einführung zu Excursionen der Deutschen Mineralogischen Gesellschaft zu den Vulkanschloten der Schwäbischen Alb und in den Hegau vom 11–17 Sept. 1961. *Fortschr. Miner.*, **40**, 5–28.

Ernst, Th., 1936. Der Melilithbasalt des Westberges bei Hofgeismar, nördlich von Kassel, ein Assimilationsprodukt ultrabasischer Gesteine. *Chemie der Erde*, **10**, 631–66.

Ernst, Th., 1961. Folgerungen für die Entstehung der Basalte aus dem speziellen Auftreten der Pyroxene in diesen Gesteinen. *Geol. Rundschau*, **51**, 364–74.

Ernst, Th., 1968. Die Basalte des 'Maintrapps' und des Vogelsbergs. *Vortragsreferat, 3. Kolloquium der internat. Rheingraben-Forschungsgruppe in Karlsruhe*, 36–7.

Ernst, Th., Kohler, H., Schütz, D., and Schwab, R., 1968. Über die Möglichkeiten der Entstehung verschiedener Basaltmagmentypen im Vogelsberg (Hessen). *Vortragsreferat*, **46.** Jahrestagung der Deutschen Mineralogischen Gesellschaft, 14–15.

Ernst, Th., and Schwab, R., 1971. Eine neue Theorie über Bildung und Aufstieg basischer Magmen, insbesondere der untersättigten Serien. *Fortschr. Miner.*, **49**, Beih. 1, 93–5.

Ficke, B., 1961. Petrologische Untersuchungen an tertiären basaltischen bis phonolithischen Vulkaniten der Rhön. *Tschermaks miner.-petrogr. Mitt., III*, **7**, 335–436.

Frechen, J., 1947. Die Vorgänge der Sanidinitbildung im Laacher Seegebiet. *Fortschr. Miner.*, **26**, 147–66.

Frechen, J., 1948. Die Genese der Olivinausscheidungen vom Dreiser Weiher (Eifel) und Finkenberg (Siebengebirge). *Neues Jb. Miner. Abh.*, **79A**, 317–406.

Frechen, J., 1962. *Führer zu vulkanologisch-petrographischen Exkursionen im Siebengebirge am Rhein, Laacher Vulkangebiet und Maargebiet der Westeifel.* 151 pp., Schweizerbart, Stuttgart.

Frechen, J., 1963. Kristallisation, Mineralbestand, Mineralchemismus und Förderfolge der Mafitite vom Dreiser Weiher in der Eifel. *Neues Jb. Miner. Mh.*, 205–25.

Frechen, J., and Lippolt, H. J., 1965. Kalium–Argon–Daten zum Alter des Laacher Vulkanismus, der Rheinterrassen und der Eiszeiten. *Eiszeitalter und Gegenwart*, **16**, 5–30.

Frechen, J., and Vieten, K., 1970. Petrographie der Vulkanite des Siebengebirges. *Decheniana*, **122**, 337–77.

Frenzel, G., 1967. On petrochemistry and genesis of the volcanic rocks from the Katzenbuckel (Odenwald). *The Rhinegraben Progress Report, Abh. Geol. Landesamt Baden-Württemberg*, **6**, 131–3.

Glangeaud, Ph., 1913. La chaîne des Puys et la Petite Chaîne des Puys. *Bull. Serv. Carte géol. France*, **22**, 256 pp.

Glangeaud, L., and Létolle, R., 1962. Evolution géochimique des magmas du massif volcanique du Mont-Dore et de l'Auvergne. Méthodes et conséquences générales. *Bull. Soc. franç. Miner. Cristallogr.*, **85**, 296–308.

Glangeaud, L. and Létolle, R., 1965. La théorie des deux magmas fondamentaux dans le volcanisme intracontinental et l'évolution géochimique des laves du Mont-Dore (France). *Geol. Rundschau*, **55**, 316–29.

Goër de Hervé, A. de. 1968. Réflexions sur les pegmatoides bulleux des carrières de dolérites de Bouzentes (Planèze de Saint-Flour, Cantal). *C. R. Acad. Sci. Paris, Ser. D*, **267**, 2260–3.

Goër de Hervé, A. de, and Vatin-Pérignon, N., 1966. Les ordanchites et roches affines du massif du Cantal. *Bull. Soc. Géol. France*, 7ème sér., **8**, 298–307.

Grünhagen, H., 1964. *Petrographische und geochemische Untersuchungen an den tertiären Andesiten und Trachyten der Hocheifel.* Dissertation, Köln, 106 pp.

Grünhagen, H., and Mergoil, J., 1963. Découverte d'hydrocalumite et afwillite associés a l'ettringite dans les porcélanites de Boisséjour près Ceyrat (Puy-de-Dôme). *Bull. Soc. franç. Miner. Cristallogr.*, **86**, 149–57.

Gutberlett, H. G., and Huckenholz, H. G., 1965. Mineralbestand und Chemismus der Analcim–Alkalitrachyte vom Selberg und der Grader Seife bei Quiddelbach in der Hocheifel. *Neues Jb. Miner. Mh.*, 10–19.

Hentschel, G., 1964. Mayenit, 12CaO.7Al_2O_3, und Brownmillerit, 2CaO.$(Al,Fe)_2O_3$, zwei neue Minerale in den Kalksteineinschlüssen der Lava des Ettringer Bellerberges. *Neues Jb. Miner. Mh.*, 22–9.

Hentschel, H., 1958. 'Die vulkanischen Gesteine', in *Erläut. geol. Karte v. Hessen 1:25 000*, Blatt Kassel-West, 83–110.

Hentschel, H., 1968. Der Basalt des Meissner. *Aufschluss Sonderh.*, **17**, 151–65.

Hentschel, H., and Pfeffer, P., 1954. Chemisch-petrographische Untersuchungen an Basalten des Westerwaldes. *Geol. Jb.*, **69**, 361–78.

Herrmann, A. G., 1968. Die Verteilung der Lanthaniden in basaltischen Gesteinen. *Contr. Miner. Petrology*, **17**, 275–314.

Herrmann, A. G., and Wedepohl, K. H., 1966. Die Verteilung des Yttriums und der Lanthaniden in einem Olivin–Alkali–Basalt mit Peridotit–Einschlüssen. *Contr. Miner. Petrology*, **13**, 366–73.

Hibsch, J. E., 1926. Erläuterungen zur geol. Übersichtskarte des böhmischen Mittelgebirges und der unmittelbar angrenzenden Gebiete. *Tetschen, Selbstverlag des freien Lehrervereins.* 158 pp.

Hoefs, J., and Wedepohl, K. H., 1968. Strontium isotope studies on young volcanic rocks from Germany and Italy. *Contr. Miner. Petrology*, **19**, 328–38.

Huckenholz, H. G., 1965a. Der petrogenetische Werdegang der Klinopyroxene in den tertiären Vulkaniten der Hocheifel, I. *Beitr. Miner. Petrogr.*, **11**, 138–95.

Huckenholz, H. G., 1965b. Der petrogenetische Werdegang der Klinopyroxene in den tertiären Vulkaniten der Hocheifel, II. *Beitr. Miner. Petrogr.*, **11**, 415–48.

Huckenholz, H. G., 1965c. Die Verteilung des Niobs in den Gesteinen und Mineralen der Alkalibasalt-Assoziation der Hocheifel. *Geochim. Cosmochim. Acta*, **29**, 807–20.

Huckenholz, H. G., 1966. Der petrogenetische Werdegang der Klinopyroxene in den tertiären Vulkaniten der Hocheifel, III. *Beitr. Miner. Petrogr.*, **12**, 73–95.

Jasmund, K., and Hentschel, G., 1964. Seltene Mineralparagenesen in den Kalksteineinschlüssen der Lava des Ettringer Bellerberges bei Mayen (Eifel). *Beitr. Miner. Petrogr.*, **10**, 296–314.

Jasmund, K., and Schreiber, T., 1965. Geochemische und petrographische Untersuchungen an basaltischen Gesteinen der Eifel aus der Umgebung von Daun. *Chem. der Erde*, **24**, 27–66.

Jasmund, K., and Seck, H. A., 1964. Geochemische Untersuchungen an Auswürflingen (Gleesiten) des Laacher-See-Gebietes. *Beitr. Miner. Petrogr.*, **10**, 275–95.

Jasmund, K., *et al.*, 1972. 'Petrochemie und Petrologie des basischen Tertiärvulkanismus in Hocheifel und Westerwald sowie vergleichende experimentelle Untersuchungen,' Forschungsbericht *Unternehmen Erdmantel*, ed. Deutche Forschungsgemeinschaft (W. Kertz, K. v. Gehlen, F. Goerlich, G. Knetsch, H. Wolf), 256–60.

Jung, H., 1928. Die chemischen und provinzialen Verhältnisse der jungen Eruptivgesteine Deutschlands und Nordböhmens. *Chemie der Erde*, **3**, 137–340.

Jung, J., 1946. Géologie de l'Auvergne et de ses confins bourbonnais et limousins. *Mém. explic. Carte géol. France*, 372 pp.

Jung, J., and Brousse, R., 1962. Les provinces volcaniques néogènes et quaternaires de la France. *Bull. Serv. Carte géol. France*, **58**, No. 267, 1–61.

Kalb, G., 1934. Beiträge zur Kenntnis der Auswürflinge, im besonderen der Sanidinite des Laacher See-Gebietes. *Miner.-petrogr. Mitt.*, **46**, 20–55.

Kalb, G., 1935. Beiträge zur Kenntnis der Auswürflinge, im besonderen der Sanidinite des Laacher See-Gebietes. *Miner.-petrogr. Mitt.*, **47**, 185–210.

Kalb, G., 1936. Beiträge zur Kenntnis der Auswürflinge, im besonderen der Sanidinite des Laacher See-Gebietes. *Miner. -petrogr. Mitt.*, **48**, 1–26.

Kardymowicz, J., 1967. Inclusions in some Silesian basalts. *Bjul. 197 J. G., Z badan petrogr.-miner. i geochem. w Polsce I*, 451–84.

Kavka, J., 1965. Beitrag zur Kenntnis der Phonolith-Magma-Evolution im nordböhmischen Tertiär. *Acta Univ. Carolinae, Geologica*, **2**, 91–117.

Kavka, J., 1968. Tabelaro pri hemia kaj minerala konsisto de la nord-bohemiaj fonolitoj. *Geologio internacia*, **1**, 101–92 (in Esperanto).

Kettner, R., 1958. Nejmladší České sopky. *Vesmír*, **37**, 109–12.

King, B. C., and Sutherland, D. S., 1960. Alkaline rocks of eastern and southern Africa. *Science Progress*, **48**, 298–321, 504–23, 709–20.

Knorr, H., 1932. Differentiations- und Eruptionsfolge im Böhmischen Mittelgebirge. *Miner. petrogr. Mitt.*, **42**, 318–70.

Kopecký, L., 1966a. 'Tertiary volcanics', in *Regional Geology of Czechoslovakia*, **I**, 554–81, Publishing House of the Czechoslovak Acad. Sci., Prague.

Kopecký, L., 1966b. The find of fenites and alkaline rocks in the České Středohoří Mts. *Vestník Ústřed. Ústav. geol.*, **41**, 121–4.

Kopecký, L., Fiala, J., Dobeš, M., and Stovíčková, N., 1970. Fenites of the Bohemian Massif and the relations between fenitization, alkaline volcanism and deep fault tectonics. *Sborník geol. věd. (geol. Ser.)*, **16**, 51–112.

Kopecký, L., Píšová, J., and Pokorný, L., 1967. Pyrope-bearing diatremes of the České Středohoří Mountains. *Sborník geol. věd., Geologie, G.* **12**, 81–130.

Koritnig, S., 1964. Pseudodifferentiation eines basaltischen Magmas durch sekundäre Wasseraufnahme. *Beitr. Miner. Petrogr.*, **10**, 50–9.

Kushiro, J., and Kuno, H., 1963. Origin of primary basalt magmas and classification of basaltic rocks. *J. Petrology*, **4**, 75–89.

Lacroix, A., 1908. Le mode de formation du Puy-de Dôme et les roches qui le constituent. *C. R. Acad. Sci. Paris*, **147**, 826–31.

Laspeyres, H., 1901. Das Siebengebirge am Rhein. *Verh. naturhist. Ver. preuss. Rheinl. Westfalen*, **57**, 119–591.

Lehmann, E., and Flörke, W., 1950. Zweitägige Exkursion im Vogelsberg. *Fortschr. Miner.*, **27**, 71–8.

Leopold, G., 1940a. Die chemischen und physiographischen Beziehungen der basischen Eruptiva der Alb, des Hegaus und des Polzengebietes (Sudetenland) zu den Ankaratriten Madagaskars. *Jb. Mitt. Oberrhein. geol. Ver.*, **29**, 88–126.

Leopold, G., 1940b. Über das Auftreten von Monticellit in den ankaratritischen Gesteinen der schwäbischen Alb. *Zbl. Mineral. etc. Abt. A*, 36–40.

Létolle, R., and Kulbicki, G., 1969 and 1970. Geochimie des laves du massif volcanique plioquaternaire du Mont Dore. *Bull. Centre Rech. Pau—SNPA*, **3**, 401–27, and **4**, 191–233.

Lohmann, L., 1964. Ein Beitrag zur Petrographie melilithführender Olivinnephelinite aus dem Gebiet Fritzlar-Naumburg (Nordhessen). *Beitr. Miner. Petrogr.*, **9**, 533–84.

Macdonald, G. A., and Katsura, T., 1964. Chemical composition of Hawaiian lavas. *J. Petrology*, **5**, 82–133.

Macháček, V., and Shrbený, O., 1970a. The geochemistry of volcanic rocks of the central part of the České Středohoří Mts. *Sborn. geol. Věd.*, **16**, 7–47.

Macháček, V., and Shrbený, O., 1970b. Beryllium and chlorine in the volcanic rocks of the České Středohoří Mts. *Vestn. Ústred. Ústav. geol.*, **45**, 321–30.

Manson, V., 1967. 'Geochemistry of basalts: major elements', in A. Poldervaart and H. H. Hess. *Basalts*, **1**, 215–70.

Martini, H. J., Pilger, A., and Schiebel, W., 1970. Die Verbreitung der Eruptiva in der Hohen Rhön *Geol. Jb.*, **88**, 127–36.

Mergoil, J., 1968. Gisement en filons annulaires de phonolites du Velay (Massif Central française). *C. R. Acad. Sci. Paris, Ser. D*, **267**, 12–14.

Michel, R., 1953. Contribution à l'étude pétrographique des pépérites et du volcanisme tertiaire de la Grande Limagne. *Publ. Fac. Sci. Univ. Clermont*, **1**, 140 pp.

Michel-Levy, A., 1890. Situation stratigraphique des régions volcaniques de l'Auvergne. La Chaîne des Puys. Le Mont Dore et ses alentours. *Bull. Soc. géol. France*, 3ème Série, **18**, 688–814.

Overkott, E., 1961. Petrologische Untersuchungen an den Trachyttuffen des Siebengebirges, unter besonderer Berücksichtigung der trachytischen Bestandteile. *Neues Jb. Miner., Abh.*, **95**, 337–69.

Pacák, O., 1928. Čediče Jeseníku a přilehlých území. *Věst. Čes. spol. nauk, Praha, Kl. II*, 1–172.

Pacák, O., 1957. Čedičové magmatity v Královedvorském úvalu. *Sbor. Ustr. ust. geol.*, **23,** 7–78.

Pietzsch, K., 1963. *Geologie von Sachsen*, 870 pp., Berlin (Deutscher Verl. Wiss.).

Roth, Z. and Hanzlíková, E., 1968. The western part of the West-Carpathian flysch belt; stratigraphy. *Regional Geology of Czechoslovkia*, **2,** 374–429.

Scheumann, K. H., 1922. Zur Genese alkalisch-lamprophyrischer Ganggesteine. *Cbl. Miner., Geol., Paläont.*, 495–545.

Schneider, A., 1970. The sulphur isotope composition of basaltic rocks. *Contr. Miner. Petrol.*, **25,** 95–124.

Schorer, G., 1970. Die Pyroxene tertiärer Vulkanite des Vogelsberges. *Chem. d. Erde*, **29,** 69–138.

Schottler, Wilh., 1937. Der Vogelsberg. *Notizbl. Hess. geol. Landesanstalt Darmstadt, V. Folge*, **1,** 3–86.

Schürmann, H., 1960. Petrographische Untersuchung der Gleesite des Laacher Seegebietes. *Beitr. Miner. Petrogr.*, **7,** 104–36.

Schütz, D., 1967. Petrographisch-geochemische Untersuchungen an Olivinknollen verschiedener Vorkommen. *Neues Jb. Miner. Abh.*, **106,** 158–90.

Schuster, M., 1927. Die Basaltgänge zwischen Hofheim-Römhild, Hildburghausen und Heldburg. *Geologie von Bayern*, part VI, 140–2.

Schwarzmann, S., 1957. Über die Feldspat- und Feldspatgesteins-Fremdlinge in den Tertiärvulkanen des Oberweser-Fulda-Gebietes. *Abh. Akad. Wiss., Göttingen, math.-physik. Kl., 3. Folge*, **25,** 165 pp.

Smulikovski, K., 1957. Tertiary lavas of Lower Silesia. *Congr. geol. internat., 20th session, Mexico, Vulcanologia del Cenozoico*, **2,** 487–94.

Smulikowski, K., 1960. 'Tertiary volcanites', in *Regional Geology of Poland: Geol. Soc. Poland*, **3,** 321–34.

Smulikowski, K., 1960. Wulkanity trzeciorzędowe. *Regionalnej Geologii Polski*, **3,** 321–34.

Strunz, H., 1967. Die Basalte der Oberpfalz: Zur Mineralogie und Geologie der Oberpfalz. *Aufschluß, Sonderheft*, **16,** 315–25.

Strnad, J., 1962. Souvislost části europského neogenníko vulkanismu s kimberlitovým magmatismen. *Casopis pro Mineral. a Geol.*, **7,** 434–42.

Sutherland, D. S., 1967. A note on the occurrence of potassium-rich trachytes in the Kaiserstuhl carbonatite complex, West Germany. *Miner. Mag.*, **36,** 334–41.

Szpila, K., 1962. Trace elements in basic volcanic rocks of Lower Silesia. *Arch. miner.*, **23,** 431–52.

Taylor, H. P., Frechen, J., and Degens, E. T., 1967. Oxygen and carbon isotope studies of carbonatites from the Laacher See District, West Germany and the Alnö District, Sweden. *Geochim. cosmochim. Acta*, **31,** 407–30.

Tilley, C. E., 1958, The leucite nepheline dolerite of Meiches, Vogelsberg, Hessen. *Am. Miner.*, **43,** 758–61.

Tröger, W. E., 1939. Über Theralith und Monchiquit. *Zbl. Miner., Geol., Paläont., Abt. A*, 80–94.

Varet, J., 1967. *Les trachytes et les phonolites du Cantal Septentrional.* Thèse 3ème cycle, Orsay, 354 pp.

Varet, J., 1969a. Les phonolites agpaïtiques et miaskitiques du Cantal septentrional (Auvergne, France). *Bull. volcanol.*, **33,** 621–56.

Varet, J., 1969b. Les pyroxènes des phonolites du Cantal. *Neues Jb. Miner. Mh.*, 174–84.

Varne, R., 1968. The petrology of Moroto Mountains, eastern Uganda, and the origin of nephelinites. *J. Petrology*, **9,** 169–90.

Vatin-Pérignon, N., 1968. Les formations eruptives et la structure de l'édifice volcanique au centre du Cantal (Massif Central Français). *Bull. volcanol.*, **32,** 207–51.

Velde, D., 1967. Sur le caractère alcalin des basaltes tertiaires et quaternaires de France. *C.R. Acad. Sci. Paris.* **264,** sér. D, 1141–4.

Vieten, K., 1961. *Die Trachyt-Latit-Alkalibasalt-assoziation des Siebengebirges am Rhein.* Dissertation, Bonn.

Wambeke, L. van, *et al.*, 1964. Les roches alcalines et les carbonatites du Kaiserstuhl. *Euratom Rapport* **1827,** d,f,e, 232 pp.

Wedepohl, K. H., 1961, Geochemische und petrographische Untersuchungen an einigen jungen Eruptivgesteinen Nordwestdeutschlands. *Fortschr. Miner.*, **39,** 142–8.

Wedepohl, K. H., 1963a. Die Nickel- und Chromgehalte von basaltischen Gesteinen und deren olivinführenden Einschlüssen. *Neues Jb. Miner., Mh.*, 237–42.

Wedepohl, K. H., 1963b. Die Untersuchung petrologischer Probleme mit geochemischen Methoden. *Fortschr. Miner.*, **41,** 99–121.

Wedepohl, K. H., 1968a. Die tertiären basaltischen Gesteine im nördlichen Hessen und südlichen Niedersachsen. *Aufschluß, Sonderheft* **17,** 112–20.

Wedepohl, K. H., 1968b. Tertiärer Vulkanismus und Magmenentwicklung im südlichen Niedersachsen und nördlichen Hessen. *Vortragsreferat, 46. Jahrestagung der Deutschen Mineralog. Ges.*, 65.

Weiskirchner, W., 1967a. Bemerkungen zum Hegau-Vulkanismus. *The Rhinegraben Progress Report, Abh. Geol. Landesamt Baden-Württemberg*, **6,** 139–41.

Weiskirchner, W., 1967b. Der Vulkanismus der Schwäbischen Alb. *The Rhinegraben Progress Report, Abh. Geol. Landesamt Baden-Württemberg*, **6,** 142–3.

Weiskirchner, W., 1967c. Über die Deckentuffe des Hegaus. *Geologie*, **16**, Beih. 58, 1–90.

Wimmenauer, W., 1963. Die Bedeutung der Olivinnephelinite und Melilithankaratrite im tertiären Vulkanismus Mitteleuropas. *Neues Jb. Miner. Mh.*, 278–82.

Wimmenauer, W., 1966. 'The eruptive rocks and carbonatites of the Kaiserstuhl, Germany', in *Carbonatites*; ed. O. F. Tuttle and J. Gittins, Interscience, New York, 183–204.

Wimmenauer, W., 1967. Igneous rocks of the Rhinegraben. *The Rhinegraben Progress Report, Abh. Geol. Landesamt Baden-Württemberg*, **6**, 144–8.

Wimmenauer, W., 1972. Gesteinsassoziationen des jungen Magmatismus in Mitteleuropa. *Tschermaks miner. petrogr. Mitteil., 3, Folge*, **18**, 56–63.

Yamasaki, M., and Brousse, R., 1963. La diversité des basaltes. *Bull. Soc. géol. France*, 7ème sér., **5**, 202–9.

Zartner, W. R., 1938. Geologie des Duppauer Gebirges, I. Nördliche Hälfte. *Abh. deutsch. Ges. Wiss. und Künste, Prag, math.-nat. Abt.*, **2**, 132 pp.

IV.5. THE MONGOL–TUVA PROVINCE OF ALKALINE ROCKS

A. S. Pavlenko

IV.5.1. Introduction

The Mongol–Tuva province of alkaline rocks occupies the enormous territory from the axial part of the East Sayans in the north to the Khangai Ridge in the south, and from the Large Lake Basin in the west to the Hamar-Daban Ridge in the east. To the north-east it extends into the Baikal belt of alkaline rocks, and still farther east into North Zabaikalje (northern part of the large area lying to the east of the Baikal lake) and South Yakutia. In the north-west the Mongol–Tuva province borders on the Minusa–Kazir province of alkaline basaltoid rocks; in the north on the belt of alkaline ultramafic massifs of the northern slope of the East Sayans.

When compared with other provinces composed of the granitoid–alkaline association (Sheynmann *et al.*, 1961) the Mongol–Tuva province is distinguished by the most complete evolution of petrographic, geochemical and genetic features.

The province is situated in the region deformed during the oldest Paleozoic folding of Cambrian age. The greater part of the alkaline province is situated within the Sangylen–Khubsugul mass, which was formed in a wide area of miogeosynclinal type (Fig. 1). In the north this mass borders on the East-Sayan anticlinorium along a sublatitudinal system of faults, in the north-west and in the south on the eugeosynclinal East-Tuva and Ider–Dzhida fold belts.

The Sangylen–Khubsugul mass is made up mostly of Late Proterozoic greenschists and marbles and of Vend (Eocambrian)–Early Cambrian effusive-terrigeneous and carbonate rocks, the latter being predominant. This is also typical for the East-Sayan anticlinorium.

The East-Tuva and Ider–Dzhida fold belts are characterized by the wide development of Vend–Early Cambrian terrigenous-effusive formations; thick carbonate sequences of this age being entirely absent.

Numerous intrusions take part in the regional structure. Gneissic leucocratic muscovite granites, plagiogranites and pegmatites belong to the Pre-Cambrian complex. Early-Cambrian ultramafic complexes are made up of periodotite, dunite, pyroxenite and serpentinite. The ultramafic massifs are clearly confined to large fracture zones, that divide the largest structural zones. The Early Paleozoic geosynclinal intrusions vary from gabbro to granodiorites and plagiogranites with predominance of the latter. The Early Paleozoic massifs are usually confined to the fields of development of Pre-Cambrian strata, and only rarely break through younger rocks.

Fig. 1 The geological structure of the Mongol–Tuva alkaline province

(1) Cenozoic rocks; (2) Middle Paleozoic, Late Paleozoic and Mesozoic effusive–terrigenous rocks.

Vend–(Eocambrian)–Early Cambrian:

(3) Terrigenous–effusive rocks of the East Tuva and the Ider–Dzhida fold zones; (4) limestone dolomite deposits of the Sangilen–Khubsugul mass; (5) effusive–terrigenous rocks of the Sangilen–Khubsugul mass.

Late Proterozoic:

(6) Greenschists; (7) crystalline schist; (8) marbles.

Archean, Middle Proterozoic:

(9) Gneisses, crystalline schists, marbles.

Paleozoic Orogenic Igneous Rocks:

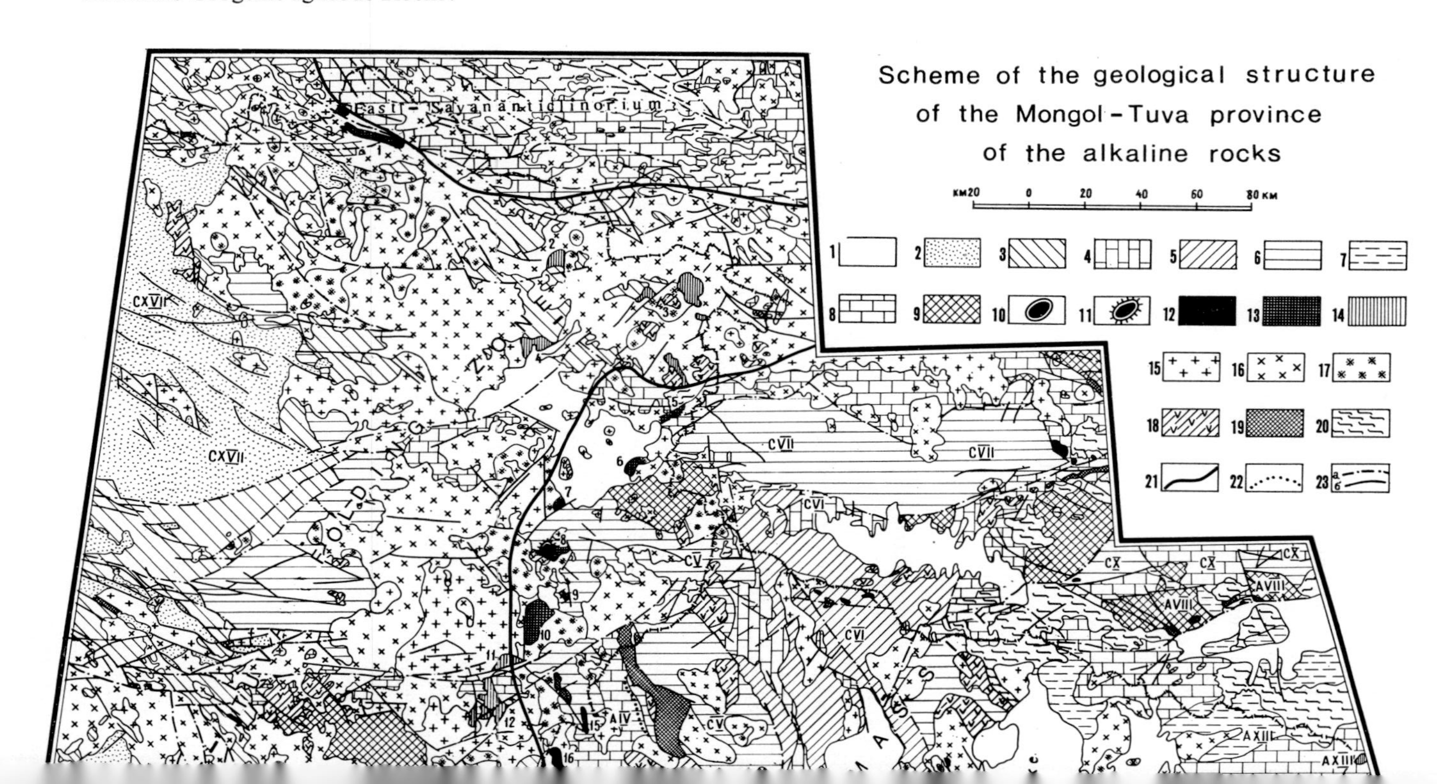

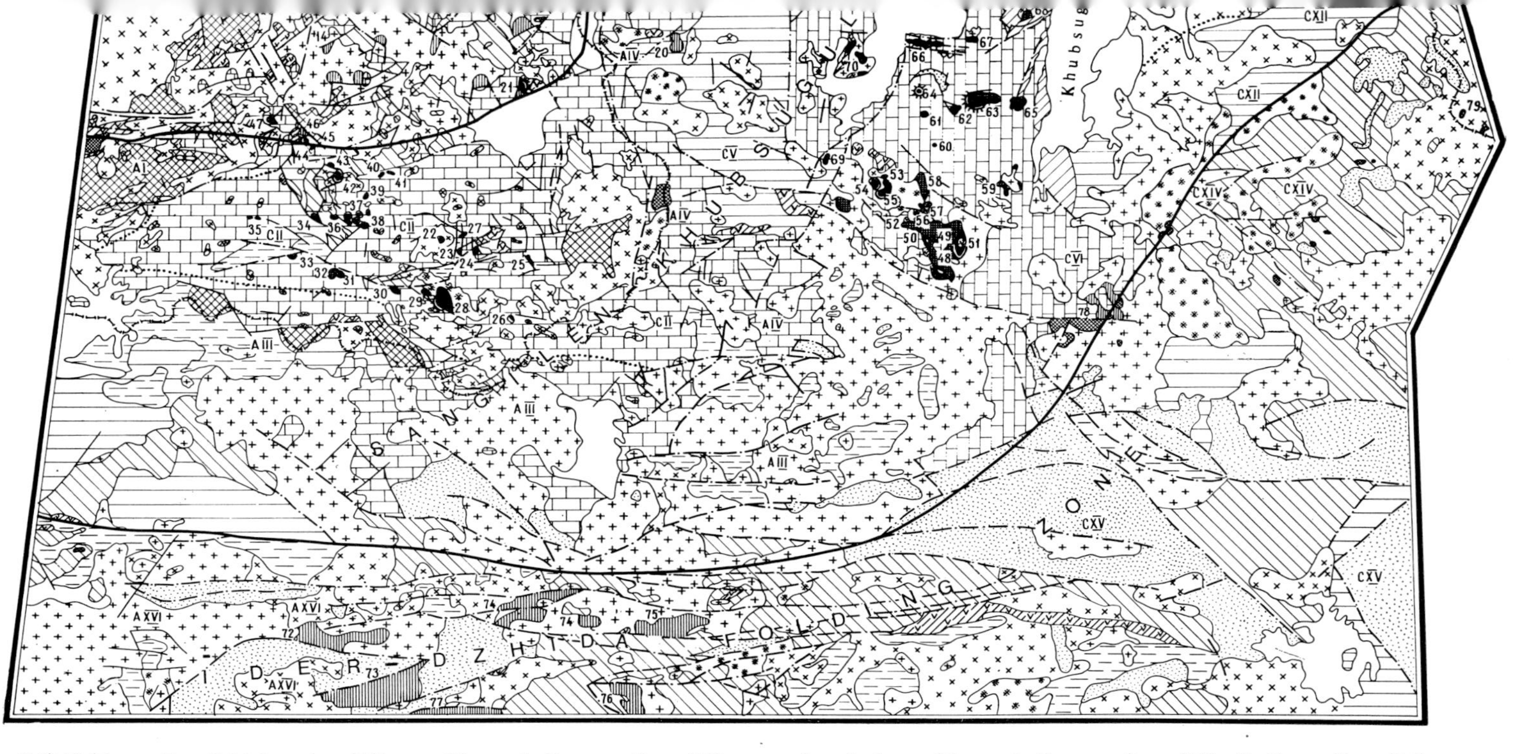

(10) Feldspar-free foidal rocks; (11) agpaitic nepheline syenites; (12) normal and plumasitic nepheline syenites; (13) alkali syenites; (14) alkali granites and apogranites; (15) granites.

Early Paleozoic Geosynclinal Complex:

(16) Granites, plagiogranites, granodiorites; (17) diorites, granodiorites; (18) gabbro, gabbro–diorites.

Earl Cambrian Ultramafic Complex:

(19) Dunites, peridotites, pyroxenites.

Proterozoic Granitoid Complex:

(20) Muscovite and two-mica granite-gneisses and pegmatoid granites; (21) boundaries of tectonic zones; (22) boundaries of individual structural units; (23) faults; (a) main, (b) minor.

Symbols on the map:

Structures: (AI–AXVI) Blocks, anticlinoria, domes; (CII–CXVII) Synclinoria, basins.

Alkaline Massifs: (1–3) East-Sayan belt; (4–21), (40–47); East-Tuva belt; (22–39) Sangilen belt; (48–71) Khubsugul belt; (72–79) Bolnaj belt.

The Paleozoic orogenic (postgeosynclinal) alkaline and granitoid intrusions, which constitute the alkaline province, are represented by a wide series of rocks from granite, through alkali granite and syenite to nepheline syenite and non-feldspathic varieties. The absolute ages (Zykov *et al.*, 1961; Pavlenko *et al.*, 1969) vary from 530 to 280 m.y. and decrease successively in the different tectonic zones as one goes from the north to the south. This interesting fact points to special conditions of generation of the orogenic magmatic complexes, as the existence during such a long period of deep-seated heterogeneous hearths 'feeding' the intrusions throughout the whole province seems to be highly improbable.

IV.5.2. Geological Position of the Alkaline Rocks of the Province

Fig. 1 shows that the belts of alkaline intrusions are confined to the miogeosynclinal zones and especially to the boundaries between mio- and eugeosynclinal zones. The main belts of alkaline rocks controlled by the most important tectonic lines are: the East Sayan belt—where the East-Sayan anticlinorium joins the East-Tuva fold zone; the East-Tuva belt where the Sangylen–Khubsugal mass meets the East-Tuva fold zone; and the Bolnaj belt—on the boundary of the Sangylen–Khubsugal mass and the Ider–Dzhida fold zone. All these belts have asymmetric structures. The massifs occurring within terrigenous-volcanic rocks and ancient granitoids of the eugeosynclinal zones consist of alkali granites and granosyenites, rarely changing to alkali syenites; while the massifs of alkali and nepheline syenites are restricted to the carbonate rocks and gabbroids of the miogeosynclines. The Sangylen and Khubsugal belts, confined to synclinal structures in the axial part of the Sangylen–Khubsugal mass, are also connected with carbonate rocks and gabbroids.

IV.5.3. Structure of the Alkaline Massifs

The alkaline massifs form small and middle-sized bodies, a few m^2 up to 250 km^2 in area. Their location and shape are often controlled by faults. The shape of the massifs is highly variable and can usually be reconstructed easily using trachytoid structures, relics and xenoliths of the surrounding rocks and the character of the contacts. The alkali granites and syenites occur mainly as slightly elongated steeply dipping stocks and sills; the nepheline syenites as more isometric stocks and ethmoliths with reverse dip of the contacts. Migmatite fields, in which rootless lens-like and vein-like bodies of igneous rocks and the deeply modified initial rocks are in close alternation, are widely distributed.

The majority of the syenite and nepheline-syenite intrusions consists of two phases. The rocks of the second phase are more siliceous than those of the first phase. The small bodies and dykes of the second phase are usually situated within the limits of the intrusions. The nepheline syenite massifs are further characterized by 'accompanying' vein rocks: dyke swarms of aplites and trachyte-porphyries, situated in the exocontact zones and having no direct connection with the intrusion. Zoning resulting from reactions with the enclosing rocks, especially with carbonate and basic rocks, is typical for the alkaline massifs. The central part of the zoned intrusions is, as a rule, more leucocratic and siliceous, while the outer parts are enriched in alkalis and bases and impoverished in silica. A typical example of a zoned two-phase intrusion is given in Fig. 3. The exocontacts are characterized by aureoles of fenite and alkaline gneisses, especially extensive in the migmatite fields, where they considerably exceed the dimensions of the magmatic bodies. Zones of postmagmatic alterations, which attack the intrusions and their exocontact zones, are widely developed. The intrusions are frequently completely altered.

IV.5.4. Petrographic Features

The Mongol–Tuva province exhibits considerable petrographic variation. Granites and alkali syenites predominate, together with meso- and leucocratic nepheline syenites of normal ($2Ca + K + Na > Al > Na + K$; atomic proportions) and plumasitic types ($Al > K + Na + 2Ca$). Calc-alkalic granites and syenites, melanocratic and agpaitic ($K + Na > Al$) nepheline

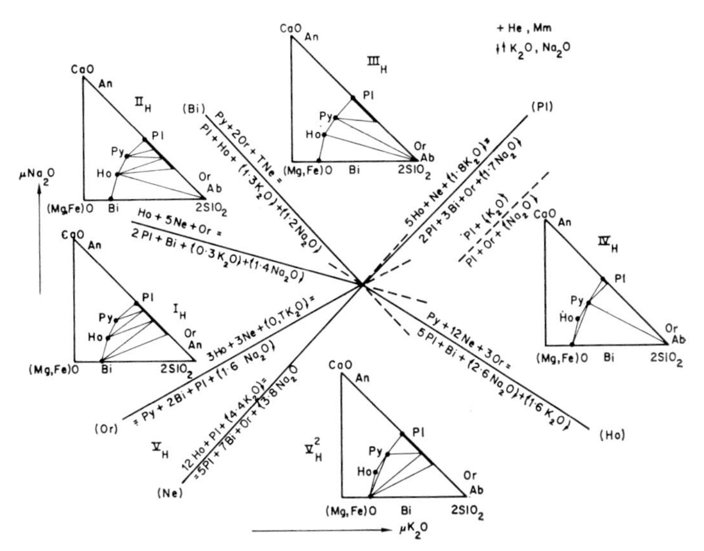

Fig. 2(a) Diagram of the dependence of the nepheline syenite paragenesis on K_2O and Na_2O potentials. Parentheses indicate minerals, which do not take part in this equilibrium

+ (plus) = abundant minerals (nepheline and magnetite); ( ) = quite mobile components; Or = orthoclase; Ab = albite; An = anorthite; Pl = plagioclase; Py = pyroxene; Ho = hornblende; Bi = biotite.

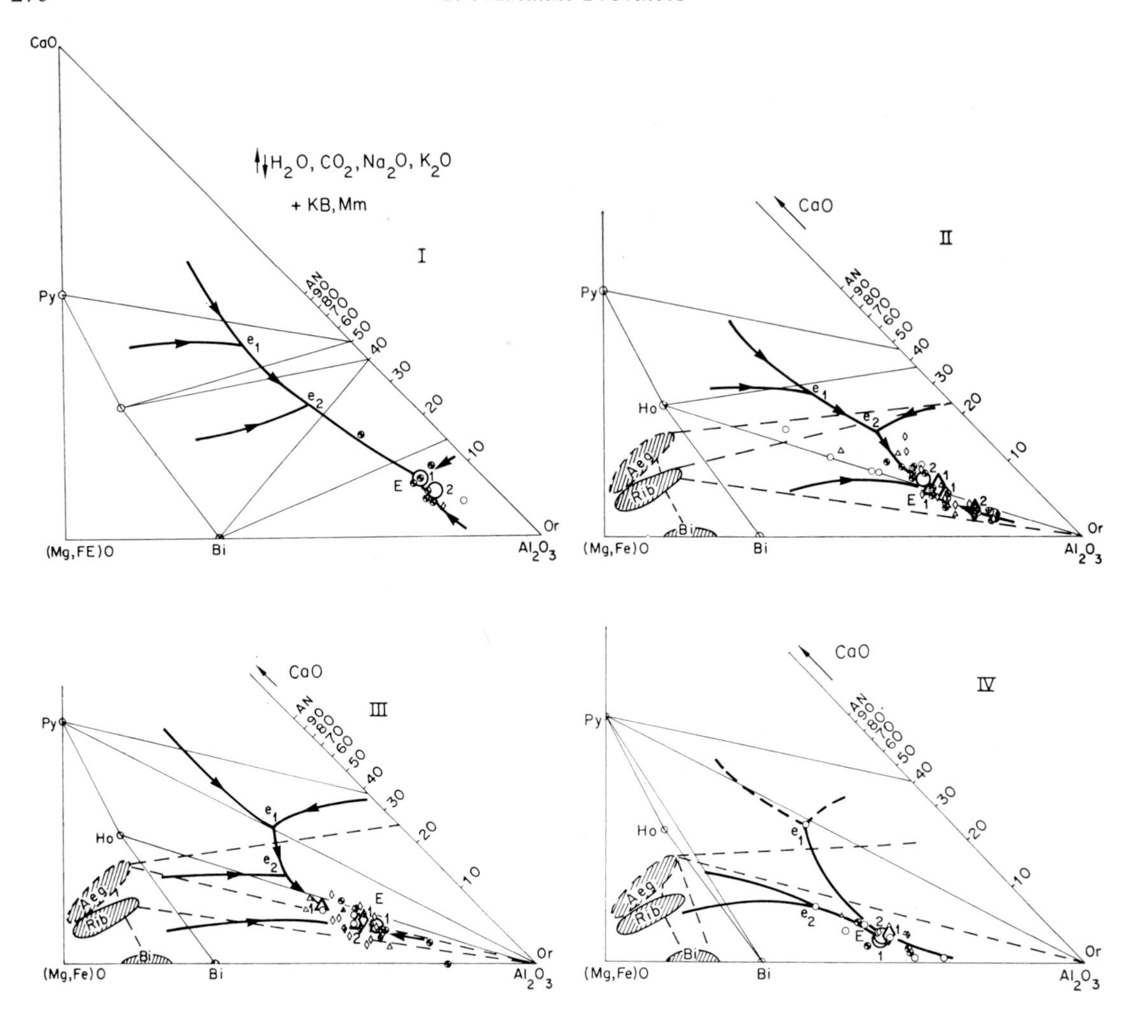

Fig. 2(b) Scheme of crystallization in fields I, II, III and IV of Korzhinsky's diagram of the dependence of granitoid parageneses on K_2O and Na_2O potentials

Continuous thin straight lines = parageneses with normal coloured minerals. Dotted straight lines = parageneses with alkaline coloured minerals. Thick continuous lines = scheme of crystallization. Small symbols = compositions of Tuva granitoids. Large symbols = their average compositions in the first and second intrusive phases. Circles = granites. Triangles = granosyenites and quartz syenites. Rhombs = syenites. Symbols showing chess-board ornamentation = supersaturated with alumina. + (plus) = abundant mineral; ↑↓ = quite mobile components; Or = alkaline feldspar; An = anorthite; Py = pyroxene, Ho = hornblende; Bi = biotite; Aeg = alkaline pyroxene; Rib = alkaline amphibole; e_1, e_2 = peritectics; E = eutectics.

The increase of alkalinity from fields I to IV results in the expansion of the fields of alkaline feldspar crystallization and the eutectics are enriched in ferruginous minerals.

syenites as well as foidal rocks occur subordinately, being formed under specific conditions. The remarkable variety of granitoid and nepheline-syenite mineral parageneses is the distinctive petrographic peculiarity of the province.

This description is based on the paragenetic diagrams of the mineral facies of granitoids developed by Korzhinsky (1957) and of nepheline syenites developed by Pavlenko and Phillipov (Kovalenko *et al.*, 1965) as well as on petrographic data (Kovalenko, 1964; Kovalenko *et al.*, 1965, Kovalenko and Popolitov, 1965, a,b; 1970; Kononova 1958, 1961, 1962; Kostin and Petrova, 1960; Kudrin, 1962, Luvsan-Danzan and Chasin, 1966; Machin and Pavlenko, 1966; Pavlenko, 1963, a,b,c; Pavlenko *et al.*, 1960; Yashina, 1957, 1962, 1963).

The above-mentioned diagrams account for the paragenetic dependence of granitoids and nepheline syenites on the chemical potentials of K_2O and Na_2O. Each field of the diagram represents a mineral facies of specific alkalinity, the mineralogical composition of which is given by the triangular diagrams of Fig. 2(a). The lines, dividing the facies fields of the bundle diagram, represent univariant equilibria and can be described by means of reactions involving exchange of alkalis, while the contents of the inert rock-forming elements remain constant. In each individual facies crystallization of the melts is dependent on the mineral associations and the chemical composition of cotectic lines and eutectics (Fig. 2 (b)).

The distribution of the petrographic types of alkaline massifs is given in Table 1. The alkalinity of the mineral facies increases from left to right—the acidity decreases from top to bottom.

IV.5.4.1. Granites

There are no plagiogranites (field V–γ) or melanocratic varieties of normal granites in the Mongol–Tuva granitoid-alkaline province. Normal microcline-plagioclase-biotite granites and alaskites (I–γ) are widespread, forming the central parts of the large massifs. They often have gradual transitions into subalkaline and alkaline granitoids and syenites.

Subalkaline biotite-hornblende-microcline granites and granosyenites (field II–γ) form most of the granitoid massifs of the East-Tuva and Bolnaj belts. They frequently constitute the bodies of phase 2 in the massifs of alkali and nepheline syenites over the whole of the province. These are coarse to middle-grained, frequently porphyritic with idiomorphic microcline-perthite phenocrysts. The main minerals are microcline-perthite, amphibole, biotite and quartz. Plagioclase (oligoclase-andesine) occurs in subordinate amounts, as grains corroded by

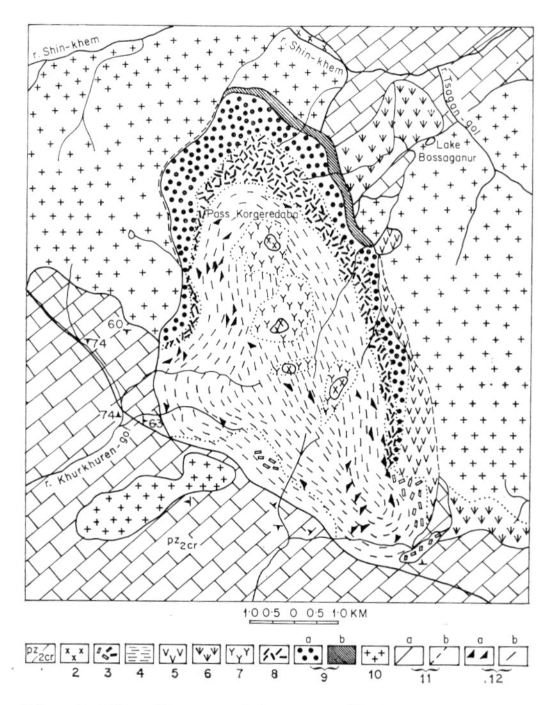

Fig. 3 The Korgeredaba massif (N 28—Fig. 1, Yachina, 1957)

(1) Marble–limestones of the Late Proterozoic.
Intrusive Rocks:
(2) Quartz and quartz-free syenites of phase 2; (3) nepheline pegmatites; (4) trachytoid nepheline pegmatites.
Ancient Intrusive Rocks:
(5) Olivine gabbro; (6) pyroxene–hornblende gabbro and diorites.
Contact–Metasomatic Alkaline Rocks:
(7) Anorthoclase quartz syenites; (8) pulaskites; (9) leucocratic essexites (a) and their hydrothermally-altered varieties (b); (10) Paleozoic orogenic granites; (11) faults traced (a), supposed (b); (12) trachytoid structure (a), layering (b).

TABLE 1 Mineral Composition of the Alkaline Rocks of the Mongol–Tuva Province and their Distribution through the Massifs

Formations		Kf–Pl–Bi Pl–Bi–Pi	Kf–Pl–Bi Pl–Bi–Hb	Kf–Bi–Hb Kf–Pl–Hb	Kf–Bi–Hb Kf–Hb–Py Kf–Pl–Py	Kf–Bi–Py Kf–Pl–Py
Granites	Q	V–γ Plagiogranite. Trondjemite. —	I–γ Alaskaite. Granite. Il, Rut, Ap, Mon, Xen, Zir, Eux, Col. 1f*-great granitoid intrusions. 2f-1, 3, 6, 10, 52†	II–γ–σγ Paisanite. Microcline granite. Granosyenite. Sph, Il, Ap, Orth, Mon, Thor, Fer, Eux. 1f-1,2,3,11,12,13, 14, 20, 48, 72–79 2f-6, 5, 10, 49, 50	III–γ Ekerite. Quartz nordmarkite. Aenig, Sph, Chev, Brith, Ce-gad, Mal, Cyr, Pyr. 1f-11,12,13,20,22, 46 2f-1, 8, 15, 23, 48, 50	IV–γ Grorudite. Quartz akerite. Astr, Chev, Brith, Thorn, Elp, Wöhl. U-Thorn 1f-12,14,73 2f-1, 8, 34
Syenites		V–γσ	I–γσ (Quartz Albitite)‡ Ilrut, Mon, Xen, Par, Bast, Thor, Mal, Lop, Fer, Eux, Col, Cryoph, Cryol, Bet. 1, 3, 6, 8, 12, 15, 20, 22, 23, 26, 27, 46, 48, 72–74, 78	II–γσ–ξσ Syenite (Plauenite). Sph, Ap, Orth, Zir. 1f-6, 7, 9, 15, 26, 32, 33, 35, 52, 66, 68 2f–10, 29, 48, 49, 52	III–γσ Nordmarkite. Timt, Sph, Ap, Orth, Zir. 1f-1, 4, 5, 10, 28, 32, 52, 59, 65, 71	IV–γσ Akerite. Sph, Chev, Astr, Brith, Pyr, Zir. 2f-8, 15, 28, 63
		V–ξσ Litchfieldite. (Nepheline Albitite). Rin, Lamp, Ram, Ap, U-Thorn, Eud, Lav, Hatch, Aesch. 8, 18, 26, 28, 29, 30, 36, 43, 44, 45, 49, 60, 63	I–ξσ Canadite. 15, 19, 21, 23, 34, 41, 42, 48		III–ξσ Toensbergite. Umptekite. Pulaskite. Sph, Ap, Orth, Zir. 1f-1, 7, 9, 10, 28, 33, 47, 62, 67	IV–ξσ Larvikite. Sph, Ap, Orth, Brith, Lav, Pyr. 2f-1, 8, 15, 34, 63
Nepheline Syenites	Ne	V–ξ (Mariupolite). Il, Ap, Zir, Cyr, Pyr. 1f-15, 26, 27, 36, 42, 45	I–ξ Miaskite. Il, Sph, Zir, Pyr, Hatch. 1f-15, 18, 26, 30, 36, 45, 49, 61	II–ξ Ditroite. Hornblende–miaskite. 1f-18, 19, 23, 27, 28, 29, 34, 38, 40, 43, 53	III–ξ Foyaite. 1f-8, 15, 16, 24, 34, 39, 50, 56, 69, 70 Essexite. Theralite. 1f-17, 28, 38, 41, 51, 70 Sph, Ap, Zir. Khibinite. Astr, Rin, Lamp, Eud. 8, 21, 28, 63, 64	IV–ξ Lardalite. 1f-1, 8, 15, 21, 34, 40, 44, 63, 69 Shonkinite. Malignite. 1f-17, 20, 21, 24, 70 Juvite. 21, 42, 44 Sph, Ap.
Peralkaline Rocks		V–νξ Congressite. Micaceous urtite. Timt, Ap. 1f-30, 42, 57		II–νξ Monmouthite. Timt, Schor, Ves. 18, 37, 55	III–νξ Urtite. Lepidomelane ijolite. Ijolite. Sph, Schor, Ap. 17, 31, 37, 44, 54, 55	
Parageneses		Bi–Hb		Hb–Py	Bi–Py	

The horizontal rows with index Q represent the granitoid rock diagram, those with index Ne the nepheline-syenite one. The vertical columns represent the fields of facies of alkalinity of these diagrams and are marked with the typomorphic paragenesis of each facies. The names of each field are arranged in the order of decreasing SiO_2 content and increasing content of coloured minerals, i.e. they correspond to definite intervals of cotectic differentiation.

microcline or in the form of relics in the cores of microcline grains. Amphibole is usually represented by common edenitic hornblende; subalkaline varieties of magnesium-hastingsitic type are also found. Accessory minerals are abundant and variable, especially in the granitoids of phase 2. As a whole, the rocks are hypidiomorphic granular with distinct xenomorphism of quartz in respect to microcline and the coloured minerals; this feature is most pronounced in the granosyenites. In the endomorphic zones there are sometimes subeffusive structures. Granophyric and pegmatoid textures occur frequently in the vein bodies of phase 2.

Alkali granites, granosyenites and quartz syenites (III–γ, III–$\gamma\sigma$, IV–γ, IV–$\gamma\sigma$) are similar to the rocks of the subalkaline group as regards geological position and origin, but they differ considerably in composition, structure and appearance. The alkaline granitoids of phase I are usually coarse-grained, frequently porphyritic with well-marked idiomorphism of the feldspar. The rocks of phase 2 are more fine-grained. They often have granular textures with idiomorphic, isometric quartz. 'White shirt' of late acid plagioclase around the potash feldspar grains and the glomeroblastic distribution of the coloured minerals make the rocks mottled, 'rapakivi-like' in appearence.

Two series of alkaline granitoids can be recognized: (1) pyroxene-hornblende granite–ekerite–quartz nordmarkite, where amphibole and pyroxene do not show reactional relations; (2) biotite-pyroxene granite–grorudite–quartz akerite characterized by the paragenesis alkali feldspar, biotite and salite (aegirine-salite). Amphibole, if present, is replaced by biotite and pyroxene. Plagioclase, with the exception of the perthitic intergrowths and late albite, are not characteristic for this group of rocks. Corroded and relic grains of plagioclase of intermediate and basic composition occur occasionally in the pyroxene-bearing varieties. The composition of the coloured minerals ranges widely from normal to alkaline varieties and there is no direct dependence between the composition of the coloured minerals and the contents of quartz or alkali feldspar.

The phase 1 rocks contain usually normal or alkali-poor calcium-magnesium mafic minerals, such as common hornblende, edenite, magnesiopargasite, magnesiohastingsite, diopside, augite and fedorovite. The granitoids of phase 2 are characterized by alkaline ferruginous varieties: ferrohastingsite, arfvedsonite, aegirine-salite. In some peralkaline granitoids unusual coloured minerals, such as astrophyllite and aenigmatite are found.

The relative amounts of quartz and alkali feldspar vary considerably, while the amount of coloured minerals remains rather constant and low (5–15%). Melanocratic rocks are formed only in contaminated and palingenetic-metasomatic facies.

The accessory minerals of the alkaline granitoids, especially of phase 2, are highly variable, but they decrease in amount and number with decreasing silica acidity.

IV.5.4.2. Syenites

Syenites form numerous massifs throughout the province. They are most widespread in the

---

Abbreviations:
Q—quartz
Kf—potash feldspar
Pl—plagioclase
Bi—biotite
Py—pyroxene
Hb—amphibole hornblende
Ne—nepheline

Aenig—aenigmatite
Astr—astrophylite
Aesch—aeschynite
Ap—apatite
Bast—bastnäsite
Brith—britholite
Bet—betafite
Cyr—cyrtolite
Col—columbite
Cryoph—cryophyllite
Cryol—cryolite
Ce-gad-Ce—gadolinite
Chev—chevkinite
Eux—euxenite
Eud—eudialyte
Elp—elpidite
Fer—fergusonite
Hatch—hatchettolite
Il—ilmenite
Ilrut—ilmenorutile
Lop—loparite
Lav—lavenite
Lamp—lamprophyllite
Mon—monazite
Mal—malakon
Orth—allanite
Par—parisite
Pyr—pyrochlore
Rut—rutile
Rin—rinkolite
Ram—ramsayite
Schor—schorlomite
Sph—titanite
Timt—Ti-magnetite
Thor—thorite
Thorn—thorianite
U-Thorn—uranothorianite
Ves—vesuvianite
Wohl—wöhlerite
Xen—xenotime
Zir—zircon

*f—phase †—numbers of alkaline plugs indicated on Fig. 1 ‡—hydrothermal metasomatic rocks

East-Sayan and in the northern part of the East-Tuva and Khubsugal belts. In the East-Tuva belt syenites occur as facies of massifs made up mainly of granite, and in the Sangilen zone as facies of nepheline syenite massifs. Besides, syenite bodies belonging to phase 2 of nepheline syenite intrusions occur within the Sangilen zone.

The syenites are as a rule coarse-grained, slightly porphyritic with well-defined idiomorphism of alkali feldspar in relation to the coloured minerals and sometimes with poorly developed trachytoid textures. *Pyroxene-amphibole umptekites and pulaskites* form the bulk of the syenites of phase 1. These syenites contain 85–90% microcline-perthite and 10–15% coloured minerals. The relative amounts of the latter vary, and pyroxene, and more rarely amphibole, may be entirely missing. Primary plagioclase is absent or occurs in markedly subordinate amounts as corroded relics. The umptekites and pulaskites are connected by gradual transitions with nepheline syenites or grano-syenites. The nordmarkites contain a total amount of 5–10% pyroxene, amphibole and biotite. The variety of the nordmarkite is tönsbergite, where up to 2% olivine (10–15% fo) often occurs together with quartz. Larvikites are also widespread. They are composed of alkali feldspar, biotite and pyroxene and occur as subordinate facies of the nepheline syenite massifs.

Lime-alkali hornblende and biotite-hornblende syenites with marked or predominant contents of oligoclase (Plauen type) occur within the intrusions emplaced in carbonate rocks. Syenites of this type, but with lower contents of plagioclase and sometimes with low amounts of quartz, occur in phase 2 of the nepheline syenite and syenite intrusions.

The compositions of the coloured minerals in the syenites vary considerably, ferruginous varieties being predominant. A distinct enrichment of the coloured minerals in Fe_2O_3 and SiO_2 is observed, approximating their compositions to those of riebeckite and aegirine. *Biotite syenite-like rocks, albitites and microclinites* with quartz or nepheline (V–γσ, I–γσ, V–ξσ, I–ξσ), occupy a particular position. These rocks are in all cases products of hydrothermal-metasomatic alteration of granitoids, syenites and nepheline syenites, as well as of the enclosing aluminosilicate rocks (Fig. 4). They are characterized by lath-like, rosette-like, diablastic and poikilitic textures and by ribbon and gneissic structures, by highly ferruginous coloured minerals (lepidomelane, riebeckite, aegirine) and by abundant and various accessory minerals.

Fig. 4 Sketch map of the Kara–Adyr area (N 34—Fig. 1, Tugarinov *et al.*, 1963)

(1) Marble; (2) feldspar–biotite–quartz schists; (3) pegmatoid nepheline syenites; (4) cummingtonite schists and fenites; (5) microclinites; (6) albitites; (7) metaquartzites; (8) boundaries of the metasomatic facies.

IV.5.4.3. Nepheline Syenites

Nepheline syenites are distinguished by the broadest variability of petrographic and chemical compositions. In contrast with the granitoids and syenites they are characterized not only by the diversity of parageneses, but also by wide variations in the relative contents of nepheline, feldspar, and coloured minerals. The series ditroit–hornblende miaskite, foyaite–essexite, and lardalite–shonkinite (II–ξ, III–ξ, IV–ξ) have the widest distribution.

The *foyaites* are coarse to medium-grained, often trachytoid,and consist of potash feldspar, usually microcline-perthite (60–70%), nepheline (20–30%) and usually 10–15% of two coloured minerals in changeable relative amounts. Sometimes three coloured minerals are present, but usually pyroxene and amphibole (magnesium, ferruginous and titanium varieties such as augite, augite-hedenbergite, titanium-aegirine-augite, hornblende, barkevikite) are associated with nepheline and K–Na-feldspar. The accessory minerals are usually represented by magnetite or titanomagnetite, titanite, apatite and zircon. The foyaites are characterized by hypidiomorphic and panidiomorphic textures, the nepheline and alkali feldspar showing idiomorphism against the coloured minerals.

Peralkaline, agpaitic varieties of foyaites, khibinites, occur more rarely. Their main features are the complete absence of primary plagioclase, higher contents of nepheline (up to 35%) and coloured minerals (up to 20%) represented by aegirine and arfvedsonite. Spatially, the khibinites usually coincide with areas of pegmatites and postmagmatic albitization. This accounts for the abundance of primary accessory minerals: astrophyllite, aenigmatite, rinkolite and eudialyte.

The melanocratic varieties of the foyaite series—*theralites and essexites*—occur subordinately as marginal facies formed by magmatic replacement of subsilicic or carbonate rocks. This group is characterized by basic plagioclases (up to An_{60}) and sometimes hypersthene.

Nepheline syenites of the ditroite–hornblende miaskite series are widely distributed in the intrusions of the Sangilen and Khubsugal belts. Hornblende miaskite consists of 30–40% microcline-perthite or K–Na orthoclase, 15–20% oligoclase, about 25% nepheline and 12–18% amphibole (hastingsite or edenite). In the ditroite primary plagioclase is absent or occurs in the form of rare corroded grains, and the coloured minerals are represented by biotite. The accessory minerals are on the whole the same as in the foyaites, but in addition ilmenite and pyrochlore are rather common. These rocks have granitoid or hypidiomorphic textures and trachytoid structures. Nepheline syenites of this type (including miaskites and mariupolites) are typical of massifs located in carbonate rocks, characterized by desilication processes. The miaskites are characterized by the association of lepidomelane with potash feldspar and oligoclase; the mariupolites by the association of biotite with oligoclase and pyroxene. It should be noted that metasomatism took part in the formation of mariupolites and some types of miaskites. This is particularly indicated by their textures: xenomorphic and -blastic, often diablastic or poikilitic.

Melanocratic rocks are rare in these series. Assimilation of marble leads to an increase in plagioclase and the formation of canaditic and litchfielditic rock types. Primary accessory minerals, such as aeschynite, cerfergusonite and uranothorite appear in the miaskites and mariupolites, as well as in their metasomatically altered varieties.

The *lardalite–shonkinite* series is characterized by alkali feldspar, biotite and pyroxene. The lardalites contain microcline-perthite (60–65%), nepheline (10–20%), pyroxene (4–10%), usually diopside or aegirine-diopside, and lepidomelane (3–6%). Meso- and melanocratic members of the series, juvites, malignites and shonkinites, which show a successive increase in the amount of coloured minerals, up to 50–60%, are usually products of magmatic replacement or palingenetic or metasomatical transformation of carbonate, subsilicic and ultrabasic rocks. The pyroxene is sharply prevalent among the coloured minerals and is represented by augite, ferroaugite and augite-hedenbergite. It is formed before the nepheline and potash feldspar, which makes these rocks different from the magmatic lardalites.

IV.5.4.4. Foidal Rocks

Nepheline is the only primary felsic mineral of the foidal peralkaline rocks. The proportion of nepheline to mafic minerals varies within wide limits. Besides, the peculiarity of the environment and of the processes of formation caused the development of minerals that are absent or rare in the other groups of rocks: calcite, garnet (andradite–grossularite series), scapolite, wollastonite, vesuvianite and graphite).

In the massifs mainly composed of peralkaline rocks, urtites and ijolites, consisting of pyroxene and nepheline, play the main role. Depending on the content of pyroxene, the colour of the rocks changes from light bluish-grey to dark green. The pyroxenes are salite, ferroaugite or augite-hedenbergite enriched in titanium. Nepheline and pyroxene are always accompanied by calcite, apatite and titanomagnetite. Titanite and hornblende (barkevikite-catophorite) are frequently present. In the ijolite-urtites developed as a result of magmatic replacement of gabbro-pyroxenites, relics of basic plagioclase are partly replaced by nepheline and carbonates with the formation of peculiar dactylotype textures. Garnet- and scapolite-ijolites form a special group, in the formation of which metasomatic processes took part.

The structures and textures of the ijolite-urtites are highly variable. Porphyroblastic (anhedral phenocrysts) and porphyraceous (euhedral phenocrysts) textures are developed in addition to hypidiomorphic ones. The contact rocks have idiomorphic porphyritic growths of pyroxene, while the exocontact rocks have nepheline. The exocontact rocks have ribbon, gneissic and migmatitic structures. Poikiloblastic and diablastic textures are widely developed in the garnet-, amphibole-, scapolite-ijolite-urtites and also in the silicate-carbonate rocks.

Other types of peralkaline rocks are rare. Lepidomelane ijolites, monmouthites and congressites are found in near-contact and marginal zones of some desilicated bodies.

IV.5.4.5. Metasomatic Rocks

Metasomatic rocks accompany all types of magmatic rocks of the province and they can be classified into two groups, the first being a product of the progressive stage accompanying the generation and injection of the melts; the second of the regressive, postmagmatic stage. The first group is restricted to the exocontacts of the massifs, the second occurs both in exo- and endocontact zones. As it is not possible to describe all the varieties of metasomatic rocks, the information available is recorded in Table 2, constructed according to the data of Machin *et al.* (1966), Pavlenko (1963c), Pavlenko and Kovalenko (1961, 1965), see also VI.7.*

IV.5.5. Petrochemical and Geochemical Characteristics of the Province

A huge amount of analytical data on the contents of the petrogenic and rare elements in the rocks of the province is available (Pavlenko *et al.*, 1957, 1958, 1959, 1960; Pavlenko and Kovalenko 1961; Pavlenko, 1963a, b; Kovalenko *et al.*, 1965 Kovalenko and Popolitov, 1970; Vajnschtejn *et al.*, 1961; Schevalejevskij *et al.*, 1960; Tugarinov *et al.*, 1969). It is therefore preferred to present the geochemical characteristics using averaged data.

Table 3 shows the composition of the individual petrographical groups of the granitoid and alkaline massifs of the province in accordance with Fig. 1 and Table 1. The following main petrochemical features can be deduced from Table 3:

1. The comparison of the data of Table 2 with those of the average types of igneous rocks of Daly (1933) and with the individual rock types of Tröger (1938) shows that most of the rocks in question are petrochemically nearer to the effusive varieties, than to the intrusive ones. This fact is in accordance with the hypabyssal conditions of formation of the rocks of the granitoid–alkaline complexes.

2. The rocks of the province as a whole have an alkaline-aluminous character with rather low contents of divalent bases and iron. In contrast to the gabbroid–alkaline association (cf. IV.1) the contents of alumina and alkalis and, to a lesser

* Readers are referred to the relevant Chapters of this book where similarly cited.

TABLE 2 Scheme of Metasomatic Processes Connected with Alkaline Intrusions of the Mongol–Tuva Province

Metasomatic stages	Magmatic and Early-Postmagmatic				Postmagmatic		
Initial rock / Intrusive rock	Carbonaceous rocks	Pyroxenites	Gabbroids	Intermediate and persilicic alumosilicate rocks	Carbonaceous rocks	Subsilicic and intermediate alumosilicate rocks	Persilicic alumosilicate rocks
Alkaline granitoids and syenites	Garnet-alkaline-feldspathic rocks/sviatonossite/	Feldspathic fenites	Alkaline feldspathic amphibolites and fenites	Alkaline-feldspathic hornfelses, alkaline migmatites	Alkaline feldspathic skarns. Wollastonite-garnet-vesuvianite rocks	Aegirine–lepidomelane albitites and microclinites	Alkaline apogranites and metaquartzites with riebeckite, astrophyllite, cryophyllite
Nepheline syenites	Diopside-feldspathic skarnoids. Nepheline-garnet-feldspathic rocks	Feldspathic ijolites. Juvites	Plagioclase-augite subsilicic schistes, metaessexites and shonkinites	Tourmaline-sillimanite and grünerite hornfelses. Nepheline fenites	Nepheline-feldspathic rocks with aegirine-diopside, hastingsite, epidote	Aegirine-lepidomelane metanephelinites	Albitites and microclinites, mainly arfvedsonite, aegirine and lepidomelane rocks
Non-feldspar peralkaline rocks	Pyroxene-nepheline-carbonaceous rocks with apatite	Melteigites	Polymineral theralite rocks	—	Calcite-cancrinite-thomsonite rocks/Tuvinites/	'Carbonatites'. Kosenites. Garnet, scapolite, cancrinite and sodalite rocks	—

TABLE 3 Average Composition of the Rocks of the Mongol–Tuva Alkaline Province

	1	2	3	4	5	6	7	8	9	10	11	12	13
Petrographic group in Table 1	I–γ	I–γ	II–γ	II–γ	II–γ	III–γ	IV–γ	II–$\gamma\sigma$ –$\xi\sigma$	III–$\gamma\sigma$	II–$\gamma\sigma$ –$\xi\sigma$	III–$\xi\sigma$	III–$\xi\sigma$	IV–$\xi\sigma$
Belts* /Fig. 1/	ES	ET	ET	S	B	ET, S	ET, S	ET, S	ET, S	ET, S	ES	ET, S	ET, S
Phase	1	1	1	1	1	1	1	1	1	1	1	1	1
Number of analyses	14	5	18	5	28	5	5	7	4	8	45	17	3
SiO_2	74·49	74·35	73·54	73·85	73·88	73·22	75·08	69·61	65·90	64·93	62·41	63·39	65·88
TiO_2	0·21	0·18	0·25	0·18	0·21	0·22	0·24	0·36	0·63	0·58	0·89	0·61	0·32
Al_2O_3	13·00	13·89	13·89	13·46	13·30	13·19	11·85	15·22	16·56	17·91	17·46	17·72	15·81
Fe_2O_3	1·15	0·57	0·94	0·98	1·45	2·68	2·73	2·01	0·82	1·25	1·79	1·80	3·54
FeO	0·92	1·18	1·46	1·04	1·19	0·63	1·49	0·99	2·99	1·89	2·62	2·98	2·00
MnO	0·09	0·05	0·04	0·07	0·06	0·12	0·06	0·06	0·13	0·04	0·15	0·14	0·09
MgO	0·42	0·26	0·34	0·62	0·35	0·42	0·38	0·43	0·98	0·33	1·14	0·58	0·72
CaO	0·95	1·42	1·09	1·32	0·83	0·89	0·41	1·17	1·98	1·45	2·02	1·34	1·07
Na_2O	4·26	3·47	4·08	4·01	4·05	4·28	4·79	4·92	5·28	6·27	6·69	5·77	5·01
K_2O	4·37	4·63	4·37	4·47	4·68	4·35	2·97	5·23	4·73	5·35	4·83	5·67	5·56

	14	15	16	17	18	19	20	21	22	23	24	25	26
Petrographic group in Table 1	I–γ	II–γ	III–γ –$\gamma\sigma$	III–γ	IV–γ –$\gamma\sigma$	II–$\gamma\sigma$ –$\xi\sigma$	V–I–ξ	II–III –ξ	III–IV –ξ	III–IV –ξ	IV–ξ	I–II –$\xi\nu$	III –$\xi\nu$
Belts* /Fig. 1/	ET, S	ET, S	ES	ET, S	ET, S	ET, S	ET, S	S	ES	H	S	S,H	S
Phase	2	2	2	2	2	2	1	1	1	1	1	1	1
Number of analyses	3	3	30	4	6	4	10	16	20	16	15	9	19
SiO_2	74·17	70·62	73·68	68·88	68·83	67·79	58·67	56·68	56·32	55·48	54·11	42·89	43·71
TiO_2	0·17	0·40	0·14	0·47	0·34	0·27	0·25	0·48	0·65	0·24	0·46	0·45	1·17
Al_2O_3	13·81	15·25	11·24	14·41	14·19	16·14	21·70	20·96	20·30	20·45	20·66	25·68	18·99
Fe_2O_3	0·67	0·68	2·95	2·27	3·64	0·62	3·13	2·90	2·53	1·66	3·43	2·64	4·01
FeO	0·76	2·54	1·77	2·03	1·90	1·76	1·48	3·06	3·40	2·89	4·50	3·97	6·99
MnO	0·06	0·07	—	0·11	0·07	0·07	0·08	0·12	0·18	0·11	0·13	0·10	0·24
MgO	0·29	0·23	0·11	0·44	0·23	0·27	0·36	0·47	0·91	0·33	0·70	1·18	1·75
CaO	1·19	1·47	0·46	1·02	0·72	0·98	0·99	2·33	2·08	2·36	3·93	8·10	13·52
Na_2O	4·46	3·75	5·19	5·12	5·13	5·84	8·17	7·93	7·42	8·06	6·37	11·49	6·69
K_2O	4·42	4·99	4·46	5·25	4·95	6·26	5·17	5·07	5·21	5·42	5·71	3·50	2·93

* ES—East-Sayan belt, ET—East-Tuva belt, S—Sangilen belt, H—Khubsugul belt, B—Bolnai belt.

degree, of divalent bases decrease sharply as the content of silica increases. With increasing SiO_2, the K_2O/Na_2O, FeO/MgO and Mg/Ca ratios increase.

3. The norms of the rocks show a slight variation of anorthite through the whole series, and a wide variation of the alkali aluminosilicates. The contents of mafic minerals increase from granite to syenite and further to nepheline syenite. Supersaturation with Al_2O_3 is typical of the granites and miaskites; supersaturation with alkalis is typical of the poorly distributed grorudites, khibinites and lardalites.

4. The compositions of the rocks of the

different facies are markedly distinct: in granitoids and nepheline syenites the content of ΣFe increases sharply from the less alkaline facies V–I to the more alkaline facies IV (Table 1, 3); the content of Ca decreases in the granites and increases in the nepheline syenites with increasing alkalinity; while with increasing alkalinity the Na content increases in the granites and decreases in the nepheline syenites.

5. The chemical compositions of the rocks of each type are practically identical in tectonic zones with different lithology, which points to their genetic closeness, in spite of significantly different absolute ages.

The distribution of rare elements is characterized by the following features: (1) Y–RE, Zr, Hf, Nb, Ta, Th and Mo are the leading rare elements, exceeding the Clarke values both in the rocks and in the minerals, especially in the granitoids of phase 2, pegmatites and postmagmatic metasomatites. Li, Be, Rb, U, Ga, Sn, Pb and As are typical elements, but they, however, do not reach considerable concentrations. The typomorphic elements of the alkaline-basaltoid rocks, such as Sr, Ba, Sc and Re, as well as B, W, Co, Ni, Cu, Zn, Cd, Te, Sb and Bi, are not typical for the complex. Among the volatiles F, CO_2 and P predominate sharply over S and Cl. (2) Table 4 shows that the distribution of rare elements differs markedly from that of the petrogenic elements. The contents of the latter in the granites and nepheline syenites of a given area are on the whole comparable. On the contrary, the dispersion of the leading rare elements in the granite and nepheline syenite complexes of lithologically different zones is great. Only the contents of Sn, as well as of Li, Be and Ga, are constant all over the province. (3) The variations in the contents of rare elements in rocks having variable amounts of mafic components (the cotectic series) are small; wide variations may be observed among the rocks of different mineral facies (horizontal rows in Table 1), especially in the more basic analogues of the geochemical pairs and groups. (4) A marked enrichment in the leading elements may be observed in the contact zones of the intrusions. This phenomenon is connected with the specifics of the infiltrational-metasomatic processes of magmatic replacement and anatexis (Fig. 5). This enrichment continues during the postmagmatic metasomatic processes, localized usually also in the contact zones.

TABLE 4 Content of Rare Elements in the Rocks of the Mongol–Tuva Province (p.p.m.)

Belts*	Rocks	RE	Zr	Th	Nb	Ce/Y	Zr/Hf	Nb/Ta	Sn
ES	Granites	460	72	16	4	4·9	21	2·2	3·6
	Nepheline syenites	500	450	11	31	6·0	49	17·2	4·1
ET	Granites	385	210	27	3	5·1	36	3·2	5·0
B	Granites	n.d.	327	n.d.	20	n.d.	n.d.	10·0	4·0
S	Granites	279	305	18	8	3·7	40	5·0	5·0
	Nepheline syenites	240	206	11	57	4·5	54	12·0	6·0
H	Granites	155	143	30	25	3·5	32	13·0	5·0
	Nepheline syenites	370	520	16	42	4·2	41	22·0	6·0

* Symbols of Table 3.

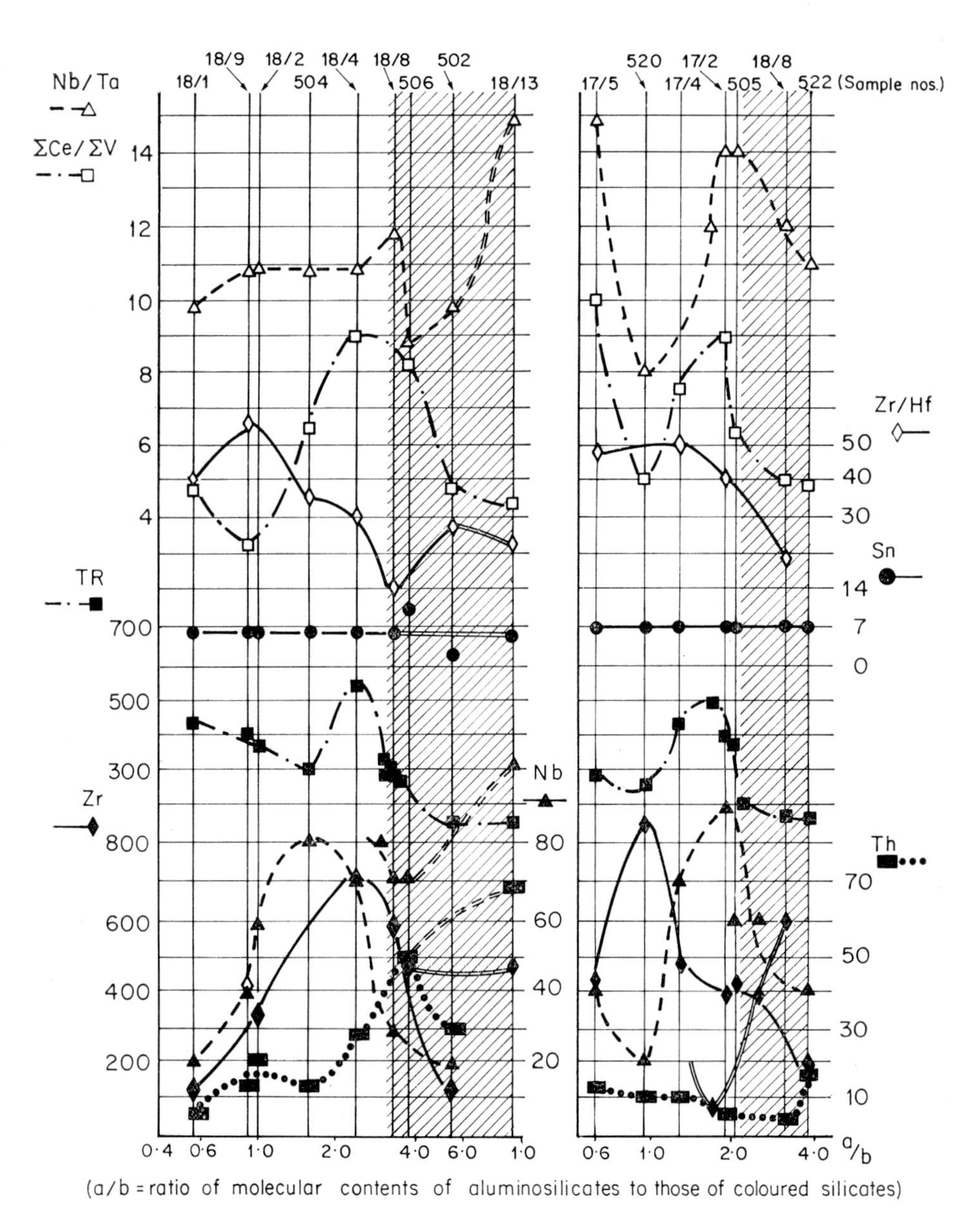

Fig. 5 Distribution of rare elements (p.p.m.) across the contact zones of granitic (18/1—18/13) and ditroitic (17/5—522) anatectic bodies. Double lines on the left are alkali-granitic variants, on the right foyaitic variants. Zone of melting is hatched

a—the content of aluminosilicates.

b—the content of coloured minerals according to A. N. Zavaritsky's system.

IV.5.6. Origin of the Alkaline Rocks of the Mongol-Tuva Province

The narrow age interval and the spatial and geochemical association of granites and alkaline rocks within the different zones of the province make it possible to combine them into a granitic-alkaline magmatic formation belonging to the post-orogenic stage of evolution of the fold belts. The origin of this magmatic formation cannot be connected with abyssal basic or ultrabasic melts, the manifestations of which are absent within the province in this interval of time. It is therefore suggested that the generation of granitic and alkaline melts and the formation of the rocks of this formation took place in the thick sialic crust produced during the previous geosynclinal stage of evolution of the region. This embraces a long interval of geological time. As a whole the processes and mechanism of generation of the melts may be grounded on the term 'palingenesis', i.e. progressive metamorphism, metasomatism and fractional fusion (anatexis). In all cases studied within the province palingenesis proceeded under amphibolite facies conditions, i.e. at temperatures not higher than 700 °C, which is confirmed by the mineralogical thermometers (Perchuk *et al.*, 1967). The series of granite-gneiss and fenite-gneiss corresponding to something like 'metasomatic cotectics' were formed during progressive metasomatism preceding anatexis. The metasomatic evolution ended with fusion at the granite or nepheline syenite eutectics and with the formation of restites. The mineral facies of the metasomatic rocks and anatectites are identical in each association and they are principally determined by the composition of the initial abyssal solutions but not by the compositions of the initial rocks. The differentiation of the rocks into silica-saturated and silica-undersaturated ones took place during the metasomatic stage, and the existence of quartz- and nepheline-bearing magmatic rocks closely connected in space does not require the crossing of any thermal barrier.

The type of contact and the morphology of the bodies of the alkaline rocks indicate their autochtonous nature (in the sense of Read). They exhibit all the structural variation known from formations such as: arterites, agmatites, nebulites, diadysites (Sederholm, 1907, Jung and Roques, 1952). It is important to note that there is here a development of a magmatic association without significant disturbance of the structural geometry of the enclosing rocks, which in the opinion of some investigators indicates a replacement origin (Wegmann, 1938; King, 1948).

A continuous transition may be traced from fenitic gneises and anatectites (migmatites) to par-autochtonous bodies with gradual contacts and again to intrusions with sharp contacts.

The intrusion of the non-overheated palingenic melt into environments chemically different from the site of magma generation may lead to differentiation of the melts: fractional apotectic crystallization (Pavlenko *et al.*, 1965, Tugarinov *et al.*, 1969). The endocontact facies of higher alkalinity is due to magmatic replacement of basic substances by weakly alkaline granitic melts. This process is accompanied by infiltrational desilication leading in turn to the effect of 'alkaline quenching', i.e. to the formation of trachytic rocks in the endocontact zones. The injection of nepheline syenitic, especially miaskitic, melts into silica-saturated rocks can give rise to selective fusion of quartz-bearing rocks constituting the 'accompanying veined series' of trachyte porphyrites, syenite-aplites and aplites in the exocontact zones of the intrusions.

Palingenesis and magmatic replacement are accompanied by a complicated wavy redistribution of the rare elements. Many rare elements accommodated in accessory and coloured minerals are concentrated in the exocontact restite zones of the anatectical bodies and in the endocontact zones of 'alkaline quenching'. The anatectic melts and the rocks formed from these melts are impoverished in most of the rare elements ('palingenic autolysis'). The low contents of rare elements and the marked dispersions of the latter, caused by their amount and phase distribution in the initial rocks, can be considered a geochemical indication of anatectical rocks.

TABLE 5 Main Massifs of Alkaline Rocks of the Mongol–Tuva Province

No. on the Map	Name of Massif	Geological Characteristics of Massif	Petrographical Variations
		North-Sayan Belt	
1	Katun–Sorug group (Katun, Sorug, Ingish, Aksuk, Irelig)	Stocks and fracture intrusions in marbles and diorites.	Granosyenite, nordmarkite, pulaskite, larvikite, lardalite. Ekerite and grorudite (second phase).
2	Oyva–Tayga	Stock in plagiogranites, diorites and tuffaceous schists.	Biotite–hastingsite microcline granites and granosyenites.
3	Dotot group (Dotot, Kharanur, Khochaar, Ostyuren)	Stocks, sills and fracture intrusions in plagiogranites, diorites and tuffaceous schists	Biotite and hastingsite microcline leucogranites and granosyenites. Quartz albitites.
		East-Tuva Belt. Biykhem Sector	
4	South Sorug	Stock in terrigenous–volcanic schists.	Quartz nordmarkite.
5–6	Ulug–Arga, Uzunoy–Tayga	Fracture intrusion in plagiogranites and limestones.	Syenite, nordmarkite, albitites.
7	Karakhol	Stock in diorites.	Quartz nordmarkite.
8	Dugda	Polyphase ethmolith in fault zone among carbonaceous shales, gabbro and diorites.	Foyaite, lardalite, khibinite. Litchfieldite and albitites. Quartz nordmarkite, ekerite, grorudite (second phase).
9	Kishtag	Stock in green shales.	Syenite.
10	Kadiross	Near-fault two-phase stock in diorites.	Nordmarkite. Microcline granite (second phase).
11	Sorligkhem	Contact alkaline facies of normal granites in contact with limestones.	Quartz nordmarkite.
12	Kadiross	Zoned stock in knot of fractures in green shales and marble.	Microcline granite, ekerite, grorudite. Albitite.
		Kaakhem Sector	
13	Karga group (Milzey, Elegteg, Byuretkin)	Alkaline facies and stock-like bodies in normal granites near xenoliths of marble and calcareous shale.	Microcline–amphibole granites and granosyenites. Albitites.
14	Ulug–Shivey group	Alkaline facies and stocks in normal granites near xenoliths of terrigenous–effusive rocks and gabbroids.	Granosyenite, ekerite, grorudite, quartz nordmarkite.
		Busingol Sector	
15	Chavach	Fracture bodies in green and calcareous shale and marble.	Syenite, akerite, larvikite, miaskite, foyaite, lardalite. Metasomatic albitites, canadite, litchfieldite.
16	Kharatugol	Fracture intrusion in limestones.	Foyaite.
17	Kaskangol	Intrusive-metasomatic stock in marble.	Miaskite and foyaite with contact facies of essexite, shonkinite, wollastonite ijolite.
18	Dukheyn–Daba group (Dukheyn–Daba, Salig, Sangalag, Arakhigol)	Intrusive-metasomatic bodies in marble.	Ditroite, syenite, Nepheline–pyroxene skarns. Monmouthite.
19	Zagan–Adar	Palingenetic-metasomatic bodies and fracture intrusives in marble and schists.	Canadite, miaskite, ditroite.

TABLE 5 *continued*

20	Uringim	Zonal stock in schists and marble with granitizited exocontacts.	Microcline granite, ekerite, quartz nordmarkite.
21	Terekhol	Zonal fracture intrusion with fenitizited contacts in limestones and gabbro.	Pulaskite, lardalite, khibinite pegmatites. Albitites, canadites.
		Sangilen Sector	
22	Khorigtig	Intrusive–metasomatic stock in zone of crushing among plagiogranites.	Ekerite. Metasomatic albitites and microclinites.
23	Burek–Kundus	Fracture intrusion in limestones.	Quartz syenite. Miaskite, ditroite, canadite.
24	Choltin	Stock in limestones and gabbro.	Foyaite, shonkinite.
25	Chekbin	Fracture intrusion in marble.	Foyaite.
26	Agash	Stock in marble.	Syenite, miaskite. Albitites.
27	Kunduss	Stock in zone of crushing in marble.	Ditroit. Metasomatic albitites.
28	Korgere–Daba	Zone intrusion with fenitizited contacts in gabbroids and marble.	Nordmarkite, pulaskite, ditroite, essexite. Akerite (second phase). Astrophyllite pegmatites. Nepheline albitites.
29	Ulan–Erge	Fracture intrusion in marble.	Ditroite, Nepheline albitite.
30	Chakhyrtoy	Metasomatic bodies in silicate–carbonate rocks.	Miaskite, congressite. Nepheline albitite.
31	Chik	Intrusive-metasomatic stocks in marble.	Ijolite. Carbonaceous-nepheline rocks.
32	Zhin–Khem	Near-fault intrusion in marble.	Syenite, nordmarkite.
33	Ak–Khem	Fracture body in marble.	Umptekite, pulaskite.
34	Pichikhol	Zoned differentiated intrusion in crush zone in marble with contact metasomatism.	Ditroite, foyaite, lardalite, Akerite (second phase). Nepheline and quartz pegmatites. Canadite. Foyaite.
35	Chartiss	Stocks in silicate marble.	Foyaite.
36	Solbeldir	Fracture body in marble.	Miaskite, mariupolite.
37	Dakhunur group	Stock-like bodies in marble and gabbro-pyroxenites.	Theralite, malignite, monmouthite, garnet ijolite, urtite, ijolite. Cancrinite pegmatites. Nepheline–vesuvianite skarns.
39	Khunchol	Stock in marble.	Foyaite, juvite.
38	Toskul	Stock in marble and gabbro-pyroxenites.	Nephelinized pyroxenites.
		Karga Sector	
40	Khonchul	Palingenetic-metasomatic bodies in silicate–carbonaceous rocks	Ditroite, essexite, theralite.
41	Khaygass		
42	Right-Bayankol	Stock in marble.	Micaceous urtite, juvite. Mariupolite.
43	Left-Bayankol	Stratified intrusion in shales.	Hedenbergite foyaite, ditroite, miaskite.
44	Kharlin	Intrusive–metasomatic near-fracture body in marble.	Hedenbergite lardalite, juvite, foyaite, miaskite. Litchfieldite. Nepheline–carbonate rocks.
45	Orugta	Fracture intrusion in marble.	Miaskite. Nepheline pegmatites. Mariupolite, litchfieldite.
46	Erzin	Near-fault stock in marble.	Albitizited riebeckite and cryophyllite granites.
47	Oleniy	Stock in marble.	Syenite, granosyenite.

TABLE 5 *continued*

		Khubsugul Belt. South Sector.	
48a	Muren group (Achitul, Dilingin, Oshigin)	Stocks and contact alkaline facies in granites on contacts with terrigenous–effusive rocks and marble.	Microcline–amphibole granites, ekerite, granosyenite, plauenite, nordmarkite.
48b	Erkhilnur group (Khukhuchulut, East-erkhinur, Sharabin, Burinkanobin, Khitagin)	Zoned intrusions in large xenoliths (relics of roof) of marble.	Quartz syenite, nordmarkite, pulaskite, foyaite, ditroite.
48c	Egiyn group (Adun, Yarkhis) Upper-Egiyn, Lower-Egiyn.	Fracture intrusion, stocks and border facies of granites on contacts with marble.	Granosyenite, plauenite, quartz nordmarkite.
	Udzhigin group		
48	Ikheerzig	Stocks and zones of fenitization and magmatic replacement in diorites and gabbro.	Granosyenite, plauenite, sodalite syenite, canadite.
49	Angarkhain	Stock in shale–calcareous series.	Miaskite, litchfieldite.
50	Khagin–Nur	Zone of fenitization and magmatic replacement in diorite on contact with marble.	Pulaskite, foyaite.
51	Lower–Udzhigin	Zoned stock and zone of magmatic replacement in gabbro and calcic–shale series.	Foyaite, essexite, foyaite, shonkinite.
51a	Burinankhan	Zoned stock in marble.	Miaskite, ditroite, monmouthite. Mariupolite.
52	Namulaulin	Zones of fenitization and palingenesis in andesitoid series.	Quartz syenite, plauenite, nordmarkite.
	Serkheulin–Middle-Beltesiin group		
53	Middle-Beltesin	Stocks and zones of magmatic replacement in diorites.	Ditroite, essexite.
54	Ijolite	Stock and metasomatic zone in marble.	Ijolite, nepheline–carbonaceous rocks.
55	Jacupirangite	Stock in contact of gabbro-pyroxenite and marble.	Monmouthite (graphitic), jacupirangite.
56	Upper-Udzhigin	Stock in marble.	Foyaite.
57	Duchin	Stocks and metasomatic zones in marble.	Urtite, calcite urtite, congressite.
58	Khorintuingol	Zone of syenitization and migmatization in diorites on contact with marble.	Plauenite, nordmarkite.
59	Khodkhal group (Berkhein, Ustugol, Dzosain)	Zones of syenitization and stocks in contact of granites and marbles.	Quartz syenite, plauenite, nordmarkite, ditroite.
		Darkhat Sector	
	East group		
60a	Aragol	Stocks in limestone.	Miaskite, ditroite.
60	Dumdakhemgol	Stocks in limestone.	Miaskite, ditroite.
61	Upper-Beltesin	Zoned stocks in limestone.	Plauenite, nordmarkite, ditroite.
62	Sharayamatugol	Zoned stocks in limestone.	Plauenite, nordmarkite, ditroite.
63	Doodkhem	Zoned differeniated intrusion in marble and dolomite.	Akerite, larvikite, lardalite. Khtibinite, eudialyte pegmatites, litchfieldite.
64	Urundush	Intrusive-metasomatic stock in marble.	Eudialyte litchfieldite.
65	Ikheul	Stock in marble.	Nordmarkite.

TABLE 5 *continued*

66–67	Alkhain, Khabtagay	Fracture intrusion in marble and calcareous shists.	Plauenite, pulaskite.
68	Khundush West group		
69	Gunain	Fracture intrusion in greenshale series.	Foyaite, shonkinite. Nepheline pegmatite.
70	Upper-Shishkid group	Intrusive–metasomatic bodies in contact of granites with gabbro and marble.	Foyaite, essexite, shonkinite.
71	Khormain	Field of fenites and migmatites in gabbro.	Nordmarkite, pulaskite.
		Bolnay Belt	
72–73	Numurga, Bederkhunur	Polyphase and polyfacies intrusions in large fault zones in volcano-genetic series (diabase–andesite) and plagigranites; alkaline facies of normal-granite.	Hastingsite and arfvedsonite granites and granosyenites. Riebeckite–aegirine and cryophyllite quartz microclinites and albitites.
74–75	Modotuin, Shumelkamish		
76–77	Toson, South-Zagannur		
78	Khangarul group	Schlieren-like bodies in normal granites.	Granosyenite, quartz nordmarkite.

ACKNOWLEDGEMENTS

The author is greatly obliged to Drs. L. P. Orlova, L. P. Philippov, A. V. Ilyin for their kind assistance.

IV.5. REFERENCES

Appleyard, E. C., 1967. Nepheline gneisses of the Wolfe Belt, Lindoch Township, Ontario. 1 Structure, stratiography, and petrography. *Can. J. Earth Sci.*, **4,** 371–96.

Appleyard, E. C., 1969. Nepheline gneisses of the Wolfe Belt, Lindoch Township, Ontario. 2 Textures and mineral paragenesis. *Can. J. Earth Sci.*, **6,** 689–718.

Barth, T. F. W., 1948. Oxygen of rocks: a basis of petrographic calculations. *J. Geology*, **56.**

Barth, T. F. W., 1955. Presentation of rock analyses. *J. Geology*, **63,** 14.

Bazorova, T. V., 1969. *Thermodynamical Condition of the Formation of some Nepheline-bearing Rocks.* Izd. 'Nauka', Moskva.

Bemmelen, R. W., van, 1952. The endogenic energy of the Earth. *Am. J. Sci.*, **250.**

Daly, R., 1933. *Igneous Rocks and the Depths of the Earth.* New York.

Härme, M., 1958. Examples of the granitization of plutonic rocks. *Bull. Comis. Geol. Finland*, **180.**

Jung, J., and Roques, M., 1952. Introduction à l'étude zonéographique des formations cristallophylliennes. *Bull. Serv. Carte Geol. Fr.*, **235.**

King, B. C., 1948. The form and structural features of aplite and pegmatite dikes and veins in the Osi area of Nigeria. *J. Geol.*, **56.**

Kononova, V. A., 1958. On the nephelinization of pyroxenites and marbles (in Russian). *Izv. Akad. Nauk SSSR*, **6,** 58–69.

Kononova, V. A., 1961. Urtit-ijolitovye intruzii Yugo-Vostochnoy Tuvy i nekotorye voprosy ikh genezisa. 109 pp. *Trudy IGEM Akad. Nauk SSSR*, vyp **60,** Izd. Akad. Nauk SSSR, Moskva. (Urtit-ijolite intrusions of the South-East Tuva and some questions of their genesis.)

Kononova, V. A., 1962. 'The Bajankol primary laminated intrusion of hedenbergite nepheline syenites' (in Russian) in Vorob'eva, O. A., *Trudy IGEM Akad. Nauk SSSR*, **76,** Izd. Akad. Nauk SSSR, Moskva, 39–70.

Korzhinsky, D. S., 1952. Granitization as the magmatic replacement. *Izv. Akad. Nauk SSSR, ser. geol.*, **2.**

Korzhinsky, D. S., 1957. *Fizico-khimicheskie osnovy analiza paragenezisa mineralov.* 184 pp., Izd- Akad. Nauk SSSR, Moskva. (Physical and chemical foundations of a paragenetic analysis of the minerals.)

Korzhinsky, D. S. 1960, The theory of metasomatic zonation (in Russian). Izd. 'Nauka', Moskva.

Kostin, N. E., and Petrova, E. A., 1960. 'Some features of the mineralogy and genesis of one of the fields of albitites in rare-metal mineralization' (in Russian), in Gliko, O. A., Shmanenko, I. O., *Mineral'noe syrie VYMS*, **1,** Gosgeoltekhizdat, Moskva, 78–85.

Kovalenko, V. I., 1964. Features of the metasomatic processes in alaskites of the Ognit complex', (in Russian), in Tauson, L. V., ed., *Geokhimiya redkikh elementov v izverzhennykh porodakh.* Izd- 'Nauka', Moskva, 63–84.

Kovalenko, V. I., Okladnikova, L. V., Pavlenko, A. S., Popolitov, E. I., Filippov, L. V., and Shmakin, B. M., 1965. 'Petrology of the Middle-Paleozoic complex of the East Tuva granitoids and alkaline rocks' (in Russian), in Shmakin, B. M., *Geokhimiya i petrologiya magmaticheskikh i metasomaticheskikh obrazovaniy. Sib. otdel. Akad. Nauk SSSR.* Izd. 'Nauka', Moskva, 5–145.

Kovalenko, V. I., and Popolitov, E. I., 1965a. On the influence of enclosing gabbro on the acidity–alkalinity of the endomorphic part of granite and nepheline-syenites massifs (in Russian). *Dokl. Akad. Nauk SSSR*, **161**, 207–9.

Kovalenko, V. I., and Popolitov, E. I., 1965b. On the origin of the North-East Tuva alkaline rocks (in Russian). *Dokl. Akad. Nauk SSSR*, **163**, 1474–6.

Kovalenko, V. I., and Popolitov, E. Y., 1970. *Petrology and geochemistry of rare elements of the alkaline and granitoid rocks of North-eastern Tuva* (in Russian). Izd. 'Nauka', Moskva.

Kudrin, V. S., 1962. The alkaline intrusions of North-East Tuva. (in Russian), *Sov. Geol.*, **4**, 40–52.

Kuznetsov, V. A., Izokh, E. P., 1969. Geological pictures of the intratelluric heat and substance flows as the agent of metamorphism and magma-formation' (in Russian), in Kuznetsov, V. A., sb., *Problemy petrologii i geneticheskoy mineralogii*, **1**. Izd. 'Nauka', Moskva, 7–20.

Lodochnikov, B. N., 1924. The simplest illustrated methods of the multicomponent systems (in Russian). *Izv. Inst. fisico-chimicheskogo analiza. Akad. Nauk SSSR*, **2**, 1926.

Luvsan-Danzan, B., and Khasin, R. A., 1966. 'New data on the West Prykhubsugal alkaline rocks' (in Russian) in Marinov, N. A. *Materialy po geologii Mongol'skoy Narodnoi Respubliki*, Izd, 'Nedra', Moskva, 111–17.

Makhin, V., and Pavlenko, A. S., 1966. 'The Sangilen complex' (in Russian), in Kudryavtsev, G. A., Kuznetsov, V. A., *Geologiya Soyuza*, **29**, Geologicheskoe opisanie, 1966, Izd 'Nedra', Moskva, 306–26.

Pavlenko, A. S., 1963a. 'Behaviour of the rock-forming and some rare elements in the formation of alkaline rocks' (in Russian) in Vinogradov, A. P., *Khimiya zemnoy kory. Trudy geokhimicheskoy konferentsii k 100-letiyu so dnya rozhdeniya akademika V.I. Vernadskogo.* Izd Akad. Nauk SSSR, Moskva, 116–29.

Pavlenko, A. S., 1963b. 'Petrology and some geochemical features of the Middle-Paleozoic complex of the East Tuva granitoids and alkaline rocks' (in Russian), in Afanas'ev, G. D., *Problemy magmy i genezica izverzhennykh gornykh porod, Trudy Yubileynogo simpoziuma posvyashchennogo F. V. Levinsone- Lessingu.* Izd Akad. Nauk SSSR, Moskva, 239–46.

Pavlenko, A. S., 1963c. 'Alkaline metasomatites of the contact type' (in Russian), in Shcherbina, V. V., *Geokhimiya shchelochnogo metasomatoza.* Izd Akad. Nauk SSSR, Moskva, 7–70.

Pavlenko, A. S., Gevorkyan, R. G., and Filippov, L. V., 1965. 'On the relation of the alkaline-earth and alkaline basaltoid series' (in Russian), in Khitarov, N. I., *Problemy geokhimii. Yubileinyi sbornik k 70-letiyu akademika A. P. Vinogradova.* Izd 'Nauka', Moskva, 350–65.

Pavlenko, A. S. and Kovalenko, V. I., 1961. Paragenetic dependence of the alkaline metasomatites in calcium-poor alumosilicate rocks on the chemical potentials of alkalis (in Russian). *Geokhimiya*, 980–7.

Pavlenko, A. S., and Kovalenko, V. I., 1965. 'The facies zoning of alkaline metasomatites and rare-metal mineralizations connected with them', (in Russian) in Kutina, V., *Trudy konferentsii*: *Problemy postmagmaticheskogo rudoobrazovaniya.* Izd Chekhoslovatskoy Akad. Nauk, Praha, 222–30.

Pavlenko, A. S., Il'in, A. V., Strizhov, V. P., and Bykhover, V. N., 1969. Absolute ages of the intrusions of the East Tuva and North Mongolian geotectonic zones (in Russian). *Geotectonika*, **3.**

Pavlenko, A. S., Syao-Chzhun-Yan, and Morozov, L. N., 1960. The comparative geochemical characteristic of granitoids enriched in tantalum-niobates (in Russian). *Geokhimiya*, 104–20.

Pavlenko, A. S., Vainshtein, E. E., and Shevaleevsky, I. D., 1957. On the zirconium and hafnium ratio in the zircons of igneous and metasomatic rocks (in Russian). *Geokhimiya*, 351–67.

Pavlenko, A. S., Vainshtein, E. E., and Kakhana, M. M., 1958. On the niobium and tantalum ratio in some minerals of igneous and metasomatic rocks (in Russian). *Geokhimiya*, 558–69.

Pavlenko, A. S., Vainshtein, E. E., and Turanskaya, N. V., 1959. On some objective laws of the behaviour of rare earths and yttrium in magmatic and postmagmatic processes (in Russian). *Geokhimiya*, 291–309.

Perchuk, L. L., 1968. 'The phase accordance in the system nepheline–alkali feldspar–aqueous solution' (in Russian) in Marakushev, A. A., *Metasomatizm i drugie voprocy fiziko-khimicheskoy petrologii.* Izd 'Nauka', Moskva, 53–95.

Perchuk, L. L., and Pavlenko, A. S., 1967. The influence of temperature on the distribution of some isomorphic components among coexisting minerals of alkaline rocks (in Russian). *Geokhimiya*, 1063–82.

Read, H. H., 1957. *The Granite Controversy*. Interscience, New York and London.

Sederholm, I. I., 1907. Om granit och gneis. *Bull. Comm. geol. Finl.*, **23.**

Sheynmann, V. M., Apel'tsin, F. R., and Nechaeva, E. A. 'Alkaline intrusions, their location and mineralization connected with them' (in Russian), in Ginzburg, A. I., *Geologiya mestorozhdeniy redkikh elementov, vyp. 12–13*. Gosgeoltekhizdat, Moskva, 1–176.

Shevaleevsky, I. D., Pavlenko, A. S., and Vainshtein, E. E., 1960. The dependence of the zirconium and hafnium behaviour on the petrochemical features of the magmatic and alkaline-metasomatic rocks (in Russian). *Geokhimiya*, 222–30.

Tröger, W. E., 1938. Eruptivgesteins-namen. *Fortschr. Min., Kryst. Petr.*, **23.**

Tugarinov, A. I., Pavlenko, A. S., and Kovalenko, V. I., 1969. The origin of apogranites according to geochemical data (in Russian). *Geokhimiya*, 1419–36.

Vainshtain, E. E., Pavlenko, A. S., Turanskaya, N. V., and Yulova, T. G., 1961. Dependence of the distribution of rare-earth elements in rocks on the petrochemical factors and its importance in the solution of petrogenetic problems (in Russian). *Geokhimiya*, 1077–86.

Vinogradov, A. P., 1962. The origin of the Earth's mantle (in Russian). *Izv. Akad. Nauk SSSR, ser. geol.*, **11.**

Wegmann, C. E., 1938. Geological investigations in Southern Greenland. Pt. 1. On the structural divisions of southern Greenland. *Meddr. Grønland*, **113(2).**

Yashina, R. M., 1957. Alkaline rocks of South-East Tuva (in Russian). *Izv. Akad. Nauk SSSR, ser. geol.*, **5**, 17–36.

Yashina, R. M., 1962. 'Kharbin concentric-zoned alkaline massif and the conditions of its formation' (in Russian), in Vorob'eva, O. A., *Trudy IGEM Akad. Nauk SSSR*, **76**, 1962. Izd. Akad. Nauk SSSR., Moskva, 7–38.

Yashina, R. M., 1963. 'On the contact-reactional interaction of nepheline syenites with xenoliths of dolomitized marbles (on the example of the East Tuva Arugta massif)' (in Russian), in Sokolov, G. A., *Fiziko-khimicheskie problemy formirovaniy gornykh porod i rud.*, **2**, Izd. Akad. Nauk SSSR, Moskva, 117–28.

Zykov, S. I., Stupnikova, N. I., Pavlenko, A. S., Tugarinov, A. I., and Orlova, L. P., 1961. Absolute age of the East-Tuva region and Yenisei ridge (in Russian). *Geokhimiya*, 547–60.

IV.6. THE MONTEREGIAN PROVINCE

A. R. Philpotts

IV.6.1. Introduction

The Monteregian province was named by Adams (1903) to include a group of alkaline intrusions that form a series of hills extending eastward from Montreal in the Province of Quebec. These hills are from west to east, Mount Royal, Bruno, St. Hilaire, Johnson, Rougemont, Yamaska, Brome and Shefford (Fig. 1). The last three lie within the Appalachian fold belt, but the remainder rise sharply from the plain of the St. Lawrence lowlands, an area underlain by flat-lying Cambrian and Ordovician sandstone, limestones and shales. Since Adams' description of these intrusions, three more have been recognized as belonging to this province; the carbonatite complex at Oka, the Iberville intrusion near Mount Johnson, and a ring complex at Mount Megantic, 110 km east of Shefford. There are, in addition, many small plug-like bodies, diatreme breccia pipes, dykes and sills, most of which occur within the east–west-trending belt of the main intrusions.

IV.6.2. Structural Settings and Age

The Monteregian province is situated at the junction of the east–west-trending Ottawa graben and the N.N.E.-trending St. Lawrence graben (Kumarapeli and Saull, 1966). Several of the intrusions are situated where the individual faults of the St. Lawrence graben intersect the

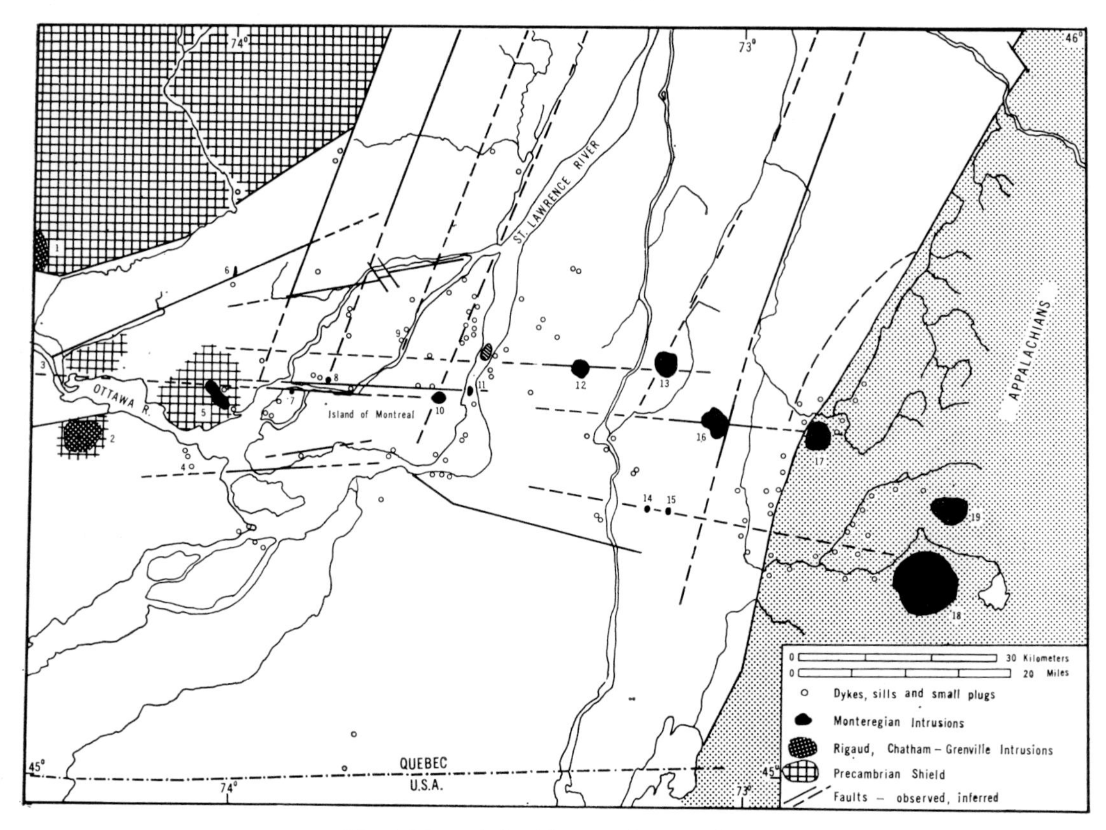

Fig. 1 Map of the Monteregian Province

(1) Chatham–Grenville; (2) Rigaud; (3) Carillon; (4) Ile Cadieux; (5) Oka; (6) St. Monique; (7) Ile Bizard; (8) St. Dorothée; (9) Visitation Island; (10) Mount Royal; (11) St. Helen's Island; (12) Mount Bruno; (13) St. Hilaire; (14) Iberville intrusion; (15) Mount Johnson; (16) Rougemont; (17) Yamaska; (18) Brome; (19) Shefford.

Information from Clark (1952) and Gold (1967). White: St. Lawrence lowlands.

faults of the Ottawa graben, as can be shown by field mapping (Clark, 1952) and by aeromagnetic evidence (Fig. 1). Radiometric dating of pseudotachylites from one of the N.N.E.-trending faults, the extension of which is shown passing through Mount Royal, indicates that this fault system has been active for at least 975 m.y. (Philpotts and Miller, 1963; Philpotts, 1964).

K–Ar dating by Doig and Barton (1968) has distinguished four main periods of alkaline igneous activity associated with the St. Lawrence rift system. During the first period, between 820 and 1000 million years, syenites, granites and carbonatites were emplaced to the north and west of the Monteregian province. The second period, 565 ± 15 m.y., involved the emplacement of carbonatites, lampophyres and syenites along the northern edge of the St. Lawrence valley and along the Ottawa graben. During the third period of activity, 450 ± 20 m.y., two alkaline syenite–granite complexes were emplaced in the Chatham–Grenville area and at Rigaud. These lie on the north and south sides respectively of the Ottawa graben at the western end of the main Monteregian province, approximately 60 km west of Mount Royal (Fig. 1). These were very near surface intrusions accompanied by diatreme breccias. The Monteregian activity at 110 ± 20 m.y. (Gold, 1967, p. 290) was the most recent period of intrusion associated with the St. Lawrence and Ottawa grabens. Dykes with ages similar to those of the Monteregian intrusions can be found extending along the Champlain valley and through New England (Zartman *et al.*, 1967) and the rocks of the White Mountain magma series of New Hampshire are only slightly older (see III.3).*

IV.6.3. General Petrology

Standard references on the petrology of the Monteregian province include Adams' original work (1903), Graham's general summary of the province (1944), and most recently Gold's coverage of the ultramafic rocks to be found in these intrusions (1967). There have been other works published on specific aspects of these rocks, but most were from the 'rock naming era' of igneous petrology and present us with such names as montrealite, yamaskite, beloeilite, rouvillite and okaite, most of which do not find common usage in modern literature. However, these works do provide excellent petrographic descriptions, and are referred to by Graham and Gold. Many recent studies have been published in a special symposium volume of the *Canadian Mineralogist* (vol. **10**, 3, 1970).

It is convenient to group the Monteregian rocks into dyke rocks, rocks of the main intrusions, ultramafic nodules and diatreme breccias.

IV.6.3.1. Dyke Rocks

There are two general types of contrasting dykes in the Monteregian province; melanocratic lamprophyres which occur throughout the province, and leucocratic, feldspar-rich rocks that are restricted to the eastern part of the province and almost entirely to the vicinity of the main intrusions.

The lamprophyres in the Monteregian province include camptonites, monchiquites and alnoites (see Table 1). The camptonites occur to the east of Mount Royal, the monchiquites between Mount Royal and St. Dorothée, and the alnoites to the west of St. Dorothée. These boundaries are relatively sharp and correspond very closely to the extensions of the north-easterly faults that pass through Mount Royal and St. Dorothée. The camptonites are all fine-grained rocks forming dykes and sills that are never more than a few metres wide. In general they strike east–west or less commonly north-easterly, paralleling the two main fracture sets in the area. They commonly contain phenocrysts of augite which have cores of endiopside and titaniferous rims, dark reddish-brown hornblende and olivine. The matrix consists of titanaugite, hornblende, plagioclase (rarely more calcic than An_{60}), and abundant oxide which is commonly ilmenite but can be magnetite. Apatite is a very common accessory and analcime occurs interstitially and in amygdules with calcite.

The monchiquites are fine-grained rocks containing titanaugite, olivine and hornblende in a turbid groundmass consisting largely of analcime, but also containing ilmenite, apatite and in some cases titanite. They commonly contain phenocrysts of augite, olivine or hornblende. With decrease in olivine content monchiquites grade to fourchites.

Approximately 20% of the camptonites and monchiquites (including fourchites) contain ocelli, small spherical bodies usually less than 1 cm in diameter, that are composed of fine-grained analcime- or nepheline-syenite (Fig. 2). In many sills they tend to segregate towards the upper part and produce lenses of syenite (see IV.6.6, and Philpotts, 1972).

The alnoites commonly occur in small plugs which were emplaced by explosive activity as indicated by the abundance of included fragments of Precambrian rocks and Potsdam sandstone, the basal member of the local Palaeozoic succession. However, some form narrow dykes, and just north of Ile Cadieux and at St. Monique they form extensive sills. Bowen (1922) described in considerable detail the alnoite of Ile Cadieux which is characteristic of the other alnoites in the Monteregian province. They are fine- to medium-grained rocks containing poikilitic crystals of biotite enclosing early formed olivine grains. Melilite and diopsidic-augite are also major components and some contain monticellite. Magnetite is an abundant accessory mineral which amounts to 10% in

* Readers are referred to the relevant Chapters of this volume where similarly cited.

TABLE 1 Compositions of Some Important Rocks of the Monteregian Province

	1	2	3	4	5	6	7	8	9	10	11	12	13	14	15	16	
SiO_2	41·18	45·51	36·66	37·56	44·45	30·85	30·27	34·32	35·19	49·96	34·14	64·62	63·30	51·71	51·53	51·44	SiO_2
TiO_2	4·32	3·84	3·40	3·72	3·05	2·87	2·84	6·58	6·58	2·40	5·27	0·27	0·58	1·41	0·25	1·65	TiO_2
Al_2O_3	16·37	15·84	12·33	14·58	14·75	8·21	10·00	14·63	9·55	18·83	13·40	17·58	18·07	19·62	20·98	19·66	Al_2O_3
Fe_2O_3	5·30	1·90	2·23	6·15	5·13	3·33	4·88	7·13	9·35	2·52	2·00	0·50	1·12	3·09	3·66	2·00	Fe_2O_3
FeO	8·25	7·59	8·19	8·20	5·41	6·52	6·95	10·47	8·86	6·64	15·86	0·86	0·99	3·52	1·17	4·50	FeO
MnO	0·52	0·40	0·09	0·23	0·29	0·21	0·16	0·20	0·33	0·20	0·19	0·09	0·13	0·28	0·21	0·27	MnO
MgO	5·87	3·85	9·98	7·47	3·78	23·16	20·11	7·68	9·20	3·52	9·63	0·33	0·43	1·22	0·14	1·50	MgO
CaO	11·58	8·35	12·20	10·08	10·58	16·46	14·73	14·44	15·01	7·42	13·05	1·07	0·82	4·46	1·11	7·25	CaO
Na_2O	2·98	5·09	2·68	4·24	4·53	1·01	1·49	1·40	1·59	5·26	2·21	6·40	7·58	9·71	13·40	7·10	Na_2O
K_2O	1·47	1·88	2·16	1·94	2·28	1·43	2·85	0·08	0·64	2·58	0·72	6·32	5·84	3·34	5·32	3·85	K_2O
P_2O_5	1·30	0·61	0·35	0·23	1·28	1·90	0·95	2·57	1·89	0·25	1·77	0·14	0·10	0·64	0·02	0·28	P_2O_5
CO_2	—	2·23	6·81	1·80	0·87	3·04	3·24	—	—	—	—	—	—	0·37	0·26	—	CO_2
H_2O^+	0·76	1·67	2·40	3·11	3·33	1·22	2·17	0·55	—	0·60	—	0·20	0·46	1·29	1·10	—	H_2O^+
H_2O^-	0·21	0·17			0·26	0·05		0·05	—		—	0·02	0·02	0·02	0·02	—	H_2O^-
Total	100·02	98·93	99·48	99·31	99·99	100·26	100·64	100·07	98·19	100·18	98·24	98·40	99·44	100·68	100·48	99·51	Total
								CIPW Norms									
Q												1·2					Q
or	8·7	11·1	12·8	11·5	13·5			0·5	3·8	15·2		37·3	34·5	19·7	31·2	22·7	or
ab	12·2	32·5	16·2	3·8	21·6			7·7	1·8	26·6		54·1	51·8	30·4	6·8	19·8	ab
an	26·9	14·8	15·2	15·0	13·2	13·6	12·2	33·4	17·0	20·1	24·2	0·6		0·1		10·4	an
lc																	lc
ne	7·0	5·7	3·5	17·4	9·0	4·6	6·8	2·2	6·3	9·7	10·3		4·7	28·0	38·7	21·8	ne
kp						4·7	9·5				2·5						kp
ac													3·2		10·6		ac
ns															5·8		ns
Ca cpx	9·2	3·6		9·3	10·7			9·0	18·9	6·3	7·1	1·6	1·4	5·3	1·6	8·2	Ca cpx
Mg cpx	6·4	2·0		6·8	8·4			7·0	16·3	3·5	3·8	0·8	0·8	3·0	0·2	3·7	Mg cpx
Fe cpx	2·0	1·4		1·5	1·1			1·0		2·6	3·0	0·8	0·6	2·1	1·6	4·4	Fe cpx
Wo														1·1		1·7	Wo
En												0·1					En
Fs												0·1					Fs
Fo	5·7	5·3	17·4	8·3	0·7	42·2	35·0	8·5	4·6	3·7	14·2		0·2		0·1		Fo
Fa	2·0	4·1	6·0	2·1	0·1	3·7	3·3	1·4		3·0	12·2		0·3		0·8		Fa
cs						13·3	10·4				3·8						cs
mt	7·7	2·7	3·2	8·9	7·4	4·8	7·1	10·3	10·5	3·6	3·0	0·7		4·5		2·9	mt
hm									2·1								hm
il	8·2	7·3	6·5	7·1	5·8	5·5	5·4	12·5	12·5	4·6	10·0	0·5	1·1	2·7	0·5	3·1	il
ap	3·1	1·5	0·8	0·5	3·1	4·5	2·3	6·2	4·5	0·6	4·0	0·3	0·2	1·5	0·1	0·7	ap
cc		5·1	15·5	4·1	2·0	6·9	7·4							0·8	0·6		cc
H_2O	1·0	1·8	2·4	3·1	3·6	1·3	2·2	0·6		0·6		0·2	0·5	1·3	1·1		H_2O
Total	100·1	98·9	99·5	99·4	100·2	105·1	101·6	100·3	98·3	100·1	98·1	98·3	99·3	100·5	99·7	99·4	Total

Legend: 1. Camptonite, Mount Johnson (Pajari, 1967) 2. Camptonite, Mount Royal (Faessler, 1962, p. 94) 3. Monchiquite, Mount Royal (Faessler, 1962, p. 95) 4. Monchiquite, Mount Royal (Faessler, 1962, p. 95) 5. Fourchite, St. Dorothée (Philpotts and Hodgson, 1968) 6. Alnoite, Ile Cadieux (Bowen, 1922) 7. Alnoite, north of Ile Cadieux (Stansfield, 1923) 8. Olivine gabbro, Brome Mountain (Valiquette and Archambault, 1970, p. 489) 9. Yamaskite, Yamaska (Gandhi, 1967, p. 188) 10. Essexite, Mount St. Hilaire (O'Neill, 1914) 11. Hornblende gabbro, Mount Royal (Woussen, 1968, p. 135) 12. Nordmarkite, Brome Mountain (Valiquette and Archambault, 1970, p. 490) 13. Pulaskite, Brome Mountain (Valiquette and Archambault, 1970, p. 490) 14. Foyaite, Mount St. Hilaire (Rajasekaran, 1967, p. 60) 15. Sodalite syenite, Mount St. Hilaire (Rajasekaran, 1967, p. 72) Analysis includes 1·31% Cl. 16. Nepheline monzonite, Mount Royal (Woussen, 1968, p. 138).

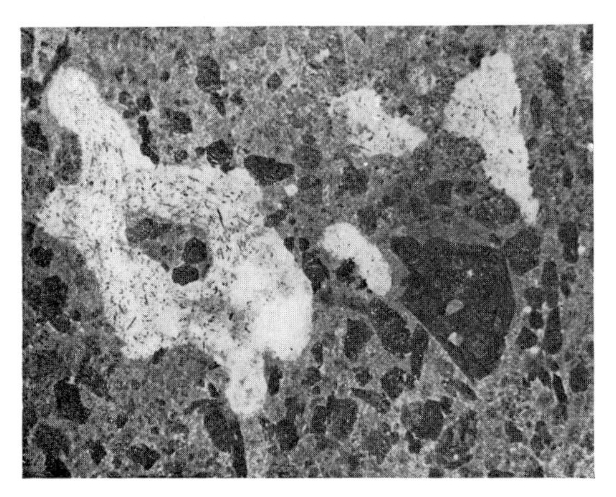

Fig. 2 Ocelli containing alkali feldspar, analcime and hornblende in a porphyritic camptonite dyke. ×2.3

some rocks. Carbonate is also very common filling interstices between the earlier formed minerals. These rocks commonly contain phenocrysts of biotite (Adams, 1892), a feature that clearly distinguishes them from the camptonites and monchiquites. Olivine and augite also form phenocrysts, and are major constituents of ultramafic nodules that are very abundant in these rocks.

A structure found in some alnoites and referred to by Gold (1967, p. 296) as lapilli, is of particular interest because of the association of these rocks with carbonatite. The lapilli consist of spheres of alnoite, rarely more than 1 cm in diameter, embedded in a carbonate rich matrix (Fig. 3). It is tempting to speculate whether this

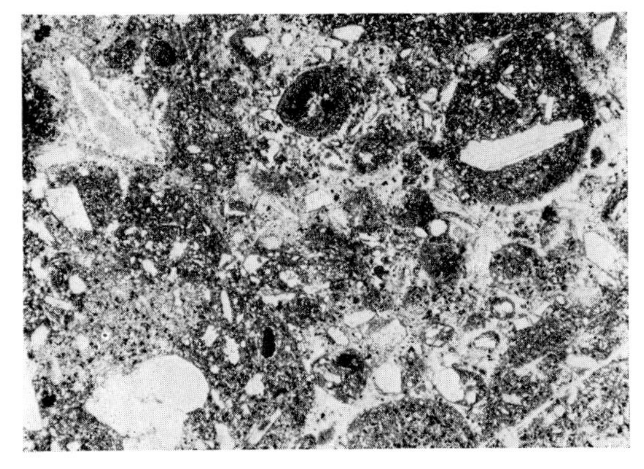

Fig. 3 Spherical bodies of alnoite in a carbonatite matrix, from Ile Bizard. ×1.2

structure could be the result of immiscibility between a carbonate magma and an alkali-rich silicate magma.

The leucocratic dykes fall into two groups, bostonites and phonolites (tinguaites) which correspond to the nordmarkites and nepheline-syenites of the main intrusions. They are composed largely of microperthitic alkali-feldspar and exhibit a trachytic texture. Quartz and brown biotite are common minor constituents of the bostonites, whereas nepheline, analcime and aegirine occur in the phonolites.

IV.6.3.2. Rocks of the Main Intrusions

The rocks of the main intrusions vary in a similar manner to the dyke rocks, with more siliceous varieties at the eastern end of the Monteregian province and undersaturated types towards the western end. For example, granite occurs at Megantic, nordmarkite and nepheline-syenite at Brome, nepheline- and sodalite-syenite at St. Hilaire and ijolite and carbonatite at Oka. In general, these rocks are more extreme in composition than the regional dykes, presumably as a result of differentiation within the intrusions (cf. Table 1). Each intrusion contains a basic phase and most have associated feldspathic or feldspathoidal differentiates. Some of the bodies are multiple intrusions showing several centres of activity. An important group of rocks that have hitherto been unrecorded from the Monteregian province were formed by the contamination of the primary basic magmas with material from the surrounding country rocks.

In discussing the rock types of the different intrusions it is important to remember that these bodies probably did not all intrude to the same level in the earth's crust, and as a result different heights in the individual magma chambers are exposed on the present surface. For example, the magnetic and gravity anomalies of the Iberville intrusion indicate a body similar in size to Mount Johnson (Kumarapeli *et al.*, 1968), even though only a few square metres of igneous rocks actually cut the present erosion surface. Similarly, an intrusion of the same general size as Mount Royal, but coming within only 180 metres of the present surface is indicated by a seismic survey in the Port of Montreal.

The basic phase of each of the intrusions varies in composition with position in the Monteregian province. At Megantic the gabbro consists of plagioclase, augite, hornblende, biotite and magnetite. At Brome and Shefford, the gabbro, in addition to containing minor olivine in places, contains plagioclase, titanaugite, hornblende, very abundant biotite and some ilmenite. In going farther west, two distinct gabbros are found at Mount Yamaska, one a very ferromagnesian rich variety containing titanaugite, hornblende and minor plagioclase, and the other an essexite. Essexites are also found at Mount Johnson and St. Hilaire. At Mount Royal much of the basic phase is a hornblendite containing some olivine, pyroxene and minor plagioclase. The common basic rock west of Mount Royal is alnoite, although kimberlite occurs on Ile Bizard, and a series of okaite-jacupirangite rocks, composed largely of melilite, nepheline and titanaugite, form the early basic phase at the Oka carbonatite complex (Gold, 1967, p. 296).

Mount Bruno and Rougemont differ from the other intrusions in being composed almost entirely of peridotite. However, these two bodies which are composed of essentially the same minerals, olivine, augite and minor plagioclase, show a significant difference. The olivine in Mount Bruno, Fo_{84-78}, coexists with plagioclase zoned from An_{60} to An_{10}, and locally with nepheline as well. In Rougemont, on the other hand, olivine ranging from Fo_{78} to Fo_{74} coexists with plagioclase between An_{98} and An_{80}. These differences are in harmony with the regional change in the compositions of the main Monteregian magmas, with nepheline and more sodic plagioclase occurring in the more westerly intrusions.

Many of the basic rocks are cumulates and show excellent layering due to the rhythmic deposition of minerals. The southern part of the Brome intrusion consists of at least 2500 metres of layered gabbros that dip at 30 to 40 degrees towards the centre of the intrusion. Layered gabbros also form a sequence of saucer-like sheets overlying the peridotite on the summit of Rougemont. However, most of the layering dips at very steep angles (Figs. 4 and 5). At Rougemont and Bruno it appears that the peridotites were formed from a crystal–liquid mush of almost the same composition as the peridotites, for surrounding these intrusions are many dykes that are composed almost entirely of phenocrysts of pyroxene and olivine.

The type of leucocratic rock occurring in each intrusion depends directly on the composition of the associated basic phase. For example, at Megantic the gabbro is not critically undersaturated and the associated leucocratic rocks include nordmarkite and granite. Although the gabbros on Brome contain no nepheline, the presence of abundant hornblende gives rise to normative nepheline and the associated leucocratic rocks are nordmarkite and nepheline-syenite. Farther west in the province nepheline-syenites are associated with essexite. Despite this apparent genetic relationship between the basic and associated feldspathic phase, in all intrusions except Mount Johnson, the leucocratic rocks are younger and transgress the basic ones with sharp contacts. This indicates that although they may be related, the leucocratic ones cannot be direct differentiates of the associated basic rocks presently exposed at the surface, but must come from some deeper source. As a result the intrusions consist of basic and leucocratic rocks with virtually no intermediate types. Only in the case of Mount Johnson is there a complete gradation from essexite to pulaskite over a distance of a few metres and no evidence of either rock type cutting the other.

The leucocratic and late stage rocks of the main intrusions can be divided into three distinct groups: quartz-bearing ones, including nordmarkites and granites; feldspathoidal varieties, such as pulaskite, nepheline-syenite and sodalite-syenite; thirdly, ijolitic rocks associated with the Oka carbonatite. These types normally occur in separate intrusions, but at Brome and Shefford both nordmarkite and nepheline-syenite occur together.

The quartz-bearing varieties which are restricted to the eastern end of the province are

Fig. 4 Vertically dipping rhythmic layering in the essexite at Mount Johnson

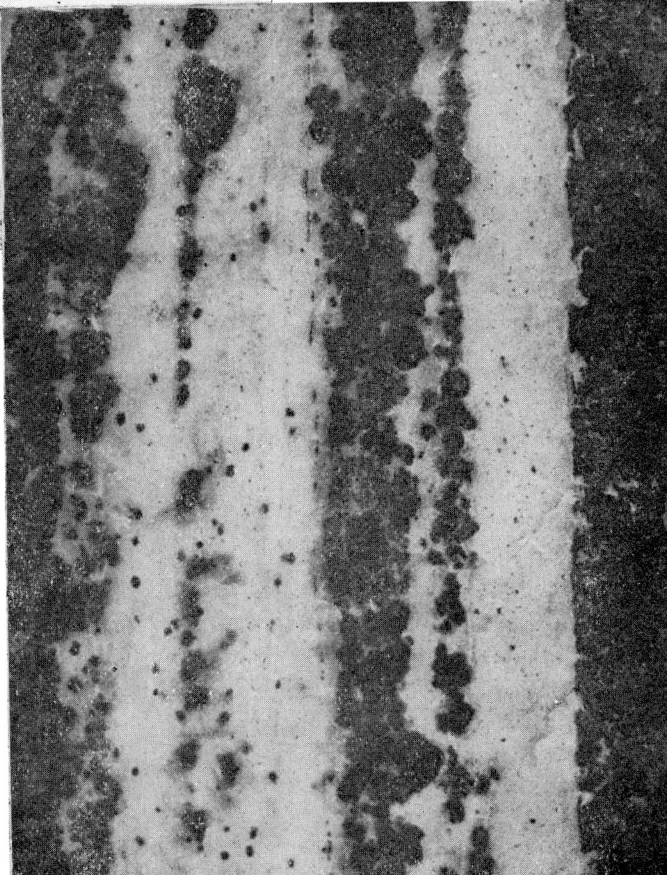

Fig. 5 Steeply dipping layering in a feldspathic part of the peridotite at Rougemont. ×1

hypersolvus rocks composed mainly of microperthitic alkali-feldspar, minor quartz and brown biotite. Sodic hornblende and aegirine-augite occur in some rocks and fayalite is present in the Megantic nordmarkite. Titanite is an extremely abundant accessory mineral. These rocks are medium- to coarse-grained and commonly exhibit a primary lamination due to the alignment of feldspar laths.

The undersaturated rocks are composed of microperthitic alkali-feldspar, nepheline, aegirine, ferrohastingsite and biotite. Locally, sodalite is abundant. At Brome, Shefford and Mount Johnson the leucocratic rocks are mainly pulaskites, being composed almost entirely of alkali feldspar with only minor nepheline. Much of the syenite at Brome contains neither quartz nor nepheline and in places, pulaskite grades into nordmarkite, but this appears to occur only around xenoliths of siliceous country rock. True nepheline-syenite occurs on Brome Mountain, but as a separate and younger intrusion than the other syenites.

Farther west nepheline is much more abundant and phenocrysts in fine-grained dykes around St. Hilaire indicate that nepheline was a primary phase on the liquidus. The feldspar is much more albitic than in the pulaskites and nordmarkites to the east, and in some pegmatites potash feldspar is entirely absent. Aegirine and aegirine-augite are the main ferromagnesian minerals and form black needles up to 5 cm long in pegmatites. Sodalite is a common accessory in nepheline-syenite and is the main constituent of a sodalite-syenite that cuts the nepheline-syenite at St. Hilaire. In this same rock miarolitic cavities are abundant and are commonly lined with many zirconium and rare earth minerals (see IV.6.5). A few cavities have been found in quarries on St. Hilaire that are large enough to walk into and are lined with siderite rhombs up to 20 cm in diameter.

Only a small amount of nepheline syenite occurs on Mount Royal and west of this there are no leucocratic rocks associated with the Monteregian intrusions. It is possible that the ijolitic rocks at Oka, composed mainly of nepheline and aegirine, bear a similar relationship to the okaite–jacupirangite series as the syenites do to the gabbros in the rest of the Monteregian province. However, the situation at Oka is more complex and as the carbonatites must also be considered, discussion of this intrusion is left to IV.6.5.

Where basic rocks come in contact with shaly or quartzo-feldspathic country rocks, for example along the southern contact of Brome, the western side of St. Hilaire, several places on Mount Royal, and the entire circumference of Mount Bruno and Rougemont, the contacts are marked by zones of breccia up to 50 metres wide. These are composed of fragments of refractory sedimentary rocks, such as quartzite and highly aluminous shale, set in an igneous matrix formed by the melting of the least refractory beds (Fig. 6). The matrix is invariably composed of quartz, feldspar and biotite, but varies in chemical composition depending on the nature of the melted country rocks. For example, at Brome it is hypersolvus quartz-syenite, whereas Rougemont and Mount Bruno it is a quartz-diorite containing abundant biotite.

The basic magmas assimilated this rheomorphic breccia to produce a variety of hybrid rocks. Augite, which is normally the most abundant ferromagnesian mineral in the basic rocks, is accompanied or replaced by hornblende and biotite in the hybrid rocks. In rocks containing only minor biotite, hypersthene is com-

Fig. 6 Rheomorphic breccia containing fragments of refractory sedimentary rocks in an igneous matrix formed by the melting of the less refractory beds, from the contact zone of Rougemont

monly abundant and rims olivine where this mineral is present. Quartz is an important mineral in some of these rocks and forms a micrographic intergrowth with plagioclase. Although these rocks occur around the margin of intrusions, at Mount Bruno they appear to have also formed a sizeable body of quartz-norite which intruded the main peridotite on the north-east side of the intrusion. It is also likely that the body of akerite that forms the northern and western margins of the Yamaska intrusion and which has been interpreted as a differentiation product of the gabbro by Gandhi (1967) is actually a hybrid rock formed by assimilation.

IV.6.3.3. Ultramafic Nodules

Ultramafic inclusions are restricted entirely to the rocks of the western end of the province, except for one camptonite dyke on St. Hilaire that contains inclusions composed of ortho- and clinopyroxene. Inclusions in the alnoites are relatively abundant and consist usually of clinopyroxene and olivine. They are commonly very coarse-grained and at Ile Cadieux olivine crystals up to 5 cm in diameter are found. More rarely orthopyroxene occurs with clinopyroxene and olivine. Inclusions are extremely abundant in the kimberlite on Ile Bizard and consist of olivine (Fo_{91-88}), orthopyroxene (En_{90-88}), clinopyroxene, garnet and spinel (Marchand, 1970). The garnet which is pyrope rich, occurs as rims around deep brown, chromium-rich spinel (see further IV.6.6). Other inclusions consist of olivine, orthopyroxene and biotite, and olivine, clinopyroxene and orthopyroxene.

IV.6.3.4. Diatreme Breccias

Finally there are the diatreme breccias which, while not being strictly igneous rocks, are important members of the Monteregian province. They are restricted to the western end of the province where they form pipes that are usually only a few tens of metres in diameter, but may be as much as 1 km, as in the case of the body at St. Helen's Island, the best known of these breccias (Osborne and Grimes-Graeme 1936). They extend vertically and in the case of St. Helen's Island, where a subway tunnel passes beneath the St. Lawrence river, there is very little change in diameter with depth. They are composed entirely of fragments from the surrounding Palaeozoic rocks and some from the Precambrian basement. At St. Helen's Island there are, in addition, several large blocks of Devonian limestones that do not occur elsewhere in the Montreal area, but are known to overlie the local rocks farther south in New York state, and hence presumably covered Montreal in Cretaceous time. The fragments range in size from approximately 1 cm up to large blocks a few metres in diameter, but most are small. They commonly show considerable rounding and alteration of their margins and are held together by smaller fragments and extremely finely comminuted material. Pyrite is extremely common with the result that these breccias are rusty weathering.

Although no igneous matrix occurs in these breccias, excavations in the St. Helen's Island body during construction for the 1967 World's Fair uncovered a small breccia pipe composed entirely of fragments of Precambrian rocks in a porphyritic camptonite matrix of a very similar composition to the basic gabbro of Mount Royal (Clark *et al.*, 1967). This suggests that the gas that caused the explosion and formation of the breccias may have emanated from such a magma chamber at depth. However, large crystals of biotite are common in these breccias, especially at the extreme western end of the province, and their presence suggests a correlation with the alnoites, a conclusion that is supported by the common association of these two rock types. It is also of interest to note that the eclogite bearing kimberlite pipe on Ile Bizard occurs within a large diatreme breccia. Regardless of the ultimate origin of these breccias, they do indicate the explosive nature of the igneous activity in the western end of the Monteregian province.

IV.6.4. Structures of the Monteregian Intrusions

All of the intrusions are vertical, plug-like bodies that are circular or lobate in plan. The contacts dip vertically even in intrusions containing gently dipping layered gabbros, and in the case of Mount Royal, under which a railway tunnel passes, the contacts are vertical over a distance of at least 550 ft (Bancroft and Howard, 1923). Gravity and aeromagnetic surveys over these intrusions also indicate that they extend downwards to considerable depths without any significant change in cross-sectional shape.

Primary layering is exhibited by almost all rocks but is most easily seen in the basic ones. In general, this layering, which is marked by alternations of feldspathic and ferromagnesian layers, parallels the contacts and dips very steeply. In the case of a small intrusion such as Mount Johnson it is concentric about the core of the body (Fig. 4), but in larger intrusions with lobate contacts the foliation parallels the contacts in each of the lobes and is commonly truncated towards the core by layering developed during later stages of crystallization. Cross-layering and trough-like structures are common (Fig. 7) and indicate that these intrusions solidified by depositing successive layers on the walls of the magma chambers. Vertical or steeply plunging mineral lineations are found in many rocks and indicate, as do the trough-like structures, that the magmas were moving in a vertical direction.

Strong convection currents might be expected to have existed in vertical plug-like bodies, and if so the magmas would have travelled down the walls of the magma chambers and up in the cores. Two lines of evidence indicate that this was the case in the Monteregian intrusions. Within the hornfels collars where partial melting of the flat-lying country rocks occurred, bedding dips in towards the intrusions at progressively steeper angles as the contacts are approached and becomes vertical at the zones of rheomorphic breccia that surround many of the intrusions. This suggests a general downward movement at the walls of the magma chambers. Also in the vertical concentric layering of the Mount Johnson essexite, plagioclase laths are oriented in such a manner as to produce an imbrication

Fig. 7 Trough-like structure in the vertically dipping layering on Mount Johnson

which indicates that the magma moved down the walls of the magma chamber (Philpotts, 1968).

In the western part of the province explosive activity was intense and it is likely that magmas were emplaced very much more rapidly than in the east. Gold (1967, p. 291) has postulated that the double ring structure of the Oka carbonatite was caused by the injection of a series of ring dykes and cone sheets.

IV.6.5. Descriptions of the Main Monteregian Intrusions

The Oka Carbonatite Complex, which contains one of the world's largest niobium deposits, forms a double ring structure in the shape of a distorted figure eight elongaged in a northwesterly direction. It is surrounded by Precambrian rocks that form an upfaulted block within the Ottawa graben. Gold *et al.* (1967) distinguish eight phases in its development:

1. Fenitization of gneissic cover rock, followed by emplacement of an early carbonatite phase as dykes and ring dykes, and ijolitization of the enclosed country rock.
2. Intrusion of okaite-jacupirangite rocks as arcuate dykes mainly in the northern ring of the complex.
3. Intrusion of a pyrochlore carbonatite, followed by a monticellite carbonatite.
4. Intrusion of ijolite dykes followed by solid-state flow in the carbonatite.
5. Hydrothermal activity along fractures, producing biotite and enrichment of the carbonatite in thorian pyrochlore.
6. Late white carbonatite dykes.
7. Lamprophyre dykes.
8. Emplacement of alnoite and alnoite breccia pipes and dykes.

Mount Royal, one of the earliest investigated of the Monteregian intrusions (Bancroft and Howard, 1923), has recently been studied by Clark (1952), Robillard (1968) and Woussen, 1970). Bancroft and Howard recognized at least seven distinct periods of injection, with each major intrusion of gabbro followed by one of syenite. However, most of the intrusion is composed of ferromagnesian-rich gabbros that are intrusive into limestone. The clinopyroxene, which generally is a titanaugite, varies in composition from $Ca_{46}Mg_{45}Fe_9$ to $Ca_{29}Mg_{26}Fe_{45}$ and the olivine from Fo_{80} to Fo_{70}. The plagioclase varies in composition from An_{80} to An_{40} with most being approximately An_{50}. A small body of essexite occurs in the western part of the intrusion and in this the plagioclase is zoned from An_{60} to An_{30}. Nepheline-syenite forms a narrow body along the northwestern side of the intrusion where it cuts and brecciates the gabbro.

Mount Bruno, which was first studied in detail by Dresser (1910) is composed almost entirely of feldspathic peridotite containing olivine (Fo_{80}), titanaugite and interstitial plagioclase (An_{60}). Along the southern and western margins the rock also contains hornblende with minor calcite and nepheline and the plagioclase is zoned from An_{60} to An_{10}. Olivines in the peridotite become noticeably enriched in iron towards the contact and are replaced by orthopyroxene in the rheomorphic contact breccia. The youngest rock forms a dyke of nordmarkite in the southern part of the intrusion.

St. Hilaire consists of two distinct bodies of rocks, an older western half that is zoned outward from a core of gabbro and peridotite, through essexite to a margin of nepheline diorite, and a younger eastern half composed largely of nepheline syenite. The syenite appears to have been emplaced forcefully and was probably associated with explosive activity, for a large part of the nepheline syenite is a breccia containing fragments of country rocks, gabbros and other syenitic rocks. On the northeastern side of the intrusion there are several small bodies of sodalite syenite and it is in this same area that pegmatite dykes occur containing many rare minerals (Chao *et al.*, 1967), the most important of which are astrophyllite, catapleiite, dawsonite, elpidite, epididymite, eudialyte-eucolite, mangan-neptunite, narsarsukite, pyrochlore, rutile, serandite and zircon.

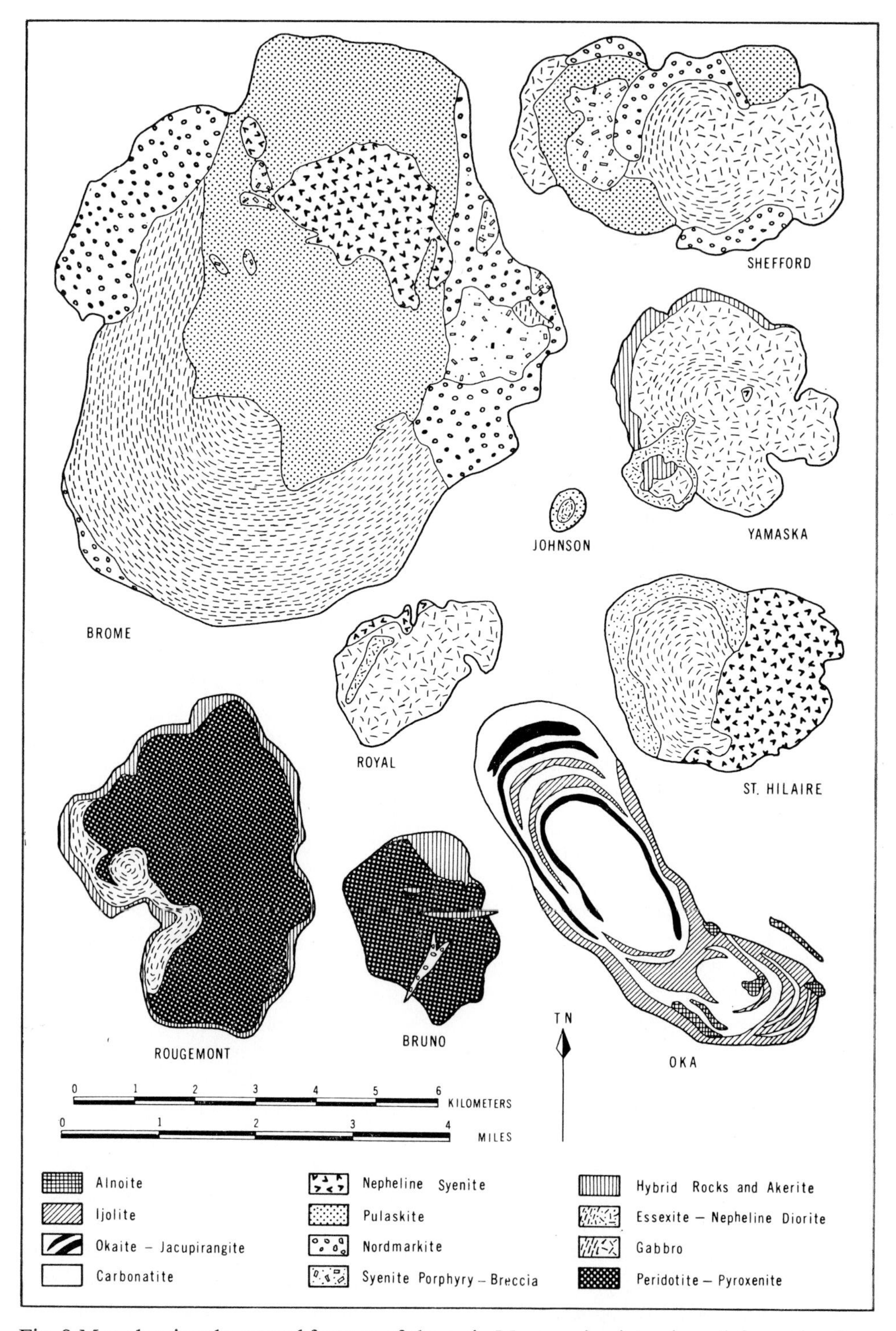

Fig. 8 Map showing the general features of the main Monteregian intrusions. Information from Pouliot (in press), Valiquette (in press), Gandhi (1967), Robillard (1968) and Gold *et al.* (1967)

Little is known of the *Iberville intrusion* as only a few outcrops of igneous rocks are exposed on the present surface (Kumarapeli *et al.*, 1968). However, these rocks appear similar to those on Mount Johnson and include gabbro and syenite.

Mount Johnson, which has been the subject of several investigations, the most recent of which are by Pajari (1967) and Philpotts (1968), consists of a nearly circular pipe-like intrusion in which the rocks grade from a core of essexite to a margin of pulaskite. At the centre of the intrusion the rock is a fine-grained oligoclase essexite containing a few per cent of olivine. There is very little mineralogical change in passing outward from the core, except for the disappearance of olivine, but the rock becomes progressively coarser grained and develops rhythmic layering due to the concentration of felsic and ferromagnesian minerals into alternating vertical layers. The transition to the pulaskite which takes place over a few metres is completely gradational and the same vertically dipping, concentric layering is present in the pulaskite (see IV.6.4).

The puzzling relationship of pulaskite having been deposited on the walls of the intrusion while molten essexite still existed in the core has been explained by thermal diffusion of the low melting components towards the cooler margins of the intrusion (Wahl, 1946), and by flowage differentiation (Bhattacharji 1966, Bhattacharji and Nehru, 1970). Philpotts (1968, 1970) has suggested that these relationships resulted from the slow upward injection of a pair of convecting cells, an upper one of pulaskitic magma, and a lower one of immiscible essexitic magma. Models indicate that with two such cells the lower one tends to penetrate into the base of the upper one, so that the pulaskitic magma would sheath the essexite one.

Rougemont is composed almost entirely of feldspathic peridotite and unlike the other intrusions has no highly feldspathic or feldspathoidal rocks associated with it. The feldspathic peridotite contains cumulus olivine and pyroxene and commonly shows a primary layering. In places the peridotite passes into pyroxenite and at the summit of the hill on the southwestern side of the intrusion it grades upward into a well layered gabbro. This material contaminated the basic magma, so that everywhere the intrusion is rimmed by a hybrid gabbro containing quartz, biotite and commonly hypersthene.

Yamaska consists of two gabbroic intrusions (Gandhi, 1967). The older one which forms the main part of the intrusion is composed of an extremely ferromagnesian-rich gabbro which, in places, grades into pyroxenite. This gabbro grades outward through a quartz bearing variety into an arcuate body of akerite which separates the basic rocks along the northern and western margins of the intrusion from the folded siltstones and shales in the country rock (cf. IV.6.3.2). The second intrusion which is composed of gabbro and essexite forms a small plug-like body on the southwestern side of the main mass. In the centre of this younger intrusion is a large body of hybrid rock containing abundant partially resorbed fragments of hornfels. Leucocratic rocks are noticeably absent from Yamaska except for one very small body of nepheline syenite that occurs in the centre of the older gabbro.

Brome Mountain is by far the largest of the Monteregian intrusions and contains the greatest complexity of rock types (Valiquette and Archambault, 1970). The oldest rocks are layered gabbros that form the southern half of the intrusion. The gabbro is truncated along its northern side by a variety of syenitic rocks, the earliest phase of which appears to have been a fine-grained syenite porphyry which in places contains considerable breccia with fragments of hornfels, gabbro and syenite. Closely related to this are two large bodies of nordmarkite that form the northwestern and eastern parts of the intrusion. Within this nordmarkite are several smaller bodies of monzonitic to dioritic rocks which appear to be related to this period of intrusion. All of these syenitic rocks contain minor quartz which distinguishes them from the younger rocks of the intrusion which contain nepheline. The northern and central parts of the intrusion are composed largely of pulaskite. Large irregular blocks of nepheline monzodiorite occur within the pulaskite and appear to represent some earlier, more basic undersaturated rock. A body of nepheline syenite in the centre of the pulaskite is the youngest member of the intrusion.

Shefford Mountain has recently been studied by Pouliot (in press). The oldest rock, which forms the eastern part of the intrusion, is a gabbro which although not as well layered as the one at Brome does exhibit some layering. Pulaskite was intruded next in the western and northern parts of the complex. This was followed by a syenite porphyry and breccia that was emplaced in the centre of the main western pulaskite body. This rock type is more volcanic in appearance than any other Monteregian igneous rock and it is quite possibly a crater filling. A ring dyke of fine-grained pulaskite was injected around the breccia and this in turn was cut by bodies of nordmarkite, the last rock to be emplaced. Although these syenitic rocks are similar to those on Brome mountain, their sequence of injection is different with undersaturated varieties being the oldest and the nordmarkites the youngest.

Mount Megantic, the most eastern member of the Monteregian province is geographically closer to the alkaline rocks of the White Mountains of New Hampshire than it is to the main Monteregian intrusions. It was first described by McGerrigle (1935) and more recently by Reid (1961). The core of the intrusion is occupied by a grey biotite granite that appears identical to the many other bodies of Devonian granite in this area. A ring dyke of nordmarkite, 2 km wide and 13 km long, extends around the northern and eastern sides of this granite core. The nordmarkite, which is a dark green rock containing microperthite, quartz, aegirine, fayalite and hornblende, contains some areas of gabbroic rocks. Although two age determinations indicate the nordmarkite to be Monteregian (Gold 1967, p. 289) further work is required to be certain that this intrusion does not belong to the slightly older White Mountain magma series (cf. III.3).

IV.6.6. Petrogenesis of the Monteregian Rocks

The compositions of the main Monteregian intrusions are controlled by the types of magma that were intruded in the different parts of the province; that is, camptonitic magma in the east, alnoitic in the west and monchiquitic in the zone between. Monchiquites could have been derived from the camptonites by fractional crystallization of calcic plagioclase. However, plagioclase phenocrysts are extremely rare in these rocks and are never found in the monchiquites where such early crystallizing plagioclase would be expected to occur. In addition, such crystal fractionation would not explain the geographical distribution of these rocks. It appears therefore, that the differences between these magmas existed in the source areas or, at least, at some considerable depth and were not produced by fractional crystallization during intrusion.

Hornblende is extremely abundant as phenocrysts in the camptonites close to the boundary of the region in which the monchiquites occur. Its presence may simply reflect the regional change in the compositions of the magmas, especially in terms of volatile content, but it is important to consider whether this mineral may, in part, be xenocrystic material from the source area and hence be of importance in determining the composition of the magmas. Some camptonites contain very large rounded hornblende phenocrysts up to 4 cm in diameter. Analyses of one of these hornblendes and its surrounding fine-grained camptonitic groundmass are given in Table 2. Apart from a difference in the iron/magnesium ratio and the sodium content, the hornblende and camptonite have very similar compositions. The barium and rare earth concentrations in this hornblende and matrix indicate equilibrium between these two phases (Schnetzler and Philpotts, 1968). The concentrations of these elements are very similar to those in the camptonite and its included phenocrystic

TABLE 2. Analyses of Hornblende Phenocrysts. (1) in a Camptonitic Groundmass (2) from the Canada Cement Quarry, Montreal. Analyst H. Ulk

	1	2
SiO_2	39·31	39·86
TiO_2	1·89	1·96
Al_2O_3	16·70	16·28
Fe_2O_3	3·91	6·67
FeO	6·55	5·82
MnO	0·09	0·25
MgO	14·61	7·32
CaO	11·99	10·86
Na_2O	1·85	3·96
K_2O	1·79	1·49
P_2O_5	0·07	1·01
CO_2	0·33	1·08
H_2O^+	0·41	3·15
H_2O^-	0·00	0·17
	99·50	99·88

or xenocrystic hornblende from the granulite-eclogite diatreme breccia at Kakanui, New Zealand (Schnetzler and Philpotts, 1968). Mason (1968, p. 1001) has drawn attention to the fact that Kakanui type hornblendes (kaersutitic) are not uncommon as xenocrysts or in xenoliths in basic rocks and suggests they are stable in the upper mantle. Hence, a possible genetic model for the generation of this particular Monteregian camptonite is very limited partial fusion of an assemblage in the lower crust or upper mantle dominated by hornblende similar to that actually occurring in the dyke (cf. VI.6).

The alnoites are drastically different in composition from the other dyke rocks and it does not appear feasible to derive them from a common magma source by any simple process of fractional crystallization. Their geographical separation from the other rock types also suggests that their composition must have been determined in the source area. The large biotite phenocrysts that are common in these rocks may, in part, be xenocrystic, for biotite is found as a primary phase in some of the peridotite nodules in the Ile Bizard kimberlite. Partial melting of a biotite peridotite could give rise to the alnoites and leave behind a refractory residue of olivine and pyroxene which could then form the ultramafic nodules that are so common in these rocks. It is reasonable to interpret these nodules as coming from some considerable depth, for the orthopyroxene that occurs in some of them is not a stable phase in the alnoitic magma at the present level.

The nodules in the Ile Bizard kimberlite (Marchand, 1970) containing clinopyroxene-orthopyroxene-spinel-forsterite-garnet are of interest because of the limited stability field of this assemblage. The rims of garnet surrounding the spinel in these nodules suggests that the following reaction was occurring;

$$\text{Diopside} + \text{Enstatite} + \text{Spinel} \rightarrow \text{Forsterite} + \text{Garnet}.$$

The experimental work pertaining to the stability of these minerals has been discussed by O'Hara (1967, p. 13) and he shows the above reaction taking place at a temperature of 1250 °C under a pressure of approximately 20 Kb, corresponding to a depth of 60 km (Fig. 9). The

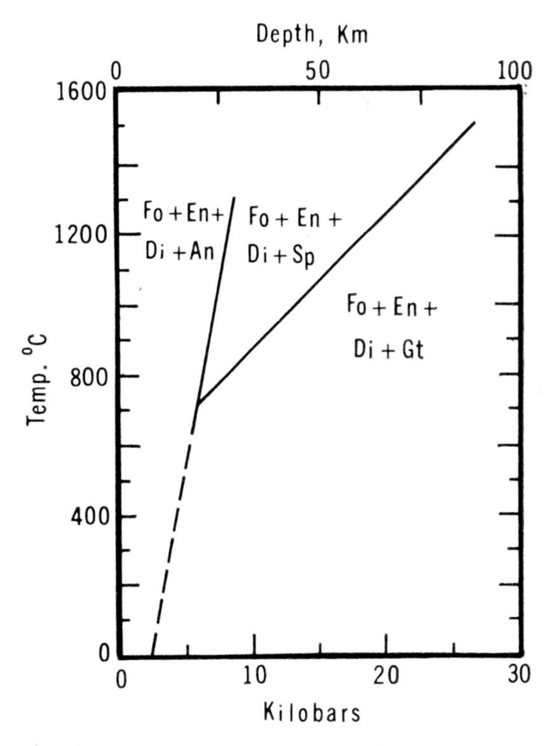

Fig. 9 Pressure–temperature relationship for the univariant reaction Diopside + Enstatite + Spinel = Forsterite + Garnet (after O'Hara, 1967, p. 13)

magmas in the other parts of the Monteregian province may have come from a similar depth, in which case their differences in composition would be due to large-scale lateral variations in the composition of the mantle. On the other hand, melting may have occurred at different depths and the compositional variations would be due to either different fractional melting products or reflect compositional layering in the mantle.

The layering exhibited by most of the basic rocks of the main intrusions indicates that crystal fractionation was, to a large extent, responsible for the differentiation within the main complexes. The steeply dipping concentric layering, channel marks and cross-layering indicate that convection cells were active and undoubtedly contributed to the differentiation. However, in almost all of the intrusions there is a lack of rocks intermediate in composition between the basic and feldspathic types and although such rocks can be postulated as occurring at depth, it strongly suggests that crystal fractionation may not have been the only process of differentiation operative.

The interpretation of the syenite ocelli in lamprophyres as droplets of immiscible liquid is based on textural and experimental evidence (Philpotts and Hodgson, 1968), a short summary of which is given here. Ocelli are found in all stages of coalescence, the larger ones having cuspate borders and containing spherical aggregates of minerals. They tend to flatten out against surfaces such as solidification fronts in dykes and sills. Surface tension appears, in general, to have prevented large phenocrysts from penetrating the ocelli, but there are cases where this happened (Fig. 2). Ocelli are not a type of amygdale, for in the upper part of many of them there is a separate amygdale, usually filled with analcime and calcite. In some ocelli there is a concentration of hornblende towards the base. Although the ocelli and surrounding lamprophyres are of very different bulk compositions, the plagioclase and hornblende have the same compositions in both phases, which is to be expected if the two liquids are in equilibrium with the crystallizing minerals. Homogeneous glasses formed by the melting of mixtures of ocelli and matrix at 1400 °C separate into two immiscible liquids with compositions similar to the original ocelli and matrix when held at temperatures below 1000 °C under low water pressures. Natural and experimentally produced ocelli have compositions that range from nepheline-syenite to quartz-syenite, making it possible that most leucocratic rocks associated with basic alkaline rocks could have formed as immiscible liquids (Philpotts, 1971, 1972).

In several thick lamprophyric sills in the Montreal area where ocelli are preserved in the chilled margins, segregations of syenitic material have formed in the upper parts. Although the syenite, in general, forms conformable lenses, small offshoots do cut the surrounding basic rock. If segregations in sills can form from the accumulation of small ocelli, it is possible that the bodies of syenite in the larger intrusions may have formed in the same way, even though the slow cooling in larger bodies would not tend to preserve the textural evidence of liquid immiscibility. Perhaps the problem of oversaturated and undersaturated syenites in the same intrusions could be explained in terms of the compositions at which differentiating basic magmas hit the postulated immiscibility field and hence overcome the difficulty of thermal barriers encountered when trying to explain the genesis of these rocks by fractional crystallization.

In discussing the origin of the oversaturated rocks of the Monteregian province it is necessary to consider the effect of assimilation, for a large number of rocks at the margins of these intrusions have clearly been contaminated by the country rocks. Although intrusion in the western end of the province appears to have been very explosive, in the rest of the area it was a rather passive event, as indicated by the lack of diatreme breccias and the scarcity of dykes. The contacts of many of the intrusions, with their inward dipping hornfels and rheomorphic breccias, indicate that emplacement was probably accomplished by partial melting and stoping of the country rocks by magmas that were convecting down the walls of the intrusions. This is substantiated by the fact that the sedimentary rocks forming the roof of the partially denuded Iberville intrusion are stratigraphically in their correct position and have not been punched upward (cf. Philpotts, 1970).

Some of the material stoped from the walls and roofs of the intrusions was incorporated into the basic magmas to produce hybrid rocks, but most of it must have been transported to depth where it is possible that it was generated into oversaturated magmas. Conflicting with the clear-cut field evidence for assimilation are the isotopic studies by Fairbairn *et al.* (1963) and Faure and Hurley (1963) on Monteregian rocks, which show that there was no crustal contamination and that the initial Sr^{87}/Sr^{86} ratio was 0·704, a value consistent with these rocks having been derived from basaltic magmas (cf. also V.4). Most of the rocks from which these data were obtained are from the cores of the intrusions where the possibility of contamination is less. Since many of the rocks have clearly been contaminated, especially those containing hypersthene, further careful isotopic studies will be necessary to determine the degree of involvement of crustal material in the generation of the alkaline rocks of the Monteregian province.

Acknowledgments

The author is grateful to the Quebec Department of Natural Resources for sponsoring much of his field work and to Dr. D. Carmichael for critically reading the manuscript. Grants from the National Research Council and the Geological Survey of Canada also aided in much of the work.

IV.6. REFERENCES

Adams, F. D., 1892. On a melilite bearing rock (alnoite) from Ste. Anne de Bellevue, near Montreal, Canada. *Am. J. Sci.*, **43,** 269–78.

Adams, F. D., 1903. The Monteregian Hills: A Canadian petrographical province. *J. Geol.*, **11,** 239–82.

Bancroft, J. A., and Howard, W. V., 1923. The essexites of Mount Royal, Montreal, P.Q. *Roy. Soc. Can.*, **17,** 13–43.

Bhattacharji, S., 1966. Flowage differentiation in Mount Johnson stock, Monteregian Hills, Canada. *Am. Geophys. Union*, **47,** 196.

Bhattacharji, S., and Nehru, C. E., 1970. Igneous structures and mechanism of emplacement of Mount Johnson, a Monteregian intrusion, Quebec: Discussion. *Can. J. Earth Sci.*, **7,** 191–4.

Bowen, N. L., 1922. Genetic features of alnoitic rocks at Ile Cadieux, Quebec. *Am. J. Sci.*, **3,** 1–34.

Dresser, J. A., 1910. Geology of St. Bruno Mountain, Province of Quebec. *Geol. Surv. Can., Mem.*, **7.**

Chao, G. Y., Harris, D. C., Hounslow, A. W., Mandarino, J. A., and Perrault, G., 1967. Minerals from the nepheline syenite, Mont St. Hilaire, Quebec. *Can. Min.*, **9,** 109–23.

Clark, T. H., 1952. Montreal area, Laval and Lachine map-areas. *Quebec Dept. Mines, Geol. Rept.*, **46,** 159 pp.

Clark, T. H., Kranck, E. H., and Philpotts, A. R., 1967. Ile Ronde breccia, Montreal. *Can. J. Earth Sci.*, **4,** 507–13.

Doig, R., and Barton, J. M., 1968. Ages of carbonatites and other alkaline rocks in Quebec. *Can. J. Earth Sci.*, **5,** 1401–7.

Faessler, C., 1962. Analyses of rocks of the Province of Quebec. *Quebec Dept. Nat. Res. Geol. Rept.* **103,** 251 pp.

Fairbairn, H. W., Faure, G., Pinson, W. H., Hurley, P. M., and Powell, J. L., 1963. Initial ratio of strontium 87 to strontium 86, whole rock age, and discordant biotite in the Monteregian igneous province, Quebec. *J. Geophys. Res.*, **68,** 6515–22.

Faure, G., and Hurley, P. M., 1963. The isotopic composition of strontium in oceanic and continental basalts: application to the origin of igneous rocks. *J. Petrology*, **4,** 31–50.

Gandhi, S. S., 1967. *Igneous petrology of Mount Yamaska, Quebec.* Unpubl. Ph.D. thesis, McGill University, Montreal.

Gold, D. P., 1967. 'Alkaline ultrabasic rocks in the Montreal area, Quebec', in Wyllie, P. J., Ed., *Ultramafic and Related Rocks*, John Wiley and Sons, New York.

Gold, D. P., Vallée, M., and Charette, J. P., 1967. Economic geology and geophysics of the Oka Alkaline Complex, Quebec. *Can. Inst. Mining Metal.*, **60,** 1131–44.

Graham, R. P. D., 1944. 'The Monteregian Hills', in Dresser, J. A., and Denis, T. C., Eds., Geology of Quebec. *Quebec Dept. Mines, Geol. Rept.*, 20, **2,** 455–82.

Kumarapeli, P. S., and Saull, V. A., 1966. The St. Lawrence valley system: A North American equivalent of the East African rift valley system. *Can. J. Earth Sci.*, **3**, 639–58.

Kumarapeli, P. S., Coates, M. E., and Gray, N. H., 1968. The Grand Bois anomaly: the magnetic expression of another Monteregian pluton. *Can. J. Earth Sci.*, **5**, 550–3.

Marchand, M., 1970. *Ultramafic Nodules from Ile Bizard.* Unpublished M.Sc. thesis, McGill University.

Mason, B., 1968. Kaersutite from San Carlos, Arizona, with comments on the paragenesis of this mineral. *Min. Mag.*, **36**, 997–1002.

McGerrigle, H. W., 1935. Mount Megantic area, Southeastern Quebec. *Quebec Dept. Mines Ann. Rept.*, Pt. D, 63–104.

O'Hara, M. J., 1967. 'Mineral facies in ultrabasic rocks', in Wyllie, P. J., Ed., *Ultramafic and Related Rocks*, John Wiley and Sons, New York.

O'Neill, J. J., 1914. St. Hilaire and Rougemont Mountains, Quebec. *Geol. Surv. Can. Mem.*, **43.**

Osborne, F. F., and Grimes-Graeme, R., 1936. The breccia on St. Helen Island, Montreal. *Am. J. Sci.*, **32,** 43–54.

Pajari, G. E., 1967. *Petrology of Mount Johnson, Quebec.* Unpub. Ph.D. thesis, University of Cambridge.

Philpotts, A. R., 1964. Origin of pseudotachylites. *Am. J. Sci.*, **262,** 1008–35.

Philpotts, A. R., 1968. Igneous structures and mechanism of emplacement of Mount Johnson, a Monteregian intrusion, Quebec. *Can. J. Earth Sci.*, **5,** 1131–7.

Philpotts, A. R., 1970. Igneous structures and mechanism of emplacement of Mount Johnson, a Monteregian intrusion, Quebec: Reply. *Can. J. Earth Sci.*, **7,** 195–7.

Philpotts, A. R., 1971. Immiscibility between feldspathic and gabbroic magmas. *Nature*, **229,** 1C7–9.

Philpotts, A. R., 1972. Density, surface tension and viscosity of the immiscible phase in a basic, alkaline magma. *Lithos*, **5,** 1–18.

Philpotts, A. R., and Hodgson, C. J., 1968. Role of liquid immiscibility in alkaline rock genesis. 23 *Int. Geol. Congress*, **2,** 175–88.

Philpotts, A. R., and Miller, J. A., 1963. A Precambrian glass from St. Alexis-des-Monts, Quebec. *Geol. Mag.*, **100,** 337–44.

Pouliot, G., in press. 'Geology of Shefford Mountain', in *The Symposium on the Monteregian Hills, Can. Min.*

Rajasekaran, K. C., 1967. *Mineralogy and petrology of nepheline syenite in Mont St. Hilaire, Quebec.* Unpublished Ph.D. thesis, McGill University.

Reid, A. M., 1961. *The petrology of Mt. Megantic igneous complex, Southern Quebec.* Unpubl. M.Sc. thesis, University of Western Ontario.

Robillard, J., 1968. *Etude des roches plutoniques mafiques du Mont Royal.* Unpubl. M.Sc. thesis, Université de Montreal.

Schnetzler, C. C., and Philpotts, J. A., 1968. 'Partition coefficients of rare earth elements and barium between igneous matrix material and rock forming mineral phenocrysts—I', in Ahrens, L. H., Ed., *Origin and Distribution of the Elements*, Pergamon.

Stansfield, J., 1923. Extension of the Monteregian petrographical province to the west and northwest. *Geol. Mag.*, **60,** 433–53.

Valiquette, G., and Archambault, G., 1970. Les gabbros et les syenites du complexe de Brome. *Can. Miner.*, **10,** 485–510.

Wahl, W. A., 1946. Thermal diffusion–convection as a cause of magmatic differentiation. *Am. J. Sci.*, **244,** 417–41.

Woussen, G., 1968. *Les monzonites du Mont Royal.* Unpublished M.Sc. thesis, Université de Montréal.

Woussen, G., 1970. La géologie du complexe igné du Mont Royal. *Can. Miner.*, **10,** 432–51.

Zartman, R. E., Brock, M. R., Heyl, A. V., and Thomas, H. H., 1967. K–Ar and Rb–Sr ages of some alkalic intrusive rocks from central and eastern United States. *Am. J. Sci.*, **265,** 848–70.

IV.7. OCEANIC ISLANDS

G. D. Borley

IV.7.1. Introduction

In view of the evidence that the most abundant oceanic rock type is of tholeiitic composition, Engel and Engel (1964), Muir and Tilley (1964), Nicholls (1965) and Aumento (1968), the small volume of alkalic rocks of the oceanic islands might seem to have attracted disproportionate attention. But the attention is understandable, for the alkalic rocks often form young 'caps' to older tholeiite successions or comprise the total exposed portions of oceanic volcanoes. However, in this short review only an outline of the distribution, chemistry and relationships of oceanic alkalic rocks can be given; more detailed discussions must be sought elsewhere.

By defining overall limits of the Thornton and Tuttle (1960) Differentiation Index (norm. $Q + ab + or + ne + leu$) it is possible to distinguish the following members of the oceanic alkalic suite:

accumulates, phyric basalts and alkalic basalts D.I. < 40, $SiO_2 < 52\%$; *trachybasalts* D.I. 40–65, SiO_2, 52–57%; *trachyandesites* D.I. 65–75, SiO_2 57–59%; *trachytes* ⟨pantellerites / phonolites⟩ D.I. > 75,

(Fig. 1). Rocks of these compositions usually plot in the field of Hawaiian alkalic rocks in the Silica v. Total Alkalis diagram of MacDonald and Katsura (1964), (Fig. 2).

IV.7.2. Selection of Islands

I have tried to include a representative selection of islands, particularly those defined as geographically oceanic by Chayes (1964). Omitted is Fernando Po of the Guinea Islands as, like Chayes, I consider it may belong geologically to the Cameroons. Similarly, in the Canary Islands the bulk of evidence, Hausen (1958), Fuster *et al.* (1968b), Gastesi (1968), Dash and Bosshard (1968), suggests that Fuerteventura and Lanzarote, the most easterly islands, have continental affinities and they are also omitted. So are the islands of the North Atlantic, Jan Mayen and Iceland, for their positions, near the merging of the Mid-Atlantic ridge with the Greenland–Iceland and Faroes–Iceland rises, makes it difficult to classify them as truly oceanic. Their omission, however, excludes few published analyses from Jan Mayen, and rather more from Iceland, from where basalts considered to be transitional or alkalic in nature have been reported by Robson and Spector (1962), Heier *et al.* (1966), Sigurdsson (1969) and by Jacobsson (1968) from the Vestmanna Islands. Included are a few islands such as Reunion, some of whose products are transitional between alkalic and tholeiitic rather than truly alkalic in nature. Details of the islands included in this review are given in Table 1.

IV.7.3. Selection of Analytical Data

Comparisons of rocks from different areas are often based on their bulk chemistry or norms. Consequently, some criteria of selection from the several hundred available analyses were necessary to ensure, as far as possible, that data used did not reflect oxidation, alteration or faulty analysis of the original rocks.

The effects of oxidation on the norms of alkalic rocks are shown by (1) the decrease or disappearance of *ol* and *hyp*; (2) the appearance of *Q* and/or *hem*. In the peralkaline rocks oxidation may result in (1) an increase in *ac*; (2) the reduction or elimination of *ns* and the appearance of *hem*, i.e. excessive oxidation can destroy evidence of peralkalinity. The only certain way of determining the oxidation state of a rock is by petrographically examining its iron–titanium

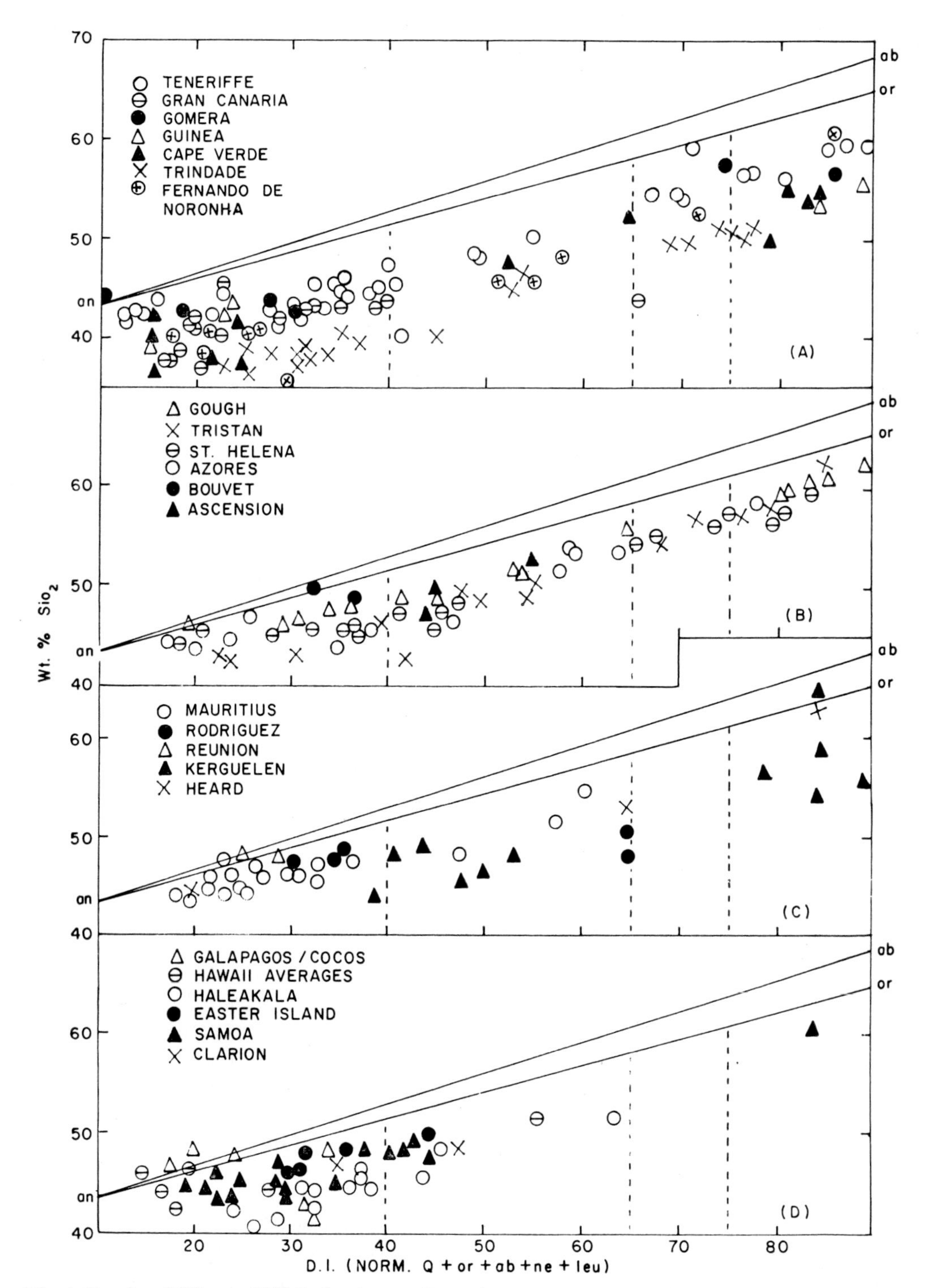

Fig. 1 Graphs of Silica *v.* Diff. Index (norm. Q + ab + or + ne + leu) for members of the alkalic suites of the oceanic islands

(a) and (b) Atlantic ocean islands.
(c) Indian ocean islands.
(d) Pacific ocean islands.

Also shown is the Thornton and Tuttle saturation line an–or which separates the field of undersaturated rocks (below) from that of saturated rocks (above), and the saturation line an–ab which separates the fields of saturated and oversaturated rocks. Dotted lines define limits of composition for: accumulative, phyric and alkalic basalts, D.I. < 40; trachybasalts, D.I. < 65; trachyandesites, D.I. < 75; trachytes, D.I. > 75.

(For sources of analyses in Figs. 1, 2 and 3 see Table 1.)

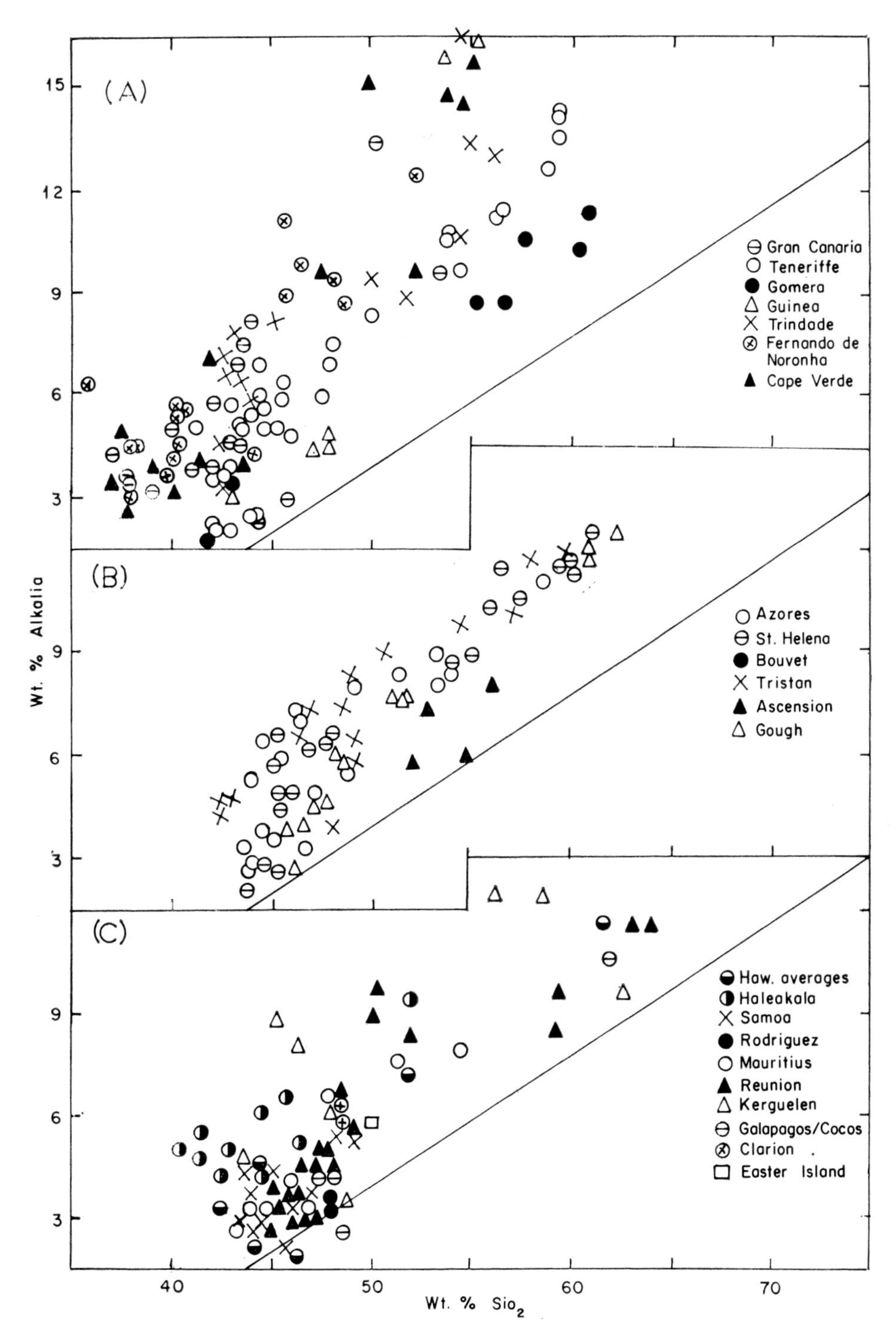

Fig. 2 Graphs of Silica *v.* Total alkalis for members of the alkalic suites of the oceanic islands

(a) and (b) Atlantic ocean islands.

(c) Pacific and Indian ocean islands.

Also shown is the boundary line separating the fields of Hawaiian alkalic and tholeiite rocks; MacDonald and Katsura (1964).

TABLE 1. Data on Oceanic Islands Discussed in this Review

Island/Group	Approx. °Lat.	°Long.	Structural or Topographic Setting	Nature of Alkalic Suite (Source, Author, Nomenclature)	Main Sources of Information
Atlantic Ocean					
Bouvet	58S	3E	Crest of Mid-Atlantic ridge	Plagioclase basalt–(pantellerite)	Broch (1946)
Gough	42S	12W	East of Mid-Atlantic ridge	Olivine basalt–trachybasalt–trachyandesite–trachyte	Le Maitre (1962)
Tristan da Cunha (Tristan, Nightingale, Inaccessible)	37S	12W	East of Mid–Atlantic ridge	Alkali basalt–trachybasalt–trachyandesite–trachyte	Baker *et al.* (1964)
Trinidade and Martin Vaz	20S	30W	Line of guyots trending 1000 km from continental slope of Brazil	Ultramafics. Basanite–grazinites–gauteites–basic phonolite. Nephelinite suite.	de Almeida (1961)
St. Helena	17S	7W	East of Mid-Atlantic ridge	Alkali basalt–trachybasalt–trachyandesite–trachyte/phonolite	Baker (1968)
Ascension	8S	14W	Crest of Mid-Atlantic ridge	Alkali basalt–trachybasalt–trachyandesite–trachyte–(pantellerite)	Bell *et al.* (1968, pers. comm.)
Fernando de Noronha	3½S	32W	Part of line of otherwise submerged volcanoes trending from continental slope of Brazil?	Ultramafics. Alkali basalts/basanites–kali gauteites–phonolites/alkali trachyte	de Almeida (1955)
Guinea Is. (Principe, Sao Tome, Annabon)	0	5E	Continuation of Cameroon fracture line?	Alkali basalt–phonolite	Fuster (1954) Cotelo Neiva (1955)
Cape Verde Is.	16N	24W	Fault zone	Ultramafics. Alkali basalt–phonolite	Part (1950)
Canary Is:	23N	17W	Fault zone (off continental slope of W. Africa)		
Gomera				Basanite–trachyandesite–trachyte/phonolite	Bravo (1964)
Gran Canaria				Basanite–hawaiite(?)–mugearite(?)–trachyte/phonolite	Fuster *et al.* (1968a)
				Alkali trachyte–alkali phonolite	Schmincke and Swanson (1967), Schmincke (1968), Rothe and Schmincke (1968)
Teneriffe				Basanite–trachybasanite–plagioclase phonolite–trachyte/phonolite	Ridley (1968) Borley (1966)
Azores	38N	30W	East–West rise of Mid-Atlantic ridge	Basanite/tephrite–alkali trachyte	Assuncao (1961)

TABLE 1 *continued*

Indian Ocean					
Reunion:	21S	56E	West of crest of Mid-Indian Ocean ridge in Rodriguez fracture zone		
Piton des Neiges				Feldsparphyric–basalt–hawaiite–mugearite–trachyte	Upton and Wadsworth (1966)
Piton de la Fournaise				Olivine basalt	Upton and Wadsworth (1967)
Rodriguez	19½S	63E	West of crest of Mid-Indian Ocean ridge in Rodriguez fracture zone	Alkali basalt	Upton *et al.* (1967)
Mauritius	20S	57E	West of crest of Mid-Indian Ocean ridge	Basalt–trachyandesite–trachyte/phonolite	Walker and Nicolaysen (1954)
Kerguelen	49S	70E	Gaussberg ridge	Basanite/tephrite–and. basalt–olig. basalt–trachyte/phonolite. Ultramafics	Edwards (1938)
Heard	53S	73E	Gaussberg ridge	Alkali basalt–trachybasalt–trachyandesite–trachyte and soda rhyolite	Tyrrell (1937)
Pacific Ocean					
Easter Is.	26S	107W	East Pacific Rise	Basalt ?–(rhyolite)	Bandy (1937)
Samoan Is.	14S	170W	System of parallel rifts		
Tutuila				Alkali basalt–hawaiite–mugearite–trachyte	Stearns (1944) MacDonald (1944b, 1968)
Manu'a				Alkali basalt–hawaiite	Stice (1968)
Upolu				Olivine basalt	MacDonald (1944b)
Savaii					
Galapagos	0	91W	East Pacific Rise	Alkali basalt	Richardson (1933) Williams (1966)
Cocos	5½N	87W	East Pacific Rise	Andesite	Chubb (1933)
Revillagigedo Is.	19N	111W	East Pacific Rise		
Clarion			(Clarion fracture zone)	Alkali basalt–trachybasalt–trachyandesite–trachyte	Bryan (1967)
Socorro				Alkali basalt–trachyte	Bryan (1966)
Hawaiian Is.	23N	170W	Hawaiian ridge	Alkali basalt–hawaiite–mugearite–trachyte. Nephelinite suite.	Yoder and Tilley (1962) MacDonald and Katsura (1964) MacDonald and Powers (1968)

oxides, as it is oxidation of these that contributes most to rock oxidation, Wilson and Watkins (1967), Watkins and Haggerty (1967), Haggerty, Borley and Abbott (1966). However, it is still rare for petrologists to do this although many undertake the tedious process of recalculating the $FeO:Fe_2O_3$ ratios of their rocks to some arbitrary figure. So in selecting data the ratio $FeO:Fe_2O_3 \leq 1$ has been taken as indicating rocks which are likely to have oxidation indices of III or more, Watkins and Haggerty (1967). Alteration of rocks may be shown by excessively high volatile contents, and faulty analysis by very high or low totals. Normative *C* or *wo* may also indicate alteration or faulty analysis, or both. To summarize criteria used in selecting data for tables and figures: (1) all analyses with totals outside the limits 99·0–101·0 were rejected, as were analyses whose reported totals differed considerably from their actual ones; (2) in the compilation of Tables 2–4 (averaged representative analyses), oxidized rocks, or rocks with normative *C* or *wo*, were rejected unless they were considered to be of unusual composition or might otherwise not have been represented, e.g. some intermediate alkalic rocks, which have high initial oxidation states and which are particularly prone to further oxidation, and some rocks from Fernando de Noronha and Trindade; (3) in Fig. 1, diagrams of Silica v. Diff. Index, rocks described under (2) were omitted; (4) in Fig. 2, diagrams of Silica v. Total Alkalis, otherwise acceptable analyses were used unless there was evidence that these oxides had been affected by alteration; (5) in compiling the F.M.A. diagrams, Fig. 3, reported $FeO + Fe_2O_3$ was recalculated as molecular FeO, thereby allowing inclusion of many otherwise satisfactory analyses of oxidized rocks.

IV.7.4. Alkalic Rocks of the Oceanic Islands

General data and representative analyses of rocks from all the oceanic islands are given in Tables 1–4 and trend-composition diagrams in Figs. 1, 2 and 3.

IV.7.4.1. The Pacific Ocean

Pacific islands have been selected to represent various topographic settings of volcanicity; (*a*) the Hawaiian chain in the north-central Pacific which lies on a submarine ridge (Jackson, 1968); (*b*) the Samoan islands of the south-west Pacific which occur in a region of parallel rifts (Stearns, 1944); and (*c*) a number of islands which emerge from the East Pacific Rise, and whose formation is almost certainly linked to that of the Rise, e.g. the Revilla Gigedo Islands of Clarion and Socorro which lie a little south of the point where the Rise is encroached on by the American continent (Bullard, 1968), the Cocos and Galapagos Islands from the equatorial area of the Rise and, much farther south and much farther removed from the continental mass, Easter Island.

No comment seems necessary on Hawaii, where the combination of active volcanicity and recognition of the intimate association of the tholeiite, alkalic and nephelinite suites led to studies that are now classics of volcanic petrology; MacDonald (1944a, 1949a), MacDonald and Katsura (1964), Muir and Tilley (1957 and 1963), Powers (1955), Tilley and Scoon (1961) and Yoder and Tilley (1962). Interesting recent papers on Hawaii are those of White (1966) and Jackson (1968) which discuss the ultramafic nodules found in the various lavas, and MacDonald and Powers (1968) on differentiation of the Haleakala alkalic lavas. Differentiation within the alkalic suite on other Pacific islands has not been as extensive as in the Hawaiian group. Among the Samoan islands, basalts dominate in Upolu, Savaii and Manu'a, whilst on Tutuila there is limited occurrence of trachytes; MacDonald (1944b and 1968) and Stice (1968). Basalts also predominate on the East Pacific Rise islands although some 'andesites' and trachytes occur in Clarion, Cocos and Easter Islands; these andesites do not resemble continental ones (cf. II.4)*. Quartz trachytes are found in Tutuila (Samoa) and rhyolitic obsidians in Easter Island. Both these rock types are believed by MacDonald (1944b and 1968) to be felsic end-members of the alkalic suite and not tholeiitic rocks. MacDonald in fact

* Readers are referred to the relevant Chapters of this book where similarly cited.

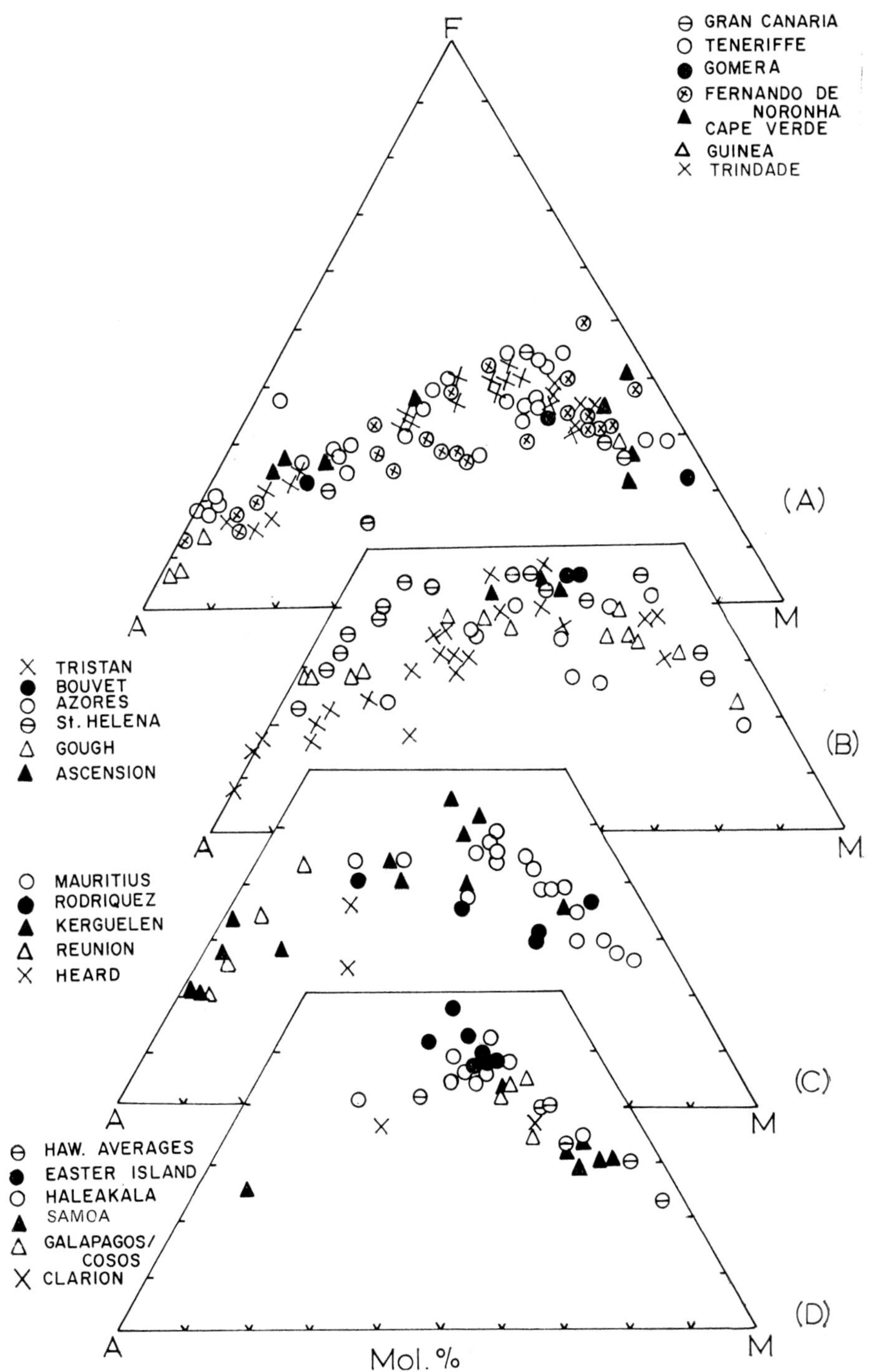

Fig. 3 F.M.A. diagrams for members of the alkalic suites of the oceanic islands

(A) Canary Islands, Guinea Islands, Cape Verde Islands, Trindade and Fernando de Noronha.

(B) Tristan, Bouvet, Gough, St. Helena, Ascension and the Azores Islands.

(C) Indian ocean islands.

(D) Pacific ocean islands.

F = total iron as mol. % FeO. M = mol. % MgO. A = mol. % Na_2O + mol. % K_2O.

TABLE 2 Averaged 'Representative' Analyses of Rocks from Pacific Ocean Islands

No.	1	2	3	4	5	6	7	8	9	10	11	12	13	14	15
Average of	10	14	28	33	13	1	3	1	3	2	5	1	1	6	3
SiO_2	44·33	46·41	46·46	48·60	51·90	51·90	41·94	46·89	47·72	46·98	46·69	43·63	60·48	45·58	48·10
TiO_2	2·65	1·98	3·01	3·16	2·57	2·11	4·06	3·03	2·15	1·70	4·11	5·30	0·39	3·97	3·58
Al_2O_3	12·80	8·53	14·64	16·49	16·65	17·10	14·26	15·78	16·60	20·03	15·11	11·70	18·16	12·87	14·91
Fe_2O_3	3·38	2·47	3·27	4·19	4·25	3·38	4·99	2·53	1·29	1·10	4·13	1·82	1·99	3·52	4·09
FeO	9·14	9·82	9·11	7·40	6·17	6·64	10·84	8·68	9·19	7·41	9·12	9·43	2·92	9·12	8·52
MnO	0·15	0·15	0·14	0·14	0·18	0·17	0·19	0·14	0·18	0·30	0·14	0·16	0·17	0·18	0·16
MgO	11·05	20·81	8·19	4·70	3·56	2·26	6·14	7·86	7·71	5·77	4·49	10·87	0·73	8·60	4·81
CaO	10·52	7·38	10·33	7·79	6·30	5·70	10·77	8·72	11·92	9·25	9·48	11·40	2·13	11·33	8·16
Na_2O	3·60	1·58	2·92	4·43	5·22	6·65	3·66	2·97	2·15	2·77	3·29	2·57	7·27	2·52	3·43
K_2O	0·99	0·32	0·84	1·60	2·01	2·72	1·51	1·78	0·41	1·52	0·83	1·63	4·12	1·10	1·68
H_2O (total)	—	—	—	—	—	0·63	0·62	0·85	0·31	2·32	2·22	1·27	0·98	0·96	1·70
P_2O_5	0·43	0·20	0·37	0·69	0·93	0·84	0·57	0·68	0·32	0·74	0·42	0·58	0·38	0·50	0·74
Totals*	99·04	99·65	99·28	99·19	99·74	100·10	99·55	99·91	99·95	99·89	100·03	100·36	99·72	100·25	99·88

* Small amounts of other constituents have been omitted.

Norms of the Analyses

	1	2	3	4	5	6	7	8	9	10	11	12	13	14	15
Q		—	—	—	—	—	—	—	—	—	0·67	—	—	—	—
or	6·12	1·67	5·00	9·45	11·68	16·12	8·89	10·56	2·46	9·01	4·89	9·63	24·35	6·58	10·00
ab	11·53	13·10	24·63	35·11	44·01	37·73	9·26	24·10	18·81	23·48	27·85	8·41	57·09	19·56	29·17
an	15·57	15·57	24·46	20·57	16·12	8·90	17·88	24·46	34·41	37·70	24·00	15·58	4·76	20·43	20·10
ne	10·22	—	—	1·14	—	9·94	11·74	0·57	—	—	—	7·23	2·39	0·98	—
di	27·18	16·15	19·08	10·75	6·96	11·54	26·08	11·18	18·52	2·74	16·49	29·67	2·81	25·60	12·69
hy	—	15·87	—	—	2·79	—		—	10·30	3·50	6·80	—	—	—	9·95
ol	17·60	29·56	14·17	8·60	4·85	4·48	5·79	16·84	9·05	14·59	2·33	14·53	2·83	12·07	1·80
mt	4·87	3·48	4·64	6·03	6·26	4·87	7·27	3·71	1·86	1·60	5·99	2·64	2·89	5·14	5·95
il	5·02	3·80	5·78	6·08	1·86	3·95	7·75	5·78	4·09	3·13	0·78	10·07	0·74	7·58	6·84
ap	1·01	0·34	1·01	1·68	2·35	2·02	1·34	1·68	0·75	1·75	1·00	1·37	0·89	1·28	1·76
Diff. Index	27·9	14·8	29·6	45·7	55·7	63·8	29·9	35·2	20·7	32·5	33·4	25·3	83·8	23·6	39·2
$Na_2O:K_2O$	3·64	4·93	3·48	2·76	2·6	2·44	2·42	1·67	5·24	1·82	3·96	1·58	1·76	2·29	2·04
$FeO:Fe_2O_3$	2·70	3·97	2·79	1·77	1·45	1·96	2·17	3·43	7·12	6·73	2·21	5·18	1·47	2·59	2·08

Averages: 1. Nepheline basanite 2. Picrite basalt of oceanite type 3. Alkali basalt 4. Hawaiite 5. Mugearite (Hawaiian averages from MacDonald and Katsura, 1964) 6. Haleakala mugearite 7. Haleakala basanitoid (from MacDonald and Powers, 1968) 8. Olivine basalt (from Clarion Island; Bryan, 1967) 9. Basalt (from Galapagos; Richardson, 1933) 10. Andesite (from Cocos Island, Chubb, 1933) 11. Basalt (from Easter Island; Bandy, 1937) 12. Limburgitic basalt 13. Trachyte (from Samoa; MacDonald, 1944b) 14. Hawaiite (from Tutuila; MacDonald, 1968) 15. Alkali basalt (from Manu'a; Stice, 1968)

considers it possible that Samoa and other similar volcanic islands consist solely of alkalic rocks produced from mantle-derived alkalic olivine basalt.

IV.7.4.2. Atlantic Ocean

Much of the attention now paid to the Atlantic Islands is due to a revival of the idea that the Afro-European and American continents drifted away from each other along a line defined by the position of the mid-Atlantic ridge. The ideas of Dietz (1961) and Wilson (1963) on ocean floor spreading and the suggestion that Atlantic islands have migrated away from the ridge crest, have also helped focus attention on their formation and growth. There is, however, no evidence of a direct association between formation of the ridge and of a number of the islands, especially those near the ocean margins. So tentative groupings of the islands can be made, as for the Pacific ones, to show the varied topographic settings of Atlantic volcanicity; (*a*) islands whose development may be connected with that of fractures off the continental slope or extending from the continent into the slope area, e.g. the Guinea and Canary Islands; (*b*) islands which are the exposed remains of chains of submarine volcanoes trending from the continental slopes into the ocean basin proper, e.g. Trindade and Fernando de Noronha; (*c*) isolated islands or island groups which may be associated with local fractures, e.g. Cape Verde and Madeira; (*d*) islands whose formation is thought to be associated with that of the ridge itself, e.g. the 'crest islands' Ascension and Bouvet, and others at varying distances from the ridge crest, e.g. Tristan, Gough, the Azores, and St. Helena. Fuster (1954) and McBirney and Gass (1967), have observed that St. Helena lies on the Cameroon–Guinea fracture trend; but it is so far removed from the Guinea Islands that its position along their trend line could be fortuitous.

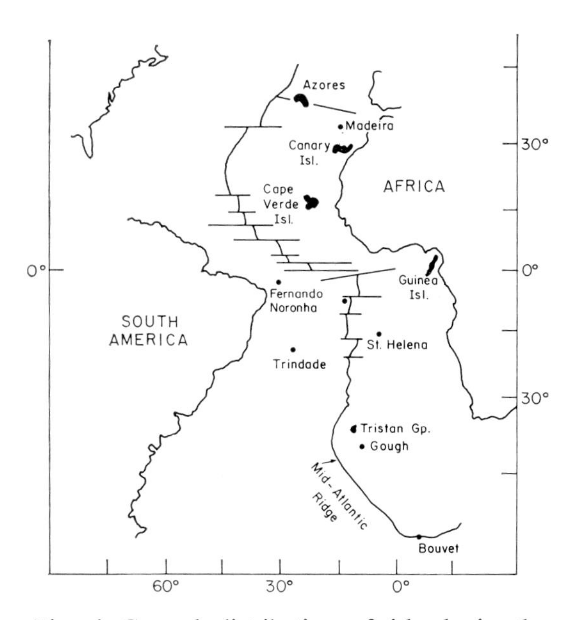

Fig. 4 General distribution of islands in the Atlantic showing their positions relative to the mid-Atlantic ridge. Iceland in the extreme north is not marked

Several points of interest emerge from the analytical data on the Atlantic Islands. (1) There is a more pronounced trend toward phonolite and trachyte derivatives in the Atlantic Islands than in those of the Indian and Pacific Oceans (Hawaii excepted), see Tables 2–4 and Figs. 1–3. (2) Careful examination of the data on which Tables 3A, B and C were based, and of the data in the Tables, does not support the suggestion of McBirney and Gass (1967) that silica saturation decreases in rocks of the Atlantic Islands with increasing distance from the crest of the Mid-Atlantic ridge. Rocks of San Miguel (Azores) and Tristan, for example, which are at similar approximate distances of 480 km from the crest, are definitely undersaturated; much more so than those of Gough, 625 km from the crest, and a little more so than most of the rocks of St. Helena, 700 km from the crest. The most highly undersaturated rocks ever reported from the Atlantic are those of Trindade and Fernando de Noronha which lie at the vastly different distances of approximately 800 km and 2000 km from the ridge crest. Rocks of these two islands have low silica, high alkali and, frequently, high volatile content and, in addition, many of them are highly oxidized; de Almeida (1955, 1961) considers their chemistry to have been considerably modified by alkali metasomatism. Data relating to trends in undersaturation obviously need to be cautiously interpreted. (3)

TABLE 3A Averaged 'Representative' analyses of Rocks from Atlantic Ocean Islands: Canary Islands: Gomera, Teneriffe, Gran Canaria

No.	1	2	3	4	5	6	7	8	9	10	11	12	13	14	15
Average of	2	1	2	2	5	1	2	1	4	3	3	2	3	3	6
SiO_2	57·10	41·90	42·73	42·30	40·69	64·05	43·30	53·44	43·36	41·18	46·39	42·37	55·70	57·43	43·88
TiO_2	0·74	0·99	5·71	3·29	3·39	0·74	3·30	1·54	2·43	3·83	3·48	3·62	1·57	1·32	3·60
Al_2O_3	19·71	4·97	17·53	12·49	12·07	14·03	14·46	19·73	9·89	12·50	15·47	10·71	19·52	19·46	15·10
Fe_2O_3	1·49	2·84	2·14	4·79	3·51	1·58	5·04	1·86	3·78	5·14	4·12	5·45	2·51	1·84	4·10
FeO	2·04	10·08	8·92	8·06	8·60	1·62	6·42	3·18	8·59	8·40	7·62	7·59	2·81	2·53	7·54
MnO	0·11	0·18	0·20	0·20	0·20	0·22	0·19	0·11	0·21	0·23	0·23	0·16	0·27	0·19	0·21
MgO	1·17	23·53	5·67	9·26	11·88	2·97	7·55	1·60	15·98	10·11	6·47	13·46	1·45	0·94	7·85
CaO	2·66	8·83	10·20	11·75	11·90	1·85	9·35	5·84	11·96	11·96	9·59	12·97	4·47	2·85	10·11
Na_2O	8·57	0·60	3·07	3·16	3·74	5·84	4·37	7·59	1·61	2·78	4·20	2·34	7·21	7·65	3·99
K_2O	3·07	1·14	1·48	1·03	1·76	3·48	2·53	2·05	0·65	1·76	1·54	0·71	3·75	4·43	1·76
H_2O (total)	3·30	4·23	1·85	2·36	0·94	3·26	2·58	2·56	1·33	1·77	0·26	0·17	0·58	1·01	1·24
P_2O_5	—	0·16	0·74	0·96	1·14	0·14	0·82	0·32	0·35	0·50	0·68	0·27	0·33	0·10	0·67
Totals*	99·96	99·45	100·24	99·65	99·82	99·78	99·91	99·82	100·14	100·16	100·05	99·82	100·17	99·75	100·05

* Small amounts of other constituents have been omitted

Norms of the Analyses

	1	2	3	4	5	6	7	8	9	10	11	12	13	14	15
Q	—	—	—	—	—	9·77	—	—	—	—	—	—	—	—	—
or	18·17	6·74	8·75	6·09	5·46	20·57	14·95	12·12	3·86	7·59	9·12	4·22	22·16	26·22	10·45
ab	49·22	1·72	16·56	11·64	—	49·41	9·41	41·56	7·32	12·15	22·54	4·42	40·70	43·09	12·24
an	6·22	7·50	29·69	16·86	10·95	1·79	12·38	13·72	14·94	8·60	18·81	16·60	9·81	5·63	18·04
leu	—	—	—	—	3·87	—	—	—	—	—	—	—	—	—	—
ne	9·08	1·82	5·10	2·42	17·15	—	14·93	12·27	3·43	11·59	7·04	8·35	11·01	11·75	11·68
di	5·64	28·09	12·94	28·00	29·69	5·09	22·60	10·65	31·22	31·49	19·46	36·25	7·10	5·53	23·06
hy	—	—	—	—	—	5·86	—	—	—	—	—	—	—	—	—
ol	1·17	42·98	9·70	11·10	16·54	—	7·57	0·57	24·27	11·08	8·66	14·66	0·87	0·67	9·96
cas	—	—	—	—	1·06	—	—	—	—	—	—	—	—	—	—
mt	2·17	4·12	3·10	6·95	5·09	3·30	7·30	2·70	5·48	7·45	5·96	7·90	3·30	2·67	5·94
il	1·41	1·88	10·84	6·25	6·45	1·40	6·27	2·92	3·59	7·27	6·60	6·87	2·97	2·52	6·85
hap	—	0·38	1·75	2·26	2·69	0·33	1·94	0·76	0·81	1·18	1·62	0·65	0·78	0·23	1·60
Diff. Index	76·5	10·3	30·4	20·2	26·5	79·8	39·3	66·0	14·6	31·3	38·7	17·0	73·9	81·1	34·4
$Na_2O:K_2O$	2·79	0·52	2·07	3·06	2·45	1·68	1·73	3·70	2·48	1·58	2·72	3·29	1·92	1·72	2·27
$FeO:Fe_2O_3$	1·36	3·55	4·16	1·68	2·13	1·03	1·27	1·71	2·27	1·63	1·84	1·39	1·11	1·37	1·83

Averages: 1. Phonolite 2. Werhlite 3. Porphyritic essexite (From Gomera, Can. Islands; Bravo, 1964) 4. Basalt 5. Ankaramite 6. Vitreous trachyte 7. Tephrite 8. Ordanchite (from Gran Canaria, Can. Islands; Fuster *et al.* 1968a) 9. Ankaramite 10. Ancient basalt 11. Basalt of Lower Canadas Series 12. Ankaramite of Lower Canadas Series 13. Phonolite from Upper Canadas Series 14. Phonolite from Recent Salic Series 15. Basalt Series III (Teneriffe: Borley, 1966)

TABLE 3B Averaged 'Representative' Analyses of Rocks from Atlantic Ocean Islands: Cape Verde, Guinea (Principe, Sao Tome, Annabon), Fernando de Noronha and Trindade

No.	1	2	3	4	5	6	7	8	9	10	11	12	13	14	15
Average of	1	2	1	2	1	1	3	1	1	2	1	1	2	3	2
SiO_2	42·08	37·31	53·76	42·77	44·96	52·23	47·52	40·80	40·40	36·51	49·40	51·00	45·90	38·04	37·50
TiO_2	4·24	3·72	0·39	3·26	1·94	2·81	1·90	1·60	2·20	3·30	0·80	0·30	1·94	3·10	2·95
Al_2O_3	15·03	7·08	22·18	10·26	18·06	16·34	18·67	13·92	13·09	12·72	20·32	20·25	19·45	12·35	14·35
Fe_2O_3	3·25	4·46	1·33	4·35	4·04	1·51	4·56	2·69	4·55	4·28	3·06	2·26	4·04	5·71	7·59
FeO	8·68	8·44	2·53	9·13	7·82	1·50	2·25	11·71	9·24	9·58	2·51	2·58	2·52	8·37	6·50
MnO	0·33	0·07	0·09	0·09	0·26	0·22	0·12	0·15	0·15	0·19	0·15	0·15	0·17	0·20	0·22
MgO	5·52	18·86	0·15	14·07	5·58	3·81	3·77	12·38	11·00	13·47	1·40	0·80	2·85	11·94	8·44
CaO	12·74	14·64	2·10	10·54	8·99	3·65	5·99	11·00	11·40	11·47	3·60	2·21	6·35	11·46	12·00
Na_2O	4·16	2·01	8·74	2·75	4·56	6·75	4·42	3·85	3·94	4·20	11·35	13·00	3·21	4·96	3·73
K_2O	2·84	0·99	5·96	1·60	2·54	5·64	4·53	0·78	1·49	1·36	5·30	5·00	6·14	1·63	0·94
H_2O (total)	0·24	1·72	2·25	1·04	0·73	5·46	5·89	0·80	2·00	1·72	2·00	1·34	7·04	1·19	3·90
P_2O_5	0·77	0·84	tr	0·25	0·62	0·43	0·36	0·65	0·80	1·02	0·29	0·18	0·30	0·91	1·37
Totals	99·88	100·17	100·03[1]	100·16	100·10	100·05	99·98	100·33	100·26	99·82	100·18	100·17[2]	100·69[3]	99·86	99·49

[1] includes 0·55 Cl, SO_3; [2,3] include 1·0 and 0·78 CO_2 for carbonate from amygdales etc.

Norms of the Analyses

	1	2	3	4	5	6	7	8	9	10	11	12	13	14	15
or	3·89	—	35·58	9·45	15·01	33·34	26·79	—	1·92	—	26·84	29·55	36·29	—	5·58
ab	—	—	21·48	3·11	10·71	21·75	18·35	—	—	—	—	6·91	4·93	—	3·44
an	13·62	7·5	5·84	10·93	21·31	—	17·73	18·40	13·64	11·84	—	—	20·53	6·61	19·63
leu	10·03	4·8	—	—	—	—	—	3·62	5·40	6·32	3·52	—	—	7·55	—
ne	19·31	8·9	26·13	10·92	15·10	16·74	10·32	17·65	18·06	19·25	40·64	37·61	11·79	22·76	15·23
ac	—	—	—	—	—	3·93	—	—	—	—	8·54	6·54	—	—	—
ns	—	—	—	—	—	—	—	—	—	—	2·55	6·11	—	—	—
di	35·75	20·9	4·14	31·58	15·66	9·49	6·53	25·82	30·26	13·06	13·22	8·32	7·01	23·34	24·22
ol	2·42	30·6	0·97	19·98	10·60	3·04	4·45	25·57	16·36	25·99	0·37	1·71	2·69	17·50	7·72
cas	—	10·3	—	—	—	—	—	0·04	—	6·81	—	—	—	4·64	—
mt	4·64	6·5	1·86	6·30	5·86	—	3·14	3·90	6·59	6·20	—	—	3·05	8·29	10·96
hem	—	—	—	—	—	0·15	2·39	—	—	—	—	—	1·93	—	0·02
il	8·06	7·1	0·76	6·20	3·68	3·64	2·94	3·03	4·18	6·27	1·52	0·57	3·68	5·89	5·60
hl	—	0·05	0·59	—	—	prv. 1·52	—	—	—	—	—	0·43	—	—	—
th	—	—	0·43	—	—	—	—	—	—	—	—	—	—	—	—
ap	2·02	2·0	—	0·59	1·46	1·01	0·85	1·53	1·88	2·41	0·68	—	0·72	2·10	3·24
Diff. Index	33·2	13·7	83·2	23·5	40·8	71·8	55·5	21·3	25·4	25·6	71·0	74·1	53·0	30·3	24·3
$Na_2O:K_2O$	1·46	2·03	1·47	1·71	1·80	1·2	0·98	4·94	2·64	3·1	2·1	2·6	0·52	3·04	3·97
$FeO:Fe_2O_3$	2·67	1·9	1·9	2·1	1·9	1·0	0·5	0·4	2·0	2·2	0·8	1·1	0·6	1·5	0·9

Averages: 1. Tephrite 2. Ankaramite 3. Phonolite (from the Cape Verde Islands; Part, 1950) 4. Masafuerite 5. Trachydolerite (from the Guinea Islands; Fuster, 1954) 6. Phonolite 7. Kali gauteite 8. Nepheline basanite 9. Olivine tephrite 10. Ankaratrite (from Fernando de Noronha; de Almeida, 1955) 11. Phonolite 12. Nosean phonolite 13. Kali gauteite 14. Ankaratrite 15. Nephelinite (from Trindade; de Almeida, 1961)

TABLE 3C Averaged 'Representative' Analyses of Rocks from Atlantic Ocean Islands: Tristan, Gough, St. Helena, Azores, Ascension and Bouvet

No.	1	2	3	4	5	6	7	8	9	10	11	12	13	14
Average of	16	1	2	3	3	3	4	4	5	1	1	2	1	2
SiO_2	46·84	48·11	55·75	47·43	51·71	60·63	45·58	55·71	60·37	49·68	58·50	52·32	52·90	49·04
TiO_2	3·36	3·27	1·74	3·33	2·49	0·37	3·36	0·63	0·18	2·28	2·40	2·49	2·24	3·11
Al_2O_3	17·34	14·29	19·19	15·64	17·79	18·50	15·76	17·50	17·99	18·38	18·40	16·95	17·15	15·24
Fe_2O_3	3·56	2·01	2·40	2·96	2·94	2·14	3·17	2·92	2·23	2·70	1·35	2·61	3·46	3·73
FeO	6·83	7·63	2·85	7·93	6·11	3·09	9·11	5·71	3·04	6·69	2·65	5·41	6·37	8·26
MnO	0·17	0·15	0·09	0·12	0·10	0·22	0·20	0·27	0·23	0·14	0·14	0·20	0·24	0·20
MgO	4·39	10·13	1·85	7·96	3·34	0·33	5·93	1·16	0·22	3·97	1·70	3·18	3·14	4·98
CaO	9·16	9·71	4·53	8·21	6·48	1·79	9·36	3·70	1·93	10·62	3·15	6·09	6·75	8·78
Na_2O	4·19	2·92	5·06	2·84	4·10	6·28	3·65	6·34	7·28	3·09	5·60	4·51	5·10	3·65
K_2O	3·16	0·98	5·03	1·65	3·59	5·83	1·33	3·25	4·58	0·86	5·50	4·21	2·22	1·25
H_2O(total)	0·30	0·55	1·25	1·48	0·93	0·62	1·99	2·46	1·86	0·84	1·34	0·94	0·51	0·60
P_2O_5	0·75	0·31	0·34	0·22	0·29	0·12	0·56	0·28	0·05	0·35	0·11	1·05	0·24	0·90
Totals*	100·05	100·06	100·08	99·77	99·87	99·92	100·00	99·93	99·96	99·60	100·84	99·96	100·32	99·74

*Small amounts of other constituents have been omitted

Norms of the Analyses

	1	2	3	4	5	6	7	8	9	10	11	12	13	14
Q	—	—	—	—	—	—	—	—	—	1·34	—	—	—	0·52
cr	18·66	5·79	29·73	9·82	21·12	34·65	7·83	19·24	27·05	5·08	32·51	24·91	13·12	7·39
ab	15·97	24·71	37·09	23·93	31·66	47·57	23·24	47·91	53·24	26·14	42·61	32·97	40·77	30·88
an	19·18	23·00	14·81	25·11	19·55	3·82	22·70	9·67	2·91	33·75	8·83	13·55	17·35	21·52
ne	10·56	—	3·10	—	1·67	2·98	4·14	3·11	4·52	—	2·59	2·83	1·29	—
di	16·78	18·49	4·26	11·27	8·85	2·82	16·34	5·89	5·49	13·59	4·74	7·94	11·92	13·12
hy	—	2·63	—	6·19	—	—	—	—	—	9·79	—	—	—	9·43
ol	5·21	15·05	2·23	10·82	6·01	2·38	11·47	5·56	1·19	—	1·46	5·89	5·53	2·88
mt	5·16	2·91	3·48	4·33	4·25	3·09	4·58	4·24	3·24	3·91	1·96	3·79	5·02	5·41
il	6·38	6·21	3·31	6·33	4·76	0·76	6·38	1·19	0·34	4·33	4·56	4·73	4·25	5·90
ap	1·94	0·73	0·81	0·56	0·67	0·34	1·33	0·65	0·12	0·83	0·26	2·47	0·57	2·12
Diff. Index	45·2	30·5	69·9	33·8	54·6	85·2	35·2	70·3	84·8	32·6	77·7	60·7	55·2	38·8
$Na_2O:K_2O$	1·32	2·97	1·0	1·72	1·14	1·07	2·74	1·95	1·58	3·59	1·0	1·0	2·29	2·92
$FeO:Fe_2O_3$	1·91	3·8	1·2	2·7	2·1	1·4	2·9	1·96	1·36	2·48	1·96	2·07	1·84	2·21

Averages: 1. Trachybasalt 2. Olivine basalt 3. Trachyandesite (from Tristan da Cunha; Baker *et al.*, 1964) 4. Basalt 5. Trachybasalt 6. Aegirine-augite trachyte (from Gough Island; LeMaitre, 1962) 7. Alkali olivine basalt 8. Trachyandesite 9. Late alkaline intrusion (from St. Helena; Baker, 1968) 10. Basalt (from Bouvet Island; Broch, 1946) 11. Trachyte 12. Andesite (from the Azores; de Assuncao, 1961) 13. Trachybasalt 14. Trachybasalt (from Ascension Island; new data Bell *et al.*, pers. comm.)

There is a little more support for the suggestion of Tilley and Muir (1964) and McBirney and Gass (1967) that rocks of Gough and Tristan represent 'potassic' trends, although the significance of such trends is not clear. $Na_2O:K_2O$ ratios for the Atlantic Islands as a whole vary from 0·5–4·9 compared with 0·4–2·1 for Gough and 0·8–3·0 for Tristan; Tables 3A, B and C and Baker *et al.* (1964), LeMaitre (1962). Within individual islands, however, the variations are very erratic, e.g. for Trindade rocks the $Na_2O:K_2O$ ratios vary from 0·4–67, but this is an extreme variation and no doubt reflects varying degrees of metasomatism of the rocks. Potash enrichment has also been noticed in rocks of the Syenite–Trachyte complex of Gran Canaria, Canary Islands, Fuster *et al.* (1968a), but some of these rocks generally are very altered and the potash enrichment could be secondary. The most notable feature of the $Na_2O:K_2O$ ratios of rocks from the Atlantic Islands is that all their more basic rocks have overlapping ratios, and obvious divergence of ratios is only shown by the more siliceous rocks. This implies that variation in $Na_2O:K_2O$ ratios among the alkalic rocks of the islands does not reflect a fundamental difference in their origin, but reflects local variations in fractionation of their parent basaltic magmas.

IV.7.4.3. The Indian Ocean

Included are the islands of Reunion, Mauritius and Rodriguez which lie west of the mid-Indian ridge in a region where fractures have displaced its crest from the main trend. Also included, far to the south, are the Kerguelen–Gaussberg ridge islands of Kerguelen and Heard (which is just within the Southern ocean). In Rodriguez, alkali basalts comprise the main eruptive sequence and they show little sign of differentiation, Upton *et al.* (1967). Similarly, in Reunion, the lavas of Piton de La Fournaise are undifferentiated basalts, but those of Piton des Neiges (the second of the island's two volcanic centres) have differentiated towards trachytes, Upton and Wadsworth (1966 and 1967). Rocks from both these islands and, to a lesser extent, those of the largely basaltic island of Mauritius, are only mildly alkaline to transitional in composition, and their analyses plot near the 'saturation line' in the Silica v. Diff. Index diagram in Fig. 1, and near the alkalic-tholeiite field boundary in the Silica v. Total Alkalis diagram in Fig. 2. Kerguelen and Heard islands, for which no published data more recent than those of Edwards (1938) and Tyrrell (1937) are available, both have highly differentiated sequences. In Kerguelen some of the late differentiates are oversaturated and peralkaline, whilst in Heard they are undersaturated and peralkaline. These trends are more like those for some of the Atlantic Islands than for those of the islands further to the north in the Indian Ocean. An alkalic basalt sequence has also been reported from Crozet Island (Gunn *et al.*, 1970).

IV.7.5. General Comments on Data for the Alkalic Rocks

The diagrams of Silica v. Total Alkalis, Silica v. Diff. Index, and F.M.A. variations (Figs. 1, 2 and 3) show that trends in the alkalic suites of the oceanic islands generally overlap, and that there is a scarcity of points representing the 'minimum frequency' compositions mentioned by Chayes (1963), see also VI.2. Points representing the basic rocks, in these diagrams, show considerable scatter which is not surprising for many of them represent accumulative rocks or phyric basalts, whose compositions are bound to vary erratically. There is less scatter than expected for points representing the siliceous end members of the alkalic suite; especially so in view of the fact that these points include some for peralkaline rocks whose position in a differentiation scheme is still uncertain, and whose compositions are frequently considered to have been modified by volatile transfer of alkalis. These comments apart, it might be unwise to give too much significance to variations in trends shown by diagrams of this sort. The results of Yoder and Tilley's (1962) experimental work on the basalt system, discussed very fully by O'Hara (1965, 1968), showed that the common basalt compositions lay close to a low pressure cotectic curve and were probably the end products of fractionation; it follows from

TABLE 4 Averaged 'Representative' Analyses of Rocks from Indian Ocean Islands

No.	1	2	3	4	5	6	7	8	9	10	11	12	13	14	15
Average of	1	1	2	2	4	33	3	1	1	1	2	2	1	3	1
SiO_2	48·12	48·14	63·75	45·94	48·64	47·98	47·77	50·48	53·05	62·40	43·71	46·21	46·01	43·95	54·49
TiO_2	3·06	3·24	0·50	1·78	2·32	2·98	1·96	2·67	1·80	tr	2·16	2·57	3·78	2·57	0·98
Al_2O_3	16·66	15·96	15·76	19·30	14·98	13·92	17·14	16·78	18·28	17·12	10·00	20·57	15·48	13·40	17·75
Fe_2O_3	3·66	4·66	1·24	2·65	3·99	2·60	2·64	4·16	3·30	2·23	3·94	4·00	2·99	2·85	3·06
FeO	8·42	8·64	4·01	7·13	7·52	9·05	6·45	5·75	4·80	2·62	8·80	5·53	9·30	9·79	5·89
MnO	—	0·17	0·17	0·25	0·18	0·13	0·10	0·14	0·30	0·30	0·08	0·21	0·19	0·17	0·30
MgO	4·29	4·40	0·03	3·08	6·28	7·67	7·73	2·39	2·30	tr	15·22	3·84	5·62	11·70	1·75
CaO	8·74	7·50	1·75	8·60	11·12	11·61	9·66	5·26	5·40	0·70	10·50	11·34	10·65	10·52	4·56
Na_2O	4·20	3·04	6·04	4·33	2·66	2·34	3·67	5·77	5·35	9·05	1·90	2·15	2·99	2·63	4·94
K_2O	1·92	2·57	5·56	4·13	0·64	0·89	1·68	3·86	4·22	4·80	0·83	1·04	1·19	1·04	3·11
H_2O (total)	0·65	1·52	1·28	1·60	1·05	1·01	1·00	0·99	0·30	0·75	2·72	1·75	1·48	1·22	2·35
P_2O_5	0·48	0·55	0·05	1·54	0·41	0·25	0·52	0·99	0·71	—	0·32	0·36	0·36	0·29	0·43
Totals*	100·20	100·19	100·14	100·33	99·79	100·43	100·32	99·24	99·81	99·29	100·18	99·57	100·04	100·13	99·61

*Small amounts of other constituents have been omitted.

Norms of the analyses

	1	2	3	4	5	6	7	8	9	10	11	12	13	14	15
Q	—	—	4·90	—	0·98	—	—	—	—	—	—	—	—	—	—
or	11·35	15·19	32·89	24·44	3·78	5·28	8·96	22·81	25·0	28·4	5·28	6·12	6·67	6·15	18·35
ab	27·18	25·72	45·94	10·19	22·48	19·75	20·65	33·83	33·5	50·1	10·87	23·05	23·45	10·35	41·92
an	20·94	22·32	2·19	21·03	27·02	24·83	26·24	8·49	13·3	—	16·12	40·42	25·58	21·40	17·24
ne	4·53	—	—	14·32	—	—	3·80	8·12	6·3	6·1	2·76	—	0·85	6·51	—
ac	—	—	0·94	—	—	—	—	—	—	6·5	—	—	—	—	—
ns	—	—	0·95	—	—	—	—	—	—	1·6	—	—	—	—	—
di	15·77	9·22	5·48	9·57	20·57	25·19	15·58	7·23	7·0	3·2	27·22	9·86	19·91	23·20	1·88
hy	—	8·90	3·27	—	12·77	8·05	—	—	—	—	—	4·44	—	—	—
ol	7·56	3·43	—	8·39	—	6·29	15·93	3·93	4·3	—	24·80	2·49	9·48	21·46	—
mt	5·31	6·47	1·32	3·85	5·78	3·78	2·80	6·03	4·9	—	5·68	5·80	4·41	4·13	4·41
il	5·81	6·15	0·96	3·38	4·41	5·67	3·77	5·07	3·5	—	4·10	4·86	7·14	4·91	1·78
hap	1·13	1·30	0·11	3·64	0·78	0·78	1·17	2·34	1·7	—	0·67	0·84	1·01	0·78	1·01
Diff. Index	43·1	40·9	83·7	49·0	27·2	25·0	33·4	64·8	64·8	84·6	18·9	29·2	31·0	23·0	60·3
$Na_2O:K_2O$	2·18	1·18	1·08	1·04	4·15	2·62	2·18	1·49	1·26	1·88	2·28	2·06	2·51	2·52	1·58
$FeO:Fe_2O_3$	2·30	1·85	3·23	2·69	1·88	3·48	2·44	1·38	1·45	1·17	2·23	1·38	3·11	3·43	1·92

Averages: 1. Andesite 2. Labradorite basalt 3. Trachyte 4. Tephrite (from Kerguelen Island; Edwards, 1938) 5. Basalt of the oceanite series 6. Basalt (from Reunion; Upton and Wadsworth, 1966) 7. Olivine basalt (from Rodriguez; Upton *et al.*, 1967) 8. Pegmatitic vein (from Rodriguez; Upton *et al.*, 1967) 9. Trachyandesite 10. Phonolite (from Heard Island; Tyrrell, 1937) 11. Ankaramite 12. Feldspar-phyric basalt 13. Olivine basalt 14. Olivine basalt 15. Trachyandesite (from Mauritius Island; Walker and Nicolaysen, 1954)

this that further low pressure fractionation of the more alkalic of these regionally 'parental' magmas might produce closely similar trends. One other general point; outside Hawaii, with few exceptions, undisputed associations of tholeiite and alkalic suites on the oceanic islands have not been demonstrated. The oversaturated rocks (including ejected blocks) of Bouvet, Ascension, Gran Canaria, Madeira, Azores (personal communication H. Schmincke), and Kerguelen, include phonolitic trachytes, comenditic rhyolites, pantellerites and peralkaline granites; all of which might be considered as end products of the alkalic sequence. The peralkaline nature of the magmas that gave rise to the ejected granite blocks of Ascension Island, was clearly demonstrated by the work of Roedder and Coombs (1967) on fluid inclusions in certain of the granite minerals (cf. V.2).

Rocks with minor amounts of normative quartz have been reported from many islands, including Cape Verde (Part, 1950), Gran Canaria (Fuster *et al.*, 1968a) and the Guinea Islands (Fuster, 1954) as well as from Samoa and Easter Island (already discussed). The Cape Verde rocks are volatile-rich trachytes and Part (1950) does not consider the normative quartz to be significant; similarly, many of the quartz normative basalts, trachytes and syenites from Gran Canaria are characterized by high oxidation states, high volatile contents, or are partly silicified. Normative quartz might be expected in rocks of this sort, but even so, the trends in rocks from this island are complicated. At the moment it can be said that much of the present evidence lends support to MacDonald's (1968), comment '... volcanoes of Samoa and some other mid-ocean islands may consist solely of alkalic rocks and thus differ fundamentally from the well established Hawaiian pattern'. This comment, which echoes that of Melson *et al.* (1967) regarding St. Paul's rock, might perhaps be true of other islands, nearer to the oceanic margins, which are believed to be of truly oceanic origin, but confirmation of the nature of the submerged bases of the oceanic islands can only be made by dredging of their lower slopes.

IV.7.6. Xenoliths from the Oceanic Alkalic Rocks

A discussion of the nature of xenoliths from alkalic rocks generally is found in chapter II.4, but a brief review of the known occurrences of xenoliths from oceanic islands may help to give an idea of their immense variety. Xenoliths have been found in:

Hawaii: Gabbro, dunite, peridotite, anorthosite, pyroxenite, troctolite, werhlite, basalt (White, 1966; Jackson 1968).

Hawaii (*Oahu*): Dunite, lherzolite, garnet websterite and garnet clinopyroxenite (Jackson and Wright, 1970).

Galapagos: Gabbro (Richardson, 1933).

Kerguelen: Gabbro, dunite, pyroxenite, amphibole-peridotite (Edwards, 1938).

Azores: Mafraite, Sanidinite (Assuncao, 1961); peralkaline syenite (Cann, 1967); dunite, gabbro, pyroxenite (LeMaitre, 1965).

Ascension: Granite (Cann, 1967); syenite, hypersthene-gabbro, peridotite (LeMaitre, 1965).

Tristan: Gabbro (Baker *et al.*, 1964). Kaersutite bearing plutonics (Le Maitre, 1969).

St. Helena: Gabbro, amphibole-rich xenoliths (Baker, 1968).

Cape Verde: Sanidinites (Part, 1950).

Gough: Hypersthene-gabbro (LeMaitre, 1965).

Canary Islands: Teneriffe: Basalt, dunite, layered pyroxenite-dunite, kaersutite-rich ultramafics (Borley *et al.*, 1971).
Gran Canaria: Clinopyroxene-amphibole xenoliths (Frisch and Schmincke, 1968).

Fernando de Noronha: Syenite, gabbro, pyroxenite (de Almeida, 1955).

Trindade: Syenite, melteigite, jacupirangite, olivinite, peridotite, glimmerite, hornblendite, kiirunavaarite (de Almeida, 1961).

Among the Atlantic islands, Trindade has a unique set of xenoliths, some of which are reminiscent of alkalic ultrabasic rocks from Africa, whilst Gough and Ascension are the only islands on which hypersthene-gabbro xenoliths have been found. Of more general interest, perhaps, is the widespread occurrence of 'brown' amphibole or kaersutite-bearing xenoliths. Experimental work by Lambert and

Wyllie (1968), suggested that amphibole bearing peridotite magma might exist beneath oceans and continents, and analytical data was presented by Mason (1968) to support his view that kaersutite-type amphibole might be a stable mantle phase (cf. II.4). More experimental work on the stability fields of amphiboles of this type has been undertaken since this volume was prepared and it should eventually help determine the significance of the presence of kaersutite bearing xenoliths in alkalic rocks.

IV.7.7. Mode of Occurrence of the Oceanic Alkalic Rocks

Oceanic alkalic rocks exist, like tholeiites, in extrusive, intrusive or clastic form: in Hawaii alkalic lavas occur as thin cappings to the tholeiite succession, but where alkalic rocks comprise the bulk of the subaerial volcano their mode of occurrence determines its overall form:

a. Islands dominated by a single cone, from whose eruptive centre lavas and interbedded pyroclastics dip radially away, for example, Tristan da Cunha. Parasitic vents and minor intrusions cut the flanks of these simple volcanoes.

b. Islands comprised of one or more shield volcanoes, e.g. St. Helena, (Fig. 5) Reunion, Gran Canaria (Canary Isl.), Tutuila (Samoa). Deep erosion of old shields may expose their cores, as in St. Helena, where dissection of the shield lavas has revealed major dyke swarms, believed by Baker (1969) to represent the original lava-feeder system. Caldera formation frequently accompanies later shield-building stages, for instance in Reunion, in the Pago Shield (Tutuila), in Gran Canaria and elsewhere.

c. Islands whose structure is complex, as are many of the Azores islands, Mauritius and Teneriffe. The volcanic stratigraphy of these islands is often complicated by the occurrence of numerous cinder-spatter cones and parasitic vents on the flanks of major volcanoes; by blanketing pumice mantles from flank vents; by the presence of intrusions; by migration of eruptive centres; and by erosion.

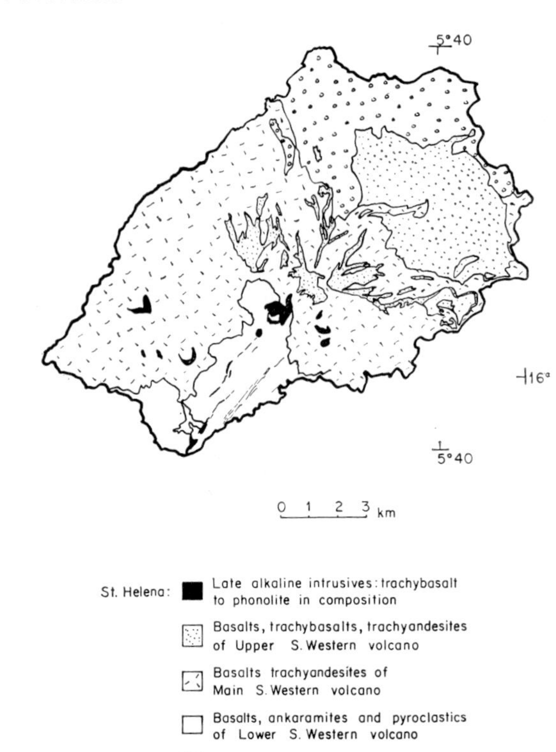

Fig. 5 St. Helena. Distribution of volcanics belonging to the two main volcanoes—north-eastern and south-western. Note the occurrence of the late alkaline intrusions in the south-western volcanics and also their arcuate shapes (after Baker, 1969)

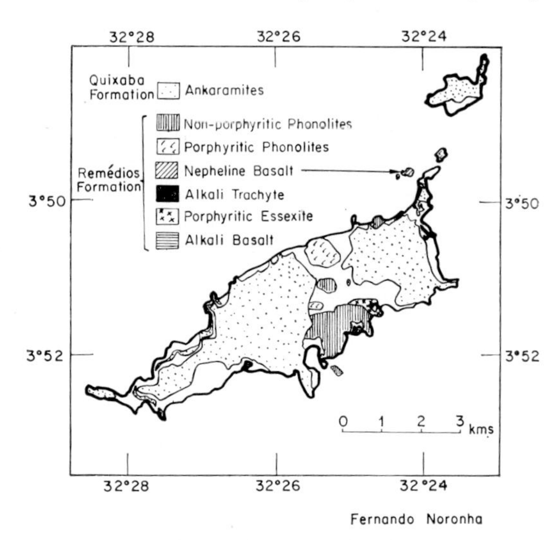

Fig. 6 The small island of Fernando Noronha with its highly alkaline volcanics. Note the extensive ankaramite formation (after Almeida, 1955)

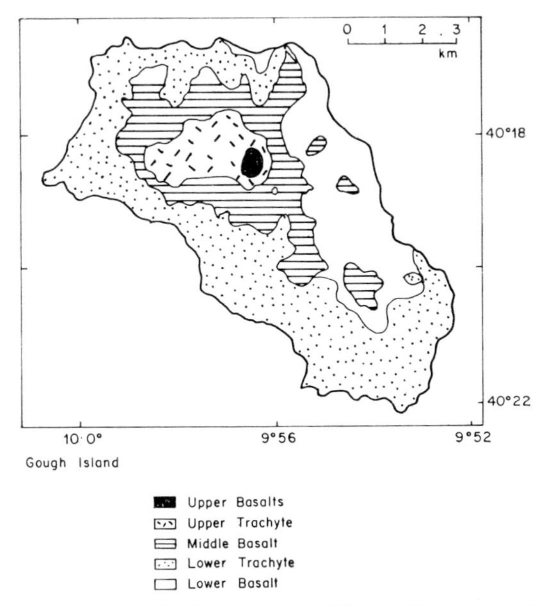

Fig. 7 Gough Island. A mildly undersaturated series trending towards trachytes. The whole series represents the so-called 'potassic' alkalic trend (after LeMaitre, 1962)

IV.7.7.1. Form of Flows, Intrusions, Pyroclastics

Flows: The extent, thickness, structure and texture of alkalic lavas is highly variable. They may be centimetres or many metres thick and they may be structureless or show marked jointing. Surface textures include aa, pahoehoe and block types, although the latter is not very common, and rubble surfaces form in lavas whose upper parts are pumiceous or clinkery. Many thick phonolite flows show massive onion skin jointing, and others develop fissility along planes of flow-oriented feldspars. Highly salic phonolite lavas are often viscous and volatile rich, and may flow only short distances from their source as have the lobate tongues of phonolite glass from Mna. Rajada, Las Canadas, Teneriffe.

Intrusive alkalic rocks whose compositions range from ultrabasic to salic occur in several modes: as basement cores in the centres of eroded volcanoes or calderas, for example the basement essexites and other plutonics of Gomera (Canary Isl.); as sheets or sills like those, which include one of differentiated mugearite, that cut the shield lavas of Piton des Neiges (Reunion); as irregular plug-like masses similar to the phonolite intrusions of Teneriffe and of St. Helena, and the trachyte plugs of Tutuila (Samoa); as dyke swarms cutting older volcanics, e.g. the dykes of the Anaga Peninsula, Teneriffe; and as coarse-grained xenoliths whose presence (in alkalic lavas) suggests possible plutonic substructures to many volcanic islands.

Clastic rocks occur as stratified or unstratified ash-lapilli deposits, pumice flows, ignimbrites and eutaxites, as cinder or spatter material.

ACKNOWLEDGMENTS

New and unpublished data made available to me include that for St. Helena (Dr. I. Baker, Centre for Volcanology, Oregon); Ascension Island (Drs. J. D. Bell, P. E. Baker, F. B. Atkins and D. G. W. Smith, of Oxford University); and Reunion Island (Dr. B. Upton of Edinburgh University. In addition, many new analyses of rocks from Teneriffe were made by my Research Assistant (Dr. M. Abbott, now of Flinders University). I am grateful to all these workers for their cooperation. Grants from NERC and the Royal Society, for equipment and transport facilities in respect of my work in Teneriffe, are also acknowledged.

IV.7. REFERENCES

Almeida, F. F. M. de, 1955. Geologia e Petrologia do Arquipelago de Fernando de Noronha. *Brazil Div. Geol. Miner., Monogr.*, **13.**

Almeida, F. F. M. de, 1961. Geologia e Petrologia do Ilha da Trindade. *Brazil Div. Geol. Miner., Monogr.*, **18.**

Assuncao, C. F. Torre de, 1961. Estudo petrografico da Ilha de S. Miguel (Acores). *Comm. dos Serv. Geol. Port.*, **65,** 81–176.

Aumento, F., 1968. The Mid-Atlantic Ridge near 45° N. 11. Basalts from the area of Confederation Peak. *Can. J. Earth Sci.*, **5,** 1–21.

Baker, I., 1968. *The Geology of St. Helena, S. Atlantic.* Unpublished Ph.D. Thesis, Univ. of London.

Baker, I., 1969. Petrology of the Volcanic rocks of St. Helena, S. Atlantic. *Bull. Geol. Soc. Am*, **80,** 1283–310.

Baker, P. E., Gass, I. G., Harris, P. G., and Le Maitre, R. W., 1964. The volcanological report of the Royal Society's Expedition to Tristan da Cunha, 1962. *Phil. Trans. Roy. Soc.*, **256,** 439–578.

Bandy, M. C., 1937. Geology and petrology of Easter Island. *Bull. Geol. Soc. Am.*, **48,** 1589–610.

Borley, G. D., 1966. The Geology of Tenerife. *Proc. Geol. Soc. Lond.*, 1635, 173–4.

Borley, G. D., Suddaby, P., and Scott, P., 1971. Some xenoliths from the alkaline rocks of Teneriffe, Can. Isl. *Contr. Miner. Petrology*, **31,** 102–14.

Bravo, T., 1964. Estudio geologico y petrografico de la Isla de la Gomera. 11. Petrologia y Quimismo de Las rocas volcanicas. *Estud. Geol.*, **20,** 23–56. Inst. Lucas Mallada. C.S.I.C. (Espana).

Broch, O. A., 1946. Two contributions to Antarctic petrography. 1. The lavas of Bouvet Island. *Sci. Res. Norweg. Antarct. Exped. 1927–8*, **25,** 3–26.

Brown, G. M., 1967. 'Mineralogy of Basaltic Rocks', in *Basalts: A treatise on Rocks of basaltic composition. 1.* (Eds. H. H. Hess and A. Poldevaart), 103–62. Interscience.

Bryan, W. B., 1966. History and mechanism of eruption of soda rhyolite and alkali basalt, Socorro Island, Mexico. *Bull. Volc.*, Ser. 2, **29,** 453–80.

Bryan, W. E., 1967. Geology and Petrology of Clarion Island, Mexico. *Bull. Geol. Soc. Am.*, **78,** 1461–76.

Bullard, E., Sir, 1968. The Bakerian Lecture 1967. Reversals of the Earth's Magnetic Field. *Phil. Trans. Roy. Soc. Lond.*, **263,** 481–524.

Cann, J. R., 1967. A second occurrence of dalyite and the petrology of some ejected syenite blocks from Sao Miguel, Azores. *Mineralog. Mag.*, **36,** 227–32.

Chayes, F., 1963. Relative abundance of Intermediate members of the oceanic basalt–trachyte association. *J. Geophys. Res.*, **68,** 1519–33.

Chayes, F. 1964. A petrographic distinction between Cenozoic volcanics in and around the open ocean. *J. Geophys. Res.*, **69,** 1573–88.

Chubb, L. J., 1933. Geology of Galapagos, Cocos and Easter Islands. *Bernice P. Bishop Mus. Bull.*, **110,** 3–43.

Dash, B. P., and Bosshard, E., 1968. Crustal studies around the Canary Islands. *Proc. Sec. 1 (Upper Mantle)*, 249–60. 23rd Inter. Geol. Congr. Prague.

Edwards, A. B., 1938. Tertiary Lavas from the Kerguelen Archipelago. *Rep. B.A.N.Z. Antart. Res. Exp. 1929–31*, **2,** 72–100.

Engel, A. E. J., and Engel, C. G., 1964. Igneous Rocks of the East Pacific Rise. *Science*, **146,** 477–85.

Ernst, T., 1967. 'Olivine nodules and the composition of the Earth's Mantle', in *Mantles of the Earth and Terrestrial Planets* (Ed. S. K. Runcorn), 321–8. Interscience.

Frisch, T., and Schmincke, H. U., 1969. Petrology of clinopyroxene-amphibole inclusions from the Roque Nublo volcanics, Gran Canaria, Canary Islands. *Bull. Volc.*, **33,** 1073–88.

Fuster, J. M., 1954. *Estudio Petrogenetico de Los Volcanes del Golfo de Guinea.* Inst. de Estudio Africanos. C.S.I.C. (Espana).

Fuster, J. M., Hernandez-Pacheco, A., Munoz, M., Rodriguez Badiola, E., and Garcia Cacho, L., 1968a. *Geologia y Volcanologia de Las Islas Canarias: Gran Canaria.* Inst. Lucas Mallada, C.S.I.C. (Espana).

Fuster, J. M., Cendrero, A., Gastesi, P., Ibarrola, E., y Lopez Ruiz, J., 1968b. *Geologia y volcanologia de Las Islas Canarias.* Fuerteventura. Inst. Lucas Mallada, C.S.I.C. (Espana).

Gastesi, P., 1969. Petrology of the ultramafic and basic rocks of Betancuria Massif, Fuerteventura Island, Canary Archipelago. *Bull. Volc.*, **33,** 1008–38.

Gunn, B. M., Coy-Yll, R., Watkins, N. D., Abramson, C. E., and Nougier, J., 1970. Geochemistry of an oceanite-ankaramite-basalt suite from East Island, Crozet Archipelago. *Contr. Miner. Petrology*, **28,** 319–39.

Haggerty, S. E., Borley, G. D., and Abbott, M. J., 1966. Iron–titanium oxides from alkaline lavas of Teneriffe. *Paper read at Inter. Mineralog. Meeting, Cambridge.*

Hausen, H., 1958. On the geology of Fuerteventura. *Sec. Sci. Fenn. Comm. Phys. Math.*, **22.**

Heier, K. S., Chappell, B. W., Arriens, P. A., and Morgan, J. W., 1966. The Geochemistry of four Icelandic basalts. *Norsk Geol. Tidss.*, **46,** 427–37.

Jackson, E. D., 1968. The Character of the Lower Crust and Upper Mantle beneath the Hawaiian Islands. *Proc. Sec. 1. (Upper Mantle)*, 135–50. 23rd Inter. Geol. Congr, Prague.

Jackson, E. D., and Wright, T. L., 1970. Xenoliths in the Honolulu Volcanic Series, Hawaii. *J. Petrology* **11,** 405–30.

Jacobsson, S., 1968. The geology and petrography of the Vestmann Islands. A preliminary report. *Surtsey Research Progress Report IV*, 113–30.

Lambert, I. B., and Wyllie, P. J., 1968. Stability of hornblende and a model for the low velocity zone. *Nature*, **219,** 1240–1.

LeMaitre, R. W., 1962. The Geology of Gough Islands, S. Atlantic. *Bull. Geol. Soc. Am.*, **73,** 1309–40.

LeMaitre, R. W., 1965. The significance of the gabroic xenolibths from Gough Islands, S. Atlantic. *Mineralog. Mag.*, **34** (Tilley Volume), 303–17.

LeMaitre, R. W., 1969. Kaersutite bearing plutonic

xenoliths from Tristan da Cunha, S. Atlantic. *Mineralog. Mag.*, **37**, 185–97.

MacDonald, G. A., 1944a. The 1840 eruption and crystal differentiation in the Kilauean magma column (Hawaii). *Am. J. Sci.*, **242**, 117–89.

MacDonald, G. A., 1944b. Petrography of the Samoan Islands. *Bull. Geol. Soc. Am.*, **55**, 1333–62.

MacDonald, G. A., 1949a. Petrography of the Island of Hawaii: *U.S. Geol. Surv. Prof. Pap.*, **214D**, 51–96.

MacDonald, G. A., 1949b. Hawaiian petrographic province. *Bull. Geol. Soc. Am.*, **60**, 1541–95.

MacDonald, G. A., 1968. Contributions to the petrology of Tutuila, American Samoa. *Geol. Rundschau*, **57**, 821–37.

MacDonald, G. A., and Katsura, T., 1964. Chemical Composition of Hawaiian Lavas. *J. Petrology*, **5**, 82–133.

MacDonald, G. A., and Powers, H. A., 1968. A further contribution to the Petrology of Haleakala Volcano, Hawaii. *Bull. Geol. Soc. Am.*, **79**, 877–88.

MacFarlane, D. J., and Ridley, W. I., 1968. An interpretation of Gravity data for Teneriffe, Canary Islands. *Earth Plan. Sci. Let.*, **4**, 481–6.

Mason, B., 1968. Kaersutite from San Carlos, Arizona with comments on the paragenesis of this mineral. *Mineralog. Mag.*, **36**, 997–1003.

McBirney, A. R., and Gass, I. G., 1967. Relations of oceanic volcanic rocks to mid-oceanic rises and heat flow. *Earth Plan. Sci. Let.*, **2**, 265–76.

Melson, W. G., Jarosewich, E., Cifelli, R., and Thompson, G., 1967. Alkali olivine basalt dredged near St. Paul's Rocks, mid-Atlantic ridge. *Nature*, **215**, 381–2.

Muir, I. D., and Tilley, C. E. 1957. The picrite basalts of Kilauea. (Pt) 1. Contributions to the Petrology of Hawaiian basalts. *Am. J. Sci.*, **255**, 241–53.

Muir, I. D., and Tilley, C. E., 1961. Mugearites and their place in Alkali igneous rock series. *J. Geol.*, **69**, 186–203.

Muir, I. D., and Tilley, C. E., 1963. The tholeiite basalts of Mauna Loa and Kilauea. (Pt) 2. Contributions to the petrology of Hawaiian basalts. *Am. J. Sci.*, **261**, 111–28.

Muir, I. D., and Tilley, C. E., 1964. Basalts from the Northern part of the rift zone of the Mid-Atlantic Ridge. *J. Petrology*, **5**, 409–34.

Neiva, J. M. Cotelo, 1955. Phonolites de L'ile de Prince. *Mem. e Not.*, No. 38, Univ. de Coimbra.

Nicholls, G. D., 1965. Basalts from the deep ocean floor. *Mineralog. Mag.*, **34** (Tilley Volume), 373–88.

O'Hara, M. J., 1965. Primary magmas and the origin of basalts. *Scot. J. Geol.*, **1**, 19–40.

O'Hara, M. J., 1968. The bearing of phase equilibria studies in synthetic and natural systems on the origin and evolution of basic and ultrabasic rocks. *Earth Sci. Rev.*, **4**, 69–133.

O'Hara, M. J., and Yoder, H. S., Jn., 1967. Formation and fractionation of basic magmas at high pressure. *Scot. J. Geol.*, **3**, 67–117.

Part, G. M., 1950. Volcanic rocks from the Cape Verde Islands. *Bull. Brit. Mus. (Nat. His.) Min.*, **1**, 27–72.

Powers, H. A., 1955. Composition and origin of basaltic magma of the Hawaiian islands. *Geoch. Cosmoch. Acta*, **7**, 77–107.

Richardson, C., 1933. Petrology of the Galapagos Islands. *Bernice P. Bishop Mus., Bull.*, **110**, 45–67.

Ridley, W. I., 1968. *The geology of Las Canadas area, Teneriffe.* Unpub. Ph.D. thesis, University of London.

Robson, G. R., and Spector, J., 1962. Crystal fractionation of the Skaergaard type in modern Icelandic magmas. *Nature*, **193**, 1277–8.

Roedder, E., and Coombs, D. S., 1967. Immiscibility in granitic melts, indicated by fluid inclusions in ejected granitic blocks from Ascension Island. *J. Petrology*, **8**, 417–52.

Rothe, P., and Schmincke, H-U., 1968. Contrasting origins of the eastern and western islands of the Canarian Archipelago. *Nature*, 218, 1152–4.

Schilling, J. G., and Winchester, J. W., 1967. 'Rare Earth fractionation and magmatic processes', in *Mantles of the Earth and Terrestrial Planets* (Ed. S. K. Runcorn). Interscience. 267–85.

Schmincke, H-U., 1968. Geologic framework and origin of alkali trachytic to alkali rhyolitic ignimbrites on Gran Canaria, Canary Islands. *Paper read at Inter. Symp. Volcan., Canary Islands, 1968.*

Schmincke, H-U., and Swanson, D. A., 1967. Ignimbrite origins of eutaxites from Teneriffe, Canary Islands. *Neus Jb. Geol. Palaont. Mh.*, 700–3.

Sigurdsson, H., 1969. Transitional and alkali basalt series in western Iceland (Abstr.). *Proc. Geol. Soc. Lon.*, **1685**, 266–7.

Stearns, H. T., 1944. Geology of the Samoan Islands. *Bull. Geol. Soc. Am.*, **55**, 1279–332.

Stice, G. D., 1968. Petrography of the Manu'a Islands, Samoa. *Contr. Miner. Petrology*, **19**, 343–57.

Thornton, C. P., and Tuttle, O. F., 1960. Chemistry of Igneous Rocks: 1. Differentiation Index. *Am. J. Sci.*, **258**, 664–84.

Tilley, C. E., and Muir, I. D., 1964. Intermediate members of the oceanic basalt–trachyte association. *Geol. Foren. Forhand.*, **85**, 434–43.

Tilley, C. E., and Scoon, J. H., 1961. Differentiation of Hawaiian basalts–trends of Mauna Loa and Kilauean historic magma. *Am. J. Sci.*, **259**, 60–8.

Tyrrell, G. W., 1937. Petrology of Heard Island. *B.A.N.Z. Ant. Res. Exp. Rep. (Ser. A)*, **2**, pt. 3.

Upton, B. G., and Wadsworth, W. J., 1966. The Basalts of Reunion Island, Indian Ocean. *Bull. Volcan.*, **29**, 7–23.

Upton, B. G., and Wadsworth, W. J., 1967. A complex basalt–mugearite sill in Piton des Neiges Volcano, Reunion. *Am. Miner.*, **52**, 1475–92.

Upton, B. G., Wadsworth, W. J., and Newman, T. C. 1967. The petrology of Rodriguez Island, Indian Ocean. *Bull. Geol. Soc. Am.*, **78**, 1495–506.

Walker, F., and Nicolaysen, L. O., 1954. The petrology of Mauritius. *Colon. Geol. Miner. Res.*, **4**, 3–44.

Watkins, N. D., and Haggerty, S. E., 1967. Primary oxidation variation and petrogenesis in a single lava. *Contr. Miner. Petrology*, **15**, 251–71.

White, R. W., 1966. Ultramafic inclusions in basaltic rocks from Hawaii. *Contr. Miner. Petrology*, **12**, 245–314.

Williams, H., 1966. Volcanic history of the Galapagos Archipelago. *Bull. Volcan.*, **39**, 27–8.

Wilson, R. L., and Watkins, N. D., 1967. Correlation of petrology and natural magnetic polarity in Columbia Plateau basalts. *Geophys. J.*, **12**, 405.

Yoder, H. S., Jn., and Tilley, C. E., 1962. Origin of basaltic magmas: an experimental study of natural and synthetic rock systems. *J. Petrology*, **3**, 342–532.

IV.8. PERALKALINE AND ASSOCIATED RING-COMPLEXES IN THE NIGERIA–NIGER PROVINCE, WEST AFRICA

P. Bowden and D. C. Turner

IV.8.1. Introduction

The Nigeria–Niger province is made up of numerous high level, nonorogenic, granitic ring-complexes confined to a narrow intrusive zone, extending for 1300 km from the northern Aïr region of the Niger Republic to the margin of the Benue valley in Nigeria. There are three geographical groupings; Aïr, southern Niger and central Nigeria, which are separate regions of uplifted Precambrian basement surrounded by younger sedimentary rocks (Fig. 1).

A description of the Jurassic younger granite province in Nigeria has been given by Jacobson, MacLeod and Black (1958) and more detailed accounts of the central and north-western parts of this region are to be published by the Geological Survey of Nigeria (MacLeod *et al.*, 1971, Buchanan *et al.*, 1971, Jacobson and MacLeod, in press). The geology of the granitic ring-complexes of Aïr has been summarized by Black, Jaujou and Pellaton (1967). In a recent review of Palaeozoic to recent igneous activity in West Africa, Black and Girod (1970) have described the tectonic setting and the structural relations of the Nigeria–Niger province in some detail. Accordingly, the present review deals only briefly with these aspects and is largely concerned with the petrography, geochemistry and petrogenesis of the province.

Nomenclature. This chapter uses the same nomenclature as that followed in Chapter II.6, with minor modifications to suit the rock types found in the Nigeria–Niger province. The peralkaline silicic rocks have a distinctive mineralogy of alkaline affinities with *ac* and sometimes *ns* appearing as CIPW normative minerals. Many such granites in the Nigeria–Niger province have been described as riebeckite granites but this term requires revision in the light of chemical and experimental studies; instead the term *riebeckic-arfvedsonite* granite is used for the sodic amphibole-bearing peralkaline granites in this chapter. The non-peralkaline granites which have *an* and/or *c* in their CIPW norm are best described by reference to their mafic mineralogy, e.g. *hastingsite granites* (metaluminous) and *biotite granites* (peraluminous).

IV.8.2. Mode of Occurrence and Associations

IV.8.2.1. Granites

Granitic ring-complexes vary greatly in size and structural complexity. The most typical ones

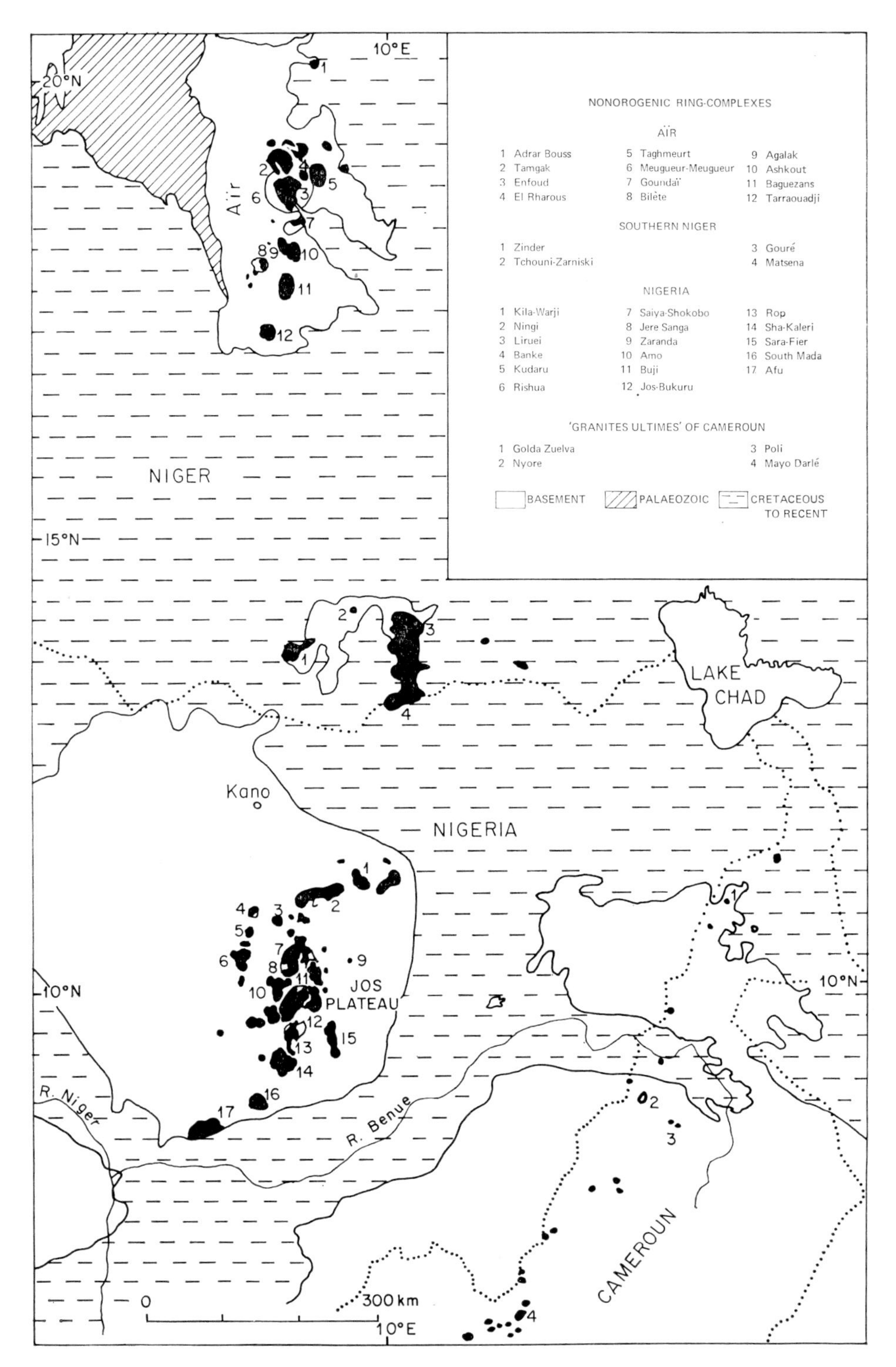

Fig. 1 Distribution of ring-complexes in the Nigeria–Niger province

are circular or elliptical in outline, 10 to 25 km in diameter and are composed of an outer ring dyke, surrounding down-faulted basement and volcanic rocks or, where deeper levels are exposed, a series of concentric granite intrusions. Many of the larger massifs are made up of a cluster of overlapping ring complexes. Such superimposed ring structures are common in the northern Aïr and Jos Plateau areas which are the two focal regions of younger granite magmatism. Sometimes overlapping ring-complexes have a linear arrangement, and alignments of separate complexes can also be distinguished along two dominant directions, N.N.E.–S.S.W. and N.N.W.–S.S.E. Such alignments, however, are not conspicuous features and are not accompanied by linear dyke swarms or other strongly directional intrusions; igneous activity appears to have been largely confined to the central complexes.

The ideal intrusive sequence in central Nigeria is illustrated by the Liruei complex (Fig. 2a) with fayalite granite (3) → peralkaline granites (4a, 4b, 4c) → biotite granites (5a, 5b). Overlapping ring-complexes (Fig. 2b) have separate centres but exhibit occasional reversals of the normal order of intrusion such as in centre IV and centre V. But these sequences are often modified in Aïr and southern Niger where the centres are dominated by peralkaline granites. For instance at Tarraouadji (Rocci, 1960) early peralkaline intrusions of aegirine granite appear instead of hastingsite granite, and biotite granite is absent from many of the Aïr complexes.

IV.8.2.2. Syenites

Syenite is an important association in several of the Aïr complexes and is abundant at Zinder (Black, 1963) and in the northern region of central Nigeria. Many of these centres contain syenite intrusions which are often preceded by rhyolite volcanics and followed by granites. For example, at Goundaï, Bilète and Jere-Sanga, syenite occupies a similar position at the beginning of an intrusive sequence as the fayalite-bearing granite in the exclusively granitic complexes. Repeated syenite–granite cycles are also found in the Nigeria–Niger province (Turner,

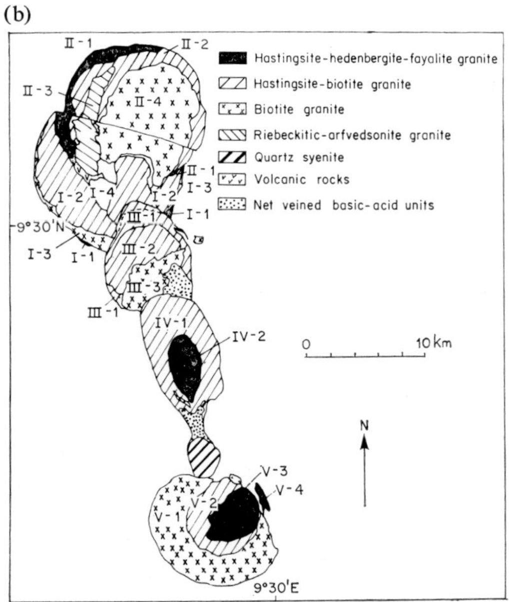

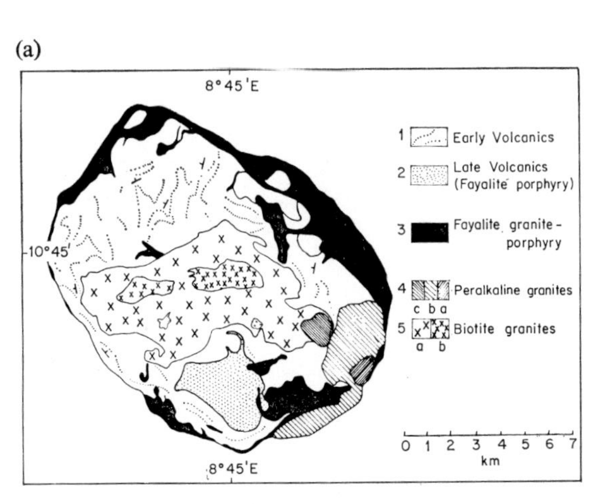

Fig. 2 (a) Liruei complex, Nigeria. Ideal magmatic sequence modified from Jacobson *et al.* (1958; p. 29); (b) Overlapping ring-complexes at Sara–Fier, Nigeria (after Turner, 1963). Roman numerals indicate the relative order of intrusion of the centres; arabic numerals give the intrusive sequence in each centre

1968). All the syenites contain small amounts of modal quartz but only the quartz-rich varieties attain peralkaline compositions.

IV.8.2.3. Anorthosites, Gabbros, Basalts and Hybrid Intrusions

Basic rocks both predate and postdate the silicic magmatic activity. The earliest types include minor basic lavas, xenoliths in certain granites, notably the early ring dykes, and small hybrid gabbroic masses such as the net veined basic–acid units of the Sara–Fier complex (Fig. 2b). However, the most abundant early basic rocks are sheet-like masses of anorthosite, found in Aïr occupying a total area of over 400 km^2 mainly in the Enfoud complex, but also in Ashkout, Tamgak and Adrar Bouss. The anorthosites represent the earliest intrusive rocks in each complex. Occasionally they are associated with small bodies of diorite, and are cut by later quartz syenites, peralkaline and non-peralkaline granites. Minor basaltic or trachybasalt dykes usually postdate the granite emplacement.

Two larger intrusions are of exceptional interest. The first is the Meugueur–Meugueur ring dyke which has a diameter of 65 km surrounding the Enfoud massif in Aïr and is composed of olivine gabbro accompanied by minor amounts of syenite. The second is a dyke cutting across the Jos–Bukuru complex which forms a 90° arc to a ring 60 km in diameter. It consists mainly of hastingsite granite-porphyry, but in places it has basic margins and a high proportion of doleritic xenoliths ranging in size from a few millimetres to blocks exceeding two feet in diameter. These large dykes are typically hybrid intrusions.

IV.8.2.4. Regional Differences Within the Nigeria–Niger Province

The main difference between the northern and southern sectors is in the relative abundance of peralkaline granites. In Aïr peralkaline (riebeckitic-arfvedsonite and aegirine) granites are common (nearly 50% by area) whilst non-peralkaline granites are very rare. In Nigeria, however, biotite granites are dominant (45% by area) and peralkaline granites relatively minor. This difference is of great practical importance because of the economic cassiterite and columbite mineralization associated with the biotite granites. Furthermore, within Nigeria, peralkaline granites become progressively less important southwards where the complexes of the Jos Plateau and near the Benue valley consist almost entirely of biotite granites and hastingsite granites.

The largest areas of volcanic rocks in Nigeria are found at the northern complexes, notably Ningi, Kila–Warji, Banke, Liruei and Saiya–Shokobo. Further south, volcanics generally are preserved only as small arcuate remnants between granite ring intrusions. This difference can be partly accounted for by regional change in depth of erosion related to north-eastward tilting towards the Chad basin and increasing erosion on approaching the Benue trough. Regional variations in post-granite erosion, however, does not in itself explain why certain complexes are preserved at high structural levels, especially as all the volcanic rocks have apparently undergone cauldron subsidence to unknown but presumably varying extents. Adjacent complexes can show contrasting structural levels. An example of this is the Amo–Buji massif (Nos. 10 and 11, Fig. 1) which is made up of two small contiguous ring-complexes, Buji consisting largely of rhyolites whilst Amo is composed of a series of concentric granite intrusions with near-vertical structures. In Aïr two complexes, Goundaï and Bilète, are distinct in having a high level volcanic character with rhyolite plateaux resting on basement within surrounding polygonal syenite ring-dykes, but neither complex has a central granite intrusion.

Regional variations are partly explained by differences in the development of two associations: rhyolite → syenite → peralkaline granite; and hastingsite granite → biotite granite. The latter group is generally found at deeper levels with many of the biotite granites forming massive irregular intrusions discordant to earlier ring structures of pyroxene granite or amphibole granite. However in Aïr large intrusions of riebeckitic-arfvedsonite granite and aegirine granite are found without associated rhyolites, for example, in the Baguezans and Taghmeurt massifs illustrating that an effusive cycle is not always present in peralkaline and associated complexes.

There are also significant differences in the distribution of non-granitic rocks; syenites are common in the Aïr and southern Niger, but in Nigeria they occupy less than 5% of the total area, most of which is in the more northerly complexes; anorthosites, quite widespread in Aïr, are absent from southern Niger and Nigeria.

IV.8.2.5. Younger Granites of the Cameroun Volcanic Line

Closely related to the Nigeria–Niger province is the *granites ultimes* province of Cameroun. About 40 small massifs ranging from 1 to 10 km in diameter, are aligned in a pronounced N.E.-S.W. direction for 1000 km, which almost coincides with the mainland part of the São Tomé–Fernando Poo–Cameroun volcanic trend. They form ring structures and are composed of rock types often identical with those of the Nigeria–Niger province, including riebeckitic-arfvedsonite granites, biotite granites, syenites and minor gabbros. Traces of tin mineralization are associated with some of the biotite granites, notably in the Mayo Darlé complex (Koch, 1959; Chaput *et al.*, 1954). These *granites ultimes* have been dated as Upper Cretaceous to Lower Tertiary (Lasserre, 1966, 1969).

IV.8.3. Petrology and Mineralogy

Acidic rocks occupy over 95% of the total area of the Nigeria–Niger province. They are extremely varied in texture and coloured mineral content, but have an overall unity through the dominance of perthitic alkali feldspars and through their small composition range in major element chemistry. Textural variations depend on grain size, the relative order of crystallization of felsic and mafic minerals and the degree of late magmatic and post magmatic recrystallization. Through the large number of rock units, both intrusive and extrusive, it is possible to trace a complete progression from originally glassy, almost aphyric rocks, to porphyries, granite-porphyries and granites. The variation in the coloured mineral content is dependent on the peralkalinity of the granites. Peralkaline granites typically contain riebeckitic-arfvedsonite and aegirine whereas the non-peralkaline granites exhibit members of the series fayalite → hedenbergite → hastingsite → biotite. The two groups differ in relative order of crystallization of felsic and mafic components. In the non-peralkaline granites fayalite, hedenbergite and hastingsite are early formed, occurring as phenocrysts in the porphyries and prismatic euhedral grains in granites. In the peralkaline granites riebeckitic-arfvedsonite and aegirine are often late formed and poikilitic, enclosing crystals of quartz and feldspar; this is the same sequence as in the agpaitic crystallization of the undersaturated rocks.

Typical modal analyses for peralkaline and non-peralkaline granites are given in Table 1.

IV.8.3.1. The Peralkaline Granites

These rocks are normally medium to coarse-grained, even-textured, white in colour; they often develop a marginal pegmatitic facies containing abundant large amphibole crystals. Astrophyllite and aenigmatite are both occasionally present as major constituents. The only constantly occurring accessory mineral is zircon, as both clear and metamict varieties, and in some granites apparently late-formed, as open clusters of grains at crystal boundaries. The albite-rich granites contain pyrochlore as a typical accessory and also have a number of interstitial and veining minerals such as fluorite, cryolite, thomsenolite and amblygonite.

Alkali amphiboles have been previously described as either riebeckite (X = very deep blue, Y = light blue, Z = light green-yellow) or arfvedsonite (X = blue-green, Y = greenish brown, Z = blue grey) with $Z = b$ instead of $Y = b$. Riebeckite appears in peralkaline granites containing 'free' albite, which are often associated with partly albitized non-peralkaline granites. But in spite of optical differences, all alkali amphiboles in peralkaline granites actually

TABLE 1 Modal Analyses of Representative Granites from the Nigeria–Niger Province

	Peralkaline Granites				Non-Peralkaline Granites			
Quartz	22·8	31·8	31·4	24·4	25·0	28·0	33·9	35·0
Perthite	55·4	57·0	22·0	54·5	66·0	51·0	41·8	37·0
'Free' albite	8·7	4·4	35·2	15·2	2·5	16·0	18·8	25·0
Riebeckitic-arfvedsonite	1·3	3·9	3·9	4·2	—	—	—	—
Aegirine	5·3	2·6	—	—	—	—	—	—
Fayalite	—	—	—	—	0·7	—	—	—
Hedenbergite	—	—	—	—	2·9	—	—	—
Hastingsite	—	—	—	—	1·1	2·0	—	—
Biotite	—	—	—	1·7	—	3·3	5·2	2·2
Iron oxides	—	—	—	—	0·8	0·3	—	0·1
Others	0·5	0·3	7·5	—	—	—	0·2	0·3

belong to a continuously varying series ranging from arfvedsonite to riebeckitic-arfvedsonite. The riebeckitic-arfvedsonites have a higher ferric iron concentration than arfvedsonite (Sundius, 1946; Miyashiro, 1957), a low MgO content, and structural formula which does not correspond to riebeckite. To illustrate important atomic variations in the Nigeria–Niger sodic amphiboles, riebeckitic-arfvedsonites from intrusive sequences at the Liruei and Tarraouadji complexes are plotted in Fig. 3. They trend with increasing Na substitution away from the Ca-rich arfvedsonite formula (AM) towards the Sundius arfvedsonite end-member composition (AS), whilst their host granites also show a parallel increase in normative *ac* and *ns*. The most sodic-rich amphibole in this series, from a strongly peralkaline albite granite, is optically a riebeckite but chemically an arfvedsonite with an excess of alkalis and ferric iron, Riebeckitic-arfvedsonites are usually found in thoroughly recrystallized granites containing a high proportion of microcline, albite and clear unstrained quartz. It is therefore suggested that riebeckitic-arfvedsonites in the Nigeria–Niger peralkaline granites crystallized initially as arfvedsonite, and were subsequently modified by magmatic hydrothermal solutions of acmitic composition (cf. II.6.3).

In partly recrystallized granites late magmatic and post magmatic changes are indicated by the occurrence of an arfvedsonite amphibole with a

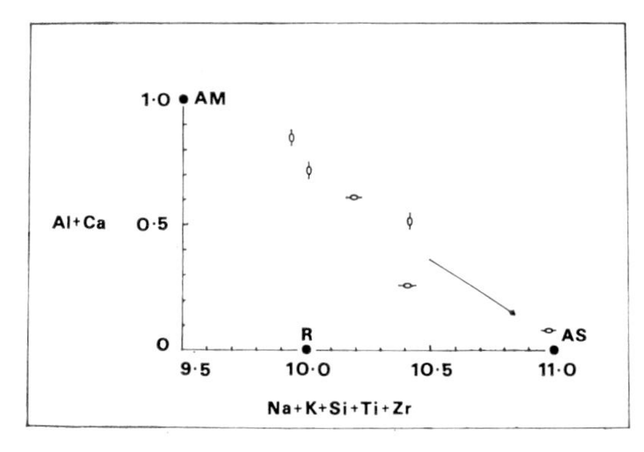

Fig. 3 Atomic variations in Al + Ca *v.* Na + K + Si + Ti + Zr for riebeckitic-arfvedsonites from Liruei and Tarraouadji, Nigeria–Niger province

Closed circles: AM natural arfvedsonite (Miyashiro, 1957); AS end-member arfvedsonite (Sundius, 1946); R end-member riebeckite.

Open circles with vertical lines: Tarraouadji, Aïr.

Open circles with horizontal lines: Liruei, Nigeria.

Arrow shows trend of amphibole compositions with increasing normative *ac* and *ns* in their host granites.

blue-green to intensely blue sodic variety penetrating the cleavage traces and fractures. It is often intergrown with aegirine and accompanied by minor amounts of 'free' albite and strained quartz. Amphiboles with a hastingsite core and a rim of arfvedsonite are common in the granite porphyries. These are of primary crystallization and represent a transition from the hastingsite syenites to the trend which is found in peralkaline granites.

Aegirine is closely associated with sodic amphibole in the peralkaline granites and often shows similar textural relationships. But occasionally aegirine may be the major mineral or even the only mafic mineral. Sometimes aegirine is clearly of late formation, appearing for example at quartz–amphibole intergrowths. However in some peralkaline granites fibrous aegirine appears to pseudomorph early euhedral crystals of hedenbergite, whilst the granite porphyries exhibit pyroxene phenocrysts with cores of hedenbergite coated with zones of aegirine. Analyses of aegirine from Nigeria presented in Table 2 closely approach the acmite end-member (Sabine, 1950). They have a high content of soda and ferric iron and may represent essentially hydrothermal pyroxenes developed at the late magmatic and post magmatic stages.

Astrophyllite has been described from a number of localities in the Nigeria–Niger province but its concentration to major modal proportions is restricted to regions of high concentration of volatiles probably near the roof of the intrusions. This effect is particularly well demonstrated in Nigeria where the Amo outer ring-dyke contains a roof pendant of basement against which astrophyllite is developed extensively. In the more normal facies of this granite, astrophyllite is only a trace constituent, frequently intergrown either with annitic mica or riebeckitic-arfvedsonite. There is also a proportional decrease of astrophyllite as the mica content increases. This relationship is compatible with the crystallographic investigation by Woodrow (1967) who has shown that in general, astrophyllite is closely related structurally to biotite but contains octahedrally co-ordinated titanium in the tetrahedral layer.

IV.8.3.2. Non-Peralkaline Granites

The most widely distributed non-peralkaline rock types are the granites and porphyries containing fayalite, hedenbergite and hastingsite, which may show transitions towards a peralkaline composition or towards hastingsite granite, as in the massive coarse-grained intrusions of the Jos Plateau. Through replacement of hastingsite by biotite there is a gradation into similar textured biotite granite. However, the most characteristic biotite granites are medium grained rocks showing no relict traces of amphibole. These granites include types with orthoclase and/or microcline perthite with only minor 'free' albite and granites with largely exsolved potash feldspar and a marked albite enrichment. The albitized biotite granites are also enriched in accessory columbite and to a lesser extent in cassiterite. A second form of local alteration of the biotite granites is greisenization, the end product of which is a quartz-mica rock, frequently cassiterite-bearing.

All of the mafic minerals in the non-peralkaline granites are iron-rich with very low magnesium (except where contamination by basic rocks has occurred). Only biotites show distinctive compositional variations.

IV.8.3.3. Biotite in Peralkaline and Non-Peralkaline Granites

The biotites in the Nigeria–Niger province fall broadly into two groups. They belong to related series of trioctahedral micas (Foster, 1960): the biotites of the perthitic peralkaline and non-peralkaline granites are iron-rich members of the phlogopite–annite–siderophyllite series; those of the albitic granites together with the somewhat similar, paler coloured micas of the greisens are lithium-poor members of the siderophyllite – protolithionite – zinnwaldite – lepidolite isomorphous series. The red-brown biotite (Table 2, analysis 10) associated with riebeckitic-arfvedsonite in granites of intermediate peralkalinity, has a low Al content and approaches most closely the annite end-member.

IV.8.3.4. Felsic Minerals in Peralkaline and Non-Peralkaline Granites

Quartz forms a textural series, appearing as isolated bipyramidal phenocrysts in the porphyries, as loose chains of phenocrysts in the granite-porphyries, and as trails and clusters of anhedral grains in the granites. Quartz is generally interstitial in the relatively minor quartz-poor granites.

Anorthoclase is a phenocryst mineral in some porphyries. In the granites alkali feldspars often show varied exsolution effects such as lamellar orthoclase perthites with a minor development of separate albite at feldspar–feldspar interfaces, patchy perthites accompanied by large late-formed albite crystals, and in a few cases separate microcline and albite in essentially two-feldspar granites. Fabriès and Rocci (1965) have found similar lamellar perthites in both biotite granites and riebeckitic-arfvedsonite granites to have a compositional range from Or_{56} to Or_{36} and a variation in $2V_x$ from 56° to 65°, making them intermediate between sanidine-high albite and orthoclase-low albite. The anorthite content of these perthites is 0·8 to 3·3% and the feldspars in almost all of the granites have a similarly low An content. Microcline appears in the perthites of some peralkaline and non-peralkaline granites and is particularly characteristic of granites that have undergone late stage albitization. Fabriès and Rocci have suggested that perthites in albitic-biotite granites contain both monoclinic and triclinic alkali feldspar. Generally the potash feldspar of the perthites is cloudy (kaolinized) while the albite is clear but in some riebeckitic-arfvedsonite granites small amounts of clear, late crystallized microcline are present with a $2V_x = 80°$, $Z{:}b = 18°$ (maximum microcline). While most feldspar relations can be attributed to recrystallization accompanying exsolution, a few generally small intrusions of peralkaline and non-peralkaline granite have a significantly higher albite content in the form of small lath-shaped crystals replacing microcline and quartz, apparently of hydrothermal origin (see II.6.3.1). These granites are the most highly mineralized in the province.

IV.8.4. Geochemistry

IV.8.4.1. Major Element Chemistry

Representative analyses of peralkaline and non-peralkaline granites, and syenites from various complexes in the Nigeria–Niger province tabulated in Table 3, are used together with other previously published data to compile Figs 4a, 4b and 4c. If only the normative salic minerals are considered the plutonic acid rocks are a closely unified group of hypersolvus alkali granites. This is illustrated in Fig. 4a, where the granites trend towards the minimum temperature composition at a water vapour pressure of 500 kg/cm² in the system $NaAlSi_3O_8$–$KAlSi_3O_8$–SiO_2–H_2O. But peralkaline albitized granites do not lie near this minimum or near the minimum for 4·5% acmite + 4·5% sodium metasilicate (at 1,000 kg/cm²) as would be expected. Instead they lie outside this magmatic system (e.g. points 2 and 3 on Fig. 4a). Similarly the albitized non-peralkaline granites (such as point 1 on Fig. 4a) do not plot near the low temperature minimum for 500 kg/cm² water vapour pressure. Perhaps these deviations may be due to the effects of hydrothermal modification.

In Fig. 4b the granites of both peralkaline and non-peralkaline compositions lie within the alkali feldspar field; the syenites lie above the boundary curve in the plagioclase field.

Probably the most significant chemical feature of the province is that slight differences in the atomic proportions of Na, K and Al can produce such striking changes in the mineralogical composition of the suite. This is depicted in Fig. 4c where the syenites and granites are separated by the bisectrix into peralkaline and non-peralkaline groups. But despite the alkaline nature of the province, the Na and K values are not particularly high, only the albite-rich peralkaline granites have a significantly high Na concentration. Thus the peralkaline character of the Nigeria–Niger province is due to a deficiency of Al relative to Na and K. Other chemical characteristics include a low Ca and Mg content: F is an important constituent in albitized granites, whilst Cl is also recorded in peralkaline ash-flow tuffs.

IV.8.4.2. Trace-Element Chemistry

It is now recognized that certain trace elements such as Zr are selectively concentrated in peralkaline rocks whilst other elements like Li and Rb exhibit a dual rôle concentrating in both

TABLE 2 Analyses of Principal Mafic Minerals in Nigeria–Niger Peralkaline Granites

	1	2 Aegirine	3	4	5	6 Riebeckitic–Arfvedsonite	7	8	9	10 Biotite
Host granite	Albite riebeckitic-arfvedsonite granite	Riebeckitic-arfvedsonite aegirine granite	Albite riebeckitic-arfvedsonite granite	Peralkaline granite		Riebeckitic-arfvedsonite microgranite	Riebeckitic-arfvedsonite aegirine granite	Aegirine riebeckitic-arfvedsonite granite	Albite riebeckitic-arfvedsonite granite	Riebeckitic-arfvedsonite biotite granite
Complex	Kigom Hills	Liruei	Liruei	Tarraouadji	Tarraouadji	Tarraouadji	Liruei	Liruei	Liruei	Amo
SiO_2	51·92	50·05	50·66	45·90	45·85	48·04	47·45	48·09	50·25	35·14
ZrO_2	n.d.	0·16	0·15	n.d.	n.d.	n.d.	0·28	0·43	0·56	n.d.
Al_2O_3	1·85	0·45	0·59	2·55	2·20	2·70	0·66	0·60	0·26	6·44
TiO_2	0·77	1·58	0·65	2·20	3·00	1·60	1·61	1·74	0·80	2·87
Fe_2O_3	31·44	29·95	33·27	14·80	13·70	13·82	12·65	10·88	13·95	4·40
FeO	0·75	2·89	0·82	21·40	21·60	20·77	21·41	25·33	17·09	34·92
MnO	n.d.	0·33	0·08	0·55	0·75	0·80	0·98	0·62	0·62	0·53
MgO	n.d.	tr.	0·05	0·24	0·48	1·06	0·03	0·05	0·04	0·43
CaO	n.d.	1·91	0·29	2·20	1·66	0·10	2·88	0·87	0·02	0·97
Na_2O	12·86	12·59	13·65	6·40	6·40	7·60	6·62	7·41	8·80	0·74
K_2O	0·19	0·28	0·23	1·18	1·07	1·46	1·82	1·01	2·20	8·92
Li_2O	n.d.	n.d.	n.d.	n.d.	n.d.	n.d.	0·42	0·53	2·20	n.d.
P_2O_5	n.d.	n.d.	n.d.	0·09	0·09	0·04	0·08	0·08	0·16	0·03
H_2O^+	0·17	n.d.	n.d.	2·98	2·52	1·67	1·70	1·30	0·40	3·12
H_2O^-	n.d.	0·06	0·03	0·28	0·28	0·36	0·22	0·15	0·09	0·06
F	n.d.	n.d.	n.d.	n.d.	n.d.	n.d.	1·40	1·03	3·31	2·97
Cl	n.d.	n.d.	n.d.	n.d.	n.d.	n.d.	0·07	0·12	0·18	n.d.
							100·61*	100·68*	102·22*	101·04
Less O=F, Cl							0·61	0·46	1·43	1·04
Total	99·95	100·25	100·47	100·77	99·60	100·02	100·00	100·22	100·79	100·00

	Greenwood (1951)	Macleod *et al.* (1971)	Macleod *et al.* (1971)	Fabriès and Rocci (1965)	Fabriès and Rocci (1965)	Fabriès and Rocci (1965)	Borley (1963)	Borley (1963)	Borley (1963)	Macleod *et al.* (1971)
α'	1·770	n.d.	n.d.	1·684	1·690	1·684	1·694	1·691	1·677	n.d.
β'	1·812	n.d.	n.d.	1·691	1·694	1·692	n.d.	n.d.	n.d.	n.d.
γ'	1·830	n.d.	n.d.	1·696	1·697	1·695	1·702	1·696	1·693	n.d.
$(\gamma'-\alpha')$	0·060	n.d.	n.d.	0·012	0·007	0·011	0·008	0·005	0·016	n.d.
$2V\alpha'$	n.d.	n.d.	n.d.	84–86°	88°	77–89°	n.d.	n.d.	n.d.	n.d.
$\alpha:z$	5–6°	n.d.	n.d.	6°	5°	0°	0°	0°	4°±1°	n.d.

Number of Ions on the Basis of 6 Oxygens

	Greenwood (1951)	Σ	Macleod *et al.* (1971)	Σ	Macleod *et al.* (1971)	Σ
Si	1·98		1·95		1·96	
Al	0·02	2·00	0·02	2·00	0·03	2·00
Ti	—		0·03		0·01	
Al	0·06		—		—	
Ti	0·02		0·02		0·01	
Fe^{III}	0·90		0·88		0·97	
Fe^{II}	0·02		0·09		0·03	
Mn	—		0·01		—	
Mg	—	1·96	—	2·04	—	2·05
Li	—		—		—	
Ca	—		0·08		0·01	
Na	0·95		0·95		1·02	
K	0·01		0·01		0·01	

Number of Ions on the Basis of 24 (O, OH, F, Cl)

	Fabriès and Rocci (1965)	Σ	Fabriès and Rocci (1965)	Σ	Fabriès and Rocci (1965)	Σ	Borley (1963)	Σ	Borley (1963)	Σ	Borley (1963)	Σ	Macleod *et al.* (1971)	Σ
Si	7·42		7·41		7·60		7·56		7·68		7·77		5·98	
Al	0·48		0·43		0·40		0·12		0·11		0·05		1·28	
Ti	0·10	8·00	0·16	8·00	—	8·00	0·19	8·00	0·18	8·00	0·09	8·00	—	8·00
Zr	—		—		—		0·02		0·03		0·04		—	
Fe^{III}	—		—		—		0·11		—		0·05		0·54	
Al	—		—		0·09		—		—		—		—	
Ti	0·17		0·21		0·19		—		0·03		—		0·36	
Fe^{III}	1·80		1·67		1·63		1·41		1·30		1·57		—	
Fe^{II}	2·88	4·98	2·91	4·99	2·73	4·98	2·85	4·66	3·37	5·13	2·21	5·18	4·91	5·44
Mn	0·06		0·09		0·10		0·13		0·08		0·08		0·07	
Mg	0·07		0·11		0·24		—		0·01		0·01		0·10	
Li	—		—		—		0·27		0·34		1·31		—	
Ca	0·37		0·29		0·02		0·49		0·15		0·03		0·17	
Na	2·00	2·62	2·00	2·52	2·33	2·65	2·05	2·91	2·29	2·64	2·64	3·11	0·24	2·32
K	0·25		0·23		0·30		0·37		0·20		0·44		1·91	
OH	1·39		1·55		1·58		1·49		1·38		0·41		3·50	
F	—		—		—		0·70	2·21	0·52	1·93	1·62	2·08	1·30	
Cl	—		—		—		0·02		0·03		0·05		—	
							* Total includes ZnO = 0·33		* Total includes ZnO = 0·43		* Total includes ZnO = 0·93 PbO = 0·18			

Tr. trace: n.d. not determined.

TABLE 3 Representative Peralkaline and Non-Peralkaline Analyses with CIPW Norms

	Peralkaline Granites				Syenites		Non-Peralkaline Granites			
Rock Type	1 Arfvedsonite granite	2 Aegirine riebeckitic arfvedsonite and aenigmatite granite	3 Albite riebeckitic arfvedsonite granite	4 Riebeckitic-arfvedsonite biotite granite	5 Quartz hastingsite syenite	6 Hastingsite augite syenite	7 Hastingsite fayalite granite	8 Hastingsite biotite granite	9 Biotite granite	10 Albite biotite granite
Complex	Sha–Kaleri	Gouré	Liruei	Amo	Jere–Sanga	Zarniski	Rop	Shere (Jos)	Tarraouadji	Jos–Bukuru
SiO_2	73·16	74·05	71·38	76·84	61·43	60·25	71·51	72·90	72·50	76·19
TiO_2	0·27	0·35	0·07	0·07	0·45	0·95	0·30	0·33	0·15	0·04
Al_2O_3	12·07	11·20	12·34	11·24	16·61	15·45	12·91	12·41	13·65	13·51
Fe_2O_3	1·86	2·10	1·96	1·06	1·09	2·80	1·14	1·80	1·25	0·43
FeO	1·99	1·70	0·91	1·27	5·49	4·37	2·87	1·71	1·40	0·61
MgO	tr.	0·40	0·16	0·11	0·68	1·20	0·15	0·34	0·08	0·08
CaO	0·50	0·45	0·17	0·54	3·64	2·50	1·16	0·89	0·65	0·10
Na_2O	4·57	4·60	7·17	4·29	4·53	5·85	4·17	4·11	4·50	4·23
K_2O	4·79	5·20	4·17	4·28	5·11	5·35	5·12	4·94	4·85	4·61
P_2O_5	0·02	0·06	0·01	0·01	0·19	0·36	0·04	0·04	0·02	tr.
MnO	0·07	0·10	0·05	0·02	0·27	0·28	0·09	0·09	0·03	0·01
F	0·06	n.d.	1·08	0·04	0·04	n.d.	0·09	0·39	n.d.	0·22
Cl	tr.	n.d.	n.d.	0·07	0·02	n.d.	0·02	n.d.	n.d.	0·02
H_2O^+	0·20	0·42	0·30	0·08	0·47	0·77	0·26	0·63	}0·95	0·13
H_2O^-	0·20	0·10	0·12	0·04	0·22	0·13	0·16	n.d.		0·15
	99·76		100·53+	99·96	100·37+		99·99	100·58		100·42+
Less O=F, Cl	0·02		0·48	0·03	0·03		0·04	0·16		0·09
Total	99·74	100·73	100·05	99·93	100·34	100·26	99·95	100·42	100·03	100·33

Specific gravity		2·61	n.d.	2·67	2·64	2·71	n.d.	2·65	n.d.	2·57	2·61
Q		27·36	28·68	22·11	35·05	5·32	—	24·39	28·25	25·80	33·52
or		28·30	30·72	24·64	25·29	30·19	31·61	30·25	29·19	28·65	27·24
ab		35·41	28·66	40·25	33·98	37·28	49·48	34·97	34·76	38·06	34·74
an		—	—	—	—	10·45	0·11	1·55	0·83	2·73	—
C		—	—	—	—	—	—	—	—	—	1·55
ac		2·85	6·07	5·67	1·11	—	—	—	—	—	—
ns		—	0·78	3·25	—	—	—	—	—	—	—
ol	fo	—	—	—	—	—	0·35	—	—	—	—
	fa	—	—	—	—	—	0·61	—	—	—	—
di	wo	0·89	0·77	—	1·03	2·59	4·15	1·51	0·79	0·15	—
	en	—	0·22	—	0·16	0·45	1·64	0·14	0·32	0·02	—
	fs	1·01	0·59	—	0·96	2·36	2·56	1·53	0·48	0·14	—
hy	en	—	0·78	0·40	0·12	1·25	0·84	0·23	0·53	0·18	0·20
	fs	1·61	2·14	1·65	0·73	6·58	1·32	2·47	0·80	1·20	0·72
mt		1·27	—	—	0·98	1·58	4·06	1·65	2·61	1·81	0·62
il		0·51	0·66	0·13	0·13	0·85	1·80	0·57	0·63	0·28	0·08
ap		0·05	0·14	—	0·02	0·45	0·85	0·09	0·09	0·05	—
fl		0·12	—	0·24	0·08	0·08	—	0·18	0·80	—	0·14
hl		—	—	—	0·12	0·12	—	0·03	—	—	0·03
				+ total includes ZrO_2 0·37, S 0·05 $(Nb, Ta)_2O_5$ 0·22		+ total includes S 0·01, CO_2 0·12					+ total includes ZrO_2 0·04 S 0·02 CO_2 0·01 $(Nb, Ta)_2O_5$ 0·02

tr. trace: n.d. not determined

Source of data

Analyses 1 and 7 taken from Jacobson *et al.* (1958, Table 11, p. 20).
Analyses 3 and 4 taken from Jacobson *et al.* (1958, Table 7, p. 17).
Analyses 2 and 6 taken from Black (1963, Tables 1 and 2, pp. 41 and 43).
Analysis 5 and 8 taken from Macleod *et al.* (1971)
Analysis 9 taken from Rocci (1960, Table 5, p. 25).
Analysis 10 taken from Jacobson *et al.* (1958, Table 4, p. 14).
All CIPW norms have been re-calculated.

12—TAR * *

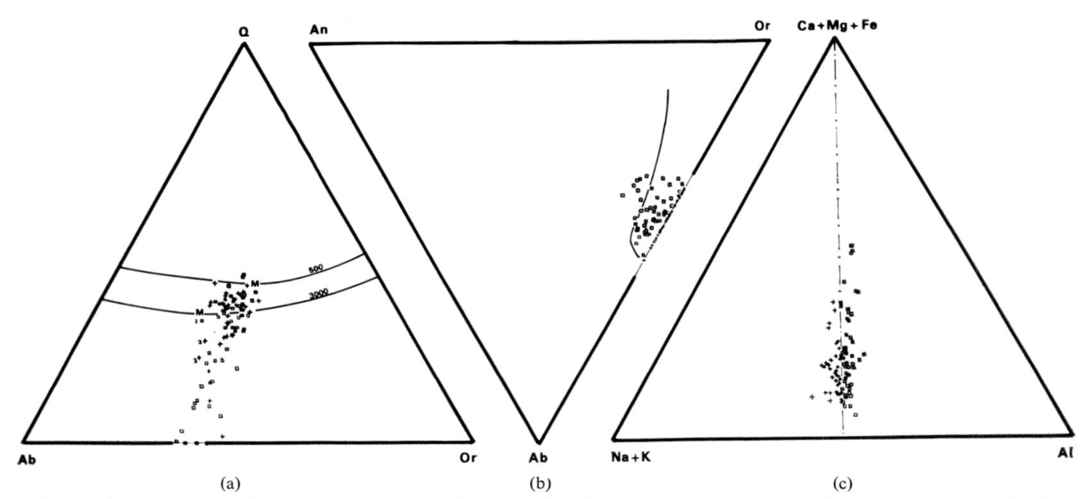

Fig. 4 (a) Projection of normative quartz, albite and orthoclase for peralkaline (+) and non-peralkaline (□) rock types from the Nigeria–Niger province, on to the anhydrous base of the system $NaAlSi_3O_8$–$KAlSi_3O_8$–SiO_2–H_2O.

M—minima at water vapour pressures of 500 kg/cm² and 3000 kg/cm² taken from Tuttle and Bowen (1958).

m—minimum in the system at 1000 kg/cm² with 4·5% acmite and 4·5% sodium metasilicate added (Carmichael and MacKenzie, 1963).

Examples of plotting positions of albitized granites given by points at 1, 2 and 3.

(b) Normative albite, anorthite and orthoclase compositions plotted in the quartz-saturated ternary feldspar system at 1000 kg/cm² water vapour pressure. Boundary curve separating fields of plagioclase and alkali feldspar taken from James and Hamilton (1969, p. 123, Fig. 7).

(c) Variation diagram showing the separation of peralkaline (+) and non-peralkaline (□) compositions, principally on the Na + K: Al atomic ratio.

Source of data: Macleod *et al.* (1971); Black (1963); Rocci (1960); Jacobson *et al.* (1958). All CIPW norms, molecular and atomic proportions have been recalculated and plotted by computer

peralkaline and non-peralkaline granites. These features are seen in the Nigeria–Niger province and in addition, significant levels of rare earths and the highly charged cations such as Nb, Hf, U and Th are also found. In contrast, there is a marked depletion in Sr and Ba and less intense depletion of Sc, Cs, and V relative to average granitic compositions (Bowden and van Breemen, 1971).

The behaviour of Zr in peralkaline liquids has been studied synthetically by Dietrich (1968) following the recognition by Bowden (1966) that Zr levels were significantly higher in peralkaline rocks from Nigeria and elsewhere, than in associated non-peralkaline types. The differences are quite dramatic and exactly match the variation in the agpaitic coefficient (cf. II.2.1). This means that the Zr concentration may be used as an index of peralkalinity. The appropriate trace element concentrations for various Nigerian complexes have been plotted against Zr in Fig. 5 to illustrate that those elements which have distinctive peralkaline tendencies are also concentrated in non-peralkaline granites.

It seems significant, in considering the origin of the non-orogenic granites, that many trace-element abundance patterns in the Nigeria–Niger province are typically peralkaline even though in Nigeria the majority of granites are non-peralkaline.

Perhaps the most important and significant element ratio to be used in igneous petrogenesis is the K/Rb ratio. Taylor *et al.* (1956) suggested that low ratios are characteristic of highly fractionated granites. The implication was that such rock types are truly magmatic, but petrographic evidence on some of the Nigerian younger granites with K/Rb ratios less than 100

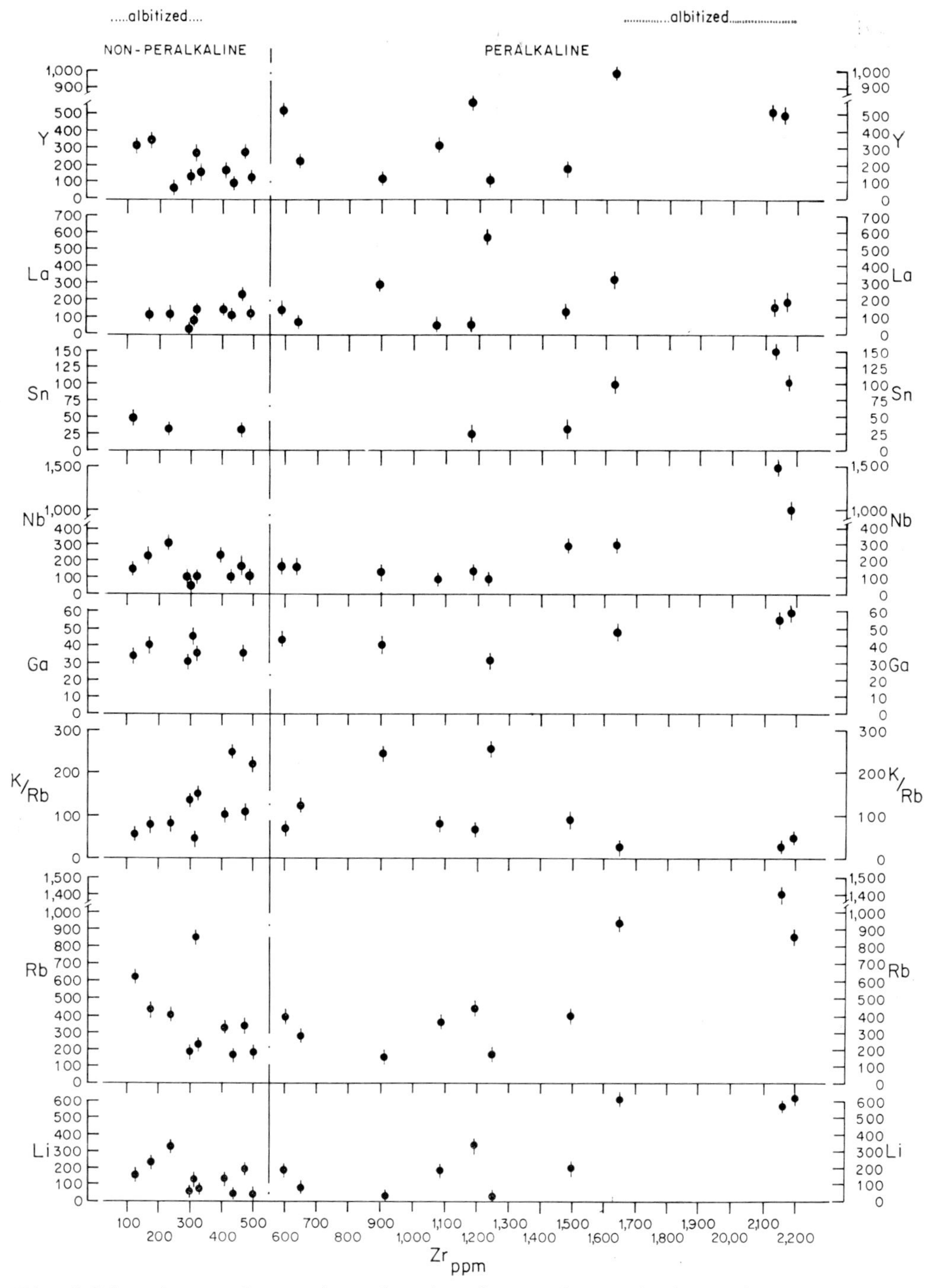

Fig. 5 Selected trace element data plotted against Zr for peralkaline and non-peralkaline granites. Zone of albitized granites indicated by dotted horizontal lines: vertical broken line separates peralkaline and non-peralkaline granites. Note that apart from Zr peralkaline and non-peralkaline granites have similar ranges of trace element concentrations and K/Rb ratios. Data taken from Table 4 and selected from Bowden and van Breemen (1971). All values in parts per million, except K/Rb

TABLE 4 Some Trace Element Data for the Liruei Complex, Nigeria

Magmatic sequence	Description and Locality	Li (ppm)	Rb (ppm)	K/Rb (wt)	Ga (ppm)	Zr (ppm)	Agpaitic coefficient (Na+K)/Al at.	La (ppm)	Yb (ppm)	Y (ppm)	Nb (ppm)	Sn (ppm)
4c	albite riebeckite-arfvedsonite granite, Kaffo valley	575	1400	25	55	2160	1·33	185	75	535	1500	150
4b	riebeckitic-arfvedsonite aegirine granite, R. Kwoya	25	160	270	33	1250	1·13	600	4	120	100	+
5b	fine-grained biotite granite near Liruei tin lode	125	860	42	45	310	0·85	70	17	280	50	+
5a	medium-grained biotite granite S. of Liruei tin lode	132	490	77	28	310	0·95	75	11	15	300	+
3	Arfvedsonite-fayalite granite porphyry, E. of R. Baba	22	140	230	42	910	0·98	300	4	120	150	+
2	quartz-hedenbergite-fayalite porphyry, S. of D. Shetu	18	270	155	29	410	1·05	575	7	110	100	+
1	comendite, N. of Korako	93	310	130	36	2300	1·14	155	18	172	225	+

Magmatic sequence modified from Jacobson *et al.* (1958, p. 29), and illustrated in Fig. 2a. + below sensitivity, 30 ppm.

TABLE 5 Rare Earth Data for Some Nigerian Granites

Sample No.	Complex	La	Ce	Pr	Nd	Sm	Eu	Gd	Tb	Dy	Ho	Er	Tm	Yb	Lu	Y	ΣREE	ΣCe/ΣY
	Non-Peralkaline Granites																	
100	Jos (Rayfield)	216	340	36	160	33	—	43	1·6	30	7·0	18	—	17	—	135	1037	3·9
102	Jos (Harwell)	58	106	13	83	7·3	—	12	2·0	35	7·6	55	6·4	98	—	175	658	0·8
113	Sara–Fier	110	200	13	90	9·2	1·9	12·5	—	2·5	1·0	3·2	—	95	—	25	563	3·4
	Peralkaline Granites																	
95	Kudaru	85	172	30	133	33	—	49	4·9	49	7·0	22	2·0	33·5	—	213	834	1·5
94	Rop	350	630	12	502	200	—	280	28	220	45	70	—	81	—	1050	3468	1·3
104	Kigom hills	258	460	51	235	54	7·3	62	2·8	42	9·5	22	—	19·5	—	165	1388	4·3

Rare earth data abstracted from Alexiev (1970, Table 1, p. 194–5).
All element values in ppm.

indicates that a hydrothermal development of albite may have been responsible for the up-grading of not only Rb but also Nb, Sn, and U. The albite-rich peralkaline granites from Nigeria have been thoroughly investigated by Mackay *et al.* (1952). It is significant that the granite from Kaffo valley, Liruei, with the lowest recorded K/Rb ratio (Table 4) was also found to be the richest potential source of Nb and U. The albite-biotite granites, although not exhibiting such extreme K/Rb ratios, are nevertheless important economic sources for Sn, Nb, Th and a number of rare earths. Rb is also enhanced considerably. It is therefore tempting to suggest that although these albite granites may have been derived originally by magmatic crystallization they have suffered extensive postmagmatic modification. Thus low K/Rb ratios should be used with extreme caution when interpreting the evolutionary history of the Nigeria–Niger granitic province. It is now considered that the albite-rich Nigerian granites were not derived by extreme differentiation as suggested by Butler *et al.* (1962), but were mainly the result of post-magmatic albitization.

Kovalenko *et al.* (1969) have inferred that a study of the rare earth patterns may substantiate whether late-stage, peralkaline granites are truly magmatic in origin, or are developed largely by hydrothermal alteration. The strong geochemical coherence and similarity of the rare earth elements is due to such features as a characteristic oxidation state (+3), similarity of cationic radii, and an essentially ionic bond formed with most ligands such as F^- OH^-, $CO^2{}_3{}^-$ and $PO^3{}_4{}^-$ (Ahrens, 1964). The rare earths have often been considered as a coherent geochemical group. However, partial separations or fractionations do occur, although very little is understood at present about how the rare earths are fractionated (Haskin and Schmitt, 1967).

Extensive data for all the rare earths in some Nigerian granites have been obtained by Alexiev (1970), who found a high total content of rare earths, and a low $\Sigma Ce/\Sigma Y$ ratio particularly for the tin-mineralized biotite granites from Jos and the albite-rich peralkaline granites (Table 5). Alexiev argued that such rare earth distribution patterns are inexplicable, if these granites are regarded solely as products of magmatic crystallization. It is believed that the increase in heavy rare earths (ΣY) is connected with trace-element overprinting of a magmatic complex by post-magmatic solutions enriched in rare earths, together with Nb, Sn, Zr, Rb etc. Indeed Alexiev considers that there is a direct correlation between the degree of albitization and the rare earth abundance patterns. If this is true then it makes a valuable tool for unravelling the petrogenesis of these rocks. It also emphasises the important role of albitization in the development of economic levels of columbite and cassiterite.

IV.8.5. Isotopic Studies

There are no reliable geochronological age dates available for the ring-complexes of Aïr and southern Niger, but stratigraphic evidence at Adrar Bouss, Goundaï and Tamgak suggests an age somewhere between post Devonian and Lower Cretaceous. However, the Nigerian region has been an active field for geochronologists who, by single whole-rock or mineral ages, have proved that the peralkaline and non-peralkaline younger granites were intruded during the Jurassic period. The radiometric age patterns illustrated in Fig. 6 suggest that there was little time difference between the emplacement of various complexes, and that there has been no subsequent extensive reheating of the region.

The only detailed Rb–Sr study on more than one sample from the same complex has been completed by Bowden and van Breemen (1971). For example, the riebeckitic-arfvedsonite-biotite granites of Nigeria are believed to represent transitional links between the peralkaline granites and non-peralkaline granites. This critical rock type forms a half mile wide circumjacent ring dyke in the Amo complex (No. 10, Fig. 1), with an outer peralkaline zone grading into an inner non-peralkaline zone. The fact that both two peralkaline and four non-peralkaline specimens of this rock type lie on the same whole-rock isochron (Fig. 7), combined with a

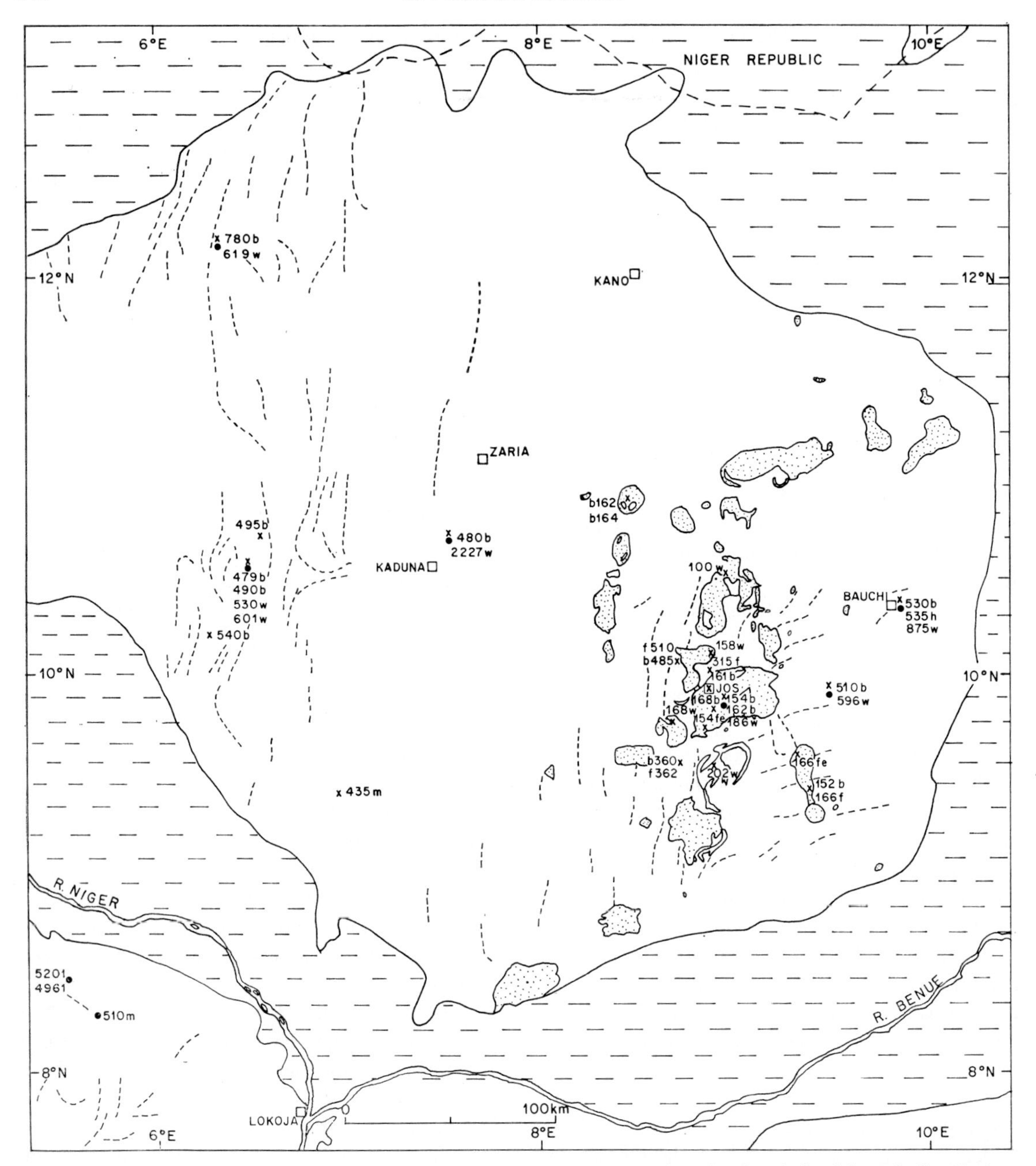

Fig. 6 Radiometric ages for the granites and surrounding basement in Nigeria, derived by Rb–Sr, K–Ar, U–Pb, Th–Pb and common lead methods

w = single whole rock; b = biotite; m = muscovite; l = lepidolite; f = feldspar; h = amphibole; fe = fergusonite.

Data obtained from: Darnley *et al.* (1962); Jacobson *et al.* (1963); Snelling (1964, 1965); Hurley *et al.* (1966); Tugarinov (1968); Bowden and van Breemen (1971).

All Rb–Sr ages standardized for a Rb^{87} decay constant of $1{\cdot}47 \times 10^{-11} yr^{-1}$.

Dashed lines indicate structural trends in the Precambrian basement: stippled areas represent Jurassic ring complexes.

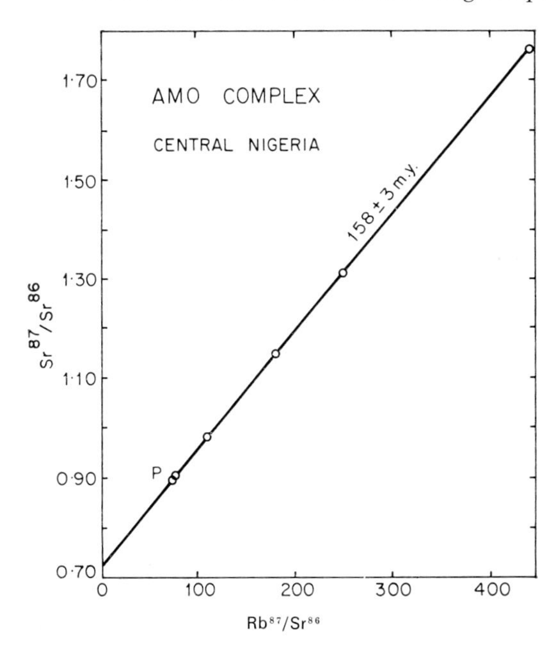

Fig. 7 Whole rock isochron for six specimens of riebeckitic-arfvedsonite biotite granite from the Amo complex. Outer peralkaline zone represented by points at P, remainder of the points are from the inner non-peralkaline zone. Age calculated using a decay constant of $1{\cdot}47 \times 10^{-11}yr^{-1}$. Initial Sr^{87}/Sr^{86} ratio 0·7212 ±0·0040

high initial Sr^{87}/Sr^{86} ratio of 0·7212 ±0·0040, suggests that a common crustal origin may be applicable for both the peralkaline and non-peralkaline granites in the Jos Plateau region. However the interpretation of strontium isotopic data is currently undergoing a critical reappraisal since the discovery, by Taylor and Forester (1971), that meteoric hydrothermal solutions enriched in radiogenic strontium may react with epizonal intrusions and lavas. The key to this interaction is a measure of the O^{18} depletion. Thus analyses for oxygen and hydrogen isotopes will be critically important for the interpretation of initial Sr^{87}/Sr^{86} ratios in the Nigeria–Niger province.

In contrast to the accurate geochronology for the Mesozoic non-orogenic igneous activity, accurate dating of the Precambrian basement has been neglected except for ages between 2000–1700 m.y. in Niger (Vachette, 1964) and a small number of determinations given in Fig. 6. The data suggest that despite the apparent dominance of the Pan African Orogeny (reflected in 500–600 m.y. ages for some orogenic granites and metasediments in Nigeria), the Nigeria–Niger province also contains remnants of only partly reworked, old cratonic continental crust.

Bowden (1970) has discussed the isotopic composition of lead in some Nigerian Precambrian, Palaeozoic, and Mesozoic rocks and minerals and concluded that the Jurassic granites with 'old lead ages' cannot be assumed to have differentiated from the upper mantle, but have been derived by fusion of mineralized crustal rocks. Although the results for the Nigerian Mesozoic rocks are few in number, the spectrum of the isotopic variations in ore-leads and rock-leads plotted in Fig. 8 are worthy of some consideration. Perhaps the most striking feature of the diagram is that none of the points lie close to the true age of emplacement (*circa* 150 m.y.).

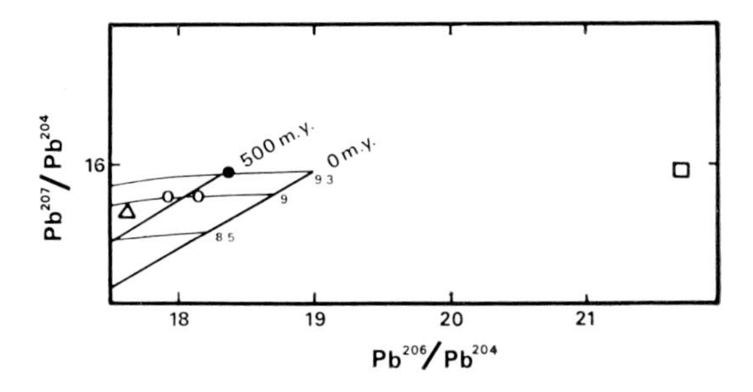

Fig. 8 Plot of Pb^{207}/Pb^{204} versus Pb^{206}/Pb^{204} for two whole rocks, potash feldspar, and two galenas from Nigerian Mesozoic ring-complexes

Open triangle = Buji rhyolite; open square = Buji quartz–fayalite hedenbergite porphyry; closed circle = potash feldspar from non-peralkaline granite, Sara–Fier; open circles = ore leads from the Liruei complex.

Data abstracted from Tugarinov (1968) and Tugarinov *et al.* (1968).

Diagram shows pertinent growth lines and isochrons for a single stage evolutionary model based on an age of the earth of 4550 m.y. and a primordial lead with a composition of Pb^{207}/Pb^{204} = 10·42 and Pb^{206}/Pb^{204} = 9·56.

Also the two whole rock determinations for

the Buji volcanic complex (No. 11. Fig. 1) are of particular interest. It is clear from the diagram that the Buji peralkaline rhyolite rock-lead is radiogenic-deficient whilst in complete contrast, the non-peralkaline quartz–fayalite–hedenbergite porphyry from the same volcanic centre as the rhyolite is radiogenically enriched.

IV.8.6. Petrogenesis

IV.8.6.1. Retrospect

In their memoir on the Nigerian younger granites, Jacobson *et al.* (1958) expressed the opinion that peralkaline and non-peralkaline granites had been derived by divergent differentiation from a more basic parent, but that the actual process by which this was achieved was not known. However a unifying theory which provided a mechanism whereby peralkaline granites developed in association with anorthosite and olivine gabbro at Aïr, was suggested by Black (1965). He concluded that tholeiitic basalt magma at deep crustal or subcrustal levels could fractionate by crystal settling to give a lower horizon of olivine gabbro and an upper concentration of anorthosite. On intrusion of the anorthosite ring dyke, the remaining liquid in the magma chamber would have been depleted in Al and hence subsequent fractionation could have led to the formation of a peralkaline acidic residuum. The non-peralkaline granites may also have been a separate differentiated product of a tholeiitic magma. Unfortunately anorthosite is not present in Nigeria or at southern Niger, and hence this theory is not generally applicable to the whole Nigeria–Niger province.

There are two current theories concerning the origin of the Nigerian younger granites. The first, developed by Bailey and Schairer (1966), assumes that peralkaline silicic magmas generated within a basic lower crust or in the upper mantle, are contaminated during their ascent into the upper crust to yield the dominant metaluminous and peraluminous granites such as those found on the Jos plateau. The other theory, proposed by Bowden (1970), involves the generation of peralkaline magmas in the upper crust from Precambrian basement rocks of granitic composition, which on further progressive melting and cooling yield the metaluminous and peraluminous trends. Both theories agree that the magmatic evolution of the younger granites began with the development of a peralkaline liquid, but the essential areas of dispute are the source regions for the generation of the peralkaline magma, and the derivation of the associated non-peralkaline silicic rocks. Unfortunately both theories have certain drawbacks. With the Bailey and Schairer model it is difficult to account for the repetition and alternation of peralkaline and non-peralkaline granites in the same ring-complex solely by infrequent contamination of repeated pulses of peralkaline magma. The main problem in Bowden's hypothesis is the quantitative transfer of the zone of high heat flow, from the arched region of the mantle to the upper crust, for progressive melting of granitic basement.

IV.8.6.2. A Petrogenic Model for the Nigeria–Niger Ring-Complexes

The overall concept that the Nigeria–Niger province provides different erosional levels of the products of the same magmatic evolution history, means that any petrogenetic model for the development of peralkaline and non-peralkaline ring-complexes must take into account the dominance of peralkaline ash-flow tuffs and early steeply-dipping mildly peralkaline ring intrusives at the beginning of the northern cycles. This should be linked with the obviously later emplacement of shallow dipping stock-like masses of non-peralkaline granites in the south of the province.

It is currently accepted from geophysical evidence that the idea of a basic lower crust in stable continental regions, which have previously undergone a complex orogenic, magmatic and metamorphic evolution, is no longer completely valid (Ringwood and Green, 1966, Bott, 1971, pp. 68–70). Indeed most workers now place the source region for any basic magmatism within the upper mantle. Thus the fact that basic rocks preceded the eruptive silicic magmatic cycle;

that they may also be interbedded with acid lavas and tuffs, or postdate the granitic cycle in the Nigeria–Niger province, means that the initial and final processes must have begun and ended with upper mantle events. There is obviously the possibility that, in areas of crustal arching, volatile concentrations and heat focussing may occur (VI.1b; Bailey, 1970) thus providing conditions for melting in the upper mantle and lower crust.

If the composition of the lower crust in the Nigeria–Niger province is a high pressure modification of 'granodiorite' (Ringwood and Green, 1966) then melting during late Palaeozoic to Mesozoic times near the crust/mantle boundary may have generated a peraluminous biotite-bearing silicic liquid, which by diapiric action rose slowly into the upper crust. It may have been preceded by an advancing wave of basaltic magma initiated during the first phase of epeirogenesis. On entering the brittle zone of the upper crust, diapiric movement would have been retarded which led to a slow build-up of fluid pressure and caused tension fractures focussed on the arrested diapirs to break through to the surface. The advancing small portion of basaltic magma had just enough superheat to initiate local partial fusion of upper crustal basement granitic rocks to form a peralkaline liquid, the temperature of which was maintained by the more massive diapir of non-peralkaline silicic liquid. The first material therefore, to be emplaced or extruded, was most likely to have been of basic composition, but concomitantly mildly peralkaline liquids generated within the upper crust were also erupted as ash flow tuffs to form shield volcanoes similar to the recent ignimbritic volcanoes described from Tibesti (Vincent, 1970, p. 307). Some intermixing of non-peralkaline and peralkaline liquids may also have occurred at this stage. Subsequently, after the newly formed upper crustal magma chamber had almost emptied, the collapse of the caldera allowed downfaulting of the lavas and ash flow tuffs which initiated the outer ring-fracture. At the same time the peraluminous biotite granite of lower crustal origin moved nearer to the surface till it intruded the volcanic cover.

During later cooling a 'boil off' of the volatiles within the biotite granite concentrated them near the roof margins, filling tension fractures with tin-rich greisens. Some of this magmatic hydrothermal fluid, which was most likely albitic-acmitic in composition and containing concentrates of Nb, Sn, Zr, Rb and the heavy rare earths, also passed along the ring-dykes of peralkaline granite causing selective but extensive recrystallization of the alkali amphibole, the generation of hydrothermal aegirine, and the development of pyrochlore.

Following the completion of acidic magma evolution, a final mantle phase in the two main centres, Aïr and Nigeria, yielding small but significant numbers of basaltic and trachybasalt dykes, gabbros and syenogabbros, brought this period of magmatism to a close.

Such a generalized petrogenetic model means that peralkaline granites are of upper crustal origin. So too may be some of the metaluminous granites. However, the majority of the peraluminous granites were probably developed at deeper levels provided the composition of the lower crust can be accepted as granodioritic. Supporting evidence for a silicic lower crust is provided by the negative Bouguer anomalies found over the Nigerian ring-complexes (Ajakaiye, 1968, 1970), which suggests that 'granite' may reach depths of the order of 12 km; such large granitic masses seem to require a broadly acidic source. The isotopic evidence gathered so far is also consistent with this model, though no biotite granites have been studied in detail.

The genesis of the syenites poses a problem. As yet insufficient work has been completed on these rock types, but since they have close associations with peralkaline granites, the syenites may also be of upper crustal origin. Of course syenites may be regarded as fractional melted products of olivine basalt (Bailey and Schairer, 1966) but the existence of large basaltic reservoirs beneath the Nigeria–Niger province seems unlikely in view of the rarity of associated basaltic magmatism, and the geophysical evidence (Ajakaiye, 1968, 1970).

It should however be obvious to the interested reader that the Nigeria–Niger province needs more detailed study before the origin of the associated peralkaline and non-peralkaline rock types can be completely understood. Meanwhile the petrogenetic model outlined above should provide a framework for subsequent modification and improvement.

IV.8. REFERENCES

Ahrens, L. H., 1964. The significance of the chemical bond for controlling the geochemical distribution of the elements. *Phys. Chem. Earth*, **5**, 1–54.

Ajakaiye, D. E., 1968. A gravity interpretation of the Liruei younger granite ring complex of Northern Nigeria. *Geol. Mag.*, **105**, 256–63.

Ajakaiye, D. E., 1970. Gravity measurements over the Nigerian younger granite province. *Nature*, **225**, 50–2.

Alexiev, E. I., 1970. Rare earth elements in younger granites of northern Nigeria and the Camerouns and their genetic significance. *Geokhimiya 1970* (in Russian), No. 2, 192–8.

Bailey, D. K., 1970. Volatile fluxing, heat-focussing and the generation of magma. *Geol. J.* (*Special Issue*), **2** 'Mechanism of igneous intrusion', Eds. Newall, G., and Rast, N., Seel House Press, Liverpool, 177–86.

Bailey, D. K., and Schairer, J. F., 1966. The system NaO_2–Al_2O_3–Fe_2O_3–SiO_2 at 1 atmosphere and the petrogenesis of alkaline rocks. *J. Petrology*, **7**, 114–70.

Black, R., 1963. Note sur les complexes annulaires de Tchouni–Zarniski et de Gouré (Niger). *Bull. Bur. Rech. géol. min.*, **1-1963**, 31–45.

Black, R., 1965. Sur la signification petrogénétique de la découverte d'anorthosites associées aux complexes annulaires subvolcaniques du Niger. *C.R. Acad. Sc. Paris*, **260**, 5829–32.

Black, R., and Girod, M., 1970. 'Late Palaeozoic to Recent igneous activity in West Africa and its relationship to basement structure', in *African Magmatism and Tectonics*, Eds. Clifford, T. N., and Gass, I. G., Oliver and Boyd, Edinburgh. 185–210.

Black, R., Jaujou, M., and Pellaton, C., 1967. Notice explicative sur la carte géologigue de l'Aïr, a l'échelle du 1:500,000. *Dir. Mines Géol.*, Niger.

Borley, G. D., 1963. Amphiboles from the younger granites of Nigeria. Part I. Chemical classification. *Mineral. Mag.*, **33**, 358–76.

Bott, M. P. H., 1971. *The Interior of the Earth.* Edward Arnold, London. 316 pp.

Bowden, P., 1966. Zirconium in younger granites of northern Nigeria. *Geochim. Cosmochim. Acta.*, **30**, 985–93.

Bowden, P., 1970. Origin of the younger granites of northern Nigeria. *Contr. Miner. Petrology.*, **25**, 153–62.

Bowden, P., and van Breemen, O., 1971. 'Isotopic and chemical studies on younger granites from northern Nigeria', in *Proceedings of the Conference on African Geology, Ibadan 1970.*

Buchanan, M. S., MacLeod, W. N., Turner, D. C., Berridge, N. G., and Black, R., (1971). The geology of the Jos Plateau Vol. 2, younger granite complexes. *Bull. geol. Surv. Nigeria*, **32.**

Butler, J. R., Bowden, P., and Smith, A. Z., 1962. K/Rb ratios in the evolution of the younger granites of northern Nigeria. *Geochim. Cosmochim. Acta.*, **26**, 89–100.

Carmichael, I. S. E., and MacKenzie, W. S., 1963. Feldspar–liquid equilibria in pantellerites: an experimental study. *Am. J. Sci.*, **261**, 382–96.

Chaput, M., Lombard, J., Lormand, J., and Michel, H., 1954. Granites et traces d'étain dans le Nord du Cameroun. *Bull. Soc. géol. Fr.*, **4**, 373–94.

Darnley, A. G., Smith, G. H., Chandler, T. R. D., and Dance, D. F., 1962. The age of fergusonite from the Jos area, northern Nigeria. Mineral. Mag., **33**, 48–51.

Dietrich, R. V., 1968. Behaviour of zirconium in certain artificial magmas under diverse P.T. conditions. *Lithos*, **1**, 20–9.

Fabriès, J., and Rocci, G., 1965. Le massif granitique du Tarraouadji (République du Niger). Etude et signification pétrogénétique des principaux minéraux. *Bull. Soc. fr. Minér. Cristallogr.*, **88**, 319–40.

Foster, M. D., 1960. Interpretation of the composition of trioctahedral micas. *Prof. pap. U.S. geol. Surv.* **354–B**, 11–49.

Greenwood, R., 1951. Younger intrusive rocks of Plateau province, Nigeria, compared with alkali rocks of New England. *Bull. geol. Soc. Am.*, **62**, 1151–78.

Haskin, L. A., and Schmitt, R. A., 1967. 'Rare earth distributions' in *Researches in Geochemistry vol. 2.*, Ed. Abelson, P. H., John Wiley and Sons, New York, 234–58.

Hurley, P. M., Rand, J. R., Pinson, W. H., Posadas, V. C., and Reid, J. B., 1966. Continental drift investigations. *14th Ann. Report. Dept. Geol. Geophys., M.I.T.* (*1966*), 3–15.

Jacobson, R. R. E., and MacLeod, W. N., (in press). Geology of the Liruei, Banke and adjacent younger granite ring-complexes. *Bull. geol. Surv. Nigeria*, **33**.

Jacobson, R. R. E., MacLeod, W. N., and Black, R., 1958. Ring-complexes in the younger granite province of northern Nigeria. *Mem. Geol. Soc. Lond.*, **1**, 71 pp.

Jacobson, R. R. E., Snelling, N. J., and Truswell, J. F., 1963. Age determinations in the geology of Nigeria, with special reference to the older and younger granites. *Overseas Geol. Mineral. Resources*, **9**, 168–82.

James, R. S., and Hamilton, D. L., 1969. Phase relations in the system $NaAlSi_3O_8$–$KAlSi_3O_8$–$CaAl_2Si_2O_8$–SiO_2 at 1 kilobar water vapour pressure. *Contr. Mineral. Petrology*, **21**, 111–41.

Koch, P., 1959. Le Précambrian de la frontiére occidental du Cameroun central. *Bull. Direct. Min. Geol. Cameroun*, **3**, 300 pp.

Kovalenko, V. I., Znamenskaya, A. S., Popolitov, E. I., and Abramova, S. R., 1969. Distribution of rare earth elements and yttrium in minerals of alkalic granitoids. *Geokhimiya 1969* (in Russian), No. 8, 997–1006.

Lasserre, M., 1966. Confirmation de l'existence d'une série de granites Tertaires en Cameroun. *Bull. Bur. Rech. géol. min.*, **3–1966**, 141–8.

Lasserre, M., 1969. Cameroun: Examen des résultats géochronologiques obtanus depuis 1967. *Ann. Fac. Sci. Clermont-Ferrand*, **41**, sér. Géol. 19, 29–33.

Mackay, R. A., Beer, K. E., and Rockingham, J. E., 1952. *The Albite-Riebeckite Granites of Nigeria.* H.M.S.O. London, 25 pp.

MacLeod, W. N., Turner, D. C., and Wright, E. P., (1971). The geology of the Jos Plateau, Vol. 1, General geology. *Bull. geol. Surv. Nigeria*, **32.**

Miyashiro, A., 1957. The chemistry, optics and genesis of the alkali amphiboles. *J. Fac. Sci. Univ. Tokyo, Sect. II*, **11**, 57–83.

Ringwood, A. E., and Green, D. H., 1966. 'Petrological nature of the stable continental crust', in *The Earth Beneath the Continents*, Eds. Steinhart, J. S., and Smith, T. J., Geophys. Monogt., No. 10, American Geophysical Union, Washington, D.C., 611–19.

Rocci, G., 1960. Le massif du Tarraouadji (République du Niger), Etude géologique et pétrographique. *Notes Bur. Rech. Géol. Min. Dakar.*, **6**, 39 pp.

Sabine, P., 1950. Optical properties and composition of acmitic pyroxenes. *Mineralog. Mag.*, **29**, 113–25.

Snelling, N. J., 1964. Age determination unit. *Ann. Rep. Overseas Geol. Survs.* (*1963*), 30–40.

Snelling, N. J., 1965. Age determination unit. *Ann. Rep. Overseas Geol. Survs.* (*1964*), 28–38.

Sundius, N., 1946. The classification of the hornblendes and the solid solution relations in the amphibole group. *Arsbok. Sveriges Geol. Undersok.*, **40**, No. 4, 36 pp.

Taylor, H. P., and Forester, R. W., 1971. Low-O^{18} igneous rocks from the intrusive complexes of Skye, Mull, and Ardnamurchan, Western Scotland. *J. Petrology*, **12**, 465–97.

Taylor, S. R., Emeleus, C. H., and Exley, C. S., 1956. Some anomalous K/Rb ratios in igneous rocks and their petrological significance. *Geochim. Cosmochim. Acta*, **10**, 224–9.

Tugarinov, A., 1968. 'Age absolu et particularités génétiques des granites du Nigeria et du Cameroun Septentrional,' in *Proceedings Symposium granites of West Africa*, 119–22.

Tugarinov, A., Kovalenko, V. I., Znamensky, E. B., Legeido, V. A., Sabatovich, E. V., Brandt, S. B., and Tsyhansky, V. D., 1968. 'Distribution of Pb-isotopes, Sn, Nb, Ta, Zr, and Hf in granitoids from Nigeria,' in *Origin and Distribution of the Elements.* Ed. Ahrens, L. H., International Series Monographs in Earth Sciences vol. 30. Pergamon Press, Oxford, 687–99.

Turner, D. C., 1963. Ring-structures in the Sara-Fier complex, Northern Nigeria. *Q. J. geol. Lond.*, **119**, 345-66.

Turner, D. C., 1968. Volcanic and intrusive structures in the Kila–Warji ring-complex, northern Nigeria. *Q. J. geol. Soc. Lond.*, **124**, 81–9.

Tuttle, O. F., and Bowen, N. L., 1958. Origin of granite in the light of experimental studies in the system $NaAlSi_3O_8$–$KAlSi_3O_8$–SiO_2–H_2O. *Mem. geol. Soc. Am.*, **74**, 153 pp.

Vachette, M., 1964. Essai de synthèse des déterminations d'âges radiométriques de formations cristallines de l'Ouest africain (Côte d'Ivoire, Mauritanie, Niger). *Ann. Fac. Sci. Clermont-Ferrand*, **25**, sér. Géol. 8, 7.

Vincent, P. M., 1970. 'The evolution of the Tibesti volcanic province, eastern Sahara,' in *African Magmatism and Tectonics.* Eds. Clifford, T. N., and Gass, I. G., Oliver and Boyd, Edinburgh, 301–19.

Woodrow, P. J., 1967. The crystal structure of astrophyllite. *Acta Cryst.*, **22**, 673–8.

V. Conditions of Formation

V.1. EXPERIMENTAL STUDIES

A. D. Edgar

V.1.1. Introduction

The same reasons that have attracted petrologists to make intensive studies of alkaline rocks, out of all proportion to their crustal abundance, have also prompted experimentalists to study systems pertinent to their genesis. The inherent difficulties involved in working out systems with the large number of components (including volatiles) found in these rocks has provided an extra challenge for experimental studies. As a result such systems are receiving ever increasing attention (Eugster, 1967).

Of all igneous rocks, those with alkaline affinities, i.e. those containing feldspathoids and/or alkali pyroxenes and amphiboles, probably show the widest range of composition, with SiO_2 contents ranging from about 35 wt. % (melilite basalt) to about 75 wt. % (alkali granite). Similar wide ranges are observed for almost all of the major oxides (cf. V.3)* In addition, many alkaline rocks are exceptionally rich in volatiles, particularly H_2O, CO_2, P, Cl, S and F, which are important in their genesis (cf. Smyth, 1927; Bailey, 1964; etc; cf. VI.4).

This wide variation in composition makes it almost impossible to delineate a system pertinent to even one-half of the different types of alkaline rocks as it would have to include the components Na_2O, K_2O, Al_2O_3, FeO, Fe_2O_3, MgO, CaO, SiO_2, H_2O, CO_2, Cl, and F in order to include the common feldspathoids, feldspars, alkali pyroxenes and amphiboles found in these rocks. Small portions of this complex system have been investigated and it is possible that we shall soon have an understanding of most of the petrologically important parts of the system.

Most experimental studies have been on systems pertinent to the commoner alkaline rocks, particularly alkali basalts and rocks of the nepheline syenite clan. These studies are considered in three categories: (*a*) systems without volatiles at atmospheric pressure, (*b*) systems done hydrothermally and under controlled partial oxygen pressures, and (*c*) melting and crystallization of alkaline rocks under controlled laboratory conditions. The last part of this chapter reviews the experimental data on the conditions of formation of the important minerals of alkaline rocks. No attempt is made in this chapter to use experimental studies in explaining the rôle of fractional crystallization in the genesis of alkaline rocks as this is covered in chapter VI.2. The following abbreviations are used to denote components of synthetic systems, and as normative minerals:

Ab—albite ($NaAlSi_3O_8$)
Ac—acmite ($NaFeSi_2O_6$)
Ak—akermanite ($Ca_2MgSi_2O_7$)
An—anorthite ($CaAl_2Si_2O_8$)
Anal—analcime ($NaAlSi_2O_6 . H_2O$)
Ct—calcite ($CaCO_3$)
Di—diopside ($CaMgSi_2O_6$)
Enst—enstatite ($MgSiO_3$)
Fo—forsterite (Mg_2SiO_4)
Geh—gehlenite ($Ca_2Al_2SiO_7$)
Hy—hypersthene ($(Mg,Fe)SiO_3$)
K-Fp—potassium feldspar ($KAlSi_3O_8$) (this will be used irrespective of the polymorph involved)
Ks—kalsilite ($KAlSiO_4$)
La—larnite (Ca_2SiO_4)
Lc—leucite ($KAlSi_2O_6$)
Ne—nepheline ($NaAlSiO_4$)
p-enst—protoenstatite ($MgSiO_3$)
Qtz—quartz (SiO_2)
Sil—silica (SiO_2)

* Readers are referred to the relevant Chapters of this book where similarly cited.

S.M.—soda melilite ($NaCaAlSi_2O_7$)
Trid—tridymite (SiO_2)
Wo—wollastonite ($CaSiO_3$)

The term 'component' is used here and throughout this chapter to denote a mineral species in a synthetic system. This word is not used in a strict thermodynamic sense since, in many cases, the phases cannot be represented by the plane or volume enclosed by these mineral species.
Other abbreviations are:

atm—atmosphere
f_{O_2}—oxygen fugacity
kb—kilobar
P_{O_2}—partial oxygen pressure
P_{H_2O}—water vapour pressure
p.s.i.—pounds per square inch
s.s.—solid solution

V.1.2. Systems Without Volatiles at Atmospheric Pressure

Alkali basalts represent an important group of 'basic' alkaline rocks. Yoder and Tilley (1962) proposed a classification of basalts based on the system Di–Ne–Fo–Sil (Fig. 1), in which normative compositions of the alkali basalts plot in the Di–Ne–Ab–Fo tetrahedron, separated from the olivine basalts by a critical plane of silica undersaturation, the system Di–Fo–Ab (cf. II.4). Assuming that Di represents clinopyroxene, Fo represents olivine and Ab represents plagioclase, this 'basalt' system can be used to explain the genesis of melilite basalts, basanites, urtites, nephelinites, olivine nephelinites, fasinites and tephrites. Iron and titanium, both important constituents of basalts, are not represented by this system.

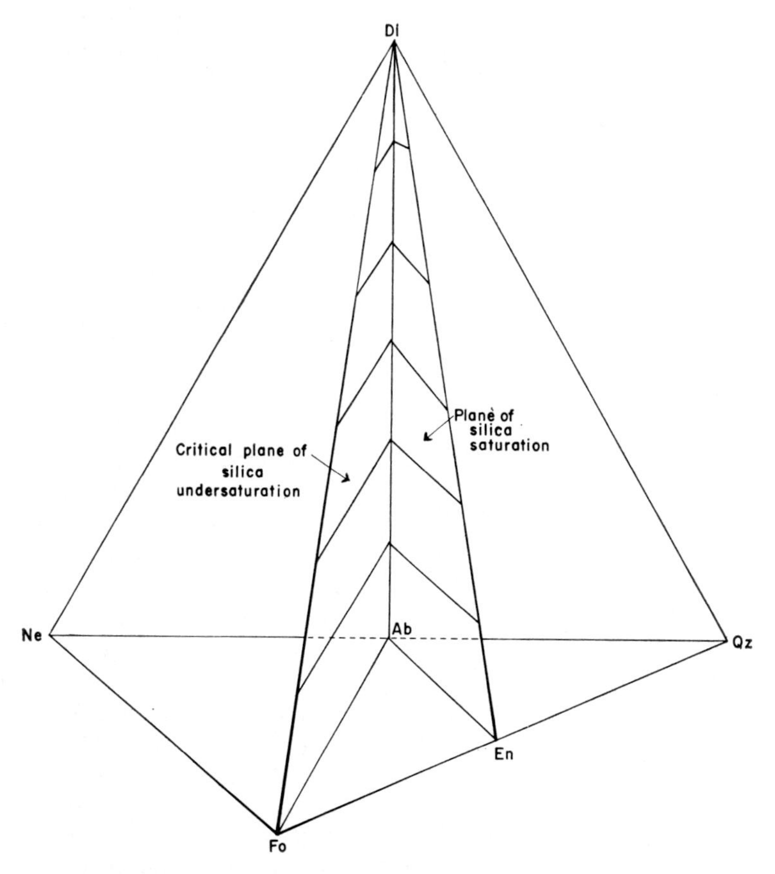

Fig. 1 The system Di–Ne–Fo–Qtz, the 'basalt' tetrahedron of Yoder and Tilley (1962)

Melilite, wollastonite, and anorthite, also occurring in alkali basalts, are not represented, although melilite is a constituent of the Ne–Di system (Bowen, 1922; Schairer *et al.*, 1962) and anorthite, as plagioclase, is present in the Di–Ne–Ab system (Schairer and Yoder, 1960a).

The influence of these components on phase relations in the Di–Ne–Ab–Fo system can be shown by examining the systems Di–Ak–Ne (Onuma and Yagi, 1967); Wo–An–Ne (Gummer, 1943); and Ne–Geh–Wo (Smalley, 1947; Schairer *et al.*, 1966). Therefore experimental data connects most of the feldspathoids, feldspars, pyroxenes and olivines found in alkali basaltic compositions. In addition, Yagi and Onuma (1969) investigated the rôle of titanium (as titanite, perovskite, iron–titanium oxides and as solid solution in pyroxene, biotite and hornblende) in alkali basalts by studying portions of the system Di–Ak–Ne–$CaTiAl_2O_6$. First we shall consider the silica undersaturated part of Yoder and Tilley's simplified 'basalt tetrahedron'.

The Di–Ab join separates the Di–Ne–Sil system into an oversaturated and undersaturated portion and is also an 'equilibrium thermal divide' (Fig. 2). Thus no single bulk composition in this system can produce both a silica-rich (or tholeiitic basalt trend) and nepheline-rich residue (alkali basalt trend) through normal fractionation processes. The formation of plagioclase in the Di–Ab system (Schairer and Yoder, 1960a) is important in the genesis of oversaturated alkaline rocks, as discussed later, while the alumina content of the pyroxene may determine whether residual liquids form an alkali basalt or tholeiitic trend. Unfortunately compositions of the plagioclase and extent of alumina substitution in the pyroxene are not completely known. The minimum melting temperature in the Di–Ne–Sil system lies along the Ne–Sil join, indicating that residual liquids in this system will be poor in the pyroxene component. Of particular petrological interest is the reaction between forsterite and liquid producing diopside and nepheline at 1138 °C similar to the reaction involving olivine and melilite in the Di–Ne join. Schairer and Yoder (1960a, p. 281) suggest that evidence of these reactions might be expected in basanites, melilite basalts and olivine nephelinites although Wilkinson (1956) believes these to be absent in alkali basalts. See also Platt and Edgar (1972).

The system Di–Fo–Ab (Schairer and Morimoto, 1959), representing the critical plane of silica undersaturation in the simplified 'basalt tetrahedron', has forsterite as the primary phase over most of its liquidus, with diopsidic pyroxene and plagioclase forming narrow fields at forsterite-poor compositions. The minimum melting temperature at $Ab_{98}Fo_2$, again indicates the lack of mafic components in residual liquids.

The third side of the alkali basalt tetrahedron is the system Di–Ne–Fo (Fig. 3) (Schairer and Yoder, 1960b) with forsterite as the largest primary phase and spinel, carnegieite s.s., nepheline s.s. and diopsidic pyroxenes occupying smaller areas. Because the system is not ternary, boundary curves between phases represent traces of boundary surfaces between primary phase volumes of the several solid phases appearing at the liquidus. Melilites occur extensively as subliquidus phases in this system.

The fourth side of the simple basalt tetrahedron is the system Fo–Ne–Sil (Fig. 4) (Schairer and Yoder, 1961). The Fo–Ab join separates this system into a silica-undersaturated portion, Fo–Ne–Ab; and a silica-saturated portion, Fo–Ab–Sil. The undersaturated part is non-ternary due to the formation of spinel at the liquidus surface, although below the temperature (point P) where spinel disappears by reaction with the liquid, the system is believed to be ternary. The last liquid in this portion of the system crystallizes at Q at a composition and temperature very close to the binary Ne–Ab eutectic (K). This system resembles the Di–Ne–Sil system (Fig. 2) in that residual liquids are enriched in alkali aluminosilicates and impoverished in ferromagnesian minerals. The 'equilibrium thermal divide' between forsterite and albite explains the absence of coexisting orthopyroxenes and nephelines in nature and also prohibits originally silica-undersaturated compositions producing oversaturated rocks, and vice-versa, by normal fractionation processes.

From consideration of the systems Di–Ne–Ab (Fig. 2), Di–Ne–Fo (Fig. 3), Di–Fo–Ab, and

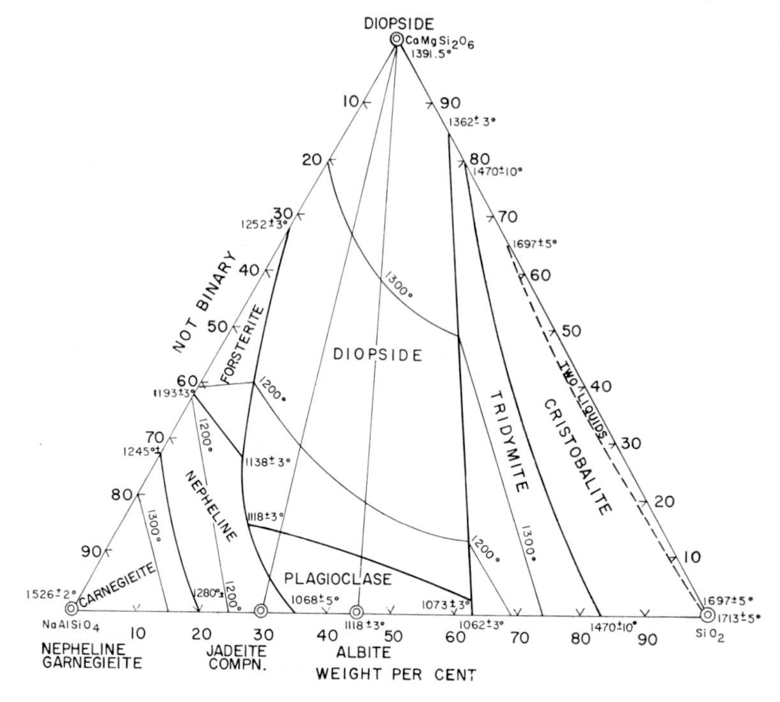

Fig. 2 The system Di–Ne–Sil (after Schairer and Yoder, 1960a)

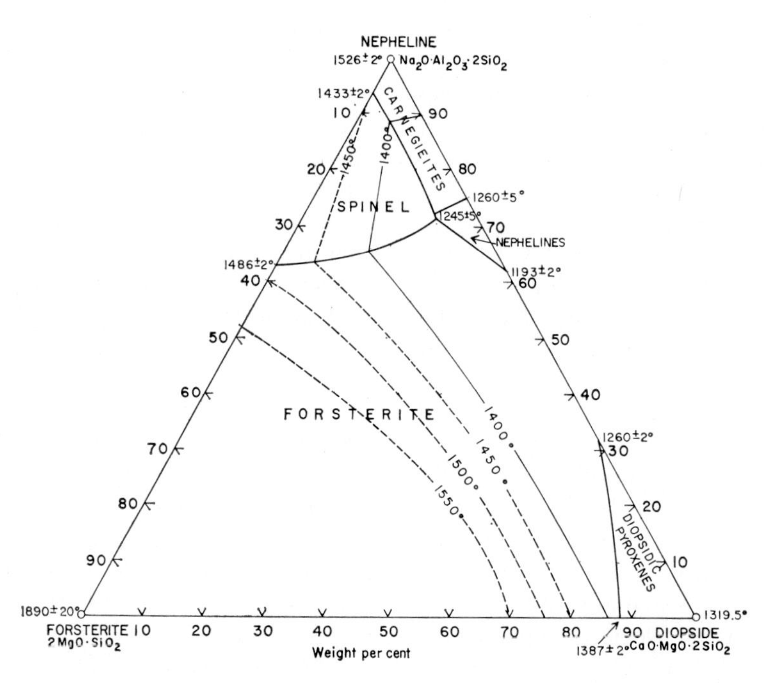

Fig. 3 The system Di–Ne–Fo (after Schairer and Yoder, 1960b)

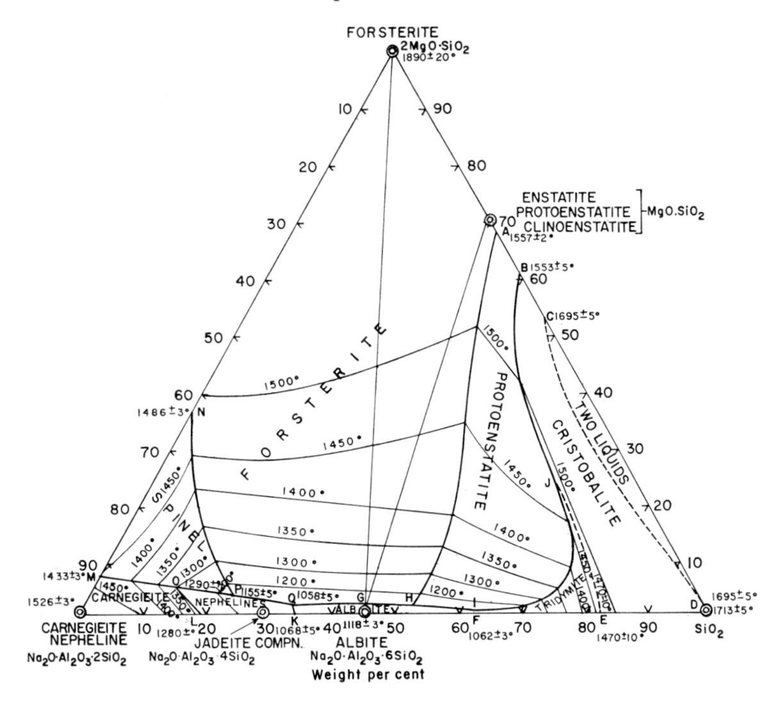

Fig. 4 The system Fo–Ne–Sil (after Schairer and Yoder, 1961). Symbols are explained in the text

Ne–Fo–Ab (Fig. 4), possible phase relations within the alkaline undersaturated part of the simple basalt tetrahedron (Di–Ne–Fo–Ab) can be deduced as shown in Fig. 5. The largest phase volume contains forsterite, a slightly smaller one diopsidic-pyroxene, and small volumes of albite (or plagioclase), nepheline s.s., carnegieite s.s. and spinel.

A number of important points arise from consideration of this tetrahedron:

1. Residual liquids trend toward the Ab apex indicating progressive enrichment in silica, soda and alumina with concomitant impoverishment of lime and magnesia. Therefore crystallization products of these liquids are rich in albite (or sodic plagioclase) and nepheline, and poor in olivine and pyroxenes.

2. The pyroxene in this system is aluminous with highest alumina contents occurring toward the Di–Ne–Ab face where the 'plagioclase effect' is operative. This enriches the coexisting liquids in soda and silica.

3. Melilites (possibly soda-rich) occur as subliquidus phases due to reaction relationships.

4. The system is not quaternary due to the formation of solid solutions and spinel toward the Ne apex.

The effect of addition of anorthite to the system Di–Ne–Fo–Ab was studied by Yoder and Tilley (1962) in the system Di–Ab–Fo–An. This tetrahedron (Fig. 6) consists of three major phase volumes (neglecting spinel)—forsterite, plagioclase and diopside. Petrologically this system is important in that, independent of the initial bulk composition, liquids reach the 4-phase curve (Di + Plag + Fo + L) and proceed along this curve in the direction of falling temperature toward a final liquid extremely rich in the albite component and impoverished in the mafic minerals, suggesting that adding anorthite does not change the general phase relations found in the system Di–Ne–Ab–Fo.

In order to discuss the trends of liquids leaving the simple basalt tetrahedron (Fig. 1), Schairer (1967) suggested an expanded basalt tetrahedron La–Ne–Fo–Qtz containing the

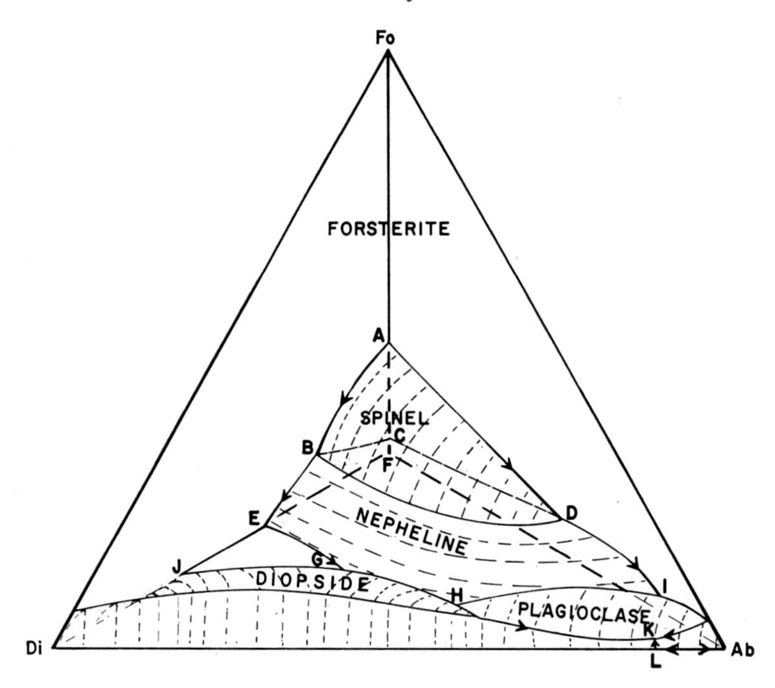

Fig. 5 Probable phase relations in the undersaturated portion of the simple basalt tetrahedron. ABCD indicates the spinel phase volume, BCDIFEGH indicates the nepheline phase volume, HIKLAb indicates the plagioclase phase volume, JGHLDi indicates the diopside phase volume. The remaining volume consists of forsterite. Arrows indicate general directions of falling temperature

simple basalt tetrahedron and the important mineral molecules akermanite, monticellite and wollastonite. Onuma and Yagi (1967) studied phase relations in one plane of this expanded system by investigating the system Di–Ak–Ne (Fig. 7).

The system Di–Ak–Ne has no invariant point but contains two piercing points at G, separating the melilite, forsterite, and diopside phase volumes in the CaO–MgO–SiO_2–$NaAlSiO_4$ system; and at H, on the boundary between the nepheline, forsterite and melilite volumes in the expanded system. Relationships of the stable phases, piercing points and temperatures are shown on the flow sheet of Fig. 8. From G (1212 °C), where forsterite, diopside and melilite coexist, temperature falls toward the Ne apex in the $NaAlSiO_4$–CaO–MgO–SiO_2 system. From H (1169 °C), temperature falls with crystallization of nepheline, forsterite and melilite. At I (1135 °C), the two quaternary univariant lines merge and nepheline, diopside, forsterite, melilite and liquid coexist. Below 1135 °C Schairer *et al.* (1962) suggest that olivine reacts with the liquid producing diopside and melilite and the temperature falls toward another quaternary invariant point.

In olivine melilitites, melilite, olivine and clinopyroxene occur in both the groundmass and phenocrysts, corresponding to G in the synthetic system (Fig. 7). Under equilibrium conditions a liquid cooling to I crystallizes nepheline, forsterite, diopside and melilite, (olivine melilite nephelinite), and at the same point forsterite reacts with liquid which, on further cooling, produces a nepheline, melilite and diopside assemblage (melilite nephelinite). From this system Onuma and Yagi infer that the differentiation trend in melilite-bearing rocks is from olivine melilitite to melilite nephelinite.

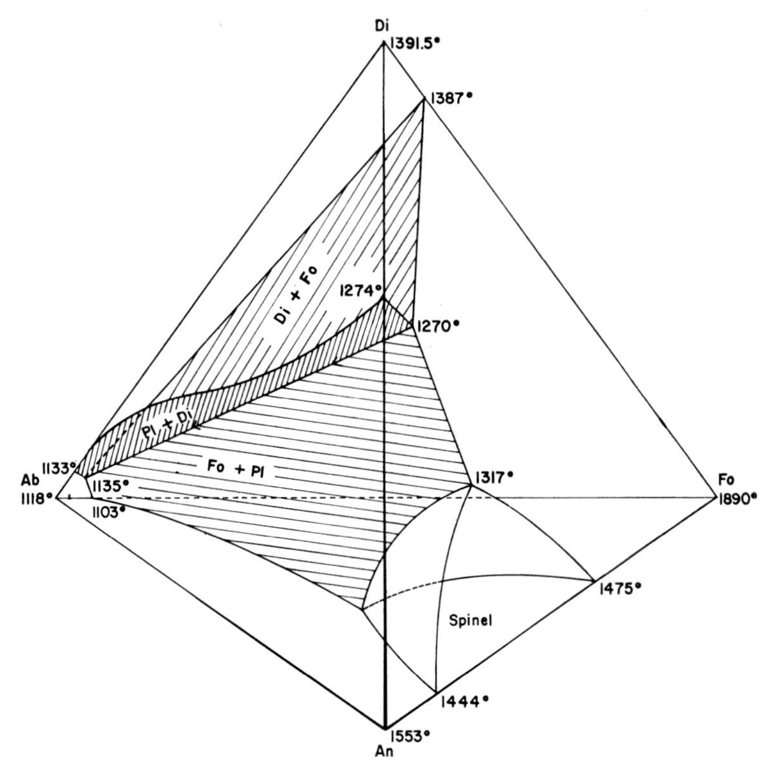

Fig. 6 The system Di–Ab–Fo–An (after Yoder and Tilley, 1962). The line joining the 1270 to 1135 °C is the 4 phase curve (Di + Plag + Fo + L)

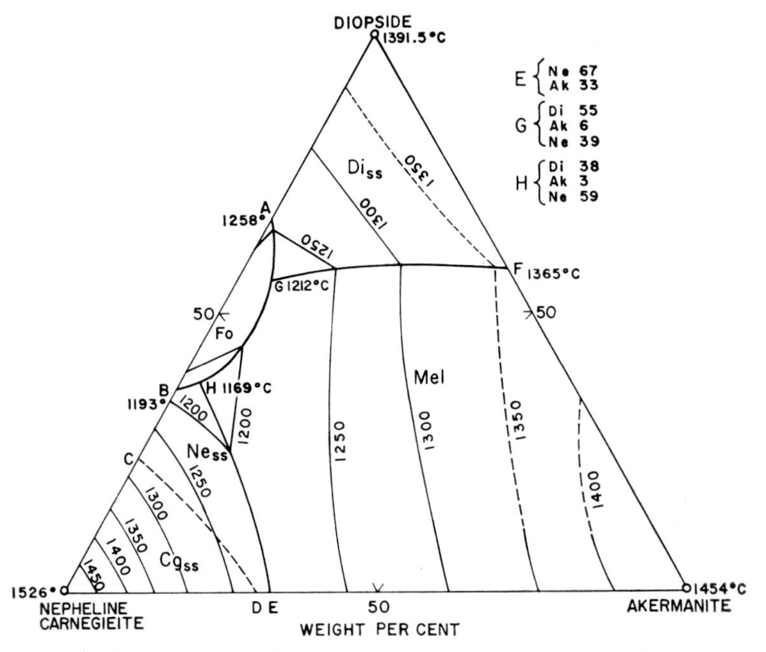

Fig. 7 The system Di–Ak–Ne (after Onuma and Yagi, 1967)

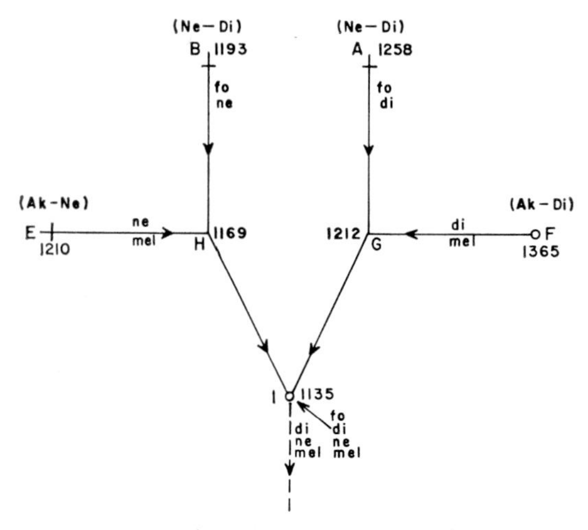

Fig. 8 Schematic flow sheet for portions of the Di–Ak–Ne system showing relationships between stable phases, piercing points and temperatures (after Onuma and Yagi, 1967)

In 1969, Yagi and Onuma (1969) investigated the system Di–Ak–Ne–$CaTiAl_2O_6$ by determining a series of slices through the tetrahedron Di–Ak–Ne–$CaTiAl_2O_6$, parallel to the Di–Ak–Ne base, at 5, 10, 15 and 20 wt. % $CaTiAl_2O_6$. Phase relations in this pseudo-quinary system are shown in Fig. 9 and the flow sheet for the system in Fig. 10. In the latter figure, points 1, 2 and 3 represent six-phase assemblages showing a reaction relationship, and point 4 a six-phase assemblage showing a eutectic relationship. Assemblages 1 and 2 (Fig. 10) have no natural analogues. The assemblage diopside, forsterite, melilite, nepheline corresponds to an olivine melilite nephelinite in which the olivine reacts with liquid at point 3 (Fig. 10) and proceeds toward point 4 (Fig. 10) producing diopside, melilite, nepheline and perovskite corresponding

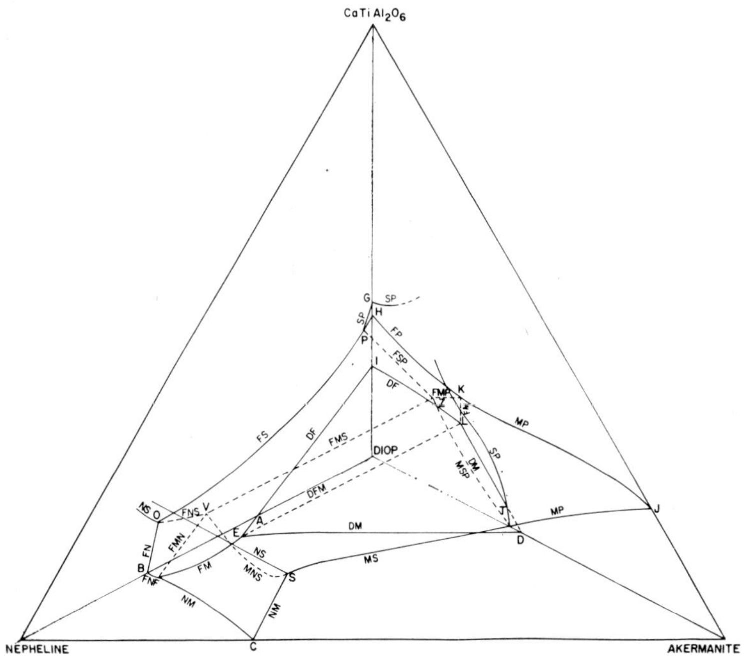

Fig. 9 The system Di–Ne–Ak–$CaTiAl_2O_6$ showing liquidus phase relations determined from the planes Di–Ne–Ak with 5, 10, 15 and 20 weight per cent $CaTiAl_2O_6$. Traces of surfaces between adjacent primary phase volumes are shown as heavy lines; heavy broken lines indicate intersections of three three-phase assemblages giving four-phase assemblages of three solids and a liquid; points represent four four-phase assemblages producing five-phase assemblages consisting of four solids and a liquid

D = diopside; F = forsterite; M = melilite; N = nepheline; P = perovskite; S = spinel; liquid omitted (after Yagi and Onuma, 1969).

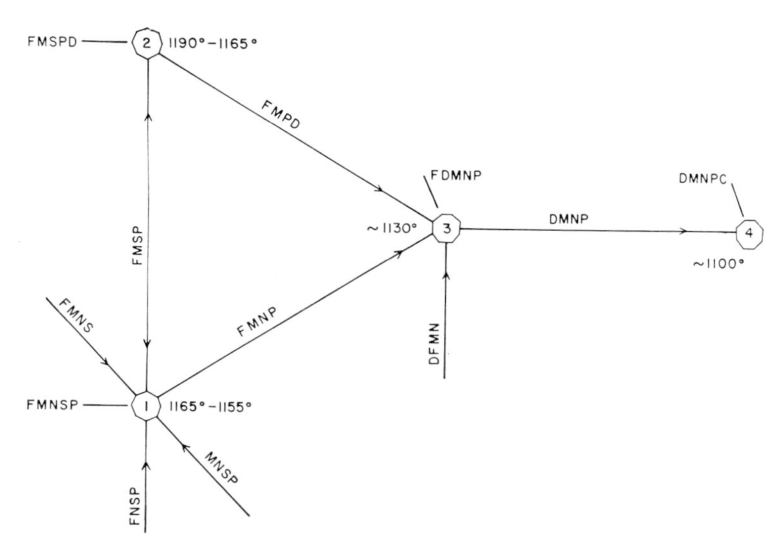

Fig. 10 Flow sheet showing relationships of five- and six-phase assemblages shown in Fig. 9. Abbreviations as for Fig. 9, C = corundum, liquid omitted (after Yagi and Onuma, 1969)

to a perovskite-bearing melilite nephelinite. On reaching point 4 corundum appears as an additional phase producing a hypothetical corundum-bearing melilite nephelinite. In nature corundum is not found in such rocks but Yagi and Onuma suggest that it may occur as hercynite or in mica or amphibole. Thus their inferred trend of differentiation is olivine melilitite → olivine melilite nephelinite → perovskite-bearing melilite nephelinite → corundum-bearing melilite nephelinite.

Yagi and Onuma also found that in compositions containing less than 4 wt. % TiO_2 the titanium occurred as solid solution in the clinopyroxene, whereas in compositions with greater than 4 wt. % TiO_2 titanium crystallized as perovskite. This is in agreement with natural occurrences of alkali basalts (Aoki, 1959, 1964, 1967; Yagi, 1953; Kushiro, 1964a; Tiba, 1966; Huckenholz, 1965a, b, 1966).

Another important experimental investigation pertaining to the differentiation of melilite nephelinite magmas is the system Na_2O–CaO–Al_2O_3–SiO_2 (Fig. 11) containing albite, anorthite, nepheline, gehlenite and wollastonite. Parts of this system have been determined by Foster (1942), Smalley (1947), Spivak (1944), Goldsmith (1949), Griffith (1944) and Gummer (1943), and its implications discussed by Bowen (1945). In Fig. 12 the portion of the system bounded by Geh–Wo–An–Ne–Ab has been redrawn in the same orientation as it occurs in the oxide system. Using this diagram Bowen discusses courses of crystallization with falling temperature from point A, representing the ternary eutectic in the Geh–Wo–An system, and the pseudo-ternary eutectics at points B, C and D in the Geh–Wo–Ne, Geh–An–Ne and Ne–Wo–Ab systems respectively. The lines AR, BR, CR and RD represent quaternary boundary curves, with the field of low CaO melilite being defined by the Geh apex and the solid angle formed from the convergence of the lines AR, BR and CR at R, an invariant point at which nepheline, wollastonite, melilite, plagioclase and liquid coexist. Thus the melilite phase volume extends beyond the Geh–An–Wo–Ne tetrahedron into the Ne–An–Wo–Ab tetrahedron. Along the line RD, nepheline, wollastonite, plagioclase (varying from An-rich at R to Ab-rich at D) and liquid coexist and temperatures decrease toward the Ab apex of the double tetrahedron.

Under perfect equilibrium conditions (corresponding to very rapid cooling), liquids lying

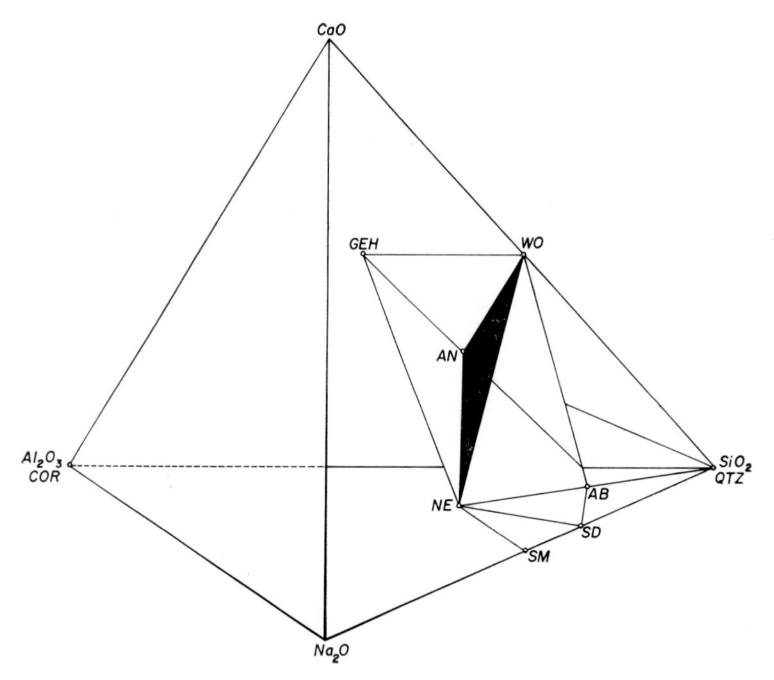

Fig. 11 The system Na_2O–CaO–Al_2O_3–SiO_2 showing compositions of the mineral molecules gehlenite, nepheline, anorthite, wollastonite, albite, and corundum. SM and SD represent sodium metasilicate and sodium disilicate respectively (after Bowen, 1945)

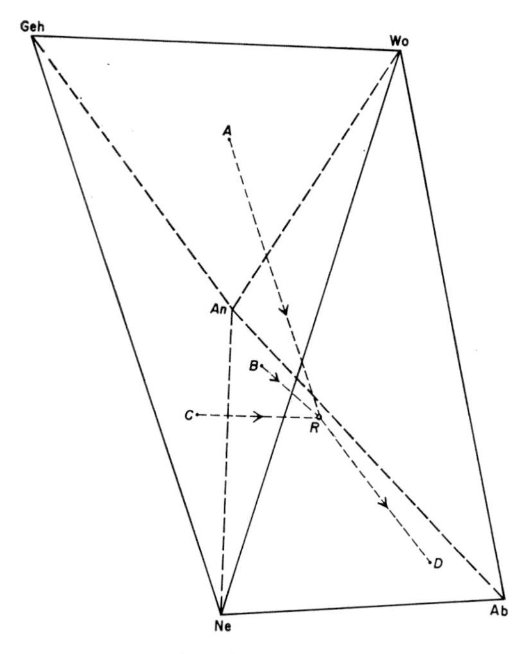

Fig. 12 Enlarged portion of the system shown in Fig. 11. Symbols are explained in the text (after Bowen, 1945)

close to the Geh–Ne–Wo system crystallize melilite, nepheline and wollastonite (a simplified melilite nephelinite) on the boundary curve BR. Under fractional crystallization conditions, liquids proceed toward R with the additional crystallization of An-rich plagioclase (a simplified tephrite). On continued fractionation, liquids move along RD crystallizing nepheline, wollastonite and a plagioclase, progressively richer in albite toward D, and approching a simplified phonolite in composition. Thus, under suitable crystallization conditions, a series of compositions, ranging from simplified melilite nephelinites to phonolites, can be produced.

This system illustrates the importance of the 'plagioclase effect' mentioned previously since, in any system with albite and a liquid containing CaO the crystallizing feldspar contains some anorthite, then residual liquids must have a component of motion away from the anorthite molecule and are enriched in Na_2O and SiO_2. Thus in the Geh–Wo–An–Ab–Ne system crystallization may proceed away from D towards another tetrahedron, one of whose faces is Ab–Wo–Ne with a fourth component being a compound such as $Na_2O.3CaO.6SiO_2$ (devitrite). Although not occurring in nature, devitrite develops at approximately the same stage as acmite which melts incongruently to hematite and a liquid enriched in Na_2SiO_3 producing, under natural conditions, a peralkaline rock containing normative Na_2SiO_3 (ns). Relations in this system suggest that under suitable conditions it is possible to produce a series of differentiates ranging in composition from a melilite nephelinite (or melilite basalt) to an alkali rhyolite through intermediate tephrites, basanites and phonolites.

Analyses of rocks with alkaline characteristics, show a general increase in alkalis and decrease in lime, magnesia and ferrous iron with progressive differentiation. Thus CaO, MgO, and FeO become less important in the genesis of rocks of the nepheline syenite type. The system Na_2O–K_2O–Al_2O_3–Fe_2O_3–SiO_2 contains virtually all of the important mineral molecules of those rocks—alkali feldspars, nepheline, kalsilite, acmite, leucite, and silica. Considerable portions of this system have been investigated both under atmospheric and P_{H_2O} conditions.

The most important portion of this system is the plane Ne–Ks–Sil (Fig. 13). Bowen (1937) termed this system 'Petrogeny's Residua System' because residual liquids from fractional crystallization will tend to be enriched in its components; thus both oversaturated and undersaturated rocks, ranging in composition from granites to nepheline syenites, can be represented by it.

Constituent joins in this system are Ne–Ks (Fig. 14) determined by Bowen (1917), Tuttle and Smith (1958); Ne–Sil (Fig. 15) determined by Greig and Barth (1938); Lc–Sil (Fig. 16) determined by Schairer and Bowen (1947); and Ab–K-Fp (shown in Fig. 17 at pressures up to 2kb P_{H_2O}) as determined by Schairer (1950) and Tuttle and Bowen (1958). The Ne–Ks system is characterized by an extensive solvus, the implications of which are discussed in section 1.5. In

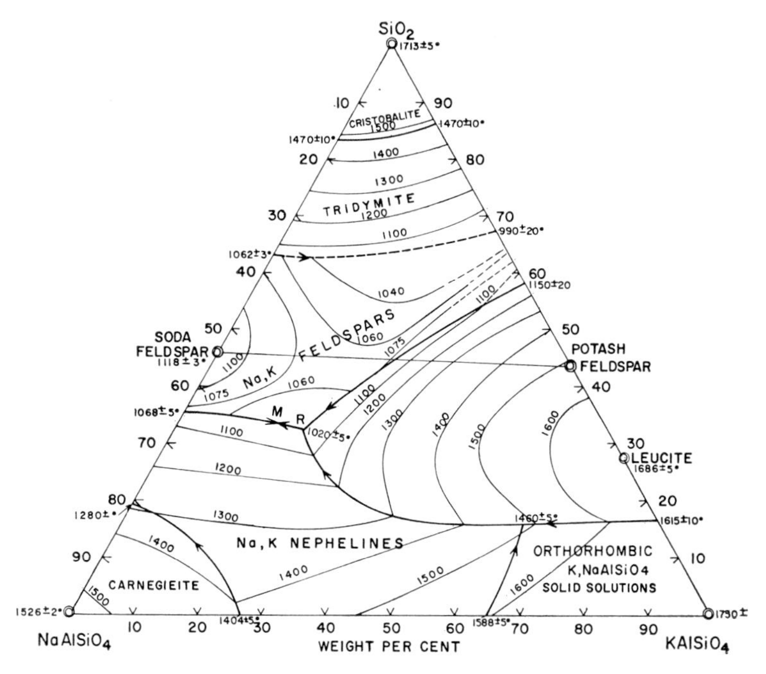

Fig. 13 The system Ne–Ks–Sil ('Petrogeny's Residua System') (modified from Bowen, 1937)

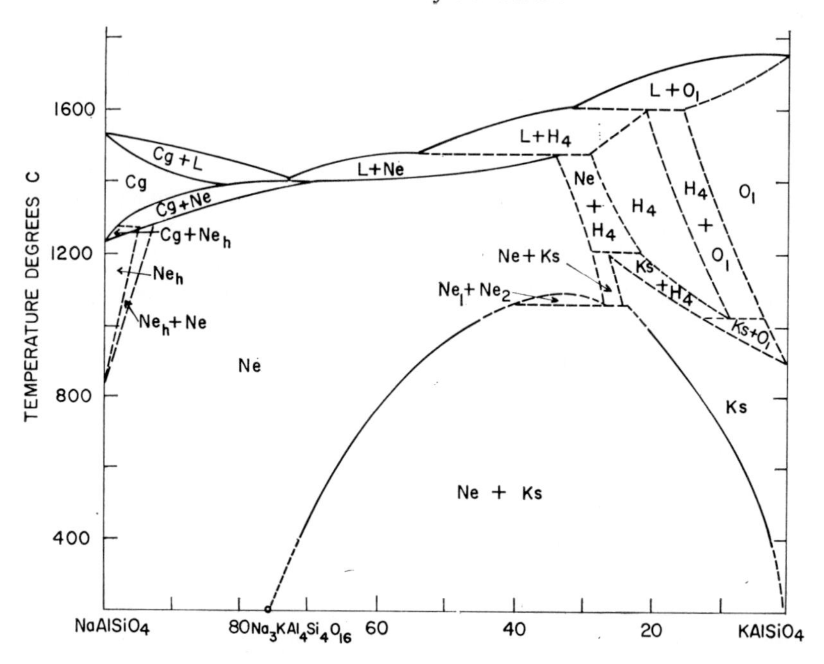

Fig. 14 The system Ne–Ks (after Bowen, 1917; Tuttle and Smith, 1958). For an explanation of the symbols see Tuttle and Smith (1958)

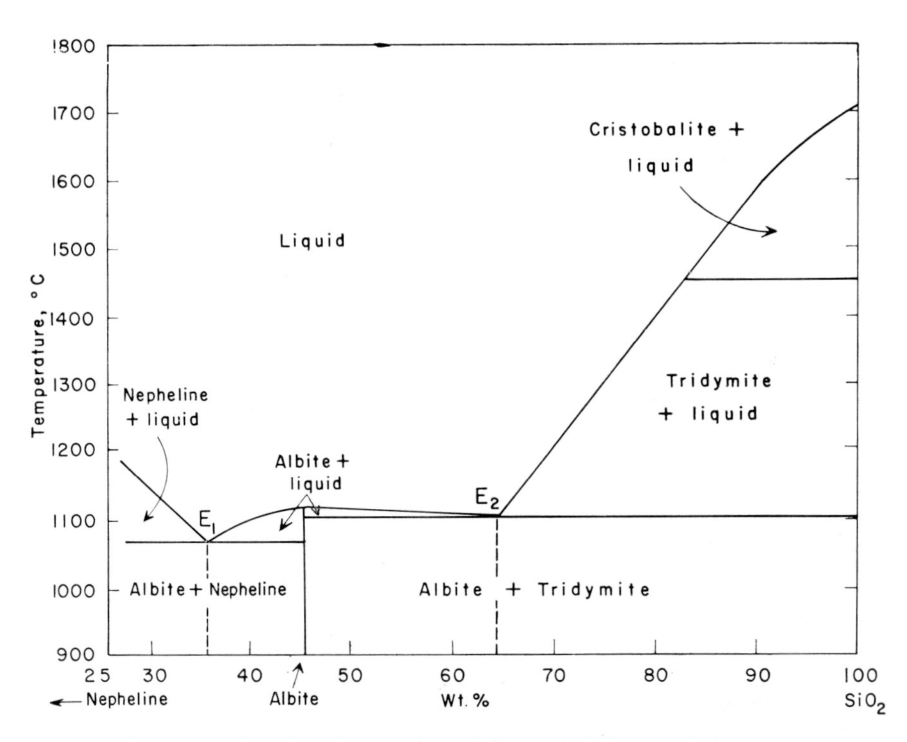

Fig. 15 The system Ne–Sil (after Greig and Barth, 1938). E_1 and E_2 represent the 'undersaturated' and 'saturated' eutectics respectively

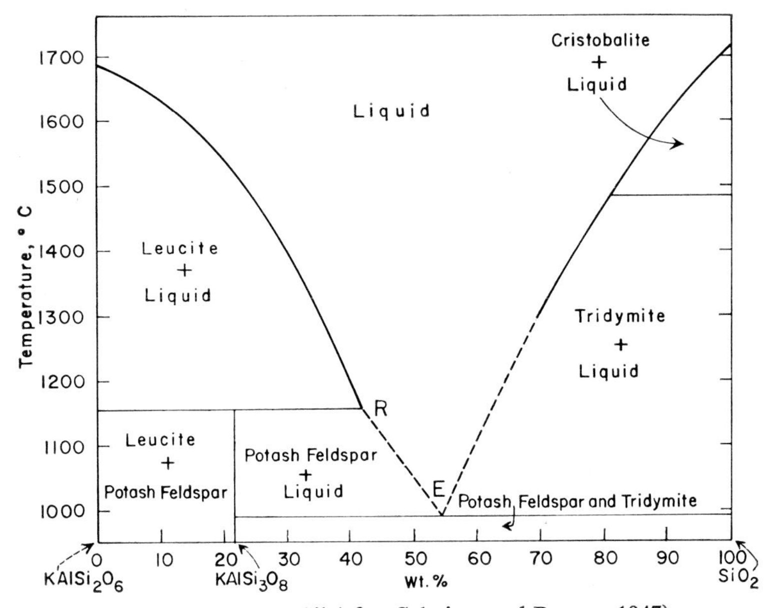

Fig. 16 The system Lc–Sil (after Schairer and Bowen, 1947)

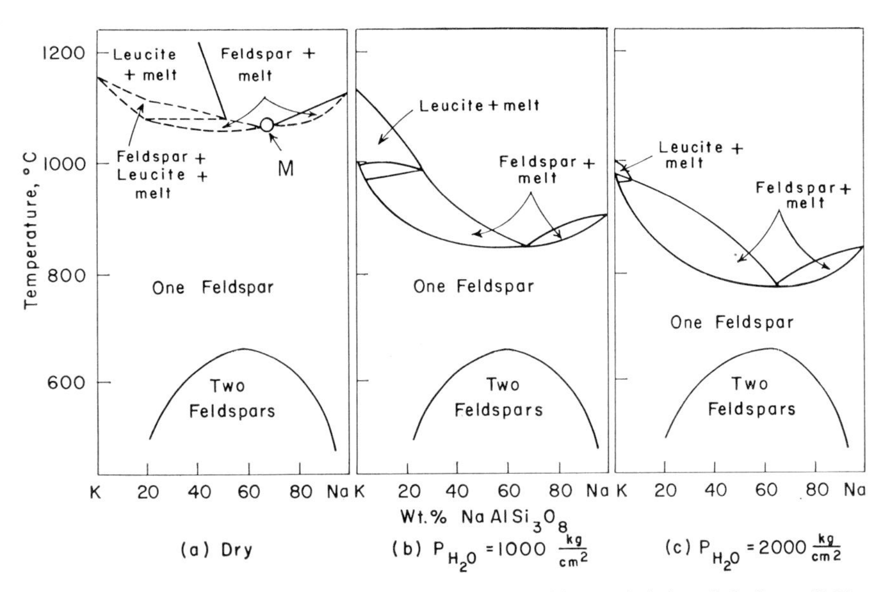

Fig. 17(a) (b) (c) The system Ab–K-Fp (up to 2 kb P_{H_2O}) (after Schairer, 1950; Tuttle and Bowen, 1958). M represents the minimum melting composition at 1 atm

the Ne–Sil join all melts initially richer in SiO_2 than $NaAlSi_3O_8$ crystallize oversaturated products, all those initially poorer in SiO_2 crystallize undersaturated products containing nepheline. The most notable feature of the Lc–Sil system is the incongruent melting of potash feldspar extending the primary leucite field to (R) containing about 20% SiO_2 in excess of the $KAlSi_3O_8$ composition. Thus leucite forms during early stages of crystallization of melts with excess silica. Fractional crystallization in this system may result in different crystalline products depending on the initial bulk composition of the liquid. The Ab–K-Fp join, divides the Ne–Ks–Sil system into silica-oversaturated and undersaturated portions. With increasing water vapour pressure the stability field of leucite contracts from about 50 wt. % $NaAlSi_3O_8$ at 1 atm. to a few per cent at 2 kb P_{H_2O}. This system also contains a solvus the crest of which remains virtually unaltered by pressure although the temperature minima at $Ab_{65}K\text{-}Fp_{35}$ decreases and changes slightly towards more Na-rich compositions with increasing P_{H_2O}.

The Ne–Ks–Sil system (Fig. 13) contains two minima, one above the feldspar join at 950 °C (the 'granite minimum'), the other below the feldspar at 1020 °C (the 'nepheline syenite minima'). Between these thermal troughs is a low-temperature, saddle-shaped area with its crest at 1063 °C forming a 'thermal barrier' between the troughs. In these low-temperature areas, residual liquids, resulting from fractional crystallization of multicomponent silicate systems, will trend toward the Na_2O–K_2O–Al_2O_3–SiO_2 plane along various thermal valleys in polydimensional space (Bowen, 1937). Thus final products of magmatic crystallization should have compositions plotting very close to the low-temperature areas of the Ne–Ks–Sil system. Normative compositions of many rocks representing the closing stages of magmatic activity, such as granites, syenites and nepheline syenites, fall within these thermal troughs in the system (Tuttle and Bowen, 1958; Hamilton and MacKenzie, 1965) strongly suggesting that a process of crystal $\rightleftharpoons$ liquid equilibria is involved in their genesis. The implication of these data are discussed in VI.2.

Crystallization courses within the residua system may be summarized as follows:

a. Compositions lying within the Sil–Ab–K-Fp area (except the leucite area) move toward the feldspar–silica boundary under equilibrium conditions and then proceed along this boundary to the minimum. Final products are silica and an alkali feldspar whose composition approaches $Ab_{65}K\text{-}Fp_{35}$.

b. Compositions lying within the Ne–Ks–Ab–K-Fp (except the leucite area) move toward the nepheline–feldspar boundary and then proceed toward the minimum on this boundary with final crystallization products being nepheline (of composition close to the 3Na:1K type) and feldspar.

c. For compositions within the leucite field, Fudali (1963) shows that the possible final assemblages are much more varied than previous authors had realized due to leucite being a solid solution rather than a stoichiometric compound. Under equilibrium conditions, and depending on the initial bulk composition, the following final crystallization products are possible, feldspar + silica, nepheline + feldspar, and nepheline + K-feldspar + leucite.

d. Under conditions of fractional crystallization, the number of possible products is greatly increased.

This system explains a number of petrographic phenomena observed in alkaline rocks.

a. By normal crystallization processes, liquids whose compositions lie in the undersaturated part of the system cannot produce differentiates containing free silica. Similarly, liquids lying in oversaturated portion cannot produce differentiates with nepheline.

b. Leucite occurs in the groundmass of rocks containing free silica whereas nepheline does not. In Fig. 13 the leucite field overlaps into the silica-oversaturated part of the system and any melt originating in this part of the leucite field can crystallize leucite in equilibrium with a melt containing excess silica. In contrast, nepheline can only crystallize from melts originally silica undersaturated.

c. Leucite is not present in plutonic rocks since on slow cooling it reacts with liquid producing alkali feldspar and nepheline at R (Fig. 13). Under plutonic conditions with high P_{H_2O}, leucite is less likely to crystallize even in the early stages of cooling due to decreasing stability of leucite with increasing P_{H_2O}.

d. Pseudoleucites are believed to form close to R (Fig. 13) where the leucite in the synthetic system is replaced by nepheline and potash feldspar. Their origin is discussed in section 1.5 of this chapter.

e. Compositions of coexisting alkali feldspars and nephelines depend on the rates of cooling. With slow cooling (plutonic or equilibrium conditions), nephelines approach the ideal 3 Na:1K composition. With rapid cooling (volcanic conditions), the feldspars in volcanic rocks will tend to be more K-rich than in their plutonic equivalents, and the nephelines more Na-rich.

Although the residua system is of fundamental importance to an understanding of the genesis of alkaline rocks, it is limited to subaluminous liquids with an alkali/alumina ratio of one. Peralkaline liquids, crystallizing acmite, normative sodium or potassium silicates (or their respective disilicates), cannot be represented by the system.

In addition to acmite, alkaline rocks contain appreciable amounts of other clinopyroxenes, principally diopside and hedenbergite. As yet there is no data on hedenbergite-alkali feldspar or hedenbergite–feldspathoid systems. Relations in the system Di–Ne–Ab (Schairer and Yoder, 1960a) have been discussed above and it has been shown that residual liquids, on equilibrium crystallization, form products enriched in salic minerals and impoverished in diopside.

In an attempt to remedy the deficiencies of petrogeny's residua system, particularly in explaining peralkaline liquids, Bailey and Schairer (1966) investigated the petrologically important part of the system Na_2O–Al_2O_3–Fe_2O_3–SiO_2, containing nepheline, carnegieite, albite, acmite, corundum, hematite, quartz, tridymite and the normative minerals sodium metasilicate (ns) and sodium disilicate (ds). From liquidus relationships, Bailey and Schairer discuss the genesis of a wide range of peralkaline rocks and suggest that the system Na_2O–Al_2O_3–Fe_2O_3–SiO_2 be termed the 'peralkaline residua system'. In order to delineate phase relations within this system, Bailey and Schairer studied the following joins through the tetrahedron Na_2O–Al_2O_3–Fe_2O_3–SiO_2: Ac–Jd–(Na_2O . $4SiO_2$), Ac–Ne–Sil, Ac–Ne–sodium disilicate, Ac–Ab–sodium disilicate, Ac–Ne–$5Na_2O$. Fe_2O_3 . $8SiO_2$, Ac–Ne–(Na_2O . $4SiO_2$). From these slices through the volume of the tetrahedron, intercepts of divariant surfaces, univariant lines, piercing points and finally quaternary reaction points and eutectics were determined. From these joins, seventeen piercing points, representing intercepts of univariant lines within the quaternary system, together with the ternary invariant points on the boundary systems were plotted on a Jäneche-type projection from the Ac apex on to the Na_2O–Al_2O_3–SiO_2 base (Fig. 18). Such a projection shows volumes in which quaternary invariant points fall and the relative positions of the univariant curves in the quaternary system. Some important relations found in the six joins are:

a. Due to the incongruent melting of acmite and the reaction of hematite with liquid producing ferriferous albites and nephelines, hematite is a primary phase over much of the liquidus.

b. Acmite may crystallize in equilibrium with nepheline + albite and quartz + albite only from liquids containing normative sodium silicate.

c. Many of the piercing points are close to ternary eutectics but solid solution (principally of Fe_2O_3) has moved them slightly away from ternary invariant points.

Using the Jäneche projection, Bailey and Schairer construct a schematic flow diagram (Fig. 19) showing how the univariant lines are linked by fifteen invariant points. Although this diagram is schematic, high-temperature relations are represented at the top of the diagram, low-temperature conditions at the bottom. Silica oversaturated assemblages are found on the

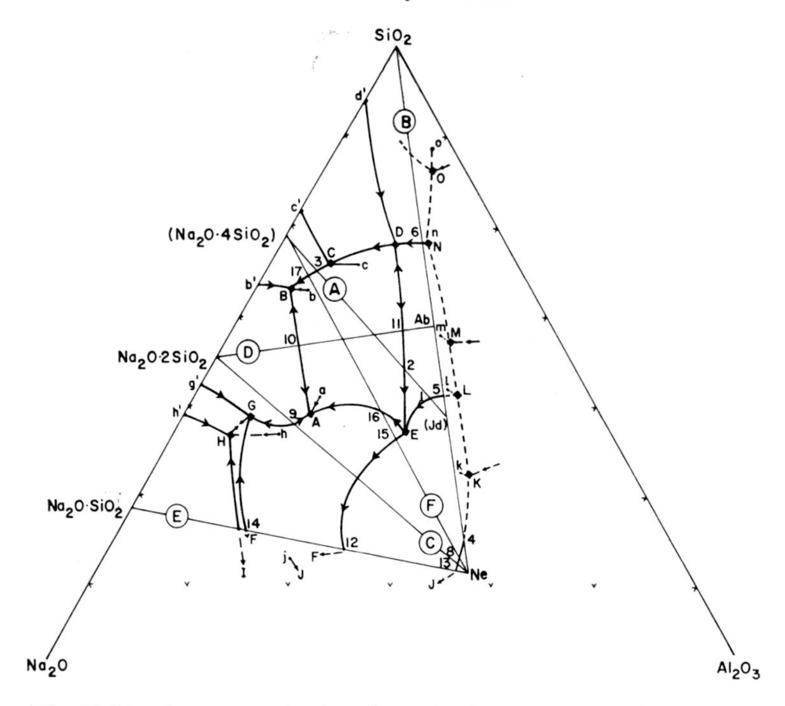

Fig. 18 Jäneche-type projection from the Ac apex on to the Na_2O–Al_2O_3–SiO_2 plane of the system Na_2O–Al_2O_3– Fe_2O_3–SiO_2 (after Bailey and Schairer, 1966). For explanation of symbols see Bailey and Schairer (1966, Fig. 14)

right-hand side of the diagram, peralkaline silica undersaturated conditions on the left-hand side. The most important points in Fig. 19 are the quaternary reaction points D and E, and the quaternary eutectics A and B. Join D, lying between the reaction points E and D and eutectics A and B, forms the boundary between silica oversaturated compositions, crystallizing tridymite (point D) and quartz (point B), and silica undersaturated compositions, crystallizing nepheline (points E and A). Between the eutectics A and B, join D forms a 'thermal barrier'.

On the basis of the experimental work summarized in Fig. 19 Bailey and Schairer make a number of petrological implications as follows:

a. At eutectic B, corresponding to peralkaline oversaturated (or pantelleritic) melts in nature, liquids are rich in potential sodium silicates and are therefore removed from the granite minimum in the residua system. The composition of this eutectic is relatively poor in silica and such compositions in nature might become undersaturated by reaction with aluminous material producing nepheline at granitic contacts without the mechanism of limestone desilication reactions. The temperature of point B also shows that pantelleritic liquids may exist at very low temperatures.

b. Eutectic A corresponds to peralkaline undersaturated (or phonolitic) melts in nature. From the quaternary reaction point E, corresponding in composition to an ijolite, liquids may become impoverished in acmite and enriched in sodium disilicate as they move to point A (a 'foyaitic' trend) or they may, under disequilibrium conditions, move to point F (a 'melteigitic' trend).

c. Liquid trends in the Na_2O–Al_2O_3–Fe_2O_3–SiO_2 system are controlled largely by the incongruent melting of acmite. If the reaction hematite + liquid→ acmite is not completed, and as a result Fe^{+3} can enter the feldspar structure, a mechanism for transition from undersaturated to

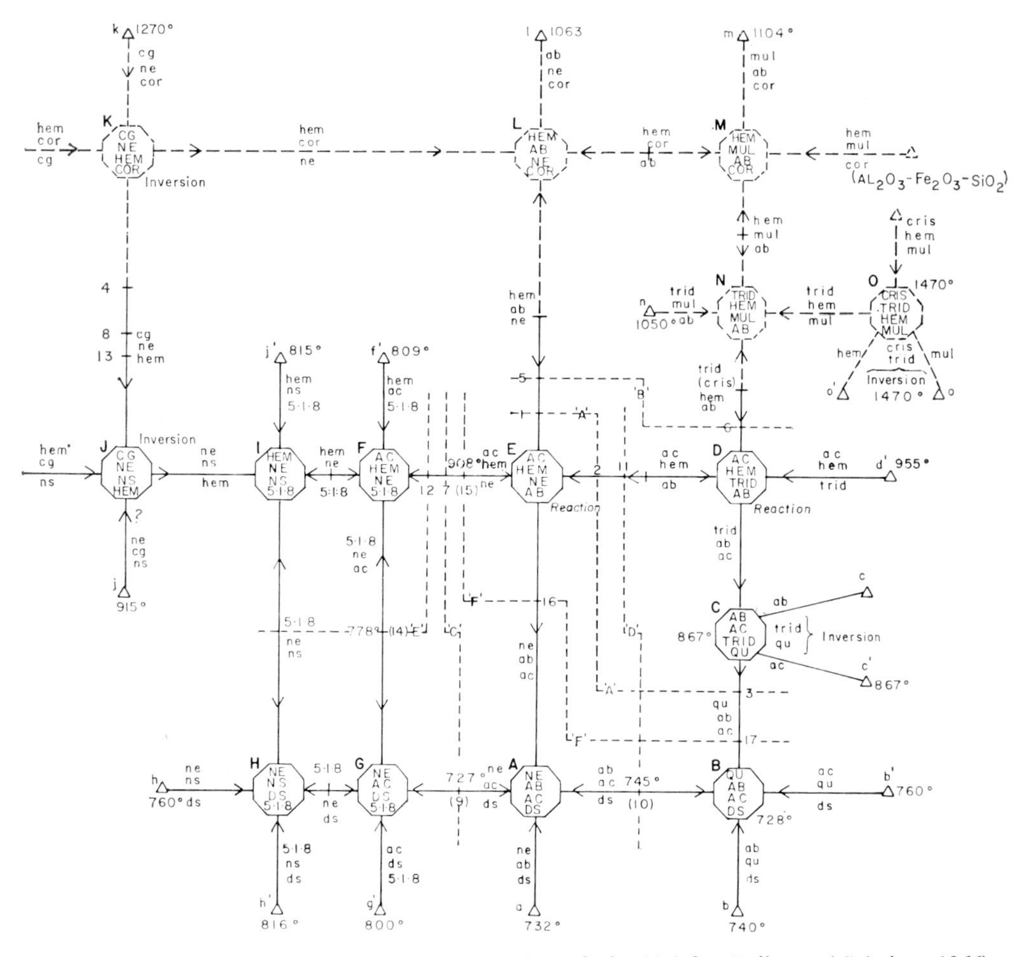

Fig. 19 Schematic flow diagram based on projection of Fig. 18 (after Bailey and Schairer, 1966). For explanation of symbols see Bailey and Schairer (1966, Fig. 15) and the text

oversaturated liquids, and vice versa, is achieved. Syenitic liquids may differentiate into oversaturated or undersaturated products by such a process.

d. Sodium silicate in the liquid is probably a prerequisite for the crystallization of acmite. Therefore Bailey and Schairer suggest that the procedure of calculating ns after ac in CIPW norms is wrong.

e. The four univariant curves leading to point E indicate that ijolitic magmas may be derived by partial melting of slightly undersaturated acmitic syenite, or as residual liquid from crystallization of a peralkaline syenite. Lack of a detectable temperature difference between points E and F (containing no feldspar) indicates that there is no significant barrier to the development of an ijolite–melteigite trend. Lime-rich ijolites may be derivatives of melilite basalts by progressive melting in the mantle, with subsequent differentiation under crustal conditions.

The first study of the addition of a basic mineral to the residua system was Sood, Platt and Edgar's (1970) study of the Di–Ne–Ks–Sil system which may be considered as the potash-rich extension of the simplified basalt tetrahedron (Di–Ne–Fo–Sil) of Yoder and Tilley (1962), thus contributing to our understanding of the genetic connection between felsic undersaturated alkaline lavas of both potassic and sodic types with their basic counterparts, the alkali basalts. From studies of the joins Di–Ne–K-Fp and Di–Ab–Lc, combined with data from the system Di–Ab–K-Fp (Morse, 1969), Sood, Platt and Edgar located a quaternary reaction point and postulated a quaternary minimum within the undersaturated volume of the system Di–Ne–Ks–Sil whose compositions and temperatures closely correspond with analogous points in the residua system (Fig. 13). Details of the phase chemistry are given in Sood, Platt and Edgar (1970). Of particular importance is the volume of primary forsterite crystallizing near the Di–Ne join in this system and which is an extension of the forsterite volume in the simple basalt tetrahedron. Within this volume, liquids containing less than approximately 10 wt. % sanidine crystallize melilite as a sub-liquidus phase over a short temperature interval, whereas liquids with greater than 10 wt. % sanidine do not crystallize melilite. On further cooling both melilite and olivine disappear by reaction with liquid which, in certain cases, lies outside the volume Di–Ne–Ks–Sil. See Platt and Edgar (1972).

By fractionation of compositions crystallizing olivine and/or melilite, Sood, Platt and Edgar deduced the sodic trend from basic to felsic lavas as shown in Fig. 20a. Although the three rock types shown at the top of this figure cannot coexist, on fractionation the differentiates of these rocks may produce a nephelinite, which, on further fractionation, may crystallize assemblages corresponding to a phonolite or wollastonite phonolite. In nature, wollastonite can form solid solutions with pyroxenes or combine with TiO_2 forming titanite.

The predominantly potash trend is produced by fractionation of liquids originating in or near the leucite volume as shown in Fig. 20b. Depending on the stage of fractionation, three trends are possible as indicated by the numbers. In addition, it is possible by suitable fractionation of liquids within a narrow compositional range to produce a silica undersaturated (leucitite) to silica-oversaturated ('rhyolite') trend.

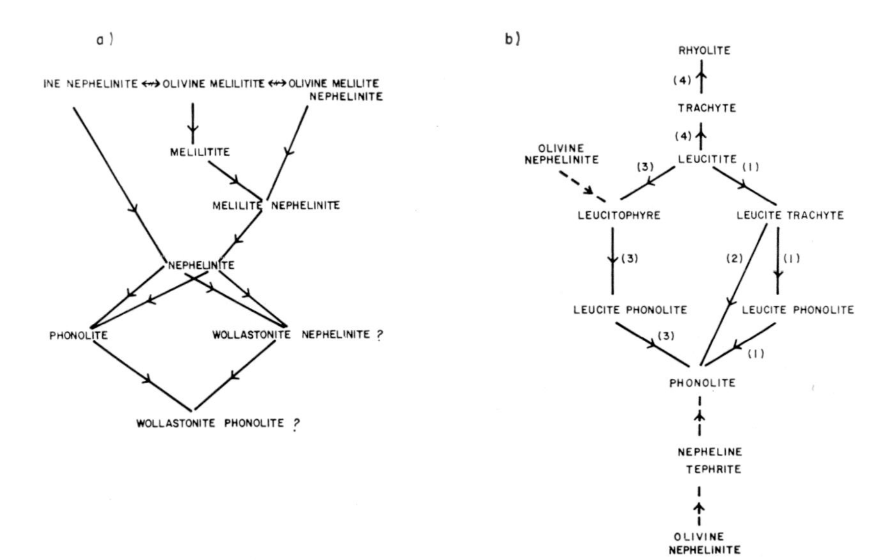

Fig. 20 (a) The sodic trend of basic to felsic alkaline lavas inferred from fractional crystallization of compositions in the system Di–Ne–Ks–Sil crystallizing olivine and/or melilite (after Sood, Platt and Edgar, 1970)
(b) The potassic trend of basic to felsic alkaline lavas inferred from fractional crystallization of potash-rich compositions in the system Di–Ne–Ks–Sil (after Sood and Edgar, 1972). Numbers refer to possible trends. Dashed lines indicate interrelationship with sodic trend

A number of examples of both of these trends have been described in rocks of the Roman Petrographic Province and from the alkaline rocks of the African Rift system (cf. Washington, 1896, 1897; Holmes and Harwood, 1937; Wright, 1963; Saggerson and Williams, 1964; King, 1965).

From known phase relations in parts of the basalt system (Di–Fo–Ne–Sil) and the Di–Ne–Sil system, Sood and Edgar (1972) postulate that the parental magmas for the sodic and potassic trends may be alkali olivine basalt (or olivine nephelinite) and olivine leucitite (ugandite) respectively. The extent of reaction between forsterite and liquid in the early stages of differentiation and leucite and liquid in the late stages may be the dominant mechanism which produces ultimate convergence of these trends resulting in phonolites.

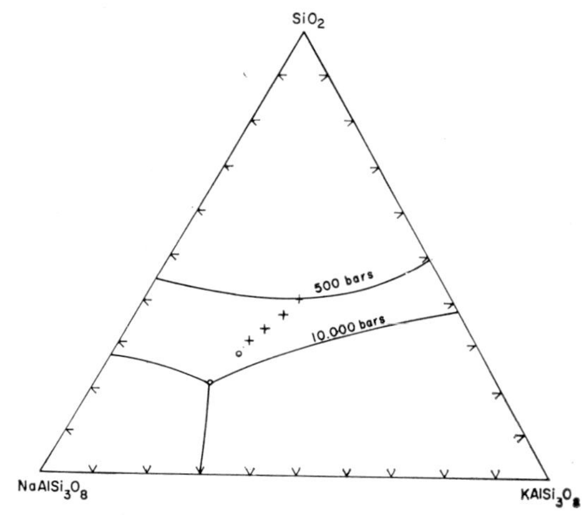

Fig. 21 The system Ab–K-Fp–Sil showing progressive shift in the minimum melting temperatures with increasing P_{H_2O} (after Luth *et al.*, 1964). Crosses represent minima, circles eutectics

V.1.3. Systems Investigated under Volatile and Controlled Oxygen Pressures

Although many systems containing volatiles have been investigated, this section will be concerned with those containing H_2O and done under controlled P_{O_2} conditions as Wyllie (1966) has recently covered systems with other volatiles (cf VI.3 and VI.4). Most systems with H_2O as an additional component are pertinent to the genesis of felsic rather than to basic alkaline rocks. Systems dealing with the hypothesis of limestone assimilation are covered in VI.3.

The importance of water in the late stages of magmatic differentiation has led to studies of the residua system up to 10 kb P_{H_2O}. In the silica oversaturated system Tuttle and Bowen (1958) investigated relations up to 4 kb P_{H_2O} and Luth *et al.* (1964) extended these to 10 kb P_{H_2O}. From 0·5 kb to 10 kb P_{H_2O} there is a progressive and almost direct shift in the minimum melting temperatures to more $NaAlSi_3O_8$-rich compositions with increasing P_{H_2O} (Fig. 21). Luth *et al.* (1964) plotted plutonic rocks containing normative Di + Ac or sodium metasilicate (corresponding to modal alkali amphiboles and pyroxenes) on the Sil–Ab–K-Fp system and showed that these rocks do not correspond in composition to the minima with increasing P_{H_2O}. This suggests that alkali-rich (or alumina undersaturated) granitic rocks may have crystallized in a relatively dry environment. Alumina undersaturated hypersolvus granites are also believed to form from fairly dry magmas. Luth *et al.* suggest that pantellerites and other alumina undersaturated lavas may be directly derived from water-poor basic magmas.

Carmichael and MacKenzie (1963), from a study of liquidus surfaces for two planes in the Ab–K-Fp–Sil with 4·5 wt. % Ac + 4·5 wt. % sodium metasilicate and 8·3 wt. % Ac + 8·3 wt. % sodium metasilicate, showed the minimum shifted with increasing Ac and sodium metasilicate contents toward the K-Fp–Sil join when projected on to the Ab–K-Fp–Sil plane. A similar shift was noted in the 'thermal valley' between the alkali feldspar minimum and the quartz-feldspar field boundary (Fig. 22). Included in Fig. 22 are recalculated analyses of 15 pantellerites and their feldspar phenocrysts, whose compositions lie parallel to the 'thermal valley' for the 8·3 wt. % Ac + 8·3 wt. % sodium metasilicate plane. Since these compositions do not correspond to rhyolitic liquids, Carmichael and MacKenzie conclude that pantellerites may have formed by fractional crystallization from a trachytic liquid by a process involving the

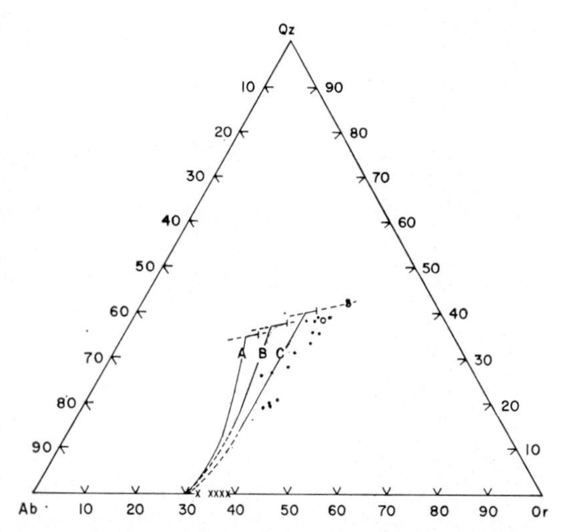

Fig. 22 Isobaric projections at 1 kb P_{H_2O} of the 'thermal valleys', their intersections with the quartz–feldspar boundaries, and the composition of the liquidus minima for the systems (A) Ab–K-Fp–Sil–H_2O; (B) the 4·5 per cent Ac + 4·5 per cent ns plane; (C) the 8·3 per cent Ac + 8·3 per cent ns plane. Solid circles represent recalculated analyses of pantellerites, open circles represent pantellerite residual glasses and crosses represent the feldspar phenocrysts from pantellerites (after Carmichael and MacKenzie, 1963)

'plagioclase effect' whereby, with decreasing temperature, residual trachytic liquids will be enriched in sodium silicate by early fractionation of plagioclase. If the amount of sodium silicate in the residual liquid is large, the resulting differentiates will be pantellerites; if small, comendites.

Carmichael and MacKenzie's method of recalculation of the pantellerite analyses and their mechanism for derivation of pantellerites from trachytic liquids has been criticized by Bailey and Schairer (1964) on three grounds: First that ns in the norm should be calculated as sodium disilicate thereby avoiding apparent excess quartz in the norm. Second, the practice of assigning all K_2O to normative orthoclase results in peralkaline characteristics being expressed only in terms of Na_2O. Third, that by ignoring part of the Na_2O content of their peralkaline liquids, Carmichael and MacKenzie failed to show the alkali balance between liquids and their feldspar phenocrysts which in general are more potassic than their parent liquids. Bailey and Schairer coin the term 'orthoclase effect' (analogous to the plagioclase effect) for the process whereby separation of alkali feldspar from a slightly alumina-deficient liquid may fractionate alumina and potash producing strongly peralkaline and sodic residual liquids.

Further evidence for the origin of peralkaline acid rocks and an evaluation of the importance of the 'orthoclase effect' (Bailey and Schairer, 1964), is provided by Thompson and MacKenzie's (1967) study of four pseudoternary planes in the system Ab–K-Fp–Sil at 1 kb P_{H_2O} by adding 5% Na_2SiO_3, 5% Ac, 5% K_2SiO_3 and 5% $NaKSiO_3$ by weight. With the addition of alkali silicates, the low temperature volume in the condensed portion of the system SiO_2–Al_2O_3–Na_2O–K_2O–H_2O (equivalent to the 'thermal valley' in the Sil–Ab–K-Fp system) moves rapidly toward the SiO_2–Al_2O_3–Na_2O plane; movement of liquid toward this volume being controlled by the compositions of the precipitating feldspar with K^+ moving preferentially into the feldspar. These results amplify Bailey and Schairer's (1964) concept of an 'orthoclase effect' and permit a wider possible parentage for pantellerites and comendites, namely sodic or potassic trachytes and rhyolites. Similarly, peralkaline granites may be derived by fractional crystallization or melting of granites or syenites. In both cases the parent magma should be either slightly peralkaline, or become peralkaline through the operation of the 'plagioclase effect' or some other mechanism.

Hamilton and MacKenzie (1965) studied the undersaturated part of the residua system at 1 kb P_{H_2O} (Fig. 23). At this pressure the leucite field is contracted and the minimum temperature lowered, although the minimum composition is very similar to that of the dry system (Fig. 13) and the smaller Lc field readily distinguishes the reaction point (R) from the minimum (M).

The following petrological implications can be made from this study:

a. The 'thermal barrier' on the feldspar join of this system is not lowered at 1 kb P_{H_2O} but becomes more prominent. Thus the presence of

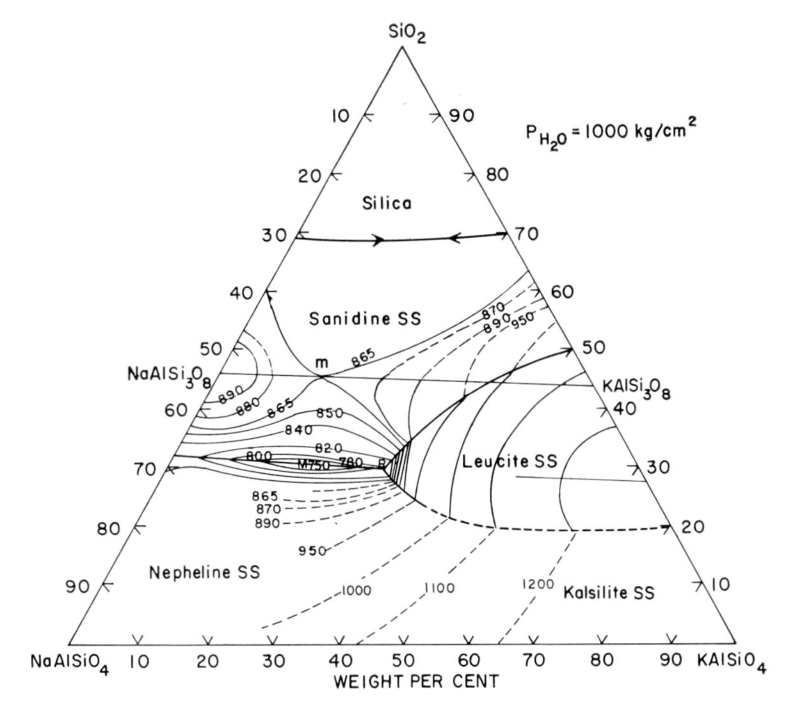

Fig. 23 The system Ne–Ks–Sil at 1 kb P_{H_2O} (after Hamilton and MacKenzie, 1965)

P_{H_2O} does not provide a mechanism for initially undersaturated compositions to produce oversaturated differentiates and vice versa.

b. The close correspondence between normative compositions of intrusive and extrusive rocks between the minimum in this system and the 'thermal valley' running from the feldspar join suggests that some process of crystal⇌liquid equilibrium has been operative in the genesis of many of these rocks. Lack of sediments of suitable composition lead Hamilton and MacKenzie to reject fractional melting as a possible process for the genesis of undersaturated alkaline rocks.

c. No difference in distribution was found for rocks with different Na_2O/K_2O ratios, although those with low ratios might have been expected to plot on the K-rich side of the minimum if they are derivatives of trachytic liquids. Those compositions plotting on this part of the system may be derived from an already undersaturated K-rich magma rather than a trachytic one, or other constituents present in natural liquids may modify the 'thermal valley' when projected onto the residua system.

d. Few liquids have approached the minimum from nepheline-rich compositions. Hamilton and MacKenzie suggest this may simply be a consequence of their method of plotting rock compositions.

Morse (1969), from studies of parts of the Ne–Ks–Ab–K-Fp system up to 5 kb P_{H_2O} (Fig. 24), found a narrow field of primary analcime s.s. lying between nepheline s.s. and Na-feldspar s.s. fields, resulting in a eutectic (E) and reaction point (R) rather than the minimum found at lower P_{H_2O}. This implies that rocks containing nepheline and albite must have formed at P_{H_2O} less than 5 kb. At this pressure, Morse finds no evidence for lowering of the 'thermal barrier' in the residua system and little change in the composition of the eutectic in comparison to the minimum determined at lower pressure. This implies that nepheline syenite-type rocks show only limited compositional variation, irrespective of the geological

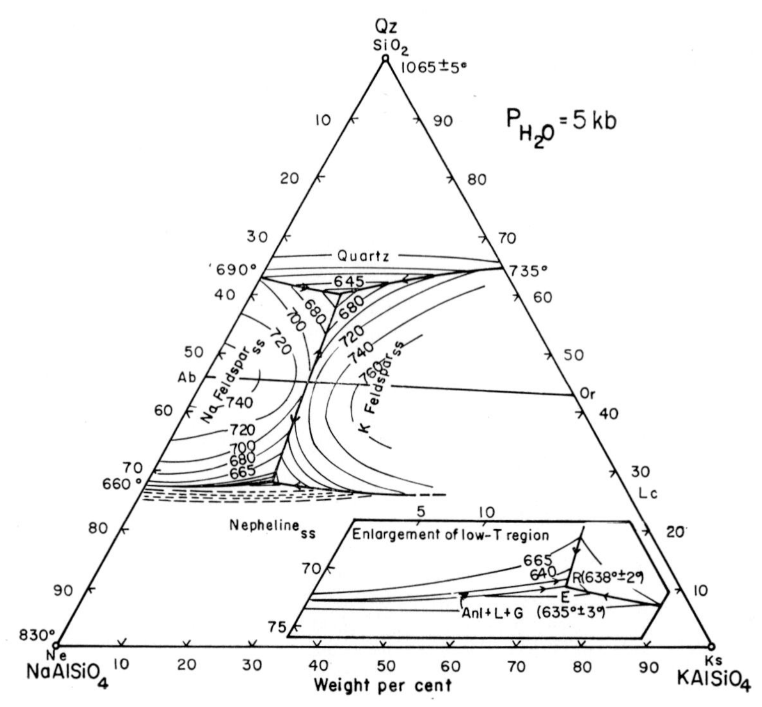

Fig. 24 The system Ne–Ks–Sil at 5 kb P_{H_2O} (after Morse, 1969)

pressure conditions under which they may have formed.

To assess the effects of adding pyroxenes to the residua system, Edgar (1964b) studied parts of the system Di–Ne–Ab–H_2O at 1 kb P_{H_2O}. In addition to lowering liquidus temperatures by about 200 °C in comparison to those at 1 atm (Schairer and Yoder, 1960b—See also Fig. 2), the field of primary plagioclase is considerably reduced at this pressure. Plagioclase synthesized in the Di–Ab–H_2O part of this system is approximately An_{10-20}, indicating the magnitude of the 'plagioclase effect'. From the close correspondence between rock compositions and the minimum melting composition, Edgar confirmed that genesis of many nepheline syenite-type rocks must have involved crystal $\rightleftharpoons$ liquid equilibria processes and that these rocks may have a syenitic parentage.

As iron-bearing pyroxenes are important constituents of alkaline rocks (Tyler and King, 1967), Nolan (1966) and Edgar and Nolan (1966) investigated portions of the system Ab–Ne–Ac–Di to 2 kb P_{H_2O} and under controlled P_{O_2} conditions. This system incorporates two important pyroxene components to the K-poor, silica undersaturated part of the residua system, and represents joins within the six-component system Na_2O–Al_2O_3–SiO_2–Fe–O–H. The position of the pyroxene composition planes studied by Nolan are shown in Fig. 25 with a composite phase diagram for relations at 1 kb P_{H_2O} being given in Fig. 26. The salient features of these relations are:

a. Lines separating primary phase volumes (unlabelled here for clarity) are traces of curved boundary surfaces within the six-component system.

b. The stippled area represents the primary phase volume (projected) of the plane Ab–Ne–Ac_{100} (the basal plane of the tetrahedron in Fig. 25). P is a piercing point ($Ab_{15}Ne_{30}Ac_{55}$ at 715 °C).

c. The ruled area represents the primary phase volume (projected) of plagioclase of the plane Ab–Ne–Di_{100} (the left-hand face of the tetra-

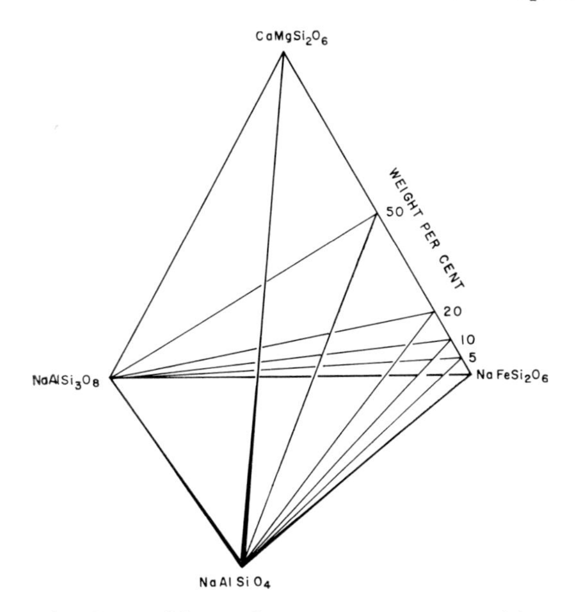

Fig. 25 Positions of the pyroxene composition planes studied by Nolan (1966) in the system Ab–Ne–Ac–Di at 1 kb P_{H_2O} (after Nolan, 1966)

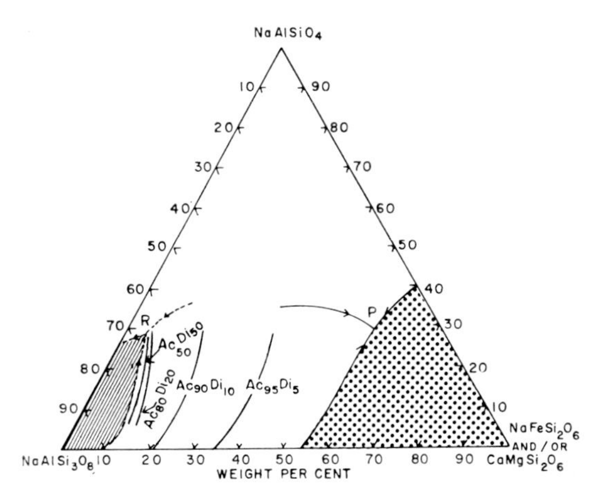

Fig. 26 Composite diagram for pyroxene composition planes given in Fig. 22 (after Nolan, 1966). Symbols are explained in the text

hedron in Fig. 25). R is a reaction point ($Ab_{66}Ne_{29}Di_5$ at 885 °C).

d. As shown by the traces of boundary surfaces (labelled $Ac_{95}Di_5$ etc.), the pyroxene phase volumes increase markedly with increments of the diopside component until the boundary surface of the plane Ab–Ne–$Ac_{50}Di_{50}$ and Ac–Ne–Di_{100} almost coincide. The exact nature of the boundary surface between points P and R is not known.

Determinations of plagioclase (see Edgar, 1964b) and pyroxene compositions led Nolan to conclude that both these phases as well as the liquids moved off the compositional planes investigated.

No correspondence was found between rock compositions and point P in the synthetic system Ab–Ne–Ac_{100} up to 2 kb P_{H_2O}, suggesting that crystal ⇌ liquid equilibrium processes were unlikely in the genesis of these rocks; but good correlation between the rock compositions and the inferred low melting area in the planes Ab–Ne–$Ac_{50}Di_{50}$, Ab–Ne–$Ac_{80}Di_{20}$, and Ab–Ne–$Ac_{90}Di_{10}$ was found. These comparisons indicate that the pyroxene composition is of major importance in controlling the position of the low-melting point for each plane studied. From the rock distribution plots (Nolan, 1966, Figs. 9–10) it is evident that, if crystal ⇌ liquid processes have been involved in their genesis, their pyroxenes would be expected to have compositions in the range $Ac_{80}Di_{20}$ to Ac_0Di_{100} (ignoring the hedenbergite molecule); values in agreement with the compositional range for pyroxenes in these rocks (Tyler and King, 1967). Agpaitic rocks (Ussing, 1912; Sørensen, 1960) plot very close to the inferred low melting point in the Ab–Ne–$Ac_{95}Di_5$ plane. In these rocks, with an excess of alkalis to alumina, the typical pyroxene is aegirine, in which Na^+ combines with Fe^{+3} rather than Al^{+3}, as in the non-agpaitic rocks. Sørensen (1969) has used this system to explain the order of crystallization in the rhythmically layered peralkaline intrusion at Ilímaussaq, Greenland.

Although pyroxene composition is the major factor controlling the sizes of the volumes of primary pyroxene in this system, increase in total P_{H_2O} and in f_{O_2} will also shift point P (Fig. 26) toward the maximum density distribution of the normative rock compositions. In natural liquids the f_{O_2} depends on original Fe_2O_3/FeO ratios; high ratios producing magnetite within its stability field. With rapid cooling, reaction between magnetite and liquid is limited and both Na_2O

and SiO_2 will probably be retained in the liquid; with slow cooling, this reaction will be completed. Increasing potential diopside in the liquid also deters the magnetite–liquid reaction since Yagi (1962) has shown that pyroxenes in the Di–Ac system only melt incongruently at more than 40 wt. % Ac. However, if early formed diopside is fractionated, a limited magnetite–liquid reaction may again produce liquids with excess Na_2O and SiO_2. Operation of the 'plagioclase effect' provides a third possibility for sodium silicate-bearing liquids. Nolan suggests that any or all of these processes can form originally undersaturated trachytic liquids. Magnetite, formed from the incongruent melting of acmitic pyroxenes, may not completely react with liquid, producing oversaturated melts crystallizing as rocks of alkali granite composition. Although unable to show petrographic evidence of the magnetite–liquid relationship, Nolan's experiments clearly show such a relationship, and thus partly verify Tilley's (1957) hypothesis that acmite might cause the 'thermal barrier' in the residua system to become inoperative.

The addition of forsterite (Mg_2SiO_4) to the K-rich portion of the residua system was investigated by Luth (1967) in the system Ks–Fo–Sil–H_2O in the range 700–1200 °C at pressures up to 3 kb. Two of the boundary curves in this complex system involve phlogopite, forsterite, liquid and vapour; and phlogopite, enstatite, liquid and vapour. With fractionation, early formed forsterite may be rimmed by phlogopite or pyroxenes may be rimmed by biotite. Such reactions, observed in basic alkaline rocks and during melting of alkali basalt (Yoder and Tilley, 1962), represent important extensions of the discontinuous reaction series (Bowen, 1928) for potassic alkaline rocks. Luth's experiments also tend to verify Bowen's (1928) hypothesis that leucite-bearing lavas can be produced by resorption of biotite sinking in a basaltic melt (cf. VI.6).

The first attempt to investigate hydrothermally a system under upper mantle conditions was Boettcher and Wyllie's (1969) investigation of $NaAlSiO_4$–SiO_2–H_2O at 35 kb P_{H_2O}. Under these conditions this system crystallizes the important minerals quartz, coesite, albite, nepheline, analcime, and the high-pressure pyroxene, jadeite. From this study no evidence was found that the slope of the solid–liquid–vapour curves changed from negative to positive, as suggested by Barth (1962), Smith (1963) and Kadik and Khitarov (1963), thus implying that the beginning of melting of silicates in the presence of excess water continues to decrease at pressures corresponding to those of the upper mantle. Abrupt changes of slope do occur however in this system at invariant points where jadeite is formed with increasing pressure. Sluggish reaction rates did not permit the determination of the reactions involved in the formation of jadeite from analcime with albite or nepheline. This study does not indicate that primary analcime can only crystallize from silica-undersaturated magmas.

Most of the systems discussed are pertinent to the igneous processes involved in alkaline rock genesis. However Saha (1961) has shown that at low temperatures and pressures, vapour may preferentially dissolve soda and alumina in the Ne–Ab–H_2O system. He suggests that this may be important in 'nephelinization' (Tilley, 1957) where volatiles, escaping from crystallizing, alkaline magma could produce 'alkalization' of surrounding country rocks. These results may be compared with experiments of MacKenzie (quoted in Nolan, 1966) and Currie (1968) where nepheline is produced from albite composition by loss of silica in an 'open' system.

V.1.4. Melting Relations of Alkaline Rocks

Recently, experiments have been made in order to obtain data on the thermal histories of igneous rocks by melting them under controlled laboratory conditions. Such experiments assume that a crystallized or partially crystallized rock can be considered as a single bulk composition in a multicomponent system whose crystallization course can be followed by available experimental methods. Although there are many assumptions and pitfalls in this method, the results, when used in conjunction with data from synthetic systems, may yield valuable petrogenetic information.

Yoder and Tilley (1962) studied melting relations of an alkali basalt, an oxidized hawaiite and an olivine nephelinite at pressures from 1 atm to 10 kb P_{H_2O}. At 1 atm for the alkali basalt and hawaiite (in common with the other basalts investigated) the appearance of the primary silicate phase, irrespective of its composition, lay within a narrow temperature range. All major silicate phases crystallized within a small temperature interval, with the total range of crystallization temperatures being short.

Comparing these results with field relations and synthetic systems, Yoder and Tilley came to the following conclusions regarding basalts:

a. Because major phases in basalts crystallize within a short temperature interval, initial basalt compositions must lie close to the 4-phase curve in the Di–Ab–An–Fo system (Fig. 6). Therefore small variations in this composition can lead to different basalt types by simple addition or subtraction of their components.

b. A single basaltic magma cannot produce both nepheline- and silica-bearing products in its early stages of fractional crystallization if only the major constituents are concerned, but minor constituents may have a profound effect on crystallization trends.

c. Oxidation processes permit passage from a Ne normative to Hy or Qtz normative rock, although the reverse is not possible.

The main objective of Yoder and Tilley's hydrothermal study was to find a possible primary magma for each of the major basalt types. Their experiments indicate a close chemical connection between basalts and hornblendites and they conclude (p. 457) that for most basalts there is a chemically equivalent hornblendite. Bowen (1928) suggested that fractional resorption of hornblende might change a basaltic liquid from a tholeiitic to an alkaline line of descent. Yoder and Tilley's experiments not only verify Bowen's hypothesis but also show that alkali basaltic magmas can be converted to tholeiitic magmas by the same mechanism.

Following publication of Yoder and Tilley's experiments in 1962, a large number of alkali basalts from many localities have been studied at the Geophysical Laboratory (Tilley *et al.*, 1964, 1965, 1966; Tilley and Yoder, 1967). Plotting various chemical parameters against liquidus temperatures indicates a direct relationship between iron enrichment

$$(FeO + Fe_2O_3)/(MgO + FeO + Fe_2O_3)$$

and temperature. Alkalic rocks of different series have consistently higher liquidus temperatures than those of the Kilauea tholeiites of comparable iron enrichment. Comparison of total iron oxides and magnesia with liquidus temperatures suggests that in tholeiites the pyroxenes and plagioclases are removed from the liquid in the correct proportions to maintain continuous iron enrichment, but keeping the value of total iron oxides and magnesia constant. In contrast, the alkali series show continuous iron enrichment with falling total iron oxide and magnesia contents. Liquidus temperatures decrease as alkali contents and iron enrichment increase for both series, with iron enrichment being a more important factor for the tholeiites. In the alkali series, olivine crystallizing at the liquidus is soon replaced by plagioclase, suggesting that a simple crystallization differentiation sequence from a parental alkali olivine basaltic liquid does not operate, and that differentiation may involve the enrichment of the residual magma in inherited crystals of plagioclase which accumulated by flotation in the fractionating liquid. Although oxidation in alkali basalts increases liquidus temperatures it does not affect the crystallizing liquidus phase.

Tilley and Yoder (1967) studied melting relations of alkali ultramafic rocks (nephelinites and alnöite) at pressures from 1 atm to 10 kb. Between 900 and 1200 °C, without excess water and at low pressures, hornblende and clinopyroxene develop. The disappearance of nepheline, melilite and most of the olivine of the unheated rock suggests that partial melting of pyroxenites at depth may yield an alkalic liquid crystallizing nephelinite at volcanic levels in the crust, and rocks such as alnöites at subvolcanic levels. This provides an alternative mechanism to that proposed by Bowen (1922) for the genesis of these rocks.

Melting relations of less basic alkaline rocks are fewer. Barker (1965) hydrothermally melted a sample of the litchfieldite from Litchfield, Maine and also a 'synthetic' litchfieldite of approximately the same composition made from analysed natural nepheline and synthetic feldspars. In the range 0·5—2 kb P_{H_2O} both samples melted at the same temperatures and had a uniformly small melting interval. Comparison of this experimentally determined curve with previously published curves based on synthetic systems, shows no evidence of nepheline syenites melting at temperatures lower than those of granites above 1·4 kb P_{H_2O}, as suggested by Yoder and Tilley (1962).

Piotrowski and Edgar (1970) and Sood and Edgar (1970), from melting and crystallization studies of both intrusive and extrusive alkaline rocks of diverse compositions under a variety of different laboratory conditions, found:

a. both P_{H_2O} and P_{O_2} altered crystallization sequences of some rocks;

b. a direct relationship between iron enrichment and liquidus temperatures;

c. good agreement between temperatures determined hydrothermally and those from the appropriate synthetic systems;

d. minor amounts of mafic minerals had a pronounced effect on liquidus temperatures;

e. fairly good agreement between experimentally determined liquidus and solidus temperatures, and crystallization sequences with those suggested by petrographic and field relations (cf. Sørensen, 1962, 1969).

A linear correlation was found between rocks with high agpaitic coefficients and melting intervals, those with highest agpaitic coefficients having much longer melting intervals than those with low coefficents. Sood and Edgar (1970) propose that the long melting intervals in agpaitic rocks are due to their enrichment in volatiles and the initially high alkali to alumina ratios in their parental liquids (Gerasimovsky, 1965, 1966). During crystallization, these volatiles, behaving more or less independently of one another, are 'fixed' in the Na-rich minerals (cf. VI.4). This independent rôle played by these volatiles thus prolongs crystallization of agpaitic rocks in comparison to miaskitic rocks containing fewer important volatile components (cf. VI.4). Sood and Edgar tentatively suggest that rocks formed under low P_{O_2} conditions may follow a miaskitic crystallization trend, whereas those formed under high P_{O_2} conditions may follow an agpaitic trend.

V.1.5. Experimental Studies on the Conditions of Formation of Alkaline Rock Minerals

Experiments on the conditions of formation of the principal minerals of alkaline rocks give indications of the genesis of these rocks and provide rapid methods of determining their compositions.

V.1.5.1. Nepheline

Many investigations have been made on various types of nepheline s.s. The most important types, using the terminology of Donnay *et al.* (1959), are:

1. Substitution type. K replacing Na. ($K_xNa_{8-x}Al_8Si_8O_{32}$). This type is represented by the nepheline s.s. region in the Ne–Ks system (Fig. 14). Smith and Tuttle (1957) found that a plot of a and c parameters versus composition produced two linear curves intersecting at a composition close to $Na_3KAl_4Si_4O_{16}$, believed to be the composition of the most stable nepheline structure and toward which natural plutonic nephelines approach on cooling (Tilley, 1954).

2. Omission and substitution type. Vacant sites replacing Na and Si replacing Al ($Na_{8-y}\square_y Al_{8-y} Si_{8+y} O_{32}$). This type, represented by the nepheline s.s region in the Ne–Sil system, is important as nearly all natural nephelines contain excess silica. Donnay *et al.* (1959) found no changes in cell dimensions of a nepheline s.s. with 24 wt. % Ab ($y = 1$ in the above formula) compared with pure sodium nepheline. For hydrothermally synthesized nepheline s.s., Edgar

(1964a) detected no changes in a or c dimensions in the range Ab_5–Ab_{20} but an abrupt change in both parameters from Ab_0–Ab_5. Refractive indices of nepheline s.s. decrease from approximately 1·538 (0) for pure $NaAlSiO_4$ to about 1·525 (0) for nephelines at 20 wt. % Ab.

3. Double substitution type. Ca replacing Na and Al replacing Si ($Na_{8-z}Ca_zAl_{8+z}Si_{8-z}O_{32}$), represented by the nepheline s.s. in the system Ne–$CaAl_2O_4$ (Goldsmith, 1949). Nepheline cell volumes increase by 2·2Å^3 per Ca ion with maximum solid solution ($z < 4{\cdot}6$ in the above formula) (Donnay *et al.*, 1959).

4. Substitution and omission type. Ca and vacant sites in equal proportion replacing Na ($Na_{8-2z}Ca_z\square_zAl_8Si_8O_{32}$) represented by nepheline s.s. in the system Ne–An (Gummer, 1943). Donnay *et al.* (1959) found no change in cell parameters of nepheline s.s. with maximum substitution in comparison to pure Na-nepheline. Natural nephelines do not approach the maximum substitution value.

Substitution of Fe^{+3} and Al is mineralogically much less important but has considerable petrogenetic significance (Bailey and Schairer, 1966).

Hamilton and MacKenzie (1960) and Hamilton (1961) determined limits of nepheline s.s. at three different temperatures at 15,000 psi P_{H_2O} in the residua system, representing the first two types of solid solution listed above. Hamilton (1961) determined probable crystallization temperatures of natural nephelines from the limits of nepheline s.s. in the Ne–Ks–Sil system. He suggests that during cooling nephelines continuously react with their residual liquids and change in composition by adjusting their Na/K and Si/Al ratios. At subsolidus temperatures, exchange of Na and K between nepheline and coexisting feldspar takes place without a change in the Si/Al ratios, producing a final nepheline of composition $Na_3KAl_4Si_4O_{16}$ as proposed by Tilley (1952, 1954, 1956). Thus the silica content of nephelines coexisting with alkali feldspars may provide an indicator of crystallization temperatures. The limitations of this geothermometer are discussed by Barth (1963).

From birefringence measurements of natural nephelines at 20 to 900 °C, Sahama (1962) distinguishes two types. Type I characterized by high birefringence at low temperature and highly negative optical Δ values (where

$$\Delta = (\omega - \epsilon)900° - (\omega - \epsilon)20°);$$

and type II with low birefringence at low temperature and a highly positive Δ value. Type I, representing nepheline with highly ordered Si and Al atoms, contained all plutonic nephelines studied; whereas type II, represented highly disordered nephelines. Volcanic nephelines range from type I to type II, indicating that there is a sluggish, order–disorder transition between the two types analogous to that found in alkali feldspars. From infrared absorption measurements, Sahama (1965) gave two equations to determine alkali ratios (100K)/(K + Na + Ca)) in nephelines with medium to high K contents, but finds no evidence of Al/Si order–disorder from these results.

Saha (1961) suggested that some of the alteration products of natural nephelines-hydronephelite, ranite, gieseckite etc., may be the natural analogues of synthetic nepheline hydrates (Saha, 1961; Peters *et al.*, 1966). From a study of nepheline alteration products, Edgar (1965a) concluded that this was not the case.

V.1.5.2. Kalsilite and Kaliophilite

Kalsilite and kaliophilite ($KAlSiO_4$) are petrological rarities. In the system Ne–Ks (Fig. 14) the only feature of interest is the solvus from $Ne_{73}Ks_{27}$ to almost Ks_{100}. Sahama (1953, 1957, 1960) described various 'perthitic' intergrowths of nepheline and kalsilite in the lavas of Mt. Nyiragongo, Zaire. Lack of knowledge of crystallography, and effects of other components do not warrant its use as a geothermometer.

V.1.5.3. Analcime

Analcime occurs as a primary and secondary mineral in alkaline rocks. Saha (1959, 1961), investigating subsolidus relations in the system

Ne–Ab–H_2O up to 45,000 psi P_{H_2O}, synthesized analcime s.s. ranging from $NaAlSiO_4$ to $NaAlSi_3O_8$, a range much more extensive than found in nature. Saha's (1961) curve for the reaction analcime s.s. ($NaAlSi_3O_8.1.5H_2O$) = albite + H_2O is about 50 °C higher than the transition for the reaction analcime + quartz = albite + quartz + H_2O (Fyfe *et al.*, 1958) suggesting that analcimes more siliceous than albite may exist, although unreported in igneous rocks. Comparison of transition temperatures for analcime s.s. of various compositions (Saha, 1961; Sand *et al.*, 1957; Yoder, 1950) indicates that decreasing silica content increases these temperatures. Although silica contents of analcime s.s. can be determined from RI's, a parameter or

$$2\theta\text{Anal}_{(639)} - 2\theta\text{Si}_{(331)}\ (\text{CuK}\alpha)$$

(Saha, 1959), the feasibility of using compositions of secondary analcimes as a geothermometer is limited because most natural analcimes are close to the 'ideal' composition

$$(NaAlSi_2O_6 . H_2O).$$

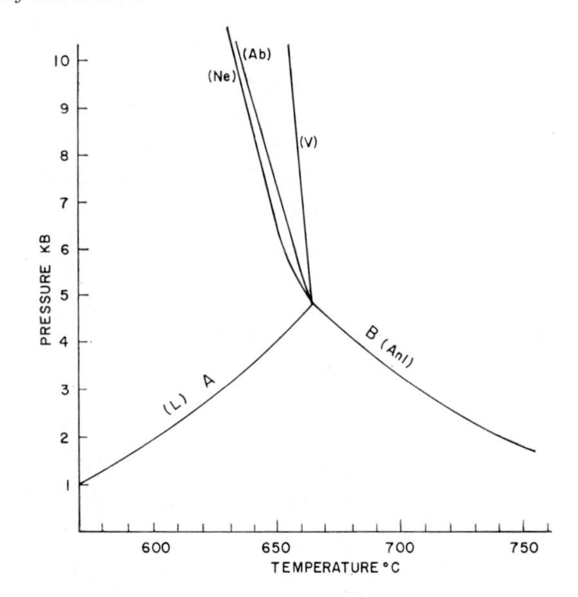

Fig. 27 Position of the univariant curves in the system Ne–Ab–H_2O, meeting at the invariant point (modified from Peters *et al.*, 1966). Bracketed symbols represent phases absent from the reactions taking place along the univariant curves. Symbols A and B are explained in the text

Peters *et al.* (1966) found that analcime of anhydrous composition ($Ab_{50}Ne_{50}$) in the system Ne–Ab–H_2O melts incongruently to albite, nepheline and liquid above an invariant point at 665 °C and 4·75 kb (Fig. 27). Boettcher and Wyllie (1968) suggest that this should be lowered to 635 °C. This point lies at the intersection of curves (A), representing the reaction analcime s.s. = nepheline + albite + vapour; and curve (B), representing the reaction albite + nepheline + vapour = liquid. Addition of potassium to this system displaces the invariant point to 2·3 kb at 650 °C with the analcime s.s. containing about 2 wt. % K_2O (comparable to natural igneous analcimes with 0·50 to 4·48 wt. % K_2O). Therefore analcime can crystallize directly from a melt at pressures above about 4·5 kb and coexist with a liquid over a small temperature range. With added potassium this pressure is considerably lower. These results are substantiated by the work of Morse (1969) mentioned previously. Peters *et al.* suggest that analcime phenocrysts in phonolites, blairmorites and trachybasalts may crystallize directly from a liquid in the small temperature range in which analcime and alkali-rich melts are in equilibrium, whereas groundmass analcime in these rocks probably crystallized within the analcime stability field. This investigation showed that increasing temperature decreased the extent of analcime s.s., implying that primary igneous analcimes should have a much narrower compositional range than those formed at lower temperatures during sedimentary processes.

Wilkinson (1962, 1963, 1965, 1968) suggests that in basic undersaturated alkaline rocks crystallized at high temperatures, Na and Al may replace Si in analcimes and that K-rich analcimes may form in rapidly cooled alkali-feldspar-free rocks which have crystallized in an hydrous environment within the analcime stability field. Natural analcimes from undersaturated rocks are generally K-poor, indicating very limited substitution of Na by K. Wilkinson (1968) shows that analcime-rich rocks plot close to the $NaAlSi_2O_6$–$KAlSi_3O_8$ join on the

Ab–Ne–Ks–K-Fp system. Compositions of coexisting feldspars and analcimes show that very Na-rich analcimes are due to the inability of the analcime structure to accommodate the larger K ion in sites normally occupied by Na ions.

V.1.5.4. Leucite

Leucite occurs principally in potash-rich, silica-deficient volcanic rocks, and occasionally in hypabyssal rocks (cf. II.5). Recent experimental work has been concerned mainly with the origin of pseudoleucites which occur in both volcanic and plutonic rocks.

Fudali (1957, 1963), investigating K-rich portions of the Ne–Ks–Sil system at 1 kb P_{H_2O}, found leucite s.s. containing up to 28 wt. % $NaAlSi_2O_6$ along the $NaAlSi_2O_6$–$KAlSi_2O_6$ join. With decreasing P_{H_2O} solid solution increased with increasing temperature, being about 40 wt. % $KAlSi_2O_6$ at 1 atm and 1000 °C. Solid solution of $KAlSiO_4$ in leucite was not found although 8 wt. % $KAlSi_3O_8$ can be taken into solid solution by leucite. No evidence of ternary solid solution was observed in this system, nor could solid solution of $NaAlSi_2O_6$ be detected by optical or X-ray methods. Sodium-rich leucites break down to nepheline and K-feldspar with falling temperature in the $NaAlSi_3O_8$–$KAlSi_2O_6$ join, providing a mechanism for the formation of plutonic pseudoleucites which supports Knight's (1906) theory for their origin but contradicts Bowen's (1928) theory that pseudoleucites form by a leucite $\rightleftharpoons$ liquid reaction in the residua system (Fig. 13).

Seki and Kennedy (1964), studying the stability of leucite from 200–850 °C at pressures up to 40 kb, found that leucite + water transformed to sanidine + kalsilite + H_2O at 450 °C and 7 kb, while increasing the pressure to 20 kb raised this temperature to about 825 °C. This suggests that pseudoleucite may form from the breakdown of leucite during magmatic cooling at a moderate to high P_{H_2O}. Since this transformation was not achieved under dry conditions, Seki and Kennedy imply that water must play an important rôle in pseudoleucite formation.

Scarfe *et al.* (1966) redetermined the reaction leucite = orthoclase + hexagonal kalsilite using a different method from Seki and Kennedy, and found the transformation took place at lower pressures. In the system Ks–Sil–H_2O, the invariant point involving K-feldspar, leucite, kalsilite, liquid and vapour occurs at 750 °C and 8·4 kb. Addition of up to 2% of the analcime molecule to leucite lowers the pressure of the transformation for a given temperature. These experiments suggest the absence of leucite in plutonic rocks is because the K-Fp–Ne join in the residua system cuts off albite from leucite and therefore these rocks have compositions which prohibit leucite as a stable phase. In contrast, under volcanic (low pressure) conditions, the stable leucite field overlaps the Ne–Ab–K-Fp portion of the residua systems and leucite is stable. In the latter case pseudoleucites may form by the mechanism proposed by Fudali (1963).

V.1.5.5. Melilites

Melilites, consisting principally of gehlenite, akermanite and soda-melilite, with minor amounts of iron-akermanite and iron-gehlenite, occur in basic alkaline rocks and in hybrid rocks formed by reaction of a basic magma with carbonates.

The system Geh–Ak shows complete solid solution with a minimum (Ferguson and Buddington, 1920). Ervin and Osborn (1949) and Neuvonen (1952) have found a linear increase in a and decrease in c parameters with increasing Ak in this system but unfortunately this relationship is not of much value in determining compositions of natural melilites containing appreciable sodium and iron.

Subsolidus relationships of various combinations of melilite molecules have been investigated by a number of authors. The reaction akermanite = monticellite + wollastonite occurs below 700 to 750 °C in the pressure range 30,000 to 60,000 psi (Harker and Tuttle, 1956). Kushiro (1964b) found akermanite to be stable from about 700 °C to less than 1500 °C at pressures under 15 kb but at pressures from about 5–8 kb it decomposed to merwinite + forsterite s.s. + wollastonite, and from 8 kb–15 kb and at high temperatures dissociated to merwinite + diopside. At pressures in excess of

5 kb, akermanite melts incongruently to merwinite + a liquid rich in the diopside molecule. These results show that akermanite is stable only under the high-temperature conditions of magma found in the upper part of the Earth's crust. In the Ak–Geh series, melilites dissociated between 500 and 650 °C at pressures from 4·8–6·7 kb (Christie, 1961), akermanitic compositions dissociating to vesuvianite + wollastonite + clinopyroxene from approximately 500 to 450 °C and clinopyroxene + xonotlite below 450 °C. With more than 20 % Geh in solid solution, the breakdown products are vesuvianite + hydrogarnet, and at lower temperatures xonotlite + hydrogarnet. From studies of the stability of akermanite and the reaction akermanite–forsterite Yoder (1967) shows that in the presence of excess water akermanite is restricted to depths within the average continental crust. In the same report, Yoder stresses the incompatibility of akermanite and albite up to 10 kb but that soda melilite and albite are compatible above 4 kb.

Soda-melilite is an important constituent of 'igneous' melilites (Yoder, 1964, Edgar, 1965b). Yoder (1964) showed that this compound is stable only at total pressures above 4 kb; below this pressure nepheline s.s. and wollastonite are the stable phases. These results indicate that soda-melilite may be an important constituent of melilites in rocks of the lower crust, whereas the equivalent low pressure rocks would have wollastonite and nepheline-bearing assemblages e.g. wollastonite-nephelinite lavas and wollastonite-ijolite dykes.

The extremely complex nature of the melilite system (Geh–Ak–SM) is evident from the studies from 1 atm to 20 kb. (Schairer and Yoder, 1964; Kushiro, 1964b; and Schairer *et al.*, 1965). Melilite is the primary phase over almost the entire liquidus in this system. Subsolidus relations in the Ak–SM and Geh–SM joins show a complex assemblage of minerals, including wollastonite, nepheline s.s. and melilites containing up to 50 wt.% of the SM molecule. Investigations of the Ak–SM join up to 20 kb (Kushiro, 1964b) show that pure akermanite is stable only at pressures up to 10·5 kb and pure soda melilite at pressures greater than 4 kb at 1000 °C. Melilites of compositions between 40 and 70 mol % are stable between 1 atm and 20 kb. With addition of gehlenite, melilites are stable up to at least 30 kb. Since most melilites in lavas contain more than 30 % of soda melilite, they may have formed at pressures in the range of 20 kb.

At 1 kb P_{H_2O}, Edgar (1965b) reported that gehlenite and akermanite could accommodate about 50 wt.% of soda-melilite in solid solution.

V.1.5.6. Sodalites and Cancrinites

Sodalites (including nosean and hauyne) and cancrinites occur both as primary and secondary minerals in felsic alkaline rocks. Experimental studies of Van Peteghem and Burley (1962) and Jarchow *et al.* (1966) support this occurrence of sodalites. Van Peteghem and Burley (1963) report complete solid solution between synthetic nosean and hauyne compositions, but only limited solid solutions between sodalite-nosean and sodalite-hauyne. Taylor (1967) has discussed the chemistry and solid solution in the sodalite group.

Synthesis of cancrinites has been done by Eitel (1922), Edgar (1964c) and Jarchow *et al.* (1966). The stability of cancrinites is strongly dependent on their compositions.

V.1.6. Summary and Conclusions

In this short review the principal systems pertinent to alkaline rock genesis have been presented. Although many investigations have not been mentioned (mainly due to space restrictions), and little attempt has been made to analyse the results of many of the systems, the contents of this chapter may indicate the contributions made by laboratory studies toward the solution of the genesis of alkaline rocks.

The majority of investigations, both of synthetic systems and of melting relations of the rocks themselves, have been done at atmospheric pressure on compositions of alkali basaltic affinities. The compositions are of fundamental importance since they are generally believed to be the precursors of many of the felsic alkaline rocks. Results of systems done under P_{H_2O} and P_{O_2} conditions clearly show the important rôle volatiles play in the formation of alkaline rocks. Considerable data are now available on the minerals of the rocks, providing a better understanding of the influence of composition, pressure and temperature on their genesis. Recent investigations have shown that mechanisms are available to bridge the so-called 'thermal barriers' between silica saturated and silica undersaturated compositions, thus providing an explanation of the close field associations of oversaturated, saturated and undersaturated rocks.

Future experimental studies pertinent to alkaline rocks could follow many lines of investigation. One of the most fruitful would be the addition of basic molecules (particularly pyroxenes, basic plagioclases and olivines) to the residua system, thus providing a physical–chemical link between the mafic and felsic rocks and giving us some understanding of the crystallization histories of residual liquids trending toward the Na_2O–K_2O–Al_2O_3–SiO_2 plane. Work has already begun on parts of this system. Although less rigorous than determinations of synthetic systems, melting and crystallization of rocks in the laboratory yields valuable information on their genesis. Another approach which should prove profitable in the future is thermodynamic treatment of experimental data (cf Perchuk and Ryabchikov, 1968) permitting the construction of equilibrium relations in systems not experimentally determined.

V.1. REFERENCES

Andrews, K. W., 1948. The lattice parameters and interplanar spacings of some artificially prepared melilites. *Mineralog. Mag.*, **28**, 374–9.

Aoki, K., 1959. Petrology of the alkalic rocks of the Iki Islands and Higaski-Matsura District, Japan. *Tohoku Univ., Sci. Repts.*, 3rd ser., **6**, 261–310.

Aoki, K., 1964. Clinopyroxenes from alkaline rocks of Japan. *Am. Miner.*, **49**, 1199–223.

Aoki, K., 1967. The role of titanium in alkali basalt magmas. *Japan. Assoc. Mineralogists, Petrologists, Econ. Geologists J.*, **57**, 188–99.

Bailey, D. K., 1964. Crustal warping—a possible tectonic control of alkaline magmatism. *J. Geophys. Res.*, **69**, 1103–11.

Bailey, D. K., and Schairer, J. F., 1964. Feldspar–liquid equilibria in peralkaline liquids—the orthoclase effect. *Am. J. Sci.*, **262**, 1198–206.

Bailey, D. K., and Schairer, J. F., 1966. The system Na_2O–Al_2O_3–Fe_2O_3–SiO_2 at 1 atmosphere, and the petrogenesis of alkaline rocks. *J. Petrology*, **7**, 114–70.

Barker, D. S., 1965. Alkalic rocks at Litchfield, Maine. *J. Petrology*, **6**, 1–27.

Barth, T. F. W., 1962. *Theoretical Petrology*. John Wiley and Sons, New York. 416 pp.

Barth, T. F. W., 1963. The composition of nepheline. *Schweiz. miner. petrogr. Mitt.*, **43**, 153–64.

Boettcher, A. L., and Wyllie, P. J., 1969. Phase relationships in the system $NaAlSiO_4$–SiO_2–H_2O to 35 kilobars pressure. *Am. J. Sci.*, **267**, 875–909.

Bowen, N. L., 1917. The sodium-potassium nephelites. *Am. J. Sci.*, 4th ser., **43**, 115–32.

Bowen, N. L., 1922. Genetic features of alnöitic rocks at Isle Cadieux, P.Q. *Am. J. Sci.*, 5th ser., **3**, 1–33.

Bowen, N. L., 1928. *Evolution of the Igneous Rocks.* 334 pp., Princeton University Press, Princeton, N.J.

Bowen, N. L., 1937. Recent high-temperature research on silicates and its significance in igneous petrology. *Am. J. Sci.*, **33**, 1–21.

Bowen, N. L., 1945. Phase equilibria bearing on the origin and differentiation of alkaline rocks. *Am. J. Sci.*, **243A**, 75–90.

Carmichael, I. S. E., and MacKenzie, W. S., 1963. Feldspar–liquid equilibria in pantellerites: an experimental study. *Am. J. Sci.*, **261**, 382–96.

Christie, O. H. J., 1961. On subsolidus relations of silicates, I. The lower breakdown temperatures of the akermanite-gehlenite mixed crystal series at moderate water pressure. *Norsk Geol. Tidsskr.*, **41**, 255–69.

Currie, K. L., 1968. On the solubility of albite in supercritical water in the range 400 to 600 °C and 750 to 3500 bars. *Am. J. Sci.*, **266**, 321–41.

Donnay, G., Schairer, J. F., and Donnay, J. D. H., 1959, Nepheline solid solutions. *Mineralog. Mag.*, **32**, 93–109.

Edgar, A. D., 1964a. Phase equilibrium relations in

the system nepheline–albite–water at 1000 kg/cm². *J. Geol.*, **72**, 448–60.

Edgar, A. D., 1964b. Phase equilibrium relations in the system $CaMgSi_2O_6$ (diopside)–$NaAlSiO_4$ (nepheline)–$NaAlSi_3O_8$ (albite)–H_2O at 1000 kg/cm² water vapor pressure. *Am. Miner.*, **49**, 573–85.

Edgar, A. D., 1964c. Studies on cancrinites: II—stability fields and cell dimensions of calcium and potassium-rich cancrinites. *Can. Miner.*, **8**, 53–67.

Edgar, A. D., 1965a. Mineralogical compositions of nepheline alteration products. *Am. Miner.*, **50**, 978–89.

Edgar, A. D., 1965b. Lattice parameters of melilite solid solutions and a reconnaissance of phase relations in the system $Ca_2Al_2SiO_7$ (gehlenite)–$Ca_2MgSi_2O_7$ (akermanite)–$NaCaAlSi_2O_7$ (soda-melilite) at 1000 kg/cm² water vapour pressure. *Can. J. Earth Sci.*, **2**, 596–621.

Edgar, A. D., and Nolan, J., 1966. Phase relations in the system $NaAlSi_3O_8$ (albite)–$NaAlSiO_4$ (nepheline) – $NaFeSi_2O_6$ (acmite) – $CaMgSi_2O_6$ (diopside)–H_2O and its importance in the genesis of alkaline undersaturated rocks. *Indian Mineralogist*, I.M.A. vol., 176–81.

Eitel, W., 1922, Über das System $CaCO_3$–$NaAlSiO_4$ (Calcit–Nephelin) und den Cancrinite. *Neues Jb. Miner.*, **2**, 45–61.

Ervin, G., and Osborn, E. F., 1949. X-ray data on synthetic melilites. *Am. Miner.*, **34**, 717–22.

Eugster, H. P., 1967. Experimental igneous petrology. *Am. Geoph. Union Trans.*, **48**, 654–61.

Ferguson, J. B., and Buddington, A. F., 1920. The binary system akermanite–gehlenite. *Am. J. Sci.*, 4th Ser., **50**, 131–40.

Foster, W. R., 1942. The system $NaAlSi_3O_8$–$CaSiO_3$–$NaAlSiO_4$. *J. Geol.*, **50**, 158–73.

Fudali, R. F., 1957. On the origin of pseudoleucite (abstract). *Am. Geophys. Union Trans.*, **38**, 391.

Fudali, R. F., 1963. Experimental studies bearing on the origin of pseudoleucite and associated problems of alkali rock systems. *Bull. geol. Soc. Am.*, **74**, 1101–26.

Fyfe, W. S., Turner, F. J., and Verhoogen, J., 1958. Metamorphic reactions and metamorphic facies. *Geol. Soc. Am. Mem.*, **73**, 259 pp.

Gerassimovsky, V. I., 1965. Role of F and other volatiles in alkaline rocks (abstract). *Geochem. Int.*, **2**, 9.

Gerassimovsky, V. I., 1966. 'Geochemical features of agpaitic nepheline syenites', in Vinogradov, A. P., Ed., *Chemistry of the Earth's Crust, Vol. 1.* Daniel Davey Co., New York, 104–18.

Goldsmith, J. R., 1949. Some aspects of the system $NaAlSiO_4$–$CaO.Al_2O_3$. *Am. Miner.*, **34**, 471–93.

Greig, J. W., and Barth, T. F. W., 1938. The System $Na_2O.Al_2O_3.5SiO_2$ (nephelite, carnegieite)–$Na_2O.Al_2O_3.6SiO_2$ (albite). *Am. J. Sci.*, **35A**, 93–112.

Griffith, S., 1944. *The System $CaSiO_3$–$NaAlSi_3O_8$–$CaAl_2Si_2O_8$.* Ph.D. thesis, University of Chicago (unpublished).

Gummer, W. K., 1943. The System $CaSiO_3$–$CaAl_2Si_2O_8$–$NaAlSiO_4$. *J. Geol.*, **51**, 503–30.

Hamilton, D. L., 1961. Nephelines as crystallization temperature indicators. *J. Geol.*, **69**, 321–9.

Hamilton, D. L., and MacKenzie, W. S., 1960. Nepheline solid solutions in the system $NaAlSiO_4$–$KAlSiO_4$–SiO_2. *J. Petrology*, **1**, 56–72.

Hamilton, D. L., and MacKenzie, W. S., 1965. Phase-equilibrium studies in the system $NaAlSiO_4$ (nepheline) – $KAlSiO_4$ (kalsilite) – SiO_2 – H_2O. *Mineralog. Mag.*, **34**, (Tilley vol.), 214–31.

Harker, R. I., and Tuttle, O. F., 1956. The lower limit of stability of akermanite ($Ca_2MgSi_2O_7$). *Am. J. Sci.*, **254**, 468–78.

Holmes, A., and Harwood, H. F., 1937. The volcanic field of Bufumbira Pt. II. The petrology of the volcanic field of Bufumbira, South-west Uganda. *Mem. geol. Surv. Uganda*, No. 3, 300 pp.

Huckenholz, H. G., 1965a. Der petrogenetische Werdegang der Klinopyroxene in den tertiären Vulkaniten der Hocheifel. I. Die Klinopyroxene der Alkali – olivin – basalt – trachyt – Assoziation. *Beitr. Mineral. Petrog.*, **11**, 138–95.

Huckenholz, H. G., 1965b. Der petrogenetische Werdegang der Klinopyroxene in den tertiären Vulkaniten der Hocheifel. II. Die Klinopyroxene der Basanitoide. *Beitr. Mineral. Petrog.*, **11**, 415–48.

Huckenholz, H. G., 1966. Der petrogenetische Werdegang der Klinopyroxene in den tertiären Vulkaniten der Hocheifel III. Die Klinopyroxene der Pikritbasalte (Ankaramite). *Beitr. Mineral. Petrog.*, **12**, 73–95.

Jarchow, O., Reese, H. H., and Saalfeld, H., 1966. Hydrothermal-synthesen von Zeolithen der Sodalith und Cancrinitgruppe. *Neues Jb. Miner.*, **66**, 289–99.

Kadik, A. A., and Khitarov, N. I., 1963. Conditions of the thermodynamic equilibria in water-silicate melt systems. *Geochemistry*, No. 10, 917–36.

King, B. C., 1965. Petrogenesis of the alkaline igneous rock suites of the volcanic and intrusive centres of Eastern Uganda. *J. Petrology*, **6**, 67–100.

Knight, C. W., 1906. A new occurrence of pseudoleucite. *Am. J. Sci.*, **21**, 286–95.

Kushiro, I., 1964a. Petrology of the Atumi dolerite, Japan. *Tokyo Univ. Fac. Sci. J.*, sec. 2, **15**, 135–202.

Kushiro, I., 1964b. The join akermanite-soda melilite at 20 kilobars. *Carnegie Inst. Wash. Year Book*, **63**, 90–2.

Luth, W. C., 1967. Studies in the system $KAlSiO_4$–Mg_2SiO_4–SiO_2–H_2O I, inferred phase relations and petrologic applications. *J. Petrology*, **8**, 372–416.

Luth, W. C., Jahns, R. H., and Tuttle, O. F., 1964. The granite system at pressures of 4 to 10 kb. *J. Geophys. Res.*, **69**, 759–73.

Morse, S. A., 1969. Nepheline–kalsilite–silica at 5 kb P_{H_2O}. *Carnegie Inst. Wash. Year Book*, **68**.

Neuvonen, K. J., 1952. Thermochemical investigation of the akermanite–gehlenite series. *Comm. Geol. Finlande Bull.*, **26**, 57 pp.

Nolan, J., 1966. Melting relations in the system $NaAlSi_3O_8$–$NaAlSiO_4$–$NaFeSi_2O_6$–$CaMgSi_2O_6$–H_2O and their bearing on the genesis of alkaline undersaturated rocks. *Q. J. geol. Soc. Lond.*, **122**, 119–57.

Onuma, K., and Yagi, K., 1967. The system diopside–akermanite–nepheline. *Am. Miner.*, **52**, 227–43.

Osborne, E. F., and Schairer, J. F., 1941. The ternary system pseudowollastonite–akermanite–gehlenite. *Am. J. Sci.*, **239**, 715–63.

Perchuk, L. L., and Ryabchikov, I. D., 1968. Mineral equilibria in the system nepheline–alkali feldspar–plagioclase and their petrological significance. *J. Petrology*, **9**, 123–67.

Peters, Tj., Luth, W. C., and Tuttle, O. F., 1966. The melting of analcite solid solutions in the system $NaAlSiO_4$–$NaAlSi_3O_8$–H_2O. *Am. Miner.*, **51**, 736–53.

Piotrowski, J. M., and Edgar, A. D., 1970. Melting relations of undersaturated alkaline rocks from South Greenland compared to those of Africa and Canada. *Medd. Grønland*, **181**, 62 pp.

Platt, R. G., and Edgar, A. D., 1972. The system nepheline–diopside–sanidine and its significance in the genesis of melilite and olivine bearing alkaline rocks. *J. Geol.*, **80**, 224–36.

Saggerson, E. P., and Williams, L. A. J., 1964. Ngurumamite from Southern Kenya and its bearing on the origin of rocks in the Northern Tanganyika alkaline district. *J. Petrology*, **5**, 40–81.

Saha, P., 1959. Geochemical and X-ray investigation of natural and synthetic analcites. *Am. Miner.*, **44**, 300–13

Saha, P., 1961. System nepheline–albite. *Am. Miner.* **46**, 859–85.

Sahama, Th. G., 1953. Parallel growth of nepheline and microperthitic kalsilite from North Kivu, Belgian Congo. *Ann. Acad. Sci. Fennia*, ser. AIII, No. 36, 1–18.

Sahama, Th. G., 1957. Complex nepheline–kalsilite phenocrysts in Kabfumu lava, Nyiragongo area, North Kivu in Belgian Congo. *J. Geol.*, **65**, 515–26.

Sahama, Th. G., 1960. Kalsilite in the lavas of Mt. Nyiragongo (Belgian Congo). *J. Petrology*, **1**, 146–71.

Sahama, Th. G., 1962. Order–disorder in natural nepheline solid solutions. *J. Petrology*, **3**, 65–81.

Sahama, Th. G., 1965. Infrared absorption of nepheline. *Comptes Rendus Soc. geol. Finlande*, **37**, 107–17.

Sand, L. B., Roy, R., and Osborn, E. F., 1957. Stability relations of some minerals in the Na_2O–Al_2O_3–SiO_2–H_2O system. *Econ. Geol.*, **52**, 169–79.

Scarfe, C. M., Luth, W. C., and Tuttle, O. F., 1966. An experimental study bearing on the absence of leucite in plutonic rocks. *Am. Miner.*, **51**, 726–35.

Schairer, J. F., 1950. The alkali–feldspar join in the system $NaAlSiO_4$–$KAlSiO_4$–SiO_2. *J. Geol.*, **58**, 512–17.

Schairer, J. F., 1967. 'Phase equilibria at one atmosphere related to tholeiitic and alkali basalts', in Abelson, P., Ed., *Researches in Geochemistry, Vol.* 2, John Wiley and Sons, 568–92.

Schairer, J. F., and Bowen, N. L., 1947. The system anorthite–leucite–silica. *Soc. geol. Finlande Bull.*, **20**, 72–5.

Schairer, J. F., and Morimoto, N., 1959. The system forsterite–diopside–silica–albite. *Carnegie Inst. Wash. Year Book*, **58**, 113–18.

Schairer, J. F., Yagi, K., and Yoder, H. S., 1962. The system nepheline–diopside. *Carnegie Inst. Wash. Year Book*, **61**, 96–8.

Schairer, J. F., and Yoder, H. S., 1960a. The nature of residual liquids from crystallization, with data on the system nepheline–diopside–silica. *Am. J. Sci.*, **258A**, 273–83.

Schairer, J. F., and Yoder, H. S., 1960b. The system forsterite–nepheline–diopside. *Carnegie Inst. Wash. Year Book*, **59**, 70–1.

Schairer, J. F. and Yoder, H. S., 1961. Crystallization in the system nepheline–forsterite–silica at one atmosphere pressure. *Carnegie Inst. Wash. Year Book*, **60**, 141–4.

Schairer, J. F., and Yoder, H. S., 1964. The join akermanite ($Ca_2MgSi_2O_7$) – Soda melilite ($NaCaAlSi_2O_7$). *Carnegie Inst. Wash. Year Book*, **63**, 89–90.

Schairer, J. F., Yoder, H. S., and Tilley, C. E., 1965. Behaviour of melilites in the join gehlenite–soda melilite–akermanite at one atmosphere pressure. *Carnegie Inst. Wash. Year Book*, **64**, 95–100.

Schairer, J. F., Yoder, H. S., and Tilley, C. E., 1966. The high temperature behaviour of synthetic melilites in the join gehlenite–soda melilite–akermanite. *Carnegie Inst. Wash. Year Book*, **65**, 217–26.

Seki, Y., and Kennedy, G. C., 1964. An experimental study on the leucite–pseudoleucite problem. *Am. Miner.*, **49**, 1267–80.

Smalley, R. G., 1947. The system $NaAlSiO_4$–$Ca_2Al_2SiO_7$. *J. Geol.*, **55**, 27–37.

Smith, F. G., 1963. *Physical Geochemistry*. 624 pp., Addison–Wesley, Reading, Mass.

Smith, J. V., and Tuttle, O. F., 1957. The nepheline–

kalsilite system: I, X-ray data for the crystalline phases. *Am. J. Sci.*, **255**, 282–305.

Smyth, C. H., 1927. The genesis of alkaline rocks. *Proc. Am. phil. Soc.*, **66**, 535–80.

Sood, M. K., and Edgar, A. D., 1970. Melting relations of undersaturated alkaline rocks from the Ilímaussaq intrusion and Grønnedal–Ika complex, south Greenland, under water vapour and controlled partial oxygen pressure. *Medd. Grønland*, **181**, 41 pp.

Sood, M. K., and Edgar, A. D., 1972. The system diopside–forsterite–nepheline–albite–leucite and its implication to the genesis of alkaline rocks. *24th Int. Geol. Congr. Montreal*, **14**, 68–74.

Sood, M. K., Platt, R. G., and Edgar, A. D., 1970. Phase relations in portions of the system diopside–nepheline–kalsilite–silica and their importance in the genesis of alkaline rocks. *Can. Miner.*, **11**.

Sørensen, H., 1960. On the agpaitic rocks. *Int. Geol. Congr., 21st sess.*, Pt. 13, 319–27.

Sørensen, H., 1962. On the occurrence of Steenstrupine in the Ilímaussaq massif, southwest Greenland. *Medd. Grønland*, **167**, 251 pp.

Sørensen, H., 1969. Rhythmic igneous layering in peralkaline intrusions. *Lithos*, **2**, 261–83.

Spivak, J., 1944. The system $NaAlSiO_4$–$CaSiO_3$–Na_2SiO_3. *J. Geol.*, **52**, 24–52.

Taylor, D., 1967. The sodalite group of minerals. *Contr. Mineral. Petrology*, **16**, 172–88.

Thompson, R. N., and MacKenzie, W. S., 1967. Feldspar–liquid equilibria in peralkaline acid liquids: an experimental study. *Am. J. Sci.*, **265**, 714–34.

Tiba, T., 1966. Petrology of the alkaline rocks of the Takakusayama district, Japan. *Tohuku Univ. Sci. Rept.*, ser. 3, **9**, 541–610.

Tilley, C. E., 1952. 'Nepheline paragenesis', in *Sir Douglas Mawson Anniversary Volume*. Hassell Press, Adelaide, 167–77.

Tilley, C. E., 1954. Nepheline–alkali feldspar paragenesis. *Am. J. Sci.*, **252**, 65–75.

Tilley, C. E., 1956. Nepheline associations. *Overdruk Uit Het Gedenbock H. A. Brouwer, Verh., Konink, Ned. Geol. Minb. Genootachap.*, **16**, 1–11.

Tilley, C. E., 1957. Problems of alkali rock genesis. *Q. J. geol. Soc. Lond.*, **113**, 323–59.

Tilley, C. E., and Yoder, H. S., 1967. The pyroxenite facies conversion of volcanic and subvolcanic, melilite-bearing and other alkali ultramafic assemblages. *Carnegie Inst. Wash. Year Book*, **66**, 457–60.

Tilley, C. E., Yoder, H. S., and Schairer, J. F., 1964. New relations on melting of basalts. *Carnegie Inst. Wash. Year Book*, **63**, 92–7.

Tilley, C. E., Yoder, H. S., and Schairer, J. F., 1965. Melting relations of volcanic tholeiite and alkali rock series. *Carnegie Inst. Wash. Year Book*, **64**, 69–82.

Tilley, C. E., Yoder, H. S., and Schairer, J. F., 1966. Melting relations of volcanic rock series. *Carnegie Inst. Wash. Year Book*, **65**, 260–9.

Tuttle, O. F., and Bowen, N. L., 1958. Origin of granite in the light of experimental studies in the system $NaAlSi_3O_8$–$KAlSi_3O_8$–SiO_2–H_2O. *Geol. Soc. Am. Mem.*, 74, 153 pp.

Tuttle, O. F., and Smith, J. V., 1958. The nepheline–kalsilite system II. phase relations. *Am. J. Sci.*, **256**, 571–89.

Tyler, R. C., and King, B. C., 1967. The pyroxenes of the alkaline igneous complexes of eastern Uganda. *Mineralog. Mag.*, **36**, 5–21.

Ussing, N. V., 1912. Geology of the country around Julianehaab, Greenland. *Medd. Grønland*, **38**, 426 pp.

Van Peteghem, J. K., and Burley, B. J., 1962. Studies on the sodalite group of minerals. *Trans. Royal Soc. Can.*, **56**, ser. III. 37–53.

Van Peteghem, J. K., and Burley, B. J., 1963. Studies on solid solution between sodalite, noseane and haüyne. *Can. Miner.*, **7**, 808–13.

Washington, H. S., 1896. Italian petrological sketches. *J. Geol.*, **4**, 541–66, 826–49.

Washington, H. S., 1897. Italian petrological sketches. *J. Geol.*, **5**, 349–77.

Wilkinson, J. F. G., 1956. Clinopyroxenes of alkali olivine–basalt magma. *Am. Miner.*, **41**, 724–43.

Wilkinson, J. F. G., 1962. Mineralogical, geochemical, and petrogenetic aspects of an analcite–basalt from the New England district of New South Wales. *J. Petrology*, **3**, 192–214.

Wilkinson, J. F. G., 1963. Some natural analcime solid solutions. *Mineralog. Mag.*, **33**, 498–505.

Wilkinson, J. F. G., 1965. Some feldspars, nephelines, and analcimes from the Square Top intrusion, Nundle, N.S.W. *J. Petrology*, **6**, 420–44.

Wilkinson, J. F. G., 1968. Analcimes from some potassic igneous rocks and aspects of analcime-rich igneous assemblages. *Contr. Miner. Petrology*, **18**, 252–69.

Wright, J. B., 1963. A note on possible differentiation trends in tertiary to Recent lavas of Kenya. *Geol. Mag.*, **100**, 164–80.

Wyllie, P. J., 1966. 'Experimental studies of carbonatite problems: The origin and differentiation of carbonatite magmas', in Tuttle, O. F., and Gittins, J., Eds., *Carbonatites*. John Wiley and Sons, New York, 311–52.

Yagi, K., 1953. Petrochemical studies on the alkalic rocks of the Morotu district, Sakhalin. *Bull. geol. Soc. Am.*, **64**, 769–810.

Yagi, K., 1962. A reconnaissance of the systems acmite–diopside and acmite–nepheline. *Carnegie Inst. Wash. Year Book*, **61**, 98–9.

Yagi, K., and Onuma, K., 1969. An experimental

study on the role of titanium in alkalic basalts in light of the system diopside–akermanite–nepheline–$CaTiAl_2O_6$. *Am. J. Sci.*, **267A**, 509–49.

Yoder, H. S., 1950. The jadeite problem. *Am. J. Sci.*, **248**, 225–48, 312–34.

Yoder, H. S., 1964. Soda melilites. *Carnegie Inst. Wash. Year Book*, **63**, 86–9.

Yoder, H. S., 1967. Akermanite and related melilite-bearing assemblages. *Carnegie Inst. Wash. Year Book*, **66**, 471–7.

Yoder, H. S., and Tilley, C. E., 1962. Origin of basalt magmas: An experimental study of natural and synthetic rock systems. *J. Petrology*, **3**, 342–532.

V.2. INCLUSIONS IN THE MINERALS OF SOME TYPES OF ALKALINE ROCKS

V. S. Sobolev, T. Yu. Bazarova and V. P. Kostyuk

V.2.1. Introduction

In this chapter it will be attempted to establish criteria on which a quantitative evaluation of the physico-chemical and thermodynamical parameters of the conditions of formation of minerals can be based.

Several methods have been suggested for the determination of the temperature of crystallization of silicates (geological thermometers).

The feldspar method proposed by Barth (1951) has later been interpreted by Ryabchikov (1963). On the basis of the synthetic nepheline–kalsilite–silica system Hamilton (1961) determined the probable relation between the chemical composition of nepheline and the temperature of crystallization when silica is in excess (see V.1)*. A similar diagram constructed by Perchuk (1965) is based on the Na_2O/K_2O ratio in coexisting nepheline and alkali feldspar.

In these methods the temperature is determined as a function of the chemical composition of minerals. Analytical inaccuracies in determining the mineral composition may, therefore, cause errors in the estimation of the temperatures of formation. A discrepancy of 300–380 °C is for instance seen when correlating the temperatures determined by means of the method of homogenization of fluid inclusions with the temperatures obtained from Perchuk's diagram for the nepheline-bearing rocks from the Botogol, Nyurgan, Ilmen and Lovozero massifs. Moreover, rather than reflecting the true temperatures of crystallization the mineralogical data may indicate the temperature conditions under which crystallization proceeded in order to maintain equilibrium during the late- and post-magmatic stage (cf. II.2.7).

A wealth of information about the temperatures of crystallization during the hydrothermal, pneumatolytic and pegmatitic stages of mineral formation has been obtained by means of the mineral-thermometric methods founded by Sorby (1898) and successfully further developed by Ermakov (1949, 1950), Smith (1953), Roedder (1958–1963), Lemmlein (1956, 1959), Dolgov (1965) and their schools. The problem of determination of two other initial parameters by means of fluid inclusions, namely pressure of formation and the chemical composition of the mineral-forming solutions, has been solved in principle (Dolgov, 1965). However, such experiments have mainly been carried out for minerals formed at relatively low temperatures, i.e. below the temperature of the α-β-inversion of quartz.

The construction of a new microthermochamber has made it possible to work at temperatures up to 1200 °C (Dolgov and Bazarov, 1965) and later on also up to 1600 °C. In this chamber both fluid–gas inclusions and quenched melt inclusions (glassy or partly crystalline) in rock-forming minerals of igneous rocks can be homogenized.

* Readers are referred to the relevant Chapters of this book where similarly cited.

V.2.2. Outline of the Geology of Some Siberian Alkaline Rocks

The present authors have especially studied the conditions of formation of the alkaline rocks developed within the folded framework of the Siberian platform (see also III.4 and IV.5). Here, intrusive alkaline rocks play an important rôle.

The majority of the examined alkaline massifs of the fold belts of Siberia belong to the miaskitic nepheline syenites and present wide petrographical variation. Ijolite, urtite, foyaite, theralite, etc., occur in some of the massifs as minor facies of the nepheline syenites, but they only rarely form independent massifs or larger masses within massifs of complicated structure.

The perpotassic alkaline pluton of Synnyr in North Pribaikal'e (Andreev, 1965; Zhidkov, 1960, 1961, 1962) is very peculiar with regard to chemical composition. In the nepheline-bearing members three fourths of the average total content of alkalis, namely 12–17%, are made up of K_2O. In the pseudoleucite-bearing 'synnyrites' (Zhidkov, 1962) the amount of K_2O reaches 18–19%. In accordance with this, the feldspar is extremely potash-rich, the nepheline contains 30 mol % or more of kalsilite, and biotite is clearly predominant among the mafic minerals.

The nepheline-free rocks are represented by alkali syenites, occasionally pulaskites (Synnyr, the East Tuva, etc.). Nordmarkite plays a notable rôle in certain regions of Siberia.

V.2.3. General Principles of the Examination of Fluid Inclusions

The first attempt to determine the lower limit of the temperature of crystallization of the nepheline syenites of East Sayan (Kerkis and Kostyuk, 1963) made it possible to confirm the magmatic nature of these rocks with sufficient certainty. Later on similar investigations have been carried out on alkaline rocks from other regions.

Most suitable for thermometric studies are the following rock-forming minerals: nepheline, leucite, the faintly coloured alkali and semi-alkali clinopyroxenes, garnet, cancrinite, sodalite, scapolite, calcite and a few other minerals. Alkali feldspars are, as a rule, almost sterile with regard to high-temperature fluid inclusions. More than 5000 quite reliable observations of homogenization of fluid inclusions have been performed. Most of these represent the high-temperature stage of crystallization.

Only solidified inclusions of melts can be regarded as definitely primary inclusions characterizing the thermal regime during crystallization of the magma. Such inclusions are relatively rare in the minerals of intrusive rocks, while they are generally present in the phenocrysts of some alkaline effusive and subvolcanic rocks. The minerals of the intrusive facies are characterized by two- and three-phase (rarely poly-phase) inclusions with predominance of either a liquid or a gas phase and with the most variable quantitative relation of these phases. The size of the inclusions is very small—generally 0·00n or rarely 0·0n mm (in effusive rocks).

Certain difficulties are involved in the determination of the primary nature of fluid inclusions, in particular of the gas–liquid inclusions which homogenize in the liquid phase between 700–980 °C. It turned out that only this type of inclusion is characterized by an important peculiarity: the modification of the shape of the inclusion (apparently as a result of redeposition of material) at temperatures close to those of homogenization. If these inclusions are really primary, then they must represent drops of the 'transmagmatic' solutions captured by a crystallizing mineral. In case these inclusions are secondary, they must represent the early post-magmatic solutions. The maximum temperature of homogenization of high-temperature inclusions in minerals of igneous rocks can be used for evaluation of the minimum temperature of crystallization.

Because of the difficulty in determining whether an inclusion is primary or not, it is necessary to stress that the absence of high-temperature inclusions in a particular rock cannot be regarded as a proof of its minerals having been formed at lower temperatures.

In these examinations it was attempted to

determine (where possible) two additional parameters of the crystallizing system: the pressure and composition of the gas phase and the chemical composition of the liquid phase of the gas–fluid inclusions.

The lower limit of the pressure of formation was determined from the density and composition of the solutions of individual high-temperature gas–liquid inclusions. In order to undertake this the true amount of CO_2 and other gases was determined. The salinity of the liquid phase was furthermore determined by crystallization during deep freezing (down to -180 °C) with a special apparatus (Bakumenko *et al.*, 1965) and by the subsequent calculation of the total pressure at the temperature of homogenization by means of physico-chemical diagrams.

The qualitative chemical composition of the inclusions and the quantitative relationship of the volatile components were determined separately, namely by analysis of water extracts and by gas analysis according to the method of Dolgov and Shugurova (1966).

V.2.4. Results of the Examination of Fluid Inclusions in Minerals of Nepheline Syenites

Table 1 records the results of the study of the temperature conditions of crystallization of the rock-forming minerals of the alkaline rocks of Siberia by means of gas–liquid inclusions (Kerkis and Kostyuk, 1963; Bazarova, 1965; Panina, 1966 a, b).

The information obtained during the examination of the Siberian rocks has been compared with the results of the study of nepheline-bearing rocks from other regions of the Soviet Union and from other countries, including Greenland (see Bazarova, 1965; Bazarova and Feigin, 1966; and Table 2).

The results of the pressure determinations are recorded in Table 3). The pressures of formation of the primary high-temperature inclusions of transmagmatic fluids in minerals from different massifs are of the order of 1100–1500 atm.

The chemical compositions of water extracts obtained from nephelines from a number of massifs, and besides, from clinopyroxene, garnet and calcite of two genetic types of nepheline syenite from the Botogol massif (Sobolev, 1947; Sobolev and Florensov, 1948; Solonenko, 1950) are presented in Table 4.

The high temperatures of homogenization of the primary (and partly the secondary) inclusions, 800–1040 °C for nepheline, 710–840 °C for alkali feldspar, and 700–800 °C for clinopyroxene, give sufficient grounds to believe that the alkaline rocks under discussion crystallized from magmatic melts.

The minimum temperature of melting of the binary mixture nepheline + alkali feldspar is 1040 °C in the dry system $NaAlSiO_4$–$KAlSiO_4$–SiO_2 (Schairer, 1950). A pressure of the order of 1 kbar will raise this temperature to 1055 °C, but if $P_{H_2O} = P_{total}$, the temperature will decrease to 855 °C (Bowen and Tuttle, 1950). The partial substitution of water by CO_2 would raise the minimum temperature, but alkali chlorides will partly or wholly balance this effect (cf. VI.4). These figures suggest that the studied rocks have been formed from magmatic melts, especially when it is recalled that the temperature of formation of the minerals is generally higher than the temperature of homogenization of fluid inclusions. The contents of the primary gas–liquid inclusions may then be treated as preserved transmagmatic or early postmagmatic solutions.

Crystallization of nepheline from magmatic melts is also indicated by the presence of recrystallized solid inclusions, which could have been captured only under magmatic conditions.

All fluid inclusions in nepheline are characterized by striking similarities with respect to morphology, phase composition, phase ratios, high concentrations of dissolved salts, behaviour during heating (redistribution of the material within the inclusions at high temperatures) and temperatures of homogenization of the primary gas–liquid inclusions in the mineral from different alkaline massifs. All this evidence emphasizes the common mechanism of formation of the primary inclusions.

In the gas phase of the primary inclusions CO_2 makes up 75–97 vol.%, the rest being made up of nitrogen and rare gases. In the secondary inclusions hydrocarbons are, as a rule,

TABLE 1 Mineral–Thermometric Examination of Alkaline Rocks from Siberia

		Types of inclusions	Interval of Temperatures of Homogenization in °C	Aggregate Condition
1	2	3	4	5
The East Sayan[1]				
		(a) The Botogol massif		
Typical nepheline syenite	nepheline	primary	850–680	liquid
	nepheline	secondary	660–120	liquid
Pegmatitic nepheline syenite	nepheline	primary	840–680	liquid
Medium-grained nepheline syenite	nepheline	primary	810–700	liquid
	nepheline	secondary	660–120	liquid
Trachytoid nepheline syenite	nepheline	primary	850–680	liquid
	nepheline	secondary	660–140	liquid
Pegmatitic pyroxene-feldspar rock	clinopyroxene	primary(?)	800–700	liquid
	clinopyroxene	secondary	660–470	liquid
	clinopyroxene	secondary	410–280	liquid
	calcite	secondary	500–400	liquid
	calcite	secondary	345–120	liquid
Nepheline syenite from contact zone against limestone	garnet	secondary	400–100	liquid
Calcite veins in zones of graphitization	calcite	primary	280–240	liquid
	calcite	secondary	240–120	liquid
Pegmatitic nepheline syenite	cancrinite	secondary	400–120	liquid
Wollastonite veinlet	wollastonite	primary–secondary	520–150	liquid
		(b) The Nyurgan massif		
Typical nepheline syenite	nepheline	primary	850–700	liquid
	nepheline	secondary	680–480	liquid
	nepheline	secondary	400–100	liquid
The North Pribaikal'e[2]				
		The Synnyr pluton		
Ditroite	nepheline	primary	950–810	liquid
	nepheline	primary	710	gas
	nepheline	secondary	610–580	liquid
	nepheline	secondary	680	liquid
	nepheline	secondary	740–680	gas
	nepheline	secondary	380–200	liquid
Pegmatitic ditroite	nepheline	secondary	630–210	liquid
	nepheline	secondary	560–160	liquid
Foyaite	nepheline	secondary	520–150	liquid
Synnyrite	nepheline	primary	860	liquid
	nepheline	secondary	610–320	liquid
Pseudoleucite syenite	nepheline	primary(?)	820–700	liquid
	nepheline	secondary	550	liquid
Foyaite	potash feldspar	primary(?)	730	liquid

TABLE 1 *Continued*

Synnyrite	potash feldspar	primary	840	liquid
Pulaskite	potash feldspar	secondary	350–280	liquid
Ditroite	sodalite	primary	630	critical phenomena
	sodalite	secondary	410–230	liquid
	cancrinite	primary	630	critical phenomena
Contact metasomatic rock	tremolite	secondary	750–600	liquid
	tremolite	secondary	550–450	gas
	tremolite	secondary	500–228	liquid
Contact skarn	scapolite	secondary	660–520	gas
	scapolite	secondary	570–470	liquid
The South Pribaikal'e[3]				
		The Borgoi massif		
nepheline syenite	nepheline	secondary	750–700	gas
	nepheline	secondary	730	liquid
	nepheline	secondary	570–500	liquid
	nepheline	secondary	480–420	liquid
	nepheline	secondary	400–130	liquid
		Western Transbaikal'e		
nepheline syenite	nepheline	primary	850	liquid
nepheline syenite	nepheline	primary	850–750	liquid
nepheline syenite	nepheline	secondary	800–780	gas
nepheline syenite	nepheline	secondary	730	gas
congressite	nepheline	secondary	800–730	gas

References:
1. Churakov (1947); Krymskii (1960); Kostyuk and Bazarova (1966); Sobolev (1947); Sobolev and Florenzov (1948); Solonenko (1950); Kerkis and Kostyuk (1963); Smirnov and Buldakov (1962).
2. Andreev (1965); Zhidkov (1960, 1961, 1962); Panina (1966a).
3. Panina (1966b).

TABLE 2 Mineral–Thermometric Examination of Alkaline Rocks from the USSR and Greenland

Rock Type	Mineral	Types of Inclusions	Interval of Temperatures of Homogenization in °C	Aggregate Condition
1	2	3	4	5
The Miaskitic Complex of the Urals (cf. II.2; Ronenson, 1964).				
		The Il'men massif		
Miaskite	nepheline	primary	950–800	liquid
	nepheline	secondary	410–180	liquid

TABLE 2 *Continued*

Miaskite pegmatites	nepheline	primary	800–700	liquid
	nepheline	secondary	500–400	gas
	nepheline	secondary	410–165	liquid
The Kola Peninsula (cf. IV.2; Bazarova and Feigin, 1966; Valyashko and Kogarko, 1966).				
		(a) The Lovozero massif		
Urtite	nepheline	primary	980–810	liquid
	nepheline	secondary	780–750	gas
	nepheline	secondary	700–650	liquid
	nepheline	secondary	700–550	liquid
	nepheline	secondary	610–445	liquid
	nepheline	secondary	400–200	liquid
	nepheline	secondary	400–300	gas
Foyaite	nepheline	primary	740–700	liquid
	nepheline	secondary	675–660	gas
	nepheline	secondary	660–430	liquid
	nepheline	secondary	410–365	liquid
	nepheline	secondary	300–200	liquid
	clinopyroxene	primary	720–700	liquid
Mesocratic lujavrite	nepheline	primary(?)	720–700	gas
	nepheline	secondary	400–350	liquid
	nepheline	secondary	300–200	gas
		(b) The Khibina massif		
Khibinite	nepheline	primary(?)	720–700	gas
	nepheline	secondary	420–380	gas
	nepheline	secondary	450–250	liquid
The Eastern Priazov'e (cf. II.2; Eliseev *et al.*, 1965).				
		The Oktjabr' massif		
Foyaite	nepheline	primary(?)	700–680	gas
	nepheline	secondary	620–600	liquid
	nepheline	secondary	570–500	gas
	nepheline	secondary	500–200	liquid
Nepheline syenite	nepheline	primary(?)	700–680	gas
	nepheline	secondary	640–600	liquid
	nepheline	secondary	550–500	gas
	nepheline	secondary	450–160	liquid
Sodalite veinlet in a mariupolite pegmatite	sodalite	primary(?)	450–390	liquid
	sodalite	secondary	240–150	liquid
Mariupolite	clinopyroxene	primary(?)	720–700	liquid

TABLE 2 *continued*

Greenland (cf. II.2; IV.3; Sobolev *et al.*, 1970).

The Ilímaussaq massif				
Green lujavrite	nepheline	primary	970–910	liquid
	nepheline	secondary	600–140	liquid
Naujaite	nepheline	primary	1040–850	liquid
Pegmatite	chkalovite	primary	980–860	liquid
	chkalovite	secondary	760–140	liquid

TABLE 3 Pressure of Formation in Fluid Inclusions in Nepheline from Botogol, Nyurgan, Lovozero, Oktjabr' and Ilmen

	Characteristics of the Inclusions at the Moment of Homogenization							
	Types of inclusions	Temperature in °C	Aggregate condition	Concentration of salts, in wt. %	Specific volume CO_2*	Pressure of solution in atm.	Pressure of dissolved CO_2 in atm.	Total pressure in atm.
The Botogol Massif								
Nepheline from	primary	760	liquid	22·8	145	1455	15	1470
nepheline syenite	primary	700	liquid	25·6	170	1255	15	1270
Nepheline from pegmatitic	primary	720	liquid	26·1	210	1340	10	1350
nepheline syenite	primary	710	liquid	26·2	125	1285	20	1300
	secondary	400	liquid	7·8	10	275	10	280
The Nyurgan massif								
Nepheline from	primary	740	liquid	46·3	120	1330	10	1340
nepheline syenite	primary	730	liquid	43·1	150	1300	10	1310
	secondary	500	liquid	41·7	100	400	20	430
The Lovozero Massif								
Nepheline from urtite	primary	810	liquid	22·0	120	>1500	30	>1530
	secondary	710	liquid	26·5	115	1280	20	1300
	secondary	710	liquid	24·7	150	1280	10	1290
	secondary	700	liquid	26·6	100	1260	20	1280
Nepheline from foyaite	primary	720	liquid	19·8	150	1330	20	1350
	secondary	450	liquid	37·2	80	310	25	335
The Oktjabr' Massif								
Nepheline from foyaite	primary	700	gas	23·0	130	1160	15	1115
	primary	680	gas	25·5	150	1240	20	1260
	secondary	620	liquid	18·3	170	990	10	1000
The Ilmen Miaskite Massif								
Nepheline from miaskite	primary	800	liquid	45·0	not determined	> 1400	—	> 1400
Nepheline from pegmatite	secondary	710	liquid	44·0	not determined	1200	—	1200

*Among the gases present in the inclusions the most important is carbon dioxide (CO_2). The effect of other gases on the value of the pressure is insignificant, owing to their great specific volumes.

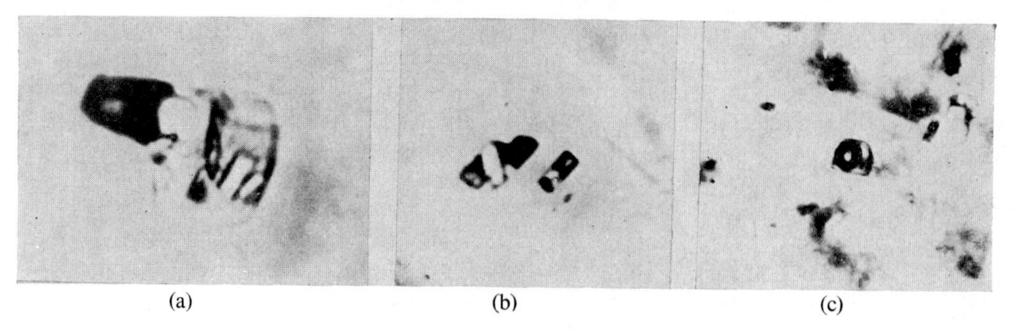

(a) (b) (c)

Fig. 1 Primary inclusions in nepheline

(a) Gaseous–liquid inclusion with high salt concentration; temperature of homogenization 1040 °C. Naujaite, Ilimaussaq, 300×.

(b) Gaseous–liquid inclusions with high salt concentrations; temperature of homogenization 1000 °C. Miaskite, the Urals, 300×.

(c) Gaseous–liquid inclusions, temperature of homogenization 950 °C. Urtite, Lovozero, 300×.

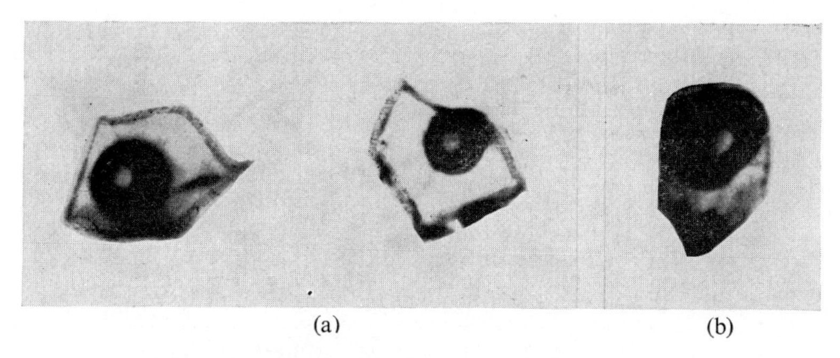

(a) (b)

Fig. 2 Secondary inclusions in nepheline

(a) Two phase, temperature of homogenization 620 °C. Foyaite Lovozero, 300×.

(b) Three-phase inclusion, temperature of homogenization 700 °C. Urtite, Lovozero, 300×.

predominant (72–91%) and the rest is made up of CO_2.

Thus, water and carbon dioxide are the predominant volatiles during the crystallization of the minerals of the examined rocks. Cl^-, F^-, SO_4^{2-} and HCO_3^- are also present and form together with Na^+, K^+, Ca^{2+} and other cations salts soluble in both water and melt. NaCl appears to be the predominant salt.

The revealed predominance of CO_2 in the high-temperature inclusions may indicate either its juvenile nature or also magmatic assimilation of carbonate material by the melt. In this connection the relatively high temperatures of homogenization of microscopically small drop-shaped poikilitic inclusions of calcite in the nepheline and feldspar of the alkaline rocks of East Sayan, Synnyr, etc., should be noted. These inclusions are often rimmed by biotite, amphibole, titanite, occasionally pyroxene, and also by cancrinite and sodalite. Due to the complete disintegration of this calcite along cleavages caused by mass bursting of low-temperature inclusions, heating of the mineral over 500 °C is

TABLE 4 Chemical Composition of Fluids from Liquid Inclusions in some Minerals from Alkaline Rocks

Mineral and Locality	Cations								Anions				Concentration in Wt. %	pH
	The Content in Wt. %													
	K	Na	Si	Al	Ca	Fe	Ti	Mg	Cl	F	SO_4	HCO_3		
Nepheline, Botogol	0·339	5·591	not found	0·292	1·123	0·377	0·037	traces	4·534	1·171	1·450	2·275	17·2	7·5
Nepheline, Nyurgan	0·766	10·157	1·236	not found	not found	0·171	not found	0·285	6·396	not determined	1·470	3·313	24·3	7·2
Nepheline, Lovozero	0·405	8·757	not found	0·865	0·554	0·229	0·070	traces	3·400	2·248	0·459	1·042	18·6	6
Nepheline, Ilmen mountains	0·724	12·966	0·230	0·836	1·381	0·177	0·190	not found	6·270	not found	4·370	4·069	26·8	6·9
Nepheline, Vishnevogorsk	0·546	5·019	0·699	0·123	0·021	0·131	0·057	not found	3·108	not found	1·881	0·624	12·2	8
Calcite, Botogol	0·256	4·028	3·661	not found	0·501	0·185	0·377	not found	4·468	not found	3·451	not determined	16·9	6·6
Calcite, Botogol	0·112	0·812	1·624	0·758	1·842	1·300	0·057	not found	1·245	not found	not found	not determined	7·7	6·3
Pyroxene, Botogol	0·308	4·726	5·193	not found	not found	not found	not found	not found	2·560	not found	1·521	0·967	15·2	6·1
Garnet, Botogol	0·549	8·340	0·584	0·115	0·134	0·260	0·080	not found	1·536	1·883	0·693	0·405	14·5	6·6

difficult. During work in progress inclusions in carbonates have been homogenized at 750 °C and higher temperatures. These figures indicate, however, the high-temperature generation of the calcite, possibly in connection with fusion of xenogenic material. This fact serves as an indirect evidence of the possible participation of calc-assimilation of limestone in the process of desilication of calc-alkaline granitic magmas.

The temperatures of homogenization of gas–fluid inclusions in cancrinite, sodalite, scapolite, wollastonite, etc. yield information about the temperature regime of the secondary post-magmatic processes. The obtained low pressure (280–400 atm.) at temperatures of homogenization of 300–500 °C reflect the minimum *P–T*-parameters for these superposed processes.

The mineral-thermometric investigations carried out on nepheline from pegmatites and alkaline rocks of pegmatoid structure from the Dakhunur intrusion in south-eastern Tuva showed that crystallization of these nephelines took place at temperatures above 750–680 °C. Hence, the formation of the ijolites enclosing these pegmatites must have taken place at temperatures higher than 750 °C.

The study of the conditions of formation of pegmatites is of particular interest. The data on the nephelines from the pegmatites of Tuva indicate high temperatures of formation. The extremely high temperatures of homogenization of fluid inclusions in chkalovite from alkaline late- or postmagmatic veins in the Ilímaussaq intrusion, Greenland (Sobolev *et al.*, 1970) emphasize the unusual composition of the inclusions which, probably, represent a water–chloride–silicate melt. This indicates that the investigation of pegmatitic minerals from alkaline rocks may yield interesting new information.

V.2.5. Inclusions of Quenched Melts in Minerals of Volcanic Rocks

While the genesis of the 'plutonic' alkaline rocks is a subject for discussion (see chapters VI.1–8), the magmatic nature of nepheline- and leucite-bearing volcanic rocks is an absolutely indisputable fact. At the same time, a knowledge of the thermal regime during crystallization of the intratelluric rock-forming minerals in poly-component magmatic systems is at present very scanty. Data concerning temperatures of homogenization of solid glass inclusions in minerals from calc-alkaline hyalodacites first appeared in 1965 (Kalyuzhnyi, 1965, Bazarova and Dmitriev, 1967; Sobolev *et al.*, 1967).

Primary inclusions of melts in nepheline, clinopyroxene and olivine from the nepheline basalts of Bakoni, Hungary (Fig. 3a) have been examined by Sobolev *et al.* (1967). The complete homogenization of such inclusions from the central zones of phenocrysts, including melting of their crystalline phases, takes place within the range of temperatures 1250–1280 °C for clinopyroxene, 1270–1290 °C for olivine, and 1250–1290 °C for nepheline. The inclusions in the nepheline are usually completely crystallized. Homogenization of inclusions from the outer zones of the crystals of these minerals takes place at 1120–1160 °C.

These minerals, besides the above mentioned inclusions, also contain what appears to be primary-secondary inclusions of melts, which are confined to fractures formed during the growth of the crystals. These inclusions are usually two-phase (gas + glass) inclusions or they contain occasionally a solid phase, namely trapped crystals. The homogenization is accomplished within the narrow temperature interval from 1120 to 1130 °C. At these temperatures there seems to be an abrupt change in the physico-chemical parameters of the crystallizing system (possibly caused by the effusion of lava flows), which is indicated by the formation of fractures and secondary inclusions.

The recrystallized inclusions of melts in leucite and clinopyroxene from the subvolcanic fergusite porphyry (Fig. 3b,d,e), which was recently discovered in Eastern Pamir, were also studied. Homogenization of the inclusions in leucite is likely to have been accomplished in a wide range of temperatures—from 1350 to 1670 °C. Similar inclusions in pyroxene from the same rocks are homogenized at 1250 to 1320 °C. It is thus possible that the data available reflect a certain

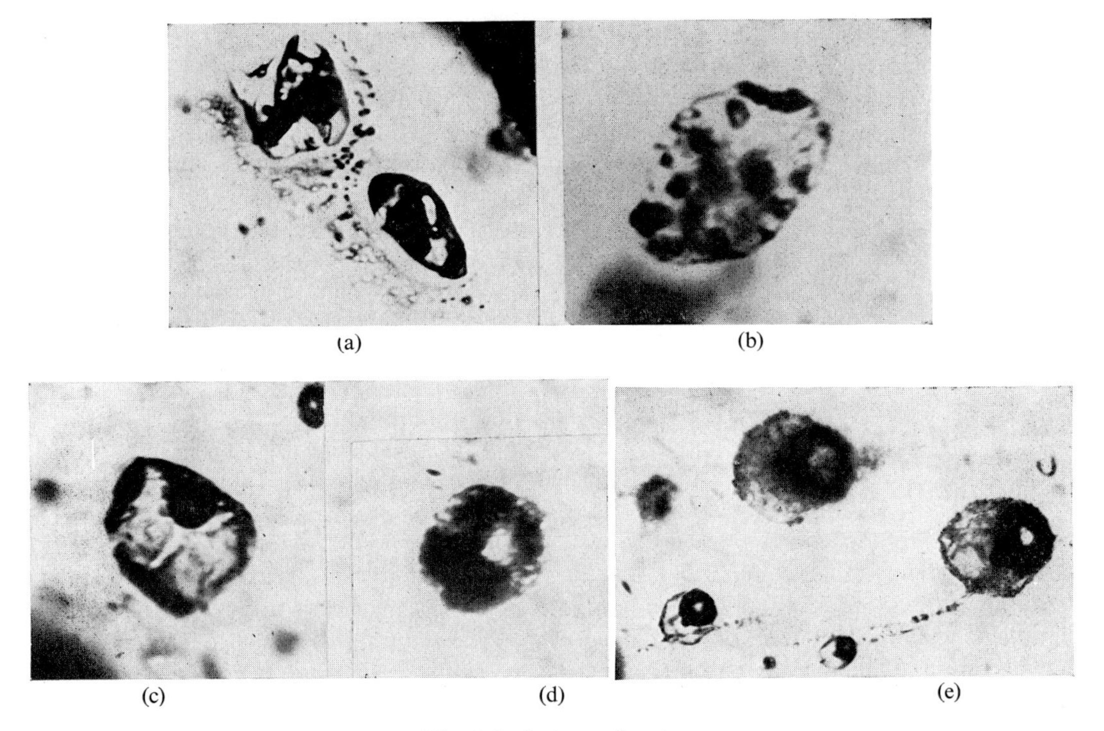

Fig. 3 Inclusions of melts

(a) Totally crystallized inclusions in nepheline; temperature of homogenization 1230 °C. Nepheline basalt, Hungary, 300×.

(b) 'Quenched' inclusion in pyroxene. Homogenization temperature 1380 °C. Fergusite–porphyry. East Pamir.

(c) Two-phase (gas + glass) inclusions in pyroxene. Temperature of homogenization 1300 °C. Leucite tephrite, Vesuvius, 300×.

(d) Inclusions in leucite (room t°).

(e) Inclusions in leucite at 1350 °C, 300×. (d) and (e) both from fergusite porphyry, East Pamir.

initial stage of intratelluric crystallization of phenocrysts.

At the same time, however, the question arises, how strict is the correlation of the temperature of homogenization of quenched inclusions of melts and the temperature of capture of the inclusions. For this purpose an experimental control (Bakumenko *et al.*, 1967) was undertaken on clinopyroxenes synthesized in the dry system or in the presence of fluorine. It is characteristic that the homogenization of inclusions of melt in such pyroxenes took place at practically the same temperatures (±10 °C) as those obtained by homogenization of melt inclusions in natural pyroxenes from volcanic and subvolcanic alkaline rocks.

Such high temperatures should be expected in the case of anhydrous or water deficient melts. This raises certain doubts as to the composition of natural melts. There remains, however, an uncertainty as to the possibility of diffusion of water through the crystal lattice at temperatures exceeding 1100 °C (and the corresponding increase of the temperature of homogenization of thus dehydrated inclusions). Repeated observations of the mode of homogenization of the same glassy and high-temperature gas–liquid inclusions do not indicate any notable diffusion of the volatiles contained in the inclusions.

If loss of water from fluid inclusions has taken place it should be most pronounced in the peripheral zones of the crystals, and the inclusions there should then have the highest temperatures of homogenization. However, this is not the case; the temperatures of homogenization in the central zones of the crystal are always the highest. This problem cannot be considered finally solved at the present time; and it is necessary to set up special investigations of material synthesized at fixed water vapour pressures.

V.2.6. Conclusions

The available information about the homogenization of fluid and glassy inclusions in minerals from alkaline rocks can be interpreted as follows:

1. The crystallization of the phenocrysts of effusive alkaline rocks began at 1250–1290 °C and in leucite rocks possibly at temperatures higher than 1600 °C. However, the extremely high temperatures of homogenization in leucites, in which one always deals with recrystallized inclusions, are questionable. These data still need experimental checking in systems containing water.

2. The crystallization of nepheline syenites proceeded within the temperature interval from 800–1040 °C. At these conditions the material must (at least partly) have been in a melted state. It was found that high-temperature calcite crystallizing above 600 °C is present. The pressure of formation exceeded 1000–1500 atm.

3. Nepheline and other minerals of nepheline syenites contain high-temperature inclusions of 'transmagmatic' or early postmagmatic solutions rich in chlorides and with predominance of CO_2 in the gas phase.

4. The pegmatitic crystallization of the nepheline syenites began at temperatures higher than 700 °C or occasionally higher than 900 °C (chkalovite) and continued to temperatures of 300–400 °C; the solutions were very rich in chlorides, being close to chloride–silicate melts.

The crystallization of secondary sodalite proceeded at temperatures of the order of 400 °C, and in some of the massifs at temperatures higher than 600 °C. For cancrinite and scapolite the temperature of crystallization is determined as 450–650 °C.

V.2. REFERENCES

Andreev, G. V., 1965. *Petrology of the Synnyr Alkaline Pluton.* 117 pp., Ulan-Ude (in Russian).

Bakumenko, I. T., Dolgov, Yu. A., and Bazarov, L. Sh., 1965. Method of determination of pressure in inclusions by combined application of homogenization and cryometry. *Reports of the II All-Union Congress on Geothermobary, Novosibirsk*, 95–7 (in Russian).

Bakumenko, I. T., Kolyago, S. S., and Sobolev, V. S., 1967. Problem of interpretation of the thermometric investigations of glassy inclusions in minerals and the initial results of the testing on artificial inclusions. *Dokl. Akad. Nauk SSSR*, **175**, 1127–30 (in Russian).

Barth, T. F. W., 1951. The feldspar geologic thermometer. *Neues Jb. Miner. Abh.*, **82**, 143–54.

Bazarova, T. Yu., 1965. Mineralothermometric investigations of inclusions in minerals of some nepheline rocks. *Dokl. Akad. Nauk SSSR*, **161**, 4 (in Russian).

Bazarova, T. Yu., and Dmitriev, E. A., 1967. Temperature conditions of fergusite–porphyry crystallization of the East Pamyr. *Dokl. Akad. Nauk SSSR*, **177**, 185–8 (in Russian).

Bazarova, T. Yu., and Feigin, Ja. M., 1966. A mineralothermometric study of nepheline from the Lovozero Massif. *Zap. Vses. Miner. Obshch.*, Ser. 2, **95**, 3 (in Russian).

Bowen, N. L., and Tuttle, O. F., 1950. The System $NaAlSi_3O_8$–$KAlSi_3O_8$–H_2O, *J. Geol.*, **58**, 489–511.

Churakov, A. N., 1947. The Upper Silurian Nepheline Belt of the East Sayan. *Review of Sci. Papers, 1945*. Dept of Geol. Geogr. Sci., Izd. AN SSSR, 84–5 (in Russian).

Dolgov, Yu. A., 1965. Formation of minerals in chamber pegmatites. *Zap. Vses. Min. Obshch.*, Ser. 2, **94**, 1, 41–8 (in Russian).

Dolgov, Yu. A., and Bazarov, L. Sh, 1965. 'A chamber for investigation of inclusions of mineral-forming solutions and melts at high-temperatures', *Mineralogical Thermometry and Barometry, I.* 118–122, Izd. 'Nauka', Moscow (in Russian).

Dolgov, Yu. A., and Shugurova, N. A., 1966. 'Investigation of the composition of individual gas inclusions', *Materials on Genetic and Experimental Mineralogy, IV.* 173–80, Izd. 'Nauka', Novosibirsk (in Russian).

Eliseev, N. A., Kushev, V. G., and Vinogradov, D. P., 1965. *The Proterozoic Intrusive Complex of the East Priazov'e.* 201 pp, Izd. 'Nauka', Moscow (in Russian).

Ermakov, N. P., 1949. Criteria of the knowledge of the genesis of minerals and the environment of their formation. *Min. Lvov Geol. Obshch.*, **3**, 6–68 (in Russian).

Ermakov, N. P., 1950. *Investigations of the Mineral-Forming Solutions.* 460 pp, Kharkov (in Russian).

Hamilton, D. L., 1961. Nephelines as crystallization temperature indicators. *J. Geol.*, **69**, 321–9.

Kalyuzhnyi, V. A., 1965. Optical and thermometric investigations of glass inclusions in the phenocrysts of the hyalodacites of Zakarpat'e, *Dokl. Akad. Nauk SSSR*, **160**, 2, 438–41 (in Russian).

Kerkis, T. Yu. and Kostyuk, V. P., 1963. The mineral-thermometric study of nepheline from Botogol (the East Sayan), *Dokl. Akad. Nauk SSSR* **150**, 5, 1125–7 (in Russian).

Kostyuk, V. P., and Bazarova, T. Yu., 1966. *Petrology of the Alkaline Rocks of the Eastern Part of the East Sayan.* 168 pp., Izd. 'Nauka', Moscow.

Krymskii, V. M., 1960. The nepheline syenites from the upper stream of the Pezo River, the East Sayans. *Geol. i Geofiz, Izd. SO AN*, **9**, 111–14 (in Russian).

Lemmlein, G. G., 1956. 'Investigation of the formation of liquid inclusions in crystals,' in *Problemi Geokhimii i Mineralogii*, Izd. AN SSSR, Moscow, 139–41 (in Russian).

Lemmlein, G. G., 1959. Classification of liquid inclusions in minerals. *Zap. Vses. Min. Obshch.*, Ser. 2, **88**, 137–43 (in Russian).

Panina, L. I., 1966a. Some data on the temperature conditions of the formation of the Synnyr alkaline massif. *Dokl. Akad. Nauk SSSR*, **170**, 6, 1411–14 (in Russian).

Panina, L. I., 1966b. On the genesis of the nepheline syenites of Borgoi. *Geol. i Geofiz., Izd. SO AN SSSR*, **8**, 114–15 (in Russian).

Perchuk, L. L., 1965. The paragenesis of nepheline and alkali feldspar—a possible indicator of the thermodynamic conditions of mineral equilibrium. *Dokl. Akad. Nauk SSSR*, **161**, 4, 932–5.

Roedder, E., 1958. Technique for the extraction and partial chemical analysis of fluid-filled inclusions from minerals. *Econ. Geol.*, **53**, 235–69.

Roedder, E., 1962. Studies of fluid inclusions. I. Low temperature application of a dual-purpose freezing and heating stage. *Econ. Geol.*, **57**, 1045–61.

Roedder, E., 1963. Studies of fluid inclusions. II. Freezing data and their interpretation. *Econ. Geol.*, **58**, 167–211.

Ronenson, B. M., 1964. 'Geology of the Vishevygory intrusion and the problem of the genesis of the alkaline rocks of the Middle Urals', in *The Origin of the Alkaline Rocks. Papers of the 3 All-Union Petrogr. Congress*, 1964. Izd. 'Nauka', Moscow, 7–11 (in Russian).

Ryabchikov, I. D. 1963. 'Improvement of the Barth feldspar geological thermometer', in *Mineralogical Thermometry and Barometry*, Izd. 'Nauka', Moscow, 49–60 (in Russian).

Schairer, D. F., 1950. The alkali–feldspar join in the system $NaAlSiO_4$–$KAlSiO_4$–SiO_2. *J. Geol.* **58**, 5, 512–17.

Smirnov, A. D., and Buldakov, V. V., 1962. *The Intrusive Complexes of the East Sayan.* Akad. Nauk SSSR, Moscow.

Smith, F. G., 1953. *Historical Development of Inclusion Thermometry.* 150 pp. Univ. of Toronto Press.

Sobolev, V. S., 1947. 'The petrology of the Botogol alkali massif,' in *The Botogol Graphite Deposit and the Perspectives of its Utilization*, OGIZ, Irkutsk, 165–218 (in Russian).

Sobolev, V. S., and Florensov, N. A., 1948. Genesis of the Botogol graphite. *Sov. Geol.*, **32**, 29–35 (in Russian).

Sobolev, V. S., Kostyuk, V. P., Bazarova, T. Yu., and Bazarov, L. Sh., 1967. Inclusions of melts in the phenocrysts of the nepheline basalts. *Dokl. Akad. Nauk SSSR*, **173**, 2. (in Russian).

Sobolev, V. S., Bazarova, T. Yu., Shugurova, N. A., Bazarov, L. Sh., Dolgov, Yu. A., and Sørensen, H., 1970. A preliminary examination of fluid inclusions in nepheline, sorensenite, tugtupite and chkalovite from the Ilímaussaq alkaline intrusion, South Greenland. *Meddr. Grønland*, **181**, 11, 1–32.

Solonenko, V. P., 1950. Genesis of the alkaline rocks and the graphite of the Botogol massif. *Izv. Akad. Nauk SSSR, Ser. geol.*, **6**, 108–18 (in Russian).

Sorby, H. G., 1898. On the microscopic structure of crystals, indicating the origin of minerals and rocks. *J. Geol. Soc. Lond.*, **14**.

Valyashko, V. M., and Kogarko, L. N., 1966. On inclusions in the apatite from the Khibina and Lovozero massifs. *Dokl. Akad. Nauk SSSR*, **166**, 1. (in Russian).

Zhidkov, A. Ya., 1960. A differentiated pluton of alkaline rocks in the North-Baikal highland. *Mater. on Geology and Mineral Deposits of the East Sibera, Izd. VSEGEI, New Ser.*, **32**, 119–25.

Zhidkov, A. Ya., 1961. A new North-Baikalian alkaline province and some feature of the nepheline-bearing rocks. *Dokl. Akad. Nauk SSSR*, **140**, 1, 181–4.

Zhidkov, A. Ya., 1962. The complex Synnyr syenite intrusion of the North-Baikalian alkaline province. *Geol. and Geophys., SO AN*, **9**,

V.3. TRACE ELEMENTS IN SELECTED GROUPS OF ALKALINE ROCKS

V. I. Gerasimovsky

V.3.1. Nepheline Syenites

The chemical composition of nepheline syenites varies within very wide ranges, not only in different regions but also within the boundaries of one massif. As, furthermore, little data is available on the contents of many elements in miaskitic nepheline syenites, it is difficult to calculate the average contents of most elements in nepheline syenites.

The average chemical compositions of some alkaline massifs composed of respectively agpaitic nepheline syenites (Ilímaussaq and Lovozero), miaskitic nepheline syenites (Vishnevogorsk, the Urals) and intermediate rocks (Khibina massif) are given in Table 1.

The distribution of trace elements in nepheline syenites is related to the geochemical history of the petrogenic elements and is in the first place determined by the proportion of alkalis and aluminium. Excess of alkalis over aluminium and considerable predominance of sodium over potassium in the magma give rise to a great number of accessory sodium-bearing minerals (numerous zirconium silicates and titanosilicates, villiaumite, etc.) in agpaitic rocks. On the contrary, accessory Na-bearing minerals are not characteristic of miaskitic rocks (cf. Table 1, II.2.1).*

The agpaitic nepheline syenites are poorer in calcium than the miaskitic ones (cf. Table 1). Ca-bearing minerals, such as apatite, titanite, fluorite and calcite, are consequently widespread in miaskitic nepheline syenites, but of subordinate importance in agpaitic rocks.

The agpaitic nepheline syenites are richer in iron than miaskitic ones, ferric iron predominating over ferrous iron. In Lovozero the average Fe^{III} content is 3·45% and the Fe^{II} content 0·83%; the corresponding values for Ilímaussaq are 4·10% and 2·93%, respectively. The average iron content in miaskitic rocks is 1·69% Fe^{III} and 2·14% Fe^{II}.

Titanium is characteristic of nepheline syenites, where it forms its proper minerals or is incorporated in mafic minerals (pyroxenes, amphiboles, etc.). In the mafic minerals titanium substitutes ferric iron isomorphously. Numerous Na-bearing titanosilicates (ramsayite, lomonosovite, murmanite, lamprophyllite, rinkolite, etc.) are characteristic of agpaitic rocks and pegmatites. In miaskitic nepheline syenites ilmenite occurs widely and titanite is the only titanosilicate.

Of the rare elements Nb, Ta, Zr, Hf, RE, Ga, Be, Li and Rb, are the most characteristic of nepheline syenites.

The contents of Nb and Ta in nepheline syenites vary within very wide ranges, Nb from 16 to 2170 ppm, Ta from 1·3 to 200 ppm. The average contents of Nb and Ta are about 200 ppm, and 18 ppm, which is considerably higher than in other types of igneous rocks. According to Borodin *et al.* (1969) the content of Ta in nepheline syenites is 8 ppm, nepheline syenites being differentiates of abyssal subcrustal magmas are richer in tantalum (on the average about 17 ppm) than rocks formed as a result of palingenic or assimilation processes (2 ppm Ta). Nb and Ta follow Ti and Zr, substituting these elements isomorphously in Ti- and Zr-bearing minerals. The contents of Nb and Ta are higher in Ti-bearing minerals than in Zr-bearing ones, since the ionic radii of Nb and Ta are almost equal to those of Ti and differ considerably from the ionic radius of Zr. Niobium and tantalum minerals are not characteristic of agpaitic nepheline syenites. They are found only in the pegmatites of these rocks. In miaskitic

* Readers are referred to the relevant Chapters of this book where similarly cited.

TABLE 1 The Chemical Composition of Nepheline Syenites of some Alkaline Massifs (in ppm)

Elements	Ilímaussaq (S.W. Greenland)	Lovozero (Kola peninsula)	Khibina (Kola peninsula)	Vishnevogorsk (the Urals)	Blue Mountain (Canada)	Stjernøy (Norway)	Sandyk (Kirgizia)	Average Composition of magmatic Rocks of the Earth's Crust
Li	310	55	20	15	54	3·5	31	32
Be	30	8·7	6·1	4	1·35	—	—	3·8
B	10	15	7	—	—	—	—	12
F	2100	1400	1230	1200	—	—	—	660
Na	86000	81180	72700	47790	68340	57500	—	250000
Mg	2200	5880	3900	3420	240	1600	—	18700
Al	93100	92160	112700	116380	—	124200	—	80500
Si	236400	250400	248600	265910	—	245600	—	295000
P	214	880	1250	400	—	—	—	930
S	910	1020	215	—	—	—	—	367
Cl	8800	1600	360	—	—	—	—	170
K	33500	48640	54100	58350	39180	66600	—	25000
Ca	7780	8700	12870	8870	4930	16100	—	32900
Ti	2070	6700	6300	2940	34	2800	—	5400
V	(90)	108	52	—	—	42	—	90
Cr	(28)	27·8	10	—	—	5–10	—	83
Mn	1900	2630	1400	690	430	550	—	1000
Fe	70300	42360	30400	27730	14600	23700	—	46500
Co	—	5	13	—	—	—	—	18
Ni	(40)	12	45	—	6·9	—	—	58
Cu	—	11	9·6	—	7·5	71	—	47
Zn	870	210	62	—	—	—	47	83
Ga	110	60	33	18	17·7	—	24	19
Rb	590	230	215	—	85	115	950	150
Sr	47	610	1070	1270	88	3500	—	340
TR	3680	2050	480	256	—	—	250	230
Zr	4735	3480	625	518	47	—	—	170
Nb	525	696	152	252	—	—	30	20
Mo	14	1·7	3·5	—	—	—	5	1·1
Sn	115	10	6·6	—	—	—	—	2·5
Cs	8·5	1·62	3·5	5	—	—	—	3·7
Ba	90	680–700	1190	1170	19	2400	—	650
Hf	97	83	15	8·5	—	—	—	1
Ta	32	60	14	21	—	—	1·7	2·5
Tl	3·05	0·86	1·2	—	—	—	3·3	6·3
Pb	225	14·6	98	10	25	—	43	16
Th	38	35	14	11	—	0·55	30	13
U	62	16·1	4·2	6·3	—	0·09	10	2·5
Agpaitic coefficient	1·39	1·40	1·09	0·84	—	—	—	—
Fe^{3+}/Fe^{2+}	1·40	4·40	1·45	0·4	—	—	—	—
References	Gerasimovsky (1968)		Kukharenko *et al.* (1965)	Eskova *et al.* (1964)	Payne (1968)	Heier (1964)	Zlobin (1963)	Vinogradov (1962)

nepheline syenites, which have no excess of alkalis over aluminium, the isomorphous entry of Nb and Ta into the crystal structures of Ti- and Zr-bearing minerals is less favourable. As a result pyrochlore is present in these rocks.

Nepheline syenites and their minerals are characterized by an approximately constant Nb/Ta ratio (on an average about 10–12). A separation of Nb and Ta occurs sometimes in rocks of this type but only in those which were formed during the final stages of magmatic evolution or during postmagmatic processes. In such cases the rocks may be characterized by higher values of the Nb/Ta ratio (up to 24).

The contents of Zr and Hf in nepheline syenites of various massifs and within the boundaries of one and the same massif are often quite variable, e.g. from 330 ppm Zr and 8 ppm Hf to 1·92% Zr and 640 ppm Hf in the rocks of Ilímaussaq and from 170 ppm Zr + Hf to 1·12% in the rocks of the Kola Peninsula.

The average Zr + Hf content in nepheline syenites is about 500–600 ppm Zr + Hf, and is higher than in other types of igneous rocks. Zr and Hf in agpaitic nepheline syenites are bound first of all in eudialyte, minor amounts are found in other zirconium silicates: lovozerite, catapleiite and in the mafic rock-forming minerals: aegirine, arfvedsonite, murmanite and other titanium-bearing minerals in which they substitute Ti isomorphously. In miaskitic nepheline syenites Zr and Hf are bound in zircon and in the rock-forming mafic minerals containing Ti.

The Zr/Hf ratio in nepheline syenites and their minerals varies from about 31 to about 70. Sometimes a slight increase of the Zr/Hf ratio occurs in the agpaitic nepheline syenites towards the end of the magmatic process. The highest values of the ratio are found in minerals having formed during postmagmatic processes (lovozerite, catapleiite, etc.). The cause of the separation of Zr and Hf is apparently some difference in chemical properties of these elements (basicity, stability of complex compounds, etc.) and crystallo-chemical factors (different types of crystal structures of Zr-bearing minerals).

The contents of the rare earth elements (RE) in nepheline syenites vary within very wide ranges, for example, in the agpaitic rocks of Ilímaussaq from 370 ppm to 1·09% (Gerasimovsky and Balashov, 1968) and in some miaskitic nepheline syenites from the Kola Peninsula from 50 to 480 ppm (according to Balashov's data, personal information). The cerium earths strongly predominate over the yttrium earths, the average ratio $\sum Ce/\sum Y$ usually ranges from 30 to 7·9. In the Lovozero nepheline syenites it ranges from 1·6 (eudialyte lujavrites) to 21 (aegirine and amphibole lujavrites, foyaites, etc. Gerasimovsky *et al.*, 1966). The distribution of RE during the magmatic stage of formation of nepheline syenites is not yet clear. The composition of the rare earth elements of nepheline syenites is given in Table 2.

The main 'concentrators' of RE in agpaitic nepheline syenites are minerals which contain both calcium and sodium, namely eudialyte, rinkite, rinkolite, steenstrupine, loparite, etc. An yttrium earth accumulation is observed in eudialyte. Thus the eudialyte of Lovozero has a $\sum Ce/\sum Y$ ratio of 0·85–2·18. In the miaskitic nepheline syenites the RE concentrate in Ca-bearing minerals, namely titanite, apatite, etc.

The content of strontium in nepheline syenites varies considerably and ranges usually from 0·001% to 0·320%, the average being about 0·085% Sr. The average content of Sr in nepheline syenites reported by Noll (1933, 1934)—0·118%—appears to be too high. Strontium substitutes Ca and K and strontium minerals are not characteristic. Only in the agpaitic nepheline syenites from Lovozero a Sr-rich mineral, lamprophyllite, is widespread. The

TABLE 2 The Composition of the Rare-Earth Elements (RE) in Nepheline Syenites (ppm)

Name of the Massif	La	Ce	Pr	Nd	Sm	Eu	Gd	Tb	Dy	Ho	Er	Tu	Yb	Lu	Y	RE	Ce/Y
Lovozero	481	860	101	298	45	—	32	6	23	5	14	1·2	10	1·8	135	2400	7·9*
Ilímaussaq	850	1395	160	495	95	12·5	84	15	60	16	41	—	34	—	390	3680	4·6†
Nepheline Syenites (average from 144 analyses)	282	547·6	64·8	252	54	12·6	47·5	12	47·9	13·1	25·3	5·0	23·8	6·0	255	1650	2·8‡

* Gerasimovsky *et al.* (1966). † Gerasimovsky and Balashov (1968). ‡ Alexiev, E. (1969).

content of strontium in miaskitic rocks may be a geochemical indicator of their origin. Rocks having low contents of Sr may be considered to be products of palingenesis, those with high contents products of crystallization of residual melts. The isotopic composition of strontium may, however, be more important in this respect (cf. V.4).

Ga, Be, Li, Cs and Tl are present in nepheline syenites in a dispersed form, besides Be, Li and Tl form their own minerals, which are usually found in pegmatitic and hydrothermal formations associated with agpaitic nepheline syenites. Chkalovite, tugtupite (beryllosodalite), eudidymite and epididymite are the most widespread Be-minerals, polylithionite and tainiolite the most common Li-minerals. The Tl-bearing mineral chalcothallite (Cu_3TlS_2) has lately been discovered in Ilímaussaq (Semenov *et al.*, 1967, Semenov, 1969).

The beryllium content of nepheline syenites varies from 2 to 50 ppm, the average being 5–7 ppm. Beryllium substitutes silicon and aluminium isomorphously in the minerals. According to Beus (1965) the main factor determining the Be isomorphism in silicates is the crystallochemical similarity of the tetrahedral complexes:

$$[BeO_4]^{6-}, [SiO_4]^{4-} \text{ and } [AlO_4]^{5-}.$$

Nepheline syenites contain from 10 ppm to 320 ppm of Li, the average being about 30 ppm. Li is mainly contained in amphiboles, pyroxenes and micas isomorphously substituting magnesium and ferrous iron.

There is very little information in the literature on the geochemistry of cesium and thallium in nepheline syenites. The cesium content varies from 0·7 to 38 ppm, the average being 6 ppm according to the data of Goldschmidt *et al.* (1934). The average Tl content in the alkaline rocks of the Soviet Union is about 0·8 ppm (Voskresenskaya, 1961). The maximum content of this element in nepheline syenites is found in the rocks from Ilímaussaq (cf. Table 1).

Rubidium in nepheline syenites is bound in microcline and nepheline in which rubidium substitutes potassium isomorphously. The K/Rb ratio of the main rock-forming minerals is very much like that of the rocks (Gerasimovsky *et al.*, 1966).

Gallium in nepheline syenites is geochemically associated with Al and Fe^{3+}. The gallium content ranges from 10 to 140 ppm, the average being about 36 ppm.

The contents of the rare lithophilic and radioactive elements (Nb, Ta, Zr, Hf, RE, Ga, Be, Li, Rb, Th, U) are considerably higher in agpaitic nepheline syenites than in miaskitic ones (Table 1). This is in accordance with the process of magmatic differentiation of melts enriched in alkalis, which first give rise to miaskitic nepheline syenites and then to agpaitic rocks enriched in volatiles (F, Cl, etc.) and trace elements (cf. Gerasimovsky *et al.*, 1966; Varet, 1969).

Agpaitic nepheline syenites are considerably enriched in thorium and uranium. The average contents of these elements are extremely high in some of the rocks of Ilímaussaq. Some of these rocks are characterized by low values of the Th/U ratio (0·6). Uranium and thorium are present in a dispersed form mainly in RE-bearing minerals. The Th- and RE-bearing mineral, steenstrupine, is found in pegmatites and hydrothermalites genetically related to agpaitic nepheline syenites, and thorite is found in pegmatites and hydrothermalites of miaskitic nepheline syenites.

Agpaitic nepheline syenites are, when compared with miaskitic ones, characterized by higher Cl, F and S contents. The rocks of Ilímaussaq are especially rich in Cl. The Cl, F and S association with Na is typical of minerals of the agpaitic rocks, so far as the geochemical history of Cl, F and S is determined by alkalis (sodium). The excess of alkalis in the melts which gave rise to the agpaitic nepheline syenites, hindered the removal of Cl, F and S into a gaseous–liquid phase (cf. VI. 4). During the formation of these rocks considerable amounts of Cl were fixed in primary sodalite; a smaller part entered eudialyte. At the final stages of rock formation Cl entered the secondary sodalite which was formed at the expense of nepheline. Sodalite is a rock-forming mineral or one of the minor minerals of most agpaitic nepheline syenites.

Fluorine in agpaitic rocks enters arfvedsonite, zirconium-titano-silicates (rinkite, rinkolite, eudialyte, etc.) and at the final stages of formation of these rocks villiaumite. Fluorine in miaskitic rocks is fixed in micas (biotite), amphiboles, fluorite and apatite and is characteristically associated with calcium.

Sulphur in agpaitic nepheline syenites is mainly fixed in sodalite and to a lesser degree in sulphides. In agpaitic rocks the sulphidic sulphur predominates over sulphate. The association of sulphidic S with sodium in sodalite is one of the most characteristic features of the geochemistry of sulphur in agpaitic nepheline syenites and hindered the removal of sulphur by postmagmatic solutions. Sulphides are therefore not widespread in pegmatites and hydrothermalites related to agpaitic rocks.

There is very little data on the geochemistry of Cl and S of miaskitic rocks.

The geochemical history of P in nepheline syenites is most variable. P-bearing minerals of agpaitic rocks, lomonosovite-murmanite, belovite, steenstrupine, 'erikite', etc., are characterized by the association of P with Na, except in some RE-phosphates. In miaskitic nepheline syenites phosphorous is bound with calcium (apatite).

The high contents of Mo, Zn and Pb in the agpaitic nepheline syenites of Ilímaussaq (14 ppm Mo, 870 ppm Zn and 225 ppm Pb) is rather unexpected. There is very little data on the contents of these elements in nepheline syenites.

The Sn content in nepheline syenites varies within very wide ranges, from 2 to 370 ppm. The average Sn content in this group of rocks is about 12 ppm; it is higher than in other types of igneous rocks. The highest average tin content is found in the rocks of Ilímaussaq (110 ppm), in which stannite and sorensenite,

$$Na_4SnBe_2O_{16}(OH)_4,$$

have been found.

Geochemical data may sometimes be used to solve geological problems. During the geochemical studies of the Lovozero alkaline massif, the rocks of which are derivatives of alkali basaltic magma (cf. IV.2), it was proved that during the process of magmatic differentiation of a magma enriched in alkalis, miaskitic nepheline syenites were formed first and then the agpaitic nepheline syenites (Gerasimovsky *et al.*, 1966). At the same time it was established that in the genetically related series of agpaitic nepheline syenites the latest rocks are considerably enriched in rare elements.

Nepheline syenites which are the end products of differentiation of alkali basaltic magma have higher contents of rare lithophilic elements than nepheline syenites formed by other processes. This thesis is especially elucidated in the Lovozero and Ilímaussaq massifs, which are supposed to have formed from strongly differentiated melts being genetically related to alkali basaltic magmas. The rocks of these massifs are substantially enriched in rare elements (Table 1) in comparison with nepheline syenites of other massifs of different genesis. For instance, in the nepheline syenites of the Mongol–Tuva province, being of a palingenic origin (cf. IV.5), the concentration of typical lithophilic elements is considerably lower: 240–500 ppm RE, 206–520 ppm Zr, 31–51 ppm Nb, 11–16 ppm Th (see IV.5). Similar data have been obtained for the miaskitic rocks of Northern Norway and of the Haliburton–Bancroft region, Ontario (Heier, 1964; Payne, 1968; cf. Table 1).

Conclusions

1. The geochemical and mineralogical peculiarities of nepheline syenites are determined in the first place by the correlation of alkalis and aluminium; of the petrochemical parameters the agpaitic coefficient most completely reflects these peculiarities.

2. The regularity of distribution of the rare and trace elements in nepheline syenites is conditioned by the geochemical history of the petrogenic elements.

3. Typomorphic elements of agpaitic nepheline syenites are Na, Li, Rb, Be, Ga, RE, Zr, Hf, Nb, Ta, Th, U, F, Cl, S and Fe. Typomorphic elements of miaskitic syenites are Na, K, Ca, Zr, Hf, Nb and Ta.

4. There is a great variation in the contents of

rare elements in nepheline syenites which is conditioned first of all by the compositions of the melts from which they have formed. Nepheline syenites, which are differentiates of alkali basaltic magmas, are enriched in rare lithophilic elements and volatile components to a greater extent than rocks formed by palingenic assimilation.

V.3.2. *Alkaline Ultramafic Rocks*

Rocks of this type have been studied in most detail in the Karelian–Kola Province (IV.2) and in the Maimecha–Kotuj Province (III.4).

The Caledonian alkaline ultramafic rocks of Karelian–Kola Province may be derived from melts corresponding to olivine melteigite or melilite basalt (IV.2; Kukharenko *et al.*, 1965). The average contents of the most abundant elements of these rocks are given in Table 3. Be, C, F, Mg, P, Cl, Ca, Sc, Ti, V, Mn, Fe, Ga, Sr, Zr, Nb, Ba, RE (the cerium group), Hf, Ta, Th and U are contained in greater amounts; Li, O, Na, Al, Si, S, K, Cr, Co, Ni, Cu, Zn, Ge, Rb, Cs, RE (the yttrium group) and Pb in lesser amounts than those of the average composition of igneous rocks of the Earth's crust. These alkaline ultramafic rocks thus differ from Alpine-type

TABLE 3 Trace Elements in the Alkaline Ultramafic Rocks of the Kola Peninsula and the Maimecha–Kotuj Province (N.W. Siberia) and in the Alkaline Effusive Association of the Maimecha–Kotuj Province (ppm)

Elements	Kola peninsula	Maimecha–Kotuj Province	
		Intrusive Rocks	Effusive Rocks
Li	2	3–30	—
Be	7·4	1–10	1–10
F	900	—	—
P	2400	60–1000	—
S	200	—	—
Cl	200	—	—
Sc	24	5–30	—
Ti	22200	4080–42720	—
V	440	10–2000	56–952
Cr	30	2–2000	34–1090
Mn	2900	100–2500	—
Co	15	1–1000	—
Ni	36	2–1000	—
Cu	34	10–1000	—
Zn	50	20–2000	—
Ga	26	3–30	—
Ge	1·3	1–3	—
Rb	80	—	—
Sr	1300	10–3000	—
Zr	340	20–300	22–1780
Nb	300	10–100	7–620
Cs	0·4	—	—
Ba	850	30–1500	—
TR	1570	28–490	160–2380
Hf	5·1	—	—
Ta	34	—	8–16
Pb	13	3–30	—
Th	90	1·4–180	—
U	15	0·3–67	—
References	Kukharenko *et al.* (1965).	Gonshakova and Egorov (1968).	Gladkikh *et al.* (1965).

ultramafic rocks and the ultramafic rocks of layered calc-alkalic intrusions in low contents of Cr, Ni and Co and in high contents of the trace elements characteristic of syenitic and granitic rocks.

The contents of rare and trace elements in alkaline ultramafic rocks vary within wide ranges (Table 4 based on Kukharenko *et al.*, 1965). The high contents of the rare lithophilic elements in these rocks, and in other alkaline rocks, is probably conditioned by the fact that post-magmatic processes of mineral formation are widely developed in these rocks. An additional supply of rare elements by the hydrothermal solutions is possible.

The intrusive alkaline ultramafic rocks of the Maimecha–Kotuj province of Permo–Triassic age are genetically related to basic and ultra-basic lavas (cf. III.4), but constitute a minor part of this province (Leontyev *et al.*, 1965) and are younger than the effusives. The intrusive rocks are characterized (Gonshakova and Egorov, 1968) by the following trace elements (Table 2) Li, Be, Ba, Ga, Sr, Cr, Sc, RE, Th, U, C, Pb, Zn, Ge, Mo, Sn, Mn, V, Cr, Ni, Co, P and B. The contents of Li, Be, Sc, Ge, Ga, Zr, Pb, Mo and Sn vary within narrow ranges, but the contents of the other elements vary substantially and increase or decrease towards the end of formation of a given complex. Decreasing contents are characteristic of the elements of the iron group (Mn, Cr, Ni, Co), increasing contents of Li, RE, Nb, Th, U and some petrogenic elements (Ba and Sr).

V.3.3. *Alkaline Effusive Rocks*

In the Maimecha–Kotuj province (III.4) the following effusive rock have been described: picritic porphyrites, basalts, tephrites, basanites, trachybasalts, trachyandesites, andesite-basalts, andesites, alkali trachytes, trachyliparites and meimechites (Leontyev *et al.*, 1965). These rocks are enriched in rare lithophilic elements: Nb, Zr, RE and Be (Table 3). An accumulation of Nb, Zr and RE is observed from basic to alkaline members of the differentiated series (Gladkikh *et al.*, 1965).

The volcanic ultrabasic potassic rocks of the western rift of East Africa (II.5) have been studied by Higazy, 1954 (Table 5). These rocks are very substantially enriched in Sr, Ba, Rb and Zr. They are according to Bell and Powell (1969) characterized by unusually high contents of K_2O (up to 12%); $K_2O > Na_2O$; relatively low contents of SiO_2; high TiO_2; relatively high contents of Ca, Mg, Ba, Sr indicating affinity to mafic rocks; and relatively high concentrations of Rb, Zr, Nb, La and Y indicating affinity to more sialic rocks.

Our unpublished investigations show that the phonolites and nephelinites of the East African rift zone are enriched in the rare lithophilic elements RE, Zr, Hf, Nb, Ta, Th and U and also in F. The phonolites contain 400 ppm RE, 758 ppm Zr + Hf, 185 ppm Nb, 16·4 ppm Ta, 6·8 ppm U and 1300 ppm F; the nephelinites have 400 ppm RE, 700 ppm Zr + Hf, 179 ppm Nb, 19·2 ppm Ta, 10·9 ppm U and 2500 ppm F. The sodic volcanic rocks of the eastern rift are, when compared with the potassic volcanics of the western rift, characterized by lower contents of Mg and Fe and of elements such as V, Cr, Ni, Co and Sc.

TABLE 4 Contents of Rare Lithophilic Elements in the Alkaline Ultramafic Rocks of the Kola Peninsula and North Karelia (ppm) (from Kukharenko *et al.*, 1965)

	Zr	Hf	Nb	Ta	RE	Be
Olivinites	58	1·2	394	45·6	900	1
Pyroxenites	340	7·2	379	47·3	2040	7
Nepheline Pyroxenites	620	7·6	202	18·7	1440	9·6
Melteigites	480	5	245	21·2	1380	10
Ijolites and Urtites	380	4				
Carbonatites	140	3·1	510	19·7	3500	—

TABLE 5 Trace Elements in Effusive Rocks of the Western Rift Zone of East Africa (in ppm) (Higazy, 1954)

Elements	Katun-gites	Mafurite	Ugandite	Ouachitite	Ankara-trites	Mela-Potash Nephelinite	Mela-Leucitite	Olivine Melilitite	Potash Nepheline Melilitite	Kivites and Leucite Trachy-basalt	Limbur-gite	Average Content in all Rock Types
Rb	220	450	450	100	170	120	180	85	300	158	50	210
Li	8	5	7	13	8	20	25	5	10	11	8	—
Ba	3370	7500	2000	1700	2900	5900	6000	2200	>1	2600	900	4100
Sr	6685	7000	1800	2500	6900	>1	>1	800	>1	4700	1200	6250
Cr	720	1300	1200	550	400	<1	5	450	<1	41	1200	535
Co	73	70	110	60	75	50	40	85	70	50	75	70
Ni	185	300	900	250	93	15	15	100	30	35	270	200
Zr	1000	900	300	1200	550	500	700	1000	1100	750	400	765
Ga	50	80	35	30	30	30	30	30	80	30	30	—
Y	<30	<30	<30	<30	<30	<30	<30	60	100	41	41	—
Cu	85	25	60	65	68	250	100	60	45	19	40	—
V	275	220	110	320	350	350	400	650	340	322	300	330
Ga	25	15	5	25	30	30	30	45	35	45	8	28
SiO_2	35·09	39·06	40·47	38·94	37·57	34·77	37·09	38·45	36·86	44·99	44·22	—
Na_2O	1·33	0·18	0·68	1·01	2·19	3·58	3·14	2·52	6·09	3·06	1·73	—
K_2O	3·56	6·98	3·46	3·96	3·80	3·88	4·13	2·46	5·47	3·37	1·05	—

Note:
Additional data on the East African Rift zones is published in Gerasimovsky, V. I., and Polyakov, A. I., 1972. Alkaline rocks of the East Africa Rift zones. *Int. Geol. Congress 24th Session Montreal*, **14**, 34–40.

V.3.4. Alkali Granites

These rocks are usually characterized by higher contents of rare lithophilic elements such as Be, RE, Zr, Hf, Nb and Ta (Table 6), than the average composition of granites. A high niobium–tantalum ratio and a substantial enrichment in yttrium ($\sum Ce/\sum Y = 0{\cdot}84$) is characteristic of the Nigerian alkali granites. Kovalenko *et al.* (1969) think that there is a dependence of the rare-earth element composition of alkali granites on alkalinity. The rôle of yttrium in the general composition of rare-earth elements diminishes with increasing alkalinity. The $\sum Ce/\sum Y$ ratio of the Siberian alkali granites varies from 2·4 to 11·6.

V.3.5. Conclusions

In all types of alkaline rocks the most typical rare elements are the lithophilic elements: Be, RE, Zr, Hf, Nb, Ta, Th and U. The concentration of these elements is often very high in alkaline rocks and is usually higher than in other types of magmatic rocks.

TABLE 6 Rare Lithophilic Elements in Alkali Granites (ppm)

Element	Region				Average Content in Acid Rocks
	Nigeria	Nigeria	Kazakhstan (USSR)	Siberia (USSR)	
$(RE + Y)_2O_3$	—	880*	700–800	500–4000	292(RE)
Nb	—	605 (195–1050)	315–490	1092	20
Ta	—	4·8–58	25–33	82	3·5
Zr	1000 (300–2200)	2390 (510–7800)	370–5700	5620	200
Hf	21–66§	8·9–110	—	—	1
Be	—	13†	20–24	18	5·5
Li	230 (54–630)	243†	—	170	40
Rb	260–840‡	410†	—	820	200
F	7200 (2700–12400)	—	1600–3000	5400	800
Nb/Ta	—	22·6 (14·8–8·39)	26	12·4	5·7
Zr/Hf	19–29§	48 (28·3–71)	—	—	200
$\Sigma Ce/\Sigma Y$	—	0·84*	—	—	—
Literary Source	Bowden (1966a, 1966b)	Tugarinov, *et al.* (1968)	Mineev (1968)	Beus *et al.* (1962)	Vinogradov (1962)

* Kovalenko *et al.* (1969).
† Kovalenko (oral information).
‡ Butler and Smith (1962).
§ Butler and Thompson (1965).

V.3. REFERENCES

Alexiev, E., 1969. *Comptes Rendus de L'Academie Bulgare des Science*, **22**, 1.

Beus, A. A., 1965. Geochemistry of beryllium. *Geokhimia*, **5** (in Russian).

Beus, A. A., Severov, E. A., Sitnin, A. A., and Subbotin, K. D., 1962. *Albitized and Greisenized Granites (Apogranites)*. Moscow, Izd. 'Nauka' (in Russian).

Bell, K., and Powell, J. L., 1969. Strontium Isotopic Studies of Alkalic Rocks: The Potassium-rich Lavas of the Birunga and Toro-Ankole Regions, East and Central Equatorial Africa. *J. Petrology*, **10**, 3, 536–72.

Borodin, L. S., Osokin, E. D., and Blum, A. A., 1969. On the average content of tantalum and some regularities of its distribution in alkaline rocks. *Geokhimia*, **11**, 1331–41 (in Russian).

Bowden, P., 1966. Lithium in younger granites of northern Nigeria. *Geochim. et Cosmochim. Acta*, **30**, 555–64.

Bowden, P., 1966. Zirconium in younger granites of northern Nigeria. *Geochim. et Cosmochim. Acta*, **30**, 985–93.

Butler, J. R., Bowden, P., and Smith, A. Z., 1961. K/Rb ratios in the evolution of the younger granites of northern Nigeria. *Geochim. et Cosmichim. Acta*, **26**, 89–100.

Butler, J. R., and Smith, A. Z., 1962. Zirconium, niobium and certain other trace elements in some alkali igneous rocks. *Geochim. et Cosmochim. Acta*, **26**, 945–53.

Butler, J. R., and Thompson, A. J., 1965. Zirconium, hafnium ratios in some igneous rocks. *Geochim. et Cosmochim. Acta*, **29**, 167–75.

Eskova, E., Zhabin, A., and Mukhitdinov, G., 1964. *Mineralogy and Geochemistry of the rare Elements of Vishnevygorsk*. Izd. 'Nauka', 1–319 (in Russian).

Gavrilin, R. D., Pevtsova, L. A., Agafonnikova, L. F., and Klassova, N. S., 1966. *Geochemistry of Variscian Intrusive Complexes of the North Tien-Shan*, Moscow, Izd. 'Nauka' (in Russian).

Gerasimovsky, V. I., 1968. Geochemistry of agpaitic nepheline syenites. *XXIII International geological Congress, vol. 6. Geochemistry*, 259–65.

Gerasimovsky, V. I., and Balashov, U. A., 1968. Some problems of the geochemistry of the rare earths in the Ilímaussaq alkaline massif. *Geokhimia*, **5**, 523–38 (in Russian).

Gerasimovsky, V. I., Volkov, V. P., Kogarko, L. N., Polyakov, A. I., Saprykina, T. V., and Balashov, U. A., 1966. *Geochemistry of the Lovozero Alkaline Massif*. Izd. 'Nauka', 1–395 (in Russian).

Gladkikh, V. S., Zhuk-Pochekutov, and Leontyev, L. N., 1965. 'Rare elements in the alkaline effusive association of the Maimecha–Kotuisk province (NW of the Siberian platform), in *Petrology and Geochemical Peculiarities of a Complex of Ultrabasites, Alkaline Rocks and Carbonatites*. Moscow, Izd. 'Nauka' (in Russian).

Goldschmidt, V. M., Bauer, H., and Witte, H., (1934). Zur Geochemie der Alkalimetalle, II. Nachr. Ges. Wissensch. Göttingen', *Math-Phys. Kl. IV. N.F.L.*, **4**, 39–55.

Gonshakova, V. I., and Egorov, L. S., 1968. *Petrochemical Peculiarities of the Ultrabasic-Alkaline Rocks of the Maimecha–Kotuj Province*. Moscow, Izd. 'Nauka' (in Russian).

Heier, K. S., 1964. Geochemistry of the nepheline syenites of Stjernøy, North Norway. *Norsk. Geol. Tidsskr.*, **42**, 205–15.

Higazy, R. A., 1954. Trace elements of volcanic ultrabasic potassic rocks of southwestern Uganda and adjoining part of the Belgian Congo. *Bull. geol. Soc. Am.*, **65**, 39–70.

Kovalenko, V. I., Legeido, V. A., Petrov, L. L., and Popolitov, E. I., 1968. Tin and beryllium in alkaline granitoids *Geokhimia*, **9**, 1078–87. (in Russian).

Kovalenko, V. I., and Popolitov, E. I., 1969. *Petrology and Geochemistry of Rare Elements of Alkaline and Granitic Rocks of Northeast Tuva*. Moscow, Izd. 'Nauka' (in Russian).

Kovalenko, V. I., Znamenskaya, A. S., Popolitov, E. I., and Abramova, S. R., 1969. Behaviour of rare-earth elements and yttrium in the process of alkaline granitoid evolution. *Geokhimia*, **5**, 541–53. (in Russian).

Kukharenko, A. A., Orlova, M. P., Bulak, A. G., Bagdasarov, E. A., Rimskaya-Korsakova, O. M., Nephedov, E. I., Ilyinsky, G. A., Sergeev, A. S., and Abakumova, N. B., 1965. *Caledonian Complex of Ultrabasic Alkaline Rocks and Carbonatites of the Kola peninsula and N. Karelia*. Izd. 'Nedra', 1–772 (in Russian).

Leontyev, L. N., Zhuk-Pochekutov, and Gladkikh, V. S., 1965. 'On the question of the so-called alkaline–ultrabasic formation'. In *Petrology and Geochemical Peculiarities of a Complex of Ultrabasites, Alkaline Rocks and Carbonatites*. Moscow, Izd. 'Nauka' (in Russian).

Mineev, D. A., 1968. *Geochemistry of Apogranites and Rare-Metal Metasomatites of North-West Tarbogatia*. Moscow, Izd. 'Nauka' (in Russian).

Noll, W., 1933/34. Geochemie des Strontium. *Chem. Erde*, **8**, 507–600.

Payne, T. G., 1968. Geology and geochemistry of the Blue Mountain nepheline syenite. *Can. J. Earth Sci.*, **5**, 259–73.

Semenov, E. I., 1969. *Mineralogy of the Ilímaussaq Alkaline Massif (Greenland).* Izd. 'Nauka' (in Russian).

Semenov, E. I., Sørensen, H., Bessmertnaja, M. S., and Novorossova, L. E., 1967. Chalcothallite—a new sulphide of copper and thallium from the Ilímaussaq alkaline intrusion, South Greenland. *Meddr. Grønland*, **181**, 5, 13–25.

Varet, J., 1969. Les phonolites agpaïtiques et miaskitiques du Cantal septentrional (Auvergne, France). *Bull. volcan.*, **33**, 621–56.

Vinogradov, A. P., 1962. Average content of chemical elements in the chief types of the igneous rocks of the Earth crust. *Geokhimia*, **7**, 555–71 (in Russian).

Voskresenskaya, N. T., 1961. Thallium content in igneous rocks of the Great Caucasus and some other regions of the USSR. *Geokhimia*, **7**, 573–83 (in Russian).

Zlobin, B. I., 1963. *A Geochemical Investigation of the Alkaline Rocks of the Sandyk Massif and their genesis.* Thesis for the degree of a candidate of geol. mineral. sciences. Moscow, Inst. of Geochem. Anal. Chem., Ac. Sc. USSR (in Russian).

V.4 ISOTOPIC COMPOSITION OF STRONTIUM IN ALKALIC ROCKS

J. L. Powell and Keith Bell

V.4.1. Introduction

Within the last few years variations in the isotopic composition of lead, oxygen, strontium and sulphur have contributed much to our understanding of petrogenetic processes. The low-mass, stable isotopes of oxygen and sulphur have been used as effective indicators of natural fractionation processes, and the radiogenic nuclides of lead and strontium have been used as natural tracers in igneous rock systems. Since relatively few measurements of the lead, oxygen, and sulphur isotopic compositions of alkalic igneous rocks have been reported, this review is restricted to the application of variations in the abundance of the strontium isotopes to the origin of alkalic rocks. Recent reviews of the isotope geology of oxygen (Epstein and Taylor, 1967; Taylor, 1968; Garlick, 1969), of lead (Slawson and Russell, 1967; Kaneswich, 1968), and of the sulphur (Thode, 1965; Jensen, 1967), are available, as is a review of the strontium isotopic composition of carbonatites (Powell, 1966). Only new, pertinent data on carbonatites are included in the present article. Faure and Powell (1972) have reviewed the general field of strontium isotope geology.

V.4.2. The Strontium-87 Tracer-Method

V.4.2.1

There are four naturally-occurring, stable isotopes of strontium: Sr^{84}, Sr^{86}, Sr^{87}, and Sr^{88}. One of these, Sr^{87}, is also formed by beta-particle emission from Rb^{87}. The relative abundance of Sr^{87} is usually expressed as the ratio of the number of atoms of Sr^{87} to the number of atoms of Sr^{86}. Because of the decay of Rb^{87}, the abundance of Sr^{87} in a system containing rubidium increases at a rate that is proportional to the Rb/Sr ratio of the system.

Rubidium and potassium are geochemically coherent elements, and both appear to have been concentrated in the Earth's crust relative to the upper mantle. Strontium, on the other hand, does not seem to have been as strongly concentrated in the crust as the alkali metals (Hurley, 1968). The crust therefore has, on the average, a substantially higher Rb/Sr ratio than the upper mantle. Estimates of the average Rb/Sr ratio of the continental crust are in the range 0·15 to 0·20 (Faure and Hurley, 1963; Armstrong, 1968). Most basaltic rocks, on the other hand, have Rb/Sr ratios of about 0·05, and this ratio is often lower for oceanic basaltic rocks. Armstrong (1968) estimated the Rb/Sr ratio of the mantle as 0·01.

If the crust and the upper mantle formed at about the same time, say three billion years ago, with the same initial ratio of Sr^{87}/Sr^{86}, and if they had Rb/Sr ratios of 0·20 and 0·05, respectively, the Sr^{87}/Sr^{86} ratio of the crust would now exceed that of the upper mantle by about 0·018. Most basaltic rocks actually are found to have Sr^{87}/Sr^{86} ratios ranging from 0·703–0·706, although somewhat higher values have been found for dolerites from Antarctica and Tasmania (Heier, Compston, and McDougall, 1965; Compston, McDougall, and Heier, 1968). The present day Sr^{87}/Sr^{86} ratios of continental crustal rocks vary widely, but most non-basaltic ones have ratios greater than 0·710, and old sialic rocks may have much higher Sr^{87}/Sr^{86} ratios. Estimates for the average present-day Sr^{87}/Sr^{86} ratio of the continental crust range from about 0·710 up to 0·725. Faure and Hurley (1963) report a figure of 0·719 for composite samples of North American shales.

V.4.2.2

Any effects of natural variations in the abundances of the strontium isotopes (fractionation) due to processes other than radioactive decay can be removed by normalizing the observed Sr^{87}/Sr^{86} ratio to an assumed value of 0·1194 for the Sr^{86}/Sr^{88} ratio (see Faure and Hurley, 1963, for details). Therefore, unless they have been contaminated, rocks derived from the same source regions as basalts should have, at the time of crystallization, the same relatively low Sr^{87}/Sr^{86} ratios that characterize basaltic rocks themselves. Rocks derived by the complete or partial melting of older continental rocks would be expected to have initial Sr^{87}/Sr^{86} ratios that are distinctly higher than those found in basaltic rocks. Thus the initial Sr^{87}/Sr^{86} ratio of a rock should give an indication of its source region. Similarly, several rocks derived from the same parent magma should have had identical Sr^{87}/Sr^{86} ratios at the time they crystallized. Rocks with significantly different initial Sr^{87}/Sr^{86} ratios either were not comagmatic, or were contaminated with strontium with a different isotopic composition.

V.4.2.3

Many possible petrogenetic processes, for example limestone syntexis or the partial melting of a eugeosynclinal sequence of sediments and volcanic rocks, may involve materials of both mantle *and* crustal origin. The products of such processes therefore could have intermediate Sr^{87}/Sr^{86} ratios that would provide little immediate petrogenetic information. Peterman *et al.* (1967), for example, have found that the initial Sr^{87}/Sr^{86} ratios of eugeosynclinal sedimentary rocks are only slightly higher than those of basaltic lavas. Granites subsequently formed from such sedimentary rocks could thus have low Sr^{87}/Sr^{86} ratios that might not reveal their crustal origin (see Hurley *et al.*, 1965; Peterman *et al.*, 1967). It may be significant in this connection that the majority of granites analyzed to date have initial Sr^{87}/Sr^{86} ratios slightly but significantly higher than those of oceanic basalts.

The strontium tracer technique is still applicable in theory even to mixtures of mantle and crustal materials, however, if the mixing processes produce characteristic isotopic patterns. In fact, Riley and Compston (1962) and Lanphere *et al.* (1964) showed that if two end members with different Sr^{87}/Sr^{86} and Rb/Sr ratios are mixed in varying proportions, the points representing the mixtures on a plot of Sr^{87}/Sr^{86} *vs.* Rb/Sr lie on a straight line that joins the positions of the two end members. Thus rocks produced by a process of mixing of mantle and crustal materials could be marked by non-time-dependent, linear relationships between initial Sr^{87}/Sr^{86} ratio and Rb/Sr ratio. Bell and Powell (1968) and Powell and Bell (1969) pointed out that plots of initial Sr^{87}/Sr^{86} ratio versus the content of another element or oxide for a series of such mixtures likewise give characteristic patterns—hyperbolas.

V.4.2.4

The hypotheses proposed for the origin of alkalic rocks can be broadly divided into three groups according to whether they involve (*a*) only mantle-derived materials, (*b*) only crustal materials, or (*c*) a mixture of the two. From the discussion above, it can be seen that as a broad generalization these three groups would have initial Sr^{87}/Sr^{86} ratios that were, respectively, (*a*) low, like those of basalt (0·702–0·706), (*b*) high, like those of average crustal rocks (>0·710), and (*c*) intermediate, perhaps marked by patterns characteristic of mixing. While this model is certainly oversimplified, it does provide a general framework for interpretation of the actual initial Sr^{87}/Sr^{86} ratios of alkalic rocks.

V.4.3. Interpretation of Strontium Isotopic Data for Alkalic Rocks

Until recently relatively few Sr^{87}/Sr^{86} ratio measurements for alkalic rocks had been reported, and most of these are from rocks of oceanic islands. Gast (1967) and Hamilton (1968) have summarized much of this information. In the following discussion we will restrict our remarks to continental alkalic rocks.

Graphs of Sr^{87}/Sr^{86} ratio *vs.* Rb/Sr ratio, *vs.* rubidium content, and *vs.* strontium content for alkalic rocks are shown as Fig. 1, 2, and 3, respectively. The data for alkalic rocks

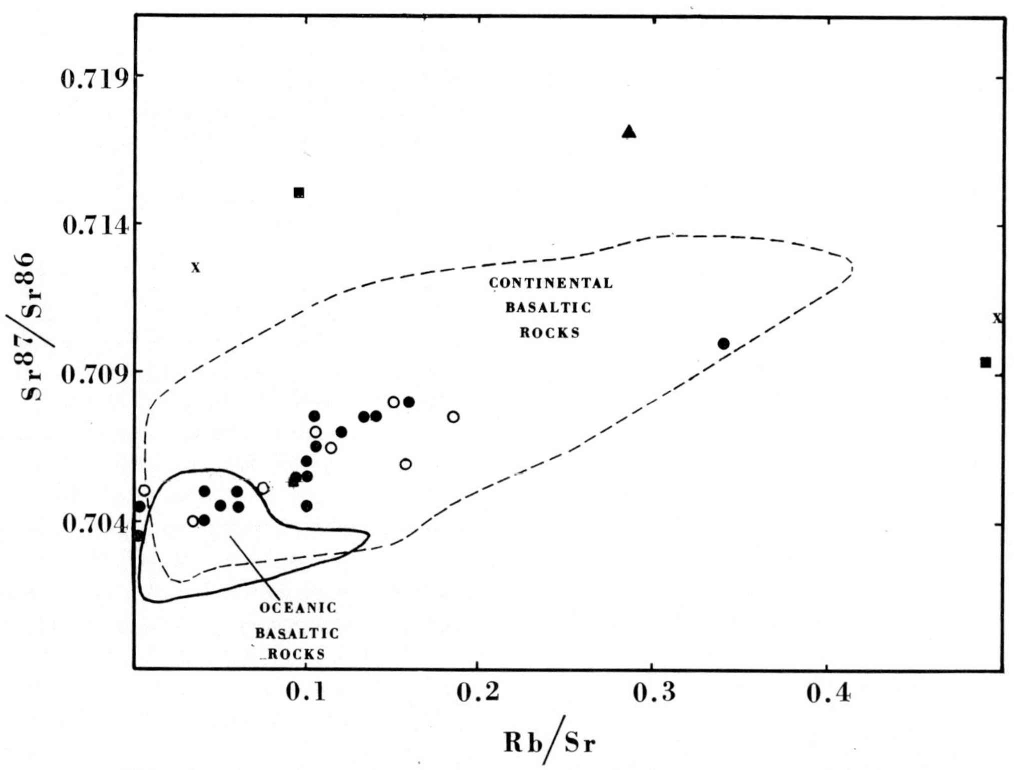

Fig. 1 Sr^{87}/Sr^{86} ratio versus Rb/Sr ratio in alkalic, oceanic basaltic and continental basaltic rocks. The data for alkalic rocks are listed in Table 1, and the data for basaltic rocks are given in Faure and Powell (1972). To maintain a reasonable scale five continental basaltic rocks with Rb/Sr > 0·5 have been omitted. Approximately 70 oceanic basaltic rocks and 90 continental basaltic rocks were used to establish the fields

Symbols used are: X = continental basaltic rocks that plot outside the main field for that group; solid circles = alkalic rocks from the Eastern and Western Rifts of Africa; triangle = alkalic suite from West Kimberley, Australia; squares = alkalic rocks from Europe; open circles = alkalic rocks from North America.

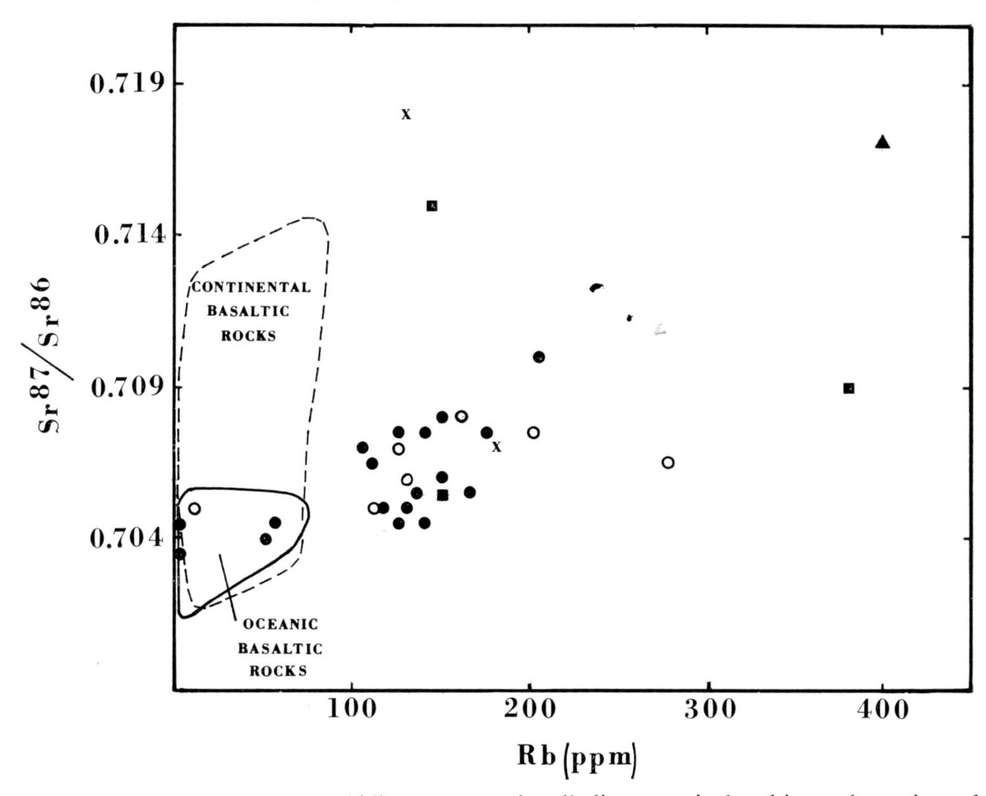

Fig 2 Sr^{87}/Sr^{86} ratio versus rubidium content in alkalic, oceanic basaltic, and continental basaltic rocks. See legend, Fig. 1

used in the construction of the diagrams are listed in Table 1. All the Sr^{87}/Sr^{86} ratios listed have been normalized to $Sr^{86}/Sr^{88} = 0{\cdot}1194$, and all represent initial Sr^{87}/Sr^{86} ratios, that is, the Sr^{87}/Sr^{86} ratio a rock possessed at the time it crystallized. For reference we have shown the fields for both oceanic and continental basaltic rocks on each graph. In some cases the points for the alkalic rocks represent the average for a particular rock type from a certain region (for example, nephelinites from the Western Rift of Africa), and in others they represent averages for a particular province (for example, the Bearpaw Mountains of Montana). It is very important for the reader to realize that because the effects of natural isotopic fractionation can be removed by normalization (see above), such processes as melting, crystallization, and magmatic differentiation will not affect the Sr^{87}/Sr^{86} ratio of either the liquid or solid phases. Since such processes can only cause chemical changes, and not changes in Sr^{87}/Sr^{86} ratio, their effects could only show up as horizontal shifts of points, parallel to the X-axis, on graphs like Fig. 1–3.

V.4.3.1. Relationship Between Alkalic Rocks and Basaltic Magma

The Sr^{87}/Sr^{86} ratios, Rb/Sr ratios, rubidium contents, and strontium contents of some alkalic rocks are similar to those of oceanic basaltic rocks, but in most cases those of the alkalic rocks are distinctly higher. The larger Sr^{87}/Sr^{86} ratios of most of the alkalic rocks compared with those of oceanic basaltic rocks suggest that alkalic rocks in general are not produced by simple differentiation of an oceanic-type basaltic magma.

The fields for continental basaltic rocks on all three graphs overlap those of oceanic basaltic rocks, but the continental basaltic rocks have a much wider range of Sr^{87}/Sr^{86} and Rb/Sr ratio.

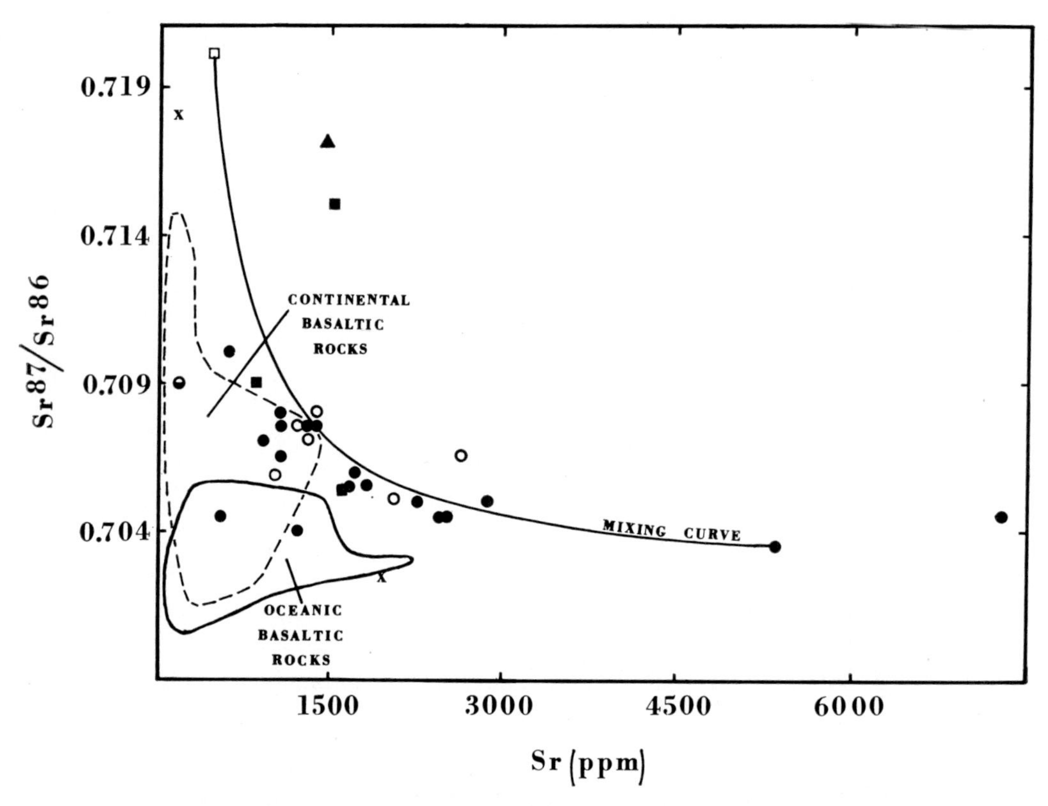

Fig. 3 Sr^{87}/Sr^{86} ratio versus strontium content in alkalic, oceanic basaltic, and continental basaltic rocks. The half-circle represents an oceanic basaltic rock that plots outside the main field for that group, and the open square represents the approximate position of average continental crust (see Faure and Hurley, 1963). See legend, Fig. 1, for general information and for the meaning of the other symbols used

The mixing curve denotes the positions of mixtures of average continental crust and average Eastern Rift carbonatite. A 1 : 1 mixture of these two end members would have $Sr^{87}/Sr^{86} = 0{\cdot}7041$ and Sr = 2955 ppm. Mixtures of the two which have the Sr^{87}/Sr^{86} ratios and strontium contents of most of the alkalic rocks shown would have to contain a large proportion of average crustal material.

The higher Sr^{87}/Sr^{86} ratios of the continental basalts may be caused by contamination with crustal radiogenic strontium, or they may indicate that these basalts come from zones with higher Rb/Sr ratios than those of the oceanic upper mantle. It should be pointed out that most of the continental basaltic rocks with Sr^{87}/Sr^{86} ratios above 0·710 are either from Antarctica or Tasmania.

The Rb/Sr and Sr^{87}/Sr^{86} ratios of alkalic rocks cover about the same range as those of continental basaltic rocks. Only the Jumilla and West Kimberley suites seem to have mean Sr^{87}/Sr^{86} ratios that are distinctly higher. Both the rubidium and strontium contents of the alkalic rocks are several times higher than those of the continental basaltic rocks. Nevertheless, from the data shown it would be possible for most alkalic rocks to have been derived from continental basaltic magmas by a process of differentiation which strongly increases the rubidium and strontium contents of the alkalic rocks. However, although the data shown in Figs. 1–3 permit the possibility that alkalic magmas are derived from continental basaltic ones, in our opinion other explanations are more con-

TABLE 1 Sr^{87}/Sr^{86} and Rb/Sr Ratios, and Rubidium and Strontium Contents of Alkalic Rocks

Locality: Rock Type	Number of Samples	Initial Sr^{87}/Sr^{86}	Rb/Sr	Rb (ppm)	Sr (ppm)	Source of Data
Africa						
(*a*) Eastern Rift						
carbonatite	6	0·7034	~0	~0	5306	1
ijolite	10	0·7045	0·10	52·5	527	1
nephelinite	9	0·7042	0·041	49·6	1211	1
(*b*) Western Rift						
absarokite	6	0·7076	0·141	138	1015	2
banakite	4	0·7074	0·134	175	1316	2
basanite	2	0·7073	0·104	127	1249	2
carbonated lava	4	0·7044	~0	~0	7240	2
katungite	5	0·7047	0·052	124	2449	2
kivite	8	0·7067	0·107	111	1047	2
latite	4	0·7099	0·339	207	613	2
leucitite	23	0·7058	0·102	152	1713	2
mafurite	6	0·7050	0·059	129	2256	2
melilitite	6	0·7049	0·040	114	2881	2
mikenite	3	0·7057	0·098	163	1670	2
murambite	5	0·7070	0·118	105	886	2
nephelinite	17	0·7047	0·058	141	2472	2
shoshonite	4	0·7081	0·158	152	1048	2
ugandite	6	0·7056	0·093	135	1808	2
Australia						
West Kimberley	8	0·7169	0·287	402	1393	3
Europe						
Jumilla, Spain	6	0·7148	0·097	144	1524	3
Laacher See Province	9	0·7054	0·093	151	1625	6, 7
Roman Province	22(18)*	0·7092	0·49*	381	839	4
North America						
Absaroka Field, Wyoming	7	0·7059	0·159	132	1018	8
Bearpaw Mountains	7	0·7076	0·183	200	1196	3
Highwood Mountains	10	0·7078	0·148	159	1370	3
Hopi Buttes	6	0·7050	0·006	10·5	1889	3
Leucite Hills	17	0·7063	0·109	268	2625	3
Montana diatremes	13	0·7050	0·075	113	2005	3
Monteregian Hills	13	0·7040	0·037			5
Navajo Province	11	0·7069	0·103	127	1305	3

Source of data
1. Bell and Powell (1970).
2. Bell and Powell (1969).
3. Powell and Bell (1970).
4. Hurley, Fairbairn and Pinson (1966).
5. Fairbairn *et al.* (1963).
6. Hoefs and Wedepohl (1968).
7. Powell, unpublished data.
8. Peterman, Doe and Prostka (1970).

* Four samples from Ischia are omitted from calculation of the average Rb/Sr ratio because of their unusually high Rb/Sr ratios.

sistent with both the geologic and the isotopic evidence.

V.4.3.2. Source Region of Alkalic Rocks

The Sr^{87}/Sr^{86} ratios of the alkalic rocks listed in Table 1, except for the Jumilla and West Kimberley suites, vary from about 0·704 to 0·710. These values appear to be lower than those of average crustal rocks, and we therefore conclude that most alkalic rocks could not have formed by anatexis of average crustal rocks, and probably not by fusion of old micas and hornblendes (Waters, 1955). The high Sr^{87}/Sr^{86} ratios of the rocks from Jumilla and West Kimberley permit the possibility that they formed entirely from crustal material or from fusion of potassic minerals.

Although we do not believe that alkalic rocks in general are formed by simple differentiation of basaltic magma, there are several ways in which partial melting or zone melting (refining) of the upper mantle could produce rocks with the required isotopic and geochemical properties.

For example, suppose that (*a*) the mantle contains a phase with a relatively high Rb/Sr ratio—phlogopite for example, (*b*) this phase has generated relatively high Sr^{87}/Sr^{86} ratios, (*c*) it does not exchange strontium isotopes with surrounding phases, (*d*) it is the first phase in the system to begin to melt. Then small amounts of partial melting will produce liquids enriched in rubidium and Sr^{87}/Sr^{86}. Further partial melting will involve other, less rubidium-rich phases, generating liquids with lower Sr^{87}/Sr^{86} ratios. This process would produce rocks with the rough correlations between Sr^{87}/Sr^{86} ratio and the other parameters shown in Figs. 1–3. However, it seems improbable that individual mineral phases in the mantle could behave as closed systems with respect to radiogenic strontium.

If the mantle is not isotopically closed on the scale of individual mineral grains, it might still be composed of relatively small, isotopically closed regions perhaps measuring several cubic kilometres in volume. The higher the average Rb/Sr ratio of one of these hypothetical regions, the higher its Sr^{87}/Sr^{86} ratio. Melting of several different regions would thus produce a series of rocks, each with a different initial Sr^{87}/Sr^{86} ratio. If the rubidium and strontium contents of the liquids produced by partial melting of these regions were proportional to those of their solid parents, rocks with a positive correlation between Sr^{87}/Sr^{86} ratio and Rb/Sr ratio, for example, would be produced.

Both of the above are *ad hoc* hypotheses designed only to explain the observed variation in Sr^{87}/Sr^{86} ratios and rubidium and strontium contents of alkalic rocks. It is clear that in terms of these three parameters and under the conditions specified, most alkalic rocks could be derived entirely from the mantle. Either an unusual type of partial melting, or areas of the mantle unusually rich in a potassic phase like phlogopite, or both, would have to be involved. Bell and Powell (1969) pointed out that since potassium content and Sr^{87}/Sr^{86} ratio do not correlate significantly in the rocks of the Western Rift, it is unlikely that they were produced by partial melting of micas or by resorption of mica by basalt magma (Bowen, 1928; Luth, 1967).

Another possible process that could give rise to alkalic rocks is zone melting (refining) (Harris, 1957). This mechanism (cf. VI.1a) can theoretically explain many of the geochemical peculiarities of alkalic rocks. But as a melt undergoing zone refining migrates slowly upward, it should continually equilibrate isotopically with the vertical section of the upper mantle through which it passes. When such melts reach the surface, therefore, their Sr^{87}/Sr^{86} ratios should be approximately the same as the average for the upper mantle—in the range 0·702–0·706. Clearly, simple zone refining of mantle material would tend to produce rather uniform Sr^{87}/Sr^{86} ratios, and not the wide range of ratios actually observed even within a single area like the Birunga province.

V.4.3.3. Origin of Alkalic Rocks by Mixing

The fact that the Sr^{87}/Sr^{86} ratios, Rb/Sr ratios, and rubidium and strontium contents of alkalic rocks overlap those of the oceanic basaltic rocks, but extend over a wider range in the direction of the composition of average continental crust (Sr^{87}/Sr^{86} = 0·720, Rb = 87 ppm, Sr = 440 ppm, Rb/Sr = 0·2; Faure and Hurley, 1963), may lend support to the view that *mixtures* of mantle and crustal materials were involved in the origin of many alkalic rocks. This interpretation is strengthened by the fact that the Sr^{87}/Sr^{86} ratios of alkalic rocks not only are variable, but may correlate with rubidium and strontium contents and with Rb/Sr ratio. Bell and Powell (1969) found similar trends to those shown in Figs. 1–3 for a single volcano, Sabinyo, in the Albert Rift section of East Africa.

If we assume (*a*) that little or no differentiation followed a hypothetical mixing event(s), and (*b*) that most of the alkalic rocks under consideration formed by the mixing of the same two materials, then the diagrams can be used to obtain information about these materials. It can be seen from the graphs that if the *silicate* rocks formed by the simple mixing of two end members more extreme in composition than any of the rocks shown, then one of the end members (A) would have had $Sr^{87}/Sr^{86} \leq 0{\cdot}704$, more than 3000 ppm strontium (more than 7000 ppm if the carbonated lavas are included), and a low rubidium content and Rb/Sr ratio. The other end member (B) would have had $Sr^{87}/Sr^{86} > 0{\cdot}717$, (or $> 0{\cdot}710$ if Jumilla and West Kimberley are excluded), less than 500 ppm strontium, more than 400 ppm of rubidium, and Rb/Sr > 0·5.

Not many materials have these properties. A is abnormally rich in strontium, and B in rubidium. Average crust and average upper mantle (as measured by basaltic rocks) will not do; neither will basaltic magma nor limestone. The only rock that actually has the properties of A is carbonatite, and B is matched best by an alkali-rich granitic rock. However, if the high concentrations of rubidium and strontium in the alkalic rocks could have been produced in part by differentiation, then the levels of rubidium and strontium in the end members need not be considered to have been so high. In this case A could have been a nephelinitic magma, for example, and B could have been a fairly typical sialic rock.

In order to further test the hypothesis that alkalic rocks form by the mixing of carbonatite magma and sial (Holmes, 1950) we have calculated the Sr^{87}/Sr^{86} ratios and the strontium contents of a series of hypothetical mixtures of average Eastern Rift carbonatite (Sr^{87}/Sr^{86} = 0·7034, Sr = 5300 ppm, Bell and Powell, 1970), and average continental crust (Sr^{87}/Sr^{86} = 0·720, Sr = 440 ppm; Faure and Hurley, 1963). We have not made similar calculations for rubidium content and Rb/Sr ratio because it is apparent without them that simple mixtures of carbonatite (Rb $\simeq$ Rb/Sr $\simeq$ 0) and average crust (Rb = 87 ppm, Rb/Sr = 0·2; Faure and Hurley, 1963) will have much lower rubidium contents and Rb/Sr ratios than those observed in many alkalic rocks.

The calculated mixing curve is shown on Fig. 3. It can be seen that in general the Sr^{87}/Sr^{86} ratios and/or strontium contents of most of the alkalic rocks represented are lower than those predicted by the calculated mixing curve. If either a less strontium-rich carbonatite end member or a lower value for the Sr^{87}/Sr^{86} ratio of average crust had been chosen, the calculated mixing line could have been made to pass more nearly through the centre of the main cluster of points. This calculation simply demonstrates that the hypothesis of carbonatite + sial of Holmes can explain the Sr^{87}/Sr^{86} ratios *and* the strontium contents of many alkalic rocks, though it fails to explain their rubidium concentrations.

V.4.4. ***Summary***

The potassic rocks from Jumilla, Spain and West Kimberley, Australia have such high initial Sr^{87}/Sr^{86} ratios that they probably contain substantial amounts of crustal material, whereas ijolites and nephelinites from the Eastern Rift of Africa have Sr^{87}/Sr^{86} ratios and strontium contents like those of oceanic basalts, and therefore could be entirely of mantle origin. Rocks from the Western Rift of Africa, the Western United States, and Italy show surprisingly similar trends of Sr^{87}/Sr^{86} and Rb/Sr ratios, and of rubidium and strontium contents. From the strontium isotopic data it seems unlikely that these rocks were formed by any of the following processes:

differentiation of basaltic magma,
resorption of mica by basaltic magma,
zone refining of mantle material,
anatexis of old micas and hornblendes,
limestone syntexis,
assimilation of sial by basalt magma.

On the other hand, the following processes can explain adequately the Sr^{87}/Sr^{86} ratios, the Rb/Sr ratios, and the strontium contents (though not the rubidium contents) of most of the alkalic rocks studied:

partial melting of an old, chemically and isotopically heterogeneous substratum, possibly a phlogopitic upper mantle,
assimilation of sial by carbonatitic magma,
assimilation of sial by a mafic, alkalic magma.

None of these hypotheses satisfactorily explain the very high potassium and rubidium contents of many alkalic rocks. In addition, it is not clear that magmas contain sufficient superheat for the rather large amounts of assimilation that would be required to produce some of the alkalic rocks. However, the very small bulk volume of the alkalic rocks found at the Earth's surface may indicate that they form only under unusual thermal conditions—conditions that could involve extensive superheating.

ACKNOWLEDGEMENTS

We are grateful to Mr. Hubert Bates, Mrs. Christine Reeves, and Mrs. Jeanne Bowman for technical assistance. The many other individuals and organizations which assisted us will be thanked elsewhere. This work was supported by N.S.F. Grants GP–3709 and GA–1329 to Oberlin College. One of us (Bell) is grateful for the support of the National Research Council of Canada, and to Prof. R. M. Farquhar for helpful discussions.

V.4. REFERENCES

Armstrong, R. L., 1968. A model for the evolution of strontium and lead isotopes in a dynamic earth. *Rev. Geophys.*, **6**, 175–99.

Bell, K., and Powell, J. L., 1968. Isotopic composition of strontium in alkalic rocks. *Program, Ann, Meeting Geol. Soc. Am., Mexico City*, 21.

Bell, K., and Powell, J. L., 1969. Strontium isotopic studies of alkalic rocks: The potassium-rich lavas of the Birunga and Toro–Ankole regions, East and Central Equatorial Africa. *J. Petrology*, **10**, 536–72.

Bell, K., and Powell, J. L., 1970. Strontium isotopic studies of alkalic rocks: The alkalic complexes of Eastern Uganda. *Bull. geol. Soc. Am.*, **81**, 3481–90.

Bowen, N. L., 1928. *Evolution of the Igneous Rocks*. Princeton University Press, 258–73.

Compston, W., McDougall, I., and Heier, K. S., 1968. Geochemical comparison of the Mesozoic basaltic rocks of Antarctica, South Africa, South America, and Tasmania. *Geochim. et Cosmochim. Acta*, **32**, 129–49.

Epstein, S., and Taylor, H. P., Jr., 1967. 'Variation in O^{18}/O^{16} in minerals and rocks', in *Researches in Geochemistry*. Wiley, New York, 29–62.

Fairbairn, H. W., Faure, G., Pinson, W. H., Hurley, P. M., and Powell, J. L., 1963. Initial ratio of strontium 87 to strontium 86 whole-rock age, and discordant biotite in the Monteregian igneous province, Quebec. *J. Geophys. Res.*, **68**, 6515–22.

Faure, G., and Hurley, P. M., 1963. The isotopic composition of strontium in oceanic and continental basalts: Application to the origin of igneous rocks. *J. Petrology*, **4**, 31–50.

Faure, G., and Powell, S. L., 1972. *Strontium Isotope Geology*. Springer-Verlag, 188 pp.

Garlick, G. D., 1969. 'The stable isotopes of oxygen', in *Handbook of Geochemistry*, H. Wedepohl, Editor, vol. 2, pt. 1, 8-B-1 to 8-B-27.

Gast, P., 1967. 'Isotopic geochemistry of volcanic rocks', in *Basalts, vol. I*. H. H. Hess, Editor, Interscience, London, New York and Sydney, 325–58.

Hamilton, E. I., 1968. 'The isotopic composition of strontium applied to the problems of the origin of the alkaline rocks', in *Radiometric Dating for Geologists*. Interscience, London, New York and Sydney, 437–63.

Harris, P. G., 1957. Zone refining and the origin of potassic basalts. *Geochim. et Cosmochim. Acta*, **12**, 195–208.

Heier, K. S., Compston, W., and McDougall, I., 1965. Thorium and uranium concentrations, and the isotopic composition of strontium in the differentiated Tasmanian dolerites. *Geochim. et Cosmochim. Acta*, **29**, 643–59.

Hoefs, J., and Wedephol, K. H., 1968. Strontium isotope studies on young volcanic rocks from Germany and Italy. *Contr. Miner. Petrology*, **19**, 328–38.

Holmes, A., 1950. Petrogenesis of katungite and its associates. *Am. Miner.*, **35**, 772–92.

Hurley, P. M., 1968. Absolute abundance and distribution of Rb, K, and Sr in the earth. *Geochim. et Cosmochim. Acta*, **32**, 273–83.

Hurley, P. M., Bateman, P. C., Fairbairn, H. W., and Pinson, W. H., Jr., 1965. Investigation of initial Sr^{87}/Sr^{86} ratios in the Sierra Nevada plutonic province. *Bull. geol. Soc. Am.*, **76**, 165–74.

Hurley, P. M., Fairbairn, H. W., and Pinson, W. H., 1966. Rb–Sr isotopic evidence in the origin of potash-rich lavas of Western Italy. *Earth Plan. Sci. Lett.*, 5, 301–6.

Jensen, M. L., 1967. 'Sulfur isotopes and mineral genesis', in *Geochemistry of Hydrothermal Ore Deposits*. Holt, Rinehart, and Winston, 143–65.

Kaneswich, E. R., 1968. 'The interpretation of lead isotopes and their geological significance', in *Radiometric Dating for Geologists*. Interscience, London, New York and Sydney, 147–223.

Lanphere, M. A., Wasserburg, G. J. F., Albee, A. L., and Tilton, G. R., 1964. 'Redistribution of strontium and rubidium during metamorphism, World Beater Complex, Panamint Range, California', in *Isotopic and Cosmic Chemistry*. North-Holland, Amsterdam, 269–320.

Luth, W. C., 1967. Studies in the system $KAlSiO_4$–Mg_2SiO_4–SiO_2–H_2O: 1, Inferred phase relations and petrologic applications. *J. Petrology*, **8**, 372–416.

Peterman, Z. E., Doe, B. R., and Prostka, H. J., 1970. Lead and strontium isotopes in rocks of of the Absaroka volcanic field, Wyoming. *Contr. Miner. Petrology*, **27**, 121–30.

Peterman, Z. E., Hedge, C., Coleman, R. G., and Snavely, P. D., Jr., 1967. Sr^{86}/Sr^{86} ratios in some eugeosynclinal sedimentary rocks and their bearing on the origin of granitic magmas in orogenic belts. *Earth Plan. Sci. Lett.*, **2**, 433–9.

Powell, J. L., 1966. Isotopic composition of strontium in carbonatites and kimberlites. *Miner. Soc. India, I.M.A. Volume*, 58–66.

Powell, J. L., and Bell, K. 1969. Recognition of contamination in igneous rocks using strontium isotopes. *Abstracts with Programs for* 1969, *Part 6, Geol. Soc. Am.*, 37.

Powell, J. L., and Bell, K., 1970. Strontium isotopic studies of alkalic rocks: Localities from Australia, Spain, and the Western United States. *Contr. Miner. Petrology*, **27**, 1–10.

Riley, G. H., and Compston, W., 1962. Theoretical and technical aspects of Rb–Sr geochronology. *Geochim. et Cosmochim. Acta*, **26**, 1255–81.

Slawson, W. F., and Russell, R. D., 1967. 'Common lead isotope abundances', in *Geochemistry of Hydrothermal Ore Deposits*. Holt, Rinehart, and Winston, 77–108.

Taylor, H. P., 1968. The oxygen isotope geochemistry of igneous rocks. *Contr. Miner. Petrology*, **19**, 1–71.

Thode, H. G., 1965. 'Sulphur isotope geochemistry', in *Studies in Analytical Geochemistry*. Univ. Toronto Press, Toronto, 25–41.

Waters, A. C., 1955. Volcanic rocks and the tectonic cycle. *Geol. Soc. Am. Spec. Paper*, **62**, 703–22.

V.5. SUMMARY

H. Sørensen

In this section of the book some of the physical and chemical parameters of the petrology of alkaline rocks have been treated.

The experimental studies of synthetic systems and of natural rocks indicate that alkaline rocks may be formed by fractionation of subalkaline magmas or by anatexis. The data give also information about the possible parental materials and magmas for different rock types (V.1).*

Volatile components and oxygen fugacity play an important rôle in the formation of many types of alkaline rocks, having influence on liquidus temperatures, order of crystallization,

* Readers are referred to the relevant Chapters of this book where similarly cited.

stability of minerals and range of consolidation. However, pantellerites and other silica-oversaturated magmas may have formed from fairly dry magmas (V.1).

The liquidus temperatures estimated from experimental data are in accordance with the temperatures obtained by homogenization of gas–liquid inclusions in the minerals of alkaline rocks (V.2). These data indicate fairly high liquidus temperatures, which explains the contact metamorphism exerted by some alkaline intrusions (II.2; IV.6).

Chemical analyses of the gases and liquids of fluid inclusions in minerals of alkaline rocks indicate that CO_2 and water are the most important gases at high temperatures, while hydrocarbon gases may play an important rôle at lower temperatures. The liquids contained in the inclusions are highly saline (V.2).

The trace element data show that alkaline rocks are generally enriched in a number of trace elements of the so-called 'incompatible' group such as Li, Be, Nb, Zr and RE, and impoverished in elements such as Cr, Ni and Co when compared with subalkaline rocks. This recalls the accumulation of rare elements in for instance granite pegmatites and indicate a formation by crystallization of melts of residual character.

There is generally a relation between the content of trace elements and the total content of alkalis (cf. V.3), the highest concentrations of the 'incompatible' group being found in peralkaline rocks such as alkali granites and agpaitic nepheline syenites. This is also seen in volcanic rocks (cf. Varet, 1969; Locardi and Mittempergher, 1967). On the other hand, the nepheline syenites of Stjernöy, Northern Norway and Blue Mountain, Ontario, are poor in rare elements in spite of high contents of potassium (Heier and Taylor, 1964; Heier, 1962; Heier, 1965). This may be a result of an anatectic origin or of a loss of a vapour phase during crystallization of the magma. The distribution of trace elements between a crystallizing magma and the gas phase in equilibrium with the magma is also reflected in the poverty of rare elements in miarolitic ekerite compared with that of massive ekerite (Dietrich and Heier, 1967). Thus the total contents of trace elements and the ratio of major elements relative to trace elements, as for instance K/Rb, may reflect conditions of crystallization as well as type of parent material and stage of crystal fractionation (cf. also IV.8 and Locardi and Mittempergher, 1970). Nevertheless ratios such as Zr/Hf appear to reflect the differentiation of magmas (cf. Gerasimovsky, 1969; Brooks, 1970).

The rare earth metals play an important rôle in alkaline rocks (cf. V.3; VI.7; VII) and may show a fractionation of heavy and light members. In the case of volatile-rich alkaline magmas complexing of the rare earth metals, partly during late- and post-magmatic processes, may have a profound influence on this fractionation and may thus conceal features inherited from the parental material. This restricts the importance of the rare earth metals as 'geochemical tracers' (cf. review by Herrmann, 1970).

The Sr^{87}/Sr^{86} and Rb/Sr ratios of alkaline rocks are discussed in V.4 and appear to indicate that many alkaline rocks are not simple differentiates of alkaline olivine basalts of oceanic type. Some alkaline rocks, as nephelinites and ijolites, may be entirely of upper mantle origin, while others contain a considerable admixture of crustal material.

The origin of the alkaline rocks will be discussed in section VI of this book in the light of the petrographical data of section II; the geological data of sections III and IV, and the laboratory data of section V.

V.5. REFERENCES

Brooks, C. K., 1970. The concentrations of zirconium and hafnium in some igneous and metamorphic rocks and minerals. *Geochim. cosmochim. Acta.*, **34**, 411–16.

Dietrich, R. V., and Heier, K. S., 1967. Differentiation of quartz-bearing syenite (nordmarkite) and riebeckitic-arfvedsonite granite (ekerite) of the Oslo series. *Geochim. cosmochim Acta*, **31**, 275–80.

Gerasimovsky, V. I., 1969. *Geochemistry of the Ilímaussaq Alkaline Massif* (in Russian). Izd. 'Nauka', Moskva 1–174.

Heier, K. S., 1962. A note on the U, Th and K contents in the nepheline syenite, and carbonatite on Stjernöy, North Norway. *Norsk Geol. Tidsskr.*, **42**, 287–92.

Heier, K. S., 1965. A geochemical comparison of the Blue Mountain (Ontario, Canada) and Stjernöy (Finmark, North Norway) nepheline syenites. *Norsk geol. Tidsskr.*, **45**, 41–52.

Heier, K. S., and Taylor, S. R., 1964. A note on the geochemistry of alkaline rocks. *Norsk Geol. Tidsskr.*, **44**, 197–204.

Herrmann, A. G., 1970. 'Yttrium and Lanthanides', in Wedepohl, K. H., ed. *Handbook of Geochemistry vol. II/2.*

Locardi, E., and Mittempergher, M., 1967. Relationship between some trace elements and magmatic processes. *Geol. Rundschau*, **57**, 313–34.

Locardi, E., and Mittempergher, M., 1970. The meaning of magmatic differentiation in some recent volcanoes of Central Italy. *Bull. volcan.*, **33**, 1089–1100.

Varet, J., 1969. Les phonolites agpaïtiques et miaskitiques du Cantal septentrional (Auvergne, France). *Bull. volcan.*, **33**, 621–56.

VI. Petrogenesis

VI.1. ORIGIN OF ALKALINE MAGMAS AS A RESULT OF ANATEXIS

a. Mantle Anatexis *by* P. G. Harris

b. Crustal Anatexis *by* D. K. Bailey

VI.1a Anatexis and Other Processes Within the Mantle (P. G. Harris)

In this discussion of the possible rôle of mantle anatexis in the origin of alkalic rocks, the writer attempts to maintain neutrality, even though, in his opinion, all mafic alkalic rocks appear to be enriched in some minor and trace elements to an extent that is difficult to explain by a simple anatectic mechanism alone.

This contribution discusses:

a. the evidence from modern volcanic rocks that might support a primary or anatectic origin,

b. the mantle environment in relation to anatexis,

c. the experimental evidence for the probable products of melting processes within the mantle,

d. other possible magmatic processes within the mantle, and criteria that might permit distinctions among these, and between primary and secondary alkalic magmas.

VI.1a.2. Evidence from Volcanic Rocks for Primary Alkalic Magmas

There can be no reliable criterion based on field or petrological evidence, for identifying primary magmas. However, one would expect that in a particular environment, a primary magma should be the dominant eruptive rock type, and there should be no obvious parental magma from which it could be derived.

Because, in oceanic regions, alkalic magmas are indisputably of mantle origin, with the minimum possibility of crustal contamination, it is desirable to look first at these.

Following McBirney and Gass (1967) it seems reasonable to make the generalization that in oceans, in the regions of high heat flow and steep thermal gradient and especially along mid-ocean ridges, the primary magma type is tholeiitic, presumably the result of shallow melting or equilibration (Oxburgh and Turcotte, 1968a; Kay *et al.*, 1970). Elsewhere in the ocean basins, in regions of normal or low geothermal gradient, the dominant magma is alkalic, the result of deeper processes.

Turning now to continental examples, although some regions, especially of flood basalts, have a dominant tholeiitic aspect, in others tholeiites are absent or uncommon and the dominant basalt type is alkalic, sometimes extremely so. The best alkalic examples are the volcanic rocks of the rift areas of East Africa, characterised by nephelinites and other highly-undersaturated types (e.g. Saggerson and Williams, 1964; Wright, 1965).

Moving away from the ocean across the circum-Pacific island arcs or continental margin, the basalts change from dominantly tholeiitic to dominantly alkalic ones (Kuno, 1959, 1967; Sugimura, 1968). Kuno relates the increasingly alkalic nature of the magma across the zone to an increasing depth of origin, corresponding to the increasing depth of earthquake foci.

VI.1a.3. The Mantle Environment

The mantle has had a complex history and is unlikely to be chemically homogeneous. Recent isotopic work suggests that some highly fractionated alkalic rocks such as kimberlites and potassic basalts may be the result of multistage processes, the final stage being the re-melting or reactivation of geochemically anomalous material within the mantle. However, the lack of data on mantle heterogeneity makes it difficult to allow for this in experimental work and in theoretical discussions of mantle behaviour. By analogy with ultramafic xenoliths from basalts and kimberlites, the uppermost mantle is thought to have the mineralogy of a spinel lherzolite, passing downward into garnet lherzolite (Ringwood, 1966, Clark and Ringwood, 1964).

Current estimates, e.g. of Clark and Ringwood (1964) and MacDonald (1965), for the average geothermal gradients in different geological environments are in agreement that these gradients are too low for melting to occur in an anhydrous upper mantle. Probable relationships between the suboceanic geothermal gradient and initial and complete melting of material of mantle composition are shown in Fig. 1, largely based on Kushiro, Syono and Akimoto (1968). In anhydrous material of probable mantle composition, melting begins at about 1200 °C and is complete at about 1650 °C, at atmospheric pressure. At high pressures, the temperature of the beginning of melting is increased by about 11 °C per kilobar.

As shown by Fig. 1, melting of anhydrous material at elevated temperatures within the mantle can occur in two ways; either by raising the local temperature above that required for initial melting (path a) or by reducing the local pressure and hence the temperature necessary for melting (path b).

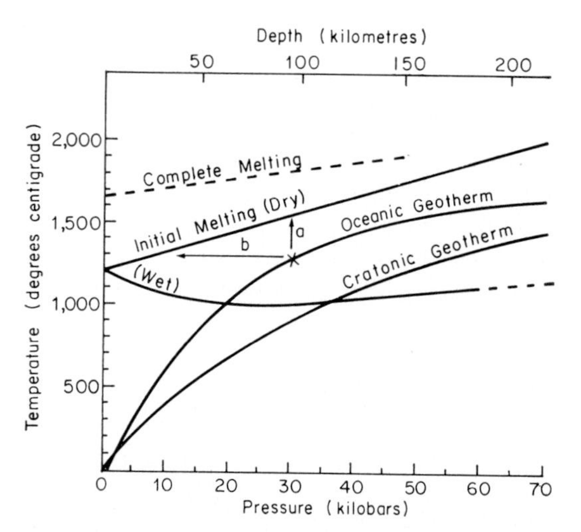

Fig. 1 The effect of pressure on the melting temperatures of material of probable upper mantle composition, based on the results of Kushiro, Syono and Akimoto (1968). The curves for the geotherms are taken from Ringwood (1966)

Increase in local temperature can result from radioactive decay of K, U and Th, but the low concentrations of these elements in the mantle make this a very slow and long-term process, requiring periods of $10^8 - 10^9$ years. Increase in temperature may result also from the release of mechanical energy. The most likely source of this is the frictional energy released during convective or gravitational movement. This energy source will occur where one crustal plate overrides another (Oxburgh and Turcotte, 1968b), for example, in the seismic areas of the circum-Pacific and the Indonesian–Himalayan–Alpine chain. The volcanism of the circum-Pacific can be attributed to this.

Release of pressure causing melting might occur during updoming or crustal tension (Yoder, 1952, see also III.2 and VI.1b).* Also, because a column of magma can be lighter than the same depth of solid mantle and crust, a magma channel opened to the surface can itself cause a reduction of pressure at depth. This effect is enhanced if the magma density is reduced by the presence within it of a separate exsolved gas phase.

Melting due to release of pressure has been invoked for the intrusive basalts of dyke swarms, and may occur in the tension zone of the mid-ocean ridges. However, it seems that the cause of magma generation in these areas is more likely to be due to the upward movement of hot mantle, as a convective plume or wedge into shallower zones of lower pressure and melting point (Oxburgh and Turcotte, 1968a), than to tension effects alone. The proportion of alkalic magma in this environment is small. In the continental rift systems, in which melting would be by the same mechanism of upward transportation of mantle material in a convective upwell, the volcanism is normally of highly alkalic type (e.g. East Africa, cf.III.2). It is tempting to assume that the difference in dominant magma type between the

* Readers are referred to the relevant Chapters of this book where similarly cited.

mid-ocean and continental rift systems is due to geothermal differences. In the relatively vigorous upwell causing ocean-floor spreading, hot mantle material rises rapidly and almost to the surface, so that near-surface geothermal gradients are very steep, and melting or final equilibration is shallow depth. Beneath continental rifts, upwell is very sluggish, too sluggish to cause crustal spreading, and penetration to shallow depths does not occur. Here the thermal effects are smaller, and magmatic processes deeper (Harris, 1969; Murray, 1970). The experimental evidence for depth control is discussed later.

Outside these regions of abnormally steepened geothermal gradient, the results of the convective heat engine, there are many areas of volcanism dominantly of alkalic mafic material. Some may be the sites of local convective upwells or diapirs, but others do not seem related to any structural features. Also, in continents, kimberlites and associated rocks seem to be geographically separated from any convective heat source. In these cases of volcanism in regions of apparently low geothermal gradient, it must be assumed that magmatism originates at greater depth, probably below the upper mantle. If one accepts that some sort of convective movement occurs within the upper mantle, this provides a cooling mechanism for the removal of excess radiogenic heat and average temperatures are well below those for melting. However, at greater depth, below the influence of convective movement, the only mechanisms for heat removal are magmatism and thermal conductivity. Therefore, one would expect thermal gradients to steepen at the interface between the convective upper and the immobile lower regions of the mantle. This region is perhaps at about 500–700 km depth. Other approaches to the Earth's thermal history and temperature distribution produce similar inflexions of the geothermal gradient (e.g. Lubimova, 1967). It seems probable that in some regions, anatexis must be much deeper than is normally realized (Harris and Middlemost, 1969).

Where a liquid phase does form in the mantle, it should be the result of partial rather than complete melting. At atmospheric pressure, the temperature interval between the beginning and completion of melting is only about 450 °C. But if we assume that the mantle contains a small proportion of water, perhaps 0·1% to 0·2%, and that this enters the first formed liquid phase, then at depths for example of 60 km, the temperature interval between solidus and water-saturated liquidus increases to 700 °C (Fig. 1). Because the energy sources are so restricted, the time required to pass from initial melting to complete melting would be enormous. Instead, it would be expected that once sufficient melting had occurred for the liquid phase to be capable of movement or escape, it would be bled off, leaving a refractory solid residue. Kay *et al.* (1970) discuss in detail the relationship between water content and degree of partial fusion.

There is no agreement on what proportion of melting must occur before the liquid phase is capable of mechanical separation. In partial fusion, the crystals would dissolve at points of stress and recrystallize elsewhere, so that the liquid cavities would enlarge and the residual solid form compacted masses. Under these conditions 10% of interstitial liquid might be capable of segregation from the solid. It is difficult to imagine 1% or 2% of liquid being capable of escape. However, where a body of liquid ascends by solution stoping (see later), adcumulus growth of the crystal phases deposited at the bottom of the liquid could almost completely displace any intergranular liquid (Wager *et al.*, 1960). In these circumstances, even 1% or 2% of an original liquid phase might be incorporated into the ascending body of liquid, the residual solid being left almost completely devoid of any liquid phase.

VI.1a.4. Experimental Evidence for the Products of Melting

The results of experimental work in the melting behaviour and fusion products of mafic and ultramafic material, in relation to basalt genesis, are reviewed by Yoder and Tilley (1962), Green and Ringwood (1967), O'Hara and Yoder (1967) and Green (1968).

At pressures below 5 kbar, partial fusion of a

peridotitic material gives a tholeiitic or silica-saturated liquid, changing to an increasingly ultramafic one with increasing degrees of fusion (Reay and Harris, 1964). The derivation of a silica-saturated liquid from an olivine-rich parental solid is a consequence of the incongruent melting of orthopyroxene. Yoder and Tilley (1962) point out that in the system forsterite–silica–nepheline–diopside there is a thermal divide between the silica-saturated and alkalic fields of composition, so that at low pressures it is not possible to pass from one field to the other by fractional crystallization or to derive an alkaline magma by the partial melting of lherzolitic upper mantle material (i.e. containing orthopyroxene). However, at higher pressures the mineral stabilities change and these thermal divides disappear or change position (O'Hara, 1968), so that it is possible to form alkalic magmas directly by fusion or indirectly by fractionation from hypersthene-normative liquids.

The initial effect of pressure is to increase the stability of orthopyroxene, so that above 5 kbar it no longer melts incongruently (Boyd, England and Davis, 1964). Green and Ringwood (1967) consider that partial fusion of mantle material between 15 to 35 km depth will yield an olivine-normative liquid corresponding probably to a high-alumina olivine tholeiite; between 35 to 70 km small degrees of fusion yield a liquid corresponding to olivine-rich alkalic basalt, while larger degrees of fusion yield an olivine tholeiite. Kushiro (1965) also indicates that the initial product will change from a silica-rich one at low pressures to a silica-poor one at high pressures, though in his case it is suggested that highly undersaturated liquids, e.g nepheline-basalts, require pressures of 30 kbar or more (90 km or deeper). At pressures of about 30–40 kbar, i.e. depths below 90–100 km, the mantle mineralogy will be that of a garnet lherzolite, and there is a marked change in the first-formed liquid which now is picritic or ultrabasic and hypersthene-normative (Davis and Schairer, 1965), with more than 30% normative olivine (Green and Ringwood, 1967).

In an alternative explanation for a primary origin of alkali magmas, Yoder and Tilley (1962, p. 507), suggested that in the melting of eclogite differences in pressure could change the ratio of garnet and omphacite in the liquid and hence the liquid composition. An alkali basalt liquid enriched in omphacite components would form at high pressures, and a tholeiitic liquid enriched in garnet components at low pressures.

All of the above experimental work and its petrogenetic implications have been based on anhydrous systems. At high water pressures the increased stability of hypersthene is inhibited so that even at 30 kbar water pressure it still melts incongruently (Kushiro, Yoder and Nishikawa, 1968). Under these conditions the liquid formed by the partial fusion of peridotite is probably silica saturated, that is, a tholeiite (Kushiro, Syono and Akimoto, 1968). It is surprising therefore that alkalic basalt magmas appear to be volatile-rich, since their volatile contents should have inhibited a fractionation or melting trend towards an alkalic magma. However, the volatiles often are very strongly CO_2 enriched. Because the solubility of CO_2 in silicate melts is so much lower than that of water, the CO_2 tends to remain in the gas phase, rather than the magma, so that the gas pressure is likely to be dominantly a CO_2 pressure. Therefore, the P_{H_2O} will be much lower than the load pressure.

Further factors in the anatexis of hydrous systems are the possible stability in the mantle of amphibole (Oxburgh, 1964; Gilbert, 1969; Varne, 1970) and phlogopite (Yoder and Kushiro, 1969). Selective fusion of an amphibole phase has been invoked for the origin of alkalic magmas (e.g. Bose, 1967), cf.VI.6.

In summary, experimental evidence indicates that a small degree of partial fusion in the mantle at pressures of about 20–30 kbar, yields a nepheline-normative liquid corresponding to an alkalic basalt. Other possible mechanisms, e.g. the selective fusion of individual minerals such as omphacite or amphibole, require experimental confirmation.

VI.1a.5. Anatexis, Fractional Crystallization and Zone Refining

The existence of mechanisms for the direct

origin by anatexis of alkalic basalts does not prove that any alkalic magmas have formed in this way. The same experimental data might equally well support an origin from a pre-existing magma by fractional crystallization or reaction with the mantle material under the appropriate pressure–temperature regime. In particular, the increased stability of enstatite at high pressure, and its preferential crystallization instead of olivine, provide a way of decreasing the silica content of a residual liquid, and of giving it an alkalic trend. For example, O'Hara, (1965) and O'Hara and Yoder (1967) would derive alkalic magmas by fractional crystallization of a parental picritic magma within the mantle, and Green and Ringwood (1967) show how fractionation of an olivine-tholeiite could give either a tholeiitic or an alkalic basalt series depending on the local pressure environment.

Criteria are required to permit the distinction between anatexis and differentiation in the origin of alkalic rocks. Such a distinction might possibly be made from the geochemistry of the rocks, and particularly from their trace element abundance patterns. So it is useful to examine the different magmatic processes in the mantle and their effects on magmatic composition.

In addition to the possible but unproven mechanisms of gaseous transfer, liquid immiscibility, ionic diffusion etc. (cf. VI.5), the main processes which control the compositions of magmas are partial fusion, fractional crystallization with or without subsequent reaction with surrounding solid phases, and zone refining or zone melting.

Although the major chemical components can be discussed in terms of phase equilibria and phase diagrams, it is convenient to discuss the behaviour of the minor and trace elements in terms of the distribution factor K (Neumann *et al.*, 1954; Schilling and Winchester, 1967; Gast, 1968; Anderson and Greenland, 1969), where

$$K = \frac{C_S}{C_L} = \frac{\text{concentration of element in solid phase}}{\text{concentration of element in liquid phase}}.$$

Although for some elements the distribution behaviour and the value of K vary widely with changes in the mineralogy and in the composition of the liquid, for others the value of K remains close to zero over a wide range of conditions.

Thus minor and trace elements, too low in concentration for their own minerals to form, and unable to substitute in any of the local crystal lattices for reason of ionic size, bond type, valency etc., are almost completely retained in the liquid phase during any liquid–solid reaction or equilibrium (Harris, 1957). These 'incompatible' (Green and Ringwood, 1967) or 'residual' (Harris, 1967) elements, such as K, Rb, Sr, Ba, Nb, U, Th, C, F and Cl, have values of K close to zero, and it is these elements that may prove to be useful and diagnostic.

VI.1a.5.1. Partial fusion. The effect of pressure in controlling the composition of the liquid during partial fusion in the mantle has already been discussed. However, this minimum-melting composition is predetermined only for the major elements with which the liquid is saturated, and their diadochs, for which there is an equilibrium distribution between liquid and solid phases. The residual or incompatible elements will go almost completely into the liquid phase where their concentrations vary inversely with the degree of fusion. If C_L and C_0 are the concentrations of an element in the newly formed liquid phase and in the total original material before melting, so that C_L/C_0 is the degree of enrichment in the liquid, and if X is the fraction of liquid phase formed during melting, then

$$\frac{C_L}{C_0} = \frac{1}{X + K(1 - X)}.$$

As K tends to zero, C_L/C_0 tends to $1/X$; as X tends to zero, C_L/C_0 tends to $1/K$, the maximum amount of enrichment (see Fig. 2).

VI.1a.5.2. Fractional crystallization. In magmatic behaviour, fractional crystallization is discussed conventionally as a process occurring within a liquid body insulated by a chilled margin from reaction with its solid environment, and in which the crystallizing phases are rapidly removed from reaction with the liquid. So, as mineral phases disappear, or new mineral phases appear during the process of fractionation, the liquid composition changes. The final liquid may be very different from the initial one or from one formed by partial fusion. At low pressures for example, fractional crystallization of an alkali basalt would lead to a final liquid close to the ternary eutectic for feldspar, leucite and nepheline in the system SiO_2–$NaAlSiO_4$–$KAlSiO_4$. The diadochs also undergo continued fractionation, so that ultimately the liquid may become almost completely denuded of the early crystallizing or high melting point component of any solid solution series. For example, fractionation in the magnesium–iron pair as illustrated by the behaviour of the system Mg_2SiO_4–Fe_2SiO_4 (Bowen and Schairer, 1935) leads to iron-rich late stages, e.g. the ferro-gabbros of layered intrusions. The residual elements will remain almost completely in the liquid phase, their final concentrations being dependant on the degree of crystallization and the proportion of residual liquid (X) remaining.

$$\frac{C_L}{C_0} = X^{K-1}.$$

As K tends to zero, C_L/C_0 tends to $1/X$; when K is close to zero then as X tends to zero, C_L/C_0 tends to an infinitely high value.

If for the different values of K the change in concentration in the residual liquid during progressive crystallization is plotted on a logarithmic scale (Fig. 2) rather than the conventional linear scale, the comparison with partial fusion is clearly seen. Partial fusion gives an upper limit of enrichment of $1/K$, while fractional crystallization does not.

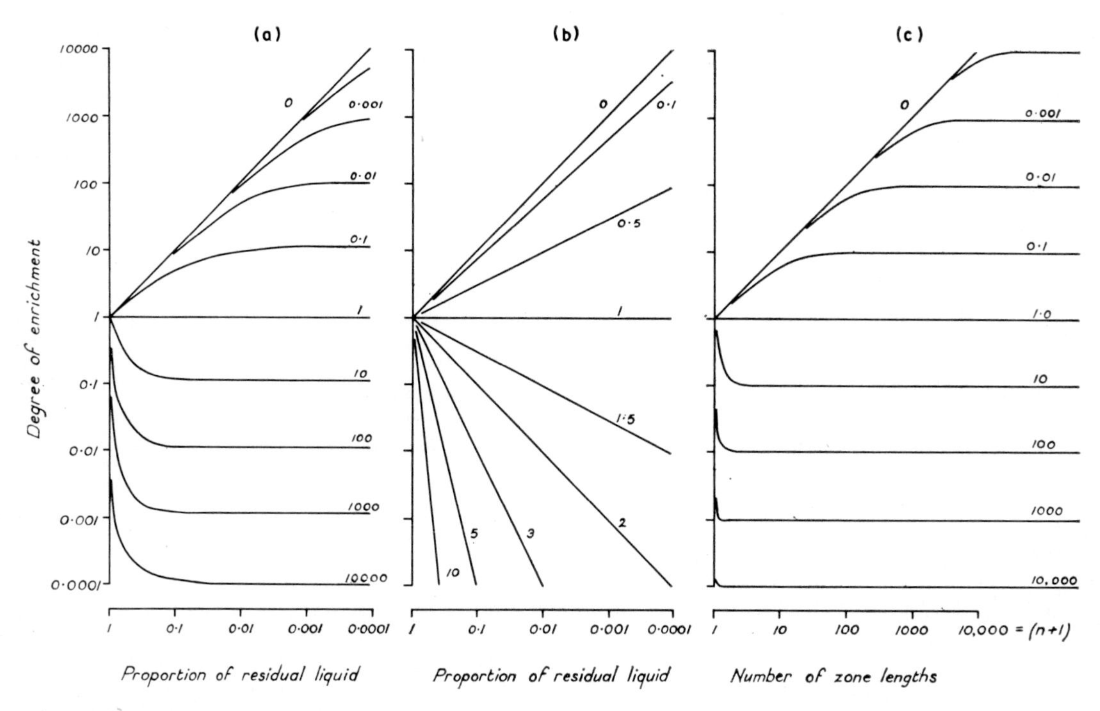

Fig. 2 The enrichment or impoverishment of a trace element in a residual liquid. for different values of K. the distribution coefficient, by the mechanisms of
(a) partial fusion,
(b) fractional crystallization,
(c) zone refining

VI.1a.5.3. Fractional crystallization plus mantle reaction. A body of liquid held in the mantle at a depth where the local temperature is only slightly lower than the solidus will cool and diminish in volume only very slowly. Under these conditions the liquid is unlikely to be chemically insulated by its own chilled margin, but instead will be in a state of continuous reaction with its crystalline environment. Even during cooling and crystallization, the magma will be dissolving roof material. In this type of environment, continued equilibration with mantle minerals will control the major element concentration in the liquid at the same level as for partial fusion at that pressure and temperature regime. Equally, the diadochs will be held at the same concentrations as in partial fusion. In other words, if there is equilibrium between liquid and solids, the liquid composition is predetermined, irrespective of whether the liquid formed *in situ* by partial fusion or was introduced from elsewhere.

The residual elements will be selectively concentrated in the liquid residue, as in the previous examples. However, the equation

$$\frac{C_L}{C_0} = X^{K-1}$$

may no longer show accurately the degree of enrichment of the residual elements. Continued liquid–solid reaction will tend to give maximum limits of enrichment of $1/K$ as in partial fusion and zone-refining. The process of mantle reaction outlined qualitatively by Harris (1957, p. 206) has been discussed in detail by Green and Ringwood (1967) under the term 'wall-rock reaction'.

VI.1a.5.4. Zone refining. The concept of zone refining as applied to geochemical problems considers that, in an environment of sufficiently high temperature (at the solidus), a liquid body may move upward through the mantle without much loss of heat or diminution in volume. Ascent will be by solution-stoping, solution of the roof and crystallization and deposition at the floor of the magma body. This will occur spontaneously, because in a gravitational pressure gradient solids will be more soluble in the magma at lower pressure at the top of body than at higher pressure at the bottom. At any point in this upward ascent the liquid will have concentrations of the major elements and their diadochs determined by the local pressure–temperature equilibrium, so that the concentrations are the same as in a liquid formed *in situ* by partial fusion. The residual elements, however, are swept out of the solid mantle by the liquid and carried upward within it. In this way, abnormally high concentrations of the residual elements can be expected in some liquids which otherwise appear as the normal products of partial fusion (Harris, 1957; Harris and Middlemost, 1969). This process has been particularly invoked by Russian geochemists as a mechanism for the chemical segregation of the mantle and crust (e.g. Vinogradov, 1968; Yaroshevskii, 1967; Vinogradov and Yaroskevskii, 1965), cf. IV.2.

The enrichment of the residual elements in the liquid can be related to n the number of zone lengths processed (i.e. the volume of solid that has reacted with a given volume of liquid).

$$\frac{C_L}{C_0} = \frac{1}{K} - \left(\frac{1}{K} - 1\right) e^{-Kn}$$

As n tends to infinity, C_L/C_0 tends to $1/K$, the same ultimate limit as for partial fusion. To this extent the products of zone refining are indistinguishable from those of very small degrees of partial fusion, in the same pressure–temperature regime (Fig. 2).

VI.1a.5.5. The nature of the liquid phase produced by these four processes can be summarized in a tabular form (Table 1).

In magmas derived from the mantle, only fractional crystallization gives a clearly distinguishable product, a feldspar- or feldspathoid-rich leucocratic rock such as phonolite or trachyte, impoverished in Ni and Mg relative to Fe. Even then the process can be ambiguous, because phonolites and trachytes might equally form by the partial melting of mafic alkalic rock.

A magma with major element contents and diadoch ratios similar to those expected for a

TABLE 1

	Major Elements	Diadochs	Residual Elements
(1). Partial fusion	Saturated	Equilibrium concentrations	$\frac{C_L}{C_0} = \frac{1}{X + K(1 - X)}$
(2). Fractional crystallization	Final feldspar-rich eutectic	Extreme fractionation	$\frac{C_L}{C_0} = X^{K-1}$
(3). Fractional crystallization plus mantle reaction	As for partial fusion	As for partial fusion	
(4). Zone refining	As for partial fusion	As for partial fusion	$\frac{C_L}{C_0} = \frac{1}{K} - \left(\frac{1}{K} - 1\right)e^{-Kn}$

partial fusion liquid within the mantle, i.e. a basic or basaltic magma, could have formed by any or all of the processes (1), (3) and (4). The difficulty is to distinguish between these processes. Only the residual elements offer any prospect of this, with two possible criteria.

One is the total concentration level of residual elements. If an alkali basalt contains a hundred times as much potassium, etc., as original upper mantle this would be equivalent to less than 1% partial fusion. It is difficult to visualize how such a small proportion of interstitial liquid could become segregated. There is no difficulty with mechanisms (3) and (4). On the other hand, if the upper mantle is already markedly heterogeneous, the result of earlier episodes of partial fusion, zone refining etc., with some regions strongly enriched in residual trace elements, then a greater degree of fusion can produce the same final degree of enrichment, and the difficulty of separating the liquid no longer occurs. The anomalous initial $^{87}Sr/^{86}Sr$ ratios of many suites of potassic rocks suggests such a multistage origin.

The other future approach lies in a better knowledge of distribution factors. Where $K = 0$, all processes produce the same effect. But actual distribution factors will be greater than zero. If, for an element, $K = 0{\cdot}1$, the maximum enrichment permissible by partial fusion or zone refining is only ten times the background, whereas fractional crystallization theoretically would permit a much greater degree of enrichment (compare a, b and c in Fig. 2).

At present the lack of data makes any distinctions uncertain, but in the future the comparison of the behaviour of several elements may permit a clear-cut distinction between each process (though probably most magmas have undergone all processes to some degrce). Already, Gast (1968) and Hubbard (1969), using trace element criteria, have suggested that alkali basalts are the result of partial fusion processes rather than fractional crystallization. In particular, the rare earth element distribution may provide a real criterion of origin, while Sr and Pb isotopic ratios may indicate the times at which the segregation processes began.

IV.1a. REFERENCES

Anderson, A. T., and Greenland, L. P., 1969. Phosphorus fractionation diagram as a quantitative indicator of crystallization differentiation of basaltic liquids. *Geochim. et Cosmochim. Acta*, **33,** 493–505.

Bose, M. K., 1967. The upper mantle and alkalic magmas. *Norsk Geol. Tidsskr.*, **47,** 121–9.

Bowen, N. L., and Schairer, J. F., 1935. The system MgO–FeO–SiO_2. *Am. J. Sci.*, **29,** 151–217.

Boyd, F. R., England, J. L., and Davis, B. T. C., 1964. Effects of pressure on the melting and polymorphism of enstatite, $MgSiO_3$. *J. Geophys. Res.* **69,** 2101–9.

Clark, S. P., and Ringwood, A. E., 1964. Density

distribution and constitution of the mantle. *Rev. Geophys.*, **2**, 35–88.

Davis, B. T. C., and Schairer, J. F. 1965 Melting relations in the join diopside–forsterite–pyrope at 40 kilobars and at one atmosphere. *Carnegie Inst. Washington Yearbook*, **64**, 123–6.

Gast, P. W., 1968. Trace element fractionation and the origin of tholeiitic and alkaline magma types. *Geochim. et Cosmochim. Acta*, **32**, 1057–86.

Gilbert, M. C., 1969. Reconnaissance study of the stability of amphiboles at high pressure. *Carnegie Inst. Washington Yearbook*, **67**, 167–70.

Green, D. H., 1968. 'Origin of basaltic magmas', in Hess, H. H., and Poldervaart, A., Eds., *Basalts, vol. 2*. Interscience, New York, 835–62.

Green, D. H., and Ringwood, A. E., 1967. The genesis of basaltic magmas. *Contr. Miner. Petrology*, **15**, 103–90.

Harris, P. G., 1957. Zone refining and the origin of potassic basalts. *Geochim. et Cosmochim. Acta*, **12**, 195–208.

Harris, P. G., 1967. 'Segregation processes in the upper mantle', in Runcorn, S. K., Ed., *Mantles of the Earth and Terrestrial Planets*. Interscience, London, 305–17.

Harris, P. G., 1969. Basalt type and African rift valley tectonism. *Tectonophysics*, **8**, 427–36.

Harris, P. G., and Middlemost, E. A. K., 1969. The evolution of kimberlites. *Lithos*, **3**, 77–88.

Hubbard, N. J., 1969. A chemical comparison of oceanic ridge, Hawaiian tholeiitic and Hawaiian alkalic basalts. *Earth Planet. Sci. Lett.*, **5**, 346–52.

Kay, R., Hubbard, N. J., and Gast, P. W., 1970. Chemical characteristics and origin of oceanic ridge volcanic rocks. *J. Geophys. Res.*, **75**, 1585–613.

Kuno, H., 1959. Origin of Cenozoic petrographic provinces of Japan and surrounding areas. *Bull. Volcan.*, **20**, 37–76.

Kuno, H., 1967. 'Volcanological and petrological evidences regarding the nature of the upper mantle', in Gaskell, T. F., Ed., *The Earth's Mantle*. Academic Press, London, 89–110.

Kushiro, I., 1965. The liquidus relations in the systems forsterite–$CaAl_2SiO_6$–silica and forsterite–nepheline–silica at high pressures. *Carnegie Inst. Washington Yearbook*, **64**, 103–9.

Kushiro, I., Syono, Y., and Akimoto, S., 1968. Melting of a peridotite nodule at high pressures and high water pressures. *J. Geophys. Res.*, **73**, 6023–9.

Kushiro, I., Yoder, H. S., and Nishikawa, M., 1968. Effect of water on the melting of enstatite. *Bull. geol. Soc. Am.*, **79**, 1685–92.

Lubimova, E. A., 1967. 'Theory of thermal state of the Earth's mantle', in Gaskell, T. F., Ed., *The Earth's Mantle*, Academic Press, London, 231–323.

McBirney, A. R., and Gass, I. G., 1967. Relations of oceanic volcanic rocks to mid-oceanic rises and heat flow. *Earth Planet. Sci. Lett.*, **2**, 265–76.

MacDonald, G. J. F., 1965. 'Geophysical deductions from observations of heat flow' in Lee, W. H. K., Ed., *Terrestrial Heat Flow*. Geophys. Monogr., **8**, Am. Geophys. Union, 191–210.

Murray, C. G., 1970. Magma genesis and heat flow; differences between mid-oceanic ridges and African rift valleys. *Earth Planet. Sci. Lett.*, **9**, 34–8.

Neumann, H., Mead, J., and Vitaliano, C. J., 1954. Trace element variation during fractional crystallization as calculated from the distribution law. *Geochim. et Cosmochim. Acta*, **6**, 90–9.

O'Hara, M. J., 1965. Primary magmas and the origin of basalts. *Scot. J. Geol.*, **1**, 19–40.

O'Hara, M. J., 1968. The bearing of phase equilibria studies in synthetic and natural systems on the origin and evolution of basic and ultrabasic rocks. *Earth Sci. Rev.*, **4**, 69–133.

O'Hara, M. J., and Yoder, H. S., 1967. Formation and fractionation of basic magmas at high pressures. *Scot. J. Geol.*, **3**, 67–117.

Oxburgh, E. R., 1964. Petrological evidence for the presence of amphibole in the upper mantle and its petrogenetic and geophysical implications. *Geol. Mag.*, **101**, 1–19.

Oxburgh, E. R., and Turcotte, D. L., 1968a. Mid-ocean ridges and geotherm distribution during mantle convection. *J. Geophys. Res.*, **73**, 2643–61.

Oxburgh, E. R., and Turcotte, D. L., 1968b. Problem of high heat flow and volcanism associated with zones of descending mantle convective flow. *Nature*, **218**, 1041–3.

Reay, A., and Harris, P. G., 1964. The partial fusion of peridotite. *Bull. Volcan.*, **27**, 115–27.

Ringwood, A. E., 1966. 'Mineralogy of the mantle', in Hurley, P. M., Ed., *Advances in Earth Science*. M.I.T. Press, Cambridge, Mass., 357–99.

Saggerson, E. P., and Williams, L. A. J., 1964. Ngurumanite from Southern Kenya and its bearing on the origin of rocks in the Northern Tanganyika alkaline district. *J. Petrology*, **5**, 40–81.

Schilling, J.-G., and Winchester, J. W., 1967. 'Rare-earth fractionation and magmatic processes', in Runcorn, S. K., Ed., *Mantles of the Earth and Terrestrial Planets*. Interscience, London, 267–83.

Sugimura, A., 1968. 'Spatial relations of basaltic magmas in island arcs', in Hess, H. H., and Poldervaart, A., Eds., *Basalts, vol. 2*. Interscience, New York, 537–71.

Varne, R., 1970. Hornblende lherzolite and the upper mantle. *Contr. Miner. Petrology*, **27**, 45–51.

Vinogradov, A. P., 1968. Geochemical problems in the evolution of the ocean. *Lithos.*, **1**, 169–78.

Vinogradov, A. P., and Yaroshevskii, A. A., 1965. Physical conditions of zone melting in the Earth's Mantle. *Geochem. Internat.*, **2**, 607.

Wager, L. R., Brown, G. M., and Wadsworth, W. J., 1960. Types of igneous cumulates. *J. Petrology*, **1**, 73–85.

Wright, J. B., 1965. Petrographic sub-provinces in the Tertiary to Recent Volcanics of Kenya. *Geol. Mag.*, **102**, 541–57.

Yaroshevskii, A. A., 1967. 'The principle of zone melting and its application to certain geochemical problems', in Vinogradov, A. P., Ed., *Chemistry of the Earth's Crust*. Israel Prog. for Scientific Translations, Jerusalem, **2**, 54–65.

Yoder, H. S., 1952. Change of melting point of diopside with pressure. *J. Geol.*, **60**, 364–74.

Yoder, H. S., and Kushiro, I., 1969. Melting of a hydrous phase: phlogopite. *Carnegie Inst. Washington Yearbook*, **67**, 161–7.

Yoder, H. S., and Tilley, C. E., 1962. Origin of basalt magmas: an experimental study of natural and synthetic rock systems. *J. Petrology*, **3**, 342–532.

VI.1b Melting in the Deep Crust (D. K. Bailey)

Rock-melting, the change of state that is the ultimate origin of all igneous rocks, has been a strangely neglected aspect of petrology. Igneous petrogenesis has been dominated by considerations of crystallization of a melt, whose origin is rarely questioned or examined, if it is mentioned at all! It is as if someone gave a description of wine, which started with the fermentation and never referred to crushing the grapes. As a consequence, this section on anatexis is not so much a review as an examination of the possibility that alkaline rocks may be formed directly by melting.

To examine such a possibility, four factors must be considered:

a. the types of melt, and any special characteristics they possess;

b. the possible source materials in the deep crust;

c. the probable physical conditions prevailing;

d. possible mechanisms by which melting may be induced.

Only within this framework can we judge the feasibility of deep crustal anatexis giving rise to the alkaline magmas seen at higher crustal levels. It should be plainly stated at the outset that the discussion will be restricted to stable continental conditions. If anatexis does occur in orogenic or mobile belts then alkaline magmas are not characteristic products.

VI.1b.1. Types of Melt

It is safe to assume that alkali basalts, basanites, and nephelinites are primarily of mantle origin, and hence outside the scope of this discussion (cf. VI.1a). The alkaline magmas in question are the trachytes, phonolites, pantellerites and comendites, and their plutonic equivalents. These appear in small amounts in alkali basalt provinces, and are characteristic of Petrogeny's Residua System (nepheline–kalsilite–silica: Bowen, 1937). They are low temperature melts, possibly end-products of fractional crystallization. When they appear on a large scale in continental alkaline provinces the question arises—are they products of partial melting of the crust?

Two special features characterize these magmas: (1) a pronounced tendency towards peralkalinity; and, (2) association with strong volatile emission.

VI.1b.2. Source Material in the Deep Crust

Composition of the deep crust below 20 km is possibly the largest area of uncertainty in a discussion of crustal anatexis. It has been known for many years that seismic velocities, and hence densities, increase with depth in the crust, with a discontinuity (the Conrad) being detectable in some regions. From these observations grew the concept of an upper crustal layer, of observed granitic composition (*sial*), underlain by a denser layer of *sima*. Doubts about the two-layer structure were eased by the discovery that the oceanic crust gave similar seismic velocities to the lower continental crust, and the latter has accordingly been sometimes referred to as the 'basaltic' layer (see Hodgson, 1964, for a review). There is no direct evidence of its composition, and the prevalent, if somewhat tentative view that the lower crust is 'gabbroic' was challenged by Ringwood and Green (1966). They argued that in the lower crust a gabbroic mineralogy would be unstable under anhydrous conditions, and any basic material would be represented by granulite and eclogite. The higher densities of granulite and eclogite would give seismic velocities significantly greater than those observed, and Ringwood and Green concluded that in general the lower crust must consist of acid to intermediate granulites, and possibly, in special areas, amphibolite. The argument hinges on a somewhat questionable representation and extrapolation of their high temperature/high pressure results to lower temperatures and pressures. A more recent study of the gabbro–eclogite transition indicates that the gabbroic mineralogy would be more stable under the conditions expected in the deep continental crust (Ito and Kennedy, 1971).

One of the most intensive geophysical enquiries into the composition of the lower crust has been made by James, *et al.* (1968). From an analysis of all the detailed seismological and gravity data on the Middle Atlantic States they conclude that the required atomic composition would best fit a 'mafic (possibly amphibolite) lower crust'. Their reference to 'amphibolite' is in deference to the Ringwood and Green hypothesis, but an anhydrous mafic mineralogy would be in perfect accord with the more recent experiments, and would avoid the serious objections to 'amphibolite' on other grounds, outlined below.

Another approach to the problem of deep crust composition is to consider the melting relations of the rock types that might be present. Fig. 3 shows the PT melting curves for a range of rock-types under different conditions, and in Fig. 4 the beginning of melting of alkali basalt composition ($P_{H_2O} = P_{Total}$) is compared with the geothermal gradient. It is clear from these data that in regions where heat flow is normal, or above average, basalts would be partly molten under hydrous conditions. Granites and syenites would be extensively melted under these conditions. It can be assumed from the absence of widespread volcanism that either the rocks of the

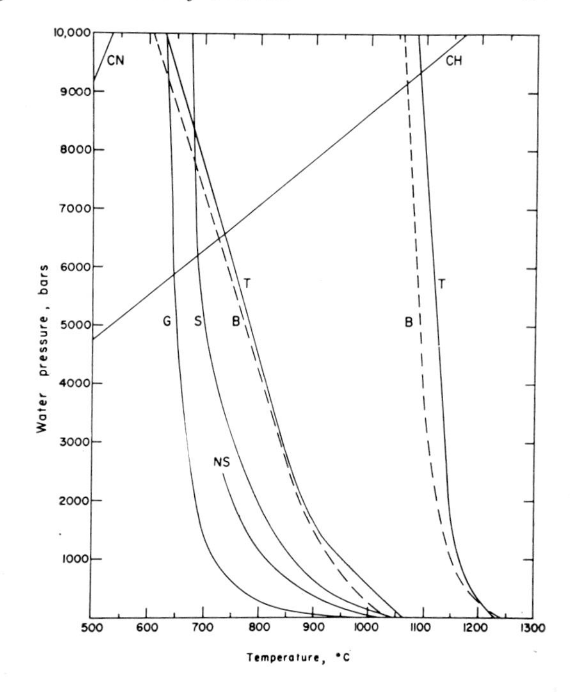

Fig. 3 Melting curves for common igneous assemblages. Curves B and T, alkali basalt and tholeiite; the upper being the hydrous liquidus curves, the lower the hydrous beginning of melting (Yoder and Tilley, 1962). Curves G and S are 'granite' and 'syenite' minimum melting curves (Luth, Jahns, and Tuttle, 1964), and NS, nepheline syenite from Barker (1965). The lines CN and CH are the geothermal gradients in continental areas of normal and high heat flow as shown in Fig. 4

deep crust are not of basaltic to granitic composition, or high partial pressures of H_2O are not normal. The former alternative is unlikely because more refractory compositions would require densities incompatible with the seismic velocities. The conclusion must be that normally the deep crust is sensibly anhydrous. Ringwood and Green (1966) arrived at the same conclusion using a different line of reasoning. They argue that the formation of granite and metamorphic assemblages in the higher crustal levels implies that temperatures in the deeper parts of the crust were probably in the region of 600–1000 °C, and that therefore 'amphibolites would have been converted into granulites and it is probable that the lower crust would have become rather thoroughly dehydrated'. This is not the whole story, however, because it neglects the experimental evidence (Yoder and Tilley, 1962) that under these very conditions amphibolite would be partly melted, and certainly any sialic rocks would have been extensively melted (see Fig. 3). The changes taking place in the deep crust, at these temperatures, would thus involve more than dehydration: the dewatering process should be accompanied by expulsion of low-temperature melts. Since the exposed areas of continental basement bear evidence of multiple episodes of granite activity and metamorphism it is reasonable to suppose that the deep crust is largely depleted in sial, and is generally mafic with only small amounts of residual felsic material.

At the present time, therefore, all the available evidence seems to indicate that the deep crust is mafic and largely anhydrous. This mafic material has suffered a long and complex history, and is presumably inhomogeneous. A lower crust of this nature imposes a crucial condition on the process of alkaline magma-genesis by deep crustal melting: the process must be capable of yielding similar end-products from somewhat heterogeneous starting materials. Trachytic, rhyolitic and phonolitic melts, similar to those in Petrogeny's Residua System, are the kind of low-temperature fractions that would be expected from a wide range of basic compositions. But the question is, what additional factors are necessary to stamp such melts with the special characteristics possessed by continental alkaline magmas?

VI.1b.3. Physical Conditions in the Deep Crust

Total pressure due to lithostatic load, in the depth range 20–40 km, must be in the range 6–12 kbar. The temperature range will vary according to the thermal model adopted to describe the observed heat flow. In Fig. 4 the thermal gradients for areas of normal (CN) and high heat flow (CH) are taken from the models proposed by von Herzen (1967): they do not differ substantially from those proposed earlier by other authors, such as Clark and Ringwood (1964).

VI.1b.4. Mechanisms of Melting

Fig. 4 shows, in addition to the geothermal gradients, the melting and liquidus curves for granitic and basaltic compositions under various hydrous and anhydrous conditions. These form a convenient basis for a discussion of the possible mechanisms that might induce anatexis in the crust.

Three possibilities, operating singly or in conjunction, can be considered:

a. Addition of heat by radiation, conduction or radioactivity. (Convection at this level would seem out of the question prior to melting.)

b. Decompression of materials in such a condition that they would melt isothermally if pressure were reduced (dT/dP of melting is positive).

c. Change of total composition by introduction of new chemical components into the lower crust.

(*a*). *Addition of heat.* Most discussions of melting assume, explicitly or tacitly, that this is the prime mechanism, but it is not particularly apt for crustal anatexis in a stable continental environment. Firstly, the melting process would be exceedingly slow by radiative–conductive transfer or radioactive decay. Secondly, the age-old problem of localizing or focussing the heat would remain. Some special heating process in the underlying mantle would be called for, such as convection, and it is questionable if this could be adequate to give much crustal anatexis without first causing extensive melting in the mantle.

(*b*) *Decompression melting*. In an earlier paper (Bailey, 1964), I suggested that alkaline magmas may be formed by relief of lithostatic load, leading to direct partial melting in the deep crust and withdrawal of fugitive constituents from the underlying mantle. The discussion was primarily concerned with the tectonic control, and the proposed mechanism of magma genesis was oversimplified. It is clear from Fig. 4 that, given an anhydrous lower crust, the mechanism of direct partial melting solely by decompression is unlikely except in regions of high heat flow, e.g. conditions close to curve LG. Such conditions are possible, but are probably found only where

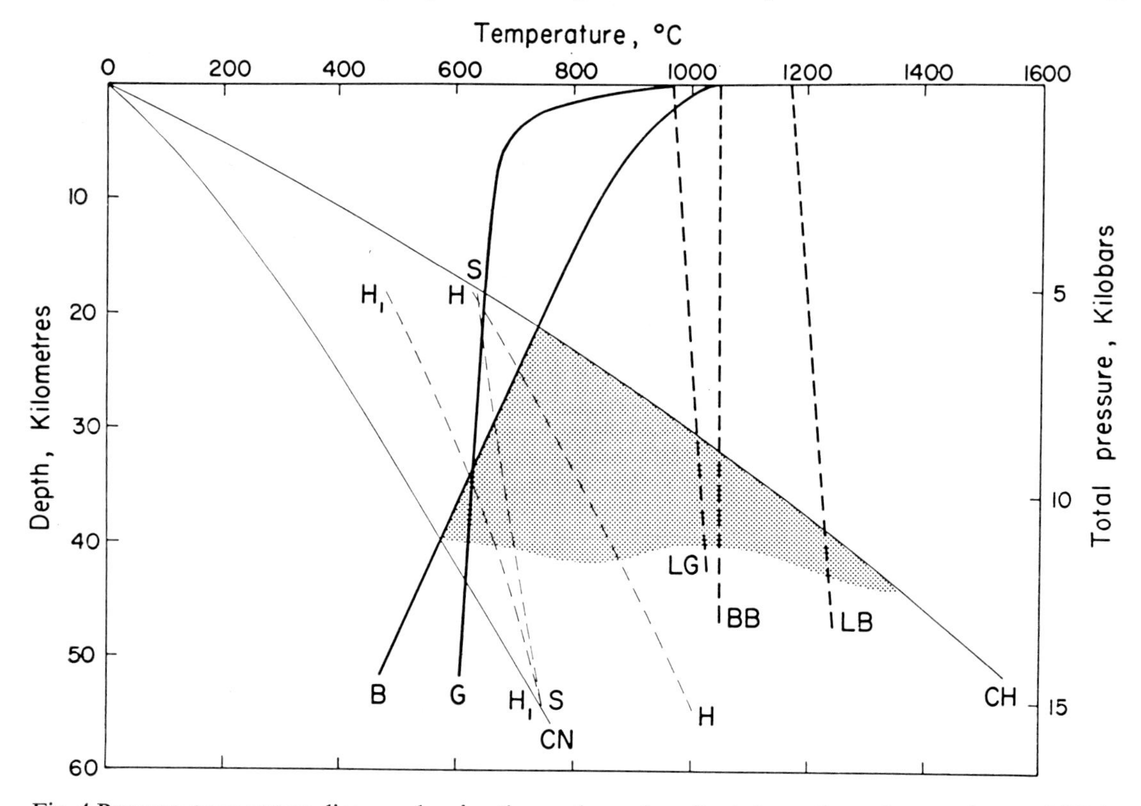

Fig. 4 Pressure–temperature diagram showing the geothermal gradients in continental areas of normal (CN) and high (CH) heat flows (after von Herzen, 1967). Curve B is the hydrous beginning-of-melting of alkali basalt (Yoder and Tilley, 1962). The stippled zone is the PT region in which anhydrous basaltic compositions would begin to melt if H_2O moved into regions of above-average heat flow. The curve SS represents an adiabatic reversible transformation ($\triangle S = 0$) of H_2O moving from mantle into lower crust. Curves H-H and H′-H′ represent constant enthalpy transformations of H_2O from mantle, at 1000 °C and 750 °C respectively, to lower crust.

Curve G is the hydrous 'granite' minimum. Curve LG is a hypothetical melting curve for the anhydrous granite minimum. Curve BB is the approximate anhydrous beginning of melting for various basalts taken from Yoder and Tilley (1962; Tables 15, 43 and 44).

Curve LB is the anhydrous liquidus of Glenelg eclogite (Yoder and Tilley, 1962; Table 42)

some external heating agency has been operating, for instance, in an already established igneous cycle. The second part of the mechanism, withdrawal of fugitive constituents from the underlying mantle, may be the crucial factor in deep crustal anatexis.

(*c*) *Introduction of volatiles into the lower crust.* Examination of the hydrous melting curves for granitic and basaltic compositions (Fig. 4, curves G and B) shows that there is a broad *PT* region in the lower crust in which granite would melt, and gabbro would partially melt in the presence of H_2O. In crustal regions of normal, or above normal, heat flow, therefore, an influx of volatiles from the underlying mantle could bring about partial melting in the lower crust. Not only would the volatiles induce melting by critically changing the overall chemical composition, they would actually carry heat into the lower crust and raise the temperature. The curves SS, HH and H_1H_1, are a selection of adiabatic transformation curves, which show the change in temperature in H_2O passing through the relevant pressure range (Bailey, 1970). Degassing of the underlying mantle would thus seem to be a mechanism worthy of consideration in any hypothesis of crustal anatexis—it has the double function of reducing the melting range of the rocks *and* focussing the heat where it will have the maximum effect. It also provides a heat-transfer process that is much more rapid than any that are normally invoked.

On the basis of the above considerations it is concluded that deep crustal anatexis is possible in the stable continents, especially in the presence of volatiles.

VI.1b.5. Crustal Anatexis in an Open System

In most discussions, melting appears to be regarded as an isochemical process, largely dependent on an increase in the heat content of the system. But on consideration this must be implausible, because the heating process would inevitably cause tectonic and chemical disturbances that would affect adjacent regions. The processes of magma generation must be complex, and involve perturbations in the pre-existing thermal, pressure and chemical regimes. It is impossible to visualize a natural system, subject to temperature and pressure disturbances, that is closed to the movement of volatiles and mobile elements. All three factors, heat-content, pressure, and chemical changes, must be present in varying degrees in a concept of magma genesis. The question that should be asked is, what are their relatives rôles in the generation of alkaline magmas by crustal anatexis? These rôles can only be assessed from the characteristics of the alkaline magmas:

1. Alkaline magmas are localized—frequently in crustal strips such as rift zones. Heating alone cannot explain this localization.

2. Alkaline magmatism shows a strong correlation with areas of continental uplift (cf. III.2 and III.4). Whatever forces are involved there must be pressure relief below the upwarp—either during uplift or as the uplift forces relax. Simple decompression melting is possible, but is unlikely to be the sole means of magma generation becase volatile influx will also modify the bulk chemistry.

3. Alkaline magmatism is accompanied by abundant volatile activity (cf. VI.4). Volatile fluxing is therefore indicated as an important factor in magma genesis.

4. High concentrations of alkalis are indicated by magma peralkalinity and metasomatism around intrusions. Volatiles accompanied by mobile elements, especially alkalis, are again indicated as an important factor.

5. Despite the probable heterogeneity of the lower crust the magmas have *consistent*, yet unusual compositions—characterized especially by high concentrations of alkalis and rare elements. This would not be possible if melting were controlled *only* by changes in temperature and pressure—low melting fractions would certainly be felsic, but *subaluminous–peraluminous* types would be possible, and might even be expected to predominate, because the source rocks would commonly have this characteristic. It is postulated, therefore, that only an external controlling influence such as fluxing by volatiles and alkalis could consistently give strongly alkaline magmas from heterogeneous source rocks.

The powerful effects of mobile alkalis can be amply demonstrated in the country rocks around the alkaline intrusions themselves, and detailed discussions of the fenitization process are available (McKie, 1966; Verwoerd, 1966). The process could be described as leading to 'metasomatic convergence'. Around the high-level Rufunsa carbonatites (Bailey, 1966), granitic rocks, sandstones and mudstones are all progressively altered to felspar rock ('syenite'). In the case of the metasomatized mudstone a highly refractory rock has been converted into a composition that would melt more readily, and more completely. In deeper complexes the metasomatic convergence is towards 'nepheline syenite', sometimes extending to the point of partial melting (von Eckermann, 1948). These are examples on a local scale, at high crustal levels, of the homogenizing influence of alkali transfer, producing rocks with the compositions of typical continental alkaline magmas. It is not difficult to see the same powerful process exerting the essential control over anatectic magma-generation in the deep crust.

The above discussion relates particularly to transfer of mobile elements in an active channel situation. Over longer periods of time it might be expected that the broad, undisturbed cratons would be subject to pervasive infiltration by mobile elements escaping from the underlying mantle. This would lead to a steady replenishment of the felsic components of the deep crust. Subsequent heating or metamorphic episodes would mobilize these lighter constituents to produce granitic rocks at higher levels in the crust. This model of 'sialic underplating' has been discussed elsewhere (Bailey, 1972) and the need for such a process has been indicated by Shackleton (1970).

Recent experimental studies on the melting of acmite composition (Bailey, 1969) show that reduction of the P_{O_2} can lower the melting range by over 170 °C, and yield alkali rich liquid in the breakdown from acmite to sodic amphibole. The indications are that reduction of P_{O_2} in the natural environment would greatly facilitate the generation of low-temperature peralkaline melts, illustrating in yet another way the crucial rôle of volatiles, and volatile composition in this process.

V.1b.6. Conclusion

From a consideration of the nature and condition of the lower crust, and the typical features of the magmatism, it is concluded that alkaline felsic magmas are generated on a *large scale* by partial melting in the deep crust. A major controlling factor is relief of pressure at depth leading to an influx of volatiles and mobile elements from the underlying mantle. The volatiles and mobile elements reduce the melting range of the deep crustal rocks, focus heat in the active zone, and give the felsic magmas their special alkaline character. At the deepest crustal levels the typical magma is probably phonolitic. As the melting regime moves to higher levels, and generally more silicic source rocks, the characteristic magmas are more trachytic; in some circumstances these may be oversaturated and associated with pantelleritic and comenditic 'residua'. For an example of the application of this concept the reader is referred to III.2.

The thick continental plate has a three-fold function:

1. Over broad areas it impedes, or muffles, the general evolutionary escape of volatiles from the underlying mantle.
2. It can be upwarped and fractured, thus channelling volatile release and focussing the melting processes.
3. Deep crustal materials are available as a source of felsic components.

It is not intended to suggest that similar felsic magmas may not form by other processes, such as fractional crystallization of basalt magma. Indeed, some may form in the mantle (Wright, 1969). But these will necessarily be limited in volume. The process of deep crustal anatexis, suggested above, is however in accord with the large scale generation of alkaline magmas in the major continental provinces.

VI.1b. REFERENCES

Bailey, D. K., 1964. Crustal warping—a possible tectonic control of alkaline magmatism. *J. Geophys. Res.*, **69**, 1103–11.

Bailey, D. K., 1966. 'Carbonatite volcanoes and shallow intrusions in Zambia,' in Tuttle, O. F., and Gittins, J., Eds., *The Carbonatites*. John Wiley and Sons, New York.

Bailey, D. K., 1969. The stability of acmite in the presence of H_2O. *Am. J. Sci.*, **267–A**, Schairer Volume, 1–18.

Bailey, D. K., 1970. Volatile flux, heat focussing and the generation of magma. *Geol. J. Spec. Iss.*, **2**, 177–86.

Bailey, D. K., 1972. Uplift, rifting and magmatism in continental plates. *Leeds University J. Earth Sciences*.

Barker, D. S., 1965. Alkalic rocks of Litchfield, Maine. *J. Petrology*, **6**, 1–27.

Bowen, N. L., 1937. Recent high-temperature research on silicates and its significance in igneous geology. *Am. J. Sci.*, **33**, 1–21.

Clark, S. P., and Ringwood, A. E., 1964. Density

distribution and constitution of the mantle. *Rev. Geophys.*, **2**, 35–88.

von Eckermann, H., 1948. The alkaline district of Alnö Island. *Sverig. Geol. Undersök.*, **36**.

von Herzen, R. P., 1967. Surface heat flow and some implications for the mantle. *The Earth's Mantle*, T. F. Gaskell, Ed., Academic Press, London, 197–230.

Hodgson, J. H., 1964. *Earthquakes and Earth Structure*. Prentice-Hall, New Jersey.

Ito, K., and Kennedy, G. C., 1971. 'An experimental study of the basalt–garnet granulite–eclogite transition', in Heacock, J. G., Ed., The structure and physical properties of the Earth's crust. *Am. Geophys. Union, Geophys. Mon.*, **14**, 383–427.

James, D. E., Smith, T. J., and Steinhart, J. S., 1968. Crustal structure of the Middle Atlantic States. *J. Geophys. Res.*, **73**, 1983–2007.

Luth, W. C., Jahns, R. H., and Tuttle, O. F., 1964. The granite system at pressures of 4 to 10 kb. *J. Geophys. Res.*, **69**, 759–73.

McKie, D., 1966. 'Fenitization', in Tuttle, O. F., and Gittins, J., Eds., *The Carbonatites*. John Wiley and Sons, New York, 261–94.

Ringwood, A. E., and Green, D. H., 1966. An experimental investigation of the gabbro-eclogite transformation and some geophysical implications. *Dept. Geophys. Geochem. Austral. Nat. Univ. Publn.*, **444**, 61–103.

Shackleton, R. M., 1970. On the origin of some African granites. *Proc. Geol. Ass.*, **81**, 549–59.

Verwoerd, W. J., 1966. 'Fenitization of basic igneous rocks', in Tuttle, O. F., and Gittins, J., Eds., *The Carbonatites*. John Wiley and Sons, New York, 295–308.

Wright, J. B., 1969. Olivine nodules in trachyte from the Jos Plateau, Nigeria. (in press).

Yoder, H. S., Jr., and Tilley, C. E., 1962. Origin of basalt magmas. *J. Petrology*, **3**, 342–532.

VI.2. THE RÔLE OF FRACTIONAL CRYSTALLIZATION IN THE FORMATION OF THE ALKALINE ROCKS

R. Macdonald

VI.2.1. Introduction

Bowen (1937) showed that when the normative salic compositions of certain East African alkaline lavas were projected into the system Quartz–Nepheline–Kalsilite (Q–Ne–Ks) they were restricted to the thermal valley on the liquidus surface, and interpreted this as evidence that crystal-liquid equilibria and hence fractional crystallization, had been the dominant process in the origin of these rocks. There is now a large body of experimental data to show that the residual liquids of fractionation of basic magma are enriched in the alkali–alumino silicates. This, and the direct evidence of differentiated intrusions, glassy residua and groundmass assemblages have convinced perhaps the majority of petrologists that fractional crystallization of basaltic magmas is the major cause of differentiation of alkaline rocks. In this chapter some aspects of recent work on the nature and evolution by fractional crystallization of the more common alkaline salic magmas are reviewed.

VI.2.2. Basic Parental Magmas

Coombs (1963) has shown that the fractional crystallization paths of alkali basaltic magma are dependent mainly on its initial composition. A mildly alkaline, transitional group showing neither appreciable hypersthene or nepheline in the norm, differentiates close to the critical plane of SiO_2-undersaturation (Yoder and Tilley, 1962 and II.4*) and leads via hawaiitic and mugearitic types to trachytic residua which straddle the feldspar join in the system Q–Ne–Ks. Under conditions of strong fractionation, the trachytic magma may itself yield phonolitic or alkali rhyolitic residua. Basanitic and nephelinitic magmas follow more strongly undersaturated

* Readers are referred to the relevant Chapters of this book where similarly cited.

crystallization paths, and may produce phonolitic derivatives without an intermediate trachytic stage. In southern Kenya and northern Tanzania two contrasted volcanic associations were supposed by Wright (1963) and Saggerson and Williams (1964) to have followed different fractionation paths: melanephelinite–nephelinite–phonolite and alkali basalt–trachyte (–comendite). Wilkinson (1966) has presented analytical data on the glassy residua of Tertiary basic lavas from New South Wales; the residua of an alkali olivine basalt and a nepheline basanite were trachyte and phonolite respectively. A similar relationship between degrees of SiO_2^- undersaturation of parent and daughter magmas has been described from the Atlantic Ocean by de Almeida (1961), who divided the islands into two groups, a hyperalkaline group comprising Trinidade, Fernando de Noronha, Cape Verde, Sao Tomé, Principe and the Canaries, and a mioalkaline group consisting of the islands on the Mid-Atlantic Ridge and Madeira (see IV.7). The hyperalkaline group with the exception of part of the Canaries succession is strongly undersaturated and is approximately equivalent to the continental nephelinite–phonolite series, whereas the mioalkaline group is comparable to Coombs' transitional group of alkali basalts.

Kuno (1968) and Coombs and Wilkinson (1969) have recently provided comprehensive reviews of the possible fractionation trends of alkaline basic magmas. Expanding on their earlier work (see above), Coombs and Wilkinson have shown that there is a continuous spectrum of lineages from basic → salic liquids, depending on the composition of the parental magma. They distinguish the following general series.

Sodic

1. Alkali basalt–hawaiite–mugearite–benmoreite–trachyte.
2. Basanite–nepheline hawaiite–nepheline mugearite–nepheline benmoreite–phonolite.
3. Nephelinite series—? (insufficient data available).

Potassic

4. Trachybasalt–trachyandesite–tristanite–trachyte.
5. Sanidine basanite–nepheline trachyandesite–nepheline tristanite–phonolite.

Before discussing the nature of the residual liquids of these various basic magmas, it is necessary to review a current point of controversy in igneous petrology which bears directly on the rôle of fractional crystallization of basalt magma in the origin of alkaline rocks.

VI.2.3. Silica Gaps

Chayes (1963) has convincingly demonstrated the scarcity on oceanic islands of published analyses of lavas in the range $53\% < SiO_2 < 57\%$, the so-called 'Daly gap'. Although statistical data are not available the Daly gap also appears to exist in continental provinces, such as the Gardar (Upton, IV.3) and Kenya (Wright, 1965). The existence of this silica-gap was considered by Chayes to be of considerable petrogenetic importance since it casts doubt on the origin of basalt → trachyte series by conventional methods of crystal fractionation, where more evolved rocks should continuously occur in progressively smaller bulk.

In reply to Chayes (1963), Harris (1963), Baker (1968) and Cann (1968) have suggested that oceanic trachytes have been oversampled relative to trachyandesites due to their mode of occurrence as erosion-resistant domes and plugs, and to their distinctive, striking appearance in hand-specimen, leading to a bias in the geologist's collecting. Trachyandesites are commonly of dull and uninteresting appearance, may not be distinguishable from basalt in the field, and are more readily eroded than trachyte. These factors may have led to an undersampling of trachyandesites, especially on islands where the volume of intermediate and salic rocks is low. Recent detailed work on Hawaii (Macdonald, 1963) the Tristan da Cunha group (Harris, 1963) and St. Helena (Baker, 1968), the latter long regarded as a classic example of an island showing the Daly gap, has shown that intermediate lavas are actually more abundant than trachytic and has prompted Baker (1968) to doubt whether the Daly gap would be substantiated during study of other islands. On the other hand, Bryan (1964) has found a distinctly bimodal distribution of lavas on the island of Socorro, with a minimum in the intermediate composition range. Further quantitative data are evidently required to determine whether the apparent scarcity of intermediate lavas is real or a result of sampling error.

Le Maitre (1968, p. 240) has pointed out that the paucity of intermediate lavas in basalt → trachyte sequences may be due to a discrimination in the eruptive process, and not to the scarcity of intermediate magmas, i.e. that trachyandesitic magma formed at depth has not been erupted at the surface as readily as more basic and more salic magma. For any given temperature and pressure acid magmas are more viscous than basic. Conversely, viscosity will decrease in the fractionation sequence basalt → trachyte due to volatile build-up. Intermediate magmas may therefore actually be more viscous than basaltic or trachytic magmas, and during the volcanic cycle may not reach the surface. A bimodal distribution of lavas, though not of magmas, would result.

A striking example of discrimination by a volcano has been given by Gass and Mallick (1968), from the Jebel Khariz volcano west of Aden, which consists of an older cone series of rhyolites, trachytes and basalts, and a younger caldera sequence of dominantly intermediate lavas. A similar relationship has been noted in the Aden and Little Aden volcanoes. Gass and Mallick (1968, p. 79) suggest that the acid magmas, being spatially nearest the surface in the differentiating magma chamber, and the relatively non-viscous basalts, were most readily extruded during the formation of the cone series, whereas the viscous intermediate magma, mainly as pyroclastic material, was erupted on caldera collapse. Had the volcanic cycle not produced caldera collapse, a perfectly good Daly gap would have been reported from the cone series.

A further explanation of silica-gaps may lie in the mechanics of crystallization. Chayes (1963) supposed that the existence of the Daly gap was incompatible with the formation of the basalt → trachyte association by simple fractional crystallization, arguing that successive residual liquids should occur in progressively smaller amounts. Bryan (1964) and Wilkinson (1966) have pointed out, however, that silica-gaps are characteristic of differentiated intrusions which provide some of the best evidence for fractional crystallization. Wyllie (1963) has presented models where apparent silica-gaps can be produced in a crystallization sequence as a result of the geometry of the phase equilibria. A continuous series of compositions from basalt → trachyte may not therefore be a requisite of fractional crystallization, and a combination of this effect, plus some sort of physical control and poor sampling as outlined above, may account for the oceanic Daly gap.

The scarcity of intermediate rocks in continental alkaline provinces is linked with the high ratio of salic: basic lavas, the salic lavas being too voluminous to be considered residual liquids of fractional crystallization of the observed volumes of basalt. This problem will be discussed in the next section.

VI.2.4. Fractionation of Salic Magmas

Alkali feldspar is the dominant mineral in trachytes, rhyolites and phonolites derived by fractionation of alkali basalt, accompanied by lesser amounts of quartz or feldspathoid. Generally the felsic minerals constitute more than 80% of the rocks and obviously their crystallization must have a strong influence on the nature of residual liquids. Fractionation of the other phenocrysts which are present in much smaller amounts, will be less influential. Among this latter group of minerals fayalitic olivine, Fe–Ti oxides, clinopyroxene, apatite and fluorite are common cumulus minerals in layered syenites, and have the effect of removing Mg, Ca, Fe, Ti, Mn, P and F from the melts. The status of biotite and calc-sodic amphiboles as fractionating phases has yet to be established, though both, being relatively undersaturated, could initiate a trend towards SiO_2-saturation in mildly undersaturated magmas. Both minerals, for example, tend to be absent from the mafic cumulates of the Gardar complexes. In trachytic and phonolitic magmas, Na–Fe minerals may be fractionated: cumulitic aegirine, riebeckite-arfvedsonite and aenigmatite have been recorded in kakortokites from the Ilímaussaq intrusion (Ferguson, 1964), accompanied by the Na–Ca–Zr-silicate, eudialyte. Though these minerals are present in much smaller amounts than alkali feldspar and feldspathoid, their crystallization has influenced the alkali:alumina balance and the Na_2O/K_2O ratios of the melts to some degree. Generally, aegirine is a phenocryst in volcanic salic rocks only rarely, if ever, the primary pyroxene being aegirine-augite or diopside-hedenbergite (Yagi, 1966). The late crystallization of the femic minerals in strongly undersaturated peralkaline liquids is the basis of the 'agpaitic order of crystallization'.

Similarly, the author is unaware of any unequivocal evidence that fractionation of riebeckite-arfvedsonite has occurred in *acid* alkaline magma. The demonstration of Noble (1968), Nicholls and Carmichael (1969) and Bailey and Macdonald (1969) that more evolved pantelleritic liquids have increasingly higher Na/K ratios (cf. Fig. 6) may suggest that crystallization of primary aenigmatite in pantellerites has had little effect in influencing the evolutionary trends of the magmas. It is suggested that the alkali: alumina ratio in alkaline *acid* rocks is determined almost entirely by crystallization of alkali feldspar (assuming fractional crystallization as the only operative process).

The salic residual liquids of fractionation of basaltic magma are conveniently considered in terms of the system Q–Ne–Ks, petrogeny's residua system, the main features of which have been discussed by Tuttle and Bowen (1958), Fudali (1963) and Hamilton and MacKenzie (1965). Fractionation of alkali feldspar from trachytic liquids lying on either side of the alkali feldspar thermal divide must drive the residual liquids away from the divide into the so-called low-temperature trough or thermal valley on the liquidus surface. Slight SiO_2-deficiency or excess in the initial magmas is accentuated on progressive fractionation, with the ultimate production of phonolitic and rhyolitic liquids whose compositions are similar to the under- and over-saturated minima respectively.

Although this system has been useful in describing trends in salic magmas, it is not strictly a residua system for peraluminous or peralkaline rocks, i.e. those where molar $Na_2O + K_2O \neq Al_2O_3$ (Bailey and Schairer, 1964). Peralkaline magmas do not fractionate, for example, towards the minima in Q–Ne–Ks, but towards natural eutectics considerably richer in excess alkali (expressed as normative acmite and sodium metasilicate), analogous to those found experimentally in Na_2O–Al_2O_3–Fe_2O_3–SiO_2 by Bailey and Schairer (1966) (q+ab+ac+sodium disilicate and ne+ab+ac +sodium disilicate). Strongly peralkaline liquids such as pantellerites and agpaites cannot be plotted in terms of Q, Ne and Ks without gross distortion, especially in the Na_2O/K_2O ratios and feldspar/liquid relationships. The salic end members of the alkali basalt–trachyte association are typically peralkaline, carrying normative acmite, or have peralkaline tendencies as shown by the presence of alkali amphiboles and pyroxenes. To examine composition trends in these rocks, Bailey and Macdonald (1969) have devised two simple plots based on the molecular proportions of SiO_2, Al_2O_3, Na_2O and K_2O. The derivation of these projections is summarized in Fig. 1, and they are used in subsequent diagrams to illustrate several features of magmatic evolution. Though the projection has been used successfully for oversaturated salic rocks, success in plotting undersaturated rocks has been limited mainly because nepheline and feldspar are the common felsic phenocrysts of phonolites. The composition of both minerals cannot lie in any one plane whose alkali:alumina ratio is greater than 1. Fractionation trends involving crystallization of both feldspar and nepheline (or any feldspathoid) cannot therefore be shown without distortion in the projection. There is an urgent need for more sophisticated methods of plotting the compositions of undersaturated, peralkaline rocks, because until these are available, we will be unable to describe quantitatively

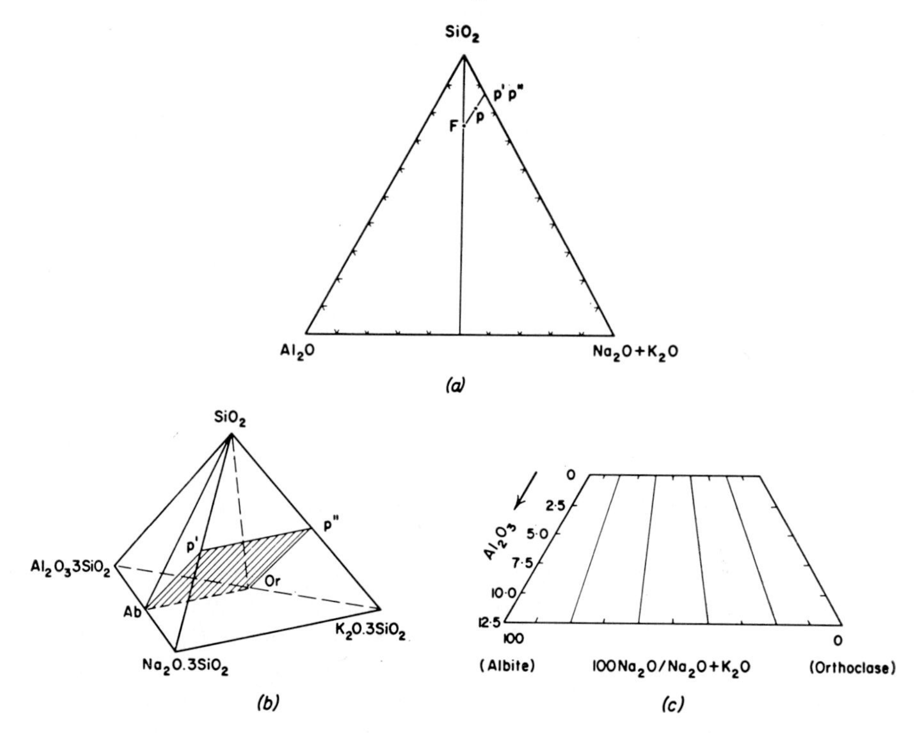

Fig. 1 (a) SiO_2–Al_2O_3–($Na_2O + K_2O$) diagram. Alkali feldspar projects as a point, F. P—hypothetical peralkaline composition. P′P″, the intersection of line Feldspar–P on alkali–silica sideline, measures the alkali–silica index of composition P. An infinite number of these planes may be drawn radiating from F.

(b) The oversaturated volume of the system Na_2O–K_2O–Al_2O_3–SiO_2, showing the plane Ab–Or–P″–P′. This plane is equivalent to F–P–P′P″ in (a).

(c) Quadrilateral representing any plane in the peralkaline volume of (b), such as Ab–Or–P″P′

the crystallization histories of the magmas.

The development of peralkaline residua is possible in any suite where feldspar is fractionating and where there is a deficiency of Al_2O_3 with respect to Na_2O, K_2O and CaO (expressed as normative diopside and/or wollastonite). Molar deficiency of this type will permit the 'plagioclase effect' of Bowen (1945) to operate. This effect depends on the fact that pure albite cannot precipitate from a liquid containing Ca, and means that precipitation of Ca-bearing plagioclase from suitable melts will tend to leave potential alkali silicate in the liquid. A kenyte from South Victoria Land, Antarctica analysed by Carmichael (1964) carries phenocrysts of anorthoclase (20%), olivine (0·6%), pyroxene (0·8%), iron ore (0·6%) and nepheline (trace). The feldspar phenocrysts have the composition $Or_{17}Ab_{63\cdot2}An_{19\cdot8}$ while the normative feldspar is $Or_{36\cdot2}Ab_{55\cdot6}An_{8\cdot2}$. The rock has no normative acmite whereas the residual liquid (groundmass) carries 4·16% of that component. Carmichael (1964, p. 56) has pointed out that this is a natural example of the plagioclase effect operating to produce peralkaline residua. The degree of peralkalinity achieved by a given magma series will depend on the initial composition of the basaltic parents, but also on the magma crystallization history. For example, fractionation from basic melts of unusually large amounts of calcic plagioclase relative to olivine and clinopyroxene would rapidly deplete Al relative to alkalis and promote the passage to peralkaline residual liquids. In this respect it is tempting to link the quite strongly peralkaline nature of several continental provinces, e.g. Gardar, New Hampshire, Oslo and Niger, with the occurrence in them of large bodies of anorthosite (cf. IV.3 and IV.8).

A peralkaline trend could also be initiated by fractionation of biotite (Carmichael, 1967), aluminous pyroxene (Schairer and Yoder, 1960), spinel, garnet or Al-bearing iron oxide (Bailey and Schairer, 1966), or of non-stoichiometric alkali feldspar (Luth and Tuttle, 1966). When a condition of Al_2O_3-deficiency has been achieved, further feldspar fractionation must accentuate this tendency, and successive residual liquids will be increasingly peralkaline. Similarly, if the trachytic liquid has even a slight excess of Al_2O_3, feldspar crystallization will promote a miaskitic trend. It was pointed out earlier that SiO_2-saturation or undersaturation in trachytic magma is also accentuated by feldspar fractionation. Bailey and Schairer (1964) have therefore suggested that trachytes lie near a 'cross-roads', which may lead to four kinds of residua, peralkaline or peraluminous oversaturated and peralkaline or peraluminous undersaturated liquids.

VI.2.4.1. Oversaturated Liquids

An excellent demonstration of the strong fractionation of saturated trachytic magma towards a rhyolitic end-point is provided by the western lower layered series of the Kûngnât complex, South Greenland (Upton, 1960). This thick series of syenites and quartz syenites showing rhythmic and cryptic layering, as well as igneous lamination, is thought to have developed by the gravitative settling of alkali feldspars in the range Or_{34-42} wt. %, Fe-olivines and clinopyroxenes, with subordinate amounts of apatite and Fe–Ti oxides. During fractionation the feldspars, originally sanidines which subsequently unmixed to perthites, showed an increase in Or relative to Ab while An decreased. Olivines became increasingly Fe-rich, and the pyroxenes showed a trend from ferroaugites to hedenbergites containing some aegirine-augite in solid solution. The other minerals present in the syenites, including quartz, hastingsitic amphiboles and biotite are of intercumulus origin.

The syenites are cut by a series of microgranite sheets which Upton (1960) has interpreted as the low-temperature residues from the layered succession. These peralkaline sheets consist essentially of alkali feldspar, quartz, riebeckite-arfvedsonite, aegirine and astrophyllite. Relative to the syenites they are depleted in Ti, Al, Fe, Mn, Mg, Ca and P, but enriched in Si, F, H_2O and in alkalis relative to Al_2O_3.

The geological relationships at Kûngnât make an origin of the microgranites by fractionation of trachytic magma seem very convincing. In the volcanic environment, however, attempts to describe the origin of peralkaline rhyolitic liquids have had mixed results, as might be expected. Most recent studies have tried to show that peralkaline rhyolites (i.e. the pantellerites and the comendites) are derived from peralkaline or peraluminous trachytes dominantly by fractionation of alkali feldspar (e.g. Carmichael and MacKenzie, 1963, Thompson and MacKenzie, 1967, Noble, 1968, Abbott, 1969, Macdonald, 1969). A simplified example of this process is shown diagrammatically in Fig. 2, where it can be seen that removal of feldspar (F) must restrict liquids (such as X) to a plane (Feldspar-X-X′) until a second phase crystallizes. This plane has a unique value of the ratio alkali:silica measured as the intercept on the SiO_2–alkalis side-line, dependent on the initial composition of the trachyte (Bailey and Macdonald, 1969) and a series of peralkaline liquids connected solely by alkali feldspar fractionation should have constant alkali:silica index.

Liquids evolving in a plane such as Feldspar-X-X′ will eventually reach the quartz–feldspar cotectic curve, and quartz becomes a phenocryst phase. Composition trends in that zone are controlled by removal of feldspar + quartz, and are approximately at right angles to any earlier feldspar trend. Their goal is presumably a natural eutectic analogous to the oversaturated eutectic (q + ab + ac + Na-disilicate) determined at 728 °C at 1 atmosphere in the system Na_2O–Al_2O_3–Fe_2O_3–SiO_2 by Bailey and Schairer (1966).

An example of a series which may have

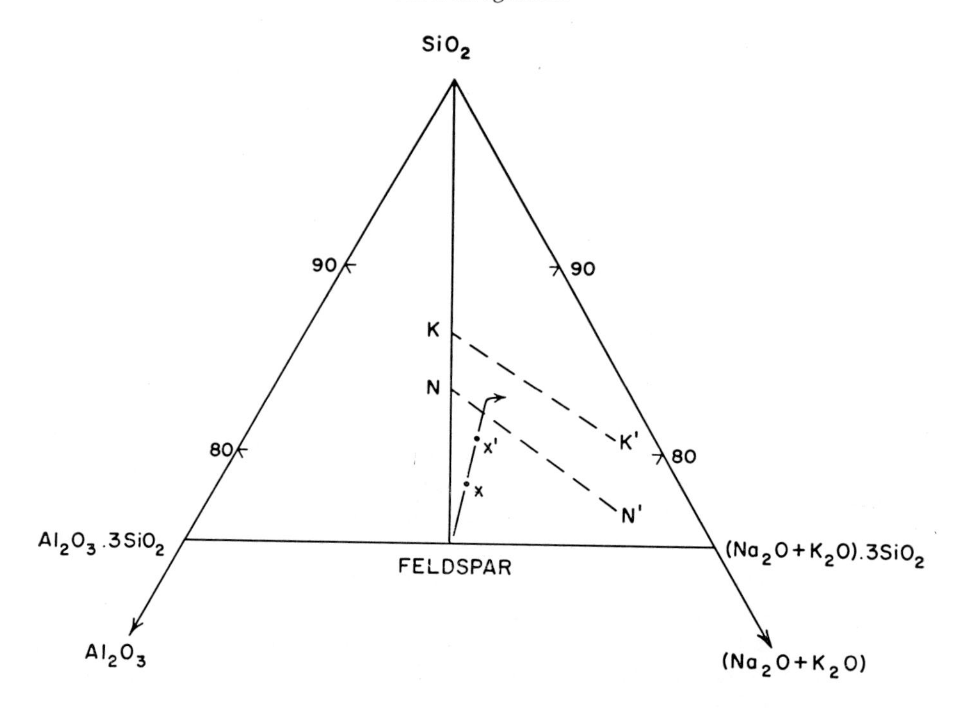

Fig. 2 SiO_2–Al_2O_3–(Na_2O + K_2O) diagram. X—hypothetical peralkaline trachyte, X′—peralkaline rhyolite. K–K′ is the quartz–potash feldspar cotectic and N–N′ the quartz-albite cotectic from the systems K_2O–Al_2O_3–SiO_2 and Na_2O–Al_2O_3–SiO_2 at 1 atmosphere (Schairer and Bowen, 1955, 1956). The zone defined by these lines is termed the quartz–feldspar cotectic zone. A possible fractionation trend of magmas of composition comparable to X is shown

evolved in this way is a peralkaline dyke swarm of Gardar age from the Tugtutôq area, South Greenland (Macdonald, 1969), which is a series passing from hastingsite microsyenites with small amounts of normative acmite to aegirine rhyolites and microgranites of comenditic affinity (Upton, 1964a). The phenocryst assemblages are:

Hastingsite microsyenites: Alkali feldspar, minor amounts of olivine, clinopyroxene, apatite, ore.

Riebeckite microsyenites: Alkali feldspar, minor amounts of ore.

Riebeckite microgranites: Alkali feldspar, quartz, minor amounts of riebeckite-arfvedsonite (or aegirine pseudomorphs), ore.

Aegirine microgranites: Alkali feldspar, quartz, minor hedenbergite, fayalite.

Analyses of the dykes have been plotted in Fig. 3; the trend is consistent with fractionation of the most abundant phenocryst minerals, i.e. feldspar from the trachytic magmas, and feldspar + quartz from the rhyolitic magmas. The analyses in the quartz–feldspar cotectic zone are rather scattered, a feature due at least partly to devitrification of the acid dykes (Macdonald, 1969).

Carmichael (1962), Carmichael and MacKenzie (1963) and Thompson and MacKenzie (1967) have specifically seen the evolution of peralkaline acid magmas as along a natural low-temperature zone analogous to the thermal valley in petrogeny's residua system. As determined experimentally the axis of this zone projects from the unique fractionation curve in Q–Or–Ab into the peralkaline volume of the system Na_2O–K_2O–Al_2O_3–SiO_2 and is strongly

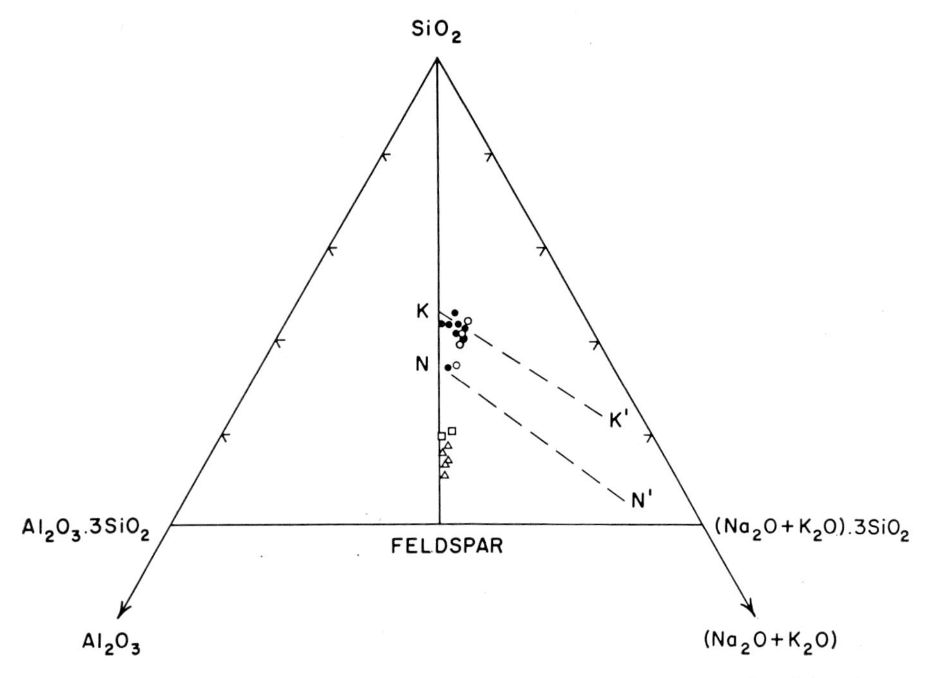

Fig. 3 Analyses of peralkaline microsyenite, rhyolite and microgranite dykes from Tugtutôq, South Greenland plotted in a SiO_2–Al_2O_3–($Na_2O + K_2O$) diagram. Triangles—hastingsite microsyenites, squares—riebeckite microsyenites, closed circles—riebeckite microgranites, open circles—aegirine microgranites

inclined towards the NaO_2–Al_2O_3–SiO_2 plane (Fig. 4, from Thompson and MacKenzie, 1967). Depending on the initial composition, the equilibrium feldspar can be more or less sodic than the trachytic liquids, and its separation will drive residual liquids towards the low-temperature zone, when further separation of a feldspar of narrow composition range (Or_{30-35}) will restrict the residual liquids to the zone. Once the liquid composition projects into the zone, this liquid must, on further feldspar fractionation, become progressively more sodic (cf. the orthoclase effect of Bailey and Schairer, 1964. See Fig. 5).

The concept of the 'low-temperature zone', based on the experimental study of synthetic material, is not so easily applied to natural peralkaline rock assemblages. Macdonald, Bailey and Sutherland (1970) were unable to relate a series of pantelleritic trachyte and pantellerite obsidians from the Kenyan Rift Valley to a natural analogue of the zone, and could not explain the chemical features of the series by simple crystal fractionation. Nicholls and Carmichael (1969, p. 273) have shown that the feldspar-residual glass relationships in comendites (transitional to pantellerites) from Mayor Island, New Zealand, are rather different from those of rocks from Pantelleria and from Kenya and stated that 'The comenditic feldspar phenocrysts cannot be so readily incorporated into Thompson and MacKenzie's picture of feldspar crystallization.' By recalculation of the experimental data of Carmichael and MacKenzie (1963), Macdonald, Bailey and Sutherland (1970) found that the axis of the proposed low-temperature zone could not be straight or even gently curved, unless some complex feldspar–liquid–vapour-relationship had not been recognized during the experiments. Until many more

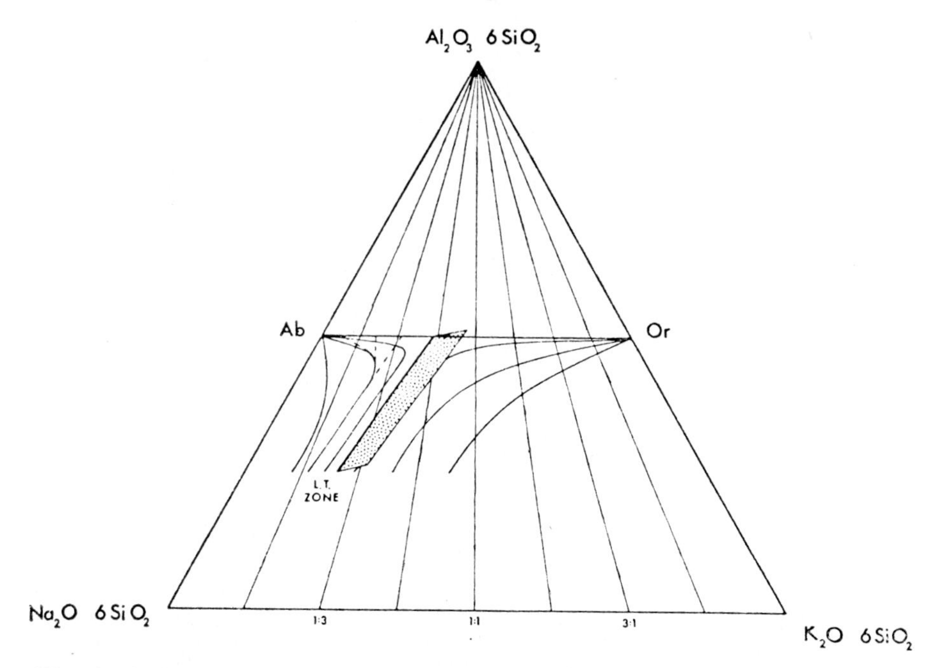

Fig. 4 The low-temperature zone (stippled) and associated fractionation curves in part of the system Na_2O–K_2O–Al_2O_3–SiO_2–H_2O, from Thompson and MacKenzie (1967, Fig. 7). The relations have been projected from SiO_2, on to the plane Al_2O_3 . $6SiO_2$–Na_2O . $6SiO_2$–K_2O . $6SiO_2$; lines radiating from the apex represent equal Na_2O/K_2O ratio. The hypothetical fractionation curves are shown as full lines

data are available on phenocryst–peralkaline siliceous glass pairs, the problem as to the existence of the natural analogue of the zone will persist. If, however, it does exist, it must be considerably more complex than envisaged from the experimental studies.

Using various continental alkaline suites as examples, Bailey and Schairer (1966), Wright (1966, 1969), Macdonald and Gibson (1969) and Bailey and Macdonald (1970) have also questioned the series basalt → peralkaline or peraluminous trachyte → peralkaline rhyolite as a product of continuous fractional crystallization. The disproportionately large volumes of salic relative to basic lavas in certain parts of the African Rifts caused Bailey and Schairer (1966) to suggest that alkaline trachytes and rhyolites are formed by partial melting in the deep crust or upper mantle, while Wright (1966, 1969) has found ultrabasic nodules in certain phonolites and trachytes from New Zealand and Nigeria which he has interpreted as indicating a direct mantle origin of these rocks. Chabbi volcano, Ethiopia, appears to have erupted only aphyric pantellerites of almost identical composition for several thousands of years, and Macdonald and Gibson (1969) found it difficult to see this as a result of recurrent fractional crystallization of basic magma. The mildly peralkaline comendites and comenditic trachytes have recently been reviewed by Bailey and Macdonald (1970). Whereas the oceanic rhyolites show chemical features consistent with their derivation from trachytic magmas dominantly by alkali feldspar fractionation, the continental comendites lie on a trend parallel to the quartz–feldspar minima in the peralkaline oversaturated volume of the system Na_2O–K_2O–Al_2O_3–SiO_2 as determined experimentally by Carmichael and MacKenzie (1963), rather than a distribution away from the minimum on the Or–Ab join (i.e. simplified alkali trachyte composition). Bailey and Mac-

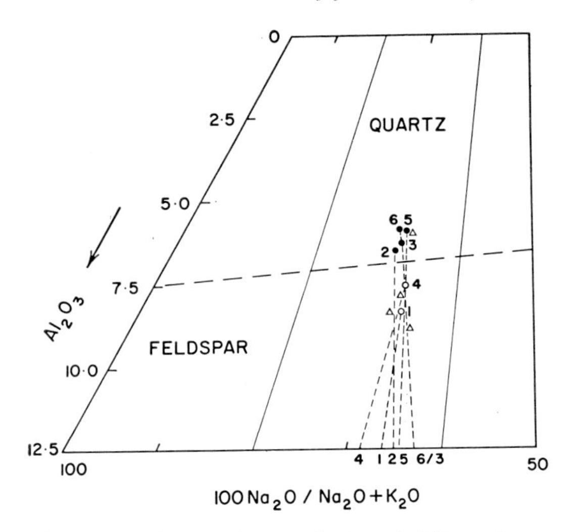

Fig. 5 Analyses of pantellerite obsidians from Pantelleria plotted in a quadrilateral representing the plane of alkali : silica index 7 : 93 (Bailey and Macdonald, 1969, Fig. 5). The quartz–feldspar cotectic has been interpolated between the intercepts of the Na_2O side-line with the quartz–feldspar cotectic, and the K_2O side-line with the quartz–potash feldspar side-line. Lavas on which there are petrographic data are shown by circles (filled where quartz–phyric) and those on which there are no data by triangles. Tie-lines connect rocks and their feldspar phenocrysts (in terms of mol. per cent Or/Or + Ab). With one exception, the feldspar phenocrysts are more potassic than the rocks, thus demonstrating the 'orthoclase effect' of Bailey and Schairer (1964). Data from Washington (1914), Zies (1960), Carmichael (1962), Chayes and Zies (1962, 1964)

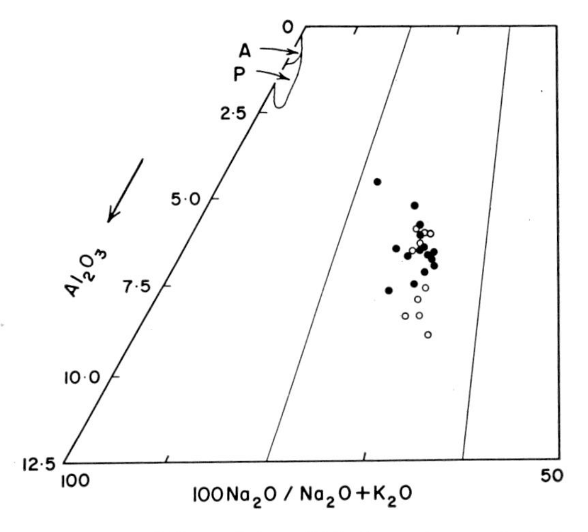

Fig. 6 Quadrilateral in which have been plotted pantellerite analyses taken from the literature. Note that the overall trend is towards increase in Na_2O/K_2O ratio with decreasing Al_2O_3. A = field of aenigmatite phenocrysts, P = field of pyroxene phenocrysts from pantellerite lavas, recalculated from data in Carmichael (1962), Zies (1966) and Ewart *et al.* (1968)

donald suggested that the continental comendites may represent partial fusion products, rather than fractional crystallization products, of basic material.

In summary, it is reasonable to assume that saturated alkaline trachytes and rhyolites may be formed by two or more processes. Where the volumes of basic and salic magmas are of the right order, and where the chemical trends are explicable by fractionation of the observed phenocryst assemblages, there is no strong case for doubting a derivation of the salic liquids by fractionation of basic magma, e.g. in the Gardar province (Upton, Chapter IV.3), in the majority of the oceanic islands, and in such volcanoes as Nandewar (Abbott, 1969) and those of the Aden Peninsula (Gass and Mallick, 1968). Where either of these conditions is unfulfilled, as in central Kenya (Macdonald *et al.*, 1970), it may be necessary to critically re-examine our petrogenetic theories. In any specific case, fractional crystallization must be proven, rather than assumed, to be the operative process.

VI.2.4.2. Undersaturated Liquids

Undersaturated trachytic magma may evolve in a rather similar way to oversaturated trachyte, i.e. dominantly by fractionation of feldspar, until the nepheline–feldspar cotectic is reached. The alkaline intrusions of the Gardar province, S. Greenland, where highly efficient crystal fractionation has produced rhythmically and cryptically layered syenites, provide excellent examples of this trend (see IV.3). The Hviddal giant dyke, a 550 m broad composite dyke in the Tugtutôq area, has a central intrusion of syenite and a narrow marginal zone of syeno-gabbro (Upton, 1964b). The syenite varies continually along its length from a larvikitic augite syenite to a soda-rich foyaite, and this differentiation is believed by Upton to be due to cryptic and phase layering in an accumulitic suite of syenites, of which some 2000 m are exposed. Details of the mineralogical and chemical changes may be found in the original paper; of particular interest here is the sequence of cumulus phases. In the lower 800 m or so, the principal cumulus minerals were alkali feldspar, clinopyroxene, olivine, ilmenomagnetite and minor fluorapatite. Upton stresses the similarity of this assemblage to that in the more basic syenites of the Kûngnât complex, an intrusion which has fractionated from syenite → peralkaline granite. Olivine disappeared in the next 2000 m and from 1000 to ~ 1400 m cumulitic pyroxene, ore and apatite decreased in amount. Feldspar was possibly the only cumulate mineral for some 100 m. Nepheline eventually joined feldspar and the ratio of cumulus feldspar:nepheline was approximately 3:1 (Upton, 1964b, p. 77). The fractionation trend of the Hviddal dyke is shown in Fig. 7 a, b. The trachytic magma evolved in a metaluminous condition and became peralkaline only in the later stages of the exposed succession. With further fractionation, involving removal of both feldspar and nepheline the residual liquids would have moved towards a composition enriched in alkali silicate, similar to the quaternary invariant point (ne+ab+ac+sodium disilicate) found experimentally at 715 °C and 1 atm by Bailey and Schairer (1966) in the system Na_2O–Al_2O_3–Fe_2O_3–SiO_2.

Strong fractionation of undersaturated trachytic magma is displayed in the classic agpaitic complex of Ilímaussaq, some 15 km east of the eastern outcrops of the Hviddal dyke, but considerably younger than it. The following outline of the petrochemical evolution of the agpaites is based on the relationship demonstrated by a Si_2O–Al_2O_3–(Na_2O + K_2O) diagram (Fig. 8) and incorporates suggestions made by Ussing (1912), Sørensen (1958), Ferguson (1964), and Hamilton (1964). The parental magma was a larvikitic augite syenite closely comparable to that of the Hviddal dyke (Fig. 7), and it differentiated *in situ*, mainly by fractionation of feldspar with lesser amounts of clinopyroxene, olivine, ore, apatite, along a similar trend to that of the Hviddal syenites, to give a foyaite. *In situ* differentiation of this foyaite led to a gravity accumulated series of layered nepheline syenites (kakortokites) and to a flotation accumulitic series of sodalite-rich syenites (naujaites, cf. II.2).

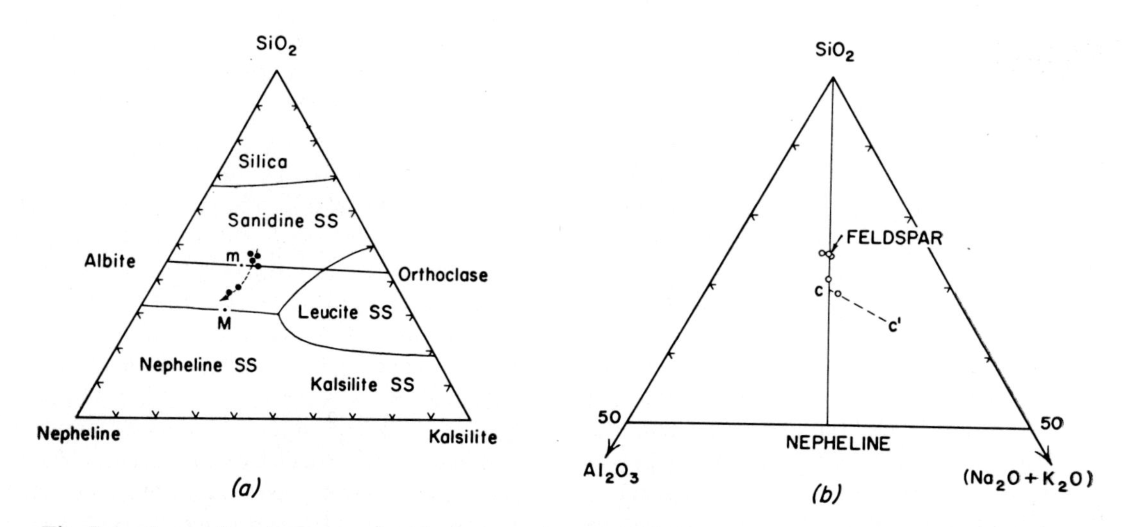

Fig. 7 Analyses of nepheline syenites from the Hviddal dyke, Tugtutôq, South Greenland plotted (a) in the system Q–Ne–Ks (after Upton, 1964b). m = alkali feldspar minimum, M = temperature minimum in Ab–Or–Ks–Ne (from Hamilton and MacKenzie, 1965). (b) in SiO_2–Al_2O_3–(Na_2O + K_2O) diagram. C–C′, albite–nepheline cotectic from Schairer and Bowen (1956)

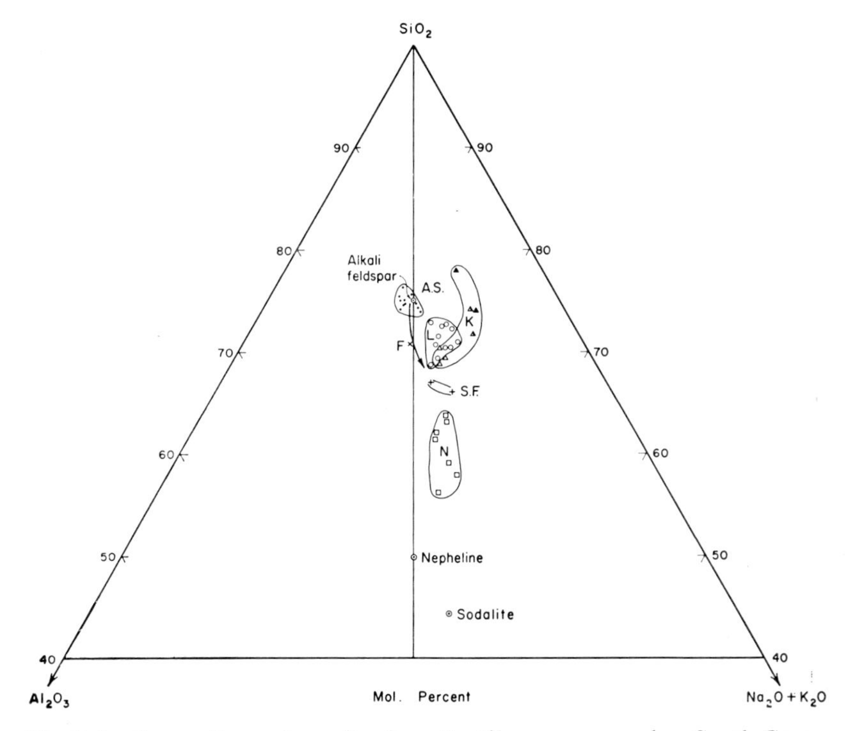

Fig. 8 Augite syenites and agpaites from the Ilímaussaq complex, South Greenland. A.S. = field of augite syenites (analyses given by solid circles), F = foyaite multiplication sign), S.F. = sodalite foyaites (crosses), N = naujaites (squares), K = kakortokites (triangles, filled for black kakortokites, half-filled for red, open for white), L = lujavrites (open circles). Data from Ussing (1912), Ferguson (1964), Hamilton (1964), Gerasimovsky and Kuznetsova (1967). Heavy line—trend of Hviddal nepheline syenites from Fig. 7

Three major types of kakortokites have been distinguished, consisting of various proportions of cumulus perthite, aegirine, riebeckite-arfvedsonite, eudialyte and nepheline (Ferguson, 1964, 1970; Sørensen, 1969).

The naujaites have formed by accumulation of floated sodalitic crystals from a sodalite foyaite, and their trend in Fig. 8, showing a range in composition between the sodalite composition point and the sodalite foyaites is consistent with this proposal. The residual liquids trapped between the downward crystallizing naujaites and the kakortokites were the lujavrites, highly peralkaline nepheline syenites consisting essentially of microcline, albite, arfvedsonite, aegirine, nepheline and eudialyte. The lujavrites show no distinct trend in Fig. 8 since their composition has been determined by fractionation of at least six cumulus phases from the immediately parental sodalite foyaite.

Agpaitic rocks such as those of Ilímaussaq are very scarce (Sørensen, 1960) and extrusives of lujavrite compositions, for example, have not been recorded. This possibly reflects the absence of the special conditions especially high volatile content necessary for the strong fractionation of foyaitic magma. Certain phonolites from the island of Trinidade have compositions comparable to the Ilímaussaq sodalite foyaites, and also carry phenocrysts of sodalite (de Almeida, 1961). Further fractionation of these magmas may eventually have led to lujavrite-like liquids, but the situation is complicated by the fact that various heteromorphic felsic assemblages may be developed in these salic liquids (Wilkinson, 1965), and by the possibility that other alkali rich

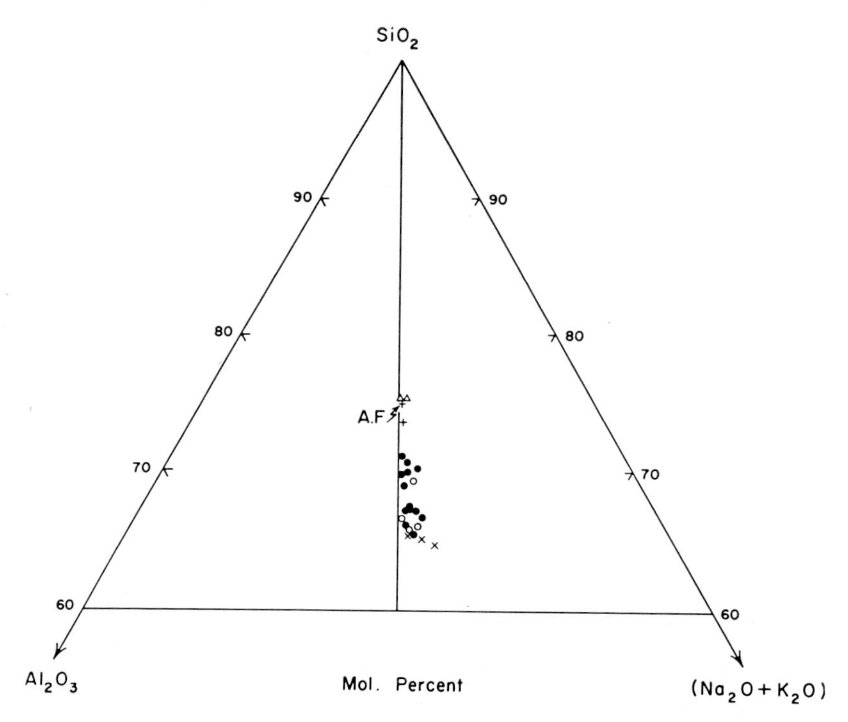

Fig. 9 Analyses of peralkaline phonolites from Atlantic islands. Triangles—St. Helena, crosses—Azores, solid circles—Principe, open circles—S. Tomé, multiplication signs—Trinidade

minerals such as aegirine or alkali amphibole may be fractionated.

The Trinidade phonolites (de Almeida, 1961) are more strongly undersaturated and more peralkaline than the phonolites of other Atlantic islands, some of which are plotted in Fig. 9. Those from St. Helena are mildly undersaturated and plot near the feldspar composition, whereas those on Principe, associated with basanites, are more strongly undersaturated and peralkaline. On Trinidade very strongly undersaturated and peralkaline phonolites occur with nephelinites and though seemingly stemming from different parental types, the phonolites of Principe and Trinidade overlap slightly in terms of SiO_2, Al_2O_3 and alkalis. It seems probable that continued fractionation of the Principe magmas would eventually produce residua comparable to those of Trinidade, though it should be possible to distinguish the two types of phonolite by their bulk chemistry, e.g. FeO, MgO and CaO contents, as they have evolved along different trends. In the general case, nepheline in strongly undersaturated series may have been crystallizing as phenocrysts in the most basic magmas, whereas in mildly alkaline basalt–phonolite suites, primary nepheline follows alkali feldspar in the crystallization sequence.

It seems clear that undersaturated basic magmas can evolve towards a phonolitic composition along several magmatic trends. One lineage descends near the critical plane of silica-undersaturation to the trachyte stage, but by continued feldspar fractionation can evolve towards a more undersaturated end-point, e.g. Ilímaussaq, St. Helena (Baker, 1969), Gough (Le Maitre, 1962) and Mt. Suswa, Kenya (Nash, Carmichael and Johnson, 1969). At the other extreme, phonolitic liquids may form from nephelinitic parents with only a small intermediate crystallization history, e.g. Trinidade. In the intermediate case, basanites may eventually fractionate towards phonolite, via feldspathoidal trachybasalts (Coombs and Wilkinson, 1969).

VI.2.5. Transitions between Undersaturated and Oversaturated Compositions

Since the discovery that the alkali feldspar join in Q–Ne–Ks is a barrier preventing the passage of quartz- or nepheline-normative liquids to the other side, several mechanisms for crossing the barrier plane have been suggested. As mentioned earlier, the degree of SiO_2-saturation of residual liquids of fractional crystallization of basalt seems to be mainly dependent on the composition of the parent. Strongly oversaturated (i.e. with appreciable normative hypersthene) and strongly undersaturated (high in normative nepheline) basalts cannot normally give rise on fractionation to undersaturated or oversaturated residua, respectively. It is apparent too that the critical plane of SiO_2 undersaturation in basalts, Fo–Di–Pl, Yoder and Tilley (1962), is generally effective in controlling fractionation trends in the mildly alkaline, transitional group of basalts (Coombs, 1963). On the islands of the Mid-Atlantic Ridge two (Ascension and Bouvet) have hypersthene-normative basalts and rhyolitic derivatives, and three (St. Helena, Tristan de Cunha and Gough) have nepheline-normative basalts and phonolitic derivatives. However in the Azores, where the average basalt is nepheline-normative, both phonolites and rhyolites are found (Baker *et al.*, 1964). In some provinces such as the Gardar and Oslo provinces, larvikitic syenite (with small amounts of normative nepheline) was parental to oversaturated and undersaturated trends, leading to quartz syenite and granite on one hand, and to nepheline syenite and foyaite on the other (Barth, 1954). If it can be assumed that the thermal barrier Or–Ab has an equivalent crystallization barrier in these more complex natural compositions, then it might be crossed by any of several mechanisms.

VI.2.5.1. Undersaturated → oversaturated

Acmite melts incongruently to hematite plus a liquid enriched in NaO_2 and SiO_2; failure of the hematite to be resorbed from a suitable range of nepheline-normative liquids could result in passage of the liquids to an oversaturated condition, (Tilley, 1958). However, acmite crystallizes only from a liquid with excess sodium silicate (Bailey and Schairer, 1966), i.e. in the natural case where $Na_2O+K_2O > Al_2O_3+Fe_2O_3$. The effect could work only in strongly peralkaline liquids, yet in the majority of cases, magmas are thoroughly committed to either an under- or oversaturated trend before sodium metasilicate becomes a normative component. The incongruent melting of acmite probably is of minor importance in petrology, and it is significant that Nolan (1966) was unable to find any petrographic evidence of its occurrence in nature.

Bowen (1928) suggested from experience in Q–Ne–Ks at 1 atmosphere that the incongruent melting of orthoclase to leucite plus SiO_2-rich liquid could permit passage of undersaturated magmas over the join Ab–Or. The field of leucite is restricted at higher water vapour pressures, however, and incongruent melting of Or is eliminated at 2·6 kb and 900 °C, (Goranson, 1938). The field of compositions which could show this effect is very small under normal magmatic conditions and Tilley (1958) and Fudali (1963) have suggested that the importance of this mechanism has probably been overemphasized.

Alkali feldspar is the most abundant phenocryst in the majority of trachytic rocks, and its composition strongly controls the fractionation paths that these rocks can follow. Even small departures from stoichiometry might influence the evolution of magma critically balanced with respect to SiO_2. For example, slight deficiency of SiO_2 from the ideal 1:1:6 formula might initiate an oversaturated from an undersaturated trend. In this respect it is interesting that the analyses of feldspars from Oslo larvikite reported by Oftedahl (1948) and Muir and Smith (1956) contain SiO_2 deficiences equivalent to 1–7% by wt. nepheline solid solution. Fractionation of SiO_2-deficient hydrous minerals such as biotite or certain amphiboles may also drive mildly undersaturated trachytic liquids through the feldspar divide. Biotite, for example, is a phenocryst in microsyenites from the Gardar and in trachytes from

Gough Island (Le Maitre, 1962) but its status as a cumulus phase has not yet been established.

VI2.5.2. Oversaturated→ undersaturated

This transition is considerably more difficult to achieve than the reverse, though again it must be questioned whether the process has more than theoretical significance in petrology. Bailey and Schairer (1966, p. 154) have suggested a possible mechanism, based on observations in the synthetic system NaO_2–Al_2O_3–Fe_2O_3–SiO_2, namely substitution of Fe^{3+} for Al in feldspar. The effect is to use Fe to make the molecule $NaFe^{3+}Si_3O_8$ rather than the acmite molecule $NaFe^{3+}Si_2O_6$, thereby using more silica. A trachytic magma plotting just on the silica-rich side of a natural crystallization barrier, such as Ab–Or–Ac–Alkali Disil., might be pushed through the plane towards the SiO_2 undersaturated side by separation of alkali feldspar carrying small amounts of Fe in its structure.

VI.2.6. Concluding Remarks

The fractionation of alkali feldspar from trachytic magma showing only a small range in the degree of SiO_2-saturation and alkali:alumina ratio can produce at least four series of residual liquids. Crystallization of feldspar is accompanied in peralkaline rhyolites by quartz, and in phonolites by a feldspathoid, and though the effect of feldspar fractionation alone is to make the slightly different parental peralkaline magmas diverge, crystallization of the second phase tends to make the different trends converge towards common eutectic compositions. In the phonolitic end-members of strongly undersaturated volcanic series, feldspathoid may be the dominant fractionating phase, and there may be an increase in the ratio feldspar:feldspathoid in residual liquids.

So far as is known the alumina:alkalis balance is not seriously affected by the crystallization of other phenocrysts in acid alkaline magmas. In undersaturated parageneses, other alkali-bearing minerals may be cumulus phases, but as they occur in relatively minor amounts the effect of fractionating them seems to be minor compared to the crystallization of the felsic minerals.

The demonstration by Coombs (1963) that the degree of undersaturation of salic alkaline magmas varies closely with that of the associated basic rocks is strong evidence that fractional crystallization has played the major rôle in the origin of the alkaline salic rocks. However, where the volume relationships of basic, intermediate and salic rocks can be shown to be totally inconsistent with such an origin, as is probable in parts of the East African Rifts, then partial melting of basic material may have contributed some part of the observed alkaline magma. One task of petrologists is to devise methods of distinguishing liquids formed by partial melting from those produced by fractionation (cf. VI. 1a. 5).

Further work should include chemical and modal analyses of phenocrysts and glass of alkaline obsidians, especially undersaturated types, coupled with relevant experimental studies, in order to establish the various evolutionary paths followed by these magmas. The direct evidence afforded by layered syenite intrusions must also be employed. Particularly large gaps exist in our knowledge of the trachytes and phonolites associated with strongly undersaturated basic rocks.

Acknowledgement

I would like to thank Dr. D. K. Bailey for reading the manuscript, and for many lively discussions on the origin of alkaline rocks.

VI.2. REFERENCES

Abbott, M. J., 1969. Petrology of the Nandewar volcano, N.S.W., Australia. *Contr. Miner. Petrology*, **20**, 115–34.

Almeida, F. F. M. de, 1961. *Brazil, Div. de Geol. e Miner., Dept. Nac. de Prod. Mineral*, Monograph 18.

Bailey, D. K., 1964. Crustal warping—a possible tectonic control of alkaline magmatism. *J. Geophys. Res.*, **69**, 1103–11.

Bailey, D. K., and Macdonald R., 1969. Alkali feldspar fractionation trends and the derivation of peralkaline liquids. *Am. J. Sci.*, **267**, 242–8.

Bailey, D. K., and Macdonald, R., 1970. Petrochemical features of mildly peralkaline (comenditic) rhyolite glasses from the continents and ocean basins. *Contr. Miner. Petrology*, **28**, 340–51.

Bailey, D. K., and Schairer, J. F., 1964. Feldspar–liquid equilibria in peralkaline liquids—the orthoclase effect. *Am. J. Sci.*, **262**, 1198–206.

Bailey, D. K., and Schairer, J. F., 1966. The system Na_2O–Al_2O_3–Fe_2O_3–SiO_2 at 1 atmosphere, and the petrogenesis of alkaline rocks. *J. Petrology*, **7**, 114–70.

Baker, I., 1968. Intermediate oceanic volcanic rocks and the 'Daly gap'. *Earth Planet. Sci. Let.*, **4**, 103–6.

Baker, I., 1969. Petrology of the volcanic rocks of Saint Helena Island, South Atlantic. *Bull. geol. Soc. Am.*, **80**, 1283–310.

Baker, P. E., Gass, I. G., Harris, P. G., and Le Maitre, R. W., 1964. The volcanological report of the Royal Society expedition to Tristan da Cunha, 1962. *Phil. Trans. Roy. Soc. Lond.*, **256**, 439–578.

Barth, T. F. W., 1954. Studies on the igneous rock complex of the Oslo region 14: Provenance of the Oslo magmas. *Skr. norske Vidensk Akad., Mat.–naturv. Kl.*, **4**.

Bowen, N. L., 1928. *The Evolution of the Igneous Rocks*. 332 pp. Princeton Univ. Press.

Bowen, N. L., 1937. Recent high-temperature research on silicates and its significance in igneous geology. *Am. J. Sci.*, **33**, 1–21.

Bowen, N. L., 1945. Phase equilibria bearing on the origin and differentiation of the alkaline rocks. *Am. J. Sci.*, **243A**, 75–89.

Bryan, W. B., 1964. Relative abundance of intermediate members of the oceanic basalt–trachyte association: evidence from Clarion and Socorro Islands, Revillagigedo Islands, Mexico. *J. Geophys. Res.*, **69**, 3047–9.

Cann, J. R., 1968. Bimodal distribution of rocks from volcanic islands. *Earth Planet. Sci., Lett.*, **4**, 479–80.

Carmichael, I. S. E., 1962. Pantelleritic liquids and their phenocrysts. *Mineralog. Mag.*, **33**, 86–113.

Carmichael, I. S. E., 1964. Natural liquids and the phonolitic minimum. *Geol. J.*, **4**, 55–60.

Carmichael, I. S. E., 1967. The iron–titanium oxides of salic volcanic rocks and their associated ferromagnesian silicates. *Contr. Miner. Petrology*, **14**, 36–64.

Carmichael, I. S. E., and MacKenzie, W. S., 1963. Feldspar–liquid equilibria in pantellerites: an experimental study. *Am. J. Sci.*, **261**, 382–96.

Chayes, F., 1963. Relative abundance of intermediate members of the oceanic basalt–trachyte association. *J. Geophys. Res.*, **68**, 1519–34.

Chayes, F., and Zies, E. G., 1962. Sanidine phenocrysts in some peralkaline volcanic rocks. *Carnegie Inst. Wash. Yearbook*, **61**, 112–18.

Chayes, F., and Zies, E. G., 1964. Notes on some Mediterranean comendite and pantellerite specimens. *Carnegie Inst. Wash. Yearbook*, **63**, 186–90.

Coombs, D. S., 1963. Trends and affinities of basaltic magmas and pyroxenes as illustrated on the diopside–olivine–silica diagram. *Min. Soc. Am. Spec. Pap.*, **1**, 227–50.

Coombs, D. S., and Wilkinson, J. F. G., 1969. Lineages and fractionation trends in undersaturated volcanic rocks from the East Otago Volcanic Province (New Zealand) and related rocks. *J. Petrology*, **10**, 440–501.

Ewart, A., Taylor, S. R., and Capp, A. C., 1968. Geochemistry of the pantellerites of Mayor Island, New Zealand. *Contr. Miner. Petrology*, **17**, 116–40.

Ferguson, J., 1964. Geology of the Ilímaussaq intrusion, South Greenland. *Meddr. Grønland*, **172**, 4.

Ferguson, J., 1970. The significance of the kakortokite in the evolution of the Ilímaussaq intrusion, South Greenland. *Meddr. Grønland*, **190**, 1, 193 pp.

Fudali, R. F., 1963. Experimental studies bearing on the origin of pseudoleucite and associated problems of alkalic rock systems. *Bull. geol. Soc. Am.*, **74**, 1101–26.

Gass, I. G., and Mallick, D. I. J., 1968. Jebel Khariz: an Upper Miocene strato-volcano of comenditic affinity on the South Arabian coast. *Bull. Volcan.*, **32–1**, 33–88.

Gerassimovsky, V. I., and Kuznetsova, S. Ya., 1967. On the petrochemistry of the Ilímaussaq intrusion, South Greenland. *Geochem. Int.*, **4**, 236–46.

Goranson, R. W., 1938. Silicate–water systems. Phase equilibria in the $NaAlSi_3O_8$–H_2O and $KAlSi_3O_8$–H_2O systems at high temperatures and pressures. *Am. J. Sci.*, **35–A**, 71–91.

Hamilton, E. I., 1964. The geochemistry of the northern part of the Ilímaussaq intrusion, S.W. Greenland. *Meddr. Grønland*, **162**, 10.

Hamilton, D. L., and MacKenzie, W. S., 1965. Phase-equilibrium studies in the system $NaAlSiO_4$ (nepheline)–$KAlSiO_4$ (Kalsilite)–SiO_2–H_2O. *Mineralog. Mag.*, **34**, 214–31.

Harris, P. G., 1963. Comments on a paper by F. Chayes, 'Relative abundances of intermediate members of the oceanic basalt–trachyte association'. *J. Geophys. Res.*, **68**, 5103–7.

Kuno, H., 1968. 'Differentiation of basalt magmas', in Hess H. H. and Poldervaart, A, Eds., *Basalts: the Poldervaart Treatise on Rocks of Basaltic Composition, vol. 2*, 623–88. Interscience, New York.

Le Maitre, R. W., 1962. Petrology of volcanic rocks, Gough Islands, South Atlantic. *Bull. geol. Soc.*, **73**, 1309–40.

Le Maitre, R. W., 1968. Chemical variations between and within volcanic rock series—a statistical approach. *J. Petrology*, **9**, 220–52.

Luth, W. C., and Tuttle, O. F., 1966. The alkali feldspar solvus in the system Na_2O–K_2O–Al_2O_3–SiO_2–H_2O. *Am. Miner.*, **51**, 1359–73.

Macdonald, G. A., 1963. Relative abundance of intermediate members of the oceanic basalt–trachyte association—a discussion. *J. Geophys. Res.*, **68**, 5100–2.

Macdonald, R., 1969. The petrology of alkaline dykes from the Tugtutôq area, South Greenland. *Bull. geol. Soc. Denmark*, **19**, 257–82.

Macdonald, R., Bailey, D. K., and Sutherland, D. S., 1970. Oversaturated peralkaline glassy trachytes from Kenya. *J. Petrology*, **11**, 507–17.

Macdonald, R., and Gibson, I. L., 1969. Pantelleritic obsidians from the Volcano Chabbi (Ethiopia). *Contr. Miner. Petrology*, **24**, 239–44.

Muir, I. D., and Smith, J. V., 1956. Crystallisation of feldspars in larvikite. *Zeit. Krist.*, **107**, 182–95.

Nash, W. P., Carmichael, I. S. E., and Johnson, R. W., 1969. The mineralogy and petrology of Mount Suswa, Kenya. *J. Petrology*, **10**, 409–39.

Nicholls, J., and Carmichael, I. S. E., 1969. Peralkaline acid liquids: a petrological study. *Contr. Miner. Petrology*, **20**, 268–94.

Noble, D. C., 1965. Gold Flat member of the Thirsty Canyon Tuff—a pantellerite ash-flow sheet in southern Nevada. *U.S. geol. Survey, Prof. Paper.*, **525–B**, 85–90.

Noble, D. C., 1967. Sodium, potassium, and ferrous iron contents of some secondarily hydrated natural silicic glasses. *Am. Miner.*, **52**, 280–6.

Noble, D. C., 1968. Systematic variation of major elements in comendite and pantellerite glasses. *Earth Planet. Sci. Let.*, **4**, 167–72.

Noble, D. C., Smith, V. C., and Peck, L. C., 1967. Loss of halogens from crystallised and glassy silicic volcanic rocks. *Geochim. cosmochim. Acta*, **31**, 215–24.

Nolan, J., 1966. Melting relations in the system $NaAlSi_3O_8$–$NaAlSiO_4$–$NaFeSi_2O_6$–$CaMgSi_2O_6$–H_2O and their bearing on the genesis of alkaline undersaturated rocks. *Q. J. geol. Soc. Lond.*, **122**, 119–57.

Oftedahl, C., 1948. Studies on the igneous rock complex of the Oslo region. 9: The feldspars. *Skr. norske Vidensk Akad., Mat.-naturv. Kl*, **3**.

Romano, R., 1968. New petrochemical data of volcanites from the Island of Pantelleria (Channel of Sicily). *Geol. Rundschau*, **57**, 773–83.

Saggerson, E. P., and Williams, L. A. J., 1964. Ngurumanite from Southern Kenya and its bearing on the origin of rocks in the northern Tanganyika alkaline district. *J. Petrology*, **5**, 40–81.

Schairer, J. F., and Bowen, N. L., 1955. The system K_2O–Al_2O_3–SiO_2, *Am. J. Sci.*, **253**, 681–746.

Schairer, J. F., and Bowen, N. L., 1956. The system Na_2O–Al_2O_3–SiO_2. *Am. J. Sci.*, **254**, 129–195.

Schairer, J. F., and Yoder, H. S., 1960. The nature of residual liquids from crystallisation, with data on the system nepheline–diopside–silica. *Am. J. Sci.*, **258A**, 273–83.

Sørensen, H., 1958. The Ilímaussaq batholith. A review and discussion. *Meddr. Grønland*, **162**, 3.

Sørensen, H., 1960. On the agpaitic rocks. *Rep. Int. geol. Congr. 21st Session, Norden*, Pt. 13, 319–27.

Sørensen, H., 1969. Rythmic igneous layering in peralkaline intrusions. *Lithos*, **2**, 261–83.

Thompson, R. N., and MacKenzie, W. S., 1967. Feldspar–liquid equilibria in peralkaline acid liquids: an experimental study. *Am. J. Sci.*, **265**, 714–34.

Tilley, C. E., 1958. Problems of alkali rock genesis: *Q.J. geol. Soc. Lond.*, **113**, 323–60.

Tuttle, O. F., and Bowen, N. L., 1958. Origin of granite in the light of experimental studies in the system $NaAlSi_3O_8$–$KAlSi_3O_8$–SiO_2–H_2O. *Geol. Soc. Am. Mem.*, **74**, 153 pp.

Upton, B. G. J., 1960. The alkaline igneous complex of Kûngnât Fjeld, South Greenland. *Bull. Grønlands geol. Unders.*, **27** (also *Meddr Grønland*, **169**, 8).

Upton, B. G. J., 1964a. The geology of Tugtutôq and neighbouring islands, South Greenland. Part 2. Nordmarkitic syenites and related alkaline rocks. *Meddr. Grønland*, **169**, 2.

Upton, B. G. J., 1964b. The geology of Tugtutôq and neighbouring islands, South Greenland. Part 4. The nepheline syenites of the Hviddal composite dyke. *Meddr. Grønland*, **169**, 3.

Ussing, N. V., 1912. The geology of the country around Julianehaab, Greenland. *Meddr. Grønland*, **38**.

Villari, L., 1968. On the geovolcanological and morphological evolution of an endogenous dome (Pantelleria, Mt. Gelkhamar). *Geol. Rundschau*, **57**, 784–95.
Washington, H. S., 1914. The volcanoes and rocks of Pantelleria. *J. Geol.*, **22**, 16–27.
Wilkinson, J. F. G., 1965. Some feldspars, nepheline and analcimes from the Square Top Intrusion, Nundle, N. S. W. *J. Petrology*, **6**, 420–44.
Wilkinson, J. F. G., 1966. Residual glasses from some alkali basaltic lavas from New South Wales. *Mineralog. Mag.*, **35**, 847–60.
Wright, J. B., 1963. A note on possible differentiation trends in Tertiary to Recent lavas of Kenya. *Geol. Mag.*, **100**, 164–80.
Wright, J. B., 1965. Petrographic sub-provinces in the Tertiary to Recent volcanics of Kenya. *Geol. Mag.*, **102**, 541–57.
Wright, J. B., 1966. Olivine nodules in a phonolite of the East Otago alkaline province, New Zealand. *Nature*, **210**, 519.
Wright, J. B., 1969. Olivine nodules and related inclusions in trachyte from the Jos Plateau, Nigeria. *Mineralog. Mag.*, **37**, 370–4.
Wyllie, P. J., 1963. Effects of the changes of slope on liquidus and solidus paths in the system diopside–anorthite–albite. *Min. Soc. Am. Spec. Pap.*, **1**, 204–12.
Yagi, K., 1966. The system acmite–diopside and its bearing on the stability relations of natural pyroxenes of the acmite–hedenbergite–diopside series. *Am. Miner.*, **15**, 976–1000.
Yoder, H. S., and Tilley, C. E., 1962. Origin of basalt magmas: An experimental study of natural and synthetic rock systems. *J. Petrology*, **3**, 342–532.
Zies, E. G., 1960. Chemical analysis of two pantellerites. *J. Petrology*, **1**, 304–8.
Zies, E. G., 1966. A new analysis of cossyrite from the island of Pantelleria. *Am. Miner.*, **51**, 200–5.

VI.3. LIMESTONE ASSIMILATION

Peter J. Wyllie

VI.3.1. Introduction

Limestone assimilation by subalkaline magma as a process for generating desilicated alkaline magma was proposed by Daly in 1910. It has remained on the petrological scene as a healthy hypothesis ever since, although in the past twenty years its strength has been waning. From the very beginning, the idea has been avidly supported by ardent advocates and vigorously contested by sceptical adversaries.

The two opposing views persist into the sixth decade. Schuiling (1964b) considered the limestone assimilation theory to be 'the only theory which really has been demonstrated in the field with a degree of certainty, scarcely ever achieved by petrological observation'. On the other hand, Turner and Verhoogen (1960, p. 396) concluded that 'Several decades of investigation have failed to confirm the efficacy of limestone assimilation as a significant factor in the development of nepheline-syenite magma.'

In the following account I have adopted a historical approach, in which I have attempted to trace certain aspects of the controversy through successive time periods. The three selected time periods of twenty-five, twenty, and fifteen years do appear to represent rather distinct stages. The first, 1910–35, is the period during which the stimulating hypothesis was launched. Arguments for and against were based mainly on interpretation of field data. The second period, 1936–55, was one during which the lines of attack and defense became deeply entrenched, and experimental study of silicate systems revealed the thermal barrier on the liquidus of silicate systems, rearing its ugly hump between subalkaline and feldspathoidal liquids. During the third period, 1956–70, the study of carbonatites disrupted the field evidence for limestone assimilation, and additional experimental data made it even more difficult to support the hypothesis.

VI.3.2. 1910-35: The Limestone Syntexis Hypothesis

VI.3.2.1. The Proposition

Daly (1910) estimated that alkaline igneous rocks amounted to about 1% of all exposed igneous rocks, and he concluded on this basis that their derivation from primeval reservoirs of alkaline magma was unlikely. They must be derived in some way from the abundant subalkaline magmas. Daly's field experience led him to develop a general theory of assimilation–differentiation.

Daly compiled a comprehensive list of all known alkaline associations, and found that for 107 of the 155 districts, there was a clear rule connecting alkaline igneous rocks and carbonate rocks. In other districts, there was insufficient evidence found. He therefore proposed the following points:

1. Solution of limestone in basalt dissociates at least part of the carbonate, and the lime thus released combines with silica, and possibly with iron oxides and magnesia, to form pyroxene. This desilicates the magma.
2. Limestone or dolomite would flux basaltic magma, improving the efficiency of gravitative differentiation. The sinking of femic and cafemic constituents, in solid or liquid phases, leaves the upper residual part of the magma richer in alkalis than the original basalt, as well as being desilicated. Nepheline may form instead of plagioclase. The amount of carbonate required is small compared to a given volume of subalkaline magma.
3. The CO_2 released must have a profoundly disturbing effect on the chemical equilibrium of the basalt. Possibly the CO_2 could form compounds with alkalis and rise to the top, where CO_2 would be gradually replaced by silica. The process may be effected by the whole carbonate or by the carbon dioxide driven from the carbonate. Juvenile CO_2 and other gases may be agents in the formation and segregation of alkaline magmas from subalkaline magmas in some examples.
4. Alkaline rocks are differentiates of the syntectic magmas produced by limestone assimilation. This explains the extreme mineralogical variability of alkaline bodies, and the occurrence of lime-rich minerals.

Daly felt confident on the basis of geological observations that future experimental results would favour the idea of limestone control, or at least CO_2 control, in the development of alkaline rock masses.

VI.3.2.2. The Opposition

Bowen (1922), in a paper which was perhaps the clearest exposition of views opposing the limestone assimilation hypothesis during this period, agreed with Harker (1909, p. 339) that there is no petrographic or theoretical evidence to indicate that magmas entered the crust with superheat, and that solution of crustal rocks would require large amounts of heat. After considering the available thermochemical data, Bowen concluded that the heat required for assimilation is probably of the order of magnitude of the heat of melting, and that this could only be supplied by partial crystallization of the magma. Therefore, assimilation must be accompanied by crystallization.

Bowen developed a simple model to illustrate the effects of limestone assimilation by considering modifications of the paths of crystallization of liquids in the system $CaO–MgO–SiO_2$ when CaO was added. Solution of CaO causes the proportions of minerals to change, and the composition of the pyroxene to change compared to the pyroxenes that would otherwise have been precipitated, but there is no fundamental change in the sequence of crystallization. Bowen maintained that addition of sedimentary inclusions to a basaltic magma adds nothing that is not already present, and therefore the same crystalline phases appear, slightly modified in composition. He added that if enough material is dissolved, then the path of crystallization might be extended sufficiently for new phases to appear, such as melilite in the system $CaO–MgO–SiO_2$ when CaO is added, and then the course of subsequent crystallization could be fundamentally modified; the derivatives of basaltic liquids precipitating melilite could conceivably include some alkaline types. Daly and others have quoted the last remark as support from Bowen for the limestone assimilation hypothesis, but they omitted to note Bowen's introduction to the remark that: it is questionable that enough heat could be available for an amount of material to be added sufficient to produce this effect.

Bowen discussed the difficulties of producing feldspathoidal liquids through desilication of granitic magma by assimilation of limestone inclusions. The amount of desilication required, first exhausting the free silica, and then desilicating the feldspar components, is very considerable. He contended that assimilation would be incapable of changing the liquid composition so far, before it was used up by the crystallization concomitant with solution and desilication. Furthermore, a review of some associations of granite with limestone indicated that the normal reaction product includes basic silicates such as pyroxene or hornblende without the formation of feldspathoids.

Other objections to the limestone assimilation hypothesis discussed during this period include the following. Limestones are so widely distributed in the crust that the observed association of alkaline rocks and limestones is accidental. Many subalkaline rocks in contact with limestones have no associated alkaline rocks.

VI.3.2.3. Counter-Arguments

Twenty-three years after first presenting the limestone syntexis hypothesis, Daly (1933) reviewed the origin of alkaline rocks in some detail in his book *Igneous Rocks and the Depths of the Earth*. He emphasized that some of the objections to the limestone syntexis theory resulted through failure of critics to understand the theory, and he therefore restated it in somewhat more precise terms than in 1910, acknowledging Shand's (1930) analysis of the processes involved. Daly maintained that the majority of feldspathoidal rock types are produced from a parent subalkaline magma, which may range in composition from basaltic to granitic, by the following processes:

1. Desilication of the subalkaline magma by: (a) assimilated CO_2 from limestone, and H_2O from other sediments; (b) assimilated CaO possibly with MgO, from limestones; (c) concentration of juvenile CO_2, H_2O, and other volatile components in magmas.

2. Differentiation of: (a) syntectic magmas corresponding to those desilicated by assimilation under (1). (b) magmas enriched in juvenile CO_2, H_2O, and other volatiles.

He emphasized the operation of three distinct processes listed as (1), (2) and (3) in section VI.3.2.1.

Shand (1930) emphasized that critics of the syntexis hypothesis had often confined their attentions to the single aspect of simple solution of limestone in basaltic or granitic magma, whereas there are three distinct factors, listed as (1), (2) and (3) in section VI.3.2.1. He proposed that reaction between limestone and albite components in the magma would produce Na_2CO_3 and Na_2SiO_3 molecules, and that these light and very soluble sodium salts must tend to rise towards the top of the magma chamber, along with juvenile and resurgent fluids. Migration of material in this way could generate alkaline magmas at sites removed from the position where assimilation was occurring; this could explain the association of subalkaline rocks and limestone without alkaline rocks, and the occurrence of alkaline rocks with no visible limestone.

Shand also provided an answer for those petrologists concerned about the source of heat for assimilation (see sections VI.3.2.1. and VI.3.2.2). Alkaline igneous bodies are usually local facies of much larger bodies of subalkaline rocks, and Shand stated that the temperature of the small body could be maintained by that of the large parent body, and in these circumstances there is no difficulty of heat supply.

VI.3.2.4. The Experimental Approach

In 1928, Stansfield published a monograph describing an extensive series of experiments which were aimed directly at testing the limestone assimilation hypothesis. His experimental programme arose from his interest in the origin of the okaite and associated rocks at St. Jospeh du Lac, Quebec (see IV.6)*. He advanced field and chemical evidence to support his conclusion that the okaite had originated by assimilation of Grenville limestone by a magma of unknown type (possibly peridotite), and he suggested that, in general, rock magmas which yield melilite and monticellite on crystallization have dissolved considerable amounts of limestone (Stansfield, 1923).

Stansfield (1928) conducted a series of fusion experiments in which ten different igneous rocks varying from peridotite to granite were made to assimilate materials representing pure sediments (kaolin, and glass sand). Most of the products of the experimental assimilation of calcite contained glass, indicating that melting had occurred, and all products were characterized by the presence of melilites. No feldspathoids were developed in any experiments. These experiments thus support the hypothesis that melilitic rocks can be derived from subalkaline magmas by limestone assimilation, provided that sufficient heat is available, but they provide no evidence for the hypothesis that feldspathoidal magmas or rocks can be produced in the same way. Stansfield concluded tentatively that 10 or 15% may be the limit of foreign rock matter which can be assimilated by natural magmas.

VI.3.3. 1936-55: Experimental and Field Interpretations

VI.3.3.1. The Thermal Barrier Between Granitic and Feldspathoidal Liquids

The significance of the system Ne–Ks–Qz for theories of origin of granitic, syenitic, and feldspathoidal magmas was discussed by Bowen in 1937. The preliminary phase diagram (Schairer and Bowen, 1935) was confirmed and revised by Schairer (1950) fifteen years later. Fig. 13 of Chapter V.1 shows that the temperature maximum on the feldspar join through the system constitutes a real thermal barrier which can be crossed only if leucite appears during the course of crystallization. The implications of this barrier for desilication of a granitic magma by limestone assimilation are illustrated by the isothermal section in Fig. 1.

Fig. 1 shows the phase fields at 1030 °C and 1 atmosphere pressure. The diagram is schematic because the tie-lines have not been determined, but petrographic studies and later experimental studies in the presence of water under pressure (Fudali, 1963; Hamilton and Mackenzie, 1965) confirm that the phase relationships must have this general pattern. Consider the granitic liquid l_1 which is just saturated with quartz and feldspar f_1. If silica is removed under isothermal conditions, the bulk composition of the system is changed to some point M, and re-adjustment then produces the assemblage liquid + feldspar, $l_2 + f_2$. If the liquid l_2 is desilicated, the bulk composition migrates further away from SiO_2, and more feldspar is precipitated. If equilibrium is maintained, the bulk composition of the liquid–feldspar system follows the line l_1Mf_3, until eventually the last trace of liquid l_3 coexists with feldspar f_3. No liquid then remains for further desilication. Even if equilibrium is not maintained, the liquid composition remains in the region of $l_1l_2l_3$ while SiO_2 is subtracted from the system, and a range of feldspar compositions is precipitated. The existence of the feldspar barrier prevents the development of liquids capable of precipitating feldspathoids, unless the temperature is greater than that of the liquidus surface on the feldspar join; but then we would be dealing with liquids that do not coexist with crystals at any stage during the process, and this is not equivalent to magmatic conditions.

If the desilication of liquid l_1 is caused by limestone assimilation, the liquid becomes enriched in CaO as feldspar is precipitated, and the feldspar is concomitantly enriched in the anorthite component. The process has to be re-examined in the system Ne–Ks–Qz–An, and here we find that the ternary feldspar join, An–Ab–Or, is also a thermal barrier, playing the same rôle as the binary feldspar join Ab–Or in Fig. 1. Looking ahead to experimental results of Yoder and Tilley in 1962, it is appropriate to

* Readers are referred to the relevant Chapters of this book where similarly cited.

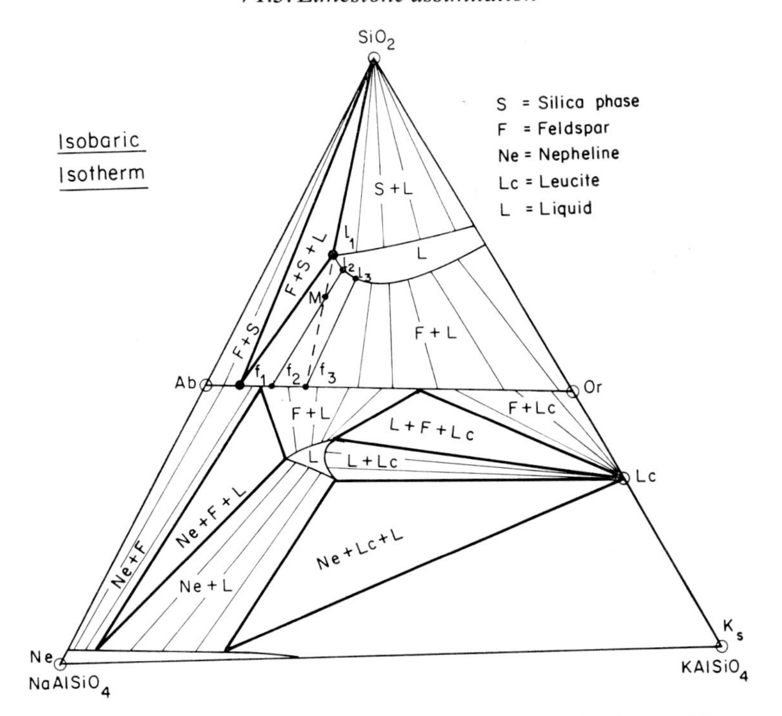

Fig. 1 Schematic isobaric isothermal section through the Residua System, based on various sources. Consider the liquid l_1. Desilication at constant pressure and temperature, and under equilibrium conditions, causes the bulk composition of the system to move along the dashed line, as shown by point M. The liquid composition changes to l_3, while feldspar is precipitated changing composition from f_1 to f_3. As explained in the text, although the system is desilicated, the liquid composition hardly changes as crystallization goes to completion and the last trace of l_3 is used up

note at this point that there exists a similar thermal barrier in basaltic systems between saturated liquids, and liquids capable of precipitating feldspathoids. These thermal barriers persist to very high pressures until albite becomes unstable and is replaced by jadeite (Yoder and Tilley, 1962, pp. 500–4). At magmatic temperatures, the pressures required correspond to positions beneath the earth's crust (Boettcher and Wyllie, 1968, 1969). The anorthite component of the plagioclase becomes unstable at comparable depths.

Those convinced that the geological evidence indicates that desilication of a silica-saturated magma does produce a feldspathoidal magma have to find a way across the thermal barrier depicted in Fig. 1. Shand (1943) suggested that perhaps dissolved H_2O and the CO_2 released from the limestone affected the liquidus so that the feldspar thermal barrier was suppressed. Towards the end of this period, experimental results on the effect of H_2O on the melting relationships of the feldspar join became available (Bowen and Tuttle, 1950), but it was two decades after Shand's suggestion before sufficient experimental data became available to show that the feldspar barrier persisted in the presence of H_2O vapour under pressure (Fudali, 1963; Hamilton and Mackenzie, 1965), and that the effect of CO_2 on the liquidus relationships in granitic systems was rather small (Wyllie and Tuttle, 1959).

In several papers, Bowen (e.g. 1928) has shown how the barrier can be crossed if leucite is precipitated from the liquid, but advocates of the limestone syntexis hypothesis do not regard this as a significant general process (see also VI.3.4.2).

VI.3.3.2. Limestones or Carbonatites?

Tilley (1952) reviewed the results of several investigations of basalt–limestone contacts in the British Tertiary province. Detailed mineralogical and chemical studies of the reaction products formed in basalt by lime absorption confirmed that residual liquids yielded mineral assemblages including melilite, and small amounts of nepheline. However, the trend of differentiation of the reaction products does not correspond to the trend of common mafic feldspathoidal rocks. Tilley therefore concluded that this evidence does not favour the limestone assimilation hypothesis for the origin of mafic alkaline rocks. He also reviewed detailed studies of granite–limestone contacts from the same province, and concluded that although desilication did occur with the production of mildly alkaline assemblages, there are no convincing examples of an overstep of the feldspar thermal barrier, with the appearance of nepheline.

Shand (1945) referred to the same evidence reviewed by Tilley, and claimed that these indisputable examples of the formation of nepheline-bearing rocks by the action of basaltic magma provided strong support for the limestone syntexis hypothesis. He asserted that all stages in the sequence from granite, to foyaite, to ijolite were supported by field evidence which hardly admits of any other interpretation.

Chayes (1942), working from Shand's department at Columbia University, published a detailed study of the Bancroft area (cf. II.7; VI.7), which had figured largely in earlier discussions about the origin of alkaline igneous rocks. There are two distinct groups of alkali rocks lying along the contact of the 'Laurentian' granites and the Grenville marble, one a series of saturated or nearly saturated rocks ranging from syenite through perknite, and the other an undersaturated group including members of the ijolite family. Chayes concluded that the nearly saturated granite–syenite–amphibolite association is best explained by the limestone syntexis hypothesis, and nothing in the field relationships of the ijolite family contradicts the theory. 'The special explanation which places least strain upon the field facts is limestone syntexis.'

During this period, increasing attention was being paid to carbonatites. Shand (1945) reviewed some of the discussions about limestones associated with alkaline rocks, and he noted that 'some petrologists preferred to put their faith even in that most strange thing, a carbonate magma'. In the second edition of his textbook Shand (1943) stated that carbonatites are either sedimentary limestone inclusions, or intrusions of mobilized limestone emplaced in the solid state rather like salt-domes.

Holmes (1950), on the other hand, assigned a major rôle to carbonatite magmas in attempting to explain the origin of the strongly alkaline mafic and ultramafic lavas of the volcanic fields near Ruwenzori. The geochemical peculiarity which requires explanation is the combination of high potash with cafemic constituents. He reasoned that it was quite impossible to explain these rocks by any variation of the limestone-syntexis hypothesis, and proposed that magmatic (largely liquid) carbonatite from the mantle had reacted with granitic rocks to produce the alkaline lavas.

A primary carbonatite magma was proposed by von Eckermann (1948, pp. 157 and 161) following his detailed study of the Alnö Island alkaline complex.

VI.3.4. 1956-70: Carbonatites, and Experimental Data on Syntexis

VI.3.4.1. Some Carbonatites are Magmatic Rocks

The growing interest in the occurrence and genesis of carbonatites, largely because of their economic potential, was brought to the attention of petrologists in general by the publication, almost simultaneously in three countries, of reviews by Pecora (1956), Campbell Smith (1956) and Agard (1956).

Shand (1945) had pointed out that a carbonate magma would be an explosive, high-temperature liquid, although he noted that the presence of sufficient H_2O might bring the temperature down to a more reasonable level: even then, he added, it would still be necessary to find a source for the strange brew. Wyllie and Tuttle (1960a) confirmed Shand's prediction by demonstrating that melts in the system $CaO–CO_2–H_2O$ precipitate calcite at temperatures down to 640 °C at pressures up to at least 4 kbar. These results have since been revised and extended to 40 kbar (Wyllie and Boettcher, 1969).

One of the major items of evidence cited for the limestone assimilation hypothesis is the association of limestone with most known alkaline complexes. Re-examination of the lists compiled by Daly shows that in many associations the supposed limestone is now interpreted as a carbonatite. Consequently, according to Heinrich (1966, p. 286), most investigators of alkalic–carbonatitic complexes have rejected the hypothesis.

Schuiling (1964a) noted that from a physicochemical viewpoint there is no essential difference between the limestone-syntexis hypothesis, and sial-syntexis by carbonatite magma.

Some remarkable evidence reported by Roedder (1965) indicates that CO_2 may play a more significant rôle in magma genesis in the mantle than petrologists and geophysicists have heretofore imagined. Roedder studied many fluid inclusions in olivine phenocrysts and olivine-bearing nodules from basalts, and in some eclogite xenoliths and kimberlite from a diamond pipe. In each of the seventy-two olivine nodules examined, Roedder found inclusions of compressed, generally liquified, nearly pure CO_2. It seems from his observations that immiscible globules of dense CO_2 fluid existed at high pressures in the basaltic liquid. The density of filling corresponds to the pressure of 10–15 km of basaltic liquid when the inclusions were trapped in the growing olivine crystals. Inclusions could have formed at even greater depths, and decrepitated on eruption. For the samples from oceanic regions, this depth places the source of the CO_2 within the mantle.

In a recent study of oxygen and carbon isotopes in the calcite and dolomite inclusions and groundmass within a mica peridotite dyke from Pennsylvania, Deines (1968) reported that the carbonate inclusions contain the heaviest terrestrial carbon discovered so far. δC^{13} values range from 12‰ to 24·8‰, with respect to the PDB standard. It is difficult to explain this enrichment in C^{13} by fractionation processes, or by assimilation or partial reaction of limestone or coal. The alternative possibility that the heavy carbon was derived from the lower crust or deep mantle appears to be preferable. A primary source of carbon or CO_2 is again implied.

Daly (1910, 1933) added to the limestone syntexis hypothesis the possibility that juvenile CO_2 and other volatile components could help the processes of desilication (item (1c) in section VI.3.2.3), and segregation of alkaline magmas from subalkaline magmas (item (3) in section VI.3.2.1). The weight of present field evidence appears to support the view that juvenile CO_2 may well contribute towards these effects, and to the production of carbonatites, with no requirement that limestone be assimilated.

VI.3.4.2. Experimental Data: Effect of Carbon Dioxide

It is well known that H_2O under pressure greatly speeds reaction rates in silicate systems, but in the presence of dry CO_2, reactions in feldspathic and granitic materials were extremely sluggish. On the basis of qualitative experiments using CO_2 in the presence of some H_2O, Wyllie and Tuttle (1959, 1960b) concluded that the phase relationships in the system albite–CO_2–H_2O would be similar to Fig. 2a.

Consider first the isobaric crystallization of liquid M in Fig. 2a. Upon cooling under equilibrium conditions, crystals of albite are precipitated. With decreasing temperature, the three-phase triangle $Ab + L + V$ moves towards the binary system $Ab–H_2O$, until the composition M just enters through the Ab–L side of the triangle. Then a vapour phase is evolved with composition given by the corner of the triangle. Note that the vapour is greatly enriched in CO_2/H_2O compared to the liquid.

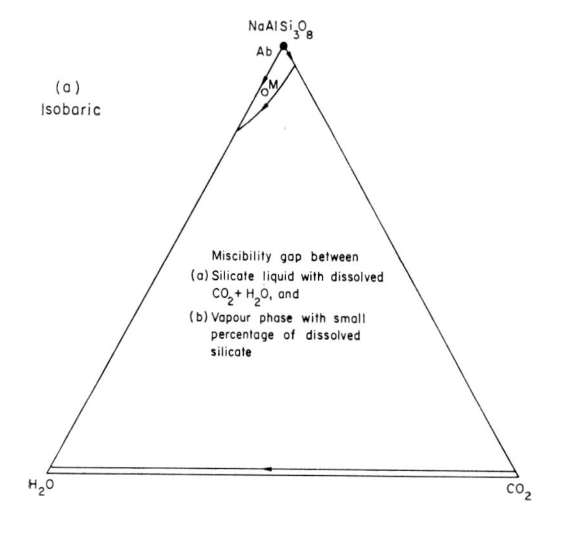

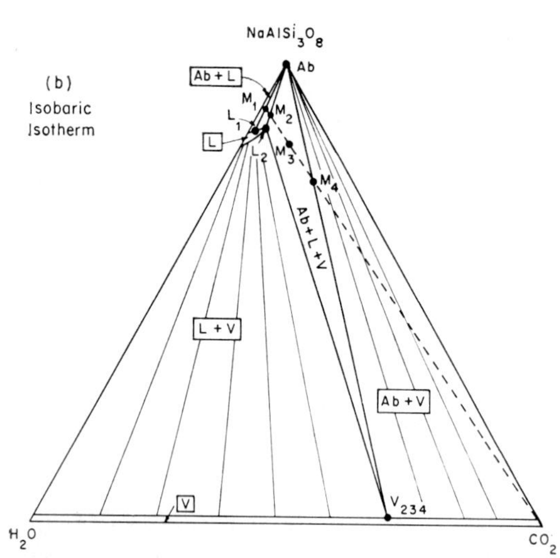

Fig. 2 Schematic phase relationships for the system albite–CO_2–H_2O, based on results of Wyllie and Tuttle (1959, 1960b). Solubility of H_2O–CO_2 in the silicate liquid is exaggerated for clarity

(a) Isobaric equilibrium diagram, with field boundaries for liquids and vapours coexisting with each other and with albite. Note the low solubility of CO_2 in the silicate liquid, compared to that of H_2O.

(b) Isobaric, isothermal section. M_1 represents a magma consisting of about 65 per cent liquid L_1, and 35 per cent albite. Addition of CO_2 to the magma at constant pressure and temperature causes its bulk composition to change along the line M_1–CO_2, with change of liquid composition to L_2, evolution of vapour V_{234}, and complete crystallization of the liquid L_2 by the time the bulk composition reaches M_4.

With further cooling (equilibrium maintained) precipitation of albite and evolution of vapour continue, while the liquid and vapour change composition towards Ab–H_2O, with partition of CO_2 between liquid and vapour continuing to be strongly in favour of the vapour. Finally, the side Ab–V of the three-phase triangle reaches the mixture M, at which temperature the last trace of liquid is used up, and there remains only albite and vapour. From this, we can conclude that a magma containing dissolved CO_2 and H_2O would give off most of the CO_2, along with some H_2O, at an early stage of crystallization, despite the fact that the amount of H_2O present may be well below the amount required for saturation in the system silicate–H_2O.

Fig. 2b is a schematic, isobaric, isothermal section through the system Ab–CO_2–H_2O. Under these conditions, the mixture M_1 consists of liquid L_1 and crystals of albite, with about 65% liquid (by weight). Let this represent a magma coming into contact with a limestone and dissociating the carbonate at least partially. Suppose that this and subsequent processes occur at constant pressure and temperature, for then the effect of the CO_2 released can be seen directly from Fig. 2b. Addition of CO_2 to the magma changes the bulk composition of the system along the line M_1–CO_2. The first CO_2 released dissolves in the liquid, as the bulk composition changes towards M_2, while the liquid changes towards L_2. Addition of more CO_2 moves the bulk composition of the system into the three-phase triangle, and the CO_2 then forms a separate vapour phase. The H_2O in the system has to be distributed between the liquid and vapour phases, in proportions controlled by the attitude of the line L_2–V_{234}. The three phases albite, L_2, and V_{234} coexist as the bulk composition M_3 changes from M_2 to M_4, but addition of CO_2 causes crystallization of the liquid L_2; the last trace of liquid is used up when sufficient CO_2 has been added for the bulk composition to reach M_4.

Thus, addition of CO_2 to magma M_1 causes extraction of H_2O from the liquid, and complete crystallization at constant temperature and pressure. For this process to go to completion in the schematic Fig. 2b a rather large quantity of CO_2 would be required. In nature, of course, such a process would not occur at constant pressure and temperature, and it is difficult to imagine the addition of sufficient CO_2 to change the composition of a body of granitic magma by so large an amount. Nevertheless, it is quite likely that near the margins of a granitic body, and in dykes emplaced in limestones, dehydration and rapid crystallization of the magma could occur in this way. This would offset any tendency of the assimilation process to yield a desilicated magma.

In the available experimental results, limited as they still are, there is no indication that the solubility of CO_2 in granitic and feldspathic liquids could preferentially lower the liquidus temperatures of the feldspar join sufficiently to destroy the thermal barrier (cf. VI.3.3.1). Tilley (1958) considered possible ways for silicate liquids to by-pass the thermal barrier, and he suggested that the incongruent melting relationship of aegirine might make this possible. Reactions involving minerals certainly appear to be more promising than the effects of volatile components on liquidus temperatures, and Bailey and Schairer (1966) have since investigated the liquidus relationships involving albite, quartz, nepheline, hematite, and aegirine; these results are reviewed in V.1 and VI.2

Finally, we have the proposal of Daly (1910) that CO_2 would form compounds with alkalis and transport them upwards (item (3) in section VI.3.2.1., and Shand's elaboration of the process as described in section VI.3.2.3). The available experimental data suggest that CO_2 is only slightly soluble in granitic magmas, and that probably very little solid material dissolves in a CO_2-rich vapour phase coexisting with a granitic magma. Under these conditions, it does not seem likely that CO_2 would be a very effective transport agent. However, the results of Morey and Fleischer (1940) in the system K_2O–SiO_2–CO_2–H_2O indicate that the solubility is greater in a more alkaline system. Koster van Groos and Wyllie (1966, 1968) studied the phase fields intersected by the composition join $NaAlSi_3O_8$–Na_2CO_3–H_2O, and discovered a compositional range characterized by the coexistence of three fluid phases: a silicate liquid containing dissolved Na_2CO_3 and H_2O, a carbonate liquid containing dissolved H_2O and very little silicate, and an aqueous phase containing dissolved CO_2 and other components. If similar phases persist in bulk compositions closer to natural rock systems, then there are many processes possible involving the separation of alkali-rich fluid phases, with their subsequent upward migration and concentration. For this to occur, however, it appears that the liquid must have excess alkali present compared to the feldspar composition. Therefore, there is some experimental evidence that gas or fluid transfer of alkalis could be effective in alkaline magmas (cf. VI.4; VI.5); but this does not help the limestone syntexis hypothesis.

VI.3.4.3. Experimental Data: Subsolidus Decarbonation and Desilication Reactions

It is well established that subsolidus reactions between feldspars and carbonates yield feldspathoids. Schuiling (1964a) described experiments at 1,000 °C at 1 atmosphere, supposedly involving the reaction of calcite or dolomite with feldspars to yield feldspathoids, but which actually involved reactions with CaO because the carbonates would be dissociated at this temperature. Philpotts *et al.* (1967) described a single experiment at 900 °C and 0·72 kbar in which they reacted microcline and dolomite in the presence of excess H_2O to form diopside and kalsilite. This result they compared with kalsilite-bearing assemblages in a sedimentary xenolith within the alkaline gabbro of Brome Mountain. There is a whole series of similar reactions awaiting experimental investigation (see Fig. 4; Watkinson, 1965; Watkinson and Wyllie, 1964, 1969).

Watkinson and Wyllie (1964) questioned Schuiling's interpretation of the experimental data, with specific reference to the difference between the formation of an undersilicated melt and the subsolidus formation of an undersilicated mineral assemblage, and they outlined additional experimental results with more direct bearing on the limestone syntexis hypothesis (see Fig. 3 and

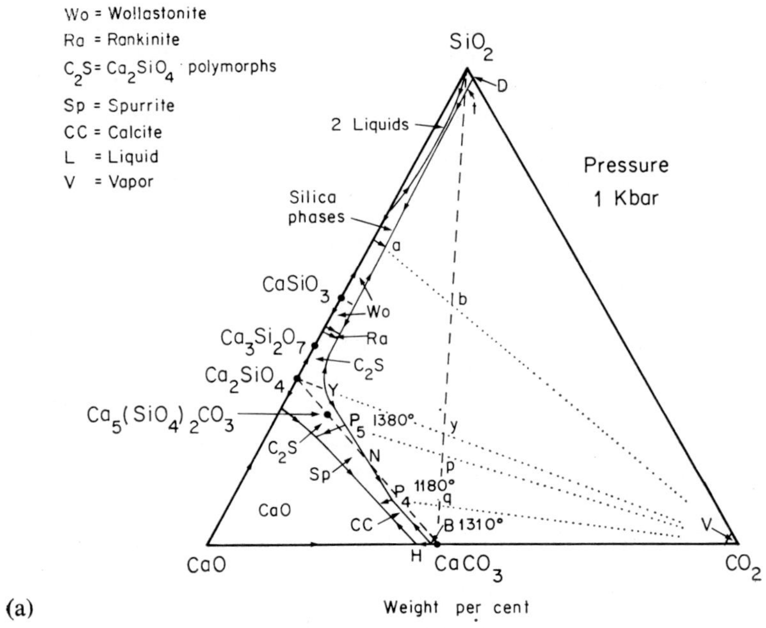

(a)

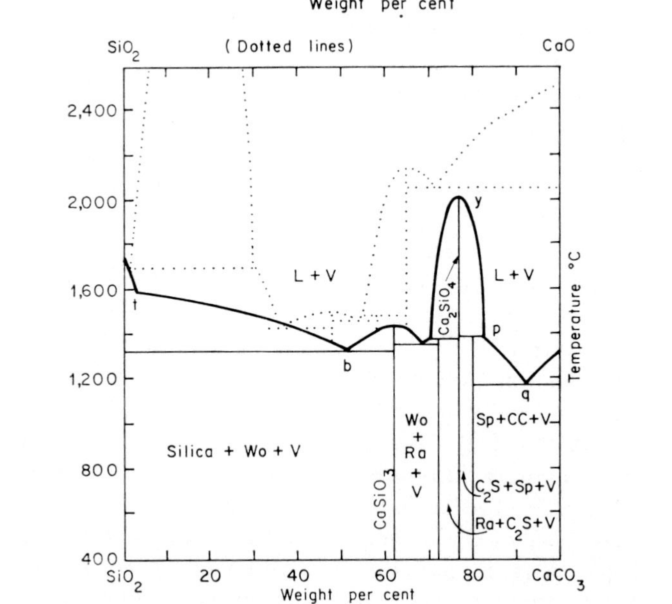

(b)

Fig. 3 (a) Schematic phase relationships at 1 kbar in the system CaO–SiO_2–CO_2 after Wyllie and Haas (1965), based on various sources.

(b) CO_2-saturated liquidus phase relationships from (a) projected on to the join SiO_2–$CaCO_3$. CO_2-vapour is present except for the portion t–SiO_2. This diagram illustrates the temperature variation along the ternay field boundary BNYD of (a). Temperatures between $CaCO_3$ and P_5 are based on experimental data; other temperatures are assumed to be lowered by about 100 °C in the presence of CO_2 at 1 kbar pressure, compared to the system CaO–SiO_2. The binary liquidus is shown for comparison by the dotted line

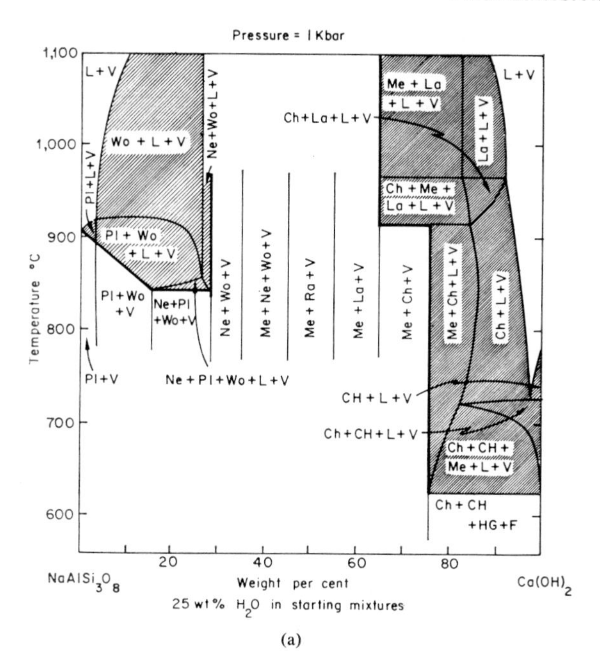

(a)

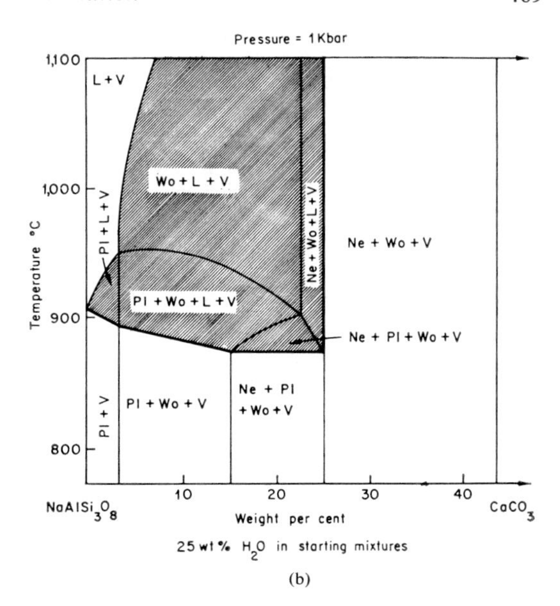

(b)

Fig. 4 Phase fields intersected by the composition join $NaAlSi_3O_8$–$CaCO_3$–$Ca(OH)_2$ with 25 weight per cent H_2O present at 1 kbar pressure (after Watkinson, 1965; Wyllie and Watkinson, 1969). The shaded areas are phase fields for crystals + liquid + vapour

(a) The 25% H_2O section of the join Ab–$Ca(OH)_2$–H_2O. The intersection of the solidus is shown by the heavy line. Note that it rises to high temperatures in the central portion of the section.

(b) The albite-rich portion of the join Ab–$CaCO_3$–H_2O. The phase fields intersected are the same as for (a), although the liquid and vapour phase composition will differ, and temperatures also differ slightly.

Key to phase symbols:
Pl = plagioclase; Wo = wollastonite; Ne = nepheline; Me = melilite; Ra = rankinite; La = = larnite (or other polymorphs of Ca_2SiO_4); Ch = calciochondrodite; CH = portlandite; HG = hydrogarnet; L = liquid; V = vapour; F = fluid; indeterminate between liquid and vapour.

4). Schuiling (1964b), in a brief reply, maintained that he was 'still in favour of the limestone-syntexis theory, or its alternative, sial-syntexis by carbonatite magma.'

VI.3.4.4. Experimental Data: Limestone-Syntexis and Sial-Syntexis

Recent phase equilibrium studies of liquidus relationships in silicate–carbonate systems provide some basis for evaluation of the processes of limestone-syntexis, and sial-syntexis by carbonatite. Wyllie and Haas (1965) described the system CaO–SiO_2–CO_2–H_2O, which is the simplest system involving a silicate melt and the 'synthetic carbonatite magma', but even this is rather complicated for convenient graphical representation. Fortunately, their schematic diagram for the system CaO–SiO_2–CO_2, which is shown in Fig. 3a, serves as a model for the quaternary system, as well as for the more complex systems shown in Fig. 4.

The field boundary BNYD in Fig. 3a gives the compositions of liquids which coexist with one crystalline phase and the vapour V (almost pure CO_2) at 1 kbar pressure. Liquid compositions such as P_4 and P_5 coexist with vapour and two crystalline phases. Fig. 3b shows the vapour-saturated liquidus boundaries projected towards CO_2 on to the composition join SiO_2–$CaCO_3$. The method of projection is shown by the lines *ab*, *Yy*, P_5p, and P_4q. The liquidus temperatures are only estimated between SiO_2 and Ca_2SiO_4, because the effect of CO_2 under pressure on the melting temperatures of the

Ca-silicate compounds has not been determined experimentally, but the results of Wyllie and Tuttle (1959) suggest that the liquidus depression would be relatively small. The liquidus for the system $CaO–SiO_2$ is shown for comparison, by dotted lines. Temperatures between Ca_2SiO_4 and $CaCO_3$ are based on experimental measurements. In Fig. 3, it is assumed that addition of CO_2 closes the liquid miscibility gap of the binary system, to simplify the discussion.

Now consider the effect of $CaCO_3$ on the liquidus temperature of molten silica at constant pressure. About 5 wt. % dissolves, lowering the liquidus to point *t*; solution of additional $CaCO_3$ is accompanied by decarbonation, evolution of CO_2, precipitation of a crystalline silica phase (S), and change in the liquid composition along the saturated liquidus field boundary *ta*. The liquid *a* precipitates wollastonite as well as silica, and continued solution of $CaCO_3$ at constant temperature would cause precipitation of wollastonite alone, while the liquid composition remained at *a*. Solution of $CaCO_3$ in the silica liquid can be examined under various conditions using Fig. 3 as a guide. The striking feature is the high liquidus temperature corresponding to the fusion of Ca_2SiO_4, and this is an effective thermal barrier between the silicate liquids and the carbonate liquids. Note also that the liquidus for wollastonite is a temperature maximum, and this is another thermal barrier, although a minor one compared to that for Ca_2SiO_4. An extrapolated conclusion from this system is that desilication of a silicate magma by limestone assimilation would rarely proceed further than the development of pyroxenes, with a 1:1 ratio of SiO_2: cafemic oxides, and could not reach a composition characterized by a 1:2 ratio of SiO_2 cafemic oxides.

The effect of SiO_2 on the carbonate liquid is also shown in Fig. 3. Solution of SiO_2 until the bulk composition reached *q* would change the liquid composition from B to P_4, with precipitation of calcite and evolution of CO_2. Continued addition of SiO_2 at constant temperature causes precipitation of spurrite until all of the liquid of composition P_4 is used up. The temperature would have to be increased for the liquid to become richer in SiO_2 than P_4. Solution of a few per cent of silica in the carbonate liquid is thus sufficient to cause precipitation of crystalline silicates; much higher temperatures are required to dissolve additional silicates, because the liquidus then climbs towards the thermal barrier corresponding to Ca_2SiO_4. An extrapolated conclusion is that sial-syntexis by carbonatite magmas (cafemic variety) can only produce silicate magmas which are less silicated than a ratio of 1:2 for SiO_2: cafemic oxides.

The same general pattern of liquidus phase relationships persists in the system $CaO–SiO_2–CO_2–H_2O$ (Wyllie and Haas, 1965), with a strongly developed thermal barrier corresponding to the fusion of Ca_2SiO_4 in the presence of $CO_2–H_2O$ mixtures. In fact, the system $CaO–SiO_2–H_2O$ provides a satisfactory model for the quaternary system. This observation permits application of the results in the CO_2-free system shown in Fig. 4a to the corresponding carbonate system of Fig. 4b.

One approach adopted for elucidation of the relationships between carbonate rocks and associated alkaline igneous rocks has been to add 'mineral molecules' such as nepheline, albite, and orthoclase to synthetic carbonatite magmas with compositions on the join $CaCO_3–Ca(OH)_2$ (Watkinson, 1965; Wyllie, 1966). The phase fields intersected by the composition join $CaCO_3–Ca(OH)_2–NaAlSi_3O_8$ in the presence of 25 wt. % H_2O at 1 kbar pressure were determined by Watkinson (1965). This is a triangular slice through a six-component system, and geometrical representation of the phase relationships is difficult. Fortunately, all significant results obtained can be illustrated by the phase fields intersected by the limiting joins Ab–$CaCO_3$ and Ab–$Ca(OH)_2$, in the presence of 25% H_2O, which are illustrated in Fig. 4. These results have been reviewed by Watkinson and Wyllie (1964, 1969).

Fig. 4a shows the phase fields intersected by the CO_2-free join and this provides a satisfactory model for the complete join Ab–$CaCO_3$–$Ca(OH)_2$–25% H_2O. Fig. 4b shows that the same fields are intersected at the albite end of the $CaCO_3$

join. The phase relationships at the $CaCO_3$ end of Fig. 4b were not determined in detail because liquidus temperatures were too high, and the presence of H_2O meant that the liquid compositions had to be represented in terms of $CaCO_3$–$Ca(OH)_2$. The results shown in Fig. 4, which are similar to those obtained using orthoclase (Watkinson, 1965; Watkinson and Wyllie, 1969), show that the system is divided into three parts. At the feldspathic end, solidus temperatures in the presence of H_2O under pressure are moderate, and at the synthetic carbonatite end melting temperatures are even lower. In the central portions, however, solidus temperatures are higher than at either end of the system. Note in Fig. 4a that the liquidus temperature is rising steeply with larnite as the primary phase, and this indicates that the carbonatite liquids are separated from the feldspathic liquids by a thermal barrier corresponding to that shown for Ca_2SiO_4 in Fig. 3b.

The subsolidus central portions of the feldspathic systems intersect a series of phase fields demonstrating the existence of a sequence of subsolidus decarbonation reactions, which cause successive desilication of the mineral assemblages, and in particular the formation of feldspathoids at the expense of feldspar. Note the wide distribution of melilite.

Fig. 4 provides a model for the assimilation of limestone by a feldspathic (syenitic) magma. Consider the addition of lime-rich material (limestone) to the $NaAlSi_3O_8$–H_2O liquid at 1 kbar pressure, with temperature maintained constant at 915 °C. Progressive solution of the limestone under equilibrium conditions causes: (1) precipitation of plagioclase feldspar, (2) precipitation of wollastonite alongside the plagioclase, (3) reaction of plagioclase, leaving wollastonite, (4) the precipitation of nepheline in small quantities through a narrow composition interval, (5) complete crystallization of the silicate liquid leaving wollastonite, nepheline, and vapour, (6) subsolidus decarbonation reactions producing a series of silica-undersaturated assemblages. These results confirm that the series of chemical and mineralogical changes required by the limestone syntexis hypothesis can occur in syenite magmas, in the presence of excess vapour ($H_2O + CO_2$), with the formation of small amounts of undersilicated alkalic liquids capable of precipitating feldspathoids. This only occurs provided that the temperature is maintained at a high level during the assimilation of up to about 20 wt. % of limestone. The same changes would *not* occur in a granitic liquid represented by albite–quartz liquids, because the join albite–H_2O is a thermal barrier separating the granitic liquids from the feldspathoidal liquids (section VI.3.3.1).

$Ca(OH)_2$–H_2O in Fig. 4a provides a model for the synthetic carbonatite magma $CaCO_3$–$Ca(OH)_2$–H_2O, and the figure therefore illustrates the effect of sial-syntexis by a carbonatite magma. Less than 5 wt. % albite dissolves in the Ca-rich liquid, with depression of liquidus temperatures, before primary silicate phases appear and the liquidus temperatures rise sharply. Continued addition of albite causes precipitation of silicate minerals with little change in liquid composition; with addition of more than 20 wt. % albite below 900 °C, or 30% above 900 °C, the liquid is used up completely leaving mineral assemblages including calcichondrodite, melilite, and larnite, with calcite and spurrite when CO_2 is present (see the system CaO–SiO_2–CO_2–H_2O; Wyllie and Haas, 1965). It appears that the composition of a carbonatite magma (cafemic variety) assimilating feldspathic or granitic material is limited to low silica contents by the thermal barrier for the orthosilicate ratio of 1:2 for SiO_2:cafemic oxides. There is no evidence in these experiments to suggest that this kind of process could produce either alkaline silicate liquids, or feldspathoidal rocks.

The results for the corresponding join nepheline–$CaCO_3$–$Ca(OH)_2$–H_2O (Watkinson, 1965; Watkinson and Wyllie, 1965) are quite different from those for the feldspathic joins. There is no thermal barrier, and fractional crystallization of a nepheline-rich magma yields a residual carbonatite magma. The feldspathoidal liquids are thus capable of dissolving large quantities of limestone, provided the temperature can be maintained. Similarly, the carbonatite magmas are capable of dissolving large quantities of feldspathoidal rocks.

VI.3.5. Discussion

The limestone syntexis hypothesis still has enthusiastic supporters, but the realization that the 'limestones' at so many of the carbonate–alkaline rock associations listed by Daly (1910) are in fact carbonatites has led most petrologists working with carbonatites to abandon the hypothesis. Tilley (1958) concluded that 'Though sweeping claims for the origin of alkali rocks by limestone syntexis continue to be made, demonstrable field evidence that feldspathoidal assemblages have been so generated is confined to contact zones of basic and intermediate rocks (Tilley, 1952)'.

The experimental evidence reviewed in this contribution confirms the persistence of a thermal barrier between SiO_2-oversaturated liquids and liquids capable of precipitating feldspathoids. This precludes the generation of alkaline magmas by desilication processes involving crystal–liquid reactions in granitic magmas. Other experimental data indicate that transfer of alkalis in vapour or fluid phases could be an effective process given an alkaline magma to begin with, but it is less likely to be significant in subalkaline magmas (cf. VI.5). There is no experimental evidence to indicate that assimilation of sialic crustal rocks by a cafemic carbonatite magma could produce a feldspathoidal magma; excess alkalis would be required.

Assimilation of limestone requires a considerable amount of heat, which is not contained as superheat in a subalkaline magma, and there is experimental evidence that the release of CO_2 contributes to the crystallization of the magma (Fig. 2). For several different reasons, based on different sets of experiments, it appears that assimilation of limestone by subalkaline magma must lead to crystallization of the magma, with concomitant development of some lime-silicates and the formation of slightly undersilicated rocks only locally, in the vicinity of the limestone undergoing assimilation. This is confirmed by field evidence, and this process must be clearly distinguished from the generation of feldspathoidal magmas capable of subsequent differentiation which is required by the limestone assimilation hypothesis.

I conclude that there is no definitive field evidence for the limestone syntexis hypothesis for the generation of feldspathoidal magmas from subalkaline magmas, and that there is considerable experimental evidence against the hypothesis. Readers are referred to other contributions in this volume for alternative petrogenetic schemes.

ACKNOWLEDGEMENT

My work in this topic has been supported by the National Science Foundation.

VI.3. REFERENCES

Agard, J., 1956. Les gites minéraux associés aux roches carbonatites. *Sci. de la Terre, Univ. Nancy*, **4**, 105–51.

Bailey, D. K., and Schairer, J. F., 1966. The system Na_2O–Al_2O_3–Fe_2O_3–SiO_2 at 1 atmosphere, and the petrogenesis of alkaline rocks. *J. Petrology*, **7**, 114–70.

Boettcher, A. L., and Wyllie, P. J., 1968. Jadeite stability measured in the presence of silicate liquids in the system $NaAlSiO_4$–SiO_2–H_2O. *Geochim. Cosmochim. Acta*, **32**, 999–1012.

Boettcher, A. L., and Wyllie, P. J., 1969. Phase relationships in the system $NaAlSiO_4$–SiO_2–H_2O to 35 kbars pressure. *Am. J. Sci.*, **267**, 875–909.

Bowen, N. L., 1922. The behavior of inclusions in igneous magmas. *J. Geol.*, **30**, 513–70.

Bowen, N. L., 1928. *The Evolution of the Igneous Rocks*, Princeton University Press. Reprinted by Dover Publications, Inc., 1956.

Bowen, N. L., 1937. Recent high-temperature research on silicates and its significance in igneous geology. *Am. J. Sci.*, **33**, 1–21.

Bowen, N. L., and Tuttle, O. F., 1950. The system $NaAlSi_3O_8$–$KAlSi_3O_8$–H_2O. *J. Geol.*, **58**, 489–511.

Chayes, F., 1942. Alkaline and carbonate intrusives near Bancroft, Ontario. *Bull. geol. Soc. Am.*, **53**, 449–512.

Daly, R. A., 1910. Origin of alkaline rocks. *Bull. geol. Soc. Am.*, **21**, 87–118.

Daly, R. A., 1933. *Igneous Rocks and the Depths of the Earth*. McGraw-Hill, New York.

Deines, P., 1968. The carbon and isotopic composition of carbonates from a mica peridotite dike near Dixonville, Pennsylvania. *Geochim. Cosmochim. Acta*, **32**, 613–25.

Eckermann, H. von, 1948. The alkaline district of Alnö Island. *Sverig. geol. Unders. Ser. Ca*, **36**, 176 pp.

Fudali, R. F., 1963. Experimental studies bearing on the origin of pseudoleucite and associated problems of alkalic rock systems. *Bull. geol. Soc. Am.*, **74**, 1101–26.

Hamilton, D. L., and Mackenzie, W. S., 1965. Phase equilibrium studies in the system $NaAlSiO_4$ (nepheline)–$KAlSiO_4$ (kalsilite)–SiO_2–H_2O. *Mineral. Mag.*, **34**, 214–31.

Harker, A., 1909. *The Natural History of Igneous Rocks*. reprinted in 1965 by Hafner Publishing Co., New York. 384 pp.

Heinrich, E. W., 1966. *The Geology of Carbonatites*. Rand McNally and Co., Chicago. 607 pp.

Holmes, A., 1950. Petrogenesis of katungite and its associates. *Am. Miner.*, **35**, 772–92.

Koster van Groos, A. F., and Wyllie, P. J., 1966. Liquid immiscibility in the system Na_2O–Al_2O_3–SiO_2–CO_2 at pressures to 1 kilobar. *Am. J. Sci.*, **264**, 234–55.

Koster van Groos, A. F., and Wyllie, P. J., 1968. Liquid immiscibility in the join $NaAlSi_3O_8$–Na_2CO_3–H_2O and its bearing on the genesis of carbonatites. *Am. J. Sci.*, **266**, 932–67.

Morey, G. W., and Fleischer, M., 1940. Equilibrium between vapor and liquid phases in the system CO_2–H_2O–K_2O–SiO_2. *Bull. geol. Soc. Am.*, **51**, 1035–58.

Naidu, P. R. J., 1966. Editor of *Papers and Proceedings, 4th general meeting International Mineralogical Association, I.M.A. vol.* Mineral Soc. India, 252 pp.

Pecora, W. T., 1956. Carbonatites: A review. *Bull. geol. Soc. Am.*, **67**, 1537–56.

Philpotts, A. R., Pattison, E. F., and Fox, J. S., 1967. Kalsilite, diopside and melilite in a sedimentary xenolith from Brome Mountain, Quebec. *Nature*, **214**, 1322–3.

Roedder, E., 1965. Liquid CO_2 inclusions in olivine-bearing nodules and phenocrysts from basalts. *Am. Miner.*, **50**, 1746–82.

Schairer, J. F., 1950. The alkali–feldspar join in the system $NaAlSiO_4$–$KAlSiO_4$–SiO_2. *J. Geol.*, **58**, 512–17.

Schairer, J. F., and Bowen, N. L., 1935. Preliminary report on equilibrium relations between feldspars, and silica. *Trans. Am. Geophys. Un.*, 16th annual meeting, 325–8.

Schuiling, R. D., 1964a. Dry synthesis of feldspathoids by feldspar–carbonate reactions. *Nature*, **201**, 1115.

Schuiling, R. D., 1964b. The limestone assimilation hypothesis. *Nature*, **204**, 1054–5.

Shand, S. J., 1930. Limestone and the origin of feldspathoidal rocks: an aftermath of the Geological Congress. *Geol. Mag.*, **67**, 415–27.

Shand, S. J., 1943. *Eruptive Rocks*. John Wiley and Sons, New York. Second edition, 444 pp.

Shand, S. J., 1945. The present status of Daly's hypothesis of the alkaline rocks. *Am. J. Sci.*, **243–A**, 495–507.

Smith, W. Campbell, 1956. A review of some problems of African carbonatites. *Q. J. geol. Soc. Lond.*, **112**, 189–219.

Stansfield, J., 1923. Extensions of the Monteregian petrographical province to the west and northwest. *Geol. Mag.*, **60**, 433–53.

Stansfield, J., 1928. *Assimilation and Petrogenesis: Separation of Ores from Magmas*. Valley Publishing Co., Urbana, Illinois, 197 pp.

Tilley, C. E., 1952. Some trends of basaltic magma in limestone syntexis. *Am. J. Sci.*, Bowen volume, 529–45.

Tilley, C. E., 1958. Problems of alkali rock genesis. Quartz. *J. geol. Soc. Lond.*, **113**, 323–60.

Turner, F. J., and Verhoogen, J., 1960. *Igneous and Metamorphic Petrology*. 2nd Edition, McGraw–Hill, New York, 694 pp.

Watkinson, D. H., 1965. *Melting relations in parts of the system Na_2O–K_2O–CaO–Al_2O_3–SiO_2–CO_2–H_2O with applications to carbonate and alkalic rocks*. Ph.D. thesis, The Pennsylvania State University.

Watkinson, D. H., and Wyllie, P. J., 1964. The limestone assimilation hypothesis. *Nature*, **204**, 1053–4.

Watkinson, D. H., and Wyllie, P. J., 1965. Phase-relations in the join $NaAlSiO_4$–$CaCO_3$–(25%) H_2O and the origin of some carbonatite–alkalic rock complexes (abstract). *Can. Miner.*, **8**, 402–3.

Watkinson, D. H., and Wyllie, P. J., 1969. Phase equilibrium studies bearing on the limestone assimilation hypothesis. *Bull. geol. Soc. Am.*, **80**, 1565–76.

Wyllie, P. J., 1966, 'Experimental studies of carbonatite problems: the origin and differentiation of carbonatite magmas', in Tuttle, O. F., and Gittins, J., Eds., *Carbonatites*, 311–52. Interscience-Wiley Publishers, New York, 591 pp.

Wyllie, P. J., and Boettcher, A. L., 1969. Liquidus phase relationships in the system CaO–CO_2–H_2O to 40 kilobars pressure, with petrological applications. *Am. J. Sci.*, Schairer vol. **267A**, 489–508.

Wyllie, P. J., and Haas, J. L., 1965. The system CaO–SiO_2–CO_2–H_2O: I. Melting relationships with excess vapor at 1 kilobar pressure. *Geochim. Cosmochim. Acta*, **29**, 871–92.

Wyllie, P. J., and Tuttle, O. F., 1959. Effect of carbon dioxide on the melting of granite and feldspars. *Am. J. Sci.*, **257**, 648–55.

Wyllie, P. J., and Tuttle, O. F., 1960a. The system $CaO–CO_2–H_2O$ and the origin of carbonatites. *J. Petrology*, **1**, 1–46.

Wyllie, P. J., and Tuttle, O. F., 1960b. Experimental investigation of silicate systems containing two volatile components. Part I. Geometrical considerations. *Am. J. Sci.*, **258**, 498–517.

Yoder, H. S., and Tilley, C. E., 1962. Origin of basalt magmas: an experimental study of natural and synthetic rock systems. *J. Petrology*, **3**, 342–532.

VI.4. RÔLE OF VOLATILES

L. N. Kogarko

VI.4.1. Introduction

In petrological hypotheses bearing on the genesis of alkaline rocks much attention is paid to the rôle of volatile components. According to Smyth (1927) alkaline rocks are products of differentiation of basalt magma under conditions of retention and accumulation of fugitive constituents up to the latest stages of this process. This mechanism may operate only in tectonically quiet zones. At the latest stages of basalt magma differentiation residual liquids saturated with respect to volatile components and alkalis may separate from the subalkaline solid phases and lead to the formation of alkaline rocks. Bowen (1928) discussing the crystallization of granitic melts arrived at a similar conclusion. Fenner (1926) and Gilson (1928) suggested that alkaline rocks originate as the result of accumulation of gaseous emanations in apical parts of completely or partially crystallized intrusions. During the metasomatic action of these emanations minerals undersaturated with respect to silica are formed: nepheline, sodalite, cancrinite, etc. Zavaritsky (1937) explained the origin of alkaline rocks in a similar way, though he assumed that the transfer of volatile components took place in the magmatic melt, and not in a gaseous phase. The possibility of transportation of fugitive constituents alongside with alkalis was recognized by Shand (1945) in his assimilation hypothesis.

Kennedy (1955) proposed, on the basis of the data of Morey and Hesselgesser (1952) on silicate–water equilibria, that alkaline rocks are formed by conjugate diffusion of volatile components and alkalis under conditions of variable temperature and pressure.

Using the calculations carried out by Verhoogen (1949) on the distribution of components in the gravitational field, Barth (1961) assumed, that alkali basalts and some alkalic lavas may have formed during diffusional differentiation of tholeiitic magma. Alkalis and volatiles migrate in the melt to the upper parts of magma chambers of great vertical extension.

All these hypotheses consider the idea of a close association of volatiles and alkalis in the processes of crystallization, gaseous transport, diffusional differentiation, etc. However, the various investigators are not unanimous concerning the aggregate state of the volatile components. One group of geologists relates the activity of volatiles with the regime of magmatic gaseous emanations while the other group considers the volatiles as ordinary components capable of migration in a condensed phase—the melt.

When discussing the rôle of volatile components we have consequently to consider the following problems: (1) the conditions of separation of volatiles into a gas phase from magmatic melts of diverse chemistry; and the cause of association of alkalis and volatile components; (2) the influence of fugitive constituents upon the properties of magmatic melts; (3) the rôle of volatiles in the origin of magmas of high alkalinity.

VI.4.2. Aggregate State of Volatile Components of Alkaline Magmas

It is generally recognized that within the range of temperatures of crystallization of magmas the volatile components are characterized by high vapour pressures. The processes in which volatiles participate depend therefore to a large extent upon pressure which in this case is an important geological factor.

At equal external pressure crystallization in different silicate–volatile systems may proceed with or without separation of a gas phase depending on the maximum pressure of univariant equilibria of the type melt + crystals + vapour. The magnitude of the maximum pressure of a saturated melt is a function of the chemistry of the magmatic system: the higher the mutual solubility of melt and gas phase, the lower is the extreme pressure of the univariant equilibrium crystals + melt + vapour.

VI.4.2.1. Water

According to the majority of petrologists water is a predominant component of magmatic gases. The separation of a gas phase depends on the solubility of water in the magma. A number of investigators have demonstrated that the solubility of water in alumino–silicate melts increases with the rise in mole fractions of alkaline cations (Ostrovsky, Orlova and Rudnitskaya, 1964). The experiments of Bowen and Tuttle (1950) showed that the solubility of water in a syenitic melt at $T = 854$ °C and $P_{H_2O} = 1000$ kg/cm^2 is equal to 7·5% H_2O which is much higher than the solubility of water in a granitic melt (2·79%) at $T =$ 900 °C, $P_{H_2O} =$ 1000 kg/cm^2 (Khitarov *et al.*, 1959). The solubility of water in molten nepheline syenite is likely to be still higher. The considerable mutual solubility of water vapour and alkaline silicate melts suggests that the limiting pressures of equilibria involving crystals + melt + vapour must be notably lower for alkalic magmatic systems than for normal magmas. This may in the case of alkaline rocks result in the gradual change of a magmatic melt into a hydrothermal solution. This is in accordance with the experimental study of the system K_2O–Na_2O–Al_2O_3–SiO_2–H_2O (Tuttle, 1961) which has shown that during the crystallization of alkalic alumino–silicate melts with an excess of alkalis over alumina the gradual change from melt into aqueous liquid solution is possible at comparatively low pressure without the separation of a gas phase. Rocks with normative sodium silicate, e.g. alkali granites, may probably crystallize in a similar way (Tuttle, 1961).

The crystallization of a part of the alkaline rocks, such as the agpaitic nepheline syenites, is likely to follow this scheme, because these rocks are characterized by a significant excess of alkalis over aluminium. This excess is still higher at the end of the process due to the agpaitic order of mineral separation (Kogarko and Gulyaeva, 1965). The crystallization of alkaline rocks which are not oversaturated with respect to alkalis (agpaitic index less than one) must have proceeded with the separation of an aqueous vapour phase.

VI.4.2.2. Fluorine, Chlorine, Sulphur and Carbon Dioxide

The relationship between the contents of fluorine, chlorine and sulphur in water vapour and the composition of the condensed phase in equilibrium with the vapour phase has been examined by Kogarko and Ryabchikov (1961). For this purpose exchange equilibria involving these components were computed thermodynamically. These calculations pretend by no means to describe complicated natural processes quantitatively. They allow, however, to establish the general tendencies in the shift of equilibria with changes in magma composition. The equilibrium constants of the following reactions were calculated at $T = 1000°$ K and $P = 1$ and 1000 atm (Figs. 1–3):

(1). $MeF + 1/2\ H_2O = 1/2\ Me_2O + HF$;

$$K_{I} = \frac{[HF]\ [Me_2O]^{1/2}}{[MeF]\ [H_2O]^{1/2}},$$

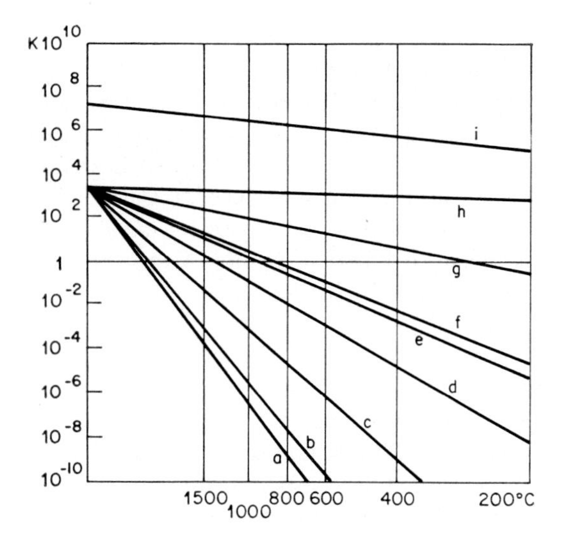

Fig. 1 Equilibrium constants for the following reactions:

(a) $KF + 1/2H_2O = 1/2K_2O + HF$;
(b) $NaF + 1/2H_2O = 1/2Na_2O + HF$;
(c) $1/2CaF_2 + 1/2H_2O = 1/2CaO + HF$;
(d) $1/2MgF_2 + 1/2H_2O = 1/2MgO + HF$;
(e) $1/2FeF_2 + 1/2H_2O = 1/2FeO + HF$;
(f) $1/3AlF_3 + 1/2H_2O = 1/6Al_2O_3 + HF$;
(g) $1/2BeF_2 + 1/2H_2O = 1/2BeO + HF$;
(h) $1/4TiF_4 + 1/2H_2O = 1/4TiO_2 + HF$;
(i) $1/4SiF_4 + 1/2H_2O = 1/4SiO_2 + HF$

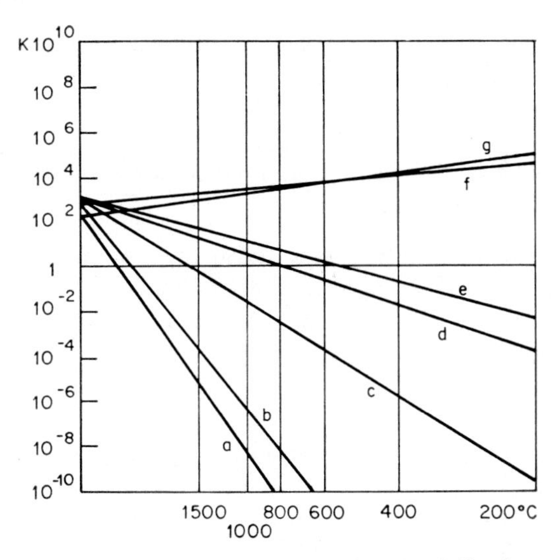

Fig. 2 Equilibrium constants for the following reactions:

(a) $KCl + 1/2H_2O = 1/2K_2O + HCl$;
(b) $NaCl + 1/2H_2O = 1/2Na_2O + HCl$;
(c) $1/2CaCl_2 + 1/2H_2O = 1/2CaO + HCl$;
(d) $1/2MgCl_2 + 1/2H_2O = 1/2\,MgO + HCl$;
(e) $1/2FeCl_2 + 1/2H_2O = 1/2FeO + HCl$;
(f) $1/3AlCl_3 + 1/2H_2O = 1/6Al_2O_3 + HCl$;
(g) $1/4SiCl_4 + 1/2H_2O = 1/4SiO_2 + HCl$

(2). $MeCl + 1/2\,H_2O = 1/2\,Me_2O + HCl$;

$$K_{II} = \frac{[HCl]\,[Me_2O]^{1/2}}{[MeCl]\,[H_2O]^{1/2}},$$

(3). $4MeF + SiO_2 = SiF_4 + 2Me_2O$;

$$K_{III} = \frac{[SiF_4]\,[Me_2O]^2}{[MeF]^4\,[SiO_2]},$$

where Me are various rock-forming elements (K, Na, Ca, Mg, etc.) and K are equilibrium constants. The values in square brackets are the activities of the components in the corresponding phases. The activities of H_2O, HCl, HF and SiF_4 are those of the gas phase, while those of all the remaining compounds relate to the condensed phases—melt or crystals. Equilibrium constants at a total pressure of 1 atm were calculated at various temperatures from the formula

$$-RT\ln K = \Delta G$$

(ΔG is the change in Gibbs free energy). As is evident from Figs. 1–3 the minimum yield of halogens into the gas phase is found in systems with alkali ions. Increasing concentrations of less basic cations leads to a significant increase in the content of halogens in the gas phase. The calculation of similar equilibria with silicates demonstrated that increasing amounts of SiO_2 leads to an intensification of the separation of halogens into the gas phase (Kogarko and Ryabchikov, 1961).

In order to approach natural conditions more closely and to investigate the action of melt composition on the content of halogens in the vapour phase at equilibrium conditions, the magnitudes of the activity of sodium oxide in melts close in composition to granite ($\lg a_{Na_2O} = -14{\cdot}06$), nepheline syenite ($\lg a_{Na_2O} = -12{\cdot}45$), and agpaitic nepheline syenite ($\lg a_{Na_2O} = -11{\cdot}84$) have been estimated by means of the EMF method by Evtiukhina *et al.* (1967). The calculation of the equilibrium P_{HF} in the vapour phase

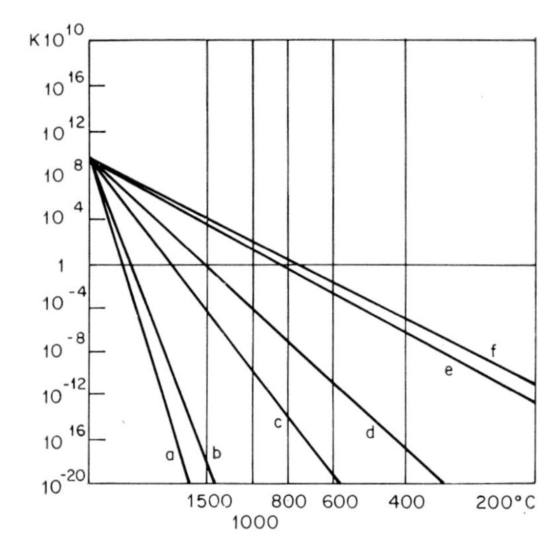

Fig. 3 Equilibrium constants for the following reactions:

(a) $4KF + SiO_2 = 2K_2O + SiF_4$;
(b) $4NaF + SiO_2 = 2Na_2O + SiF_4$;
(c) $2CaF_2 + SiO_2 = 2CaO + SiF_4$;
(d) $2MgF_2 + SiO_2 = 2MgO + SiF_4$;
(e) $2FeF_2 + SiO_2 = 2FeO + SiF_4$;
(f) $4/3AlF_3 + SiO_2 = 2/3Al_2O_3 + SiF_4$

was based on the consideration of the hydrolysis reaction: $2NaF + H_2O = Na_2O + 2HF$ at $T = 1050°K$. Other conditions being equal the HF/H_2O ratio would fall from granite to agpaitic nepheline syenite proportionally to the square root of a_{Na_2O}. Taking the P_{H_2O} of the magmatic gas phase to be equal to 1000 atm, the partial pressure of HF in equilibrium with the granitic melt was found to be 2·76 atm, that in equilibrium with nepheline syenite to be 0·4 atm, but only 0·16 atm for the agpaitic nepheline syenite (Evtiukhina *et al.*, 1967). It should be pointed out that the value for the granitic melt is less precise.

Analogous results were obtained during the investigation of the forms of fluorine separation from melts in the system Na_2O–SiO_2–NaF, in which the separation of fluorine increases regularly with the rise in silica content and the decrease of sodium oxide in the melt (Kogarko *et al.*, 1968).

The calculation of hydrolysis constants for sulphides ($MeS + H_2O = MeO + H_2S$) shows that other conditions being equal the activity of H_2S in a gas phase decreases with the rise in alkalinity of the condensed phases (Kogarko, 1970). It is proven experimentally that the capacity of slags to extract sulphur increases markedly with rising activities of strong bases (Kozheurov, 1959).

Niggli (1937) studying the systems K_2O–SiO_2–CO_2 and Na_2O–SiO_2–CO_2 at P_{CO_2} = const. established that the concentration of CO_2 in a silicate melt increases with rising content of alkalis. Considerably higher solubilities of volatile components in silicate melts enriched in alkalis is evident from silicate–fluorine, silicate–chloride, silicate–carbonate, and silicate–sulphide systems displaying immiscibility in the liquid state. The concentrations of F, Cl, S and CO_2 are always much higher in the more alkalic of two immiscible melts. On going to systems with more basic cations the partition of volatile components (enrichment in the more basic melt) increases. Thus, in the system MgF_2–SiO_2–Al_2O_3 the concentration of F in the more basic melt (with the higher Mg/Si ratio) is approximately 26 times higher than that in the more acid liquid. For the system CaF_2–SiO_2–Al_2O_3 this value approaches 40, and for the system SrF_2–SiO_2–Al_2O_3, 45 (in each case the more acid melt contained about 2% metal fluoride). These calculations are based upon experimental data of Ershova (1957). Taking into account that the activities of the anions F^-, Cl^-, S^{2-} and $CO_3{}^{2-}$ must be equal in two coexisting liquids at equilibrium one may conclude that in the more siliceous melts the activity coefficients of these anions must be higher than in more basic melts.

Thus, the above calculations and the survey of experimental data for systems with H_2O, F, Cl, S and CO_2 attest to the high capacity of alkaline melts to dissolve and retain volatile components, which leads to a decrease in their concentration in the coexisting gas phase. This is probably the reason for the absence of greisenization phenomena and pneumatolytic mobilization of ore and volatile components in alkaline rocks (these phenomena are usually associated with the most acid rocks—alaskitic granites). This statement is especially well supported by the nepheline syenites of Lovozero where the mobilization of ore and volatile components to the endo- and exocontact zones is not typical in spite of high average abundances. Minerals containing volatile components (villiaumite, sodalite, alkali amphibole, etc) are present in the rocks as accessory or rock-forming primary minerals. The accumulation of volatile components in the melt and not in the gas phase results in the parallel behaviour of fugitive constituents, e.g. F and Cl and 'nonvolatile' lithophilic elements (Na, Si, Fe) which is shown by the close correlation between halogens and petrogenic elements (Kogarko and Gulyaeva, 1965).

The increase in the solubility of volatile components with rising alkalinity of silicate melts is consistent with the distribution of these elements in igneous rocks. The data on the geochemistry of the alkaline rocks show their significant enrichment in volatiles. Of all types of igneous rocks syenites and nepheline syenites are characterized by the highest abundances of fluorine and chlorine (Turekian and Wedepohl, 1961, Gerasimovsky, 1963) reaching up to 0·14 and 0·16% by weight, respectively, in agpaitic nepheline syenites. The latter rocks also display the highest concentrations of sulphur—up to 0·24% recalculated as SO_3 (cf. Gerasimovsky, 1969). Carbonatites which concentrate enormous amounts of CO_2 are also related to alkaline complexes (the ultramafic alkaline association). The extensive development of water-containing minerals, such as amphiboles, analcime, natrolite, etc., in alkaline rocks attests to the high concentration of water in alkaline magmas. For instance, according to Daly (1914), the content of water in teschenites is 4·92% by weight.

VI.4.2.3. Migration of Alkalis and Rare Metals

The accumulation of volatiles in alkaline rocks results, on the one hand, from the fact that this type of rock is formed at the latest stages of crystallization differentiation and bears a residual character. On the other hand, any processes leading to the formation of alkaline rocks apparently included equilibria of the type crystals + liquid + gas. During these processes the alkaline liquids, characterized by high capacity to absorb F, Cl, H_2O, etc., so to say 'extracted' these components from the other phases which were in equilibrium with the melt. A significant lowering of the activities of the volatile components with the rise in alkalinity of the liquid phase caused the conjugate migration of alkalis with volatiles in various geological processes (thermo-diffusion, diffusion in the gravitational field, under the gradient of pressure, etc.).

Analogous views have been proposed concerning the migration of some rare metals (Zr, Nb, RE, etc) alongside with alkalis and volatile components. The existence of strong complexes of rare metals with volatiles in silicate melts or in the gaseous phase have been assumed. However, the presence of such complexes in alkaline melts is doubtful; on the contrary a high alkalinity must result in the decomposition of these complexes and increase the stability of oxygen complexes of the rare elements. For example it has been demonstrated experimentally that the alumino–fluoride complex AlF_6^{3-} decomposes in silicate melts oversaturated with respect to alkalis, instead aluminium tends to be linked with oxygen while sodium forms bonds with fluorine (Kogarko, 1966). Because of this fact cryolite is unstable in agpaitic nepheline syenites, and in spite of the high concentrations of Na, Al, and F, villiaumite is formed instead (Kogarko, 1966). In less alkalic silicate melts the complexing of rare elements with halogens may be expected. The absence of experimental data precludes a detailed discussion of this problem. The existence of fluoride, chloride, and carbonate complexes of rare metals in aqueous solutions at elevated temperatures and pressures and high pH values has been demonstrated experimentally (Beus and Sobolev, 1962).

From the extensive geological literature on metasomatic and hydrothermal deposits related to alkaline rocks the close association of minerals containing rare metals with minerals concentrating volatiles (F, Cl, CO_2, etc.) is well known. From this we may conclude that the rôle of volatiles in the transport of ore components was rather important in hydrothermal and metasomatic processes genetically related to alkaline rocks.

VI.4.2.4. Rôle of Organic Compounds

It has been proposed recently that the ore elements were transported in aqueous solutions in the form of organic and silicon–organic complexes (Germanov, 1961; Ermolaev, 1963). One is tempted to apply these views to alkaline rocks considerably enriched in hydrocarbons—gases and bitumens. Agpaitic nepheline syenites are especially rich in hydrocarbon gases (from 4·05 to 60·10 cm^3 per kg rock) (Petersilie, 1964; Petersilie and Sørensen, 1970). The gas fraction of these rocks is dominated by methane and in some cases by hydrogen. Carbon monoxide and carbon dioxide, heavy hydrocarbons, and helium are of subordinate significance.

The composition of such gas phases is determined by the equilibria in the system C–O–H which are primarily governed by the partial pressure of oxygen. According to the experimental data of Bailey (1969) on the stability of aegirine and arfvedsonite one may assume that the fugacity of oxygen in agpaitic nepheline syenite magmas corresponds approximately to that of the buffer QFM. The calculation of the equilibrium constants of the reactions $C+2H_2 = CH_4$; $H_2+1/2O_2=H_2O$; $C+O_2=CO_2$; $2CO+O_2=2CO_2$ (Kogarko, 1970) at 525, 400 and 300 °C and at water vapour pressures of 1000, 200 and 100 atm demonstrated that the vapour phase of agpaitic nepheline syenites should be characterized by a very high fugacity of methane which rises with decreasing temperature ($\lg fCH_4$ (525 °C) = 2·54; $\lg fCH_4$(400 °C) = 3·07; $\lg fCH_4$ (300 °C) = 5·60). The fugacity of hydrogen also increases while the fugacities of the oxidized forms of carbon (CO and CO_2) are markedly reduced ($\lg fH_2$ (525 °C) = 1·27; $\lg fH_2$ (400 °C) = 0·94; $\lg fH_2$ (300 °C) = 1·64; $\lg fCO_2$ (525 °C) = 1·99; $\lg fCO_2$(400 °C) = 0·51; $\lg fCO_2$ (300 °C) = − 1·28; $\lg fCO$ (525 °C) = − 0·02; $\lg fCO$ (400 °C) = − 1·81; $\lg fCO$ (300 °C) = − 3·90).

The comparison of the results of these calculations with the data on the composition of the gas phases of the Khibina, Lovozero and Ilímaussaq massifs shows that gas phases of this composition are very likely to be products of the evolution of nepheline syenites and formed under conditions of relatively low temperatures.

VI.4.3. Influence of Volatiles on Properties of Magmas

The high concentrations of H_2O, Cl, F, CO_2 and other 'mineralizers' in alkaline rocks and their retention in the melts must lead to noticeable changes in the physico-chemical properties of alkaline magmas, such as reduction of viscosity, lowering of crystallization temperatures and increase of rates of diffusion of components from a theoretical point of view. These problems have been discussed by Grigoriev (1935), Buerger, (1948), Belov (1950), and others.

VI.4.3.1. Temperature

The works of Volarovich *et al.* (1939), Winkler (1947) and Nikol'skaya (1955) demonstrated that the introduction of CaF_2 and LiF into melts of nepheline and diopside compositions reduces the melting temperatures of these minerals by several tens of degrees. Introduction of water leads to similar results and may, for example, lower the melting temperature of nepheline from 1548 °C under dry conditions to 1050 °C at P_{H_2O} = 2000 atm (Yoder, 1958). Wyllie and Tuttle (1960) established that the addition of HF lowers the melting temperatures of albite and granite even more than water does. These authors calculated on the basis of their experimental data that in a granite composed of 96% silicates by weight and 4% pore solution and having a total of 0·16% fluorine, this small concentration of F will reduce the beginning of melting from 635 °C (in case of granite plus pure water) to 570 °C. High concentrations of volatile components in agpaitic nepheline syenites may have lowered their crystallization temperatures to 400 °C according to Sørensen (1962). Based upon rather approximate calculations of the relative amounts of residual melt and earlier precipitated alumino-silicate phases, we have shown that the concentration of fluorine may in the course of crystallization of foyaite from the Lovozero alkaline massif have increased to 10–15% by weight. This amount of fluorine may lower the temperature of crystallization of the residual melt by 150–200 °C (calculated from the melting diagrams of silicate–fluoride systems). It may lead to a noticeable increase in the crystallization range of these rocks.

Experimental data on the melting of alkaline rocks have demonstrated a correlation between the relative amounts of minerals concentrating volatiles and the extension of the melting ranges of the rocks (Piotrowski and Edgar, 1970; Sood and Edgar, 1970).

The significant reductions of the melting temperatures of rocks caused by the volatile components is probably of great importance for palingenesis, which is likely to be the major process in the genesis of alkali granites. The presence of volatile mineralizers helps to solve the problem of heat sources for the selective melting of anchieutectic magmas in the Earth's crust.

VI.4.3.2. Viscosity

Another very important property of silicate melts, viscosity, also changes markedly by the addition of volatile components, especially fluorine. It is established experimentally that the introduction of 13% of CaF_2 into a dioritic melt reduces its viscosity 17 times at 1200 °C (Volarovich *et al.*, 1939). Valiashikhina (1953) found that a maximum decrease of viscosity takes place by adding negligible quantities of volatile mineralizers (1–3% by weight) to the melt. The reduction of viscosity may facilitate the processes of magmatic and crystallization differentiation which may proceed more completely, particularly in case of filter-pressing and separation of melt from crystalline phases during palingenesis. Mineralizers also cause an increase in the rates of diffusion in silicate melts (Grigoriev, 1935).

The formation of layered rocks from initially homogeneous melts in some alkaline complexes (the Lovozero, Khibina and Ilímaussaq massifs) may be due to sufficiently high rates of diffusion in alkaline magmas containing considerable amounts of volatiles. In particular, high concentrations of volatiles evidently facilitated the process of rhythmic eutectic crystallization in the rocks of the Lovozero alkaline massif (Kogarko, 1964), see also IV.2.*

VI.4.3.3. Growth of Minerals

The extensive experimental data on the artificial production of minerals demonstrate the exclusive rôle of volatile compounds for the growth of crystals. Mineralizers may not enter the composition of minerals but exert catalytic action on the process of their growth. Belov (1950) discussed this problem from crystal chemical positions. Ions of fluorine or hydroxyl substituting for divalent oxygen protect the surfaces of growing crystals against resorption. It is well known that many alumino-silicates fail to grow at all without admixture of mineralizers. The experiments carried out by Winkler (1947) established that the ability of nepheline to crystallize increases significantly when adding lithium fluoride.The formation of coarse-grained pegmatite-like textures in many alkaline rocks is apparently the result of volatile mineralizers.

VI.4.3.4. Acid–Base Interaction

The rôle of volatile components is not restricted to the influence on the physical properties of silicate melts. Their dissolution in a magma results in changes in its chemical characteristics by affecting the activities of other components.

It has been demonstrated experimentally (Niggli, 1937; Kurkjian and Russell, 1958) that the dissolution of H_2O, F, Cl, CO_2 and other volatiles in silicate melts follows the following schemes:

$$H_2O_{gas} + O^{2-}_{melt} = 2OH^-_{melt},$$
$$HF_{gas} + O^{2-}_{melt} = F^-_{melt} + OH^-_{melt},$$
$$HCl_{gas} + O^{2-}_{melt} = Cl^-_{melt} + OH^-_{melt},$$
$$CO_{2gas} + O^{2-}_{melt} = CO_3^{2-}_{melt}.$$

This process is thus accompanied by the substitution of a part of the O^{2-} of silicate melts by OH^-, F^-, Cl^-, etc., which causes an increase in the general acidity of the system. According to the conception of acid–base interaction (Lingafelter, 1941; Usanovich, 1953; Lewis, 1938) based on the theory of hydrogen-free acids the ions of

* Readers are referred to the relevant Chapters of this book where similarly cited.

melts and solutions may be arranged in the following sequences of increasing acidity:

rise in acid properties →

$O^{2-} < S^{2-} < OH^- < CO_3^{2-} < F^- < SO_4^{2-} < Cl^- < Br^-$,
$K^+ < Na^+ < Ca^{2+} < Mg^{2+} < Fe^{2+} < Fe^{3+} < Ti^{4+} < Al^{3+} < Si^{4+}$.

The acidity of cations is determined by their affinity to oxygen. The sequence of increasing acidity of anions coincides with decreasing strengths of their bonds with hydrogen. In relation to rock-forming cations the succession of increasing acidity of anions is the same as in aqueous solutions (Ryabchikov and Kogarko, 1963).

The acid properties of silicate and aluminosilicate anions increase with progressive polymerization:

$SiO_4^{4-} < Si_2O_7^{6-} < SiO_3^{2-} < Si_2O_5^{2-} < Si_4O_{11}^{6-} < (Si, Al)O_2^{n-}$

(Ramberg, 1952; Shcherbina, 1953). It is interesting to determine the place of anions of volatile components (OH^-, CO_3^{2-}, F^-, Cl^-) in this succession. This problem may be solved with the assistance of thermochemical calculations of reactions such as:

$$Mg_2SiO_4 + 2CaCO_3 \leftarrow Ca_2SiO_4 + 2MgCO_3,$$
$$CaSiO_3 + MgCO_3 \rightleftharpoons MgSiO_3 + CaCO_3,$$
$$CaSiO_3 + MgF_2 \rightleftharpoons MgSiO_3 + CaF_2,$$
$$CaSiO_3 + MgCl_2 \rightarrow MgSiO_3 + CaCl_2,$$
$$2NaAlSiO_4 + CaF_2 \leftarrow CaAl_2Si_2O_8 + 2NaF,$$
$$2NaAlSiO_4 + CaCl_2 \rightarrow CaAl_2Si_2O_8 + 2NaCl,$$
$$NaAlSiO_4 + KF \rightarrow KAlSi_3O_8 + NaCl,$$
$$NaAlSi_3O_8 + KCl \rightarrow KAlSi_3O_8 + NaCl.$$

Calculation of ΔG° of these reactions and analysis of natural parageneses allowed one to establish the directions in which these equilibria are shifted. According to the rule of polarity, as formulated by Ramberg (1952), Nekrasov and Bochvar (1940) and others, the equilibria of exchange reactions are always displaced in such a way as to realize the bonds of maximum and minimum polarity, i.e. to provide combination of the more acid anion with the more basic cation. Taking this into consideration one may deduce that the anions CO_3^{2-} and F^- are close in acidity to SiO_3^{2-} but lower than $AlSiO_4^-$ and $AlSi_3O_8^-$. Apparently Cl^- is more acid than $AlSiO_4^-$ and similar to $AlSi_3O_8^-$.

The application of the rule of polarity explains the relationship between mineral-concentrators of volatile components and the chemical composition of alkaline rocks. In alkaline rocks where $(Na + K)/Al \leqslant 1$, sodium and potassium as the strongest bases are completely consumed in the formation of nepheline and alkali feldspar. The anions CO_3^{2-} and F^- are bound with calcium which is the strongest cation after Na and K; calcite, fluorite and apatite being formed. The more acid anion chlorine may be bound with Na. In some miaskitic nepheline syenites sodalite is present (Gerasimovsky and Tuzova, 1964) while villiaumite has not been found in spite of high contents of fluorine in miaskitic nepheline syenites. In alkaline rocks having $(Na + K)/Al > 1$, the stronger base potassium is retained in alkali feldspar while the part of sodium excessive with respect to aluminium may be combined with F^-, CO_3^{2-} and S^{2-} with the formation of villiaumite, sodalite, etc., O^{2-}, which is the only simple anion of dry silicate melts, is less acid than F^-, Cl^-, and CO_3^{2-}. The introduction of these anions into a melt results therefore in a rise of its acidity.

Starting from the theory of acid–base interaction and analysing melting diagrams of silicate systems, Korzhinsky (1959) established that an increase in the acidity of a system results in the rise of the activity coefficients of acid components and in a reduction of the activity coefficients of all the bases. This takes place to a larger extent when the basic properties are stronger, e.g. the activities of cations in the less oxidized state decrease in comparison to that of the more oxidized forms. The increase in acidity of silicate melts caused by the dissolution of volatile components must result in a relative increase of the activities of the more acid components. This is the cause of an extension of the primary fields of crystallization of the more acid phases.

This property of volatile components was first noted by Niggli (1937). In his opinion volatiles form acid anions in silicate melts, which interact actively with the cations. As a result the rearrangement in silicon–oxygen complexes takes place and more polymerized silicate radicals are formed. The 'silicification' of the silicon–oxygen network proceeds, and the polymerized anions are separated into solid phases.

The acid–basic properties of water may be elucidated by considering equilibria in some water–silicate systems. The solubility of water at constant temperature and P_{H_2O} in melts in the systems K_2O–SiO_2, Na_2O–SiO_2, and Li_2O–SiO_2 passes through a minimum at 25 mol % of alkali oxide (Kurkjian and Russel, 1958), and, consequently, the activity coefficient of water passes at this point through a maximum. This was interpreted by Ryabchikov and Kogarko (1963) as an indication of the dual rôle of water in silicate melts: in more acid silicate liquids water acts as a base and in more basic melts it plays the rôle of an acid. In the point of minimum solubility of water, changes in its concentration do not affect the activity coefficients of SiO_2 and alkalis, i.e. water plays the rôle of an inert solvent. Similar views were proposed later for the solubility of water in melts composed of alkali oxide and B_2O_3 (Franz, 1966).

However, in melts of natural rocks water plays in the majority of cases the rôle of a weak acid. In the system $KAlSiO_4$–$NaAlSiO_4$–SiO_2–H_2O at pressures of water sufficiently high for the direct crystallization of quartz from the melt, an increase in water vapour pressure (and the consequent rise of the concentration of OH^- in the melt) will shift the cotectic line quartz–feldspars into areas of more alkalic melts (Tuttle and Bowen, 1958). The cotectic line feldspars–nephelines is also displaced by rising water vapour pressure towards compositions more undersaturated in silica and more alkalic (Hamilton and Mackenzie, 1965). Consequently, the fields of crystallization of the more acid phases are expanded: that of quartz at the expense of alkali feldspar and that of feldspar at the expense of nepheline. The increase in the chemical potential of water in melts in the system $KAlSi_2O_6$–SiO_2–H_2O leads to such a significant contraction of the field of the more alkalic phase—leucite (i.e. an expansion of that of the more acid phase—K-feldspar) that at P_{H_2O} = 2700 kg/cm^2 K-feldspar begins to melt congruently (Goranson, 1938).

The investigations of Seki and Kennedy (1965) demonstrated that at very high water pressures (*c.* 10 kbar) the join K-feldspar–water is intersected by the field of the water-containing phase —muscovite. Burnham (1967) suggests that crystallization of muscovite in granites at high P_{H_2O} may give rise to residual liquids oversaturated with alkalis in respect to alumina.

Experimental data of Bultitude and Green (1968) showed noticeable expansions of the fields of crystallization of orthopyroxenes with rising partial pressures of water in the systems olivine nephelinite–water and picritic nephelinite–water. They concluded that undersaturated alkaline magmas may originate by fractional crystallization of normal basalts or by partial melting of mantle substance under wet conditions. However, these data are preliminary and have been criticized by O'Hara (1968).

The crystallization of amphiboles and biotite, which may proceed only in the presence of water, was also used by some investigators to explain the genesis of alkaline rocks. Bowen (1928) suggested that the accumulation of biotite and hornblende at the bottom of magma chambers and subsequent remelting due to the action of hot basaltic magma may lead to the formation of liquids belonging to the nephelinite clan. According to Varne (1968) differential melting of mantle substance enriched in volatiles and containing pargasite may result in the generation of nephelinite magmas.

It is necessary to point out that the action of water dissolved in silicate melts depends upon numerous factors. Because of the similarity of the acid–basic properties of the anions O^{2-} and OH^- the effect of water is not in all cases controlled only by acid–base relations.

The dissolution of HF in silicate melts causes even greater increase in acidity than dissolution of H_2O. This is confirmed by the experimental data on the system $NaAlSi_3O_8$–HF–H_2O (Wyllie and Tuttle, 1961). When HF makes up more than 5 % of the volatile components and at a pressure of 2750 bar (gas phase present in excess) the first solid phase precipitating from the albitic melt is quartz; albite appears only at lower temperatures. This may probably be explained by such a significant enlargement of the quartz field that albite commences to melt incongruently. The presence of quartz amongst the crystalline phases after complete consumption of the melt testifies to the incongruent solubility of albite in HF-containing vapour. Increasing contents of HF in the system 'granite'–H_2O–HF also result in expansion of the quartz field at the expense of feldspars (Wyllie and Tuttle, 1961).

In the dry system MgO–MgF_2–SiO_2 studied by Hinz and Kunth (1960) increasing contents of fluorine also expand the field of crystallization of the more acid phase—quartz. Applying the ionic model of Temkin (1946) for silicate melts on this system it is calculated that when increasing the F^-/O^{2-} ratio of the melt from 0 to 0·22 and 0·39, the activity coefficient of SiO_2 increases 1·6 and 2·2 times, respectively (Kogarko, 1964).

Chlorine is apparently the most acid mineralizer compared to other volatile components. The dissolution of HCl in albitic melts expands the quartz field even more than does HF (Wyllie and Tuttle, 1961). Because of very strong displacement of the exchange equilibrium $NaCl + SiO_2 = Na_2O + SiCl_4$ in the molten state the solubility of chlorine in silicate melts is quite low and the introduction of higher concentrations of Cl leads to the formation of two immiscible liquids.

Due to its extremely low solubility chlorine does not influence the physico-chemical properties of silicate melts significantly as do F^- and H_2O.

As is evident from data on silicate–carbonate equilibria, CO_2 also is characterized by distinct acid properties, though to a lesser degree than Cl. The increase in P_{CO_2} in the system $NaAlSi_3O_8$–Na_2O–SiO_2–CO_2 leads to the formation of Na_2CO_3 at the expense of the sodium excessive with respect to aluminium, but at moderate P_{CO_2} the bonding of sodium with aluminium in the albite molecule is not broken and the quartz field does not reach the compositional point of albite (Koster van Groos, 1966).

The accumulation of volatile components during the crystallization of alkaline rocks may lead to noticeable increases in the acidity of silicate melts. For instance, in some rocks from the Lovozero alkaline massif the atomic ratio F/O of the residual melt has increased approximately seven times (Kogarko, 1964). This results in the rise of the activity of SiO_2 and in reactions of replacement of minerals less saturated in SiO_2 by more saturated ones. The development of albite and analcime after nepheline, arfvedsonite after aegirine, and so on, is related to this stage of mineral formation. High contents of F and Cl have caused deviations from the agpaitic order of crystallization in some rocks from the Lovozero massif: ilmenite and aenigmatite, minerals enriched in weak bases, separated before nepheline and K-feldspar (Kogarko, 1964).

VI.4.3.5. Oxidation-Reduction Reactions

During the formation of the Lovozero massif the volatile components apparently affected the oxidation–reduction equilibria of the magma. Oxidation–reduction reactions are primarily controlled by temperature and by the partial pressure of oxygen.

$$K \cdot P_{O_2} = \frac{\gamma^2_{Fe_2O_3} \cdot N^2_{Fe_2O_3}}{\gamma^4_{FeO} \cdot N^4_{FeO}},$$

where γ are activity coefficients and N mole fractions. However, other conditions being equal P_{O_2} = const; T = const) an increase in the acidity of the environment will be accompanied by a decrease of the activity coefficients of the less oxidized forms ($\gamma_{FeO}/\gamma_{Fe_2O_3}$ decreases) thus leading to a rise in concentration of the less oxidized ions, e.g. ferrous iron as compared with ferric. In many rocks from Lovozero there is a correlation between the content of volatiles and the FeO/Fe_2O_3 ratio. This is consistent with experimental studies (Belyaev, 1959) which demonstrated that the addition of fluorine to silicate melts results in the rise of the Fe^{2+}/Fe^{3+} ratio. It is quite probable that the genesis of amphibole lujavrites in the Lovozero and Ilímaussaq massifs containing high concentrations of ferrous iron is related to the action of volatile components. The origin of these rocks as a consequence of a sharp fall of the partial oxygen pressure, in our opinion, is not likely, because gradual transitions between amphibole and aegirine lujavrites are observed and amphibole lujavrites form occasionally schlieren within aegirine lujavrites (Sørensen, 1962).

VI.4.3.6. Concluding Remarks

High concentrations of volatile components in alkaline magmas may lead to the formation of immiscible liquids. The separation by liquation probably controlled the origin of some carbonatites, apatite-magnetite deposits, sodalite and analcime rocks (cf.VI.5).

Thus, the most pronounced action of volatile components manifests itself at the latest stages of crystallization of alkaline rocks when the concentration of fugitive constituents is especially high. The volatile mineralizers are the cause of changes in the physical properties of silicate melts which results in the greater completeness of crystallization and magmatic differentiation. In alkaline rocks the chemical rôle of the volatile components is first of all to cause an increase in the acid properties of the melts, 'silicification', and increasing the activities of weaker bases in relation to stronger ones.

VI.4.4. Origin of Alkaline Magmas

Because of the enormous complexity of the problem of the origin of the alkaline rocks the conception presented below is suggested only as one of the numerous mechanisms leading to the formation of magmas of high alkalinity.

The consideration of the regularities of crystallization of granitic magmas in the light of experimental data on the system $NaAlSiO_4$–$KAlSiO_4$–SiO_2 shows that melts, whose compositional points lie in the triangle $NaAlSi_3O_8$–$KAlSi_3O_8$–SiO_2, in the course of fractional crystallization in the 'dry' system will reach the eutectic (or more exactly minimum of melting) point which is close in composition to granitic rocks. These compositions are separated from the nepheline–feldspar end point (simplified nepheline syenite) of crystallization of alkaline melts (eutectic or minimum melting) by the feldspar line of maximum melting temperatures. An increase in the chemical potentials of acid volatile components must, as it was stated above, lead to the displacement of the cotectic lines towards more alkalic compositions. This may cause, as is demonstrated by the experiments of Wyllie and Tuttle (1961), the transition of the line of simultaneous crystallization of quartz and feldspars from eutectical to peritectical, i.e. it may result in incongruent melting of the feldspars. In this case the join orthoclase–albite is no longer a thermal divide and any composition in this system will in the course of fractional crystallization reach the nepheline–feldspar eutectic corresponding to a simplified nepheline syenite (Fig. 4).

Significant expansion of the fields of crystallization of the more acid phases, that of quartz at the expense of feldspars and that of feldspars at the expense of nepheline, in the system $NaAlSiO_4$–$KAlSiO_4$–SiO_2 under the influence of volatile components must lead to desilication

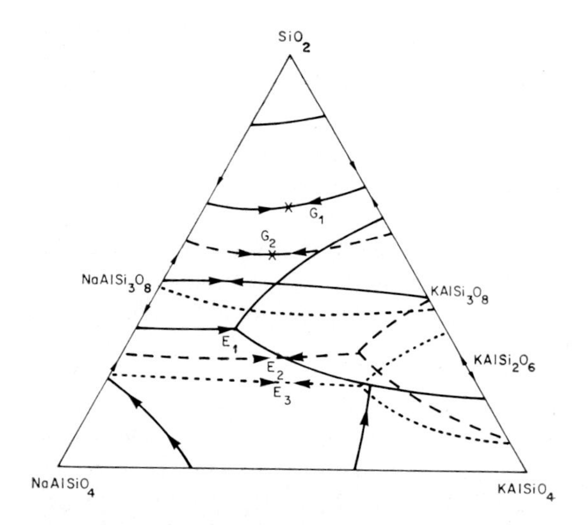

Fig. 4 Hypothetical phase diagram for the system $NaAlSiO_4$–$KAlSiO_4$–SiO_2 at various pressures of acid volatile components

——1, – – –2, ... 3

1—phase boundaries for dry system; 2—boundary lines at intermediate partial pressures of acid volatile components; 3—boundary lines at high partial pressures of acid volatile components.

of the melt at the latest stages of crystallization differentiation. The changes in the fields of crystallization of nepheline and feldspars play a great rôle during the differentiation of nepheline syenite melts. At high pressures of acid volatile components these melts are located within the crystallization field of feldspars and would change their composition towards rocks more undersaturated in silica. At lower volatile pressures the same compositions are located in the field of nepheline which results in the appearance of more silica saturated residual liquids.

The factual data obtained from liquidus diagrams of polycomponent silicate systems show that basic magma after separation of the mafic components during fractional crystallization may reach compositions close to the system $NaAlSiO_4$–$KAlSiO_4$–SiO_2. If such fractional crystallization proceeds under heightened potentials of acid volatile components, the expansion of the pyroxene field (the more acid crystalline phase) at the expense of the olivine field would lead to essential desilication of the residual melt. This may in the system $KAlSiO_4$–$NaAlSiO_4$–Mg_2SiO_4–Fe_2SiO_4–SiO_2 result in the possible intersection of the thermal divide olivine–feldspar and in a transition from melts containing normative enstatite to more undersaturated nepheline-normative melts. In other words the situation may arise that in a dry system the residual liquid would reach the eutectic point quartz + coloured mineral + feldspars while under the heightened chemical potential of acid volatile components it may reach the eutectic point nepheline + coloured minerals + feldspars. If the chemical potentials of the volatile components are so high that amphiboles crystallize instead of pyroxenes, desilication of the residual melts may take place to an even greater extent. This will, of course, depend upon the character of the crystallizing amphibole.

An increase in the amount of volatile components in a magma may thus result in the rise of the acidity of the melt. Fractional crystallization under conditions of high chemical potentials of acid volatile components may then lead to the generation of residual alkaline magma due to the expansion of the fields of crystallization of the more acid c ystalline phases. Other processes, which take place under high chemical potentials of acid volatile components, e.g. partial melting of the rocks, may also contribute to the generation of alkaline magmas.

VI.4. REFERENCES

Barth, T. F., 1961. 'Composition and evolution of magma in the southern Mid-Atlantic Ridge' (in Russian), in *Physical-Chemical Problems of Genesis of Igneous Rocks and Ores*. I, 31–55. Izd. AN SSSR, Moscow.

Bailey, D. K., 1969. The stability of acmite in the presence of H_2O. *Am. J. Sci.*, **267–A,** 1–18.

Belyaeev, G. I., 1959. The action of fluorine on some properties of ground enamel (in Russian). *Steklo i Keramika*, **16**, 3, 30–2.

Belov, N. V., 1950. Crystallochemistry of mineralizers (in Russian). *Dokl. AN SSSR*, **71**, 61–4.

Beus, A. A., and Sobolev, B. P., 1962. 'On the halogen transportation of elements in endogenous processes', in *Experimental Investigations in the Field of Deep-Seated Processes* (in Russian). Izd. AN SSSR, Moscow, 67–76.

Bowen, N. L., 1928. *The Evolution of the Igneous Rocks*. Princeton Univ. Press, Princeton.

Bowen, N. L., and Tuttle, O. F., 1950. The system $NaAlSi_3O_8$–$KAlSi_3O_8$–H_2O. *J. Geol.*, **58**, 489–511.

Buerger, M. I., 1948. The structural nature of the mineralizer action of fluorine and hydroxyl. *Am. Miner.*, **33**, 744–7.

Bultitude, R. S., and Green, D. H., 1968. Experimental study at high pressures on the origin of olivine nephelinite and olivine melilite nephelinite magmas. *Earth Planet. Sci. Lett.*, **3**, 325–37.

Burnham, C. W., 1967. 'Hydrothermal fluids at the magmatic stage', in Barnes, H. L., Ed. *Geochemistry of Hydrothermal Ore Deposits*, Holt, Rinehart and Winstone, New York, 34–76.

Daly, R. A., 1914. *Igneous Rocks and their Origin*. McGraw-Hill, New York, 1–563.

Ermolaev, N. P., 1963. 'The geochemistry of uranium in the pegmatite process' (in Russian), in *Principal Features of Uranium Geochemistry*. Izd. AN SSSR, Moscow, 70–109.

Ershova, Z. P., 1957. Equilibria of immiscible liquids in systems of the type MeF_2–Al_2O_3–SiO_2 (in Russian, English summary). *Geokhimia*, **4**, 296–303.

Evtiukhina, I. A., Kogarko, L. N., Kunin, L. L., Malkin, V. I., and Rudchenko, L. N., 1967. Acid–base properties of some aluminosilicate melts, simplified analogues of natural rocks (in Russian). *Dokl. AN SSSR*, **175**, 1369–71.

Fenner, C. H., 1926. The Katmai Magmatic Province. *J. Geol.*, **34**, 700–3.

Franz, H., 1966. Solubility of water vapour in alkali borate melts. *J. Am. Ceram. Soc.*, **49**, 473–7.

Germanov, A. I., 1961. On the possible part of organic matter in geochemical processes in the regions of recent and modern volcanic activity (in Russian). *Tr. Lab. Vulk. AN SSSR*, **19**.

Gerasimovsky, V. I., 1963. 'Geochemical features of agpaitic nepheline syenites', in *Chemistry of the Earth's Crust, vol. 1* (in Russian). 102–15. Izd. AN SSSR, Moscow.

Gerasimovsky, V. I., 1969. *Geochemistry of the Ilímaussaq Massif* (in Russian). Izd. Nauka, Moscow.

Gerasimovsky, V. I., and Tuzova, A. M., 1964. On the geochemistry of chlorine in nepheline syenites (in Russian, English summary). *Geokhimia*, 872–85.

Gillson, I. C., 1928. On the origin of the alkaline rocks. *J. Geol.*, **26**, 471–4.

Goranson, 1938. Silicate–Water Systems: phase equilibria in the $NaAlSi_3O_8$ and $KAlSi_3O_8$–H_2O systems at high temperatures and pressures. *Am. J. Sci.*, 5th Ser., **35A**, 71–91.

Grigoriev, D. P., 1935. On the action of fluorine, chlorine, and wolframite compounds in the artificial production of magnesian micas (in Russian). *Zap. Vseross. Miner. Obshch.*, 2 ser., **64**, 2, 347–54.

Hamilton, D. L., and Mackenzie, W. S., 1965. Phase equilibrium studies in the system $NaAlSiO_4$ (nepheline)–$KAlSiO_4$ (kalsilite)–SiO_2–H_2O. *Mineralog. Mag.*, **34**, 214–31.

Hinz, W., and Kunth, P. O., 1960. Phase equilibrium data for the system MgO–MgF_2–SiO_2. *Am. Miner.*, **45**, 1198–210.

Kennedy, G. C., 1955. Some aspects of the role of water in rock melts. *Geol. Soc. Am. Spec. Pap.*, **62**, 489–504.

Khitarov, N. I., Lebedev, E. B., Rengarten, E. V., and Arsenieva, R. V., 1959. Comparative characteristics of the solubility of water in basaltic and granitic melts (in Russian). *Geokhimia*, 387–96.

Kogarko, L. N., 1964. The geochemistry of fluorine in alkaline rocks, exemplified by the Lovozero alkaline massif (in Russian, English summary). *Geokhimia*, 119–27.

Kogarko, L. N., 1966. Physico-chemical analysis of cryolite paragenesis (in Russian, English summary). *Geokhimia*, 1300–10.

Kogarko, L. N., 1970. 'Magmatic equilibria in some natural systems of heightened alkalinity', in *Geochemistry, Petrology and Mineralogy of Alkaline Rocks* (in Russian). Izd. Nauka, Moscow.

Kogarko, L. N., 1970. Thermodynamic activities of components in agpaitic nepheline syenites and their application for geological problems (in Russian, English summary). *Geokhimia*, **4**.

Kogarko, L. N., and Gulyaeva, L. A., 1965. Geochemistry of halogens in alkaline rocks on the example of the Lovozero Massif (Kola Peninsula) (in Russian, English summary). *Geokhimia*, 1011–23.

Kogarko, L. N., Krigman, L. D., and Sharudilo, N. S., 1968. Experimental investigations of the forms of fluorine separation from melts of varying alkalinity (in Russian, English summary). *Geokhimia*, 948–56.

Kogarko, L. N., and Ryabchikov, I. D., 1961. The content of halogen compounds in the gas phase as a function of the chemistry of magmatic melts (in Russian, English summary). *Geokhimia*, 1068–76.

Korzhinsky, D. S., 1959. Acid–base interaction of components in silicate melts and the direction of cotectic lines (in Russian). *Dokl. AN SSSR*, **128**, 383–6.

Koster van Groos, A. F., 1966. *The effect of NaF, NaCl and Na_2CO_3 on the phase relationships in selected joins of the system Na_2O–CaO–Al_2O_3–SiO_2–H_2O at elevated temperatures and pressures.* Ph.D. thesis, Leiden University.

Koster van Groos, A. F., and Wyllie, P. J., 1966. Liquid immiscibility in the system Na_2O–Al_2O_3–SiO_2–CO_2 at pressures to 1 kilobar. *Am. J. Sci.*, **264**, 234–55.

Kozheurov, V. A., 1959. 'On the thermodynamics of ionic solutions with an arbitrary number of anions' (in Russian), in *Thermodynamics and Structure of Solutions*. Izd. AN SSSR, Moscow, 186–90.

Kurkjian, C. R., and Russell, L. E., 1958. Solubility of water in molten alkali silicates. *J. Soc. Glass Tech.*, **42**, 130–44.

Lewis, G. N., 1938. Acids and bases. *J. Franklin Inst.*, **226**, 293–313.

Lingafelter, E. C., 1941. The relative strengths of acids and bases. *J. Am. Chem. Soc.*, **63**, 1999–2000.

Morey, G. W., and Hesselgesser, J. M., 1952. The System H_2O–Na_2O–SiO_2 at 400 °C. *Am. J. Sci.*, Bowen vol., 343–71.

Nekrasov, B., and Bochvar, A., 1940. The properties of ions. IV Ionic radii and exchange reactions among alkali halides (in Russian). *Zhurn. Obshch. Khim.*, **10**, 1218–19.

Niggli, P., 1937. *Das Magma und seine Produkte.* Akademische Verlagsgesellschaft, M.B.H., Leipzig.

Nikol'skaya, T. L., 1955. To the problem of mineralizer action upon the melting points, crystallization capacity and viscosity in the system diopside–anorthite (in Russian). *Uch. Zap. LGU. ser. Geol.*, **5**, 188, 129–46.

Nolan, J., 1966. Melting-relations in the system $NaAlSi_3O_8$–$NaAlSiO_4$–$NaFeSi_2O_6$–$CaMgSi_2O_6$–H_2O and their bearing on the genesis of alkaline undersaturated rocks. *Q. J. geol. Soc. Lond.*, **122**, 119–57.

O'Hara, M. J., 1968. The bearing of phase equilibria studies in synthetic and natural systems on the origin and evolution of basic and ultrabasic rocks. *Earth Sci. Rev.*, **4**, 69–133.

Ostrovsky, I. A., Orlova, G. P., and Rudnitskaya, E. S., 1964. On the stoichiometry of the dissolution of water in alkali aluminosilicate melts (in Russian). *Dokl. AN SSSR*, 157, 1146–8.

Petersilie, I. A., 1964. *Geology and Geochemistry of Natural Gases and Disperse Bitumens of Some Geological Formations from the Kola Peninsula* (in Russian). Izd. Nauka, Moscow, 171 pp.

Petersilie, I. A., and Sørensen, H., 1970. 'Hydrocarbon gases and bitumens in the rocks of the Ilímaussaq Massif (Greenland)' (in Russian), in *Geochemistry, Petrology and Mineralogy of Alkaline Rocks* . Izd. Nauka, Moscow.

Piotrowski, J. M., and Edgar, A. D., 1970. Melting relations of undersaturated alkaline rocks from South Greenland. *Meddr. Grønland*, **181**, 9, 1–62.

Ramberg, H., 1952. Chemical bonds and distribution of cations in silicates. *J. Geol.*, **60**, 331–5.

Ryabchikov, I. D., and Kogarko, L. N., 1963. The action of anion substitution on the acidity of magmatic melts (in Russian, English summary). *Geokhimia*, 305–11.

Seki, Y., and Kennedy, G. C., 1965. Muscovite and its melting relations in the system $KAlSi_3O_8$–H_2O. *Geochim. cosmochim. Acta*, **29**, 1077–84.

Shand, S. J., 1945. The present status of Daly's hypothesis of the alkaline rocks. *Am. J. Sci.*, **234 A**, 495–507.

Shcherbina, V. V., 1953. 'On the form of existence of chemical elements in magmatic melts', in *Problems of Petrography and Mineralogy, vol. 1* (in Russian). Izd. AN SSSR, Moscow, 48–52.

Smyth, C. H., 1913. Composition of the alkaline rocks and its significance as to their origin. *Am. J. Sci.*, **36**, 1–36.

Smyth, C. H., 1927. The genesis of alkaline rocks. *Am. Phil. Soc. Proc.*, **66**, 535–80.

Soød, M. K., and Edgar, A. D., 1970. Melting relations of undersaturated alkaline rocks from the Ilímaussaq intrusion and Grønnedal–Ika complex, South Greenland, under water vapour and controlled partial oxygen pressure. *Meddr. Grønland*, **181**, 12, 41pp.

Sørensen, H., 1962. On the occurrence of steenstrupine in the Ilímaussaq massif, Southwest Greenland. *Meddr. Grønland*, **167**, 1, 251 pp.

Temkin, N. I., 1946. Molten salt mixtures as ionic solutions (in Russian). *Zhurn. Fiz. Khim.*, **20**, 105–10.

Turekian, K. K., and Wedepohl, K. H., 1961. Distribution of the elements in some major units of the Earth's crust. *Bull. geol. Soc. Am.*, **72**, 172–91.

Tuttle, O. F., 1961. 'Residual solution formed by crystallization', in *Physico-chemical Problems of the Genesis of Ores and Rocks* (in Russian, English summary). Izd. AN SSSR, Moscow 647–53.

Tuttle, O. F., and Bowen, N. L., 1958. Origin of granite in the light of experimental studies. *Geol. Soc. Am., Mem.*, **74**, 1–153.

Usanovich, M. I., 1953. *What is Acids and Bases?* (in Russian). Alma-Ata.

Valyashikhina, E. P., 1953. The action of mineralizers on some properties of the system diopside–lithium disilicate (in Russian). *Trans. IV. Conf. Experim. Mineralogy and Petrography, vol. II*, Moscow. Izd. AN SSSR 201–13.

Varne, R., 1968. The petrology of Moroto Mountain, Eastern Uganda, and the origin of nephelinites. *J. Petrology*, **9**, 168–90.

Verhoogen, J., 1949. Thermodynamics of a magmatic gas phase. *Univ. California Publications, Bull. Dep. Geol. Sci.*, **28**, 5, 91–135.

Volarovich, M. P., Leont'eva, A. A., Korchiomkin, L. I., and Fridman, R. S., 1939. On the action of fluorine on the density and viscosity of molten diorite (in Russian). *Tr. Inst. Geol. Nauk.*, **20**, 6, 51–7.

Winkler, H. G. F., 1947. On the synthesis of nepheline. *Am. Miner.*, **32**, 131–6.

Wyllie, P. J., and Tuttle, O. F., 1960. Melting in the Earth's crust. *Rep. 21st Sess. Int. Geol. Congress, Copenhagen, part XVIII*, 227–35.

Wyllie, P. J., and Tuttle, O. F., 1961. Experimental investigation of silicate systems containing two volatile components. Part II. The effects of NH_3 and HF, in addition to H_2O on the melting temperatures of albite and granite. *Am. J. Sci.*, **259**, 128–43.

Yoder, H. S., 1958. Effect of water on melting of silicates. *Carnegie Inst. Washington, Year Book*, **57**, 189–91.

Zavaritsky, A. N., 1937. Petrography of the Berdiansh pluton (in Russian). *Tr. Tsentr. nauch. issl. geol. gazv. inst.*, **96**, 1–316.

VI.5. LIQUID FRACTIONATION

L. N. Kogarko, I. D. Ryabchikov and H. Sørensen

VI.5.1. Introduction

In a number of igneous provinces differentiation processes, such as fractional crystallization, filter pressing or assimilation, fail to explain the observed serial relationships. Instead the investigators refer to processes involving enrichment in fugitive components, such as gaseous transfer, pneumatolytic, filtrational or diffusional differentiation and thermodiffusion (the Soret effect), or to liquid immiscibility.

Hamilton (1965) has coined the term 'liquid fractionation' in order to describe the processes by which a magma is differentiated into parts of graded or contrasted compositions without involving crystallization of anhydrous minerals. This migration and separation of the components take place in the liquid state by diffusion of mobile components in response to temperature–pressure gradients, by gaseous transfer of volatile material, or by liquid immiscibility (liquation).

These processes may be described under two main headings, pneumatolytic differentiation, resulting in upward migration of volatile constituents towards the apical parts of magma reservoirs, and liquid immiscibility, resulting in a splitting up of the magma during or prior to crystallization.

In cases it may be difficult to distinguish a build-up of volatiles in residual melts due to crystallization of anhydrous minerals from concentration of volatiles brought about by migration of material in the liquid magma.

VI.5.2. Examples of the Application of the Pneumatolytic Differentiation Hypothesis

VI.5.2.1. Volcanic Provinces

In Chapter II.5 pneumatolytic differentiation is invoked to explain the petrology of the Fitzroy Basin, Western Australia, the Nyiragongo volcanic field, Kivu; and the Roman volcanic province. In Chapter IV.4 this process is considered in order to explain the relations at Mont Dore, and in Cantal, Eifel and Bohemia.

In the Roman volcanic region the first emitted portion of a magma, the most superficial part, is richest in pneumatolytic elements (Locardi and Mittempergher, 1969, p. 10). This is demonstrated very clearly in the distribution of U and Th and it appears that this effect is best developed in volcanoes fed by magma reservoirs of great vertical extension, as for instance the Vico volcano, while magma chambers of slight vertical extension do not display this phenomenon.

In the Vico complex there is a progressive enrichment in potassium but when pneumatolytic differentiation dominates in the apical part of the magma there is a sharp decrease in potassium which is not balanced by an increase in sodium (Locardi and Mittempergher, 1967, p. 330). This may be due to a loss of K from the pyromagma and is displayed in low K/Rb ratios and high U/K and Th/K ratios indicating that there is no simultaneous enrichment in K, Rb, U and Th.

Upton (1969, p. 5) has pointed out that in the Midland Valley of Scotland the uprising magma bodies have experienced some form of differentiation such that their upper portions were enriched in silica and alkalis. This process was minimal in the fast rising magmas, which formed the plateau lavas, but more important in slower rising or stagnant magma columns.

VI.5.2.2. Differentiated Sills and Other Intrusions

In differentiated sills and laccoliths there often is a silicified (in tholeiitic intrusions, cf. Hamilton, 1965) or a zeolitized (in alkali basaltic intrusions, cf. Wilshire, 1967) zone immediately beneath the upper chilled zone. This volatile enrichment is partly caused by a build-up of volatiles in the residual melt, but also by liquid fractionation (Hamilton, 1965; Wilshire, 1967). A famous example is the Shonkin Sag laccolith (Hurlbut and Griggs, 1939; Nash and Wilkinson, 1970) in which the upward increase in the content of volatiles is seen in increasing grain size and zeolite content, and in the change of mafic minerals from olivine-augite to sodic pyroxenes and amphiboles.

In the Gardar province (IV.3)* syenitic and gabbroic rocks are commonly associated within the same dyke fissure. In the giant dykes gabbro is intruded by syenite, in thin dykes the opposite relation is seen. Fractional crystallization cannot alone account for this evolution, but it is suggested that the parent alkali gabbroic magma was split into an upper syenitic and a lower gabbroic part (Bridgwater and Harry, 1968) by means of diffusion of alkalis etc. in the liquid. This differentiation may have taken place during a slow ascend of the magma in the dyke fissure.

The order of succession of the phases of the composite Kûngnât intrusion, the Gardar province (IV.3) suggests an emplacement of magma from an underlying graded magma reservoir consisting of a syenitic top and gabbroic bottom. Each body of syenite in this intrusion displays enrichment in alkalis and volatiles in its upper parts.

This enrichment of volatiles in the uppermost parts of magma bodies has a pronounced influence on the mineralogy and petrochemistry of the intrusions, and also on their internal structures. The density gradients established in this way create stagnant conditions and promote crystallization in the lower parts of the magma body, while crystallization is retarded in the upper parts.

VI.5.3. Mechanism of Pneumatolytic Differentiation

In Chapters I.2.7. and VI.4.1. reference is made to some of the more important papers discussing the rôle of pneumatolytic differentiation.

As discussed in Chapter VI.4.1, the different authors have reached no unanimous opinion concerning the aggregate state of the volatile components migrating towards the apical parts of magma chambers.

Saether (1948) and Kennedy (1955) emphasized the rôle of diffusion along pressure and temperature gradients resulting in a concentration of gases in the upper parts of magma chambers, characterized by the lower temperatures and pressures, in order to maintain the physico-chemical equilibrium in the magma. Alkalis accompany the volatiles towards the apical parts. The Soret effect is thus not the only cause of differentiation but pressure gradients also provoke diffusion (cf. Wilshire, 1967, p. 153).

The importance of diffusion of volatiles in granitic and granite pegmatitic magmas has been contested by Burnham (1967) and Jahns and Burnham (1969) who point out that the equilibrium water pressure gradients are much smaller than considered by Kennedy (1955) and that this driving force for diffusion of water to establish osmotic equilibrium is thus fairly small. They favour transportation of volatiles in vapour bubbles rising through the magma. The importance of transportation of material in such high-temperature fluid phases has been demonstrated experimentally, for instance by Orville (1963) and Burnham (1967).

However, Kogarko (Chapter VI.4.2) has pointed out that water is more soluble in syenitic than in granitic magmas and that most volatiles are easily soluble in alkaline or peralkaline melts. Also granitic melts may dissolve larger quantities of water when they, as demonstrated by Tuttle and Bowen (1958) and Luth and Tuttle (1969), are enriched in sodium or potassium. In these cases crystallization of a granitic melt may take place without separation of a water-rich phase and there may be a continuous passage from granitic melts into water-rich low-temperature liquids.

The authors therefore maintain that liquid fractionation by diffusion of volatile components

* Readers are referred to the relevant Chapters of this book where similarly cited.

in very fluid alkaline and peralkaline melts is an important process in the genesis of alkaline rocks and that a water-rich vapour phase mainly separates from alkaline melts by boiling at near-surface conditions (cf. McCall, 1964 and Locardi and Mittempergher, 1967), or by filter pressing, cf. the Ilímaussaq lujavrites (Sørensen, 1962).

VI.5.4. Examples of the Application of the Liquation Hypothesis

Liquid immiscibility has by a number of geologists (e.g. Loewinson-Lessing, 1884, 1935; Fenner, 1948; Holgate, 1954) been recognized as a possible mechanism in the origin of diverse magma types. These investigators suggested that an initially homogeneous silicate melt at a certain stage prior to crystallization splits into two immiscible melts: one approaching rhyolites in composition and the other close to basalts. A number of workers (Afanasiev, ed., 1963) considered some structural features of acid lavas as the results of immiscibility in the liquid state.

Marshall (1914) explained the sequential extrusion of basic, acid and alkaline lavas (including trachytes, phonolites and basanites) within the Cenozoic petrographic provinces of New Zealand by liquation in the essexitic magma of the source reservoir.

The splitting up of alkali basaltic magma into two melts, one of which markedly enriched in water, has been proposed to account for the origin of rounded bodies of analcime trachybasalt in phonolite (Tomkeieff, 1952), for natrolite bearing globules in a picritic sheet at Igdlorssuit, West Greenland (Drever, 1960), and for analcime syenitic ocelli in lamprophyres from the Monteregian alkaline province, Canada (Philpotts and Hodgson, 1968). The relations in the last named example correspond to the common association of nepheline syenite and essexite in this province.

Liquid immiscibility may therefore have played an important petrogenetic rôle in this province (IV.6).

Vugs of primary calcite, analcime and zeolite in ngurumanite (melteigite with iron-rich mesostasis) may be products of liquid immiscibility (Saggerson and Williams, 1964).

The separation of aqueous–saline liquid solutions immiscible with silicate melts may also take place at the latest stages of crystallization of alkali granites (Roedder and Coombs, 1967).

The formation of immiscible silicate and chloride melts is likely to take place when silicate magma intrudes strata of evaporites (Pavlov and Ryabchikov, 1968). The formation of 'anatectic' salt melts immiscible with alkaline magma at the time of intrusion may be expected at such contacts (cf. Jones and Madsen, 1959).

Analcime-rich spheroids in lujavrites, and 'dense' analcime rocks in naujaites from the Ilímaussaq massif may have originated by separation of two immiscible liquids, one of which extremely rich in water (Sørensen, 1962). An aqueous–saline liquid immiscible with the silicate magma could be separated during the crystallization of agpaitic nepheline syenites. This second liquid phase, depending on the relative enrichment in fluorine, chlorine, or water, could form segregations of respectively villiaumite, sodalite or analcime rocks. Villiaumite is distributed rather irregularly in the Lovozero foyaites forming separate patches, occasionally its content rises to 2–3%. Studies of thin sections suggest that this mineral could crystallize from separate liquids enriched in fluorine, chlorine and water. This is corroborated by the close paragenesis of villiaumite with sodalite and analcime (Fig. 1).

Another possible example of liquid immiscibility in alkaline magma is the formation of carbonatites, which may be correlated with two-liquid equilibria in alkali carbonate–silicate systems. This problem is discussed in detail elsewhere (VI.3; Koster van Groos and Wyllie, 1963, 1966; Wyllie, 1966).

The origin of magnetite (titanomagnetite)–apatite deposits associated with alkali gabbro intrusions is often ascribed to the process of separation of an immiscible liquid. The ore body of magnetite–apatite at Kiruna, Sweden (Geijer, 1931; Asklund, 1949) is made up of magnetite with an admixture of apatite and negligible

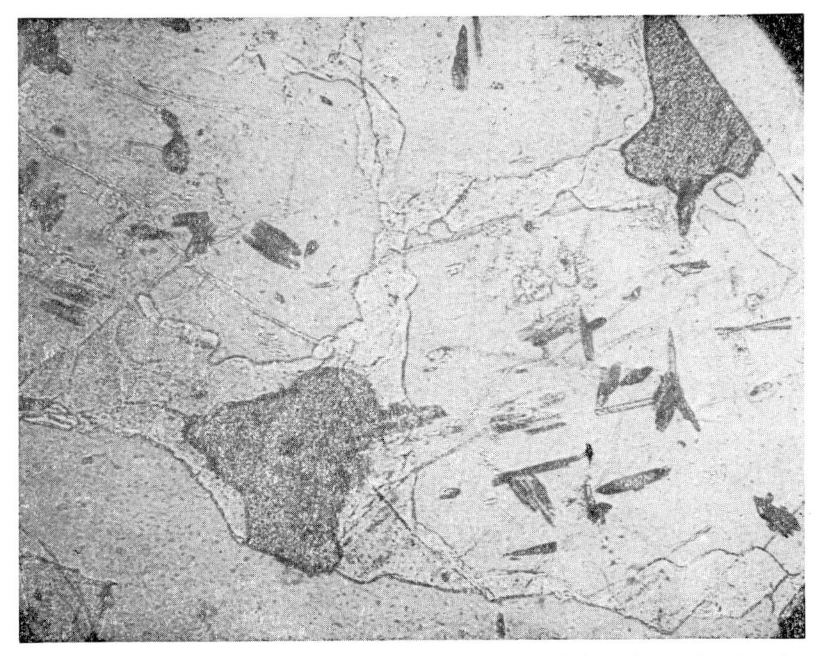

Fig. 1 Villiaumite dark and sodalite (network around villiaumite grains showing 'relief') aggregates in the mesostasis of foyaite from the Lovozero Massif. ×30, 1 nicol

amounts of hornblende. The wall-rocks are porphyries of syenitic and quartz–syenitic composition. Fischer (1950), analyzing the phase-diagram of the system

$$Fe_xO_y–Ca_5(PO_4)_3F–SiO_2–Na_2O$$

concluded that experimentally obtained immiscible melts correspond respectively to the ores and wall-rocks (cf. VI.5.5.3). Bogatikov (1966) also demonstrated that the compositions of gabbro-syenites from the Sayans containing apatite–titanomagnetite ores correspond to the two-liquid field.

The Khibina apatite deposits (see IV.2 and VII) may be products of immiscibility in an alkaline magma with the formation of two liquids: silicate and phosphate melts (Ivanova, 1968). The rich apatite deposits form sheet-like bodies along the hanging wall of an apatite–nepheline intrusion. These bodies are underlain by massive urtites. The majority of investigators consider the apatite ores and the underlying urtites to be syngenetic formations (Dudkin *et al.*, 1964; Minakov *et al.*, 1967).

VI.5.5. Experimental Data on Liquid Immiscibility

Liquid immiscibility in silicate–oxide systems (i.e. containing O^{2-} as the only simple anion) was demonstrated experimentally in the classical work of Greig (1927). However, Greig has shown that two liquids may be in equilibrium in these systems only in a rather narrow compositional range, characterized by extremely high concentrations of silica and low amounts of alkalis and alumina. According to Greig (1927) even the most acid of the natural igneous rocks are located outside the immiscibility gap. The compositions of alkaline rocks rich in alkalis and alumina are situated still farther from the two-liquid region.

The addition of volatile components (F, Cl, S, P, etc.) leads to the expansion of the two-liquid regions in silicate systems (Kogarko and Ryabchikov, 1969). This is related to the fact that the dissolution of fugitive constituents in silicate melts results in the replacement of O^{2-} anions by salt-forming anions such as F^-, Cl^-,

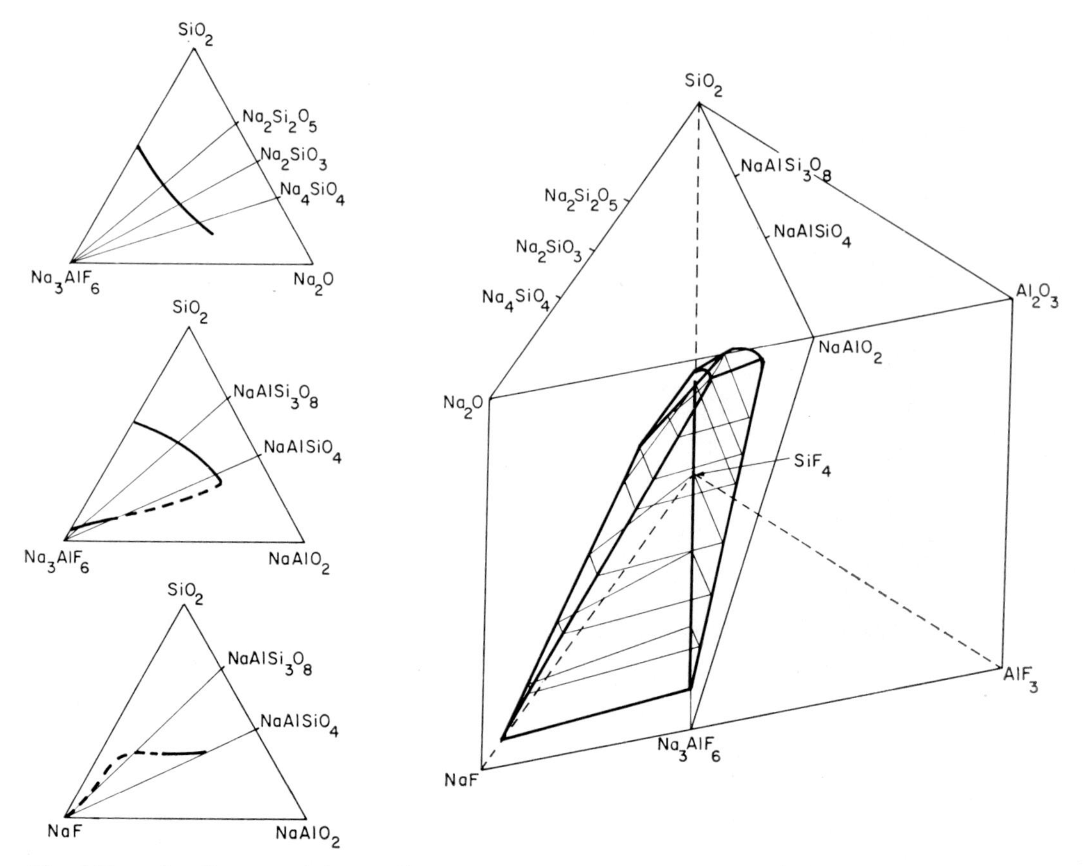

Fig. 2 Tentative diagram of the two liquid region in the system Na, Al, Si|O, F after Kogarko (1967) and Kogarko and Ryabchikov (1969)

S^{2-}, and SO_4^{2-}, which in turn leads to the appearance of exchange equilibria of the type

$$2Na_2O + SiCl_4 \rightarrow 4NaCl + SiO_2.$$

These equilibria are shifted in such a way that salt-forming anions are bound with basic cations—Na, K, Ca, Mg, etc., while silicon is surrounded by the polarizable oxygen. This leads to a microheterogeneity of the silicate melt, and in the case of strong displacement of equilibria (when ΔG of these reactions is large, see Blander and Topol, 1966) it causes a heterogeneity in a megascopic scale—i.e. two separate melts are formed—one rich in salt components (ionic) and the other rich in silicates (polymerized).

VI.5.5.1. Systems Containing Fluorine

Ol'shansky (1957); Ershova (1957, 1962); Ershova and Ol'shansky (1958) have demonstrated significant increases in the dimensions of the immiscibility gaps in the systems CaO–CaF_2–SiO_2, MgO–MgF_2–SiO_2, CaO–CaF_2–Al_2O_3–AlF_3–SiO_2, etc., as a result of the substitution of oxygen by fluorine. Later it was shown (Kogarko, 1967; Kogarko and Ryabchikov, 1969) that immiscibility exists in the system Na–Al–Si|O–F (Fig. 2). This field of liquid immiscibility includes the petrologically important joins $NaAlSiO_4$–NaF, Na_3AlF_6–$NaAlSiO_4$, Na_3AlF_6–$NaAlSi_3O_8$, Na_3AlF_6–SiO_2. The join $NaAlSi_3O_8$–NaF is probably not far from the critical point of the immiscibility field.

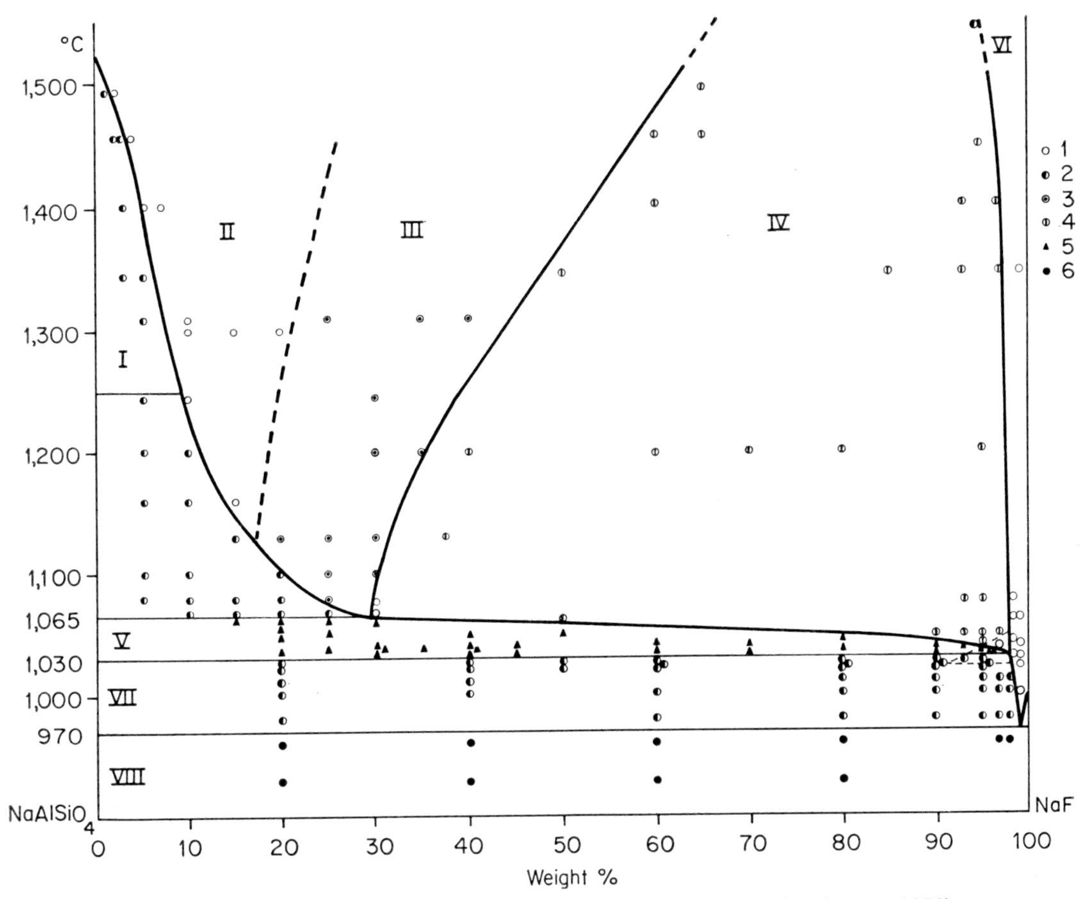

Fig. 3 Phase diagram for the join nepheline–NaF (Kogarko and Krigman, 1970)

I. nepheline or carnegieite and silicate melt;
II. silicate melt;
III. liquid immiscibility in microscopic scale;
IV. two liquids;
V. two liquids and nepheline;
VI. fluoride-rich liquid;
VII. nepheline and fluoride-rich liquid;
VIII. nepheline and villiaumite;
△ field of unknown crystalline phase.

(1) homogeneous melt; (2) one crystalline phase + liquid; (3) microliquation; (4) macroliquation; (5) one crystalline phase + two immiscible liquids; (6) below solidus.

In the system $NaAlSiO_4$–NaF (Fig. 3) there is, besides the field of stability of two liquid phases, a field in which nepheline coexists with the two immiscible liquids. This indicates a slightly non-binary behaviour of the join nepheline–NaF.

VI.5.5.2. Systems Containing NaCl

Kotlova *et al.* (1960) established an almost complete immiscibility of liquids in the system SiO_2–NaCl. The immiscibility gap in the molten state was shown to extend from this boundary join into the region of considerably more alkalic compositions in the system NaCl–Na_2O–Al_2O_3–SiO_2 (Ryabchikov, 1963). With increasing alkalinity the mutual miscibility of silicate and chloride melts rises, but the addition of even

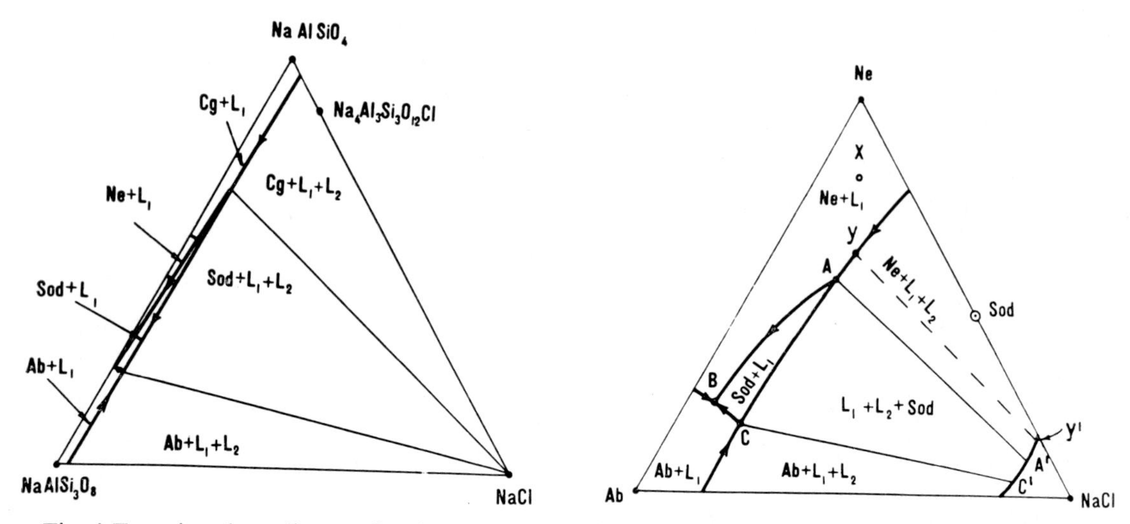

Fig. 4 Tentative phase diagram for the system $NaAlSiO_4$–$NaAlSi_3O_8$–NaCl. (a) Semi-quantitative diagram drawn in scale (the composition of the cloride-rich melt coincides with the NaCl corner); (b) Qualitative diagram for the convenience of discussion. The transition nepheline ⇌ carnegieite is not shown

Legend: Cg = carnegieite; Ne = nepheline; Sod = sodalite; Ab = albite; L_1 = silicate-rich melt; L_2 = chloride-rich melt.

small amounts of NaCl to silicate melts which are close in composition to alkaline magmas leads to the appearance of two immiscible melts. In particular, in the presence of as little as 1% chlorine by weight a melt of natural lujavrite (from the Lovozero massif) splits into two liquids (Pavlov and Ryabchikov, 1968).

The solubility of NaCl in silicate melts in equilibrium with sodium chloride melts in the joins albite–NaCl and the albite + nepheline eutectic–NaCl is within 2–3 wt. % at various temperatures (Ryabchikov, 1963; Koster van Groos and Wyllie, 1969). Even at 1550 °C the solubility of NaCl in nepheline melts saturated with respect to sodium chloride melt is only 2·2 wt. % (Kogarko and Ryabchikov, 1969).

VI.5.5.3. Systems Containing Phosphorus, Sulphur or CO_2

Wide fields of immiscibility between melts of almost pure SiO_2 and silicate–phosphate liquids in the system CaO–P_2O_5–SiO_2 were demonstrated by the experiments of Trömel (1943). The immiscibility of fused calcium phosphate and ferrous oxide in the system CaO–FeO–P_2O_5 was reported by Olsen and Metz (1945–46). Fischer (1950) continued the investigation of iron oxide–phosphate systems. He found an extensive two-liquid field in the system Fe_xO_y–$Ca_5(PO_4)_3F$, which is closer to natural compositions.

In the system Na_2O–SiO_2–$Ca_3(PO_4)_2$–Al_2O_3 investigated by Melentiev and Ol'shansky (1952) the liquid immiscibility field includes the petrologically important joins: $Ca_3(PO_4)_2$–albite, the albite + silica eutectic–$Ca_3(PO_4)_2$, and nepheline + albite eutectic–$Ca_3(PO_4)_2$. With increase in the contents of alumina and sodium oxide the immiscibility gap contracts (Fig. 5).

Numerous silicate–sulphide mixtures give two liquids after melting. These systems are reviewed by Ol'shansky (1950), Smith (1961) and MacLean (1969).

The equilibrium of two liquid phases (silicate and carbonate) was also observed at elevated pressures and temperatures in the system $Na_2O - CaO - Al_2O_3 - SiO_2 - CO_2 - H_2O$ (VI.3; Koster van Groos and Wyllie, 1963, 1965, 1966; Koster van Groos, 1966).

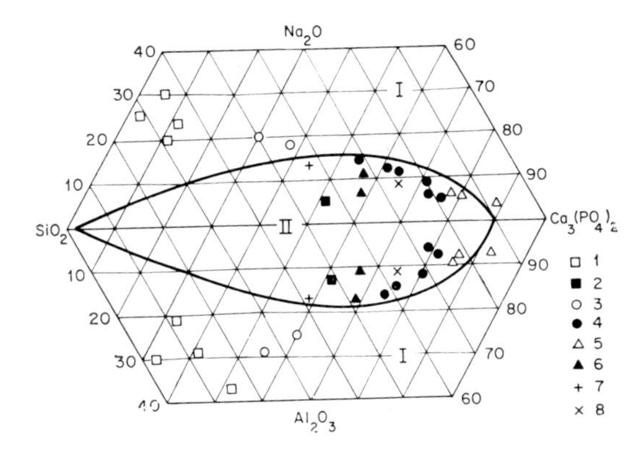

Fig. 5 The two-liquid field in the system Na_2O–Al_2O_3–SiO_2–$Ca_3(PO_4)_2$ (after Melentiev and Ol'shansky, 1952) with the plotted compositions of various rocks from the Khibina apatite deposit after Dudkin *et al.*, 1964)

I—one liquid; II—two melts; (1) ijolite–urtite; (2) apatite ijolite; (3) reticular apatite–nepheline rocks; (4) lense-like and banded apatite-nepheline rocks; (5) spotted nepheline–apatite rocks; (6) lense-like and banded apatite–titanite–nepheline rocks; (7) spotted apatite–nepheline rock from the Poach–Vumchorr deposit; (8) banded rock from the same deposit.

VI.5.5.4. Systems Containing Aqueous Salt Solutions

Tuttle and Friedman (1948) and Friedman (1951) described equilibria of two immiscible liquids (silicate-rich and water-rich) and vapour in the systems Na_2O–SiO_2–H_2O and Na_2O–Al_2O_3–SiO_2–H_2O. The appearance of such equilibria in these systems is due to the considerable decrease of silicate liquidus temperatures caused by the addition of water. Above certain critical temperatures (*c.* 390 °C in the experiments of Tuttle and Friedman) aqueous liquid and gas phases become identical.

An analogous situation arises in systems containing silicates, salts and water at much higher temperatures and pressures, because in the presence of salts the region of coexistence of gaseous and liquid aqueous phases extends to higher parameters than for pure water (i.e. addition of salt results in the rise of critical temperatures and pressures, Sourirajan and Kennedy, 1962; Ravich, 1966). With decreasing water content the equilibrium silicate melt + aqueous saline solution may gradually merge into equilibria of silicate melt + salt melt in boundary systems of silicates–salts.

Due to the immiscibility between silicate melts and concentrated aqueous saline solutions the systems silicate–salt–water must below certain pressures be characterized by the presence of univariant phase assemblages including crystalline silicates, silicate melt, aqueous saline liquid, and gas. This problem is discussed in detail elsewhere (Koster van Groos and Wyllie, 1969; Ryabchikov, 1967).

VI.5.6. Discussion of the Petrological Importance of Liquid Immiscibility

From the brief survey of experimental data for silicate systems displaying immiscibility in the molten state one may conclude that the substitution of the O^{2-} of the silicate melt by such acid anions as F^-, Cl^-, PO_4^{3-}, CO_3^{2-}, SO_4^{2-}, etc. may lead to the formation of immiscible liquids. As has been pointed out in Chapter VI.4, alkaline rocks are characterized by maximum abundances of F, Cl, S and other volatiles. However, the concentrations of volatiles are not sufficient for the separation of liquid phases prior to crystallization. During the crystallization of alkaline magmas the concentrations of volatile components are continuously increasing and the conditions of the separation of immiscible liquid may be realized.

VI.5.6.1. Sodalite Rocks

Among silicate–salt systems the most extensive two-liquid fields are characteristic for the silicate–chloride systems.

The maximum abundance of chlorine is reached in certain varieties of peralkaline nepheline syenites: tawites and poikilitic sodalite syenites from the Lovozero massif, Kola Peninsula (IV.2; Gerasimovsky *et al.*, 1966), naujaites from the Ilímaussaq massif, Greenland (IV.3; Gerasimovsky, 1969). They contain in average about 2·3% of chlorine and sometimes up to 5%. These compositions fall into the two-liquid regions of the experimentally examined systems.

Polyakov and Kostetskaya (1965) suggested on the basis of geological data that the sodalite-rich rocks and the lujavrites of the Lovozero massif were products of two immiscible liquids (cf. IV.2). We shall consider this hypothesis in the light of physico-chemical and geochemical evidence.

For this purpose we constructed the semi-quantitative melting diagram (Fig. 4) for the system $NaAlSiO_4$–$NaAlSi_3O_8$–$NaCl$ based upon the experimental data of Koster van Groos (1966), Wellman (1968) and Kogarko and Ryabchikov (1969). The topology of the more complicated phase diagram $(Na, K)AlSiO_4$–$(K, Na)AlSi_3O_8$–$(Na, K)Cl$, more closely corresponding to natural magmas, must be similar. According to this diagram crystallization of certain compositions situated in the field of primary carnegieite (or nepheline under hydrous conditions) will result in the separation of the second chloride-rich liquid phase before any sodalite crystals appear. This liquid is in equilibrium with silicate melt and nepheline crystals. The chloride-rich liquid will either be disseminated in the mixture of silicate melt and crystals or ascend and accumulate in structurally favourable parts of the magma chamber. With further cooling the primary nepheline will react with the chloride-rich liquid under the formation of sodalite. This results in the consumption of the chloride-rich liquid in the major part of the intrusion, while near the places where the chloride-rich liquid is accumulated the crystals of nepheline will disappear. The crystallization of this system is discussed in more detail elsewhere (Kogarko and Ryabchikov, 1969). The residual chloride-rich liquid (if it is preserved) has an extremely low viscosity, it may migrate to places where nepheline is present and be responsible for the metasomatic sodalitization of the latter.

This scheme based upon a tentative phase diagram fits many features of the natural systems: the presence of both primary and metasomatic sodalite and the occurrence of irregularly shaped bodies of sodalite-rich rocks surrounded by lujavrites with lesser amounts of sodalite in the Lovozero massif. These sodalite-rich bodies may be products of the interaction of NaCl-rich liquid with nepheline and silicate melt near spots of accumulation of the chloride-rich liquid (Kogarko and Ryabchikov, 1969). The data on the distribution of Cl and Br among the poikilitic sodalite syenite and the surrounding lujavrite of the Lovozero massif corroborates the hypothesis on the leading rôle of liquid immiscibility in the origin of the sodalite-rich bodies (Kogarko and Gulyaeva, 1965).

The poikilitic sodalite syenites of the Ilímaussaq intrusion (naujaites, cf. II.2.2.5) are nearly identical to those of Lovozero. It should, however, be pointed out that all investigators of the Ilímaussaq intrusion favour the view that the naujaites, which form a zone several hundred metres thick in the upper part of this intrusion, are flotation cumulates formed by flotation of sodalite crystals in a very fluid gas-rich magma. This interpretation is in agreement with field, as well as experimental evidence (Sørensen, 1969). It may for instance be pointed out that the naujaite upwards grades into sodalite foyaite having interstitial sodalite and that it is intruded by the clearly younger lujavrite. Sørensen (1970) has therefore pointed at the possibility that the Lovozero naujaites similarly may, originally, have formed a continuous upper zone which is now partly engulfed by lujavrites.

Sulphide deposits, whose genesis is ascribed to the formation of immiscible sulphide melts, are not related to alkaline magmas: they occur with basic–ultrabasic complexes. However, the presence of S^{2-} alongside with Cl^- in agpaitic nepheline syenites may contribute to the formation of a second immiscible liquid in the case of sodalite-rich rocks which contain significant concentrations of S^{2-} (0·16 wt. %) as contrasted with the surrounding lujavrites (0·08 wt. %) (Gerasimovsky *et al.*, 1966).

VI.5.6.2. Fluorine-Rich Nepheline Syenites

Experimental factual knowledge shows that fluorine expands the two-liquid fields in silicate systems, though to a lesser degree than chlorine. Agpaitic nepheline syenites (Lovozero, Ilímaussaq, Islands of Los) contain exceptional high concentrations of fluorine, which continuously rise to the end of crystallization. According to our estimates the residual melt (taken as mesostasis among euhedral nepheline and feldspar) may have contained above 30 wt. % of NaF.

When comparing this value with the data from the system Na_2O–Al_2O_3–SiO_2–NaF and in particular the join nepheline–NaF (cf. VI.5.5.1), one may deduce that under dry conditions the composition of the residual melt is close to the immiscibility gap. (Unfortunately, a detailed knowledge of this system under hydrous conditions is still absent.)

VI.5.6.3. The Apatite–Nepheline Ores of Khibina

Dudkin *et al.* (1964) have plotted the compositions of the apatite–nepheline ores of Khibina (cf. VI.5.4) and the underlying urtites on the melting diagram of the system Na_2O–SiO_2–Al_2O_3–$Ca_3(PO_4)_2$ (cf. VI.5.5.3). The bulk compositions of numerous ores (silicate + phosphate components) fall into the two-liquid region (Fig. 5). The compositions of the very rich ores and the underlying urtites fall beyond the limits of the experimentally determined immiscibility gap. Hence Dudkin *et al.*, concluded, that the major process in the formation of ores was the separation of an immiscible second liquid. However, the extraordinary high melting points of F-apatite—1650 °C (rich ores are almost monomineralic apatite)—makes this hypothesis less plausible.

Our data on the system apatite–nepheline–H_2O (Kogarko and Lebedev, 1968) show negligible lowering of the apatite liquidus by the addition of nepheline and by increase in water vapour pressure. There is no basis for assuming that the melting temperatures of the apatite–nepheline rocks were significantly suppressed by the presence of sodium chloride or sodium fluoride, because villiaumite and sodalite are scarce in these rocks. Due to the very extensive field of crystallization of primary apatite (Kogarko and Lebedev, 1968; Kogarko, 1971) only rather high concentrations of NaF and NaCl may noticeably lower the liquidus temperature of apatite.

VI.5.6.4. Conclusions

The students of alkaline igneous provinces often have to invoke pneumatolytic differentiation in order to explain the geological relations observed in the field, and the geochemical trends observed in the laboratory. As most volatiles, including water, are easily soluble in alkaline and peralkaline melts, processes involving migration of volatiles along pressure and temperature gradients in alkaline magmas are likely to be of petrological importance.

The separation by liquation is, however, not likely to be an important factor in the genesis of alkaline magmas. However, at the late stages of crystallization of alkaline magmas the separation of a second immiscible liquid is possible, which consists predominantly of salts and volatile components. This liquid phase differs profoundly in its chemistry from silicate melts, and its crystallization or metasomatic action on the earlier solidified minerals will consequently lead to the appearance of exotic rocks, such as carbonatites, apatite–magnetite ores, sodalite and analcime rocks, and agpaitic nepheline syenites strongly enriched in villiaumite.

VI.5. REFERENCES

Afanasiev, G. D., ed., 1963. Petrographic criteria of liquation in acid lavas (in Russian). *Trudy IGEM Ac. Sci. USSR*, **90**, 1–99.

Asklund, B., 1949. Apatitjärnmalmernas differentiation. *Geol. För. Stockh. Förh.*, **71**, 127–76.

Blander, M., and Topol, L. E., 1966. The topology of phase diagrams of reciprocal molten salt systems. *Inorg. Chem.*, **5**, 1641–5.

Bogatikov, O. A., 1966. *Petrology and Metallogeny of Gabbro–Syenite Complexes from the Altai–Sayan Region* (in Russian). Izd. Nauka, Moscow, 240 pp.

Bridgwater, D., and Harry, W. T., 1968. Anorthosite xenoliths and plagioclase megacrysts in Precambrian intrusions of South Greenland. *Meddr. Grønland*, **185**, 2, 1–66.

Burnham, C. W., 1967. 'Hydrothermal fluids at the magmatic stage', in Barnes, H. Ed. *Geochemistry of Hydrothermal Ore Deposits*, 36–76. Holt, Reinhart and Winston, Inc.

Drever, H. I., 1960. Immiscibility in the picritic intrusion at Igdlorssuit, West Greenland. *Rep. 21st Int. Geol. Congr. Norden*, **13**, 47–58.

Dudkin, O. B., Kozyreva, L. V., and Pomerantseva, N. G., 1964. *The Mineralogy of the Apatite Deposits from the Khibina Tundras* (in Russian). Nauka, Moscow and Leningrad, 235 pp.

Ershova, Z. P., 1957. The equilibrium of two immiscible liquids in systems of the type MeF_2–Al_2O_3–SiO_2 (in Russian, English summary). *Geokhimia*, 296–303,

Ershova, Z. P., 1962. Some regularities of immiscibility in fluor–silicate melts (in Russian). *Trudy VI Soveshchania po Experimental'noy Tekhnicheskoy Mineralogii i Petrographii*, Moscow, 176–8.

Ershova, Z. P., and Ol'shansky, Ya. I., 1957. Equilibrium of two liquids in systems of the type MeF_2–MeO–SiO_2 (in Russian, English summary). *Geokhimia*, 214–21.

Ershova, Z. P., and Ol'shansky, Ya. I., 1958. Equilibrium of two liquid phases in fluor–silicate systems, containing alkaline metals (in Russian, English summary). *Geokhimia*, 144–54.

Eugster, H. P., and Prostka, H. J., 1960. Synthetic scapolites (Abs.). *Bull. geol. Soc. Am.*, **51**, 1859.

Fenner, C. N., 1948. Immiscibility of igneous magmas. *Am. J. Sci.*, **246**, 465–502.

Fischer, R., 1950. Entmischungen in Schmelzen aus Schwermetalloxyden, Silikaten und Phosphaten, Ihre geochemische und lagerstättenkundliche Bedeutung. *Neues Jb. Miner.*, **81**, 315–64.

Friedman, I. I., 1951. Some aspects of the system H_2O–Na_2O–SiO_2–Al_2O_3. *J. Geol.*, **59**, 19–31.

Geijer, P., 1931. The iron ores of the Kiruna type, geographical distribution, geological characters and origin. *Sver. Geol. Unders.*, 367.

Gerasimovsky, V. I., 1969. *Geochemistry of the Ilímaussaq Massif.* Izd. 'Nauka', Moscow, 174 pp. (in Russian).

Gerasimovsky, V. I., Volkov, V. P., Kogarko, L. N., Polyakov, A. I., Saprykina, T. V., and Balashov, Yu. A., 1966. *The Geochemistry of Lovozero Alkaline Massif* (in Russian). Izd. Nauka, Moscow, 393 pp.

Greig, J. W., 1927. Immiscibility in silicate melts. *Am. J. Sci.*, ser. 5, **13**, 1–44, 133–54.

Hamilton, W., 1965. Diabase sheets of the Taylor Glacier region Victoria Land, Antarctica. *Prof. Pap. U.S. geol. Surv.*, **456-B**, 1–71.

Holgate, N., 1954. The role of igneous immiscibility in igneous petrogenesis. *J. Geol.*, **62**, 439–80.

Hurlbut, C. S., jr., and Griggs, D., 1939. Igneous rocks of the Highwood Mountains, Montana. Part I. The laccoliths. *Bull. geol. Soc. Am.*, **50**, 1043–112.

Ivanova, T. N., 1968. 'The results of long-term investigation of apatite ores and the tasks for further studies', in *Geological Structure, Development, and Ore-Resources of Kola Peninsula* (in Russian). Kol. Filial AN SSSR, Apatity, 86–96.

Jahns, R. H., and Burnham, C. W., 1969. Experimental studies of pegmatite genesis: I. A model for the derivation and crystallization of granitic pegmatites. *Econ. Geol.*, **64**, 843–63.

Jones, C. L., and Madsen, B. M., 1959. Observations on igneous intrusions in late Permian evaporites, south eastern New Mexico. *Bull. geol. Soc. Am.*, **70**, 1625–6.

Kennedy, G. C., 1955. 'Some aspects of the role of water in rock melts', in Poldervaart, A., Ed,. Crust of the Earth—a Symposium. *Spec. Pap. geol. Soc. Am.*, **62**, 489–503.

Kogarko, L. N., 1967. The field of immiscibility in the system Na, Al, Si/O, F (in Russian). *Dokl. AN SSSR*, **176**, 918–20.

Kogarko, L. N., 1971. Phase equilibria in the system nepheline–fluor–apatite (in Russian, English summary). *Geokhimia*, 160–8.

Kogarko, L. N., and Gulyaeva, L. A., 1965. Geochemistry of halogens in alkaline rocks on the example of Lovozero Massif (Kola Peninsula) (in Russian, English summary). *Geohkimia*, 1011–23.

Kogarko, L. N., and Krigman, L. D., 1970. Phase equilibria in the system nepheline–sodium fluoride (in Russian, English summary). *Geokhimia*, 162–7.

Kogarko, L. N., and Lebedev, E. B., 1968. Equilibria in the system nepheline–apatite–water (in Russian). *Geokhimia*, 375–7.

Kogarko, L. N., and Ryabchikov, I. D., 1969. Peculiarities of differentiation of alkaline magmas rich in volatiles (in Russian, English summary). *Geokhimia*, 1439–50.

Koster van Groos, A. F., 1966. *The effect of NaF, NaCl, and Na_2CO_3 on the phase relationships in selected joins of the system Na_2O–CaO–Al_2O_3–SiO_2–H_2O at elevated temperatures and pressures.* Ph.D. Thesis, Leiden.

Koster van Groos, A. F., and Wyllie, P. J., 1963. Experimental data bearing on the role of liquid immiscibility in the genesis of carbonatites. *Nature*, 4895, 801–2.

Koster van Groos, A. F., and Wyllie, P. J., 1965. The system $NaAlSi_3O_8$–NaCl–H_2O at 1 kb pressure (Abstract): *Trans. Am. Geophys. Un.*, **46**, 179–80.

Koster van Groos, A. F., and Wyllie, P. J., 1966. Liquid immiscibility in the system Na_2O–Al_2O_3–SiO_2–CO_2 at pressures to 1 kilobar. *Am. J. Sci.*, **264**, 234–55.

Koster van Groos, A. F., and Wyllie, P. J., 1969. Melting relationships in the system $NaAlSi_3O_8$–NaCl–H_2O at one kilobar pressure with petrological applications, *J. Geol.*, **77**, 581–605.

Kotlova, A. G., Ol'shansky, Ya. I., and Tsvetkov, A. I., 1960. Some regularities of liquid immiscibility in binary silicate and borate systems (in Russian). *Trudy IGEM*, **42**, 3–20.

Locardi, E., and Mittempergher, M., 1967. Relationship between some trace elements and magmatic processes. *Geol. Rundschau*, **57**, 313–34.

Locardi, E., and Mittempergher, M., 1969. The meaning of magmatic differentiation in some recent volcanoes of Central Italy. *Bull. volcan.*, **33**, 1–12.

Loewinson-Lessing, F., 1884. Die Variolite von Jalguba im Gouvernement Olonez. *Tsch. Min. Pet. Mitt.*, **6**, 281–300.

Loewinson-Lessing, F. J., 1935. On a peculiar type of differentiation represented by variolites of Yalguba, Karelia (in Russian, English summary). *Trav. Instn. Petr. Ac. Sci. USSR*, **5**, 21–7.

Luth, W. C., and Tuttle, O. F., 1969. 'The hydrous vapour phase in equilibrium with granite and granite magmas', in Larsen, L. H., Ed., Poldervaart Memorial volume, *Geol. Soc. Amer. Mem.*, **115**, 513–48.

McCall, G. J. H., 1964. Froth flows in Kenya. *Geol. Rundschau*, **54**, 1148–95.

MacLean, W. H., 1969. Liquidus phase relations in the FeS–FeO–Fe_3O_4–SiO_2 system and their application to geology. *Econ. Geol.*, **64**, 865–84.

Marshall, P., 1914. The sequence of lavas at North Head, Otago. *Q. J. Geol. Soc.*, **70**, 382–406.

Melentiev, B. N., and Ol'shansky, Ya. I., 1952. Equilibrium of immiscible liquids in the system Na_2O–Al_2O_3–SiO_2–$Ca_3(PO_4)_2$ (in Russian). *Dokl. AN SSSR*, **86**, 1125–8.

Minakov, F. V., Kamenev, E. A., and Kalinkin, M. M., 1967. On the original composition and evolution of the ijolite–urtite magma from the Khibina Alkaline Massif (in Russian, English summary). *Geokhimia*, 901–15.

Nash, W. P., and Wilkinson, J. F. G., 1970. Shonkin, Sag laccolith, Montana. I. Mafic minerals and estimates of temperature, pressure, oxygen fugacity and silica activity. *Contr. miner. petrology*, **25**, 241–69.

Olsen, W., and Metz, H., 1945–6. Zur Metallurgie des Thomasverfahrens. *Archiv. Eisenhüttenw.*, **19**, 111–17.

Ol'shansky, Ya. I., 1950. The results of experimental investigation of sulphide–silicate systems (in Russian). *Trudy Inst Geol. Nauk*, **121**, ser. petr., 36, 12–38.

Ol'shansky, Ya. I., 1957. The equilibrium of two liquid phases in the simplest fluor–silicate systems (in Russian). *Dokl. AN SSSR*, **114**, 1246–9.

Orville, P. M., 1963. Alkali ion exchange between vapor and feldspar phases. *Am. J. Sci.*, **261**, 201–37.

Pavlov, D. I., and Ryabchikov, I. D., 1968. On the dolerites solidified in salts (in Russian). *Izv. AN SSSR*, ser. geol., **2**, 52–63.

Philpotts, A. R., and Hodgson, C. J., 1968. Role of liquid immiscibility in alkaline rock genesis. *Rep. 23rd Intern. geol. Congr. Czechoslovakia*, **2**, 175–88.

Polyakov, A. I., and Kostetskaya, E. V., 1965. Poikilitic sodalite syenites of the Lovozero Massif (some problems of petrology and geochemistry) (in Russian). *Izv. AN SSSR*, ser. geol., **6**, 16–25.

Ravich, M. I., 1966. Phase equilibria in supercritical regions of some water–salt systems of the type P–Q. (in Russian). *Geokhimia*, 1275–85.

Roedder, E., and Coombs, D. S., 1967. Immiscibility in granitic melts, indicated by fluid inclusions in ejected granitic blocks from Ascension Island. *J. Petrology*, **8**, 417–51.

Ryabchikov, I. D., 1963. Experimental investigation of the distribution of alkalis between immiscible silicate and chloride melts (in Russian). *Dokl. AN SSSR*, **142**, 1174–7.

Ryabchikov, I. D., 1967. Possible rôle of concentrated saline solutions for the mobilization of ore components from magma (Abstract). *Appl. Earth Sci.*, **76**, 14.

Saether, E., 1948. On the genesis of peralkaline rock provinces. *Rep. 18th Intern. geol. Congr. Great Britain*, **2**, 123–30.

Smith, F. G., 1961. Metallic sulphide melts as igneous differentiates. *Can. Miner.*, **6**, 663–9.

Sourirajan, S., and Kennedy, G. C., 1962. The system H_2O–NaCl at elevated temperatures and pressures. *Am. J. Sci.*, **260**, 115–41.

Sørensen, H., 1962. On the occurrence of steenstrupine in the Ilímaussaq Massif, Southwest Greenland. *Meddr. Grønland*, **167**, 1, 251 pp.

Sørensen, H., 1969. Rhythmic igneous layering in peralkaline intrusions. *Lithos*, **2**, 261–83.

Sørensen, H., 1970. Internal structures and geological setting of the three agpaitic intrusions—Khibina and Lovozero of the Kola Peninsula and Ilímaussaq, South Greenland. *Can. Miner.*, **10**, 299–334.

Tomkeieff, S. I., 1952. Analcite-trachybasalt inclusions in the phonolite of Traprain Law. *Trans. geol. Soc. Edinb.*, **15**, 360–73.

Trömel, G., 1943. Untersuchungen im Dreistoffsystem CaO–P_2O_5–SiO_2 und ihre Bedeutung für die Erzeugung der Thomasschlacke. *Stahl und Eisen*, **63**, 21–30.

Tuttle, O. F., and Bowen, N. L., 1958. Origin of granite in the light of experimental studies in the system $NaAlSi_3O_8$–$KAlSi_3O_8$–SiO_2–H_2O. *Mem. geol. Soc. Am.*, **74**, 153 pp.

Tuttle, O. F., and Friedman, I. I., 1948. Liquid immiscibility in the system H_2O–Na_2O–SiO_2. *Am. Chem. Soc. J.*, **70**, 919–26.

Upton, B. G. J., 1969. Field excursion guide to the carboniferous volcanic rocks of the Midland Valley of Scotland. *Int. symp. volcan. Oxford*, 1–46.

Wellman, T., 1968. Stability of sodalite in the system $NaAlSiO_4$–NaCl–H_2O. *Am. Geoph. Un. Trans.*, **49**, 342–3.

Wilshire, H. G., 1967. The Prospect alkaline diabase–picrite intrusion, New South Wales, Australia. *J. Petrology*, **8**, 97–162.

Wyllie, P. J., 1966. 'Experimental studies of carbonatite problems: the origin and differentiation of carbonatite magmas', in Tuttle, O. F., and Gittins, J., Eds. *Carbonatites*. Interscience, New York, London and Sydney, 311–52.

VI.6. RESORPTION OF SILICATE MINERALS

W. C. Luth

VI.6.1. Basic Principles

When attempting to evaluate the rôle of resorption of silicate minerals by a magma the overall mass, energy and volume relations in the system must be considered. A key factor is related to the thermal energy, or heat, balance in the resorption process. As Bowen (1928) pointed out, in general the heat of mixing in the liquid term will be small in comparison with the heat of melting. This feature, in combination with the increase of solubility of silicates with temperature, led him to conclude that the solution of a (crystalline) silicate results in absorption of heat, an endothermic process. However, if as a consequence of the solution of a crystalline silicate in a saturated magma, crystallization ensues then some heat (of crystallization) will be liberated, an exothermic reaction.

In the analysis of resorption processes we will be concerned with the interdependence of several thermodynamic variables. Rather than treat the relationships in terms of energy we shall examine these phenomena in terms of the more familiar, and at least potentially measurable, parameters pressure (P, units of bars, $= 10^6$ dynes/cm^2), temperature (T, units of °C, or °K), Volume (V, units of cm^3), and mass (units of grammes). In this sense P and T are intensive parameters, V and mass are extensive parameters. It is often useful to represent mass reduced parameters derived from the extensive parameters such as specific volume ($\tilde{V} = 1/\rho$) and mass fraction (or wt. %) in order to evaluate the behaviour of the system in terms of intensive parameters.

A pertinent question relating to an analysis of resorption could be formulated as follows. What are the consequences of adding X g of crystalline material, or mineral, to Y g of a liquid, or silicate melt? Intuitively we would believe this to be dependent on the condition of the crystalline material, as well as on the value of X relative to Y, and whether or not the silicate melt is saturated with respect to the same, or a different, crystalline phase as the one which is being added. Thus we see that the relatively simple question, as stated above, is not capable of a straightforward simple answer, other than to say that changes will, in general, be the result. In order to express the question in an answerable form it must be modified. It is, I think, obvious that the nature and mass of both crystalline material and liquid must be stated.

A modified question would then be: What are the effects on P, T, V, $\tilde{V}$, mass, relative proportions and composition of the phases if 10 g of diopside are added to 100 g of a liquid of the composition 60% $Ca_2MgSi_2O_7$ 40% $CaMgSi_2O_6$ at 1405 °C and 1 bar which is just saturated (at the liquidus) with akermanite? (Fig. 1a). A portion of the question can be answered directly in that the total mass of the system is now 110 g and the bulk composition is composed of 50 g $CaMgSi_2O_6$ and

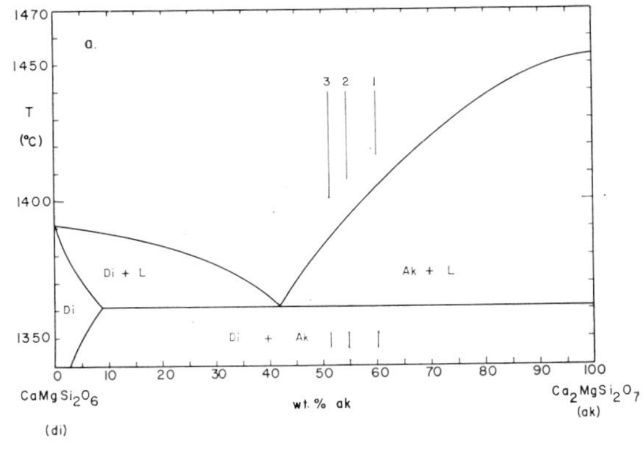

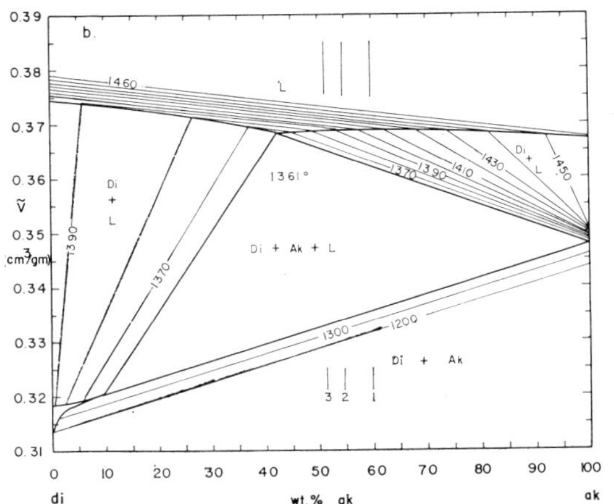

Fig. 1 The system $CaMgSi_2O_6$(di)–$Ca_2MgSi_2O_7$(ak) at 1 bar

(a) Temperature–composition diagram. From Kushiro and Schairer (1964).

(b) Specific volume–composition diagram. Constructed on the basis of data listed in Clark (1966). Isothermal two-phase tie lines shown at 10 °C intervals where liquid is present.

60 g of $Ca_2MgSi_2O_7$, or = 45·4 wt. % $CaMgSi_2O_6$ and 54·6 wt. % $Ca_2MgSi_2O_7$. However there is still no answer to the portion of the question dealing with the effects on P, T, V, $\tilde{V}$, composition and relative proportions of the phases when equlibrium is re-established. Let us consider the resorption process to be at constant pressure then we can evaluate the case in terms of a series of limiting situations. We are then concerned with the interplay of T, V, $\tilde{V}$, composition and proportions of the coexisting phases. From Fig. 1a we see that the primary crystalline phase for both the initial liquid (1) and the bulk composition resulting from the addition of 10 g of crystalline diopside (2) will be akermanite. If the resorption process is isothermal then it appears from Fig. 1a that the new bulk composition (2) will be undersaturated with respect to akermanite. That is, it will exist at 1405 °C, some 11 °C above the liquidus value for that composition, and would thus be characterized as containing the equivalent of 11 °C of 'superheat'. In this sense we could answer the portion of the question dealing with the composition and proportions of the phases. If the liquid of composition 2 is to exist as a single phase at 1405 °C and 1 bar the $\tilde{V}$ for that liquid will be given by the intersection of the 1405 °C isotherm and the line of constant composition (isopleth) 2 of Fig. 1b or $\tilde{V} = 0{\cdot}3694$, and V will be given by $(\tilde{V})(m)$ or 40·634 cm³. Thus the effects of adding 10 g of diopside to the particular liquid at the constant pressure and temperature are:

1. To produce a liquid of composition 2 undersaturated with akermanite, which has $\tilde{V} = 0{\cdot}3694$.
2. To increase the mass of the system to 110 g and to change the volume of the system to 40·634 cm³.

However, the net change in temperature and V cannot be specified unless the initial condition of the diopside is given. If the diopside was initially at 1200 °C ($\tilde{V} = 0{\cdot}314$) the results will not be the same as if the diopside was added as the stable phase at 1405 °C, a liquid ($\tilde{V} = 0{\cdot}3755$). In Fig. 1 the geometry requires in the first case that the mean temperature for the initial composition 2 (of 10 g diopside at 1200 °C and 100 g of liquid 1 at 1405 °C) to be

$$\frac{(10)\,(1200) + (100)\,(1405)}{110} = 1386\ ^\circ\mathrm{C}.$$

Thus in this case the process of resorption requires an increase in temperature of 21 °C, or in terms of percentage increase over initial mean temperature (in Kelvin) of 1·28 %.

The $\tilde{V}$ relations may be considered in a similar fashion. 10 g of diopside at 1200 °C at

V of 3·14 cm^3 is added to 100 g of liquid 1 at $\tilde{V}$ of 0·3694 cm^3 (V = 36·94) yields a total volume of 40·08 cm^3 of initial composition 2 or a net initial $\tilde{V}$ = 0·3644. The final values are $\tilde{V}$ = 0·3694 and V = 40·62, on an increase in both $\tilde{V}$ and V on completion of the resorption process. If the diopside is added as a liquid phase at 1405 °C there will be no net change in $\tilde{V}$ or V since we have assumed that the liquids exhibit no volume of mixing.

Returning to the question: What are the effects on $P,T,V,\tilde{V}$, mass, relative proportions and composition of the phases if 10 g diopside are added to 100 g of a liquid of the composition 60% $Ca_2MgSi_2O_7$, 40% $CaMgSi_2O_6$ at 1405 °C and 1 bar which is just saturated with akermanite? In the previous section we assumed that the resorption process was isothermal. The resulting phase assemblage, a liquid, was 'superheated'. If we remove the restriction that the resorption process be isothermal and instead impose a restriction such that the resulting phase assemblage contains no 'superheat' the effects of isobaric resorption can again be predicted from Fig. 1a and 1b. Composition 2 has a liquidus value of 1394 °C, and $\tilde{V}$ = 0·3688 thus the total volume 40·57 cm^3 represents the system volume on completion of resorption of 10 g diopside by 100 g of liquid 1. Again we assume that the diopside which is added is at a temperature of 1200 °C ($\tilde{V}$ = 0·314), resulting in an initial mean temperature of 1386 °C and an initial mean $\tilde{V}$ of 0·3644 and initial V of 40·08 cm^3. Thus the final assemblage in this case has a reduced increase in $\tilde{V}$, V and T, as compared to the similar isothermal case. We may modify the question still further by removing the restriction that the resulting liquid be saturated with akermanite but impose a restriction that the temperature is not increased over the initial value of 1386 °C. Composition 2 is a two phase assemblage akermanite (8%)–liquid 3 (92%) at 1386 °C. The liquid coexisting with akermanite has a composition of 49% $CaMgSi_2O_6$ 51% $Ca_2MgSi_2O_7$. The effective bulk composition 2 resulting on completion of the resorption process is 45·4% $CaMgSi_2O_6$, 54·6% $Ca_2MgSi_2O_7$. The final assemblage then consists of 8·8 g akermanite and 101·2 g liquid 3. From Fig. 1b the $\tilde{V}$ of the two-phase final assemblage is 0·3674 giving a total volume of 40·414. This represents a net increase in volume from the initial condition of 40·08 cm^3. Thus although no temperature increase over the initial mean temperature was observed there was still work, of PV nature, done by (or on) the system. In this case, the resorption of 10 g of diopside results in the precipitation of 8·8 g of akermanite.

We may consider the effects to be expected on T, $\tilde{V}$, composition, and proportions of the coexisting phases on completion of resorption if the isobaric constant volume process is considered. Again we consider 10 g of diopside at 1200 °C added to liquid 1 at 1405 °C as given in Fig. 1a and 1b. However in this case the final volume on completion of resorption will be the same as the initial volume occupied by 100 g of liquid 1. Physically this process is somewhat difficult to visualize unless a series of infinitesimal steps are conceived, although we can easily envisage such a process taking place in a deep magma chamber. 100 g of liquid 1 at $\tilde{V}$ = 0·3694 (Fig. 1b) occupies a volume V of 36·94 cm^3. To this we add 10 g of diopside, at 1200 °C ($\tilde{V}$ = 0·314) and impose the restrictions that the resorption process is isobaric, that total mass remains constant at 110 g and that the resulting phase assemblage occupies a net volume of 36·94 cm^3. Since the final specific volume is given by 36·94/110 = 0·336, examination of Fig. 1b demonstrates that the stable phase assemblage at this $\tilde{V}$ for composition 2 is diopside, akermanite and liquid at ~ 1361 °C. The weight proportions of the three phases are given by the application of the 'lever-rule' as: 46·6% Diopside, 48·6% Akermanite and 4·8% liquid. In terms of masses of the participating phases we would have 51·3 g of diopside 53·4 g akermanite. Thus under these specific conditions the addition of a small amount of diopside to a large amount of liquid saturated with respect to akermanite has resulted in nearly complete crystallization of the liquid. Further analysis in the same manner demonstrates that the addition of only 1 g of diopside (1200 °C) to 100 g of liquid 1 would result in the precipitation of about 19 g of akermanite. Consequently, at least in this case, the restrictions of constant volume and constant pressure result in dramatic changes.

It is frequently implied, or assumed, that the phase diagrams so useful in the classical description of heterogeneous equilibria are not applicable in the discussion of open system equilibria. In the previous analysis we have been concerned with an analysis in terms of a system which is open in the sense that the total mass changes as a consequence of the resorption process. We can imagine the converse case in which the total mass remains constant and the composition varies during resorption.

A simple illustration of this type of process can be framed in terms of Figs. 1a and 1b. Suppose that 10 g of diopside (at 1200 °C) are allowed to interact with 100 g of L_1 at 1405 °C, with the restrictions that the total mass of the system remains constant at 100 g and the liquid remains saturated with akermanite. The addition of 10 g diopside to 100 g of liquid 1 would change the effective bulk composition to 2 (60 g, 54·6% $Ca_2MgSi_2O_7$; 50 g, 45·4% $CaMgSi_2O_6$). However in order to obtain a mass of 100 g some 10 g of material must be extracted from the system. If the liquid (2) is to remain saturated with akermanite it must be on the liquidus curve of Fig. 1a and at a temperature of 1394 °C. Since 100 g of the liquid are required the liquid (2) must be composed of 54·6 g $Ca_2MgSi_2O_7$ and 45·4 g $CaMgSi_2O_6$. Consequently 5·4 g $Ca_2MgSi_2O_7$ and 4·6 g $CaMgSi_2O_6$ are removed from the system. Conceivably this material could be removed either as liquid or as crystalline akermanite + diopside. However there is no necessity for the liquid to remain composition 2. Given the restrictions stated above the 100 g of liquid could vary between 50% $Ca_2MgSi_2O_7$ 50% $CaMgSi_2O_6$ and 60% $Ca_2MgSi_2O_7$, 40% $CaMgSi_2O_6$. If the first case holds the addition of 10 g diopside results in the removal of 10 g akermanite, and the resulting akermanite saturated liquid would be composition 3 of Fig. 1 at a temperature of 1386 °C. In this case the $\tilde{V}$ of the liquid is 0·3694 and a final system volume is 36·94 cm³ as is the initial system volume of liquid 1 alone. Thus the net volume of the system remains constant but there are changes involving the total volume when both the diopside added and the akermanite extracted are considered. Initially we have 10 g of diopside (3·14 cm³) and 36·94 cm³ of liquid 1. Upon completion of the resorption process we have 37 cm³ of liquid 3 and 3·55 cm³ of akermanite. Although the volume of the system does not change the volume of the 'environment' increases significantly.

In the preceding discussion we have been concerned with the interaction of intensive and extensive parameters of importance in dealing with isobaric resorption in a simple, idealized binary silicate system. In fact only one composition has been seriously considered. However the principles involved are general and serve to demonstrate the necessity for specifying fully the conditions under which resorption is being considered. The modifying effects of pressure variation are readily visualized in terms of the volume relationships. In view of the lack of data at elevated pressure and temperature the pressure effects will not be considered further.

The isobaric $\tilde{V}$–composition diagram (Fig. 1b) may well be unfamiliar, however its usefulness should be apparent. The general form of the isobaric $\tilde{V}$–X diagram will be as given in Fig. 1b if the P–T univariant curves pertaining to the unary and binary melting reactions have $dP/dT > 0$. Several other cases may arise which prohibit general application of the form of Fig. 1b. These cases, as well as the relationships involving the entropy function, will be discussed elsewhere.

Resorption may be contrasted genetically with assimilation in that assimilation refers to interaction of the magma with 'foreign' material while resorption refers to a similar interaction with material coexisting with that magma or its parent under different conditions. If resorption is complete, all constituents of the particular phase are dissolved in the liquid. However if resorption is partial, or fractional, then there is a partitioning of the various constituents of the phase between the magma and the mineral phases with which it is saturated.

Both complete and fractional resorption are

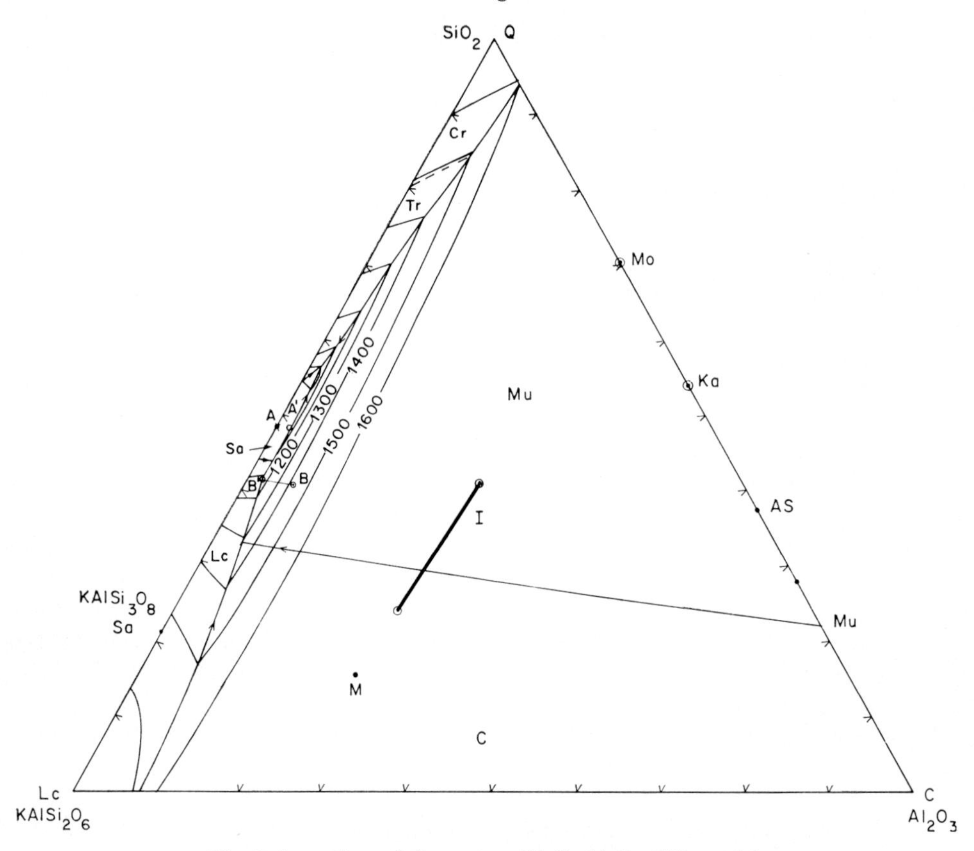

Fig. 2 A portion of the system K_2O–Al_2O_3–SiO_2 at 1 bar

The symbols on the liquidus diagram refer to the idealized and anhydrous compositions of: Q = quartz; Cr = cristobalite; Tr = tridymite; Sa = sanidine; Lc = leucite; Mo = montmorillonite; Ka = kaolinite; AS = aluminosilicates kyanite, andalusite and sillimanite; Mu = mullite; M = muscovite; I = illite; C = corundum. The letters A, A′, B, and B′ refer to specific compositions discussed in the text.

well illustrated in many of the synthetic anhydrous silicate systems studied at one atmosphere, although the process is treated without invoking the word resorption. For example we may consider the 'resorption' of 'early' olivine in the system $CaMgSi_2O_6$–Mg_2SiO_4–SiO_2 in which crystallization was discussed so elegantly by Bowen (1914). Boyd and Schairer (1964) have provided new and important data on the system.

In general, resorption of silicates in a melt can produce changes in: (1) liquid composition, (2) composition or kind of crystals with which the melt is saturated, (3) the proportion of crystals to liquid, (4) the path followed by the liquid on further equilibrium (or fractional) crystallization, and (5) the final equilibrium crystalline assemblage.

VI.6.2. Muscovite

In terms of a slightly more complex situation the results of resorption of muscovite by a granitic melt may be considered. Fig. 2 is a portion of the liquidus diagram of the system K_2O–Al_2O_3–SiO_2 at one atmosphere, (Schairer and Bowen, 1955). In view of the lack of data concerning liquidus relations in the presence of a separate aqueous vapour phase under pressure we will assume that this diagram represents a reasonable model, for near surface conditions. In addition we will assume the process to be taking place in the presence of excess H_2O so that this component can be ignored in the following discussion. In terms of anhydrous composition of the phases we are able to show muscovite (M), illite (I), kaolinite (Ka) and montmorillonite (Mo) in Fig. 2.

Consider the equilibrium resorption of muscovite in a liquid (A) composed of 64·91% KAl Si_3O_8 and 35·09% SiO_2. Assume that 30 g of muscovite (73·19% or 21·957 g $KAlSi_3O_8$ and 26·81% or 8·043 g Al_2O_3) interacts with 100 g of this liquid (A). This muscovite–liquid assemblage might correspond to the stable phase assemblage at high pressure and the problem is to evaluate the change in assemblage on emplacement near the surface.

Three of the possible cases may be evaluated:

a. All the muscovite is resorbed, and the liquid remains just saturated with the appropriate phase, that is the assemblage is all liquid, but is not 'superheated'.

b. The liquid is just saturated with a second crystalline phase, and coexists with its primary phase.

c. The resorption process is isothermal.

In all three of these cases the temperature, and phase assemblage are of key importance, as is the subsequent crystallization history of the liquid.

The effective bulk composition produced by the equilibrium resorption of 30 g muscovite by 100 g of liquid A is given by composition B (Table 1, Fig. 2). This composition is in the primary field of mullite and would be just saturated with that phase at the liquidus temperature shown (1430 °C). Thus resorption under the conditions given in (*a*) requires an increase in temperature of 380 °C.

If the resorption is to proceed with the results expected in (*b*) then the liquid would lie on a boundary curve, and be colinear with mullite and composition B; composition B′. Of considerable importance is the fact that the second phase with which the liquid is saturated is leucite.

TABLE 1 Resorption of Muscovite

wt. %	A	M	B	B_1	A_1
$KAlSi_3O_8$	64·91	73·19	66·82	72·17	67·75
SiO_2	35·09	—	26.99	26·89	31·61
Al_2O_3	—	26·81	6·19	0·94	0·64
grammes					
$KAlSi_3O_8$	64·91	21·96	86·87	86·87	69·00
SiO_2	35·09	—	35·09	32·37	32·19
Al_2O_3	—	8·04	8·04	1·13	0·65
Total	100·00	30·00	130·00	120·37	101·84

In this case the liquid has increased in mass by 20·37 g relative to A, and has increased its $KAlSi_3O_8$ and Al_2O_3 content at the expense of SiO_2. The temperature has increased from 1050 °C to approximately 1150 °C. If this liquid cools further to the 1140 °C peritectic, leucite and mullite will coprecipitate to give an assemblage composed of 117·435 g (90·335%) liquid, 10·243 g (7·879%) mullite, and 2·322 g (1·786%) leucite. If the liquid is allowed to fractionally crystallize from this point, leucite and mullite are effectively removed from the system. Thus given these special conditions we see that resorption of muscovite by a liquid of 65 **or**–35 Q composition at low pressure can result in precipitation of leucite.

If the resorption is to be considered as an isothermal process the liquid produced will lie on the mullite-potassium feldspar boundary curve, near A in composition, at A′. At the completion of this process the assemblage will consist of 10·293 g (7·918%) mullite, 17·874 g (13·749%) potassium feldspar and 101·833 g (78·333%) liquid A′. In this case the amount (in grammes) of the liquid has remained essentially constant, and composition has changed only slightly as the muscovite has been resorbed.

It is apparent that the production of silica-undersaturated phase assemblages by the resorption of muscovite, paragonite, illite, kaolinite or mixtures of these phases with quartz by a 'granitic' magma would be a relatively inefficient mechanism. Several key factors provide limiting conditions. These include the diminished field of leucite when a free aqueous vapour phase is present under modest pressure; the fact that either an aluminosilicate or corundum rather than mullite are the typical phases, and these only in the nepheline rich rocks; and the relatively large amounts of the peraluminous hydrates required.

VI.6.3. Biotite

Luth (1967) considered some aspect of the fractional resorption of biotite in connection with studies in the system $KAlSiO_4$–Mg_2SiO_4–SiO_2–H_2O, using phlogopite as a model for biotite. Obviously the model is simplified, yet it permits illustration of some features of the resorption process.

At pressures below 500 bars phlogopite does not coexist with a water-saturated silicate liquid, and at pressures greater than 1200 bars phlogopite coexists with a wide variety of water saturated liquids.

If we visualize the resorption process as taking place as the magma moves from a position deep within the crust to the near surface region then a simplified analogous case involves the readjustment of a phase assemblage on isothermal-polybaric crystallization. Luth (1967) constructed the isothermal polybaric diagram showing the appropriate boundary curves at 900 °C. This diagram and the corresponding isothermal polybaric diagrams at 950 °C and 1000 °C are given in Fig. 3. The calculated abundance of the various phases present along each isotherm with decreasing pressure for a bulk composition 65% $KAlSi_3O_8$ 35% $MgSiO_3$ (weight) are shown in Fig. 4. The path defined by the liquid with decreasing pressure is given by the heavy dashed line for each of the three cases in Fig. 3. The double-dash (=) line in each of these three cases represents that portion of the crystallization history when phlogopite is being resorbed, with enstatite or forsterite being precipitated. Figs. 3 and 4 illustrate several key features of the polybaric–isothermal resorption process. At 900 °C the major part of the resorption takes place simultaneously with the disappearance of the liquid, while at higher temperatures resorption is completed in the presence of significant amounts of liquid. Although the liquid change is toward an 'alkaline' state in all three cases, it does not reach a silica undersaturated condition. Since the liquid is always silica oversaturated the crystalline assemblage must be silica undersaturated when a liquid is present. Indeed we do see assemblages reminiscent of the alkaline ultramafic rocks. Fig. 3 may be used to illustrate isothermal resorption (polybaric) for a wide variety of bulk compositions. This type of resorption is not, at these temperatures, capable of generating a silica undersaturated alkaline liquid for any silica saturated or oversaturated bulk composition.

Luth (1967, p. 407) pointed out that the resorption process need not be of an equilibrium nature, and that this would tend to be more effective in producing a potassic magma. This can be modeled in terms of the 1000 °C isotherm (Fig. 3c). If the enstatite produced by the isothermal resorption of phlogopite from 3400 to 2450 bars is removed from the system, the effective bulk composition has changed from A to B and is composed of 71% L, 29% Ph (91% of A). Further resorption of Ph will then take place on decrease in pressure as Fo is produced. If this Fo is removed from the system as it is produced, then the effective bulk composition will change from B to C with the complete resorption of Ph. However at C the precipitation of Fo continues across the Fo field, then follows a path similar to that given by the original bulk composition. Thus resorption of this type is only slightly more effective than equilibrium resorption for silica saturated bulk compositions. Many other variants on the schemes outlined above can be given, however the general conclusion which can be obtained is that resorption of phlogopite caused by decrease in pressure alone requires very special circumstances in order to generate alkaline *magmas*, though mineral assemblages typical of alkaline ultramafic and mafic rocks are produced with ease.

The preceding discussion has involved resorption in the presence of a free aqueous vapour phase. The general relationships in the absence

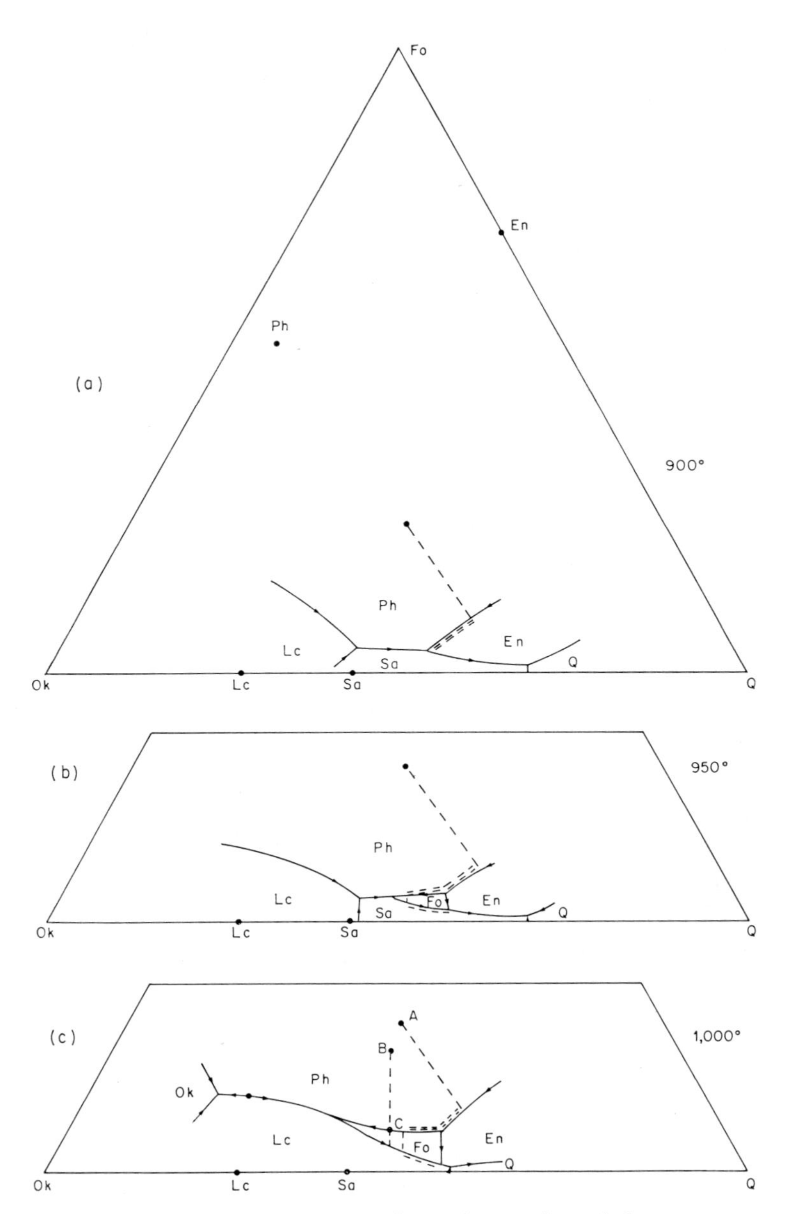

Fig. 3 Isothermal–polybaric projections of a portion of the vapour-saturated liquidus in the system $KAlSiO_4$–Mg_2SiO_4–SiO_2–H_2O. Compositions and primary phase fields on the liquidus symbolized by:

Fo = forsterite, Mg_2SiO_4; En = enstatite, $MgSiO_3$; Q = quartz, SiO_2; Sa = sanidine, $KAlSi_3O_8$; Lc = leucite, $KAlSi_2O_6$; Ok = ortho-kalsilite, $KAlSiO_4$; Ph = phlogopite, $KMg_3AlSi_3O_{10}(OH)_2$.

Dashed and double dashed curves refer to liquid compositions discussed in the text.

(a) 900 °C; (b) 950 °C; (c) 1000 °C.

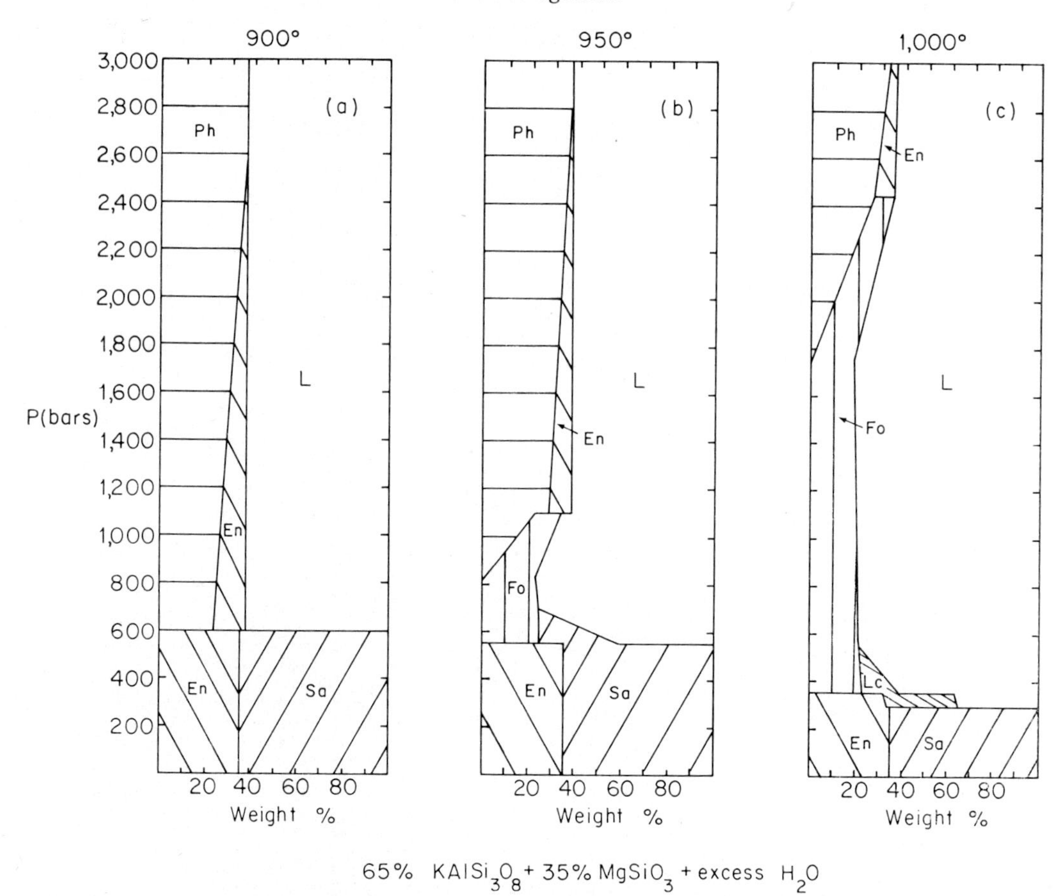

Fig. 4 Proportions of the phases present under isothermal–polybaric conditions at equilibrium for the composition 65% $KAlSi_3O_8$, 35% $MgSiO_3$ with excess H_2O. The amount of the vapour phase has been ignored in the calculations. Symbols are defined in the legend to Fig. 3

of a free vapour phase will differ in the relative stability of phlogopite bearing assemblages. In the absence of data in the water undersaturated portion of the system it is not possible to make adequate predictions.

Yoder and Kushiro (1969) have modified the author's postulated reaction governing the ultimate stability of phlogopite. The reaction indicated by their study is Phlogopite = forsterite + liquid, at pressures greater than 1·7 Kb. As noted by Yoder and Kushiro (1969, p. 568, 571) the interpretation of the experimental products is complicated by the fact that phlogopite, leucite and forsterite all form from the glass on the quench, under various *P–T* conditions. The interpretation of textural relations implying phase stability when quench phenomena are present is difficult (compare Yoder and Eugster, 1954) and Yoder and Kushiro, 1969, on the topic) particularly when only one bulk composition is studied using short duration experiments. The interpretation given by Yoder and Kushiro (1969) does not represent a unique solution, but is possible and may be accepted for the purposes of this discussion. Considering Fig. 4 of Yoder and Kushiro (1969, p. 576) we see that the addition of phlogopite to a liquid saturated with respect to phlogopite and forsterite, but undersaturated with respect to H_2O isothermally at 10 Kb, 1250 °C, may result in either the precipitation of phlogopite, or the solution of phlogopite. The consequences depend on whether the liquid of concern is rich or poor in H_2O.

The resorption of biotite can not be considered adequately in terms of phlogopite due to the iron-free nature of phlogopite. Eugster and Wones (1962), Wones (1963) and Wones and Eugster (1965) have discussed the theoretical and experimental basis of experiments at controlled oxygen fugacity (f_{O_2}) pertaining to biotite equilibria.

A most useful approach for our purposes is to consider the effects of P, T and f_{O_2} on the equilibrium composition of biotite coexisting with sanidine, magnetite and gas (Wones and Eugster, 1965). We may consider the effects of each of the three variables separately by holding the other two constant. Reduction of f_{O_2} at constant P, T results in a more annite rich biotite. However a reduction of temperature at constant f_{O_2} and P may result in either a more annite rich biotite (for magnesian biotites) or a more phlogopite rich biotite (for ferroan biotites). A reduction of pressure at constant f_{O_2} and T results in a decrease in the annite content of the biotite coexisting with Sa, Mt, and V, at least for iron rich biotites.

Several aspects of the biotite decomposition sequence are of particular interest in the analysis of resorption effects. The essential reactions as given by Wones and Eugster (1965) are:

1. biotite + gas = biotite* + sanidine + hematite + gas,
2. biotite + gas = biotite* + sanidine + magnetite + gas,
3. biotite + gas = biotite* + leucite + olivine + magnetite + gas,
4. biotite + gas = biotite* + leucite + olivine + kalsilite + gas,

where biotite* is more magnesian than biotite for each reaction. Since we may consider the reactions in terms of the decomposition of a portion of the annite component of the biotite an alternative form would be:

1a. $KFe_3AlSi_3O_{10}(OH)_2 + 3/4O_2 = KAlSi_3O_8 + 3/2\ Fe_2O_3 + H_2O$,

2a. $KFe_3AlSi_3O_{10}(OH)_2 + 1/2O_2 = KAlSi_3O_8 + Fe_3O_4 + H_2O$,

3a. $KFe_3AlSi_3O_{10}(OH)_2 + 1/6O_2 = KAlSi_2O_6 + Fe_2SiO_4 + 1/3\ Fe_3O_4 + H_2O$,

4a. $KFe_3AlSi_3O_{10}(OH)_2 = 1/2\ KAlSi_2O_6 + 1/2\ KAlSiO_4 + 3/2\ Fe_2SiO_4 + H_2O$.

Thus it is apparent that at values of f_{O_2} and T below the quartz–fayalite–magnetite buffer curve oxidation of biotite by a magma will result in tendency toward potassium enrichment and silica depletion of the remaining liquid. The fayalite freed by this reaction would interact with the crystalline material with which the magma is saturated producing a more iron rich olivine or pyroxene. By the same token the leucite and kalsilite produced by the decomposition would be expected to dissolve in the liquid. However oxidation of a biotite by a magma in the f_{O_2}–T region where la or 2a are pertinent would be expected only to add sanidine to the liquid. The situation is complicated by the complex rôles played by the oxybiotite and ferriannite end-members in the crystalline solution series.

Both Bowen (1928) and Holmes and Harwood (1937) placed considerable emphasis on the rôle of resorption of biotite in the generation of potassic alkaline magmas. This concept is certainly supported by the available experimental data, however we have seen that this mechanism is relatively inefficient in terms of the production of potassium rich silica undersaturated magmas. It seems appropriate to note that in many of these rocks the glass, or mesostasis, is silica oversaturated (Carmichael, 1967). Thus it seems possible that the postulated mechanism may be extremely efficient in terms of the production of the observed *rocks*.

VI.6.4. Hornblende

Yoder and Tilley (1962) were particularly concerned with the effects of amphibole and biotite resorption in governing the observed trends shown in the basalt magma types. In terms of the generalized simple basalt tetrahedron (nepheline–silica–olivine–diopside) they

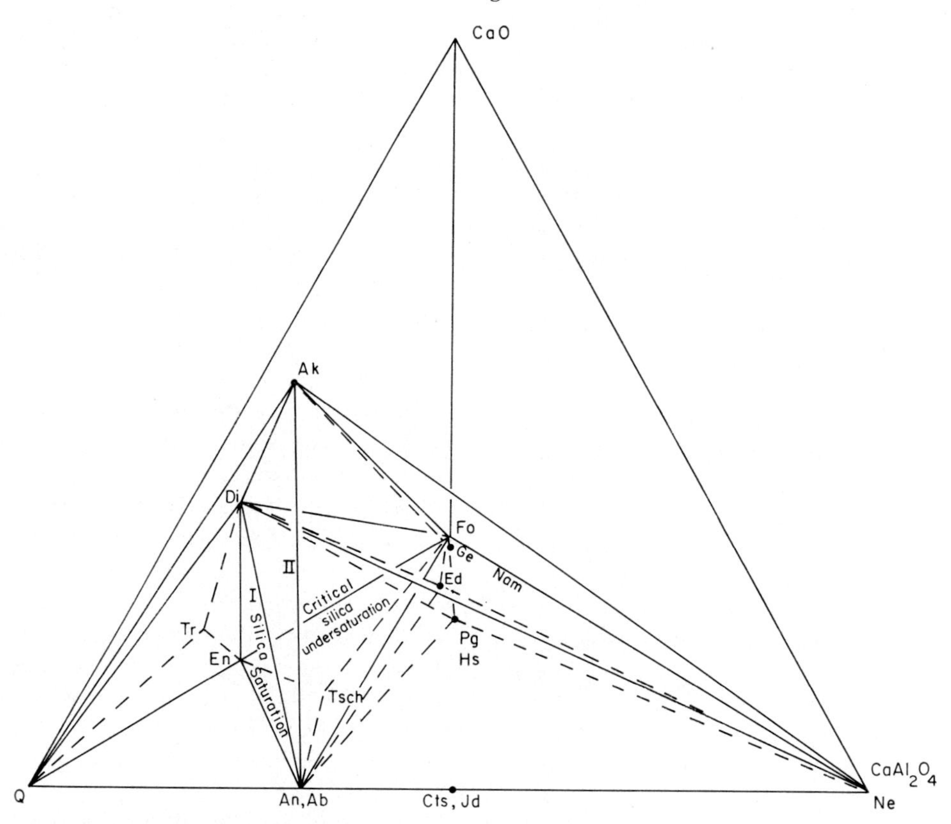

Fig. 5 A generalized simplified basalt tetrahedron. The diagram represents the superposition of two tetrahedra; SiO_2–$NaAlSiO_4$–Mg_2SiO_4–CaO and SiO_2–$CaAl_2O_4$–Mg_2SiO_4–CaO. The symbols; Tr, Pg, Hs, Ed, Tsch (Ts) are defined in Table 2. Other symbols are: Q = quartz; An = anorthite; Ab = albite; Cts = Ca-Tschermak's molecule; Jd = jadeite; Ne = nepheline; En = enstatite (or ortho-pyroxene, opx, if Fe—Mg); Fo = forsterite (or olivine, ol, if Fe—Mg); Di = diopside (or clinopyroxene, cpx, if Fe—Mg); Ak = akermanite; Ge = gehlenite; Nam = soda-melilite.

Plane I: plagioclase (An,Ab)–orthopyroxene (En)–clinopyroxene (Di).
Plane II: plagioclase–orthopyroxene–olivine (Fo)

noted that the major constituents of hornblende: hastingsite (we will use pargasite for the Al^{3+} end-member and reserve the name hastingsite for the Fe^{3+} end-member), edenite, tschermakite and tremolite, are compositionally in different portions of the tetrahedron. In Fig. 5 a modified basalt tetrahedron is shown with the anhydrous composition of the various amphiboles plotted. Note that in order to consider these compositions two separate tetrahedra are superimposed: $NaAlSiO_4$–SiO_2–olivine–diopside and $CaAl_2O_4$–SiO_2–olivine–diopside. In this tetrahedron the planes Pl–Opx–Cpx (plagioclase–orthopyroxene–clinopyroxene) and Pl–Cpx–Ol (plagioclase–clinopyroxene–olivine) represent planes of saturation and critical undersaturation with respect to SiO_2.

A crucial problem in relating possible trends in the derivation of one type of basaltic magma from another is that the Pl–Opx–Cpx and Pl–Cpx–Ol planes are apparent thermal barriers in terms of liquid compositions as noted by Yoder and Tilley (1962). They believed that resorption of primary amphiboles offered a very plausible

TABLE 2 Nomenclature and Chemical Relationships of the Amphiboles

Pg	pargasite	$NaCa_2(Mg,Fe)_4AlSi_6Al_2O_{22}(OH)_2$	ferropargasite
Hs	magnesiohastingsite	$NaCa_2(Mg,Fe)_4Fe^{III}Si_6Al_2O_{22}(OH)_2$	hastingsite
Ed	edenite	$NaCa_2(Mg,Fe)_5Si_7AlO_{22}(OH)_2$	ferroedenite
Ts	tschermakite	$Ca_2(Mg,Fe)_3Al_2Si_6Al_2O_{22}(OH)_2$	ferrotschermakite
Tr	tremolite	$Ca_2(Mg,Fe)_5Si_8O_{22}(OH)_2$	ferrotremolite
Gl	glaucophane	$Na_2(Mg,Fe)_3Al_2Si_8O_{22}(OH)_2$	ferroglaucophane
Eck	eckermanite	$NaNa_2(Mg,Fe)_4AlSi_8O_{22}(OH)_2$	ferroeckermanite
Rch	richterite	$Na(CaNa)(Mg,Fe)_5Si_8O_{22}(OH)_2$	ferrorichterite
Rb	magnesioriebeckite	$Na_2(Mg,Fe)_3Fe_2^{III}Si_8O_{22}(OH)_2$	riebeckite
Arfv	arfvedsonite	$NaNa_2(Mg,Fe)_4Fe^{III}Si_8O_{22}(OH)_2$	ferroarfvedsonite

4Pg = 4Ne + 2An + 6Di + 4Fo + 2Sp
4Hs = 4Ne + 2An + 6Di + 4Fo + 2Mf
4EEd = 3Ne + 1Ab + 8Di + 6Fo
4Ts = 8An + 4Fo + 4En
4Tr = 8Di + 12En + 4Q
4Gl = 8Ab + 4En + 4Fo
4Eck = 4Ne + 8En + 4Fo + $4Na_2O.2SiO_2$
4Rch = 16En + $4Na_2O.2SiO_2$ + 4Di
4Rb = 16En + $4Na_2O.2SiO_2$ + 16Q + 4Mf
4Arfv = 14En + $6Na_2O.2SiO_2$ + 6Q + 2Mf

All reactions given above are for the magnesian end members of the series.

means by which magmas could cross these apparent thermal barriers.

Although Yoder and Tilley rationalized the formula of hastingsite (our pargasite) as:

$$3NaCa_2Mg_4AlSi_6Al_2O_{22}(OH)_2 = Di + 3An + 5Fo + 3Ne + Ak(+3H_2O)$$

there are several alternative forms. In particular, and with reference to the Cpx–Ol–Ne plane we find 3Pg = 3Ne + 6Di + 3Fo + $3Al_2O_3$ and 2Pg = 2Ne + An + 3Di + Sp + 2Fo. The suggestion that the peraluminous rather than the melilite bearing assemblages may be more appropriate, receives confirmation in Boyd's (1959) experimental study of pargasite. The importance of the Cpx–Pl–Ne plane is also demonstrated by Gilbert's (1966) study of ferropargasite.

If we are to consider the effects on magma composition of resorption of early-formed amphibole (hornblende) it is apparent that the composition of the amphibole is of crucial importance. Pargasitic and tschermakitic amphiboles seem to be most characteristic of primary amphiboles in basic rocks, while hastingsitic amphiboles are typical of the diorite–granodiorite–granite suite (Ernst, 1968). Thus the region of amphibole compositions of particular importance in the expanded simple basalt tetrahedron intersects the plane of critical silica undersaturation. From Table 2, a mixture of 1Pg (or Hs) + 4Tsch contains neither En or Ne, thus lies on the plane of critical silica undersaturation

$$2Pg(Hs) + 8Tsch = 2Ab + 17An + 3Di + 14Fo + Sp(MF).$$

In a similar sense a mixture of

$$4Pg \text{ (or Hs)} + 3{\cdot}2Tr$$

lies on the plane of critical silica undersaturation:

$$[4Pg \text{ (or Hs)} + 3{\cdot}2Tr = 4Ab + 2An + 12{\cdot}4Di + 8{\cdot}8Fo + 2Sp(Mf)].$$

This suggests that many 'early' hornblendes will be nepheline normative. Bowen (1928, p. 271) suggested the normative nepheline, albite, leucite, potassium feldspar, and diopside would become constituents of the liquid, while the anorthite, hypersthene, and olivine would be effectively precipitated through reaction with plagioclase and pyroxene. Boyd's results (1959) on pargasite suggest that at the maximum stability of Pg both An and Ne are contributed to the liquid, while diopside (aluminous), forsterite and spinel coexist with the liquid. This indicates that this liquid is on or near the Pl–Ne join of the basalt tetrahedron. We note that decomposition of 100 g of pargasite would produce approximately 40 g liquid,

$$(34{\cdot}4 \text{ g } (\text{Ne} + \text{An}) + 5 \text{ g others})$$

and 60 g remaining Di + Ol + Sp. Thus using this model, interaction of relatively small amounts of pargasite with olivine tholeiite liquid would be expected to produce a ne normative composition. Varne (1968) invoked partial melting of pargasite to generate the olivine melanephelinites and nephelinites of Moroto, Uganda. However resorption of early pargasite by an alkaline olivine basalt, or olivine basalt, magma might well produce the same result.

In their study of the melting relations of natural basalts Yoder and Tilley (1962) observed that the plagioclase was the first phase to disappear on an increase in temperature over a wide range of pressure above the solidus. This indicates that the early liquids in melting, or the later liquids in crystallization are enriched in plagioclase.

It appears that the primary field of amphibole (Pg–Ed–Tsch) crosses into the Ol–Cpx–Opx–Pl volume of the simple basalt tetrahedron at modest to high pressures in the presence of a free vapour phase. Thus under these conditions resorption of such an amphibole would not be able to produce a silica undersaturated residuum. However if the amphibole–liquid assemblage is transported to a *P–T* region nearer the surface, this amphibole could then be resorbed to produce the desired alkaline magma, particularly if a portion of the liquid had been removed.

Of interest in the case of the amphiboles is the fact that the substitution of Fe^{3+} for Al^{3+} in the octahedral sites results in little change in the upper stability limit of the amphibole (compare glaucophane and magnesio-riebeckite) though the phase assemblage produced on decomposition may be quite different. However the substitution of Fe^{2+} for Mg^{2+} in the octahedral site results in a marked depression of the upper temperature stability limit of the amphibole. In this case we may consider pargasite–ferropargasite (Boyd, 1959, Gilbert, 1966) and magnesioriebeckite–riebeckite (Ernst, 1960, 1962). The $Fe^{2+} = Mg^{2+}$ case is similar to that observed in annite–phlogopite. The $Fe^{3+} = Al^{3+}$ case is not analogous to that observed in annite–ferriannite (Eugster and Wones, 1962, Wones, 1963) where the interchange is in a tetrahedral site. This tetrahedral Fe^{3+} substitution results in a higher thermal stability of ferriannite than annite. Unfortunately there is little experimental information available on crystalline solutions between the amphibole end-members cited above, comparable to the phlogopite–annite relations.

Members of the pargasite–ferropargasite series are important constituents of hornblendes to be expected coexisting with basaltic magmas (see previous discussion of study by Yoder and Tilley, 1962). Hamilton *et al.* (1964) have concluded that a wide variety of f_{O_2} values are possible when considering the generation of a basaltic magma at depth. At 14Kb the limiting conditions are: 1400 °C, dry, $f_{O_2} = 10^{-0{\cdot}5}$ and 900 °C, saturated with respect to H_2O, 10^{-18} bars f_{O_2}. Fudali (1965) arrived at an estimated f_{O_2} of 10^{-9} to $10^{-6{\cdot}5}$ atm at 1200 °C for basaltic and andesitic magmas of the Cascades, presumably under near-surface conditions of equilibration. Hamilton *et al.* (1964) suggest that in general as a basaltic magma moves to the surface from depth, an increase in f_{O_2} will occur.

On the basis of an analogous relationship between phlogopite–pargasite and annite–ferropargasite it is possible to predict the same *general* effects on the Fe/(Fe + Mg) ratio for the amphibole series pargasite–ferropargasite coexisting with the decomposition products as for the phlogopite–annite series.

1. A reduction of f_{O_2} at constant *P*, *T* results in a more iron rich crystalline solution.

2. A reduction in T at constant f_{O_2}, P may produce a more magnesian (for pargasite rich members) composition, or a more ferroan (for ferropargasite rich members) composition.

3. A reduction of P at constant f_{O_2} and T will cause a decrease in the ferropargasite content.

The reactions analogous to those in the phlogopite–annite series are:

1. amphibole + xO_2 = amphibole* + magnetite-magnesioferrite + garnet + plagioclase + fluid,
2. amphibole + yO_2 = amphibole* + spinel phase + garnet + plagioclase + aluminous clinopyroxene + nepheline + fluid,
3. amphibole + zO_2 = amphibole* + spinel phase + aluminous clinopyroxene + olivine + plagioclase + nepheline + fluid,

and several others which are not entirely predictable. The reactions in the limiting case of pargasite (Boyd, 1959) are

4. pargasite = aluminous diopside + forsterite + nepheline + spinel + anorthite + fluid (low pressure),
5. pargasite = aluminous diopside + forsterite + spinel + melt + fluid ($P > 800$ bars).

Using these estimates of the phase relations it is apparent that the oxygen fugacity when resorption takes place may control to a very large degree which constituents of the amphibole are contributed to the melt.

The amphiboles of the granitic and intermediate rocks tend to be riebeckite–arfvedsonite or hastingsite rich. The barkevikitic amphiboles characteristic of the alkaline rocks may be approximated by an equimolar mixture of edenite and hastingsite (Table 2).

Compositionally we may consider:

arfvedsonite = sodium disilicate + acmite + 4opx,

riebeckite = 2acmite + 3opx + quartz.

Following the basic concept of Bowen (1928) we would anticipate that arfvedsonite would contribute sodium disilicate and acmite to the liquid which presumably would be saturated with opx. In a similar sense we would expect the resorption of riebeckite to contribute quartz and acmite to the liquid. Indeed this is the case as shown by Ernst's studies of magnesio riebeckite (1960) and riebeckite–arfvedsonite crystalline solutions (1962).

The major effect to be expected by the resorption of riebeckite-arfvedsonite amphiboles by a granitic magma would be a tendency to produce a peralkaline (pantellerite–comendite) suite. Carmichael and Mackenzie's (1963) study on the effects of simultaneous addition of acmite and sodium metasilicate to the Ab–Or–Q–H_2O system studied by Tuttle and Bowen (1958) are of particular importance in this case. As Tuttle and Bowen (1958) noted the system Ab–Or–Q–H_2O is a residua system only for compositions on the plane Ab–Or–Q in the system Na_2O–K_2O–Al_2O_3–SiO_2. Bailey and Schairer (1964) have presented a vigorous critique of the interpretation of the data given by Carmichael and Mackenzie (1963) but not of the data itself. Tuttle and Bowen (1958) and Luth and Tuttle (1969) have shown that the addition of excess alkali to compositions which are on the Ab–Or–Q plane has two major effects. First, the solubility of water in the silicate melt is increased significantly, and second that both the 'solidus' and 'liquidus' temperatures are decreased markedly. Both of these features are similar to those observed by Carmichael and Mackenzie. Since resorption of arfvedsonite-riebeckite crystalline solutions would contribute sodium disilicate + acmite + quartz to the liquid in addition to H_2O, it seems quite possible that one result of such resorption would be to produce a water-undersaturated magma.

The major effect of resorption of hastingsite-barkevikite amphiboles by a granitic magma would be to contribute Ne, Ab, An to the liquid. This then would tend to 'desilicate' the granitic melt.

The situation changes considerably when we consider the resorption of sodic amphiboles by granodioritic or dioritic melts. In this case we must consider the relations involving calcic plagioclase (either potential or already precipitated). If the amphibole contributes sodium disilicate to the liquid then the plagioclase in equilibrium with that liquid should be converted to a more sodic plagioclase. Simultaneously one would expect the 'free' Ca to interact with opx to generate cpx. Similar phenomena would be expected in terms of the acmite, though possibly magnetite would be an important crystalline product. Thus resorption of a sodic amphibole by an intermediate magma offers a possible mechanism for the generation of alkaline magmas. However the available data suggests that 'normal' intermediate magmas coexist with a hastingsite-barkevikitic amphibole, rather than riebeckite-arfvedsonite crystalline solutions. Ernst (1968) and Borley and Frost (1963) have noted the bulk composition control (Ab/An) on this feature. Consequently we might expect the more common case to involve resorption of a hastingsite amphibole by an intermediate magma. In general the consequence of such interaction would be to desilicate the magma, due to the normative ne and ol of hastingsite. Presumably the greatest effects, in terms of producing alkaline magmas would be expected of dioritic, monzonitic and andesitic magmas.

VI.6.5. Conclusions

This brief review of some aspects of resorption demonstrates that this process may play a rôle significant in the generation of alkaline magmas and rocks, as suggested by Bowen (1928), Holmes and Harwood (1937) and others.

The treatment given here is largely qualitative, yet we find little to deny the importance of the process. Though more experimental data will help in understanding the details of the resorption process, the evidence concerning the extent to which the process is active in nature must come from field and petrographic studies.

Resorption of complex minerals such as amphiboles and biotites is phenomenologically inseparable from the process normally thought of as magmatic reaction, such as the common olivine–pyroxene reaction observed in many igneous rocks, and believed to have taken place in the presence of the magma. In both cases the subject matter is the influence of changing environmental parameters (pressure, temperature, f_{H_2O}, f_{O_2}, etc.) on the mineral assemblage which stably coexists with the melt. Thus the term resorption is equally appropriate (or inappropriate) to such contrasted relationships as the olivine–pyroxene, leucite–potassium feldspar reaction relationships as to the more complicated (and less well understood) reaction relationships involving the biotites and amphiboles.

VI.6 REFERENCES

Bailey, D. K., and Schairer, J. F., 1964. Feldspar–liquid equilibria in peralkaline liquids—the orthoclase effect. *Am. J. Sci.*, **262**, 1198–206.

Borley, G. D., and Frost, M. T., 1963. Some observations on igneous ferrohastingsites, *Mineralog. Mag.*, **33**, 646–62.

Bowen, N. L., 1914. The ternary system diopside–forsterite–silica, *Am. J. Sci.*, **38**, 207–64.

Bowen, N. L., 1928. *The Evolution of the Igneous Rocks*. Dover Publications, 332 pp.

Boyd, F. R., 1959. 'Hydrothermal investigations of amphiboles,' in Abelson, P. H., *Researches in Geochemistry*, John Wiley and Sons, 377–96.

Boyd, F. R., and Schairer, J. F., 1964. The system $MgSiO_3$–$CaMgSi_2O_6$. *J. Petrology*, **5**, 275–309.

Carmichael, I. S. E., 1967. The mineralogy and petrology of the volcanic rocks from the Leucite Hills, Wyoming. *Contr. Miner. Petrology*, **15**.

Carmichael, I. S. E., and MacKenzie, W. S., 1963. Feldspar–liquid equilibria in pantellerites; an experimental study. *Am. J. Sci.*, **261**, 382–96.

Clark, S. P., Jr., 1966. Handbook of physical constants, Geol. Soc. Mem., **97**.

Ernst, W. G., 1960. Stability relations of magnesioriebeckite, *Geochim. et Cosmochim. Acta*, **19**, 10–40.

Ernst, W. G., 1962. Synthesis, stability relations, and occurrence of riebeckite and riebeckite-arfvedsonite solid solutions. *J. Geol.*, **70**, 689–736.

Ernst, W. G., 1968. *Amphiboles, Crystal Chemistry, Phase Relations, and Occurrence*. Springe-Verlag New York Inc., 59 fig., 125 pp.

Eugster, H. P., and Wones, D. R., 1962. Stability relations of the ferruginous biotite, annite. *J. Petrology*, **3**, 82–125.

Fudali, R. F., 1965. Oxygen fugacites of basaltic and andesitic magmas. *Geochim. et Cosmichim. Acta*, **29**, 1063–75.
Gilbert, M. C., 1966. Synthesis and stability relationships of ferropargasite. *Am. J. Sci.*, **264**, 698–742.
Hamilton, D. L., Burnham, C. W., and Osborn, E. F., 1964. The solubility of water and effects of oxygen fugacity and water content on crystallization in mafic magmas. *J. Petrology*, **5**, 21–39.
Holmes, A., and Harwood, H. F., 1937. The volcanic area of Bufumbira, Part II, The petrology of the volcanic field of Bufumbira, South west Uganda and of other parts of the Birunga field. *Geol. Surv. Uganda, Mem.*, **3**, 300 pp.
Kushiro, I., and Schairer, J. F., 1964. The join diopside–akermanite. *Carnegie Inst. of Wash. Year Book*, **63**, 132–3.
Luth, W. C., 1967. Studies in the system $KAlSiO_4$–Mg_2SiO_2–H_2O: I, Inferred phase relations and petrologic applications. *J. Petrology*, **8**, 372–416.
Luth, W. C., and Tuttle, O. F., 1969. 'The hydrous vapor phase in equilibrium with granite, and granite magmas,' in Larsen, L. H., Ed., Poldervaart Memorial volume, *Geol. Soc. Am. Mem.*, 513–48.
O'Hara, M. J., 1965. Primary magmas and the origin of basalts. *Scot. J. Geol.*, **1**, part 1, 19–40.
Schairer, J. F., and Bowen, N. L., 1955. The system K_2O–Al_2O_3–SiO_2. *Am. J. Sci.*, **253**, 681–746.
Tuttle, O. F., and Bowen, N. L., 1958. Origin of granite in the light of experimental studies in the system $NaAlSi_3O_8$–$KAlSi_3O_8$–SiO_2–H_2O. *Mem. Geol. Soc. Am.*, **74**, 153 pp.
Varne, R., 1968. The petrology of Moroto Mountain, eastern Uganda, and the origin of nephelinites. *J. Petrology*, **9**, 169–90.
Wones, D. R., 1963. Phase equilibria of ferriannite, $KFe_3^{2+}Fe^{3+}Si_3O_{10}(OH)_2$. *Am. J. Sci.*, **261**, 581–96
Wones, D. R., and Eugster, H. P., 1965. Stability of biotite; experiment, theory and application. *Am. Miner.*, **50**, 1228–72.
Yoder, H. S., and Eugster, H. P., 1954. Phlogopite synthesis and stability range. *Geochim. et Cosmochim. Acta*, **6**, 157–85.
Yoder, H. S., and Kushiro, I., 1969. Melting of a hydrous phase: Phlogopite. *Am. J. Sci.*, **267a**, Schairer vol., 558–82.
Yoder, H. S., and Tilley, C. E., 1962. Origin of basaltic magmas: an experimental study of natural and synthetic systems. *J. Petrology*, **3**, 342–532.

VI. 7. THE RÔLE OF METASOMATIC PROCESSES IN THE FORMATION OF ALKALINE ROCKS

L. S. Borodin and A. S. Pavlenko

VI.7.1. Introduction

Metasomatic processes are considered to have been essential or predominant in the formation of certain types of alkaline rocks. According to geological and experimental data one may distinguish progressive and regressive stages of alkaline metasomatism. The progressive stage embraces regional metasomatism preceding anatexis, and metasomatism at the contacts of alkaline intrusions and is caused by increasing temperature. Postmagmatic alkaline metasomatism belongs to the regressive stage. The products of progressive metasomatism occur only at the exocontacts of intrusions and anatectic bodies. A special case is the regionally metasomatized alkaline rocks which appear to be without direct connection with magmatic rocks (Glagolev, 1966; Tugarinov *et al.*, 1963; Pavlenko, 1959). In this case progressive or regressive trends of metasomatism may be distinguished by the temperature sequence of mineral generation. This chapter deals with metasomatic processes connected with the alkaline ultramafic, nepheline syenitic and alkaline granitoid formations, since metasomatism appears to have been most widely developed in these associations.

VI.7.2. Alkaline Ultramafic Complexes

This formation is characterized by nepheline-pyroxene rocks belonging to the series alkali pyroxenite (jacupirangite) – melteigite – ijolite. These rocks are often associated with ultramafic rocks and carbonatites, thus forming complex massifs (cf. IV.2)*.

In shallow intrusions such as stocks and dykes, metasomatic processes are not important in the formation of alkaline rocks. Such rocks are in these cases interpreted as differentiates of primary alkaline magmas. Many deep-seated massifs represent, however, another extreme case, metasomatic rocks dominating over magmatic ones. The most typical examples are found in complex massifs, where cores of pyroxenite and olivinite are surrounded by concentric zones of ijolite-melteigite and other alkaline rocks. In such massifs the central bodies (cores) are built up of the oldest rocks contrary to intrusions of the so-called central type where the central stock represents the youngest rocks (Anderson, 1936).

In some massifs relics of olivinites and pyroxenites are distributed over most of the area occupied by alkaline ultramafic rocks, the latter being characterized by rather variable modal compositions and by corrosive metasomatic textures. It is believed, therefore, that such alkaline ultramafic rocks are formed by metasomatic transformation of pyroxenites and olivinites which originally made up stocks and other types of intrusive bodies. Alkaline solutions and magmas, which caused the high-temperature alkaline metasomatism (nephelinization, phlogopitization, etc.), were introduced along the contacts of the ultramafic bodies (stocks) with the country rocks. As a result of the irregular development of these processes the minerals of the primary ultramafic rocks are partly dissolved and replaced by alkaline minerals—nepheline, phlogopite, etc.—and partly undergo recrystallization. In this way, olivine-phlogopite rocks, nepheline-pyroxene rocks, and other metasomatites of very variable composition arise. The zones of metasomatites precede magmatic ijolites and other intrusive rocks. Dykes and veins of ijolite quite often intersect the metasomatic rocks. The above-mentioned processes may therefore be related directly to a magmatic stage (progressive metasomatism).

* Readers are referred to the relevant Chapters of this book where similarly cited.

VI.7.2.1. Nephelinization and Phlogopitization

Nephelinization is a major metasomatic process in deep-seated complexes of alkaline ultramafic rocks. The great importance of this process in the formation of ijolites was first established at Spitskop (Strauss and Truter, 1951), where alkaline emanations appear to have ascended in front of the intruding magma along the zone of contact between a pyroxenite intrusion and the surrounding country rocks. These emanations attacked both types of rocks. Thus, an aureole of fenites arose around the intrusion, while the intrusive pyroxenites were nephelinized and transformed into nepheline-pyroxene rocks of ijolitic type. The poor degree of exposure at Spitskop and the lack of comparative data on similar complexes prevented a detailed evaluation of the rôle and the degree of nephelinization in this massif. In this respect, the carbonatite provinces of the Kola peninsula (IV.2) and Polar Siberia (III.4) are more illustrative (Borodin, 1957, 1958a, b, 1961, 1962, 1965; Egorov, 1960; Lapin, 1963). In these provinces there are fine examples of a successive transition from ultramafic massifs, in which alkaline rocks are poorly developed, into ijolite-melteigite ones, where, by contrast, primary ultramafic rocks are represented only by irregular relic inclusions in the main alkaline rocks (Fig. 1).

In the initial stage of nephelinization of pyroxenite separate veinlets and dispersed spots of nepheline impregnation occur. In this case metacrystals of nepheline corrode the pyroxene and are filled with small irregular remnants of the latter. In subsequent stages the pyroxene is partly dissolved and the rock becomes increasingly saturated with nepheline. The residual pyroxene recrystallizes at the same time into alkali pyroxene and is partly replaced by phlogopite and garnet. Thus, ijolite-like metasomatites arise in place of pyroxenite. Similar processes have been described by King and Sutherland (1967) from Napak.

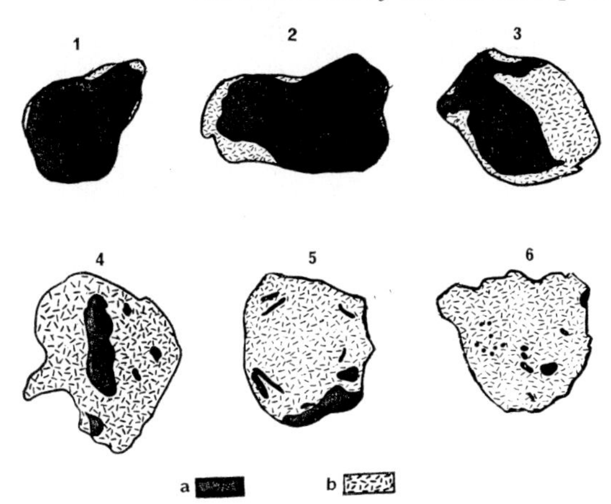

Fig. 1 Stages of transformation of ultramafic rocks into alkaline–ultramafic metasomatites in the following massifs: (1) Bor–Urjakh; (2) Vuorijarvy; (3) Kugda; (4) Kovdor; (5) Salma; (6) Odikincha. 1,3,6 are situated in Polar Siberia; 2,4,5 in Karelia and the Kola Peninsula

(a) Pimary ultramafic rocks.

(b) Alkaline ultramafic and other metasomatic rocks.

These rocks correspond to any member of the urtite–jacupirangite series, but under closer petrographic examination a number of features are found which make it possible to distinguish them from their magmatic equivalents.

1. Textural heterogeneity and irregularity of composition in one and the same specimen due to uneven distribution of nepheline. Nepheline veinlets as well as bands and pockets enriched in nepheline are typical.
2. General presence of textures indicating corrosion and blastic growth. Numerous relics of earlier, often partly modified ultramafic rocks.
3. Rims of phlogopite and garnet or melilite around porphyroblasts and veinlets of nepheline as a result of local garnetization, phlogopitization and melilitization of pyroxene and its partial aegirinization. The latter process is marked by changes of colour from pale to green when the pyroxenes are studied in thin sections.

In a number of fairly large massifs, which have preserved cores of the initial ultramafic rocks, ijolite-like metasomatites and magmatic ijolites most often form a ring zone along the periphery of the massifs. A series of intermediate metasomatic zones occurs between this marginal zone and the ultramafic core. Phlogopite and sometimes melilite are the main minerals in these intermediate zones, while nepheline often plays a minor rôle or is absent. The reactional metasomatic zoning is best noted where remnants of olivine cores are present. In such cases it is evident that most of the 'pyroxenites' are of non-intrusive metasomatic origin. Such 'ultrabasites' are essentially distinguished from their intrusive counterparts by the pattern of distribution of the minor elements Nb, RE, Cr and Ni (Borodin, 1965).

It is thus possible to establish the combined appearance of the following metasomatic processes (from periphery to centre of massifs): nephelinization $\rightarrow$ phlogopitization $\rightarrow$ pyroxenization.

The whole complex of simultaneous metasomatic zones forms a single metasomatic column of nephelinization. In this column the rear zones gradually replace the frontal ones. In the extreme rear zone of nepheline pyroxenite metasomatites metasomatism may be succeeded by melting of rocks and crystallization of magmatic ijolites. The most typical column of metasomatites may be represented by the following scheme (cf. Table 1):

	III	II	I	
Magmatic ijolite	Nepheline-pyroxene rocks (Gr, Phl, Mel, Mag)	Phlogopite-pyroxene (Gr, Per, Mag)	Pyroxene rocks (Phl, Per, Mag)	Olivinites (pyroxenites)

The most typical minor minerals are Gr—garnet–schorlomite; Phl—mica from phlogopite–biotite group; Mel—melilite; Mag—titanomagnetite and magnetite; Per—perovskite.

This concentric zonal structure may cover areas of tens of square kilometres and has previously been considered to be a result of several successive stages of intrusive activity. This may be illustrated by the Kovdor massif, the Kola peninsula (Fig. 2). The same sequence of zones also exists *en miniature*, i.e. around relics of olivinite or on the contact of olivinite and

TABLE 1 Average Chemical Analyses of the Metasomatic Zones of the Kovdor Massif, the Kola Peninsula (Kukharenko *et* al., 1965; Volotovskaja, 1958; Rimskaja-Korsakova, 1968)

	Olivinite		I Pyroxenite			II Phlogopite–Pyroxene Rocks			IIIa Melilite–Pyroxene Rocks			IIIb Nepheline–Pyroxene Rocks		
	Wt %	Cation %	Wt %	Cation %	Gain (+) Loss (−)	Wt %	Cation %	Gain (+) Loss (−)	Wt %	Cation %	Gain (+) Loss (−)	Wt %	Cation %	Gain (+) Loss (−)
SiO_2	37·05	32	43·29	42	+10	43·30	41	−1(+9)	39·79	37	+5(−4)	42·18	39	+7(−2)
TiO_2	0·06		1·59	2	+2	1·09	1	−1(+1)	1·37	1	+1(0)	2·13	1	+1(0)
Al_2O_3	1·50	2	2·59	3	+1	5·20	6	+3(+4)	10·61	11	+9(+5)	11·36	13	+11(7)
Fe_2O_3	7·16	5	10·00	7	+2	8·60	7	0(+2)	4·02	3	−2(−4)	6·34	5	0(−2)
FeO	6·50	5	6·01	5	0	5·22	4	−1(−1)	4·18	3	−2(−1)	5·26	4	−1(0)
MnO	0·36		0·14			0·12			0·24			0·14		
MgO	42·86	54	14·58	20	−34	16·00	22	+2(−32)	12·76	17	−37(−5)	9·30	13	−41(−9)
CaO	2·15	2	20·19	21	+19	18·40	18	−3(+16)	19·94	20	+18(+2)	15·27	15	+13(−3)
Na_2O			0·25			0·32			3·48	6	+8(+7)	4·03	7	
K_2O	0·23		0·35			1·50	1	+1(+1)	2·09	2		2·17	3	+10(+9)
P_2O_5	0·35		0·11			—			0·14			0·49		
$H_2O(CO_2)$	2·05		0·88			0·40			1·06			0·94		
	100·27	100	99·96	100		100·00	100			100		99·61	100	

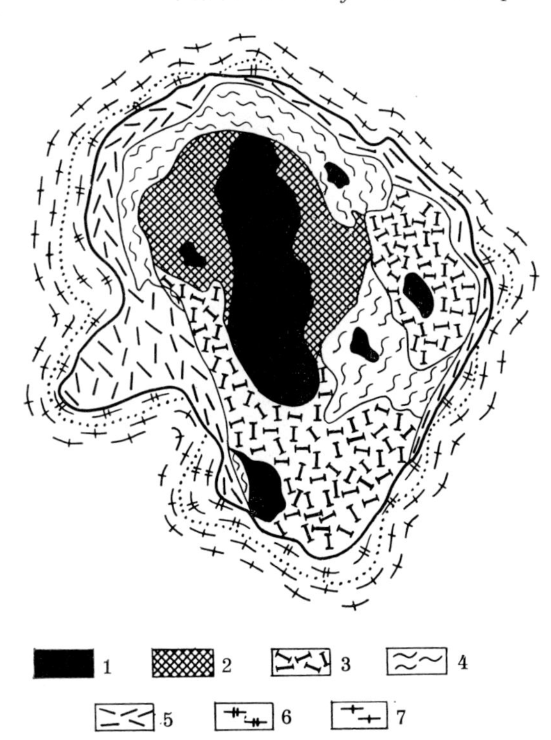

Fig 2 Schematic geological map of the Kovdor massif. (1) Olivinites; (2) pyroxenized and phlogopitized olivinites; (3) nepheline-pyroxene rocks (jacupirangite–melteigite); (4) turjaites and other melilite rocks; (5) ijolites and other nepheline–pyroxene rocks; (6) fenites; (7) country rocks (granite–gneisses, a.o.)

nepheline-pyroxene veinlets (Fig. 3). Anchimonomineralic compositions are found locally in zones of polymineralic metasomatites (patches of glimmerite among phlogopite-pyroxene rocks, 'urtite' lenses in nepheline-pyroxene rocks, etc.).

In view of this tendency towards a monomineralic composition of the metasomatites, the chemical features of an alkaline metasomatic process may be represented by the following principal reactions:

1. Nephelinization of pyroxene

$2Ca(Mg,Fe)Si_2O_6 \rightarrow 3(Na,K)AlSiO_4$
or when Mg : Fe = 4 : 1 and Na : K = 4 : 1
$Ca_2Mg_{1\cdot 6}Fe_{0\cdot 4}Si_4O_{12} + 2{\cdot}4Na + 0{\cdot}6K + 3Al = Na_{2\cdot 4}Ka_{0\cdot 6}Al_3Si_3O_{12} + 2Ca + 1{\cdot}6Mg + 0{\cdot}4Fe + Si.$

2. Phlogopitization of pyroxene

$2Ca(Mg,Fe)Si_2O_6 \rightarrow K(Mg,Fe)_3AlSi_3(O,OH)_{12}$
or $Ca_2Mg_{1\cdot 6}Fe_{0\cdot 4}Si_4O_{12} + K + Al + 0{\cdot}65Mg + 0{\cdot}35Fe = KMg_{2\cdot 25}Fe_{0\cdot 75}AlSi_3(O,OH)_{12} + 2Ca + Si.$

3. Pyroxenization of olivine

$3(Mg,Fe)_2SiO_4 \rightarrow 2Ca(Mg,Fe)Si_2O_6$
or when olivine and pyroxene have the same Mg/Fe ratio $Mg_6Si_3O_{12} + 2Ca + Si = Ca_2Mg_2Si_4O_{12} + 4Mg.$

These reactions are calculated for the volume of twelve oxygen atoms. When making more accurate calculations it is necessary to take the differences in density into consideration (Borodin, 1961).

These reactions indicate that the above mentioned combination of simultaneous metasomatic processes depends to a great extent on migration of Ca, Mg and Fe, which are successively released in the interaction between alkaline solutions and ultramafic rocks, and on the high activity of the bases in the alkaline solutions. The enrichment of these components as well as of Ti and rare elements causes the development of minerals like titanomagnetite, perovskite and garnet (melanite) in the partly transformed olivinites and pyroxenites. Zones I and II in the metasomatic column of nephelinization, which are composed of pyroxene (augite-diopside, salite), garnet, iron ores, perovskite and 'alkaline' minerals (phlogopite, melilite, nepheline), may be considered as alkaline skarns.

Table 1 shows a comparison of chemical analyses of the main metasomatic petrographic zones of the Kovdor massif (Fig. 2). Such zones are also typical of other complex massifs of Karelia and the Kola peninsula. The pyroxenites correspond to the frontal zone of the metasomatic column (zone I), the phlogopite-pyroxene rocks to the intermediate one (zone II), turjaites (IIIa) and nepheline-pyroxene rocks of ijolite-melteigite type (IIIb) to the rear zones. As the rocks have massive textures and fairly similar densities the nature of the change in composition is evident from direct comparison

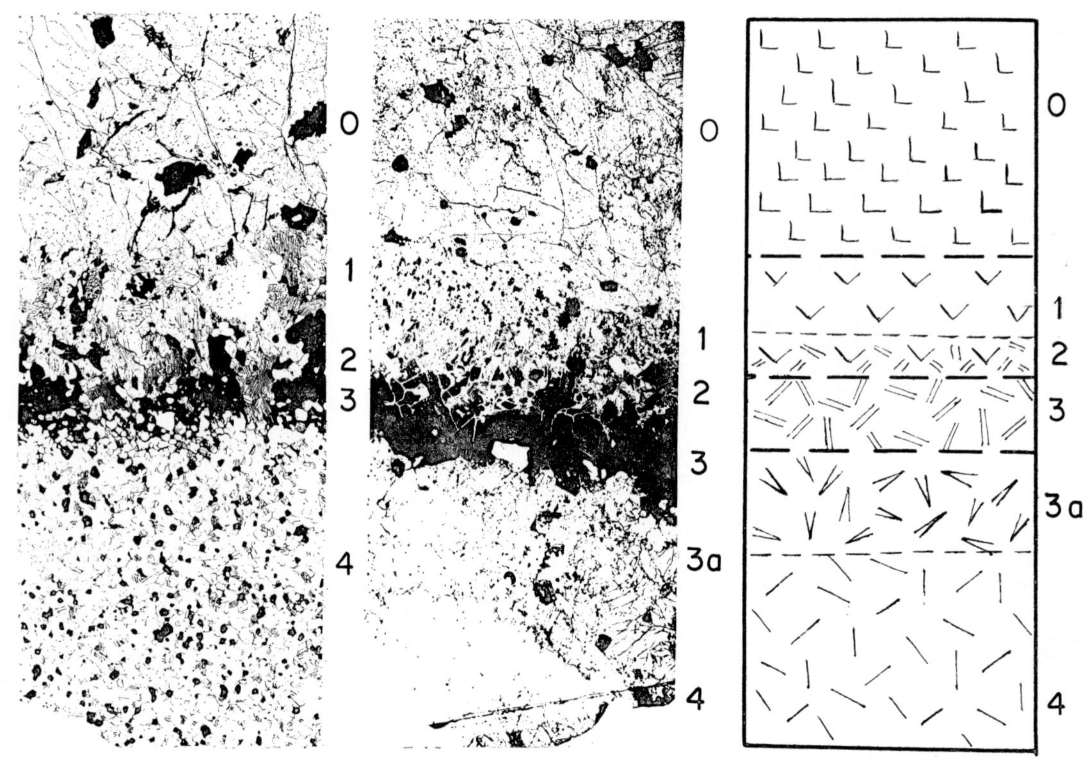

Fig. 3 Metasomatic zoning of the column of nephelinization along the contact of olivinites and ijolites at Odikincha, Polar Siberia. O—olivinite. (1) pyroxene zone with impregnations of titano-magnetite and perovskite ('ore pyroxenite'); (2) phlogopite–pyroxene zone ('micatized pyroxenite'); (3) phlogopitized zone ('glimmerite'); (3a) melilite zone ('turjaite'); (4) ijolite

of analytical data. It is seen from Table 1, that up to 30–40% of the amount of cations, essentially Mg, Ca, Al and alkalis as well as Fe and Ti is involved in the process of replacement of the primary olivinites. The chemical composition changes to a lesser degree when passing from one zone of the metasomatic column to the next (the figures in brackets).

The above considerations lead to the conclusion that the nephelinization of ultramafic rocks, as well as the transformation of such rocks into alkaline skarns, is characterized by a number of features which are also typical of other high-temperature infiltrative metasomatic processes (large scale of metasomatism and of introduction of material, variable composition of rock-forming minerals, etc.).

This conclusion based on geological observations is also in good agreement with the results of experiments performed by us to obtain some physical and chemical verification of the reactions of nephelinization and phlogopitization. These reactions were performed during hydrothermal synthesis at 550 °C and pressures of 1000–1200 atm. The main purpose of the experiments was to synthesize nepheline and phlogopite without supply of silica, i.e. at the cost of pyroxene and olivine during their interaction with alkaline solutions (pH = 11–12). Nephelinization of pyroxene (diopside–augite) was performed according to the scheme:

pyroxene + NaOH + $KAlO_2$ + H_2O → nepheline (+ phlogopite + garnet)

Nepheline was synthesized as prismatic crystals up to 0·8 mm long. It contains 11·8% of Na and 6·9% of K.

'Phlogopitization' was performed according to the scheme:

olivine + $KAlO_2$ + NaOH + CaO + H_2O → phlogopite (+ nepheline, perovskite).

The initial olivine contained 10–12% fayalite. Phlogopite was synthesized as light green leaflets (Ng = Nm from 1·570 to 1·610) and has the following composition:

SiO_2—32·5%; Al_2O_3—15·4%; Fe_2O_3—1·9%; FeO—4·5%; MgO—23·1%; Na_2O—3·8%; K_2O—5·8%; H_2O—13·4%.

VI.7.2.2. Fenitization Around Alkaline Ultramafic Rocks

As stated above, fenitization (alkaline syenitization) of the country rocks (granites, granite gneisses, a.o.) is closely associated with nephelinization. In the process of fenitization quartz and primary coloured minerals, first of all biotite, are progressively replaced by alkali amphiboles, pyroxenes, acid plagioclase or potassic orthoclase. In the process of intense fenitization the granites, gneisses, etc. enclosing alkaline massifs may be replaced by an aureole composed of the following zones: an outer zone of syenitic fenites; an intermediate zone of pyroxene syenite fenites and an inner zone of nepheline syenitic fenites. Melting of these alkali and nepheline syenites is a culmination of this process. Some authors suggest that such rheomorphic rocks can intrude into overlying horizons giving foyaite and sometimes carbonatite (King and Sutherland, 1960). Aureoles of fenitization may arise around plugs and dykes of carbonatites also when ijolites are absent (von Eckermann, 1948; Garson, 1962, 1967).

Postmagmatic (regressive) metasomatism is of minor importance in complexes of ultramafic alkaline rocks. Local phlogopitization and garnetization of the alkali ultramafic and the ultramafic rocks may be associated with residual solutions of the post-ijolitic stage. The hydrothermal stage of formation of carbonatites is characterized by carbonatization, amphibolization and phlogopitization.

VI.7.3. Nepheline Syenitic and Alkaline-Granitoid Complexes

Two types of progressive alkaline metasomatism—one leading to anatexis and the other manifested at the contacts of alkaline intrusions, give rise to similar results. This is supported by mineralogical data, e.g. the information about the rock-forming minerals of alkaline rocks of different genesis summarized by Perchuk (1968a, b).

A problem is whether the source of the agent, which causes the metasomatism, is of subcrustal origin (juvenile) or magmatic (derived from crustal magma chambers). Regressive metasomatic processes are displayed mainly as postmagmatic alteration of igneous rocks.

VI.7.3.1. Nepheline Gneisses as Products of Progressive Alkali Metasomatism

In the Haliburton–Bancroft district, Ontario (cf. II.2 and II.7) numerous small bodies of alkaline rocks of miaskitic type (canadite, congressite, monmouthite, craigmontite, and others, which are usually confined to carbonate rocks), are developed among Pre-Cambrian gneisses, amphibolites, dolomites and marbles. A number of investigators have explained their appearance by effect of granitic emanations. Gummer and Burr (1946) suggested that nephelinization could in general be caused by adding Na_2O, K_2O and Al_2O_3 to calcic paragneisses with the appropriate removal of CaO and CO_2. Moyd (1949) explained the nephelinization as a result of successive reactions of granitic emanations, which were previously desilicated by limestone, with the dark gneisses of the Grenville series. On the other hand, Tilley and Gittins (1961) and Gittins (1961) associated the nephelinization with emanations from nepheline syenitic and theralitic intrusions. Tilley (1958) suggested the following series of replacement in the York-River area:

1. Limestone → calcite–biotite–plagioclase → biotite–plagioclase–nepheline → biotite–nepheline

TABLE 2 Gains and Losses Involved in the Transformation of Marble into Nepheline-Bearing Rocks (Wt %)

	1	2	3	4	5	6	7	8	9	10	11	12	13	14
SiO_2	6·60	47·2	45·11	40·0	41·25	0·66	+13·8	4·03	+136	41·80	+4·0	44·16	44·09	54·63
TiO_2	0·08	0·3	0·43	0·4	0·48	—		—	+0·6	0·17	+4·3	0·95	2·04	0·58
Al_2O_3	2·38	17·0	26·42	17·3	28·42				+14·3	4·81	+4·2	6·94	9·76	20·94
Fe_2O_3	0·11	0·6	1·31	0·9	2·37	0·38		1·00	+6·9	1·89	+26·0	7·37	5·85	2·52
FeO	0·96	4·0	4·78	1·9	2·84				+5·8	3·99	+30·4	9·73	11·01	3·21
MnO	0·02	—	0·04	—	0·01	—		—		0·25		0·50	0·41	0·18
MgO	4·00	−3·0	1·09	−3·4	0·04	23·27	+16·8	21·47	−8·4	11·31	−20·3	5·94	3·53	0·78
CaO	47·22	−34·9	5·51	−34·9	5·84	35·68	−9·0	37·87	+4·0	28·44	−18·6	19·43	14·04	1·87
Na_2O	0·19	4·4	4·08	11·0	10·48	0·08		—		0·96		2·0	4·32	7·70
K_2O	1·56	2·6	4·49	2·0	4·07	0·06		—		0·33		0·55	1·87	5·79
P_2O_5	0·01	0·1	0·10		0·01	—		—		—		0·04	—	—
CO_2	35·50	−36·6	2·08	−36·6	2·29			33·62		6·78		2·37	0·94	1·84
H_2O^+	1·66	1·6	4·58	0·3	1·32	39·87		—		0·11		—	0·05	0·23
H_2O^-	0·13		0·11		0·05	—		—		0·39		—	0·89	1·46
	100·42		100·13		100·37	100·00		97·89		100·73		99·98	99·02	100·15

Legend

1–5. Haliburton–Bancroft District. (Gittins, 1961).
1. Marble with phlogopite and some diopside, York River.
2. Gains and losses calculated on a unit-volume basis to convert marble (1) into nepheline-bearing gneiss (3).
3. Biotite–plagioclase–nepheline–(calcite) gneiss.
4. Gains and losses calculated on a unit-volume basis to convert marble (1) into nepheline-bearing gneiss (5).
5. Nepheline–plagioclase–biotite–hornblende–(calcite) gneiss.

6–14. North-Eastern Tuva. (Kovalenko *et al.*, 1968).
6. Dolomite marble.
8. Forsterite calciphyre.
10. Fassaite skarn.
12. Endomorphic ijolite.
7, 9, 11. Gains and losses in g/100 cm^3 in the corresponding transformation.
13. Melanocratic nepheline syenite.
14. Foyaite.

2. Limestone → calcite–hastingsite–plagioclase → hastingsite–plagioclase–nepheline → hastingsite–clinopyroxene–nepheline

Gittins (1961) calculated the material balance in the transformation of skarn-marbles into nepheline-bearing gneisses according to the above mentioned types of reactions (Table 2). Besides, Gittins noted the microcline, albite, scapolite, grossularite, apatite and magnetite in the nepheline-bearing gneisses associated with nepheline-free gneisses.

Appleyard (1967, 1969) has interpreted the nepheline gneisses of the Wolfe Belt as products of metamorphism under upper amphibolite facies conditions of sedimentary rocks during a static interval prior to the second of three periods of folding.

A similar geological situation may be observed in the Vishnevogorsk and Ilmenogorsk alkaline massifs of the middle Urals (cf. I and II.2), which were investigated in detail by Ronenson (1966). The initial rocks were greenschists, which were metamorphosed into gneisses containing plagioclase, biotite, sillimanite and garnet. According to reliable geological data, the progressive alkaline metasomatism was caused by juvenile alkaline solutions and controlled by tectonic zones. In the course of metasomatism the mineralogical and chemical compositions of the rocks approached those of granitic or miaskitic eutectics and fusion of these compositions completed the process. According to Ronenson (1966) the development of nepheline gneisses and melts, when compared with granitic ones, require more time for the metasomatic action under stress conditions.

Table 3 shows the metasomatic changes involved in the formation of miaskitic rocks and also the accompanying physical changes (volumetric weight, porosity). As seen from the table the progressive metasomatism took place not only by displacement of the main elements, but also by the rare ones: Ti, Zr, RE and P. A similar phenomenon has been mentioned in Chapter IV.5, Fig. 4.

In the Mongol–Tuva province palingenetic and metasomatic formations of alkaline rocks are widespread. Among the effusive-sedimentary intermediate and subsilicic rocks there are linear zones of gneisses and fenitic gneisses with rootless lens-shaped and vein-like, anatectic bodies of alkali and nepheline syenites (Tugarinov *et al.*, 1968). These bodies have close spatial and temporal relations. The palingenetic metasomatic series is characterized by a great variety of mineral facies which each display a special line of petrochemical evolution, (see Fig. 4, where foyaitic (6), miaskitic (7) and granosyenitic (8) series are shown).

VI.7.3.2. Alkali Granitic Gneisses as Products of Progressive Alkali Metasomatism

The alkaline-granitoid palingenetic and metasomatic formations (alkali gneisses, cf. II.7) are more widespread than those of nepheline syenites. Many authors give convincing evidence that juvenile solutions are responsible for the genesis of such rocks: Chumakov (1958)—the Keivy formation, Kola Peninsula; King (1955)—the granites of Semarule, Bechuanaland; Korikovsky (1967)—the Aldan Shield and the Charsky block, Yakutia.

VI.7.3.3. Products of Progressive Metasomatism in Exocontact Zones of Alkaline Intrusions (Fenitization)

These metasomatic rocks are extremely variable (cf. IV.5, Table 2) and like palingenic rocks they appear to approach the mineralogical and chemical compositions of igneous rocks towards the intrusive contacts. They differ from palingenic rocks in the reaction relation displayed by endocontact igneous rocks with the country rocks resulting in differences between the marginal and central parts of intrusions. This effect, described by many authors (Zharikov, 1959; Kovalenko and Popolitov, 1965 and 1969; Kovalenko *et al.*, 1969; Marakuschev *et al.*, 1968; Perchuk, 1963, 1964; Pavlenko, 1963a, b) is a result of acid–basic interactions of the components of melts and solutions, as stated by Korzhinsky (1959, 1960). Infiltrative magmatic replacement may be generalized as follows: the more basic the enclosing rock, the more alkaline and less

TABLE 3 The Mineral Reactions and Chemical Changes Involved in the Transformation of Greenschists into Miaskite, Vishnevogorsk, the Urals (Ronenson, 1966)

N	Mineral Reactions	v	p	O	Si	
0.	$0{\cdot}2Mu + Hl + 8{\cdot}1Q + 0{\cdot}2Ka + 5{\cdot}7Hm + (1{\cdot}3K + 0{\cdot}3Na)\rightarrow$	2·58	3·48	125·02	83·40	C_0
1.	$\rightarrow 1{\cdot}1Bi_1 + 0{\cdot}4Mu + 0{\cdot}3Pl_{20} + 5{\cdot}0Q + 2{\cdot}1Mt + 0{\cdot}2Sl + 0{\cdot}2Alm + 5{\cdot}5H_2O + 0{\cdot}2CO_2$;					
	$5{\cdot}8Pl_{30} + 1{\cdot}35Bi_2 + 6{\cdot}65Q + 0{\cdot}6Mt + (1{\cdot}35K + 1{\cdot}05Na + 0{\cdot}3H_2O)\rightarrow$	2·61	3·52	125·89	81·60	C_1
		—	—	+0·87	−1·8	Δ^{1}_{2-3}
2.	$\rightarrow 5{\cdot}4Pl_{10} + 1{\cdot}3Kf_1 + 1{\cdot}65Bi_3 + 2{\cdot}0Q + 0{\cdot}3Mt$;					
	$2{\cdot}0Pl_{10} + Kf_1 + Bi_3 + Q + 0{\cdot}85Mt + (1{\cdot}5Ca + 0{\cdot}1Na + 2{\cdot}450)\rightarrow$					
		2·56	5·89	121·96	80·20	C_{2-3}
3.	$\rightarrow 1{\cdot}05Pl_{30} + Kf_1 + Gs_1 + 1{\cdot}1Q + 0{\cdot}2Mt + (0{\cdot}35K + 0{\cdot}6H_2O)$;					
	$18{\cdot}5Pl_{30} + 15Kf_1 + 5{\cdot}0Hs_1 + 15{\cdot}4Q + 2{\cdot}35Mt + (0{\cdot}37H_2O + 2{\cdot}7Mg + 6{\cdot}0Na + 0{\cdot}9K)\rightarrow$					
		—	—	−3·06	−3·20	Δ^{2-3}_{4-5}
4.	$\rightarrow 15Kf_2 + 7{\cdot}0Hs_2 + 18{\cdot}5Pl_{10} + (0{\cdot}9Ca)$;					
	$Kf_2 + 0{\cdot}52Hs_2 + 0{\cdot}5Pl_{10} + (0{\cdot}55Mg + 1{\cdot}22K + 0{\cdot}84H_2O)\rightarrow$					
		2·50	6·72	117·64	73·20	C_{4-5}
5.	$\rightarrow 1{\cdot}51Kf_2 + 1{\cdot}04Bi_4 + (0{\cdot}78Ca + 0{\cdot}95Na + 0{\cdot}680)$;					
	$8{\cdot}2Kf_2 + 1{\cdot}0Bi_4 + 3{\cdot}6Pl_{10} + (1{\cdot}33K + 2{\cdot}90)\rightarrow$					
		—	—	−7·38	−10·20	Δ^{4-6}_{6}
6.	$\rightarrow 12{\cdot}3Kf_3 + 0{\cdot}9Bi_5 + (0{\cdot}3Mg + 0{\cdot}36Ca + 0{\cdot}58Na + 0{\cdot}1H_2O)$;					
	$16{\cdot}05Kf_3 + 0{\cdot}9Bi_5\rightarrow$	2·52	3·82	118·30	72·50	C_6
		—	—	−6·72	−10·90	Δ^{6}_{7}
7.	$\rightarrow 14{\cdot}65Kf_4 + 1{\cdot}0Bi_6 + 0{\cdot}2Nf + (0{\cdot}5Na + 0{\cdot}45H_2O + 7{\cdot}10)$;	2·54	5·87	120·45	66·70	C_7
		—	—	−4·57	−16·70	Δ^{1}_{7}

1.

Al	Fe	K	Na	Ca	Sr	Ba	Mg	Mn	Ti	Nb	Zr	RE	P	H	*k*	
18·20	13·52	3·42	5·13	3·24	0·13		4·36	0·20	0·97	—	—	—	0·12	0·28		C_0
18·60	11·31	9·34	4·97	3·04	0·07	0·01	4·59	0·16	1·14	0·011	0·077	0·032	0·09	0·10		C_1
+0·4	−2·21	+6·08	−0·16	−0·20	−0·06	+0·01	+0·23	−0·04	+0·17	+0·011	+0·077	+0·032	−0·03	−0·18	97·0	Δ^{1}_{2-3}
20·02	10·11	8·50	5·87	4·55	0·03	0·15	2·42	0·18	1·33	0·008	0·098	0·148	0·35	0·07		C_{2-3}
+1·82	−5·41	+5·08	+0·74	+1·31	−0·10	+0·14	−1·94	−0·02	+0·36	−0·003	+0·021	+0·116	+0·23	−0·21	95·1	Δ^{2-3}_{4-5}
26·56	6·24	11·80	8·33	3·44	0·12	0·16	0·90	0·26	0·88	0·024	0·121	0·075	0·17	0·08		C_{4-5}
+8·36	−7·28	+8·38	+3·20	+0·20	−0·01	+0·15	−3·46	+0·06	−0·09	+0·013	+0·044	+0·043	+0·05	−0·20	91·8	Δ^{4-5}_{6}
27·40	5·51	11·73	12·63	0·93	0·28	0·62	0·93	0·08	0·58	0·006	0·125	0·077	0·15	0·15		C_6
+9·20	−8·01	+8·31	+7·50	−2·31	+0·15	+0·61	−3·43	−0·12	−0·39	−0·005	+0·048	+0·045	+0·03	−0·13	89·6	Δ^{6}_{7}
29·40	5·67	14·47	12·02	2·55	0·32	0·30	0·77	0·17	0·63	0·075	0·126	0·075	0·10	0·17		C_7
+11·20	−7·85	+11·05	+6·89	−0·69	+0·19	+0·29	−3·59	−0·03	−0·36	+0·064	+0·049	+0·043	+0·02	−0·11	88·3	Δ^{1}_{7}

N—zones.
O—greenschist.
1. staurolite–quartz and sillimanite–almandine gneisses.
2. biotite migmatite.
3. amphibole migmatite.
4. amphibole fenite.
5. biotite fenite.
6. biotite–feldspar rock.
7. miaskite.
Alm—almandine.
Biotites:
Bi_1—$K(Mg_2Fe'')[AlSi_3O_{10}](OH)_2$,
Bi_2—$K(Mg_{1\cdot5}Fe''_{1\cdot5})[AlSi_3O_{10}](OH)_2$,
Bi_3—$K(Mg_{1\cdot35}Fe''_{1\cdot65})[AlSi_3O_{10}](OH_2)$,
Bi_4—$K(Mg_{1\cdot2}Fe''_{1\cdot8})[AlSi_3O_{10}](OH)_2$,
Bi_5—$K(Mg_{1\cdot0}Fe''_{2\cdot0})[AlSi_3O_{10}](OH)_2$,
Bi_6—$K_{0\cdot8}Na_{0\cdot1}(Mg_{0\cdot9}Fe''_{1\cdot3})(Fe'''_{0\cdot5}Al_{0\cdot2})[Al_{1\cdot2}Si_{2\cdot6}O_{10}](OH)_{0\cdot9}O_{0\cdot8})$.
Hl—chlorite.

Hm—hematite. $Hs_1 = Hs_2$ — $Ca_{1\cdot4}(Na_{1\cdot5}K_{0\cdot45})(Fe_{1\cdot7}Mg_{1\cdot35})Fe_{1\cdot65}[(Al_{1\cdot85}Fe_{0\cdot20})Si_{5\cdot85}O_{22}][(OH)_{0\cdot75}O_{1\cdot1})]$—hastingsite.
Ka—calcite.
K—feldspar.
$Kf_1 = Kf_2$—$(K_{0\cdot8}Na_{0\cdot2})[AlSi_{2\cdot9}O_8]$,
Kf_3—$(K_{0\cdot65}Na_{0\cdot35})[AlSi_{2\cdot8}O_8]$,
Kf_4—$(K_{0\cdot65}Na_{0\cdot35})[AlSi_3O_8]$
Mt—magnetite.
Mu—muscovite.
Nf—nepheline.
$Pl_{10,20,30}$—plagioclases and their contents of anorthite.
Q—quartz.
Sl—sillimanite.
v—volumetric weight.
p—porosity.
k—coefficient of survival of the composition.
$c_0 \ldots c_7$—the contents of components in zones 0–7.
$\Delta^{1}_{2-3} \ldots \Delta^{6}_{7}$—gains and losses in g/100 cm³ in reaction between zones.

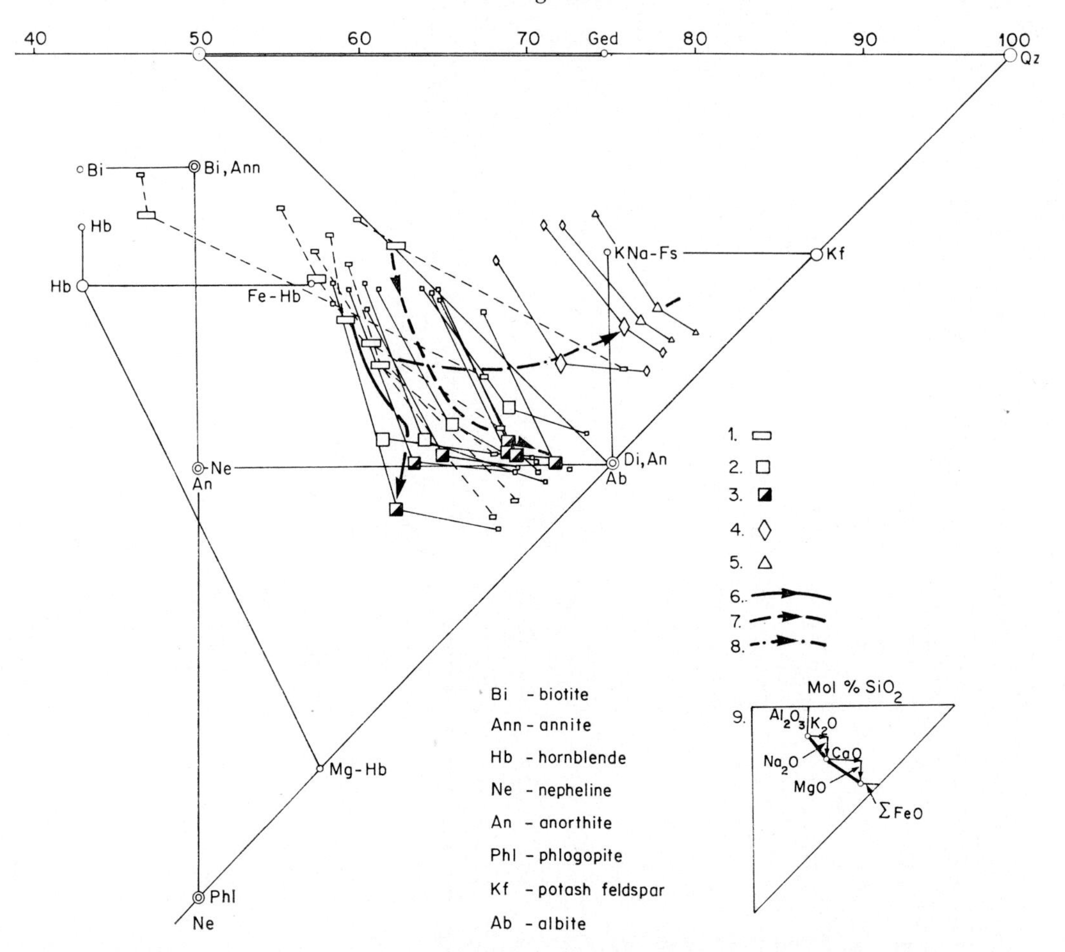

Fig. 4 Petrochemical evolution of the metasomatic–anatectic series of the West Khubsugul region, Mongolia. (6) foyaite series; (7) miaskite series; (8) granosyenite series. (1) initial rocks of gabbroic and dioritic compositions; (2) nepheline–fenite–gneisses; (3) anatectic nepheline syenite; (4) nepheline-free syenites; (5) anatectic quartz syenites; (9) the mode of construction of the diagram. (The molecular proportions of SiO_2, Al_2O_3, K_2O, Na_2O, CaO, MgO and $\Sigma(FeO + Fe_2O_3)$ recalculated to a sum of 100 are plotted on a Lodochnikov barycentrical diagram (Lodochnikov, 1926, 1927; Korzhinsky, 1957). Each value is represented by three figurative points or a broken vector. The arrows numbered 6, 7 and 8 connect the figurative points of sequential rock members of the metasomatic–anatectic series of the different mineral facies represented by the fields III-ξ, I-ξ and III-$\gamma\delta$, III-γ of Table 1, Chapter IV.5)

siliceous the endomorphic facies, and vice versa. Perchuk (1963) interpreted the formation of nepheline rocks in the endocontact zone of the Dezhnev granitoid massif (Chukotka) by replacement of the enclosing carbonate sequences. On the other hand in the Matchin massif (Turkestan–Alay, Soviet Middle Asia) aegirine-hedenbergite-biotite-nepheline syenites pass in the direction of the enclosing biotite-quartz schists into pyroxene-hastingsite alkali syenites, hornblende-biotite-quartz syenites and, finally, biotite granites (Perchuk, 1964). It is interesting to note, that in contact with gabbroids, both nepheline syenites and granites are in their marginal portions transformed into pyroxene-amphibole alkali syenites (Kovalenko

TABLE 4 Chemical Compositions of Khibinite, Fenites and Country Rocks of the Khibina massif Recalculated by Barth's Method (Tikhonenkova, 1967)

Rocks	Number of ions in the standard cell								
	K	Na	Ca	Mg	Fe^{2+}	Fe^{3+}	Al	Ti	Si
g. Biotite gneiss	26	108	26	16	20	17	155	—	606
2g. Feldspathic gneiss	56	92	20	12	19	14	218	—	563
1g. Nepheline–feldspathic gneiss	104	76	34	21	23	33	184	10	542
O. Khibinite	76	167	15	10	17	26	239	5	513
1d. Feldspathic fenite with nepheline	77	113	13	15	16	8	224	8	536
2d. Malignite–fenite	32	135	60	123	53	35	121	24	492
3d. Theralite–fenite	38	89	56	163	62	27	171	72	407
4d. Pyroxene–plagioclase hornfels	8	53	133	102	55	25	140	4	511
d. Metadiabase	30	62	63	21	91	32	147	9	546

and Popolitov, 1965, 1969; Kovalenko *et al.*, 1969).

The exocontact alkaline-metasomatic rocks may be subdivided into three principal groups corresponding to the initial rocks that are replaced: alumino-silicate, carbonate and ultrabasic.

The formation of fenites, i.e. syenite-like rocks, sometimes with quartz or nepheline, takes place in alumino-silicate rocks, the composition of which is equivalent to the range from granite to gabbro. One of the essential reactions of fenitization is:

$$\underset{\text{biotite}}{HK(Fe,Mg)_4Ab_3Si_6O_{24}} + \underset{\text{quartz}}{7SiO_2} + \langle 4Na_2O \rangle =$$
$$\underset{\text{potash feldspar}}{KAlSi_3O_8} + \underset{\text{albite}}{2NaAlSi_3O_8} + \underset{\text{aegirine}}{2NaFeSi_2O_6} + \langle H_2O + 2MgO + 0{\cdot}5K_2O \rangle.$$

Arrow brackets indicate mobile components (Korzhinsky, 1957). If quartz is lacking nepheline is formed.

The production of nepheline and potash feldspar at the expense of the anorthite component of the plagioclase with the loss of CaO or precipitation of calcite is also typical:

$$\underset{\text{anorthite}}{CaAl_2Si_2O_8} + \underset{\text{quartz}}{2SiO_2} + \langle 0{\cdot}5K_2O + 0{\cdot}5Na_2O \rangle =$$
$$\underset{\text{nepheline}}{NaAlSiO_4} + \underset{\text{potash feldspar}}{KAlSi_3O_8} + \langle CaO \rangle.$$

If quartz is lacking in such a reaction corundum appears, especially in the presence of calcite and scapolite (Moyd, 1949).

A typical example of the chemical changes involved in the formation of fenites at the expense of gneisses and greenstones is given by Tichonenkova (1967) for the Khibina massif, Kola peninsula (Table 4). Attention is drawn to the formation of the zone of basification which coincides with theralite fenites, in which Mg, Fe^{2+}, Ti and Al are at a maximum and SiO_2 is at a minimum. Such zones are very typical of fenitization of intermediate and basic rocks.

Magnesian skarns containing spinel, forsterite and pyroxene and analogous to those in the contacts of granite intrusions are usually formed in the contact zones between carbonate (calcite-dolomite) rocks and alkaline intrusions. Perchuk (1964) describes the following zonation of magnesian skarns connected with the alkaline intrusions of Turkestan–Alay (Soviet Middle Asia):

(oo) dolomite–calcite; (V) spinel–calcite; (IV) spinel–forsterite–calcite; (III) spinel–forsterite–pyroxene–calcite; (II) spinel–pyroxene–calcite; (I) pyroxene–calcite; (O) melanocratic syenite. The interaction of nepheline syenitic melts and carbonate rocks in different regions is also considered by Yashina (1962, 1963), Kononova (1961), Kovalenko *et al.* (1968). The last-named work reports the chemical changes when passing from the initial dolomite marbles through the forsterite and fassaite rocks, the endocontact ijolites and the melanocratic nepheline syenites into the central foyaite (Table 2).

VI.7.3.4. Postmagmatic Alkali Metasomatism

Postmagmatic alkali metasomatism is displayed in the replacement of rocks of any composition by alkali feldspar, alkaline mafic minerals and other alkali-rich minerals at temperatures lower than those of the granite and nepheline syenite eutectics, i.e. after crystallization of the magmatic rocks.

These metasomatic products appear often as diffuse patches, particularly in the endocontact zones of intrusions, while the metasomatic products of the progressive stage often reflect the primary structures of the country rocks. The postmagmatic metasomatic rocks are further characterized by two main features: their location is generally determined by post-intrusive fractures and the appearance of zoned metasomatic columns with a varying number of minerals and displaying sharp mutual boundaries (Korzhinsky, 1953, 1959). Large bodies of anchimonomineralic rocks (microclinites and albitites) are characteristic for the most transformed zones.

Postmagmatic minerals differ markedly in composition from the minerals of the progressive stage. The postmagmatic feldspars approximate pure microcline and pure albite; the coloured minerals are distinguished by high contents of iron and alkalis. Lithium appears in riebeckite and micas.

The chemical evolution of the postmagmatic metasomatism is the reverse to that of the progressive metasomatism. With decreasing temperature the acidity of the solutions increases, which is manifested by the succession of microclinization, albitization and silicification processes. At the final stages the alkalinity increases again and the late generation of albite, sodalite, cancrinite, natrolite and analcime is formed. The general temperature interval of the post-magmatic alkaline metasomatism may be estimated to vary from 600–300 °C to 200 °C, as indicated by the two-feldspar and nepheline–feldspar thermometers (Perchuk, 1968a, b; Perchuk and Pavlenko, 1967) and by the method of fluid inclusions (V.2; Kerkis and Kostyuk, 1963; Bazarova, 1969).

The mafic minerals, which are important carriers of a number of trace elements (cf. V.3), are often replaced by magnetite during potassium metasomatism which results in a release of elements such as RE–Y, Zr, Nb, Be and Th and in the formation of accessory minerals such as zircon, columbite, xenotime and thorite (Pavlenko, 1972). Steenstrupine is formed in this way in agpaitic nepheline syenites (cf. II.2).

By means of the escape of iron, potassium metasomatism can lead to the formation of sviatonossite, i.e. a garnetiferous syenite-like rock:

$$\underset{\text{aegirine–hedenbergite}}{6[(CaFe^{2+})_{0\cdot5}(NaFe^{3+})_{0\cdot5}Si_2O_6]} + \underset{\text{nepheline}}{6NaAlSiO_4} + \lceil 2\cdot5K_2O\rfloor = \underset{\text{microcline}}{5KAlSi_3O_8} + \underset{\text{garnet}}{Ca_3AlFe^{3+}Si_3O_{12}} + \lceil 4\cdot5Na_2O + 3FeO + Fe_2O_3\rfloor.$$

In accordance with the reaction

$$2\text{biotite} + (3\cdot1 + n)\text{ microcline} + \text{nepheline} + \lceil[5\cdot45 + 0\cdot5n]\ Na_2O\rfloor = 4\cdot4\text{aegirine} + (6\cdot5 + n)\text{ nepheline} + \text{albite} + \lceil[2\cdot55 + 0\cdot5n]K_2O\rfloor$$

albitization of biotite syenite with or without nepheline leads to the formation of mariupolite.

VI.7.3.5. Decisive Factors in Alkali Metasomatism

The relation between the concentrations or chemical potentials of K_2O and Na_2O is the determining factor in the alkaline–metasomatic process, as one can see from the above-mentioned reactions. This is also in agreement with experimental data (Wyart and Sabatier, 1962; Ivanov, 1962a, b, 1966).

Bundle and multy-bundle diagrams (Schreinemakers, 1948; Korzhinsky, 1957) are appropriate for the representation of the different mineral associations originating at variable values and relations of the chemical potentials (μ) of K_2O and Na_2O. An example of such a diagram is given in Fig. 5 (Pavlenko and Kovalenko, 1961). The inclinations of the univariant lines are determined by the relation of K_2O and Na_2O added and/or discharged in the reaction; the direction of stable and metastable rays is taken from the rule of Le Chatelier; the scale is conventional. Each of the fields of the diagram corresponds to a facies of alkalinity with a definite mineral paragenesis, reflected in the triangular concentrational diagrams. Fields C1–C3 belong to meta-nephelinites; B1–B3 to albitites; B2–B4 to microclinites; A1–A4 to meta-quartzites.

In nature the facies occur jointly, comprising a zonation that reflects the temporal and spatial sequence of their formation (Pavlenko and Kovalenko, 1965). Broken arrows in Fig. 5 show the common facies sequence from early facies to late ones, i.e. from peripheral facies to central ones (see Chapter IV.5, Fig. 3). They indicate the tendency of the increase in acidity in the course of a metasomatic process associated with decreasing temperatures.

Within the limits of certain facies, columns of metasomatic zones displaying a decreasing number of minerals may result from a differential mobility of the components (Korzhinsky, 1953, 1968, 1969).

As the components of the rocks passed into the mobile state without saturating the metasomatic solutions, the minerals of the mobile components disappeared successively leaving monomineralic residual zones behind. In the alkaline metasomatites Al_2O_3 is usually less mobile. Monomineralic microclinite or albitite zones are consequently formed during maximal metasomatic alteration. The replacement of mafic minerals, the carriers of rare elements, by microclinite or albitite caused the accumulation of various accessory minerals.

The favourable geochemical situation (the increased content of rare elements in the initial magmatic rocks, abundance of mineralizers as F, CO_2 and alkalis, the considerable temperature interval) determines the appearance in the alkaline metasomatites of numerous rare minerals containing Y, RE, Zr, Hf, Nb, Ta, Th, U, Ti, Be and Li. Some of them are given in Fig. 5. There is a number of peculiar minerals in the metasomatites of agpaitic massifs, for example in the Lovozero intrusion: apophyllite, rhabdophanite, catapleiite, elpidite, vlasovite, narsarsukite, neptunite, nenadkevichite, labuntsovite, epididymite (Tichonenkov and Tichonenkova, 1962).

Minerals of Y, RE, Ta, Be and Li—gagarinite, thalenite, yttrialite, gadolinite, tengerite, synchysite, cenosite, microlite, milarite, bertrandite, phenacite and cryophyllite (Mineev, 1968) develop in apogranites, i.e. granites, subjected to alkali metasomatism (Beus *et al.*, 1962), along with cryolite and thomsenolite.

The rare elements form their own accessory minerals according to the mineralogical phase rule of Goldschmidt. The compositions of these minerals, like those of the rock-forming minerals, are determined by the alkalinity of the prevailing mineral facies: the more alkaline facies, the higher content of strong bases in the mineral. The basicity of minerals is most conveniently evaluated by means of their average geometrical electronegativity (EN) which rises with decreasing mineral basicity. EN-values calculated according to the data of Pauling are given in Table 5 (cf. Pavlenko, 1972).

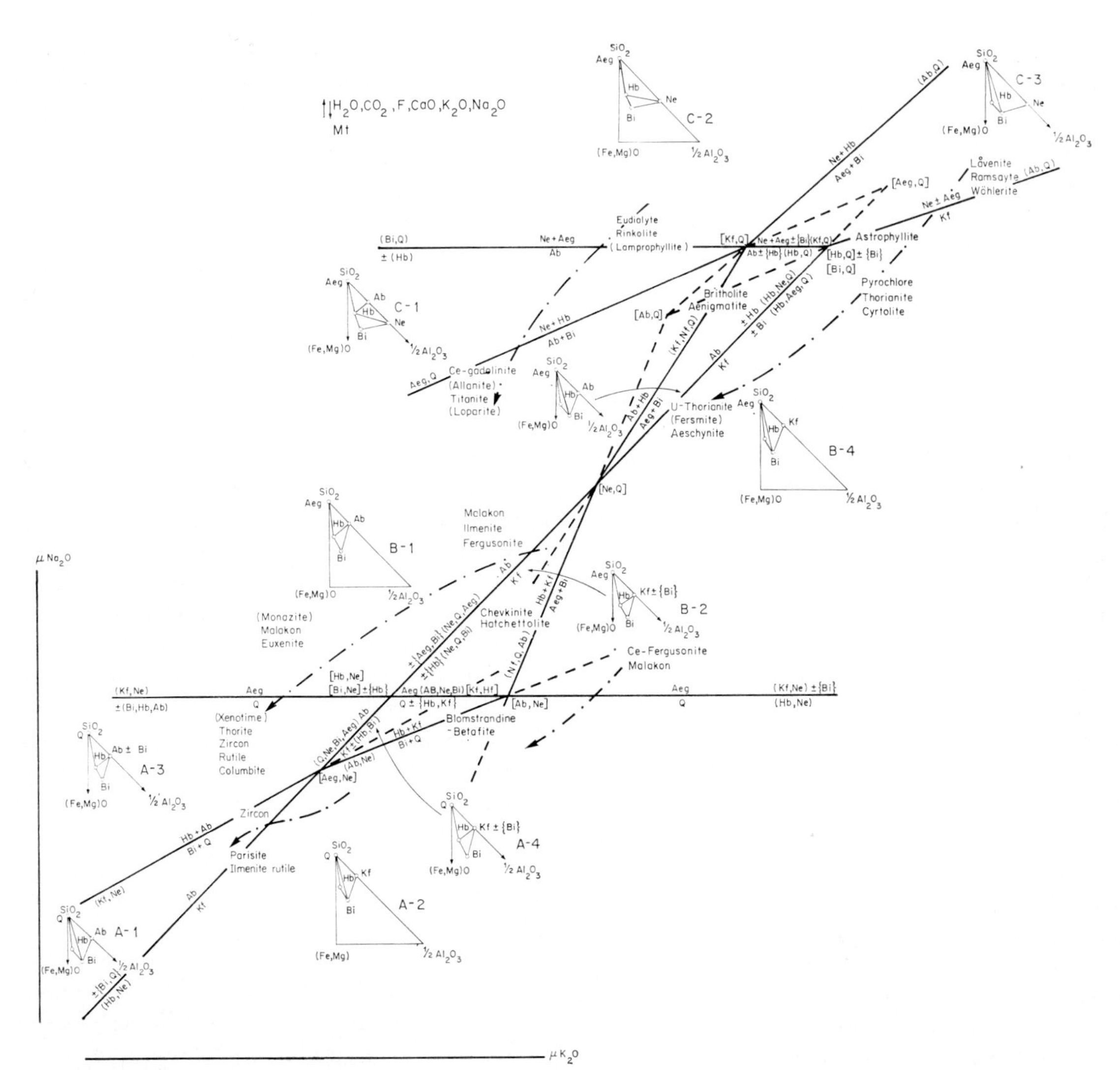

Fig. 5 Paragenetic diagram of a metasomatic alkaline complex formed from aluminosilicate rocks impoverished in calcium. The parageneses are shown as functions of the chemical potentials of K_2O and Na_2O. Round brackets: minerals, that are absent in univariant equilibria; square brackets: minerals, that are absent in invariant equilibria; { }: indifferent phases—minerals. Ne = nepheline; Kf = microcline; Ab = albite; Aeg = aegirine; Hb = amphibole; Bi = biotite; Mt = magnetite; C–1 . . . A–4—facies of rocks. Solid lines: rays of stable equilibria; dashed lines: rays of metastable equilibria; dashed and dotted arrows: the typical facies sequence; ↑↓ components, for which the system is open, the completely mobile components

TABLE 5 Electronegativity (EN) of the Accessory Minerals of the Alkaline Metasomatites (EN after Pauling; the figure 2 omitted: 39 read 2.39)

Facies	RE, Y	EN	Th, U	EN	Zr, Hf	EN	Ti	EN	Nb, Ta	EN
C C–2 (C–1)					Eudialyte $(Na,Ca)_5ZrSi_6O_{16}(OH)Cl_2$	39	Rinkolite $Ca_4Na_2Cl(Ti,Nb)[Si_2O_7](OH,F)_4$	33		
Ne EN 32							(Lamprophyllite $Na_4Sr_2Fe_2Ti_4Si_6O_{25}F_2$	36		
C–3	(Apatite) $Ca_5[PO_4]_3(F,OH)$	42			Låvenite $(Na,Ca,Mn'')_3Zr[SiO_4]_2F$	42	Ramsayite $Na_2Ti_2Si_2O_9$	43	Wöhlerite $(Ca,Na)_3(Zr,Nb,Ti)[SiO_4]_2F$	43
							Astrophyllite $(K,Na)_2(Fe,Mn)_4Ti[Si_2O_7]_2(OH,F)_2$	47		
B B–4a									Pyrochlore $(Na,Ca)_2(Nb,Ti)_2(F,O)_7$	47
B–3 Kf EN 54	Britholite $(Na,Ca,Ce)_5[(Si,P)O_4]_3(OH,F)_2$	50					Aenigmatite $Na_2(Fe,Mn)_6Ti[Si_2O_7]_3$	51		
Ab EN 56	Ce-Gadolinite $(Ce,Y)_2Fe''BeSi_2O_{10}$	56					Titanite $CaTi[SiO_4]O$	53	(Loparite) $(Ce,Na,CO)_2(Ti,Nb)_2O_6$	56
B–1a Aeg EN 53 Rib EN 52 Bi EN 55	(Allanite) $(Ca,Ce)_2(Al,Fe)_3Si_3O_{12}(OH)$	58								
B–2							Chevkinite $(Ca,Ce)Ti[SiO_4]O$	55	Hatchettolite $(Ce,U,Ca)_{1\cdot3}(Ti,Nb)_2O_7$	56
Mt EN 61			Thorianite ThO_2	58					Fersmite $CaNb_2O_6$	59
B–4b			Uranothorianite $(U,Th)O_2$	60			Ilmenite $FeTiO_3$	59	Aeschynite $(Ca,Ce,Th)(Ti,Nb)_2O_6$	63
					Cyrtolite $(Zr,RE)SiO_4\ nH_2O$	(70)			Fergusonite $Y_2Nb_2O_8$	63
B–1b	(Monazite) $CePO_4$	69			Malacon $(Zr,U)SiO_4\ nH_2O$	(73)			Euxenite $(Ca,Y,U)(Nb,Ti)_2O_6$	66
A A–3									Blomstrandite Betafite $(Ca,RE,U)_{0\cdot7}(Ti,Nb)_2O_7$	72
A–2	Xenotime YPO_4	72	Thorite $ThSiO_4$	72	Zircon $ZrSiO_4$	75	Rutile(Anatase) TiO_2	70	Columbite $(Fe,Mn)(Ta,Nb)_2O_5$	76
Q EN 77 A–1	Parasite $Ca(La,Ce)_2[CO_3]_3F_2$	75					(Ilmenorutile) $(Ti,Nb,Fe)O_2$	73		

Ne—Nepheline, Kf—Potash Feldspar, Ab—Albite, Aeg—Aegirine, Rib—Riebeckite, Bi—Biotite, Mt—Magnetite, Q—Quartz. C–1 . . . A–1—Facies of Fig. 5.

VI.7. REFERENCES

Anderson, E. M., 1936. The dynamics of the formation of cone sheets, ring dykes and cauldron subsidence. *Proc. Roy. Soc. Edinburgh*, **56**.

Appleyard, E. C., 1967. Nepheline gneisses of the Wolfe Belt, Lyndoch Township, Ontario. I. Structure, stratigraphy, and petrography. *Can. J. Earth Sci.*, **4**, 371–95.

Appleyard, E. C., 1969. Nepheline gneisses of the Wolfe Belt, Lyndoch Township, Ontario. II. Textures and mineral paragenesis. *Can. J. Earth Sci.*, **6**, 689–717.

Bazarova, T. Yu., 1969. *Thermodynamic Conditions of Formation of Some Nepheline-Bearing Rocks.* (in Russian). Izd. 'Nauka', Moskva.

Beus, A. A., Severov, E. A., Sitnin, A. A., and Subbotin, K. D., 1962. *Albitized and greisenized Granites (Apogranites).* (in Russian). 198 pp., Izd-vo Akad. Nauk SSSR, Moskva.

Borodin, L. S., 1957. Types of carbonatite deposits and their relationship to massifs of ultrabasic alkaline rocks (in Russian). *Izv. Akad. Nauk SSSR*, Geol. Ser., **5**, 3–16. French transl. BRGM No. 2048.

Borodin, L. S., 1958a. The chemistry of aegirinization and nephelinization of pyroxene in the formation of metasomatic nepheline-pyroxene rocks (ijolites); *Geokhimiya*, **5**, 501–2. English transl. *Geochemistry*, **5**, 637–40.

Borodin, L. S., 1958b. On the process of nephelinization and aegirinization of pyroxenites in connection with the problem of the genesis of alkaline rocks of ijolite-melteigite type (in Russian). *Izv. Akad. Nauk SSSR*, Geol. Ser., **6**, 48–57.

Borodin, L. S., 1961. 'Nephelinization of pyroxenites and paragenesis of rock-forming ijolite minerals from massifs of ultrabasic alkaline rocks', in *Physico-Chemical Problems of the Formation of Rocks and Ores*, 1. (in Russian) Akad. Nauk SSSR, Moskva.

Borodin, L. S., 1962. 'The petrography and genesis of the Vuori-Yarvi massif', in Rare elements in massifs of alkaline rocks. (in Russian) *Trans. Inst. Min., Geochem., Crystal Chemistry of Rare Elem.*, **9**, 161–205.

Borodin, L. S., 1965. 'Rare elements in ultrabasites from complex massifs of ultrabasic-alkaline rocks (to the problem of genesis of carbonatites and their relation to alkaline magma' (in Russian), in *Problems of Geochemistry*, Nauka, Moskva, 396–405.

Chumakov, A. A., 1958. 'On the origin of the Keivi alkali granites', in Vorob'eva, O. A., Ed., *Schelochnye Granity Kol'skogo Poluostrova* (in Russian). Izd. Akad. Nauk SSSR, Moskva–Leningrad, 308–68.

Eckermann, H., 1948. The alkaline district of Alnö-Island. *Sver. geol. undersökn*, **36**.

Egorov, L. S., 1960. On the problem of nephelinization and iron–magnesium–calcium metasomatism in intrusions of alkaline and ultrabasic rocks (in Russian). *Trudy NIIGA*, **114**, 14, 102–18.

Garson, M. S., 1962. The Tundulu carbonatite ring complex in southern Nyasaland. Nyasaland Protectorate. *Geol. Surv. Dept. Mem.*, **2**.

Garson, M. S., 1967. 'Carbonatites in Malawi', in Tuttle, O. and Gittins, J., Eds., *Carbonatites.* John Wiley and Sons, 33–71.

Gittins, J., 1961. Nephelinisation in the Haliburton–Bancroft district, Ontario. *J. Geol.*, **69**, 291–308.

Glagolev, A. A., 1966. *Metamorphism of Pre-Cambrian rocks of KMA—Kursk magnetic anomaly* (in Russian). 158 pp. Izd. 'Nauka', Moskva.

Gummer, W. K., and Burr, S. V., 1946. Nephelinised paragneisses in the Bancroft Area, Ontario. *J. Geol.*, **3**, 137–68.

Ivanov, I. P., 1962a. 'On the nature of the albitizing solutions', in Khitarov, N. I., Ed. *Experimental Investigations in the Field of Abyssal Processes* (in Russian). Izd. Akad. Nauk SSSR. Moskva, 92–103.

Ivanov, I. P., 1962b. 'Experiments on hydrothermal metamorphism of mica schists under dynamic conditions', in Tsvetkov, A. I., Ed. *Trudy 4-go Soveskchaniya po Eksperimental'noy i Tekhnicheskoi Mineralogii i petrografii* (in Russian). Izd. Akad. Nauk SSSR, Moskva, 53–60.

Ivanov, I. P., 1966. 'On the question of experimental research of open systems', in Tsvetkov, A. I., Ed., *Issledovaniya Prirodnogo i Teknicheskogo Mineraloobrasovaniya* (in Russian). Izd. 'Nauka', Moskva 90–6.

Kerkis, T. Yu., and Kostyuk, V. P., 1963. Mineralogical investigation of the Botogol nepheline (the East Sayan) (in Russian). *Dokl. Akad. Nauk SSSR*, 85, **150**, 1125–7.

King, B. C., 1955. Syenitisation des granites à Semarule, pres de Molepolole prefectorat du Bechuanaland. *Science de la Terre, Nhors serie*, 1–16.

King, B. C., 1967. 'The nature and genesis of migmatites, metasomatism or anatexis', in Pitcher, U.S., and Flin, G. U., *Priroda metamorfisma* (in Russian). Izd 'Mir', Moskva., 227–42.

King, B. C., and Sutherland, D. S., 1960. Alkaline rocks of Eastern and Southern Africa, 1, 2, 3. *Science Progress*, **48**, 298–321, 504–24, 709–20.

King, B. C., and Sutherland, D. S., 1967. 'The carbonatite complexes of Eastern Uganda', in Tuttle, O. F., and Gittins, J., Eds., *Carbonatites*. John Wiley and Sons, 73–126.

Kononova, V. A., 1961. Urtite–ijolite intrusions of South-East Tuva and some questions of their genesis (in Russian). 119 pp., *Trudy IGEM*, **60**. Izd Akad. Nauk SSSR, Moskva.

Korikovsky, S. P. 1967. *Metamorphism, granitization and Postmagmatic Processes in the Pre-Cambrian of the Udakano-Stanovaya Zone* (in Russian). 298 pp., Izd 'Nauka', Moskva.

Korzhinsky, D. S., 1953. 'Sketch of the metasomatic processes', in Sokolov, G. A., Ed., *Osnovnye Problemy v Uchenii Magmatogenykh Rudnykh Mestorozhdeniyakh* (in Russian). Izd. Akad. Nauk SSSR, Moskva., 332–450.

Korzhinsky, D. S., 1957. *Physico-Chemical Principles of Mineral Paragenesis* (in Russian). 182 pp., Izd Akad. Nauk SSSR, Moskva.

Korzhinsky, D. S., 1959. Acid–basic interaction of the components in silicate melts and the direction of the cotectic lines (in Russian). *Dokl. Akad. Nauk SSSR*, **128**, 383–6.

Korzhinsky, D. S., 1960. 'Acidity–alkalinity as the most principal factor in magmatic and post-magmatic processes', in Antropov, P. I., Ed., *Magmatism i Svyas s Nim Polesnykh Iskopaemykh. Trudy II Vsesoyznogo Petrograficheskogo Soveshchyaniya* (in Russian). Izd Akad. Nauk SSSR, Moskva. 21–31.

Korzhinsky, D. S., 1968. 'New derivation of an equation of metasomatic zoning', in Marakushev, A. A., Ed., *Metasomatizm i Drugie Voprosy Fiziko-Khimicheskoy Petrologii* (in Russian). Izd 'Nauka', Moskva, 3–8.

Korzhinsky, D. S., 1969. *The Theory of Metasomatic Zonation* (in Russian). Izd. 'Nauka', Moskva.

Kovalenko, V. I., Brandt, S. B., Pisarskaya, V. A., Popolitov, E. I., and Frolova, L., 1969. 'On the granitization of gabbro and the mechanism of formation of endomorphic syenite rocks from alaskitic magma', in Tauson, L. V. Ed., *Primenenie Metodov Fisitsheskoi Khimii v Petrologii* (in Russian). Izd 'Nauka', Moskva.

Kovalenko, V. I., Okladnikova, L. V., Pavlenko, A. S., Popolitov, E. I., and Philippov, L. V., 1965. 'Petrology of the Middle Paleozoic complex of East Tuva granitoids and alkaline rocks. The magmatic rocks of the East Tuva Middle-Paleozoic complex', in Shmakin, B. M. Ed., *Geokhimiya i Petrologiya Magmaticheskikh i Metasomaticheskikh Obrazovaniy*. Izd 'Nauk', Moskva, 5–75.

Kovalenko, V. I., Okladnikova, L. V., and Popolitov, E. I., 1968. 'The features of interaction of nepheline syenite magma with dolomite marble (on the example of the North-East Tuva alkaline massifs)', in Marakushev, A. A., Ed., *Metasomatizm i Drugie Voprosy Fiziko-Khimicheskoy Petrologii* (in Russian). Izd 'Nauka', Moskva, 165–80.

Kovalenko, V. I., and Popolitov, E. I., 1965. On the influence of the enclosing gabbro on the acidity-alkalinity of the endomorphic parts of granite and nepheline-syenite intrusions (in Russian). *Dokl. Akad. Nauk SSSR*, **161**, 207–9.

Kovalenko, V. I., and Popolitov, E. I., 1969. *Petrology and Geochemistry of Rare Elements of the Alkaline and Granitoid Rocks of the North-East Tuva*. Izd 'Nauka', Moskva.

Kukharenko, A. A., *et al.*, 1965. *Kaledonian Complex of Ultrabasic Alkaline Rocks and Carbonatites of Kola Peninsula and North Karelia* (in Russian). Moskva, 550 pp.

Lapin, A. W., 1963. Nephelinization and ijolite dykes in the Kovdor massif of ultrabasic alkaline rocks (in Russian). *Izv Akad. Nauk SSSR, Geol. Ser.*, **5**, 9–23.

Marakushev, A. A., Tararin, I. A., and Zalischak, V. L., 1968. 'Mineral facies of acidity-alkalinity of calcium poor granitoids', in Govorov, I. I., and Tararin, I. A., Ed. *Mineralnye Fatsii i ikh Rudonosnost* (in Russian). Izd 'Nauka', Moskva, 5–72.

Mineev, D. A., 1968. *Geochemistry of Apogranites and Metasomatites of the North-East Tarbagatay* (in Russian). 185 pp., Izd 'Nauka', Moskva.

Moyd, L., 1949. Petrology of the nepheline and corundum rocks of Southeastern Ontario. *Am. Miner.*, **34**, 736–51.

Pavlenko, A. S., 1959. The features of metasomatism in one of the regions of the North-Krivoy Rog (in Russian). *Izv. Akad. Nauk SSSR, ser, geol.*, **1**, 81–101.

Pavlenko, A. S., 1963a. Behaviour of rock-forming and some rare elements in the processes of formation of alkaline rocks', in Vinogradov, A. P., Ed. *Khimiya Zemnoy Kory, Trudy Geokhim. Konferentsii. Posvyashchennoy Stoletiyu so Dnya Rozhdeniya Akad. V. I. Vernadskogo, I* (in Russian). Izd. Akad. Nauk SSSR, Moskva, 116–28.

Pavlenko, A. S., 1963b. 'Petrology and some geochemical features of the Middle-Paleozoic complex of the East Tuva granitoids and alkaline rocks', in Afanas'ev, G. D., Ed., *Trudy Yubileinogo Sympoziuma, Posvyashchennogo F. O. Levinsone-Lessingy* (in Russian). Izd. Akad. Nauk SSSR, Moskva, 239–47.

Pavlenko, A. S., 1972. 'The effect of alkalinity on the behaviour of the rare-metal accessory minerals of magmatic and metasomatic rocks', in Tauson, L. V., Ed. *Primenenie Metodov Fizicheskoy Khimii v Petrologii* (in Russian). Izd. 'Nauka, Moskva.

Pavlenko, A. S., Il'in, A. V., Strizhov, V. P., and Bykhover, V. N., 1969. Absolute age of the intrusions of the main tectonic zones of the East-Tuva and North Mongolia (in Russian). *Geotectonika*, **3**.

Pavlenko, A. S., and Kovalenko, V. I., 1961. The paragenetic dependence of alkaline metasomatites on the chemical potential of alkalis in calcium-poor alumosilicate rocks (in Russian). *Geokhimya* **11**, Moskva, 980–7.

Pavlenko, A. S., and Kovalenko, V. I., 1965. 'Facies zoning of alkaline metasomatites and the associated rare-metal mineralization', in Kutina, V., Ed. *Trudy Konferentsii Problemy Postmagmaticheskogo Rudoobrasovaniya, 2, 1965 Chekhoslovakiya*. Izd Chekhoslov. Akad. Nauk, Praga, 179–85.

Perchuk, L. L., 1963. 'Magmatic replacement of the carbonate sequence with the formation of nepheline syenites and other alkaline rocks on the example of the Dezhnev massif', in Sokolov, G. A., Ed. *Fisiko-Khimicheskie Problemy Formirovaniya Gornykh Porod i Rud, t. 2* (in Russian). Izd. Akad. Nauk SSSR, Moskva, 160–82.

Perchuk, L. L., 1964. *Physico-Chemical Petrology of the Central Turkestan–Alay Granitoid and Alkaline Intrusions* (in Russian). 241 pp., Izd. 'Nauka', Moskva.

Perchuk, L. L., 1968a. 'The phase accordance in the system nepheline–alkali feldspar–aqueous solution', in Marakushev, A. A., Ed. *Metasomatism i Drugie Voprosy Fiziko-Khimicheskoy Petrologii* (in Russian). Izd. 'Nauka', Moskva, 53–95.

Perchuk, L. L., 1968b. The composition variations of calcium-poor Fe–Mo–Mn minerals of alkaline rocks', in Marakushev, A. A., Ed. *Metasomatizm i Drugie Voprosy Fiziko-Khimicheskoy Petrologii* (in Russian). Izd. 'Nauka', Moskva, 96–135.

Perchuk, L. L., and Pavlenko, A. S., 1967. The effect of temperature on the distribution of some isomorphic components among coexisting minerals of alkaline rocks (in Russian). *Geokhimya*, **9**, 1063–82.

Rimskaya-Korsakova, 1968. 'Apatite mineralization in the Kovdor massif, in *Apatites*, Nauka, Moskva, 191–8.

Ronenson, B. M., 1966. 'The origin of miaskites and their associated rare-metallic mineralization', in Ginzburg, A. I., Ed. *Geologiya Mestorozhdeniy Redkikh Elementov, 28* (in Russian). Izd. 'Nedra', Moskva, 1–174.

Schreinemakers, F. A., 1948. *Nonvariant, monovariant and divariant equilibria* (in Russian). 212 pp., Izd. Inostr. Lit-ry, Moskva.

Strauss, C. A., and Truter, F. C., 1951. The alkaline complex at Spitskop Sekukuniland, Eastern Transvaal. *Trans geol. Soc. S. Africa.*, **53**, 169–90.

Tikhonenkov, I. P., and Tikhonenkova, R. P., 1962. On the mineralogy of the contact zone of the Lovozero massif', in Vlasov, K. A., *Redkie Elementy v Massivakh Shchelochnykh Porod. Trudy IMGRE, 9* Izd. Akad. Nauk SSSR, 3–34.

Tikhonenkova, R. P., 1967. 'Fenites of the Khibina alkaline massif,' in Borodin, L. S., Ed. *Redkometal'nye metasomatity shchelochnykh massivov* (in Russian). Izd. 'Nauka', Moskva, 5–94.

Tilley, C. E., 1958. Problems of alkali rocks genesis. *Q. J. geol. Soc. Lond.*, **113**, 323–60.

Tilley, C. E., and Gittins, J., 1961. Igneous nepheline-bearing rocks of the Haliburton–Bancroft province of Ontario. *J. Petrology*, **2**, 38–48.

Tugarinov, A. I., Pavlenko, A. S., and Aleksandrov, T. V., 1963. *Geochemistry of alkali Metasomatism* (in Russian). 202 pp., Izd. Akad. Nauk SSSR, Moskva.

Tugarinov, A. I., Pavlenko, A. S., and Kovalenko, V. I., 1968. Origin of apogranites according to geochemical data (in Russian). *Geokhimya*, **12**, 1419–36.

Volotovskaja, N. A., 1958. The Kovdor-massif (in Russian). *Geologiya SSSR*, **27**, 419–28.

Wyart, I., Sabatier, G., 1962. Sur le probleme de l'équilibre des feldspathes et des plagioclases. *C. R. Acad. Sci.*, **255**, 1551–6.

Yashina, R. M., 1962. 'The Kharlin concentric-zonal alkaline massif and the conditions of its formation', in Vorob'eva, O. A., Ed., *Trudy IGEM*, **76**, (in Russian). Izd. Akad. Nauk SSSR, Moskva, 7–38.

Yashina, R. M., 1963. 'On the contact-reactional interaction of nepheline syenites with xenoliths of dolomite-bearing marbles (on the example of the Orukhta alkaline massif of the South-East Tuva)', in Sokolov, G. A., Ed., *Fiziko-khimicheskie problemy formirovaniya gornykh porod i rud.*, 11, (in Russian). Izd. Akad. Nauk SSSR, Moskva, 104–17.

Zharikov, V. A., 1959. Geology and metasomatic features of skarn–polymetallic deposits of East Karamazar (in Russian). *Trudy IGEM*, **14**, Izd. Akad. Nauk SSSR, Moskva.

VI.8. ORIGIN OF THE ALKALINE ROCKS—A SUMMARY AND RETROSPECT

Henning Sørensen
(with information supplied by V. P. Volkov)

The origin of the various types of alkaline rocks is considered in Chapters II.1–7; III.1–5; IV.1–8 and V.1–4. The suggested mechanisms of formation are discussed in Chapters VI.1–7. In the present chapter the major groups of hypotheses are reviewed with emphasis on the historical aspects (see also Chapter I).

VI.8.1. Primary Alkaline Magmas—Rôle of Anatexis

There is now general agreement that no single parent magma is responsible for the formation of alkaline rocks, since alkaline rocks occur in many petrological associations (IV.1).*

This makes the old ideas of Daubrée of a primary magma rich in alkalis and 'mineralizers', of Becke (1903) of a primary Atlantic magma, and of Rosenbusch (1890) of a foyaitic 'Kern'—(Na,K)$AlSiO_2$—obsolete. According to Rosenbusch a homogeneous primary 'Urmagma' was ever persistent at depth. This metallic magma split into 'Metallkerne' according to chemical affinities. These metal magmas were oxidized and took up water during ascent to the surface.

It should however be mentioned here that large volumes of alkali basalt, syenite, trachyte, alkali granite etc. may be regarded as products of anatexis at depth (see VI.1a and b). In this case one may speak of primary alkaline magmas. This explanation is also given for the formation of perpotassic rocks (II.5).

VI.8.2. Alkaline Rocks as Products of Magmatic Differentiation and of Resorption of Minerals

Many authors consider alkaline rocks as products of consolidation of residual liquids from common subalkaline magmas (cf. VI.2).

* Readers are referred to the relevant Chapters of this book where similarly cited.

Harker (1909) suggested an origin by squeezing out of interstitial (normally more alkaline) fluids from the partly consolidated magma ('wine press mechanism'). Such mixtures of crystals and liquid must exist underneath any region of long-continued igneous activity. The interstitial liquid may be displaced in response to the type of stresses which affected the North Atlantic basin in Tertiary times.

Brøgger (1890) concluded from his examination of the Oslo Province that the successive eruptions originated in a common subjacent magma fairly rich in sodium. The sequence of crystallization and intrusion provides clues to the development of the magma. The 'Krystallisationsfolge' (crystallization sequence) in a cooling magma reflects the 'Differentiationsfolge' (sequence of differentiation) and he introduced the term 'Gesteinsserie' (rock series). The late members of the rock series of Oslo are enriched in alkalis (lardalite and natron granite).

Bowen (1928) suggested that nepheline syenite may originate as a residual liquid from granitic magma which crystallized 'in the appropriate manner'.

Daly (1933) emphasized the rôle of crystal fractionation, but (p. 402) also that no single-course-fractionation of saturated basalt magma could be responsible for the formation of alkaline magmas. He envisaged an introduction of alkalis by gaseous transfer, of feldspar, or of siliceous alkaline liquids.

According to Bowen (1928) undersaturated alkaline melts could be formed by fractionation of pyroxene from basaltic magmas. Also Larsen (1940) and Barth (1936) have advocated this view (see VI.2).

Fractional separation of amphibole or settling and resorption of amphibole and biotite may contribute to the formation of undersaturated alkaline rocks (Bowen, 1928; Waters, 1955; Yoder and Tilley, 1962; see also VI.6. and V.4).

Eliseev (1941) and Polkanov *et al.* (1967) explained the formation of nepheline syenites associated with alkali gabbroids, syenites and ijolites-urtites in stratified intrusions by means of a modified principle of crystallization differentiation. They emphasized the importance of gravitational flotation of crystals of plagioclase or nepheline during emplacement of the alkaline magma.

The incongruent melting of orthoclase and leucite accumulation may be the cause of derivation of feldspathoidal rocks from saturated magmas (Bowen, 1928).

Most authors have pointed at basaltic parent magmas. Barth (1936 and 1962) suggested that minor differences in chemical composition of a primary basaltic magma may determine the subsequent trend of differentiation.

Peridotitic or kimberlitic–carbonatitic magmas have also been claimed as primary magmas for alkaline rocks (Holmes, 1932; von Eckermann, 1948; Backlund, 1932; Strauss and Truter, 1950; Schröcke, 1955).

Alkali granite may develop as a residual product from saturated syenite magmas (see IV.3 and VI.2).

VI.8.3. Assimilation

Nearly identical alkaline rocks occur in widely different country rocks in different provinces. This is an indication that alkalinity was acquired before emplacement of the magma. Assimilation of country rocks (limestone-syntexis) as advocated by Daly (1910), Shand (1945), and Belijankin (1909–10, 1926) cannot therefore explain the origin of large masses of undersaturated alkaline rocks by desilication as discussed by Wyllie (VI.3). Assimilation of carbonate rocks, alkaline argillitic rocks and saline sediments may, however, be of local importance on a smaller scale (Jensen, 1908; Brouwer, 1946; van Bemmelen, 1947; Knopf, 1957; Tilley, 1957; Kostyuk and Bazarova, 1966; Barker and Long, 1969). The rôle of interaction of granitoid melts with carbonate and basic rocks in the formation of alkaline rocks is discussed by Korzhinsky and his collaborators (see VI.8.6).

VI.8.4. Rôle of Volatiles

Volatiles or 'mineralizers' have repeatedly been made responsible for the development of alkaline trends in subalkaline magmas as discussed in VI.4 and VI.5, in which reference is made to the classical papers of Smyth, Gilson, Fenner and Zavaritsky. Volatiles appear to be of outstanding importance in the genesis of many types of alkaline rocks. A few examples from the literature are listed here.

Daly (1933, p. 474) believed that water in basalt came from the invaded sediments and could contribute to the development of trachytic differentiates. Lindgren (1933) demonstrated that the Cordilleran igneous activity in the U.S.A. moved eastward, the magmas becoming gradually more differentiated and alkaline. The rôle of volatiles in this process was emphasized.

Tomkeieff (1937) considered a differential movement of alkalis in an olivine basaltic magma towards the upper part of the magma giving upper teschenitic and lower quartz–doleritic magma portions. Saether (1948) has developed this idea further (see VI.5).

Edwards (1938) regarded gas streaming to be active in the development of trachytic and phonolitic lavas at Kerguelen. Shand (1949, p. 143) explained the occasional outpouring of trachytic and phonolitic lavas from volcanoes normally yielding basalt as a result of expulsion of residual interstitial liquid by streams of gas bubbles.

Desilication of silica-saturated magmas may be a result of a flux of an aqueous phase through the magma (Currie, 1970).

VI.8.5. Liquid Immiscibility

Liquid immiscibility (liquation) has been considered by Loewinson-Lessing (1899) to be an important agent in the development of magmas. Bilibin (1939, 1940; V. P. Volkov, personal information) advanced the hypothesis of magmatic differentiation of primary basaltic magmas during the precrystallization period, the cause of this differentiation being the association of the ions of the melt into complex molecular structures. It is suggested that the different rate of diffusion of these associated molecules affected the formation of facies of higher alkalinity in the peripheric zones of the magma chamber.

The hypothesis of liquid immiscibility has until recently been disregarded in the interpretation of the origin of alkaline rocks but is now again considered important, especially in the later stages of crystallization, see IV.6 and VI.5.

VI.8.6. Metasomatism

'Nephelinization', has been considered by a number of authors (see VI.7). The formation of alkaline rocks by fenitization is of widespread importance and mobilization of fenites may give rise to intrusive nepheline syenites (von Eckermann, 1948; Kopeck ′ *et al.*, 1970). A number of occurrences of alkali gneisses are also regarded to be results of metasomatic processes (see II.2 and II.7).

The rôle of 'transmagmatic' solutions enriched in volatile components during the regional process of granitization and magmatic replacement is emphasized by Korzhinsky (1952, 1955, 1960, V. P. Volkov, personal information). One of the main statements of his theory is the 'principle of acid–basic interaction': alteration of the alkalinity of a magma expressed in terms of the chemical potentials of the mobile components (water, alkalis) expands the fields of crystallization of minerals containing these components and results in changes of the eutectic compositions and the crystallization trends of the magma (Korzhinsky, 1955). The source of matter and energy is assumed to be subcrustal 'transmagmatic' solutions. If such solutions are infiltrated through sialic rocks granitoid calc-alkaline melts should be generated within the Earth's crust at high temperatures. The successive infiltration of the solutions results in the replacement of the wall-rocks during the magmatic stage of the process. If the wall-rocks are enriched in bases (basic and ultrabasic rocks, dolomites) subalkaline and alkaline magmas could be generated. This mechanism is advocated by several investigators of alkalic–granitoid and alkalic–gabbroid associations of rocks located in regions of completed folding (Omel'iyanenko, 1959; Perchuk, 1964; Pavlenko, 1963 etc., see IV.5).

The idea of Korzhinsky (1960) that the formation of basic and ultrabasic magmas is accompanied by 'transmagmatic' solutions primarily enriched in alkalis was evolved by Borodin (1961, 1963, VI.7) and Kuznetsov (1964) in order to explain the genesis of alkaline ultrabasic rocks.

VI.8. REFERENCES

Backlund, H. G., 1932. On the mode of intrusion of deep-seated alkaline bodies. *Bull. Geol. Inst. Uppsala*, **24**, 1–24.

Barker, D. S., and Long, L. E., 1969. Feldspathoidal syenite in a quartz diabase sill, Brookville, New Jersey. *J. Petrology*, **10**, 202–21.

Barth, T. F. W., 1936. The crystallization process of basalt. *Am. J. Sci.*, **31**, 321–51.

Barth, T. F. W., 1962. *Theoretical Petrology*. John Wiley and Sons, 2nd edition, 1–416.

Becke, F., 1903. Die Eruptivgebiete des böhmischen Mittelgebirges und der amerikanischen Andes. *Tschermaks miner. petrogr. Mitt. Neue Folge*, **22**, 209–65.

Belijankin, D. S., 1909, 1910. Outlines of the petrography of the Ilmen Mountains (in Russian). *Izv. S-Peterb. Politechn. Inst.*, **12**, and **13**.

Belijankin, D. S., 1926. On the interpretation of the Ilmen complex (in Russian). *Geol. Vestnik*, **5**.

Bemmelen, R. W., van, 1947. The Muriah volcano (central Java) and the origin of its leucite-bearing rocks. *Proc. K. ned. Akad. Wet.*, **50**, 653–8.

Bilibin, Yu. A., 1939. Dissociation of molecules in the magmatic melt as a factor of magma differentiation (in Russian). *Dokl. Akad. Nauk SSSR*, **24**, 783–5.

Bilibin, Yu. A., 1940. On the genesis of alkaline rocks (in Russian). *Zap. Vses. min. Ob-va*, **69**, 2–3, 228–48.

Borodin, L. S., 1961. 'Nephelinization of pyroxenites and the parageneses of rock-forming minerals of ijolites in the ultrabasic–alkaline rocks' (in Russian), in *Fiziko-chimicheskie Problemy Formirovaniya Gornýkh Porod i Rud. I.* Izd. Akad. Nauk SSSR.

Borodin, L. S., 1963. Carbonatites and nepheline syenites (on the general petrology of ultrabasic and carbonatite massifs) (in Russian). *Izv. Akad. Nauk SSSR, ser. geol.*, **8**.

Bowen, N. L., 1928. *The Evolution of the Igneous Rocks*. Princeton Univ. Press., 1–332.

Brøgger, W. C., 1890. Die Mineralien der Syenitpegmatitgänge der südnorwegischen Augit- und Nephelinsyenite. *Z. Kristallogr. Miner.*, **16**, 1–663.

Brouwer, H. A., 1946. The association of alkali rocks and metamorphic limestone in a block ejected by the volcano Merapi (Java). *Proc. K. ned. Akad. Wet.*, **48**, 166–89.

Currie, K. L., 1970. An hypothesis on the origin of alkaline rocks suggested by the tectonic setting of the Monteregian Hills. *Can. Miner.* **10**, 411–20.

Daly, R. A., 1910. Origin of alkaline rocks. *Bull. geol. Soc. Am.*, **21**, 87–118.

Daly, R. A., 1933. *Igneous Rocks and the Depths of the Earth.* McGraw-Hill, New York.

Eckermann, H. von, 1948. The alkaline district of Alnö Island. *Sver. geol. Unders. Ser. Ca.*, **36**, 1–176.

Edwards, A. B., 1938. Tertiary lavas from the Kerguelen Archipelago, B.A.N.Z. Antarctic Expedition (D. Mawson). 1929–31. *Rept. ser. A*, **5**, 72–100.

Eliseev, N. A., 1941. On the origin of primary layering in the Lovozero pluton. *Zap. Vses. min. Ob-va*, **70**, 1, 86–105.

Harker, A., 1909. *The Natural History of the Igneous Rocks*. Macmillan, New York.

Holmes, A., 1932. The origin of igneous rocks. *Geol. Mag.*, **69**, 543–58.

Jensen, H. I., 1908. The distribution, origin and relationships of alkaline rocks. *Proc. Linn. Soc. N.S.W.*, **33**, 491–588.

Knopf, A., 1957. The Boulder batholith of central Montana. *Am. J. Sci.*, **255**, 81–103.

Kopecký, L. Dobes, M., Fiala, J., and Stovíčkova, N., 1970. Fenites of the Bohemian Massif and the relations between fenitization, alkaline volcanism and deep fault tectonics. *Sbornik Geol. Věd. Geologie*, rada G. **16**, 51–112.

Korzhinsky, D. S., 1952. Granitization as a factor of magmatic replacement (in Russian). *Izv. Akad. Nauk SSSR, ser. geol.*, **2**.

Korzhinsky, D. S., 1955. 'Problems of petrography of the magmatic rocks related to transmagmatic solutions and granitization' (in Russian), in *Magmatizm i svyaz' s nim poleznykh iskopaemykh.* Izd. Akad. Nauk SSSR.

Korzhinsky, D. S., 1960: 'Acidity–basicity as the most important factor in magmatic and postmagmatic processes' (in Russian), in *Magmatizm i svyaz' s nim poleznykh iskopaemykh.* Gosgeoltekhizdat.

Kosty'uk, V. P., and Bazarova, T. Yu., 1966. *The Petrology of the Alkaline rocks of the Eastern part of the Eastern Sayan* (in Russian). Izd. 'Nauka', Moskva, 1–168.

Kuznetsov, Yu. A., 1964. *Chief Types of Magmatic Formations* (in Russian). Izd. 'Nedra'.

Larsen, E. S., 1940. Petrographic province of central Montana. *Bull. geol. Soc. Am.*, **51**, 887–948.

Lindgren, W., 1933: 'Differentiation and ore deposition, Cordilleran region of the United States', in *Ore deposits of the Western States (Lindgren volume)*. Amer. Inst. Min. Met. Eng., 152–80.

Loewinson-Lessing, F., 1899, 1900. Kritische Beiträge zur Systematik der Eruptivgesteine I, II, III. *Tschermaks miner. petrogr. Mitt. Neue Folge*, **18**, 519–24; **19**, 169–81; **19**, 291–307.

Omel'iyanenko, B. I., 1959. Possible trends of alkaline magma formation in geosynclinal regions with Turkestan–Alay as an example (in Russian). *Izv. Akad. Nauk SSSR, ser. geol.*, **12**.

Pavlenko, A. S., 1963. 'Petrology and some geochemical peculiarities of Middle-Palaeozoic complex of granitoids and alkaline rocks of the Eastern Tuva' (in Russian), in *Problemy magmy i genezisa izverzhennykh gornykh porod.* Izd. Akad. Nauk SSSR, Moskva, 239–46.

Perchuk, L. L., 1964. *Physico-Chemical Petrology of Granitoid and Alkaline Intrusions of the Central Turkestan–Alay* (in Russian). Izd. 'Nauka'.

Polkanov, A. A., Eliseev, N. A., Eliseev, E. N., and Kavardin, G. I., 1967. *The Gremijakha-Vyrmes Massif of the Kola Peninsula* (in Russian). Izd. 'Nauka', Leningrad, 1–236.

Ronenson, B. M., 1966. *The Origin of Miaskites and the Associated Rare-Metal Mineralization* (in Russian). 'Geologiya mestorozhdeniy redkikh elementov' 28, 1–174. Izd. 'Nedra', Moskva.

Rosenbusch, H., 1890. Über die chemischen Beziehungen der Eruptivgesteine. *Tschermaks miner. petrogr. Mitt.*, **11**, 144–78.

Saether, E., 1948. On the genesis of peralkaline rock provinces. *Rep. 18th Intern geol. Congr. Great Britain*, **2**, 123–30.

Schröcke, H., 1955. Über Alkaligesteine und deren Lagerstätten. *Neues Jb. Miner. Mh.*, 169–89.

Shand, S. J., 1945. The present status of Daly's hypothesis of the alkaline rocks. *Am. J. Sci.*, **243A**, 495–507.

Shand, S. J., 1949. *Eruptive Rocks*. Thomas Murby and Co., 1–488.

Strauss, C. A., and Truter, F. C., 1950. The alkali complex at Spitskop, Sekukuniland, eastern Transvaal. *Trans. geol. Soc. S. Afr.*, **53**, 81–125.

Tilley, C. E., 1957. Problems of alkali rock genesis. *Q. Jl. geol. Soc. Lond.*, **113**, 323–60.

Tomkeieff, S. J., 1937. Petrochemistry of the Scottish Carboniferous–Permian igneous rocks. *Bull. volcan.*, **2**, 59–87.

Waters, A. C., 1955. Volcanic rocks and the tectonic cycle. *Geol. Soc. Am. Spec. Paper*, **62**, 703–22.

Yoder, H. S., and Tilley, C. E., 1962. Origin of basalt magmas; an experimental study of natural and synthetic rocks systems. *J. Petrology*, **3**, 342–532.

VII. Economical Geology

VII.1. ECONOMIC MINERALOGY OF ALKALINE ROCKS

E. I. Semenov

VII.1.1. Introduction

The minerals of alkaline rocks which are of demonstrated or potential economic importance are mainly oxides or silicates of Nb, Ti, Zr, RE, Al, Be and Na. Phosphates of Ca and RE are also important.

Some minerals, as for instance pyrochlore, occur in several types of alkaline rocks, each characterized by its own specific variety of the mineral. Thus the pyrochlore of alkali granites is rich in Ta, Pb, Bi and Y; that of agpaitic pegmatites rich in RE, and that of carbonatites rich in Ba, Ti and Zr. Other minerals, as for instance steenstrupine and loparite, are restricted to rocks of agpaitic type (cf. II.2.1;* Semenov, 1967).

Alkaline rocks and their minerals are exploited in occurrences such as Khibina, the Kola peninsula (apatite, nepheline), Kovdor, the Kola peninsula (vermiculite, apatite, magnetite), and Jos, Nigeria (columbite, cassiterite). Nepheline rocks are furthermore important sources of ceramic raw materials, Blue Mountain, Ontario and Stjernøy, Northern Norway. The bauxitic crust of weathering of nepheline syenites is exploited in Arkansas, U.S.A., and in the Los Islands, Republic of Guinea, and nepheline is produced as a source of aluminium in the U.S.S.R. Furthermore, carbonatitic deposits of pyrochlore, apatite, etc. associated with alkaline rocks are, or have been, mined in a number of places such as Oka, Quebec; Arasha, Brazil; Kaiserstuhl, Germany; and Palabora, South Africa. Some types of alkaline rocks are potential sources of uranium.

Reviews of the economic geology of alkaline rocks have been published by Schröcke (1955); Agard (1956); Sheynmann *et al.* (1961); Semenov (1967), and in a general way in the three volume work by Vlasov and collaborators (Vlasov, 1966–8).

The economic geology of carbonatites has been exhaustively treated by Ginzburg *et al.*, (1958); Deans (1966); Heinrich (1966) and van Wambeke *et al.*, (1964). The mineral deposits of these rocks will consequently be mentioned only briefly in this paper.

VII.1.2. Types of Deposits and Genetical Associations

Most economically important minerals of alkaline rocks are formed at high temperature conditions, either as primary accumulations in the magma, as for instance layers and lenses of eudialyte, or disseminated mineralizations, as for instance perovskite and titanomagnetite in alkaline ultramafic rocks.

Pneumatolytic processes are important in the formation of deposits of vermiculite, albitized pyrochlore-bearing zones in alkali granites, etc.

A number of mineral deposits are of pneumatolytic–hydrothermal type, as for instance rare-earth–fluorite–carbonate veins.

Fenitic zones around alkaline rocks may be enriched in beryllium (Seal Lake, Labrador), zirconium (Lovozero) and uranium (Ilímaussaq).

The crust of weathering of alkaline rocks contain important mineral deposits. Examples are silicate bauxites formed at the expense of nepheline syenites and sometimes enriched in Nb and Ti; and the weathering crusts of carbonatitic derivatives of alkaline rocks enriched in earthy rare earth minerals (bastnaesite, rhabdophanite, leucoxene ores of Ce, Th, U, Ti, Nb).

* Readers are referred to the relevant Chapters of this book where similarly cited.

TABLE 1 Types of Deposits Genetically Connected with Alkaline Rocks

Rock Type	Stage	Characteristic Minerals	Characteristic Elements	Examples of Deposits
Sodalite syenites	pneumatolytic	*steenstrupine*, *epistolite*, *chkalovite*, sorensenite, analcime, monazite, aegirine	Ce, Th, *U*, *Be*, *Nb*	Ilímaussaq, Lovozero
Eudialyte nepheline syenites	magmatic	*eudialyte*, microcline aegirine, rinkite	*Zr*, Nb, Ce	—
Urtites	magmatic	*nepheline*, *apatite*, *NbCe-perovskite*, *titanite*, eudialyte	*P*, *Al*, *Nb*, *Ti*, *Ce*, *Zr*, *F*	Khibina, Lovozero
Ultramafic alkaline rocks	pneumatolytic	*perovskite*, *titano-magnetite* diopside	*Ti*, *Fe*, Mg, Ca, *Ce*	Kovdor
		a) *phlogopite* (*vermiculite*) diopside, titanite	K, Mg, Ca, Ti	Kovdor, Palabora
		b) *magnetite*, *apatite*, baddeleyite, pyrochlore	*Fe*, Mg, Ca, P, Zr, *Nb*, *Ta*	—
		c) *chalcopyrite*	*Cu*	Palabora
		d) calcite, *pyrochlore*, apatite, perovskite, magnetite	*Nb*, P, Ti, Ca	Oka
	hydrothermal	e) ankerite, rodochrosite *pyrochlore*, *bastnaesite* strontianite, *barite*, *fluorite*	*Nb*, *Ce*, *Ba*, *Sr* Fe, Mn	Kangankunde Arasha
Nepheline and alkali syenites	pneumatolytic	a) albite, *zircon*, *pyrochlore*, titanite, apatite, britholite	*Zr*, *Nb*, Ti, P, Ce, Ca, Na	Miask
	hydrothermal	b) calcite, *pyrochlore*, apatite	*Nb*, P, Ca	Miask
	supergenic	bauxite (*gibbsite*) leucoxene (*anatase*)	*Al*, Fe, *Ti*	Los
Potassium-rich leucite rocks	magmatic	*leucite*, *kaliophillite*, microcline, nepheline, apatite	*K*, *Al*, Ca, P	Siberia Italy
	hydrothermal	barite, fluorite, *brannerite*, thorite	K, Ba, Tl, *U*, Th, F	—
alkali syenites	pneumatolytic	albite, microcline, *barylite*, epididymite, aegirine, niobophyllite, joaquinite	*Be*, Nb, Ce, Ba	Seal Lake
	hydrothermal	fluorite, *leucophanite*, bastnaesite	*Be*, F, Ce	Siberia
Alkali granites	magmatic–pneumatolytic	albite, riebeckite, *columbite*, *cassiterite*, zinnwaldite, helvite, *pyrochlore*, xenotime, cryolite	*Nb*, *Ta*, *Sn*, Zr, Li, Be, F	Jos plateau
	hydrothermal	*hematite*, siderite, *fluorite*, barite, *bastnaesite*, *monazite*, *cryolite*	*Fe*, *Ba*, *Ce*, *Y*, Sc, Nb, *F*	Mountain Pass, Mongolia
	supergenic	halloysite, limonite, *bastnaesite*, rhabdophanite	Al, Fe, *Ce*	S. Siberia Poços-de-Caldas

Placer deposits formed at the expense of alkali granites contain columbite, cassiterite and pyrochlore (Nigeria); the urtites of Khibina give rise to placers rich in apatite and nepheline; the miaskites of the Urals deposits of pyrochlore and zircon; and placers formed at the expense of carbonatites may be rich in pyrochlore, apatite and perovskite.

Some other types of deposits may be genetically connected with alkaline magmatism. This includes concentrations of bitumens, hydrocarbon gases and hot mineralized waters, which were recently found in Khibina (cf. II.2 and VI.4). Deposits of oilshales in the Green River region, U.S.A. are also associated with alkaline volcano–sedimentary rocks.

Deposits of soda, dawsonite, analcime and other zeolites appear to be products of volcano–sedimentary processes in alkaline volcanic provinces. Nepheline syenites, ijolites, alkali granites (and carbonatites) are the economically most important alkaline rocks (cf. Table 1). These rocks are characterized by high concentrations of sodium, fluorine and other volatile components and often contain F-arfvedsonite (Semenov, 1967).

VII.1.2.1. Agpaitic Nepheline Syenites

The agpaitic nepheline and sodalite syenites of Ilímaussaq are accompanied by pegmatites and pneumatolytic–hydrothermal derivatives rich in Ce–Th–U–Be–Nb–Zn mineralizations (steenstrupine, monazite, epistolite, pyrochlore, chkalovite, epididymite, sphalerite, etc.). The rare alkalis Rb, Cs and Li are concentrated in rocks and pegmatites containing astrophyllite and lithium-mica (cf. Sørensen, 1962; Semenov, 1969).

Besides agpaitic meso- or melanocratic rocks, such as lujavrites and kakortokites, are rich in eudialyte. The agpaitic rocks of Ilímaussaq and Lovozero present huge deposits of complex type in the technology of the future, and may by combined exploitation yield elements such as Zr, Hf, U, Th, Ti, Sn, Nb, Ta, V, Mo, Tl, RE, Ga, Al, Li, Na, K, Rb, Cs, F, Be, Zn, Fe, etc.

Nepheline syenites of intermediate type (Semenov, 1967, cf. II.2.1), as for instance Khibina, contain deposits of apatite–nepheline–titanite ores.

VII.1.2.2. Miaskitic Nepheline Syenites and Alkali Syenites

Nepheline syenites of miaskitic type can carry niobium and zirconium (pyrochlore, zircon and sometimes britholite) in pneumatolytic–hydrothermal albite and calcite mineralizations. These derivatives are known from the Urals (Miask) and Ukraine. In Canada (Lake Nipissing, Ontario) there is a similar type of deposit, but without nepheline rocks (Rowe 1958). Carbonate derivatives containing uranium-rich pyrochlore are known in alkali syenite (without nepheline) also in Bancroft, Ontario; in the Baikal area; etc.

Potassium-rich alkaline rocks contain sometimes big quantities of leucite and kaliophilite and may as in Siberia be accompanied by uraniferous hydrothermal veins containing brannerite and thorite. Late derivatives of potassic alkaline rocks may also be enriched in barium, rubidium and thallium, elements which replace potassium isomorphously.

VII.1.2.3. Ijolites and Alkaline Ultramafic Rocks

Rocks belonging to the series urtite–ijolite–jacupirangite contain in addition to the above-mentioned apatite–nepheline ore of Khibina, concentrations of the perovskite-group minerals rich in Nb and RE and sometimes associated with apatite.

The ijolite–melteigites of alkaline ultramafic massifs carry perovskite, titanomagnetite and apatite; phlogopitization (and secondary vermiculite) is widely distributed (cf. VI.7).

VII.1.2.4. Alkali Granites and Granosyenites

Alkali granites are characterized by deposits of pyrochlore, columbite and cassiterite and may also be enriched in xenotime, gagarinite (Y-rare earths), zircon and cryolite (Archangelskaya, 1968). The cryolite deposit at Ivigtut, South Greenland (IV.3) is associated with a stock of alkali granite.

Alkali barkevikite granites and granosyenites of more basic type (Ca, Mg) are accompanied by more important hydrothermal veins and bodies rich in RE, Ba, F and Fe-ores. Such deposits are now mined in California (Mountain Pass) and Mongolia. The presence of scandium and rare alkalis (Li, Cs) can be used to distinguish this type of deposit from late carbonatites accompanying alkaline ultramafic rocks.

VII.1.3. Economically Important Elements, Minerals and Rocks, and Examples of Deposits

Phosphorus: Apatite is associated with many types of alkaline rocks (Deans, 1967). The Khibina massif of the Kola peninsula (IV.2) contains an annular zone of apatite–nepheline ore (cf. IV.5; Ivanova, 1963) sandwiched between a lower zone of urtite and an upper zone of poikilitic nepheline syenite (rischorrite). The ore body is up to 200 m thick and several km long. It dips towards the centre of the intrusion and has been traced to depths of several hundred metres. The apatite–nepheline ore contains 40–90% apatite corresponding to 20–45% P_2O_5. The apatite reserve equals 2000 million tons, the yearly production is about 10 million tons and is planned to increase to 18 million tons/year in the near future. It has also proved economical to mine ore containing less than 20% P_2O_5. It is possible to extract fluorine (about 1·5% in apatite), rare earth metals (about 0·8%) and strontium (about 1%) from the apatite ore. Nepheline is obtained as a by-product.

Apatite is also produced from alkaline ultramafic massifs such as Kovdor, the Kola peninsula (IV.2, VI.7) where it is associated with magnetite; and from the carbonatite complex at Palabora, South Africa (Heinrich, 1970). The apatite of Kovdor contains RE and Sr, but in lower concentrations than the Khibina apatite.

It is proposed to mine apatite from alkaline ultramafic rocks and their crusts of weathering also in Brazil and Africa (Leonardos, 1956).

Niobium: The important minerals are pyrochlore, Nb-perovskite, columbite and epistolite (Rowe, 1958).

Pyrochlore deposits are especially connected with carbonatitic derivatives of ultramafic alkaline massifs (Oka, Quebec, cf. IV.6; Gold, 1966; Arasha, Brazil; Lueshe, Congo; etc.) and less commonly with miaskitic nepheline syenites and alkali syenites (Miask, the Urals, Eskova *et al.*, 1964; Lake Nipissing, Ontario, Rowe, 1958). The Lake Nipissing deposit is associated with nepheline-free syenites, the ore deposit holds 0·7% Nb_2O_5 and measures 27 million tons. Niobium minerals of the perovskite group occur not only in carbonatites, but also in ijolite–urtites and in alkaline ultramafic rocks.

Loparite is a characteristic niobium mineral in the differentiated complex of the Lovozero massif (IV.2; Vlasov *et al.*, 1959; Vlasov, 1968). In the agpaitic nepheline and sodalite syenites of the Ilímaussaq intrusion, South Greenland (IV.3; Hansen, 1968; Semenov, 1969) pyrochlore is of widespread occurrence in pneumatolytic hydrothermal analcime veins (cf. II.2.9), while minerals of the epistolite–murmanite group and niobium astrophyllite (in part niobophyllite) occur in fenitized basaltic rocks in the roof of the intrusion (Sørensen, *et al.*, 1969).

Alkali granites in Northern Nigeria (Williams *et al.*, 1956) and in Southern Siberia contain huge amounts of niobium and tantalum. Aegirine- and riebeckite-albite granites carry pyrochlore, the biotite varieties—columbite and cassiterite (cf. II.6; IV.8). Eluvial placers of Northern Nigeria have been mined for a long time for columbite and cassiterite. The unweathered rocks contain in places about 0·1% $(Nb,Ta)_2O_5$.

Tantalum does not form its own minerals: but columbite from certain zones in alkali granites and pyrochlore from carbonatites may be enriched in tantalum, up to 10–20% Ta_2O_5. The Nb/Ta ratio may in such cases reach a value of 2, while, on the contrary, this ratio is 100 in epistolite.

Molybdenum: Molybdenite forms a deposit of economic importance in a syenite massif at Werner Bjerge, East Greenland (Bearth, 1959). Disseminated molybdenite occurs in albitized rocks in Khibina.

Titanium: Perovskite and titano-magnetite are greatly concentrated in nepheline–pyroxene rocks of alkaline ultramafic massifs. Examples are Afrikanda, Kola Peninsula (IV.2); Libby, Montana (Larsen and Pardee, 1929) and Jacupiranga, Brazil (Melcher, 1966) The hanging wall of some Khibina apatite–nepheline ore bodies contains rather big quantities of titanite (Ivanova, 1963). In genetic connection with nepheline and alkali syenites of the Urals (Ilmen) and Arkansas there are deposits of rutile (and ilmeno-rutile).

The bauxite crusts of weathering of some alkaline massifs (Arkansas, the Enissey Range) are enriched in leucoxene and sometimes also in niobium.

Zirconium: Eudialyte forms huge accumulations in some agpaitic nepheline syenites: the kakortokites of Ilímaussaq (IV.3) have up to 30% eudialyte (Bohse, Brooks and Kunzendorf, 1971) and the lujavrites of Lovozero (IV.2) have up to 10% eudialyte and contain lenses of monomineralic eudialytite (Vlasov *et al.*, 1959). The concentration of ZrO_2 is, however, generally lower than 2% in agpaitic rocks. The eudialyte contains interesting concentrations of niobium (about 1% Nb_2O_5) and about 2% rare earth metals (enriched in the Y-group metals).

The fenitic zone around Lovozero is in places rich in zirconium, partly occurring in vlasovite (Vlasov, 1968).

Albitized miaskites and their crusts of weathering may be richer still, containing big quantities of zircon (the Urals).

Cryptocrystalline baddeleyite is mined in hydrothermal veins at Poços de Caldas, Brazil (Tolbert, 1966). Baddeleyite and zirkelite can be concentrated also in ultramafic massifs (Kola, S. Africa). From the Palabora copper ore 2000 tons of baddeleyite is extracted annually.

In the above-mentioned minerals zirconium is isomorphously substituted by hafnium (up to 6% HfO_2 in zircon from Nigerian granites). In some occurrences, e.g. hydrothermal veins, the hafnium content falls abruptly, resulting in zirconium practically free of hafnium. This is important for the use of zirconium for nuclear purposes which must be purified for hafnium.

Tin: Cassiterite is mined together with columbite in placers formed from the crust of weathering of alkali granites in N. Nigeria. Some agpaitic rocks from the Ilímaussaq intrusion are also rich in tin (stannite, sorensenite, cf. V.3).

Uranium: Agpaitic nepheline syenites (and corresponding lavas) and peralkaline granites are richer in disseminated uranium than other igneous rocks, and alkalic types of igneous rocks are generally richer in uranium than their subalkalic equivalents (cf. review by Sørensen, 1970b). The highest concentrations reported so far are alkali granites of the Bokan Mountain, Alaska (MacKevett, 1963) which contain up to 200 ppm U (uranothorite, xenotime, uranothorianite, uraninite and brannerite) and up to 0·5% U_3O_8 in albitized and hydrothermally altered zones. Some lujavritic rocks of the Ilímaussaq intrusion have up to 0·3% U_3O_8, and average values of 200–600 ppm U, this is especially the case of hydrothermally altered rocks in the roof of the intrusion (Sørensen, 1970a). The radioactive minerals are steenstrupine, monazite and thorite.

Albitized and otherwise altered alkaline rocks are especially rich in uranium (Sørensen, 1970b). Mixtures of baddeleyite and zircon ('caldasite') form veins and lenses in altered eudialyte-bearing rocks at Poços de Caldas (Wedow, 1967) and have up to 0·6% U_3O_8. These veins are probably formed by leaching and redeposition of the rare elements of the primary alkaline rocks.

Pitchblende occurs in alkali metasomatic rocks containing albite and aegirine in some quartz–magnetite deposits of the USSR.

Uraniferous hydrothermal veins containing brannerite and thorite are associated with some Siberian potassic alkaline rocks (Bilibina *et al.*, 1963).

Carbonatites may contain uranium as a constituent of pyrochlore, zirkelite and baddeleyite.

Thorium: Thorium minerals and rare earth minerals enriched in thorium are found in agpaitic nepheline syenites (steenstrupine, britholite) and their vein facies (Sørensen, 1962; Poljakov, 1970); and in veins of miaskitic nepheline syenites (thorite), for instance in the Urals (Eskova *et al*., 1962). Hydrothermal carbonate–hematite–fluorite veins containing thorite are found in a number of regions (cf. Olson and Overstreet, 1964; Vlasov, 1968; Sørensen, 1970b).

Thorium mineralizations are found in quartz, baryte and carbonate-bearing veins associated with a stock and veins of albite-rich syenite in the Wet Mountains, Colorado (Christman *et al*., 1959).

The rare earth metals (*Semenov, 1963*): Concentrations of 1–2% RE_2O_3 are found in agpaitic rocks, and their hydrothermal derivatives, for instance those of Ilímaussaq (Sørensen, 1962; Semenov, 1969; Gerasimovsky, 1969). The most prominent minerals are steenstrupine, rinkite and monazite. The eudialyte of these rocks contains 1–2% RE_2O_3.

Pegmatites rich in lovchorrite-rinkolite have been exploited in the Khibina massif (Slepnev, 1962).

Derivatives of ijolite-urtite (Kola) contain RE in apatite and perovskite and concentrations of britholite are known in miaskitic massifs (the Urals).

Late hydrothermal carbonatites are often enriched in bastnaesite, parisite, monazite and florencite. Carbonatites such as Kangankunde, Malawi (Holt, 1965) contain about 5% RE_2O_3. Huge hydrothermal deposits of bastnaesite, monazite, fluorite, baryte, hematite, magnetite and siderite are known from California (Mountain Pass) and Mongolia in genetic connection with alkaline grano-syenites (Olson *et al*., 1954; Tugarinov, Pavlenko and Aleksandrov, 1963). The Mountain Pass ore contains nearly 5% of RE, the reserve is 5 million tons. These rare earths are enriched in europium.

The crusts of chemical weathering of alkaline massifs (Poços de Caldas, the Enissey Range) contain secondary RE-hydrophosphates, carbonates and oxides (rhabdophanite, bastnaesite, cerianite). Deposits of this type in southern Siberia have crusts of pyrolusite and limonite rich in earthy bastnaesite, rhabdophanite and monazite overlying a more than 100 m thick zone of halloysite.

All the above-mentioned minerals (with the exception of xenotime from alkali granites) contain rare earths of the cerium group. Eudialyte and rinkite are relatively enriched in the heavy yttrium lanthanoids (including europium). Steenstrupine, monazite, perovskite and bastnaesite are rich in lanthanum. Enrichment in europium (when compared to samarium) is characteristic for fluorite, bastnaesite and monazite from carbonatites. Green monazite may contain strontium, which can be isomorphously replaced by europium.

Aluminium: Aluminium is produced on a large scale from gibbsite-bauxite from the crusts of weathering of nepheline syenites (Arkansas, Gordon *et al*., 1958; Guinea). Gibbsite is also widely distributed in the zeolitized urtites from Lovozero. Nepheline is now an important raw material for the production of aluminium (Khibina, W. Siberia) and leucite is a potential source (Italy, Siberia). It has been proposed to produce alumina from its carbonate—dawsonite—which is concentrated in the volcano–sedimentary soda-deposits of Green River, U.S.A. Similar in origin are the alumina-rich analcime strata in sandstones from the Caucasus, etc.

Gallium: Nepheline contains about 0·00n% Ga and gallium is consequently produced in nepheline plants. Tugtupite and natrolite from agpaitic pneumatolytic-hydrothermal deposits are still richer in gallium (up to 0·03% Ga).

Iron: Magnetite is mined from alkaline ultramafic massifs (Kovdor, Palabora). Siderite and its alteration product, hematite, are widely distributed in late carbonatites (as are manganese carbonates and oxides). The magnetite of Kovdor is rich in Mg, but free from titanium.

The ore contains 35% Fe, the reserves amount to 350 million tons.

Beryllium: Rather rich deposits are found in pneumatolytic derivatives of alkali syenites (Seal Lake, Labrador; E. Siberia). The most important minerals are leucophane, barylite and epididymite (Heinrich and Deane, 1962). The fenitic aegirine–feldspar rocks of the Seal Lake region contain 0·4% BeO (Mulligan, 1968). The hydrothermal Siberian occurrences are enriched in leucophane and fluorite.

Pneumatolytic–hydrothermal veins in naujaite and augite syenite in the Ilímaussaq intrusion contain important quantities of chkalovite, stockwork deposits of thin ussingite–analcime veins have average contents of 0·1% Be (Engell *et al.*, 1971).

Strontium: Strontium can be separated from calcium in the apatite technology (Khibina, Kovdor). Strontianite is widely distributed in some late rare-earth carbonatites. The carbonatites of Kangankunde, Malawi (Holt, 1965) contain about 15% of this mineral.

Barium: Baryte occurs in rare-earth carbonatites associated with grano-syenites (Mountain Pass, cf. Olson *et al.*, 1954).

Sodium and potassium: These metals are obtained together with aluminium from nepheline (Khibina, etc.). Volcano–sedimentary soda-deposits are known in the Green River Belt, U.S.A. (Bradley, 1964) and from Lake Magadi, Tanzania, the latter in genetic connection with alkaline ultrabasic volcanism.

The rare alkalis: Concentrations of up to 12% of Cs + Rb are known in astrophyllite. Polylithionite contains similar amounts of Li + Rb. This has been found in the nepheline and alkali syenites and in the alkali granites of Ilímaussaq, the Emissey range, the Turkestan ridge, the Baykal area and Nigeria. Amphiboles (arfvedsonite and riebeckite) from the Lovozero nepheline syenites and the Nigeria alkali granites are often rich in lithium (1–2% Li_2O).

Fluorine: Many rocks in Ilímaussaq and Lovozero are rich in villiaumite, NaF, and are potential sources of fluorine. Besides, fluorine is obtained as a biproduct in the apatite industry.

Fluorite is concentrated in some late hydrothermal carbonatites (Sayan, S.W. Africa, etc.). It is now mined in Amba Dongar (W. India)

The cryolite body at Ivigtut, Greenland (Berthelsen, 1962) is found in the apical part of a stock of alkali granite which is transformed into greisen adjacent to the cryolite. The riebeckite–albite granites of Nigeria have accessory cryolite.

Copper: Chalcopyrite is exploited in the Palabora alkaline ultramafic massif. The ore contains 0·69% Cu.

Pyrrhotite: Pyrrhotite concentrations in the contact zone of Khibina have been mined (Fersman, 1935).

Gold and silver: The auriferous (Au–Te) pipes of Cripple Creek, the Front Ranges of Colorado (Lovering and Goddard, 1950) are associated with phonolites.

Silver is a minor accessory in some ussingite veins of Ilímaussaq (Semenov, 1969).

Lead and zinc: Galena and sphalerite are widely distributed minor accessories in alkaline rocks. The lujavrites of Ilímaussaq contain more than 0·1% Zn over large areas (cf. Sørensen *et al.*, 1969; Gerasimovsky, 1969).

VII.1.4. Technically Important Minerals and Rocks

Graphite: High-grade graphite deposits occur in the nepheline and cancrinite syenites of the Botogol massif, the Sayans (Sobolev, 1947; Sobolev and Florensov, 1948).

Corundum: Corundum for abrasives has been mined in the nepheline gneiss belts of the Haliburton–Bancroft region, Ontario (Adams and Barlow, 1910).

Hexagonal enantiomorphic crystals of nepheline and cancrinite display stronger piezo effects than the usual piezo-material quartz. But natural monocrystals of nepheline and cancrinite without fractures have not been found up to now. Natural material can, however, be used as seeds in the synthesis of monocrystals.

Vermiculite (and phlogopite) is exploited from alkaline ultramafic rocks in the Kola peninsula (Ternovoy *et al.*, 1969), Siberia and Montana.

The olivinites of alkaline ultramafic massifs in cases consist of very pure magnesian olivine of refractory grade.

Iron-poor nepheline syenites at Blue Mountain, Ontario (Hewitt, 1960) and Stjernøy, Northern Norway (Heier, 1962) are mined for ceramical purposes (Allen and Charlsley, 1968). In some parts of the U.S.S.R. rich in electrical power nepheline-rich rocks, especially urtites, are raw materials for the production of aluminium, soda, potash, cement. Examples are the Kija–Shaltyr deposit in the Krasnojarsk area, Pambak in Armenia (Andreeva, 1968) etc. Titanite, garnet and rare-metal minerals can be obtained as biproducts. The nepheline obtained as a biproduct in the apatite production at Khibina is also used in the production of alkali-metals, gallium, ceramics, cement and as a filler material.

Considerable concentrations of zeolites (natrolite, etc.) in rocks, pegmatites and hydrothermal veins of Lovozero, Ilímaussaq, and other nepheline-sodalite syenite massifs may be of interest as molecular sieves in the chemical technology.

Leucite (and kaliophilite-kalsilite) from potassic alkaline volcanic rocks represent potential sources of potassium fertilizers. Rocks of this type are widely distributed in Italy, the Baikal area, Armenia etc.

Gemstones: Transparent or semitransparent varieties of pink or red tugtupite, and of ussingite, eudialyte, green ekanite, apatite, etc. are occasionally used as semi-precious gemstones. The same applies to transparent olivine (crysolite) from the alkaline ultramafic rocks of Taimyr.

Building and ornamental stones: Blue sodalite in the Haliburton–Bancroft region and lazurite deposits in Afghanistan, Pamir and the Baikal area (near Sludanka, Voskoboinokova, 1938) are genetically connected with alkaline rocks and are used for ornamental purposes.

The nepheline and sodalite syenites of Saint-Hilaire, the Monteregian province (IV.6) are used as building stones, road material and in the production of concrete.

Many alkaline rocks of striking appearance may be used as decorative building material. The larvikite of the Oslo province and the pulaskite of the Fourche Mountains, Arkansas, are prominent examples. Some kakortokites and naujaites of Ilímaussaq with bright red eudialyte, green sodalite and white feldspar; greenish-grey khibinite from Khibina with red eudialyte; dense fine-grained rocks (tinguaite etc.) from Khibina and Lovozero with gold-yellow needles of lamprophyllite, astrophyllite, green aegirine, red eudialyte, violet plates of murmanite, etc., are potential ornamental stones.

VII.1. REFERENCES

Adams, F. D., and Barlow, A. E., 1910. Geology of the Haliburton and Bancroft areas, Province of Ontario. *Mem. Geol. Surv. Can.*, **6**, 419 pp.

Agard, J., 1956. Les gites minéraux associés aux roches alcalines et aux carbonatites. *Sciences Terre*, **4**, 105–51.

Allen, J. B., and Charlsley, T. J., 1968. *Nepheline-Syenite and Phonolite.* Institute of Geological Sciences, London, 1–169.

Andreeva, E. D., 1968. *Alkaline Magmatism of Kuznezk Alatau* (in Russian). Izd. 'Nauka', Moskva, 168 pp.

Archangelskaia, V. V., 1968. Tantal–niobium ores in ancient metasomatic rocks of E. Siberia (in Russian). *Geol. Rud. Mestor.*, **5**, 29–40.

Bearth, P., 1959. On the alkali massif of the Werner Bjerge in East Greenland. *Meddr Grønland*, **153**, 4, 1–63.

Bilibina, T. V., Donakov, V. I., and Titov, V. K., 1963. Hydrothermal uranium ores connected with alkaline intrusive complexes (in Russian). *Geol. Rud. Mestor.*, **5**, 35–54.

Berthelsen, A., 1962. On the geology of the country around Ivigtut, SW-Greenland. *Geol. Rundschau*, **52**, 269–80.

Bohse, H., Brooks, C. K., and Kunzendorf, H., 1971. Field observations on the kakortokites of the Ilímaussaq intrusion, South Greenland, including mapping and analyses by portable X-ray fluorescence equipment for zirconium and niobium. *Rapp Grønl. geol. Unders.*, **38**.

Bradley, W. H., 1964. Geology of the Green River formation and associated Eocene rocks. *Prof. Pap. U.S. Geol. Surv.*, **496A**, 86 pp.

Christman, R. A., Brock, M. R., Pearson, R. C., and Singewald, Q. D., 1959. Geology and thorium deposits of the Wet Mountains, Colorado. *Bull. U.S. Geol. Surv.*, **1072-H**, 491–535.

Deans, T., 1966. 'Economic mineralogy of African carbonatites', in Tuttle, O. F., and Gittins, J., Eds., *Carbonatites.* Interscience Publishers, 385–413.

Deans, T., 1967. Exploration for apatite deposits associated with carbonatites and pyroxenites. *Seminar on Sources of Mineral Raw Materials for the Fertilizer Industry in Asia and the Far East, Bangkok, Thailand*, 1–10.

Engell, J., Hansen, J., Jensen, M., Kunzendorf, H., and Løvborg, L., 1971. Beryllium mineralization in the Ilímaussaq intrusion, South Greenland, with description of a field beryllometer and chemical methods. *Rapp. Grønl. Geol. Unders.*, **33**.

Eskova, E. M., Mineyev, D. A., and Mineyeva, I. G., 1962. Uranium and thorium in the alkalic rocks of the Urals. *Geochemistry*, **9**, 885–94.

Eskova, E. M., Zabin, A. G., and Muchitdinov, G. N., 1964. *The Mineralogy and Geochemistry of Rare-Metals in Vishnevogorsk* (in Russian). Izd. 'Nauka', Moskva, 318 pp.

Fersman, A. E., 1935. *The Scientific Study of Soviet Mineral Resources.* Martin Lawrence Ltd., London, 1–149.

Gerasimovsky, V. I., 1969. *The Geochemistry of the Ilímaussaq Alkali Massif* (in Russian). Izd. 'Nauka' Moskva, 174 pp.

Ginzburg, A. I., *et al.*, 1958. *Rare-Metal Carbonatites* (in Russian). Gosgeoltekhizdat, Moskva, 127 pp.

Gold, D. P., 1966. The minerals of the Oka carbonatite and alkaline complex, Oka, Quebec. *Min. Soc. India*, IMA volume, 83–91.

Gordon, M., Tracey, J. I., and Ellis, M. W., 1958. Geology of the Arkansas Bauxite region. *Prof. Pap. U.S. Geol. Surv.*, **299**, 268 pp.

Hansen, J., 1968. Niobium mineralization in the Ilímaussaq alkaline complex, South-West Greenland. *Rep. 23rd Intern. Geol. Congr. Czechoslovakia, section 7*, 263–73.

Heier, K. S., 1962. A note on the U, Th and K contents in the nepheline syenite, and carbonatite on Stjernöy, North Norway. *Norsk. Geol. Tidsskr.*, **42**, 287–92.

Heinrich, E. Wm., 1966. *The Geology of Carbonatites.* Rand McNally and Company, Chicago, 1–555.

Heinrich, E. Wm., 1970. The Palabora carbonatitic complex—A unique copper deposit. *Can. Miner.* **10**, 585–98.

Heinrich, E. Wm., and Deane, R. W., 1962. An occurrence of barylite near Seal Lake, Labrador. *Am. Miner.*, **47**, 758–63.

Hewitt, D. F., 1960. Nepheline Syenite Deposits of southern Ontario. *Rep. Ontario Dep. Mines.*, **69**, 8, 194 pp.

Holt, D. N., 1965. 'The Kangankunde Hill rare-earth prospect. *Bull. Geol. Surv. Malawi*, **20**.

Ivanova, T. N., 1963. *The Apatite Deposits of Khibina* (in Russian). Gosgeoltekhizdat, Moskva, 288 pp.

Larsen, E. S., and Pardee, J. T., 1929. The stock of alkaline rocks near Libby, Montana. *J. Geol.*, **37**, 97–112.

Leonardos, O. H., 1956. Carbonatitos com apatita e pirocloro no estrangeiro e no Brazil. *Engen. Miner. Metal*, **23**, 136.

Lovering, T. S., and Goddard, E. N., 1950. Geology and ore deposits of the Front Range Colorado. *Prof. Pap. U.S. geol. Surv.*, **223**, 1–319.

MacKevett, E. M., jr., 1963. Geology and ore deposits of the Bokan Mountain uranium–thorium area, Southeastern Alaska. *Bull. U.S. geol. Surv.*, **1154**, 1–125.

Melcher, G. C., 1966. 'The carbonatites of Jacupiranga, São Paulo, Brazil', in Tuttle, O. F., and Gittins, J., Eds. *Carbonatites.* Interscience Publishers, 169–82.

Mulligan, R., 1968. Geology of Canadian beryllium deposits. *Geol. Surv. Can., Econ. Geol. Rep.*, **23**, 1–109.

Olson, J. C., and Overstreet, W. C., 1964. Geologic distribution and resources of thorium. *Bull. U.S. Geol. Surv.*, **1204**, 1–61.

Olson, J. C., Shawe, D. R., Pray, L. C., and Sharp, W. N., 1954. Rare-earth mineral deposits of the Mountain Pass district, San Bernardino County, California. *Prof. Pap. U.S. Geol. Surv.*, **261**, 1–75.

Poljakov, A. I., 1970 *Geochemistry of Thorium in the Alkaline Rocks of Kola Peninsula* (in Russian). Izd. 'Nauka', Moskva, 166 pp.

Rowe, R. B., 1958. Niobium (columbium) deposits of Canada. *Geol. Surv. Can., Econ. Geol. Ser.*, **18**, 1–108.

Schröcke, H., 1955. Über Alkaligesteine und deren Lagerstätten. *Neues Jb. Miner. Mh.*, 169–89.

Semenov, E. I., 1963. *The Mineralogy of the Rare-Earth Metals* (in Russian). Izd. Akad. Nauk, USSR, 412 pp.

Semenov, E. I., 1967. 'The mineralogical–geochemical types of derivatives of nepheline syenites. The activity of alkalis and volatiles' (in Russian), in Tichonenkova R. P. and Semenov, E. I., Eds. *Mineralogy of Pegmatites and Hydrothermalites from Alkaline Massifs*, 52–71.

Semenov, E. I., 1969. *The Mineralogy of the Ilímaussaq Alkali Massif* (in Russian). Izd. 'Nauka', Moskva, 165 pp.

Sheynmann, Yu. M., Apel'tsin, F. R., and Nechayeva, Ye. A., 1961. Alkalic Intrusions, Their Mode of Occurrence and Associated Mineralization (in Russian). *Geologiya Mestorozhdenii Redikkh Elementov*, **12-13**, Moscow (Gosgeoltekhizdat), 177 pp. (Review in *Int. Geol. Rev.*, **5**, 451–8).

Slepnev, Yu. S., 1962. *Lovchorrite-Rinkolite Pegmatites* (in Russian). Izd. Akad. Nauk USSR, Moskva, 150 pp.

Sobolev, V. S., 1947. 'The petrology of the Botogol alkali massif', in *The Botogol Graphite Deposit and the Perspectives of its Utilization*. OGIZ, Irkutsk, 165–218 (in Russian).

Sobolev, V. S., and Florensov, N. A., 1948. Genesis of the Botogol Graphite. *Sovietskaiya Geologiya*, **32**, 29–35 (in Russian).

Sørensen, H., 1962. On the occurrence of steenstrupine in the Ilímaussaq massif, Southwest Greenland. *Meddr. Grønland*, **167**, 1, 1–251.

Sørensen, H., 1970a 'Occurrence of uranium in alkaline igneous rocks', in *Uranium Exploration Geology*. International Atomic Energy Agency, Vienna, 161–8.

Sørensen, H., 1970b. 'Low-grade uranium deposits in agpaitic nepheline syenites, South Greenland', in *Uranium Exploration Geology*. International Atomic Energy Agency, Vienna, 151–9.

Sørensen, H., Hansen, J., and Bondesen, E., 1969. Preliminary account of the geology of the Kvanefjeld area of the Ilímaussaq intrusion, South Greenland. *Rapp. Grønl. geol. Unders.*, **18**, 1–40.

Ternovoy, V. I., Afanasiev, B. V., and Sulimov, B. I., 1969. *Geology and Prospecting of the Kovdor Vermiculite-Phlogopite Deposit* (in Russian). Izd. 'Nedra', Moskva, 287 pp.

Tolbert, G. E., 1966. The uraniferous zirconium deposits of the Poços de Caldas Plateau, Brazil. *Bull. U.S. Geol. Surv.*, **1185-C**, 1–28.

Tugarinov, A. I., Pavlenko, A. S., and Alexandrov, I. V., 1963. *Geochemistry of Alkali Metasomatism* (in Russian). Izd. Akad. Nauk, Moskva, 202 pp.

Vlasov, K. A., (ed.), 1966–8. *Genetic Types of Rare-Element Deposits*. Israel Program for Scientific Translations, Jerusalem, 916 pp.

Vlasov, K. A., Kuzmenko, M. V., and Eskova, E. M., 1959. *The Lovozero Alkali Massif* (in Russian). Izd. Akad. Nauk. Moskva, 623 pp.

Voskoboinikova, N., 1938. On the mineralogy of the Slyudyanka lazurite deposit (in Russian). *Zap. Vseros. Miner. Obsh.*, **67**, 4.

Wambeke, L. van, Brinck, J. W., Deutzmann, W., Gonfiantini, R., Hubaux, A., Métais, D., Omenetto, P., Tongiorgi, E., Verfaillie, G., Weber, K., and Wimmenauer, W., 1964. Les roches alcalines et les carbonatites du Kaiserstuhl. *Euratom*, **1827,d,f,e**, 1–232.

Wedow, H., jr., 1967. The Morro do Ferro thorium and rare-earth ore deposit, Poços de Caldas district, Brazil. *Bull. U.S. Geol. Surv.*, **1185-D**, 1–34.

Williams, F. A., Meehan, I. A., Paulo, K. L., John, T. U., and Rushton, H. G., 1956. Economic geology of the decomposed columbite-bearing granites, Jos Plateau, Nigeria. *Econ. Geol.*, **51**.

ADDITIONAL REFERENCES

Chapter I

Marshall, P., 1906. The geology of Dunedin, New Zealand. *Q. J. geol. Soc. Lond.*, **62**, 381–424.

Chapter II.2

Shand, S. J., 1922. The problem of the alkaline rocks. *Proc. geol. Soc. S. Afr.*, **25**, xix–xxxiii.

Chapter II.7

Teixeira, C., 1954, not 1965 as in reference list.

Chapter IV.3

Macdonald, R., 1968. *Petrological studies of some peralkaline acid rocks from the Gardar province, South Greenland.* Unpubl. Ph.D. thesis, Edinburgh.

O'Hara, M. J., and Yoder, H. S. (Jr.), 1967. Formation and fractionation of basic magmas at high pressures. *Scot. J. Geol.*, **3**, 67–117.

Pauly, H., 1960. Paragenetic relations in the main cryolite ore of Ivigtut, South Greenland. *Neues Jb. Miner. Abh.*, **94** (Fesband Ramdohr), 121–39.

Poulsen, V., 1964. The sandstones of the Precambrian Eriksfjord Formation in South Greenland. *Geol. Surv. Greenland Rep.*, **2**, 1–16.

Pulvertaft, T. C. R., 1961. The Puklen intrusion, Nunarssuit, S. W. Greenland. *Bull. Grønlands geol. Unders.*, **29**, 33–50 (also *Meddr. Grønland*, **123**, 6).

Pulvertaft, T. C. R., 1965. The Eqaloqarfia layered dyke, Nunarssuit, South Greenland. *Bull. Grønlands geol. Unders.* **55**, 39 pp. (also *Meddr. Grønland* **169**, 10).

Scharbert, H. G., 1968. Microsyenitic dykes from the northern part of the Ilímaussaq peninsula, southern Greenland. *Tschermaks miner. petrogr. Mitt.*, **12**, 443–62.

Chapter IV.4

Bultitude, R. J., and Green, D. H., 1968. Experimental studies at high pressures on the origin of olivine nephelinite and olivine melilite nephelinite magma. *Earth Planet. Sci. Lett.*, **3**, 325–37.

Klominský, J., 1963. Cancrinitführende Alkalisyenite aus dem Čistá–Massiv, Westböhmen, Tschechoslovakei. *Neues Jb. Miner. Abh.*, **99**, 295–306.

Schneider, A., 1970. The sulphur isotope composition of basaltic rocks. *Contr. Miner. Petrol.*, **25**, 95–124.

Shrbený, O., 1969. Tertiary magmatic differentiation in the central part of the České středohoří Mountains. *Časopis pro Miner. a Geol.*, **14**, 285–98.

Spencer, A. B., 1969. Alcalic igneous rocks of the Balcones Province, Texas. *J. Petrology*, **10**, 272–306.

Tröger, W. E., 1935. *Spezielle Petrographie der Eruptivgesteine.* 360 pp. Schweizerbart, Stuttgart.

Chapter IV.7

Dietz, R. S., 1961. Continent and ocean basin evolution by spreading of the sea floor. *Nature*, **190**, 854–57.

Wilson, J. T., 1963. Evidence from Islands on the spreading of ocean floors. *Nature*, **197**, 536–538.

Chapter V.3

Tugarinov, A., Kovalenko, V. I., Znamensky, E. B., Legeido, V. A., Sabatovich, E. V., Brandt, S. B., and Tsyhansky, V. D., 1968. 'Distribution of Pb-isotopes, Sn, Nb, Ta, Zr and Hf in granitoids from Nigeria', in *Origin and Distribution of the Elements.* Ed. Ahrens, L. H., International Series Monographs in Earth Sciences vol. **30**, Pergamon Press, Oxford 687–99.

Chapter VI.5

Saggerson, E. P., and Williams, L. A. J., 1964. Ngurumanite from southern Kenya and its bearing on the origin of rocks in Northern Tanganyika alkaline district. *J. Petrology*, **5**, 40–81.

Chapter VI.7

Lodochnikov, B. N., 1926. The simplest methods of illustrating multicomponent systems (in Russian). *Izv. Inst. Fisicochimicheskogo Analiza. Akad. Nauk SSSR*, **1924**, 2.

Appendix

1. CLASSIFICATION AND NOMENCLATURE OF PLUTONIC ROCKS

Recommended by the Subcommision on the Systematics of Igneous Rocks (the International Union of Geological Sciences). Reviewed by Henning Sørensen

The Subcommission on the Systematics of Igneous Rocks (chairman: Professor A. Streckeisen) has at its meeting in Montreal in August 1972, and after deliberations carried out during a number of years, agreed on a system of classification and nomenclature for the plutonic rocks (except charnockitic rocks, carbonatites, and melilite-rich rocks). The system, which is based on the proposals of A. Streckeisen (1967), is presented in *Neues Jahrbuch für Mineralogie*, Monatshefte 4, 1973 and in *Geological News Letter*, 1973, no. 2. The future work of the Subcommission will deal with the volcanic rocks and with the plutonic rocks which are not covered by the system agreed on in Montreal.

Principles of classification

1. The rocks in question are classified and named according to their actual (modal) mineral content (measured in volume percent).

2. Rocks with colour index lower than 90 are classified primarily according to their light-coloured minerals, rocks with colour index 90–100 according to their mafic minerals.

3. Rocks with colour index lower than 90 are classified and named according to their position in the *QAPF* (see 4 below) double triangle, the light-coloured minerals being calculated to the sum of 100 (i.e. $Q + A + P = 100$, or $F + A + P = 100$), see Fig. 1.

The names given to each field are 'root names' for larger groups of rocks, subsidiary diagrams should be used in order to give each specific rock its proper name (see II.1, Table 2)*.

4. The following minerals and mineral groups are used:

*Readers are referred to the relevant Chapters of this book where similarly cited.

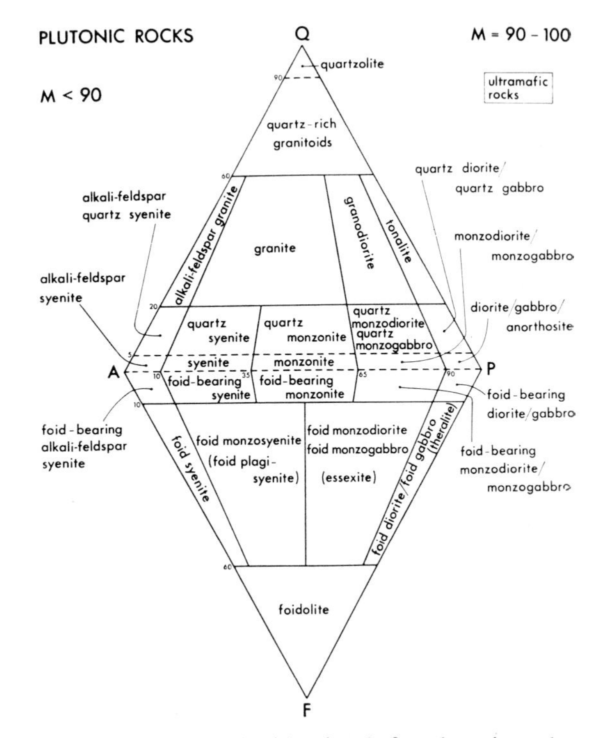

Fig. 1 The *QAPF* double triangle for plutonic rocks having less than 90% dark minerals

- *Q* quartz (or: silica minerals, mainly quartz);
- *A* alkali feldspars: orthoclase, microcline, perthite, anorthoclase, albite (An_{00-05});
- *P* plagioclase (An_{05-100}), scapolite;
- *F* feldspathoids (foids): leucite and pseudoleucite, nepheline, sodalite, noseane, hauyne, analcime, cancrinite, etc.;
- *M* mafic minerals (micas, amphiboles, pyroxenes, olivines, opaque minerals, accessories (zircon, apatite, titanite, etc.), epidote, allanite, garnets, melilites, monticellite, primary carbonates, etc.).

5. It is suggested to use the prefixes *mela-* and

leuco- to designate the more mafic and felsic types of each rock group. The term melasyenite applies, for instance, to syenites darker then the average syenite.

If rock suites are to be divided into colour index groups, the following division is proposed: colour index 0–35 (leucocratic); 35–65 (mesocratic); 65–90 (melanocratic), and 90–100 (ultramafic.)

6. It is recommended that the minerals in composite rock names are arranged in order of increasing amounts, the most abundant mineral being placed closest to the root name. Thus, in a eudialyte–aegirine–nepheline syenite, aegirine is more abundant than eudialyte.

Alkaline rocks

The terms *alkali* and *alkaline* are regarded as chemical terms which should not be used in general rock names or 'root names' in a classification based on modal compositions. It is recommended to use the terms in a chemical sense to indicate that the rocks in question are enriched in alkalis with respect to silica and/or alumina. This is expressed by the presence modally of feldspathoids and/or alkali pyroxenes and/or amphiboles. See also discussion on pp. 3–7.

The term *alkaline* is a general term, while *alkali* is used to indicate the presence of feldspathoids, alkali pyroxenes/amphiboles in a specific rock, as in the names alkali granite, alkali syenite and alkali gabbro. Thus, all rocks plotting in the *FAP*-triangle are alkaline, by definition.

The root name feldspathoidal, foidal or foid syenite is regarded as a general name for all rocks plotted in this field of the *FAP*-triangle. Each specific rock should be given its specific name: nepheline syenite, nepheline–sodalite syenite, aegirine–nepheline syenite, etc. The detailed classification and naming should be based on nature of feldspathoid, nature of mafic minerals, colour index and in cases also on textural relations. Such a system of classification is presented in Chapter II.1, Table 2.

Rocks plotting in the field named foidolite are named according to nature of feldspathoid, kind of mafic mineral and colour index (see Chapter II.1, Table 2).

Rocks plotting in the fields named foid monzosyenite (foid plagisyenite), foid monzodiorite/monzogabbro (essexite) and foid diorite/gabbro are also named according to mineralogy and colour index. Theralite may be used synonymously for nepheline gabbro.

2. GLOSSARY OF ALKALINE AND RELATED ROCKS

Compiled by
Henning Sørensen

A glossary of rock names is being prepared by the Commission on Systematics in Petrology. This glossary will contain definitions of names and also recommendations for terms to be retained or abandoned.

Furthermore, as stated on p. 17, the works of Johannsen (1939, etc.), Tröger (1935 and 1938), Jung and Brousse (1959) and Ronner (1963) contain easily accessible information on terminology. To this list should be added Holmes (1928) and the main Soviet reference book by Loewinson-Lessing and Struve (1963).

The editor of this book therefore found it untimely and unnecessary to bring a list of rock names and definitions. However, a number of colleagues have pointed out that the alkaline rocks comprise so many rock names that a glossary of these names might be useful to the reader of the book. For this reason the index of rock names and the succeeding glossary of alkaline and related rocks have been compiled.

With regard to the plutonic rocks the principles of classification listed in the first part of the appendix have been followed. It should at this place be emphasized that the *QAPF*-double triangle gives only the 'root names' or 'group names' for major groups or classes of rocks, while a specific rock should be given a more precise name. Thus, an alkali syenite is an alkali feldspar syenite having sodic pyroxene/amphibole. The specific rock should be named according to its characteristic minerals, e.g. aegirine syenite or aegirine–albite syenite. Such self-evident terms are only included in the glossary in special cases.

With regard to the volcanic rocks the definitions given are based on Streckeisen (1967) and it has been attempted to follow the terminology used for the plutonic rocks as closely as possible. However, the Commission on Systematics in Petrology has not yet accomplished the revision of the classification of volcanic rocks, not yet, for example, having decided on what principles this classification should be based.

Similarly, lamprophyres and melilite-bearing rocks have not been discussed by the Commission, for which reason the definitions given in the glossary are provisional.

A number of subalkaline rocks, which commonly occur in alkaline provinces, have been included in the glossary.

The editor finds, as stated in Chapter II.1 (p. 16), that many of the names given to alkaline rocks are entirely unnecessary, a simple set of names, such as that presented in Table 2 (Chapter II.1), would serve most descriptive purposes. It should, however, be admitted that, even if many of the names listed in the glossary appear to be obsolete, they may nevertheless be of practical use in specific cases. The editor would, for instance, find it cumbersome to substitute brief names such as naujaite and kakortokite with more precise (and complicated) names when discussing the petrology of the Ilímaussaq intrusion. The use of a local name is furthermore justified when it is given to a rock unit which covers several fields in the *QAPF*-double triangle. The name larvikite is an example of this.

It is recommended that petrologists using local names always give a precise definition of the rock, as for instance the position in the *QAPF*-double triangle or as in Table 2, Chapter II.1.

The compiler of the glossary has endeavoured to make the list of names as comprehensive as possible, but would be grateful for information regarding possible omissions.

Finally, the compiler would like to stress that in a number of cases it has been impossible to give a precise and unambiguous definition of a rock. This is caused partly by the fact that some, undoubted rock units cover a certain scope in petrographical variation, partly that the original descriptions of the rock types are incomplete or erroneous. In some cases rock names have not been used in a consistent way over the years.

It has not always been possible to give the type localities in modern geographical terms.

The attached list of references contains the papers consulted when compiling this glossary. For references to the original descriptions of the various rocks see Johannsen (vols. II–IV) and Tröger (1935 and 1938).

GLOSSARY

Absarokite (Iddings, 1895; Absaroka Range, Yellowstone, U.S.A.)
Trachybasalt, mesocratic; alkali feldspar, labradorite, and sometimes leucite in matrix. Phenocrysts of olivine and augite.

Aegiapite (Belijankin and Vlodavetz, 1932; Turij Mis, Kola Peninsula, U.S.S.R.)
Calcite–apatite–aegirine–augite rock.

Aegirinolith (Kretschmer, 1917)
Aegirine augite–titanite–magnetite pyroxenite.

Afrikandite (Afrikanda, Kola Peninsula, U.S.S.R.) Ultramafic rock composed of knopite, titanomagnetite, melilite.

Agpaitic nepheline syenites (and phonolites) (Ussing, 1912; Ilímaussaq, Greenland), see pp. 6, 22–24.
Peralkaline nepheline syenites and phono-

lites, Na + K > Al, leuco- to mesocratic. Aegirine, arfvedsonite, complex zirconium–titanium silicates such as eudialyte characteristic. No Fe–Ti-oxides. Villiaumite common.

Ailsyte (Heddle, 1897; Ailsa Craig, Scotland)
Alkali quartz microsyenite, riebeckite-bearing.

Aiounite (Duparc, 1925; Mestigmer, Morocco)
Melteigite containing titanaugite.

Akerite (Brøgger, 1890, Oslo Province, Norway)
Microsyenite or -monzonite, quartz-bearing, characterized by rectangular grains of oligoclase. Augite, hornblende, biotite, sometimes aegirine augite. Hypersthene may be present.

Alaskite (Spurr, 1900; Alaska)
Leucogranite composed almost entirely of alkali feldspar and quartz.

Albanite (Washington, 1920)
Melaleucitite containing phlogopite.

Albite–nepheline syenite, see Mariupolite.

Albitite (Turner, 1896; Plumas Co., California, U.S.A.)
Albite syenite, often aplitic.

Algarvite (Lacroix, 1922; Caldeira de Monchique, Portugal)
Melteigite, biotite-rich.

Alkali basalt (Tilley, 1950), see pp. 20, 67, 70–71
Basalt containing normative nepheline and often modal nepheline, analcime or alkali feldspar. Also used for basalt containing alkali feldspar, devoid of foids and with less than 52% SiO_2 (U.S.S.R.).

Alkali basaltoids
In the U.S.S.R. used for basaltic rocks containing foids; with or without plagioclase.

Alkali gabbro (see p. 70)
Foid- and/or alkali feldspar-bearing gabbro, less than 10% foids in rock, alkali feldspar less than 10% of total feldspar.

Alkali granite and -rhyolite
Granites and rhyolites containing sodic pyroxenes and/or -amphiboles.

Alkali syenite, -monzonite, -diorite, etc.
Rocks containing sodic pyroxenes and/or -amphiboles, and/or foids (less than 10% of total of felsic minerals).

Alkali pyroxenite
Pyroxenite containing sodic pyroxenes or minor foids.

Allochetite (Doelter, 1902; Monzoni, Tyrol)
Micro-nepheline-monzosyenite, porphyritic.

Alnöite (Rosenbusch, 1887; Alnö, Sweden)
Alkali lamprophyre containing melilite, biotite, olivine, titanaugite, nepheline, perovskite, garnet, calcite, etc.; olivine–biotite–melilitolite.

Alvikite (von Eckermann, 1948; Alvik, Sweden)
Quartz-bearing melanocratic lamprophyre rich in calcite and biotite, sometimes nepheline in matrix.

Ampasimenite (Lacroix, 1922; Ampasimena, Madagascar)
Ijolite porphyry or nephelinite porphyry with glass matrix.

Amphigenite (Cordier, 1868)
= Leucite tephrite

Analcime basalt (Lindgren, 1883; Lacroix, 1924)
= Olivine analcimite.

Analcime dolerite
= Crinanite, microteschenite.

Analcimite (Gemmellaro, 1845; Cyclop Isl.; Pirsson, 1896)
Volcanic rock containing analcime as only or predominant felsic mineral. Mesocratic. Olivine subordinate; olivine analcimite is sometimes termed analcime basalt.

Analcimolith (Johannsen, 1938)
Volcanic rock composed entirely of analcime, = leucoanalcimite.

Analcitite
= analcimite.

Ankaramite (Lacroix, 1916; Ankaramy, Madagascar)
Augite-rich olivine melabasalt = augite-rich picrite basalt.

Ankaratrite (Lacroix, 1916; Ankaratra, Madagascar), see p. 55
Biotite-bearing olivine melanephelinite.

Anorthoclasite (Loewinson-Lessing, 1901)
Alkali feldspar syenite composed of anorthoclase.

Apachite (Osann, 1896; Apache Mts., Texas, U.S.A.)
Nepheline phonolite rich in sodic amphiboles, besides sodic pyroxene and aenigmatite.

Apaneite (Vlodavetz, 1930; Khibina, Kola Peninsula, U.S.S.R.)
Nepheline-bearing apatitite.

Arkite (Washington, 1901; Magnet Cove, Arkansas, U.S.A.)
Nepheline fergusite or leucite/pseudoleucite ijolite, leucocratic, containing aegirine augite, melanite-rich.

Arsoite (Rheinisch, 1912; Ischia, Italy)
Leucotrachyte, olivine-, andesine- and sometimes nepheline-bearing. Related to rhomb porphyry and kenyte. Arso-type trachyte of Rosenbusch.

Assyntite (Shand, 1909; Assynt, Scotland)
Titanite-rich sodalite syenite containing augite.

Atlantite (Lehmann, 1924; Rungwe, Tanzania)
Nepheline tephrite or nepheline basanite.

Augite syenite
= Hypersolvus syenite (or monzonite) carrying augite, titanaugite, or ferroaugite and with alkaline affinity. Sometimes ternary feldspar, and sometimes nepheline- or quartz-bearing. See also larvikite.

Augitite (Doelter, 1883; Cap Verde Isl.)
Hyalo–nepheline–tephrite with foids in matrix or 'concealed' in glass.

Baldite (Johannsen, 1938; Little Belt Mts., Montana, U.S.A.)
Melamonchiquite.

Banakite (Iddings, 1895; Absaroka, Yellowstone, U.S.A.)
Trachybasalt. Similar to absarokite, but poorer in olivine and augite.

Barshawite (Johannsen, 1938; Barshaw, Scotland)
Micro-analcime–nepheline monzosyenite or orthoclase-bearing andesine lugarite, with barkevikite and titanaugite.

Basalt, see p. 67

Basanite (Pliny (Plinius), Brongniart, 1813), see p. 70
= Olivine tephrite, that is olivine basalt in which foids make up more than 10% of felsic minerals.

Basanitoid (Bücking, 1880)
The term has been used in different meanings: originally for rocks intermediate between olivine basalt and basanite. On p. 71 the term is used for alkali basalts in which modal foid is not apparent.

Batukite (Iddings and Morley, 1917; Celebes)
= Olivine melaleucitite.

Bebedourite (Tröger, 1928; Salitre Mts., Brazil)
Sodic pyroxenite containing biotite (= alkali pyroxenite or biotite jacupirangite).

Bekinkinite (Rosenbusch, 1907; Bekinkina Mt., Madagascar)
Melatheralite, rich in analcime (secondary after nepheline), barkevikite-bearing.

Beloeilite (Johannsen, 1920; St. Hilaire, Quebec, Canada)
Sodalite syenite or feldspar-bearing tawite.

Benmoreite (Tilley and Muir, 1964; Mull, Scotland), see p. 88
Mugearite–trachyte having anorthoclase and normative plagioclase; $Na_2O > K_2O$.

Bergalite (Söllner, 1913; Kaiserstuhl, W. Germany)
Hauyne–micromelilitolite, biotite-rich = melilite-rich foidolite; distinguished from polzenite by absence of olivine.

Beringite (Starzynski, 1913; Bering Isl., U.S.S.R.)
Alkali feldspar melatrachyte containing barkevikite.

Bermudite (Pirsson, 1914; Bermuda Isl.)
= Biotite nephelinite.

Berondrite (Lacroix, 1920; Berondra, Madagascar)
Theralite rich in barkevikite, melanocratic.

Bigwoodite (Quirke, 1936; French River, Ontario, Canada)
Alkali syenite (orthoclase, microcline, albite) with traces of foids and sodic pyroxene and/or amphibole.

Bjerezite (Bereshite) (Erdmannsdörfer, 1928; Bjerez, U.S.S.R.)
Micro-leucoessexite rich in nepheline, containing analcime, titanaugite, aegirine, etc.

Blairmorite (Knight, 1904; Blairmore, Alberta, Canada)

Leucoanalcimite with aegirine augite and minor nepheline, sanidine, melanite, etc.

Boderite (Frechen, 1967; Eifel, W. Germany)
= Noseane–sanidine–biotite–augitite (= noseane shonkinite), subvolcanic, granular or cumulate.

Bogusite (Johannsen, 1938; Boguschowitz, Moravia, Czechoslovakia)
Teschenite.

Borolanite (Teall, 1892; Loch Borolan, Scotland)
Biotite–pseudoleucite–syenite rich in melanite.

Bostonite (Hunter and Rosenbusch, 1890; Brazil)
Micro-alkali feldspar syenite having poorly oriented rectangular laths of microcline and albite (bostonitic texture), hololeucocratic.

Bowralite (Mawson, 1906; Bowral, Wales)
Alkali syenite pegmatite, rich in aegirine augite.

Braccianite (Lacroix, 1917; Bracciano Lake, Italy)
= Leucite–leucotrachybasalt or tephritic leucitite.

Brandbergite (Chudoba, 1930; Brandberg, S.W. Africa)
Alkali granite with minor arfvedsonite.

Buchonite (Sandberger, 1872; Buchonia, W. Germany)
Tephrite to phonolitic tephrite; mesocratic; phenocrysts of brown hornblende and biotite.

Caltonite (Johannsen, 1938; Calton Hill, England)
Basanite rich in analcime, mesocratic.

Campanite (Lacroix, 1917; Vesuvius, Italy)
Leucoleucite–trachybasalt or tephritic leucoleucitite containing aegirine augite, sanidine, hauyne, nepheline.

Camptonite (Rosenbusch, 1887; Campton, New Hampshire, U.S.A.)
Gabbroic lamprophyre, mesocratic, containing barkevikite, titanaugite, biotite and sometimes olivine.

Canadite (Quensel, 1913; Ontario, Canada)
Potash feldspar–albite–nepheline syenite, rich in biotite, with normative anorthite.

Cancalites (Fuster *et al.*, 1967; Cancarix and Calasparra, Spain), see p. 98
Lamproites with potential olivine, devoid of potential leucite (normative).

Cancarixite (Parga-Pondal, 1935; Sierra de las Cabras, Spain)
Aegirine augite mela-alkali quartz-syenite, lamprophyric.

Cascadite (Pirsson, 1905; Highwood Mts., Montana, U.S.A.)
Lamprophyre having phenocrysts of biotite, olivine and augite in matrix of alkali feldspar, leucite, etc.

Cecilite (Cordier, 1868; Capo di Bove, Italy)
Melilite–leucoleucitite, nepheline- and anorthite-bearing.

Cedricite (Wade and Prider, 1940; Kimberley, Australia), see p. 97
Leucite–diopside–lamproite = melaleucitite containing phlogopite, magnophorite, etc.

Chibinite, see Khibinite.

Ciminite (Washington, 1896; Cimino, Italy)
Leucotrachybasalt or sanidine absarokite containing labradorite and olivine.

Cocite (Lacroix, 1933; Coc-Pia, Haut-Tonkin)
Leucite-rich lamprophyre having augite, olivine, biotite in matrix of alkali feldspar, leucite and mafic minerals.

Columbretite (Johannsen, 1938; Columbrete Isl., Spain)
Tephritic–leucite phonolite.

Comendite (Bertolio, 1895; Comende Isl., Sardinia, Italy), see p. 109
Alkali leucorhyolite, see also pantellerite.

Congressite (Adams and Barlow, 1913; Congress Bluff, Ontario, Canada)
Feldspar–biotite–urtite.

Coppaelite (Sabatini, 1903; Coppaeli di Sotto, Italy)
Biotite–melilitite, olivine-free.

Covite (Washington, 1901; Magnet Cove, Arkansas, U.S.A.)
Mela-nepheline syenite (= malignite).

Craigmontite (Adams and Barlow, 1913; Craigmont Hill, Ontario, Canada)
Nepheline-rich leuco-nepheline diorite, corundum-bearing.

Crinanite (Flett, 1909; Loch Crinan, Scotland)

Olivine–analcime dolerite = ophitic analcime-poor olivine teschenite.

Cromaltite (Shand, 1906; Cromalt Hills, Assynt, Scotland)
Melanite–alkali pyroxenite.

Cuyamite (Johannsen, 1938; Cuyamas Valley, California, U.S.A.)
Teschenite (microbogusite).

Dahamite (Pelikan, 1902; Sokotra Isl., Red Sea)
Micro-alkali granite, rich in albite.

Damkjernite (Brøgger, 1921; Fen, Norway)
Nepheline-rich lamprophyre, melanocratic. Phenocrysts of titanaugite, biotite, in base of nepheline, alkali feldspar, calcite, perovskite, etc.

Dancalite (De Angelis, 1925; Assakoma, Ethiopia)
Analcime leucotephrite, alkali feldspar-bearing (= analcime trachyandesite). Augite with aegirine rims, barkevikite.

Deldoradoite (Johannsen, 1938; Deldorado Bay, Colorado, U.S.A.)
Cancrinite syenite containing biotite, aegirine, etc., leucocratic.

Ditroite (Zirkel, 1866; Ditro, Roumania)
Sodalite-bearing nepheline syenite with some cancrinite, but sodalite is of secondary origin. Name therefore not suitable as synonym for sodalite–nepheline syenites. Brøgger (1890) suggested that the name should be used for nepheline syenites having granular texture (in contrast to foyaites). This is not to be recommended, either.

Domite (von Buch, 1809; Puy-de-Dôme, Auvergne, France)
Biotite trachyte to trachyandesite, subalkaline. Tridymite and cristobalite in matrix.

Doreite (Lacroix, 1923; Mont Dore, Auvergne, France)
Olivine-bearing trachyandesite, subalkaline.

Drachenfels trachyte
= Subalkaline (calc-alkaline) trachyte.

Drakontite (Rheinish, 1912; Drachenfels, W. Germany)
= biotite syenite containing oligoclase to labradorite, sometimes minor sodalite, aegirine augite, etc.

Dumalite (Loewinson-Lessing, 1905; Caucasus)
= Sodic trachyandesite.

Dungannonite (Adams and Barlow, 1908; Dungannon, Ontario, Canada)
Nepheline-bearing corundum diorite, containing scapolite.

Durbachite (Sauer, 1891; Schwarzwald, W. Germany)
Micromelasyenite, subalkaline, with hornblende and biotite.

Ekerite (Brøgger, 1906; Oslo Province, Norway)
Alkali granite, hypersolvus.

Elaeolite syenite (Blum, 1861)
= Old name for nepheline syenite.

Elkhornite (Johannsen, 1937; Elkhorn, Montana, U.S.A.)
Alkali feldspar syenite, carrying some labradorite.

Espichellite (Souza-Brandao, 1907; Cape Espichel, Portugal)
Analcime lamprophyre, mesocratic.

Essexibasalt (Lehmann, 1924; Utanjilua, Africa)
Nepheline basanite containing bytownite.

Essexite (Sears, 1891; Salem Neck, Essex County, Massachusetts, U.S.A.), see p. 70
Term used as synonym for foidal monzodiorite and -gabbro having nepheline as predominant foid. Glenmuirite is the term for the corresponding analcime rock.

Essexite gabbro
Term sometimes used for essexites with more than An_{50} in plagioclase.

Etindite (Lacroix, 1923; Etinde, Cameroun)
Leucite nephelinite, mesocratic.

Euktolite (Rosenbusch, 1899; San Venanzo, Umbria, Italy)
Same as venanzite.

Eustratite (Ktenas, 1928; Haghios Eustratios Isl., Aegean Isl.)
= Nepheline monchiquite.

Evergreenite (Ritter, 1908; Evergreen Mine, Oregon, U.S.A.)
Alkali granite or nordmarkite with wollastonite.

Evisite (Niggli, 1923; Evisa, Corsica)
Alkali granites and syenites of Evisa area.

Farrisite (Brøgger, 1898; Oslo Province, Norway)

Melilite lamprophyre, melanocratic, containing augite, barkevikite, biotite, olivine. No feldspar.

Fasibitikite (Lacroix, 1915; Ampasibitika, Madagascar)
Mela-alkali microgranite rich in aegirine. Contains zircon and eucolite.

Fasinite (Lacroix, 1916; Ampasindava, Madagascar)
Melteigite containing titanaugite. Chemically equivalent to berondrite.

Fenite (Brøgger, 1921; Fen, Norway)
Alkali syenite formed by contact metasomatic processes around alkaline intrusions.

Fergusite (Pirsson, 1905; Highwood Mts., Montana, U.S.A.)
Pseudoleucitolite, leuco- to mesocratic.

Fiasconite (Johannsen, 1938; Mt. Vulsini, Italy)
Leucite melabasanite; plagioclase rich in anorthite.

Finandranite (Lacroix, 1922; Madagascar)
Microcline syenite.

Fitzroydite (Wade and Prider, 1940; Fitzroy Basin, Australia)
Leucite lamproite containing phenocrysts of leucite and phlogopite.

Foidites and foidolites
Feldspar-free (and feldspar-poor) rocks composed of foids and mafic minerals. Foidite is now recommended as group name for the respective volcanic rocks, foidolite for plutonic equivalents.

Fortunite (De Yarza, 1893; Fortuna, Spain), see p. 98
Trachyte containing bronzite and phlogopite; silica-oversaturated glass base. Subalkaline.

Fourchite (Williams, 1891; Fourche Mts., Arkansas, U.S.A.)
Analcime lamprophyre, melanocratic, olivine- and feldspar-free (= augite monchiquite.

Foyaite (Blum, 1861; Serra de Monchique, Portugal), see p. 25
Nepheline syenite, agpaitic to intermediate between agpaitic and miaskitic. Foyaitic (intergranular) texture because of arrangement of plates and tables of microperthite. Often used as group name for nepheline syenites. It is recommended here to restrict name to hypersolvus rocks having foyaitic texture (cf. Brøgger, 1890).

Gabbroproterobase (Brøgger, 1894)
= Essexite.

Gabbrosyenite (Tarasenko, 1896)
= Monzogabbro.

Gaussbergite (Lacroix, 1926; Gaussberg, Antarctis)
Trachyleucitite having phenocrysts of leucite, glass matrix of sanidine composition.

Gauteite (Hibsch, 1898; Gaute (= Kout, Bohemia, Czechoslovakia)
Syenite porphyry or trachyandesite, analcime may be present.

Ghizite (Washington, 1914; Ghizo, Sardinia)
Analcime andesite, mesocratic, having analcime phenocrysts, glass matrix of andesine composition.

Gibelite (Washington, 1913; Pantelleria, Italy)
Alkali trachyte having aegirine augite, diopside, olivine and cossyrite, but no biotite. Quartz-bearing.

Giumarrite (Viola, 1901; Giumarra, Sicily)
= Amphibole monchiquite.

Gleesites (Kalb, 1934; Laacher See, W. Germany)
Collective name for subvolcanic hauyne-rocks.

Glenmuirite (Johannsen, 1938; Lugar, Scotland)
Olivine teschenite, alkali feldspar-bearing, = olivine–analcime monzogabbro.

Glimmerite (Larsen and Pardee, 1929)
Biotite rock = biotitite.

Gooderite (Johannsen, 1938; Haliburton–Bancroft, Ontario, Canada)
Albite syenite with some potash feldspar and nepheline.

Granosyenite
= Quartz syenite.

Grazinite (de Almeida, 1961; Trindade)
Phonolitic nephelinite, mesocratic; contains analcime, devoid of olivine (= olivine-free murite).

Grorudite (Brøgger, 1890; Oslo Province, Norway)

Mela-alkali microgranite having tinguaitic texture.

Guardiaite (Bieber, 1924; Guardia, Italy)
Nepheline leucobasanite with content of alkali feldspar (= nepheline–trachybasalt to –trachyandesite).

Grennaite (Adamson, 1944; Norra Kärr, Sweden)
Catapleiite–eudialyte–nepheline syenite, agpaitic, porphyritic with fine-grained matrix (= micro-leucolujavrite).

Hakutoite (Yamari, 1925; Hakuto-San, Manchuria)
Alkali rhyolite, quartz-poor, similar to taurite.

Hatherlite (Henderson, 1898; Leeuwfontein, S. Africa)
= Perthite or anorthoclase syenite (= leeuwfonteinite).

Hauynophyre (Abich, 1839)
Name equivalent to leucitophyre for rocks rich in hauyne and noseane.

Hawaiite (Daly, 1911; Iddings, 1913; Hawaii), see p. 88
Andesine basalt which is member of the alkali basalt–alkali trachyte series; = olivine andesite. The rock is characterized by basaltic texture.

Hedrumite (Brøgger, 1890; Oslo Province, Norway)
Nepheline-bearing alkali feldspar microsyenite. Groundmass has a trachytoid (foyaitic) texture. Biotite-rich, minor sodic amphibole.

Helsinkite (Laitakari, 1918; Helsinki, Finland)
Subalkaline albite–epidote syenite.

Heptorite (Busz, 1904; Siebengebirge, W. Germany)
Hauyne-rich lamprophyre or hauyne basanite having phenocrysts of barkevikite, titanaugite and hauyne in glassy or analcime matrix containing labradorite, olivine, etc. Mesocratic (= hauyne monchiquite).

Heronite (Coleman, 1899; Heron Bay, Lake Superior, Canada)
Analcime plagisyenite containing labradorite, potash feldspar, aegirine, etc.

Heumite (Brøgger, 1898; Oslo Province, Norway)
Nepheline lamprophyre, mesocratic, containing barkevikite, biotite, alkali feldspar, sodalite, etc.

Highwoodite (Johannsen, 1938; Highwood Mts., Montana, U.S.A.)
Nepheline-bearing monzonite with labradorite, mesocratic (biotite and diopside).

Hilairite (Johannsen, 1938; St. Hilaire, Quebec, Canada)
Sodalite–nepheline syenite, agpaitic.

Hollaite (Brøgger, 1921; Fen, Norway)
Sövitic ijolite–melteigite.

Holmite (Johannsen, 1938; Orkney Isl., U.K.)
Biotite-poor, pyroxene-rich alnöite.

Holyokeite (Emerson, 1902; Holyoke, Massachusetts, U.S.A.)
Albite syenite, subalkaline, rich in calcite.

Hovlandite (Barth, 1945; Oslo Province, Norway)
Biotite–hypersthene–olivine monzogabbro.

Hurumite (Brøgger, 1931, Oslo Province, Norway)
Quartz micromonzonite, subalkaline.

Husebyite (Brøgger, 1933; Oslo Province, Norway)
Nepheline syenite, containing aegirine augite and barkevikite.

Ijolite (Ramsay and Berghell, 1891; Iivaara, Finland), see p. 53
Nepheline–pyroxene rock, mesocratic, feldspar-free.

Ijussite (Rakovski, 1911; Ijuss River, Siberia, U.S.S.R.)
Melateschenite with titanaugite, barkevikite, analcime, bytownite and alkali feldspar.

Isenite (Bertels, 1874; Nassau, W. Germany)
Noseane trachyandesite, hornblende- and biotite-bearing. According to Tröger (1935) obsolete term, since apatite has originally been misidentified as hauyne.

Italite (Washington, 1920; Albano Hills, Italy)
Leucite-rich rock containing minor mafic minerals. No feldspar. Leucocratic (= leucofergusite or subvolcanic leucoleucitite).

Itsindrite (Lacroix, 1922; Itsindra, Madagascar)

Nepheline microsyenite containing aegirine, biotite, melanite, etc., potash-rich.

Jacupirangite (Derby, 1891; Jacupiranga, Brazil)
Alkali pyroxenite: titanaugite, titanomagnetite, nepheline, apatite, perovskite, melanite, etc.

Jumillite (Osann, 1906; Jumilla, Spain, see p. 98.
Leucite phonolite, mesocratic, having olivine, augite, sanidine, phlogopite in matrix of leucite, sanidine, aegirine augite, catophorite, etc. (= phonolitic olivine leucitite).

Juvite (Brøgger, 1921; Fen, Norway)
Nepheline syenite having orthoclase as only feldspar.

Kaiwekite (Marshall, 1906; Kaiweke, New Zealand)
Olivine trachyandesite or trachyte, phenocrysts of anorthoclase (volcanic equivalent of larvikite), titanaugite with aegirine rims, barkevikite, etc.

Kajanite (Lacroix, 1926; Kajan Isl., Borneo)
Melaleucitite rich in biotite.

Kakortokite (Ussing, 1912; Ilímaussaq, Greenland)
Eudialyte–arfvedsonite–nepheline syenite, agpaitic. Leuco- to mesocratic, igneous lamination, distinct gravitative layering with black, red and white layers.

Kalmafite (Hatch, Wells and Wells, 1961)
Volcanic rock composed of kalsilite plus mafic minerals (= mafurite, katungite).

Kamafugite (Sahama, this volume), see pp. 96, 100
Collective name derived from KAtungite–MAfurite–UGandite.

Karite (Karpinsky, 1903; Kara Valley, Transbaikalia, U.S.S.R.)
Alkali microgranite, quartz- and aegirine-rich (equivalent of grorudite but richer in quartz).

Karlsteinite (Waldmann, 1935; Karlstein, Austria)
Alkali granite, potash-rich (microcline).

Kåsenite (Brøgger, 1921; Fen, Norway)
Carbonatite, nepheline-bearing, and rich in pyroxene.

Kassaite (Lacroix, 1918; Iles de Los, Guinea)
Hauyne monzonite to theralite, carrying plagioclase (An_{55-20}), sanidine, mafic minerals.

Katungite (Holmes, 1937; Katunga, Uganda), see p. 100
Olivine melilitite containing kalsilite, mesocratic, devoid of pyroxene (= kalmafite).

Katzenbuckelite (Osann, 1903; Katzenbuckel, Germany)
Hauyne–nepheline microsyenite, leucocratic, carrying biotite, aegirine augite, etc.; tinguaitic.

Kauaiite (Iddings, 1913; Hawaii)
Monzodiorite (syenodiorite), mesocratic. Titanaugite, oligoclase, sanidine, olivine, etc.

Kaxtorpite (Adamson, 1944; Norra Kärr, Sweden)
Pectolite–eckermannite–aegirine–nepheline syenite.

Kenyte/kenyite (Gregory, 1900; Mt. Kenya, Kenya)
Phonolite, olivine-bearing, leucocratic, with aegirine augite, cossyrite, etc. Rhomb-shaped anorthoclase.

Keratophyre (Gümbel, 1874)
Albite trachyte, subalkaline (= albitized felsic extrusives).

Khagiarite (Washington, 1913)
= Hyalo-pantellerite.

Khibinite (Ramsay, 1898; Khibina (Umptek), Kola, U.S.S.R.)
Nepheline syenite, leucocratic, agpaitic to intermediate between agpaitic and miaskitic nepheline syenites. Granular or trachytoid texture. Eudialyte-bearing.

Kirunavaarite (Rinne, 1921; Kirunavaara, Sweden)
= Magnetitite.

Kivite (Lacroix, 1923; Kivu Lake, Zaire)
Leucite basanite, mesocratic, olivine-poor (or olivine-bearing leucite tephrite. Leucite basanite having $Na_2O \sim K_2O$, while $K_2O > Na_2O$ in general case).

Kjelsåsite (Brøgger, 1933; Oslo Province, Norway)
Monzodiorite, or monzonite, leucocratic,

having brown hornblende, aegirine augite, biotite, etc. (= basic larvikite).

Kosenite
= Kåsenite

Kristianite (Brøgger, 1893)
Local name for the biotite granite (Drammen granite) of the Oslo Province (Kristiania is old name for Oslo).

Kulaite (Washington, 1894; Kula, Lydia, Asia Minor)
Nepheline basanite, mesocratic, rich in hornblende, sanidine-bearing (= nepheline trachybasalt).

Kvellite (Brøgger, 1898; Oslo Province, Norway)
Ultramafic porphyrite containing olivine, barkevikite, biotite in matrix of alkali feldspar and nepheline.

Kylite (Tyrell, 1912; Kyle, Scotland)
Nepheline-bearing gabbro, melanocratic, rich in olivine (olivine-rich melatheralite).

Labradorite (Senft, 1857)
Leucobasalt.

Lakarpite (Törnebohm, 1906; Norra Kärr, Sweden)
Nepheline syenite, arfvedsonite-rich, subsolvus with microcline and albite, redefined by Adamson (1944): Arfvedsonite–albite–nepheline syenite.

Lamproite (Niggli, 1923), see p. 96
Group name for K- and Mg-rich extrusive rocks = lamprophyric extrusives, rich in biotite (phlogopite).

Lardalite/laurdalite (Brøgger, 1890; Oslo Province, Norway)
Nepheline syenite, hypersolvus, leucocratic, with rhomb-shaped alkali feldspar or ternary feldspar; a 'nepheline-rich larvikite.'

Larvikite/laurvikite (Brøgger, 1890; Oslo Province, Norway)
Augite monzonite or hypersolvus syenite, leucocratic, rhomb-shaped ternary feldspar, which is schillerizing. Nepheline or quartz may be present. Sometimes olivine-bearing (= augite syenite or monzonite).

Ledmorite (Shand, 1910; Ledmore River, Scotland)
= Melanite malignite having aegirine augite and biotite.

Leeuwfonteinite (Brouwer, 1903; Leeuwfontein, S. Africa)
Alkali syenite, anorthoclase-rich, with andesine, barkevikite, diopside, biotite, etc., (= hatherlite).

Lestiwarite (Rosenbusch, 1896; Khibina, Kola Peninsula, U.S.S.R.)
Alkali microsyenite, with eudialyte, quartz-poor or quartz-free. Aplitic texture.

Leucite basalt (Zirkel, 1870)
= Olivine leucitite.

Leucite basanite
= Olivine basalt containing leucite.

Leucite kentallenite (Lacroix, 1917; Vesuvius, Italy)
Olivine–diopside–leucite melamonzosyenite, (= dark sommaite).

Leucite lamproite
Lamprophyric rock containing leucite (see p. 96)

Leucite monchiquite
Ultrabasic rock having olivine and augite in groundmass of leucite.

Leucite phonolite and leucite trachyte
See discussion on p. 30. It is now recommended to use the term leucite phonolite for foidal alkali feldspar-rich extrusives having leucite as predominant feldspathoid. The name leucite trachyte is then to be discarded.

Leucite tephrite
Tephrite with leucite as the predominant foid. If nepheline is present: nepheline–leucite tephrite.

Leucitite (Rosenbusch, 1877; Orvieto, Italy)
Volcanic rock having leucite and augite as predominant components. Feldspar is lacking. When olivine is present the rock is termed olivine leucitite (or sometimes leucite basalt). Generally mesocratic (= leumafite).

Leucitonephelinite (Jung and Brousse, 1959)
Nephelinite containing leucite.

Leucitophyre (von Humboldt, 1837; Eifel, W. Germany)
Variety of leucite phonolite or leucito-

nephelinite, that is leucocratic volcanic rock containing leucite, nepheline, sodalite group minerals, some alkali feldspar and mafic minerals, such as aegirine augite, melanite, mica. According to Hatch, Wells and Wells leucitophyres have subordinate nepheline and are thus alkali–feldspar leucite rocks.

Leucotephrite, -leucitite, -syenite, etc.
Rock more leucocratic than generally in that rock group.

Limburgite (Rosenbusch, 1872; Kaiserstuhl, W. Germany), see p. 72
Phenocrysts of olivine and augite in alkali-rich glass (chemically corresponding to nepheline + sanidine + plagioclase); = glass-rich picrite.

Leumafite (Hatch, Wells and Wells, 1961)
Volcanic rock composed of leucite + mafic minerals (= leucitite).

Lindinosite (Lacroix, 1922; Lindinosa, Corsica)
Alkali melagranite (= riebeckite-rich granite).

Lindöite (Brøgger, 1894; Oslo Province, Norway)
Leucosölvsbergite = alkali microgranite, leucocratic, bostonitic texture.

Linosaite (Johannsen, 1938; Linosa Isl., Italy)
Nepheline-bearing basalt.

Litchfieldite (Bailey, 1892; Litchfield, Maine, U.S.A.)
Nepheline syenite, miaskitic, leucocratic, subsolvus, with albite, potash feldspar, cancrinite, biotite, etc.

Lugarite (Tyrrell, 1912; Lugar, Scotland)
Labradorite–analcime ijolite (= analcime-rich microteschenite), mesocratic, with some labradorite and no alkali feldspar. Titan-augite and barkevikite.

Luhite (Scheumann, 1922; Luh, Bohemia, Czechoslovakia)
Hauyne–nepheline–calcite–micromelilitolite = melilite damkjernite = hauyne–nepheline–alnöite; melanocratic, phenocrysts of titanaugite and olivine in matrix of melilite, titanaugite, calcite, nepheline, hauyne, biotite, perovskite, etc.

Lujavrite (Luijaurite) (Brøgger, 1890; Lovozero (Lujavr-Urt), Kola Peninsula, U.S.S.R.), see p. 27
Agpaitic nepheline syenite, mesocratic (aegirine and/or arfvedsonite), having trachytoid texture (igneous lamination). Fine- to coarse-grained. Rich in rare minerals such as eudialyte. Hypersolvus with microperthite or subsolvus with separate microcline and albite.

Lujavritite (Antonov, 1934; Khibina, Kola Peninsula, U.S.S R.)
Feldspar-poor lujavrite or feldspar-bearing ijolite.

Lundyite (Hall, 1914; Lundy Isl., England)
Catophorite-quartz orthophyre (= quartz nordmarkite).

Luscladite (Lacroix, 1916; Alter Pedroso, Portugal)
Theralite, nepheline-poor, mesocratic, with labradorite rimmed by alkali feldspar. Transitional into foidal monzogabbro (melaessexite). Hornblende is missing. Olivine and biotite occur. Kylite is a melanocratic type.

Lusitanite (Lacroix, 1916; Alter Pedroso, Portugal)
Mela-alkali syenite.

Lutalite (Holmes, 1937; Bufumbira, Uganda)
Olivine leucitite with higher Na_2O/K_2O than ordinary olivine leucitite. Glass-rich.

Macedonite (Skeats, 1909; Macedon, Australia)
Olivine–biotite trachyte to trachyandesite, subalkaline. (Early reports of melilite and noseane have not been substantiated).

Madeirite (Gagel, 1913; Madeira)
Micro-melagabbro.

Madupite (Cross, 1897; Leucite Hills, Wyoming, U.S.A.), see p. 98
Hyaloleucitonephelinite; phenocrysts of diopside and phlogopite in glass having composition of leucite + nepheline. Melanocratic.

Maenaite (Brøgger, 1898; Oslo Province, Norway)
= Plagioclase-bearing bostonite = microsyenite with separate albite and orthoclase.

Mafraite (Lacroix, 1920; Mafra, Portugal)
Hornblende (barkevikite) monzogabbro.

Differs from berondrite by the absence of nepheline.

Mafurite (Holmes, 1945; Mafura, Uganda), see p. 100
Olivine kalsilitite, melanocratic (= kalmafite).

Malignite (Lawson, 1896; Maligna River, Ontario, Canada)
Mela-nepheline syenite, mesocratic.

Mamilite (Wade and Prider, 1940; Mamilu Hill, Western Australia), see p. 97
Magnophorite–leucite lamproite.

Mandchurite (Lacroix, 1923; Tsao-Shih-Err, Manchuria)
Hyalobasanite, mesocratic, to nepheline–hyalotrachybasalt.

Mareugite (Lacroix, 1917; Mont Dore, Auvergne France)
Hauyne-bearing alkali gabbro, with bytownite.

Marienbergite (Johannsen, 1938; Marienberg (= Mariánská hora) Bohemia, Czechoslovakia)
Natrolite phonolite.

Mariupolite (Morozewicz, 1902; Mariupol (Oktj'abr), Ukraine, U.S.S.R.)
Albite–nepheline syenite, leucocratic, with aegirine, biotite, etc., subordinate potash feldspar.

Marosite (Iddings, 1913; Maros, Celebes)
Nepheline melamonzogabbro (melamonzonite), rich in biotite, with bytownite = biotite–nepheline melamonzogabbro.

Masafuerite (Johannsen, 1937)
Picrite basalt.

Martinite (Johannsen, 1936; Vico, Italy)
Phonolitic leucite tephrite (basanite).

Meimechite/Meymechite (Kotulsky, 1943; Maimecha–Kotuj, U.S.S.R.)
Ultramafic volcanic rock composed of abundant (50–60 %) olivine phenocrysts in matrix of glass with microlites of titanaugite.

Melmafite (Hatch, Wells and Wells, 1961)
Volcanic rock composed of melilite plus mafic minerals (= melilitite).

Melilite basalt (v. Rath, 1866; Stelzner, 1882)
= Olivine melilitite.

Melilitite (Lacroix, 1896)
Volcanic rock mainly composed of melilite and augite. Olivine-free, feldspar-free.
Varieties: nepheline melilitite, leucite melilitite (= melmafite).

Melilitolite (Lacroix, 1933)
Intrusive equivalent of melilitite.

Melteigite (Brøgger, 1921; Fen, Norway), see p. 53
Melanocratic rock composed mainly of alkali pyroxene and nepheline (= 'melaijolite').

Mestigmerite (Duparc, 1926; Mestigmer, Morocco)
Mesocratic hornblende–augite ijolite or pyroxene malignite (perhaps altered malignite).

Miaskite (Rose, 1839; Miask, the Urals, U.S.S.R.), see pp. 22–23
Biotite–nepheline monzosyenite, miaskitic, leucocratic, with albite–oligoclase and potash feldspar (= nepheline plagisyenite).

Mikenite (Lacroix, 1933; Birunga, Uganda)
Leucitite having higher Na_2O/K_2O than ordinary leucitite.

Missourite (Weed and Pirsson, 1896; Shonkin Creek, Montana, U.S.A.)
Melafergusite; phanerocrystalline, melanocratic equivalent of leucitite and olivine leucitite (= leucite melteigite).

Modlibovite (Scheumann, 1922; Modlibov, Bohemia, Czechoslovakia)
Foidal olivine–biotite–melilitolite, mesocratic (= melilite–nepheline lamprophyre = biotite polzenite).

Modumite (Brøgger, 1933; Oslo Province, Norway)
Anorthosite of bytownitic composition with some pyroxene and barkevikite.

Monchiquite (Hunter and Rosenbusch, 1890; Sierra de Monchique, Portugal)
Analcime lamprophyre having olivine, titanaugite and often biotite in matrix of analcime. May be nepheline- and leucite-bearing.

Mondhaldeite (Gruss, 1900; Kaiserstuhl, W. Germany)
Micromonzonite with alkali-rich glass.

Monmouthite (Adams, 1904; Haliburton–Bancroft, Ontario, Canada)
Hastingsite urtite with minor albite.

Montrealite (Adams, 1913; Montreal, Canada)
Nepheline-bearing olivine–titanaugite melagabbro, also kaersutitic amphibole.

Mugearite (Harker, 1904; Skye, Scotland), see p. 88
Olivine leucotrachyandesite; member of alkali basalt–alkali trachyte series.

Muniongite (David, 1901; Kosciusco, New South Wales, Australia)
Nepheline-rich tinguaite, leucocratic.

Murambite (Holmes, 1936; Bufumbira, Uganda)
= Olivine-rich kivite = leucite basanite.

Murite (Lacroix, 1927; Cook Isl., Pacific Ocean)
Olivine melaphonolite, rich in nepheline.

Natronshonkinite (Nieland, 1931)
Mela-nepheline–syenite.

Naujaite (Ussing, 1912; Ilímaussaq, Greenland)
Poikilitic sodalite–nepheline syenite and sodalite syenite, agpaitic, leucocratic. Crystals of sodalite are poikilitically enclosed in other rock-forming minerals. Up to 70% sodalite. Eudialyte-rich.

Neapite (Vlodavetz, 1930; Khibina, the Kola Peninsula, U.S.S.R.)
Apatite-rich ijolite–urtite.

Nemafite (Hatch, Wells and Wells, 1961)
Volcanic rock composed of nepheline and mafic minerals (= nephelinite).

Nemite (Lacroix, 1933; Lake Nemi, Italy)
Melaleucitite.

Nepheline basalt (Naumann, 1850)
= Olivine nephelinite.

Nepheline andesite, basanite, diorite, monzonite, etc.
Andesite, etc., with nepheline making up more than 10% of volume of felsic minerals.

Nepheline syenite (Rosenbusch, 1878)
Group name for foidal syenites with nepheline as predominant foid. Divided into agpaitic and miaskitic types which are connected by transitional forms (see pp. 23–24). See under names of varieties such as foyaite, lujavrite, mariupolite, miaskite, etc.

Nephelinite (Naumann, 1849)
Volcanic rock composed mainly of nepheline and augite. Devoid of feldspar and olivine (= nemafite).

Nephelinolite (Loewinson-Lessing, 1901)
Hololeucocratic urtite.

Ngurumanite (Saggerson and Williams, 1964; Lake Magadi, Kenya)
Hypabyssal ankaratrite (or melteigite): nepheline–pyroxene rock having iron-rich mesostasis (serpentine–chlorite) and vugs of calcite and analcime.

Niligongite (Lacroix, 1933; Kivu Lake, Zaire)
Phanerocrystalline equivalent of leucite nephelinite and nepheline leucitite = melilite-bearing leucite ijolite. Equal amounts of leucite and nepheline.

Nordmarkite (Brøgger, 1890; Oslo Province, Norway)
Hypersolvus, quartz-bearing alkali feldspar syenite, both alkaline and subalkaline.

Nordsjöite (Johannsen, 1938; Fen, Norway)
Calcite-rich nepheline syenite having more nepheline than orthoclase, melanite-bearing.

Noseanite (Boricky, 1873)
Nosean-bearing equivalent of nephelinite.

Nosykombite (Niggli, 1923; Nossi-Komba Isl., Madagascar)
Barkevikite–nepheline monzosyenite or plagisyenite (= nepheline covite).

Oceanite (Lacroix, 1923; Reunion, Indian Ocean)
Olivine melabasalt; picrite basalt with more than 50% modal olivine.

Okaite (Stansfield, 1923; Oka, Quebec, Canada)
Hauyne melilitolite (hauyne turjaite).

Okawaite (Nemoto, 1934; Okawa River, Hokkaido, Japan)
Aegirine augite pitchstone.

Olivinite (Eichstädt, 1884)
Formerly used as group name for hornblende dunites and peridotites. Now used in the U.S.S.R. for the olivine rocks of alkaline ultramafic massifs. They consist mainly of olivine (Fa_{12-15}) with varying amounts of phlogopite, perovskite and titanomagnetite (rocks have up to 20% TiO_2). Augite, hornblende, melilite and monticellite may occur. Rocks are poor in Cr, Ni compared

to common dunites. See Johannsen (1938, vol. 4, p. 410).

Onkilonite (Backlund, 1915; Wilkitski Isl., U.S.S.R.)
Olivine–leucite melanephelinite.

Ordanchite (Lacroix, 1917; Mont-Dore, Auvergne, France)
Hauyne trachyandesite, leucocratic. Volcanic equivalent of essexite and glenmuirite.

Ordosite (Lacroix, 1925; Hoangho, Ordos Province, China)
Mela-alkali syenite.

Orendite (Cross, 1897; Leucite Hills, Wyoming, U.S.A.), see p. 96
Leucite phonolite, phlogopite-bearing (= phonolitic leucoleucitite = leucite lamproite).

Orotvite (Streckeisen, 1938; Ditro, Roumania)
Nepheline-bearing hornblende diorite to nepheline diorite of Ditro intrusion.

Orthophyre (Coquand, 1857)
Orthoclase porphyry, paleovolcanic.

Orthosite (Turner, 1900)
Orthoclase syenite, hololeucocratic.

Orthosyenite (Hatch, Wells and Wells, 1972)
= Silica saturated syenite. Also used as synonym for calc-alkaline or subalkaline syenites.

Orthotrachyte (Hatch, Wells and Wells, 1961)
Silica saturated trachyte.

Orvietite (Niggli, 1923; Orvieto, Italy)
Leucite leucotrachybasalt, olivine-bearing.

Ottajanite (Lacroix, 1917; Vesuvius, Italy)
Leucite leucobasanite (or leucotrachybasalt) (richer in plagioclase and poorer in leucite than vesuvite).

Ouachitite (Kemp, 1891; Ouachita River, Arkansas, U.S.A.)
Biotite monchiquite, olivine-free, base of glass or analcime, Melanocratic.

Paisanite (Osann, 1893; Paisano, Texas, U.S.A.)
Alkali microgranite, riebeckite-bearing.

Pantellerite (Förstner, 1881; Pantelleria, Italy), see p. 109
Alkali rhyolite, leucocratic, more mafic minerals than comendite. Macdonald and Bailey (personal information) have proposed a chemical distinction of comendite and pantellerite, the latter having a higher iron content (calculated as FeO) in a $FeO–Al_2O_3$ rectilinear diagram, the separation line having the equation $Al_2O_3 = 1{\cdot}33\ FeO + 4{\cdot}40$.

Parchettite (Johannsen, 1938; San Martin, Italy)
Leucite trachyandesite, mesocratic.

Pedrosite (Osann, 1923; Pedrosa, Portugal)
Hornblendite, made up of osannite.

Peridotite
Peridotites of alkaline ultramafic massifs are dominated by titanaugite, olivine (more than 40%), titanomagnetite, perovskite and apatite. May contain nepheline.

Perthosite (Phemister, 1926; Creag na Fearna, Scotland)
Hololeucocratic alkali syenite with minor aegirine augite.

Phonolite (Haüy, in Cordier, 1816; Klaproth, 1801), see p. 30
Volcanic equivalent of foidal syenite, agpaitic to miaskitic, leucocratic. Phonolite s.s = nepheline phonolite; in leucite phonolite leucite is only foid; leucite–nepheline phonolite contains less leucite than nepheline.

Picrite (Tschermak, 1866; N. Freiberg, Czechoslovakia), see p. 86
Olivine melagabbro, or olivine melateschenite, ultramafic (alkali peridotite and labradorite peridotite). Titanaugite is characteristic.

Picrite basalt (Quensel, 1912; Juan Fernandez Isl.,), see p. 86
Olivine melabasalt, rich in olivine (cf. limburgite). If phenocrysts are mainly olivine: oceanite; if olivine and augite: ankaramite.

Pienaarite (Brouwer, 1910; Leeuwfontein, S. Africa)
Titanite mela-nepheline syenite (= titanite shonkinite or malignite).

Pilandite (Henderson, 1898; Pilanesberg, S. Africa)
Alkali microsyenite containing anorthoclase, aegirine augite, etc., (hatherlite porphyry).

Plauenite (Brøgger, 1895; Dresden, E. Germany)
Hornblende–quartz syenite to quartz monzonite, subalkaline (original syenite of Werner).

Pollenite (Lacroix, 1907; Vesuvius, Italy)
Hyalophonolite, olivine-bearing, content of labradorite, glass composition corresponds to sanidine + nepheline + oligoclase; = tephritic phonolite.

Polzenite (Scheumann, 1922; Polzen, Czechoslovakia), see p. 262
Melilite lamprophyres, mesocratic with nepheline, olivine and devoid of augite. The group includes modlibovite, vesecite, luhite and wesselite.

Ponzite (Washington, 1913; Ponza Isl., Italy)
Sodalite phonolite, containing aegirine augite, låvenite, etc., (= sodalite-bearing alkali trachyte, trachyte of Ponza type: Rosenbusch).

Puglianite (Lacroix, 1917; Vesuvius, Italy)
Leucite melatheralite or melamonzogabbro; contains anorthite.

Pulaskite (Williams, 1891; Pulaski County, Arkansas, U.S.A.)
Nepheline-bearing hypersolvus alkali syenite.

Raabsite (Waldmann, 1935; Raabs, Austria)
Minette containing sodic amphibole.

Rafaelite (Johannsen, 1938; San Rafael Swell, Utah, U.S.A.)
Analcime-bearing amphibole microsyenite containing labradorite (= analcime plagisyenite).

Raglanite (Adams and Barlow, 1908; Haliburton–Bancroft, Ontario, Canada)
Corundum leuco-nepheline–diorite.

Rauhaugite (Brøgger, 1921; Fen, Norway)
Dolomite carbonatite containing alkali feldspar (= beforsite).

Rhombporphyry (von Buch, 1810; Oslo Province, Norway)
Porphyritic trachyte to trachyandesite with rhomb-shaped phenocrysts of ternary feldspar (oligoclase + anorthoclase). Volcanic equivalent of larvikite. Subalkaline to weakly alkaline (aegirine augite).

Riedenite (Brauns, 1922; Laacher See, W. Germany)
Noseane melteigite, melanocratic, biotite-rich (= noseane–biotite–augitite).

Ringite (Brøgger, 1921; Fen, Norway)
Carbonatite containing aegirine, alkali feldspar, etc., a sövite–fenite hybrid.

Rischorrite (Kupletsky, 1932; Khibina, Kola Peninsula, U.S.S.R.), see p. 216
Nepheline syenite having nepheline crystals poikilitically enclosed in microcline perthite. Biotite, aegirine augite, etc.

Rizzonite (Doelter, 1903; Mt. Rizzoni, Tyrol)
Variety of limburgite.

Rockallite (Judd, 1897; Rockall, Atlantic Ocean)
Mela-alkali granite rich in aegirine and albite.

Rodderite (Frechen, 1967; Eifel, W. Germany)
Sanidine–noseane–biotite–augitite (= foyaitic noseane melteigite), subvolcanic.

Rongstockite (Tröger, 1935, Rongstock (Roztoky), Bohemia, Czechoslovakia)
Foid-bearing monzogabbro/monzodiorite (= nepheline-poor essexite enriched in sodium compared to calcium). Containing titanaugite, biotite, brown hornblende.

Rougemontite (O'Neill, 1914; Rougemont, Quebec, Canada)
Anorthite–olivine gabbro, rich in titanaugite.

Rouvillite (O'Neill, 1914; St. Hilaire, Quebec, Canada)
Leucotheralite rich in nepheline and bytownite.

Rutterite (Quirke, 1936; Sudbury, Ontario, Canada)
Alkali feldspar syenite, subsolvus with microcline and albite, nepheline-bearing.

Salitrite (Tröger, 1928; Salitre Mts., Brazil)
Alkali pyroxenite (aegirine augite), rich in titanite. (= titanite-rich jacupirangite).

Sancyite (Lacroix, 1923; Mont-Dore, Auvergne, France)
Leucotrachyandesite containing tridymite, subalkaline.

Sanidinite (Nose, in Nöggerath, 1808)
Sanidine syenite.

Sannaite (Brøgger, 1921; Fen, Norway)
Lamprophyre containing phenocrysts of

augite with aegirine rims, barkevikite rimmed by biotite, and biotite in base of alkali feldspars (orthoclase and albite), augite, aegirine, nepheline, etc.

Särnaite (Brøgger, 1890; Särna, Sweden)
Nepheline–cancrinite syenite, leucocratic, containing aegirine augite, orthoclase rimmed by albite, etc.

Scanoite (Lacroix, 1924, Sardinia)
Analcime hyalobasanite, mesocratic.

Schorenbergite (Brauns, 1922; Laacher See, W. Germany)
Noseane–nepheline leucoleucitite containing aegirine augite (tinguaitic texture) (= micro-noseane–niligongite, or noseane–nepheline–fergusite to leucocratic noseane–leucitophyre).

Selbergite (Brauns, 1922; Laacher See, W. Germany)
Noseane–leucite–alkali microsyenite–leucocratic, containing aegirine augite.

Shackanite (Daly, 1912; Shackan, Brit. Columbia, Canada)
Analcime hyalophonolite.

Semejtavite (Gornostaev, 1933)
Quartz–alkali feldspar syenite with anorthoclase, ilmenite, augite and riebeckite.

Shonkinite (Weed and Pirsson, 1895; Highwood Mts., Montana, U.S.A.)
Melanocratic rock composed of alkali feldspar, foids and more than 60% mafic minerals (pyroxene, olivine, biotite).

Shoshonite (Iddings, 1895; Yellowstone Park, U.S.A.)
Olivine leucotrachyandesite, subalkaline.

Sodalitite (Ussing, 1912; Ilímaussaq, Greenland)
Sodalite-rich sodalite syenite (= sodalite urtite).

Sölvsbergite (Brøgger, 1894; Oslo Province, Norway)
Micro-alkali syenite. Richer in mafic minerals than bostonite.

Sommaite (Lacroix, 1902; Vesuvius, Italy)
Leucite monzosyenite to monzonite containing labradorite.

Sörkedalite (Brøgger, 1933; Oslo Province, Norway)
Olivine monzodiorite or troctolitic monzonite, partly leucocratic (apotroctolite); rich in apatite and Fe–Ti-oxides. Plagioclase antiperthitic andesine. (Bose, 1969).

Sövite (Brøgger, 1921; Fen, Norway)
Biotite–calcite carbonatite (= alvikite).

Sphenetite (Allen, 1914)
Titanite pyroxenite, nepheline-bearing.

Sumacoite (Johannsen, 1938; Sumaco, Equador)
Nepheline trachyandesite, leucocratic; (= andesitic tephrite).

Sussexite (Kemp, 1892; Sussex County, New Jersey, U.S.A.) Micro-nepheline syenite with tinguaitic texture; containing aegirine augite, catophorite, biotite. Rich in nepheline, poor in feldspar (= feldspar-bearing urtite).

Sviatonossite (Eskola, 1920; Sviatoy Noss, Transbaikalia, U.S.S.R.)
Andradite alkali syenite containing orthoclase perthite and albite.

Syenite (Pliny, (Syene), Werner, 1788 (Plauensche Grund, Dresden, E. Germany))
Alkali feldspar rock, often with minor oligoclase. Sometimes quartz- or nepheline-bearing. Leucocratic with biotite, hornblende and augite. When sodic pyroxenes and amphiboles (and foids) are present: alkali syenite.

Syenitite (Loewinson-Lessing, 1900)
Biotite syenite.

Syenodiorite (Evans, 1916)
Feldspar-rich rock having both alkali feldspar and plagioclase (with less than An_{50}) = monzonite + monzodiorite.

Syenogabbro (Johannsen, 1917)
Feldspar-rich rock, having alkali feldspar and plagioclase (More than An_{50}) = monzogabbro + monzonite.

Synnyrite (Zhidkov, 1962; Synnyr, Northern Baikal Region, U.S.S.R.)
Pseudoleucite leuco-nepheline syenite with micropegmatitic structure composed of potash feldspar and kalsilite. Minor biotite and ore minerals.

Syenoide (Shand, 1910)
Abbreviation for foidal syenite.

Tahitite (Lacroix, 1917; Tahiti)
Hauyne leucotrachyandesite, often glass-

bearing (= hauynophyre or hauyne benmoreite).

Taimyrite (Chrustschoff, 1894; Taimyr, U.S.S.R.)
Noseane phonolite.

Talzastite (Termier and Jouravsky, 1948; Talzast, Morocco)
Titanaugite ijolite.

Tamaraite (Lacroix, 1918; Iles de Los, Guinea)
Nepheline lamprophyre, melanocratic, with titanaugite, biotite, olivine in base of augite, hornblende, nepheline, analcime, plagioclase, potash feldspar, biotite, cancrinite, etc. (= nepheline-rich camptonite).

Tannbuschite (Johannsen, 1938; Tannbusch, Bohemia, Czechoslovakia)
Augite ankaratrite or olivine melanephelinite.

Tasmanite (Johannsen, 1938; Shannon Tier, Tasmania)
Melilite–olivine ijolite, zeolite-rich.

Taurite (Lagorio, 1897; Crimea, U.S.S.R.)
Alkali rhyolite with spherulitic groundmass.

Tautirite (Iddings, 1918; Tahiti)
Nepheline trachyandesite, leucocratic.

Tavolatite (Washington, 1906; Tavolato, Italy)
Foid-rich phonolite containing leucite, nepheline and hauyne, aegirine augite, biotite, melanite, etc. Phenocrysts of labradorite (= tephritic phonolite to phonolitic leucitite).

Tawite (Ramsay, 1894; Lovozero, Kola Peninsula, U.S.S.R.)
Leuco-sodalite–ijolite, agpaitic.

Tephrite (Pliny, Cordier, 1816; Rosenbusch, 1877), see p. 70
Alkali basaltic volcanic rock composed essentially of plagioclase, augite and feldspathoid.

Tephritoid (Bücking)
A glass-bearing rock having the chemical composition of tephrite.

Teschenite (Hohenegger, 1861; Teschen, Czechoslovakia), see p. 70
Analcime theralite or analcime gabbro (cf. crinanite).

Theralite (Rosenbusch, 1887; Doupov, Bohemia, Czechoslovakia), see p. 70
Nepheline gabbro.

Thuresite (Waldmann, 1935; Thures, Austria)
Alkali syenite with sodic amphibole.

Tinguaite (Rosenbusch, 1887; Serra de Tingua, Brazil)
Nepheline microsyenite, having tinguaitic texture: needles of aegirine occur interstitially in mosaic of alkali feldspar and foids, mainly nepheline.

Titanolite (Kretschmer, 1917)
Alkali pyroxenite rich in titanite and magnetite.

Tjosite (Brøgger, 1906; Oslo Province, Norway)
Micromelteigite, anorthoclase-bearing, shonkinitic; microjacupirangite, lamprophyric.

Tönsbergite (Brøgger, 1898; Oslo Province, Norway)
Alkali syenite with rhomb-shaped grains of alkali feldspar = red, altered variety of larvikite.

Topsailite (Lacroix, 1911; Iles de Los, Guinea)
Nepheline microessexite containing augite and biotite, lamprophyric.

Toryhillite (Johannsen, 1920; Haliburton–Bancroft, Ontario, Canada)
Nepheline-rich albite–nepheline syenite, containing aegirine augite, devoid of potash feldspar (= albite-rich urtite).

Trachyandesite (Michel-Levý, 1894), see p. 88
= Latite, plagioclase with less than 50% An (also benmoreite, tristanite, toscanite), potassic equivalent of mugearite.

Trachybasalt (Boricky, 1874), see p. 70
= Latite–basalt, plagioclase with more than 50% An.

Trachylabradorite (Jung and Brousse, 1959)
Leucotrachybasalt, plagioclase with more than 50% An.

Trachyte (Haüy in Brongniart, 1813), see p. 30
Alkali feldspar-rich volcanic rock; often with subordinate oligoclase. May be quartz- or nepheline-bearing. Extrusive equivalent of syenite, When sodic pyroxene and amphibole present: alkali trachyte. Often trachytic texture.

Trass
An Italian name for tuffs widely distributed in central Italy and in Eifel, W. Germany.

Tristanite (Tilley and Muir, 1964; Tristan da Cunha), see p. 88
Sanidine- and anorthoclase-bearing trachyandesite. $K_2O > Na_2O$.

Turjaite (Ramsay, 1921; Turij Mis (= Turja), Kola, U.S.S.R.)
Biotite–nepheline melilitolite containing melanite, perovskite, etc.

Turjite (Belijankin, 1924; Turij Mis (= Turja), Kola, U.S.S.R.)
Melanocratic lamprophyre rich in analcime, biotite, calcite and melanite (= analcime–calcite–alnöite).

Tusculite (Cordier, 1868)
= Leuco-melilite–leucitite.

Tutvetite (Johannsen, 1938; Oslo Province, Norway)
Microcline–albite microsyenite with trachytoid texture.

Tveitåsite (Brøgger, 1921; Fen, Norway)
Mela-alkali–syenite = lusitanite, rich in aegirine augite, may be nepheline-bearing (= melafenite).

Ugandite (Holmes, 1945; Bufumbira, Uganda), see p. 100
Olivine melaleucitite, glass base corresponds chemically to plagioclase + nepheline (= kalmafite).

Ulrichite (Marshall, 1906; Dunedin, New Zealand)
Nepheline microsyenite rich in aegirine augite and barkevikite (= tinguaite porphyry).

Umptekite (Ramsay, 1894; Umptek (Khibina), Kola, U.S.S.R.)
Alkali syenite, rich in arfvedsonite, with aegirine, biotite, etc., hypersolvus with microperthite. Nepheline-poor or nepheline-free.

Uncompahgrite (Larsen and Hunter, 1914; Uncompahgre Mt., Colorado, U.S.A.)
Pyroxene melilitolite.

Urtite (Ramsay, 1896; Lovozero (Lujaur-Urt), Kola, U.S.S.R.), see p. 53
Nephelinolite with more than 70% nepheline; devoid of feldspar, differs from ijolite in smaller content of mafic minerals.

Venanzite (Sabatini, 1898; Umbria, Italy)
Olivine–leucite–kalsilite–melilitite or mela-melilite–olivine–leucitite.

Verite (Osann, 1889; Cabo de Gata, Spain)
Hyalotrachyte, subalkaline, containing biotite, augite and olivine in glass.

Vesbite (Washington, 1920; Vesuvius, Italy)
Melilite fergusite or melilite italite (Kalsilite-bearing nepheline–leucite foidite).

Vesecite (Scheumann, 1922; Vesec Svetla, Bohemia, Czechoslovakia)
Monticellite polzenite
= monticellite–phlogopite–nepheline–olivine melilitite devoid of augite.

Vesuvite (Lacroix, 1917; Vesuvius, Italy)
Basanitic leucitite.

Vetrallite (Johannsen, 1938; Vico, Italy)
Tephritic phonolite containing labradorite.

Vibetoite (Brøgger, 1921; Fen, Norway)
Biotite–amphibole pyroxenite, calcite-bearing.

Vicoite (Washington, 1906; Vico, Italy)
Leucite leucotrachybasalt, very leucite-rich = trachybasaltic leucitite or phonolitic leucite tephrite.

Viterbite (Washington, 1906; Viterbo, Italy)
Leucite-rich tephritic phonolite, leucocratic, containing labradorite.

Vogesite (Rosenbusch, 1887; Vosges, France)
Monzonitic lamprophyre rich in hornblende.

Vulsinite (Washington, 1896; Bolsena, Italy)
Labradorite–trachyte, foid-bearing.

Wesselite (Scheumann, 1922; Wesseln, Bohemia, Czechoslovakia)
Hauyne–nepheline–biotite ankaratrite or hauyne–biotite–melanephelinite.

Windsorite (Daly, 1903; Windsor, Vermont, U.S.A.)
Alkali feldspar–quartz microsyenite to quartz micromonzonite, subalkaline.

Wolgidite (Wade and Prider, 1940; Kimberley, Western Australia), see p. 97
Phlogopite–diopside–magnophorite–leucite lamproite.

Wyomingite (Cross, 1897; Leucite Hills, Wyoming, U.S.A.), see p. 97
Phlogopite–leucite lamproite or phonolitic pholgopite leucitite, glass-rich.

Yamaskite (Young, 1904; Mt. Yamaska, Quebec, Canada)
Pyroxenite dominated by titanaugite and hornblende minor bytownite (= amphibole-bearing jacupirangite).

LIST OF REFERENCES USED IN COMPILING THE GLOSSARY

Adamson, O. J., 1944. The petrology of the Norra Kärr district. An occurrence of alkaline rocks in southern Sweden. *Geol. För. Stockh. Förh.*, **66**, 113–255.

Barth, T. F. W., 1945. Studies on the igneous rock complex of the Oslo region. II. Systematic petrography of the plutonic rocks. *Skr. norske Vidensk-Akad. Mat.-naturv. Kl.*, **1944**, No. 9, 1–104.

Bose, M. K., 1969. Studies on the igneous rock complex of the Oslo region. XXI. Petrology of the sørkedalite—a primitive rock from the alkali igneous province of Oslo. *Skr. norske Vidensk-Akad. Mat.-naturv. Kl. Ny Serie*, No. 27, 1–28.

Commission on Systematics in Petrology, 1973. Classification and nomenclature of plutonic rocks—recommendations (Subcommission on Systematics of Igneous Rocks). *Geological Newsletter*, **1973**, no. 2, pp. 110–27.

Frechen, J., 1967. (In Taylor, H. P., Frechen, J., and Degens, E. T., 1967.) Oxygen and carbon isotope studies of carbonatites from the Laacher See District, West Germany and the Alnö District, Sweden. *Geoch. Cosmochim. Acta*, **31**, 407–30.

Fúster, J. M., Gastesi, P., Sagredo, J., and Fermoso, M. L., 1967. Las rocas lamproíticas del S.E. de España. *Estud. Geol.*, **23**, 35–69.

Hatch, F. H., Wells, A. K., and Wells, M. K., 1961. *Petrology of the Igneous Rocks*, 12th ed. Thomas Murby and Co., London, 1–515.

Hatch, F. H., Wells, A. K., and Wells, M. K., 1972. *Petrology of the Igneous Rocks*, 13th ed. Thomas Murby and Co., London, 1–551.

Holmes, A., 1928. *The Nomenclature of Petrology*, 2nd. ed. Thomas Murby and Co., London, 1–284.

Holmes, A., 1937. The petrology of katungite. *Geol. Mag.*, **74**, 200–19.

Holmes, A., 1945. Leucitized granite xenoliths from the potash-rich lavas of Bunyaruguru, South-West Uganda. *Am. J. Sci.*, **243-A**, Daly Volume, 313–32.

Holmes, A., 1950. Petrogenesis of katungite and its associates. *Am. Miner.*, **35**, 772–92.

Irvine, T. N., and Baragar, W. R. A., 1971. A guide to the chemical classification of the common volcanic rocks. *Can. J. Earth Sci.*, **8**, 523–48.

IUGS Subcommission on the Systematics of Igneous Rocks, 1973. Classification and nomenclature of plutonic rocks; recommendations. *Neues Jb. Miner. Mh.*, **4**, 149–64.

Johannsen, A., 1932, 1937, 1938, 1939. *A Descriptive Petrography of the Igneous Rocks, vols. I–IV.* Chicago University Press.

Joplin, G. A., 1964. *A Petrography of Australian Igneous Rocks.* Angus and Robertson, Sydney, 1–210.

Jung, J., and Brousse, R., 1959. *Classification Modale des Roches Eruptives.* Masson & Cie., Paris, 1–122.

Kotulsky, V. K., 1943. *Bull. Teschn. Inf. Norilsk Gorno-obogat. Kombinat.*

Lacroix, A., 1933. Contribution à la connaissance de la composition chimique et minéralogique des roches éruptives de l'Indochine. *Bull. Serv. géol. Indochine*, **20**, 3, 1–208.

Loewinson-Lessing, F.Yu., and Struve, E. A., 1963. *Petrographic Dictionary* (in Russian).

Muir, I. D., 1973. *A Mineralogical Classification of Igneous Rocks and Glossary of Rock Names.* Department of Mineralogy and Petrology, Cambridge, 1–21.

Ronner, F., 1963. *Systematische Klassifikation der Massengesteine.* Springer-Verlag, Wien, 1–380.

Saggerson, E. P., and Williams, L. A. J., 1964. Ngurumanite from Southern Kenya and its bearing on the origin of rocks in the Northern Tanganyika alkaline district. *J. Petrology*, **5**, 40–81.

Streckeisen, A., 1967. Classification and nomenclature of igneous rocks. *Neues Jb. Miner. Abh.*, **107**, 144–240.

Tilley, C. E., 1950. Some aspects of magmatic evolution. *Q. J. geol. Soc. Lond.*, **106**, 37–61.

Tilley, C. E., and Muir, I. D., 1964. Intermediate members of the oceanic basalt–trachyte association. *Geol. För. Stockh. Förh.*, **85**, 434–43.

Tröger, W. E., 1935. *Spezielle Petrographie der Eruptivgesteine. Ein Nomenklatur-Kompendium.* Schweizerbart'sche Verlagsbuchhandlung, Stuttgart, 1–360, (reprinted 1969).

Tröger, W. E., 1938. Spezielle Petrographie der Eruptivgesteine. Eruptivgesteinsnamen. 1. Nachtrag. *Fortsch. Miner., Kristall., und Petrogr.*, **23**, 41–90, (reprinted 1969).

Wade, A., and Prider, R. T., 1940. The leucite-bearing rocks of the West Kimberley area, Western Australia. *Q. J. geol. Soc. Lond.*, **96,** 39–98.

Zhidkov, A.Ya. (= Jidkov, A.Iy.), 1962. Complex Synnyr intrusion of syenites of the North-Baikal province (in Russian). *Geologia i Geofisika*, No. 9.

INDEX OF ROCK NAMES

Abbreviations:

def = definition
min = mineralogical composition
chem = chemical composition
occ = occurrence
gen = genesis
ec.min = economic mineralogy

SUBJECT INDEX

GEOGRAPHICAL INDEX

(*Italics* indicate that localities in a country, province, etc. have been compiled into one list)

AUTHOR INDEX